Water and Wastewater Engineering

Water and Wastewater Engineering

Hydraulics, Hydrology and Management

Fourth Edition

Volume 1

Lawrence K. Wang (PhD)
Rutgers University, New Brunswick, NJ, USA

Mu-Hao Sung Wang (PhD)
Rutgers University, New Brunswick, NJ, USA

Nazih K. Shammas (PhD)
University of Michigan, Ann Arbor, MI, USA

Published by John Wiley & Sons, Inc., Hoboken, New Jersey.
Published simultaneously in Canada.

For general information on our other products and services or for technical support, please contact our Customer Care Department within the United States at (800) 762-2974, outside the United States at (317) 572-3993 or fax (317) 572-4002.

Wiley also publishes its books in a variety of electronic formats. Some content that appears in print may not be available in electronic formats. For more information about Wiley products, visit our website at www.wiley.com.

Library of Congress Cataloging-in-Publication Data:

Names: Wang, Lawrence K., author. | Wang, Mu Hao Sung, author. |
Shammas, Nazih K., author.
Title: Water and wastewater engineering. Volume 1 : hydraulics, hydrology
and management / Lawrence K. Wang, Mu-Hao Sung Wang, Nazih K. Shammas.
Description: Fourth edition. | Hoboken, New Jersey : Wiley, [2024] |
Includes index.
Identifiers: LCCN 2023058678 (print) | LCCN 2023058679 (ebook) | ISBN
9781394179107 (cloth) | ISBN 9781394179145 (adobe pdf) | ISBN
9781394179138 (epub)
Subjects: LCSH: Water–Purification. | Hydrology. | Hydraulics.
Classification: LCC TD430 .W3537 2024 (print) | LCC TD430 (ebook) | DDC
628.1–dc23
LC record available at https://lccn.loc.gov/2023058678
LC ebook record available at https://lccn.loc.gov/2023058679

Cover Design: Wiley
Cover Image: © Am Balducci/Getty Images

Set in 9.5/12.5pt STIXTwoText by Straive, Pondicherry, India
Printed and bound by CPI Group (UK) Ltd, Croydon, CR0 4YY

C9781394179107_090424

Contents

About the Authors

Lawrence K. Wang has worked for decades in industrial, academic, and policy roles related to water and wastewater engineering, including as a Senior Advisor to the United Nations Industrial Development Organization (UNIDO), an Advisor to the US Environmental Protection Agency (USEPA), a Delegate Leader of the People to People International Foundation (PPIF), and a retired Acting President/Professor of the Lenox Institute of Water Technology (LIWT). Dr. Wang is a licensed Professional Engineer, a certified Laboratory Director, an OSHA Train-the-Trainers Instructor, a vastly prolific author, and the holder of 29 invention patents.

Mu-Hao Sung Wang has served the society for decades as a New York State government official, a university professor, a science researcher, and a book writer. She is a licensed Professional Engineer, a Diplomate of the American Academy of Environmental Engineers (AAEE), and an AWWA-APHA-WEF Committee Member of the Standard Methods for the Examination of Water and Wastewater. Dr. Wang has published over 20 books and over 200 scientific articles and has had 14 patented inventions in the field of environmental engineering.

Nazih K. Shammas was an environmental consultant and professor for over 45 years. He was an Dean/Director of the Lenox Institute of Water Technology (LIWT) and an advisor to Krofta Engineering Corporation, USA. Dr. Shammas was the author of over 250 publications and 19 books in the field of environmental engineering. He had expertise in water quality control, wastewater reclamation, pollution prevention, waste management, water purification, distribution, sludge treatment, hydraulics, hydrology, computer science, and environmental engineering education.

Preface

History of This Book

This text is a revision of the classic text *Water and Wastewater Engineering*, originally authored by Gordon M. Fair, John C. Geyer, and Daniel A. Okun. All three original authors were good professional friends of Professor Marvin L. Granstrom of Rutgers University where Professor Lawrence K. Wang and Professor Mu-Hao Sung Wang received their PhD degrees under Professor Granstrom's supervision. The last surviving member of the three original authors, Professor Daniel A. Okun, asked his student, Professor Nazih K. Shammas, to collaborate with Professor Lawrence K. Wang on publishing the third edition, *Water and Wastewater Engineering: Water Supply and Wastewater Removal*, before Professor Okun's passing in 2007. Just before beginning the preparation of this fourth edition of *Water and Wastewater Engineering: Hydraulics, Hydrology and Management*, Professor Shammas died in 2020. The current surviving authors (Lawrence K. Wang and Mu-Hao Sung Wang) are dedicating this book and all remaining books in the *Water and Wastewater Engineering* series to the three original authors (Professors Fair, Geyer and Okun), co-author Professor Shammas and the book series coordinator Professor Granstrom for their outstanding lifetime academic accomplishments and their lifetime friendship with the surviving authors.

Goals of This Book

Today, effective design and efficient operation of engineering works ask, above all, for a fuller understanding and application of scientific principles. Thus, the results of scientific research are being incorporated with remarkable success into new designs and new operating procedures. Like other fields of engineering, water and wastewater engineering has its science and its art. To reach the audience to which this book is addressed, the science of water and wastewater engineering is given principal emphasis. However, the art of water and wastewater engineering is not neglected. Enough elements of engineering practice, experience, common sense, and rules of thumb are included to keep the reader aware of the water environment and constructions that place water at the service of cities, towns, villages, and homesteads, and collect wastewater for treatment, recycle or disposal.

Further Study in Addition to Classroom Education

The study of scientific principles is best accomplished in the classroom; the application of these principles follows as a matter of practice. To further bridge the way from principle to practice, we suggest that the study of this textbook be supplemented by (a) visits to water and wastewater works, (b) examination of plans and specifications of existing water and wastewater systems, (c) readings in the environmental science and engineering periodicals, (d) study of the data and handbook editions of trade journals, (e) examination of the catalogs and bulletins of equipment manufacturers, and (f) searching for the latest water and wastewater engineering developments from the internet.

Intended Audience

Like its forerunners, the new work is intended for engineers of civil and environmental engineering, regardless of whether they belong to the student body of a university or are already established in their profession. Specifically, the target audience is engineering students who have had introductory calculus, chemistry, and fluid mechanics, typically civil and environmental engineering majors. Several chapters of this book contain introductory material appropriate for juniors as well as more advanced material that might only be appropriate for upper-level undergraduate engineering students. Applied hydrology and hydraulics are reviewed and included in this book. The inclusion of this material makes this book important also to physical and investment planners of urban and regional developments.

Course Suggestions

The book is comprehensive and covers all aspects of water supply, water sources, water distribution, sanitary sewerage, and urban stormwater drainage. This comprehensive coverage gives faculty the flexibility of choosing the material they find fit for their courses, and this wide coverage is helpful to engineers in their everyday practice.

Courses where this book may be used include

1) water and wastewater engineering;
2) **hydraulics and hydrology engineering**;
3) water supply and sewerage;
4) civil and sanitary engineering design;
5) environmental engineering design;
6) supply, transmission, and distribution system;
7) design of sewage and stormwater collection systems; and
8) water supply and wastewater collection systems.

Key Features of This Book

Several items unique to this textbook include

- *Solved problems*: A reliable problem-solving experience for students is carried throughout the text and demonstrated in every example problem to reinforce the best practices.
- *Photos and illustrations*: Photos are used throughout the text to clarify infrastructure systems and show examples of built and constructed water supply and waste-water collection features.
- *Current water and wastewater infrastructure issues*: Current infrastructure and global issues are addressed in the text. Examples of such issues include (a) water quality in distribution systems; (b) groundwater under the direct influence of surface water; (c) dual water systems; (d) cross-connection control and back-flow prevention; (e) computer-aided design; (f) design nomograms for fast water infrastructure analysis; (g) trenchless technology and rehabilitation of sewers; (h) computer-aided water distribution system modeling and analysis; (i) computer-aided gravity sewer system modeling and design; (j) modern alternative wastewater collection systems, including low-pressure sewer system design; and (k) urban runoff and overflow management
- *Equations and example problems with both US customary and SI units suitable for engineers with international water and wastewater engineering projects:* The text has a multitude of example problems. Such equations and problems incorporate both SI and the more customary English (or US) unit systems. We feel that many introductory texts fall short in both of these areas by not providing students with equations and examples that help explain difficult technical concepts and by only focusing on one system of units.
- *Applied hydraulics and hydrology*: Hydraulics and hydrology concepts are critical for the civil and environmental engineering professional, and thus the reader. Applied hydraulics topics such as pumps, weirs, pressurized pipe flow, gravity flow, head losses water transmission, water distribution, sanitary sewer analysis, and storm sewer analysis are reviewed in this book. Applied hydrology topics such as water cycle, precipitation, runoff, groundwater, surface water, evaporation, transpiration, and percolation are also reviewed, both for the practical design of water and wastewater handling facilities.
- *Prevention through design and system safety*: A complete chapter (20) is dedicated to prevention through design (PtD), because it is important for readers to learn about this new strategy. The National Institute for Occupational Safety and Health (NIOSH) is promoting the inclusion of PtD in undergraduate engineering education, has reviewed this chapter, and provided the illustrative examples described in Chapter 6, Water Distribution Systems: Components, Design, and Operation, and Chapter 14, Design of Sewer Systems. In addition, many examples and applications of these safety issues are given for actual situations in systems design chapters. Other water and wastewater engineering texts do not address this important topic.

Instructor Resources

The following resources are available to instructors on the book's website at www.wiley.com/go/Wang/Waterandwaste water4e visit the Instructor section of the website to register for access to these password-protected resources:

- *Solutions manual*: Complete solutions for every homework problem and answers to all discussion questions in the text are available to instructors.
- *Image gallery*: Images from the text in electronic form can be provided to instructors, upon request, for preparation of lecture PowerPoint slides.

Haestad Methods Water Solutions Software by Bentley

The following software modules are discussed in the text:

- **WaterGEMS** is used to illustrate the application of various available software programs that can help civil and environmental engineers design and analyze water distribution systems. It is also used by water utility managers as a tool for the efficient operation of distribution systems. See Chapter 7, Water Distribution Systems: Modeling and Computer Applications.
- **SewerCAD** is used as a demonstration for the application of modeling and computer techniques in the sanitary sewer design process. See Chapter 15, Sewerage Systems: Modeling and Computer Applications.
- **StormCAD** is used as a demonstration for the application of modeling and computer techniques in the stormwater street inlets and storm sewer design process. See Chapter 15, Sewerage Systems: Modeling and Computer Applications.

  Available at: selectserver.bentley.com

Dedication and Acknowledgments

The authors are dedicating this book and all remaining books in the *Water and Wastewater Engineering* series to both Professor Nazih K. Shammas of the Lenox Institute of Water Technology and Professor Marvin L. Granstrom of Rutgers University for their outstanding lifetime academic accomplishments and their lifetime friendship with the authors. Professor Shammas contributed significantly to the old editions of this "Water and Wastewater Engineering series", so his name is listed as a co-author of all related future books in the series. Professor Granstrom was the Academic Advisor of both Author Lawrence K. Wang and Author Mu-Hao Sung Wang when both authors were PhD students at Rutgers University.

Books do not come off the press through the efforts of authors alone. Their preparation passes successively through the editorial and production stages. The authors would like to thank our colleagues, mentors, Wiley editors/reviewers, and the resource providers and designers who have guided our efforts along the way. Sincere appreciation is extended to the families of Professors Gordon M. Fair (Harvard University), John C. Geyer (Johns Hopkins University), Daniel A. Okun (University of North Carolina), Marvin L. Granstrom (Rutgers University), and Nazih K. Shammas (Lenox Institute of Water Technology) and the reviewers whose suggestions and comments have significantly improved the overall quality of this book series. Special thanks are due to Dr. Richard Reinhart of NIOSH for his review of Chapter 20 PtD and to Dr. Carolyn M. Jones of the San Francisco Public Utilities Commission for her contribution of PtD engineering examples.

A book is not written in long evenings and on holidays without the consent, encouragement, and cooperation of the writers' families. This, too, should be a matter of record.

Lawrence K. Wang
National Cheng Kung University, Taiwan, ROC, BSCE
Missouri University of Science and Technology, MSCE
University of Rhode Island, MS
Rutgers University, PhD

Mu-Hao Sung Wang
National Cheng Kung University, Taiwan, ROC, BSCE
University of Rhode Island, MS
Rutgers University, PhD

About the Companion Website

This book is accompanied by a companion website:

www.wiley.com/go/Wang/Waterandwastewater4e

This website includes:

- Solutions Manuals
- Image Gallery

1

Introduction to Water Systems

The right to water is an implicit part of the right to an adequate standard of living and the right to the highest attainable standard of physical and mental health, both of which are protected by the United Nations' *International Covenant on Economic, Social and Cultural Rights,* which was established in 1976. However, some countries continue to deny the legitimacy of this right. In light of this fact and because of the widespread noncompliance of states with their obligations regarding the right to water, the United Nations' Committee on Economic, Social and Cultural Rights confirmed and further defined the right to water in its General Comment No. 15 in 2002. The comment clearly states that water is indispensable for an adequate standard of living and is one of the most fundamental conditions for survival:

> *The human right to water entitles everyone to sufficient, safe, acceptable, physically accessible and affordable water for personal and domestic uses. An adequate amount of safe water is necessary to prevent death from dehydration, reduce the risk of water-related disease and provide for consumption, cooking, personal and domestic hygienic requirements.*

According to the World Health Organization (WHO), 1.1 billion people (17% of the global population) lack access to safe drinking water, meaning that they have to revert to unprotected wells or springs, canals, lakes, or rivers to fetch water; 2.6 billion people lack adequate sanitation; and 1.8 million people die every year from diarrheal diseases, including 90% of children under age 5. This situation is no longer bearable. To meet the WHO's *Water for Life Decade (2005–2015),* an additional 260,000 people per day need to gain access to improved water sources.

In 2004 about 3.5 billion people worldwide (54% of the global population) had access to piped water supply through house connections. Another 1.3 billion (20%) had access to safe water through other means than house connections, including standpipes, "water kiosks," protected springs, and protected wells.

In the United States 95% of the population that is served by community water systems receives drinking water that meets all applicable health-based drinking water standards through effective treatment and source water protection. In 2007, approximately 156,000 US public drinking water systems served more than 306 million people. Each of these systems regularly supplied drinking water to at least 25 people or 15 service connections. Beyond their common purpose, the 156,000 systems vary widely. Table 1.1 groups water systems into categories that show their similarities and differences. For example, the table shows that most people in the United States (286 million) get their water from a community water system. Of the approximately 52,000 community water systems, only 8% of those systems (4048) serve 82% of the people.

Water is used in population centers for many purposes: (a) for drinking and culinary uses; (b) for washing, bathing, and laundering; (c) for cleaning windows, walls, and floors; (d) for heating and air conditioning; (e) for watering lawns and gardens; (f) for sprinkling and cleaning streets; (g) for filling swimming and wading pools; (h) for display in fountains and cascades; (i) for producing hydraulic and steam power; (j) for employment in numerous and varied industrial processes; (k) for protecting life and property against fire; and (l) for removing offensive and potentially dangerous wastes from households, commercial establishments, and industries. To provide for these varying uses, which total about 100 gallons per capita per day (gpcd) or 378 liters per capita per day (Lpcd) in average North American *residential* communities and 150 gpcd (568 Lpcd) or more in large *industrial* cities, the supply of water must be satisfactory in quality and adequate in quantity, readily available to the user, relatively cheap, and easily disposed of after it has served its many purposes. Necessary engineering works are waterworks, or water supply systems, and wastewater works, or wastewater management systems.

Water and Wastewater Engineering: Hydraulics, Hydrology and Management, Volume 1, Fourth Edition.
Lawrence K. Wang, Mu-Hao Sung Wang, and Nazih K. Shammas.

Companion website: www.wiley.com/go/Wang/Waterandwastewater4e

Table 1.1 US Public Water Systems Size by Population Served in 2007

Water System		Very Small (500 or less)	Small (501–3300)	Medium (3301–10,000)	Large (10,001–100,000)	Very Large (>100,000)	Total
Community water system[a]	No. of systems	29,282	13,906	4822	3702	398	52,110
	Population served	4,857,007	19,848,329	27,942,486	105,195,727	128,607,655	286,451,204
	Percentage of systems	56%	27%	9%	7%	1%	100%
	Percentage of population	2%	7%	10%	37%	45%	100%
Nontransient noncommunity water system[b]	No. of systems	16,034	2662	120	22	1	18,839
	Population served	2,247,556	2,710,330	639,561	533,845	203,000	6,334,292
	Percentage of systems	85%	14%	1%	0%	0%	100%
	Percentage of population	35%	43%	10%	8%	3%	100%
Transient noncommunity water system[c]	No. of systems	81,873	2751	102	15	3	84,744
	Population served	7,230,344	2,681,373	546,481	424,662	2,869,000	13,751,860
	Percentage of systems	97%	3%	0%	0%	0%	100%
	Percentage of population	53%	19%	4%	3%	21%	100%
Total no. of systems		127,189	19,319	5044	3739	402	155,693

[a] Community water system: a public water system that supplies water to the same population year-round.
[b] Nontransient noncommunity water system: a public water system that regularly supplies water to at least 25 of the same people at least 6 months per year, but not year-round. Some examples are schools, factories, office buildings, and hospitals that have their own water systems.
[c] Transient noncommunity water system: a public water system that provides water in a place such as a gas station or campground where people do not remain for long periods of time.
Source: Courtesy U.S. Environmental Protection Agency.

Waterworks withdraw water from natural sources of supply, purify it if necessary, and deliver it to the consumer. Wastewater works collect the spent water of the community—about 70% of the water supplied—together with varying amounts of entering ground and surface waters. The collected wastewaters are treated and reused or discharged, usually into a natural water body (more rarely onto land). Often the receiving body of water continues to serve also as a source of important water supplies for many purposes. It is this multiple use of natural waters that creates the most impelling reasons for sound water quality management.

1.1 Components of Water Systems

Each section of this chapter offers, in a sense, a preview of matters discussed at length in later parts of this book. There they are dealt with as isolated topics to be mastered in detail. Here they appear in sequence as parts of the whole so that their general purpose and significance in the scheme of things may be understood and may give a reason for closer study.

Municipal water systems generally comprise (a) *collection works,* (b) *purification works,* (c) *transmission works,* and (d) *distribution works.* The relative functions and positions of these components in a surface water supply are sketched in Fig. 1.1. Collection works either tap a source continuously adequate in volume for present and reasonable future demands or convert an intermittently insufficient source into a continuously adequate supply. To ensure adequacy, seasonal and, in large developments, even annual surpluses must be stored for use in times of insufficiency. When the quality of the water

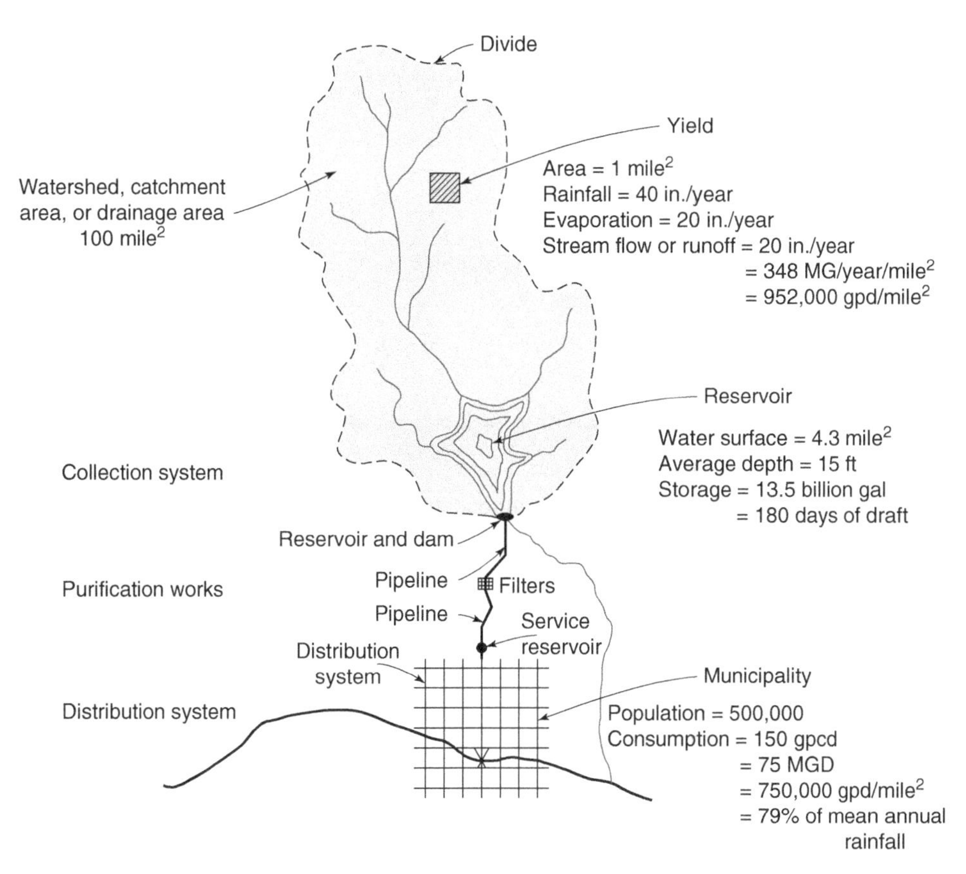

Figure 1.1 Rainfall, Runoff, Storage, and Draft Relations in the Development of Surface Water. Conversion factors: 1 mi^2 = 2.59 km^2; 1 in. /yr = 25.4 mm/yr; 1 ft = 0.3048 m; 1 MG/yr/mi^2 = 1.46 ML/yr/km^2; 1 gpd/mi^2 = 1.461 L/d/km^2; 1 billion gal = 1 BG = 3.785 billion L = 3.785 BL; 1 gpcd = 3.785 Lpcd; 1 MGD = 3.785 MLD.

collected is not satisfactory, purification works are introduced to render it suitable for the purposes it must serve: Contaminated water is disinfected, aesthetically displeasing water made attractive and palatable, water containing iron or manganese deferrized or demanganized, corrosive water deactivated, and hard water softened. Transmission works convey the collected and purified supply to the community, where distribution works dispense it to consumers in wanted volume at adequate pressure. Ordinarily, the water delivered is metered so that an equitable charge can be made for its use and, often, also for its disposal after use.

1.2 Required Capacity

Water supply systems are designed to meet population needs for a reasonable number of years in the future. The rate of consumption is normally expressed as the mean annual use in gallons per capita daily (gpcd) or liters per capita daily (Lpcd), and seasonal, monthly, daily, and hourly departures in rate are given in percentages of the mean. In North America, the spread in consumption is large: from 35 to 500 gpcd (132–1890 Lpcd), varying radically with industrial water demands. Average rates between 100 and 200 gpcd (378–757 Lpcd) are common, and a generalized average of 150 gpcd (568 Lpcd) is a useful guide to normal requirements.

The capacity of individual system components is set by what is expected of them. Distribution systems, for example, must be large enough to combat and control serious conflagrations without failing to supply maximum *coincident* domestic and industrial drafts. Fire demands vary with size and value of properties to be protected and are normally a function of the gross size of the community. The distribution system leading to the high-value district of an average American city of 100,000 people, for example, must have an excess of *fire standby* capacity equal in itself to the average rate of draft. For smaller or larger American communities, the standby capacity falls or rises, within certain limits, more or less in proportion to the square root of the population.

1.3 Sources of Water Supply

The source of water commonly determines the nature of the collection, purification, transmission, and distribution works. Common sources of freshwater and their development are as follows:

1) Rainwater:
   a) From roofs, stored in cisterns, for small individual supplies.
   b) From larger, prepared watersheds, or catches, stored in reservoirs, for large communal supplies.
2) Surface water:
   a) From streams, natural ponds, and lakes of sufficient size, by continuous draft.
   b) From streams with adequate flood flows, by intermittent, seasonal, or selective draft of clean floodwaters, and their storage in reservoirs adjacent to the streams, or otherwise readily accessible from them.
   c) From streams with low dry-weather flows but sufficient annual discharge, by continuous draft through storage of necessary flows in excess of daily use in one or more reservoirs impounded by dams thrown across the stream valleys.
   d) From brackish and seawater by desalination. Desalination is an artificial process by which saline water is converted to freshwater. The most common desalination processes are distillation and reverse osmosis. Desalination is currently expensive compared to most alternative sources of water, and only a small fraction of total human use is satisfied by desalination. It is only economically practical for high-valued uses (such as household and industrial uses) in arid areas. The most extensive use is in the Persian (Arabian) Gulf. Mildly saline waters (brackish) are desalted most economically by reverse osmosis, and strongly saline waters by evaporation and condensation.
3) Groundwater:
   a) From natural springs.
   b) From wells.
   c) From infiltration galleries, basins, or cribs.
   d) From wells, galleries, and, possibly, springs, with flows augmented from some other sources (i) spread on the surface of the gathering ground, (ii) carried into charging basins or ditches, or (iii) led into diffusion galleries or wells.
   e) From wells or galleries with flows maintained by returning to the ground the water previously withdrawn from the same aquifer for cooling or similar purposes.

Several schemes have been proposed to make use of *icebergs* as a water source; to date, however, this has only been done for novelty purposes. One of the serious moves toward the practical use of icebergs is the formation of an Arabian-American investment group to search for the optimal way to transport and melt icebergs for use as a source of drinking water supply without the need for on-land storage. Glacier runoff is considered to be surface water.

An iceberg is a large piece of freshwater ice that has broken off from a snow-formed glacier or ice shelf and is floating in open water. Because the density of pure ice is about 920 kg/m^3, and that of sea water is about 1025 kg/m^3, typically only one-tenth of the volume of an iceberg is above water. The shape of the rest of the iceberg under the water can be difficult to surmise from looking at what is visible above the surface. Icebergs generally range from 1 to 75 m (about 3–250 ft) above sea level and weigh 100,000–200,000 metric tons (about 110,000–220,000 short tons). The tallest known iceberg in the North Atlantic was 168 m (about 551 ft) above sea level, making it the height of a 55-story building. Despite their size, icebergs move an average of 17 km (about 10 mi) a day. These icebergs originate from glaciers and may have an interior temperature of −15 to −20°C (5 to − 4°F).

Municipal supplies may be derived from more than one source, the yields of available sources ordinarily being combined before distribution. *Dual public water supplies* (see Chapter 8) of unequal quality are unusual in North America. However, they do exist, for example, as a high-grade supply for general municipal uses and a low-grade supply for specific industrial purposes or firefighting. Unless the low-grade (nonpotable) supply is rigorously disinfected, its existence is frowned on by health authorities because it may be cross-connected, wittingly or unwittingly, with the high-grade (potable) supply. A *cross-connection* is defined as a junction between water supply systems through which water from doubtful or unsafe sources may enter an otherwise safe supply.

1.4 Rainwater

Rain is rarely the immediate provenance of municipal water supplies. Instead, the capture of rainwater is confined to farms and rural settlements usually in semiarid regions devoid of satisfactory ground or surface waters. On homesteads, rainwater running off roofs is led through gutters and downspouts to rain barrels or cisterns situated on or in the ground. Storage transforms the intermittent rainfall into a continuous supply. For municipal service, sheds or catches on ground that is naturally impervious or made tight by grouting, cementing, paving, or similar means must usually be added.

The gross yield of rainwater is proportional to the receiving area and the amount of precipitation. However, some rain is blown off the roof, evaporated, or lost in wetting the collecting surfaces and conduits and in filling depressions or improperly pitched gutters. Also, the first flush of water may have to be wasted because it contains dust, bird droppings, and other unwanted materials. The combined loss may be high. A cutoff, switch, or deflector in the downspout permits selective diversion of unwanted water from the system. Sand filters will cleanse the water as it enters the cistern and prevent its deterioration via the growth of undesirable organisms and consequent tastes, odors, and other changes in attractiveness and palatability.

The storage to be provided in *cisterns* depends on the distribution of rainfall. Storage varies with the length of dry spells and commonly approximates one-third to one-half the annual consumption. If rainfalls of high intensity are to be captured, standby capacity must exist in advance of filtration. Because their area is small, roofs seldom yield much water. A careful analysis of storm rainfalls and seasonal variations in precipitation is, therefore, required.

Example 1.1 Calculating the volume of rainfall that can be collected from a building roof

Make a rough estimate of the volume in gallons or liters of water that can be caught by 3000 ft^2 (278.7 m^2) of horizontally projected roof area (the average area of American farm buildings) in a region where the mean annual rainfall is 15 in. (38.1 cm).

Solution 1 (US Customary System):

$$\text{Gross yield} = 3{,}000\ \text{ft}^2 \times (15/12\ \text{ft}) \times 7.48\ \text{gal/ft}^3 = 28{,}100\ \text{gal annually} = 28{,}100\ \text{gal}/365\ \text{days} = 77\ \text{gpd}.$$

Net yield approximates two-thirds gross yield = 18,800 gal annually = **51 gpd**.

About half the net annual yield, or 9,400 gal = 1,250 ft^3, must normally be stored to tide the supply over dry spells.

Solution 2 (SI System):

$$\text{Gross yield} = \left(278.7\ \text{m}^2\right)(38.1/100\ \text{m})\left(1{,}000\ \text{L/m}^3\right) = 106{,}178\ \text{L annually} = 291\ \text{L/day} = 291\ \text{L/d}.$$

Net yield approximates two-thirds gross yield = 291 L/d (2/3) = **194 L/d** = 70,790 L/year.

About half the net annual yield = 0.5 (70,790 L/year) = 35,395 L = 35.4 m^3 stored to tide the supply over dry spells.

1.5 Surface Water

In North America, by far the largest volumes of municipal water are collected from surface sources. The quantities that can be gathered vary directly with the size of the catchment area, or watershed, and with the difference between the amounts of water falling on it and the amounts lost by evapotranspiration. The significance of these relationships to water supply is illustrated in Fig. 1.1. Where surface water and groundwater sheds do not coincide, some groundwater may enter from neighboring catchment areas or escape to them.

1.5.1 Continuous Draft

Communities on or near streams, ponds, or lakes may take their supplies from them by continuous draft if stream flow and pond or lake capacity are high enough at all seasons of the year to furnish requisite water volumes. Collecting works ordinarily include (a) an intake crib, gatehouse, or tower; (b) an intake conduit; and (c) in many places, a pumping station. On small streams serving communities of moderate size, an intake or diversion dam may create sufficient depth of water to submerge the intake pipe and protect it against ice. From intakes close to the community the water must generally be lifted to purification works and thence to the distribution system.

Most large streams are polluted by wastes from upstream communities and industries. Purification of their waters is then a necessity. Cities on large lakes usually guard their supplies against their own and their neighbor's wastewater and spent industrial-process waters by moving their intakes far away from shore and purifying both their water and wastewater. Diversion of wastewater from lakes will retard the lakes' eutrophication.

1.5.2 Selective Draft

Low stream flows are often left untouched. They may be wanted for other downstream purposes, or they may be too highly polluted for reasonable use. Only clean floodwaters are then diverted into reservoirs constructed in meadow lands adjacent to the stream or otherwise conveniently available. The amount of water so stored must supply demands during seasons of unavailable stream flow. If draft is confined to a quarter year, for example, the reservoir must hold at least three-fourths of the annual supply. In spite of its selection and long storage, the water may have to be purified.

1.5.3 Impoundage

In their search for clean water and water that can be brought and distributed to the community by gravity, engineers have developed supplies from upland streams. Most of them are tapped near their source in high and sparsely settled regions. To be of use, their annual discharge must equal or exceed the demands of the community they serve for a reasonable number of years in the future. Because their dry season flows generally fall short of concurrent municipal requirements, their floodwaters must usually be stored in sufficient volume to ensure an adequate supply. Necessary reservoirs are impounded by throwing dams across the stream valley. In this way, amounts up to the mean annual flow can be utilized. The area draining to an impoundment is known as the catchment area or watershed. Its economic development depends on the value of water in the region, but it is a function, too, of runoff and its variation, accessibility of catchment areas, interference with existing water rights, and costs of construction. Allowances must be made for evaporation from new water surfaces generated by the impoundage (Fig. 1.2) and also often for release of agreed-on flows to the valley below the dam (compensating water). Increased ground storage in the flooded area and the gradual diminution of reservoir volumes by siltation must also be considered.

Figure 1.2 A Watershed Lake in Western Missouri that Provides Water Supply (*Source:* Courtesy of the National Resources Conservation Service and USDA)

Intake structures are incorporated in impounding dams or kept separate. Other important components of impounding reservoirs are (a) spillways safely passing floods in excess of reservoir capacity and (b) diversion conduits safely carrying the stream past the construction site until the reservoir has been completed and its spillway can go into action. Analysis of flood records enters into the design of these ancillary structures.

Some impounded supplies are sufficiently safe, attractive, and palatable to be used without treatment other than protective disinfection. However, it may be necessary to remove high color imparted to the stored water by the decomposition of organic matter in swamps and on the flooded valley bottom; odors and tastes generated in the decomposition or growth of algae, especially during the first years after filling; and turbidity (finely divided clay or silt) carried into streams or reservoirs by surface wash, wave action, or bank erosion. Recreational uses of watersheds and reservoirs may call for treatment of the flows withdrawn from storage.

Much of the water in streams, ponds, lakes, and reservoirs in times of drought, or when precipitation is frozen, is seepage from the soil. Nevertheless, it is classified as surface runoff rather than groundwater. Water seeps *from* the ground when surface streams are low and *to* the ground when surface streams are high. Release of water from ground storage or from accumulations of snow in high mountains is a determining factor in the yield of some catchment areas. Although surface waters are derived ultimately from precipitation, the relations between precipitation, runoff, infiltration, evaporation, and transpiration are so complex that engineers rightly prefer to base calculations of yield on available stream gaugings. For adequate information, gaugings must extend over a considerable number of years.

Example 1.2 Estimates of Yields from Water Sheds and Storage Requirements
Certain rough estimates of the yield of surface water sheds and storage requirements are shown in Fig. 1.1. Rainfall is used as the point of departure, merely to identify the dimensions of possible rainfall–runoff relationships. Determine

1) the yields from the water sheds,
2) the storage requirements,
3) the number of people who can be supported by a drainage area of 100 mile^2 (259 km^2) if there is adequate impoundage for water storage, and
4) the number of people who can be supported by a drainage area of 100 mile^2 (259 km^2) if there is no impoundage for water storage.

The following assumptions are made: (a) Rainfall=20 in. /mile^2 annually = 19.6 cm/km^2, (b) a stream flow of about 1 MGD/mile^2 (million gallons per day per square mile) or (1.547 ft^3/s)/mile^2 [or 1.46 MLD/km^2 (million liters per day per square kilometer)] is a good average for the well-watered sections of North America, (c) for 75% development (0.75 × 1 MGD/mile^2 or 0.75 × 1.46 MLD/km^2), about half a year's supply must generally be stored. In semiarid regions, storage of three times the mean annual stream flow is not uncommon, that is, water is held over from wet years to supply demands during dry years, (d) average water consumption =150 gpcd = 567.8 Lpcd, (e) for water supply by continuous draft, low water flows rather than average annual yields govern. In well-watered sections of North America, these approximate 0.1 ft^3/s or 64,600 gpd/mile^2 (or 28.32 L/s, or 0.094316 MLD/km^2).

Solution 1 (US Customary System):

1) The following conversion factors and approximations are being employed:

   1 in. rainfall/mile^2 = 17.378 MG

   Hence, 20 in. /mile^2 annually = 20 × 17.378 = 348 MG or 348/365 = **0.952 MGD**.

2) A stream flow of about 1 MGD/mile^2 is a good average for the well-watered sections of North America. Not all of it can be adduced economically by storage.For 75% development (0.75 MGD/mile^2, or 750,000 gpd/mile^2), about half a year's supply must generally be stored. For a catchment area of 100 mile^2, therefore,

   Storage = (0.75 MGD/mile^2)(100 mile^2) × (0.5 × 365 days) = 13,688 MG = **13.5 BG** (billion gallons) approximately.

   In semiarid regions storage of three times the mean annual stream flow is not uncommon, that is, water is held over from wet years to supply demands during dry years.
3) For an average consumption of 150 gpcd, the drainage area of 100 mile^2 and impoundage of 13.5 BG will supply a population of 100 × 750,000/150 = **500,000 persons**.
4) For water supply by continuous draft, low water flows rather than average annual yields govern.
   In well-watered sections of North America, these approximate 0.1 ft^3/s or 64,600 gpd/mile^2.
   A catchment area of 100 mile^2, therefore, can supply without storage:

   100 × 64,600/150 = **43,000 people.**

   This is compared against 500,000 people in the presence of proper storage.

Solution 2 (SI System):

1) The following conversion factors and approximations are being employed:

   1 cm/km^2 = 67.12 ML (million liters).

   Hence, 19.6 cm/km^2 annually =19.6 × 67.12 = 1,315.6 ML annually = **3.6 MLD**.

2) A stream flow of about 1.46 MLD/km^2 is a good average for the well-watered sections of North America. Not all of it can be adduced economically by storage.

For 75% development (0.75×1.46 MLD/km^2), about half a year's supply must generally be stored.
For a catchment area of 259 km^2, therefore,

$$\text{Storage} = 0.75(1.46\ \text{MLD/km}^2)(259\ \text{km}^2)(0.5 \times 365) = 51{,}758\ \text{ML} = \mathbf{51.758\ BL}\ \text{(billion liters)}.$$

In semiarid regions storage of three times the mean annual stream flow is not uncommon, that is, water is held over from wet years to supply demands during dry years.

3) For an average consumption of 567.8 Lpcd, a drainage area of 259 km^2 and impoundage of 51.758 BL will supply a population of

$$\left(0.75 \times 1.46\ \text{MLD/km}^2\right)\left(259\ \text{km}^2\right)(1{,}000{,}000\ \text{L/ML})/(567.8\ \text{Lpcd})$$
$$= \mathbf{500{,}000\ persons}.$$

4) For water supply by continuous draft, low water flows rather than average annual yields govern. In well-watered sections of North America these approximate 28.32 L/s, or 0.094316 MLD/km^2.
A catchment area of 259 km^2, therefore, can supply without storage:

$$(259\ \text{km}^2)(0.094316\ \text{MLD/km}^2)(1{,}000{,}000\ \text{L/ML})/(567.8\ \text{Lpcd}) = \mathbf{43{,}000\ people.}$$

This is compared against 500,000 people in the presence of proper storage.

1.6 Groundwater

Smaller in daily delivery, but many times more numerous than surface water supplies, are the municipal and private groundwater supplies of North America. Groundwater is drawn from many different geological formations: (a) from the pores of alluvial (water-borne), glacial, or aeolian (windblown) deposits of granular, unconsolidated materials such as sand and gravel, and from consolidated materials such as sandstone; (b) from the solution passages, caverns, and cleavage planes of sedimentary rocks such as limestone, slate, and shale; (c) from the fractures and fissures of igneous rocks; and (d) from combinations of these unconsolidated and consolidated geological formations. Groundwater sources, too, have an intake or catchment area, but the catch, or recharge, is by infiltration into soil openings rather than by runoff over its surface. The intake area may be nearby or a considerable distance away, especially when flow is confined within a water-bearing stratum or *aquifer* (from the Latin *aqua,* "water," and *ferre,* "to bear") underlying an impervious stratum or *aquiclude* (from the Latin *aqua,* "water," and *cludere,* "to shut" or "to close out").

The maximum yield of groundwater is directly proportional to the size of the intake area and to the difference between precipitation and the sum of evapotranspiration and storm runoff. Laterally, flow extends across the width of the aquifer; vertically, it is as deep as the zone of open pores and passages in Earth's crust and as shallow as the *groundwater table.* When the water surface rises and falls with seasonal changes in recharge, flow is unconfined or free, and the groundwater table slopes downward more or less parallel to the ground surface. Flow then moves at right angles to the water table contours. If a porous stratum dips beneath an impervious layer, flow is confined as in a pipe dropping below the hydraulic grade line. When this kind of aquifer is tapped, *artesian water* rises from it under pressure, in some geological situations, even in free-flowing fountains. In other geological formations, water is perched on a lens of impervious material above the true groundwater table.

Groundwater reaches daylight through springs: (a) when the ground surface drops sharply below the normal groundwater table (depression springs), (b) when a geological obstruction impounds soil water behind it and forces it to the surface (contact springs), and (c) when a fault in an impervious stratum lets artesian water escape from confinement (also contact springs). A cutoff wall carried to bedrock will hold back subsurface and surface flows behind an impounding dam and so put the full capacity of the catchment area to use unless there is lateral leakage through the sides of the reservoir or around the abutments of the dam.

The rate of flow through the substantially vertical cross-section of ground at right angles to the direction of flow is not great. Because of the high resistance of the normally narrow pores of the soil, the water moves forward only slowly, traveling about as far in a year as stream flow does in an hour. Natural rates of flow are seldom more than a few feet per hour (or meters per hour); nor are they less than a few feet per day (or meters per day) in aquifers delivering useful water supplies. However, if a well is sunk into the ground and the level of water in it is lowered by pumping, water is discharged into the well not only from the direction of natural flow but from all directions. That is why wells can be spaced many times their own diameter apart and yet intercept most of the water once escaping through the intervening space.

1.6.1 Springs

Springs are usually developed to capture the natural flow of an aquifer. In favorable circumstances their yield can be increased by driving collecting pipes or galleries, more or less horizontally, into the water-bearing formations that feed them. Pollution generally originates close to the point of capture. It is prevented (a) by excluding shallow seepage waters through encircling the spring with a watertight chamber penetrating a safe distance into the aquifer and (b) by diverting surface runoff away from the immediate vicinity. Some springs yield less than 1 gpm (3.78 L/min); a few yield more than 50 MGD (189 MLD). Some are perennial; others are periodically or seasonally intermittent.

1.6.2 Wells

Depending on the geological formations through which they pass and on their depth, wells are *dug, driven, bored,* or *drilled* into the ground. A well and its pumping equipment are shown in Fig. 1.3. Dug and driven wells are usually confined to soft ground, sand, and gravel at depths normally less than 100 ft (30 m). Hard ground and rock generally call for bored and drilled wells sunk to depths of hundreds and even thousands of feet. In well-watered regions successful wells of moderate depth and diameter yield 1–50 gpm (4–190 L/min) in hard rock and 50–500 gpm (190–1900 L/min) in coarse sand and gravel as well as coarse sandstone. Wells in deep aquifers may yield 100 gpm (400 L/min) or more.

Except in hard rock, particularly limestone, without sand or gravel cover, wells are generally not polluted by lateral seepage but by vertical entrance of pollution at or near the ground surface. Pollution is excluded by watertight casings or seals extending into the aquifer and at least 10 ft (3 m) below the ground surface, together with diversion of surface runoff from the well area and its protection against inundation by nearby streams.

1.6.3 Infiltration Galleries

Groundwater traveling toward streams or lakes from neighboring uplands can be intercepted by infiltration galleries laid more or less at right angles to the direction of flow and carrying entrant water to pumping stations. Water is drawn into more or less horizontal conduits from both sides, or the riverside is blanked off to exclude the often less satisfactory water seeping in from the river itself. Infiltration basins and trenches are similar in conception. They are, in essence, large, or long, shallow, open wells. Filter cribs built into alluvial deposits of streams intercept the underflow. Groundwater can also be collected from the driftways and slopes of mines, galleries driven into mountainsides specifically for this purpose, or abandoned mines. Some infiltration galleries yield as much as a 1 MGD/1000 ft (12.4 MLD/1000 m) of gallery. They are particularly useful in tapping aquifers of shallow depth or where deep saline waters are to be excluded.

Figure 1.3 A Well Providing about 1.5 MGD (5.68 MLD) of Water to Central Maui, Hawaii (*Source:* Courtesy of the Department of Water Supply, Maui County, Hawaii, USA)

1.6.4 Recharging Devices

As outlined earlier, the yield of groundwater works can be augmented or maintained at high level by water spreading or diffusion. The necessary structures are built close to the collecting works within the groundwater shed. Charging ditches or basins are filled with river or lake water by gravity or pumping. In the flooding method, water diverted from streams by check dams is led onto a suitable area of pervious soils. The applied waters soak into the ground and increase its natural flows. The incentive is either augmentation of a dwindling or inadequate supply or taking advantage of natural filtration as a means of water purification. Gathering a more uniformly cool water is also a consideration. Badly polluted surface water may be partially purified before it is introduced into the charging structure. Some diffusion galleries and wells return waters abstracted earlier from the ground for cooling and other purposes.

Groundwater collection works usually include pumps. To them, water flows from all or much of the well field either by gravity through deep-lying conduits or under negative pressure through suction mains. Individual pumping units are often used instead, especially when the water table lies at considerable depths.

Most natural groundwaters are clean, palatable, and cool. However, passage through some soils may make them unpalatable, unattractive, corrosive, or hard (soap-consuming). Their treatment must be varied according to needs.

To determine the yield of groundwater areas, the engineer must know the geology and the hydrology of the region. He can learn much from existing supplies in nearby areas, but his ultimate judgment must generally rest on the behavior of test wells.

Example 1.3 Determination of Aquifer Yield

Make a rough estimate of the yield of an aquifer 20 ft (6.10 m) deep through which water moves at a rate of 3 ft (0.91 m) a day, (1) if all of the groundwater laterally within 500 ft (152 m) of the well comes fully within its influence and (2) if a gallery 1000 ft (305 m) long collects water from both sides.

Solution 1 (US Customary System):

1) $20\text{ ft} \times 500\text{ ft} \times 2 \times 3\text{ ft/d} \times (7.5\text{ gal/ft}^3)/(1{,}440\text{ min/d}) = \mathbf{310\ gpm}$.
2) $20\text{ ft} \times 1{,}000\text{ ft} \times 2 \times 3\text{ ft/d} \times (7.5\text{ gal/ft}^3)/(1{,}000{,}000) = \mathbf{0.90\ MGD}$.

Solution 2 (SI System):

1) $(6.10\text{ m})(305\text{ m})(0.91\text{ m})(1000\text{ L/m}^3)(1/1{,}440\text{ min}) = 1{,}176\text{ L/min}$.
2) $2(6.10\text{ m})(305\text{ m})(0.91\text{ m})/\text{d} = 3397\text{ m}^3/\text{d} = 3{,}396{,}500\text{ L/d} = \mathbf{3.4\ MLD}$.

1.7 Purification Works

The quality of some waters from surface or ground sources is naturally satisfactory for all common uses. *Disinfection* may be the only required safeguard. Other waters contain objectionable substances that must be removed, reduced to tolerable limits, destroyed, or otherwise altered in character before the water is sent to the consumer. Impurities are acquired in the passage of water through the atmosphere, over the earth's surface, or through the pores of the earth. Their pollution is associated with man's activities, in particular, with his own use of water in household and industry and the return of spent water to natural water courses. Some of the *heavy metals* (lead, copper, zinc, and iron) come from the corrosion of metallic water pipes. Contamination of distribution systems through cross-connections with impure water supplies and through *backflow* in plumbing systems is another hazard. (Backflow permits water drawn into a fixture, tank, or similar device to flow back into the supply line by gravity or by siphonage.)

How to treat a given supply depends on its inherent traits and on accepted water quality standards. Municipal works must deliver water that is (a) hygienically safe, (b) aesthetically attractive and palatable, and (c) economically satisfactory for its intended uses. The most common classes of municipal water purification works and their principal functions are as follows:

1) *Filtration plants* remove objectionable color, turbidity, and bacteria, as well as other potentially harmful organisms by filtration through sand or other granular substances after necessary preparation of the water by coagulation and sedimentation (Fig. 1.4a).

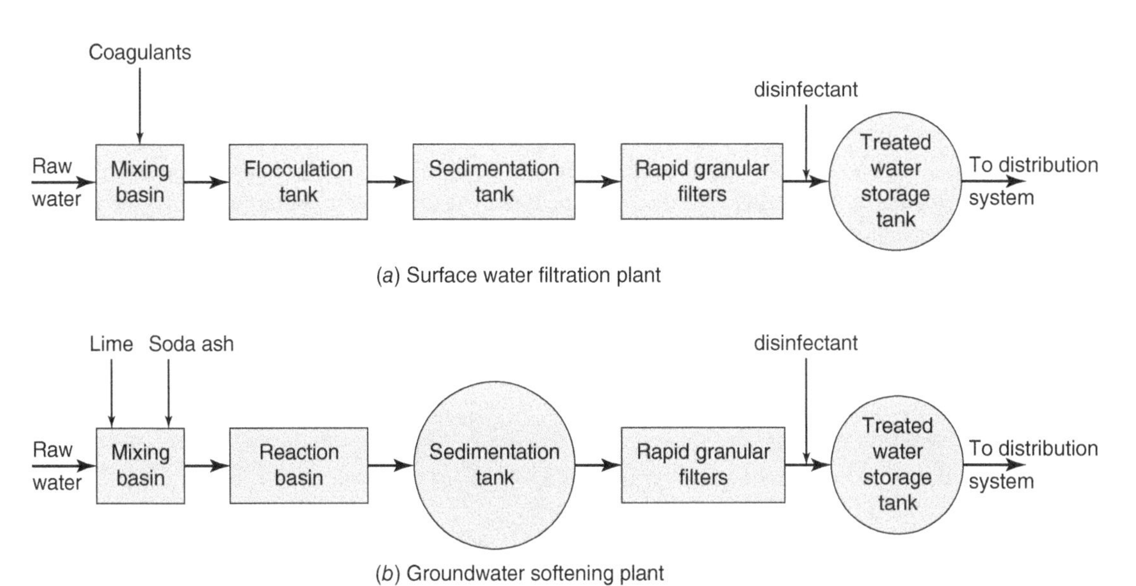

Figure 1.4 Common Types of Water Treatment Plants. (*Note:* A sedimentation tank may be replaced by a dissolved air flotation tank.)

2) *Iron and manganese treatment plants* remove excessive amounts of iron and manganese by oxidizing the dissolved metals and converting them into insoluble flocs removable by sedimentation and filtration.
3) *Softening plants* remove excessive amounts of scale-forming, soap-consuming ingredients, chiefly calcium and magnesium ions (a) by the addition of lime and soda ash, which precipitate calcium as a carbonate and magnesium as a hydrate (Fig. 1.4b), or (b) by passage of the water through cation exchange media that substitute sodium for calcium and magnesium ions and are themselves regenerated by brine.

Today most water supplies are either chlorinated or ozonated to ensure their disinfection. Lime or other chemicals are often added to reduce the corrosiveness of water to iron and other metals and so to preserve water quality during distribution and ensure a longer life for metallic pipes in particular. Odor- or taste-producing substances are adsorbed onto activated carbon, or destroyed by high doses of chlorine, chlorine dioxide, or other oxidants. Numerous other treatment methods serve special needs. The perspective of a water treatment plant in northern Portugal is shown in Fig. 1.5.

Figure 1.5 A Water Treatment Plant in Northern Portugal (*Source:* http://en.wikipedia.org/wiki/Image:Bragan%C3%A7a43.jpg)

Water purification plants must take into consideration the following design functions:

1) *Process design:* An understanding of unit operations that bring about the removal or modification of objectionable substances.
2) *Hydraulic design:* A knowledge of how water flows through the structures composing water purification plants: channels, pipes including perforated pipes, gates, measuring devices, basins, beds of sand and other granular materials, and pumps.
3) *Structural design:* A comprehension of the behavior of needed structures under load.
4) *Economic design:* An appreciation of treatment costs and associated benefits.

The following normally applicable requirements provide the reader with a concept of the sizing of principal structures:

1) *Mixing basins* hold a few minutes of flow.
2) *Flocculating and reaction basins* hold about half an hour's flow.
3) *Sedimentation basins* hold an hour or more of flow and are rated at about 0.50 gpm/ft^2 (20 L/min/m^2) of water surface area. It should be noted that dissolved air flotation (DAF) clarification with much higher clarification rate can now replace sedimentation clarification in modern water treatment plants.
4) *Slow sand filters* pass water at rates of about 3 MGD/acre (28 MLD/ha) in surface water filtration, stepping up to about 10 MGD/acre (94 MLD/ha) in groundwater treatment for iron and manganese removal or when they are preceded by roughing filters.
5) *Rapid filters* operate at rates of 125 MGD/acre or 2 gpm/ft^2 (1170 MLD/ha or 81 L/min/m^2), but rates run higher in modern works that include flocculating chambers.
6) *Coke tricklers* for aeration are rated at about 75 MGD/acre or 1.2 gpm/ft^2 (700 MLD/ha or 50 L/min/m^2).

Example 1.4 Determination of the Capacity of Water Treatment Plant Units
Estimate the capacity of the components of a rapid sand filtration plant (Fig. 1.4 top) that is to deliver 10 MGD or 6940 gpm (37.85 MLD or 26,268 L/min) of water to a city of 67,000 people.

Solution 1 (US Customary System):

1) Two mixing basins, H = 10 ft deep; number of mixing basins N = 2.
   a) Assumed detention period t = 2 min.
   b) Volume V = Qt/N = 6,940 × 2/2 = 6,940 gal = 928 ft^{3}each.
   c) Surface area A = V/H = 928/10 = 92.8 ft^2 = 0.785 D^2.
   d) Diameter $D = \sqrt{\frac{A \times 4}{\pi}} = \sqrt{\frac{92.8 \times 4}{\pi}} = \mathbf{10.9\ ft}$.

2) Two flocculating basins, H = 10 ft deep.
   a) Assumed detention period t = 30 min; number of flocculating basins N = 2.
   b) Volume V = Qt/N = 6,940 × 30/2 = 104,000 gal = 13,900 ft^3.
   c) Surface area A = V/H = (13,900 ft^3)/(10 ft) = **1,390 ft^2** each (such as 20 ft by 70 ft).

3) Two settling basins, H = 10 ft deep, but allow for 2 ft of sludge; number of settling basins N = 2.
   a) Assumed detention period t = 2 h.
   b) Effective volume V = Qt/N = 6,940 × 2 × 60/2 = 416,000 gal = 55,700 ft^3.
   c) Surface area A = V/H = 55,700/(10 – 2)ft = **6,960 ft^2** (such as 35 ft by 200 ft).
   d) Surface rating SR = Q/A = 6,940/6,960 = 1.0 gpm/ft^2.

4) Six rapid sand filters.
   a) Assumed rating SR = Q/A = 3 gpm/ft^2; number of filters N = 6.
   b) Area A = Q/(N × SR) = 6,940/(6 × 3) = **385 ft^2** (such as 15 ft by 26 ft).

Solution 2 (SI System):

1) Two mixing basins, H = 3.05 m deep; number of mixing basins N = 2.
   a) Assumed detention period t = 2 min .
   b) Volume V = Qt/N = (26,268 × 2)/2 = 26,268 L = 26.27 m^3 each.

c) Surface area $A = V/H = (26.27/3.048) = 8.62\ m^2$.

d) Diameter $D = \sqrt{\frac{A \times 4}{\pi}} = \sqrt{\frac{8.62 \times 4}{\pi}} = \mathbf{3.31\ m}$.

2) Two flocculating basins, $H = 3.05$ m deep.
   a) Assumed detention period $t = 30$ min; number of flocculating basins $N = 2$.
   b) Volume $V = Qt/N = (26{,}268\ L/\min \times 30\ \min)/2 = 394{,}020\ L = 394\ m^3$.
   c) Surface area $A = V/H = (394\ m^3)/(3.05\ m) = \mathbf{129.27\ m^2}$ each (such as $\mathbf{6.1\ m \times 21.3\ m}$).

3) Two settling basins, $H = 3.05$ m deep, but allow for 0.61 m of sludge; number of settling basins $N = 2$.
   a) Assumed detention period $t = 2$ h.
   b) Effective volume $V = Qt/N = (26{,}268\ L/\min \times 2 \times 60\ \min)/2 = 1{,}576{,}080\ L = 1{,}576\ m^3$.
   c) Surface area $A = V/H = (1{,}576\ m^3)/(3.05 - 0.61\ m) = \mathbf{646\ m^2}$ (such as 10.7 m by 61 m).
   d) Surface rating $SR = Q/A = (26{,}268\ L/min)/(646\ m^2) = 40.7\ L/\min/m^2$.

4) Six rapid sand filters.
   a) Assumed rating $SR = Q/A = 122.1\ L/\min/m^2$; number of filters $N = 6$.
   b) Area $A = Q/(N \times SR) = (26{,}268\ L/min)/(6 \times 122.1) = \mathbf{35.86\ m^2}$ (such as 4.6 m by 7.9 m).

1.8 Transmission Works

Supply conduits, or aqueducts, transport water from the source of supply to the community and so form the connecting link between collection works and distribution systems. Source location determines whether conduits are short or long, and whether transport is by gravity or pumping. Depending on topography and available materials, conduits are designed for open-channel or pressure flow. They may follow the hydraulic grade line as canals dug through the ground, flumes elevated above the ground, grade aqueducts laid in balanced cut and cover at the ground surface, and grade tunnels penetrating hills; or they may depart from the hydraulic grade line as pressure aqueducts laid in balanced cut and cover at the ground surface, pressure tunnels dipping beneath valleys or hills, and pipelines of fabricated materials following the ground surface, if necessary over hill and through dale, sometimes even rising above the hydraulic grade line. The 336-mile (541 km)-long Central Arizona Project aqueduct shown in Fig. 1.6 is the largest and most expensive aqueduct system ever constructed in the United States. The Colorado River aqueduct of the Metropolitan Water District of Southern California is 242 miles (389 km) long and includes 92 miles (148 km) of grade tunnel, 63 miles (101 km) of canal, 54 miles (87 km) of grade aqueduct, 29 miles (47 km) of inverted siphons, and 4 miles (6.4 km) of force main. The Delaware aqueduct of New York City comprises 85 miles (137 km) of pressure tunnel in three sections. Pressure tunnels 25 miles (40 km) in length supply the metropolitan districts of Boston and San Francisco. The supply conduits of Springfield, Massachusetts, are made of steel pipe and reinforced concrete pipe; those of Albany, New York, of cast-iron pipe (CIP).

Figure 1.6 Central Arizona Project Aqueduct in the USA (*Source:* http://en.wikipedia.org/wiki/Image:Arizona_cap_canal.jpg)

The size and shape of supply conduits are determined by hydraulic, structural, and economic considerations. Velocities of flow ordinarily lie between 3 and 5 ft/s (0.91 and 1.52 m/s). Requisite capacities depend on the inclusion and size of supporting *service* or *distributing reservoirs*. If these store enough water to (a) care for hourly variations in water consumption in excess of inflow, (b) deliver water needed to fight serious fires, and (c)

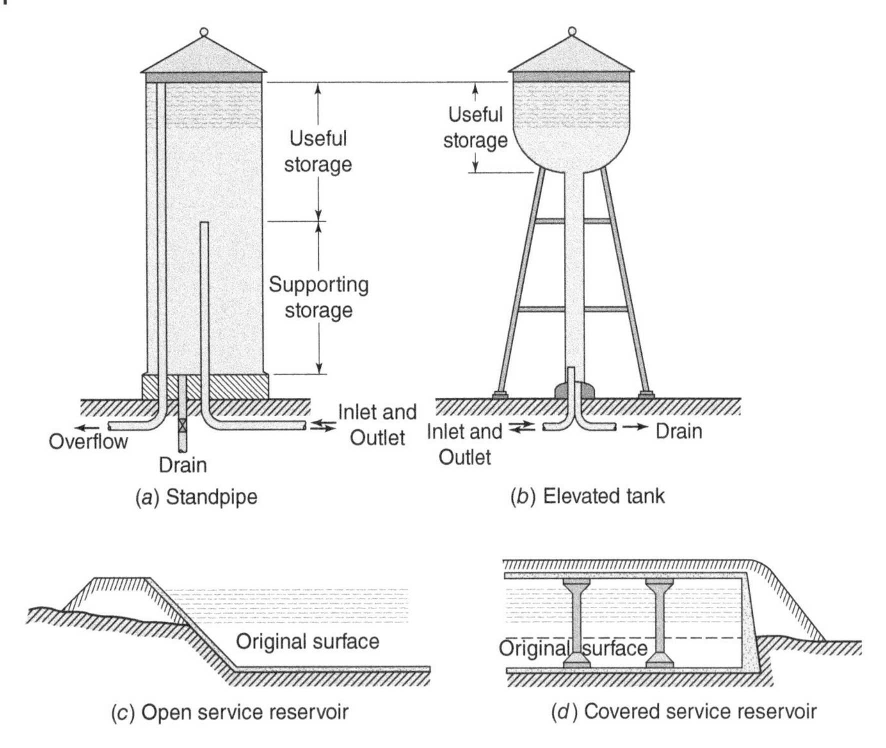

Figure 1.7 Four Types of Service, or Distribution, Reservoirs

permit incoming lines to be shut down for inspection and minor repairs, the supply conduits need to operate only at the maximum daily rate, about 50% in excess of the average daily rate. Ordinarily, required storage approximates a day's consumption. Distribution reservoirs are open or covered basins in balanced cut and fill, standpipes, or elevated tanks. Selection depends on size and location in particular reference to available elevations above the area served (Fig. 1.7). More than one reservoir may be needed in large systems. Open reservoirs are troubled by soot and dust falls, by algal growths, and, in seacoast cities, by sea gulls. Today, covered reservoirs are preferred.

Example 1.5 Estimation of the Size of a Water Conduit

Estimate roughly the size of a supply conduit leading to an adequate distributing reservoir serving (i) relatively small residential community of 10,000 people and (ii) a relatively large industrial community of 400,000 people.

The following are the design conditions specifically for a North America region:

a) Average daily water consumption for small communities with populations of 10,000 or less = 100 gpcd (378.5 Lpcd).
b) Average daily water consumption for communities with populations of greater than 10,000 = 150 gpcd (567.8 Lpcd).
c) Maximum daily water consumption is about 50% greater than average daily water consumption.
d) Design water velocity in the circular conduit when flowing full =4 ft/s = 1.22 m/s.

Solution 1 (US Customary System):

1) Average daily water consumption at (a) 100 gpcd and (b) 150 gpcd for the 10,000 people community and the 400,000 people community, respectively:
   i) $10{,}000 \times 100/1{,}000{,}000 =$ **1.0 MGD**.
   ii) $400{,}000 \times 150/1{,}000{,}000 =$ **60 MGD**.

2) Maximum daily water consumption is 50% greater than the average:
   i) $1.0 \times 1.5 = 1.5$ MGD $= 1.5 \times 1{,}000{,}000/(7.5 \times 24 \times 60 \times 60) =$ **2.32 ft^3/s**.
   ii) $60 \times 1.5 = 90$ MGD $= 90 \times 1{,}000{,}000/(7.5 \times 24 \times 60 \times 60) =$ **139 ft^3/s**.

3) Diameter of circular conduit flowing at 4 ft/s:
   i) Cross-sectional area A = Q/v = 2.32/4 = $\pi D^2/4 = 0.785\ D^2$.
      Diameter D = 0.833 ft = 10 in. for the small 10,000 people community.
   ii) Cross-sectional area A = Q/v = 139/4 = $\pi D^2/4 = 0.785\ D^2$.
      Diameter D = 6.667 ft = 80 in. for the large 400,000 people community.

Solution 2 (SI System):

1) Average daily water consumption =378.5 Lpcd for the 10,000 people community and average daily water consumption = 567.8 Lpcd for the 400,000 people community.
   i) 10,000 × 378.5/1,000,000 = **3.785 MLD**.
   ii) 400,000 × 567.8/1,000,000 = **227.1 MLD**.
2) Maximum daily water consumption is 50% greater than the average:
   i) (3.785 MLD) × 1.5 = 5.6775 MLD = 5,677.5 m^3/d = (5,677.5 m^3)/(1,440 × 60)s = 0.066 m^3/s = **66 L/s**.
   ii) (227.1 MLD) × 1.5 = 340.65 MLD = 340,650 m^3/d = (340,650 m^3)/(1,440 × 60)s = **3.94 m^3/s**.
3) Diameter of circular conduit flowing at 1.22 m/s:
   i) Cross-sectional area A = Q/v = (0.066 m^3/s)/(1.22 m/s) = 0.054 m^2 = 0.785 D^2.
      Diameter D = 0.26 m = 260 mm for the small 10,000 people community.
   ii) Cross-sectional area A = Q/v = (3.94 m^3/s)/(1.22 m/s) = 3.23 m^2 = 0.785 D^2.
      Diameter D = 2.03 m = 2030 mm for the large 400,000 people community.

1.9 Distribution Works

Supply conduits (Fig. 1.8) feed their waters into the distribution system that eventually serves each individual property—household, mercantile establishment, public building, or factory . Street plan, topography, and location of supply works and service storage establish the type of distribution system and its character of flow. In accordance with the street plan, two distribution patterns emerge: (a) a *branching pattern* on the outskirts of the community, in which ribbon development follows the primary arteries of roads and streets (Fig. 1.9a), and (b) a *gridiron pattern* within the built-up portions of the community where streets crisscross and water mains are interconnected (Fig. 1.9b).

Hydraulically, the gridiron system has the advantage of delivering water to any spot from more than one direction and of avoiding dead ends. Gridiron systems are strengthened by substituting for a central feeder a loop or belt of feeders that supplies water to the *congested,* or *high-value,* district from at least two directions. This more or less doubles the delivery of the grid. In large systems feeders are constructed as pressure tunnels, pressure aqueducts, steel pipes, or reinforced concrete pipes. In smaller communities the entire distribution system may consist of cast iron pipes (CIP)s. Cast iron is, indeed, the most common material for water mains, but asbestos-cement, in general, and plastics, in the case of small supplies, are also important.

Figure 1.8 A Pipeline in the Goldfields Water Supply Scheme, Perth, Australia (*Source:* http://en.wikipedia.org/wiki/Image:Goldfields_Pipeline_SMC.JPG)

1.9.1 High and Low Services

Sections of the community too high to be supplied directly from the principal, or *low-service,* works are generally incorporated into separate distribution systems with independent piping and service storage. The resulting *high services* are normally fed by pumps that take water from the main supply and boost its pressure as required. Areas varying widely in elevation may be formed into intermediate districts or zones. Gated connections

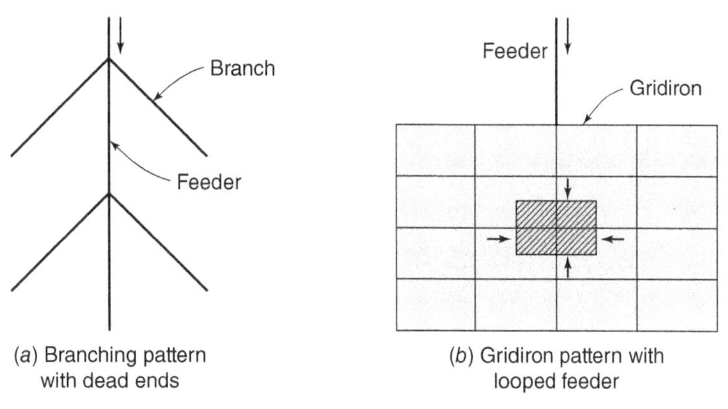

Figure 1.9 Patterns of Water Distribution Systems

between the different systems are opened by hand during emergencies or go into operation automatically by means of pressure-regulating valves.

1.9.2 Fire Supplies

Before the days of high-capacity, high-pressure, motorized fire engines, conflagrations in the congested central, or *high-value,* district of some large cities were fought through independent high-pressure systems of pipes and hydrants. Taking water from the public supply and boosting its pressure by pumps in power stations whenever an alarm was rung in, these systems performed well. For extreme emergencies, rigorously protected connections usually led to independent sources of water: rivers, lakes, or tidal estuaries. Large industrial establishments, with heavy investments in plant, equipment, raw materials, and finished products, concentrated in a small area, are generally equipped with high-pressure fire supplies and distribution networks of their own. Because such supplies may be drawn from sources of questionable quality, some regulatory agencies enforce rigid separation of private fire supplies and public systems. Others prescribe protected cross-connections that are regularly inspected for tightness. Ground-level storage and pumping are less advantageous.

1.9.3 Pressures

In normal municipal practice, pressures of 60–75 psig (416–520 kPa) are maintained in business blocks and 40 psig (278 kPa) in residential areas. Higher pressures, such as 100 psig (694 kPa) or more, delivering adequate amounts of water for firefighting through hoses attached directly to fire *hydrants* are no longer important. Instead, modern motor pumpers can discharge thousands of gallons per minute at even greater pressures. Moreover, low operating pressures make for low *leakage* from mains and reduce the amount of water that is *unaccounted* for. To supply their upper stories, tall buildings boost water to tanks at various elevations and on their roofs or in towers. In individual industrial complexes, the water pressure may be raised during fires by fixed installations of fire pumps.

1.9.4 Capacity

The capacity of distribution systems is dictated by domestic, industrial, and other normal water uses and by the *standby* or *ready-to-serve* requirements for firefighting. Pipes should be able to carry the maximum *coincident* draft at velocities that do not produce high pressure drops and water hammer. Velocities of 2–4 ft/s (0.60–1.2 m/s) and minimum pipe diameters of 6 in. (150 mm) are common in North American municipalities.

1.9.5 Service to Premises

Water reaches individual premises from the street main through one or more service pipes tapping the distribution system. The building supply between the public main and the take-offs to the various plumbing fixtures or other points of water use is illustrated in Fig. 1.10. Small services are made of cement-lined iron or steel, brass of varying copper content,

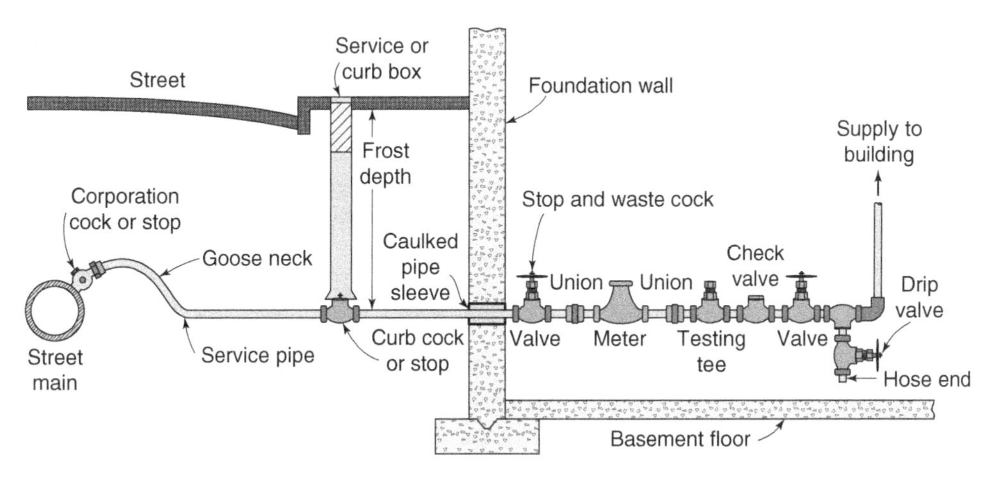

Figure 1.10 Service Pipe, Fittings, and Accessories. There are many possible modifications, both inside and outside the building. In many instances, the meter is conveniently placed in a vault outside the building

admiralty metal, copper, and plastics such as polyethylene (PE), high-density polyethylene (HDPE), or polyvinyl chloride (PVC). Because lead and lead-lined pipes may corrode and release lead to the water, they are no longer installed afresh. For large services, coated or lined CIP is often employed. For dwellings and similar buildings, the minimum desirable size of service is $^3/_4$ in. (19 mm). *Pipe-tapping machines* connect services to the main without shutting off the water. They also make larger connections within water distribution systems.

1.10 Water Systems Management

Construction of water supplies from the ground up, or their improvement and extension, progresses from preliminary investigations or planning through financing, design, and construction to operation, maintenance, and repair. Political and financial procedures are involved, as well as engineering.

1.10.1 Municipal Supplies

The cost of public water supplies in the United States provides the reader with some concept of the magnitude of engineering activity and responsibility associated with their design and construction. Per capita investment in physical plant depends on many factors: nature, proximity, and abundance of suitable water sources; need for water treatment; availability and price of labor and materials; size and construction conditions of the system; habits of the people; and characteristics of the areas served. Wide differences in these factors make for much variation in initial costs. For communities in excess of 10,000 population, replacement costs in North America lie in the vicinity of $1860 per capita (for 2022 price levels; for other years multiply by the ratio of an applicable *utilities price index;* see Appendix 16), with much of the investment in small communities chargeable to fire protection.

Of the various system components, collection and transportation works cost about a fourth, distribution works slightly less than a half, purification and pumping works about a tenth, and service lines and meters nearly a sixth of the total. The initial 2022 cost of conventional water filtration plants is about $12,400,000 per MGD ($3,280,000 per MLD) capacity, varying with plant size as the two-thirds power of the capacity. The 2022 cost of water treatment, excluding fixed charges, lies in the vicinity of $521 per MG ($138 per ML), varying with plant output capacity inversely as the two-fifths power of the daily production. Including interest and depreciation as well as charges against operation and maintenance, water costs $372–$3,720 per million gallons ($98.3–$983 per million liters) and is charged for accordingly. As one of our most prized commodities, water is nevertheless remarkably cheap—as low as 15 cents a ton delivered to the premises of large consumers and as little as 30 cents a ton to the taps of small consumers.

Example 1.6 Estimation of Waterworks Cost

Roughly, what is the replacement cost of a conventional filtration plant and other waterworks for a city of 100,000 people and what is the average plant flow?

The following conditions are assumed:

a) A per capita first cost of \$1860 in 2022.
b) Thirty percent of the first cost is to be invested in the collection works, 10% in the purification works, and 60% in the distribution works.
c) A water consumption rate of 150 gpcd (568 Lpcd) for the city.

Solution 1 (US Customary System):

1) Assuming a per capita cost of \$1860, the total first cost is \$1860 × 100,000 = \$186,000,000.
2) Assuming that 30% of this amount is invested in the collection works, 10% in the purification works, and 60% in the distribution works, the breakdown is as follows:

- Collection works = 0.3 × \$186,000,000 = \$55,800,000
- Purification works = 0.1 × \$186,000,000 = \$18,600,000
- Distribution works 0.6 × \$186,000,000 = \$111,600,000.

3) Assuming a water consumption rate of 150 gpcd, the total water consumption of the city is 150 × 100,000 gpd = 15 MGD.
4) Per MGD cost = \$186,000,000/(15 MGD) = \$12.4 × 10^6 per MGD.

Solution 2 (SI System):

1) Assuming a per capita cost of \$1860, the total first cost is \$1860 × 100,000 = \$186,000,000.
2) Assuming that 30% of this amount is invested in the collection works, 10% in the purification works, and 60% in the distribution works, the breakdown is as follows:

- Collection works = 0.3 × \$186,000,000 = \$55,800,000
- Purification works = 0.1 × \$186,000,000 = \$18,600,000
- Distribution works 0.6 × \$186,000,000 = \$111,600,000.

3) Assuming a water consumption of 567.75 Lpcd, the total water consumption of the city is (567.75 Lpcd) × 100,000 people = 56.775 MLD.
4) Per MLD cost = \$186,000,000/(56.775 MLD) = \$3.276 × 10^6 per MLD.

1.10.2 Individual Small Supplies

The term *individual* describes those situations in which the needs and amenities of water supply and wastewater disposal are normally satisfied by relatively small and compact systems individually owned, developed, and operated, and kept within the property lines of the owner. Normally, this implies construction of wanted or required systems through individual rather than community effort. But there have been developments for villages and communities with scattered buildings in which local government has taken the initiative and assumed responsibility for construction and care of individualized systems. Property owners, as well as the community, then enjoy the benefits of adequate planning, design, construction, management, and supervision. Otherwise, unfortunately, necessary works are rarely designed by qualified engineers and often end up not satisfying their purposes, both in a sanitary and an economic sense.

Reasonably good results can be obtained (a) if engineering departments of central health authorities publish manuals of design, construction, and operation that fit local conditions, and (b) if they give needed advice and supervision, as well as provide for regulation. Nevertheless, villages and fringe areas are best served, in the long run, by the extension of central water lines and sewers, or by incorporation of *water* and *sewer districts* comprising more than a single unit of local government.

1.11 Individual Water Systems

Because of the natural purifying capacity and protection of the soil, individual and *rural* water supplies are generally drawn from springs, infiltration galleries, and wells. Where groundwater is highly mineralized or unavailable, rainwater is next

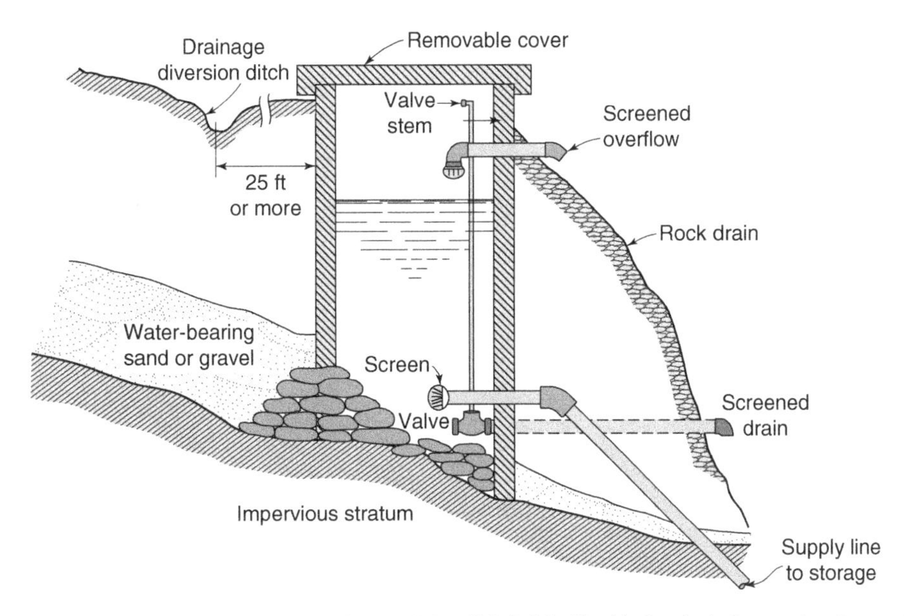

Figure 1.11 Water Supply from Spring (After U.S. Public Health Service). Conversion factor: 1 ft = 0.3048 m

best in general safety and quality. Only in uninhabited and well-protected upland areas should ponds and streams be tapped without purifying the waters drawn.

Some of the safeguards for groundwater works are illustrated in Fig. 1.11–1.13. They share the following features in common:

1) Diversion of surface water from intake structures
2) Drainage of overflow or spillage waters away from intake structures
3) Water tightness of intake works for at least 10 ft (3 m) below the ground surface and, if necessary, until the aquifer is reached
4) Prevention of backflow into intakes; where there is no electric power, water is pumped by hand, wind, water, or gasoline engines.

Individual and rural water supplies are not without their purification problems. Gravity and pressure filters are employed to improve waters of doubtful purity, and zeolite softeners and other ion-exchange units are used for the removal of unwanted hardness. Iron-bearing groundwaters that issue from their source sparklingly clear but become rusty on exposure to air (by oxidation and precipitation of iron) are best treated in manganese cation exchange units. Hexametaphosphates may keep iron from precipitating, but this requires skillful management. It may be advisable to seek an iron-free source instead.

Some soft groundwaters containing much carbon dioxide are highly corrosive. Passage through marble or limestone chips takes calcium into solution and reduces

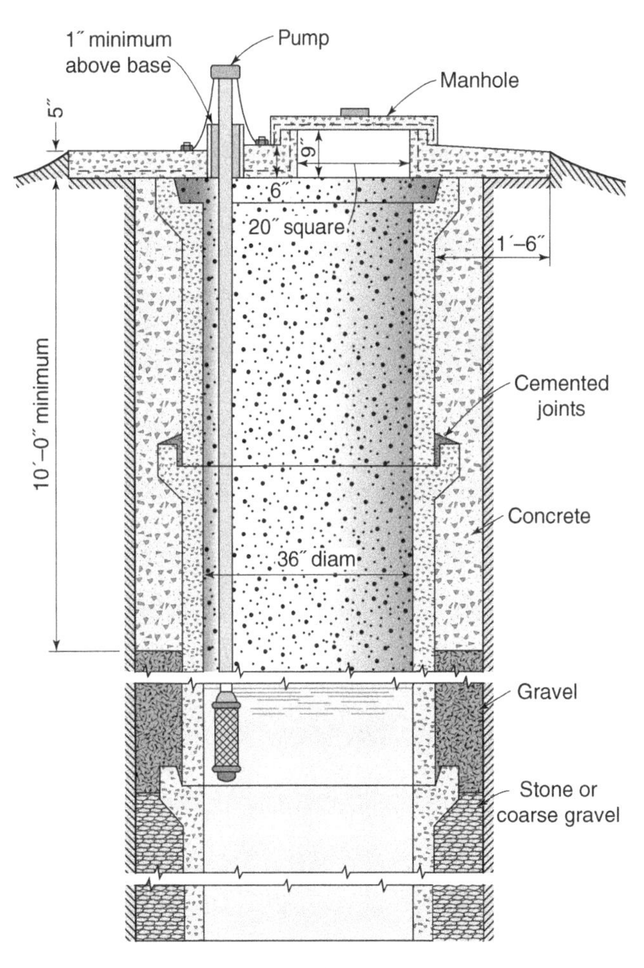

Figure 1.12 Water Supply from Dug Well (After U.S. Department of Agriculture). Conversion factors: 1′ = 1 ft = 0.3048 m; 1″ = 1 in. = 2.54 cm

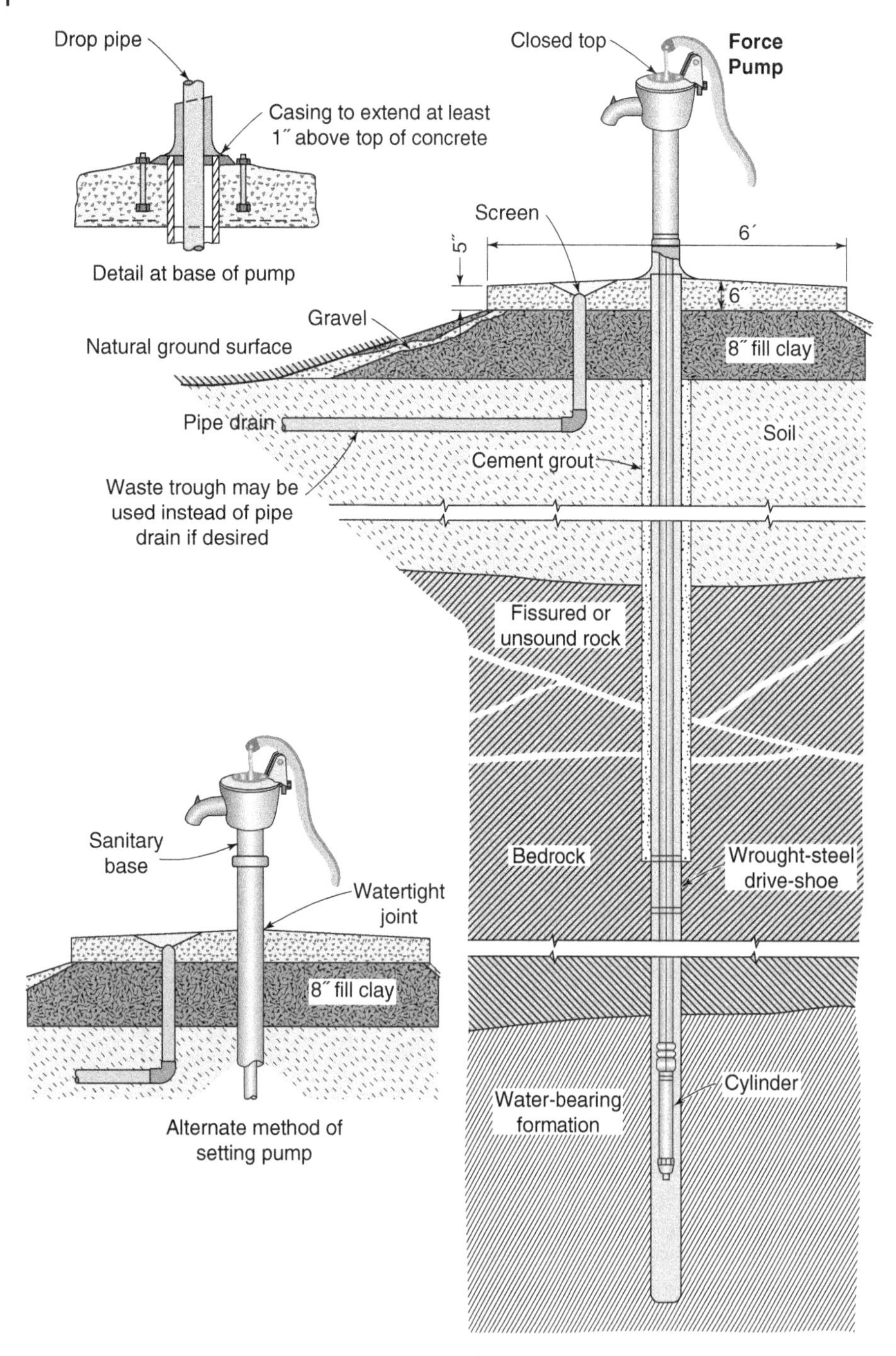

Figure 1.13 Water Supply from Driven Well (After Virginia State Department of Health). Conversion factors: 1′ = 1 ft = 0.3048 m; 1″ = 1 in. = 2.54 cm

the carbon dioxide proportionately. Hardness is increased, but corrosiveness is decreased. For the chlorination of polluted rural supplies, there are solution-feed dosing devices that proportion the amount of added chlorine to flow. Instead, the householder may prefer to boil his drinking and culinary water. Investment in an inherently safe and satisfactory supply, however, is usually wisest in the long run.

Problems/Questions

1.1 What is the stream flow in MGD (MLD) for a catchment area of 80 mile^2 (207.2 km^2) where rainfall rate is 45 in./year (114.3 cm/year) and evaporation rate is 20 in./year (50.8 cm/year)?

1.2 A city is served by a raw water reservoir that has a water surface of 5.8 mile^2 (15.02 km^2) and an average effective depth of 18 ft (5.49 m). Determine the water storage volume.

1.3 The population of a city is 400,000, and the average daily per capita water demand is 150 gpcd (567.75 Lpcd). Determine the city's average daily water demand.

1.4 How many days of draft can a raw water reservoir support for a city of 400,000 people? The reservoir has a water surface of 5.8 mile^2 = 15.02 km^{2}and an average effective depth of 18 ft = 5.49 m.

1.5 What percentage of mean annual runoff is to be consumed by a city of 400,000 people in an area with (a) rainfall rate = 45 in./year = 114.3 cm/year, (b) evaporation rate = 20 in./year = 50.8 cm/year, and (c) watershed catchment or drainage area = 80 mile^2 = 207.2 km^2?

1.6 Determine the net yield and water storage volume of a rainwater system assuming that (a) the net yield of a rain collection facility approximates two-thirds of its gross yield, (b) the mean annual rainfall = 25 in./year = 63.5 cm/year, (c) the mean annual evaporation rate is 8 in./year = 20.32 cm/year, (d) the rain collection roof area equals 3,200 ft^2 = 297.28 m^2, and (e) the water storage volume equals 50% of annual net yield.

1.7 Determine the storage volume of a new raw water reservoir in accordance with the following given technical information: (a) city population = 400,000, (b) water consumption = 150 gpcd = 568 Lpcd, (c) watershed or catchment area = 80 mile^2 = 207.2 km^2, (d) rainfall rate = 45 in./year = 114.3 cm/year, (e) evaporation rate = 20 in./year = 50.8 cm/year, (f) minimum reservoir volume = 50% annual net yield, or half of a year's water supply, whichever is greater, and (g) 75% water resources development.

1.8 Determine the number of people who can be sustainably supported by a watershed under the following conditions: (a) watershed area = 80 mile^2 = 207.2 km^2, (b) annual rainfall rate = 45 in. /year = 114.3 cm/year, (c) annual evaporation rate = 20 in. /year = 50.8 cm/year, (d) water resource development = 75 %, (e) raw water reservoir volume to store 50% net annual yield or provide half of a year's water supply, whichever is higher = 13 BG = 49.205 BL, and (f) water consumption rate = 150 gpcd = 568 Lpcd.

1.9 Determine the number of people who can be adequately supported by a watershed under the following conditions: (a) watershed area = 80 mile^2 = 207.2 km^2, (b) water supply system with no reservoir for water storage, (c) low-water flow = 0.1 ft^3/s = 64,600 gpd per mile^2 = 2.83 L/s = 0.00283 m^3/s, and (d) water consumption rate = 150 gpcd = 568 Lpcd.

1.10 Make a rough estimate of the groundwater movement velocity (ft/day or m/day), if (a) all of the groundwater laterally within 400 ft (122 m) of the well comes fully within its influence and (b) the yield of the aquifer is 258 gpm (gallon per minute) = 976.53 L/ min = 16.28 L/s. The aquifer through which the groundwater moves is 25 ft (7.62 m) deep.

1.11 Estimate the surface area (ft^2, or m^2) of a slow sand filter that is to deliver water to a village of 1000 people assuming that (a) the average daily water demand = 100 gpcd = 378.5 Lpcd, (b) the slow sand filter's filtration rate is 3 million gallons per acre per day (MGAD) =3 MGD/acre = 28.08 MLD/ha = 2, 808 MLD/km^2, and (c) two slow sand filters are required. Each filter is able to treat the full water flow and one of the two filters is a standby unit.

1.12 Estimate roughly the size of a water supply pipe leading to a water-distributing reservoir serving a small village of 2000 people assuming that (a) the water consumption rate is 100 gpcd = 378.5 Lpcd and (b) water velocity in the pipe = 3 ft/s = 0.91 m/s.

1.13 Determine the diameter of a water main to serve a residential area, assuming (a) an average water demand of 150 gpcd = 568 Lpcd, (b) population = 30,000, (c) fire flow requirement = 500 gpm = 2,082 L/ min = 32 L/s, and (d) recommended water velocity = 3.5 ft/s = 1.07 m/s.

1.14 Roughly, what is the replacement cost of the waterworks of a city of 10,000 people?

Bibliography

Al-Dhowalia, K., and Shammas, N. K., Leak detection and quantification of losses in a water network, *Int. J. Water Resour. Develop.*, *7*, 1, 30–38, 1991.

Department of Water Supply Web Site, County of Maui, Hawaii, http://mauiwater.org/, 2010.

Fair, G. M., Geyer, J. C., and Okun, D. A., Water and Wastewater Engineering, in *Water Supply and Wastewater Removal*, vol. *1*, John Wiley & Sons, New York, USA, 1966.

Fair, G. M., Geyer, J. C., and Okun, D. A., *Elements of Water Supply and Wastewater Disposal*, John Wiley & Sons, New York, USA, 1971.

National Resources Conservation Service, U.S. Department of Agriculture, Water Supply Forecasting, http://www.wcc.nrcs.usda.gov/wsf/, 2009.

Orlob, G. T., and Lindorf, M. R., Cost of water treatment in California, *J. Am. Water Works Ass.*, *50*, 45, 1958.

Pankivskyi, Y. I., and Oshurkevych-Pankivska, O. Y., Water supply and sewerage systems, in *Environmental Science, Technology, Engineering, and Mathematics*, Lenox Institute Press, MA, USA, 261 pp, 2023. https://doi.org/10.17613/x5qn-d460.

Quraishi, A., Shammas, N. K., and Kadi, H., Analysis of per capita household water demand for the City of Riyadh, Saudi Arabia, *Arabian J. Sci. Technol.*, *15*, 4, 539–552, 1990.

Shammas, N. K., Wastewater Management and Reuse in Housing Projects, *Water Reuse Symposium IV, Implementing Water Reuse*, August 2–7, 1987, AWWA Research Foundation, Denver, CO, pp. 1363–1378.

Shammas, N. K., and Al-Dhowalia, K., Effect of pressure on leakage rate in water distribution networks, *J. Eng. Sci.*, *5*, 2, 155–312, 1993.

Shammas, N. K., and El-Rehaili, A., *Wastewater Engineering, Textbook on Wastewater Treatment Works and Maintenance of Sewers and Pumping Stations*, General Directorate of Technical Education and Professional Training, Institute of Technical Superintendents, Riyadh, Kingdom of Saudi Arabia, 1988.

U. S. Army Corps of Engineers, *Yearly Average Cost Index for Utilities, Civil Works Construction Cost Index System Manual, 110-2-1304*, U.S. Army Corps of Engineers, Washington, DC, 2020, March 2020.

U.S. Environmental Protection Agency, *Drinking Water and Ground Water Statistics for 2008*, EPA 816-K-08-004, Washington, DC, USA, December 2008.

U.S. Environmental Protection Agency, Drinking Water Data, 2010. Drinking water data tables, Available at http://www.epa.gov/safewater/data/getdata.html.

Wang, L. K., Hung, Y. T., and Shammas, N. K., *Physicochemical Treatment Processes*, Humana Press, Totowa, NJ, USA, 2005, p. 723.

Wang, L. K., Hung, Y. T., and Shammas, N. K., *Advanced Physicochemical Treatment Processes*, Humana Press, Totowa, NJ, USA, 2006, p. 690.

Wang, L. K., Hung, Y. T., and Shammas, N. K., *Advanced Physicochemical Treatment Technologies*, Humana Press, Totowa, NJ, USA, 2007, p. 710.

Wang, L. K., Wang, M. H. S., Shammas, N. K., and Aulenbach, D. B., *Environmental Flotation Engineering*, Springer Nature Switzerland, 433 pp., 2021.

World Health Organization, 2004, *Water, Sanitation and Health, Facts and Figures*, http://www.who.int/water_sanitation_health/publications/facts2004/en/index.html.

World Water Council Web Site, 2010. http://www.worldwatercouncil.org/index.php?id=23.

2

Water Sources: Surface Water

Figure 2.1 shows where water is and how it is distributed on Earth. The bar on the left shows where the water on Earth exists; about 97% of all water is in the oceans. The middle bar shows the distribution of that 3% of all Earth's water that is freshwater. The majority, about 69%, is locked up in glaciers and ice caps, mainly in Greenland and Antarctica. You might be surprised that of the remaining freshwater, almost all of it is below our feet, as groundwater. No matter where on Earth one is standing, chances are that, at some depth, the ground below is saturated with water. Of all the freshwater on Earth, only about 0.3% is contained in rivers and lakes—yet rivers and lakes are not only the water we are most familiar with, they are also where most of the water we use in our everyday lives exists.

For a detailed explanation of where Earth's water is, look at the data in Table 2.1. Notice how, of the world's total water supply of about 326 million mile^3 (1,360 million km^3), more than 97% of it is saline. Also, of the total freshwater, about 7 million mile^3 (29 million km^3) is locked up in ice and glaciers. Another 30% of freshwater is in the ground. Thus, surface water sources only constitute about 30,300 mile^3 (126,300 km^3), which is about 0.009% of total water.

What interconnects groundwater and surface water is the *water cycle* (Fig. 2.2). The water cycle has no starting point, but we will begin in the oceans, because that is where most of Earth's water exists. The sun, which drives the water cycle, heats water in the oceans. Some of it evaporates as vapor into the air. Ice and snow can sublimate directly into water vapor. Rising air currents take the vapor up into the atmosphere, along with water from *evapotranspiration*, which is water transpired from plants and evaporated from the soil. The vapor rises into the air where cooler temperatures cause it to condense into clouds. Air currents move clouds around the globe; cloud particles collide, grow, and fall out of the sky as precipitation. Some precipitation falls as snow and can accumulate as ice caps and glaciers, which can store frozen water for thousands of years. Snowpacks in warmer climates often thaw and melt when spring arrives, and the melted water flows overland as snowmelt. Most precipitation falls back into the oceans or onto land, where, due to gravity, the precipitation flows over the ground as *surface runoff*. A portion of runoff enters rivers in valleys in the landscape, with stream flow moving water toward the oceans. Runoff and groundwater seepage accumulate and are stored as freshwater in lakes.

Not all runoff flows into rivers, however. Much of it soaks into the ground as *infiltration*. Some water infiltrates deep into the ground and replenishes *aquifers* (saturated subsurface formations), which store huge amounts of freshwater for long periods of time. Some infiltration stays close to the land surface and can seep back into surface water bodies (and the ocean) as groundwater discharge, and some groundwater finds openings in the land surface and emerges as freshwater springs. Over time, however, all of this water keeps moving; some reenters the oceans, where the water cycle "ends"—or, where it "begins."

2.1 Sources of Surface Water

In the United States, by far the largest volumes of municipal water are collected from surface sources. Possible yields vary directly with the size of the catchment area, or watershed, and with the difference between the amount of water falling on it and the amount lost by evapotranspiration. The significance of these relations to water supply is illustrated in Fig. 1.1. Where surface water and groundwater sheds do not coincide, some groundwater may enter from neighboring catchment areas or escape to them.

Communities on or near streams, ponds, or lakes may withdraw their supplies by *continuous draft* if stream flow and pond or lake capacity are high enough at all seasons of the year to furnish requisite water volumes. Collecting works include

Water and Wastewater Engineering: Hydraulics, Hydrology and Management, Volume 1, Fourth Edition.
Lawrence K. Wang, Mu-Hao Sung Wang, and Nazih K. Shammas.

Companion website: www.wiley.com/go/Wang/Waterandwastewater4e

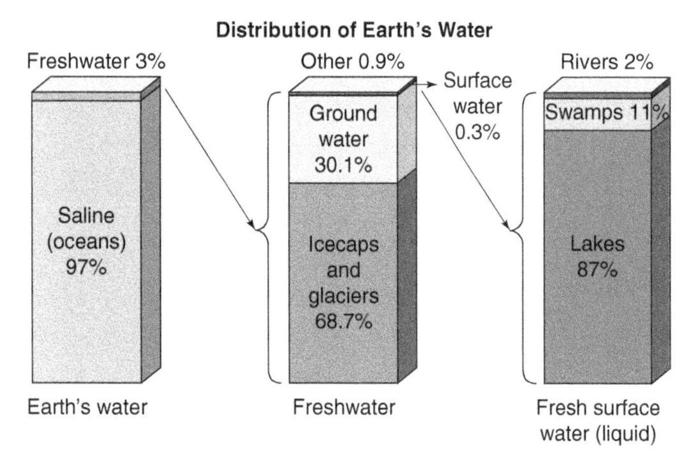

Figure 2.1 Distribution of Earth's Water (*Source:* Courtesy of USGS)

Table 2.1 Estimated Water Distribution on Earth

	Surface Area (mi²)	Volume (mi³)	% of Total
Salt water			
The oceans	139,500,000	317,000,000	97.2
Inland seas and saline lakes	270,000	25,000	0.008
Freshwater			
Freshwater lakes	330,000	30,000	0.009
All rivers (average level)	–	300	0.0001
Antarctic ice cap	6,000,000	6,300,000	1.9
Arctic ice cap and glaciers	900,000	680,000	0.21
Water in the atmosphere	197,000,000	3,100	0.001
Groundwater within half a mile from surface	–	1,000,000	0.31
Deep-lying groundwater	–	1,000,000	0.31
Total (rounded)	–	326,000,000	100.00

Conversion factors: 1 mi^2 = 2.59 km^2; 1 mi^3 = 4.17 km^3.

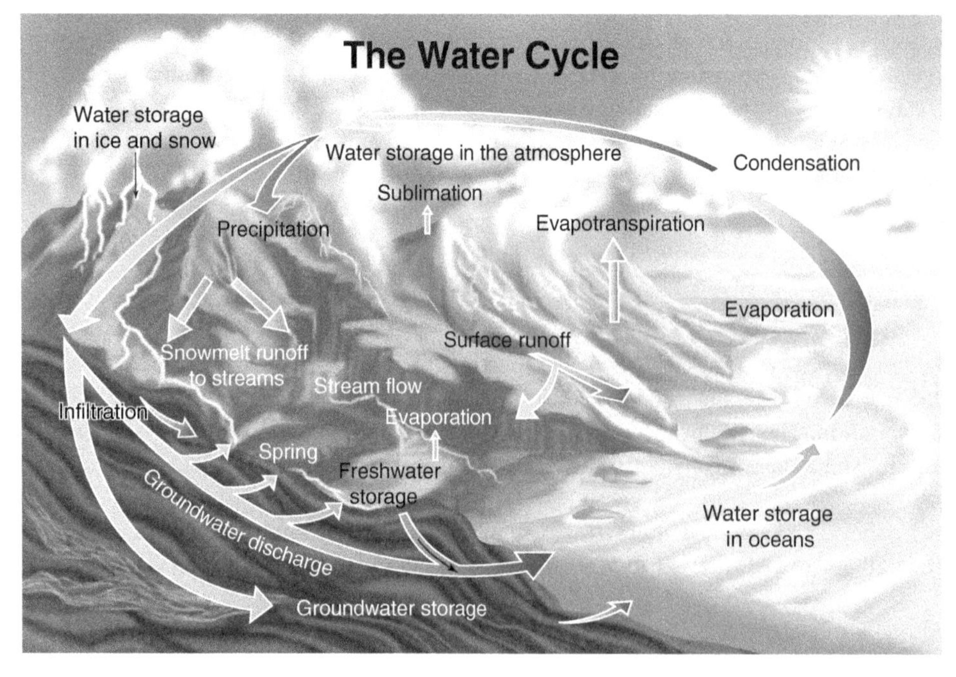

Figure 2.2 The Water Cycle (*Source:* Courtesy of USGS)

ordinarily (a) an intake crib, gatehouse, or tower; (b) an intake conduit; and, (c) in many places, a pumping station. On small streams serving communities of moderate size, intake or diversion dams can create a sufficient depth of water to submerge the intake pipe and protect it against ice. From intakes close to the community the water must generally be lifted to purification works and thence to the distribution system (Fig. 2.3).

Because most large streams are polluted by wastes from upstream communities and industries, their waters must be purified before use. Cities on large lakes must usually guard their supplies against their own and their neighbors' wastewaters and spent industrial process waters by moving their intakes far away from shore and purifying both their water and wastewater. Diversion of wastewaters and other plant nutrients from lakes will retard lake eutrophication.

Low stream flows are left untouched when they are wanted for other valley purposes or are too highly polluted for reasonable use. Only clean floodwaters are then diverted into reservoirs constructed in meadowlands adjacent to the stream or otherwise conveniently available. The amount of water so stored must supply demands during seasons of unavailable stream flow. If draft is confined to a quarter year, for example, the reservoir must hold at least three-fourths of a community's annual supply. In spite of its selection and long storage, the water may still have to be purified.

In search of clean water and water that can be brought and distributed to the community by gravity, engineers have developed supplies from upland streams. Most of them are tapped near their source in high and sparsely settled regions. To be of use, their annual discharge must equal or exceed the demands of the community they serve for a reasonable number of years in the future. Because their dry-season flows generally fall short of concurrent municipal requirements, their floodwaters must usually be stored in sufficient volume to ensure an adequate supply. Necessary reservoirs are impounded by throwing dams across the stream valley (Fig. 2.4). In this way, amounts up to about 70% or 80% of the mean annual flow can be utilized. The area draining to impounded reservoirs is known as the *catchment area* or *watershed*. Its economic development depends on the value of water in the region, but it is a function, too, of runoff and its variation, accessibility of catchment

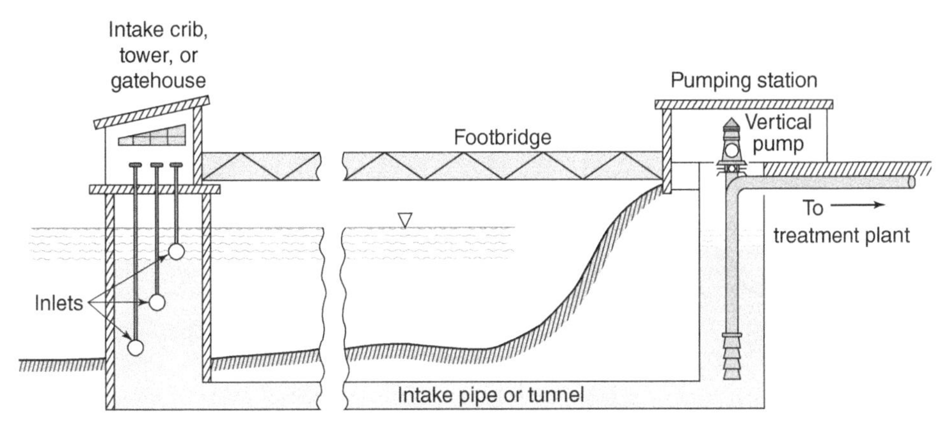

Figure 2.3 Continuous Draft of Water from Large Lakes and Streams

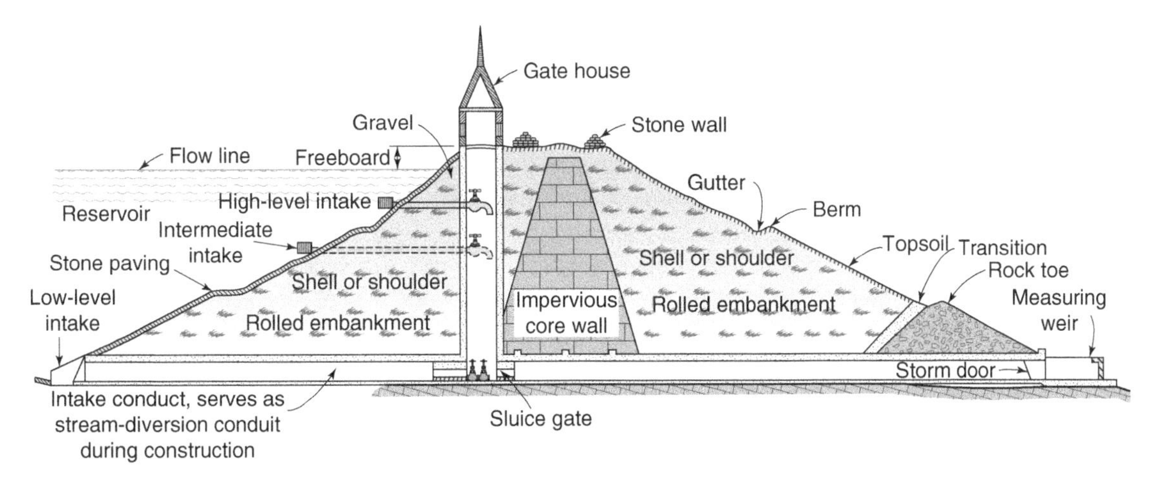

Figure 2.4 Dam and Intake Towers for an Impounded Surface Water Supply

Figure 2.5 Concrete Spillway in Dam (*Source:* Courtesy of the Department of Water Supply, Maui County, Hawaii)

Figure 2.6 Hoover Dam, Clark County, NV (*Source:* Courtesy of the National Resources Conservation Service and USDA)

areas, interference with existing water rights, and costs of construction. Allowances must be made for evaporation from new water surfaces generated by the impoundage, and often, too, for the release of agreed-on flows to the valley below the dam (compensating water). Increased ground storage in the flooded area and the gradual diminution of reservoir volumes by siltation must also be considered.

Intake structures are incorporated in impounding dams or kept separate. Other important components of impounding reservoirs are (a) spillways (Fig. 2.5) safely passing floods in excess of reservoir capacity and (b) diversion conduits safely carrying the stream past the construction site until the reservoir has been completed and its spillway can go into action. Analysis of flood records enters into the design of these ancillary structures. Some impounded supplies are sufficiently safe, attractive, and palatable to be used without treatment other than protective disinfection. However, it may be necessary to remove (a) high *color* imparted to the stored water by the decomposition of organic matter in swamps and on the flooded valley floor, (b) *odors and tastes* generated in the decomposition or growth of algae, especially during the first years after filling, and (c) *turbidity* (finely divided clay or silt) carried into streams or reservoirs by surface wash, wave action, or bank erosion. Recreational uses of watersheds and reservoirs may endanger the water's safety and call for treatment of the flows withdrawn from storage.

Much of the water entering streams, ponds, lakes, and reservoirs in times of drought, or when precipitation is frozen, is seepage from the soil. Nevertheless, it is classified as surface runoff rather than groundwater. Water seeps *from* the ground when surface streams are low and *to* the ground when surface streams are high. Release of water from ground storage or from accumulations of snow in high mountains is a determining factor in the yield of some catchment areas. Although surface waters are derived ultimately from precipitation, the relations between precipitation, runoff, infiltration, evaporation, and transpiration are so complex that engineers rightly prefer to base calculations of *yield* on available *stream gaugings*. For adequate information, gaugings must extend over a considerable number of years.

In the absence of adequate natural storage, engineers construct impounding reservoirs (Fig. 2.6). More rarely they excavate storage basins in lowlands adjacent to streams. Natural storage, too, can be regulated. Control works (gates and weirs or sills) at the outlets to lakes and ponds are examples. Some storage works are designed to serve a single purpose only; others are planned to perform a number of different functions and to preserve the broader economy of natural resources. Common purposes include the following:

1) Water supply for household, farm, community, and industry
2) Dilution and natural purification of municipal and industrial wastewaters
3) Irrigation of arable land

4) Harnessing water power
5) Low-water regulation for navigation
6) Preservation and cultivation of useful aquatic life
7) Recreation, for example, fishing, boating, and bathing
8) Control of destructive floods.

The greatest net benefit may accrue from a judicious combination of reservoir functions in *multipurpose* developments. The choice of single-purpose storage systems should indeed be justified fully.

Storage is provided when stream flow is inadequate or rendered unsatisfactory by heavy pollution. Release of stored waters then swells flows and dilutes pollution. Storage itself also affects the quality of the waters impounded. Both desirable and undesirable changes may take place. Their identification is the responsibility of *limnology*, the science of lakes or, more broadly, of inland waters.

If they must receive wastewaters, stream flows should be adjusted to the pollution load imposed on them. Low-water regulation, as such, is made possible by headwater or upstream storage, but lowland reservoirs, too, may aid dilution and play an active part in the natural purification of river systems. Whether overall results are helpful depends on the volume and nature of wastewater flows and the chosen regimen of the stream.

2.2 Safe Yield of Streams

In the absence of storage, the safe yield of a river system is its lowest *dry-weather flow*; with full development of storage, the safe yield approaches the *mean annual flow*. The economical yield generally lies somewhere in between. The attainable yield is modified by (a) evaporation, (b) bank storage, (c) seepage out of the catchment area, and (d) silting. Storage–yield relations are illustrated in this chapter by calculations of storage to be provided in impounding reservoirs for water supply. However, the principles demonstrated are also applicable to other purposes and uses of storage.

2.3 Storage as a Function of Draft and Runoff

A dam thrown across a river valley impounds the waters of the valley. Once the reservoir has filled, the water drawn from storage is eventually replenished by the stream, provided runoff, storage, and draft are kept in proper balance. The balance is struck graphically or analytically on the basis of historical records or replications generated by suitable statistical procedures of operational hydrology.

Assuming that the reservoir is full at the beginning of a dry period, the maximum amount of water S (MG/mi^2 or ML/km^2) that must be withdrawn from storage to maintain a given average draft D (MG/mi^2 or ML/km^2) equals the maximum cumulative difference between the draft D and the runoff Q (MG/mi^2 or ML/km^2) in a given dry period, or

$$S = \text{maximum value of} \sum (D - Q) \tag{2.1}$$

To find S, $\sum (D - Q)$ is summed arithmetically or graphically. The *mass diagram method* illustrated in Fig. 2.7 is a useful demonstration of finding $\sum(D - Q) = \sum D - \sum Q$. The shorter the interval of time for which runoff is recorded, the more exact the result. As the maximum value is approached, therefore, it may be worthwhile to shift to short intervals of time—from monthly to daily values, for example. The additional storage identified by such a shift may be as much as 10 days of draft.

Assuming that inflow and drafts are repeated cyclically, in successive sets of T years, a procedure called *sequent peak* was developed for determining minimum storage for no shortage in draft based on two needed cycles. Example 2.1 illustrates the procedure.

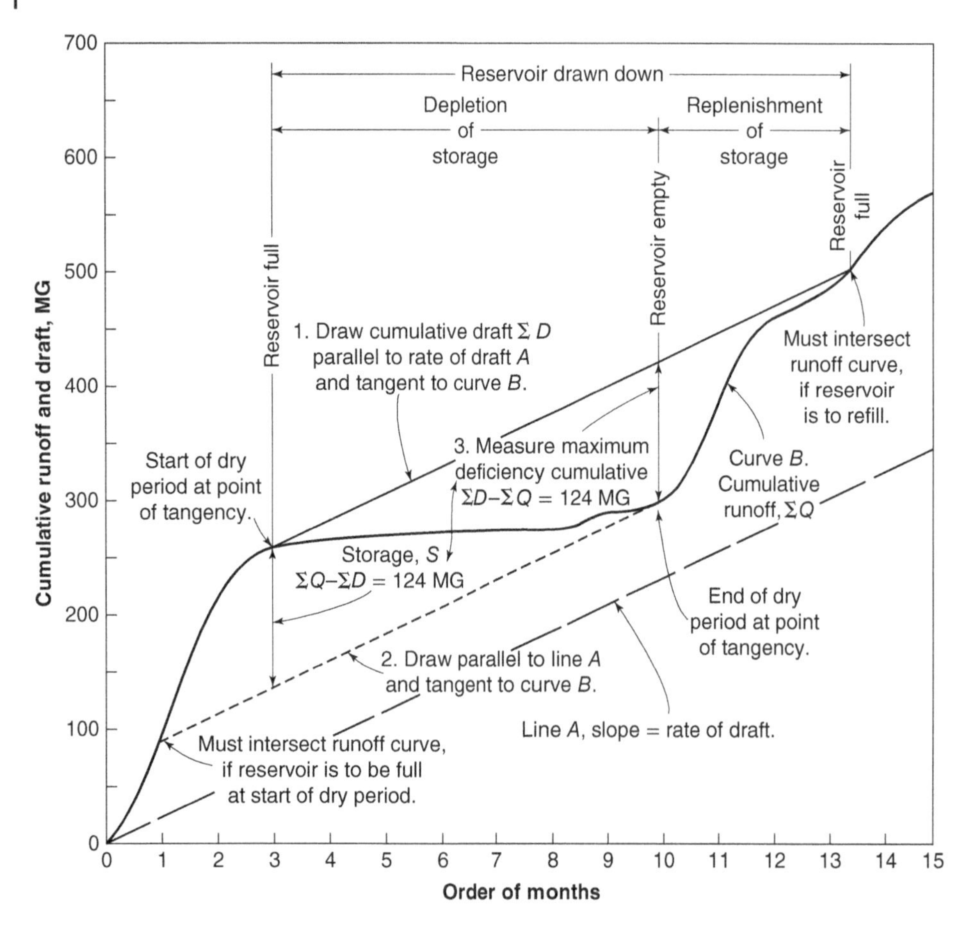

Figure 2.7 Mass Diagram Method for the Determination of Storage Required in Impounding Reservoirs (a constant draft of 750,000 gpd/mile^2 = 23 MG/mile^2 for a month of 30.4 days is assumed). Conversion factors: 1 *MG* = 1,000,000 *gal* = 3.785 *ML* = 3,785,000 *L*; 1 MG/mile^2 = 1.461 ML/km^2; 1 gpd/mile^2 = 1.461 L/d/km^2

Example 2.1 Calculation of Required Storage from Runoff Records

From the recorded monthly mean runoff values shown in column 2, Table 2.2, find the required storage for the estimated rates of draft listed in column 3, Table 2.2.

Solution 1 (US Customary System):

Conversion factors: 1 MG/mi^2 = 1.461 ML/km^2; 1 MGD/mi^2 = 1.461 MLD/km^2

Column 2: These are observed flows for the Westfield Little River, near Springfield, MA, USA.

Column 3: The values 27, 30, 33, and 36 MG/mi^2 = 0.89, 1.1, 1.09, and 1.18 MGD/mi^2, respectively, for 30.4 day/month. For a total flow of 462 MG/mi^2 in 12 months, the average flow is 462/365 = 1.27 MGD/mi^2, and for a total draft of 360 MG/mi^2, the development is $100 \times 360/462 = 78\,\%$.

Column 4: Positive values are surpluses; negative values, deficiencies.

Column 5: P_1 is the first peak, and T_1 is the first trough in the range P_1P_2, where P_2 is the second higher peak; similarly, T_2 is the second trough in the range P_2P_3.

Column 6: The required maximum storage $S_m = \max(P_j - T_j) = P_m - T_m = \boldsymbol{P_1 - T_1} = 177 - 1 = \mathbf{176\ MG/mi^2}$ in this case. The fact that $\boldsymbol{P_2 - T_2} = 279 - 103 = \mathbf{176\ MG/mi^2}$ also implies that there is seasonal rather than over-year storage. Storage at the end of month i is $S_i = \min\{S_m, [S_{i-1} + (Q_i - D_j)]\}$; for example, in line 2, $S_m - 176$ and $[S_{i-1} + (Q_i - D_j)] = 67 + 95 = 162$, or $S_i = 162$; in line 3, however, $S_m = 176$ and $[S_{i-1} + (Q_i - D_j)] = 162 + 95 = 257$ or $S_i = S_m = 176$.

Column 7: The flow wasted $W_i = \max\{0, [(Q_i - D_i) - (S_m - S_{i-1})]\}$; for example, line 3, $(Q_i - D_i) - (S_m - S_{i-1}) = 15 - (176 - 162) = 1$ or $W_i = 1$; in line 3 of the second series, however, $(Q_i - D_i) - (S_m - S_{i-1}) = 15 - (176 - 176) = 15$. There is no negative waste.

Table 2.2 Calculation of Required Storage (Example 2.1)[a]

Order of Months	Recorded Runoff, Q	Estimated Draft, D	$Q-D$	$\sum(Q-D)$	Storage[b] $S = P - T$	Waste, W	Reservoir State[c]
(1)	(2)	(3)	(4)	(5)	(6)	(7)	(8)
1	94	27	+67	67	67	0	R
2	122	27	+95	162	162	0	R
3	45	30	+15	$177P_1$	176	1	L
4	5	30	−25	152	151	0	F
5	5	33	−28	124	123	0	F
6	2	30	−28	96	95	0	F
7	0	27	−27	69	68	0	F
8	2	27	−25	44	43	0	F
9	16	30	−14	30	29	0	F
10	7	36	−29	$1T_1$	0	0	E
11	72	33	+39	40	39	0	R
12	92	30	+62	102	101	0	R
1	94	27	+67	169	168	0	R
2	122	27	+95	264	176	87	L
3	45	30	+15	$279P_2$	176	15	L
4	5	30	−25	254	151	0	F
5	5	33	−28	226	123	0	F
6	2	30	−28	198	95	0	F
7	0	27	−27	171	68	0	F
8	2	27	−25	146	43	0	F
9	16	30	−14	132	29	0	F
10	7	36	−29	$103T_2$	0	0	E
11	72	33	+39	142	39	0	R
12	92	30	+62	$204P_3$	101	0	R

[a] Runoff Q, draft D, and storage S are expressed in MG/mi^2.
[b] P = peak; T = trough.
[c] R = rising; F = falling; L = spilling; E = empty.
Conversion factor: 1 MG/mi^2 = 1.461 ML/km^2.

Solution 2 (SI System):

An SI or metric system solution can be obtained using these conversation factors: 1 MG = 3,785 m^3 = 3.785 ML; 1 ML = 1,000 m^3; 1 MGD = 3,785 m^3/day = 3.785 MLD = 0.0438 m^3/s = 43.8 L/s; 1 mi^2 = 2.59 km^2; 1 MG/mi^2 = 1.461 ML/km^2; 1 MGD/mi^2 = 1.461 MLD/km^2.

For variable drafts and inclusion of varying allowances for evaporation from the water surface created by the impoundage, the analytical method possesses distinct advantages over the graphical method. The principal value of the mass diagram method, indeed, is not for the estimation of storage requirements, but for determining the yield of catchment areas on which storage reservoirs are already established.

2.4 Design Storage

Except for occasional series of dry years and very high-density housing and industrial developments, *seasonal storage* generally suffices in the well-watered regions of the United States. Water is plentiful, stream flows do not vary greatly from year to year, reservoirs generally refill within the annual hydrologic cycle, and it does not pay to go in for advanced or complete development of catchment areas. In semiarid regions, on the other hand, water is scarce, stream flows fluctuate widely from year to year, runoff of wet years must be conserved for use during dry years, and it pays to store a large proportion of the mean annual flow. In these circumstances, operational records of adequate length become important, along with computational aids.

Given a series of storage values for the flows observed or generated statistically, the engineer must decide which value he will use. Will it be the highest on record, or the second, third, or fourth highest? Obviously, the choice depends on the degree of protection to be afforded against water shortage. This must also be considered in terms of drought experience, which is a function of the length of record examined. To arrive at a reasonable answer and an economically justifiable storage design, the engineer may resort to (a) a *statistical analysis* of the arrayed storage values and (b) estimates of the difficulties and costs associated with shortage in supply. Storage values equaled or exceeded but once in 20, 50, or 100 years, that is, 5%, 2%, and 1% of the years, respectively, are often considered. For water supply, Hazen (1956) suggested employing the 5% value in ordinary circumstances. In other words, design storage should be adequate to compensate for a drought of a severity not expected to occur more often than once in 20 years. In still drier years, it may be necessary to *curtail the use of water* by limiting or prohibiting, for example, lawn sprinkling and car washing.

Restricting water use is irksome to the public and a poor way to run a public utility. As a practical matter, moreover, use must be cut down well in advance of anticipated exhaustion of the supply. It would seem logical to consider not only the frequency of curtailment but also the depletion point at which conservation should begin. In practice, the iron ration generally lies between 20% and 50% of the total water stored. Requiring a 25% reserve for the drought that occurs about once in 20 years is reasonable. An alternative is a storage allowance for the drought to be expected once in 100 years. This is slightly less in magnitude than the combination of a 25% reserve with a once-in-20-years risk.

In undeveloped areas, few records are even as long as 20 years. Thus, estimation of 5%, 2%, and 1% frequencies, or of recurrence intervals of 20, 50, and 100 years, respectively, requires extrapolation from available data. *Probability plots* lend themselves well to this purpose. However, they must be used with discretion. Where severe droughts in the record extend over several years and require annual rather than seasonal storage values to be used, the resulting series of storage values becomes nonhomogeneous and is no longer strictly subject to ordinary statistical interpretations. They can be made reasonably homogeneous by including, besides all truly seasonal storage values, not only all true annual storage values, but also any seasonal storage values that would have been identified within the periods of annual storage if the drought of the preceding year or years had not been measured. Plots of recurrence intervals should include minor storage capacities as well as major ones. The results of these statistical analyses are then conveniently reduced to a set of *draft–storage–frequency* curves.

Example 2.2 Design of Storage Requirement for Various Frequencies

Examination of the 25-year record of runoff from an eastern stream shows that the storage amounts listed in Table 2.3 are needed in successive years to maintain a draft of 750,000 gpd/mi^2 (1,096,000 L/d/km^2). Estimate the design storage requirement that is probably reached or exceeded but once in 20, 50, and 100 years.

Solution:

1) The 25 calculated storage values arrayed in order of magnitude are plotted on arithmetic-probability paper in Fig. 2.8 at $100 \times k/26$; therefore, $100 \times 1/26 = 3.8$, $100 \times 2/26 = 7.7$, $100 \times 3/26 = 11.5\,\%$, and so forth. A straight line of best fit is identified in this instance, but not necessarily others, the arithmetic mean storage being μ = **67 MG** (254 ML) and the standard deviation σ = 33 MG (125 ML).
2) The storage requirements reached or exceeded once in 20, 50, and 100 years, or 5%, 2%, and 1% of the time, are read as **123, 137**, and **146 MG**, respectively (466, 519, and 553 ML, respectively). Probability paper is used because it offers a rational basis for projecting the information beyond the period of experience. The once-in-20-years requirement with 25% reserve suggests a design storage of 123/0.75 = **164 MG/mi^2 (230 ML/km^2) of drainage area**.

Table 2.3 Storage Requirements (Example 2.2)

Order of year	1	2	3	4	5	6	7	8	9	10	11	12	13
Calculated storage (MG)	47	39	104	110	115	35	74	81	124	29	37	82	78
Order of year	14	15	16	17	18	19	20	21	22	23	24	25	
Calculated storage (MG)	72	10	117	51	61	8	102	65	73	20	53	88	

Conversion factors: 1 MG = 1,000,000 gal = 3.785 ML = 3,785,000 L.

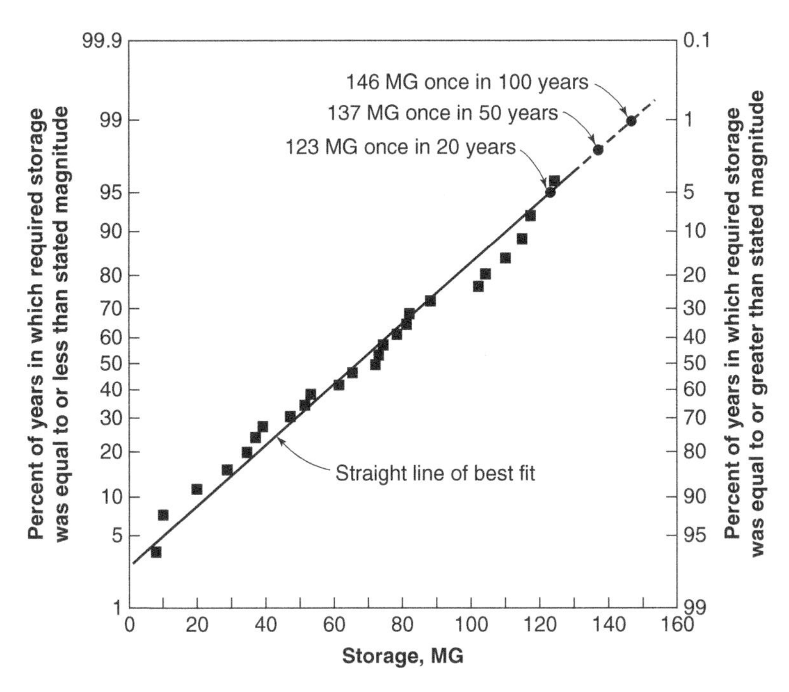

Figure 2.8 Frequency Distribution of Required Storage Plotted on Arithmetic-Probability Paper. Conversion factors: 1 MG = 1,000,000 gal = 3.785 ML = 3,785,000 L

2.5 Loss by Evaporation, Seepage, and Silting

When an impounding reservoir is filled, the hydrology of the inundated area and its immediate surroundings is changed in a number of respects: (a) The reservoir loses water by evaporation to the atmosphere and gains water by direct reception of rainfall, (b) rising and falling water levels alter the pattern of groundwater storage and movement into and out of the surrounding reservoir banks, (c) at high stages, water may seep from the reservoir through permeable soils into neighboring catchment areas and so be lost to the area of origin, and (d) quiescence encourages subsidence of settleable suspended solids and silting of the reservoir.

2.5.1 Water Surface Response

The response of the new water surface is to establish new hydrologic equilibria (a) through loss of the runoff once coming from precipitation on the land area flooded by the reservoir Qa (closely), where Q is the areal rate of runoff of the original watershed and a is the water surface area of the reservoir, and through evaporation from the water surface Ea, where E is the areal rate of evaporation; and (b) through gain of rainfall on the water surface Ra, where R is the areal rate of rainfall. The net rate of loss or gain is $[R-(Q+E)]a$; a negative value records a net loss and a positive value a net gain.

Individual factors vary within the annual hydrologic cycle and from year to year. They can be measured. Exact calculations, however, are commonly handicapped by inadequate data on evaporation. Required hydrological information should come from local or nearby observation stations, areas of water surface being determined from contour maps of the reservoir site. The mean annual water surface as a fraction of the reservoir area at the spillway, f, is normally about 0.90 or 90%.

For convenience, the water surface response is expressed in one of the following ways:

(1) Revised runoff: $Q_r = Q-(Q+E-R)(\text{f}a/A)$ (2.2)

(2) Equivalent draft: $D_e = (Q+E-R)(\text{f}a/A)$ (2.3)

(3) Effective catchment area: $A_e = A-\text{f}a[1-(R-E)/Q]$ (2.4)

(4) Adjusted flow line: $\text{F} = Q+E-R$ (2.5)

(5) Effective draft: $D_{ed} = D_{md} + D_e(A)$ (2.6)

where Q_r is revised runoff, in./year or cm/year; Q is mean annual runoff, in./year or cm/year; R is mean annual runoff, in./year or cm/year; E is mean annual evaporation, in./year or cm/year; a is reservoir area, mi^2 or km^2; A is catchment area, mi^2 or km^2; f = 90 % = 0.9 = effective factor for reservoir area; D_e is equivalent draft, in./year or cm/year (MGD/mi^2 or MLD/km^2); F is adjusted flow line at spillway level, in. or cm; A_e is effective catchment area, mi^2 or km^2; D_{ed} is effective draft, MGD or MLD; and D_{md} is mean draft, MGD or MLD.

The values of A (total catchment area) and a (reservoir surface area) obtained are used as follows in recalculating storage requirements: Q_r replaces Q; $D + D_e$ replaces D; and A_e replaces A. A fourth allowance calls for raising the flow line of the reservoir by $Q + E - R$, expressed in units of length yearly. In rough approximation, the spillway level is raised by a foot or two (0.3048 m or 0.6096 m) in the eastern United States.

Example 2.3 Calculations of Mean Annual Runoff and Draft

A mean draft of 30.0 MGD (113.6 MLD) is to be developed from a catchment area of 40.0 mi^2 (103.6 km^2). First calculations ask for a reservoir area of 1,500 acres (6.07 km^2) at flow line. The mean annual rainfall is 47.0 in./year (119.4 cm/year), the mean annual runoff is 27.0 in./year (68.6 cm/year), and the mean annual evaporation is 40.0 in./year (101.6 cm/year). Find the following:

1) The revised mean annual runoff
2) The equivalent mean draft
3) The equivalent land area
4) The adjusted flow line.

Solution 1 (US Customary System):

1) By Eq. (2.2), the revised annual runoff is

$$Q_r = Q - (Q + E - R)(\mathrm{f}a/A)$$

$$Q_r = 27.0 - (27.0 + 40.0 - 47.0)[(0.9 \times 1{,}500/640)/(40.0)] = 27.0 - 1.1 = \mathbf{25.9\ in./year}.$$

Here, 1 mi^2 = 640 acres.

2) By Eq. (2.3), the equivalent mean draft is

$$D_e = (Q + E - R)(\mathrm{f}a/A)$$

$$D_e = (27.0 + 40.0 - 47.0)[(0.9 \times 1{,}500/640)/(40.0)] = 1.1\ \text{in./year}$$

$$D_e = 1.1\ \text{in./year} = 52{,}360\ \text{gpd/mi}^2 = 0.052\ \text{MGD/mi}^2$$

and the effective draft D_{ed} is

$$D_{ed} = D_{md} + D_e(A) = 30.0\ \text{MGD} + \left(0.052\ \text{MGD/mi}^2\right)\left(40.0\ \text{mi}^2\right) = \mathbf{32.1\ MGD}.$$

3) By Eq. (2.4), the equivalent land area is

$$A_e = A - \mathrm{f}a[1 - (R - E)/Q]$$

$$A_e = 40.0 - (0.9 \times 1{,}500/640)[1 - (47.0 - 40.0)/27.0] = 40.0 - 1.6 = \mathbf{38.4\ mi^2}.$$

4) By Eq. (2.5), the adjusted flow line is

$$F = Q + E - R$$

$$F = 27.0 + 40.0 - 47.0 = 20\ \text{in., equaling } 20 \times 0.9 = \mathbf{18\ in.\ at\ spillway\ level}$$

Solution 2 (SI System):

5) By Eq. (2.2), the revised annual runoff is

$$Q_r = Q - (Q + E - R)\ (\mathrm{fa}/A)$$

$$Q_r = 68.58 - (68.58 + 101.60 - 119.38)[(0.9 \times 6.07)/(103.6)]$$
$$= 68.58 - 2.68 = \mathbf{65.9\ cm/year}.$$

6) By Eq. (2.3), the equivalent mean draft is

$$D_e = (Q - E + R)(\mathrm{fa}/A)$$

$$D_e = (68.58 + 101.60 - 119.38)[(0.9 \times 6.07)/(103.6)] = 2.68\ \mathrm{cm/year}.$$

$D_e = 2.68$ cm/year = 0.07337 MLD/km^2 and the effective draft D_{ed} is

$$D_{ed} = D_{md} + D_e(A) = 113.55\ \mathrm{MLD} + \left(0.07337\ \mathrm{MLD/km^2}\right)\left(103.6\ \mathrm{km^2}\right) = 121.2\ \mathrm{MLD}.$$

Here, 1 cm/year = 0.0273793 MLD/km^2.

7) By Eq. (2.4), the equivalent land area is

$$A_e = A - \mathrm{fa}[1 - (R - E)/Q]$$

$$A_e = 103.6 - (0.9 \times 6.07)[1 - (119.38 - 101.60)/68.58] = 103.6 - 4.47 = \mathbf{99.13\ km^2}.$$

8) By Eq. (2.5), the adjusted flow line is

$$F = Q + E - R$$

$$F = 68.58 + 101.60 - 119.38 = 50.8\ \mathrm{cm},\ \text{equaling}\ (50.8\ \mathrm{cm}) \times 0.9 = \mathbf{45.7\ cm\ at\ spillway\ level}.$$

2.5.2 Seepage

If the valley enclosing a reservoir is underlain by porous strata, water may be lost by seepage. Subsurface exploration alone can foretell how much. Seepage is not necessarily confined to the dam site. It may occur wherever the sides and bottom of the reservoir are sufficiently permeable to permit water to escape through the surrounding hills.

2.5.3 Silting

Soil erosion on the watershed causes reservoir silting. Both are undesirable. Erosion destroys arable lands. Silting destroys useful storage (see Fig. 2.9). How bad conditions are in a given catchment area depends principally on soil and rock types, ground surface slopes, vegetal cover, methods of cultivation, and storm rainfall intensities.

Silt accumulations cannot be removed economically from reservoirs by any means so far devised. Dredging is expensive, and attempts to flush out deposited silt by opening scour valves in dams are fruitless. Scour only produces gullies in the silt. In favorable circumstances, however, much of the heaviest load of suspended silt can be steered through the reservoir by opening large sluices installed for this purpose. Flood flows are thereby selected for storage in accordance with their quality and their volume.

Reduction of soil erosion is generally a long-range undertaking. Involved are proper farming methods, such as contour plowing, terracing of hillsides, reforestation, cultivation of permanent pastures, prevention of gully formation through construction of check dams or debris barriers, and revetment of stream banks.

In the design of impounding reservoirs for silt-bearing streams, suitable allowance must be made for loss of capacity by silting. Rates of deposition are especially high in impoundments on flashy streams draining easily eroded catchment areas. The proportion of sediment retained is called its *trap efficiency*. A simple calculation will show that 2,000 mg/L of suspended

Figure 2.9 A Watershed Dam in Northwest Iowa that Is Completely Silted In. The dam no longer functions to store water, since its capacity has been lost (*Source:* Courtesy of the National Resources Conservation Service and USDA)

solids equals 8.3 ton/MG (1.989 ton/ML) and that an acre-ft of silt weighs almost 1,500 tons (1,360.8 metric tons) if its unit weight is 70 lb/ft^3 (1,211.4 kg/m^3). In some parts of the United States, the volume of silt V_s in acre-ft (1 acre - ft = 0.32585 MG = 43,560 ft^3 = 1,233.5 m^3) deposited annually can be approximated by the following equation:

$$V_s = cA^n \quad \text{(US customary units)}, \tag{2.7}$$

where V_s is the volume of silt deposited annually, acre-ft, `. A is the size of the drainage area in mi^2 (note: 1 mi^2 = 2.59 km^2), and c and n are coefficients with a value of n = 0.77 for the US southwestern streams and values of c varying from 0.43 through 1.7 to 4.8 for low, average, and high deposition, respectively, and the corresponding values for southeastern streams being c = 0.44 only and n = 1.0. Understandably, the magnitudes of c and n, here reported, apply only to the regions for which they were developed.

The volume of silt deposited annually can also be approximated by the following equation using metric units:

$$V_s = 1{,}233.5c\left[A/\left(2.59 \times 10^6\right)\right]^n \quad \text{(SI units)}, \tag{2.7a}$$

where V_s is the volume of silt deposited annually, m^3; A is the size of the drainage area, m^2; c is a coefficient with a value varying from 0.43 to 1.7 to 4.8 for low, average, and high deposition, respectively; andn is a coefficient to be determined specifically for a target drainage area 0.77 for southwestern steams in the United States.

A plot of trap efficiency against the proportion of the mean annual flow stored in a reservoir traces curves quite similar to curves for the expected performance of settling basins of varying effectiveness. Close to 100% of the sediment transported by influent streams may be retained in reservoirs storing a full year's tributary flow. Trap efficiency drops to a point between 65% and 85% when the storage ratio is reduced to 0.5 (half a year's inflow) and to 30% and 60% when the storage ratio is lowered to 0.1 (5 weeks' inflow). Silting is often fast when reservoirs are first placed in service and may be expected to drop off and reach a steady state as delta building goes on and shores become stabilized. An annual silting rate of 1.0 acre-ft/mi^2 (note: 1 acre - ft/mi^2 = 476.25 m^3/km^2 = 0.47625 ML/km^2) of watershed corresponds roughly to a yearly reduction in storage of 0.32585 MG/mi^2 (0.476 ML/km^2) because an acre 3 ft deep (4,047 m^2 area × 0.9144 m depth) is about 1 MG (3.785 ML).

Example 2.4 Volume of silting

Determine the volume of silt accumulations for a drainage area having the following characteristics:

Area = 100 mi^2 = 259 km^2 = 2.59 × 10^8 m^2 = 259,000,000 m^2

Average deposition of silt: c = 1.7

Area located in the Southwestern United States: n = 0.77.

Solution 1 (US Customary System):

$$V_s = cA^n, \tag{2.7}$$

where V_s is the volume of silt deposited annually, acre - ft; and A is the size of the drainage area, mi^2.

$$\begin{aligned} V_s &= 1.7 \times (100)^{0.77} \\ &= 1.7 \times 34.7 \\ &= \mathbf{59\ acre\text{-}ft}. \end{aligned}$$

Solution 2 (SI System):

$$V_s = 1{,}233.5c\left[A/\left(2.59 \times 10^6\right)\right]^n, \tag{2.7a}$$

where V_s is the volume of silt deposited annually, m^3; and A is the size of the drainage area in m^2

$$V_s = 1{,}233.5c\left[A/\left(2.59 \times 10^6\right)\right]^n$$

$$V_s = 1{,}233.5(1.7)\left[\left(2.59 \times 10^8\right)/\left(2.59 \times 10^6\right)\right]^{0.77}$$

$$= 1{,}233.5(1.7)(34.6737)$$

$$= \mathbf{72{,}708.98\ m^3}.$$

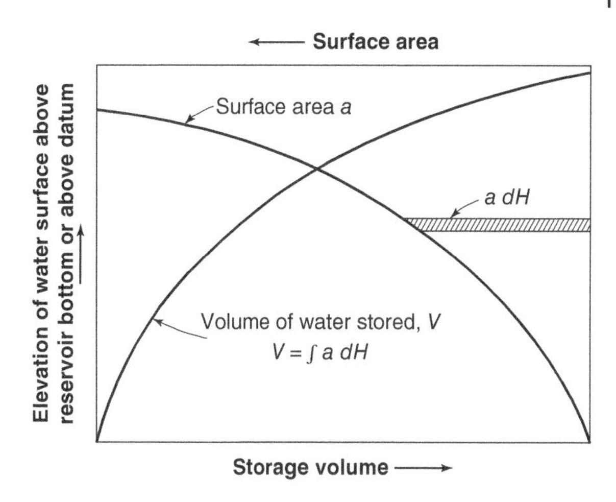

Figure 2.10 Surface Area of a Reservoir and Volume of Water Stored

2.6 Area and Volume of Reservoirs

The surface areas and volumes of water at given horizons are found from a contour map of the reservoir site. Areas enclosed by each contour line are planimetered, and volumes between contour lines are calculated. The *average-end-area method* is generally good enough for the attainable precision of measurements.

For uniform contour intervals h (ft or m) and successive contour areas $a_0, a_1, \ldots, a_n$,(acre or m^2), the volume V of water (acre-ft or m^3) stored up to the nth contour is

$$V = 1/2\,h\left[(a_0 + a_1) + (a_1 + a_2) + \ldots + (a_{n-1} + a_n)\right]$$

$$V = 1/2\,h\left(a_0 + a_n + 2\sum_{1}^{n-1} a\right). \tag{2.8}$$

For general use, surface areas and volumes are commonly plotted against contour elevations as in Fig. 2.10. Note that volumes must be determined from the surface area curve by planimetering the area enclosed between the curve and its ordinate.

In reservoir operation, a small amount of water lies below the invert of the reservoir outlet. Constituting the dregs of the impoundage, this water is of poor quality. The associated reduction in *useful storage* is offset, in general, by bank storage released from the soil as the reservoir is drawn down. Moreover, the water below the outlet sill does form a conservation pool for fish and wildlife.

Surface areas and volumes enter not only into the solution of hydrologic problems but also into the management of water quality, such as the control of algae by copper sulfate and destratification by pumping or aeration.

2.7 Management of Catchment Areas

The comparative advantage of developing surface rather than underground waters is offset, in large measure, by the unsteadiness of surface runoff, both in quantity and quality, and the recurrence of flow extremes. Those hydrologic factors that enter strongly into the development of surface water supplies must, therefore, be kept clearly in mind in their design and operation, with special reference to

1) The principles of selecting, preparing, and controlling catchment areas
2) The choice and treatment of reservoir areas and the management of natural ponds and lakes, as well as impounding reservoirs
3) The siting, dimensioning, construction, and maintenance of necessary engineering works, including dams and dikes, intake structures, spillways, and diversion works. Also keep in mind that river systems may have to be developed for multiple purposes, not just for municipal uses.

The gathering grounds for public water supplies vary *in size* from a few hundred acres to thousands of square miles, and in *character* from sparsely inhabited uplands to densely populated river valleys. The less developed they are, the better, relatively, they lend themselves to exploitation for steady yields and the production of water of high quality.

2.7.1 Upland Areas

Occasionally, a water utility can, with economic justification, acquire the entire watershed of its source and manage solely for water supply purposes, *excluding habitations and factories* to keep the water safe and attractive; *letting arable lands lie fallow* to prevent wasteful runoff and high turbidities; *draining swamps* to reduce evaporation and eliminate odors, tastes, and color; and *cultivating woodlots* to hold back winter snows and storm runoff and help to preserve the even tenor of stream flow. As competition for water and land increases, land holdings of water utilities are understandably confined to the marginal lands of water courses, especially those closest to water intakes themselves. Yet water quality management need not be neglected. Scattered habitations can be equipped with acceptable sanitary facilities, wastewaters can be adequately treated or, possibly, diverted into neighboring drainage areas not used for water supply, swamps can be drained, and soil erosion can be controlled. *Intelligent land management* of this kind can normally be exercised most economically when water is drawn from *upland sources* where small streams traverse land of little value and small area. However, some upland watersheds are big enough to satisfy the demands of great cities. The water supplies of Boston, New York, and San Francisco are examples.

2.7.2 Lowland Areas

When water is drawn from large lakes and wide rivers that, without additional storage, yield an abundance of water, management of their catchments ordinarily becomes the concern of more than one community (examples are the Ohio and Mississippi rivers) and sometimes of more than a single state (the Delaware River is a notable example) and even of a single country (for example, the Great Lakes are shared with Canada and the Colorado River is shared with Mexico). Regional, interstate, and international authorities must be set up to manage and protect land and water resources of this kind.

2.7.3 Quality Control

To safeguard their sources, water utilities can fence and post their lands, patrol watersheds, and obtain legislative authority for enforcing reasonable rules and regulations for the environmental management of the catchment area. When the cost of policing the area outweighs the cost of purifying its waters in suitable treatment works, purification is often preferred. It is likewise preferred when lakes, reservoirs, and streams become important recreational assets and their enjoyment can be encouraged without endangering their quality. It goes without saying that recreation must be properly supervised and recreational areas suitably located and adequately equipped with sanitary facilities.

2.7.4 Swamp Drainage

Three types of swamps may occur on catchment areas:

1) Rainwater swamps where precipitation accumulates on flat lands or where rivers overflow their banks in times of flood
2) Backwater swamps or reaches of shallow flowage in sluggish, often meandering streams where bends or other obstructions can hamper flow
3) Seepage-outcrop swamps where hillside meets the plain or where sand and gravel overlie clay or other impervious formations.

Rainwater swamps can be drained by ditches cut into the floodplain; backwater swamps by channel regulation; and seepage-outcrop swamps by marginal interception of seepage waters along hillsides sometimes supplemented by the construction of central surface and subsurface drains.

2.8 Reservoir Siting

In the absence of natural ponds and lakes, intensive development of upland waters requires the construction of impounding reservoirs. Suitable siting is governed by interrelated considerations of adequacy, economy, safety, and palatability of the supply. Desirable factors include the following:

1) *Surface topography* that generates a low ratio of dam volume to volume of water stored; for example, a narrow gorge for the dam, opening into a broad and branching upstream valley for the reservoir. In addition, a favorable site for a stream diversion conduit and a spillway, and a suitable route for an aqueduct or pipeline to the city are desirable.

2) *Subsurface geology* that ensures (a) safe foundations for the dam and other structures; (b) tightness against seepage through abutments and beneath the dam; and (c) materials, such as sand, gravel, and clay, for the construction of dam and appurtenant structures.
3) A *reservoir valley* that is sparsely inhabited, neither marshy nor heavily wooded, and not traversed by important roads or railroads; the valley being *so shaped* that waters pouring into the reservoir are not short-circuited to the outlet, and *so sloped* that there is little shallow flowage around the margins. Natural purification by storage can be an important asset. Narrow reservoirs stretching in the direction of prevailing winds are easily short-circuited and may be plagued by high waves. Areas of shallow flowage often support heavy growths of water plants while they are submerged and of land plants while they are uncovered. Shoreline vegetation encourages mosquito breeding; decaying vegetation imparts odors, tastes, and color to the water.
4) *Reservoir flowage* that interferes as little as possible with established property rights, proximity to the intake to the community served, and location at such an elevation that supply can be by gravity.

Large reservoirs may inundate villages, including their dwellings, stores, and public buildings; mills and manufacturing establishments; farms and farmlands, stables, barns, and other outhouses; and gardens, playgrounds, and graveyards. Although such properties can be seized by the *right of eminent domain*, a wise water authority will proceed with patience and understanding. To be humane and foster goodwill, the authority will transport dwellings and other wanted and salvable buildings to favorable new sites, establish new cemeteries or remove remains and headstones to grounds chosen by surviving relatives, and assist in reconstituting civil administration and the regional economy.

When reservoir sites are flooded, land plants die and organic residues of all kinds begin to decompose below the rising waters; nutrients are released; algae and other microorganisms flourish in the eutrophying environment; and odors, tastes, and color are intensified. Ten to fifteen years normally elapse before the biodegradable substances are minimized and the reservoir is more or less stabilized.

In modern practice, reservoir sites are cleared only in limited measure as follows:

1) *Within the entire reservoir area:* (a) Dwellings and other structures are removed or razed; (b) barnyards, cesspools, and privies are cleaned, and manure piles are carted away; (c) trees and brush are cut close to the ground, usable timber is salvaged, and slash, weeds, and grass are burned; (d) swamp muck is dug out to reasonable depths, and residual muck is covered with clean gravel, the gravel, in turn, being covered with clean sand; and (e) channels are cut to pockets that would not drain when the water level of the reservoir is lowered.
2) *Within a marginal strip between the high-water mark reached by waves and a contour line about 20 ft (6.1 m) below reservoir level:* (a) Stumps, roots, and topsoil are removed; (b) marginal swamps are drained or filled; and (c) banks are steepened to produce depths near the shore that are close to 8 ft (2.44 m) during much of the growing season of aquatic plants—to do this, upper reservoir reaches may have to be improved by excavation or fill or by building auxiliary dams across shallow arms of the impoundage.

Soil stripping, namely, the removal of all topsoil containing more than 1% or 2% organic matter from the entire reservoir area, is no longer economical.

In malarious regions, impounding reservoirs should be so constructed and managed that they will not breed dangerous numbers of anopheline mosquitoes. To this purpose, banks should be clean and reasonably steep. To keep them so, they may have to be protected by riprap.

2.9 Reservoir Management

The introduction of impounding reservoirs into a river system or the existence of natural lakes and ponds within it raises questions of quality control. Limnological factors are important not only in the management of ponds, lakes, and reservoirs but also in reservoir design.

2.9.1 Quality Control

Of concern in the quality management of reservoirs is the control of water weeds and algal blooms; the bleaching of color; the settling of turbidity; destratification by mixing or aeration; and, in the absence of destratification, the selection of water of optimal quality and temperature by shifting intake depths in order to suit withdrawals to water uses or to downstream quality requirements.

2.9.2 Evaporation Control

The thought that oil spread on water will suppress evaporation is not new. It is well known that

1) Certain chemicals spread spontaneously on water as layers no more than a molecule thick.
2) These substances include alcohol (hydroxyl) or fatty acid (carboxyl) groups attached to a saturated paraffin chain of carbon atoms.
3) The resulting *monolayers* consist of molecules oriented in the same direction and thereby offering more resistance to the passage of water molecules than do thick layers of oil composed of multilayers of haphazardly oriented molecules.
4) The hydrophilic radicals (OH or COOH) at one end of the paraffin chain move down into the water phase, while the hydrophobic paraffin chains themselves stretch up into the gaseous phase. Examples of suitable chemicals are alcohols and corresponding fatty acids.

The cost and difficulty of maintaining adequate coverage of the water surface have operated against the widespread use of such substances. Small and light plastic balls have also been used to retard evaporation from water surfaces of reservoirs.

2.10 Dams and Dikes

Generally speaking, the great dams and barrages of the world are the most massive structures built by man. To block river channels carved through mountains in geologic time periods, many of them are wedged between high valley walls and impound days and months of flow in deep reservoirs. Occasionally, water reservoirs reach such levels that their waters would spill over low saddles of the divide into neighboring watersheds if saddle dams or dikes were not built to complement the main structure. In other ways, too, surface topography and subsurface geology are of controlling influence. Hydraulically, they determine the siting of dams; volumes of storage, including subsurface storage in glacial and alluvial deposits; and spillway and diversion arrangements. Structurally, they identify the nature and usefulness of foundations and the location and economic availability of suitable construction materials. Soils and rock of many kinds can go into the building of dams and dikes. Timber and steel have found more limited applications. Like most other civil engineering constructions, therefore, dams and their reservoirs are derived largely from their own environment.

Structurally, dams resist the pressure of waters against their upstream face by gravity, arch action, or both. Hydraulically, they stem the tides of water by their tightness as a whole and the relative imperviousness of their foundations and abutments. They combine these hydraulic and structural properties to keep seepage within tolerable limits and channeled such that the working structures remain safe. Various materials and methods of construction are used to create dams of many types. The following are the most common types: (a) embankment dams of Earth, rock, or both, and (b) masonry dams (today largely concrete dams) built as gravity, arched, or buttressed structures.

2.10.1 Embankment Dams

Rock, sand, clay, and silt are the principal materials of construction for rock and Earth embankments. Permeables provide weight, impermeables watertightness. Optimal excavation, handling, placement, distribution, and compaction with special reference to selective placement of available materials challenge the ingenuity of the designer and constructor. Permeables form the shells or shoulders, impermeables the core or blanket of the finished embankment. Depending in some measure on the abundance or scarcity of clays, relatively thick cores are centered in a substantially vertical position, or relatively thin cores are displaced toward the upstream face in an inclined position. Common features of an earth dam with a central clay core wall are illustrated in Fig. 2.4. Concrete walls can take the place of clay cores, but they do not adjust well to the movements of newly placed, consolidating embankments and foundations; by contrast, clay is plastic enough to do so. If materials are properly dispatched from borrow pits, earth shells can be ideally graded from fine at the watertight core to coarse and well draining at the upstream and downstream faces. In rock fills, too, there must be effective transition from core to shell, the required change in particle size ranging from a fraction of a millimeter for fine sand through coarse sand (about 1 mm) and gravel (about 10 mm) to rock of large dimensions.

Within the range of destructive wave action, stone placed either as paving or as *riprap* wards off erosion of the upstream face. Concrete aprons are not as satisfactory, sharing as they do most of the disadvantages of concrete core walls. A wide *berm* at the foot of the protected slope helps to keep riprap in place. To prevent the downstream face from washing away, it is commonly seeded with grass or covering vines and provided with a system of surface and subsurface drains. Berms break up

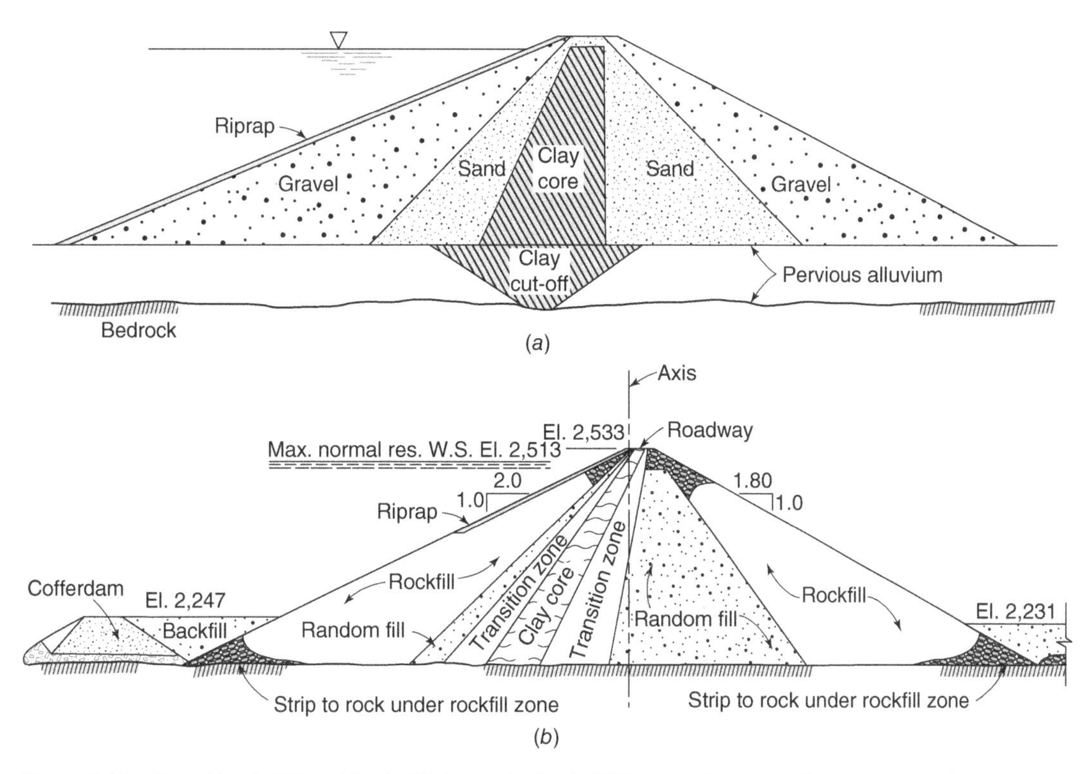

Figure 2.11 Zoned Earth-Fill and Rock-Fill Dams: (a) Earth-fill Dam on Pervious Alluvium and (b) Rock-Fill Dam on Bedrock

the face into manageable drainage areas and give access to slopes for mowing and maintenance. Although they are more or less horizontal, berms do slope inward to gutters; moreover, they are pitched lengthwise for the gutters to conduct runoff to surface or subsurface main drains and through them safely down the face or abutment of the dam, eventually into the stream channel.

Earth embankments are constructed either as *rolled fills* or *hydraulic fills*; rock embankments are built as *uncompacted* (dumped) or *compacted* fills. In rolled earth fills, successive layers of earth 4–12 in. (100–300 mm) thick are spread, rolled, and consolidated. Sheep's foot rollers do the compacting, but they are helped in their work by heavy earth-moving vehicles bringing fill to the dam or bulldozing it into place. Portions of embankment that cannot be rolled in this way are compacted by hand or power tampers. Strips adjacent to concrete core walls, the walls of outlet structures, and the wingwalls of spillway sections are examples.

In *hydraulic fills* water-carried soil is deposited differentially to form an embankment graded from coarse at the two faces of the dam to fine in the central core.

Methods as well as materials of construction determine the strength, tightness, and stability of embankment dams. Whether their axis should be straight or curved depends largely on topographic conditions. Whether upstream curves are in fact useful is open to question. The intention is to provide axial compression in the core and prevent cracks as the dam settles. Spillways are incorporated into some embankment dams and divorced from others in separate constructions.

Where rock outcrops on canyon walls can be blasted into the streambed or where spillways or stream diversion tunnels are constructed in rock, rock embankment becomes particularly economical. In modern construction, rock fills are given internal clay cores or membranes in somewhat the same fashion as earth fills (Fig. 2.11). Concrete slabs or timber sheathing once much used on the upstream face can be dangerously stressed and fail as the fill itself, or its foundation, settles. They are no longer in favor.

2.10.2 Masonry Dams

In the construction of gravity dams, *cyclopean* masonry and mass concrete embedding great boulders have, in the course of time, given way to poured concrete; in the case of arched dams rubble has also ceded the field to concrete. Gravity dams are designed to be in compression under all conditions of loading. They will fit into almost any site with a suitable foundation.

Some arched dams are designed to resist water pressures and other forces by acting as vertical cantilevers and horizontal arches simultaneously; for others, arch action alone is assumed, thrust being transmitted laterally to both sides of the valley, which must be strong enough to serve as abutments. In constant-radius dams, the upstream face is vertical or, at most, slanted steeply near the bottom; the downstream face is projected as a series of concentric, circular contours in plan. Dams of this kind fit well into U-shaped valleys, where cantilever action is expected to respond favorably to the high-intensity bottom loads. In constant-angle dams, the upstream face bulges up valley; the downstream face curves inward like the small of a man's back. Dams of this kind fit well into V-shaped valleys, where arch action becomes their main source of strength at all horizons.

Concrete buttresses are designed to support flat slabs or multiple arches in buttress dams. Here and there, wood and steel structures have taken the place of reinforced concrete. Their upstream face is normally sloped one on one and may terminate in a vertical cutoff wall.

All masonry dams must rest on solid rock. Foundation pressures are high in gravity dams; abutment pressures are intense in arched dams. Buttress dams are light on their foundations. Making foundations tight by sealing contained pockets or cavities and seams or faults with cement or cement-and-sand grout under pressure is an important responsibility. Low-pressure grouting (up to 40 psig or 278 kPa) may be followed by high-pressure grouting (200 psig or 1,390 kPa) from permanent galleries in the dam itself, and a curtain of grout may be forced into the foundation at the heel of gravity dams to obstruct seepage. Vertical drainage holes just downstream from the grout curtain help reduce uplift.

2.11 Spillways

Spillways have been built into the immediate structure of both embankment and masonry dams, in each instance as masonry sections (see Fig. 2.5). Masonry dams may indeed serve as spillways over their entire length. In general, however, spillways are placed at a distance from the dam itself to divert flow and direct possible destructive forces—generated, for example, by ice and debris, wave action, and the onward rush of waters—away from the structure rather than toward it. Saddle dams or dikes may be built to a lower elevation than the main impounding dam in order to serve as emergency floodways.

The head on the spillway crest at time of maximum discharge is the principal component of the *freeboard*, namely, the vertical distance between maximum reservoir level and elevation of dam crest. Other factors are wave height (trough to crest), wave runup on sloping upstream faces, wind setup or tilting of the reservoir surface by the drag exerted in the direction of persistent winds in common with differences in barometric pressure, and (for earth embankments only) depth of frost.

Overflow sections of masonry and embankment dams are designed as masonry structures and separate spillways as *saddle*, *side channel*, and *drop inlet* or *shaft* structures. Spillways constructed through a *saddle* normally discharge into a natural floodway leading back to the stream below the dam. Usually they take the form of open channels and may include a relatively low overflow weir in the approach to the floodway proper. Overflow sections and overflow weirs must be calibrated if weir heads are to record flood discharges accurately, but their performance can be approximated from known calculations of similar structures. If their profile conforms to the ventilated lower nappe of a sharp-crested weir of the same relative height d/h (Fig. 2.12), under the design head, h, the rate of discharge, Q, becomes

$$Q = 2/3\mathrm{c}\ \sqrt{2g}\, lh^{3/2} = \mathrm{C}lh^{3/2}, \tag{2.9}$$

where C = 2/3 c is the coefficient of discharge, g the gravity constant, and l the unobstructed crest length of the weir. For a crest height d above the channel bottom, the magnitude of C is approximately

$$\mathrm{C} = 4.15 + 0.65h/d \text{ for } h/d < 4 \text{ or } \mathrm{C} = 4.15 \text{ to } 6.75. \tag{2.10}$$

Under heads h' other than the design head h, C approximates as follows:

$$\mathrm{C} = 4.15(h'/h)^{7/5} \text{ up to a ratio of } h'/h = 3.0. \tag{2.11}$$

If the entrance to the spillway is streamlined, little if any energy is lost—certainly no more than $0.05v^2/2g$. As suggested in Fig. 2.12, substantial quiescence within the reservoir must be translated into full channel velocity. Discharge is greatest when flow becomes critical. The velocity head h then equals one-third the height H of the reservoir surface above the entrance sill to a rectangular channel, and the rate of discharge Q becomes

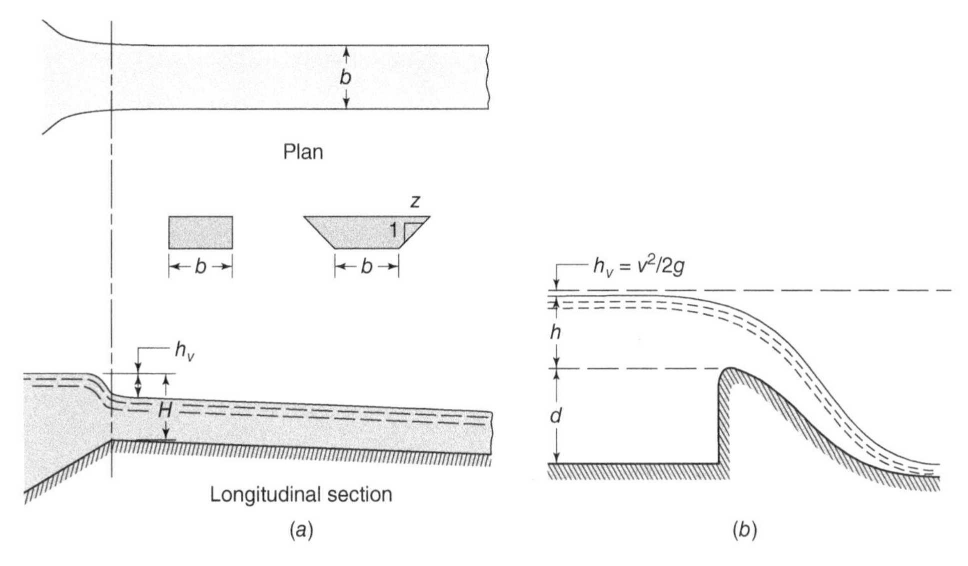

Figure 2.12 Spillways: (a) Channel Spillway and (b) Ogee Spillway (Kindness of Arthur Casagrande)

$$Q = 2/3CbH\sqrt{2gH/3} = 3.087CbH^{3/2}, \tag{2.12}$$

where b is the width of the channel and C is an entrance coefficient varying from 1.0 for a smooth entrance to 0.8 for an abrupt one. A trapezoidal channel with side slopes of 1:2 discharges

$$Q = 8.03Ch_v{}^{1/2}\,(H - h_v)[b + z(H - h_v)]. \tag{2.13}$$

where

$$h_v = \frac{3(2zH + b) - \left(16z^2H^2 + 16zbH + 9b^2\right)^{1/2}}{10z}. \tag{2.14}$$

Best hydraulic but not necessarily best economic efficiency is obtained when a semicircle can be inscribed in the cross-section.

Flow is uniform below the entrance when friction and channel slope are in balance. Otherwise, flow becomes nonuniform and channel cross-section must be adjusted accordingly. A weir within the channel produces a backwater curve.

Side-channel spillways occupy relatively little space in the cross-section of a valley. The crest more or less parallels one abutting hillside and can be made as long as wanted. The channel, into which the spillway pours its waters, skirts the end of the dam and delivers its flows safely past the toe. If it is blasted out of tight rock, the channel can be left unlined. Crest length and channel size are determined in much the same way as for washwater gutters in rapid sand filters.

As shown in Fig. 2.13, *shaft* or *drop-inlet spillways* consist of an overflow lip supported on a shaft rising from an outlet conduit, often the original stream-diversion tunnel. The lip can be of any desired configuration. A circular-lip and trumpet-like transition to the shaft form a *morning-glory spillway* that must lie far enough from shore to be fully effective. By contrast, a three-sided semicircular lip can be placed in direct contact with the shore; accessibility is its advantage. The capacity of shaft spillways is governed by their constituent parts and by flow conditions including air entrainment and hydraulic submersion. Hydraulic efficiency and capacity are greatest when the conduit flows full. Model studies are useful in arriving at suitable dimensions.

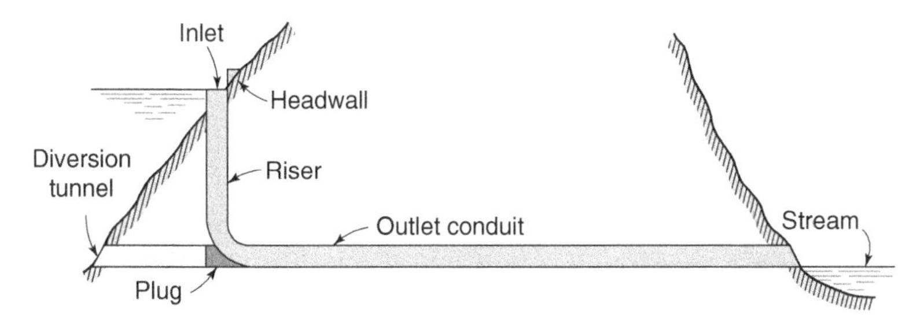

Figure 2.13 Shaft Spillway and Diversion Tunnel

Flashboards or stop logs and gates of many kinds are added to spillways to take advantage of storage above crest level. They must be so designed and operated that the dam itself is not endangered in times of flood.

2.12 Intakes

Depending on the size and nature of the installation, water is drawn from rivers, lakes, and reservoirs through relatively simple submerged intake pipes, or through fairly elaborate tower-like structures that rise above the water surface and may house intake gates; openings controlled by stop logs; racks and screens, including mechanical screens, pumps, and compressors; chlorinators and other chemical feeders; venturi meters and other measuring devices; and even living quarters and shops for operating personnel (Fig. 2.3). Important in the design and operation of intakes is that the water they draw be as clean, palatable, and safe as the source of supply can provide.

2.12.1 River Intakes

Understandably, river intakes are constructed well upstream from points of discharge of wastewater and industrial wastes. An optimal location will take advantage of deep water, a stable bottom, and favorable water quality (if pollution hugs one shore of the stream, for example), all with proper reference to protection against floods, debris, ice, and river traffic (Fig. 2.14). Small streams may have to be dammed up by *diversion* or *intake dams* to keep intake pipes submerged and preclude hydraulically wasteful air entrainment. The resulting intake pool will also work well as a settling basin for coarse silt and allow a protective sheet of ice to form in winter.

2.12.2 Lake and Reservoir Intakes

Lake intakes are sited with due reference to sources of pollution, prevailing winds, surface and subsurface currents, and shipping lanes. As shown in Fig. 2.15, shifting the depth of draft makes it possible to collect clean bottom water when the wind is offshore and, conversely, clean surface water when the wind is onshore. If the surrounding water is deep enough, bottom sediments will not be stirred up by wave action, and ice troubles will be few.

Reservoir intakes resemble lake intakes but generally lie closer to shore in the deepest part of the reservoir (see Fig. 2.6). They are often incorporated into the impounding structure itself (see Fig. 2.4). Where a reservoir serves many purposes, the intake structure is equipped with gates, conduits, and machinery not only for water supply but also for the regulation of low

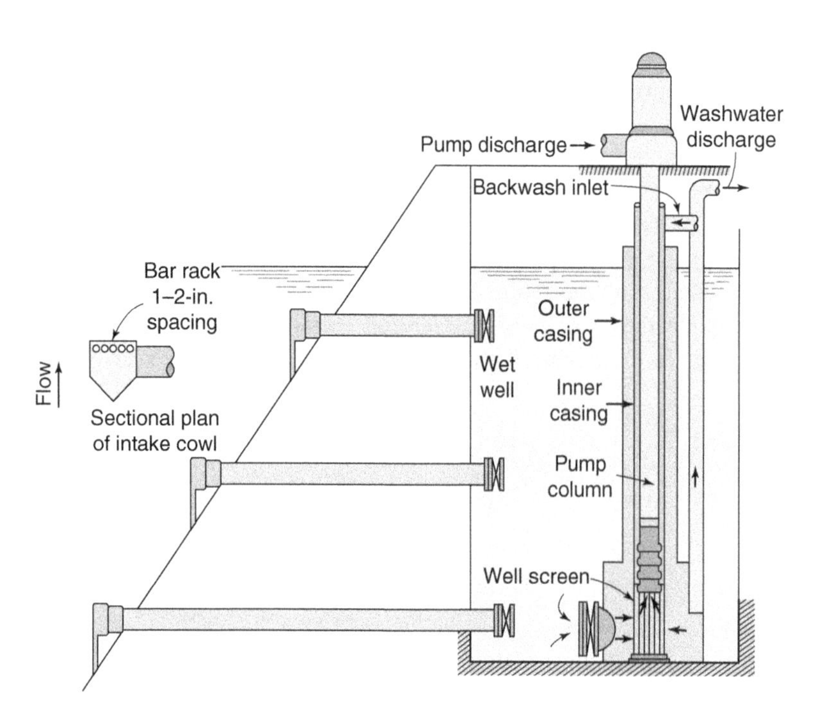

Figure 2.14 River or Lake Intake with Vertical Pump and Backwashed Well-Type Screen

water flows (including compensating water), generation of hydroelectric power, release of irrigation waters, and control of floods. Navigation locks and fish ladders or elevators complete the list of possible control works.

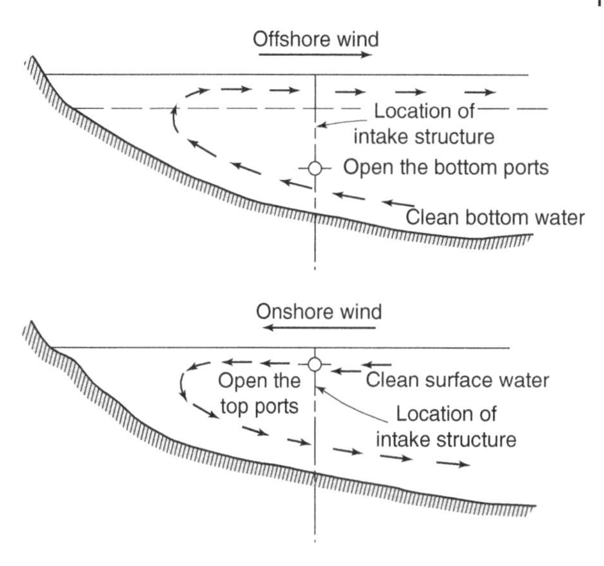

Figure 2.15 Effect of Onshore and Offshore Winds on Water Quality at Water Intake

2.12.3 Submerged and Exposed Intakes

Submerged intakes are constructed as *cribs* or *screened bell-mouths*. Cribs are built of heavy timber weighted down with rocks to protect the intake conduit against damage by waves and ice and to support a grating that will keep large objects out of the central intake pipe.

Exposed intake gatehouses, often still misnamed cribs, are tower-like structures built (a) into dams, (b) on banks of streams and lakes, (c) sufficiently near the shore to be connected to it by a bridge or causeway, and (d) at such distance from shore that they can be reached only by boat (see Figs. 2.3 and 2.4). In *dry intakes*, ports in the outer wall admit water to gated pipes that bridge a circumferential dry well and open into a central wet well comprising the entrance to the intake conduit. In *wet intakes*, water fills both wells. Open ports lead to the outer well, whence needed flows are drawn through gated openings into the inner well.

2.12.4 Intake Velocities and Depths

In cold climates, ice troubles are reduced in frequency and intensity if intake ports lie as much as 25 ft (7.5 m) below the water surface and entrance velocities are kept down to 0.30 ft/s (0.9 m/s). At such low velocity, ice spicules, leaves, and debris are not entrained in the flowing water and fish are well able to escape from the intake current.

Bottom sediments are kept out of intakes by raising entrance ports 4–6 ft (1.2–1.8 m) above the lake or reservoir floor. Ports controlled at numerous horizons permit water quality selection and optimization. A vertical interval of 15 ft (4.5 m) is common. Submerged gratings are given openings of 2–3 in. (5–7.5 cm). Specifications for screens commonly call for 2–8 meshes to the inch and face (approach) velocity of 0.30 ft/s (0.9 m/s). Wet wells should contain blow-off gates for cleaning and repairs.

2.12.5 Intake Conduits and Pumping Stations

Intakes are connected to the shores of lakes and reservoirs (a) by pipelines (often laid with flexible joints) or (b) by tunnels blasted through rock beneath the lake or reservoir floor. Pipelines are generally laid in a trench on the floor and covered after completion. This protects them against disturbance by waves and ice. Except in rock, conduits passing through the foundations of dams are subjected to heavy loads and to stresses caused by consolidation of the foundation.

Intake conduits are designed to operate at self-cleansing velocities of 3–4 ft/s (0.9–1.2 m/s). Flow may be by gravity or suction. Pump wells are generally located on shore. Suction lift, including friction, should not exceed 15–20 ft (4.5–6 m). Accordingly, pump wells or rooms are often quite deep. The determining factor is the elevation of the river, lake, or reservoir in times of drought. Placing pumping units in dry wells introduces problems of hydrostatic uplift and seepage in times of flood. Wet wells and deep-well pumps may be used instead.

2.13 Diversion Works

Depending on the geology and topography of the dam site and its immediate surroundings, streams are diverted from the construction area in two principal ways:

1) The entire flow is carried around the site in a diversion conduit or tunnel. An upstream cofferdam and, if necessary, a downstream cofferdam lay the site dry. After fulfilling its duty of bypassing the stream and protecting the valley during construction, the diversion conduit is usually incorporated in the intake or regulatory system of the reservoir (see Figs. 2.4 and 2.13).

2) The stream is diverted to one side of its valley, the other side being laid dry by a more or less semicircular cofferdam. After construction has progressed far enough in the protected zone, stream flow is rediverted through a sluiceway in the completed section of the dam, and a new cofferdam is built to pump out the remaining portion of the construction site.

Diversion conduits are built as grade aqueducts and tunnels, or as pressure conduits and tunnels. As a matter of safety, however, it should be impossible for any conduit passing through an earth embankment dam to be put under pressure; a leak might bring disaster. Accordingly, gates should be installed only at the inlet portal, never at the outlet portal. If a pipe must work under pressure, it should be laid within a larger access conduit. To discourage seepage along their outer walls, conduits passing through earth dams or earth foundations are often given projecting fins or collars that increase the length of path of seepage (by, say, 20% or more) and force flow in the direction of minimum as well as maximum permeability. At their terminus near the toe of the dam, moreover, emerging conduits should be surrounded by rock, through which residual seepage waters can escape safely.

The capacity of diversion conduits is determined by flood-flow requirements. Variations in the head and volume of floodwater impounded behind the rising dam are important factors in this connection. Rising heads normally increase the capacity of diversion conduits, and increasing storage reduces the intensity of floods. At the same time, however, dangers to the construction site and the valley below mount higher.

2.14 Collection of Rainwater

Rain is rarely the immediate source of municipal water supplies—a notable example is the water supply of the communities in the islands of Bermuda, on which streams are lacking and groundwater is brackish. The use of rainwater is generally confined (a) to farms and towns in semiarid regions devoid of satisfactory groundwater or surface water supplies, and (b) to some hard-water communities in which, because of its softness, roof drainage is employed principally for household laundry work and general washing purposes, while the public supply satisfies all other requirements. In most hard-water communities, the installation and operation of municipal water-softening plants can ordinarily be justified economically. Their introduction is desirable and does away with the need for supplementary rainwater supplies and the associated objection of their possible cross-connection with the public supply.

For individual homesteads, rainwater running off the roof is led through gutters and downspouts to a rain barrel or cistern situated on the ground or below it (Fig. 2.16). *Barrel* or *cistern* storage converts the intermittent rainfall into a continuous supply. For municipal service, roof water may be combined with water collected from sheds or catches on the surface of ground that is naturally impervious or rendered so by grouting, cementing, paving, or similar means.

The *gross yield* of rainwater supplies is proportional to the receiving area and the amount of precipitation. Some rain, however, is blown off the roof by wind, evaporated, or lost in wetting the collecting area and conduits and in filling depressions or improperly pitched gutters. Also, the first flush of water contains most of the dust and other undesirable washings from the catchment surfaces and may have to be wasted. The combined loss is particularly great during the dry season of the year. A cutoff, switch, or deflector in the downspout permits selection of the quality of water to be stored. *Sand filters* (Fig. 2.17) are successfully employed to cleanse the water and prevent its deterioration (a) by growth of undesirable organisms and (b) by the bacterial decomposition of organic materials, both of which may give rise to tastes, odors, and other changes in the attractiveness and palatability of the water.

Storage to be provided in cisterns depends on seasonal rainfall characteristics and commonly approximates one-third to one-half of the annual needs in accordance with the length of dry spells. If the water is to be filtered before storage, standby capacity in advance of filtration must be provided if rainfalls of high intensity are not to escape. Because of the relatively small catchment area available, roof drainage cannot be expected to yield an abundant supply of water, and a close analysis of storm rainfalls and seasonal variations in precipitation must be made if catchment areas, standby tanks, filters, and cisterns are to be proportioned and developed properly.

A properly located and constructed controlled catchment and cistern, augmented by satisfactory treatment facilities, will provide safe water. A *controlled catchment* is a defined surface area from which rainfall runoff is collected. For these controlled catchments, simple guidelines to determine water yield from rainfall totals can be established. When the controlled catchment area has a smooth surface or is paved and the runoff is collected in a cistern, water loss due to evaporation, replacement of soil moisture deficit, and infiltration is small. As a general rule, losses from smooth concrete or asphalt-covered ground catchments average less than 10%. For shingled roofs or tar and gravel surfaces, losses should not exceed 15%; and for sheet metal roofs, the

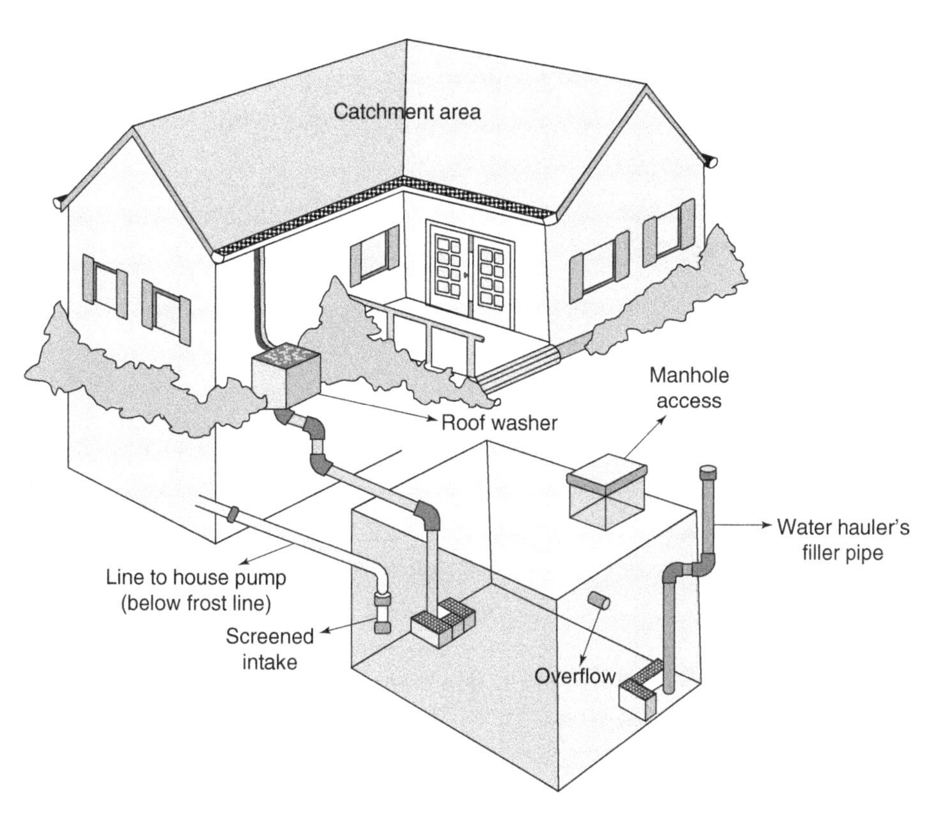

Figure 2.16 Rainwater Collection

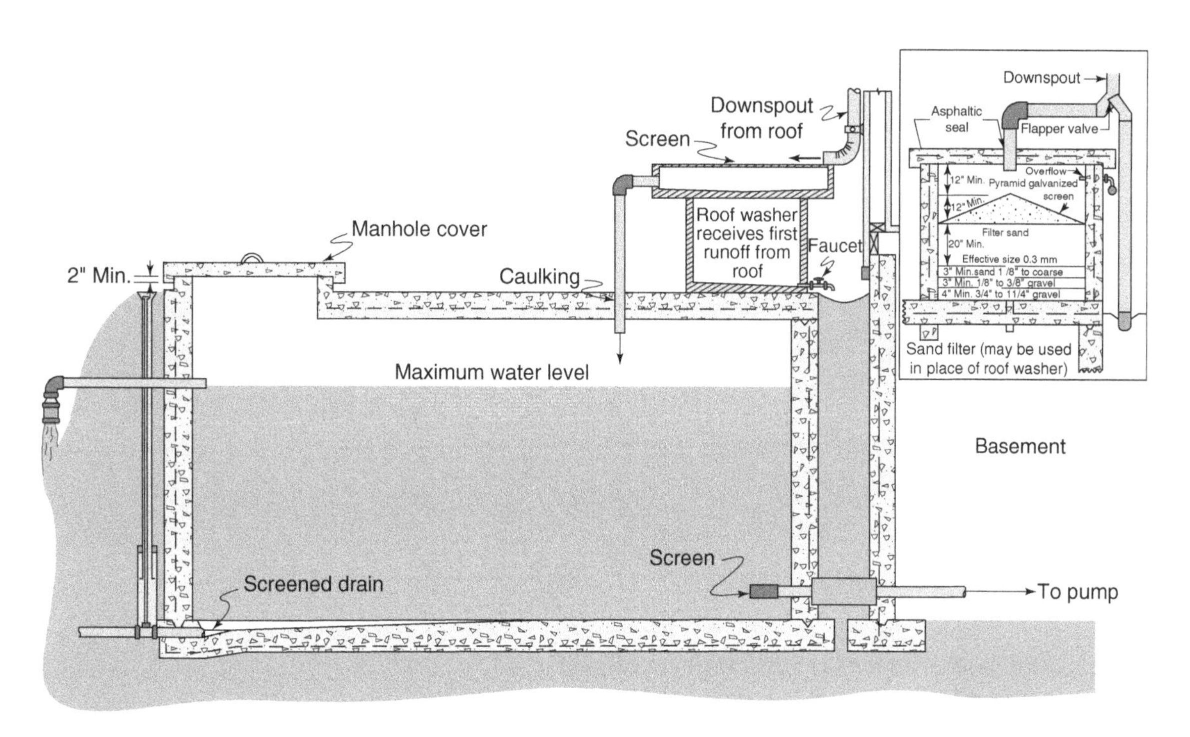

Figure 2.17 Cistern Equipped with Sand Filtration for Collection of Surface Water in Rural Area for Drinking Conversion factor: 1″ = 1 in. = 2.54 cm

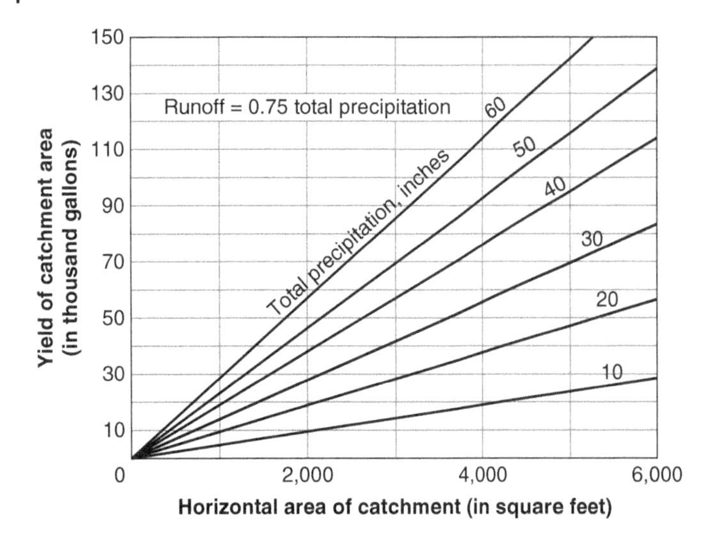

Figure 2.18 Yield of Impervious Catchment Area. Conversion factors: 1 in. = 25.4 mm; 1 gal = 3.785 L; 1 ft^2 = 0.0929 m^2

loss is negligible. A conservative design can be based on the assumption that the amount of water that can be recovered for use is three-fourths of the total annual rainfall. The location of the cistern should be governed by both convenience and quality protection. A cistern should be as close to the point of ultimate use as practical. A cistern should not be placed closer than 50 ft (15 m) to any part of a sewage-disposal installation and should be on higher ground.

Cisterns collecting water from roof surfaces should be located adjacent to the building, but not in basements subject to flooding. They may be placed below the surface of the ground for protection against freezing in cold climates and to keep water temperatures low in warm climates, but should be situated on the highest ground practicable, with the surrounding area graded to provide good drainage.

The size of cistern needed will depend on the size of the family and the length of time between periods of heavy rainfall. The size of the catchment area will depend on the amount of rainfall and the character of the surface. The safety factor allowed should be for lower-than-normal rainfall levels. Designing for two-thirds of the mean annual rainfall will result usually in a catchment area of adequate capacity (Fig. 2.18).

Example 2.5 Sizing of Cistern and Required Catchment Area

A farmhouse for a family of four people has minimum drinking and culinary requirements of 120 gpd (454 L/d). The mean annual rainfall is 45 in. (114 cm), and the effective period between rainy periods is 100 days. Determine

1) the size of the required rainfall collection cistern and
2) the size of the required catchment area.

Solution 1 (US Customary System):

1) The size of the required rainfall collection cistern:

The minimum volume of the cistern required will be

$$120\,\text{gpd} \times 100\,\text{days} = \mathbf{12{,}000\ gal} = 12{,}000/7.48 = 1{,}604\,\text{ft}^3.$$

Say, a **10-ft deep, 13-ft-square cistern**.

2) The size of the required catchment area:

Because the mean annual rainfall is 45 in., then the total design rainfall is

$$45 \times 2/3 = 30\,\text{in.}$$

The total year's requirement =365 days × 120 gpd = 43,800 gal.

Referring to Fig. 2.18, the catchment area required to produce 43,800 gal from a design storm of 30 in. is **3,300 ft^2**.

Say, a **100 - ft × 33 - ft catchment area**.

Solution 2 (SI System):

1) The size of the required rainfall collection cistern:
The minimum volume of the cistern required will be

$$(454.2\ \text{L/d}) \times 100\ \text{days} = \mathbf{45{,}420\ L} = 45.42\ \text{m}^3.$$

Say, a **3-m-deep, 4-m-square cistern (3 m × 4 m × 4 m)**.

2) The size of the required catchment area:
Because the mean annual rainfall is 114 cm, then the total design rainfall is

$$(114\ \text{cm}) \times 2/3 = 76.2\ \text{cm}.$$

The total year's requirement =(365 day)(454 L/d) = 165,800 L.
Referring to Fig. 2.18, the catchment area required to produce 165,800 L (43,800 gal) from a design storm of 76 cm (30 in.) is 307 m^2 (**3,300 ft^2**).
Say, a **31-m × 10-m catchment area.**

Problems/Questions

2.1 The response of a new water surface will establish new hydrologic equilibria. Determine the equivalent mean draft under the following conditions: (a) areal rate of runoff of the original watershed Q = 27.5 in. /year (69.85 cm/year), (b) areal rate of evaporation E = 40.5 in. /year (102.87 cm/year), (c) areal rate of rainfall R = 47.5 in. /year (120.65 cm/year), (d) reservoir area a = 1,500 acres (6.07 km^2), (e) catchment area A = 25,600 acres (103.60 km^2), and (f) mean annual water surface = 90% of reservoir area at spillway level.

2.2 Determine the revised mean annual runoff, Q_r, assuming (a) the original mean annual runoff Q = 27.5 in. /year (69.85 cm/year) and (b) the equivalent mean draft D_e = 1.08 in. /year (2.74 cm/year).

2.3 Determine the effective catchment area assuming (a) original catchment area A = 25,600 acres (103.60 km^2), (b) reservoir area a = 1,500 acres (6.07 km^2), (c) mean annual water surface = 90% of reservoir area at spillway level, (d) original watershed's runoff rate Q = 27.5 in. /year (69.85 cm/year), (e) evaporation rate E = 40.5 in. /year (102.87 cm/year), and (f) rainfall rate R = 47.5 in. /year (120.65 cm/year).

2.4 In some parts of the United States, the volume of silt V_s in acre-ft deposited annually can be approximated by the following equation:

$$V_s = c(A)^n,$$

where A is drainage area, mi^2; c = 0.43 for low deposition, 1.7 for average deposition, 4.8 for high deposition; and n = 0.77 for southwestern streams.
Determine the volume of silt accumulations for drainage area, where A, c, and n are 40 mi^2 (103.60 km^2), 1.7, and 0.77, respectively.

2.5 The volume of silt deposited annually can be approximated by the following equation:

$$V_s = 1{,}233.5c\left[\left(A/\left(2.59 \times 10^6\right)\right]^n\right.$$

where V_s is the volume of silt deposited annually, m^3; A is the size of the drainage area, m^2; c is a coefficient with a value varying from 0.43 through 1.7 to 4.8 for low, average, and high deposition, respectively; and n is a coefficient to be determined specifically for a target drainage area.
Determine the volume of silts deposited annually V_s in m^3, assuming it is known that c = 1.7, n = 0.77, and $A = 648 \times 10^6\ m^2$ (160,119 acre).

2.6 Using the same metric equation from Problem 2.5 of $V_s = 1{,}233.5c[(A/(2.59 \times 10^6)]^n$ and assuming c = 1.7, n = 0.77, and drainage area = 984×10^6 m^2 (243,143 acre), determine the volume of silt deposited annually.

2.7 A research team has investigated two drainage areas in the southwestern region of the United States and has measured the volumes of their annually deposited silts. The following are the field data collected by the researchers using the metric units.

For drainage area 1:

$$A_1 = 984 \times 10^6 \text{m}^2 = 243{,}143 \text{ acre} = 379.8 \text{ mi}^2$$

$$V_{s1} = 203{,}000 \text{ m}^3 = 164.6 \text{ acre-ft.}$$

For drainage area 2:

$$A_2 = 648 \times 10^6 \text{m}^2 = 160{,}119 \text{ acre} = 250 \text{ mi}^2$$

$$V_{s2} = 147{,}000 \text{ m}^3 = 119.17 \text{ acre-ft.}$$

Determine the coefficients of c and n for the southwestern region of the United States.

2.8 The volume of silt deposited annually can be approximated by the following equation using English units:

$$V_s = c(A)^n,$$

where V_s is the volume of silt deposited annually, acre - ft; A is the size of the drainage area, mi^2; c is a coefficient with a value varying from 0.43 through 1.7 to 4.8 for low, average, and high deposition, respectively; and n is a coefficient to be determined specifically for a target drainage area.

A research team has investigated two drainage areas in the southwestern region of the United States and has measured the volumes of their annually deposited silts. The following are the field data collected by the researchers using English units.

For drainage area 1:

$$A_1 = 380 \text{ mi}^2 = 984.2 \times 10^6 \text{ m}^2$$

$$V_{s1} = 165 \text{ acre-ft} = 203{,}528 \text{ m}^3.$$

For drainage area 2:

$$A_2 = 250 \text{ mi}^2 = 647.5 \times 10^6 \text{ m}^2$$

$$V_{s1} = 120 \text{ acre-ft} = 148{,}020 \text{ m}^3.$$

Determine the coefficients of c and n for the southwestern region of the United States.

2.9 A spillway weir has an unobstructed crest length of 20 ft (6.1 m) and a coefficient of discharge equal to 5.50. If the design water head is 0.75 ft (0.23 m), what is the rate of discharge?

2.10 A house for a family of four people has minimum drinking and culinary requirements of 25 gpcd (95 Lpcd). The mean annual rainfall is 50 in./year (127 cm/year) and the effective period between rainy periods is 150 days. Determine (a) the size of the required rainfall collection cistern and (b) the size of the required catchment area.

Bibliography

American Society of Civil Engineers, *Design and Construction of Dams*, Reston, VA, USA 1967.

American Water Works Association, *Water Sources*, AWWA, Denver, CO. 202 pages, 1995.

Bradley, J. N., Wagner, W. E., and Peterka, A. J., Morning-Glory Shaft Spillways: A Symposium, *Trans. Am. Soc. Civil Eng.*, *121*, 312, 1956.

Department of Water Supply website, County of Maui, Hawaii, http://mauiwater.org/, 2009.

Fair, G. M., Geyer, J. C., and Okun, D. A., *Elements of Water Supply and Wastewater Disposal*, John Wiley & Sons, New York, USA, 1971.

Fair, G. M., Geyer, J. C., and Okun, D. A., Water and Wastewater Engineering: *Water Supply and Wastewater Removal*, vol. *1*, John Wiley & Sons, New York, USA, 1966.

Hazen, A., as revised by Richard Hazen, in Abbott, R. W. (Ed.), *American Civil Engineering Practice*, vol. *2*, John Wiley & Sons, New York, 1956, pp. 18–09.

Ippen, A. T., Channel Transitions and Controls, in Rouse, H. (Ed.), *Engineering Hydraulics*, John Wiley & Sons, New York, pp. 534, 535, 1950.

Jenkins, J. E., Moak, C. E., and Okun, D. A., Sedimentation in Reservoirs in the Southeast, *Trans. Am. Soc. Civil Engrs. III*, *68*, 3133, 1961.

La Mer, V. K. (Ed.), *Retardation of Evaporation by Monolayer: Transport Processes*, Academic Press, New York, 1962.

Lischer, V. C., and Hartung, H. O., Intakes on Variable Streams, *J. Am. Water Works Ass.*, *33*, 873, 1952.

National Resources Conservation Service, U.S. Department of Agriculture, *Water Supply Forecasting*, http://www.wcc.nrcs.usda.gov/wsf/, 2009.

Sherard, J. L., Woodward, R. J., Gizienski, S. F., and Clevenger, W. A., *Earth and Earth-Rock Dams*, John Wiley & Sons, New York, 1963.

Wang, L. K., and Wang, M. H. S., Understanding Evaporation, Transpiration, Evapotranspiration, Precipitation and Runoff Volume for Agricultural Waste Management, *Evolutionary Progress in Science, Technology, Engineering, Arts, and Mathematics*, vol. *4(5)*, May 2022. 81 pp. https://doi.org/10.17613/m8tf-zd10.

U.S. Environmental Protection Agency, Drinking Water Data, 2008. Drinking Water Data Tables. http://www.epa.gov/safewater/data/getdata.html.

U.S. Environmental Protection Agency, *Drinking Water and Ground Water Statistics for 2008*, EPA 816-K-08-004, Washington, DC, USA, December 2008.

U.S. Geological Survey, U.S. Department of the Interior, 2009, *Science for a Changing World*, http://ga.water.usgs.gov/edu/earthwherewater.html.

U.S. Geological Survey, U.S. Department of the Interior, *The Water Cycle*, http://ga.water.usgs.gov/edu/watercycle.html, 2009.

3

Water Sources: Groundwater

Groundwater from wells and springs has served as a source of domestic water supply since antiquity. Table 3.1 shows that in the United States more water systems have groundwater than surface water as a source—but more people drink from a surface-water system. Thirty-five percent (107 million) of the total population of 306 million served by public water systems depend on groundwater. The groundwater works installations are 10 times more than surface-water installations (about 141,648 versus 14,006; see Table 3.1); the average capacity of groundwater facilities is, however, much smaller. Contributions from groundwater also play a major role in the supplies depending on surface sources. It is the discharge of groundwater that sustains the dry-season flow of most streams.

Groundwater is more widely distributed than surface water. Its nearly universal, albeit uneven, occurrence and other desirable characteristics make it an attractive source of water supply. Groundwater offers a naturally purer, cheaper, and more satisfactory supply than does surface water. It is generally available at the point of use and obviates the need to incur substantial transmission costs. It occurs as an underground reservoir, thus eliminating the necessity of impoundment works. It is economical even when produced in small quantities.

To an increasing degree, engineers are being called on to investigate the possibility of developing groundwater as a usable resource. The following factors need to be considered:

1) The effective water content, that is, the maximum volume of water that can be withdrawn from a body of groundwater through engineering works. The effective *porosity* and *storage coefficient* of the water-bearing material control the useful storage.
2) The ability of the aquifer to transmit water in requisite quantities to wells or other engineering installations. *Permeability* and *transmissivity* are the indicators of this capability.
3) The suitability of the quality of water for the intended use, after treatment if necessary.
4) The reliability and permanence of the available supply with respect to both the quantity and the quality of water.

As a source of a permanent and reliable water supply, only that portion of the subsurface water that is in the zone of saturation need be considered. In this zone almost all the interstices are completely filled with water under hydrostatic pressure (atmospheric pressure or greater). That water is free to move in accordance with the laws of saturated flow from places where it enters the zone of saturation (recharge areas) to places where it is discharged. The main features of the groundwater phase of the hydrologic cycle are depicted in Fig. 3.1.

3.1 Porosity and Effective Porosity

The amount of groundwater stored in saturated materials depends on the material's *porosity,* the ratio of the aggregate volume of interstices in a rock or soil to its total volume. It is usually expressed as a percentage. The concept of porosity involves all types of interstices, both primary (original) and secondary. Primary interstices were created at the time of the rock's origin. In granular unconsolidated sediments, they coincide with intergranular spaces. In volcanic rocks, they include tubular and vesicular openings. Secondary interstices result from the action of geologic, mechanical, and chemical forces on the original rock. They include joints, faults, fissures, solution channels, and bedding planes in hard rocks. The extent of

Water and Wastewater Engineering: Hydraulics, Hydrology and Management, Volume 1, Fourth Edition.
Lawrence K. Wang, Mu-Hao Sung Wang, and Nazih K. Shammas.

Companion website: www.wiley.com/go/Wang/Waterandwastewater4e

Table 3.1 US Water Sources in 2007

Water System		Groundwater	Surface Water	Totals
Community water system[a]	No. of systems	40,646	11,449	52,095
	Population served	90,549,995	195,887,109	286,437,104
	Percentage of systems	78%	22%	100%
	Percentage of population	32%	68%	100%
Nontransient noncommunity water system[b]	No. of systems	18,151	679	18,830
	Population served	5,503,282	787,555	6,290,837
	Percentage of systems	96%	4%	100%
	Percentage of population	87%	13%	100%
Transient noncommunity water system[c]	No. of systems	82,851	1,878	84,729
	Population served	11,077,369	2,668,985	13,746,354
	Percentage of systems	98%	2%	100%
	Percentage of population	81%	19%	100%
Total no. of systems		141,648	14,006	155,654

[a] Community water system: a public water system that supplies water to the same population year-round.
[b] Nontransient noncommunity water system: a public water system that regularly supplies water to at least 25 of the same people at least 6 months per year, but not year-round. Some examples are schools, factories, office buildings, and hospitals that have their own water systems.
[c] Transient noncommunity water system: a public water system that provides water in a place such as a gas station or campground where people do not remain for long periods of time.
Source: Courtesy U.S. Environmental Protection Agency.

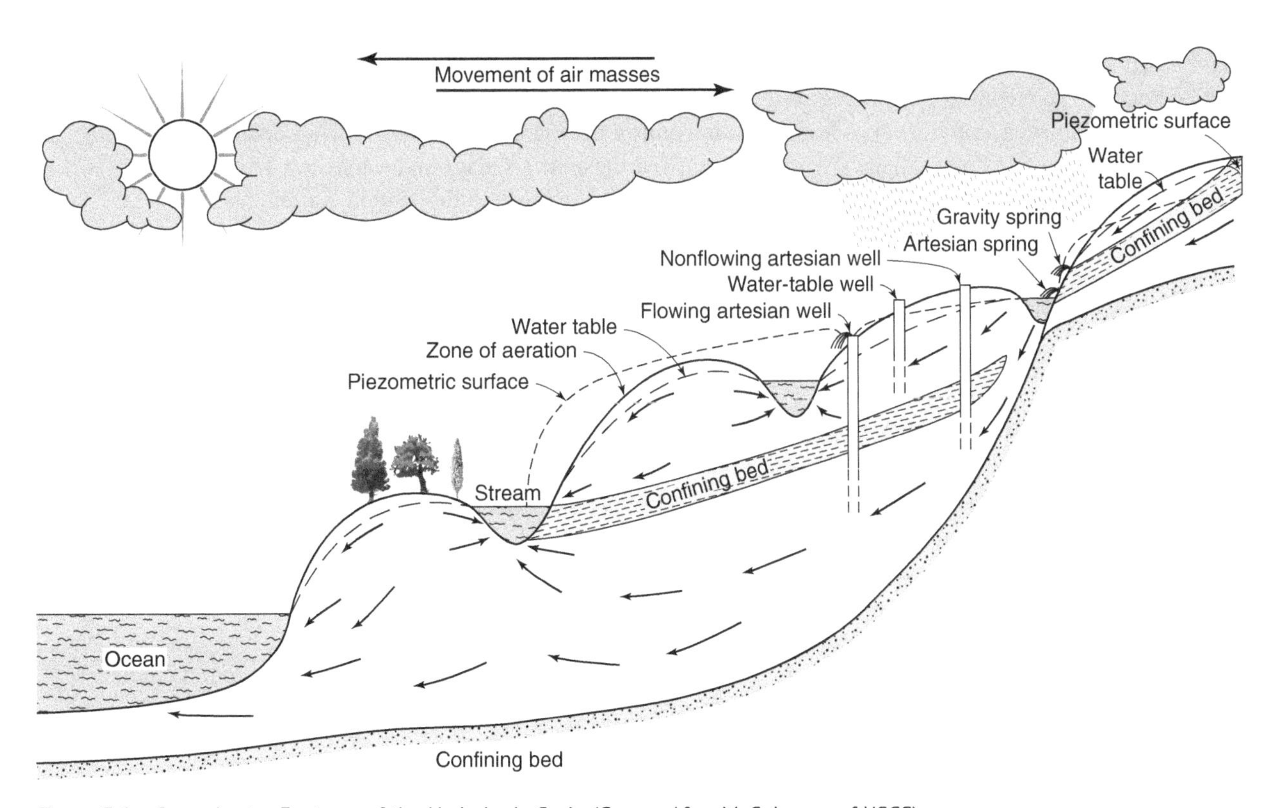

Figure 3.1 Groundwater Features of the Hydrologic Cycle (*Source:* After McGuinness of USGS)

fracturing and intensity of weathering exert a profound influence on the distribution of larger interstices. The importance of secondary porosity in determining the amount of water that can be obtained from a formation is often great in those hard rocks that lack intergrain porosity. This type of porosity is dependent on local conditions and gives water-bearing formations a heterogeneous character. The distribution of secondary porosity varies markedly with depth.

Porosity is a static quality of rocks and soils. It is not itself a measure of *perviousness* or *permeability,* which are dynamic quantities controlling the flow. Not all the water stored in a saturated material is available for movement; only the interconnected interstices can participate in flow. Water in isolated openings is held immobile. Furthermore, water in a part of the interconnected pore space is held in place by molecular and surface tension forces. This is the dead storage and is called *specific retention*. Thus not all the water stored in a geologic formation can be withdrawn by normal engineering operations. Accordingly, there is a difference between total storage and useful storage. That portion of the pore space in which flow takes place is called *effective porosity,* or *specific yield* of the material, defined as the proportion of water in the pores that is free to drain away or be withdrawn under the influence of gravity. Specific yields vary from zero for plastic clays to 30% or more for uniform sands and gravels. Most aquifers have yields of 10%–20%.

3.2 Permeability

The *permeability* or *perviousness* of a rock is its capacity for transmitting a fluid under the influence of a hydraulic gradient. An important factor affecting the permeability is the geometry of the pore spaces and of the rock particles. The nature of the system of pores, rather than their relative volume, determines the resistance to flow at given velocities. There is no simple and direct relationship between permeability and porosity. Clays with porosities of 50% or more have extremely low permeability; sandstones with porosities of 15% or less may be quite pervious.

A standard unit of intrinsic permeability, dependent only on the properties of the medium, is the *darcy,* D. It is expressed as flow, in cubic centimeters per second, of a fluid of one centipoise viscosity, through a cross-sectional area of 1 cm^2 of the porous medium under a pressure gradient of 1 atm/cm. It is equivalent to a water flow of 18.2 gpd/ft^2 (0.743 $m^3/d/m^2$) under a hydraulic gradient of 1 ft/ft (1 m/m) at a temperature of 60°F (15.5°C).

The *homogeneity* and *isotropy* of a medium refer to the spatial distribution of permeability. A porous medium is isotropic if its permeability is the same in all directions. It is called *anisotropic* if the permeability varies with the direction. Anisotropy is common in sedimentary deposits where the permeability across the bedding plane may be only a fraction of that parallel to the bedding plane. The medium is *homogeneous* if the permeability is constant from point to point over the medium. It is *nonhomogeneous* if the permeability varies from point to point in the medium. Aquifers with secondary porosity are nonhomogeneous. Isotropy and homogeneity are often assumed in the analysis of groundwater problems. The effects of nonhomogeneity and anisotropy can, however, be incorporated into an analysis under certain conditions.

3.3 Groundwater Geology

The geologic framework of an area provides the most valuable guide to the occurrence and availability of groundwater. Rocks, the solid matter forming Earth's crust, are an assemblage of minerals. In the geologic sense, the term *rock* includes both the hard, consolidated formations and loose, unconsolidated materials. With respect to their origin, they fall into three broad categories: *igneous, metamorphic,* and *sedimentary*.

The two classes of igneous rocks, intrusive and extrusive, differ appreciably in their hydrologic properties. Fresh intrusive rocks are compact and, in general, not water-bearing. They have very low porosities (less than 1%) and are almost impermeable. When fractured and jointed, they may develop appreciable porosity and permeability within a few hundred feet of the surface. Permeability produced by fracturing of unweathered rocks generally ranges from 0.001 to 10.0 D, where D is darcy, which is the unit of permeability, named after Henry Darcy.[1] Extrusive or volcanic rocks can be good aquifers.

Metamorphic rocks are generally compact and highly crystalline. They are impervious and make poor aquifers.

Rocks may be grouped into hydrologic units on the basis of their ability to store and transmit water. An *aquifer* is a body of rock that acts as a hydrologic unit and is capable of transmitting significant quantities of water. An *aquiclude* is a rock

1 One Darcy is equal to the passage of 1 cm^3 of fluid of 1 centipose viscosity in 1 second (i.e., 1 mL/s) under a pressure differential of 1 atmosphere having an area of cross-section of 1 cm^2 and a length of 1 cm.

formation that contains water but is not capable of transmitting it in significant amounts. Aquicludes usually form the boundaries of aquifers, although they are seldom absolute barriers to groundwater movement. They often contain considerable water in storage, and there is frequently some interchange between the free groundwater above an aquiclude and the confined aquifer below. Materials that have permeabilities intermediate between those of aquifers and aquicludes have been termed *aquitards*.

The boundaries of a geologic rock unit and the dimensions of an aquifer often do not correspond precisely. The latter are arrived at from the considerations of the degree of hydraulic continuity and from the position and character of hydrologic boundaries. An aquifer can thus be a geologic formation, a group of formations, or part of a formation.

Sedimentary formations include both consolidated, hard rocks (shale, sandstone, and limestone) and loose, unconsolidated materials (clay, gravel, and sand). Some sandstones may be almost impermeable, and others highly pervious. The degree of cementation plays a crucial role. Partially cemented or fractured sandstones have very high yields. Porosity of sandstones ranges from less than 5% to a maximum of about 30%. Permeability of medium-range sandstones generally varies from 1 to 500 mD (millidarcy).

Limestones vary widely in density, porosity, and permeability. When not deformed, they are usually dense and impervious. From the standpoint of water yield, secondary porosity produced as a result of fracturing and solution is more important than density and permeability. The nonuniform distribution of interstices in limestones over even short distances results because of marked differences in secondary porosity, which depends on local conditions. They are second only to sandstones as a source of groundwater. Limestones are prolific producers under suitable conditions.

Although consolidated rocks are important sources of water, the areas served by them in the United States are relatively small. Most developments lie in granular, unconsolidated sediments. Unconsolidated, sedimentary aquifers include (a) marine deposits, (b) river valleys, (c) alluvial fans, (d) coastal plains, (e) glacial outwash, and, to a much smaller degree, (f) dune sand. Materials deposited in seas are often extensive; sediments deposited on land by streams, ice, and wind are less extensive and are usually discontinuous.

Sands and gravels are by far the best water-producing sediments. They have excellent water storage and transmission characteristics and are ordinarily so situated that replenishment is rapid, although extremely fine sands are of little value. Porosity, specific yield, and permeability depend on particle size, size distribution, packing configuration, and shape. Uniform or well-sorted sands and gravels are the most productive; mixed materials containing clay are least so. Boulder clay deposited beneath ice sheets is an example. Typical porosities lie between 25% and 65%. Gravel and coarse sands usually have specific yields greater than 20%.

Clays and silts are poor aquifers. They are highly porous but have very low permeabilities. However, the permeability is seldom zero. They are significant only when they (a) confine or impede the movement of water through more pervious soils and (b) supply water to aquifers through leakage by consolidation.

3.4 Groundwater Situation in the United States

Geologic and hydrologic conditions vary greatly in various parts of the United States. To permit useful generalizations about the occurrence and availability of groundwater, Thomas (1952) divided the United States into 10 major groundwater regions (Fig. 3.2). McGuiness (1963) provided an updated assessment of the groundwater situation in each of Thomas's regions and has also described the occurrence and development of groundwater in each of the states. The Water Resources Division (WRD) of the U.S. Geological Survey is the principal agency of the federal government engaged in groundwater investigations. The published reports and the unpublished data of the WRD are indispensable to any groundwater investigation. In addition, many states have agencies responsible for activities in groundwater.

3.5 Types of Aquifers

Because of the differences in the mechanism of flow, three types of aquifers are distinguished: (1) *unconfined* or *water table*, (2) *confined* or *artesian*, and (3) *semiconfined* or *leaky*.

Unconfined aquifers (also known as water table, *phreatic*, or *free* aquifers) are those in which the upper surface of the zone of saturation is under atmospheric pressure. This surface is free to rise and fall in response to the changes of storage in the saturated zone. The flow under such conditions is said to be unconfined. An imaginary surface connecting all rest or static

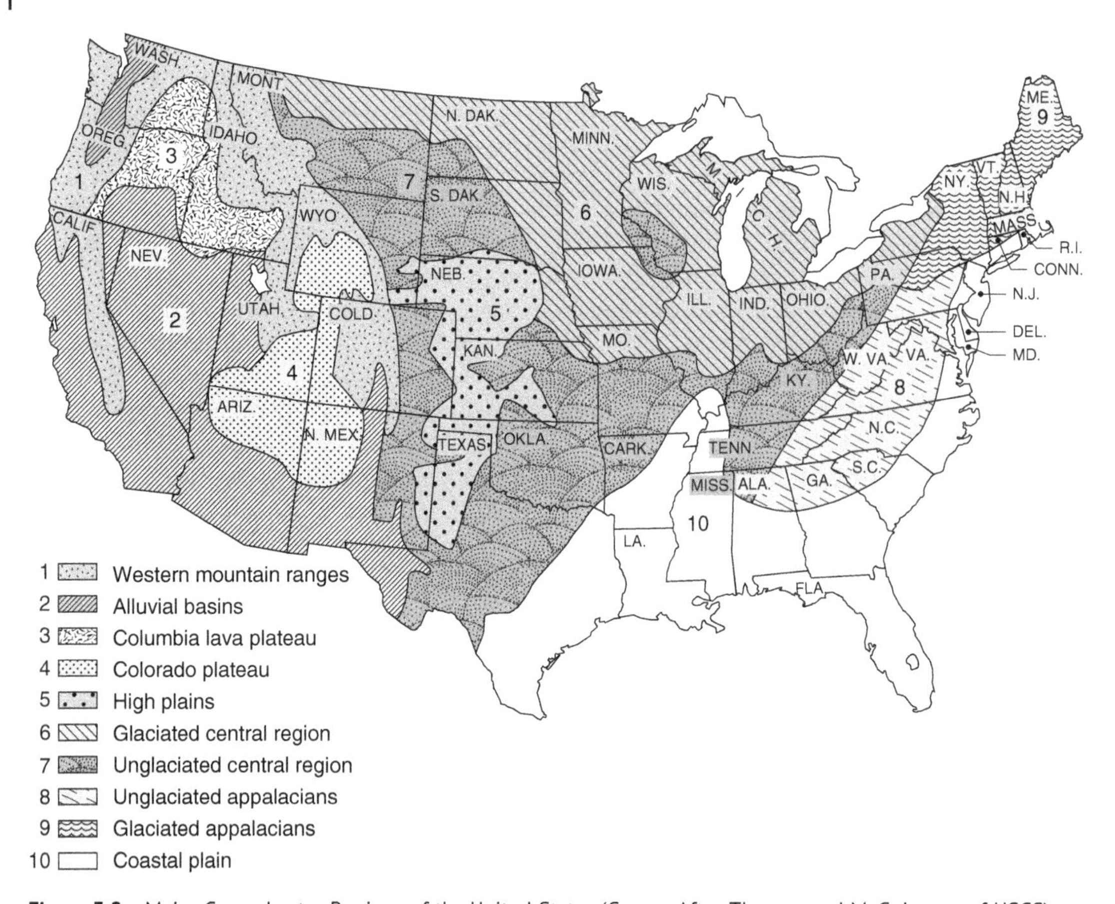

Figure 3.2 Major Groundwater Regions of the United States (*Source:* After Thomas and McGuinness of USGS)

levels in wells in an unconfined aquifer is its *water table* or *phreatic surface*. This defines the level in the zone of saturation, which is at atmospheric pressure. The water held by capillary attraction at less than atmospheric pressure may fully saturate the interstices to levels above those observed in wells. Thus the upper limit of the zone of saturation and water table are not coincident. The capillary fringe may be significant for sediments with small interstices and low permeability, such as clay.

More than one zone of saturation occurs when an impervious or semipervious layer or lens in the zone of aeration supports a less-extensive zone of saturation above the main water table, giving rise to the so-called *perched water table*.

If a porous stratum in the zone of saturation dips beneath an impervious layer, the flow is confined in much the same way as in a pipe that drops below the hydraulic grade line. There is no free surface in contact with the atmosphere in the area of confinement. The water level in a well tapping this confined or artesian aquifer will rise, under pressure, above the base of the confining layer to an elevation that defines the piezometric level. If the recharge areas are at a sufficiently high elevation, the pressure may be great enough to result in free-flowing wells or springs. An imaginary surface connecting the piezometric levels at all points in an artesian aquifer is called the *piezometric surface* (Fig. 3.1 depicts some of these terms). The rise and fall of water levels in artesian wells result primarily from changes in pressure rather than from changes in storage volume. The seasonal fluctuations are usually small compared with unconfined conditions.

Aquifers that are overlain or underlain by aquitards are called *leaky aquifers*. In natural materials, confining layers seldom form an absolute barrier to groundwater movement. The magnitude of flow through the semipervious layer is called *leakage*. Although the vertical permeability of the aquitard is very low and the movement of water through it extremely slow, leakage can be significant because of the large horizontal areas involved.

3.6 Groundwater Movement

Groundwater in the natural state is constantly in motion. Its rate of movement under the force of gravity is governed by the frictional resistance to flow offered by the porous medium. The difference in head between any two points provides the driving force. Water moves from levels of higher energy potential (or head) to levels of lower energy potential, the loss in head being dissipated as heat. Because the magnitudes of discharge, recharge, and storage fluctuate with time, the head distribution at various locations is not stationary. Groundwater flow is both unsteady and nonuniform. Compared with surface water, the rate of groundwater movement is generally very slow. Low velocities and the small size of passageways give rise to very low *Reynolds numbers* and consequently the flow is almost always *laminar. Turbulent* flow may occur in cavernous limestones and volcanic rocks, where the passageways may be large, or in coarse gravels, particularly in the vicinity of a discharging well. Depending on the intrinsic permeability, the rate of movement can vary considerably within the same geologic formation. Flow tends to be concentrated in zones of higher permeability, that is, where the interstices are larger in size and have a better interconnection.

In aquifers of high yield, velocities of 5–60 ft/d (1.5–18.3 m/d) are associated with hydraulic gradients of 10–20 ft/mi (1.89–3.78 m/km). Underflow through gravel deposits may travel several hundred ft/d (m/d). Depending on requirements, flows as low as a few ft/year (m/year) may also be economically useful.

In homogeneous, isotropic aquifers, the dominant movement is in the direction of greatest slope of the water table or piezometric surface. Where there are marked nonhomogeneities and anisotropies in permeability, the direction of groundwater movement can be highly variable.

3.7 Darcy's Law

Although other scientists were the first to propose that the velocity of flow of water and other liquids through capillary tubes is proportional to the first power of the hydraulic gradient, credit for verification of this observation and for its application to the flow of water through natural materials or, more specifically, its filtration through sand, must go to Darcy. The relationship known as *Darcy's law* may be written as follows:

$$v = K(dh/dl) = KI \tag{3.1}$$

where v is the hypothetical, superficial or face velocity (darcy; not the actual velocity through the interstices of the soil) through the gross cross-sectional area of the porous medium; $I = dh/dl$ is the *hydraulic gradient,* or the loss of head per unit length in the direction of flow; and K is a constant of proportionality known as *hydraulic conductivity,* or the *coefficient of permeability*. The actual velocity, known as *effective velocity,* varies from point to point. The average velocity through pore space is given by

$$v_e = KI/\theta \tag{3.2}$$

where θ is the effective porosity. Because I is a dimensionless ratio, K has the dimensions of velocity and is in fact the velocity of flow associated with a hydraulic gradient of unity.

The proportionality coefficient in Darcy's law, K, refers to the characteristics of both the porous medium and the fluid. By dimensional analysis,

$$K = Cd^2\gamma/\mu \tag{3.3}$$

where C is a dimensionless constant summarizing the geometric properties of the medium affecting flow, d is a representative pore diameter, μ is the viscosity, and γ is the specific weight of fluid. The product Cd^2 depends on the properties of the medium alone and is called the intrinsic or specific permeability of a water-bearing medium, $k = Cd^2$. It has the dimensions of area.

Hence,

$$K = k\gamma/\mu = k\rho g/\mu = kg/\nu \tag{3.4}$$

where ρ is the specific density and ν the kinematic viscosity.

The fluid properties that affect the flow are viscosity and specific weight. The value of K varies inversely with the kinematic viscosity, ν, of the flowing fluid. The ratio of specific weight to viscosity is affected by changes in the temperature and salinity of groundwater. Measurements of K are generally referred to a standard water temperature such as 60°F or 15.5°C. The necessary correction factor for field temperatures other than standard is provided by the relationship:

$$K_1/K_2 = \nu_1/\nu_2 \tag{3.5}$$

Most groundwaters have relatively constant temperatures, and this correction is usually ignored in practice and K is stated in terms of the prevailing water temperature. Special circumstances in which correction may be important include influent seepage into an aquifer from a surface-water body where temperature varies seasonally.

Darcy's law is applicable only to laminar flow, and there is no perceptible lower limit to the validity of the law. The volume rate of flow is the product of the velocity given by Darcy's law and the cross-sectional area A normal to the direction of motion. Thus,

$$Q = KA(dh/dl) \tag{3.6}$$

and solving for K,

$$K = Q/[A(dh/dl)] \tag{3.7}$$

Hydraulic conductivity may thus be defined as the volume of water per unit time flowing through a medium of unit cross-sectional area under a unit hydraulic gradient. In the standard coefficient used by the U.S. Geological Survey, the rate of flow is expressed in gpd/ft^2 (m^3/d/m^2) under a hydraulic gradient of 1 ft/ft (m/m) at a temperature of 60°F (15.5°C). This unit is called *meinzer*. For most natural aquifer materials, values of K fall in the range of 10–5,000 meinzers.

3.8 Aquifer Characteristics

The ability of an aquifer to transmit water is characterized by its *coefficient of transmissivity*. It is the product of the saturated thickness of the aquifer, b, and the average value of the hydraulic conductivities in a vertical section of the aquifer, K. The transmissivity, $T = Kb$, gives the rate of flow of water through a vertical strip of an aquifer 1 ft wide extending the full saturated thickness of the aquifer under a unit of hydraulic gradient. It has the dimensions of (length)^2/time, that is, ft^2/d or gpd/ft. Equation 3.6 can be rewritten as

$$Q = TW(dh/dl) \tag{3.8}$$

where W is the width of flow.

The *coefficient of storage* is defined as the volume of water that a unit decline in head releases from storage in a vertical prism of the aquifer of unit cross-sectional area (Fig. 3.3).

The physical processes involved when the water is released from (or taken into) storage in response to head changes are quite different in cases in which free surface is present from those in which it is not. A confined aquifer remains saturated during the withdrawal of water. In the case of a confined aquifer the water is released from storage by virtue of two

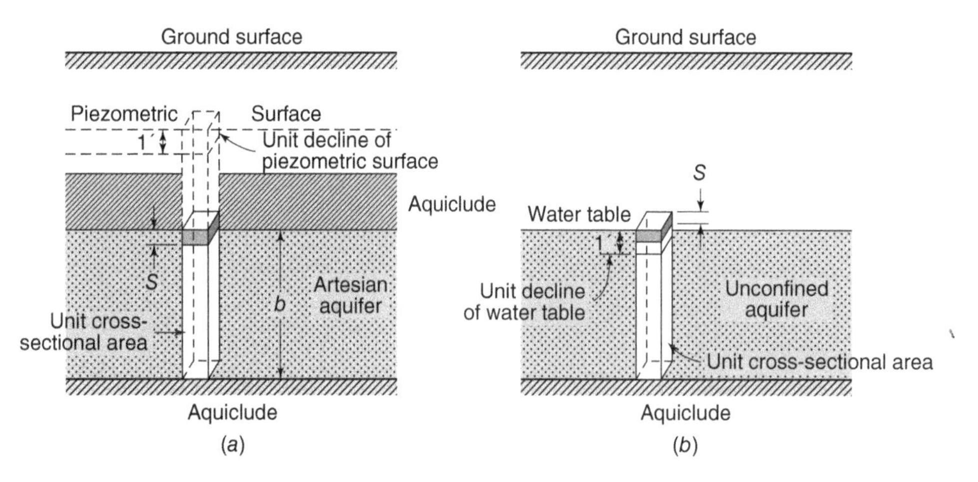

Figure 3.3 Graphical Representation of Storage Coefficient. The volume of water that a unit decline in head releases from storage in a vertical prism of the aquifer of unit cross-sectional area. (a) Confined aquifer and (b) unconfined aquifer. Conversion factor: 1′ = 1 ft = 0.3048 m

processes: (a) lowering of the water table in the recharge or intake area of the aquifer and (b) elastic response to pressure changes in the aquifer and its confining beds induced by the withdrawal of water. For this the storage coefficient is expressed as

$$S = \theta\gamma b[\beta + (\alpha/\theta)] \tag{3.9}$$

in which θ is the average porosity of the aquifer, γ is the specific weight of water, β is the compressibility of water, and α is the vertical compressibility of aquifer material. In most confined aquifers, storage coefficient values lie in the range of 0.00005–0.0005. These values are small and thus large pressure changes over extensive areas are required to develop substantial quantities of water.

A confined aquifer for which S in Eq. 3.9 is 3×10^{-4} will release from 1 square mile, 64,125 gal (93,711 L/km^2) by lowering the piezometric surface by 1 ft (0.3048 m).

A water table aquifer also releases water from storage by two processes: (a) dewatering or drainage of material at the free surface as it moves downward and (b) elastic response of the material below the free surface. In general, the quantity released by elastic response is very small compared to the dewatering of the saturated material at the water table. Thus the storage coefficient is virtually equal to the specific yield of the material. In unconfined aquifers, the full complement of storage is usually not released instantaneously. The speed of drainage depends on the types of aquifer materials. Thus, in water table aquifers, the storage coefficient varies with time, increasing at a diminishing rate. Ultimately it is equal to specific yield. Furthermore, since the dewatered portion of the aquifer cannot transmit water, transmissivity of the aquifer decreases with the lowering of the water table. Transmissivity is thus a function of head in an unconfined aquifer. The storage coefficient of unconfined aquifers may range from 0.01 to 0.3. A water table aquifer with a storage coefficient of 0.15 will release from a 1-mi^2 (2.59 km^2) area with an average decline in head of 1 ft (0.3048 m), $209 \times 10^6 \times 0.15$ gal = 31.30 MG (118.4 ML).

Hydraulic diffusivity is the ratio of transmissivity, T, to storage coefficient, S, or of permeability, K, to unit storage, S'. Where D is hydraulic diffusivity,

$$D = T/S = K/S' \tag{3.10}$$

In an unconfined aquifer, even if S is assumed constant, the diffusivity will vary with transmissivity, which varies with the position of the free surface.

The conductivity, the transmissivity, the storage coefficient, and the specific yield are usually referred to as *formation constants,* and provide measures of the hydraulic properties of aquifers.

The capacity of an aquifer to transmit water can be measured by several methods:

1) Laboratory tests of aquifer samples
2) Tracer techniques
3) Analysis of water level maps
4) Aquifer tests.

Laboratory measurements of hydraulic conductivity are obtained by using samples of aquifer material in either a constant-head or a falling-head permeameter. Undisturbed core samples are used in the case of well-consolidated materials, and repacked samples in the case of unconsolidated materials. Observations are made of the time taken for a known quantity of water under a given head to pass through the sample. The application of Darcy's law enables hydraulic conductivity to be determined. The main disadvantage of this method arises from the fact that the values obtained are point measurements. Aquifers are seldom, if ever, truly homogeneous throughout their extent, and laboratory measurements are not representative of actual "in-place" values. Most samples of the material are taken in a vertical direction, whereas the dominant movement of water in the aquifer is nearly horizontal, and horizontal and vertical permeabilities differ markedly. Also, some disturbance is inevitable when the sample is removed from its environment. This method cannot, therefore, be used to give a reliable quantitative measure of hydraulic conductivity.

The measurement of hydraulic conductivity in undisturbed natural materials can be made by measurement of hydraulic gradient and determination of the speed of groundwater movement through the use of tracers. A tracer (dye, electrolyte, or radioactive substance) is introduced into the groundwater through an injection well at an upstream location, and measurements are made of the time taken by the tracer to appear in one or more downstream wells. Uranin, a sodium salt of fluorescein, is an especially useful dye because it remains visible in dilutions of 1:(14×10^7) without a fluoroscope and 1:10^{10} with one. Tritium has been used as a radioactive tracer.

The time of arrival is determined by visual observation or colorimetry when dyes are added, by titration or electrical conductivity when salt solutions are injected, or by a Geiger or scintillation counter when radioactive tracers are used. The distance between the wells divided by the time required for half the recovered substance to appear is the median velocity. The observed velocity is the actual average rate of motion through the interstices of the aquifer material. The face velocity can be calculated, if effective porosity is known. The application of Darcy's law enables the hydraulic conductivity to be computed. The problems of direction of motion, dispersion and molecular diffusion, and the slow movement of groundwater limit the applicability of this method. The method is impractical for a heterogeneous aquifer that has large variations in horizontal and vertical hydraulic conductivity.

The drop in head between two equipotential lines in an aquifer divided by the distance traversed by a particle of water moving from a higher to a lower potential determines the hydraulic gradient. Changes in the hydraulic gradient may arise from either a change in flow rate, Q, hydraulic conductivity, K, or aquifer thickness, b (Eq. 3.6). If no water is being added to or lost from an aquifer, the steepening of the gradient must be due to lower transmissivity, reflecting either a lower permeability, a reduction in thickness, or both (Eq. 3.8).

Of the currently available methods for the estimation of formation constants, aquifer tests (also called pumping tests) are the most reliable. The mechanics of a test involve the pumping of water from a well at a constant discharge rate and the observation of water levels in observation wells at various distances from the pumping well at different time intervals after pumping commences. The analysis of a pumping test comprises the graphical fitting of the various theoretical equations of groundwater flow to the observed data. The mathematical model giving the best fit is used for the estimation of formation constants. The main advantages of this method are that the sample used is large and remains undisturbed in its natural surroundings. The time and expense are reasonable. The main disadvantage of the method concerns the number of assumptions that must be made when applying the theory to the observed data. Despite the restrictive assumptions, pumping tests have been successfully applied under a wide range of conditions actually encountered.

3.9 Well Hydraulics

Well hydraulics deals with predicting yields from wells and in forecasting the effects of pumping on groundwater flow and on the distribution of potential in an aquifer. The response of an aquifer to pumping depends on the type of aquifer (confined, unconfined, or leaky), aquifer characteristics (transmissivity, storage coefficient, and leakage), aquifer boundaries, and well construction (size, type, whether fully or partially penetrating) and well operation (constant or variable discharge, continuous or intermittent pumping).

The first water pumped from a well is derived from aquifer storage in the immediate vicinity of the well. Water level (i.e., piezometric surface or water table) is lowered and a *cone of depression* is created. The shape of the cone is determined by the hydraulic gradients required to transmit water through aquifer material toward the pumping well. The distance through which the water level is lowered is called the *drawdown*. The outer boundary of the drawdown curve defines the *area of influence* of the well. As pumping is continued, the shape of the cone changes as it travels outward from the well. This is the dynamic phase, in which the flow is *time-dependent* (*nonsteady*), and both the velocities and water levels are changing. With continued withdrawals, the shape of the cone of depression stabilizes near the well and, with time, this condition progresses to greater distances. Thereafter the cone of depression moves parallel to itself in this area. This is the depletion phase. Eventually the drawdown curve may extend to the areas of natural discharge or recharge. A new state of equilibrium is reached if the natural discharge is decreased or the natural recharge is increased by an amount equal to the rate of withdrawal from the well. A *steady state* is then reached and the water level ceases to decline.

3.10 Nonsteady Radial Flow

Solutions have been developed for nonsteady radial flow toward a discharging well. Pumping test analyses for the determination of aquifer constants are based on solutions of unsteady radial flow equations.

3.10.1 Confined Aquifers

In an effectively infinite artesian aquifer, the discharge of a well can only be supplied through a reduction of storage within the aquifer. The propagation of the area of influence and the rate of decline of head depend on the hydraulic diffusivity of the aquifer. The differential equation governing nonsteady radial flow to a well in a confined aquifer is given by

$$\frac{\partial^2 h}{\partial r^2} + \frac{1}{r}\frac{\partial h}{\partial r} = \frac{S}{T}\frac{\partial h}{\partial t} \tag{3.11}$$

Using an analogy to the flow of heat to a sink, Theis (1963) derived an expression for the drawdown in a confined aquifer due to the discharge of a well at a constant rate. His equation is really a solution of the Eq. 3.11 based on the following assumptions: (a) The aquifer is homogeneous, isotropic, and of infinite areal extent; (b) transmissivity is constant with respect to time and space; (c) water is derived entirely from storage, being released instantaneously with the decline in head; (d) storage coefficient remains constant with time; and (e) the well penetrates, and receives water from, the entire thickness of the aquifer. The Theis equation may be written as follows:

$$s = h_0 - h = \frac{Q}{4\pi T}\int_{\frac{r^2 S}{4Tt}}^{\infty} \frac{e^{-u}}{u}du = \frac{Q}{4\pi T}W(u) \quad \text{(SI units)} \tag{3.12}$$

where h is the head at a distance r from the well at a time t after the start of pumping, h_0 is the initial head in the aquifer prior to pumping, Q is the constant discharge of the well, S is the storage coefficient of the aquifer, and T is the transmissivity of the aquifer. The integral in the above expression is known as the exponential integral and is a function of its lower limit. In groundwater literature, it is written symbolically as $W(u)$, which is read "well function of u," where

$$u = \left(r^2 S\right)/(4Tt) \quad \text{(SI units)} \tag{3.13}$$

The drawdown s (m) at a distance r (m) at time t (days) after the start of pumping for a constant discharge Q (m^3/d) under the transmissivity T (m^3/d/m) is given by Eq. 3.12 and Eq. 3.13.

Its value can be approximated by a convergent infinite series:

$$W(u) = -0.5772 - \ln u + u - u^2/2 \times 2! + u^3/3 \times 3! \ldots \tag{3.14}$$

Values of $W(u)$ for a given value of u are tabulated in numerous publications. A partial listing is given in Table 3.2.

The drawdown s (ft), at a distance r (ft), at time t (days) after the start of pumping for a constant discharge Q (gpm), is given by

$$s = h_0 - h = 1{,}440\, QW(u)/(4\pi T) = 114.6\, QW(u)/T \quad \text{(US customary units)} \tag{3.15}$$

where

$$u = 1.87(S/T)\left(r^2/t\right) \quad \text{(US customary units)} \tag{3.16}$$

and T is transmissivity in gpd/ft.

The equation can be solved for any one of the quantities involved if other parameters are given. The solution for drawdown, discharge, distance from the well, or time is straightforward. The solution for transmissivity T is difficult, because it occurs both inside and outside the integral. Theis (1963) devised a graphical method of superposition to obtain a solution of the equation for T and S.

If the discharge Q is known, the formation constants of an aquifer can be obtained as follows:

1) Plot the field or data curve with drawdown, s, as the ordinate and r^2/t as the abscissa on logarithmic coordinates on a transparent paper.
2) Plot a "type curve" with the well function, $W(u)$, as the ordinate and its argument u as the abscissa on logarithmic coordinates using the same scale as the field curve.
3) Superimpose the curves shifting vertically and laterally, keeping the coordinate axes parallel until most of the plotted points of the observed data fall on a segment of the type curve.
4) Select a convenient matching point anywhere on the overlapping portion of the sheets and record the coordinates of this common point on both graphs.
5) Use the two ordinates, s and $W(u)$, to obtain the solution for transmissivity T from Eq. 3.15.
6) Use the two abscissas, r^2/t and u, together with the value of T, to obtain the solution for the storage coefficient S from Eq. 3.16.

Table 3.2 Values of the Well Function *W*(*u*) for Various Values of *u*

	u							
N	$N \times 10^{-15}$	$N \times 10^{-14}$	$N \times 10^{-13}$	$N \times 10^{-12}$	$N \times 10^{-11}$	$N \times 10^{-10}$	$N \times 10^{-9}$	$N \times 10^{-8}$
1.0	33.96	31.66	29.36	27.05	24.75	22.45	20.15	17.84
1.5	33.56	31.25	28.95	26.65	24.35	22.04	19.74	17.44
2.0	33.27	30.97	28.66	26.36	24.06	21.67	19.45	17.15
2.5	33.05	30.74	28.44	26.14	23.83	21.53	19.23	16.93
3.0	32.86	30.56	28.26	25.96	23.65	21.35	19.05	16.75
3.5	32.71	30.41	28.10	25.80	23.50	21.20	18.89	16.59
4.0	32.56	30.27	27.97	25.67	23.36	21.06	18.76	16.46
4.5	32.47	30.15	27.85	25.55	23.25	20.94	18.64	16.34
5.0	32.35	30.05	27.75	25.44	23.14	20.84	18.54	16.23
5.5	32.26	29.95	27.65	25.35	23.05	20.74	18.44	16.14
6.0	32.17	29.87	27.56	25.26	22.96	20.66	18.35	16.05
6.5	32.09	29.79	27.48	25.18	22.88	20.58	18.27	15.97
7.0	32.02	29.71	27.41	25.11	22.81	20.50	18.20	15.90
7.5	31.95	29.64	27.34	25.04	22.74	20.43	18.13	15.83
8.0	31.88	29.58	27.28	24.97	22.67	20.37	18.07	15.76
8.5	31.82	29.52	27.22	24.91	22.61	20.31	18.01	15.70
9.0	31.76	29.46	27.16	24.86	22.55	20.25	17.95	15.65
9.5	31.71	29.41	27.11	24.80	22.50	20.20	17.89	15.59
	u							
N	$N \times 10^{-7}$	$N \times 10^{-6}$	$N \times 10^{-5}$	$N \times 10^{-4}$	$N \times 10^{-3}$	$N \times 10^{-2}$	$N \times 10^{-1}$	*N*
1.0	15.54	13.24	10.94	8.633	6.332	4.038	1.823	2.194×10^{-1}
1.5	15.14	12.83	10.53	8.228	5.927	3.637	1.465	1.000×10^{-1}
2.0	14.85	12.55	10.24	7.940	5.639	3.355	1.223	4.890×10^{-2}
2.5	14.62	12.32	10.02	7.717	5.417	3.137	1.044	2.491×10^{-2}
3.0	14.44	12.14	9.837	7.535	5.235	2.959	0.9057	1.305×10^{-2}
3.5	14.29	11.99	9.683	7.381	5.081	2.810	0.7942	6.970×10^{-3}
4.0	14.15	11.85	9.550	7.247	4.948	2.681	0.7024	3.779×10^{-3}
4.5	14.04	11.73	9.432	7.130	4.831	2.568	0.6253	2.073×10^{-3}
5.0	13.93	11.63	9.326	7.024	4.726	2.468	0.5598	1.148×10^{-3}
5.5	13.84	11.53	9.231	6.929	4.631	2.378	0.5034	6.409×10^{-4}
6.0	13.75	11.45	9.144	6.842	4.545	2.295	0.4544	3.601×10^{-4}
6.5	13.67	11.37	9.064	6.762	4.465	2.220	0.4115	2.034×10^{-4}
7.0	13.60	11.29	8.990	6.688	4.392	2.151	0.3738	1.155×10^{-4}
7.5	13.53	11.22	8.921	6.619	4.323	2.087	0.3403	6.583×10^{-5}
8.0	13.46	11.16	8.856	6.555	4.259	2.027	0.3106	3.767×10^{-5}
8.5	13.40	11.10	8.796	6.494	4.199	1.971	0.2840	2.162×10^{-5}
9.0	13.34	11.04	8.739	6.437	4.142	1.919	0.2602	1.245×10^{-5}
9.5	13.29	10.99	8.685	6.383	4.089	1.870	0.2387	7.185×10^{-6}

Source: After U.S. Geological Survey.

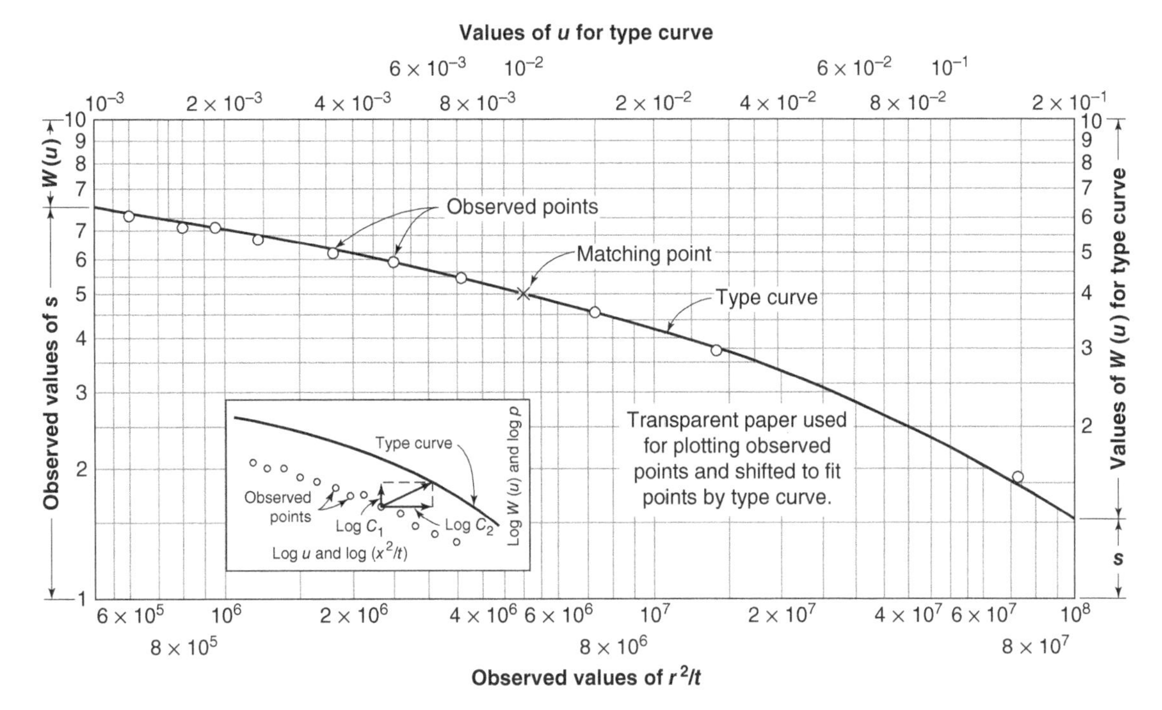

Figure 3.4 Theis Type-Curve Determination of the Formation Constants of a Well Field Using the US Customary System (Data by courtesy of the U.S. Geological Survey). Conversion factor: (r^2/t of US customary units) × 0.0929 = (r^2/t of SI units)

Values of s are related to the corresponding values of $W(u)$ by the constant factor $114.6Q/T$, whereas values of r^2/t are related to corresponding values of u by the constant factor $T/(1.87S)$. Thus, when the two curves are superimposed, corresponding vertical axes are separated by a constant distance proportional to $\log[114.6(Q/T)] = \log C_1$, whereas the corresponding horizontal axes are separated by a constant distance proportional to $\log[T/(1.87S)] = \log C_2$ as shown in Fig. 3.4.

Example 3.1 Calculation of the Formation Constants T and S Using the Theis Method

The observed data from a pumping test are shown plotted in Fig. 3.4 along with a Theis-type curve, as if the transparency of the observed data had been moved into place over the type curve. The observation well represented by the data is 225 ft (68.58 m) from a pumping well, where the rate of discharge is 350 gpm (1324.75 L/min; 1908 m^3/d). Calculate the formation constants T and S.

Solution 1 (US Customary System):

The match-point coordinates are as follows:

$$W(u) = 4.0$$

$$s = 5.0\,\text{ft}\,(1.524\,\text{m})$$

$$u = 10^{-2}$$

$$r^2/t = 5 \times 10^6\ (4.645 \times 10^5 \text{ for SI units}).$$

Compute the formation constants:

$$T = 114.6QW(u)/s$$

$$T = 114.6 \times 350 \times 4.0/5.0$$

$$\mathbf{T = 3.2 \times 10^4\ gpd/ft}$$

$$S = uT/\left(1.87\,r^2/t\right)$$

$$S = 10^{-2}\left(3.2 \times 10^4\right)/\left[1.87\left(5 \times 10^6\right)\right]$$

$$\mathbf{S = 3.4 \times 10^{-5}}$$

Solution 2 (SI System):

$$T = [Q/(4\pi s)]W(u)$$
$$T = [1808/4 \times 3.14 \times 1.524)] \times 4$$
$$\mathbf{T = 398.72\ m^3/d/m}$$
$$S = 4Tu/\left(r^2/t\right)$$
$$S = 4 \times 398.72 \times 10^{-2}/\left(4.645 \times 10^5\right)$$
$$\mathbf{S = 3.43 \times 10^{-5}}$$

Example 3.2 Calculations of Drawdown with Time in a Well

In the aquifer represented by the pumping test in Example 3.1, a gravel-packed well with an effective diameter of 24 in. (610 mm) is to be constructed. The design flow of the well is 700 gpm (3815 m^3/d). Calculate the drawdown at the well with total withdrawals from storage (i.e., with no recharge or leakages) after (a) 1 minute, (b) 1 hour, (c) 8 hours, (d) 24 hours, (e) 30 days, and (f) 6 months of continuous pumping, at design capacity.

Solution 1 (US Customary System):

From Eq. 3.15: $s = [114.6\ Q/T][W(u)]$

$$= \left[114.6 \times 700/\left(3.2 \times 10^4\right)\right][W(u)]$$
$$= 2.51\ \text{ft}\ [W(u)]$$

From Eq. 3.16: $u = 1.87\ r^2S/Tt$

$$= 1.87 \times 1^2 \times 3.4 \times 10^{-5}/\left(3.2 \times 10^4\right)t$$
$$= \left(2.0 \times 10^{-9}\right)/t$$

For various values of t, compute u, then from Table 3.2 obtain the well function, $W(u)$, for the calculation of drawdown. **The values of drawdown for various values of time are given in Table 3.3.**

Solution 2 (SI Units):

From Eq. 3.12: $s = [Q/(4\pi T)][W(u)]$

$$= [3,815/(4 \times 3.14 \times 398.72)][W(u)]$$
$$= 0.72\ \text{m}[W(u)]$$

From Eq. 3.13: $u = r^2S/(4Tt)$

$$= (0.3048)^2 \times 3.43 \times 10^{-5}/(4 \times 398.72t)$$
$$= \left(2.0 \times 10^{-9}\right)/t$$

For various values of t, compute u, then from Table 3.2, obtain the well function, $W(u)$, for the calculation of drawdown. **The values of drawdown for various values of time are given in Table 3.3.**

Table 3.3 Variation of Drawdown with Time for Example 3.2

	Time, days	u	$W(u)$	Drawdown, s, ft (m)
(a)	1/1440	2.86×10^{-6}	12.19	30.6 (9.33)
(b)	1/24	4.8×10^{-8}	16.27	40.8 (12.44)
(c)	1/3	6.0×10^{-9}	18.35	46.0 (14.02)
(d)	1	2.0×10^{-9}	19.45	48.8 (14.87)
(e)	30	6.6×10^{-11}	22.86	57.3 (17.46)
(f)	180	1.1×10^{-11}	24.66	61.8 (18.84)

3.10.2 Semilogarithmic Approximation

It was recognized that when u is small, the sum of the terms beyond ln u in the series expansion of $W(u)$ (Eq. 3.14) is relatively insignificant. The Theis equation (Eq. 3.12) then reduces to

$$s = [Q/(4\pi T)]\{\ln[(4Tt)/(r^2S)] - 0.5772\}$$

$$s = Q/(4\pi T)\ \ln[(2.25Tt)/(r^2S)] \quad \text{(SI units)} \tag{3.17}$$

where Q is in m^3/d, T in m^3/d/m, t in days, and r in m.

When Q is in gpm, T in gpd/ft, t in days, and r in ft, the equation becomes

$$s = 264(Q/T)\ \log[(0.3Tt)/(r^2S)] \quad \text{(US customary units)} \tag{3.18}$$

A graphical solution was proposed for this equation. If the drawdown is measured in a particular observation well (fixed r) at several values of t, the equation becomes

$$s = 264(Q/T)\ \log(Ct)$$

where

$$C = 0.3T/(r^2S).$$

If, on semilogarithmic paper, the values of drawdown are plotted on the arithmetic scale and time on the logarithmic scale, the resulting graph should be a straight line for higher values of t where the approximation is valid. The graph is referred to as the time–drawdown curve. On this straight line an arbitrary choice of times t_1 and t_2 can be made and the corresponding values of s_1 and s_2 recorded. Inserting these values in Eq. 3.18, we obtain

$$s_2 - s_1 = 264(Q/T)\ \log(t_2/t_1) \tag{3.19}$$

Solving for T,

$$T = 264\,Q\ \log(t_2/t_1)/(s_2 - s_1) \tag{3.20}$$

Thus transmissivity is inversely proportional to the slope of the time–drawdown curve. For convenience, t_1 and t_2 are usually chosen one log cycle apart. Eq. 3.20 then reduces to

$$T = 264\,Q/\Delta s \quad \text{(US customary units)} \tag{3.21a}$$

where T is the transmissivity, in gpd/ft; Q is the well flow in gpm; and Δs is the change in drawdown, in ft, over one log cycle of time.

An equivalent equation using the SI units is

$$T = 0.1833\,Q/\Delta s \quad \text{(SI units)} \tag{3.21b}$$

where T is the transmissivity, in m^3/d/m; Q is the well flow, in m^3/d; and Δs is the change in drawdown, in m, over one log cycle of time.

The coefficient of storage of the aquifer can be calculated from the intercept of the straight line on the time axis at zero drawdown, provided that time is converted to days. For zero drawdown, Eq. 3.18 gives

$$0 = 264(Q/T)\ \log[0.3\,Tt_0/(r^2S)]$$

that is,

$$0.3\,Tt_0/(r^2S) = 1$$

which gives

$$S = 0.3\,Tt_0/r^2 \quad \text{(US customary units)} \tag{3.22a}$$

where S is the coefficient of storage of an aquifier, dimensionless; T is the transmissivity, gpd/ft; t_0 is the time at zero drawdown, d; and r is the distance between an observation well and a pumping well, ft.

An equivalent equation using the SI units is

$$S = 2.24\,Tt_0/r^2 \quad \text{(SI units)} \tag{3.22b}$$

where S is the coefficient of storage of an aquifier, dimensionless; T is the transmissivity, $m^3/d/m$; t_0 is the time at zero drawdown, d; and r is the distance between an observation well and a pumping well, m.

Example 3.3 Determination of the T and S Coefficients of an Aquifer using the Approximation Method
A time–drawdown curve for an observation well at a distance of 225 ft (68.6 m) from a pumping well discharging at a constant rate of 350 gpm (1907.5 m^3/d) is shown in Fig. 3.5. Determine the transmissivity and storage coefficient of the aquifer.

Solution 1 (US Customary System):

To determine the slope of the straight-line portion, select two points one log cycle apart, namely:

$t_1 = 1$ min $\quad s_1 = 1.6$ ft (0.49 m)
$t_2 = 10$ min $\quad s_2 = 4.5$ ft (1.37 m)

The slope of the line per log cycle is $\Delta s = 4.5 - 1.6 = 2.9$ ft (0.88 m). The line intersects the zero-drawdown axis at $t_0 = 0.3$ min . The transmissivity and the storage coefficient of the aquifers are

$$\begin{aligned} T &= 264\, Q/\Delta s \\ &= 264 \times 350/2.9 \\ &= \mathbf{3.2 \times 10^4\ gpd/ft\ \left(397\ m^3/d/m\right)} \end{aligned}$$

$$\begin{aligned} S &= 0.3\, Tt_0/r^2 \\ &= 0.3\left(3.2 \times 10^4\right)[0.3/(60 \times 24)]/(225)^2 \\ &= \mathbf{4.0 \times 10^{-5}} \end{aligned}$$

Solution 2 (SI System):

A time–drawdown curve shown in Fig. 3.5 is used, except that the drawdown s (m) versus the time (min) should be plotted instead.

To determine the slope of the straight-line portion, select two points one log cycle apart, namely:

$t_1 = 1$ min $\quad s_1 = (0.49$ m)
$t_2 = 10$ min $\quad s_2 = (1.37$ m)

The slope of the line per log cycle is $\Delta s = 1.37$ m $- 0.49$ m $= (0.88$ m). The line intersects the zero-drawdown axis at $t_0 = 0.3$ min $= 0.3/(60 \times 24)d$. The transmissivity and the storage coefficient of the aquifiers are

$$\begin{aligned} T &= 0.1833\, Q/\Delta s \\ &= 0.1833 \times 1907.5/0.88 \\ &= 397.3\ m^3/d/m \end{aligned}$$

$$\begin{aligned} S &= 2.24\, Tt_0/r^2 \\ &= 2.24\,(397.3)[0.3/(60 \times 24)]/68.6^2 \\ &= 4 \times 10^{-5} \end{aligned}$$

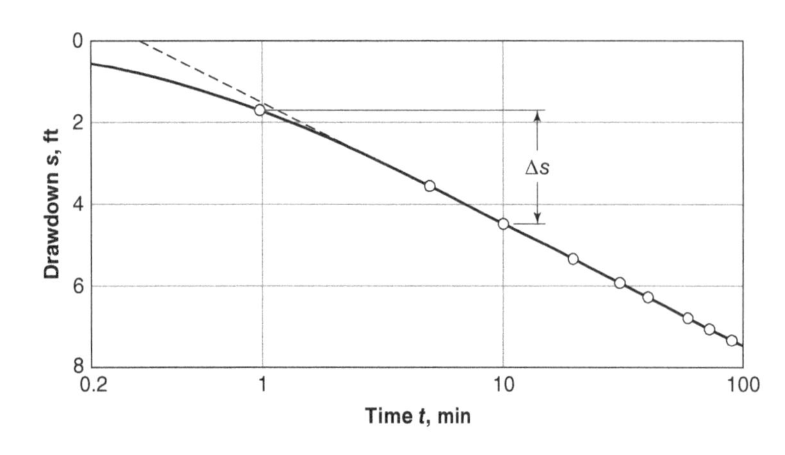

Figure 3.5 Time–Drawdown Curve (Data by courtesy of the U.S. Geological Survey). Conversion factor: 1 ft = 0.3048 m

Equation 3.18 can also be used if the drawdown is measured at several observation wells at essentially the same time, that is, from the shape of the cone of depression. Drawdowns are plotted on the arithmetic scale and distance on the log scale and the resulting straight-line graph is called the distance–drawdown curve. It can be shown that the expressions for T and S in this case are

$$T = 528Q/\Delta s \tag{3.23}$$

$$S = 0.3Tt/r^2{}_0 \tag{3.24}$$

With the formation constants T and S known, Eq. 3.18 gives the drawdown for any desired value of r and t, provided that u (Eq. 3.13) is less than 0.01. The value of u is directly proportional to the square of the distance and inversely proportional to time t. The combination of time and distance at which u passes the critical value is inversely proportional to the hydraulic diffusivity of the aquifer, $D = T/S$. The critical value of u is reached much more quickly in confined aquifers than in unconfined aquifers.

3.10.3 Recovery Method

In the absence of an observation well, transmissivity can be determined more accurately by measuring the recovery of water levels in the well under test after pumping has stopped than by measuring the drawdown in the well during pumping. For this purpose, a well is pumped for a known period of time, long enough to be drawn down appreciably. The pump is then stopped, and the rise of water level within the well (or in a nearby observation well) is observed (Fig. 3.6). The drawdown after the shutdown will be the same as if the discharge had continued at the rate of pumping and a recharge well with the same flow had been superimposed on the discharge well at the instance the discharge was shut down. The residual drawdown, s', can be found from Eq. 3.15 as follows:

$$s' = (114.6Q/T)[W(u) - W(u')] \tag{3.25}$$

where

$$u = 1.87r^2S/4Tt$$

$$u' = 1.87r^2S/(4Tt')$$

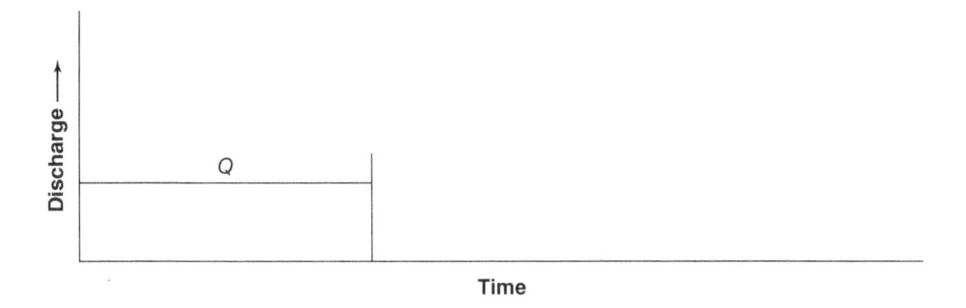

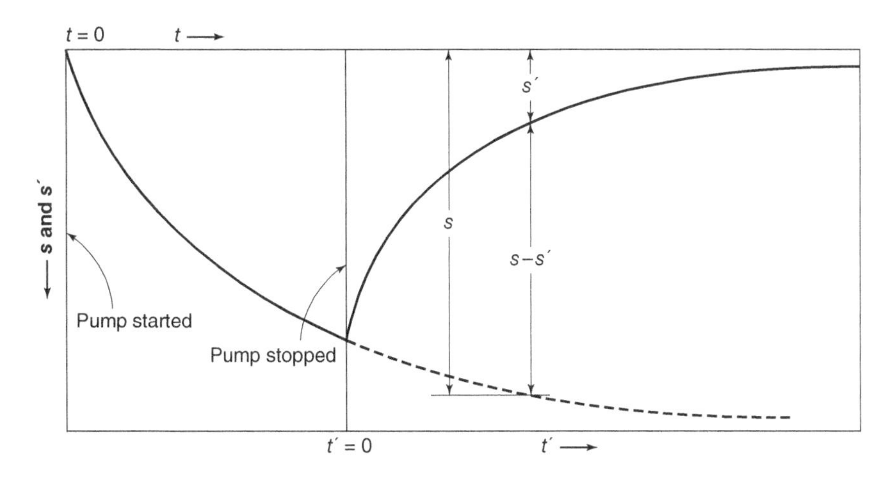

Figure 3.6 Water-Level Recovery After Pumping Has Stopped

where r is the effective radius of the well (or the distance to the observation well), t is the time since pumping started, and t' is the time since pumping stopped. For small values of r and large values of t, the residual drawdown may be obtained from Eq. 3.19:

$$s' = (264Q/T)\ \log\ (t/t') \quad \text{(US customary units)} \tag{3.26a}$$

Solving for T,

$$T = (264Q/s')\ \log(t/t') \quad \text{(US customary units)} \tag{3.27a}$$

Plotting s' on an arithmetic scale and t/t' on a logarithmic scale, a straight line is drawn through the observations. The coefficient of transmissivity can be determined from the slope of the line, or for convenience, the change of residual drawdown over one log cycle can be used as

$$T = 264Q/\Delta s' \quad \text{(US customary units)} \tag{3.28a}$$

where the US customary units are s' (ft), Q (gpm), T (gpd/ft), t (min), t' (min), and $\Delta s'$ (ft). The comparable equations using the SI units of s' (m), Q (m^3/d), T (m^3/d/m), t (min), t' (min), and $\Delta s'$ (m) are as follows:

$$s' = (0.1833Q/T)\log(t/t') \quad \text{(SI units)} \tag{3.26b}$$

$$T = (0.1833Q/s')\log(t/t') \quad \text{(SI units)} \tag{3.27b}$$

$$T = 0.1833Q/\Delta s' \quad \text{(SI units)} \tag{3.28b}$$

Strictly speaking, the Theis equation and its approximations are applicable only to situations that satisfy the assumptions used in their derivation. They undoubtedly also provide reasonable approximations in a much wider variety of conditions than their restrictive assumptions would suggest. Significant departures from the theoretical model will be reflected in the deviation of the test data from the type curves. Advances have recently been made in obtaining analytical solutions for anisotropic aquifers, for aquifers of variable thickness, and for partially penetrating wells.

3.10.4 Unconfined Aquifers

The partial differential equation governing nonsteady unconfined flow is nonlinear in h. In many cases, it is difficult or impossible to obtain analytical solutions to the problems of unsteady unconfined flow. A strategy commonly used is to investigate the conditions under which a confined flow equation would provide a reasonable approximation for the head distribution in an unconfined aquifer. These conditions are that (a) the drawdown at any point in the aquifer must be small relative to the total saturated thickness of the aquifer, and (b) the vertical head gradients must be negligible. This implies that the downward movement of the water table should be very slow, that is, that sufficient time must elapse for the flow to become stabilized in a portion of the cone of depression. The minimum duration of pumping depends on the properties of the aquifer.

The observed drawdown s, if large compared to the initial depth of flow h_0, should be reduced by a factor $s^2/2h_0$ to account for the decreased thickness of flow due to dewatering before Eq. 3.15 can be applied. For an observation well at a distance greater than $0.2h_0$, the minimum duration of pumping beyond which the approximation is valid is given as

$$t_{\min} = 37.4h_0/K \quad \text{(US customary units)} \tag{3.29a}$$

where t is time in days, h_0 is the saturated aquifer thickness in ft, and K is hydraulic conductivity in gpd/ft^2.

$$t_{\min} = 5h_0/K \quad \text{(SI units)} \tag{3.29b}$$

where $t_{\min}$ is time in days, h_0 is saturated aquifer thickness in m, and K is hydraulic conductivity in m^3/d/m^2.

3.10.5 Leaky Aquifers

The partial differential equation governing nonsteady radial flow toward a steadily discharging well in a leaky confined aquifer is

$$\frac{\partial^2 s}{\partial r^2} + \frac{1}{r}\frac{\partial s}{\partial r} - \frac{s}{B^2} = \frac{S}{T}\frac{\partial s}{\partial t} \tag{3.30a}$$

where

$$B = \sqrt{\frac{T}{K'/b'}} \tag{3.30b}$$

and s is the drawdown at a distance r from the pumping well, T and S are the transmissivity and storage coefficient of the lower aquifer, respectively, and K' and b' are the vertical permeability and thickness of the semipervious confining layer, respectively. The solution in an abbreviated form is given as

$$s = 114.6Q/T[W(u, r/B)] \tag{3.31}$$

where

$$W(u, r/B) = \int_{u}^{\infty} \frac{1}{y} \exp \frac{-y - r^2}{4B^2 y} dy$$

and

$$u = 1.87\, r^2 S/Tt \tag{3.32}$$

Here, $W(u, r/B)$ is the well function of the leaky aquifer, Q is the constant discharge of the well in gpm, T is transmissivity in gpd/ft, and t is the time in days.

In the earlier phases of the transient state, that is, at very small values of time, the system acts like an ideal elastic artesian aquifer without leakage and the drawdown pattern closely follows the Theis-type curve. As time increases, the drawdown in the leaky aquifer begins to deviate from the Theis curve. At large values of time, the solution approaches the steady-state condition. With time, the fraction of well discharge derived from storage in the lower aquifer decreases and becomes negligible at large values of time as steady state is approached.

The solution to the above equation is obtained graphically by the match-point technique described for the Theis solution. On the field curve drawdown versus time is plotted on logarithmic coordinates. On the type curve the values of $W(u, r/B)$ versus $1/u$ are plotted for various values of r/B as shown in Fig. 3.7. The curve corresponding to the value of r/B giving the best fit is selected. From the match-point coordinates s and $W(u, r/B)$, T can be calculated by substituting in Eq. 3.31. From the other two match-point coordinates and the value of T computed above, S is determined from Eq. 3.32. If b' is known, the value of the vertical permeability of the aquitard can be computed from Eq. 3.30, knowing r/B and T. Values of $W(u, r/B)$ for the practical range of u and r/B are given in Table 3.4.

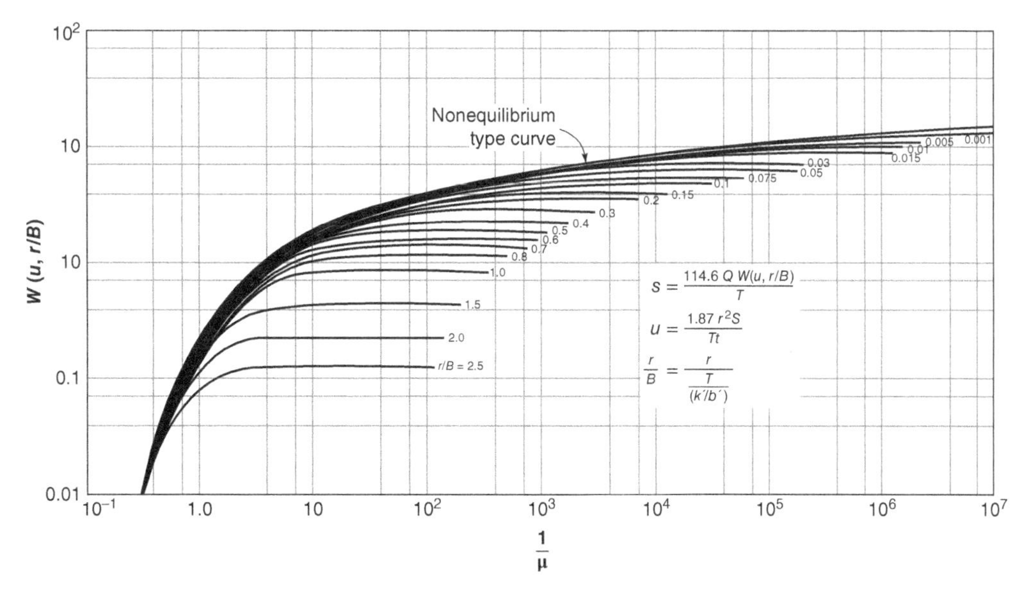

Figure 3.7 Nonsteady-State Leaky Artesian-Type Curves, Using the US Customary System (*Source:* After Walton of Illinois State Water Survey)

Table 3.4 Values of the Function $W(u, r/B) = W(u, r/B) = \int_u^\infty \frac{1}{y} \exp \frac{-y - r^2}{4B^2 y} dy$

u \ r/B	0.005	0.01	0.025	0.05	0.075	0.10	0.15	0.2	0.3	0.4	0.5	0.6	0.7	0.8	0.9	1.0	1.5	2.0
0	10.8286	9.4425	7.6111	6.2285	5.4228	4.8541	4.0601	3.5054	2.7449	2.2291	1.8488	1.5550	1.3210	1.1307	0.9735	0.8420	0.4276	0.2278
0.000001	10.8283																	
0.000005	10.6822	9.4413																
0.00001	10.3963	9.4176																
0.00005	9.2052	8.8827	7.6000															
0.0001	8.5717	8.3983	7.5199	6.2282	5.4228													
0.0005	7.0118	6.9750	6.7357	6.0821	5.4062	4.8530												
0.001	6.3253	6.3069	6.1823	5.7965	5.3078	4.8292	4.0595	3.5054										
0.005	4.7249	4.7212	4.6960	4.6084	4.4713	4.2960	3.8821	3.4567	2.7428	2.2290								
0.01	4.0373	4.0356	4.0231	3.9795	3.9091	3.8150	3.5725	3.2875	2.7104	2.2253	1.8486	1.5550	1.3210	1.1307				
0.05	2.4678	2.4675	2.4653	2.4576	2.4448	2.4271	2.3776	2.3110	2.1371	1.9283	1.7075	1.4927	1.2955	1.1210	0.9700	0.8409		
0.1	1.8229	1.8227	1.8218	1.8184	1.8128	1.8050	1.7829	1.7527	1.6704	1.5644	1.4422	1.3115	1.1791	1.0505	0.9297	0.8190	0.4271	0.2278
0.5	0.5598	0.5598	0.5597	0.5594	0.5588	0.5581	0.5561	0.5532	0.5453	0.5344	0.5206	0.5044	0.4860	0.4658	0.4440	0.4210	0.3007	0.1944
1.0	0.2194	0.2194	0.2194	0.2193	0.2191	0.2190	0.2186	0.2179	0.2161	0.2135	0.2103	0.2065	0.2020	0.1970	0.1914	0.1855	0.1509	0.1139
5.0	0.0011	0.0011	0.0011	0.0011	0.0011	0.0011	0.0011	0.0011	0.0011	0.0011	0.0011	0.0011	0.0011	0.0011	0.0011	0.0011	0.0010	0.0010

3.11 Prediction of Drawdown

Predictions of drawdowns are useful when a new well field is to be established or where new wells are added to an existing field. To predict drawdowns, T, S, and proposed pumping rates must be known. Any of the several equations can be used. The Theis equation is of quite general applicability. The approximation does not accurately show drawdowns during the first few hours or first few days of withdrawals ($u > 0.01$). Because the equations governing flow are linear, the principle of superposition is valid.

3.11.1 Constant Discharge

Example 3.3 illustrated a method that can be used to evaluate the variation in drawdown with time, whereas the following example, Example 3.4, illustrates the variation in drawdown with distance when the pumping rate is constant.

Example 3.4 Determination of the Profile of a Cone of Depression

Determine the profile of a quasi-steady-state cone of depression for a proposed 24-in. (61-cm) well pumping continuously at (a) 150, (b) 200, and (c) 250 gpm in an elastic artesian aquifer having a transmissivity of 10,000 gpd/ft and a storage coefficient of 6×10^{-4}. Assume that the discharge and recharge conditions are such that the drawdowns will be stabilized after 180 days.

Solution:

The distance at which drawdown is approaching zero, that is, the radius of cone of depression, can be obtained from Eq. 3.24:

$r^2_0 = 0.3\ Tt/S = 0.3\ Dt$, where D is the diffusivity of the aquifer $= T/S$

$= 0.3(1 \times 10^4)180/(6 \times 10^{-4}) = 9 \times 10^8$

$r_0 = (0.3\text{Dt})^{0.5} = 3 \times 10^4 \text{ft}\ (0.914 \times 10^4\ \text{m})$

This is independent of Q and depends only on the diffusivity of the aquifer. The change in drawdown per log cycle from Eq. 3.23 is

$\Delta s = 528\ Q/T$

For 150 gpm: $\Delta s_1 = 528 \times 150/1 \times 10^4 =$ **7.9 ft (2.4 m)**

For 200 gpm: $\Delta s_2 =$ **10.6 ft (3.23 m)**

For 250 gpm: $\Delta s_3 =$ **13.2 ft (4.02 m)**

Using the value of $r = 30{,}000$ ft (9, 144 m) as the starting point, straight lines having slopes of 7.9, 10.6, and 13.2 ft (2.4 m, 3.2 m, 4.0 m) are drawn in Fig. 3.8.

The contours of the piezometric surface can be drawn by subtracting the drawdowns at several points from the initial values.

3.11.2 Variable Discharge

The rate at which water is pumped from a well field in a water supply system will vary with time in response to changes in demand. The continuous rate of pumping curve can be approximated by a series of steps as shown in Fig. 3.9. Fig. 3.9a shows a planned variable discharge curve. However, Fig. 3.9b shows an actual discharge curve, and Fig. 3.9c shows an actual drawdown curve. Then each step can be analyzed, using one of the conventional equations. From the principle of superposition, the drawdown at any point at any specific time can be obtained as the sum of increments in drawdowns caused by the step increases up to that time.

$$s = \Delta s_1 + \Delta s_2 + \ldots + \Delta s_t \tag{3.33}$$

Using a semilogarithmic approximation,

$$s = (264\ \Delta Q_i/T)\ \log\left[(0.3\ Tt_i)/\left(r^2 S\right)\right] \tag{3.34}$$

where

$t_i = t - t_{i-1}$

Increments of drawdown are determined with respect to the extension of the preceding water-level curve.

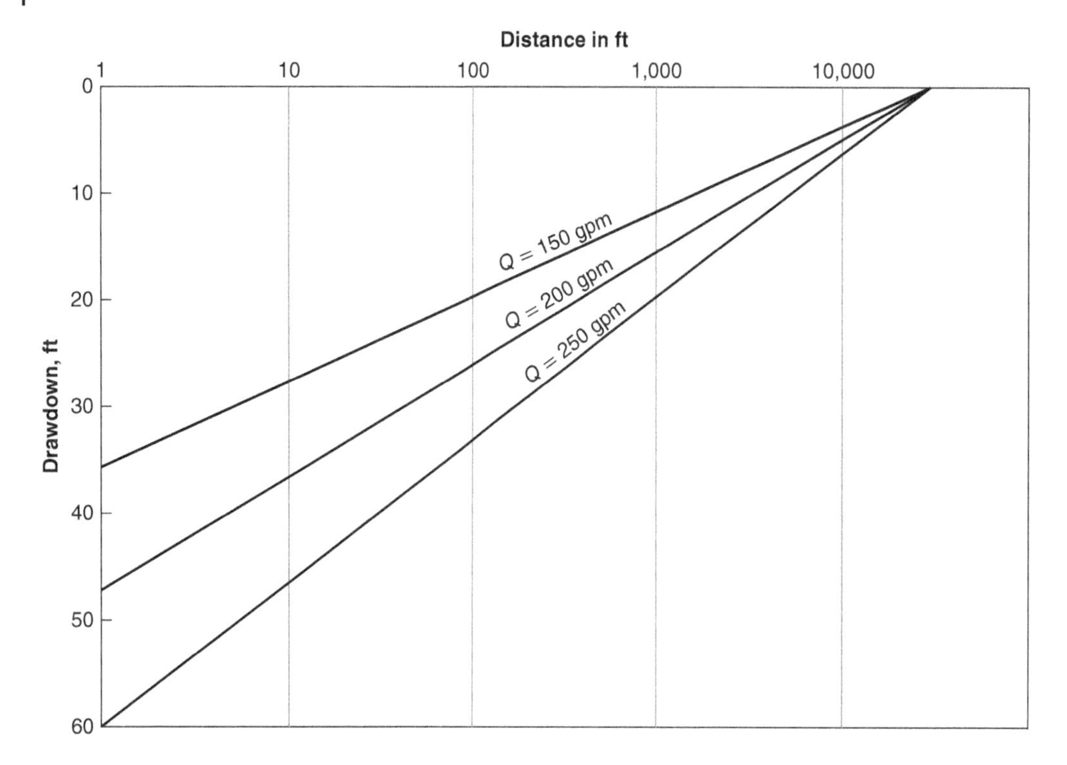

Figure 3.8 Distance–Drawdown Curves for Various Rates of Pumping (Example 3.4). Conversion factors: 1 ft = 0.3048 m; 1 gpm = 5.45 m^3/d

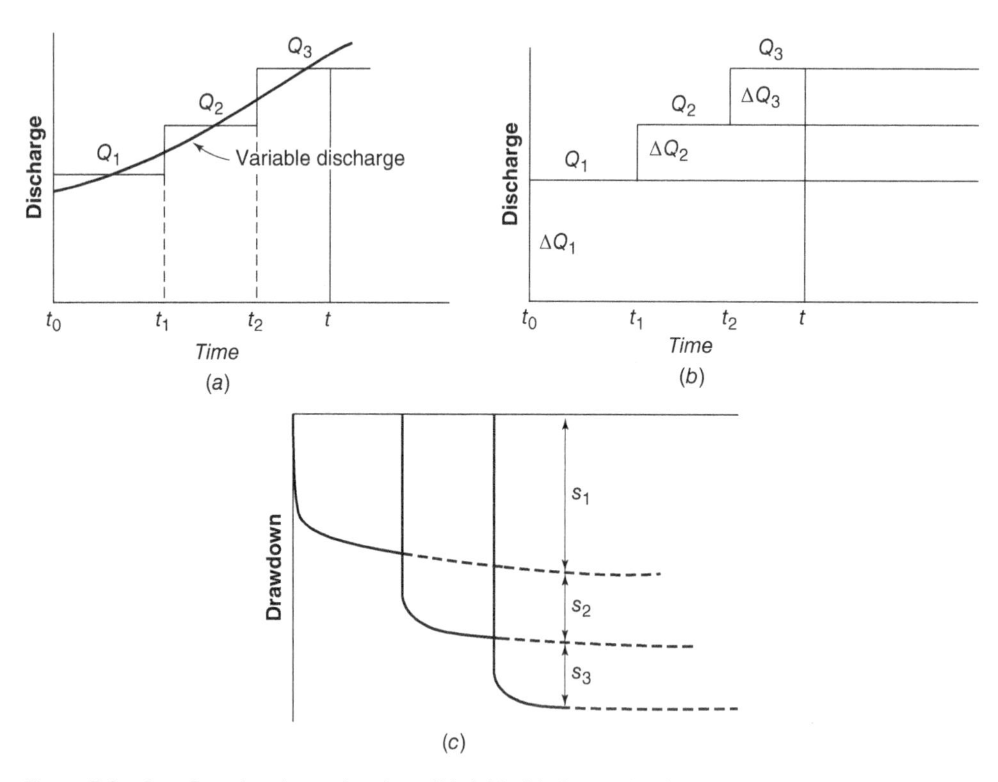

Figure 3.9 Step Function Approximation of Variable Discharge: (a) planned variable discharge curve; (b) actual discharge curve; and (c) actual drawdown curve

3.11.3 Intermittent Discharge

In a water supply system, a well (or a well field) may be operated on a regular daily cycle, pumping at a constant rate for a given time interval and remaining idle for the rest of the period. Brown (1963) gives the following expression for computing the drawdown in the pumped well after n cycles of operation:

$$s_n = (264\,Q/T)\ \log[(1.2.3\ldots n)/(1-p)(2-p)\ldots(n-p)] \quad \text{(US customary units)} \tag{3.35a}$$

where s_n is the drawdown in the pumped well after n cycles in ft, p is the fractional part of the cycle during which the well is pumped, Q is the discharge in gpm, and T is the transmissivity in gpd/ft.

$$s_n = (380{,}160\,Q/T)\ \log[(1.2.3\ldots n)/(1-p)(2-p)\ldots(n-p)] \quad \text{(SI units)} \tag{3.35b}$$

where s_n is the drawdown in the pumped well after n cycles in m, p is the fractional part of the cycle during which the well is pumped, Q is the discharge in m^3/d, and T is the transmissivity in $m^3/d/m$.

The pumping regime may involve switching on a well only during periods of peak demand. The problem of computing drawdown in a well then consists of applying one of the equations of nonsteady flow to each of the periods of pumping and recovery. The drawdown in the well, or at any other point, may be obtained by an algebraic sum of the individual values of drawdown and "buildup" resulting from each period of pumping and recovery resulting from each shutdown.

Example 3.5 Calculation of Well Drawdown

A well was pumped at a constant rate of 350 gpm (1907.5 m^3/d) between 7 and 9 a.m., 11 a.m. and 1 p.m., and 3 and 6 p.m., remaining idle the rest of the time. What will be the drawdown in the well at 7 a.m. the next day when a new cycle of pumping is to start? Assume no recharge or leakage. The transmissivity of the artesian aquifer is 3.2×10^4 gpd/ft (397.38 $m^3/d/m$).

Solution 1 (US Customary System):

The problem can be decomposed into three pumping and recovery periods, with Eq. 3.26a applied to each of the subproblems:

$$s' = (264Q/T)\log(t/t')$$
$$264\ Q/T = 264 \times 350/3.2 \times 10^4 = 2.89$$

For the first period of pumping:

t = time since pumping started = 1,440 min
t' = time since pumping stopped = 1,320 min
$\log(t/t')_1 = \log(1{,}440/1{,}320) = 0.038$

Similarly, for the second period:

$\log(t/t')_2 = \log(1{,}200/1{,}080) = 0.046$

For third period:

$\log(t/t')_3 = \log(960/780) = 0.090$

Total residual drawdown:

$s' = (264\ Q/T)\ [\sum \log t/t'] = 2.89 \times 0.174 = \mathbf{0.5\ ft\ (0.15\ m)}$

Solution 2 (SI System):

The problem can be decomposed into three pumping and recovery periods, with Eq. 3.26b applied to each of the subproblems:

$$\begin{aligned} s' &= (0.1833\ Q/T)\log(t/t') \\ &= (0.1833 \times 1907.5/397.38)\log(t/t') \\ &= 0.8798\log(t/t') \end{aligned}$$

For the first period of pumping:

t = time since pumping started = 1,440 min
t' = time since pumping stopped = 1,320 min
$\log(t/t')_1 = \log(1{,}440/1{,}320) = 0.038$

For the second period of pumping:

$\log(t/t')_2 = \log(1{,}200/1{,}080) = 0.046$

For the third period of pumping:

$\log(t/t')_3 = \log(960/780) = 0.090$

Total residual drawdown:

$$\begin{aligned} s' &= (0.1833\ Q/T)\ [\textstyle\sum \log t/t'] \\ &= 0.8798\ (0.038 + 0.046 + 0.090) \\ &= 0.15\ \text{m} \end{aligned}$$

3.12 Multiple-Well Systems

Because the equations governing steady and unsteady flow are linear, the drawdown at any point due to several wells is equal to the algebraic sum of the drawdowns caused by each individual well, that is, for n wells in a well field:

$$s = \sum_{i=1}^{n} s_i$$

where s is the drawdown at the point due to the ith well. If the location of wells, their discharges, and their formation constants are known, the combined distribution of drawdown can be determined by calculating drawdown at several points in the area of influence and drawing contours.

Example 3.6 Drawdown in Three Wells

Three 24-in. (610-mm) wells are located on a straight line 1000 ft (304.8 m) apart in an artesian aquifer with $T = 3.2 \times 10^4$ gpd/ft (397.4 m^3/d/m) and $S = 3 \times 10^{-5}$. Compute the drawdown at each well when (a) one of the outside wells is pumped at a rate of 700 gpm (3815 m^3/d) for 10 days and (b) the three wells are pumped at 700 gpm (3815 m^3/d) for 10 days.

Solution 1 (US Customary System):

From Eq. 3.16

$u = 1.87\ r^2S/Tt$

$u_{1\ \text{ft}} = (1.87 \times 1^2 \times 3.0 \times 10^{-5})/(3.2 \times 10^4 \times 10) = 1.75 \times 10^{-10},\ W(u) = 21.89$

$u_{1{,}000\ \text{ft}} = 10^6\ u_1 = 1.75 \times 10^{-4},\ W(u) = 8.08$

$u_{2{,}000\ \text{ft}} = 4 \times 10^6\ u_1 = 7 \times 10^{-4},\ W(u) = 6.69$

Drawdown constant $= (114.6\ Q)/T = (114.6 \times 700)/(3.2 \times 10^4) = 2.51$ ft (0.765 m)

From Eq. 3.15, $s = (114.6\ Q/T)\ W(u)$

a) Drawdown at face of pumping well: $s_1 = 2.51 \times 21.89 =$ **54.9 ft** (16.73 m)

   Drawdown in central well:

   $s_2 = 2.51 \times 8.08 =$ **20.3 ft** (6.19 m)

   Drawdown in the other outside well:

   $s_3 = 2.51 \times 6.69 =$ **16.8 ft** (5.12 m)

b) Drawdown in outside wells:

   $s_1 + s_2 + s_3 = 54.9 + 20.3 + 16.8 =$ **92 ft** (28.04 m)

   Drawdown in central well:

   $s_1 + s_2 + s_3 = 54.9 + 20.3 + 20.3 =$ **95.5 ft** (29.11 m)

Solution 2 (SI System):

From Eq. 3.13

1 ft = 0.3048 m, 1,000 ft = 304.8 m, 2, 000 ft = 609.6 m

$$u = r^2S/(4Tt)$$
$$u_{0.3048\ \mathrm{m}} = (0.3048)^2(3\times10^{-5})/(4\times397.4\times10) = 1.75\times10^{-10}$$
$$W(u) = 21.89$$
$$u_{304.8\ \mathrm{m}} = (304.8)^2(3\times10^{-5})/(4\times397.4\times10) = 1.75\times10^{-4}$$
$$W(u) = 8.08$$
$$u_{609.6\ \mathrm{m}} = (609.6)^2(3\times10^{-5})/(4\times397.4\times10) = 7.01\times10^{-4}$$
$$W(u) = 6.69$$

From Eq. 3.12

$$S = (Q/4\pi T)\ W(u)$$
$$= (3,815/4\times3.14\times397.4)\ W(u)$$
$$= 0.765\ W(u)$$

a) Drawdown at face of pumping well:

$$s_1 = 0.765\times21.89 = 16.73\ \mathrm{m}$$

Drawdown in central well:

$$s_2 = 0.765\times8.08 = 6.18\ \mathrm{m}$$

Drawdown in other outside well:

$$s_3 = 0.765\times6.69 = 5.12\ \mathrm{m}$$

b) Drawdown in outside wells:

$$s_1+s_2+s_3 = 16.73+6.18+5.12 = 28.03\ \mathrm{m}$$

Drawdown in central well:

$$s_1+s_2+s_3 = 16.73+6.18+6.18 = 29.09\ \mathrm{m}$$

A problem of more practical interest is to determine the discharges of the wells when their drawdowns are given. This involves simultaneous solution of linear equations, which can be undertaken by numerical methods or by trial and error.

Example 3.7 Discharge from Three Wells at a Given Drawdown

Suppose we want to restrict the drawdown in each of the wells from Example 3.6 to 60 ft (18.3 m). What will be the corresponding discharges for the individual wells?

Solution 1 (US Customary System):

Eq. 3.15 is $s = (114.6\ Q/T)\ W(u)$

$$\text{Well 1: } [114.6/T][Q_1W(u_1)+Q_2W(u_{1,000})+Q_3W(u_{2,000})] = 60\ \mathrm{ft}$$
$$\text{Well 2: } [114.6/T][Q_1W(u_{1,000})+Q_2W(u_1)+Q_3W(u_{1,000})] = 60\ \mathrm{ft}$$
$$\text{Well 3: } [114.6/T][Q_1W(u_{2,000})+Q_2W(u_{1,000})+Q_3W(u_1)] = 60\ \mathrm{ft}$$
$$114.6/(3.2\times10^4)[21.89\ Q_1+8.08\ Q_2+6.69\ Q_3] = 60\ \mathrm{ft}$$
$$114.6/(3.2\times10^4)[8.08\ Q_1+21.89\ Q_2+8.08\ Q_3] = 60\ \mathrm{ft}$$
$$114.6/(3.2\times10^4)[6.69\ Q_1+8.08\ Q_2+21.89\ Q_3] = 60\ \mathrm{ft}$$
$$21.89\ Q_1+8.08\ Q_2+6.69\ Q_3 = 16{,}800$$
$$8.08\ Q_1+21.89\ Q_2+8.08\ Q_3 = 16{,}800$$
$$6.69\ Q_1+8.08\ Q_2+21.89\ Q_3 = 16{,}800$$

Solving the three equations for the three unknown discharges Q_1, Q_2, and Q_3:

$Q_1 = Q_3 =$ **468 gpm (2, 550 m³/d)**
$Q_2 =$ **420 gpm (2, 289 m³/d)**

Solution 2 (SI System):

Eq. 3.12 is $s = (Q/4\pi T)\ W(u)$

$$\text{Well 1}: [1/4\pi T][Q_1W(u_{0.3048\ \mathrm{m}})+Q_2W(u_{304.8\ \mathrm{m}})+Q_3W(u_{609.6\ \mathrm{m}})] = 18.3\ \mathrm{m}$$
$$\text{Well 2}: [1/4\pi T][Q_1W(u_{304.8\ \mathrm{m}})+Q_2W(u_{0.3048\ \mathrm{m}})+Q_3W(u_{304.8\ \mathrm{m}})] = 18.3\ \mathrm{m}$$
$$\text{Well 3}: [1/4\pi T][Q_1W(u_{609.6\ \mathrm{m}})+Q_2W(u_{304.8\ \mathrm{m}})+Q_3W(u_{0.3048\ \mathrm{m}})] = 18.3\ \mathrm{m}$$

Then,

$[1/(4 \times 3.14 \times 397.4)][21.89\ Q_1 + 8.08\ Q_2 + 6.69\ Q_3] = 18.3$ m
$[1/(4 \times 3.14 \times 397.4)][8.08\ Q_1 + 21.89\ Q_2 + 8.08\ Q_3] = 18.3$ m
$[1/(4 \times 3.14 \times 397.4)][6.69\ Q_1 + 8.08\ Q_2 + 21.89\ Q_3] = 18.3$ m

Then,

$21.89\ Q_1 + 8.08\ Q_2 + 6.69\ Q_3 = 91{,}341.6$
$8.08\ Q_1 + 21.89\ Q_2 + 8.08\ Q_3 = 91{,}341.6$
$6.69\ Q_1 + 8.08\ Q_2 + 21.89\ Q_3 = 91{,}341.6$

Solving the three equations for the three unknowns, Q_1, Q_2, and Q_3:

$Q_1 = Q_3 = 2{,}550\ m^3/d$
$Q_2 = 2{,}289\ m^3/d$

When the areas of influence of two or more pumped wells overlap, the draft of one well affects the drawdown of all others. In closely spaced wells, interference may become so severe that a well group behaves like a single well producing a single large cone of depression. When this is the case, discharge–drawdown relationships can be studied by replacing the group of wells by an equivalent single well having the same drawdown distribution when producing water at a rate equal to the combined discharge of the group.

The effective radius of a heavily pumped well field could be a mile or more and have a circle of influence extending over many miles. By contrast, lightly pumped, shallow wells in unconfined aquifers may show no interference when placed 100 ft (30.5 m) apart or even less. The number of wells, the geometry of the well field, and its location with respect to recharge and discharge areas and aquifer boundaries are important in determining the distribution of drawdown and well discharges. An analysis of the optimum location, spacing, and discharges should be carried out when designing a well field.

3.13 Aquifer Boundaries

Most methods of analysis assume that an aquifer is infinite in extent. In practice, all aquifers have boundaries. However, unless a well is located so close to a boundary that the radial flow pattern is significantly modified, the flow equations can be applied without appreciable error. Nevertheless, in many situations definite geologic and hydraulic boundaries limit aquifer dimensions and cause the response of an aquifer to deviate substantially from that predicted from equations based on extensive aquifers. This is especially true if the cone of depression reaches streams, outcrops, or groundwater divides; geologic boundaries, such as faults and folds; and valley fills of limited extent.

The effect of aquifer boundaries can be incorporated into analysis through the method of images. The *method of images* is an artifice employed to transform a bounded aquifer into one of an infinite extent having an equivalent hydraulic flow system. The effect of a known physical boundary (in the flow system) is simulated by introducing one or more hypothetical components, called images. The solution to a problem can then be obtained by using the equations of flow developed for extensive aquifers for this hypothetical system.

3.13.1 Recharge Boundaries

The conditions along a recharge boundary can be reproduced by assuming that the aquifer is infinite and by introducing a negative image well (e.g., a recharge image well for a discharging real well) an equal distance on the opposite side of the boundary from the real well, the line joining the two being at right angles to the boundary (Fig. 3.10). The drawdown, s, at any distance, r, from the pumping well and r_i from the image well is the algebraic sum of the drawdowns due to the real well, s_r, and buildup due to the image well, s_i.

$$s = s_r + s_i = (528\,Q/T)\ \log(r_i/r) \quad \text{(US customary units)} \tag{3.36a}$$

$$s_w = (528\,Q/T)\ \log(2a/r_w) \quad \text{(US customary units)} \tag{3.37a}$$

where s is the drawdown at any distance, r, from a pumping well, ft; s_r is the drawdown due to real well, ft; s_i is the drawdown due to image well, ft; s_w is the drawdown at the well, ft; r is any distance between the drawdown location and the pumping well, ft; r_i is the distance between the drawdown location and the image well, ft; r_w is the radius of the pumping well, ft; Q is pumping discharge rate, gpm; T is transmissivity, gpd/ft; a is the distance between the stream and the pumping well ft.

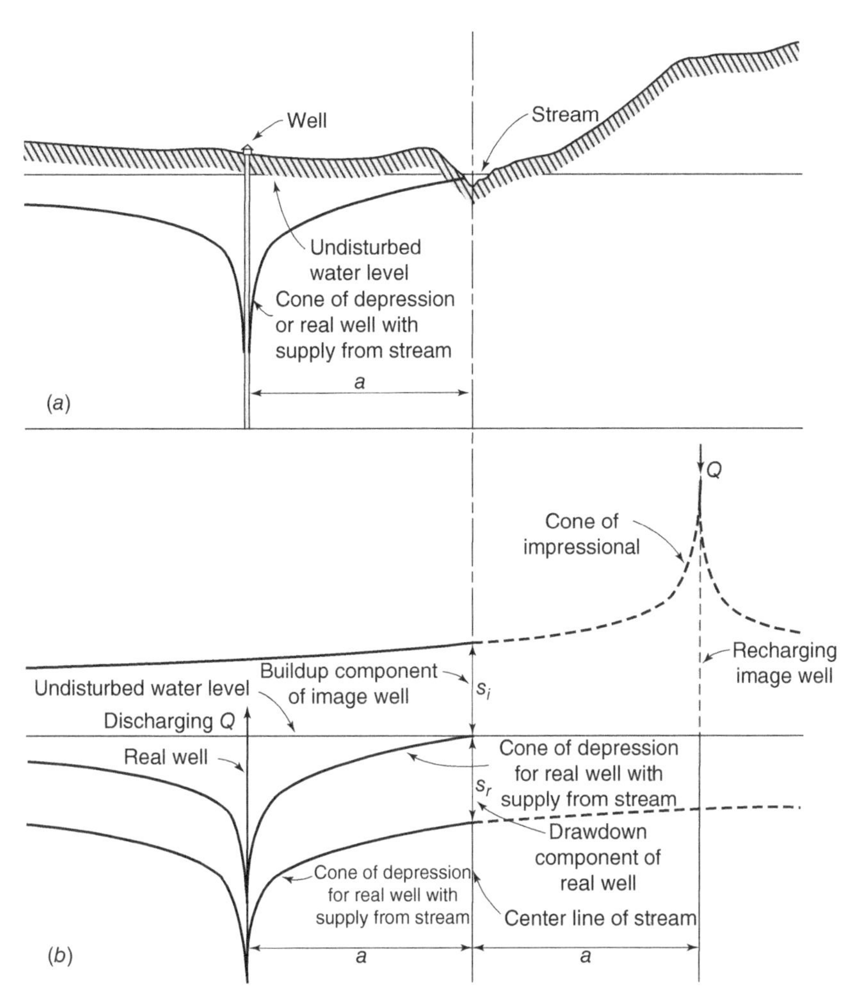

Figure 3.10 Application of the Method of Images to a Well Receiving Water from a Stream (Idealized): (a) Real System. (b) Equivalent System in an Infinite Aquifer

The following are two equivalent equations using the SI or metric units of: s (m), s_r (m), s_i (m), s_w (m), r (m), r_i (m), r_w (m), Q (m^3/d), T (m^3/d/m), and a (m):

$$s = s_r + s_i = (0.366\,Q/T)\,\log(r_i/r) \quad \text{(SI units)} \tag{3.36b}$$

$$s_w = (0.366\,Q/T)\,\log(2a/r_w) \quad \text{(SI units)} \tag{3.37b}$$

Example 3.8 Determination of a Well Drawdown and Cone of Depression Profile

A gravel-packed well with an effective diameter of 24 in. (610 mm) pumps water from an artesian aquifer having $T = 3.2 \times 10^4$ gpd/ft (397.4 m^3/d/m) and $S = 3.4 \times 10^{-5}$. The well lies at a distance of 1000 ft (304.8 m) from a stream that can supply water fast enough to maintain a constant head. Find the drawdown in the well after 10 days of pumping at 700 gpm (3815 m^3/d). Determine the profile of the cone of depression with a vertical plane through the well normal to the stream.

Solution 1 (US Customary System):

In the region of interest, the use of a semilogarithmic approximation is valid:

$$s_w = (528\;Q/T)\log(2a/r_w)$$
$$s = (528\;Q/T)\log(r_i/r)$$
$$528\;Q/T = (528 \times 700)/(3.2 \times 10^4) = 11.55 \text{ ft}$$

Drawdown at the well:

$$a = 1{,}000 \text{ ft}$$
$$r_w = 24 \text{ in.}/2 = 1 \text{ ft}$$
$$s_w = (528\;Q/T)\log(2a/r_w)$$
$$s_w = 11.55 \log 2{,}000 = 38.0 \text{ ft}$$

Drawdown at 500 ft from the stream:

$r = 500$ ft
$r_i = a + r = 1{,}000 + 500 = 1{,}500$ ft
$s = 11.55 \log(1{,}500/500) = \mathbf{5.5\ ft}$

Drawdown at 1500 ft (5 m) from the stream:

$r = 1{,}500$ ft
$r_i = a + r = 1{,}000 + 1{,}500 = 2{,}500$ ft
$s = 11.55 \log(2{,}500/1{,}500) = \mathbf{2.56\ ft}$

Drawdowns at other points can be calculated in a similar manner.
The results are shown in Fig. 3.10.

Solution 2 (SI System):

In the region of interest, the use of semilogarithmic approximation is valid:

$$s = (0.366\ Q/T) \log(r_i/r)$$
$$s_w = (0.366\ Q/T) \log(2a/r_w)$$

$0.366\ Q/T = 0.366 \times 3{,}815/397.4 = 3.5$ m

Drawdown at the well:

$a = 304.8$ m
$r_w = 610$ mm$/2 = 305$ mm $= 0.305$ m
$s_w = 3.5 \log(2 \times 304.8/0.305) = 11.55$ m

Drawdown at 152.4 m (or 500 ft) from the stream:

$r = 152.4$ m
$r_i = a + r = 304.8$ m $+ 152.4$ m $= 457.2$ m
$s = 3.5 \log(457.2/152.4) = 1.67$ m

Drawdown at 457.2 m (or 1500 ft) from the stream:

$r = 457.2$ m
$r_i = a + r = 304.8$ m $+ 457.2$ m $= 762$ m
$s = 3.5 \log(762/457.2) = 0.78$ m

The results are shown in Fig. 3.10.

For a well located near a stream, the proportion of the discharge of the well diverted directly from the source of recharge depends on the distance of the well from the recharge boundary, the aquifer characteristics, and the duration of pumping. Fig. 3.11 shows an inflatable rubber dam that is inflated seasonally (late spring to late autumn) to raise the river stage by 3 m. The higher stage allows greater pumping rates to be maintained in the collector wells that supply water to Sonoma County

Figure 3.11 Inflatable Rubber Dam (*Source:* Courtesy USGS)

residents in Maui, Hawaii. The contribution from a line source of recharge and distribution of drawdown in such a system can be evaluated and are extremely useful in determining the optimal location of well fields.

The problem of recirculation between a recharge well and a discharge well pair is of great practical importance because of the use of wells (or other devices) for underground waste disposal (or artificial recharge) and for water supply in the same area. The recirculation can be minimized by locating the recharge well directly downstream from the discharge well. The critical value of discharge and optimum spacing for no recirculation can be evaluated.

The permissible distance, r_c, between production and disposal wells in an isotropic, extensive aquifer to prevent recirculation is given by

$$r_c = 2Q/(\pi/TI) \tag{3.38}$$

where r_c is in ft or m, Q is the equal pumping and disposal rate in gpd or m^3/d, T is the transmissivity in gpd/ft (m^3/d/m), and I is the hydraulic gradient of the water table or piezometric surface.

3.13.2 Location of Aquifer Boundaries

In many instances, the location and nature of hydraulic boundaries of an aquifer can be inferred from the analysis of aquifer-test data. The effect of a boundary when it reaches an observation well causes the drawdowns to diverge from the curve or the straight-line methods. The nature of the boundary, recharge, or barrier is given by the direction of departures. An observation well closer to the boundary shows evidence of boundary effect earlier than does an observation well at a greater distance. The theory of images can be used to estimate the distance to the boundary. The analysis can be extended to locate multiple boundaries.

For the estimation of the formation constants, only those observations should be used that do not reflect boundary effects, that is, the earlier part of the time–drawdown curve. For the prediction of future drawdowns, the latter part of the curve incorporating the boundary effects is pertinent.

3.14 Characteristics of Wells

The drawdown in a well being pumped is the difference between the static water level and the pumping water level. The well drawdown consists of two components:

1) *Formation loss,* that is, the head expended in overcoming the frictional resistance of the medium from the outer boundary to the face of the well, which is directly proportional to the velocity if the flow is laminar
2) *Well loss,* which includes (a) the entrance head loss caused by the flow through the screen and (b) the head loss due to the upward axial flow of water inside the screen and the casing up to the pump intake. This loss is associated with the turbulent flow and is approximately proportional to the square of the velocity.

The well drawdown D_w can be expressed as

$$D_w = BQ + CQ^2 \tag{3.39a}$$

where B summarizes the resistance characteristics of the formation and C represents the characteristics of the well.

For unsteady flow in a confined aquifer, from Eq. 3.18,

$$B = (264/T)\ \log\left(0.3Tt/r_w{}^2S\right) \tag{3.39b}$$

This shows that the resistance of an extensive artesian aquifer increases with time as the area of influence of the well expands. For relatively low pumping rates, the well loss may be neglected, but for higher rates of discharge it can represent a sizable proportion of the well drawdown.

3.14.1 Specific Capacity of a Well

The productivity and efficiency of a well is generally expressed in terms of *specific capacity,* defined as the discharge per unit drawdown, that is, the ratio of discharge to well drawdown:

$$Q/D_w = 1/(B + CQ) \tag{3.40a}$$

The specific capacity of a well depends on the formation constants and hydrogeologic boundaries of the aquifer, on well construction and design, and on test conditions. It is sometimes useful to distinguish between *theoretical specific capacity,* which depends only on formation characteristics and ignores well losses, and actual specific capacity. The former is a measure of the productivity. The difference between the two, or their ratio, is a measure of the efficiency of the well.

For unsteady flow in a confined aquifer,

$$Q/D_w = 1/\{(264/T)\ \log[(0.3\ Tt)/(r_w{}^2S)] + CQ\} \tag{3.40b}$$

Hence the specific capacity is not a fixed quantity, but decreases with both the period of pumping and the discharge. It is important to state not only the discharge at which a value of specific capacity is obtained, but also the duration of pumping. Determination of specific capacity from a short-term acceptance test of a few hours' duration can give misleading results, particularly in aquifers having low hydraulic diffusivity, that is, low transmissivity and high storage coefficients.

3.14.2 Partial Penetration

The specific capacity of a well is affected by partial penetration. A well that is screened only opposite a part of an aquifer will have a lower discharge for the same drawdown or larger drawdown for the same discharge, that is, a smaller specific capacity. The ratio of the specific capacity of a partially penetrating well to the specific capacity of a completely penetrating well in homogeneous artesian aquifers is given by the *Kozeny formula,* which is valid for *steady-state conditions* using either the US customary units or the SI units:

$$(Q/s_p)\ (Q/s) = K_p\left\{1 + 7[r_w/(2K_pb)]^{1/2} \cos(\pi K_p/2)\right\} \tag{3.41}$$

where Q/s_p is the specific capacity of a partially penetrating well, gpm/ft or $m^3/d/m$; Q/s is the specific capacity of a completely penetrating well, gpm/ft or $m^3/d/m$; r_w is effective well radius, ft or m; b is aquifer thickness, ft or m; and K_p is the ratio of length of screen to saturated thickness of the aquifer.

If the right-hand side of Eq. 3.41 is denoted by F_p, the equation may be written as

$$Q/s_p = (Q/s)F_p \tag{3.42}$$

The formula is not valid for small b, large K_p, and large r_w.

A graph of F_p versus K_p for various values of b/r_w is given in Fig. 3.12 within the valid range of the formula.

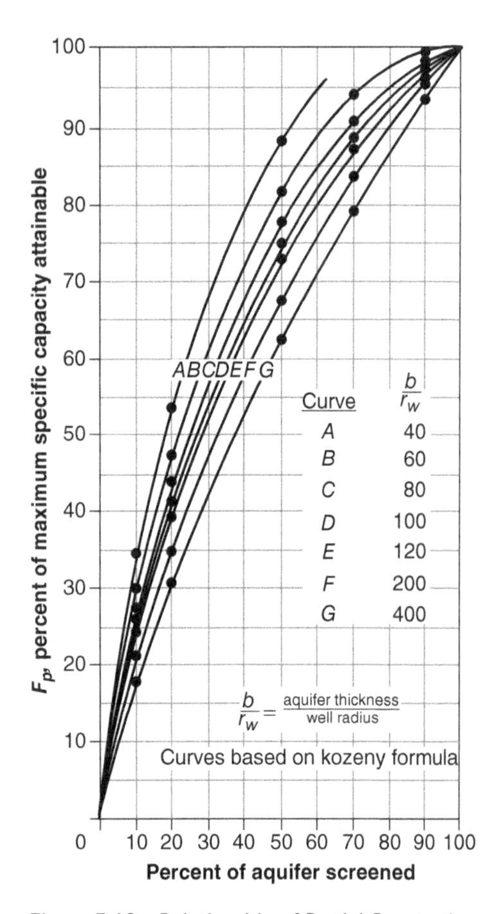

Figure 3.12 Relationship of Partial Penetration and Specific Capacity for Wells in Homogeneous Artesian Aquifers

3.14.3 Effective Well Radius

The effective radius of a well is seldom equal to its nominal radius. *Effective radius* is defined as that distance, measured radially from the axis of a well, at which the theoretical drawdown equals the actual drawdown just at the surface of the well. Depending on the method of construction and development, and the actual condition of the intake portion of the well, the effective radius may be greater than, equal to, or less than the nominal radius. The transmissivity of the material in the immediate vicinity of a well is the controlling factor. If the transmissivity of the material surrounding the well is higher than that of the aquifer, the effective well radius will be greater than the nominal radius. On the other hand, if the material around the well has a lower transmissivity due to caving or clogging because of faulty construction, the effective radius will be less than the nominal radius.

3.14.4 Measurement of Well Characteristics

The well-loss factor and the effective radius of a well can be determined by the multiple-step drawdown test. In this test, a well is pumped at a constant rate (generally for a few hours), after which the rate is increased and held constant at that rate for the same period. Three or four steps are used. The size of the steps depends on the yield of the well and the capacity of the pumping equipment. Another method for carrying out the test is to pump the well at a constant rate for a specified period, shut off the pump for an equal period, and then restart the pump at a higher rate. This is continued for three or four steps.

The analysis of the test data is similar to that discussed under variable discharge. Equation 3.34 is modified to include well losses. Increments of drawdown are determined at the same period of pumping in each step. Simultaneous solution of the equations gives the well-loss coefficient C, effective well radius r, and the values of formation constants. Usually a graphical procedure of solution is employed. Good results are obtained only if the value of C does not change appreciably with the rate of pumping. This indicates a stable well. A decrease in C for higher discharges may indicate development of a well during testing; an increase in C may denote clogging. The acceptance test for a well should include a step-drawdown test, because such a test permits the characteristics of an aquifer and of the well that govern the efficient performance of a water supply system to be evaluated.

3.15 Yield of a Well

For the optimal design of a well (or a well field), the combination of discharge and drawdown that gives the lowest-cost solution is sought. Both the capital outlays and the operation and maintenance costs need to be considered over the economic life of the structure. The interrelationship of pumping rate, drawdown, and specific yield serves as a basis for the selection of optimal design capacity. We have shown previously that the specific capacity decreases as the pumping rate is increased. Hence the earlier increments of drawdown are more effective in producing yields than the later ones; that is, each additional unit of yield is more expensive than the previous one. Increasing the yield of a well by one unit is economically justified only if the cost of developing this unit from alternate sources, such as another well or surface supplies, is higher.

The yield obtainable from a well at any site depends on (a) the hydraulic characteristics of the aquifer, which may be given in terms of a specific capacity drawdown relationship; (b) the drawdown at the pumping well; (c) the length of the intake section of the well; (d) the effective diameter of the well; and (e) the number of aquifers penetrated by the well.

3.15.1 Maximum Available Drawdown

The maximum available drawdown at a well site can be estimated by the difference in elevations between the static water level and a conservation level below which it is undesirable to let the water levels drop. The conservation level is controlled by hydrogeologic conditions (type and thickness of the aquifer and the location of the most permeable strata), maintenance of the efficiency of the well, preservation of water quality, and pumping costs. In an artesian aquifer, good design practice requires that the drawdown not result in the dewatering of any part of the aquifer. Hence the maximum allowable drawdown is the distance between the initial piezometric level and the top of the aquifer.

In a water table aquifer, the pumping level should be kept above the top of the screen. The yield–drawdown relationship of homogeneous water table aquifers indicates that optimum yields are obtained by screening the lower one-half to one-third of the aquifer. A common practice is to limit the maximum available drawdown to one-half to two-thirds of the saturated thickness. In very thick aquifers, such as artesian or water table aquifers, the limiting factor in obtaining yields is not the drawdown but the cost of pumping. In some locations, the available drawdown may be controlled by the presence of poor-quality water. The maximum drawdown should be such as to avoid drawing this poor-quality water into the pumping well.

3.15.2 Specific Capacity–Drawdown Curve

A graph of specific capacity versus drawdown is prepared from the data on existing wells in the formation if such data are available. Specific capacities should be adjusted for well losses and partial penetration and should be reduced to a common well radius and duration of pumping. If no data are available, a step-drawdown test is conducted on the production well. The curve is extended to cover the maximum available drawdown. For a well receiving water from more than one aquifer, the resultant specific capacity is the sum of the specific capacities of the aquifer penetrated, reduced appropriately for partial penetration.

3.15.3 Maximum Yield

The following procedure is carried out to estimate the maximum yield of a well:

1) Calculate the specific capacity of the fully penetrating well having the proposed diameter from Eq. 3.15 or Eq. 3.18.
2) Reduce the specific capacity obtained above for partial penetration. This can be done by using Eq. 3.41.
3) Adjust the specific capacity for the desired duration of pumping from Eq. 3.40.
4) Calculate the maximum available drawdown.
5) Compute the maximum yield of the well by multiplying the specific capacity in step 3 by the maximum available drawdown.

Example 3.9 Computation of the Specific Capacity of a Well and Its Maximum Yield

A well having an effective diameter of 12 in. (305 mm) is to be located in a relatively homogeneous artesian aquifer with a transmissivity of 10,000 gpd/ft (124.18 m^3/d/m) and a storage coefficient of 4×10^{-4}. The initial piezometric surface level is 20 ft (6.1 m) below the land surface. The depth to the top of the aquifer is 150 ft (45.72 m) and the thickness of the aquifer is 50 ft (15.24 m). The well is to be finished with a screen length of 20 ft (6.1 m). Compute the specific capacity of the well and its maximum yield after 10 days of pumping. Neglect well losses.

Solution 1 (US Customary System):

The specific capacity of a 12-in., fully penetrating well after 40 days of pumping can be calculated from Eqs. 3.15 and 3.16:

$$s = 114.6\,QW(u)/T \tag{3.15}$$

$$u = 1.87(S/T)(r^2/t) \tag{3.16}$$

Given data are as follows:

$t = 10$ d
$T = 124.18$ m^3/d/m = 10,000 gpd/ft = 10^4 gpd/ft
$S = 4 \times 10^{-4}$
$r = (305$ mm/2) = 0.1525 m = (12 in.)/2 = 1 ft/2 = 0.5 ft.

Then

$$u = 1.87(4 \times 10^{-4}/10^4)(0.5^2/10) = 1.87 \times 10^{-9}$$

From Table 3.2: $W(u) = 19.52$

$$(Q/s) = T/114.6\,W(u) = 10,000/(114.6 \times 19.52) = 4.47\text{ gpm/ft }(80\text{ m}^3/\text{d/m})$$

The percentage of aquifer screened is $Kp = 20/50 = 40\,\%$.
The slenderness of the well factor is $b/r_w = 50/0.5 = 100$.
The value of F_p from Fig. 3.12 is 0.65.
The expected **specific capacity** of the well = 0.65×4.47 = **2.9 gpm/ft** (**51.87** m^3/d/m).
The maximum available drawdown = 150 − 20 = 130 ft (39.62 m).
The maximum **yield** of the well = 130×2.9 = **377 gpm** (**2, 055 m^3/d**).
In view of the approximations involved with the evaluation, the actual maximum yield will be between perhaps 300 and 450 gpm (1635 and 2453 m^3/d).

Solution 2 (SI System):

The specific capacity of a 305-mm, fully penetrating well after 40 days of pumping can be calculated from Eqs. 3.12 and 3.13:

$$s = (Q/4\pi T)W(u) \tag{3.12}$$

$$u = (r^2 s)/(4Tt) \tag{3.13}$$

Given data are as follows:

$t = 10$ d
$T = 124.18$ m^3/d/m= 10,000 gpd/ft = 10^4 gpd/ft
$S = 4 \times 10^{-4}$
$r = (305$ mm/2$) = 0.1525$ m= (12 in.)/2 = 1 ft/2 = 0.5 ft.

Then

$$u = (0.1525)^2 \times (4 \times 10^{-4})/(4 \times 124.18 \times 10)$$
$$= 1.87 \times 10^{-9}$$

From Table 3.2, $W(u) = 19.52$

$$(Q/s) = 4\pi T/W(u) = 4 \times 3.14 \times 124.18/19.52 = 80 \text{ m}^3/\text{d/m}$$

The percentage of aquifer screened is $K_p = 6.1/15.24 = 40\,\%$.
The slenderness of the well factor is $b/r_w = 15.24/0.1525 = 100$.
The value of F_p from Fig. 3.12 is 0.65.
The expected **specific capacity** of the well $(Q/s_p) = (Q/s)\,F_p = 80 \times 0.65 = 51.9$ m^3/d/m.
The maximum available drawdown = 45.72 − 6.1 = 39.62 m.
The maximum **yield** of the well = (51.9 m^3/d/m)(39.62 m) = 2,057 m^3/d.
The actual maximum **yield** will be between 1635 and 2457 m^3/d.

3.16 Well Design

From the standpoint of well design, it is useful to think of a well as consisting of two parts: (a) the *conduit portion* of the well, which houses the pumping equipment and provides the passage for the upward flow of water to the pumping intake; and (b) the *intake portion,* where the water from the aquifer enters the well. In consolidated water-bearing materials, the conduit portion is usually cased from the surface to the top of the aquifer, and the intake portion is an uncased, open hole. In unconsolidated aquifers, a perforated casing or a screen is required to hold back the water-bearing material and to allow water to flow into the well.

The depth of a well depends on the anticipated drawdown for the design yield, the vertical position of the more permeable strata, and the length of the intake portion of the well.

The well size affects the cost of construction substantially. The well need not be of the same size from top to bottom. The diameter of a well is governed by (a) the proposed yield of the well, (b) entrance velocity and loss, and (c) the method of construction. The controlling factor is usually the size of the pump that will be required to deliver the design yield. The diameter of the casing should be two nominal sizes larger than the size of the pump bowls, to prevent the pump shaft from binding and to reduce well losses. Table 3.5 gives the casing sizes recommended for various pumping rates.

The selection of the well size may depend on the size of the open area desired to keep entrance velocities and well losses at a reasonable value. In deep-drilled wells, the minimum size of the hole may be controlled by the equipment necessary to reach the required depth. Deep wells in consolidated formations are often telescoped in size to permit drilling to required depths.

The wells are generally lined or cased with mild steel pipe, which should be grouted in place in order to prevent caving and contamination by vertical circulation and to prevent undue deterioration of the well by corrosion. If conditions are such that corrosion is unusually severe, then plastic or glass fiber pipes can be used if practicable.

The intake portion of the well should be as long as economically feasible to reduce the drawdown and the entrance velocities. In relatively homogeneous aquifers, it is not efficient to obtain more than 90% of the maximum yield. In nonhomogeneous aquifers, the best strategy is to locate the intake portion in one or more of the most permeable strata.

Perforated pipes or prefabricated screens are used in wells in unconsolidated aquifers. The width of the screen openings, called the slot size, depends on the critical particle size of the water-bearing material to be retained and on the grain-size distribution, and is chosen from a standard sieve analysis of the aquifer material. With a relatively coarse and graded material, slot sizes are selected that will permit the fine and medium-sized particles to wash into the well during development and to retain a specified portion of the aquifer material around the screen. A graded filter is thereby generated around the well, which has higher permeability than the undisturbed aquifer material.

Perforated casings are generally used in uncemented wells when relatively large openings are permissible. If the casing is slotted in place after installation, the smallest practical opening is one-eighth inch (3.18 mm). Machine-perforated casings are also available. Fabricated well screens are available in a wide variety of sizes, designs, and materials. The choice of material is governed by water quality and cost.

Table 3.5 Recommended Well Diameters

Anticipated Well Yield (gpm)	Nominal Size of Pump Bowls (in.)	Optimum Size of Well Casing (in.)	Smallest Size of Well Casing (in.)
Less than 100	4	6 ID	5 ID
75–175	5	8 ID	6 ID
150–400	6	10 ID	8 ID
350–650	8	12 ID	10 ID
600–900	10	14 OD	12 ID
850–1300	12	16 OD	14 OD
1200–1800	14	20 OD	16 OD
1600–3000	16	24 OD	20 OD

Conversion factors: 1 gpm = 5.45 m^3/d; 1 in. = 25.4 mm. Notes: ID = inside diameter; OD = outside diameter.

For maximum efficiency, the frictional loss of the screen must be small. The head loss through a screen depends on screen length L, diameter D, percentage open area A_p, coefficient of contraction of openings C_c, velocity in the screen v, and the total flow into the screen Q (ft^3/s). It has been shown that for minimum screen loss, $CL/D > 6$, where $C = 11.31C_cA_P$. The value of CL/D may be increased by increasing C_c, A_p, or L, or by decreasing D. Thus, for the screen loss to be a minimum, the percentage of open area depends on the length and diameter of the screen. The screen length is usually fixed by the considerations of hydrogeology and cost.

The screen length and diameter can be selected from the slot size and the requirement that the entrance velocity be less than that needed to move the unwanted sand particle sizes into the well. Experience has shown that, in general, a velocity of 0.1 ft/s (0.03 m/s) gives negligible friction losses and the least incrustation and corrosion.

Where the natural aquifer material is fine and uniform (effective size less than 0.01 in. [0.25 mm] and uniformity coefficient less than 3.0), it is necessary to replace it by a coarser gravel envelope next to the screen. The slot size is selected to fit this gravel pack. The gravel pack increases the effective well radius and acts as a filter and a stabilizer for the finer aquifer material. A gravel pack well is shown in Fig. 3.13. There are no universally accepted rules for the selection of slot sizes or for the design of a gravel pack. A correctly designed well should provide a virtually sand-free operation (less than 3 mg/L). The thickness of the gravel pack should not be less than 3 in. or more than 9 in. and the particle size distribution curve of the pack should approximately parallel that of the aquifer.

3.17 Well Construction

There is no one optimum method of well construction. The size and depth of the hole, the rocks to be penetrated, and the equipment and experience of local drillers control the method of well sinking and determine the cost of construction. Well sinking is a specialized art that has evolved along a number of more or less regional lines. In the United States, well drillers are generally given much latitude in the choice of a suitable method. What they undertake to do is to sink a well of specified size at a fixed price per foot. Ordinarily, therefore, the engineer gives his attention not so much to drilling operations as to the adequacy, suitability, and economics of proposed developments and the location of the works.

Well categories generally take their names from the methods by which wells are constructed. Shallow wells can be dug, driven, jetted, or bored.

3.17.1 Dug Wells

Small dug wells are generally excavated by hand. In loose overburden, they are cribbed with timber; lined with brick, rubble, or concrete; or cased with large-diameter vitrified tile or concrete pipe. In rock, they are commonly left unlined. Excavation is continued until water flows in more rapidly than it can be bailed out. Dug wells should be completed when the water table is at or near its lowest level. Otherwise, they may have to be deepened at a later date.

Figure 3.13 Gravel-Packed Well with Deep-Well Turbine Pump
(*Source:* After Wisconsin State Board of Health)

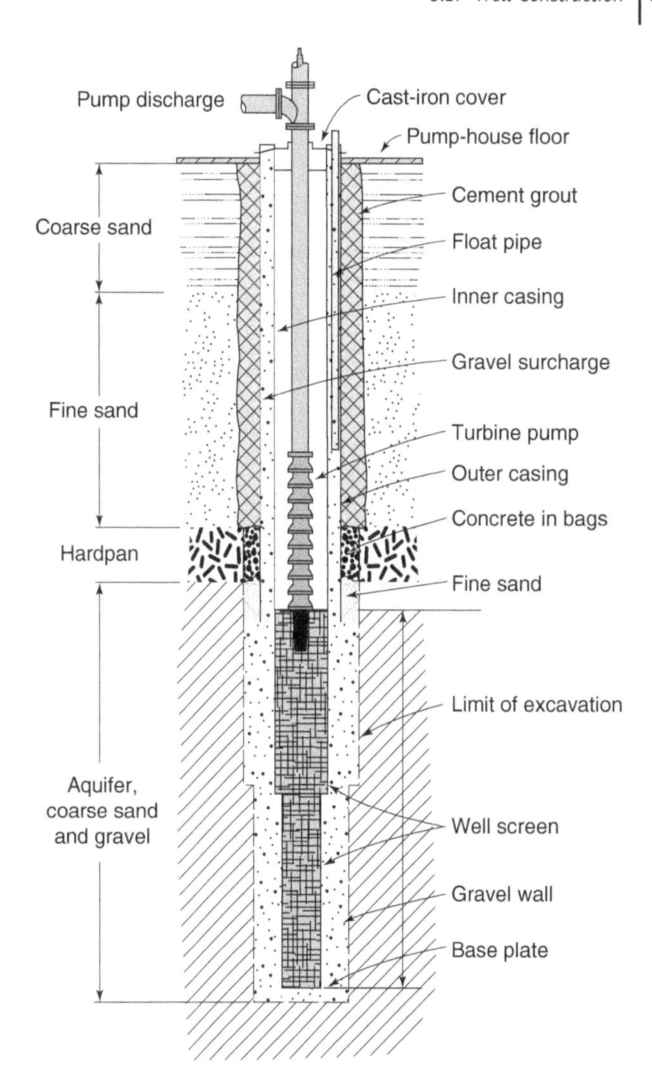

Large and deep dug wells are often constructed by sinking their liners as excavation proceeds. The lead ring has a steel cutting edge; new rings are added as excavation progresses.

3.17.2 Driven and Jetted Wells

Wells can be driven into relatively shallow sand formations. As shown in Fig. 3.14, the driving point is attached to a strainer or perforated section of pipe. To reduce friction, the point is somewhat larger than the casing. The driving weight is commonly suspended from a block attached to a tripod. In hard ground a cylindrical shoe equipped with water jets loosens the soil and washes it to the surface. Batteries of driven wells may be connected to a suction header to supply enough water.

3.17.3 Bored Wells

Wells can be bored with hand or power augers into sufficiently cohesive (noncaving) soils. Above the water table, the soil is usually held in the auger, which must then be raised from time to time to be cleaned. Below the water table, sand may wash out of the auger and have to be removed from the bore hole by a bailer or sand pump. As the well becomes deeper and deeper, sections of rod are added to the auger stem. Bits up to 36 in. (914 mm) in diameter have been used successfully, and wells have been enlarged in diameter up to 48 in. (1,219 mm) by reaming. A concrete, tile, or metal casing is inserted in the hole and cemented in place before the strainer is installed.

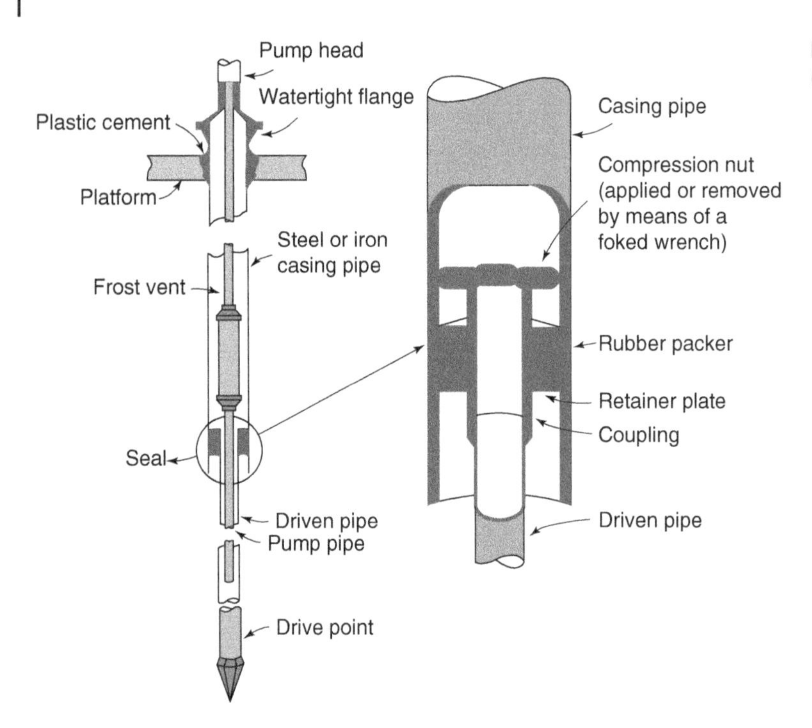

Figure 3.14 Driven Well and Its Sanitary Protection (*Source:* After Iowa State Department of Health)

3.17.4 Drilled Wells

High-capacity, deep wells are constructed by drilling. Because the water-bearing materials vary so widely, no one method of drilling can be adopted under all conditions. The method of drilling is selected to suit the particular conditions of a site. The systems of drilling used in water-well construction are based on either the percussion or the rotary principle.

3.17.5 Collector Wells

A *collector well* consists of a central shaft of concrete caisson some 15 ft (4.57 m) in internal diameter and finished off below the water table with a thick concrete plug. From this shaft, perforated radial pipes 6 or 8 in. (150 or 200 mm) in diameter and 100–250 ft (30–76 m) long are jacked horizontally into a water-bearing formation through ports near the bottom of a caisson. The collector pipes may be installed and developed in the same manner as for ordinary wells.

3.17.6 Pumps

Many types of well pumps are on the market to suit the wide variety of capacity requirements, depths to water, and sources of power. Fig. 3.15 shows a well pump and 190,000-gal (719,150-L) water tank. In the United States almost all well pumps are driven by electric motors.

Domestic systems commonly employ one of the following pumps:

1) For lifts under 25 ft (7.62 m), a small reciprocating or piston pump
2) For lifts up to 125 ft (38.10 m), a centrifugal pump to which water is lifted by recirculating part of the discharge to a jet or ejector
3) For lifts that cannot be managed by jet pumps, a cylinder pump installed in the well and driven by pump rods through a jack mounted at the well head.

Systems of choice normally incorporate pressure tanks for smooth pressure-switch operation. The well itself may provide enough storage to care for differences between demand rates in the house and flow rates from the aquifer. This is why domestic wells are seldom made less than 100 ft (30 m) deep even though the water table may lie only a few feet below the ground surface. Deep wells and pump settings maintain the supply when groundwater levels sink during severe droughts or when nearby wells are drawn down steeply.

Figure 3.15 Well and 190,000-Gallon (719,150-L) Tank, Maui, Hawaii (*Source:* Courtesy of the Department of Water Supply, Maui County, Hawaii)

Large-capacity systems are normally equipped with centrifugal or turbine pumps driven by electric motors. A sufficient number of pump bowls are mounted one above the other to provide the pressure necessary to overcome static and dynamic heads at the lowest water levels. For moderate quantities and lifts, *submersible* motors and pumps, assembled into a single unit, are lowered into the well. The water being pumped cools the compact motors normally employed. Large-capacity wells should be equipped with suitable measuring devices. Continuous records of water levels and rates of withdrawal permit the operator to check the condition of the equipment and the behavior of the source of supply. This is essential information in the study and management of the groundwater resource.

3.17.7 Development

Steps taken to open up or enlarge flow passages in the formation in the vicinity of the well are called *development*. Thorough development of the completed well is essential regardless of the method of construction used to obtain higher specific capacities, to increase effective well radius, and to promote efficient operation over a longer period of time. This can be achieved in several ways. The method selected depends on the drilling method used and on the formation in which the well is located. The most common method employed is overpumping, that is, pumping the well at a higher capacity than the design yield. Temporary equipment can provide the required pumping rates. Pumping is continued until no sand enters the well. Other methods used include flushing, surging, high-velocity jetting, and backwashing. Various chemical treatments and explosives are used in special circumstances.

3.17.8 Testing

After a well is completed, it should be tested to determine its characteristics and productivity. Constant-rate and step-drawdown pumping tests are used for this purpose. The test should be of sufficient duration; the specific capacity of a well based on a 1-hour test may be substantially higher than that based on a 1-day test. Longer duration is also required to detect the effect of hydraulic boundaries, if any. The extent to which the specific yield would decrease depends on the nature and the effectiveness of the boundaries.

3.17.9 Sanitary Protection of Wells

The design and construction of a well to supply drinking water should incorporate features to safeguard against contamination from surface and subsurface sources. The protective measures vary with the geologic formations penetrated and the site conditions. The well should be located at such a distance from the possible sources of pollution (e.g., wells used for the disposal of liquid wastes or artificial recharge; seepage pits; and septic tanks) that there is no likelihood of contaminated water reaching the well. The casing should be sufficiently long and watertight to seal off formations that have undesirable characteristics. Failure to seal off the annular space between the casing and well hole has been responsible for bacterial contamination in many instances. The casing should be sealed in place by filling the open space around the casing with

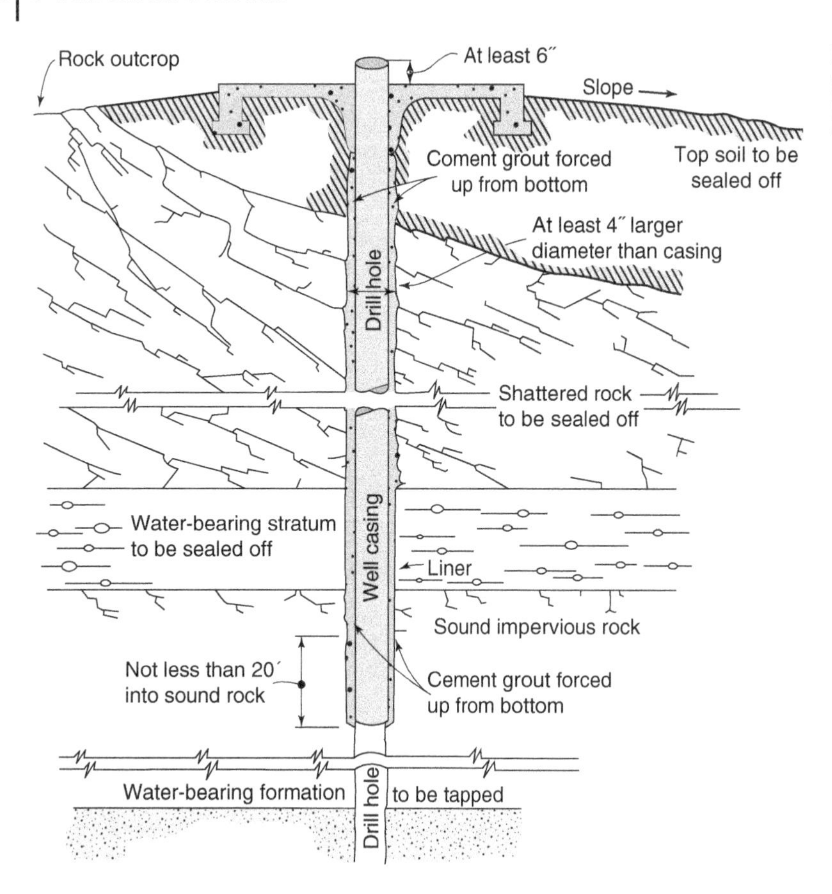

Figure 3.16 Drilled Well and Its Sanitary Protection (After Iowa State Department of Health). Conversion factor: 1′ = 1 ft = 0.3048 m; 1″ = 1 in. = 25.4 mm

cement grout or other impermeable material down to an adequate depth. This prevents seepage of water vertically along the outside of the pipe. A properly cemented well is shown in Fig. 3.16. The well casing should extend above the ground. The top of the well should contain a watertight seal; the surface drainage should be away from it in all directions.

An essential final step in well completion is the thorough disinfection of the well, the pump, and the piping system. Although the water in the aquifer itself may be of good sanitary quality, contamination can be introduced into the well system during drilling operations and the installation of other elements of the system. Periodic disinfection of the well during the drilling is a good practice and should be encouraged. In the case of an artificially gravel-packed well, all gravel-pack material should be sterilized before being placed in the well. Solution strengths of 50–200 mg/L chlorine are commonly used for sterilizing wells. The effectiveness of disinfection should be checked after the completion of the work. Disinfection of the system is also necessary after repairs of any part of the system.

3.17.10 Maintenance

Good maintenance extends the life of a well. The maintenance of the yield of a well depends on (a) the well construction, (b) the quality of water pumped (water may be corroding or encrusting), and (c) any interference from neighboring wells. If the performance of the well declines, renovation measures should be undertaken that may include mechanical cleaning, surging, and chemical treatment.

3.18 Evaluation of Aquifer Behavior

Planning for the optimum utilization of the groundwater resource in an aquifer system requires the evaluation of the merits of alternative strategies of development. The steps involved in predicting the consequences of various plans are (a) quantitative assessment of the hydraulic and hydrologic characteristics of the aquifer system and (b) elaboration of the cause-and-effect relationships between pumping, replenishment, and water levels.

3.18.1 Hydrologic Equation

The basic groundwater balance equation is an expression of material balance:

Inflow (or recharge) = outflow (or discharge) ± change in storage

$$I = O \pm \Delta S$$

This equation must be applied to a specific area for a specific period of time. When drawdowns imposed by withdrawals reduce the hydraulic gradient in the discharge areas, the rate of natural discharge is reduced. These flows become available for development and a new equilibrium condition is approached with the water table or the piezometric surface at a lower level.

The effect of drawdown on aquifer recharge is difficult to evaluate. Additional recharge may be induced into an aquifer through the former discharge areas by reversing the hydraulic gradient. In leaky aquifers, inflow may be induced by the reduction of heads, the contributions being roughly proportional to drawdown. Additional recharge may also be induced in the recharge areas if drawdown causes a dewatering in areas where recharge was limited because the aquifer was full. This is referred to as the capture of *rejected recharge*.

3.18.2 Safe Yield of an Aquifer

The yield of an aquifer depends on the following:

1) The characteristics of the aquifer
2) The dimensions of the aquifer and the hydraulic characteristics of its boundaries
3) The vertical position of each aquifer and the hydraulic characteristics of the overlying and underlying beds
4) The effect of proposed withdrawals on the recharge and discharge characteristics of the aquifer.

Thus it is evident that the safe yield of an aquifer is not necessarily a fixed quantity, and it is not strictly a characteristic of the groundwater aquifer. It is a variable quantity dependent on natural hydrogeologic conditions and on recharge and discharge regimes. Safe yield has been defined in a variety of ways, each definition placing emphasis on a particular aspect of groundwater resource development. These include, within economic limits, (a) development to the extent that withdrawals balance recharge and (b) development to the extent that change in the quality of groundwater allows.

3.19 Groundwater Quality Management

In a majority of cases when polluted water has been drawn from wells, the contamination was introduced at the well site, indicating faulty construction. There are, however, numerous examples of contamination of groundwater caused by disposal of wastes. Once groundwater is contaminated, the impairment of the groundwater resource is long-lasting, and recovery is extremely slow.

To predict where the contaminating fluids will go requires a three-dimensional geologic, hydrodynamic, and geochemical analysis. The rate and extent of the spread of pollution are controlled by (a) the characteristics of the source of pollution, (b) the nature of rock formations in the unsaturated and saturated zones, and (c) the physical and chemical properties of the contaminant. The phenomena governing the disposition of the contaminant are capillary attraction, decay, adsorption, dispersion, and diffusion.

There have been numerous examples of contamination of groundwater by wastes allowed to seep into the ground, wastes discharged into pits and ponds, and leaks from holding tanks and sewers. The safe distance from a polluting source of this type is determined to a large degree by the velocity of percolation through the unsaturated zone and by the lateral movement once the contamination reaches groundwater. Water table aquifers, being near the surface and having a direct hydrologic connection to it, are more subject to contamination than are deeper lying artesian aquifers.

The discharge of wastes into streams has had both direct and indirect effects on the quality of groundwater. The polluted rivers that cross recharge areas of artesian aquifers tapped by wells have affected the quality of their discharge. The aquifers that are replenished by infiltration from polluted streams will eventually be contaminated by soluble chemical wastes carried in the stream. Induced contamination of an aquifer can result when the cone of depression of a discharging well intersects a polluted river. This is frequently the case in coastal areas in wells located near streams containing brackish water.

Artificial recharge with river water of poorer quality than that found in the aquifer will ultimately result in the deterioration of the quality of groundwater.

3.19.1 Biological Contamination

Because of increasing numbers of septic tanks and growing use of effluents from wastewater treatment plants for artificial recharge of aquifers, the potential for contamination of groundwater by bacteria and viruses needs to be considered. Filtration through granular material improves the biological quality of water. A 10-ft (3-m) downward percolation in fine sand is capable of removing all bacteria from water. The length of time bacteria and viruses may survive and the distances they may travel through specific rock materials in different subsurface environmental conditions are uncertain. They seem to behave in a manner similar to the degradable and adsorbable contaminants. Under favorable conditions some bacteria and viruses may survive up to at least 5 years in the underground environment. However, the distances traveled in both the saturated and unsaturated media are surprisingly short when reasonable precautions are taken in disposal.

The principal determinant of the distance traveled seems to be the size of the media. Romero (1970) provides diagrams that may be used to evaluate the feasibility of disposing of biologically contaminated wastes in saturated and unsaturated granular media. The danger of bacterial pollution is greater in fractured rocks, cavernous limestones, and gravel deposits where the granular materials have no filtering capacity. The distances traveled will be higher in areas of influence of discharging or recharging wells because higher velocities are present. The higher rates of artificial recharge and greater permeability of artificial recharge basins enable bacteria to be carried to a greater depth.

3.19.2 New Wells Disinfection During Construction

During the drilling of new wells, contamination could be introduced into the well from the drilling tools and mud, make-up water, topsoil falling in or sticking to tools, and from the gravel itself. Disinfection should take place following development, testing for yield, and before the test pump is removed from the well. This will assure that the well is purged of drilling mud, dirt, and other debris that reduces the effectiveness of the disinfecting solution. The procedure described below is generally satisfactory for disinfecting a well: (a) The chlorine compounds is required to dose 100 ft (30 m) of water-filled casing at 50 mg/L for diameters ranging from 6 to 24 in. (150–600 mm). Since organic matter such as oil may need an initial concentration of 1000 mg/L before being injected into the well. This is 20 times the recommended values of 50 mg/L; (b) turn the pump on and off several times so as to thoroughly mix the disinfectant with the water in the well. Pump until the water discharged has the odor of chlorine. Repeat this procedure several times at one-hour intervals; (c) allow the well to stand without pumping for 24 hours; (d) the water should then be pumped to waste until the odor of chlorine is no longer detectable. Use a chlorine test kit to determine the absence of a chlorine residual; (e) collect a bacteriological sample in a sterile container and submit it to a laboratory for examination, and (f) if the laboratory analysis shows the water is not free of bacterial contamination, repeat the disinfection procedure and retest the water. If repeated attempts to disinfect the well are unsuccessful, a detailed investigation to determine the cause of the contamination should be undertaken.

3.19.3 New Wells Disinfection After Construction

Prior to placing a new well pumping installation in service, the well disinfection procedure previously described should be repeated. This will protect against contamination caused during the construction of the pump base and related appurtenances, and the installation of the permanent pumping unit.

3.19.4 Existing Wells Disinfection After Well or Pump Repairs

Those deposits due to well system repairs are difficult to disinfect by typical well disinfection procedures. The following special procedures are recommended: (a) Swab inside of well casing with a strong non-foaming detergent; (b) calculate amount of chlorine based on diameter of well and water depth. The chlorine dosage rate should produce a free chlorine concentration of at least 100 mg/L in the water-filled casing; (c) add chlorine solution to the well, preferably through a hose raised and lowered to reach all areas of the well, including the well casing above the water line; (d) clean and disinfect pump, pump column pipe, cable, and other equipment before the units are lowered into the well; (e) follow procedures outlined for new well disinfection from Section 3.19.2 (b) through Section 3.19.2 (f).

3.19.5 Contaminated Wells Disinfection

If the well and pumping unit are indeed contaminated by bacteria, the well pump should be left in place, and the following procedures are recommended for well system disinfection: (a) To the well add a chlorine solution, strong enough that it produces a chlorine concentration of 200 mg/L in the well casing; (b) turn the pump on and off several times to thoroughly mix the disinfectant with the water in the well. Pump until the water discharged has the odor of chlorine. Repeat this procedure several times at one-hour intervals; (c) allow the well to stand without pumping for 24 hours; (d) pump the water to waste until the odor of chlorine is no longer detectable. A chlorine test kit should be used to determine the absence of a chlorine residual; (e) take a bacteriological sample and submit it to a laboratory for examination, and (f) repeat the disinfection procedure and retest the water if the laboratory analysis shows the water is not free of bacterial contamination. If repeated attempts to disinfect the well are unsuccessful, a detailed investigation to determine the cause or source of the contamination should be undertaken.

3.19.6 Subsurface Disposal of Liquid Wastes

Subsurface space may be used to an increasing degree for the disposal of wastes. The oil industry pumps nearly 20 million barrels of salt water per day into subsurface formations from which oil has been extracted. Some highly toxic chemical wastes are disposed of underground. The use of an aquifer as a receptacle for toxic waste materials is justified only if it has little or no value as a present, or potential, source of water supply. Further, there should not be any significant risk of contaminating other aquifers or of inducing fractures in the confining formations. Recharging of groundwater by injection or spreading of reclaimed municipal wastewaters is an accepted practice that will undoubtedly be increasingly used in the future.

3.20 Groundwater Under the Direct Influence of Surface Water

The U.S. Federal Surface Water Treatment Rule (SWTR) gives the following definition for *groundwater directly under the influence of surface water (GWUDI):* "any water beneath the surface of the ground which exhibits significant and rapid shifts in water characteristics such as turbidity, temperature/conductivity or pH which closely correlates to climatological or surface water conditions and/or which contains macroorganisms, algae, large diameter (three microns or greater) pathogens or insect parts of a surface water origin." In the United States, true groundwater, which is not directly influenced by surface water, will be monitored and/or treated under the Groundwater Rule, whereas GWUDI will be monitored and treated in accordance with the SWTR and the Long-Term 2 Enhanced Surface Water Treatment Rule (LT2-ESWTR).

The purpose of regulating groundwater sources under the direct influence of surface water in the SWTR is to protect against contamination from large-diameter pathogens associated with surface waters. Groundwater sources determined to be under the direct influence of surface water must be filtered or meet filtration avoidance criteria as contained in US federal and state sanitary codes. In some cases, it will be easier to replace the source with a properly designed and constructed well or spring, or possibly to modify the source to eliminate the direct influence of surface water. Public water systems with groundwater sources under the direct influence of surface water are also subject to more stringent monitoring requirements for total coliform, turbidity, and entry point disinfection residual. The types of groundwater sources potentially regulated under the SWTR include dug wells, springs, infiltration galleries, shallow or improperly constructed wells, or other collectors in subsurface aquifers near surface waters.

Each local health department (LHD) throughout the United States is responsible for identifying which public water sources are subject to the SWTR. However, it is the responsibility of the water supplier to provide the information needed to make this determination to the LHD. The LHD is also responsible for recording and reporting the criteria used and the results of determinations. All groundwater sources used to supply public water systems must be evaluated for evidence of groundwater under the direct influence of surface-water GWUDI. This evaluation will focus on the likelihood that the groundwater source could be contaminated with large-diameter pathogens, such as *Giardia lamblia* and *Cryptosporidium,* through a hydraulic connection with surface water. If a drinking water source has been identified as GWUDI, the source must meet the criteria established under the SWTR.

3.20.1 GWUDI Determination: Source Screening Phase

Information gathered during sanitary surveys is important when making GWUDI determinations. In addition, information such as compliance monitoring data, topographic maps, geologic reports, well logs, and data on potential contaminant source(s) is useful.

A two-phase methodology is being used in the United States to determine whether or not a groundwater source is under the direct influence of surface water: The *source screening phase* is used first to separate those sources that are clearly not subject to surface-water influences from those sources in need of further evaluation. Then the second phase, the *detailed evaluation phase,* is applied to sources identified for testing to evaluate their degree of hydraulic connection with surface water.

The source screening phase should be used to separate those sources that are clearly not subject to surface-water influences from those sources in need of further evaluation. A schematic of the screening procedure is shown in Fig. 3.17, and the overall methodology for the detailed evaluation phase is presented as a flowchart in Fig. 3.18.

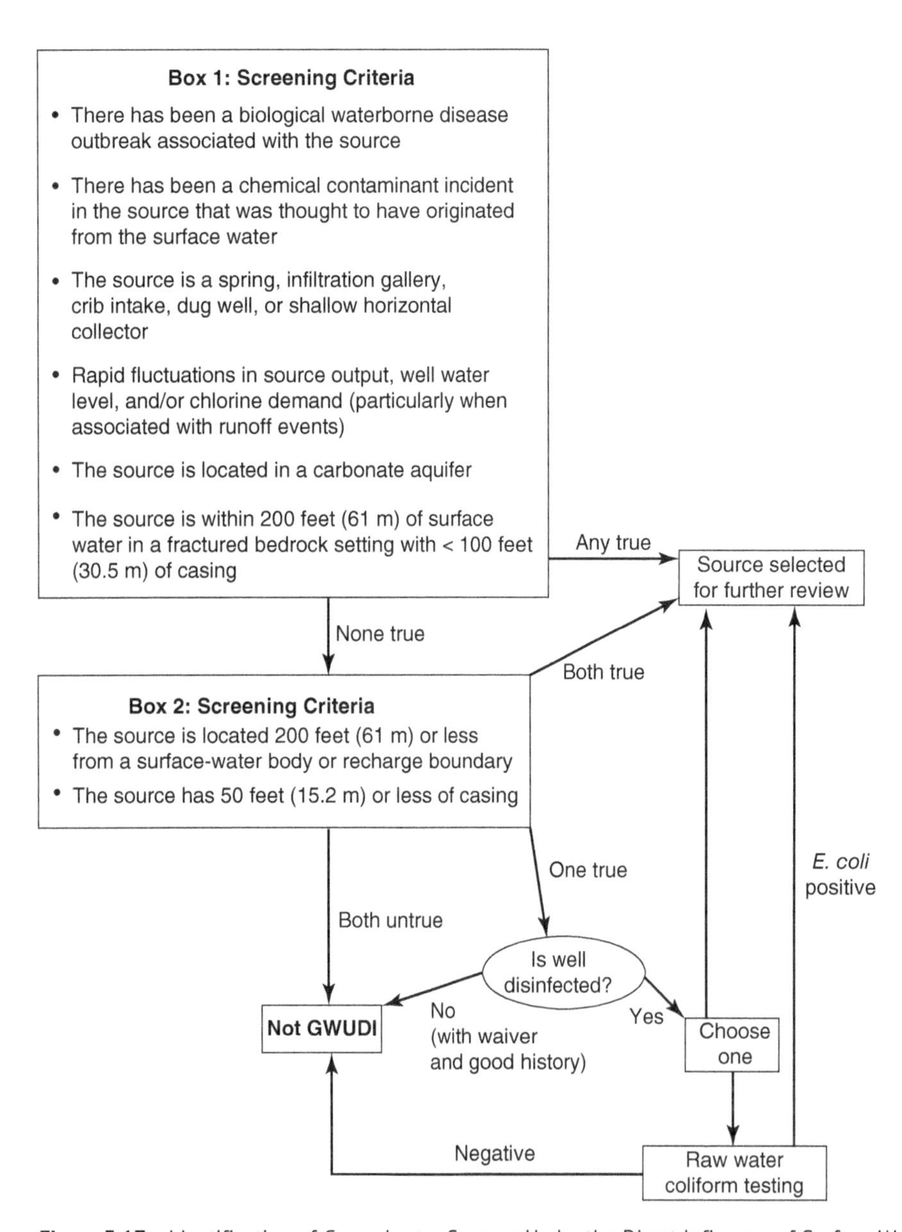

Figure 3.17 Identification of Groundwater Sources Under the Direct Influence of Surface Water (GWUDI)—Source Screening Phase Methodology

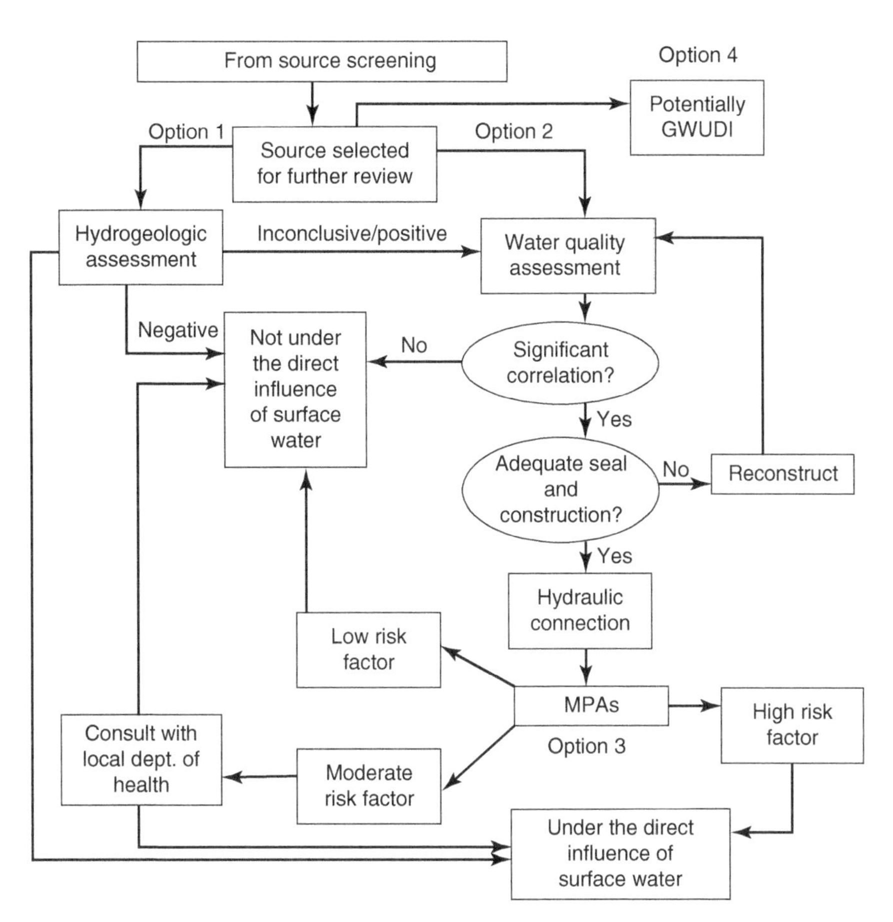

Figure 3.18 Identification of Groundwater Sources Under the Direct Influence of Surface Water—Detailed Evaluation Phase Methodology

Box 1 in Fig. 3.17 includes criteria that will immediately select a groundwater source for further review. These source-screening water criteria include the following: (a) A biological waterborne disease outbreak has been associated with the source; (b) a chemical contaminant incident has occurred in the source that was thought to have originated from the surface water; (c) the source is a spring, infiltration gallery, crib intake, dug well, or shallow horizontal collector; (d) rapid fluctuations have been noted in the source output, well water level, and/or chlorine demand (particularly when associated with runoff events); (e) the source is located in a carbonate aquifer; and (f) the source is within 200 ft (60 m) of surface water in a fractured bedrock setting with <100 ft (30 m) of casing. If none of these criteria are met, then the screening process continues in Box 2 of Fig. 3.17. If any of these criteria is met, then the screening process continue to the box "Source selected for further review."

The criteria in Box 2 of Fig. 3.17 ask if the source is located within 200 ft (60 m) of surface water, and/or if the well has 50 ft (15 m) or less of casing. When both criteria are met, the source is selected for further review under the box "Source selected for further review." If neither of these conditions is met, the source is designated as not being under the direct influence of surface water, or not-GWUDI. When only one of the criteria in Box 2 of Fig. 3.17 is met, the next step in the GWUDI determination is dependent on whether or not the water source is currently disinfected. Undisinfected wells that have met the criteria for a disinfection waiver and have an adequate coliform monitoring history (typically 5 years of quarterly monitoring) are designated as not being under the direct influence of surface water. Disinfected wells are subject to the options of either performing 1 year of monthly raw water coliform monitoring or moving directly into the detailed evaluation phase. Any raw water sample that is positive for *Escherichia coli* would require the system to perform a detailed evaluation. Conversely, once 1 year of satisfactory monthly raw water coliform samples have been obtained, the source can be designated as not being under the direct influence of surface water, or not-GWUDI.

3.20.2 GWUDI Determination: Detailed Evaluation Phase

Once a groundwater source has been selected for further review, as a result of the source screening phase, the procedure described in Fig. 3.18 should be closely followed.

The detailed evaluation phase has three components, as shown in Fig. 3.18: (a) hydrogeologic assessment, (b) water quality assessment, and (c) microscopic particulate analysis (MPA). The water supplier makes the decision about whether to begin with option 1 (hydrogeologic assessment), option 2 (water quality assessment), or option 3 (MPAs). Either the hydrogeologic assessment or water quality assessment is capable of providing the information required to determine that no surface-water influence is present. Sometimes, for example, when the project time is extremely important, the project engineer may want to choose option 3 (MPAs) directly or even option 4 (accept GWUDI determination).

If the hydrogeologic assessment option was chosen, but its results were inconclusive, the project engineer who represents the groundwater source owner/developer may request further water quality assessment (option 2) or MPAs (option 3) to complete the detailed evaluation phase, or simply accept the GWUDI determination (option 4).

Most project engineers would like to choose the straightforward option of water quality assessment instead of the complex option 1, hydrogeologic assessment, for their clients. If the option 2 test fails, then option 3 follows.

The MPA option is usually chosen, conducted during times of worst-case GWUDI conditions, as indicated by the water quality assessment or as predicted by historical information.

3.20.3 Hydrogeologic Assessment

As discussed and shown in Fig. 3.18, a water system has four evaluation options. Option 1 requires a detailed hydrogeologic assessment that addresses the potential of surface water to move quickly to the subsurface collection device. However, at any time during the hydrogeologic assessment, the water system's engineer can halt an ongoing evaluation by accepting a GWUDI designation (option 4) and making the appropriate modifications to bring the system into compliance with the SWTR, or changing it to option 2 or option 3.

If the hydrogeologic assessment is selected, and results of the assessment indicate that the aquifer supplying the source is not in hydraulic connection with surface water, no further analysis will be required. However, if the LHD determines that the hydrogeologic assessment does not contain enough information to establish whether there is a hydraulic connection between surface water and the source-water collection device, the water supplier should collect additional hydrogeologic information or proceed with a water quality assessment or MPAs.

The hydrogeologic assessment should be designed to provide the following information:

1) Well construction details, which provide (a) a well log, well construction diagram, and well description; and (b) its installation methods for comparison with current well standards
2) Aquifer characteristics, data which include (a) aquifer geometry and texture, including the unsaturated zone; (b) saturated thickness; (c) hydraulic conductivity; and (d) transmissivity
3) Prepumping and pumping groundwater flow characteristics data, which include (a) water table/potentiometric surface elevations, (b) groundwater flow directions, (c) groundwater flow velocity, and (d) zone of contribution/influence of the well
4) Degree of hydraulic connection between surface-water source and aquifer, which is indicated by (a) geology underlying surface-water body, (b) characterization of bottom sediments in surface-water body, (c) determination of vertical hydraulic gradient in surface-water body, (d) hydraulic relationship between the surface-water body and the well, and (e) calculations of travel times between the surface-water body and the well
5) Seasonal variations in hydrogeologic characteristics, such as the changes in flow patterns during seasonal fluctuations or periods of drought.

The hydrogeologic assessment should include, as a minimum, geologic logs and construction details for the pumping well and any observation wells or piezometers; aquifer pumping test(s); a survey of the elevations of water level monitoring measuring points; water level monitoring of groundwater and surface sources; and preparation of detailed maps of water table/potentiometric surface and geologic cross-sections. The owner/developer may need to install observation/monitoring wells or piezometers if these do not already exist.

In addition to requiring an assessment of hydrogeologic factors, the hydrogeologic assessment should include a description and review of the collection device (i.e., type, age) and a summary of any current or historical sanitary conditions. Any information available from previous sanitary surveys or field investigations should be included in the assessment, as appropriate. Table 3.6 and Fig. 3.19 provide examples of sanitary and field survey reports that can be used for information gathering.

To be definitive, a hydrogeologic assessment needs to include an interpretation of the information collected with respect to the potential for a hydraulic connection between a surface-water body and the aquifer. If the hydrogeologic assessment indicates a potential hydraulic connection, the water system should be required to initiate a water quality assessment.

3.20.4 Water Quality Assessment

Option 2, a water quality assessment, entails either (a) an evaluation of water quality parameters in terms of conductivity and temperature daily over a 12-month period or (b) an evaluation of water quality parameter in terms of monthly coliform levels for 12 consecutive months.

Conductivity, or specific conductance, is the measure of water's ability to carry an electric current. This ability depends on the presence of ions in the water and the water's temperature. Groundwater is generally higher in conductivity than surface water, because groundwater dissolves minerals from substrates through which it moves. Generally, the longer the contact times between groundwater and its aquifer, the higher the conductivity. However, there are exceptions to this generalization (e.g., surface-water bodies receiving large amounts of groundwater recharge or surface-water bodies contaminated with salts, clays, metals, or polar organics). Conductivity data are especially important when making determinations for springs (and other situations where large seasonal fluctuations in temperature are expected). Overall, conductivity tends to be a more sensitive parameter than temperature and more difficult to interpret.

Once water quality assessment data have been collected and analyzed, a determination must be made regarding whether there is a significant hydraulic connection between the surface-water body and the groundwater source. A significant hydraulic connection exists when water movement from the surface-water body to the groundwater source allows for the transport of *Giardia lamblia* cysts or *Cryptosporidium* oocysts. Dilution and time of travel estimates should be considered when determining the significance of the hydraulic connection. If the time of travel estimate for the source is less than 100 days, a significant hydraulic connection should be assumed and the supplier should proceed to the next step, MPAs (option 3).

3.20.5 Microscopic Particulate Analyses

When the hydrogeologic assessment (option 1) and/or water quality assessment (option 2) results suggest that the groundwater source is probably under the direct influence of surface water, then the water supplier may either accept the GWUDI determination, or choose to conduct MPAs, as shown in Fig. 3.18. Information collected as part of the characterization of hydraulic communication and time of travel will be very important in determining the correct timing of MPA sample collection. MPA samples must be collected at least twice, and the dates of sampling should represent worst-case conditions, when maximum potential recharge from the nearby surface water is taking place (usually during extremely wet or dry periods).

Basically, the use of MPA in a GWUDI determination involves the careful enumeration of microscopic organisms (and other particulates) in the raw drinking water. These data are then systematically evaluated to determine if the particles found are more indicative of surface water or groundwater. Some of the organisms that are considered to be characteristic of surface water include *Giardia, Cryptosporidium,* algae, diatoms, and rotifers.

Note that the hydrogeologic information and the results of the water quality assessment must be used in conjunction with MPAs to make the GWUDI determinations. Most importantly, MPA samples should be collected at a time when the water quality assessment and hydrogeologic data indicate the greatest probability that surface water is impacting the groundwater source, as indicated by hydrogeologic and water quality data. It is relatively safe to conclude from a "high" MPA rating that the groundwater source is under the direct influence of surface water, particularly when considered along with corroborating information collected in the earlier phases of a GWUDI evaluation. Moreover, it is more difficult to conclude from a "low" MPA rating that the groundwater source is not under the direct influence of surface water, because surface-water influences often only occur intermittently under particular hydrologic conditions (usually during very wet or dry periods).

Table 3.6 Example of Sanitary and Field Survey Report for Hydrogeologic Assessment

Well Location and Characteristics

Water System Name/Address/Fed ID# ______________________________

____

____

Name of Site ______________________________

Description of Device ______________________________

Name and Description of Nearby Surface Source(s) ______________________________

____

____

Describe Datum for Elevations ______________________________

Is there a USGS gauging station nearby, or are other flow records available?

____

Static Water Level

as a Depth ______________ Date ______________

as an Elevation ______________ Date ______________

Pumping Water Level

as a Depth ______________ Date ______________

as an Elevation ______________ Date ______________

Characterize pumping practices or spring flow (rates, seasonally, etc.)

____

____

Describe the vertical distance between the aquifer and the surface water under pumping conditions

____

____

Describe the horizontal distance between the aquifer and the surface-water source under pumping conditions

____

____

Well Construction

Construction date ______________

Does construction conform to the current standards (i.e., Ten State)? ______________________________

Construction methods/materials ______________________________

____

Casing ______________________________

Grouting ______________________________

Screening ______________________________

Well House ______________________________

Other Information ______________________________

Is a detailed plan/drawing available? ______________

Hydrogeology

Is a geologic log for this well available? ______________

What is the thickness of the unsaturated zone? ______________

What is the hydraulic conductivity of the unsaturated zone and the aquifer? ______

Is there a confining layer? At what elevation(s)? ______________

Describe ______________________________

Does the surface-water body penetrate the aquifer? ______________

Draw the aquifer(s), confining unit(s), and water table using one or more cross-sections with at least one cross-section parallel to the flow direction ______________

Summary of Sanitary Conditions

Describe the results of past sanitary surveys and give the location of written reports

____

____

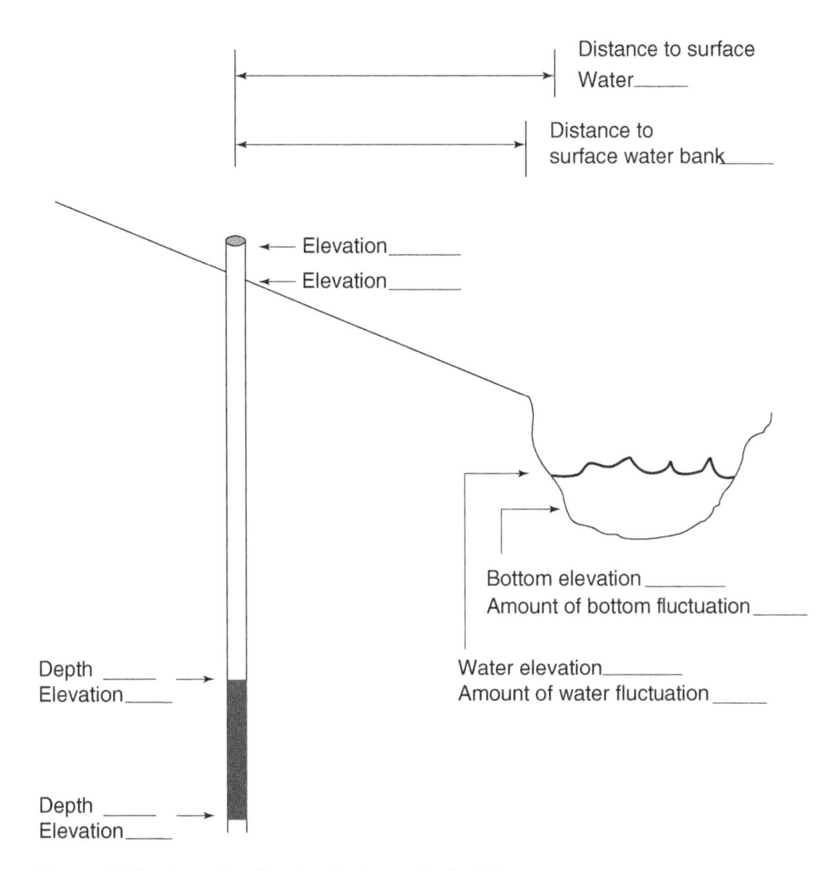

Figure 3.19 Investigation for Hydrogeological Assessment

Problems/Questions

3.1 Differentiate between porosity and permeability and between specific retention and specific yield.

3.2 Differentiate among aquifers, aquicludes, and aquitards.

3.3 Explain the classification of aquifers, and why the unconfined aquifers can be easily contaminated.

3.4 Demonstrate the relatively slow rate of groundwater movement by using Eq. 3.2 to determine the rate of movement through an aquifer and a confining bed. The following data are assumed:

- Aquifer composed of coarse sand: $K = 60$ m/d, dh/dl $= I = 1$ m/1,000 m and $\theta = 0.20$.
- Confining bed composed of clay: $K = 0.0001$ m/d, dh/dl $= I = 1/10$ m and $\theta = 0.50$.

3.5 The observed data from a pumping test were plotted in a figure similar to Fig. 3.4 along with a Theis type curve, as if the transparency of the observed data had been moved into place over the type curve. The observation well represented by the data is 400 ft (121.92 m) from a pumping well where the rate of discharge is 275 gpm (1498.75 m^3/d). Calculate the formation constants T and S using these match-point coordinates:

$W(u) = 3.0$

$s = 3.7\,\text{ft}\,(1.13\,\text{m})$

$u = 2.3 \times 10^{-2}$

$r^2/t = 1.4 \times 10^7 (\text{SI } 1.3 \times 10^6)$

3.6 A time–drawdown curve for an observation well at a distance of 400 ft (121.92 m) from a pumping well discharging at a constant rate of 275 gpm (1498.75 m^3/d) is shown in Fig. 3.5. Determine the transmissivity and storage coefficient of the aquifer.

3.7 In the aquifer represented by the pumping test in Problem 3.5 ($T = 2.6 \times 10^4$ gpd/ft = 322.87 m^3/d/m and $S = 2.3 \times 10^{-5}$), a gravel-packed well with an effective diameter of 36 in. (914 mm) is to be constructed. The design flow of the well is 1000 gpm (5450 m^3/d). Calculate the drawdown at the well with total withdrawals from storage (i.e., with no recharge or leakages) after (a) 1 min, (b) 1 h, (c) 1 day, (d) 1 month, and (e) 1 year of continuous pumping, at design capacity.

3.8 Determine the drawdown of a quasi-steady state for a proposed 36-in. (914 mm) effective diameter well pumping continuously at 350 gpm (1907 m^3/d) in an elastic artesian aquifer having a transmissivity of 5000 gpd/ft (62 m^3/d/m) and a storage coefficient of 4×10^{-4}. Assume that the discharge and recharge conditions are such that the drawdown will be stabilized after 180 days.

3.9 A well was pumped at a constant rate of 500 gpm (2725 m^3/d) between 8 a.m. and 11 a.m., 2 p.m. and 5 p.m., and remained idle the rest of the time. What will be the drawdown in the well at 8 a.m. the next day when a new cycle of pumping is to start? Assume no recharge or leakage. The transmissivity of the artesian aquifer is 4.1×10^4 gpd/ft (509.14 m^3/d/m).

3.10 Two wells with effective diameters of 36 in. (914 mm) are located 1500 ft (457.2 m) apart in an artesian aquifer with $T = 2.5 \times 10^4$ gpd/ft (310.45 m^3/d/m) and $S = 4 \times 10^{-5}$. Compute the drawdown at each well when the two wells are pumped at 500 gpm (2725 m^3/d) for 20 days.

3.11 Suppose it is desired to restrict the drawdown in each of the wells to 40 ft in Problem 3.10. What will be the corresponding discharges for individual wells?

3.12 A gravel-packed well with an effective diameter of 36 in. (914 mm) pumps water from an artesian aquifer having $T = 4.2 \times 10^4$ gpd/ft (521.56 m^3/d/m) and $S = 4.4 \times 10^{-5}$. The well lies at a distance of 2000 ft (610 m) from a stream that can supply water fast enough to maintain a constant head. Find the drawdown in the well after 20 days of pumping at 550 gpm (2725 m^3/d).

3.13 The interrelationship of pumping rate, drawdown, and specific yield serves as a basis for the selection of optimal design capacity. List the factors on which the obtainable yield from a well depends.

3.14 A well having an effective diameter of 36 in. (914 mm) is to be located in a relatively homogeneous artesian aquifer with a transmissivity of 15,000 gpd/ft (186.27 m^3/d/m) and a storage coefficient of 3×10^{-4}. The initial piezometric surface level is 30 ft (9.14 m) below the land surface. The depth to the top of the aquifer is 140 ft (42.67 m) and the thickness of the aquifer is 60 ft (18.28 m). The well is to be finished with a screen length of 30 ft (9.14 m). Compute the specific capacity of the well and its maximum yield after 10 days of pumping. Neglect well losses.

3.15 Discuss the sanitary protection of wells constructed for supplying drinking water.

3.16 Visit the U.S. Environmental Protection Agency website (www.epa.org) and find the Groundwater Rule (GWR). Explain the background of the GWR and outline briefly the most recent final GWR final requirements.

3.17 About 25% of Earth's freshwater supply is stored beneath the surface of the land, where it remains for thousands of years. Only a relatively small proportion of this groundwater is located in soil formations (aquifers) from which it can be withdrawn in significant amounts. Explain the following briefly:
(a) The reasons why groundwater is the principal source of water for small systems
(b) The U.S. EPA's Groundwater Rule promulgation
(c) The meaning of "groundwater under the direct influence of surface water" (GWUDI) and the water monitoring and treatment requirements of GWUDI.

Bibliography

American Society of Civil Engineers, Groundwater Basin Management, in *Manual of Engineering Practice, No. 40*, Reston, VA, 1961.

Batu, V., *Aquifer Hydraulics: A Comprehensive Guide to Hydrogeologic Data Analysis*, John Wiley & Sons, New York, USA, February 1998.

Brown, R. H., Drawdowns Resulting from Cyclic Intervals of Discharge, in *Methods of Determining Permeability, Transmissibility and Drawdown*, U.S. Geological Survey Water Supply Paper 1537-1, 1963.

Brown, R. H., Selected procedures for analyzing aquifer test data, *J. Am. Water Works Ass.*, *45*, 844–866, 1953.

Department of Water Supply website, *County of Maui*, Hawaii, http://mauiwater.org, 2009.

Downing, A., and Wilkinson, W. B., *Applied Groundwater Hydrology: A British Perspective*, Oxford University Press, Oxford, UK, 1992.

Edward E. Johnson, Inc., *Groundwater and Wells*, Saint Paul, MN, 1966.

Fair, G. M., Geyer, J. C., and Okun, D. A., *Elements of Water Supply and Wastewater Disposal*, John Wiley & Sons, New York, USA, 1971.

Fair, G. M., Geyer, J. C., and Okun, D. A., Water and Wastewater Engineering, in *Water Supply and Wastewater Removal*, vol. *1*, John Wiley & Sons, New York, USA, 1966.

Hantush, M. S., and Jacob, C. E., Nonsteady radial flow in an infinite leaky aquifer, *Trans. Am. Geophys. Union*, 1955.

Hantush, M. S., Hydraulics of wells, *Adv. Hydro-sci.*, *1*, 281, 1964.

Heath, R. C., *Basic Ground-Water Hydrology*, U.S. Geological Survey Water-Supply Paper 2220, revised 2004, 1983.

Jacob, C. E., Determining the Permeability of Water-Table Aquifers, in *Methods of Determining Permeability, Transmissibility and Drawdown*, U.S. Geological Survey Water Supply Paper 1536-I, 245, 1963.

Jacob, C. E., Flow of ground water, in Rouse, H. (Ed.), *Engineering Hydraulics*, John Wiley & Sons, New York, 1950, p. 321.

McGuiness, C. L., *The Role of Groundwater in the National Water Situation*, U.S. Geological Survey Water Supply Paper 1800, 1963.

New York State Department of Health, Identification of Ground Water Sources Under the Direct Influence of Surface Water, in *Environmental Health Manual*, October 2001.

Romero, J. C., The movement of bacteria and viruses through porous media, *Ground Water*, *8*, 2, 1970.

Theis, C. V., Drawdowns Caused by a Well Discharging Under Equilibrium Conditions from an Aquifer Bounded on a Straight-Line Source, in *Short Cuts and Special Problems in Aquifer Tests, U.S. Geological Survey Water Supply Paper 1545-C*, 1963.

Thomas, H. E., Groundwater Regions of the United States—Their Storage Facilities, in *U.S. 83rd Congress, House Interior and Insular Affairs Comm., The Physical and Economic Foundation of Natural Resources*, vol. *3*, 1952.

Todd, D. K., and Mays, L. W., *Groundwater Hydrology*, John Wiley and Sons, New York, December 2008.

U.S. Army Corps of Engineers, *Engineering and Design—Groundwater Hydrology*, Engineering Manual M 1110-2-1421, Washington, DC, February 1999.

U.S. Environmental Protection Agency and California Department of Health Services, Water Distribution System Operation and Maintenance, 2002.

U.S. Environmental Protection Agency, Drinking Water Data, 2009. Drinking water data tables, Available at http://www.epa.gov/safewater/data/getdata.html.

U.S. Environmental Protection Agency, *Factoids: Drinking Water and Ground Water Statistics for 2007*, EPA 816-K-08-004, Washington, DC, March 2008.

U.S. Geological Survey, *Basic Ground-Water Hydrology*, Water Supply Paper 2220, 1983, http://pubs.usgs.gov/wsp/wsp2220, 2009a.

U.S. Geological Survey, *Sonoma County Pictures*, http://ca.water.usgs.gov/user_projects/sonoma/sonomapics.html, 2009b.

Vasconcelos, J., and Hams, S., *Consensus Method for Determining Groundwater Under the Direct Influence of Surface Water Using Microscopic Particulate Analysis*, U.S. Environmental Protection Agency, Port Orchard, WA, 1992.

Walton, W. C., Leaky Artesian Aquifer Conditions in Illinois, *Illinois Slate Water Survey Report Invest.*, *39*, 1960.

Wang, L. K. and Yang, C. T. *Modern Water Resources Engineering*, Humana Press, New Jersey, USA, 866 pages, 2014.

4

Quantities of Water and Wastewater Flows

Knowledge of the required quantities of water and wastewater flows is fundamental to systems design and management. In the United States, volumes of water supplied to cities and towns or removed from them are expressed in US gallons per year (gal/yr), month (gal/mo), day (gpd), or minute (gpm). The US gallon occupies a volume of 0.1337 cubic feet (ft^3) or 3.785 liters (L) and weighs 8.344 pounds (lb) or 3.785 kilograms (kg). The fundamental metric (SI) unit in engineering work is the cubic meter (m^3), weighing 1,000 kilograms (kg) or 1 metric ton (T) and equaling 1,000 liters (L). In the United States, annual water or wastewater volumes are conveniently recorded in million gallons (MG) or million liters (ML). Daily volumes are generally expressed in million gallons per day (MGD) or million liters per day (MLD) if more than 100,000 gpd (378,500 L/d). The gallons per capita daily volumes are stated in gpcd, while the liters per capita daily volumes are stated in Lpcd. Connected or tributary populations and numbers of services or dwelling units may take the place of total populations.

Per capita and related figures generalize the experience. They permit comparison of the experience and practices of different communities and are helpful in estimating future requirements of specific communities. Fluctuations in flow are usefully expressed as ratios of maximum or minimum annual, seasonal, monthly, weekly, daily, hourly, and peak rates of flow to corresponding average rates of flow.

Most water and wastewater systems include massive structures (dams, reservoirs, and treatment works) that have long construction timelines and are not readily expanded; they also include pipes and other conduits sunk into city streets, which disrupts traffic while they are being laid. Accordingly, the principal system components are purposely made large enough to satisfy community needs for a reasonable number of years. For this reason, selecting the initial or design capacity is not simple. It calls for skill in interpreting social and economic trends and sound judgment in analyzing past experience and predicting future requirements. Among needed estimates are the following:

1) The number of years, or *design period*, for which the proposed system and its component structures and equipment are to be adequate
2) The number of people, or *design population*, to be served
3) The rates of water use and wastewater release, or *design flows*, in terms of per capita water consumption and wastewater discharge as well as industrial and commercial requirements
4) The area to be served, or *design area*, and the allowances to be made for population density and areal water consumption as well as water supply and wastewater release from residential, commercial, and industrial districts
5) The rates of rainfall and runoff, or *design hydrology*, for storm and combined systems.

4.1 Design Period

New water and wastewater works are normally made large enough to meet the needs and wants of growing communities for an economically justifiable number of years in the future. Choice of a relevant design period is generally based on the following:

1) The useful life of component structures and equipment, taking into account obsolescence and wear and tear
2) The ease or difficulty of enlarging contemplated works, including consideration of their location
3) The anticipated rate of population growth and water use by the community and its industries

Water and Wastewater Engineering: Hydraulics, Hydrology and Management, Volume 1, Fourth Edition.
Lawrence K. Wang, Mu-Hao Sung Wang, and Nazih K. Shammas.

Companion website: www.wiley.com/go/Wang/Waterandwastewater4e

Table 4.1 Design Periods for Water and Wastewater Structures

Type of Structure	Special Characteristics	Design Period, Years
Water supply		
Large dams and conduits	Hard and costly to enlarge	25–50
	Easy to extend	
Wells, distribution systems, and filter plants	When growth and interest rates are low[a]	20–25
	When growth and interest rates are high[a]	10–15
Pipes more than 12 in.in diameter	Replacement of smaller pipes is more costly in long run	20–25
Laterals and secondary mainsless than 12 in. in diameter	Requirements may change fast in limited areas	Full development
Sewerage		
Laterals and submains less than 15 in. in diameter	Requirements may change fast in limited areas	Full development
Main sewers, outfalls, andinterceptors	Hard and costly to enlarge	40–50
Treatment works	When growth and interest rates are low[a]	20–25
	When growth and interest rates and high[a]	10–15

Conversion factor: 1 in. = 25.4 mm.
[a] The dividing line is in the vicinity of 3% per annum.

4) The going rate of interest on bonded indebtedness
5) The performance of contemplated works during their early years when they are expected to be under minimum load.

Design periods often employed in practice are shown in Table 4.1.

4.2 Design Population

4.2.1 Population Data

For information on the population of given communities or regions at a given time, engineers turn to the records of official censuses or enumerations. The US government has conducted a decennial census since 1,790. Some state and local enumerations provide additional information, usually for years ending in 5, and results of special surveys sponsored by public authorities or private agencies for political, social, or commercial purposes may also be available. United State census dates and intervals between censuses are listed in Table 4.2.

The information obtained in the decennial censuses is published by the U.S. Bureau of the Census, Department of Commerce. Political or geographic subdivisions for which population data are collated vary downward in size from the country

Table 4.2 US Census Dates and Intervals Between Censuses

Year	Date	Census Interval, Years
1790–1820	First Monday in August	Approximately 10
1830–1900	June 1	Exactly 10, except between 1,820 and 1,830
1910	April 15	9.875
1920	January 1	9.708
1930	April 1	10.250
1940–2010	April 1	Exactly 10

as a whole, to its coterminous portion only, individual states and counties, metropolitan districts, cities and wards, townships and towns, and—in large communities—census tracts. The tracts are areas of substantially the same size and large enough to house 3,000–6,000 people.

4.2.2 Population Growth

Populations increase by births, decrease by deaths, and change with migration. Communities also grow by annexation. Urbanization and industrialization bring about social and economic changes and growth. Educational and employment opportunities and medical care are among the desirable changes. Among unwanted changes are the creation of slums and the pollution of air, water, and soil. Least predictable of the effects on growth are changes in commercial and industrial activity. Examples are furnished in Table 4.3 (a) for Detroit, Michigan, where the automobile industry was responsible for a rapid rise in population between 1910 and 1950; (b) for Providence, Rhode Island, where competition with southern textile mills was reflected in low rates of population growth after 1910; and (c) for Miami, Florida, where recreation added a new and important element to prosperity from 1910 onward.

Were it not for industrial vagaries of the Providence type, human population kinetics would trace an S-shaped growth curve in much the same way as spatially constrained microbial populations. As shown in Fig. 4.1, the trend of seed populations is progressively faster at the beginning and progressively slower toward the end as a saturation value or upper limit is approached. What the future holds for a given community, therefore, is seen to depend on where on the growth curve the community happens to be at a given time.

The growth of cities and towns and characteristic portions of their growth curves can be approximated by relatively simple equations that derive historically from chemical kinetics. The equation of a first-order chemical reaction, possibly catalyzed by its own reaction products, is a recurring example. It identifies also the kinetics of biological growth and other biological reactions including population growth, kinetics, or dynamics. This widely useful equation can be written as follows:

$$dy/dt = ky(L - y), \tag{4.1}$$

where y is the population at time t, L is the saturation or maximum population, and k is a growth or rate constant with the dimension $1/t$. It is pictured in Fig. 4.1 together with its integral, Eq. (4.6).

Three related equations apply closely to characteristic portions of this growth curve: (a) a first-order progression for the terminal arc *ec* of Fig. 4.1, (b) a logarithmic or geometric progression for the initial arc *ad*, and (c) an arithmetic progression for the transitional intercept, *de*, or,

Table 4.3 Census Populations of Detroit, MI, Providence, RI, and Miami, FL, in the USA in the period 1910–2006

	City		
Census Year	**Detroit**	**Providence**	**Miami**
1910	466,000	224,000	5,500
1920	994,000	235,000	30,000
1930	1,569,000	253,000	111,000
1940	1,623,000	254,000	172,000
1950	1,850,000	249,000	249,000
1960	1,670,000	207,000	292,000
1970	1,493,000	177,000	332,000
1980	1,203,000	157,000	347,000
1990	1,028,000	161,000	359,000
2000	951,000	174,000	363,000
2006 (estimated)	871,000	175,000	404,000

All population values have been rounded to the nearest thousand.
Source: After U.S. Bureau of the Census.

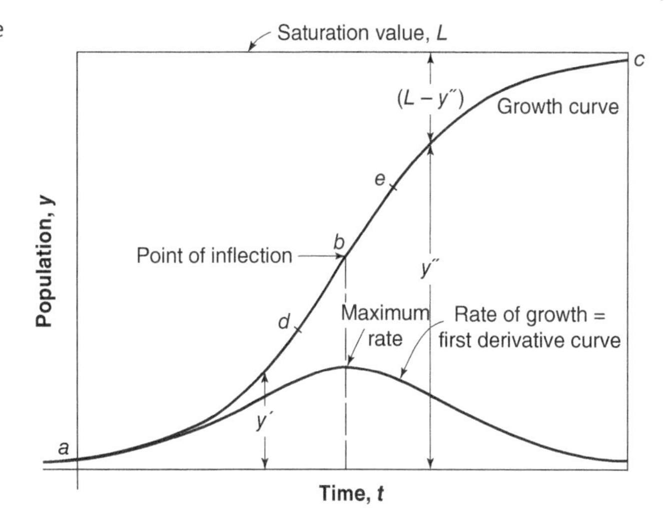

Figure 4.1 Population Growth Idealized. Note Geometric Increase from *a* to *d*, Straight-line Increase from *d* to *e*, and First-order Increase from *e* to *c*

For arc *ec*: $$dy/dt = k(L - y) \quad (4.2)$$

For arc *ad*: $$dy/dt = ky \quad (4.3)$$

For arc *de*: $$dy/dt = k. \quad (4.4)$$

If it is assumed that the initial value of k, namely k_0, decreases in magnitude with time or population growth rather than remaining constant, k can be assigned the following value:

$$k = k_o/(1 + nk_ot) \quad (4.5)$$

in which n, as a coefficient of retardance, adds a useful concept to Eqs. (4.2)–(4.4).

On integrating Eqs. (4.1)–(4.4) between the limits $y = y_0$ at $t = 0$ and $y = y$ at $t = t$ for unchanging k values, they change as shown next.

For autocatalytic first-order progression (arc *ac* in Fig. 4.1):

$$\ln[(L - y)/y] - \ln[(L - y_0)/y_0] = -kLt$$

or

$$y = L/\{1 + [(L - y_0)/y_0] \exp(-kLt)\}. \quad (4.6)$$

For first-order progression without catalysis (arc *ec* in Fig. 4.1):

$$\ln[(L - y)/(L - y_0)] = -kt$$

or

$$y = L - (L - y_0) \exp(-kt). \quad (4.7)$$

For geometric progression (arc *ad* in Fig. 4.1):

$$\ln(y/y_0) = kt$$

or

$$y = y_0 \exp(kt). \quad (4.8)$$

For arithmetic progression (arc *de* in Fig. 4.1):

$$y - y_0 = kt. \quad (4.9)$$

Substituting Eq. (4.5) into Eqs. (4.2)–(4.4) yields the retardant expressions shown next.

For retardant first-order progression:

$$y = L - (L - y)(1 + nk_0t)^{-1/n}. \tag{4.10}$$

For retardant, geometric progression:

$$\ln(y/y_0) = (1/n)\ \ln(1 + nk_0t)$$

or

$$y = y_0(1 + nk_0t)^{-1/n}. \tag{4.11}$$

For retardant, arithmetic progression:

$$y - y_0 = (1/n)\ \ln(1 + nk_0t). \tag{4.12}$$

These and similar equations are useful in water and wastewater practice, especially in water and wastewater treatment kinetics.

4.2.3 Short-Term Population Estimates

Estimates of midyear populations for current years and the recent past are normally derived by arithmetic from census data. They are needed perhaps most often for (a) computing per capita water consumption and wastewater release and (b) for calculating the annual birth and general death rates per 1,000 inhabitants, or specific disease and death rates per 100,000 inhabitants.

Understandably, morbidity and mortality rates from waterborne and otherwise water-related diseases are of deep concern to sanitary engineers.

For years between censuses or after the last census, estimates are usually interpolated or extrapolated as arithmetic or geometric progressions. If t_i and t_j are the dates of two sequent censuses and t_m is the midyear date of the year for which a population estimate is wanted, the rate of arithmetic growth is given by Eq. (4.9) as

$$k_{\text{arithmetic}} = y_j - y_i/(t_j - t_i)$$

and the midyear populations, y_m, of intercensal and postcensal years are as follows:

Intercensal:

$$y_m = y_i + (t_m - t_i)\left(y_j - y_i\right)/(t_j - t_i) \tag{4.13}$$

Postcensal:

$$y_m = y_j + (t_m - t_j)\left(y_j - y_i\right)/(t_j - t_i). \tag{4.14}$$

In similar fashion, Eq. (4.8) states that

$$k_{\text{arithmetic}} = \left(\log\, y_i - \log\, y_j\right)/(t_j - t_i)$$

and the logarithms of the midyear populations, $\log y_m$, for intercensal and postcensal years are as follows:

Intercensal:

$$\log\, y_m = \log y_i + (t_m - t_i)\left(\log y_j - \log y_i\right)/(t_j - t_i) \tag{4.15}$$

Postcensal:

$$\log\, y_m = \log y_j + (t_m - t_j)\left(\log y_j - \log y_i\right)/(t_j - t_i). \tag{4.16}$$

Geometric estimates, therefore, use the logarithms of the population parameters in the same way as the population parameters themselves are employed in arithmetic estimates; moreover, the arithmetic increase corresponds to capital growth by simple interest, and the geometric increase to capital growth by compound interest. Graphically, arithmetic progression is characterized by a straight-line plot against arithmetic scales for both population and time on double-arithmetic coordinate paper and, thus, geometric as well as first-order progression by a straight-line plot against a geometric (logarithmic) population scale and an arithmetic timescale on semilogarithmic paper. The suitable equation and method of plotting are best determined by inspection from a basic arithmetic plot of available historic population information.

Example 4.1 Estimation of Population

As shown in Table 4.3, the rounded census population of Miami, Florida, USA, was 249,000 in 1950 and 292,000 in 1960. Estimate the midyear population (1) for the fifth intercensal year and (2) for the ninth postcensal year by (a) arithmetic and (b) geometric progression. The two census dates were both April 1.

Solution:

Intercensal estimates for 1955:

$$t_m = 1{,}955.25\ (\text{note : there are 3 months} = 0.25\ \text{yr, from April 1 to midyear, June 30})$$

$$t_m - t_i = 1{,}955.25 - 1950 = 5.25\ \text{yr}$$

$$t_j - t_i = 1960 - 1950 = 10.00\ \text{yr}$$

$$(t_m - t_i)/(t_j - t_i) = 5.25/10.00 = 0.525$$

	(a) Arithmetic	(b) Geometric
1960	$y_j = 292{,}000$	$\log y_j = 5.4654$
1950	$y_i = 249{,}000$	$\log y_i = 5.3962$
	$y_j - y_i = 43{,}000$	$\log y_j - \log y_i = 0.0692$
	$0.525(y_j - y_i) = 23{,}000$	$0.525(\log y_j - \log y_i) = 0.03633$
1955	$\mathbf{y_m = 272{,}000}$	$\mathbf{y_m = 268{,}000}$

Postcensal estimate for 1969:

$$t_m - t_i = 9.25\ \text{yr}$$

$$t_j - t_i = 10.00\ \text{yr}$$

$$(t_m - t_i)/(t_j - t_i) = 0.925$$

	(a) Arithmetic	(b) Geometric
1960	$y_j = 292{,}000$	$\log y_j = 5.4654$
1950	$y_i = 249{,}000$	$\log y_i = 5.3962$
	$y_j - y_i = 43{,}000$	$\log y_j - \log y_i = 0.0692$
	$0.925(y_j - y_i) = 40{,}000$	$0.925(\log y_j - \log y_i) = 0.0620$
1969	$\mathbf{y_m = 332{,}000}$	$\mathbf{y_m = 337{,}000}$

Geometric estimates are seen to be lower than arithmetic estimates for intercensal years and higher for postcensal years.

The U.S. Bureau of the Census estimates the current population of the whole nation by adding to the last census population the intervening differences (a) between births and deaths, that is, the natural increases; and (b) between immigration and emigration. For states and other large population groups, postcensal estimates can be based on the apportionment method, which postulates that local increases will equal the national increase times the ratio of the local to the national intercensal population increase. Intercensal losses in population are normally disregarded in postcensal estimates; the last census figures are used instead.

Supporting data for short-term estimates can be derived from sources that reflect population growth in ways different from, yet related to, population enumeration. Examples are records of school enrollments; house connections for water, electricity, gas, and telephones; commercial transactions; building permits; and health and welfare services. These are translated into population values by ratios derived for the recent past. The following ratios are not uncommon:

1) Population/(school enrollment) = 5:1
2) Population/(number of water, gas, or electricity services) = 3:1
3) Population/(number of land-line telephone services) = 4:1.

4.2.4 Long-Range Population Forecasts

Long-range forecasts, covering design periods of 10–50 years, make use of available and pertinent records of population growth. Again dependence is placed on mathematical curve fitting and graphical studies. The logistic growth curve is an example.

The logistic growth equation is derived from the autocatalytic, first-order equation (Eq. 4.6) by letting

$$p = (L - y_0)/y_0 \quad \text{and} \quad q = kL$$

or

$$y = L/[1 + p\ \exp(-qt)]$$
$$y = L/[1 + \exp(\ln p - qt)]$$

and equating the first derivative of Eq. (4.1) to zero, or

$$d(dy/dt)/dt = kL - 2ky = 0.$$

It follows that the maximum rate of growth dy/dt is obtained when

$$y = 1/2L \text{ and}$$
$$t = (-\ln p)/q = (-2.303 \log p)/q.$$

It is possible to develop a logistic scale for fitting a straight line to pairs of observations as in Fig. 4.2. For general use of this scale, populations are expressed in terms of successive saturation estimates L, which are eventually verified graphically by lying closely in a straight line on a logistic-arithmetic plot. The percentage saturation P is

$$P = 100y/L = 100/[1 + p\ \exp(-qt)] \text{ and}$$
$$\ln[(100 - P)/P] = \ln p - qt.$$

The straight line of best fit by eye has an ordinate intercept $\ln p$ and a slope $-q$, when $\ln[(100 - P)/P]$ is plotted against t or values of $n[(100 - P)/P]$ are scaled in either direction from a 50th percentile or middle ordinate.

Graphical forecasts offer a means of escape from mathematical forecasting. However, even when mathematical forecasting appears to give meaningful results, most engineers seek support for their estimates from plots of experienced and

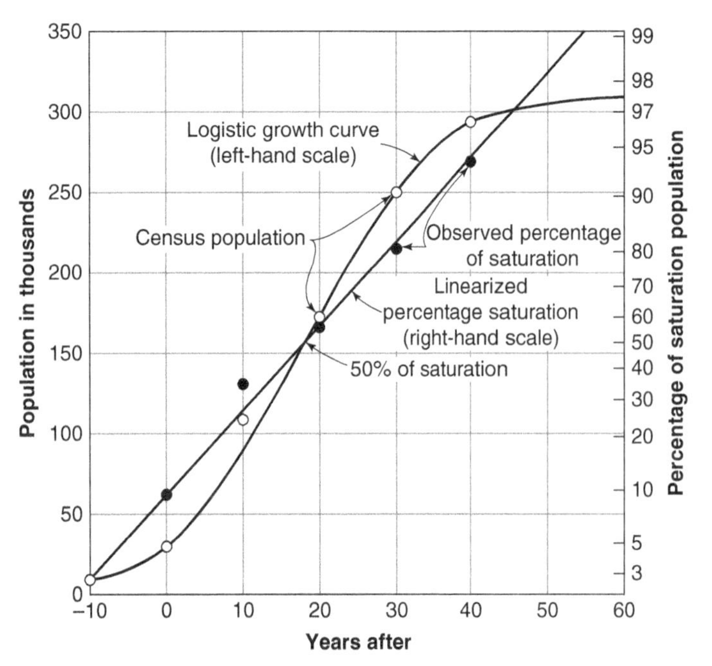

Figure 4.2 Logistic Growth of a City. Calculated Saturation Population, Confirmed by Graphical Good Straight-line Fit, is 313,000. Right-hand Scale is Plotted as log [(100 − P)/P] about 50% at the Center

projected population growth on arithmetic or semilogarithmic scales. Trends in rates of growth rather than growth itself may be examined arithmetically, geometrically, or graphically with fair promise of success. Estimates of arithmetic and semilogarithmic straight-line growth of populations and population trends can be developed analytically by applying least squares procedures, including the determination of the coefficient of correlation and its standard error.

At best, since forecasts of population involve great uncertainties, the probability that the estimated values turn out to be correct can be quite low. Nevertheless, the engineer must select values in order to proceed with planning and design of works. To use uncertainty as a reason for low estimates and short design periods can lead to capacities that are even less adequate than they otherwise frequently turn out to be. Because of the uncertainties involved, populations are sometimes projected at three rates—high, medium, and low. The economic and other consequences of designing for one rate and having the population grow at another can then be examined.

4.2.5 Simplified Method for Population Forecasts

The following are two simple equations that are used by consulting engineers for computing the rate of population increase and the future population forecast:

$$P_n = P_o(1 + R)^n \tag{4.17}$$

$$P_o = P_p(1 + R)^n, \tag{4.17a}$$

where P_n is future population, P_o is present population, P_p is past population, R is the probable rate of population increase per year, and n is the number of years considered.

When population data (P_p, P_o, and n) for the past are available, the value of R in the preceding equations can be computed. In case the rate of population increase per year is changing, several R values can be calculated in order to obtain an average for the future population forecast (P_n) using the known data (R, P_o, and n).

4.2.6 Population Distribution and Area Density

Capacities of water collection, purification, and transmission works and of wastewater outfall and treatment works are a matter of areal and population size. Within communities, their individual service areas, populations, and occupancy are the determinants. A classification of areas by use and of expected population densities in persons per acre is shown in Table 4.4. Values of this kind are founded on analyses of present and planned future subdivisions of typical blocks. Helpful, in this connection, are census tract data; land-office, property, zoning, fire insurance, and aerial maps; and other information collected by planning agencies.

Table 4.4 Common Population Densities

	Persons Per Acre
1) Residential areas	
(a) Single-family dwellings, large lots	5–15
(b) Single-family dwellings, small lots	15–35
(c) Multiple-family dwellings, small lots	35–100
(d) Apartment or tenement houses	100–1,000 or more
2) Mercantile and commercial areas	15–30
3) Industrial areas	5–15
4) Total, exclusive of parks, playgrounds, and cemeteries	10–50

Conversion factors: 1 acre = 0.4046 ha = 4,047 m^2 = 0.004047 km^2; 1 person/acre = 2.4716 person/ha.

4.3 Water Consumption

Although the draft of water from distribution systems is commonly referred to as *water consumption,* little of it is, strictly speaking, consumed; most of it is discharged as spent or wastewater. *Use of water* is a more exact term. True consumptive use refers to the volume of water evaporated or transpired in the course of use—principally in sprinkling lawns and gardens, in raising and condensing steam, and in bottling, canning, and other industrial operations.

Service pipes introduce water into dwellings, mercantile and commercial properties, industrial complexes, and public buildings. The water delivered is classified accordingly. Table 4.5 shows approximate per capita daily uses in the United States. Wide variations in these figures must be expected because of differences in the following:

1) Climate
2) Standards of living
3) Type of residences
4) Extent of sewerage
5) Type of mercantile, commercial, and industrial activity
6) Water pricing
7) Use of private supplies
8) Water quality for domestic and industrial purposes
9) Distribution-system pressure
10) Completeness of metering
11) Systems management.

4.3.1 Domestic Consumption

Although domestic water use is about 50% of the water drawn in urban areas, 90% of the consumers are domestic. A breakdown of household flows apportions the various uses as follows:

1) 41% to flushing toilets
2) 37% to washing and bathing
3) 6% to kitchen use
4) 5% to drinking water
5) 4% to washing clothes
6) 3% to general household cleaning
7) 3% to watering lawns and gardens
8) 1% to washing family cars.

Although domestic use is commonly expressed in gpcd (or Lpcd), the daily draft per dwelling unit, gpud (or Lpud), may offer more meaningful information.

Table 4.5 Water Consumption

	Quantity, gpcd	
Class of Consumption	**Normal Range**	**Average**
Domestic or residential	20–90	55
Commercial	10–130	20
Industrial	22–80	50
Public	5–20	10
Water unaccounted for	5–30	15
Total	60–350	150

Conversion factor: 1 gpcd = 3.785 Lpcd.

Extremes of heat and cold increase water consumption: hot and arid climates by frequent bathing, air conditioning, and heavy sprinkling; and cold climates by bleeding water through faucets to keep service pipes and internal water piping from freezing during cold spells. In metered and sewered residential areas, the observed average daily use of water for lawns and gardens, $Q_{sprinkling}$, in gpd during the growing season is about 60% of the estimated average potential evapotranspiration E, reduced by the average daily precipitation, P, effective in satisfying evapotranspiration during the period, or

$$Q_{sprinkling} = 1.63 \times 10^4 A(E - P) \quad \text{(US customary units)}, \tag{4.18}$$

where $Q_{sprinkling}$ is the average daily use of water for lawns and gardens (gpd), 1.63×10^4 is the number of gallons in an acre-inch considering a 60% factor, A is the average lawn and garden acreage per dwelling unit, E is the average potential evapotranspiration expressed (in./d), and$P$ is the average daily precipitation (in./d).

The following is a sister equation using SI units:

$$Q_{sprinkling} = 6003\, A(E - P) \quad \text{(SI units)}, \tag{4.18a}$$

where $Q_{sprinkling}$ is the average daily use of water for lawns and gardens (L/d), A is average lawn and garden area per dwelling unit (ha), E is average potential evapotranspiration expressed (mm/d), and$P$ is average daily precipitation also expressed (mm/d).

The average lawn and garden area per dwelling unit A is given by the observational relationship:

$$A = 0.803D^{-1.26} \quad \text{(US customary units)}, \tag{4.19}$$

where A is the average lawn and garden area, in acres per dwelling unit, and D is the gross housing density, in dwelling units per acre. The following is its sister equation using SI units:

$$A = 1.016D^{-1.26} \quad \text{(SI units)}, \tag{4.19a}$$

where A is the average lawn and garden area, in ha per dwelling unit, and D is the gross housing density, in dwelling units per ha.

In arid climates, Shammas (1991) studied the relationship between actual water application and evapotranspiration rates for various categories of green areas in Riyadh, Saudi Arabia (Table 4.6). As expected, irrigation water application rates increased with rising evapotranspiration rates for every category of green areas. For private residences, as evapotranspiration rates rose from 3.4 to 5.3 to 8.0 mm/d (L/m^2/d), the corresponding water application rates increased from 7 to 10 to 23 mm/d (L/m^2/d). Similar behavior can be noticed for the other categories of planted areas.

High standards of cleanliness, large numbers of water-connected appliances, oversized plumbing fixtures, and frequent lawn and garden sprinkling—all of which are associated with wealth—result in heavy drafts. For sewered properties, the average domestic use of water $Q_{domestic}$, in gpud or Lpud for each dwelling unit, is related to the average market value M of the units in thousands of current U.S. dollars by the following observational equations:

$$Q_{domestic} = 157 + 3.46M \quad \text{(US customary units)} \tag{4.20}$$

$$Q_{domestic} = 594 + 13.1M \quad \text{(SI units)}. \tag{4.20a}$$

Table 4.6 Relationship of Irrigation Water Application Rates to Evapotranspiration in Arid Areas

Period	December–January			February–March			April–May			Overall
Category	I[a] (L/m^2/d)	E[b] (mm/d)	I/E Ratio	I[a] (L/m^2/d)	E[b] (mm/d)	I/E Ratio	I[a] (L/m^2/d)	E[b] (mm/d)	I/E Ratio	I/E Ratio
Private residences	7.0	3.4	2.1	10	5.3	1.9	23	8.0	2.9	2.3
Road medians	15	3.4	4.4	15	5.3	2.8	33	8.0	4.1	3.8
Public parks	3.0	3.4	0.9	7.0	5.3	1.3	12	8.0	1.5	1.2
Special developments	1.0	3.4	0.3	3.0	5.3	0.6	6.0	8.0	0.8	0.6

[a] I is the irrigation rate (L/m^2/d).
[b] E is the evapotranspiration rate (1 mm/d equivalent to L/m^2/d).
Conversion factors: 1 gpd/ft^2 = 0.0408 m^3/m^2/d = 40.8 L/m^2/d; 1 in./d = 25.4 mm/d.

It is important to note that these mathematical models do vary with time and locations; therefore, they must be modified and updated for a specific site/time situations.

4.3.2 General Urban Water Demands

Some commercial enterprises—hotels and restaurants, for instance—draw much water; so do industries such as breweries, canneries, laundries, paper mills, and steel mills. Industries, in particular, draw larger volumes of water when it is cheap than when it is dear. Industrial draft varies roughly inversely as the manufacturing rate and is likely to drop by about half the percentage increase in cost when rates are raised. Hospitals, too, have high demands. Although the rate of draft in firefighting is high, the time and annual volume of water consumed in extinguishing fires are small and seldom identified separately for this reason.

Water of poor quality may drive consumers to resort to uncontrolled, sometimes dangerous, sources, but the public supply remains the preferred source when the product water is clean, palatable, and of unquestioned safety; soft for washing and cool for drinking; and generally useful to industry. The availability of groundwater and nearby surface sources may persuade large industries and commercial enterprises to develop their own process and cooling waters.

Hydraulically, leaks from mains and plumbing systems and flows from faucets and other regulated openings behave like orifices. Their rate of flow varies as the square root of the pressure head, and high distribution pressures raise the rate of discharge and with it the waste of water from fixtures and leaks. Ordinarily, systems pressures are not raised above 60 psig (lb/in.^{2} gage), or 416 kPa, in American practice, even though it is impossible to employ direct hydrant streams in firefighting when hydrant pressures are below 75 psig (520 kPa).

Metering encourages thrift and normalizes the demand. The costs of metering and the running expense of reading and repairing meters, however, are substantial. They may be justified in part by accompanying reductions in waste and possible postponement of otherwise needed extensions. Under study and on trial there is an

- encouragement of off-peak-hour draft of water by large users. To this purpose, rates charged for water drawn during off-peak hours are lowered preferentially. The objective is to reap the economic benefits of a relatively steady flow of water within the system and the resulting proportionately reduced capacity requirements of systems components. The water drawn during off-peak hours is generally stored by the user at ground level even when this entails repumping.

Distribution networks are seldom perfectly tight. Mains (see Fig. 4.3), valves, hydrants, and services of well-managed systems are therefore regularly checked for leaks. Superficial signs of controllable leakage are as follows:

1) High night flows in mains
2) Water running in street gutters
3) Moist pavements
4) Persistent seepage
5) Excessive flows in sewers
6) Abnormal pressure drops
7) Unusually green vegetation (in dry climates).

Figure 4.3 Water Leakage from a Crack in a Water Pipe

Table 4.7 Water Losses in Various Cities, 1980–1990

City	Water Loss (%)
Toronto, Canada	18
Munich, Germany	12
Hamburg, Germany	2
Hong Kong, China	30
Bombay, India	33
Delhi, India	18
Daegu, Korea	37
Manila, Philippines	51
Riyadh, Saudi Arabia	30
Colombo, Sri Lanka	30
Stockholm, Sweden	19
Kalmar, Sweden	5
Bangkok, Thailand	49
United Kingdom	24
Boston, MA, USA	33
Cambridge, MA, USA	18.5
Springfield, IL, USA	25
Westchester, NY, USA	16
46 Communities in Massachusetts, USA	2–50

Leakage is detected by (a) driving rods into the ground to test for moist earth, (b) using a sounding method in which listening devices amplify the sound of running water, (c) conducting a leak noise correlation that is based on a sonic technique, (d) tracing leaks by injecting a gas into the network and detecting its escape by a suitable instrument to determine the location of the leak, and (e) inspecting premises for leaky plumbing and fixtures. Leakage detection of well-managed waterworks may be complemented by periodic and intensive but, preferably, routine and extensive water-waste surveys. Generally involved is the isolation of comparatively small sections of the distribution system by closing valves on most or all feeder mains and measuring the water entering the section at night through one or more open valves or added piping on fire hoses. Common means of measurement are pitot tubes, bypass meters around controlling valves, or meters on one or more hose lines between hydrants that straddle closed valves. Table 4.7 summarizes reported water losses ranging from 2% to 50% in several cities including developed and developing countries.

4.3.3 Industrial Water Consumption

The amounts of water used by industry vary widely. Some industries draw in excess of 50 MGD (190 MLD); others, no more than comparably sized mercantile establishments. On average, US industry satisfies more than 60% of its water requirements by internal reuse and less than 40% by draft through plant intakes from its own water sources or through service connections from public water systems. Only about 7% of the water taken in is consumed; 93% is returned to open waterways or to the ground, whence it may be removed again by downstream users. On balance, industry's consumptive use is kept down to 2% of the draft of all water users in the United States. Table 4.8 shows the relative amounts of water consumed by different industries. Not brought out is the fact that once-through cooling, particularly by the power industry, is by far the biggest use component and the principal contributor to the thermal pollution of receiving waters.

To draw comparisons between the water uses of different industries and of plants within the same industrial category, it is customary to express plant or process use in volumes of water—gallons, for instance—per unit of production (Table 4.9). For the chemical industry, however, this may not be meaningful, because of the diversity of chemicals produced.

Table 4.8 Water Percentage of Total Water Intake Consumed by Industry

Industry	Percent of Intake	Industry	Percent of Intake
Automobile	6.2	Meat	3.2
Beet sugar	10.5	Petroleum	7.2
Chemicals	5.9	Poultry processing	5.3
Coal preparation	18.2	Pulp and paper	4.3
Corn and wheat milling	20.6	Salt	27.6
Distillation	10.4	Soap and detergents	8.5
Food processing	33.6	Steel	7.3
Machinery	21.4	Sugar, cane	15.9
		Textiles	6.7

Table 4.9 Water Requirements of Selected Industries

Industry	Unity of Production	Gal per Unit
Food products		
Beet sugar	ton of beets	7,000[a]
Beverage alcohol	Proof gal	125–170
Meat	1000 lb live weight	600–3500[b]
Vegetables, canned	Case	3–250
Manufactured products		
Automobiles	Vehicle	10,000
Cotton goods	1000 lb	20,000–100,000
Leather	1,000 ft^2 of hide	200–64,000
Paper	ton	2,000–100,000
Paper pulp	ton	4,000–60,000
Mineral products		
Aluminum (electrolytic smelting)	ton	56,000 (max)
Copper		
Smelting	ton	10,000[c]
Refining	ton	4,000
Fabricating	ton	200–1,000
Petroleum	Barrel of crude oil	800–3,000[d]
Steel	ton	1,500–50,000

[a] Includes 2,600 gal (9,841 L) of flume water and 2,000 gal (7,570 L) of barometric condenser water.
[b] Lower values for slaughterhouses; higher for slaughtering and packing.
[c] Total, including recycled water; water consumed is 1,400 gal (5,390 L).
[d] Total, including recycled water; water consumed is 30–60 gal (113.55–227.1 L).
Conversion factors: 1 gal = 3.785 L; 1 ton = 2,000 lb = 0.9072 metric ton; 1 lb = 0.4536 kg; 1 ft^2 = 0.0929 m^2; 1 barrel = 5.615 ft^3 = 0.159 m^3 = 42 gal.

Rising water use can be arrested by conserving plant supplies and introducing efficient processes and operations. Most important, perhaps, are the economies of multiple reuse through countercurrent rinsing of products, recirculation of cooling and condensing waters, and reuse of otherwise spent water for secondary purposes after their partial purification or reunification.

About two-thirds of the total water intake of US manufacturing plants is put to use for cooling. In electric power generation, the proportion is nearly 100%; in manufacturing industries, it ranges from 10% in textile mills to 95% in beet sugar refineries. It averages 66% in industries as reported by the National Association of Manufacturers.

Industry often develops its own supply. Chemical plants, petroleum refineries, and steel mills, for example, draw on public or private utilities for less than 10% of their needs. Food processors, by contrast, purchase about half of their water from public supplies, largely because the bacterial quality of drinking water makes it *de facto* acceptable.

About 90% of the industrial draft is taken from surface sources. Groundwaters may be called into use in the summer because their temperature is then seasonally low. They may be prized, too, for their clarity and their freedom from color, odor, and taste.

Available sources may be drawn on selectively: municipal water for drinking, sanitary purposes, and delicate processes, for example; and river water for rugged processes and cooling and for emergency uses such as fire protection. Treatment costs and economic benefits are the determinants.

4.3.4 Rural Water Consumption

The minimum use of piped water in rural dwellings is about 20 gpcd (75 Lpcd); the average about 50 gpcd (190 Lpcd). Approximate drafts of rural schools, overnight camps, and rural factories (exclusive of manufacturing uses) are 25 gpcd (95 Lpcd); of wayside restaurants, 10 gpcd (38 Lpcd) on a patronage basis; and of work or construction camps, 45 gpcd (170 Lpcd). Resort hotels need about 100 gpcd (380 Lpcd), and rural hospitals and the like, nearly twice this amount.

Farm animals have the following approximate requirements: dairy cows, 20 gpcd (75 Lpcd); horses, mules, and steers, 12 gpcd (45 Lpcd); hogs, 4 gpcd (15 Lpcd); sheep, 2 gpcd (8 Lpcd); turkeys, 0.07 gpcd (0.26 Lpcd); and chickens, 0.04 gpcd (0.15 Lpcd). Cleansing and cooling water add about 15 gpcd (57 Lpcd) for cows to the water budget of dairies. Greenhouses may use as much as 70 gpd per 1,000 ft^2 (2.856×10^{-3} $m^3/d/m^2$) and garden crops about half this amount.

Military requirements vary from an absolute minimum of 0.5 gpcd (1.9 Lpcd) for troops in combat through 2–5 gpcd (7.6–19 Lpcd) for soldiers on the march or in bivouac, and 15 gpcd (57 Lpcd) for temporary camps, up to 50 gpcd (190 Lpcd) or more for permanent military installations.

4.4 Variations or Patterns of Water Demand

A pattern is a function relating water use to time of day. Patterns allow the user to apply automatic time-variable changes within the system. Most patterns are based on a multiplication factor versus time relationship, whereby a multiplication factor of 1.0 represents the base value (often the average value, Q_{avg}). This relationship is written as $Q_t = kQ_{avg}$, where Q_t is demand at time t and k is multiplier for time.

Water consumption changes with seasons, days of the week, and hours of the day. Fluctuations are greater in small rather than large communities, and during short rather than long periods of time. Variations are usually expressed as ratios to the average demand. Estimates for the United States are as follows:

Ratio of Rates, k	Normal Range	Average
Maximum day: average day	(1.5–3.5):1	2.0:1
Maximum hour: average day	(2.0–7.0):1	4.5:1

Where existing water quantity data are not available to accurately determine the instantaneous peak demand for the design year, the following criteria may be used as a minimum for estimating the instantaneous peak demand:

For 220 people or less: $Q_{ins\ peak} = 9(Q_{avg\ day})$

For more than 220 people: $Q_{ins\ peak} = 7(Q_{avg\ day})/P_k^{0.167}$,

where $Q_{ins\ peak}$ is instantaneous peak water demand (gpm or L/min), $Q_{avg\ day}$ is average daily water demand (gpm or L/min), and P_k is the design year population (thousands).

4.4.1 Domestic Variations

Observations in a Johns Hopkins University study and standards of the Federal Housing Administration for domestic drafts are brought together in Table 4.10. Damping effects produced by network size and phasing of commercial, industrial, and domestic drafts explain the differences between community-wide and domestic demands. Observational values for peak hourly demands $Q_{\text{peak h}}$ in gpud for US customary units and Lpud for SI units are given by the regression equation:

$$Q_{\text{peak h}} = 334 + 2.02\,Q_{\text{max day}} \quad \text{(US customary units)} \tag{4.21}$$

$$Q_{\text{peak h}} = 1264 + 2.02\,Q_{\text{max day}} \quad \text{(SI units)}, \tag{4.21a}$$

where $Q_{\text{max day}}$ is the maximum daily water demand in gpud for US customary units and Lpud for SI units.

Calculation of the confidence limit of expected demands makes it possible to attach suitably higher limits to these values. For design, this may be the 95% confidence limit, which lies above the expected rate of demand by twice the variance, that is, by $2\sigma^2$, where σ is the standard deviation of the observed demands. Approximate values of this variance are shown in Table 4.11 for gross housing densities of 1, 3, and 10 dwelling units per acre (1 acre = 0.4046 ha) and a potential daily evapotranspiration of 0.28 in. (7.1 mm) of water.

In another study in Riyadh, Saudi Arabia, Quraishi et al. (1990) reported similar daily variations in water consumption, as well as variations in average daily water consumption as a function of type of residence and household income (see Table 4.12).

Table 4.10 Variations in Average Daily Rates of Water Drawn by Dwelling Units

	gpd per Dwelling Unit and (Ratios to Average Day)		
	Average Day	**Maximum Day**	**Peak Hour**
Federal Housing Administration standards	400 (1.0)	800 (2.0)	2,000 (5.0)
	Observed drafts—metered dwellings		
National average	400 (1.0)	870 (2.2)	2,120 (5.3)
West	460 (1.0)	980 (2.1)	2,480 (5.2)
East	310 (1.0)	790 (2.5)	1,830 (5.9)
	Unmetered dwellings		
National average	690 (1.0)	2,350 (3.4)	5,170 (7.6)
	Unsewered dwellings		
National average	250 (1.0)	730 (2.9)	1,840 (7.5)

Conversion factor: 1 gpd per dwelling unit = 1 gpud = 3.785 Lpd per dwelling unit = 3.785 Lpud.

Table 4.11 Increase in Design Demands for Dwelling Units Above Expected Maximum Day and Peak Hour

	Maximum Demand (1,000 gpud) for Stated Housing Density in Dwelling Units/Acre			Peak Demand (1,000 gpud) for Stated Housing Density in Dwelling Units/Acre		
Number of Dwelling Units	**1**	**3**	**10**	**1**	**3**	**10**
1	5.0	4.8	4.6	7.0	7.0	7.0
10	1.6	1.5	1.5	3.5	2.3	2.2
10^2	0.7	0.5	0.5	1.5	0.8	0.7
10^3	0.6	0.3	0.2	1.5	0.4	0.3
10^4	0.5	0.2	0.1	1.4	0.3	0.2

Example: For 100 dwelling units, the maximum water demand for the housing density of three dwelling units/acre is 0.5 × 1,000 gpud = 500 gpud. The peak water demand for the housing density of three dwelling units/acre is 0.8 × 1,000 gpud = 800 gpud.
Conversion factors: 1,000 gpud = 3,785 Lpud; 1 dwelling unit/acre = 2.4716 dwelling units/ha; 1 acre = 0.4046 ha.

Table 4.12 Variations in Per Capita Domestic Water Consumption in Riyadh, Saudi Arabia

Description	Consumption (Lpcd)	Ratio to Average Day
Overall population		
Average day	310	1.00
Maximum day	1,323	4.27
Minimum day	41	0.13
Income level		
Low income (average day)	339	1.09
Medium income (average day)	219	0.71
High income (average day)	432	1.39
House occupants (average day)	343	1.11
Building occupants (average day)	221	0.71

Conversion factor: 1 gpcd = 3.785 Lpcd.

Rates of rural water use and wastewater production are generally functions of the water requirements and discharge capacities of existing fixtures.

Design demands determine the size of hydraulic components of the system; expected demands determine the rate structure and operation of the system.

4.4.2 Fire Demands

Height, bulk, area, congestion, fire resistance, type of construction, and building occupancy determine the rate at which water should be made available at neighboring hydrants, either as hydrant or engine streams, to extinguish localized fires and prevent their spread into areal or citywide conflagrations. Needed fire flow (NFF) is defined as the water flow rate, measured at a residual pressure of 20 psi (138 kPa) and for a given duration, which is required for fighting a fire in a specific building. Analysis of water demands actually experienced during fires in communities of different size underlies the formulation of the general standards. The American Water Works Association (1998) in its M31 Manual describes three methods for calculating the fire flow requirements that were developed by the following organizations:

1) Insurance Services Office Inc. (ISO)
2) Illinois Institute of Technology Research Institute (IITRI)
3) Iowa State University (ISU).

The following general information is taken from the 2006 edition of the ISO's *Guide for Determination of Needed Fire Flow*:

1) For one- and two-family dwellings not exceeding two stories in height, the following NFFs should be used:

Distance Between Buildings, ft (m)	Fire Flow, gpm (L/min)
Over 100 (over 30)	500 (1,893)
31–100 (9.1–30)	750 (2,839)
11–30 (3.1–9)	1,000 (3,785)
10 or less (3 or less)	1,500 (5,678)

2) For wood-shingle roof coverings on the building or on exposed buildings, add 500 gpm (1,893 Lpm) to the above flows.
3) High-risk areas. Multifamily, commercial, and industrial areas are considered high-risk areas. The fire flows available for these areas require special consideration. The distribution and arterial mains in the high-risk areas are to accommodate the required fire flows in those areas.

4) Typically the water main system must be able to meet the flow requirements of (a) peak day demands plus fire flow demands as a minimum for any water main design and (b) instantaneous peak demands for special water mains from water source, water treatment plant (WTP), and/or water storage facilities.
5) Water storage facilities should have sufficient capacity, as required by the *Recommended Standards for Water Works* (2007 edition), to meet domestic average daily water demands and, where fire protection is provided, fire flow demands.

Fire demand can also be estimated according to the community size and realty subdivision development by using the following empirical equation from the National Board of Fire Underwriters:

$$Q_{\text{fire}} = 1,020\,(P_k)^{0.5}\left[1 - 0.01\,(P_k)^{0.5}\right] \quad \text{(US customary units)} \tag{4.22}$$

$$Q_{\text{fire}} = 3,860.7\,(P_k)^{0.5}\left[1 - 0.01\,(P_k)^{0.5}\right] \quad \text{(SI units)}, \tag{4.22a}$$

where Q_{fire} is the fire demand, in gpm for US customary units and L/min for SI units, and P_k is the population, in thousands.

The National Board of Fire Underwriters requires provision for a 5-hour fire flow in places with populations of less than 2,500 and provision for a 10-hour flow in larger places.

The ISU method is relatively simple and quick to use but yields low fire flow requirements, whereas the IITRI method produces excessively high rates. The ISO methodology calculates values in between the other two. The most recent *International Fire Code* (International Code Council 2006) recommends a minimum fire flow of 1,000 gpm (3,785 L/min) for one- and two-family dwellings having an area that does not exceed 3,600 ft^2 (344 m^2). Fire flow and flow durations for larger buildings having areas in excess of 3,600 ft^2 (344 m^2) are not to be less than what is shown in Table 4.13. Type I-A structures are typically concrete frame buildings made of noncombustible materials. All of the building elements—structural frame, bearing walls, floors, and roofs—are fire-resistance-rated. Type V-B construction is typically wood-frame construction, which is very common because it does not require a fire rating.

Table 4.13 Required Fire Flow for Buildings

Area[a] (ft^2) Type of Building Construction			
Type I-A[b] Noncombustible, Protected	**Type V-B[c] Combustible, Unprotected**	**Fire Flow (gpm)**	**Duration (h)**
0–22,700	0–3,600	1,500	2
22,701–30,200	3,601–4,800	1,750	2
30,201–38,700	4,801–6,200	2,000	2
38,701–48,300	6,201–7,700	2,250	2
48,301–59,000	7,701–9,400	2,500	2
59,001–70,900	9,401–11,300	2,750	2
70,901–83,700	11,301–13,400	3,000	3
83,701–97,700	13,401–15,600	3,250	3
97,701–112,700	15,601–18,000	3,500	3
112,701–128,700	18,001–20,600	3,750	3
128,701–145,900	20,601–23,300	4,000	4
145,901–164,200	23,301–26,300	4,250	4
164,201–183,400	26,301–29,300	4,500	4
183,401–203,700	29,301–32,600	4,750	4
203,701–225,200	32,601–36,000	5,000	4
225,201–247,700	36,001–39,600	5,250	4
247,701–271,200	39,601–43,400	5,500	4
271,201–295,900	43,401–47,400	5,750	4
295,901–greater	47,401–51,500	6,000	4
—	51,501–55,700	6,250	4
—	55,701–60,200	6,500	4

Table 4.13 (Continued)

Area[a] (ft^2) Type of Building Construction			
Type I-A[b] Noncombustible, Protected	**Type V-B[c] Combustible, Unprotected**	**Fire Flow (gpm)**	**Duration (h)**
—	60,201–64,800	6,750	4
—	64,801–69,600	7,000	4
—	69,601–74,600	7,250	4
—	74,601–79,800	7,500	4
—	79,801–85,100	7,750	4
—	85,101–greater	8,000	4

[a] The specified area shall be the area of the three largest successive floors.
[b] Typically these are concrete-frame buildings made of noncombustible materials. The "A" stands for protected.
[c] Typically these are wood-frame buildings. The "B" stands for unprotected.
Conversion factors: 1 ft^2 = 0.0929 m^2; 1 gpm = 3.785 L/ min.
Source: Data from International Fire Code.

The benefits of early fire suppression are acknowledged by the acceptance that the firefighting water requirement can be reduced by 50% for one- and two-family dwellings and by up to 75% for larger buildings (exceeding 3,600 ft^2 or 344 m^2) when buildings are fitted with automatic sprinkler systems.

Standard fire requirements take into account probable loss of water from connections broken in the excitement of a serious fire. Coincident draft of water for purposes other than firefighting is rarely assumed to equal the maximum hourly rate. Depending on local conditions, the maximum daily rate may be a reasonably safe assumption instead.

Example 4.2 Determination of Capacities of Waterworks Systems

The four typical waterworks systems shown in Fig. 4.4 supply a community with an estimated future population of 120,000. Determine the required capacities of the constituent structures for an average consumption of 150 gpcd (568 Lpcd) and a distributing reservoir so sized that it can balance out differences between hourly and daily flows, fire demands, and emergency requirements. Assume a requirement for fire protection that is needed to control two simultaneous fires in two unprotected wood-frame construction, multistory buildings having a 12,000-ft^2 (1,115-m^2) area per floor.

Solution 1 (US Customary System):

Required capacities for waterworks systems of Fig. 4.4:

Average daily draft $= 150 \times 1.2 \times 10^5/10^6 = 18$ MGD
Maximum daily draft = coincident draft $= 2 \times 18 = 36$ MGD
Maximum hourly draft $= 4.5 \times 18 = 81$ MGD

Fire flow from Table 4.13:

Area of three floors $= 3 \times 12{,}000$ ft^2/floor $= 36{,}000$ ft^2
Flow for one fire = 5,000 gpm
Flow or two simultaneous fires $= 2 \times 5{,}000 = 10{,}000$ gpm = 14.4 MGD

Coincident draft plus fire flow $= 36 + 14.4 = 50.4$ MGD
Provision for breakdowns and repair of pumps and water purification units by installing one reserve unit, as shown in Table 4.14, assuming total units $= 3 + 1 = 4$:

Low-lift pumps: 4/3× maximum daily draft $= (4/3) \times 36 = 48$ MGD
High-lift pumps: 4/3 × maximum hourly draft $= (4/3) \times 81 = 108$ MGD
Treatment works: 4/3 × maximum daily draft = 48 MGD.

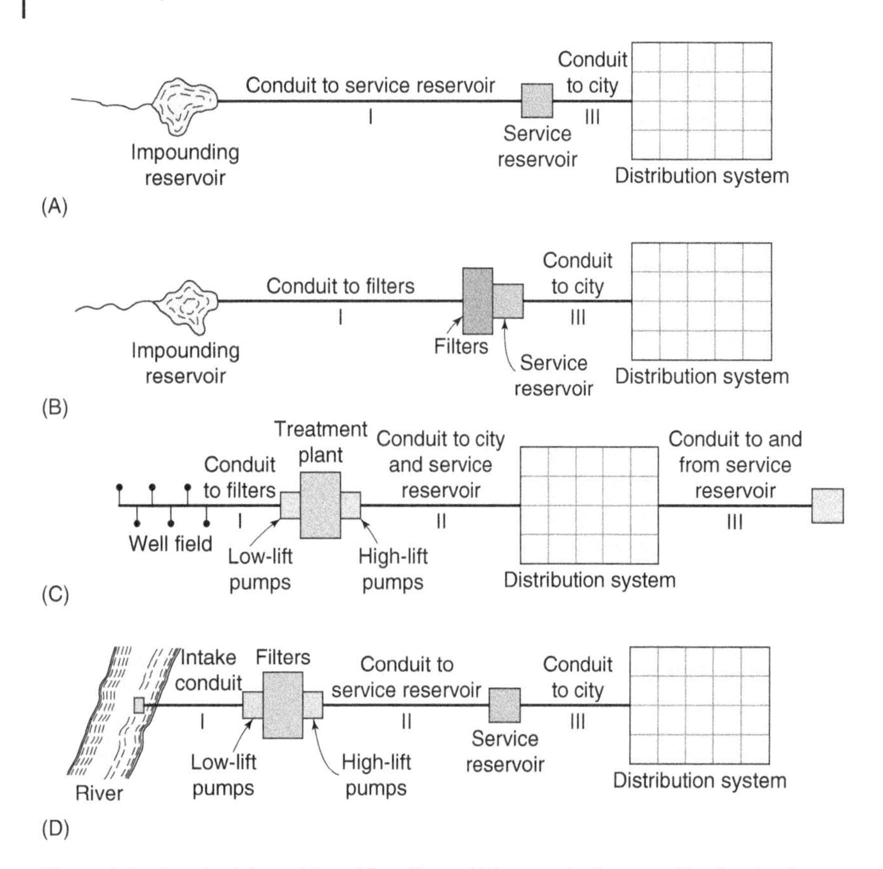

Figure 4.4 Required Capacities of Four Typical Waterworks Systems. The Service Reservoir is Assumed to Compensate for Fluctuations in Draft and Fire Drafts, and to Hold an Emergency Reserve. **System A:** Treating high quality lake water from high elevation with simple disinfection (no water filtration), transmitting water by gravity (no pumping), and regulating water drafts with a service reservoir between the lake and the distribution system; **System B:** Treating regular quality lake water from high elevation with a complete filtration plant, transmitting water by gravity (no pumping), and regulating water drafts with a service reservoir at the filtration plant; **System C:** Treating groundwater from low elevation with a complete filtration plant, low-lift pumping low elevation groundwater to filtration plant for treatment, high-lift pumping the finished water to the water distribution system and a service reservoir, and regulating the water drafts with the service reservoir located far away; and **System D:** Treating river water from low elevation with a complete filtration plant, low-lift pumping low elevation river water to filtration plant for treatment, high-lift pumping the finished water to the service reservoir and the water distribution system, and regulating the water drafts with the service reservoir located between the filtration plant and the water distribution system.

The resultant capacities of systems components are summarized in Table 4.14.

Table 4.14 Capacities of the Four Systems of Example 4.2

		Capacity of System, MGD			
Structure	**Required Capacity**	**A**	**B**	**C**	**D**
1. River or well field	Maximum day	–	–	36.0	36.0
2. Conduit I	Maximum day	36.0	36.0	36.0	36.0
3. Conduit II	Maximum day	None	None	36.0	36.0
4. Conduit III	Maximum hour	81.1	81.0	81.0	81.0
5. Low-life pumps	Maximum day plus reserve	None	None	48.0	48.0
6. High-life pumps	Maximum hour plus reserve	None	None	108.0	108.0
7. Treatment plant	Maximum day plus reserve	None	48.0	48.0	48.0
8. Distribution system high-value district	Maximum hour	81.0	81.0	81.0	81.0

Conversion factor: 1 MGD = 3.785 MLD.

Solution 2 (SI system):

Required capacities for waterworks systems of Fig. 4.4:

Average daily draft = $Q_{\text{avg day}}$ = (568 Lpcd) (120,000)/10^6 = 68.2 MLD
Maximum daily draft = $Q_{\text{max day}}$ = coincident draft = $2\times Q_{\text{avg day}} = 2\times 68.2$ = 136.4 MLD
Maximum hourly draft = $Q_{\text{max h}} = 4.5\times Q_{\text{avg day}} = 4.5\times 68.2$ = 306.9 MLD

Fire flow from Table 4.13:

Area of three floors = $3\times 12{,}000$ ft^2/floor = 36,000 ft^2 = 3,344.4 m^2
Flow for one fire = 5,000 gpm = 18,925 Lpm
Flow or two simultaneous fires = $2\times 18{,}925$ = 37,850 Lpm = 54.5 MLD

Coincident draft plus fire flow = 136.4 + 54.5 = 190.9 MLD

Provision for breakdowns and repair of pumps and water purification units by installing one reserve unit, as shown in Table 4.14 assuming total units = 3 + 1 = 4:

Low-lift pumps: 4/3× maximum daily draft = (4/3) × 136.4 = 181.9 MLD
High-lift pumps: 4/3× maximum hourly draft = (4/3) × 306.9 = 409.2 MLD
Treatment works: 4/3× maximum daily draft = 181.9 MLD.

The resultant capacities of systems components are summarized in Table 4.14.

Example 4.3 Estimation of Water Demand to a Residential Area
For management and design purposes, find the expected maximum day and peak hour demand and daily rates of water to be supplied to a residential area of 200 houses with a gross housing density of three dwellings per acre (7.415 unit/ha), an average market value of USD 125,000 per dwelling unit, and a potential evaporation of 0.28 in./d (7.11 mm/d) on the maximum day.

Solution 1 (US Customary System):

Average maximum and peak daily domestic demands per dwelling unit:

By Eq. (4.20): $Q_{\text{domestic}} = 157 + 3.46\ M = 157 + 3.46\times 125 = 590$ gpud
By Eq. (4.19): $A = 0.803\ D^{-1.26} = 0.803/3^{1.26} = 0.20$ acre per dwelling unit
By Eq. (4.18): $Q_{\text{sprinkling}} = 1.63\times 10^4\ A\ (E-P) = 1.63\times 10^4\times 0.20\times(0.28-0) = 913$ gpud excluding precipitation

For management:

$$Q_{\text{max day}} = Q_{\text{domestic}} + Q_{\text{sprinkling}} = 590 + 913 = 1{,}503 \text{ gpud}$$

By Eq. (4.21): $Q_{\text{peak h}} = 334 + 2.02\ Q_{\text{max day}} = 334 + 2.02\times 1{,}503 = 3{,}640$ gpud

For design based on 200 dwellings:

From Table 4.11: $Q_{\text{max day}} = 1{,}503 + 0.4\ (1{,}000) = 1{,}903$ gpud
From Table 4.11: $Q_{\text{peak h}} = 3{,}640 + 0.7\ (1{,}000) = 4{,}340$ gpud.

Solution 2 (SI System):

Average maximum and peak daily domestic demands per dwelling unit:

By Eq. (4.20a): $Q_{\text{domestic}} = 594 + 13.1\ M = 594 + 13.1\times 125 = 2{,}232$ Lpud
By Eq. (4.19a): $A = 1.016\ D^{-1.26} = 1.016/(7.415)^{1.26} = 1.016/12.4836 = 0.08138$ ha/unit
By Eq. (4.18a): $Q_{\text{sprinkling}} = 6{,}003\ A\ (E-P) = 6{,}003\times 0.08138\times(7.11-0) = 3{,}473$ Lpud excluding precipitation

For management:

$$Q_{\text{max day}} = Q_{\text{domestic}} + Q_{\text{sprinkling}} = 2{,}232 + 3{,}473 = 5{,}705 \text{ Lpud}$$

By Eq. (4.21a): $Q_{\text{peak h}} = 1{,}264 + 2.02\ Q_{\text{max day}} = 1{,}264 + 2.02\times 5{,}705.2 = 12{,}789$ Lpud

For design based on 200 dwellings:

From Table 4.11: $Q_{\text{max day}} = 5{,}705 + 0.4\ (1{,}000)\times 3.785 = 7{,}219$ Lpud
From Table 4.11: $Q_{\text{peak h}} = 12{,}789 + 0.7\ (1{,}000)\times 3.785 = 15{,}438$ Lpud.

4.5 Demand and Drainage Loads of Buildings

The demand load of a building depends on the number and kinds of fixtures and the probability of their simultaneous operation. Different fixtures are furnished with water at different rates as a matter of convenience and purpose. Expressed in cubic feet per minute (ft^3/min), or liter per second (L/s) for fixture units, these rates become whole numbers of small size. Common demand rates are shown in Table 4.15.

It is quite unlikely that all fixtures in a building system will draw water or discharge it at the same time. A probability study of draft demands leads to the relationships plotted in Fig. 4.5. In practice, the values shown are modified as follows:

1) Demands for service sinks are ignored in calculating the total fixture demand.
2) Demands of supply outlets, such as sill cocks, hose connections, and air conditioners, through which water flows more or less continuously for a considerable length of time, are added to the probable flow rather than the fixture demand.
3) Fixtures supplied with both hot and cold water exert reduced demands on main hot-water and cold-water branches (not fixture branches). An allowance of three-fourths of the demand shown in Table 4.15 for individual fixtures is suggested.

Example 4.4 Finding the Demand Load

A two-story dwelling contains the following fixtures: a pair of bathroom groups, an additional water closet, an additional washbasin, a kitchen sink, a laundry tray, and a sill cock. All water closets are served by flush tanks. Find the demand load.

Solution:

From Table 4.15: 2 bathroom groups at 6 = 12; 1 water closet = 3; 1 washbasin = 1; 1 kitchen sink = 2; 1 laundry tray = 3; total = 21 $\text{ft}^3/\text{ min}$, or 0.5943 m^3/min.

From Fig. 4.5, the probable flow in the building main will be 15 gpm plus ($2/3\ \text{ft}^3/\text{ min} = (2/3) \times 7.48 = 5$ gpm for the sill cock, or a total of 20 gpm, or 75.7 L/min.

Table 4.15 Fixture Rates

Type of Fixture and Supply (1) and Discharge (2)	Rates of Supply and Discharge, ft^3/min	
	Private Buildings	Public and Office Buildings
Wash basin, faucet—(1) and (2)	1	2
Water closet, (2) only	6	8
Flush tank, (1) only	3	5
Flush valve, (1) only	6	10
Urinal		
(Stall or wall), flush tank (1) only	...	3
(Stall or wall), flush valve	...	5
(Pedestal), flush valve	...	10
Bathtub or shower, faucet or mixing valve	2	4
Bathroom group (2) only	8	...
Flush tank for closet, (1) only	6	...
Flush valve for closet, (1) only	8	...
Separate shower head, (1) only	2	...
Separate shower stall, (2) only	10	...
Kitchen sink, faucet 2	2	4
Laundry trays (1–3), faucet	3	...
Combination fixture, faucet	3	...
Service sink, faucet (hotel or restaurant)	...	3
Sill cock	$\frac{2}{3}$	$\frac{2}{3}$

Conversion factor: $1\ \text{ft}^3/\text{ min} = 28.32\ \text{L}/\text{ min} = 0.02832\ \text{m}^3/\text{ min}$.

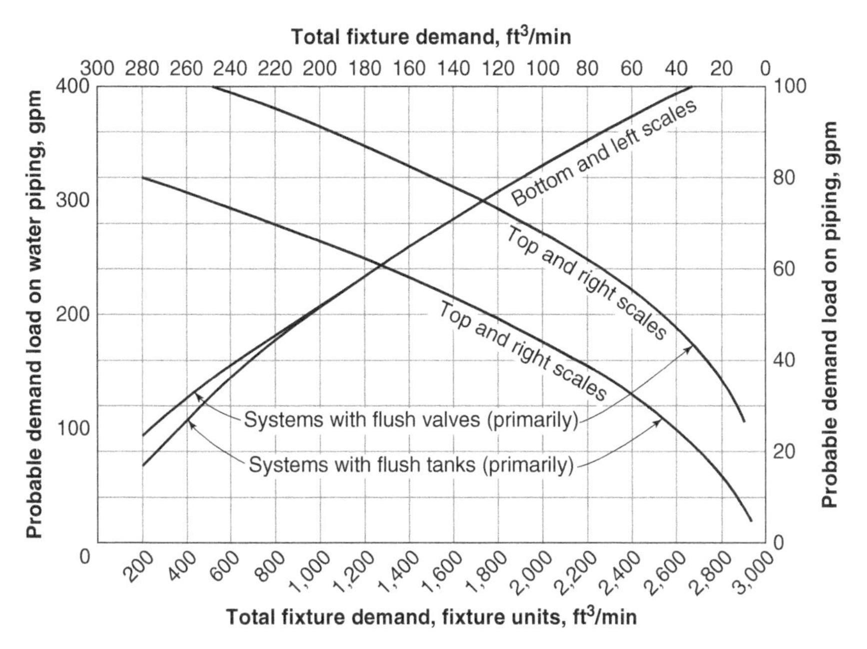

Figure 4.5 Demand Load on Water Piping. Conversion factor: 1 ft^3/min = 7.48 gpm = 0.0283 m^3/min.1 gpm = 3.785 L/min

Like demand load, the drainage load placed on different parts of the system does not equal the sum of the discharge rates of individual fixtures. Not all of them are likely to be operated at the same instant or in such manner as to concentrate the full flow at a given point in the system. The larger the system, the steadier the flow and the smaller the intensity of flow. In this respect, building-drainage systems behave, on a small scale, like wastewater systems. Moreover, the time factor varies with differences in fixture distribution and complicates the estimation of a reasonable load. Probable discharges are greater for an individual horizontal branch than for the stack into which it empties, and greater, too, for an individual stack than for the building drain in which it terminates. As a consequence, allowable ratios of fixture load to design discharge are relatively small for horizontal drains, larger for stacks, and greatest for building drains. These complicating factors make it impossible to show the relation between fixture discharge and drainage load by simple curves as is done for water piping.

4.6 Wastewater Flows

Public sewers receive and transport one or more of the following liquids: spent water, groundwater seepage or infiltration, and runoff from rainfall.

4.6.1 Spent Water or Domestic Wastewater

Spent waters are primarily portions of the public water supply discharged into sewers through the drain pipes of buildings, and secondarily waters drawn from private or secondary sources for air conditioning, industrial processing, and similar uses. Of the water introduced into dwellings and similar buildings, 60%–70% becomes wastewater, which is discharged into the sewerage system (see Table 4.16). The remainder is used consumptively. Commercial areas discharge about 20,000 gpd per acre (= 0.02 MGD/acre = 0.187 MLD/ha).

4.6.2 Groundwater Seepage

Well-laid street sewers equipped with modern preformed joints and tight manholes carry little groundwater. This cannot be said of house sewers unless they are constructed of cast iron or other materials normally laid with tight joints.

The following seepage may be expected when the sewers are laid above the groundwater table: 500–5,000 gpd per acre, average 2,000 (or 4,680–46,795 L/d/ha, average 18,718); 5,000–100,000 gpd per mile of sewer and house connection, average 20,000 (or 11,762–235,239 L/d/km, average 47,040); and 500–5,000 gpd per mile and inch diameter of sewer and house

Table 4.16 Residential Wastewater Flows

City	Average Flow (gpcd)
Denver, Colorado	69
Los Angeles, California	90
Milwaukee, Wisconsin	64
Phoenix, Arizona	78
Riyadh, Saudi Arabia	70
San Diego, California	58
Seattle, Washington	57
Stamford, Connecticut	80
Tampa, Florida	66
USA	65

Conversion factor: 1 gpcd = 3.785 Lpcd.

connection, average 2,500 (or 46–463 L/d/km-mm, average 232), plus 500 gpd per manhole (1,893 L/d/manhole). These ranges in seepage were made so broad because of the great uncertainty regarding the magnitude of groundwater flows in leaky sewers. Where poor construction permits entrance of groundwater, the flows may be expected to vary with rainfall, thawing of frozen ground, changes in groundwater levels, and nature of the soil drained. As suggested by the diurnal flow variations shown in Fig. 4.6, seepage makes up most of the early morning flows. Tightness of sewers may be measured by blocking off a portion of the system, usually with balloons designed for the purpose, then subjecting the tested portion to an appropriate pressure and observing the rate of water loss. Various other tests have been devised. If contracts call for such tests and they are made, even occasionally, the estimates of the rate of groundwater seepage into street sewers can be greatly reduced and will have validity.

4.6.3 Stormwater

In well-watered regions, runoff from rainfall, snow, and ice normally outstrips spent water in intensity and annual volume. Spent water reaching combined sewers is indeed so small a fraction of the design capacity that it may be omitted from calculations for combined systems. Relatively large volumes of raw wastewaters and large amounts of pollutional solids are, therefore, discharged in times of heavy rain into receiving waters through overflows from combined sewers.

Although stormwater runoff is seldom allowed to be introduced into spent-water systems, it is difficult to enforce the necessary ordinances. Even when connections to the wastewater system can be made only by licensed drain layers, illicit connections to sanitary sewers must be expected to add to them some of the runoff from roof, yard, basement entrance, and

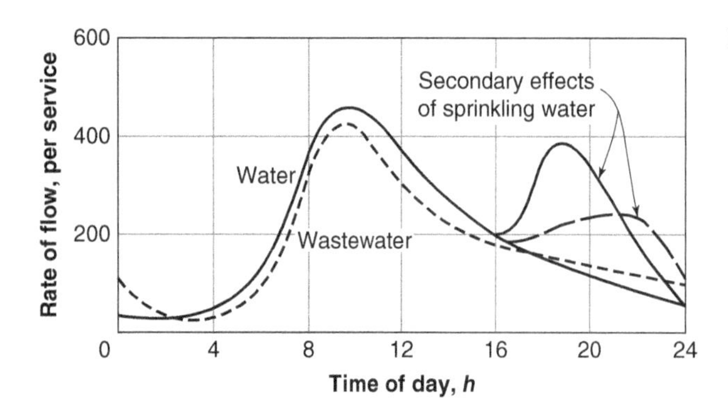

Figure 4.6 Flow Variations of Water and Wastewater

foundation drains. Poorly sealed manhole covers permit further entrance of runoff. Total amounts vary with the effectiveness of enforcing regulations and conducting countermeasures. Allowances for illicit stormwater flow are as high as 70 gpcd (265 Lpcd) and average 30 gpcd (114 Lpcd). A rainfall of 1 in./h (25.4 mm/h) may shed water at a rate of 12.5 gpm (47.3 L/min) from 1,200 ft^2 (111.5 m^2) of roof area, or 1.008 ft^3/s from an acre (0.0705 m^3/s from a hectare) of impervious surfaces. Leaky manhole covers may admit 20–70 gpm (75.7–265 L/min) when streets are under an inch (25.4 mm) of water. The volume of illicit stormwater approximates the difference between normal dry-weather flows and flows during intense rainfalls.

4.6.4 Industrial, Commercial, and Institutional Wastewaters

Industrial wastewaters may be discharged into municipal wastewater systems at convenient points, provided they do not overload them or damage the collecting and treatment works. Nevertheless, it may be advantageous to lead spent process and cooling waters into separate disposal systems where they exist or can conveniently be built or otherwise provided. Pretreatment before discharge to the municipal sewer is also a matter for decision.

Within manufacturing plants themselves there may be rigid separation of different process waters and other wastewaters that can be isolated as such. A metal-finishing shop, for example, may install separate piping for each of the following: (a) strong chromic acid; (b) other strong acids; (c) weak acid wastes, including chromium; (d) strong alkalis, including cyanide; (e) weak alkalis; and (f) sanitary wastes. In addition, separate lines may carry copper rinses and nickel rinses for their individual recovery. Not all lines need be laid as underground gravity-flow conduits. Relatively small volumes of wastewaters may be collected in sumps and pumped through overhead lines instead.

When there is good promise of reasonable recovery of water or waste matters or of treatment simplification, spent industrial waters may be segregated even if collecting lines must be duplicated. Examples are the pretreatment of strong wastewaters before admixture with similar dilute wastewaters and also the separation of cyanide wastewaters for destruction by chlorine before mixing them with wastewaters containing reaction-inhibiting nickel. However, it may also pay to blend wastewaters in order to (a) dilute strong wastes, (b) equalize wastewater flows and composition, (c) permit self-neutralization to take place, (d) foster other beneficial reactions, and (e) improve the overall economy. As a rule, it pays to separate wastewaters while significant benefits can still accrue. After that they may well be blended to advantage into a single waste stream.

Wash waters may require special collection and treatment when they differ from other process waters. Thus, most wastewaters from food processing contain nutrients that are amenable as such to biological treatment. However, they may no longer be so after strong alkalis; soaps; or synthetic detergents, sanitizers, and germicides have been added to them along with the wash waters of the industry.

4.7 Variations in Wastewater Flows

Imprinted on flows in storm and combined sewers is the pattern of rainfall and snow and ice melt. Fluctuations may be sharp and high for the storm rainfall itself and as protracted and low as the melting of snow and ice without the benefit of spring thaws. As shown in Fig. 4.6, (a) flows of spent water normally lie below and lag behind the flows of supplied water, and (b) some of the water sprayed onto lawns and gardens is bound to escape into yard and street drains.

The open-channel hydraulics of sewers allows their levels to rise and fall with the volume rate of entrant waters. Rising levels store flows; falling levels release them. The damping effect of storage is reinforced by the compositing of flows from successive upstream areas for which shape and size are governing factors. Low flows edge upward and high flows move downward. Indeed, wastewater and stormwater flows are quite like stream and flood flows and hence subject also to analysis by flood-routing procedures.

Gifft (1945) has evaluated damping effects by the following observational relationships:

$$Q_{\text{max day}}/Q_{\text{avg day}} = 5.0P^{-1/6} \tag{4.23}$$

$$Q_{\text{min day}}/Q_{\text{avg day}} = 0.2P^{1/6} \tag{4.24}$$

$$Q_{\text{max day}}/Q_{\text{min day}} = 25.0P^{-1/3}. \tag{4.25}$$

Here $Q_{\text{max day}}$, $Q_{\text{avg day}}$, and $Q_{\text{min day}}$ are, respectively, the maximum, average, and minimum daily rates of flow of spent water and P is the population in thousands.

The most commonly used ratio is $Q_{\text{max day}}/Q_{\text{avg day}}$, which is known as the peaking factor, *PF*. In this way the maximum flow is calculated by multiplying the average flow by the PF:

$$Q_{\text{max day}} = PF \times Q_{\text{avg day}}. \tag{4.26}$$

Other common variable PF calculation methods include the following:

$$\text{Babbitt:} \quad PF = \frac{5.0}{P^{0.2}} \tag{4.27}$$

$$\text{Harmon:} \quad PF = 1.0 + \frac{14.0}{4.0 + P^{0.5}} \tag{4.28}$$

$$\text{Ten} - \text{States Standards:} \quad PF = \frac{18 + \sqrt{P}}{4 + \sqrt{P}} \tag{4.29}$$

$$\text{Federov:} \quad PF = \frac{2.69}{Q_{\text{avg day}}^{0.121}}, \tag{4.30}$$

where P is the population, in thousands; and $Q_{\text{avg day}}$ is the average daily flow, in L/s.

For population-based PF methods, the peaking factor decreases as the population increases. A larger population means that peak loads from different sources are likely to occur in a more staggered manner. Thus, the peak loads are likely to be less pronounced compared to the average loading rate. In systems servicing smaller populations, peak loads from different sources are more likely to coincide, causing more pronounced differences in average loading rates. The graph in Fig. 4.7 shows how the extreme flow factor decreases with increasing population when applying the Ten-States Standards.

Eight commonly used sets of ratios are compared in *Manuals of Practice*, ASCE No. 60 and WEF No. FD-5 (2007). Such variations of ratios with population are shown in Table 4.17. Table 4.18 presents the expected wastewater hydraulic loading rates of various residential, commercial, and industrial installations compiled by the authors based on data from the U.S. Environmental Protection Agency and the New York State Department of Environmental Conservation.

Expected flow rates from sewered areas of moderate size (tens of square miles) in industrial countries are as follows:

$$Q_{\text{max day}} = 2\,Q_{\text{avg day}} \tag{4.31}$$

$$Q_{\text{min day}} = (2/3)\,Q_{\text{avg day}} \tag{4.32}$$

$$Q_{\text{peak h}} = (3/2)\,Q_{\text{max day}} = 3\,Q_{\text{avg day}} \tag{4.33}$$

$$Q_{\text{min h}} = (1/2)\,Q_{\text{min day}} = (1/3)\,Q_{\text{avg day}}. \tag{4.34}$$

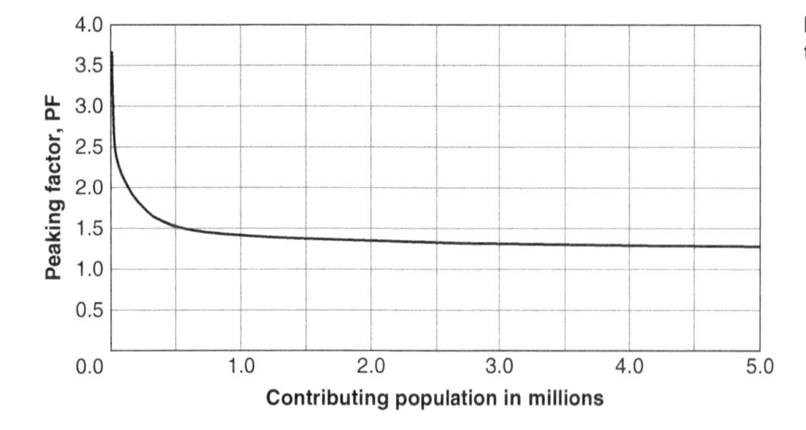

Figure 4.7 PF Versus Contributing Population from the Ten-States Standards

Table 4.17 Variations of Wastewater Flow Ratios with Population

Contributing Population	Ratio of Maximum to Average Daily Flow	Ratio of Minimum to Average Daily Flow
1,000	5.0	0.20
5,000	4.0	0.28
10,000	3.5	0.32
25,000	3.0	0.37
50,000	2.6	0.40
75,000	2.4	0.44
100,000	2.3	0.46
250,000	1.8	0.55
500,000	1.7	0.60
1,000,000	1.6	0.70

Table 4.18 Design Flows Recommended by the U.S. Environmental Protection Agency and the New York State Department of Environmental Conservation for Various Infrastructure Developments

Type of Facility or Development	Flow Rate Per Person (gpcd)	Flow Rate PerUnit (gpud)
Airports		
(Per passenger)	3	
(Per employee)	15	
Apartments	75	
Average per apartment		300
One Bedroom		150
Two Bedrooms		300
ThreeBedrooms		400
Bathhouse	10	
(Per swimmer)		
Boarding house	75	
Bowling alley		
(Per lane no food)		75
(With food-add food service value)		
Campgrounds (recreational vehicle per site)		
Sewered sites		100
Central facilities, served sites, 300-ft (91 m) radius		100
Peripheral sites, 500-ft (152 m) radius		75
Subtractions from above		
No showers		25
Dual service (central facilities and sewered facilities overlapping the central)		25
Campground (summer camp)		
Central facilities	50	
Separate facilities		

(Continued)

Table 4.18 (Continued)

Type of Facility or Development	Flow Rate Per Person (gpcd)	Flow Rate PerUnit (gpud)
Separate facilities		
toilet	10	
shower	25	
kitchen	10	
Campground dumping stations		
(Per unsewered site)		10
(Per sewered site)		5
Camps, day	13	
Add for lunch	3	
Add for showers	5	
Carwashes, assuming no recycle		
Tunnel, per car		80
Rollover, per car		40
Handwash, per 5-minute cycle		20
Churches—per seat		3
(With catering—add food service value)		
Clubs, country		
Per resident member		75
Per non-resident member		25
Racquet (per court per hour)		80
Commercial, general (per acre)		2,000
Factories		
Per person/shift	25	
Add for showers	10	
Food service operations (per seat)		
Ordinary restaurant		35
24-hour restaurant		50
Restaurant along freeway		70
Tavern (little food service)		20
Curb Service (drive-in, per car space)		50
Catering, or banquet facilities	20	
Hair dresser (per station)		170
Hospitals (per bed)		175
Hotels (per room)		120
Add for banquet facilities, theater, night club, as applicable		
Homes		
One Bedroom		150
TwoBedrooms		300
Three Bedrooms		400
Four Bedrooms		475
FiveBedrooms		550
Industrial, general (per acre)		10,000

Table 4.18 (Continued)

Type of Facility or Development	Flow Rate Per Person (gpcd)	Flow Rate PerUnit (gpud)
Institutions (other than hospitals)	125	
Laundromats (per machine)		580
Mobile home parks		
Less than 5 units: use flow rates for homes		
Twenty or more units, per trailer, double wide		200
Five to twenty units, use prorated scale		300
Motels		
Per living unit		100–130
With kitchen		150
Office, dental buildings		
Per employee	15–20	
Per square foot		0.1–0.2
Dentist per chair		750
Parks (per picnicker)		
Restroom only	5	
Showers and restroom	10	
Residential		
General	100	
Single Family, unit		370
Townhouse, unit		300
Schools (per student)	16	
General	75	
Boarding	10	
Day, cafeteria add	5	
Day, showers add	5	
Service stations		
Per toilet (not including car wash)		400
Shopping lefts		
Per ft^2—food extra		0.1
Per employee	15	
Per toilet		400
Swimming pools (per swimmer)	10	
Sports stadium	5	
Theater		
Drive-in (per space)		20
Movie (per seat)		20
Dinner theater, individual (per seat)		20
With hotel		10
Warehouse, general (per acre)		600

Conversion factors: 1 gal/day/person = 1 gpcd = 3.785 Lpcd; 1 gal/day/unit = 1 gpud = 3.785 Lpud; 1 ft = 0.3048 m; 1 acre = 0.4046 ha; 1 square foot = 1 sq. ft. = 0.929 m^2.

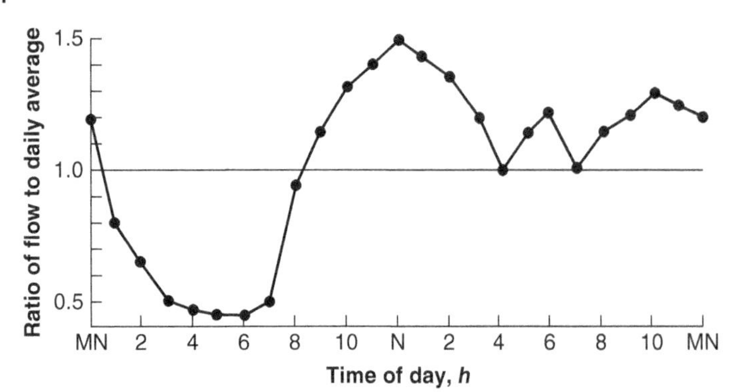

Figure 4.8 Hourly Variations in Wastewater Flow

For developing countries, Shammas (1984) has evaluated the variations in wastewater flows from sewered areas of moderate size and population in Riyadh, Saudi Arabia, to be as follows:

$$Q_{\text{max mo}} = 1.19\, Q_{\text{avg day}};\quad Q_{\text{min mo}} = 0.89\, Q_{\text{avg day}} \tag{4.35}$$

$$Q_{\text{max day}} = 1.36\, Q_{\text{avg day}};\quad Q_{\text{min day}} = 0.64\, Q_{\text{avg day}} \tag{4.36}$$

$$Q_{\text{max h}} = 1.50\, Q_{\text{avg day}};\quad Q_{\text{min h}} = 0.45\, Q_{\text{avg day}} \tag{4.37}$$

$$Q_{\text{peak h}} = 1.36 \times 1.50\, Q_{\text{avg day}} = 2.04\, Q_{\text{avg day}};\quad Q_{\text{extreme min h}} = 0.64 \times 0.45\, Q_{\text{avg day}} = 0.29\, Q_{\text{avg day}}. \tag{4.38}$$

The hourly flow variations are shown in Fig. 4.8. The flow varies continuously throughout the day, with extremely low flows of 0.45 of the daily average occurring between 4:00 and 6:00 a.m., and maximum flows during the daylight hours culminating at noon with a peak of 1.5 times the daily average. In essence, three peaks are characteristic of the Riyadh wastewater flow: one absolute peak occurring at noon, with the other two relatively smaller peaks occurring around 6:00 to 10:00 p.m. Note that this pattern of variation is different from those of Western communities (Fig. 4.6) where such secondary peaks are absent. Taking into account the time taken by the wastewater to reach the treatment plant, the occurrence of such secondary peaks in Riyadh is attributed to the religious practice of ablution before prayers.

Example 4.5 Estimation of Flow Rates in Sewers

Estimate the average, peak, and low rates of flow in a spent-water sewer of a community of 9000 inhabitants with an average rate of water consumption of 150 gpcd (568 Lpcd), using Table 4.17 and Griff equations for the purpose of comparison.

Solution 1 (US Customary System):

1) Average daily spent water: 0.7×150 gpcd = 105 gpcd = 0.945 MGD
2) Maximum daily: $3.5 \times 105 = 367.5$ gpcd = 3.308 MGD with Eq. (4.23): $Q_{\text{max day}} = 105 \times 5(9)^{-1/6} = 360$ gpcd = 3.24 MGD
3) Minimum daily: $0.32 \times 105 = 33.6$ gpcd = 0.302 MGD with Eq. (4.24): $Q_{\text{min day}} = 105 \times 0.2(9)^{1/6} = 30$ gpcd = 0.27 MGD.

Solution 2 (SI System):

1) Average daily spent water: 0.7×568 Lpcd = 397 Lpcd = 3.573 MLD
2) Maximum daily: 3.5×397 Lpcd = 1,389.5 Lpcd = 12.505 MLD with Eq. (4.23): $Q_{\text{max day}} = 397 \times 5\,(9)^{-1/6} = 1,377$ Lpcd = 12.393 MLD
3) Minimum daily: 0.32×397 Lpcd = 127.04 Lpcd = 1.14 MLD with Eq. (4.24): $Q_{\text{min day}} = 397 \times 0.2(9)^{1/6} = 115$ Lpcd = 1.035 MLD.

Expected low flows in spent-water and combined sewers are as meaningful as expected high flows because suspended solids are deposited or stranded as flows and velocities decline. Flow may be obstructed and malodorous, and dangerous gases may be released from the accumulating detritus.

4.8 Design Flows for Wastewater Infrastructure Development

In the state of New York, USA, Table 4.18 can be used as a basis for the design of sewage treatment and disposal facilities for new developments, and for existing establishments when the hydraulic loading cannot be measured. Alternatively, water usage data can be used to estimate wastewater flow, if it is available for an establishment. Adjustments should be made for infiltration and for water that will not reach the sewer (such as boiler water). For commercial establishments variations in flow may be extreme; in these cases it is necessary to examine the significant delivery period of the wastewater and base the peak design flow on this information to prevent an excessive rate of flow through the treatment system. It may be desirable to include an equalization basin prior to the treatment system.

New York State's environmental conservation law mandates the use of water-saving plumbing facilities in new and renovated buildings. Hydraulic loading, as determined from Table 4.18, may be decreased by 20% in those installations serving premises equipped with certified water-saving plumbing fixtures. A combination of new and old fixtures can be considered on a pro rata basis.

New toilets, which use as little as 0.5 gal (1.89 L) of water per flush, are becoming available on the market and the reduction of wastewater flow attributable to these and other new technologies is considered on a case-by-case basis. The reduction allowance depends in part on the ability of the builder or owner to ensure adequate maintenance and/or replacement in kind when necessary.

Once the design wastewater flow is chosen, an environmental engineer may design the sanitary sewers to carry peak residential, commercial, institutional, and industrial flows, as well as infiltration and inflow. Gravity sewers are designed to flow full at the design peak flow, in accordance with U.S. Environmental Protection Agency rules. As discussed previously, design flows are based on various types of developments. Table 4.18 provides a list of design flows for various development types. The design for a long-lived wastewater infrastructure should consider serviceability factors, such as ease of installation, design period, useful life of the conduit, resistance to infiltration and corrosion, and maintenance requirements. The design period should be based on the ultimate tributary population and usually ranges from 25 to 30 years.

4.9 Design Flows for Water Infrastructure Development

Of 50 municipalities surveyed in 1993 by the Water Environment Association of Ontario (WEAO), 45 reported ratios of annual wastewater treated to water produced ranging from 0.5 to 2.0, and 5 reported ratios from 2.0 to 4.0. Simple ratios such as these fail to provide a perspective on real life. Ratios tend to obscure the reality of our less than perfect world—a world with leaky sewers, water losses, inaccuracies in metering, and the need for large volumes of water for irrigation—and should be largely ignored. The path from water treatment plant to wastewater treatment plant (WWTP) is far from predictable.

It is only by disaggregating daily water demand and daily wastewater flow that their relationship can be understood. Geerts et al. (1966) disaggregated several years of daily water production and wastewater treatment records for three Ontario communities, referred to as C1, C2, and C3, shown in Table 4.19. By estimating dry-weather wastewater flow and normal water use and comparing these with yearly averages (that is, average over 365 days, including the "extremes"), first cut estimates on the contribution of inflow/infiltration (I/I) to the WWTP and the amount of water used for irrigation could be made.

In the sample communities, the ratio of annual sewage flow to water produced was 0.88, 1.09, and 1.22 (Table 4.19). The relationship between water and wastewater appeared reasonably close to 1:1 in all communities. However, when the statistics on these three municipalities were analyzed, it was found that C1, instead of having what appeared to be a good water

Table 4.19 Comparison of Water Usage Versus Wastewater Volumes of Three Municipalities in Ontario, Canada

Municipality	C1	C2	C3
Population	19,000	71,000	45,000
Annual sewage to water ratio	0.88	1.09	1.22
Actual percentage of water to WWTP	61%	75%	51%
Percentage flow to WWTP from I/I	39%	25%	49%
Total I/I (dry weather)	15.6%	22.3%	23.7%
Total I/I (wet weather)	23.4%	2.7%	25.3%

to sewage ratio, in fact, had a system of sewers that contributed an excess flow of 39% from the inflow/infiltration of groundwater and stormwater sources (I/I). Community C2 was found to have little flow from wet-weather events but contributed groundwater flows to the WWTP to the tune of 22% of total flows over the year. It is important to appreciate that the methodology presented is intended to generally characterize water and wastewater systems for planning purposes.

The analysis using daily water and wastewater records has provided the answer to the question "How much treated water ends up in the sewer?" helping consulting engineers to design new water infrastructure and helping municipalities to anticipate the impact of water conservation on servicing requirements.

The recommendation then is that the ratio of wastewater flow to water demand should be roughly 1.0, given that the volume of water consumed is approximately equal to the wastewater generated. Table 4.18 may then be used for determination of both water design flows and wastewater design flows of various residential, industrial, commercial, and institutional facilities.

Problems/Questions

4.1 As shown in Table 4.3, the April census population of Detroit, Michigan, was 1,028,000 in 1990 and decreased to 951,000 in 2000. Estimate the population for the following periods:

1) For the fifth intercensal year in April using (a) arithmetic and (b) geometric progression.
2) For the sixth postcensal year in April using (a) arithmetic and (b) geometric progression.

4.2 As shown in Table 4.3, the census population of Providence, Rhode Island, was 161,000 in 1990 and 174,000 in 2000. Estimate the 2009 population by (a) arithmetic and (b) geometric progression.

4.3 As shown in Table 4.3, the city of Miami recorded a population of 111,000 in its 1930 and 172,000 in its 1940 consecutive censuses. Estimate the midyear (July 1) populations for the following periods:

1) For the fifth intercensal year by (a) arithmetic increase and (b) geometric increase.
2) For the ninth postcensal year by (a) arithmetic increase and (b) geometric increase.

Assume a census date of April 1.

4.4 Each of the four waterworks systems shown in Fig. 4.4 serves a community with an estimated future population of 100,000. Estimate the required capacities of their constituent structures for an average water consumption of 150 gpcd (568 Lpcd), a fire demand of 9160 gpm (34,671 L/min), and a distributing reservoir so sized that it can provide enough water to care for differences between hourly and daily flows and for fire demands and emergency water requirements.

4.5 An old town in an established agricultural area with no prospects for extensive development is located along a river that is used as the source of water supply for the community. The conditions of the river are such that it receives its water from a drainage area of 1000 mi^2 (2590 km^2) and its low water flow is 0.1 $ft^3/s/mi^2$ (0.00109 $m^3/s/km^2$). The regulations are that only 10% of the river flow can be used at any time for water supply.

Past records reveal that the population increased as follows:

Year	Population
1940	10,000
1950	12,500
1960	15,200
1970	17,500
1980	19,700
1990	22,000
2000	24,400

If you are asked to design a water supply system for this community,

1) What would be the expected population in the year 2040? Justify your method of choice for estimating the future population. (Give the figure to the nearest 1000.)
2) Do you think that the amount of water available in the river would be enough of a water supply and how much water can be supplied to each person in the community?
3) Assuming that you decide to supply the community at an annual average demand of 70 gpcd, or 264 Lpcd, and that the maximum daily demand is two times the annual average and the maximum hourly demand is 1.5 times the daily demand, would you be able to supply this demand directly from the river? If not, what would you recommend in order to be able to supply the peak demand?

4.6 A small village is located near a stream that is to be used as the source of water supply for the community. The minimum flow in the stream is 550 m^3/h. The population analysis reveals that the population is expected to increase linearly from a population of 10,000 in 2008 to 30,000 in 2058.

The water consumption of the community is estimated to reach an annual average demand of 300 Lpcd, or 79.26 gpcd, a maximum daily demand of 1.8 times the annual average, and peak hourly demand of 1.5 times the daily average.

Would you be able to supply these demands directly from the stream? If not, what would you recommend (1) in the year 2030 and (2) in the year 2058?

4.7 A well that is the source of water supply for a community has a uniform production capacity of 350 m^3/h (79.52 gpm). Currently the community has a population of 5000 people with an anticipated linear growth rate of 10% per year.

If the average water consumption for the community is 400 Lpcd (105.68 gpcd), would you be able to supply the water demand (a) 5 years, (b) 10 years, and (c) 20 years from now, directly from this well? If not, what would you recommend for each of the above periods? Show all of your calculations and explain the reasons behind your assumptions and recommendations.

4.8 Estimate the number of people who can be supplied with water from (a) 12-in. (304.8-mm), and (b) 24-in. (609.6-mm) water main (1) in the absence of fire service for a maximum draft of 200 gpcd (757 Lpcd) and (2) with a residential fire-flow requirement of 500 gpm (1892.5 Lpm) and a coincident draft of 150 gpcd (567.75 Lpcd). Also, find the hydraulic gradient.

Note: Consider the most economical design velocity in the system piping to be 3 ft/s (0.91 m/s).

4.9 Estimate the number of people who can be supplied with water from (a) a 300-mm water main and (b) a 600-mm water main.

Case 1: In the absence of fire service for a maximum draft of 800 Lpcd.

Case 2: With a residential fire-flow requirement of 1890 Lpm and a simultaneous domestic draft of 600 Lpcd.

Note: Consider the most economical design velocity in the system piping to be 1 m/s.

Bibliography

Al-Dhowalia, K. H., and Shammas, N. K., Leak Detection and Quantification of Losses in a Water Network, *Int. J. Water Resour. Develop.*, 7, 1, 30–38, 1991.

American Society of Civil Engineers and Water Environment Federation, *Gravity Sanitary Sewer, Design and Construction*, 2nd ed., ASC Manuals and Reports on Engineering Practice #60, WEF Manual of Practice #FD-5, Reston, VA, 2007.

American Water Works Association, *Distribution System Requirements for Fire Protection*, 3rd ed., Manual of Water Supply Practices M31, Denver, CO, 1998.

Aziz, H. A., Wastewater Engineering, in *Environmental Science, Technology, Engineering, and Mathematics*, 209 p, Lenox Institute Press, MA, 2023. https://doi.org/10.17613/9g1r-6886.

Dufor, C. N., and Becker, E., *Public Water Supplies of the 100 Largest Cities in the United States*, U.S. Geological Survey Water Supply Paper, 1812, 5, 1964.

Fair, G. M., Geyer, J. C., and Okun, D. A., *Water and Wastewater Engineering*, vol. *1*, John Wiley & Sons, New York, USA, 1966.

Fair, G. M., Geyer, J. C., and Okun, D. A., *Elements of Water Supply and Wastewater Disposal*, John Wiley & Sons, New York, USA, 1971.

Geerts, H., Goodings, R. A., and Davidson, R., How Much Treated Water Ends in Sewers? *Environ. Sci. Eng.*, 1966. Also available online at http://esemag.net/1196/treated.html.

Geyer, J. C., and Lentz, J. J., An Evaluation of Problems of Sanitary Sewer System Design, *J. Water Pollution Control Federat.*, *38*, 1138, 1966.

Gifft, H. M., Estimating Variations in Domestic Sewage Flows, *Waterworks Sewerage*, *92*, 175, 1945.

Great Lakes-Upper Mississippi River Board of State and Provincial Public Health and Environmental Managers, *Recommended Standards for Water Works*, Health Research Inc., Albany, NY, USA, 2018a.

Great Lakes-Upper Mississippi River Board of State and Provincial Public Health and Environmental Managers, *Recommended Standards for Wastewater facilities*, Health Research Inc., Albany, NY, 2018b.

Insurance Services Office, *Guide for Determination of Needed Fire Flow*, Insurance Services Office, Jersey City, NJ, 2006.

International Code Council, *International Fire Code*, Country Club Hills, IL, USA, 2006.

Khadam, M., Shammas, N. K., and Al-Feraiheedi, Y., Water Losses from Municipal Utilities and their Impacts, *Water Int.*, *13*, 1, 254–261, 1991.

Linaweaver, F. P. Jr., Geyer, J. C., and Wolff, J. B., Summary Report on the Residential Water Use Research Project, *J. Am. Water Works Ass.*, *59*, 267, 1967.

National Association of Manufacturers, *Water in Industry*, 1965, http://www.NAM.org, 2009.

New York State Department of Environmental Conservation, *Design Standards for Wastewater Treatment Works*, Albany, NY, USA, 1988.

Quraishi, A. A., Shammas, N. K., and Kadi, H. M., Analysis of Per Capita Household Water Demand for the City of Riyadh, Saudi Arabia, *Arabian J. Sci. Technol.*, *15*, 4, 539–552, 1990.

Shammas, N. K., Characteristics of the Wastewater of the City of Riyadh, *Int. J. Develop. Technol.*, *2*, 141–150, 1984.

Shammas, N. K., Investigation of Irrigation Water Application Rates to Landscaped Areas in Riyadh, *J. King Saud Univ., Eng. Sci.*, *3*, 2, 147–165, 1991.

Shammas, N. K., and Al-Dhowalia, K., Effect of Pressure on Leakage Rate in Water Distribution Networks, *J. Eng. Sci.*, *5*, 2, 155–312, 1993.

U.S. Bureau of the Census, Department of Commerce, 2009. http://www.census.gov/population/www/documentation/twps0027.html.

U.S. Environmental Protection Agency, *Collection Systems Technology Fact Sheet—Sewers*, Conventional Gravity, Washington, DC, 2009.

5

Water Hydraulics, Transmission, and Appurtenances

5.1 Fluid Mechanics, Hydraulics, and Water Transmission

5.1.1 Fluid Mechanics and Hydraulics

Fluids are substances that are capable of flowing and that conform to the shape of the vessel in which they are contained. Fluids may be divided into liquids and gases. Liquids are practically incompressible, whereas gases are compressible. Liquids occupy definite volumes and have free surfaces, whereas gases expand until they occupy all portions of any containing vessels. Water, wastewater, oil, and so on are liquids.

Fluid mechanics is a branch of applied mechanics dealing with the behavior of fluids (liquids and/or gases) at rest and in motion. The principles of thermodynamics must be included when an appreciable compressibility does occur. Hydraulics is a branch of fluid mechanics dealing with the behavior of particularly incompressible water, wastewater, liquid sludge, and so on.

The characteristics of gases are determined by Boyle's and Charles's laws:

$$P_a V_s / T_a = R_g \tag{5.1}$$

where P_a is absolute pressure $= P_{gauge} + P_{atm}$, lb/ft^2; $V_s = 1/\gamma =$ specific volume, ft^3/lb; T_a is absolute temperature $= T + 460$°F; R_g is gas constant, ft/°F; T is temperature, °F; γ is specific weight, lb/ft^3; and

$$\gamma = 1/V_s = P_a / R_g T_a \tag{5.2}$$

The specific weight, γ, of a substance is the weight of a unit volume of the substance. The specific (unit) weight of water is 62.4 lb/ft^3 (or 9.81 kN/m^3) for ordinary temperature variations.

The specific gravity of a substance is that pure number, which denotes the ratio of the weight of a substance to the weight of an equal volume of a substance taken as a standard. Solids and liquids are referred to water (at 39.2°F = 4°C) as standard, whereas gases are referred to air free of hydrogen and carbon dioxide (at 32°F = 0°C and at 1 atmosphere 14.7 lb/in.^{2} pressure) as standard. For example, if the specific gravity of a given oil is 0.755, its specific weight is 0.755 × (62.4 lb/ft^3) = 47.11 lb/ft^3.

The mass of a body ρ (rho), which is termed density, can be expressed by the following equation:

$$\rho = \frac{\gamma}{g} \tag{5.3}$$

In the US customary system of units, the density of water is (62.4 lb/ft^3)/(32.2 ft/s^2) = 1.94 slug/ft^3. In the metric system, the density of water is 1 g/cm^3 = 1 kg/L = 1,000 kg/m^3 = 1 tonne/m^3 at 4°C.

Fluid flow may be steady or unsteady, uniform or nonuniform. Steady flow occurs at any point if the velocity of successive fluid particles is the same at successive instants, that is, if the fluid velocity is constant with time. The uniform flow occurs when the velocity of successive fluid particles does not change with distance. This book introduces mainly the fundamentals of the steady and uniform flows involving water and wastewater.

In steady flow, the mass of fluid passing any and all sections in a stream of fluid per unit of time is the same. For incompressible fluids, such as water and wastewater, the following principle of conservation of mass governs:

$$A_1 v_1 = A_2 v_2 = \text{Constant} = Q \tag{5.4}$$

Water and Wastewater Engineering: Hydraulics, Hydrology and Management, Volume 1, Fourth Edition.
Lawrence K. Wang, Mu-Hao Sung Wang, and Nazih K. Shammas.

Companion website: www.wiley.com/go/Wang/Waterandwastewater4e

where A_1 and A_2 are, respectively, the cross-sectional areas (ft^2 or m^2) at section 1 and section 2; v_1 and v_2 are, respectively, the average velocity of the stream (ft/s or m/s) at section 1 and section 2; and Q is the flow rate (ft^3/s or m^3/s).

The Bernoulli equation results from application of the principle of conservation of energy to fluid flow, and is written between two points in a hydraulic system:

$$H_A + H_a - H_1 - H_e = H_B \tag{5.5a}$$

$$H_A + H_a = H_B + (H_1 + H_e) = H_B + h_f \tag{5.5b}$$

$$H_A = P_A/\gamma + v_A^2/2g + Z_A \tag{5.6a}$$

$$H_B = P_B/\gamma + v_B^2/2g + Z_B \tag{5.6b}$$

where

H_A = energy at section A;
H_B = energy at section B;
H_a = energy added;
H_1 = energy lost;
H_e = energy extracted (The units of H_A, H_B, H_a, H_1, and H_e are feet or meters of the fluid.);
P_A, P_B = pressures at sections A and B $\left(\text{lb/ft}^2 \text{ or kN/m}^2 \left[1{,}000 \text{ Newton/m}^2 = 1 \text{ kN/m}^2\right]\right)$;
γ = specific weight of water $\left(62.4 \text{ lb/ft}^3 \text{ or } 9.81 \text{ kN/m}^3\right)$;
g = acceleration due to gravity $\left(32.2 \text{ ft/s}^2 \text{or } 9.81 \text{ m/s}^2\right)$;
V_A, V_B = velocities at sections A and B (ft/s or m/s);
Z_A, Z_B = heights of stream tube above any assumed datum plane at sections A and B (ft or m); and
h_f = head loss (ft or m).

It is important to determine the magnitude, direction, and sense of the hydraulic force exerted by fluids in order to design the constraints of the structures satisfactorily. The force P_f(lb or kg) exerted by a fluid on a plane area A (ft^2 or m^2) is equal to the product of the specific weight γ (lb/ft^3 or kN/m^3) of the liquid, the depth (ft and m) of the center of gravity h_{cg} of the area, and the area A (ft^2 or m^2).

The engineering equation is

$$P_f = \left(\gamma h_{cg}\right)A = I_p A \tag{5.7}$$

where I_p is the intensity of pressure at center of gravity of the area. The line of action of the force passes through the center of pressure, which can be determined by the following equation:

$$y_{cp} = \left[I_{cg}/y_{cg}A\right] + y_{cg} \tag{5.8}$$

where I_{cg} is the moment of inertia of the area about its center of gravity axis, y_{cp} is the distance of the center of pressure measured along the plane from an axis located at the intersection of the plane and the water surface, extended if necessary, and y_{cg} is the distance of the center of gravity measured along the plane from an axis located at the intersection of the plane and the liquid surface, extended if necessary.

The moment of inertia of the most common rectangular shape (width b and height h) can be calculated by the following equation:

$$\begin{aligned} I_{cg} &= (1/12)Ah^2 \\ &= (1/12)(bh)h^2 \\ &= (1/12)bh^3 \end{aligned} \tag{5.9}$$

Example 5.1 Equation of Continuity

For incompressible fluids, such as water, the equation of continuity shown in Eq. (5.4) can be used for all practical purposes. Derive the equations of continuity for compressible fluids, such as air, carbon dioxide, and chlorine, which are frequently used in water and wastewater treatment plants.

Solution:

The following equations of continuity should be used:

$$\rho_1 A_1 v_1 = \rho_2 A_2 v_2 \quad (5.4a)$$

$$\gamma_1 A_1 v_1 / g = \gamma_2 A_2 v_2 / g \quad (5.4b)$$

$$\gamma_1 A_1 v_1 = \gamma_2 A_2 v_2, \quad (5.4c)$$

where $\rho_1, \rho_2, A_1, A_2, v_1, v_2, \gamma_1, \gamma_2$, and g have been defined in Eqs. (5.2)–(5.4). The above equations of continuity result from the basic principle of conservation of mass. For a steady flow, the mass of fluid passing any and all sections in a stream of fluid per unit of time is the same.

Example 5.2 Sizing Air Pipe Diameter
Determine the minimum diameter of an air pipe to carry 0.55 lb/s (0.25 kg/s) of air with a maximum velocity of 19 ft/s (5.8 m/s). The air is at 82°F (27.8°C) and under an absolute pressure of 34.5 lb/in.^{2} (239.4 kPa). Gas constant for air is 53.3 ft/°R (0.287 kJ/kg-K).

Solution 1 (US Customary System):

$$P_a = P_{gauge} + P_{atm} = 34.5\ \text{lb/in.}^2 = 34.5 \times 144\ \text{lb/ft}^2.$$
$$R_g = 53.3\ \text{ft/}^\circ\text{R}.$$
$$T_a = \text{absolute temperature} = 82 + 460 = 542^\circ\text{R}.$$
$$\gamma_a = P_a/[R_g T_a] = (34.5 \times 144\ \text{lb/ft}^2)/[(53.3\ \text{ft/}^\circ\text{R})(542^\circ\text{R})] = 0.172\ \text{lb/ft}^3.$$
$$W = 0.55\ \text{lb/s} = \gamma_a Q = (0.172\ \text{lb/ft}^3)(Q)$$
$$Q = 3.19\ \text{ft}^3\text{/s} = Av = A(19\ \text{ft/s}).$$
$$A = 3.19/19 = 0.168\ \text{ft}^2 = D^2 \times 0.785$$
$$\mathbf{D} = (0.168/0.785)^{0.5} = \mathbf{0.46\ ft = 5.55\ in.}$$

Solution 2 (SI System):

$$P_a = 239.4\ \text{kPa} = 239.4\ \text{kN/m}^2.$$
$$R_g = 0.287\ \text{kJ/kg-K} = 29.26\ \text{m/K}.$$
$$\gamma_a = P_a/[R_g T_a] = (239.4\ \text{kN/m}^2)/[(29.26\ \text{m/K})(27.8 + 273)\ \text{K}].$$
$$\gamma_a = 0.027\ \text{kN/m}^3 = 0.027 \times 1{,}000/9.81\ \text{kg/m}^3 = 2.752\ \text{kg/m}^3.$$

Here 1 kg = 9.81 N.

$$W = 0.25\ \text{kg/s} = \gamma_a Q = (2.752\ \text{kg/m}^3)Q.$$
$$Q = 0.09084\ \text{m}^3\text{/s} = Av = A(5.8\ \text{m/s}).$$
$$A = 0.01788\ \text{m}^2 = D^2 \times 0.785.$$
$$\mathbf{D} = (0.01788/0.785)^{0.5} = \mathbf{0.15\ m = 15\ cm}.$$

Example 5.3 Compressible Carbon Dioxide Gas Flow Calculations
Carbon dioxide passes point A in a 3.5 in. (0.0889 m) pipe at a velocity of 16 ft/s (4.88 m/s). The pressure at point A is 32 psi (222 kPa) and the temperature is 70°F (21°C). At point B downstream the pressure is 22 psi (152.7 kPa) and the temperature is 85°F (29.4°C). For a barometric reading of 14.7 psi (102 kPa), determine the gas velocity at point B, and the gas flows at points A and B. The gas constant of carbon dioxide is 38.86 ft/°R (0.209 kJ/kg-K).

Solution 1 (US Customary System):

$$P_a = \text{absolute pressure} = P_{gauge} + P_{atm} = 32 + 14.7 = 46.7\ \text{psi}.$$
$$P_a = (46.7\ \text{lb/in.}^2)(144\ \text{in.}^2\text{/ft}^2) = 6{,}724.8\ \text{lb/ft}^2.$$
$$T_a = \text{absolute temperature} = T + 460 = 70 + 460 = 530^\circ\text{R}.$$
$$\gamma_A = P_a/[R_g T_a] = [(32 + 14.7) \times 144]/[38.86(70 + 460)].$$
$$\gamma_A = 6{,}724.8\ \text{lb/ft}^2/[(38.86\ \text{ft/}^\circ\text{R})(530^\circ\text{R})] = 0.326\ \text{lb/ft}^3.$$

Similarly γ_B can be determined:

$$\gamma_B = P_a/[R_g T_a] = [(22 + 14.7) \times 144]/[38.86(85 + 460)] = 0.249\ \text{lb/ft}^3$$

$$W(\text{lb/s}) = (\gamma_A)(A_A)(v_A) = (\gamma_B)(A_B)(v_B) = (\gamma_A)(Q_A) = (\gamma_B)(Q_B)$$

Since $A_A = A_B$, then $(\gamma_A)(v_A) = (\gamma_B)(v_B)$.

$$(0.326\ \text{lb/ft}^3)(16\ \text{ft/s}) = (0.249\ \text{lb/ft}^3)(v_B)$$

$$\mathbf{v_B = 20.9\ ft/s}.$$

$$A_A = A_B = (3.5/12)^2(3.14/4) = 0.0668\ \text{ft}^2.$$

$$\mathbf{Q_A} = (A_A)(v_A) = (0.0668\ \text{ft}^2)(16\ \text{ft/s}) = \mathbf{1.068\ ft^3/s}.$$

$$\mathbf{Q_B} = (A_B)(v_B) = (0.0668\ \text{ft}^2)(20.9\ \text{ft/s}) = \mathbf{1.396\ ft^3/s}.$$

Solution 2 (SI System):

$$P_a = \text{absolute pressure} = P_{\text{gauge}} + P_{\text{atm}} = 222 + 102 = 324\ \text{kPa} = 324\ \text{kN/m}^2.$$

$$R_g = 0.209\ \text{kJ/kg-K} = 209\ \text{J/kg-K} = 209 \times 0.101972\ \text{kg-m/kg-K} = 21.31\ \text{m/K} = 38.86\ \text{ft-lb/lb-}^{\circ}\text{R}.$$

$$T_a = \text{absolute temperature} = T + 273 = 21 + 273 = 294\ \text{K}.$$

$$\gamma_A = P_a/[R_g T_a] = (324\ \text{kN/m}^2)/[(21.31\ \text{m/K})(294\ \text{K})] = 0.0517\ \text{kN/m}^3.$$

$$\gamma_A = (0.0517 \times 1{,}000)/9.81\ \text{kg/m}^3 = 5.27\ \text{kg/m}^3.$$

Similarly γ_B can also be determined:

$$\gamma_B = P_a/[R_g T_a] = (152.7 + 102)/[21.31(29.4 + 273)] = 0.0395\ \text{kN/m}^3.$$

$$\gamma_B = (0.0395 \times 1{,}000)/9.81\ \text{kg/m}^3 = 4.029\ \text{kg/m}^3.$$

$$W(\text{kg/s}) = (\gamma_A)(A_A)(v_A) = (\gamma_B)(A_B)(v_B) = (\gamma_A)(Q_A) = (\gamma_B)(Q_B).$$

Since $A_A = A_B$, then $(\gamma_A)(v_A) = (\gamma_B)(v_B)$.

$$(5.27\ \text{kg/m}^3)(4.88\ \text{m/s}) = (4.029\ \text{kg/m}^3)(v_B).$$

$$\mathbf{v_B = 6.38\ m/s}.$$

$$A_A = A_B = (0.0889)^2(3.14/4) = 0.0062\ \text{m}^2.$$

$$\mathbf{Q_A} = (A_A)(v_A) = (0.0062\ \text{m}^2)(4.88\ \text{m/s}) = 0.03\ \text{m}^3/\text{s}.$$

$$\mathbf{Q_B} = (A_B)(v_B) = (0.0062\ \text{m}^2)(6.38\ \text{m/s}) = 0.04\ \text{m}^3/\text{s}.$$

Example 5.4 Compressible Air Flow Calculations

Air flows in a 6.5 in. (165 mm) pipe at a pressure of 35 psi (242.9 kPa) gauge and a temperature of 100°F (37.78°C). If the barometric pressure is 14.7 psi (102 kPa) and the velocity is 10.8 ft/s (3.29 m/s), determine the air mass flow (lb/s and kg/s) when air is flowing. Gas constant for air is 53.3 ft/°R absolute (0.287 kJ/kg-K).

Solution 1 (US Customary System):

The gas laws require absolute units for temperature and pressure, when using Eq. (5.2).

$$P_a = \text{absolute pressure} = P_{\text{gauge}} + P_{\text{atm}} = 35 + 14.7 = 49.7\ \text{psi}.$$

$$P_a = (49.7\ \text{lb/in.}^2)(144\ \text{in.}^2/\text{ft}^2) = 7{,}156.8\ \text{lb/ft}^2.$$

$$R_g = 53.3\ \text{ft/}^{\circ}\text{R absolute}.$$

$$T_a = \text{absolute temperature} = T + 460 = 100 + 460 = 560^{\circ}\text{R absolute}.$$

$$\gamma = P_a/[R_g T_a] = (7{,}156.8\ \text{lb/ft}^2)/[(53.3\ \text{ft/}^{\circ}\text{R})(560^{\circ}\text{R})] = 0.24\ \text{lb/ft}^3\ \text{for air}.$$

$$W = \text{mass flow} = (\gamma)(A)(v) = (\gamma)(Q).$$

$$W = (0.24\ \text{lb/ft}^3)[(6.5/12)^2(3.14/4)\text{ft}^2](10.8\ \text{ft/s}) = \mathbf{0.597\ lb/s}.$$

Solution 2 (SI System):

$$P_a = \text{absolute pressure} = P_{\text{gauge}} + P_{\text{atm}} = 242.9 + 102 = 344.9\ \text{kPa} = 344.9\ \text{kN/m}^2.$$

$$R_g = 53.3\ \text{ft/}^{\circ}\text{R absolute} = 0.287\ \text{kJ/kg-K} = 287 \times 0.101972\ \text{kg-m/kg-K} = 29.26\ \text{m/K}.$$

Here 1 J = 0.101972 kg-m.

T_a = absolute temperature = $T + 273 = 37.78 + 273 = 310.8$ K absolute.

$\gamma = P_a/[R_g T_a] = (344.9\ \text{kN/m}^2)/[(29.26\ \text{m/K})(310.8\ \text{K})] = 0.0379\ \text{kN/m}^3$ for air $= 3.87\ \text{kg/m}^3$.

Here 1 kg = 9.81 N.

$$W = \text{mass flow} = (\gamma)(A)(v) = (\gamma)(Q).$$
$$W = (3.87\ \text{kg/m}^3)[(0.165\ \text{m})^2(3.14/4)](3.29\ \text{m/s}) = \mathbf{0.272\ kg/s}.$$

Example 5.5 Calculation of Velocities in Water Pipes

When 550 gpm (2,082 Lpm) water flows through a 12 in. (30.48 cm) pipe, which later reduces to a 6 in. (15.24 cm) pipe, calculate the average velocities in the two pipes.

Solution 1 (US Customary System):

$$Q = (A_1)(v_1) = (A_2)(v_2) = 550\ \text{gpm} = 1.224\ \text{ft}^3/\text{s}.$$
$$A_1 = (12/12)^2 \times 0.785\ \text{ft}^2 = 0.785\ \text{ft}^2.$$
$$A_2 = (6/12)^2 \times 0.785\ \text{ft}^2 = 0.196\ \text{ft}^2.$$
$$Q = 1.224\ \text{ft}^3/\text{s} = (A_1)(v_1) = (A_2)(v_2).$$
$$1.224\ \text{ft}^3/\text{s} = (0.785\ \text{ft}^2)(v_1) = (0.196\ \text{ft}^2)(v_2).$$
$$\mathbf{v_1 = 1.6\ ft/s}.$$
$$\mathbf{v_2 = 6.2\ ft/s}.$$

Solution 2 (SI System):

$$Q = (A_1)(v_1) = (A_2)(v_2) = 2,082\ \text{Lpm} = 0.0347\ \text{m}^3/\text{s}.$$
$$A_1 = (30.7/100)^2 \times 0.785\ \text{m}^2 = 0.0725\ \text{m}^2.$$
$$A_2 = (15.24/100)^2 \times 0.785\ \text{m}^2 = 0.0182\ \text{m}^2.$$
$$Q = 0.0347\ \text{m}^3/\text{s} = (A_1)(v_1) = (A_2)(v_2).$$
$$0.0347\ \text{m}^3/\text{s} = (0.0725\ \text{m}^2)(v_1) = (0.0182\ \text{m}^2)(v_2).$$
$$\mathbf{v_1 = 0.479\ m/s}.$$
$$\mathbf{v_2 = 1.907\ m/s}.$$

Example 5.6 Determination of Hydraulic Pressure in a Water Pipe

Water flows through the nozzle shown in Fig. 5.1 and deflects the mercury in the U-tube gauge at a treatment plant. Determine the value of h if the pressure at point D is 20 psi (138.8 kN/m^2), and m is 2.75 ft (0.84 m).

Solution 1 (US Customary System):

Method A. Pressure balance method: (See Fig. 5.1)

$$20\ \text{lb/in.}^2 + [(62.4\ \text{lb/ft}^3)(2.75 + h)\text{ft}]/(144\ \text{in.}^2/\text{ft}^2) = [(13.57 \times 62.4\ \text{lb/ft}^3)(h)\text{ft}]/(144\ \text{in.}^2/\text{ft}^2) + P_A.$$

$P_A = 0\ \text{psi} = 0\ \text{lb/in.}^2$.

$144 \times 20 + 62.4(2.75 + h) = 13.57 \times 62.4h$.

$\boldsymbol{h} = \mathbf{3.9\ ft}$.

Method B. Water column length method:

1 psi = 2.307 ft of water.

1 ft Hg = 13.57 ft of water.

$20 \times 2.307 + (2.75 + h) = 13.57h$.

$\boldsymbol{h} = \mathbf{3.9\ ft}$.

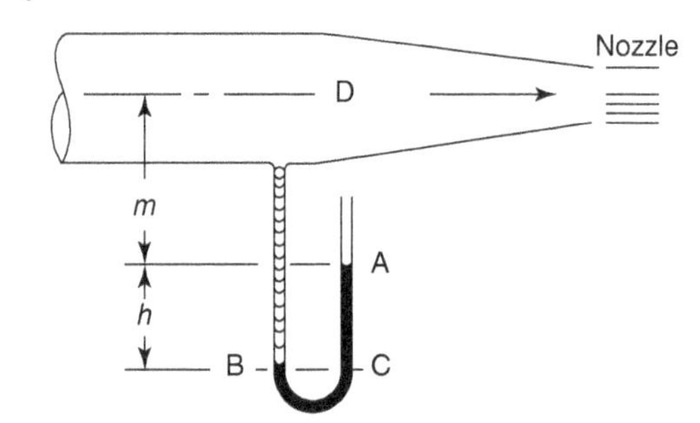

Figure 5.1 Fluid Flow Through a Nozzle

Solution 2 (SI System):

Method A. Pressure balance method: (See Fig. 5.1)

$$138.8\ \text{kN/m}^2 + (9.8\ \text{kN/m}^3)(0.84\ \text{m} + h)$$
$$= (13.57 \times 9.8\ \text{kN/m}^3)h + P_A.$$

$P_A = 0\ \text{kN/m}^3$ assumed.

$138.8 + 9.8 \times 0.84 + 9.8h = 13.57 \times 9.8h.$

$\boldsymbol{h = 1.19\ \text{m}}$.

Method B. Water column length method:

$1\ \text{psi} = 2.307\ \text{ft of water} = 0.703\ \text{m of water} = 6.94\ \text{kN/m}^2.$

$1\ \text{m Hg} = 13.57\ \text{m of water}$

$1\ \text{kN/m}^2 = (0.703/6.94)\ \text{m}\ H_2O = 0.1013\ \text{m}\ H_2O$

$138.8 \times 0.1013\ \text{m} + 0.84\ \text{m} + h = 13.57h.$

$\boldsymbol{h = 1.1\ \text{m}}$.

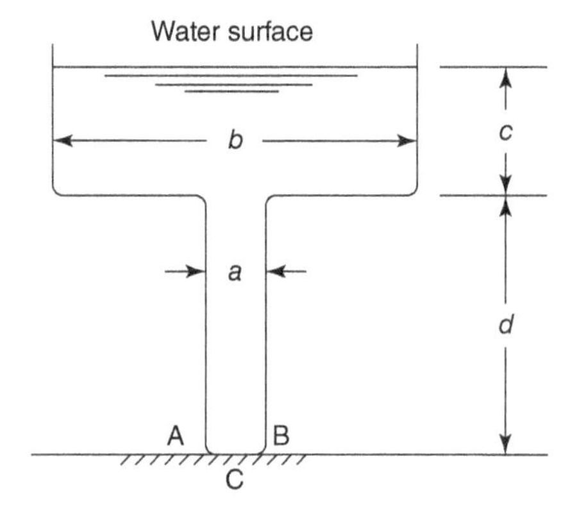

Figure 5.2 Riser Pipe Analysis

Example 5.7 Determination of Water Force on Riser Pipe
Referring to Fig. 5.2, determine (a) the force exerted by water on the bottom plate AB of the 2-ft (0.61-m)-diameter riser pipe for water storage and (b) the total force on plane C. The dimensions given are a = 2 ft (0.61 m), b = 12 ft (3.66 m), c = 6 ft (1.83 m) and d = 10 ft (3.05 m).

Solution 1 (US Customary System):

a. $P_{fAB} = \left(16 \times 2^2 \times 0.785\ \text{ft}^3\right)\left(62.4\ \text{lb/ft}^3\right) = 3{,}135\ \text{lb}.$

$p_{AB} = P_{fAB}/A = 3{,}135\ \text{lb}/\left[\left(2^2 \times 0.785\ \text{ft}^2\right)\left(144\ \text{in.}^2/\text{ft}^2\right) = \mathbf{6.93\ psi}.$

b. $P_{fC} = \left[\left(12^2 \times 0.785 \times 6\ \text{ft}^3\right) + \left(2^2 \times 0.785 \times 10\ \text{ft}^3\right)\right]\left(62.4\ \text{lb/ft}^3\right) = 44{,}282\ \text{lb}.$

$p_C = 44{,}282\ \text{lb}/\left[\left(2^2 \times 0.785\ \text{ft}^2\right)\left(144\ \text{in.}^2/\text{ft}^2\right)\right] = \mathbf{97.93\ psi}.$

Solution 2 (SI System):

a. $P_{fAB} = (1.83 + 3.05) \times 0.61^2 \times 0.785\ \text{m}^3\left(1{,}000\ \text{kg/m}^3\right) = 1{,}425.4\ \text{kg}.$

$p_{AB} = P_{fAB}/A = 1{,}425.4\ \text{kg}/\left(0.61^2 \times 0.785\ \text{m}^2\right) = \mathbf{4{,}880\ kg/m^2}.$

b. $P_{fC} = \left[\left(3.66^2 \times 0.785 \times 1.83\ \text{m}^3\right) + \left(0.61^2 \times 0.785 \times 3.05\ \text{m}^3\right)\right]\left(1{,}000\ \text{kg/m}^3\right) = 20{,}131\ \text{kg}.$

$p_C = 20{,}131\ \text{kg}/\left(0.61^2 \times 0.785\ \text{m}^2\right) = \mathbf{68{,}919\ kg/m^2}.$

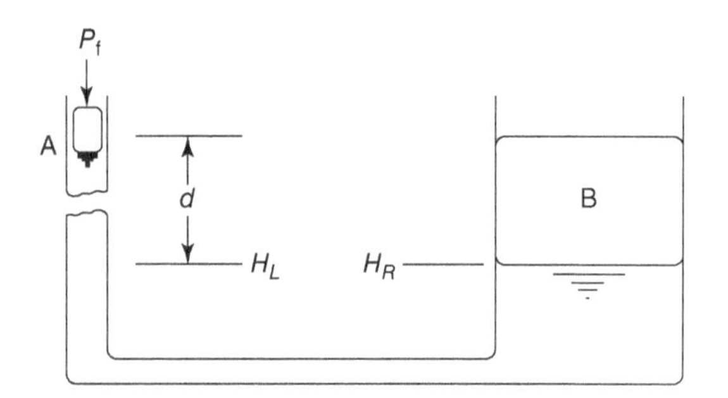

Figure 5.3 Force Analysis for a Vessel with a Plunger and a Cylinder

Example 5.8 Hydraulic Force Equilibrium
In Fig. 5.3, the areas of the hydraulic plunger A and cylinder B are 6.5 in.^{2} (42.52 cm^2) and 650 in.^{2} (4,193.15 cm^2), respectively. The weight of B is 9,500 lb (4,313 kg). The vessel and the connecting passages are filled with water of specific gravity 1.0. What force P_f is required for hydraulic force equilibrium, neglecting the weight of A? Distance d = 16 ft = 4.88 m.

Solution 1 (US Customary System):

Determine the unit pressure acting on plunger A first. Since H_L and H_R are at the same level in water, then

Pressure at H_L(psi) = pressure at H_R(psi).

Pressure at H_L = pressure under A + 16 ft water.

Pressure at H_R = 9,500 lb/650 in.^{2} = 14.65 psi.

$P_f/(6.5\text{ in.}^2) + (62.4\text{ lb/ft}^3)(16\text{ft})/(144\text{ in.}^2/\text{ft}^2) = 9{,}500\text{ lb}/(650\text{ in.}^2)$.

$\boldsymbol{P_f}$ **= 50.18 lb**.

Solution 2 (SI System):

Determine the unit pressure acting on plunger A first. Since H_L and H_R are at the same level in water, then

Pressure at $H_L(\text{g/cm}^2)$ = pressure at $H_R(\text{g/cm}^2)$.

Pressure at H_L = pressure under A + 4.88 m water.

Pressure at H_R = 4,313,000 g/4,193.15 cm^2.

$\gamma = 1\text{ g/cm}^3 = 1\text{ kg/L} = 1\text{ kg}/1{,}000\text{ cm}^3$.

$P_f/(42.52\text{ cm}^2) + (1\text{ g/cm}^3)(488\text{ cm}) = 4{,}313{,}000\text{ g}/4{,}193.15\text{ cm}^2$.

$\boldsymbol{P_f}$ **= 22,985 g = 22.99 kg**.

Example 5.9 Determination of Hydraulic Force Exerted by Fluid in a Tank

Determine the hydraulic force developed on one of the vertical sides of a cubical tank with 4 ft (1.22 m) sides filled with water. Also determine the locations of the center of gravity C_g and the center of pressure C_p, as shown in Fig. 5.4.

Solution 1 (US Customary System):

Average pressure exerted on one side is pressure existent at C_g of the side times area A of the side. Since pressure varies linearly with depth, as shown in Fig. 5.4, the C_p (center of pressure) is located below the C_g (center of gravity). The C_p is the location where a single concentrated force P_f can be used to replace the triangle-shaped distributed pressure forces for moment purposes. The pressure existent at the area centroid times area equals force exerted normal to the area acting through the C_p.

$h_{cg} = 0.5 \times d = 0.5 \times 4\text{ ft} = 2\text{ ft}$.

$\boldsymbol{C_g}$ **= 2 ft down from water surface**.

$\boldsymbol{P_f} = (\gamma h_{cg})A = I_pA = (62.4\text{ lb/ft}^3)(2\text{ ft})(4\text{ ft} \times 4\text{ ft}) = \mathbf{2{,}000\text{ lb}}$.

$y_{cg} = h_{cg} = 2\text{ ft}$ in this particular case.

$b = 4\text{ ft}$.

$h = 4\text{ ft}$.

$I_{cg} = bh^3/12 = 4 \times 4^3/12 = 21.33\text{ ft}^4$.

$y_{cp} = [I_{cg}/y_{cg}A] + y_{cg} = [21.33\text{ ft}^4/(2\text{ ft} \times 4\text{ ft} \times 4\text{ ft})] + 2\text{ ft} = 2.67\text{ ft}$.

It is noted that 2.67/4 = 2/3.

$\boldsymbol{C_p}$ **= 2.67 ft down from water surface**.

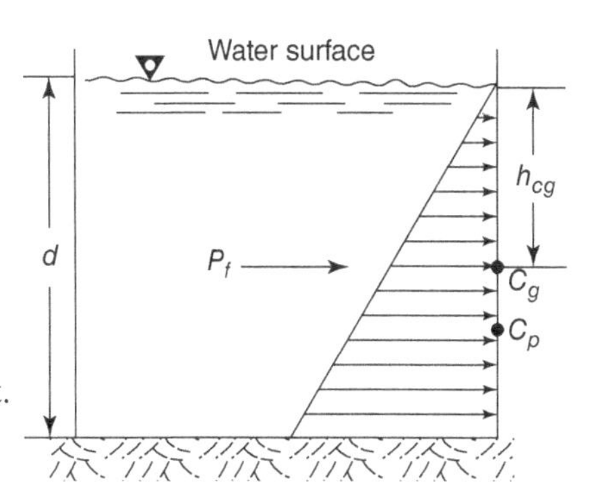

Figure 5.4 Cubical Tank with Four Equal Sides

Solution 2 (SI System):

Read the explanations in Solution 1.

$$h_{cg} = 0.5 \times d = 0.5 \times 1.22\ \text{m} = 0.61\ \text{m}.$$

$$\boldsymbol{C_g} = \textbf{0.61 m downward from water surface}.$$

$$\boldsymbol{P_f} = (\gamma h_{cg})A = I_p A = (9.8\ \text{kN/m}^3)(0.5 \times 1.22\ \text{m})(1.22\ \text{m} \times 1.22\ \text{m}) = \textbf{8.9 kN}.$$

$$y_{cg} = h_{cg} = 0.5 \times 1.22\ \text{m} = 0.61\ \text{m in this particular case.}$$

$$b = 1.22\ \text{m}.$$

$$h = 1.22\ \text{m}.$$

$$I_{cg} = bh^3/12 = 1.22 \times 1.22^3/12 = 0.185\ \text{m}^4.$$

$$y_{cp} = \left[I_{cg}/y_{cg}A\right] + y_{cg} = \left[0.185\ \text{m}^4/(0.61\ \text{m} \times 1.22\ \text{m} \times 1.22\ \text{m})\right] + 0.61\ \text{m} = 0.81\ \text{m}.$$

$$\boldsymbol{C_p} = \textbf{0.81 m downward from water surface}.$$

Example 5.10 Determination of Depth of Center of Pressure

Determine the depth to the center of pressure y_{cp} for a rectangular area $(b \times h)$ vertically submerged with the long side $(h = 8.2\ \text{ft} = 2.5\ \text{m})$ at the water surface.

Solution 1 (US Customary System):

Using Eqs. (5.8) and (5.9), $y_{cp} = [I_{cg}/y_{cg}A] + y_{cg}$, where $y_{cg} = h/2$, $A = b \times h$, and $I_{cg} = bh^3/12$.

$$y_{cp} = (bh^3/12)/[(h/2)(bh)] + (h/2) = 2\,h/3.$$

$$\boldsymbol{y_{cp}} = (2 \times 8.2\ \text{ft})/3 = \textbf{5.47 ft}.$$

Solution 2 (SI System):

Using Eqs. (5.8) and (5.9), $y_{cp} = [I_{cg}/y_{cg}A] + y_{cg}$, where $y_{cg} = h/2$, $A = b \times h$, and $I_{cg} = bh^3/12$.

$$y_{cp} = (bh^3/12)/[(h/2)(bh)] + (h/2) = 2h/3.$$

$$\boldsymbol{y_{cp}} = (2 \times 2.5\ \text{m})/3 = \textbf{1.67 m}.$$

Example 5.11 Determination of Hydrostatic Force and Location of Center of Pressure

Determine the hydrostatic force and the location of the center of pressure on the 92-ft (28-m)-long dam shown in Fig. 5.5. The face of the dam is at an angle of 60°. The water temperature is 77°F (25°C) and the depth is 26.25 ft (8 m).

Solution 1 (US Customary System):

From Appendix 3, $\gamma = 62.28\ \text{lb/ft}^3$ at 77°F.

$$\boldsymbol{P_f} = (\gamma h_{cg})A = (62.28\ \text{lb/ft}^3)(26.25\ \text{ft}/2)(92\ \text{ft})(26.25\ \text{ft}/\sin 60^\circ) = \textbf{2,279.54 lb}.$$

The center of pressure is at two-thirds of the total water depth, (2/3)(26.25 ft) = **17.5 ft.**

Solution 2 (SI System):

From Appendix 3, $\gamma = 9.779\ \text{kN/m}^3$ at 25°C.

$$\boldsymbol{P_f} = (\gamma h_{cg})A = (9.779\ \text{kN/m}^3)(8\ \text{m}/2)(28\ \text{m})(8\ \text{m}/\sin 60^\circ) = \textbf{10,118 kN}.$$

The center of pressure is at two-thirds of the total water depth, (2/3)(8 m) = **5.33 m.**

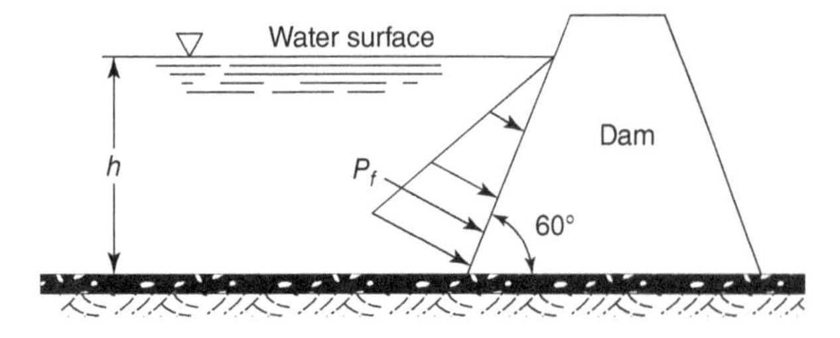

Figure 5.5 Hydrostatic Force on a Dam (Example 5.11)

Example 5.12 Determination of Resultant Force Due to Water Pressure Acting on Vertical Gate
Determine the resultant force and its location due to the water acting on a 4 ft by 8 ft (1.22 m by 2.44 m) rectangular gate AB shown in Fig. 5.6. The top of the gate is 16 ft (4.88 m) below the water surface. The dimensions in Fig. 5.6 are a = 16 ft = 4.88 m, h = 8 ft = 2.44 m, and b = 4 ft = 1.22 m.

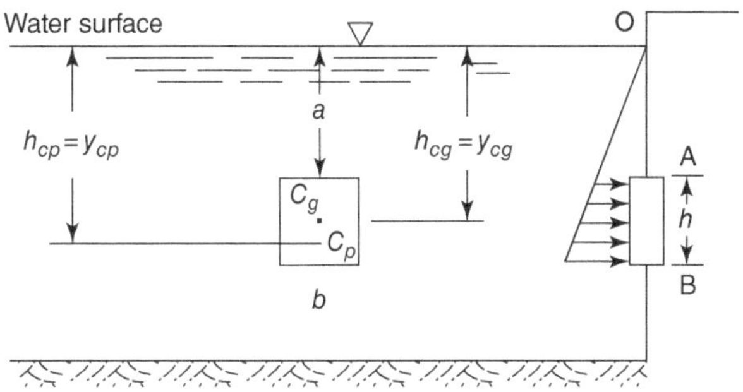

Figure 5.6 Vertical Rectangular Gate in a Water Tank (Example 5.12)

Solution 1 (US Customary System):

$$\boldsymbol{P_f} = (\gamma h_{cg})A = (62.4\ \text{lb/ft}^3)(16\ \text{ft} + 8\ \text{ft} \times 0.5)(4\ \text{ft} \times 8\ \text{ft}) = \mathbf{39{,}936\ lb}.$$

The above resultant force acts at the center of pressure, which is at a distance y_{cp} from water surface O.
At vertical position, $h_{cg} = y_{cg}$.

$$y_{cg} = 16\ \text{ft} + 8\ \text{ft} \times 0.5 = 20\ \text{ft}.$$
$$A = 4\ \text{ft} \times 8\ \text{ft} = 32\ \text{ft}^2.$$
$$y_{cp} = \left[I_{cg}/y_{cg}A\right] + y_{cg},\ \text{where}\ A = b \times h,\ \text{and}\ I_{cg} = bh^3/12.$$
$$y_{cp} = \left[I_{cg}/y_{cg}A\right] + y_{cg} = \left[(4 \times 8^3\ \text{ft}^4/12)/(20\ \text{ft} \times 32\ \text{ft}^2)\right] + 20\ \text{ft} = \mathbf{20.27\ ft\ from\ point}\ \text{O}.$$

Solution 2 (SI System):

$$P_f = (\gamma h_{cg})A = (9.8\ \text{kN/m}^3)(4.88\ \text{m} + 2.44\ \text{m} \times 0.5)(1.22\ \text{m} \times 2.44\ \text{m}) = \mathbf{177.95\ kN}.$$

The above resultant force acts at the center of pressure, which is at a distance y_{cp} from water surface O.
At vertical position, $h_{cg} = y_{cg}$.

$$y_{cg} = 4.88\ \text{m} + 2.44\ \text{m} \times 0.5 = 6.1\ \text{m}.$$
$$A = 1.22\ \text{m} \times 2.44\ \text{m} = 2.98\ \text{m}^2.$$
$$y_{cp} = \left[I_{cg}/y_{cg}A\right] + y_{cg},\ \text{where,}\ A = b \times h,\ \text{and}\ I_{cg} = bh^3/12.$$
$$\boldsymbol{y_{cp}} = \left[I_{cg}/y_{cg}A\right] + y_{cg} = \left[(1.22 \times 2.44^3\ \text{m}^4/12)/(6.1\ \text{m} \times 2.98\ \text{m}^2)\right] + 6.1\ \text{m} = \mathbf{6.18\ m\ from\ point}\ \text{O}.$$

Example 5.13 Determination of Resultant Force Due to Water Acting on an Inclined Rectangular Gate
Determine the total resultant force and its location acting on the gate shown in Fig. 5.7. The gate is 4.5 ft × 6.5 ft (1.37 m × 1.98 mm) in dimension. The apex of the triangle is at C. The gate top is 5.5 ft (1.68 m) below the water surface. The related dimensions in the figure are a = 5.5 ft = 1.68 m, h = 4.5 ft = 1.37 m, and b = 6.5 ft = 1.98 m.

Solution 1 (US Customary System):

$$h_{cg} = 5.5\ \text{ft} + 0.5(4.5\ \text{Cos}\ 60^\circ) = 6.625\ \text{ft}.$$
$$\boldsymbol{P_f} = (\gamma h_{cg})A = (62.4\ \text{lb/ft}^3)(6.625\ \text{ft})(4.5\ \text{ft} \times 6.5\ \text{ft}) = \mathbf{12{,}092\ lb}.$$
$$y_{cg} = h_{cg}/\text{Sin}\ 30^\circ = 6.625\ \text{ft}/0.5 = 13.25\ \text{ft, or}$$
$$y_{cg} = (5.5\ \text{ft}/\text{Cos}\ 60^\circ) + (4.5\ \text{ft}/2) = 13.25\ \text{ft}.$$
$$I_{cg} = bh^3/12 = (6.5 \times 4.5^3\ \text{ft}^4)/12 = 49.35\ \text{ft}^4.$$
$$y_{cp} = \left[I_{cg}/y_{cg}A\right] + y_{cg},\ \text{where}\ A = b \times h,\ \text{and}\ I_{cg} = bh^3/12.$$
$$y_{cp} = \left[I_{cg}/y_{cg}A\right] + y_{cg} = \left[49.35\ \text{ft}^4/(13.25\ \text{ft} \times 4.5\ \text{ft} \times 6.5\ \text{ft})\right] + 13.25\ \text{ft} = 13.38\ \text{ft}.$$
$$\boldsymbol{h_{cp}} = y_{cp}\ \text{Sin}\ 30^\circ = 13.38\ \text{ft} \times 0.5 = \mathbf{6.69\ ft}.$$

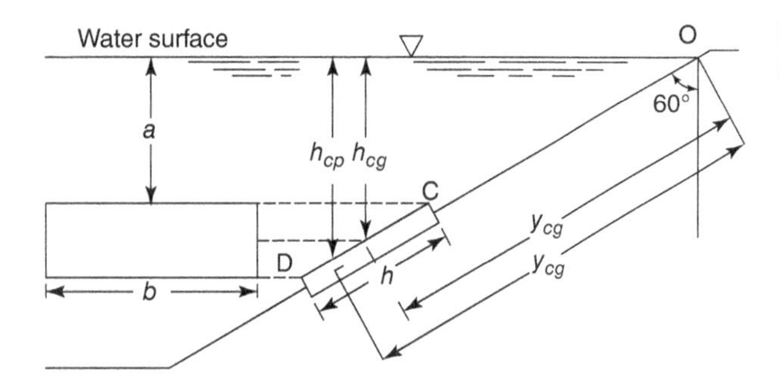

Figure 5.7 Inclined Rectangular Gate in a Water Tank (Example 5.13)

Solution 2 (SI System):

$$h_{cg} = 1.68\text{ m} + 0.5(1.37\text{ m Cos }60^{o}) = 2.02\text{ m}.$$
$$\boldsymbol{P_f} = (\gamma h_{cg})A = (9.79\text{ kN/m}^3)(2.02\text{ m})(1.37\text{ m} \times 1.98\text{ m}) = \mathbf{53.64\ kN}.$$
$$y_{cg} = h_{cg}/\text{Sin }30^{o} = 2.02\text{ m}/0.5 = 4.04\text{ m, or}$$
$$y_{cg} = (1.68\text{ m/Cos }60^{o}) + (1.37\text{ m}/2) = 4.04\text{ m}.$$
$$I_{cg} = bh^3/12 = (1.98 \times 1.37^3\text{ m}^4)/12 = 0.4243\text{ m}^4.$$
$$y_{cp} = \left[I_{cg}/y_{cg}A\right] + y_{cg}, \text{where } A = b \times h, \text{and } I_{cg} = bh^3/12.$$
$$y_{cp} = \left[I_{cg}/y_{cg}\text{A}\right] + y_{cg} = \left[0.4243\text{ m}^4/(4.04\text{ m} \times 1.37\text{ m} \times 1.98\text{ m})\right] + 4.04\text{ m} = 4.08\text{ m}.$$
$$h_{cp} = y_{cp}\text{ Sin }30^{o} = 4.08\text{ m} \times 0.5 = \mathbf{2.04\ m}.$$

Example 5.14 Water, Specific Weight, Force, and Moment of Inertia

Water rises to level E in the pipe attached to water storage tank ABCD in Fig. 5.8. Neglecting the weight of the tank and riser pipe, determine (a) the resultant force acting on area AB, which is 8 ft wide (2.4 m); (b) the location of the resultant force acting on area AB; (c) the pressure on the bottom BC; and (d) the total weight of the water.

Solution 1 (US Customary System):

a) The resultant force, P_f, acting on area AB:

$$\begin{aligned}\boldsymbol{P_f} &= (\gamma h_{cg})A \\ &= (62.4\text{ lb/ft}^3)(12\text{ ft} + 3\text{ ft})(6\text{ ft} \times 8\text{ ft}) \\ &= \mathbf{45{,}000\ lb}.\end{aligned}$$

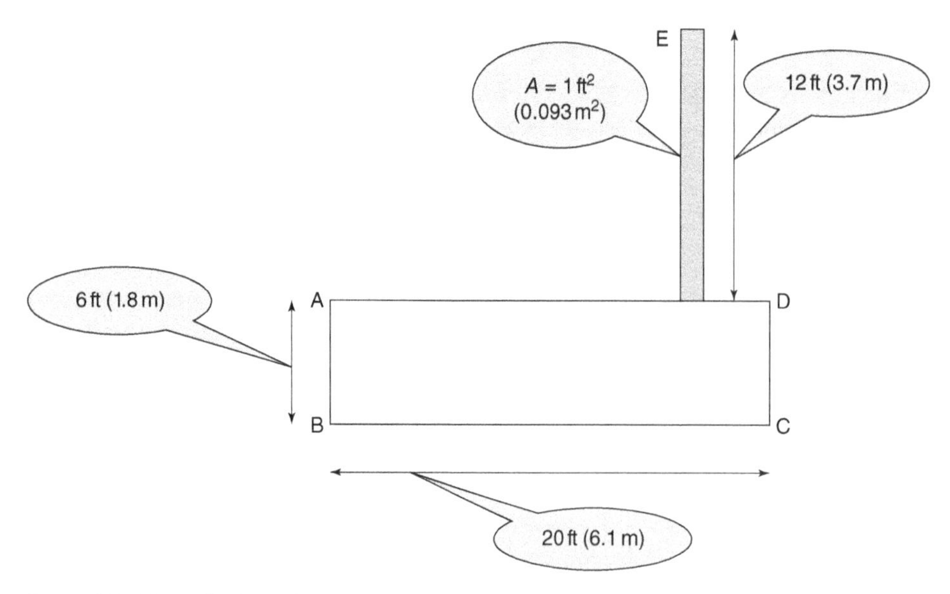

Figure 5.8 Water Tank for Example 5.14 (Example 5.14)

b) The location of the resultant force acting on area AB:

$$
\begin{aligned}
y_{cp} &= \left[I_{cg}/y_{cg}A\right] + y_{cg} \\
I_{cg} &= (1/12)bh^3 \\
y_{cp} &= \left[8(6)^3/12\right]/[15(6 \times 8)] + 15 \\
&= 0.20 + 15 \\
&= \mathbf{15.2\ ft.}
\end{aligned}
$$

c) The pressure, p, on the bottom BC:

$$
\begin{aligned}
p &= \gamma h \\
&= \left(62.4\ \text{lb/ft}^3\right)(6\ \text{ft} + 12\ \text{ft})/\left(144\ \text{in.}^2/\text{ft}^2\right) \\
&= \mathbf{7.8\ psi.}
\end{aligned}
$$

d) The total weight, W, of the water:

$$
\begin{aligned}
\boldsymbol{W} &= \gamma V \\
&= (62.4\ \text{lb/ft}^3)(20 \times 6 \times 8 + 12 \times 1)\text{ft}^3 \\
&= \mathbf{60{,}700\ lb.}
\end{aligned}
$$

Solution 2 (SI System):

a) Specific weight for water at 39°F (4°C) is 62.4 lb/ft^3 (9.81 kN/m^3). Specific weight in SI unit can be calculated as

$$
\begin{aligned}
\gamma &= (\rho)(g) = \left(1{,}000\ \text{kg/m}^3\right)\left(9.81\ \text{m/s}^2\right) \\
&= 9{,}810\ \text{N/m}^3 = 9.81\ \text{kN/m}^3.
\end{aligned}
$$

The resultant force, P_f, acting on area AB:

$$
\begin{aligned}
P_f &= (\gamma)(hcg)(A) \\
&= \left(9.81\ \text{kN/m}^3\right)(3.7\ \text{m} + 1.8\ \text{m}/2)(1.8\ \text{m} \times 2.44\ \text{m}) \\
&= \mathbf{198.2\ kN.}
\end{aligned}
$$

b) The location of the resultant force acting on area AB:

$$
\begin{aligned}
y_{cp} &= \left[I_{cg}/y_{cg}A\right] + y_{cg} \\
I_{cg} &= (1/12)bh^3 \\
y_{cp} &= \left[\left(bh^3/12\right)/y_{cg}A\right] + y_{cg} \\
&= \left(2.44 \times 1.8^3/12\right)/[(4.56)(1.8 \times 2.44)] + 4.56 \\
&= \mathbf{4.62\ m.}
\end{aligned}
$$

c) The pressure, p, on the bottom BC:

$$
\begin{aligned}
p &= \gamma h = \left(9.81\ \text{kN/m}^3\right)(1.8\ \text{m} + 3.7\ \text{m}) = 53.9\ \text{kN/m}^2 \\
&= \mathbf{53.9\ kPa.}
\end{aligned}
$$

d) The total weight, W, of the water:

$$
\begin{aligned}
W &= \rho V \\
&= \left(1{,}000\ \text{kg/m}^3\right)\left(6.1\ \text{m} \times 1.8\ \text{m} \times 2.44\ \text{m} + 3.7\ \text{m} \times 0.093\ \text{m}^2\right) \\
&= \mathbf{26{,}865\ kg.}
\end{aligned}
$$

5.1.2 Transmission Systems

Supply *conduits*, or *aqueducts*, transport water from the source of supply to the community and, hence, form the connecting link between collection works and distribution systems. The location of the source determines whether conduits are short or long, and whether transport is by *gravity* or *pumping*. Depending on topography and available materials, conduits are designed for *open channel* or *pressure flow*. They may follow the hydraulic grade line (HGL) as canals dug through the ground, flumes elevated above the ground, grade aqueducts laid in balanced cut and cover at the ground surface, and grade tunnels penetrating hills; or they may depart from the HGL as pressure aqueducts laid in balanced cut and cover at the ground surface, pressure tunnels dipping beneath valleys or hills, and pipelines of fabricated materials following the ground surface, if necessary over hill and through dale, sometimes even rising above the HGL. Illustrative examples of supply conduits and aqueducts include the following:

1) The *Central Arizona aqueduct* is a multipurpose water resource project that delivers Colorado River water into central and southern Arizona (see Fig. 1.6). The aqueduct diversion canal runs about 336 mi (541 km) from Lake Havasu to a terminus 14 mi (22.5 km) southwest of Tucson. The final extension to Tucson includes a tunnel through the mountains.
2) The *Colorado River aqueduct* of the Metropolitan Water District of Southern California is 242 mi (389 km) long and includes 92 mi (148 km) of grade tunnel, 63 mi (101 km) of canal, 54 mi (86.9 mi) of grade aqueduct, 29 mi (46.7 km) of inverted siphons, and 4 mi (6.4 km) of force main.
3) The *Delaware aqueduct* of New York City comprises 85 mi (137 km) of pressure tunnel in three sections.
4) *Pressure tunnels* 25 mi (40.2 km) long supply the metropolitan districts of Boston and San Francisco.
5) The *supply conduits* of Springfield, MA, are made of steel pipe and reinforced concrete pipe; those of Albany, NY, of cast iron pipe.
6) The *Nanzenji aqueduct* water channel of Kyoto, Japan (Fig. 5.9).

The profile and typical cross-sections of a supply conduit are shown in Fig. 5.10. Static heads and HGLs are indicated for pressure conduits.

Figure 5.9 Water Channel of the Nanzenji Aqueduct, Kyoto, Japan (*Source:* Tx-re/Wikimedia Commons)

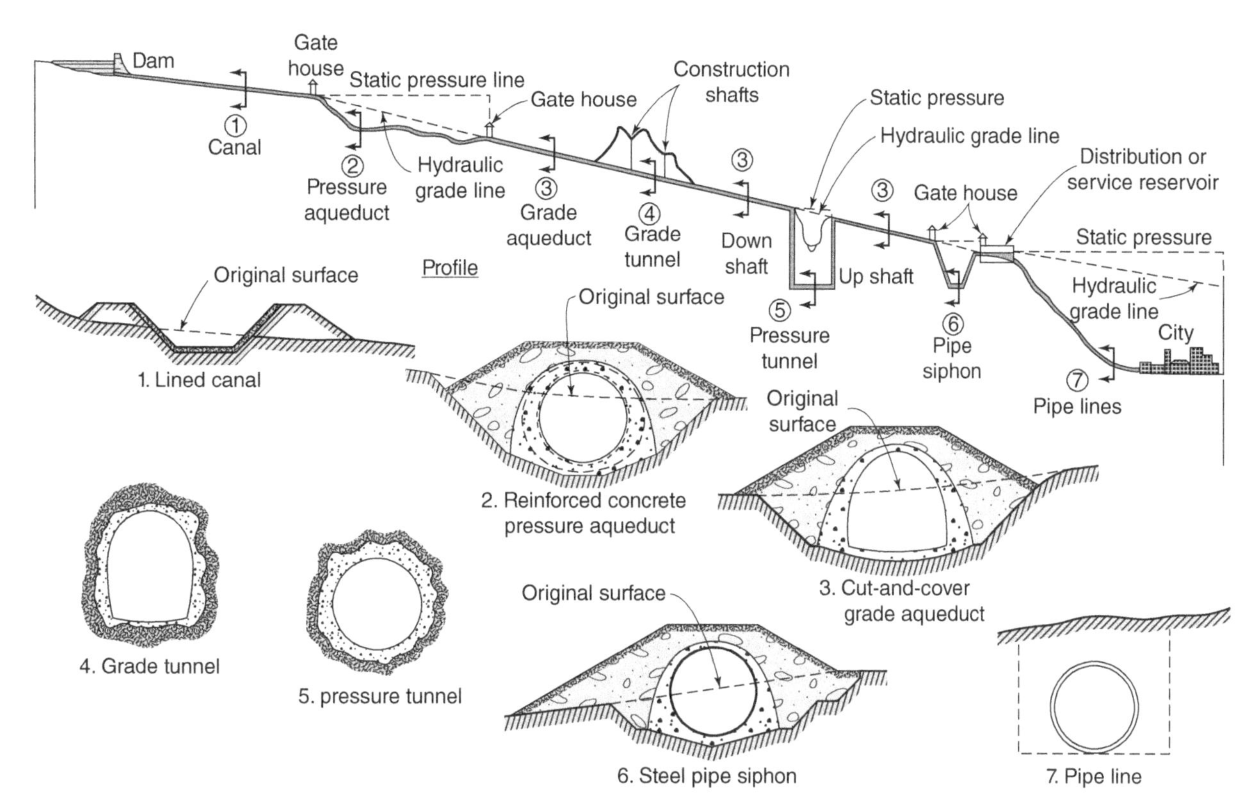

Figure 5.10 Profile and Typical Cross-sections of a Water Supply Conduit

5.2 Fluid Transport

The hydraulic design of supply conduits is concerned chiefly with (a) resistance to flow in relation to available and needed heads or pressures and (b) required and allowable velocities of flow relative to cost, scour, and sediment transport. In long supply lines, *frictional* or *surface resistance* offered by the pipe interior is the dominant element. *Form resistance* responsible for losses in transitions and appurtenances is often negligible. In short transport systems, on the other hand, form resistance may be of controlling importance.

5.2.1 Rational Equation for Surface Resistance

The most nearly rational relationship between velocity of flow and head loss in a conduit is also one of the earliest. Generally referred to as the *Darcy–Weisbach formula*, it is actually written in the form suggested by Weisbach, rather than Darcy, namely,

$$h_f = f(L/d)\left(v^2/2g\right) \tag{5.10a}$$

$$h_f = KQ^2 \tag{5.10b}$$

where h_f is the head loss in ft (m) (energy loss because of surface resistance) in a pipe of length L in ft (m) and diameter d in ft (m) through which a fluid is transported at a mean velocity v in ft/s (m/s) and flow rate Q in ft^3/s (m^3/s); g is the acceleration of gravity, 32.2 ft/s^2 (9.81 m/s^2); f is a dimensionless friction factor (see Fig. 5.11); and $K = 8fL/\pi^2 gd^5$. In the more than 100 years of its existence, use, and study, this formulation has been foremost in the minds of engineers concerned with the transmission of water as well as other fluids. That this has often been so in a conceptual rather than a practical sense does not detract from its importance.

Within Eq. (5.10a), the dimensionless friction factor f is both its strength and its weakness in applications—(a) its strength as a function of the *Reynolds number R*,

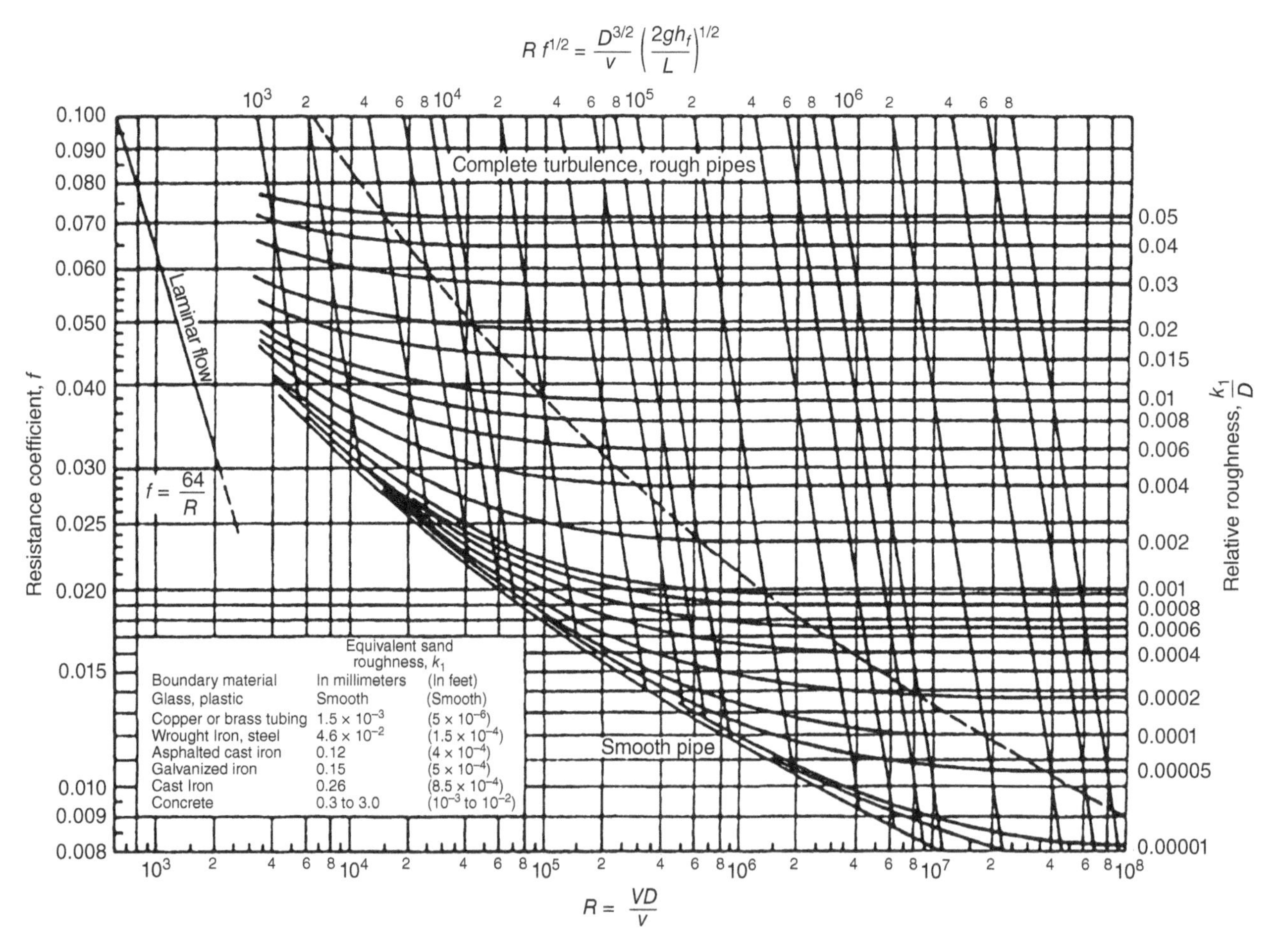

Figure 5.11 Moody Diagram with the Friction Factor f, as a Function of Reynolds Number R, in Darcy–Weisbach Formula for Flow in Conduits

$$\mathbf{R} = vd\rho/\mu = vd/\upsilon \tag{5.11}$$

where μ is the absolute viscosity, $\upsilon = \mu/\rho$ is the kinematic viscosity of the fluid, and ρ is its density; and (b) its weakness as a function of relative roughness ε/r, where ε is a measure of *absolute roughness* and r is the inside radius of the pipe ($2\varepsilon/d = \varepsilon/r$). The $f:R$ relationship is shown in Fig. 5.11, a general resistance diagram for flow in uniform conduits. This diagram evolves from a logarithmic plot of $1/\sqrt{f}$ against $\mathbf{R}/\sqrt{f}$ with scales for f and R added for convenience in finding f for use in Eq. (5.10a).

In reference to R and ε/r:

1) Laminar flow persists until $R = 2{,}000$, and the $f{:}R$ relationship is quite simply as follows:

$$f = 64/\mathbf{R} \tag{5.12a}$$

Reynolds number R and friction factor f are dimensionless.

2) Above $R = 4{,}000$, turbulent flow is fully established, and the single trace for laminar flow branches into a family of curves for increasing values of ε/r above a lower boundary that identifies the $f{:}R$ relationship for smooth pipes as

$$1/\sqrt{f} = 2\log \mathbf{R}\sqrt{f} - 0.8 \tag{5.13}$$

3) For rough pipes, the relative roughness ε/r takes command and

$$1/\sqrt{f} = 2 \log \varepsilon/r + 1.74 \tag{5.14}$$

where r is the radius of the pipe (ft or m) and ε/r is the relative roughness, dimensionless.

4) In the critical zone between $\mathbf{R} = 2{,}000$ and $\mathbf{R} = 4{,}000$, both $\mathbf{R}$ and ε/r make their appearance in the semiempirical equation of Colebrook and White (1937–1938):

$$1/\sqrt{f} = 1.74 - 2\log\left(\varepsilon/r + 18.7/R\sqrt{f}\right) \tag{5.15}$$

The magnitudes of absolute roughness ε in the $f:\mathbf{R}$ and $f:(\varepsilon/r)$ relationships depend on the angularity, height, and other geometrical properties of the roughness element and its distribution. Common magnitudes of ε/r have been evaluated for large pipes by Bradley and Thompson (1951).

Despite the logic and inherent conceptual simplicity of the combination of friction–factor diagram and Weisbach formulation, there are important reasons why water engineers do not make use of them for the routine solution of fluid transport problems encountered in water transmission lines and pipe networks. Among the reasons are the following:

1) Because the relative roughness ε/r is a key to f, it is not possible to find r (or d) directly when h_f, v or Q, ε, and water temperature (or ν) are given. A trial-and-error solution is required.
2) Because transmission lines may include noncircular as well as circular conduits, additional $f:\mathbf{R}$ diagrams are needed.
3) Because entry 2 also often applies to partially filled sections, additional diagrams are also necessary for them. For such sections, moreover, trial-and-error solutions must be performed whenever the depth of flow is unknown.

Example 5.15 Head Loss for Laminar Flow

This example introduces Poiseuille's equation for laminar flow when Reynolds number R is smaller than or equal to 2,000.

$$h_f = (32\mu Lv)/\left(\gamma d^2\right) = (32\upsilon Lv)/\left(gd^2\right) \tag{5.12b}$$

where

h_f = head loss, ft, m;

μ = absolute viscosity;

d = pipe diameter, ft, m;

γ = specific weight, 62.4 lb/ft^3, 9.8 kN/m^3 for water;

L = pipe length, ft, m;

v = average velocity, ft/s, m/s;

$\upsilon = \mu/\rho$ = kinematic viscosity; and

g = acceleration due to gravity, 32.2 ft/s^2, 9.81 m/s^2.

Develop Eqs. (5.10a) and (5.12a) using Poiseuille's equation (Eq. 5.12b) shown above.

Solution:

$h_f = (32\mu Lv)/\left(\gamma d^2\right) = (32\upsilon Lv)/\left(gd^2\right) = (32\upsilon Lv)/\left(gd^2\right)(2v/2v)$.

$h_f = (64\upsilon/vd)(L/d)\left(v^2/2g\right) = (64/\mathbf{R})(L/d)\left(v^2/2g\right) = f(L/d)\left(v^2/2g\right)$.

$h_f = f(L/d)\left(v^2/2g\right)$, which is Eq.(5.10a).

$f = 64/R$, which is Eq.(5.12a).

f and R are the resistance coefficient and the Reynolds number, respectively.

Example 5.16 Velocity, Reynolds Number, and Head Loss Under Laminar Flow Conditions

The elevation of reservoir A's water surface is 210 m (688.98 ft), while the elevation of reservoir B's water surface is 200 m (656.17 ft). The two reservoirs are connected by a pipe which is 610 m (2,001.31 ft) long and 5 mm (0.0164 ft) in diameter. Assume (a) water flows from reservoir A to reservoir B under laminar flow conditions; (b) $\gamma = 9.8\ \text{kN/m}^3 = 62.4\ \text{lb/ft}^3$; and (c) $\mu = 1.003 \times 10^{-3}\ \text{N-s/m}^2 = 2.05 \times 10^{-5}\ \text{lb-s/ft}^2$ at 20°C (68°F). Determine the velocity, the Reynolds number, and head loss per unit length.

Solution 1 (US Customary System):

$$h_f = 688.98\ \text{ft} - 656.17\ \text{ft} = 32.81\ \text{ft}.$$
$$d = 0.0164\ \text{ft}.$$
$$L = 2{,}001.31\ \text{ft}.$$
$$\mu = 2.05 \times 10^{-5}\ \text{lb-s/ft}^2\ \text{at}\ 68^{\circ}\text{F}.$$
$$\gamma = 62.4\ \text{lb/ft}^3$$
$$h_f = (32\ \mu L v)/(\gamma d^2).$$
$$32.81\ \text{ft} = \left[32(2.05 \times 10^{-5}\ \text{lb-s/ft}^2)(2{,}001.31\ \text{ft})(v\ \text{ft/s})\right]/\left[(62.4\ \text{lb/ft}^3)(0.0164\ \text{ft})^2\right].$$
$$\mathbf{v\ (velocity) = 0.41\ ft/s}.$$

Check the Reynolds number.

$$R = vd/\upsilon = vd\rho/\mu = vd\gamma/g\mu.$$
$$R = \left[(0.41\ \text{ft/s})(0.0164\ \text{ft})(62.4\ \text{lb/ft}^3)\right]/\left[(32.2\ \text{ft/s}^2)(2.05 \times 10^{-5}\ \text{lb-s/ft}^2)\right] = \mathbf{630}.$$

Flow is laminar because R is below 2,000.

$$Q = Av = \left[(0.0164\ \text{ft})^2 \times 0.785\right](0.41\ \text{ft/s}) = 8.66 \times 10^{-5}\ \text{ft}^3/\text{s}.$$
$$\boldsymbol{H_f} = h_f/L = 32.81\ \text{ft}/2{,}001.31\ \text{ft} = \mathbf{0.0164\ ft/ft}.$$

Solution 2 (SI System):

$$h_f = 210\ \text{m} - 200\ \text{m} = 10\ \text{m}.$$
$$d = 5\ \text{mm} = 0.005\ \text{m}.$$
$$L = 610\ \text{m}.$$
$$\mu = 1.003 \times 10^{-3}\ \text{N-s/m}^2\ \text{at}\ 20^{\circ}\text{C}. (\text{Note}: 1\ \text{N-s/m}^2 = 0.021\ \text{lb-s/ft}^2)$$
$$\gamma = 9.8\ \text{kN/m}^3 = 9.8 \times 1{,}000\ \text{N/m}^3.$$
$$\rho = 1{,}000\ \text{kg/m}^3.$$
$$h_f = (32\ \mu L v)/(\gamma d^2).$$
$$10\ \text{m} = \left[32(1.003 \times 10^{-3}\ \text{N-s/m}^2)(610\ \text{m})(v\ \text{m/s})\right]/\left[(9.8 \times 1{,}000\ \text{N/m}^3)(0.005\ \text{m})^2\right].$$
$$\mathbf{v\ (velocity) = 0.125\ m/s}.$$

Check the Reynolds number.

$$R = vd/\upsilon = vd\rho/\mu.$$
$$R = \left[(0.125\ \text{m/s})(0.005\ \text{m})(1{,}000\ \text{kg/m}^3)\right]/(1.003 \times 10^{-3}\ \text{N-s/m}^2) = \mathbf{630}.$$

Flow is laminar because R is below 2,000.

$$Q = Av = \left[(0.005\ \text{m})^2 \times 0.785\right](0.125\ \text{m/s}) = 2.5 \times 10^{-6}\ \text{m}^3/\text{s}.$$
$$\boldsymbol{H_f} = h_f/L = 10\ \text{m}/610\ \text{m} = \mathbf{0.0164\ m/m}.$$

Example 5.17 Head Loss Under Turbulent Flow Conditions

Compare Poiseuille's equation with Darcy–Weisbach formula, and discuss their applications for determining the head loss under turbulent and laminar flow conditions.

Solution:

From previous example, it can be seen that Poiseuille's equation (Eq. 5.12b) can be rearranged to have a form:

$$h_f = f(L/d)(v^2/2g) \quad (5.10a)$$

in which $f = 64/R$, which is Eq. (5.12a).

Poiseuille's equation is used for hydraulic analyses under laminar flow conditions, $R < 2{,}000$.

Darcy–Weisbach formula has the same general form:

$$h_f = f(L/d)(v^2/2g) \quad (5.10a)$$

in which $f = a/R^b$, and a and b are numerical constants. Darcy–Weisbach formula with $f = a/R^b$ is mainly used for determining the head loss of turbulent flows, $R > 4{,}000$. The condition of the pipe wall causes complex degrees of turbulence and different values of f for the same R. Equations (5.13) and (5.14) show the complexity of f values.

In the critical zone between $R = 2{,}000$ and $R = 4{,}000$, Eq. (5.15) represents the complexity of f values.

A Moody diagram shown in Fig. 5.11 can be used for determination of f values, in turn, for calculation of h_f using Eq. (5.10a), when the pipe roughness, and Reynolds number, are known.

Example 5.18 Head Loss Determination Using Moody Diagram

A 3,200-ft (975.4-m)-long cast iron pipeline with 12 in. (304.8 mm) diameter carries a water flow at a velocity of 15 ft/s (4.572 m/s) and temperature of 68°F (20°C). Determine the Reynolds number, the resistance coefficient f, and the head loss using Moody diagram.

Solution 1 (US Customary System):

v = kinematic viscosity of water at 68°F = 1.059×10^{-5} ft²/s.

$R = vd/v = (15\ \text{ft/s})(12/12\ \text{ft})/(1.059 \times 10^{-5}\ \text{ft}^2/\text{s}) = \mathbf{1.4 \times 10^6 > 2{,}000}$ **turbulent flow**.

$K_s = 8.5 \times 10^{-4}$ ft for cast iron pipe (from Fig.5.11).

$K_s/d = 8.5 \times 10^{-4}\ \text{ft}/(12/12)\text{ft} = 0.0008$.

From Moody diagram (Fig. 5.11), $\boldsymbol{f} = \mathbf{0.02}$.

$\boldsymbol{h_f} = f(L/d)(v^2/2g) = 0.02(3{,}200\ \text{ft}/1\ \text{ft})(15\ \text{ft/s})^2/[2(32.2\ \text{ft/s}^2)] = \mathbf{223.6\ ft}$.

Solution 2 (SI System):

v = kinematic viscosity of water at 20°C = 1.007×10^{-6} m²/s.

$\mathbf{R} = vd/v = (4.572\ \text{m/s})(0.3048\ \text{m})/(1.007 \times 10^{-6}\ \text{m}^2/\text{s}) = \mathbf{1.4 \times 10^6 > 2{,}000}$ **turbulent flow**.

$K_s = 0.26\ \text{mm} = 0.00026$ m for cast iron pipe (from Fig.5.11).

$K_s/d = 0.00026\ \text{m}/0.3048\ \text{m} = 0.00085$.

From Moody diagram (Fig. 5.11), $\boldsymbol{f} = \mathbf{0.02}$.

$\boldsymbol{h_f} = f(L/d)(v^2/2g) = 0.02(975.4\ \text{m}/0.3048\ \text{m})(4.572\ \text{m/s})^2/[2(9.81\ \text{m/s}^2)] = \mathbf{68.18\ m}$.

Example 5.19 Laminar or Turbulent Flow

Determine the type of flow occurring in a 16 in. (40.64 cm) pipe when (a) water at 60°F (15.55°C) flows at a velocity of 3.8 ft/s (1.1582 m/s); and (b) heavy fuel oil at 60°F (15.55°C) flows at the same velocity. At this temperature, kinematic viscosity is 1.217×10^{-5} ft²/s (1.1306×10^{-6} m²/s) for water, and 221×10^{-5} ft²/s (20.53×10^{-5} m²/s) for heavy oil.

Solution 1 (US Customary System):

$R = vd/v$ = (velocity) (diameter)/(kinematic viscosity).

$R = (3.8\ \text{ft/s})(16/12)\ \text{ft}/(1.217 \times 10^{-5}\ \text{ft}^2/\text{s}) = 416{,}324 > 2{,}000$ for water.

So the water flow is under turbulent flow condition.

$$R = (3.8\text{ ft/s})(16/12)\text{ ft}/\left(221 \times 10^{-5}\text{ ft}^2/\text{s}\right) = 2{,}292 > 2{,}000 \text{ for heavy oil.}$$

So the heavy oil is under turbulent flow condition, but very close to laminar flow condition.

Solution 2 (SI System):

$$R = vd/\upsilon = (\text{velocity})(\text{diameter})/(\text{kinematic viscosity}).$$
$$R = (1.1582\text{ m/s})(0.4064\text{ m})/\left(1.1306 \times 10^{-6}\text{ m}^2/\text{s}\right) = 416{,}324 > 2{,}000 \text{ for water.}$$

So the water flow is under turbulent flow condition.

$$R = (1.1582\text{ m/s})(0.4064\text{ m})/\left(20.53 \times 10^{-5}\text{ m}^2/\text{s}\right) = 2{,}292 > 2{,}000 \text{ for heavy oil.}$$

So the heavy oil is under turbulent flow condition, but very close to laminar flow condition.

Example 5.20 Pipe Sizing Under Laminar Flow Conditions

Determine the pipe diameter for delivering 96 gpm (363.3 L/min) of medium fuel oil at 41°F (5°C), assuming that the kinematic viscosity = 6.55×10^{-5} ft²/s = 6.085×10^{-6} m²/s.

Solution 1 (US Customary System):

$$Q = Av = (\text{area})\ (\text{velocity}) = \left(d^2 \times 0.785\text{ ft}^2\right)(v\text{ ft/s}) = 96\text{ gpm} = 0.2139\text{ ft}^3/\text{s}.$$
$$v = Q/A = \left(0.2139\text{ ft}^3/\text{s}\right)/\left(0.785\, d^2\text{ ft}^2\right) = 0.272/d^2\text{ ft/s}.$$
$$R = vd/\upsilon = (\text{velocity})\ (\text{diameter})/(\text{kinematic viscosity}).$$
$$R = \left(0.272/d^2\text{ ft/s}\right)(d\text{ ft})/\left(6.55 \times 10^{-5}\text{ ft}^2/\text{s}\right) = 2{,}000.$$

Pipe diameter = 2.07 ft.

Solution 2 (SI System):

$$Q = Av = (\text{area})\ (\text{velocity}) = \left(d^2 \times 0.785\text{ m}^2\right)(v\text{ m/s}) = 363.3\text{ L/min} = 0.3633/60\text{ m}^3/\text{s}.$$
$$v = Q/A = (0.0.3633/60)\text{m}^3/\text{s}/\left(0.785 d^2\text{ m}^2\right) = 0.0077/d^2\text{ m/s}.$$
$$R = vd/\upsilon = (\text{velocity})\ (\text{diameter})/(\text{kinematic viscosity}).$$
$$R = \left(0.0077/d^2\text{ m/s}\right)(d\text{ m})/\left(6.085 \times 10^{-6}\text{ m}^2/\text{s}\right) = 2{,}000.$$

Pipe diameter = 0.63 m.

Example 5.21 Head Loss Using Darcy–Weisbach Formula

Oil of absolute viscosity 0.0021 lb-s/ft² (0.1 N-s/m²) and specific gravity 0.851 flows through 10,000 ft (3,048 m) of 12 in. (300 mm) stainless-steel pipe at a flow rate of 1.57 ft³/s (0.0445 m³/s). Determine the Reynolds number, the friction factor, and the head loss in the pipe.

Solution 1 (US Customary System):

$$\begin{aligned}\text{Velocity}, v &= Q/A\\ &= \left(1.57\text{ ft}^3/\text{s}\right)/\left[(1\text{ ft})^2 \times 0.785\right]\\ &= \mathbf{2\text{ ft/s}}.\end{aligned}$$

$$\begin{aligned}\text{Reynolds number}, R &= vd\rho/\mu = vd\gamma/\mu g\\ &= (2\text{ ft/s})(1\text{ ft})\left(0.851 \times 62.4\text{ lb/ft}^3\right)/\left[\left(0.0021\text{ lb-s/ft}^2\right) \times \left(32.2\text{ ft/s}^2\right)\right]\\ &= \mathbf{1{,}570} < 2{,}000, \text{ which indicates a laminar flow.}\end{aligned}$$

$$\begin{aligned}\textbf{Friction factor}, f &= 64/\mathbf{R} \text{ under laminar flow conditions}\\ &= 64/1{,}570\\ &= \mathbf{0.041}.\end{aligned}$$

$$\textbf{Head loss, } h_f = f(L/d)(v^2/2g)$$
$$= (0.041)(10{,}000\ \text{ft}/1\ \text{ft})(2\text{ft/s})^2/(2 \times 32.2\ \text{ft/s}^2)$$
$$= \mathbf{25.5\ ft}.$$

Solution 2 (SI System):

$$\text{Velocity, } v = Q/A$$
$$= (0.0445\ \text{m}^3/\text{s})/[(0.300\ \text{m})^2 \times 0.785]$$
$$= \mathbf{0.62\ m/s}.$$

$$\textbf{Reynolds number, } \mathbf{R} = vd\rho/\mu$$
$$= (0.62\ \text{m/s})(0.300\ \text{m})(0.851 \times 1{,}000\ \text{kg/m}^3)/(0.1\ \text{N-s/m}^2)$$
$$= \mathbf{1{,}580} < 2{,}000, \text{which indicates a laminar flow.}$$

$$\textbf{Friction factor, } f = 64/\mathbf{R} \text{ under laminar flow conditions}$$
$$= 64/1{,}580$$
$$= \mathbf{0.041}.$$

$$\textbf{Head loss, } h_f = f(L/d)(v^2/2g)$$
$$= (0.041)(3{,}048\ \text{m}/0.300\ \text{m})(0.62\ \text{m/s})^2/(2 \times 9.81\ \text{m/s}^2)$$
$$= \mathbf{8.2\ m}.$$

Example 5.22 Ratio of Head Loss in a Pipe to that in a Perforated Pipe

Show that the head loss h_f in a pipe is equal to three times the head loss h_f in a perforated pipe, having the same length, diameter, and friction factor.

Take the flow in the imperforated pipe as Q_0; assume a straight-line variation of the flow Q with distance in the perforated pipe, with $Q = Q_0$ at the inlet of the pipe and $Q = 0$ at the end of the line. Consider only losses to pipe friction, and assume no variation in the value of f with a changing Q.

Solution:

$$h_f = f(L/d)(v^2/2g) \tag{5.10a}$$

Since

$$v = Q/A \tag{5.4}$$

then

$$h_f = f(L/D)(Q^2/2gA^2) \tag{5.16}$$

$$h_f = (f/D)(1/2gA^2)Q^2L$$
$$h_f = KQ^2L \tag{5.17}$$

where

$$K = (f/D)(1/2gA^2)$$

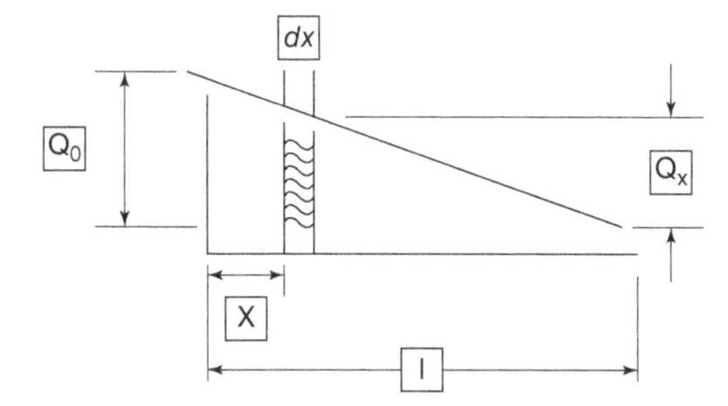

Taking a small length along the perforated pipe (see sketch) = dx, then

$$h_f = K\int_0^L Q_x^2\,dx$$

But

$$\frac{Q_x}{Q_0} = \frac{L-x}{L}$$

$$Q_x = Q_0\frac{L-x}{L}$$

So

$$h_f = K\frac{Q_0^2}{L^2}\int_0^L (L-x)^2\,dx.$$

$$h_f = K\frac{Q_0^2}{L^2}\int_0^L \left[\left(L^2 - 2xL + x^2\right)dx\right].$$

$$h_f = K\frac{Q_0^2}{L^2}\left[L^2x - x^2L + \frac{x^3}{3}\right]_0^L.$$

$$h_f = K\frac{Q_0^2}{L^2}\int_0^L\left[L^3 - L^3 + \frac{L^3}{3}\right].$$

$$h_f = KQ_0^2\frac{L}{3}.$$

Hence,

$$h_f = 1/3\,(h_f)_0.$$
$$(h_f)_0 = 3h_f.$$

That is, **the head loss h_f in a pipe is equal to three times the head loss h_f in a perforated pipe.**

Example 5.23 Head Loss Between Two Pressure Gauges

Water flows in a horizontal pipe of constant cross-sectional area. The pipe has a pressure gauge at location A with a pressure of 120 psig (832.8 kPa) and a pressure gauge at location B with a pressure of 90 psig (624.6 kPa). Determine the head loss between the two pressure gauges.

Solution 1 (US Customary System):

Use the energy equations (5.5a and 5.6b):

$$\left(P_A/\gamma + v_A^2/2g + Z_A\right) + H_a = \left(P_B/\gamma + v_B^2/2g + Z_B\right) + h_f.$$

Here $Z_A = Z_B$, and $H_a = 0$

Since the pipe size is constant, $v_A^2/2g = v_B^2/2g$.

The above energy equation reduces to

$$\boldsymbol{h_f} = P_A/\gamma - P_B/\gamma = \left(120\text{ lb/in.}^2 - 90\text{ lb/in.}^2\right)\left(144\text{ in.}^2/\text{ft}^2\right)/\left(62.4\text{ lb/ft}^3\right) = \mathbf{69.23\ ft}.$$

Solution 2 (SI System):

$$1\text{ kPa} = 1\text{ kN/m}^2.$$

Use the energy equations (5.5a and 5.6b),

$$\left(P_A/\gamma + v_A^2/2g + Z_A\right) + H_a = \left(P_B/\gamma + v_B^2/2g + Z_B\right) + h_f.$$

Here $Z_A = Z_B$, and $H_2 = 0$.
Since the pipe size is constant, $v_A^2/2g = v_B^2/2g$.
The above energy equation reduces to

$$h_f = P_A/\gamma - P_B/\gamma = (832.8\ \text{kN/m}^2 - 624.6\ \text{kN/m}^2)/(9.81\ \text{kN/m}^3)$$
$$= \mathbf{21.22\ m}.$$

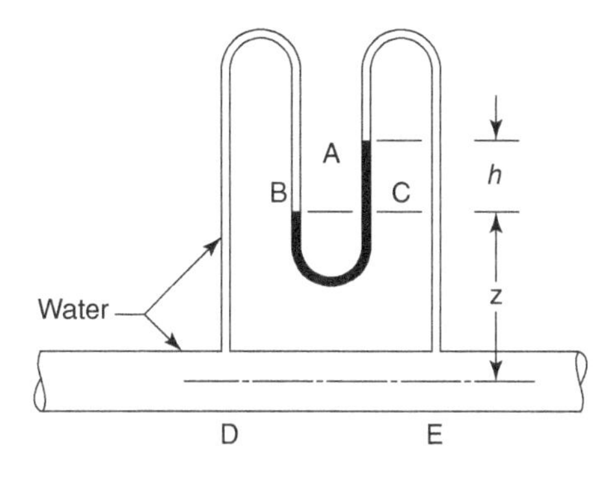

Figure 5.12 Differential Gauge Attached to a Horizontal Pipe (Example 5.24)

Example 5.24 Differential Gauge Measurement

A differential gauge shown in Fig. 5.12 is attached to two cross-sections D and E in a horizontal water pipe in which water is flowing. The deflection of the mercury h in the gauge is 2 ft (0.61 m), the level nearer D being the lower one. Determine the difference in lb/in.^{2} (kN/m^2) between sections D and E.

Solution 1 (US Customary System):

Pressure head at B = pressure head at C

$$P_D/\gamma - Z = P_E/\gamma - Z - 2\ \text{ft} + 13.57 \times 2\ \text{ft}.$$
$$P_D/\gamma - P_E/\gamma = \mathbf{25.4\ ft\ H_2O}.$$

Or $25.4 \times 0.433 =$ **11 psi**.

Solution 2 (SI System)

Pressure head at B = pressure head at C.

$$P_D/\gamma - Z = P_E/\gamma - Z - 0.61\ \text{m} + 13.57 \times 0.61\ \text{m}.$$
$$P_D/\gamma - P_E/\gamma = \mathbf{7.66\ m\ H_2O}.$$

$$7.66\ \text{m H}_2\text{O} = 7.66/0.1013 = \mathbf{75.69\ kN/m^2}.$$

Note: Since $(P_D/\gamma - P_E/\gamma)$ is positive, the pressure at D is larger than the pressure at E. For more accurate measurement, the differential gauges should have the air extracted from all tubing before gauge readings are taken. 13.57 is the specific gravity of mercury.

Example 5.25 Determination of Unbalanced Moment

Gate EBC shown in Fig. 5.13 is hinged at B and is 4.5 ft (1.37 m) wide. Neglecting the weight of the gate, determine the unbalanced moment due to water acting on the gate EBC. It is assumed that angle of the gate is 60°, $d = 8.5$ ft $= 2.59$ m, and $a = 3.5$ ft $= 1.07$ m.

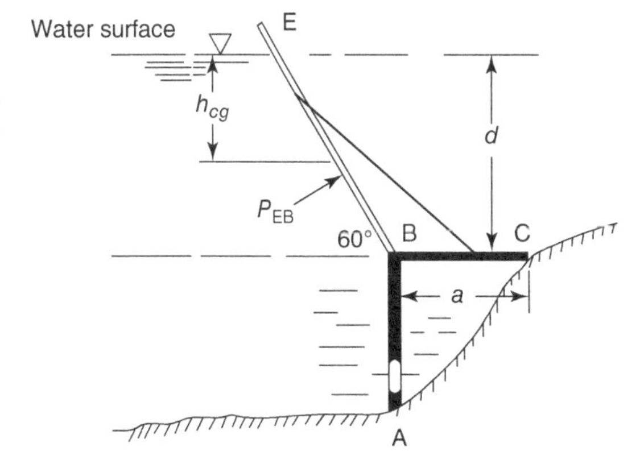

Figure 5.13 Analysis of Moment for a Hinged Gate (Example 5.25)

Solution 1 (US Customary System):

$$P_f = (\gamma h_{cg})A = P_{EB}.$$
$$EB = 8.5\ \text{ft}/\sin(60^\circ) = 9.81\ \text{ft}.$$
$$h_{cg} = 8.5/2 = 4.25\ \text{ft}.$$
$$A = 9.81\ \text{ft} \times 4.5\ \text{ft} = 44.2\ \text{ft}^2.$$
$$P_{EB} = (62.4\ \text{lb/ft}^3)(4.25\ \text{ft})(44.2\ \text{ft}^2) = 11{,}722\ \text{lb}.$$

Force $P_{EB} = 11{,}722$ lb acting at $(2/3) \times 9.81$ ft $= 6.54$ ft from E, and perpendicular to gate EB. $P_{BC} = (62.4\ \text{lb/ft}^3)(8.5\ \text{ft})(3.5\ \text{ft} \times 4.5\ \text{ft}) = 8{,}354$ lb, acting at the center of gravity of BC since the pressure on BC is uniform. Taking moment about B (clockwise plus)

$$\textbf{Unbalanced moment } M = 11{,}722\ \text{lb} \times (1/3) \times 9.81\ \text{ft} - 8{,}354\ \text{lb} \times (1/2) \times 3.5\ \text{ft}$$
$$= 38{,}331\ \text{lb-ft} - 14{,}620\ \text{lb-ft}$$
$$= \mathbf{23{,}711\ lb\text{-}ft\ clockwise}.$$

Solution 2 (SI System):

$P_f = (\gamma h_{cg})A = P_{EB}$.
$EB = 2.59\ \text{m}/\sin(60^{o}) = 2.99\ \text{m}$.
$h_{cg} = 2.59\ \text{m}/2 = 1.3\ \text{m}$.
$A = 2.99\ \text{m} \times 1.37\ \text{m} = 4.1\ \text{m}^2$.
$P_{EB} = (9.81\ \text{kN/m}^3)(1.3\ \text{m})(4.1\ \text{m}^2) = 52.23\ \text{kN}$.

Force $P_{EB} = 52.23$ kN acting at $(2/3) \times 2.99$ m $= 1.99$ m from E, and perpendicular to gate EB. $P_{BC} = (9.81\ \text{kN/m}^3)(2.59\ \text{m})(1.07\ \text{m} \times 1.37\ \text{m}) = 37.25$ kN, acting at the center of gravity of BC since the pressure on BC is uniform. Taking moment about B (clockwise plus)

Unbalanced moment $M = 52.23\ \text{kN} \times (1/3) \times 2.99\ \text{m} - 37.25\ \text{kN} \times (1/2) \times 1.07\ \text{m}$
$= (52.05 - 19.93)$ kN-m
$=$ **32.12 kN-m clockwise**.

Note: 1 kN-m = 737.6 lb-ft.

Example 5.26 Hydraulic Grade Line and Energy Grade Line
In Fig. 5.14, water flows from A to B at a flow rate of 15 ft^3/s (0.4245 m^3/s) and the pressure head at A is 25 ft (7.62 m). Determine the pressure head at B assuming no loss of energy from A to B, $d_A = 12$ in. (30.48 cm), $d_B = 24$ in. (60.96 cm), $Z_A = 10$ ft (3.048 m), $Z_B = 25$ ft (7.62 m).

Solution 1 (US Customary System):

$Q = (A_A)(v_A) = (A_B)(v_B)$.
$15\ \text{ft}^3/\text{s} = [(12/12)^2 \times 0.785\ \text{ft}^2](v_A\ \text{ft/s}) = [(24/12)^2 \times 0.785\ \text{ft}^2](v_B\ \text{ft/s})$.
$v_A = 19.1$ ft/s.
$v_B = 4.78$ ft/s.
$(v_A)^2/2g = (19.1\ \text{ft/s})^2/(2 \times 32.2\ \text{ft/s}^2) = 5.6$ ft.
$(v_B)^2/2g = (4.78\ \text{ft/s})^2/(2 \times 32.2\ \text{ft/s}^2) = 0.35$ ft.

Apply the Bernoulli equation from A to B.
Use Eq. (5.5a), in which H_a, H_1, and H_e are all zero.

$$[P_A/\gamma + (v_A)^2/2g + Z_A] + 0 - 0 - 0 = P_B/\gamma + (v_B)^2/2g + Z_B.$$

$$[25 + 5.6 + 10] = P_B/\gamma + 0.35 + 25 \text{ assuming datum D} - \text{D}.$$

$$[25 + 5.6 + 0] = P_B/\gamma + 0.35 + 15 \text{ assuming datum A}.$$

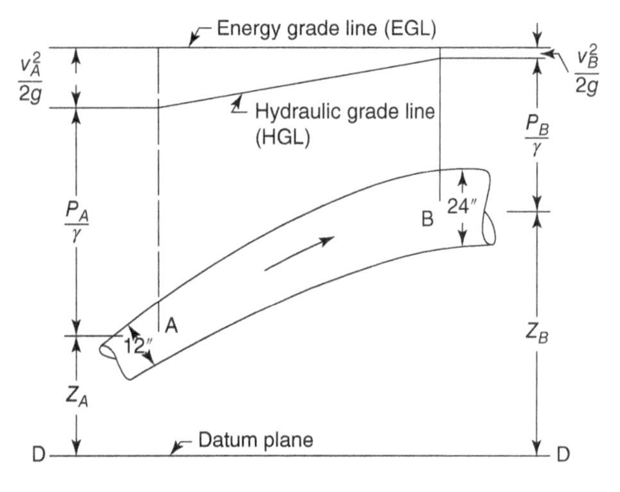

Figure 5.14 Hydraulic and Energy Grade Lines. (Example 5.26) Conversion Factor: 1″ = 1 in. = 2.54 cm

For either datum, we obtain $P_B/\gamma = 15.25$ ft of water.The total energy at any section can be plotted based on a chosen datum plane. Using D–D in this case,

$$\text{Energy at A} = P_A/\gamma + (v_A)^2/2g + Z_A = 25 + 5.6 + 10 = 40.6\ \text{ft}.$$

$$\textbf{Energy at B} = P_B/\gamma + (v_B)^2/2g + Z_B = 15.25 + 0.35 + 25 = \mathbf{40.6\ ft}.$$

It is noted that transformation from one form of energy to another occurs during flow. In this case a portion of both the pressure energy and the kinetic energy at A is transformed to potential energy at B.

Solution 2 (SI System):

$$Q = (A_A)(v_A) = (A_B)(v_B).$$
$$0.4245\ \text{m}^3/\text{s} = \left[(0.3048)^2 \times 0.785\ \text{m}^2\right](v_A\ \text{m/s}) = \left[(0.6096)^2 \times 0.785\ m^2\right](v_B\ \text{m/s}).$$
$$v_A = 5.82\ \text{m/s}.$$
$$v_B = 1.46\ \text{m/s}.$$
$$(v_A)^2/2g = (5.82)^2/(2 \times 9.8) = 1.72\ \text{m}.$$
$$(v_B)^2/2g = (1.46)^2/(2 \times 9.8) = 0.109\ \text{m}.$$

Apply the Bernoulli equation from A to B.
Use Eq. (5.5a), in which H_a, H_1, and H_e are all zero.

$$\left[P_A/\gamma + (v_A)^2/2g + Z_A\right] + 0 - 0 - 0 = P_B/\gamma + (v_B)^2/2g + Z_B.$$

$$[7.62 + 1.72 + 3.048] = P_B/\gamma + 0.109 + 7.62\ \text{assuming datum D – D}.$$

$$[7.62 + 1.72 + 0] = P_B/\gamma + 0.109 + 4.57\ \text{assuming datum A}.$$

For either datum, we obtain $P_B/\gamma = 4.66$ m of water.
The total energy at any section can be plotted based on a chosen datum plane. Using D–D in this case,

$$\text{Energy at A} = P_A/\gamma + (v_A)^2/2g + Z_A = 7.62 + 1.72 + 3.048 = 12.39\ \text{m}.$$

$$\textbf{Energy at B} = P_B/\gamma + (v_B)^2/2g + Z_B = 4.66 + 0.109 + 7.62 = \mathbf{12.39\ m}.$$

It is noted that transformation from one form of energy to another occurs during flow. In this case a portion of both the pressure energy and the kinetic energy at A is transformed to potential energy at B.

Example 5.27 Venturi Meter Analysis

The deflection of the mercury in the differential gauge of a venturi meter is $y = 16.5$ in. (41.91 cm) shown in Fig. 5.15.

Determine the water flow through the venturi meter if no energy is lost between A and B, and the following are known: $d_A = 12$ in. (30.48 cm), $d_B = 6$ in. (15.24 cm), $x = 36$ in. (91.44 cm), $y = 16.5$ in. (41.91 cm), and $z =$ unknown.

Solution 1 (US Customary System):

Apply the Bernoulli equation from A to B and use A as the datum:

$$\frac{P_A}{\gamma} + \frac{v_A^2}{2g} + 0 = \frac{P_B}{\gamma} + \frac{v_B^2}{2g} + \left(\frac{36}{12}\right).$$
$$\frac{P_A}{\gamma} - \frac{P_B}{\gamma} = \frac{v_B^2}{2g} - \frac{v_A^2}{2g} + 3.$$

From continuity equation,

$$Q = A_A v_A = A_B v_B.$$

$$(\pi/4)(12/12)^2 v_A = (\pi/4)(6/12)^2\ v_B.$$

$$4v_A = v_B.$$

$$v_A = 0.25\ v_B.$$

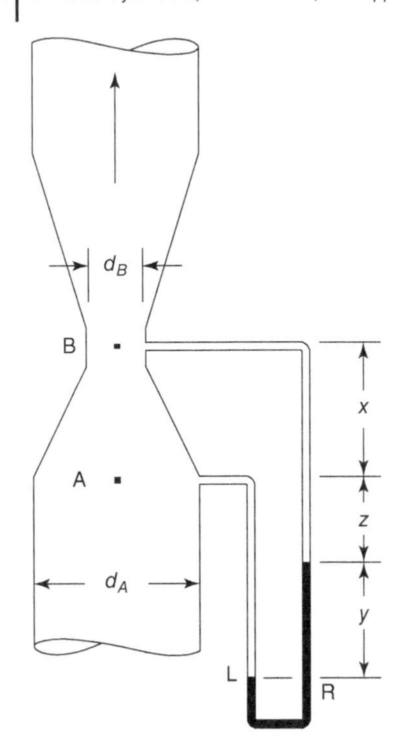

Figure 5.15 Venturi Meter Analysis for Example 5.27

Apply the Bernoulli equation for L to R (s.g. mercury, Hg = 13.6):

$$\frac{P_A}{\gamma} + z + \frac{16.5}{12} = \frac{P_B}{\gamma} + \frac{36}{12} + z + \frac{16.5}{12} \times 13.6.$$

$$\frac{P_A}{\gamma} - \frac{P_B}{\gamma} = 20.3 \text{ ft of water.}$$

$$20.3 = \frac{v_B^2}{2g} - \frac{(0.25\, v_B)^2}{2g} + 3.$$

Since g = 32.2 ft/s^2

$$v_B = 34.49 \text{ ft/s}.$$

$$Q = A_B v_B = (6/12)^2(\pi/4)(34.49) = \mathbf{6.77\ ft^3/s}.$$

Solution 2 (SI System):

Apply the Bernoulli equation from A to B and use A as the datum:

$$\frac{P_A}{\gamma} + \frac{v_A^2}{2g} + 0 = \frac{P_B}{\gamma} + \frac{v_B^2}{2g} + 0.9144.$$

$$\frac{P_A}{\gamma} - \frac{P_B}{\gamma} = \frac{v_B^2}{2g} - \frac{v_A^2}{2g} + 0.9144.$$

From continuity equation

$$Q = A_A v_A = A_B v_B.$$

$$(\pi/4)(0.3048)^2 v_A = (\pi/4)(0.1524)^2 v_B.$$

$$4\, v_A = v_B.$$

$$v_A = 0.25 v_B.$$

Apply the Bernoulli equation for L to R (s.g. mercury, Hg = 13.6):

$$\frac{P_A}{\gamma} + z + 0.4191 = \frac{P_B}{\gamma} + 0.9144 + z + 0.4191 \times 13.6.$$

$$\frac{P_A}{\gamma} - \frac{P_B}{\gamma} = 6.195 \text{ m of water.}$$

$$6.195 = \frac{v_B^2}{2g} - \frac{(0.25 v_B)^2}{2g} + 0.9144.$$

Since g = 9.81 m/s^2

$$v_B = 10.51 \text{ m/s}.$$

$$Q = A_B v_B = (0.1524)^2(\pi/4)(10.51) = \mathbf{0.192\ m^3/s}.$$

Example 5.28 Buoyancy and Flotation

A body, floating or immersed in a liquid, is acted upon by a buoyant force equal to the weight of the liquid displaced. The buoyancy center is the point through which this buoyant force acts. It is located at the center of gravity of the displaced liquid. If the center of gravity of the body lies directly below the center of buoyancy (gravity) of the displaced liquid, the submerged body is stable. If the two points coincide, the submerged body is in neutral equilibrium position.

Determine the volume of a metal body and its specific gravity if the body weighs 100 lb (45.4 kg) in air and weighs 60 lb (27.2 kg) when immersed in water.

Solution 1 (US Customary System):

Buoyant force = (Weight in air) – (Weight in water)
= 100 lb – 60 lb = 40 lb.

Buoyant force = Weight of displaced liquid.

$40 \text{ lb} = \gamma V = (62.4 \text{ lb/ft}^3)V.$

$\mathbf{V = 0.641 \text{ ft}^3}$.

Specific gravity = (Weight of metal body)/(Weight of an equal volume of water)
= 100 lb/40 lb
= **2.5**.

Solution 2 (SI System):

Buoyant force = (Weight in air) – (Weight in water)
= 45.4 kg – 27.2 kg = 18.2 kg.

Buoyant force = Weight of displaced liquid.

$18.2 \text{ kg} = \gamma V = (1 \text{ kg/L})V.$

$\mathbf{V = 18.2 \text{ L}}$.

Specific gravity = (Weight of metal body)/(Weight of an equal volume of water)
= 45.4 kg/18.2 kg
= 2.5.

Example 5.29 Effect of a Floating Object on Water Depth

A floating cylinder 8 cm in diameter and weighing 960 g is placed in a cylindrical container 26 cm in diameter partially full of water. Determine the increase in the depth of water in the container due to placing the float in it.

Solution:

A 960 g cylinder will displace 960 g of water. Since 1 g of water occupies 1 cm^3 volume, the cylinder will displace 960 cm^3 of water. The change in total volume beneath the water surface ΔV equals the area of the cylindrical container A, times the change in water level Δh, or

$$\Delta V = A\Delta h$$
$$\Delta h = \Delta V/A$$
$$= 960 \text{ cm}^3/\left[\pi(26)^2 \text{ cm}^2\right]$$
$$= \mathbf{1.81 \text{ cm}}.$$

5.2.2 Exponential Equation for Surface Resistance

Because of practical shortcomings of the Weisbach formula, engineers have resorted to the so-called exponential equations in flow calculations. Among them the Chezy formula is the basic for all:

$$v = C\sqrt{rs} \tag{5.18a}$$

where v is average velocity, ft/s; C is coefficient; r is hydraulic radius, which is defined as the cross-section area divided by the wetted perimeter, ft; and s is the slope of water surface or energy gradient:

$$r = A/P_w \tag{5.19}$$

$$s = h_f/L \tag{5.20}$$

where r is hydraulic radius, ft or m; A is cross-section area, ft^2 or m^2; P_w is wetted perimeter, ft or m; s = slope of water surface, dimensionless; and L is pipe length, ft or m.

The coefficient C can be obtained by using one of the following expressions:

$$\text{Chezy expression: } C = (8g/f)^{0.5} \tag{5.21}$$

$$\text{Manning expression: } C = (1.486/n)(r)^{1/6} \tag{5.22}$$

$$\text{Bazin expression: } C = 157.6/\left(1 + mr^{-0.5}\right) \tag{5.23}$$

$$\text{Kutter expression: } C = (41.65 + 0.00281/s + 1.811/n)/\left[1 + \left(n/r^{0.5}\right)(41.65 + 0.00281/s)\right] \tag{5.24}$$

In the preceding expressions, f, n, and m are the friction or roughness factors determined by hydraulic experiments using water only. Of the above hydraulic equations, Robert Manning's expression is commonly used for both open channels and closed conduits. Combining Eqs. (5.18a) and (5.22) will give the following Manning formula using US customary units:

$$v = (1.486/n)(r)^{0.67}(s)^{0.5} \quad \text{(US customary units)} \tag{5.25}$$

where v is velocity, ft/s; n is the coefficient of roughness, dimensionless; r is hydraulic radius, ft; and s is the slope of energy grade line, ft/ft.

Equation (5.26) is the equivalent Manning formula using SI units:

$$v = (1/n)(r)^{0.67}(s)^{0.5} \quad \text{(SI units)} \tag{5.26}$$

where v is velocity, m/s; n is the coefficient of roughness, dimensionless; r is hydraulic radius, m; and s is the slope of energy grade line, m/m.

For a pipe flowing full, the hydraulic radius, Eq. (5.19) becomes

$$r = \left[(\pi/4)D^2\right]/(\pi D) = D/4 \tag{5.27}$$

where r is hydraulic radius, ft or m; and D is pipe diameter, ft or m.

Substituting for r into Eqs. (5.25) and (5.26), the following Manning equations are obtained for practical engineering designs for circular pipes flowing full:

$$v = (0.59/n)(D)^{0.67}(s)^{0.5} \quad \text{(US customary units)} \tag{5.28}$$

$$Q = (0.46/n)(D)^{2.67}(s)^{0.5} \quad \text{(US customary units)} \tag{5.29}$$

where Q is flow rate, ft^3/s; v is velocity, ft/s; D is pipe diameter, ft; s is the slope of energy grade line, ft/ft and n is roughness coefficient, dimensionless. For SI measurements:

$$v = (0.40/n)(D)^{0.67}(s)^{0.5} \text{ (SI units)} \tag{5.30}$$

$$Q = (0.31/n)(D)^{2.67}(s)^{0.5} \text{ (SI units)} \tag{5.31}$$

where Q is flow rate, m^3/s; v is velocity, m/s; D is pipe diameter, m; s is the slope of energy grade line, m/m; and n is roughness coefficient, dimensionless.

The *Hazen–Williams formula* is most widely used in the United States to express flow relations in pressure conduits or conduits flowing full, the *Manning formula* in free-flow conduits or conduits not flowing full. The Hazen–Williams formula, which was proposed in 1905, is discussed in another author's sister book entitled *Water Supply and Wastewater Removal*, 3rd ed., John Wiley & Sons (2011).

The following notation is used for the US customary units: Q is the rate of discharge, gpm, gpd, MGD, or ft^3/s as needed; d is the diameter of small circular conduits, in.; D is the diameter of large circular conduits, ft; v is mean velocity, ft/s; $a = \pi D^2/4 = \pi d^2/576$ is the cross-sectional area of conduit, ft^2; $r = a$/wetted perimeter $= D/4 = d/48 =$ hydraulic radius, ft; h_f is the loss of head, ft; L is conduit length, ft; and $s = h_f/L$ is hydraulic gradient, dimensionless.

There is another set of notation to be used for the SI units: Q is the rate of discharge, m^3/s; d is the diameter of small circular conduits, mm; D is the diameter of large circular conduits, m; v is mean velocity, m/s; a is the cross-sectional area of conduit, m^2; r is hydraulic radius, m; h_f is the loss of head, m; L is conduit length, m; and $s = h_f/L$ is hydraulic gradient, dimensionless.

As written by the authors, the Hazen–Williams formula

$$v = Cr^{0.63}s^{0.54}\left(0.001^{-0.04}\right) \tag{5.31}$$

$$v = 1.318\ Cr^{0.63}s^{0.54} \quad \text{(US customary units)} \tag{5.32}$$

$$v = 0.849\ Cr^{0.63}s^{0.54} \quad \text{(SI units)} \tag{5.33}$$

where C is a coefficient known as the Hazen–Williams coefficient, and the factor $(0.001^{-0.04}) = 1.318$ makes C conform in general magnitude with established values of a similar coefficient in the more-than-a-century-older Chezy formula.

For circular conduits, the Hazen–Williams formulation can take one of the following forms:

$$\begin{aligned} v &= 0.115\ Cd^{0.63}s^{0.54} \quad \text{(US customary units)} \\ v &= 0.550\ CD^{0.63}s^{0.54} \quad \text{(US customary units)} \end{aligned} \tag{5.34}$$

$$\begin{aligned} v &= 0.3545\ CD^{0.63}s^{0.54} \quad \text{(SI units)} \\ h_f &= 5.47\ (v/C)^{1.85}L/d^{1.17} \quad \text{(US customary units)} \end{aligned} \tag{5.35}$$

$$\begin{aligned} h_f &= 3.02\ (v/C)^{1.85}L/D^{1.17} \quad \text{(US customary units)} \\ h_f &= 6.81\ (v/C)^{1.85}L/D^{1.17} \quad \text{(SI units)} \\ Q_{gpd} &= 405\ Cd^{2.63}s^{0.54} \quad \text{(US customary units)} \end{aligned} \tag{5.36}$$

$$\begin{aligned} Q_{MGD} &= 0.279\ CD^{2.63}s^{0.54} \quad \text{(US customary units)} \\ Q_{\mathrm{ft^3/s}} &= 0.432\ CD^{2.63}s^{0.54} \quad \text{(US customary units)} \\ Q_{\mathrm{m^3/s}} &= 0.278\ CD^{2.63}s^{0.54} \quad \text{(SI units)} \\ h_f &= 1.50 \times 10^{-5}\left(Q_{gpd}/C\right)^{1.85}L/d^{4.87} \quad \text{(US customary units)} \end{aligned} \tag{5.37}$$

$$\begin{aligned} h_f &= 10.6\ (Q_{MGD}/C)^{1.85}L/D^{4.87} \quad \text{(US customary units)} \\ h_f &= 4.72\left(Q_{\mathrm{ft^3/s}}/C\right)^{1.85}L/D^{4.87} \quad \text{(US customary units)} \\ h_f &= 10.67\left(Q_{\mathrm{m^3/s}}/C\right)^{1.85}L/D^{4.87} \quad \text{(SI units)} \\ h_f &= KQ^{1.85} \quad \text{(US customary units or SI units).} \end{aligned}$$

Note that in this Hazen–Williams relationship, the head loss is proportional to the flow raised to the power of 1.85, whereas in Darcy–Weisbach relationship (Eq. 5.10b) the head loss is proportional to the square of the flow.

Solutions of Eqs. (5.32) and (5.34)–(5.37) for Q, v, r, D, d, s, h_f, L, or C requires the use of logarithms, a log–log slide rule, tables, a diagram with logarithmic scales (see Figs. 5.16a and b), an alignment chart (Fig. 5.17 and Appendix 14) or a computer program. For C other than 100, one can multiply given Q or v in Fig. 5.17 by $(100/C)$ to find s, or can multiply found value of Q or v in Fig. 5.17 by C/100 for given s. Use of the Hazen–Williams pipe flow diagram (Fig. 5.16) is explained in Fig. 5.16c. The weakest element in the Hazen–Williams formula is the estimate of C in the absence of measurements of loss of head and discharge or velocity. Values of C vary for different conduit materials and their relative deterioration in service. They vary somewhat also with size and shape. The values listed in Table 5.1 reflect more or less general experience.

For purposes of comparison, the size of a noncircular conduit can be stated in terms of the diameter of a circular conduit of equal carrying capacity. For identical values of C and s, multiplication of Eq. (5.32) by the conduit area a and equating the resulting expression to Eq. (5.35), the diameter of the equivalent conduit becomes

$$D = 1.53a^{0.38}r^{0.24}. \tag{5.38}$$

Equation (5.38) is applicable to both the US customary system and the SI system. The units for the above equation are D (ft or m), a (ft^2 or m^2), and r (ft or m).

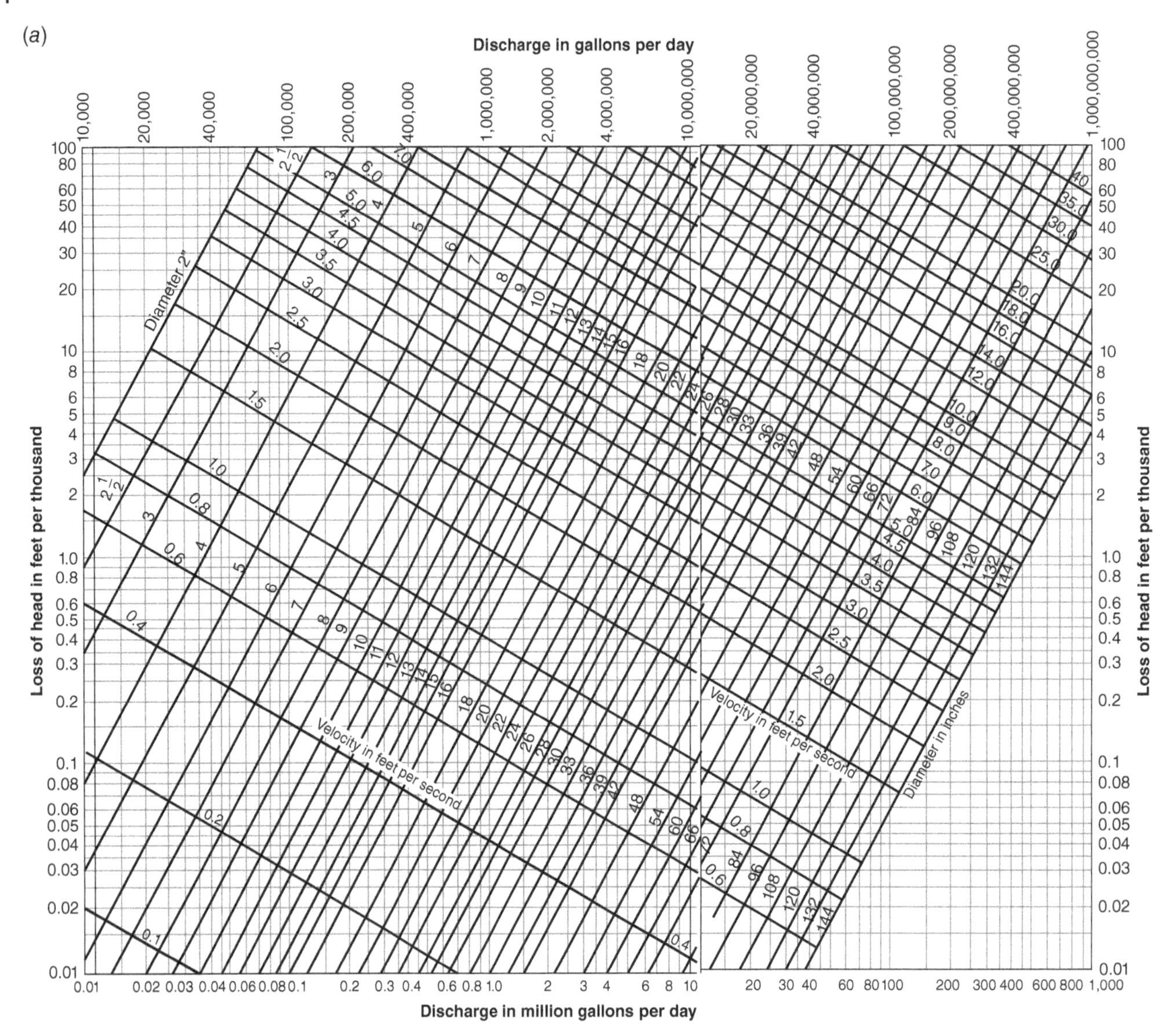

Figure 5.16 (a) Hazen–Williams Pipe Flow Diagram for Discharge of 0.01–10 MGD for $C = 100$. Conversion Factors: 1 MGD = 3.785 MLD = 43.8 L/s; 1 gpd = 3.785 L/d; 1 ft/1,000 ft = 1 m/1,000 m = 1‰; 1 in. = 25.4 mm; 1 ft/s = 0.3048 m/s. (b) Use of Hazen–Williams pipe flow diagram (a): (i) given Q and d, to find s; (ii) given d and s, to find Q; (iii) given d and s, to find v; (iv) given Q and s, to find d; (v) given Q and h, to find Q for different h; (vi) given Q and h, to find h for different Q. For C other than 100: (a) Multiply given Q or v by (100/C) to find s; (b) Multiply found value of Q or v by (C/100) for given s. Conversion factors: 1″ = 1 in. = 25.4 mm; 1 ft/s = 0.3408 m/s.

Example 5.30 Metric Open Channel Equations

Compare the US customary system's open channel equations with the SI system's equations.

Solution:

Chezy formula, in terms of the US customary units, is presented below:

$$v = C\sqrt{rs} \tag{5.18a}$$

where v is average velocity, ft/s; C is coefficient, dimensionless; r is hydraulic radius, ft; and s is the slope of water surface or energy gradient, dimensionless.

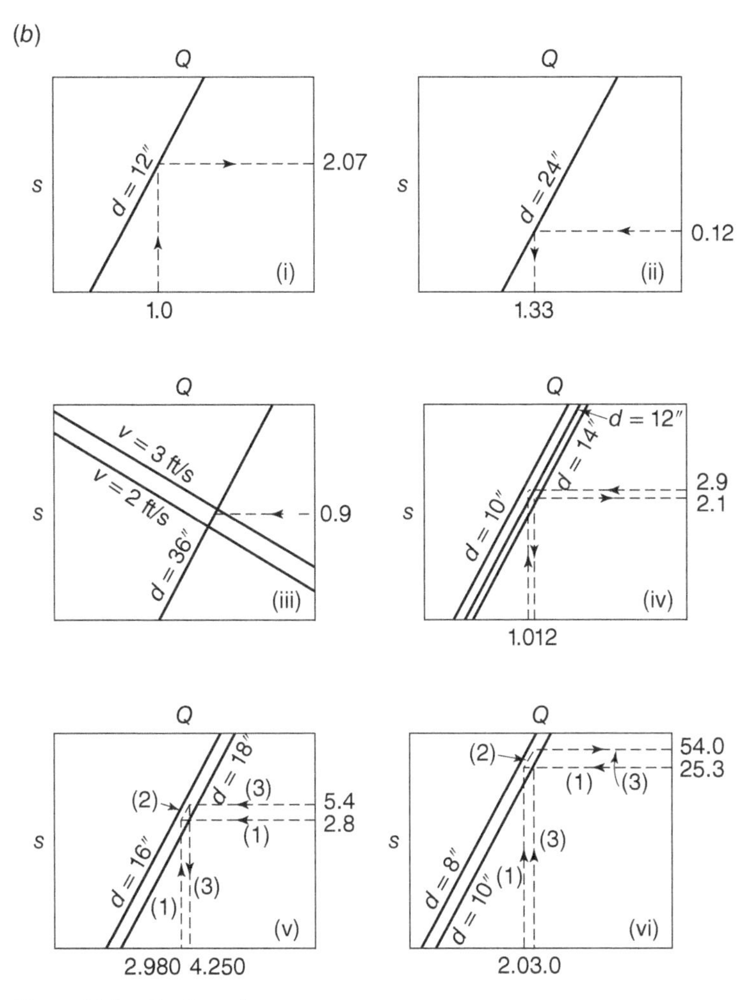

Figure 5.16 (Continued)

The equivalent Chezy formula using the SI system is

$$v = 0.552C\sqrt{rs} \tag{5.18b}$$

where v is average velocity, m/s; C is coefficient, dimensionless; r is hydraulic radius, m; and s is the slope of water surface or energy gradient, dimensionless.

Substituting Eq. (5.22) into Eq. (5.18a), one obtains the Manning's US customary equation:

$$v = \frac{1.486}{n}(r)^{0.67}(s)^{0.5} \tag{5.25}$$

Its equivalent SI equation is

$$v = \frac{1}{n}(r)^{0.67}(s)^{0.5} \tag{5.26}$$

where v is average velocity, m/s; C is coefficient, dimensionless; r is hydraulic radius, m; and s is the slope of water surface or energy gradient, dimensionless.

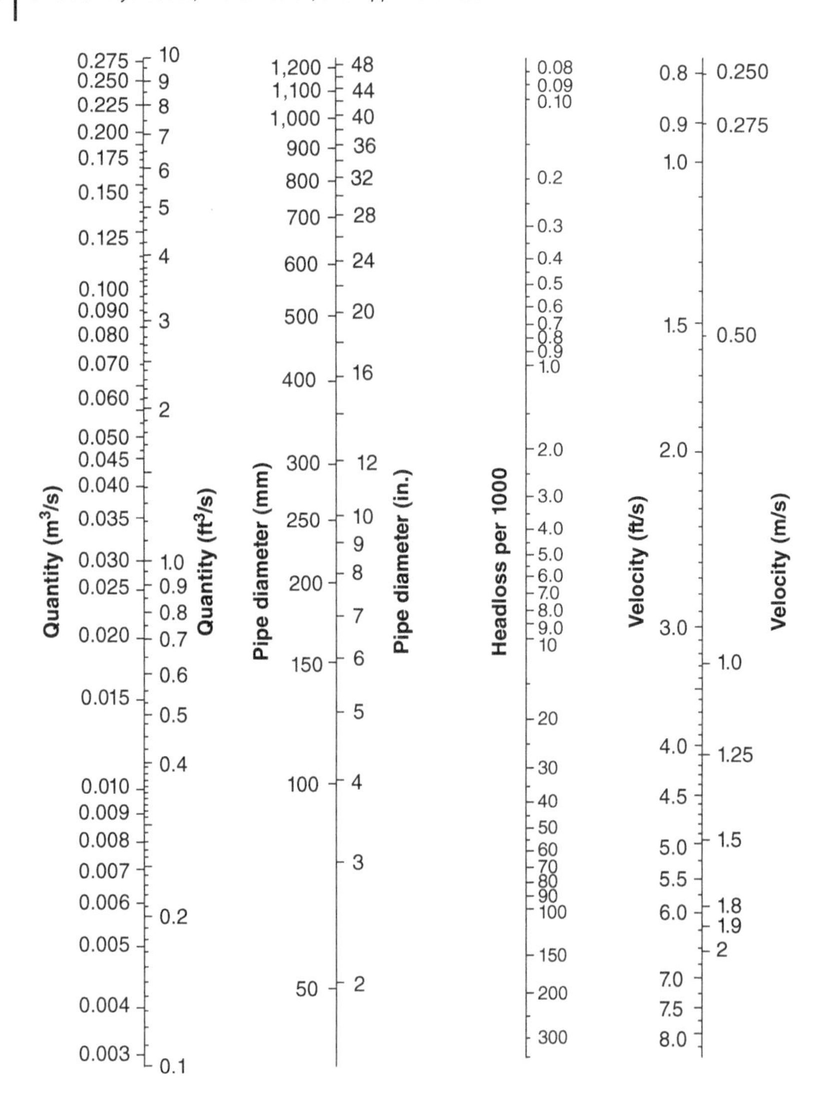

Figure 5.17 Nomogram for Solving the Hazen–Williams Equation (C = 100)

Table 5.1 Values of the Hazen–Williams Coefficient, C, for Different Conduit Materials and Age of Conduit

	Age	
Conduit Material	New	Uncertain
Cast iron pipe, coated (inside and outside)	130	100
Cast iron pipe, lined with cement or bituminous enamel	130[a]	130[a]
Steel, riveted joints, coated	110	90
Steel, welded joints, coated	140	100
Steel, welded joints, lined with cement or bituminous enamel	140[a]	130[a]
Concrete	140	130[a]
Wood stave	130	130
Cement-asbestos and plastic pipe	140	130

[a] For use with the nominal diameter, that is, the diameter of unlined pipe.

Example 5.31 Roughness Coefficient
Discuss the applications of the roughness coefficient. What is the roughness coefficient for an average cement-lined open channel?

Solution:

The following are the average values of the roughness coefficient n, for use in the application of the Kutter's and Manning's equations:

0.010 Smooth cement lining; best planed timber
0.012 New wood flumes; lined cast iron
0.013 Vitrified sewer pipe; concrete pipe; metal flumes
0.015 Average clay pipe; average cast iron pipe; average cement-lined pipe
0.023 Well-maintained earth canals
0.027 Dredged earth canals
0.040 Canals cut in rocks
0.030 Rivers in good condition

Therefore, selection of $n = 0.015$ for an average cement-lined open channel appears to be appropriate.

Example 5.32 Open Channel Flow
Determine the expected flow in a 4.2-ft (1.28-m)-wide rectangular cement-lined open channel. The channel is laid on a slope of 4.8 ft in 10,000 ft (4.8 m in 10,000 m) and water flows 2.2 ft (0.67 m) deep. Use the Kutter's C assuming the roughness coefficient n equals 0.015.

Solution 1 (US Customary System):

$$v = C\sqrt{rs} \quad (5.18a)$$

$$r = A/P_w = (4.2 \text{ ft} \times 2.2 \text{ ft})/(2.2 \text{ ft} \times 2 + 4.2 \text{ ft}) = 1.074 \text{ ft}.$$

$$s = 4.8 \text{ ft}/10{,}000 \text{ ft} = 0.00048.$$

$$C = (41.65 + 0.00281/s + 1.811/n)/\left[1 + \left(n/r^{0.5}\right)(41.65 + 0.00281/s)\right] \quad (5.24)$$

$$C = (41.65 + 0.00281/0.00048 + 1.811/0.015)/\left[1 + \left(0.015/1.074^{0.5}\right)(41.65 + 0.00281/0.00048)\right].$$

$$C = 100.$$

$$\begin{aligned} Q &= Av = AC\sqrt{rs} \\ &= (4.2 \times 2.2)(100)\sqrt{1.074 \times 0.00048} \\ &= \mathbf{21.0\ ft^3/s}. \end{aligned}$$

Solution 2 (SI System):

$$v = 0.552C\sqrt{rs} \quad (5.18b)$$

$$r = A/P_w = (1.28 \text{ m} \times 0.67 \text{ m})/(0.67 \text{ m} \times 2 + 1.28 \text{ m}) = 0.327 \text{ m}.$$

$$s = 4.8 \text{ m}/10{,}000 \text{ m} = 0.00048.$$

From Solution 1, $C = 100$.

$$\begin{aligned} Q &= Av = 0.552AC\sqrt{rs} \\ &= 0.552(1.28 \times 0.67)(100)\sqrt{0.327 \times 0.00048} \\ &= \mathbf{0.59\ m^3/s}. \end{aligned}$$

Example 5.33 Open-Channel Field Investigation—Kutter Method
A water flow of 21.0 ft^3/s (0.59 m^3/s) was measured in a rectangular open channel 4.2 ft (1.28 m) wide with water flowing 2.2 ft (0.67 m) deep. Determine the roughness factor n for the channel lining, if the channel slope is 0.00048. Use the Kutter's method.

Solution 1 (US Customary System):

$$v = C\sqrt{rs} \tag{5.18a}$$

$r = A/P_w = (4.2\text{ ft} \times 2.2\text{ ft})/(2.2\text{ ft} \times 2 + 4.2\text{ ft}) = 1.074\text{ ft}.$
$Q = Av = AC\sqrt{rs}.$
$21.0 = (4.2 \times 2.2)(C)\sqrt{1.074 \times 0.00048}.$
$C = (21.0)/(4.2 \times 2.2)\sqrt{1.074 \times 0.00048}$
$= 100.$

Substitute $C = 100$, $r = 1.074$ ft, and $s = 0.00048$ into Eq. (5.24):

$$C = (41.65 + 0.00281/s + 1.811/n)/\left[1 + \left(n/r^{0.5}\right)(41.65 + 0.00281/s)\right] \tag{5.24}$$

$$100 = (41.65 + 0.00281/0.00048 + 1.811/n)/\left[1 + \left(n/1.074^{0.5}\right)(41.65 + 0.000281/0.00048)\right].$$

By the trial-and-error method, assuming $n = 0.011,0.013,0.015,0.017$, and so on, ***n* is found to be 0.015.**

Solution 2 (SI System):

$$v = 0.552AC\sqrt{rs} \tag{5.18b}$$

$r = A/P_w = (1.28\text{ m} \times 0.67\text{ m})/(0.67\text{ m} \times 2 + 1.28\text{ m}) = 0.327\text{ m}(1.074\text{ ft}).$

$Q = Av = 0.552AC\sqrt{rs}.$

$0.59 = 0.552(1.28 \times 0.67)(C)\sqrt{0.327 \times 0.00048}.$

$C = 100.$

Substitute $C = 100$, $r = 1.074$, and $s = 0.00048$ into Eq. (5.24):

$$C = (41.65 + 0.00281/s + 1.811/n)/\left[1 + \left(n/r^{0.5}\right)(41.65 + 0.00281/s)\right] \tag{5.24}$$

$$100 = (41.65 + 0.00281/0.00048 + 1.811/n)/\left[1 + \left(n/1.074^{0.5}\right)(41.65 + 0.000281/0.00048)\right].$$

By the trial-and-error method, assuming $n = 0.011, 0.013, 0.015, 0.017$, and so on, ***n* is found to be 0.015.**

Example 5.34 Open-Channel Field Investigation—Manning Method

A water flow of 21.0 ft^3/s (0.59 m^3/s) was measured in a rectangular open channel 4.2 ft (1.28 m) wide with water flowing 2.2 ft (0.67 m) deep. Determine the roughness factor n for the channel lining, if the channel slope is 0.00048. Use the Manning's method.

Solution 1 (US Customary System):

$$v = \frac{1.486}{n}(r)^{0.67}(s)^{0.5} \tag{5.25}$$

$A = 4.2 \times 2.2 = 9.24\text{ ft}^2.$

$r = A/P_w = (9.24)/(2.2\text{ ft} \times 2 + 4.2\text{ ft}) = 1.074\text{ ft}.$

$v = (1.486/n)(1.074)^{0.67}(0.00048)^{0.5}.$

$Q = Av.$

$21.0 = 9.24\left[(1.486/n)(1.074)^{0.67}(0.00048)^{0.5}\right].$

$n = 9.24\left[(1.486)(1.074)^{0.67}(0.00048)^{0.5}\right]/21.0.$

$\boldsymbol{n = 0.015}.$

Solution 2 (SI System):

$$v = \frac{1}{n}(r)^{0.67}(s)^{0.5} \tag{5.26}$$

$A = 1.28 \times 0.67 = 0.858.$

$r = A/P_w = (0.858\ \text{m}^2)/(0.67\ \text{m} \times 2 + 1.28\ \text{m}) = 0.327\ \text{m}.$

$v = (1/n)(r)^{0.67}(s)^{0.5}$

$Q = Av.$

$0.59 = (0.858)\left[(1/n)(0.327)^{0.67}(0.00048)^{0.5}\right].$

$n = (0.858)\left[(0.327)^{0.67}(0.00048)^{0.5}\right]/0.59.$

$\boldsymbol{n = 0.015}.$

Example 5.35 Application of Manning Formula for a Pipe Flowing Full

Determine the velocity and flow of a circular pipe 24 in. (061 m) in diameter, which is flowing full. The pipe roughness coefficient is 0.013 and the slope of its energy line is $s = 0.0004$.

Solution 1 (US Customary System):

For a pipe flowing full $r = A/P_w = (\pi D^2/4)/(\pi D) = D/4$.

$$v = (0.59/n)(D)^{0.67}(s)^{0.5} \tag{5.28}$$

$$v = (0.59/0.013)(24/12)^{0.67}(0.0004)^{0.5}.$$
$$= 1.44\ \text{ft/s}.$$

$$Q = (0.46/n)(D)^{2.67}(s)^{0.5} \tag{5.29}$$

$$Q = (0.46/0.013)(24/12)^{2.67}(0.0004)^{0.5}.$$
$$\boldsymbol{Q = 4.5\ \text{ft}^3/\text{s} = 2{,}020\ \text{gpm}}.$$

Using the nomogram (Appendix 15), for $n = 0.013$:

$\boldsymbol{v = 1.45\ \text{ft/s}}.$

$\boldsymbol{Q = 2{,}000\ \text{gpm}}.$

Solution 2 (SI System):

$$v = (0.40/n)(D)^{0.67}(s)^{0.5} \tag{5.30}$$

$$v = (0.40/0.013)(0.61)^{0.67}(0.0004)^{0.5}.$$
$$\boldsymbol{v = 0.44\ \text{m/s}}.$$

$$Q = (0.31/n)(D)^{2.67}(s)^{0.5} \tag{5.31}$$

$$Q = (0.31/0.013)(0.61)^{2.57}(0.0004)^{0.5}.$$
$$\boldsymbol{Q = 0.127\ \text{m}^3/\text{s}}.$$

Using the nomogram (Appendix 15), for $n = 0.013$:

$\boldsymbol{v = 0.44\ \text{m/s}}.$

$\boldsymbol{Q = 0.126\ \text{m}^3/\text{s}}.$

Example 5.36 Head Loss Determination Using Hazen–Williams Formula

Determine the loss of head in ft/1,000 ft (m/1,000 m) for a 12 in. (0.3048 m) pipe with a flow of 1 MGD (3.785 MLD) and a $C = 100$. Assume that the pipe is flowing full.

Solution 1 (US Customary System):

Method 1—use Fig. 5.16a.
When $Q = 1$ MGD, $d = 12$ in., $C = 100$

$\boldsymbol{h_f = 2.07\ \text{ft}/1{,}000\ \text{ft}}.$

Method 2—use Fig. 5.17.
When $Q = 1\ \text{MGD} = 1.547\ \text{ft}^3/\text{s}$, $d = 12$ in., $C = 100$

$\boldsymbol{h_f = 2.07\ \text{ft}/1{,}000\ \text{ft}}.$

Method 3—use Appendix 14
When $Q = 1$ MGD $= 684.4$ gpm, $d = 12$ in., $C = 100$

$\boldsymbol{h_f = 2.07}$ **ft/1,000 ft.**

Method 4—use Eq. (5.37).

$$h_f = 10.6\,(Q_{MGD}/C)^{1.85} L/D^{4.87} \tag{5.37}$$

$$h_f = 10.6(1/100)^{1.85} 1{,}000/(12/12)^{4.87}.$$

$\boldsymbol{s = 2.07}$ **ft/1,000 ft.**

Solution 2 (SI System):

Method 1—use Fig. 5.16a.
When $Q = 3.785$ MLD, $D = 0.3048$ m, $C = 100$, after all metric units are converted to the US customary units

$\boldsymbol{h_f} = 2.07$ ft/1,000 ft $=$ **2.07 m/1,000 m.**

Method 2—use Fig. 5.17.
Read Solution 1 after SI units are converted to the US customary units.

$\boldsymbol{h_f = 2.07}$ **m/1,000 m.**

Method 3—use Appendix 14.
When $Q = 0.0438$ m^3/s $= 43.8$ L/s, $D = 0.3048$ m $= 30.48$ cm, and $C = 100$

$\boldsymbol{h_f = 2.07}$ **m/1,000 m.**

Method 4—use Eq. (5.37).
where $Q = 0.0438$ m^3/s, $D = 0.3048$ m, $L = 1{,}000$ m and $C = 100$

$$h_f = 10.67\left(Q_{m^3/s}/C\right)^{1.85} L/D^{4.87}$$

$$h_f = 10.67(0.0438/100)^{1.85} 1{,}000/(0.3048)^{4.87} \tag{5.37}$$

$$s = \mathbf{2.17\ m/1{,}000\ m}$$

Example 5.37 Mathematical and Graphical Basis of the Pipe Flow Diagram
The pipe flow diagram (Fig. 5.16) establishes the numerical relationships between Q, v, d, and s for a value of $C = 100$. Conversion to other magnitudes of C is simple because both v and Q vary directly as C. Show the mathematical and graphical basis of this diagram.

Solution:

1) Written in logarithmic form, Eq. (5.36), $Q_{gpd} = 405Cd^{2.63}\,s^{0.54}$, is
   a) $\log Q_{gpd} = 4.61 + 2.63 \log d + 0.54 \log s$, or
   b) $\log s = -8.54 - 4.87 \log d + 1.85 \log Q_{gpd}$.

   A family of straight lines of equal slope is obtained, therefore, when s is plotted against Q on log–log paper for specified diameters d. Two points define each line. Pairs of coordinates for a 12 in. pipe, for example, are as follows:
   c) $Q = 100{,}000$ gpd, $s = 0.028$ ft/1,000 ft.
   d) $Q = 1{,}000{,}000$ gpd, $s = 0.028$ ft/1,000 ft.
2) Written in logarithmic form, Eq. (5.34), $v = 0.115Cd^{0.63}s^{0.54}$, is as follows:

$$\log v = 1.0607 + 0.63 \log d + 0.54 \log s$$

   If the diameter d is eliminated from the logarithmic transforms of Eqs. (5.36) and (5.34), then
   a) $\log Q = 0.180 + 4.17 \log v - 1.71 \log s$.
   b) $\log s = 0.105 + 2.43 \log v - 0.585 \log Q$.

3) A family of straight lines of equal slope is obtained when s is plotted against Q on log–log paper for specified velocities v. Two points define each line. Pairs of coordinates for a velocity of 1 ft/s, for example, are as follows:
   a) $Q = 100{,}000$ gpd, $s = 1.5$ ft/1,000 ft.
   b) $Q = 10{,}000{,}000$ gpd, $s = 0.10$ ft/1,000 ft.

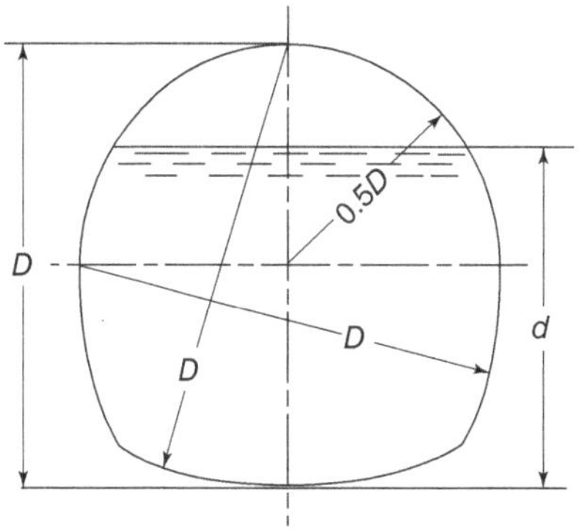

Example 5.38 Circular Conduit Equivalence to a Horseshoe Conduit
A tunnel having a horseshoe shape (Fig. 5.18) has a cross-sectional area of 27.9 ft^2 (2.59 m^2) and a hydraulic radius of 1.36 ft (0.41 m). Find the diameter, hydraulic radius, and area of the hydraulically equivalent circular conduit.

Figure 5.18 Horseshoe Conduit Section (Example 5.38)

Solution 1 (US Customary System):

$$D = 1.53\, a^{0.38} r^{0.24} \quad (5.38)$$

$D = 1.53 \times (27.9)^{0.38} \times (1.36)^{0.24} = \mathbf{5.85\ ft}\ \ (\mathbf{1.78\ m})$.

Hydraulic radius $r = D/4 = \mathbf{1.46\ ft}\ \ (\mathbf{0.45\ m})$.

Area $a = \pi D^2/4 = \mathbf{26.7\ ft^2}\ \ (\mathbf{2.48\ m^2})$.

Solution 2 (SI System):

$D = 1.53\, a^{0.38} r^{0.24}$

$D = 1.53(2.59)^{0.38}(0.41)^{0.24} = \mathbf{1.78\ m}\ \ (\mathbf{5.85\ ft})$.

Hydraulic radius $r = D/4 = \mathbf{0.45\ m}\ \ (\mathbf{1.46\ ft})$.

Area $a = \pi D^2/4 = \mathbf{2.48\ m^2}\ \ (\mathbf{926.7\ ft^2})$.

Note that neither the cross-sectional area nor the hydraulic radius of this equivalent circular conduit is the same as that of the horseshoe section proper. Equation (5.38) is applicable to both the US customary system and the SI system.

Example 5.39 Determination of Pipe Diameter Using a Nomogram for Solving The Hazen–Williams Equation Given the following:

1) If a flow of 210 L/s is to be carried from point A to point B by a 3,300-m ductile iron pipeline ($C = 100$) without exceeding a head loss of 43 m, what must be the pipe diameter?
2) If the elevation of point A is 580 m and the level of point B is 600 m and a minimum residual pressure of 3 bars (30 m of water) must be maintained at point B, what then must be the minimum actual pressure at point A?

Solution:

1) Pipe diameter:

   $C = 100$.
   $s = h_f/L = 43/3{,}300 = 13‰$.
   $Q = 210\ \text{L/s} = 0.210\ \text{m}^3/\text{s}$.

   From the nomogram of Fig. 5.17, $D = 380$ mm.
   Therefore, use a nominal pipe diameter of 400 mm so as not to exceed a head loss of 43 m.

2) Actual pressure at point A:

   $$P_A/\gamma + Z_A + (v_A)^2/2g = P_B/\gamma + Z_B + (v_B)^2/2g + h_f.$$

   Knowing that

   $v_A = v_B$ since no change in D or Q,

   the nomogram ($C = 100$) of Fig. 5.17 for $D = 400$ mm and $Q = 0.210\ \text{m}^3/\text{s}$ will give $s = 11‰$.

$h_f = sL = 11‰ \times 3{,}300\ \text{m} = 36\ \text{m}.$

$P_A/\gamma + 580 = 30 + 600 + 36.$

$\mathbf{P_A/\gamma = 86\ m\ or\ 8.6\ bars\ is\ the\ actual\ pressure\ at\ point}\ A.$

Example 5.40 Applications of Hazen–Williams Formula

Determine the expected velocity, full flow, and head loss for a 24 in. (609.6 mm) circular conduit with a Hazen–Williams coefficient $C = 100$, a length $L = 1{,}000$ ft (304.8 m), and a hydraulic gradient $s = 2.5‰$.

Solution 1 (US Customary System):

1) According to the nomogram shown in Fig. 5.17,

$v = 3.3$ ft/s and $Q = 10\ \text{ft}^3/\text{s}.$

$s = 2.5‰ = 0.0025 = h_f/\text{L} = h_f/1{,}000$ ft.

$h_f = 1{,}000\ \text{ft} \times 0.0025 = 2.5\ \text{ft}.$

2) Using Eqs. (5.34) and (5.36),

$$v = 0.115Cd^{0.63}s^{0.54} = 0.115 \times 100(24)^{0.63}(0.0025)^{0.54} = 3.3\ \text{ft/s}.$$
$$v = 0.55CD^{0.63}s^{0.54} = 0.55 \times 100(24/12)^{0.63}(0.0025)^{0.54} = \mathbf{3.3\ ft/s}.$$
$$Q_{\text{ft}^3/\text{s}} = 0.432CD^{2.63}s^{0.54} = 0.432 \times 100(24/12)^{2.63}(0.0025)^{0.54} = \mathbf{10.5\ ft^3/s}.$$
$$h_f = 1{,}000\ \text{ft} \times 0.0025 = \mathbf{2.5\ ft}.$$

Solution 2 (SI System):

1) According to the nomogram shown in Fig. 5.17,

$v = 1$ m/s and Q = $0.28\ \text{m}^3/\text{s}.$

$h_f = 304.8\ \text{m} \times 0.0025 = 0.762\ \text{m}.$

2) Using Eqs. (5.34) and (5.36),

$$v = 0.3545CD^{0.63}\ s^{0.54} = 0.3545 \times 100(0.6096)^{0.63}(0.0025)^{0.54} = \mathbf{1\ m/s}$$
$$Q_{\text{m}^3/\text{s}} = 0.278CD^{2.63}s^{0.54} = 0.278 \times 100(0.6096)^{2.63}(0.0025)^{0.54} = \mathbf{0.298\ m^3/s}.$$
$$h_f = 304.8\ \text{m} \times 0.0025 = \mathbf{0.762\ m}.$$

5.2.3 Form Resistance

Pipeline transitions and appurtenances add *form* resistance to *surface* resistance. Head losses are stepped up by changes in cross-sectional geometry and changing directions of flow. Expansion and contraction exemplify geometric change; elbows and branches, directional change. Valves and meters as well as other appurtenances may create both geometrical and directional change. With rare exceptions, head losses are expressed either in terms of velocity heads, such as $Kv^2/2g$, or as equivalent lengths of straight pipe, $Le = h_f/s = Kv^2/2gs = KD/f$ (see Appendix 17). The outstanding exception is the loss on sudden expansion or enlargement $(v_1 - v_2)^2/2g$, where v_1 is the velocity in the original conduit and v_2 the velocity in the expanded conduit; even it is, however, sometimes converted, for convenience, into $Kv^2/2g$. Because continuity as $a_1v_1 = a_2v_2$ equates $k_1v_1^2/2g$ with $(v_1^2/2g)(1 - a_1/a_2)^2$, the loss at the point of discharge of a pipeline into a reservoir (making a_2 very large in comparison with a_1) equals approximately $v_1^2/2g$; consequently, there is no recovery of energy. In all but special cases like this, k must be determined experimentally. When there is no experimental information, the values of k in Table 5.2 give useful first approximations on likely losses.

Example 5.41 Form Resistance

Summarize and discuss the methods for determination of form (minor) resistances of pipe connections, contractions, expansions (enlargement), and elbows, fittings, and valves.

Table 5.2 Values of k for Form Losses

Fitting	k-value	Fitting	k-value
Pipe entrance		90° smooth bend	
Bellmouth	0.03–0.05	Bend radius/D = 4	0.16–0.18
Rounded	0.12–0.25	Bend radius/D = 2	0.19–0.25
Sharp edged	0.50	Bend radius/D = 1	0.35–0.40
Projecting	0.80	Mitered bend	
Contraction—sudden		$\theta = 15°$	0.05
$D_2/D_1 = 0.80$	0.18	$\theta = 30°$	0.10
$D_2/D_1 = 0.50$	0.37	$\theta = 45°$	0.20
$D_2/D_1 = 0.20$	0.49	$\theta = 60°$	0.35
Contraction—conical		$\theta = 90°$	0.80
$D_2/D_1 = 0.80$	0.05	Tee	
$D_2/D_1 = 0.50$	0.07	Line flow	0.30–0.40
$D_2/D_1 = 0.20$	0.08	Branch flow	0.75–1.80
Expansion—sudden		Cross	
$D_2/D_1 = 1.25$	0.16	Line flow	0.50
$D_2/D_1 = 2.00$	0.57	Branch flow	0.75
$D_2/D_1 = 5.00$	0.92	45° wye	
Expansion—conical		Line flow	0.30
$D_2/D_1 = 1.25$	0.03	Branch flow	0 50
$D_2/D_1 = 2.00$	0.08		
$D_2/D_1 = 5.00$	0.13		

Note: Subscript 1 = Upstream; Subscript 2 = Downstream.
Source: Courtesy of Haestad Methods Water Solutions, Bentley Institute Press.

Solution:

Hydraulic losses of head in pipe fittings are generally expressed as

$$h_f = K\frac{v^2}{2g}$$

or

$$h_f = K\frac{(v_1 - v_2)^2}{2g}$$

where subscript 1 refers to upstream and 2 refers to downstream.

Typical loss of head items can be listed as follows:

$$\text{From tank to pipe (entrance loss)} = K\frac{v_2^2}{2g}.$$

$$\text{From pipe to tank (exit loss)} = K\frac{v_1^2}{2g}.$$

$$\text{Expansion (enlargement)} = K\frac{(v_1 - v_2)^2}{2g}.$$

$$\text{Contraction} = K\frac{(v_2)^2}{2g}.$$

$$\text{Fittings, elbows, and valves} = K\frac{v^2}{2g}.$$

Ideally, K must be determined experimentally. If this is unavailable, the values of K can be obtained from Table 5.2 and Appendixes 17 and 19.

Example 5.42 Form and Minor Head Loss Determination
Determine the form (minor) resistance caused by a sudden contraction from a 12-in. (0.3048 m)-diameter pipe to a 6-in. (0.1504-m) diameter pipe when water flow is 0.5 ft^3/s (0.01416 m^3/s) and angle of contraction $\theta = 180^o$.

Solution 1 (US Customary System):

$$D_1 = 12 \text{ in.} = 1 \text{ ft.}$$
$$D_2 = 6 \text{ in.} = 0.5 \text{ ft.}$$
$$\theta = 180^o.$$
$$D_2/D_1 = 0.5/1 = 0.5.$$

From Appendix 19, $K_c = 0.37$.
From Table 5.2, $K = 0.37$.
From Appendix 17, equivalent length of 6 in. pipe = 5.5 ft.

$$Q_1 = A_1 v_1 = Q_2 = A_2 v_2.$$
$$0.5 = (\pi/4)(1)^2 v_1 = (\pi/4)(0.5)^2 v_2.$$
$$v_1 = 0.637 \text{ ft/s}.$$
$$v_2 = 2.548 \text{ ft/s}.$$
$$h_f = K\frac{(v_2)^2}{2g}.$$
$$\boldsymbol{h_f} = 0.37(2.548)^2/(2 \times 32.2) = \mathbf{0.037\ ft = 0.44\ in.}$$

Solution 2 (SI System):

$$D_1 = 0.3048 \text{ m}.$$
$$D_2 = 0.1504 \text{ m}.$$
$$\theta = 180^o.$$
$$D_2/D_1 = 0.1504/0.3048 = 0.5.$$

From Appendix 19, $K_c = 0.37$.
From Table 5.2, $K = 0.37$.
From Appendix 17, equivalent length of 6 in. pipe =5.5 ft = 1.676 m.

$$Q_1 = A_1 v_1 = Q_2 = A_2 v_2.$$
$$0.01416 = (\pi/4)(0.3048)^2 v_1 = (\pi/4)(0.1504)^2 v_2.$$
$$v_1 = 0.194 \text{ m/s}.$$
$$v_2 = 0.797 \text{ m/s}.$$
$$h_f = K\frac{(v_2)^2}{2g}.$$
$$\boldsymbol{h_f} = 0.37(0.797)^2/(2 \times 9.80) = \mathbf{0.012\ m = 12\ mm.}$$

Example 5.43 Plotting of Hydraulic and Energy Grade Lines
A water transmission line consists of three pipes connected in series, pipes AB, CD, and EF, to carry the required water flow from A to F (see Fig. 5.19). There is a sudden reduction BC and a sudden enlargement DE.

a) Draw the HGLs by calculating/plotting H_1, H_2, H_3, H_4, H_5, and H_6.
b) Draw the energy grade lines by calculating/plotting E_1, E_2, E_3, E_4, E_5, and E_6.

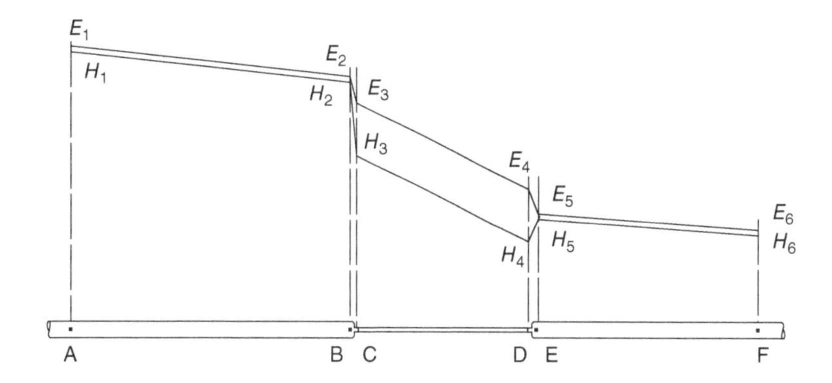

Figure 5.19 Plotting of Hydraulic and Energy Grade Lines (Example 5.43)

The following data are given:

- Pipe AB is 200 ft long (61 m) and 12 in. (30.48 cm) in diameter, $f = 0.020$.
- Pipe CD is 100 ft long (30.5 m) and 6 in. (15.24 cm) in diameter, $f = 0.015$.
- Pipe EF is 100 ft long (30.5 m) and 12 in. (30.48 cm) in diameter, $f = 0.020$.
- Velocity in pipes AB and EF is 8.025 ft/s (2.446 m/s).
- Pressure head at point A is 300 ft (91.44 m).
- Sudden reduction $K = 0.37$.

Solution 1 (US Customary System):

$$Q_{AB} = A_{AB}v_{AB} = (\pi/4)(12/12)^2(8.025) = 6.3\ \text{ft}^3/\text{s}.$$
$$Q_{CD} = Q_{AB} = 6.3 = A_{CD}v_{CD} = (\pi/4)(6/12)^2(v_{CD}).$$
$$v_{CD} = 32.1\ \text{ft/s} \quad (v_{CD})^2/2\,\text{g} = (32.1)^2/(2 \times 32.2) = 16\ \text{ft}.$$
$$v_{AB} = 8.025\ \text{ft/s} \quad (v_{AB})^2/2\,\text{g} = (8.025)^2/(2 \times 32.2) = 1\ \text{ft}.$$
$$v_{EF} = 8.025\ \text{ft/s} \quad (v_{EF})^2/2\,\text{g} = (8.025)^2/(2 \times 32.2) = 1\ \text{ft}.$$

The energy grade line drops in the direction of water flow by the amount of the lost head. The HGL is below the energy line by the amount of the velocity head at any cross-section. It is important to note that the HGL can rise where a change (an enlargement in pipe size) occurs. This is so because the enlargement decreases the velocity, hence the velocity head decreases and the pressure head increases (conservation of total energy).

$$(h_f)_{AB} = f(L/D)(v^2/2g) = 0.02(200/1)(1) = 4\ \text{ft}.$$
$$(h_f)_{BC} = K(v^2/2g) = 0.37 \times 16 = 5.9\ \text{ft (Note : } K = 0.37 \text{ from Table 5.2).}$$
$$(h_f)_{CD} = f(L/D)(v^2/2g) = 0.015(100/0.5)(16) = 48\ \text{ft}.$$
$$(h_f)_{DE} = (v_D - v_E)^2/2g = (32.1 - 8.085)^2/(2 \times 32.2) = 9\ \text{ft (see Section 5.2.3).}$$
$$(h_f)_{EF} = f(L/D)(v^2/2g) = 0.020(100/1)(1) = 2\ \text{ft}.$$

At	From	Head loss, h_f (ft)	EGL elevation (ft)		$v^2/2g$ (ft)	HGL elevation (ft)	
A	Level = 0		E_1	301.0	1.0	H_1	300.0
B	A to B	4	E_2	297.0	1.0	H_2	296.0
C	B to C	5.9	E_3	291.1	16	H_3	275.1
D	C to D	48	E_4	243.1	16	H_4	227.1
E	D to E	9	E_5	234.1	1.0	H_5	233.1
F	E to F	2	E_6	232.1	1.0	H_6	231.1

Plotting and connecting the dots of H_1 to H_6 will form the HGL. Plotting and connecting the dots of E_1 to E_6 will form the energy grade line as shown in Fig. 5.19.

Solution 2 (SI System):

$$Q_{AB} = A_{AB}v_{AB} = (\pi/4)(0.3048)^2(2.446) = 0.1784\ \text{m}^3/\text{s}.$$
$$Q_{CD} = Q_{AB} = 0.1784 = A_{CD}v_{CD} = (\pi/4)(0.1524)^2(v_{CD}).$$
$$v_{CD} = 9.784\ \text{m/s} \quad (v_{CD})^2/2g = (9.784)^2/(2 \times 9.80) = 4.884\ \text{m}.$$
$$v_{AB} = 2.446\ \text{m/s} \quad (v_{AB})^2/2g = (2.446)^2/(2 \times 9.80) = 0.305\ \text{m}.$$
$$v_{EF} = 2.446\ \text{m/s} \quad (v_{EF})^2/2g = (2.446)^2/(2 \times 9.80) = 0.305\ \text{m}.$$

Read the explanation in Solution 1.

$$(h_f)_{AB} = f(L/D)(v^2/2g) = 0.02(61/0.3048)(0.305) = 1.22\ \text{m}.$$
$$(h_f)_{BC} = K(v^2/2g) = 0.37 \times 4.884 = 1.807\ \text{m}.$$
$$(h_f)_{CD} = f(L/D)(v^2/2g) = 0.015(30.5/0.1524)(4.884) = 14.66\ \text{m}.$$
$$(h_f)_{DE} = (v_D - v_E)^2/2g = (9.784 - 2.446)^2/(2 \times 9.8) = 2.747\ \text{m.(see Section 5.2.3)}.$$
$$(h_f)_{EF} = f(L/D)(v^2/2g) = 0.020(30.5/0.3048)(0.305) = 0.61\ \text{m}.$$

At	From	Head loss h_f (m)	EGL elevation (m)		$v^2/2g$ (m)	HGL elevation (m)	
A	Level = 0		E_1	91.745	0.305	H_1	91.44
B	A to B	1.220	E_2	90.525	0.305	H_2	90.22
C	B to C	1.807	E_3	93.297	4.884	H_3	88.41
D	C to D	14.66	E_4	78.637	4.884	H_4	73.75
E	D to E	2.747	E_5	71.311	0.305	H_5	71.01
F	E to F	0.610	E_6	70.701	0.305	H_6	70.40

Note: 1 ft = 0.3048 m

Read the explanations in Solution 1 for plotting the EGL and HGL.

Example 5.44 Frictional Head Loss Determination

Given the following energy data at two sections, A and B, across a pipe transporting water in a steady-state flow, determine the head loss between the two sections.

a) Potential energy, Z

Section A = 66 ft = 20.12 m.
Section B = 136 ft = 41.45 m.

b) Kinetic energy, $v^2/2g$

Section A = 50 ft = 15.24 m.
Section B = 50 ft = 15.24 m.

c) Pressure energy, P/γ

Section A = 336 ft = 102.41 m.
Section B = 246 ft = 74.98 m.

d) Total energy, E

Section A = 452 ft = 137.77 m.
Section B = 432 ft = 131.67 m.

Solution 1 (US Customary System):

Total energy at B = total energy at A + energy input – energy loss

$$432 = 452 + 0 - h_f.$$
$$\boldsymbol{h_f = 20\,\text{ft of water}}.$$

Solution 2 (SI System):

Total energy at B = total energy at A + energy input – energy loss.

$$131.67\,\text{m} = 137.77\,\text{m} + 0 - h_f.$$
$$\boldsymbol{h_f = 6.1\,\text{m of water}}.$$

Example 5.45 Mass Flow of Incompressible Fluid

Determine the mass-flow rate at section B of a pipeline if the static pressure at section A is 102 psig (707.9 kPa = 707.0 kN/m^2) and the 4-in. (101.6-mm)-diameter pipe is flowing full with turbulent flow at an average velocity of 32 ft/s (9.75 m/s).

Solution 1 (US Customary System):

$$Q = Av$$
$$\gamma Q = \gamma Av = \text{mass flow}$$
$$n = (62.4\ \text{lb/ft}^3)\left[(\pi/4)(4/12)^2\text{ft}^2\right](32\ \text{ft/s})$$

Mass flow = 174 lb/s.

Solution 2 (SI System):

$$Q = Av$$
$$\gamma Q = \gamma Av = \text{mass flow}$$
$$n = (9.8\ \text{kN/m}^3)\left[(\pi/4)(0.1016)^2\text{m}^2\right](9.75\ \text{m/s})$$

Mass flow = 0.77 kN/s.

Mass flow rate is the same at all sections. Mass flow at section A equals mass flow at section B.

Example 5.46 Mass Flow of Compressible Fluid

Determine the mass-flow rate of air traveling in a long length of 1-in. (25.4-mm)-diameter pipe. At section A, the pressure is 31 psia (215.14 kPa = 215.14 kN/m^2), the temperature is 300°F (148.88°C), and the air velocity is 32 ft/s (9.75 m/s). At the downflow section B, the pressure has been reduced by friction and heat loss to 21 psig (145.7 kPa = 145.7 kN/m^2) (R_g, the gas constant of air = 53.34 ft-lb/lb-°R (286.9 J/kg-K); 1 J = 0.101972 kg-m; and 1 N = 0.101972 kg).

Solution 1 (US Customary System):

$$R_g = PV/T.$$
$$53.4\ \text{ft-lb/lb-°R} = (31 \times 144\ \text{lb/ft}^2)(V)/(300 + 460\text{°R}).$$
$$V = 9.081\ \text{ft}^3/\text{lb}.$$
$$\gamma = 1/V = 1/9.081\ \text{ft}^3/\text{lb} = 0.11\ \text{lb/ft}^3.$$
$$\textbf{Mass flow} = \gamma Q = \gamma Av$$
$$= (0.11\ \text{lb/ft}^3)\left[(\pi/4)(1/12)^2\text{ft}^2\right](32\ \text{ft/s})$$
$$= \mathbf{0.019\ lb/s}.$$

Solution 2 (SI System):

$$R_g = PV/T.$$
$$286.9\ \text{J/kg-K} \times (0.101972\ \text{kg-m/J}) = (215.14\ \text{kN/m}^2)V/(148.88 + 275)\text{K}.$$
$$V = 57.36\ \text{m}^3/\text{kN}.$$
$$\gamma = 1/V = 1/57.46 = 0.0174\ \text{kN/m}^3.$$

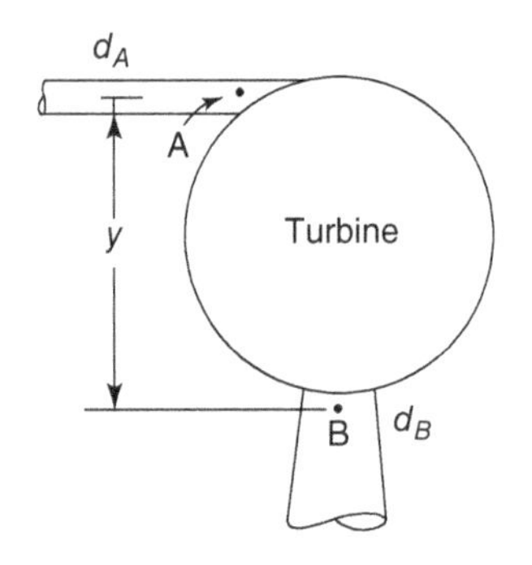

Figure 5.20 Water Flow Through a Turbine (Example 5.47)

$$\text{Mass flow} = \gamma Q = \gamma A v$$
$$= (0.0174 \text{ kN/m}^3)\left[(\pi/4)(0.0254)^2 \text{m}^2\right](9.75 \text{ m/s})$$
$$= 0.0000859 \text{ kN/s}$$
$$= \mathbf{0.00875 \text{ kg/s}}$$
$$= \mathbf{8.75 \text{ g/s}}.$$

Example 5.47 Turbine Head Loss and Horsepower Calculation
Water flows from pipe A (diameter d_A = 12 in. = 30.48 cm) to a turbine and then pipe B (diameter d_B = 24 in. = 60.96 cm) at the flow rate of 8 ft³/s (0.2264 m³/s). The pressures at A and B are 22 psi (152.68 kPa) and −5.2 psi (−36.09 kPa), respectively. The Distance (y) between A and B is 3.2 ft (0.975 m). Determine the head loss of turbine and the horsepower delivered to the turbine by the water.

Solution 1 (US Customary System):

$$Q_A = A_A V_A = \left(\frac{12}{12}\right)^2 \frac{3.14}{4} \text{ ft}^2\, V_A = 8 \text{ ft}^3/\text{s}$$
$$V_A = 10.19 \text{ ft/s}$$
$$Q_A = Q_B = A_B V_B = \left(\frac{24}{12}\right)^2 \frac{3.14}{4} \text{ ft}^2\, V_B = 8 \text{ ft}^3/\text{s}$$
$$V_B = 2.55 \text{ ft/s}$$
$$\frac{P_A}{\gamma} + \frac{V_A^2}{2g} + Z_A - H_T = \frac{P_B}{\gamma} + \frac{V_B^2}{2g} + Z_B$$
$$Z_B = 0 \text{ (datum)}; Z_A = 3.2 \text{ ft}; g = 32.2 \text{ ft/s}^2$$
$$\frac{22 \times 144 \text{ lb/ft}^2}{62.4 \text{ lb/ft}^3} + \frac{10.19^2}{2g} + 3.2 - H_T = \frac{-5.2 \times 144 \text{ lb/ft}^2}{62.4 \text{ lb/ft}^3} + \frac{2.55^2}{2g} + 0$$
$$H_T = 67.58 \text{ ft}$$
$$\text{HP} = (Q \text{ gpm})(H \text{ ft})/3957$$
$$\text{HP} = (\gamma \text{ lb/ft}^3)(Q \text{ ft}^3/\text{s})(H \text{ ft})\Big/\left(550 \frac{\text{ft-lb/s}}{\text{HP}}\right)$$
$$\text{HP} = \frac{(62.4 \text{ lb/ft}^3)(8 \text{ ft}^3/\text{s})(67.58 \text{ ft})}{550 \text{ (ft-lb/s)/hp}} = 61.33 \text{ hp to turbine}$$

Solution 2 (SI System):

$$Q_A = A_A V_A = (0.3048)^2 \times 0.785 \text{ m}^2 V_A = 0.2264 \text{ m}^3/\text{s}$$
$$V_A = 3.1 \text{ m/s}$$
$$Q_A = Q_B = A_B V_B = (0.6096)^2 \times 0.785\, V_B = 0.2264 \text{ m}^3/\text{s}$$
$$V_B = 0.77 \text{ m/s}$$
$$\frac{P_A}{\gamma} + \frac{V_A^2}{2g} + Z_A - H_T = \frac{P_B}{\gamma} + \frac{V_B^2}{2g} + Z_B$$
$$Z_B = 0 \text{ (datum)};\; Z_A = 0.975 \text{ m};\; g = 9.81 \text{ m/s}^2$$
$$\gamma = 9.8 \text{ kN/m}^3; 1 \text{ kPa} = 1 \text{ kN/m}^2$$
$$\frac{152.68 \text{ kN/m}^2}{9.80 \text{ kN/m}^3} + \frac{3.1^2 \text{ m}^2/\text{s}^2}{2 \times 9.81 \text{ m/s}^2} + 0.975 \text{ m} - H_T = \frac{-36.09 \text{ kN/m}^2}{9.8 \text{ kN/m}^3} + \frac{0.77^2 \text{ m}^2/\text{s}^2}{2 \times 9.8 \text{ m/s}^2} + 0$$
$$H_T = 20.695 \text{ m}$$

MP = 9.8066 (Q m³/s)(H m) = 9.8066 (0.2264 m³)(20.695 m) = 45.95 kW. Note: 1 hp = 0.7457 kW

5.2.4 Hydraulic Transients

Transmission lines are subjected to transient pressures when valves are opened or closed or when pumps are started or stopped. Water hammer and surge are among such transient phenomena.

Water hammer is the pressure rise accompanying a sudden change in velocity. When velocity is decreased in this way, energy of motion must be stored by elastic deformation of the system. The sequence of phenomena that follows sudden closure of a gate, for example, is quite like what would ensue if a long, rigid spring, traveling at uniform speed, were suddenly stopped and held stationary at its forward end. A pressure wave would travel back along the spring as it compressed against the point of stoppage. *Kinetic energy* would change to *elastic energy*. Then the spring would vibrate back and forth. In a pipe, compression of the water and distention of the pipe wall replace the compression of the spring. The behavior of the pressure wave and the motion of the spring and the water are identically described by the differential equations for one-dimensional waves. Both systems would vibrate indefinitely, were it not for the dissipation of energy by internal friction.

Water hammer is held within bounds in small pipelines by operating them at moderate velocities, because the *pressure* rise in psi or kPa cannot exceed about 50 times the velocity expressed in ft/s or m/s. In larger lines the pressure is held down by arresting flows at a sufficiently slow rate to allow the relief wave to return to the point of control before pressures become excessive. If this is not practicable, pressure relief or surge valves are introduced.

Very large lines, 6 ft (1.8 m) or more in equivalent diameter, operate economically at relatively high velocities. However, the cost of making them strong enough to withstand water hammer would ordinarily be prohibitive if the energy could not be dissipated slowly in surge tanks. In its simplest form, a *surge tank* is a standpipe at the end of the line next to the point of velocity control. If this control is a gate, the tank accepts water and builds up back pressure when velocities are regulated downward. When demand on the line increases, the surge tank supplies immediately needed water and generates the excess hydraulic gradient for accelerating the flow through the conduit. Following a change in the discharge rate, the water level in a surge tank oscillates slowly up and down until excess energy is dissipated by hydraulic friction in the system.

5.3 Capacity and Size of Conduits

With rates of water consumption and fire demand known, the capacity of individual supply conduits depends on their position in the waterworks system and the choice of the designer for (a) a structure of full size or (b) duplicate lines staggered in time of construction.

Minimum workable size is one controlling factor in the design of tunnels. Otherwise, size is determined by hydraulic and economic considerations. For a gravity system, that is, where pumping is not required, controlling hydraulic factors are available heads and allowable velocities. Head requirements include proper allowances for drawdown of reservoirs and maintenance of pressure in the various parts of the community, under conditions of normal as well as peak demand. Reservoir heads greater than necessary to transport water at normal velocities may be turned into power when it is economical to do so.

Allowable velocities are governed by the characteristics of the water carried and the magnitude of the hydraulic transients. For silt-bearing waters, there are both lower and upper limits of velocity; for *clear* water, only an upper limit. The minimum velocity should prevent deposition of silt; it lies in the vicinity of 2–2.5 ft/s (0.60–0.75 m/s). The maximum velocity should not cause erosion or scour, nor should it endanger the conduit by excessive water hammer when gates are closed quickly. Velocities of 4–6 ft/s (1.2–1.8 m/s) are common, but the upper limit lies between 10 and 20 ft/s (3 and 6.1 m/s) for most materials of which supply conduits are built and for most types of water carried. Unlined canals impose greater restrictions. The size of force mains and of gravity mains that include power generation is fixed by the relative cost or value of the conduit and the cost of pumping or power.

When aqueducts include more than one kind of conduit, the most economical distribution of the available head among the component classes is effected when the change in cost Δc for a given change in head Δh is the same for each kind. The proof for this statement is provided by *Lagrange's method* of undetermined multipliers. As shown in Fig. 5.21 for three components of a conduit with an allowable, or constrained, head loss H, the Lagrangian requirement of $\Delta c_1/\Delta h_1 = \Delta c_2/\Delta h_2 = \Delta c_3/\Delta h_3$ is met when parallel tangents to the three c:h curves identify, by trial, three heads h_1, h_2, and h_3 that satisfy the constraint $h_1 + h_2 + h_3 = H$.

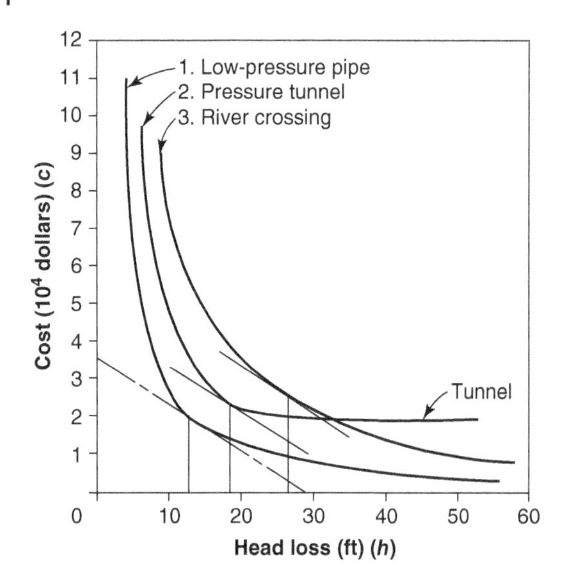

Figure 5.21 Lagrangian Optimization of Conduit Sections by Parallel Tangents. Conversion Factor: 1 ft = 0.3048 m

Example 5.48 Determination of the Most Economical Distribution of Heads for three Conduit Sections
Given the costs and losses of head shown in Fig. 5.21 for three sections of a conduit, find the most economical distribution of the available head $H = 60$ ft (18.3 m) between the three sections.

Solution:

By trial, the three heads h_1, h_2, and h_3 that satisfy the constraint $h_1 + h_2 + h_3 = H$ are as follows:

ft	m	
$h_1 = 13.5$	4.1	$c_1 = 2.0 \times 10^4$
$h_2 = 19.0$	5.8	$c_2 = 2.1 \times 10^4$
$h_3 = 27.5$	8.4	$c_3 = 2.2 \times 10^4$
$H = 60.0$	18.3	$C = 6.3 \times 10^4$.

Example 5.49 Upgrading Capacity and Size of Pipes
A gravity water supply system consists of a water pipe ($C = 100$) 150 mm in diameter and 3,000 m long that joins two reservoirs that have a difference of water surface elevation of 13 m.

Expecting a higher water demand in the near future, the flow needs to be increased between the reservoirs to three times what would be produced by the present system. Two alternatives are being investigated for tripling the flow: either place a pump on the existing line or install an additional parallel line between the two reservoirs.

1) If a pump is placed, compute the required water head for the added pump.
2) If an additional line is installed, calculate the required size of the added pipe.

Solution:

First let us determine the existing flow in the 150 mm pipe:

$$P_A/\gamma + Z_A + (v_A)^2/2g = P_B/\gamma + Z_B + (v_B)^2/2g + h_f.$$
$$0 + (Z_A - Z_B) + 0 = 0 + h_f.$$
$$h_f = 13\text{ m}.$$
$$s = h_f/L = 13/3{,}000 = 4.3‰.$$

The nomogram ($C = 100$) of Fig. 5.17 for $D = 150$ mm and $s = 4.3‰$ will give $\mathbf{Q = 0.010\ m^3/s}$.

1) The required water head for the added pump:

Future flow $Q_2 = 3Q = 3 \times 0.010 = 0.030\text{ m}^3/\text{s}$.

The nomogram ($C = 100$) of Fig. 5.17 for $D = 150$ mm and $Q_2 = 0.030\text{ m}^3/\text{s}$ will give $s = 30‰$.

$$h_f = sL = 30‰ \times 3{,}000 = 90\text{ m}. \quad (\text{Note: } 30‰ = 0.03)$$
$$P_A/\gamma + Z_A + (v_A)^2/2g + h_P = P_B/\gamma + Z_B + (v_B)^2/2g + h_f.$$
$$0 + (Z_A - Z_B) + 0 + h_P = 0 + 0 + h_f.$$
$$13 + h_P = 90.$$
$$h_P = 90 - 13 = 77\text{ m}.$$

The required water head for the pump = 77 m.

2) The required size of the added pipe:

$$Q_{pipe2} = 0.030 - 0.010 = 0.020\text{ m}^3/\text{s}.$$

The nomogram ($C = 100$) of Fig. 5.17 for $Q_{pipe2} = 0.020\text{ m}^3/\text{s}$ and $s = 4.3‰$ will give $D = 200$ mm.
The required size of the added pipe is 200 mm.

Example 5.50 Capacity and Size ff Parallel Pipes
A town is receiving its water supply from a groundwater tank located at the treatment plant through a gravity main that terminates at an elevated water tank adjacent to the town. The difference between the levels of the tanks is 100 m, and the distance between them is 50 km. The tanks were originally connected by a single pipeline designed to carry 13,000 m^3/d. It was later found necessary to increase the supply to 19,500 m^3/d and consequently the decision was made to lay another pipeline of the same diameter alongside part of the original line and cross-connected to it.
Calculate (given C for all pipes = 100)

1) The diameter of the pipes
2) The length of the second pipe, which was necessary to install.

Solution:

1) The diameter of the pipes:

$$\text{At present}: Q = 13{,}000\ \text{m}^3/\text{d} = \left(13{,}000\ \text{m}^3/\text{d}\right)/(24\ \text{h/d})(60\ \text{min/h})(60\ \text{s/min}) = 0.150\ \text{m}^3/\text{s}\ \ (150\ \text{L/s}).$$

$$S = h_f/L = 100\ \text{m}/50\ \text{km} = 2‰.$$

The nomogram (C = 100) of Fig. 5.17 for Q = 0.150 m^3/s and s = 2‰ will give a pipe diameter =500 mm.
Diameter of existing pipe no. 1 = 500 mm.

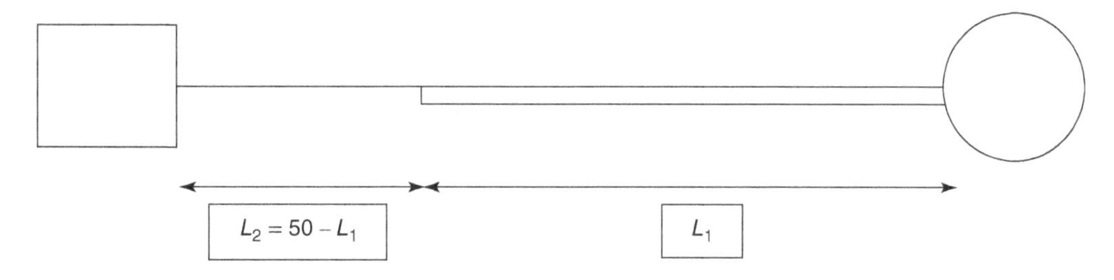

$$Q_2 = \left(19{,}500\ \text{m}^3/\text{d}\right)/(24\ \text{h/d})(60\ \text{min/h})(60\ \text{s/min})$$
$$= \mathbf{0.226\ m^3/s} = (226\ \text{L/s}).$$
$$Q_1 = 0.226/2 = \mathbf{0.113\ m^3/s = 113\ L/s}.$$

Lines are in parallel, their flows are equal.

2) The length of the second pipe, which had to be installed:

$$h_f = \left(h_f\right)_1 + \left(h_f\right)_2.$$

Lines are in series, their head losses are additive.

$$100 = L_1 s_1 + L_2 s_2.$$
$$100 = L_1 s_1 + (50 - L_1) s_2.$$

The nomogram (C = 100) of Fig. 5.17 for Q_1 = 0.113 m^3/s and D = 500 mm will give s_1 = 1.0 ‰.
The nomogram (C = 100) of Fig. 5.17 for Q_2 = 0.226 m^3/s and D = 500 mm will give s_2 = 4.3 ‰.

$$100 = L_1\ \text{km} \times 1.0‰ + (50 - L_1) \times 4.3‰.$$
$$100 = L_1 + 215 - 4.3 L_1.$$
$$3.3 L_1 = 115.$$
$$L_1 = 115/3.3 = 35\ \text{km}.$$

The length of the second pipe, which had to be installed, is 35 km.

Figure 5.22 The Dual 1,600-mm Desalinated Water Transmission Line in Abu Dhabi, UAE. The Pipes are Long. The Pipes are Ductile Iron Laid Above Ground Level with Bitumen/Zinc Coating (*Source:* http://www.water-technology.net/projects/shuweihat/shuweihat4.html)

5.4 Multiple Lines

Although masonry aqueducts and tunnels of all kinds are best designed to the full projected capacity of the system, this is not necessarily so for pipelines. Parallel lines (Fig. 5.22) built a number of years apart may prove to be more economical. Cost, furthermore, is not the only consideration. It may be expedient to lay more than one line (a) when the maximum pipe size of manufacture is exceeded; 36 in. in the case of centrifugal cast iron pipe, for example; (b) when possible failure would put the line out of commission for a long time; and (c) when pipe location presents special hazards—floods, ice, and ships' anchors endangering river crossings or submarine pipes and cave-ins rupturing pipelines in mining areas, for example.

Twin lines generally cost 30–50% more than a single line of equal capacity. If they are close enough to be interconnected at frequent intervals, gates should be installed in the bridging pipes to keep most of the system in operation during repairs to affected parts. However, if failure of one line will endanger the other, twin lines should not be laid in the same trench. Thus, cast iron pipe can fail so suddenly that a number of pipe lengths will be undermined and pulled apart before the water can be turned off. Another reason for having dual lines traverse different routes is to have them feed water into opposite ends of the distribution system.

5.5 Cross-Sections

Both hydraulic performance and structural behavior enter into the choice of cross-section. Because hydraulic capacity is a direct function of the hydraulic radius, and the circle and half circle possess the largest hydraulic radius or smallest (frictional) surface for a given volume of water, the circle is the cross-section of choice for closed conduits and the semicircle for open conduits whenever structural conditions permit. Next best are cross-sections in which circle or semicircles can be inscribed. Examples are (a) trapezoids approaching half a hexagon as nearly as maintainable slope of canals in earth permit; (b) rectangles twice as wide as they are deep for canals and flumes of masonry or wood; (c) semicircles for flumes of wood staves or steel; (d) circles for pressure aqueducts, pressure tunnels, and pipelines; and (e) horseshoe sections for grade aqueducts and grade tunnels.

Internal pressures are best resisted by cylindrical tubes and materials strong in tension; external earth and rock pressures (not counterbalanced by internal pressures) by horseshoe sections and materials strong in compression. By design, the hydraulic properties of horseshoe sections are only slightly poorer than are those of circles. Moreover, their relatively flat *invert* makes for easy transport of excavation and construction materials in and out of the aqueduct. As shown in Fig. 5.18, four circular arcs are struck to form the section: a circular arc rising from the *springing line* of the arch at half depth, two lateral arcs struck by radii equaling the height of the crown above the invert, and a circular arc of like radius establishing the bottom.

5.6 Structural Requirements

Structurally, closed conduits must resist a number of different forces singly or in combination:

1) Internal pressure equal to the full head of water to which the conduit can be subjected
2) Unbalanced pressures at bends, contractions, and closures
3) Water hammer or increased internal pressure caused by sudden reduction in the velocity of the water—by the rapid closing of a gate or shutdown of a pump, for example
4) External loads in the form of backfill and traffic
5) Their own weight between external supports (piers or hangers)
6) Temperature-induced expansion and contraction.

Internal pressure, including water hammer, creates transverse stress or *hoop tension*. Bends and closures at dead ends or gates produce unbalanced pressures and *longitudinal stress*. When conduits are not permitted to change length, variations in temperature likewise create longitudinal stress. External loads and foundation reactions (manner of support), including the weight of the full conduit, and atmospheric pressure (when the conduit is under a vacuum) produce *flexural stress*.

In jointed pipes, such as bell-and-spigot cast iron pipes, the longitudinal stresses must either be resisted by the joint or be relieved by motion. Mechanical joints offer such resistance. The resistance of joints in bell-and-spigot cast iron pipe to being pulled apart can be estimated from Prior's (1935) observational equations:

$$p = \frac{3{,}800}{d + 6} - 40 \quad \text{(US customary units)} \tag{5.39a}$$

$$P_f = \left(\frac{3{,}000}{d + 6} - 31\right) d^2 \quad \text{(US customary units)} \tag{5.40a}$$

where d is the diameter, in.; p is the intensity of pressure, psig; and P_f is the total force, lb.

The equivalent equations using the SI units are the following:

$$p = \frac{670{,}000}{d + 152} - 278 \quad \text{(SI units)} \tag{5.39b}$$

$$P_f = \left(\frac{525}{d + 152} - 0.2\right) d^2 \quad \text{(SI units)} \tag{5.40b}$$

where d is the diameter, mm; p is the intensity of pressure kPa gauge; and P_f is the total force, N.

Example 5.51 Pressure Intensity and Force of Pipe's Mechanical Joints

In jointed cast iron pipes, the longitudinal stresses can be resisted by the mechanical joints. Determine the pressure intensity and total force resisted by the mechanical joints of a 24 in. (609.6 mm) bell-and-spigot cast iron pipe.

Solution 1 (US Customary System):

$$\begin{aligned} p &= [3{,}800/(d + 6)] - 40 \\ &= [3{,}800/(24 + 6)] - 40 \\ &= 86.66 \text{ psig.} \end{aligned}$$

$$\begin{aligned} P_f &= [3{,}000/(d + 6) - 31]d^2 \\ &= [3{,}000/(24 + 6) - 31](24)^2 \\ &= 39{,}744 \text{ lb.} \end{aligned}$$

Solution 2 (SI System):

$$\begin{aligned} p &= [670{,}000/(d + 152)] - 278 \\ &= [670{,}000/(609.6 + 152)] - 278 \\ &= 601.73 \text{ kPa gauge.} \end{aligned}$$

$$\begin{aligned} P_f &= [525/(d + 152) - 0.2]d^2 \\ &= [525/(609.6 + 152) - 0.2](609.6)^2 \\ &= 181{,}844 \text{ N.} \end{aligned}$$

Figure 5.23 Concrete Water Pipe (*Source:* http://upload.wikimedia.org/wikipedia/commons/2/2e/Concrete water_pipe.jpg)

Tables of standard dimensions and laying lengths are found in professional manuals, specifications of the American Water Works Association, and publications of manufacturers and trade associations.

5.7 Location

Supply conduits are located in much the same way as railroads and highways.

5.7.1 Line and Grade

The invert of a grade aqueduct or grade tunnel is placed on the same slope as the HGL. Cut and fill, as well as cut and cover, are balanced to maintain a uniform gradient and reduce haul. Valleys and rivers that would be bridged by railroads and highways may be bridged also by aqueducts. Such indeed was the practice of ancient Rome, but modern aqueducts no longer rise above valley, stream, and hamlet except where a bridge is needed primarily to carry road or railway traffic. Pressure conduits have taken their place. Sometimes they are laid in trenches as sag pipes to traverse valleys and pass beneath streams; sometimes they strike deep below Earth's surface in pressure tunnels for which geologic exploration fixes both line and grade.

Pressure aqueducts and pipelines move freely up and down slopes. For economy they should hug the HGL in profile and a straight line in plan (Fig. 5.23). The size and thickness of conduit and difficulty of construction must be kept in balance with length. The shortest route is not necessarily the cheapest.

Air released from the water and trapped at high points reduces the waterway, increases friction, and may interrupt flow unless an air relief valve or vacuum pump is installed. True siphons should be avoided if possible. However, if the height of rise above the hydraulic grade is confined to less than 20 ft (6 m) and the velocity of flow is kept above 2 ft/s (0.60 m/s), operating troubles will be few. For best results, the line should leave the summit at a slope less than that of the hydraulic gradient.

In practice, possible locations of supply conduits are examined on available maps of the region; the topographic and geologic sheets of the US Geological Survey are useful examples. Route surveys are then carried into the field. Topography and geology are confirmed and developed in needed detail, possibly by aerial surveys, borings, and seismic exploration. Rights of way, accessibility of proposed routes, and the nature of obstructions are also identified. The use of joint rights-of-way with other utilities may generate economies.

5.7.2 Vertical and Horizontal Curves

In long supply lines, changes in direction and grade are effected gradually in order to conserve head and avoid unbalanced pressures. Masonry conduits built in place can be brought to any desired degree of curvature by proper form work. Cast iron and other sectional pipelines are limited in curvature by the maximum angular deflection of standard lengths of pipe at which joints will remain tight. The desired curve is built up by the necessary number of offsets from the tangent. Sharper curves can be formed by shorter or shortened pipes. The smaller the pipe, the sharper can be the deflection. Welded pipelines less than 15 in. (400 mm) in diameter are sufficiently flexible to be bent in the field. The ends of large steel pipe must be cut at an angle that depends on the type of transverse joint, the thickness of the steel plate, and the size of the pipe.

For sharp curves, transitions, and branches, special fittings are often built up or manufactured of the same materials as the main conduit.

5.7.3 Depth of Cover

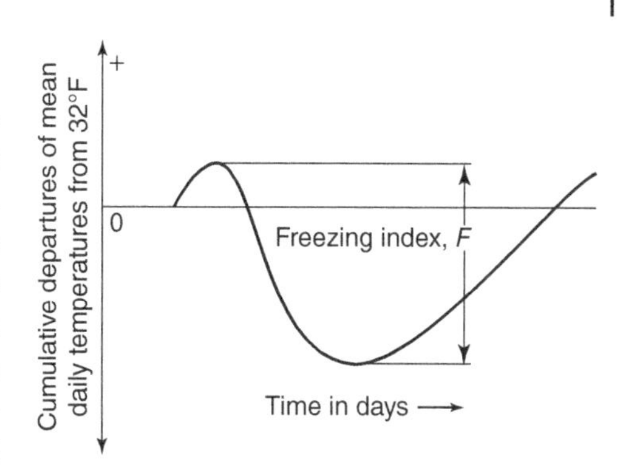

Figure 5.24 Determination of the Freezing Index of Soils as the Cumulative Departure of the Mean Daily Temperature from 32°F

Conduits that follow the surface of the ground are generally laid below the *frost line*, although the thermal capacity and latent heat of water are so great that there is little danger of freezing as long as the water remains in motion. To reduce the external load on large conduits, only the lower half may be laid below frost. Along the 42nd parallel of latitude, which describes the southern boundaries of Massachusetts, upper New York, and Michigan in the United States, frost seldom penetrates more than 5 ft (1.5 m) beneath the surface; along the 45th parallel the depth increases to 7 ft (2 m). The following equation approximates Shannon's (1945) observations of frost depth:

$$d = 1.65F^{0.468} \quad \text{(US customary units)} \tag{5.41a}$$

where d is the depth of frozen soil, in.; and F, the freezing index, is the algebraic difference between the maximum positive and maximum negative cumulative departures, $\Sigma(T_d - 32)$, of the daily mean temperatures (T_d) from 32°F. Accumulation, as shown in Fig. 5.24, begins with the first day on which a freezing temperature is recorded. In concept, the *freezing index* is analogous to the *degree day*, which describes the heat requirements of buildings during the heating season. The authors of this book have developed the following frost depth equation using the SI units:

$$d = 55.18F^{0.468} \quad \text{(SI units)} \tag{5.41b}$$

where d is the depth of frozen soil, mm; and F is the freezing index, which is the algebraic difference between the maximum positive and maximum negative cumulative departures, $\Sigma(T_d - 0)$, of the daily mean temperatures (T_d) from 0°C.

In the absence of daily readings, the value of F may be approximated, in North America, from the mean monthly temperatures as follows:

$$F = (32n - \Sigma T_m)30.2 \quad \text{(US customary units)} \tag{5.42a}$$

$$F = (0 - \Sigma T_m)30.2 \quad \text{(SI units)} \tag{5.42b}$$

Here n is the number of months during which the temperature is less than 32°F (0°C); ΣT_m is the sum of the mean temperatures °F or °C during each of these months; and 30.2 is the mean number of days in December, January, February, and March.

Pipes laid at depths of 2–3 ft (0.60 – 0.90 m) are safe from extremes of heat and ordinary mechanical damage, but it is wise to go to 5 ft (1.5 m) in streets or roads open to heavy vehicles. Otherwise, structural characteristics of conduits determine the allowable depth of cover or weight of backfill. Some conduits may have to be laid in open cut to keep the depth of backfill below the maximum allowable value.

Example 5.52 Freezing Index and Frost Depth Determination

The following are the average consecutive 2-week temperature records of a typical coldest month.

Day	1	2	3	4	5	6	7	8	9	10	11	12	13	14
T_d (°F)	35	37	37	30	28	25	29	30	24	36	40	40	42	45
T_d (°C)	1.67	2.78	2.78	−1.11	−2.22	−3.88	−1.66	−1.11	−4.44	2.22	4.44	4.44	5.55	7.22

Determine the freezing index and the required minimum frost depth.

Solution 1 (US Customary System):

Day	T_d (°F)	$T_d - 32$	$\Sigma(T_d - 32)$	
1	35	3	3	
2	37	5	8	
3	37	5	13	First maximum positive
4	30	−2	11	

(Continued)

Day	T_d (°F)	$T_d - 32$	$\Sigma(T_d - 32)$	
5	28	−4	7	
6	25	−7	0	
7	29	−3	−3	
8	30	−2	−5	
9	24	−8	−13	First maximum negative
10	36	4	−9	
11	40	8	−1	
12	40	8	7	
13	42	10	17	
14	45	13	30	

Figure 5.24 can be plotted for determining F.
The freezing index $=13 + 13 =$ **26 (US customary unit).**
The frost depth $d = 1.65F^{0.468}$.

$$d = 1.65(26)^{0.468} = \mathbf{7.58\ in.}$$

There shall be a minimum of 1 ft (0.3048 m) of fill material on the top of a sewer or water main for freezing protection. Certain US states near Canada require a minimum of 3 ft (1 m) of fill material.

Solution 2 (SI system):

Day	T_d (°C)	$T_d - 0$	$\Sigma(T_d - 0)$	
1	1.67	1.67	1.67	
2	2.78	2.78	4.45	
3	2.78	2.78	7.22	First maximum positive
4	−1.11	−1.11	6.11	
5	−2.22	−2.22	3.9	
6	−3.88	−3.88	0.02	
7	−1.66	−1.66	−1.64	
8	−1.11	−1.11	−2.75	
9	−4.44	−4.44	−7.19	First maximum negative
10	2.22	2.22	−4.97	
11	4.44	4.44	−0.53	
12	4.44	4.44	3.91	
13	5.55	5.55	9.46	
14	7.22	7.22	26.68	

The freezing index $=7.22 + 7.19 =$ **14.41 (SI unit).**
The frost depth $d = 55.18F^{0.468}$.

$$d = 55.18(14.41)^{0.468} = \mathbf{192.33\ mm.}$$

There shall be a minimum of 305 mm of fill material on the top of a sewer or water main for freezing protection. Certain US states near Canada require a minimum of 1 m (3 ft) of fill material.

Example 5.53 Freezing Index and Frost Depth Requirement in North America

In the absence of daily readings, the value of freezing index F may be approximated by Eqs. (5.42a) and (5.42b), using the US customary units and the SI units, respectively. Determine the freezing index of a northern state in the United States assuming the following coldest monthly temperatures recorded:

	T_m (°F)	T_m (°C)
December	23	−5
January	26	−3.33
February	24	−4.44
March	31	−0.55

Solution 1 (US Customary System):

$$\Sigma T_m = 23 + 26 + 24 + 31 = 104.$$

$$\begin{aligned} F &= (32n - \Sigma T_m) \times 30.2 \\ &= (32 \times 4 - 104) \times 30.2 \\ &= \mathbf{724.8}. \end{aligned}$$

$$\begin{aligned} d &= 1.65F^{0.468} \\ &= 1.65(724.8)^{0.468} \\ &= \mathbf{35.98\ in.} > 12\,\text{in. Select 36 in. of fill material.} \end{aligned}$$

Solution 2 (SI System):

$$\Sigma T_m = -5 - 3.33 - 4.44 - 0.55 = -13.32.$$

$$\begin{aligned} F &= [0 - \Sigma T_m] \times 30.2 \\ &= [0 - (-13.32)] \times 30.2 \\ &= \mathbf{402.26}. \end{aligned}$$

$$\begin{aligned} d &= 55.18F^{0.468} \\ &= 55.18(402.26)^{0.468} \\ &= \mathbf{913.46mm} > 305\text{ mm. Select 1 m of fill material.} \end{aligned}$$

5.8 Materials of Construction

Selection of pipeline materials is based on carrying capacity, strength, life or durability, ease of transportation, handling, and laying, safety, availability, cost in place, and cost of maintenance. Various types of iron, steel, reinforced concrete, and fiberglass are most used for water transmission pipes, but plastic pipes are now being made in the smaller sizes. Other materials may come into use in the future.

5.8.1 Carrying Capacity

The initial value of the Hazen–Williams coefficient *C* hovers around 140 for all types of well-laid pipelines but tends to be somewhat higher for reinforced concrete and fiberglass lines and to drop to a normal value of about 130 for unlined cast iron pipe. Cast iron and steel pipes lined with cement or with bituminous enamel possess coefficients of 130 and over on the basis of their nominal diameter; improved smoothness offsets the reduction in cross-section.

Loss of capacity with age or, more strictly, with service depends on the properties of the water carried and the characteristics of the pipe. Modern methods for controlling aggressive water promise that the corrosion of metallic pipes and the disintegration of cement linings and of reinforced concrete pipe will be held in check very largely, if not fully.

Cement and bituminous-enamel linings and reinforced concrete and fiberglass pipes do not, as a rule, deteriorate significantly with service.

5.8.2 Strength

Steel pipes can resist high internal pressures, but large lines cannot withstand heavy external loads or partial vacuums unless special measures are taken to resist these forces. Ductile iron and fiberglass pipes are good for moderately high water pressures and appreciable external loads, provided that they are properly bedded. Prestressed reinforced concrete pipe is satisfactory for high water pressures. All types of concrete pipe can be designed to support high external loads.

5.8.3 Durability

Experience with all but coated ductile iron pipe (length of life 100 years) has been too short and changes in water treatment have been too many to give us reliable values on the length of life of different pipe materials. The corrosiveness of the water, the quality of the material, and the type and thickness of protective coating all influence the useful life of the various types of water pipes. External corrosion (soil corrosion) is important, along with internal corrosion. Pipes laid in acid soils, seawater, and cinder fills may need special protection.

5.8.4 Transportation

When pipelines must be built in rugged and inaccessible locations, their size and weight become important. Ductile iron pipe is heavy in the larger sizes; steel pipe relatively much lighter. The normal laying length of cast iron pipe is 12 ft (3.7 m). Lengths of 16.4, 18, and 20 ft (5, 5.5, and 6 m) are also available in different types of bell-and-spigot pipe. The length of steel pipe is 20–30 (6–9 m). Both prestressed and cast reinforced concrete pipe are generally fabricated in the vicinity of the pipeline. The sections are 12 and 16 ft (3.6 and 4.9 m) long and very heavy in the larger sizes. A diameter smaller than 24 in. (600 mm) is unusual.

5.8.5 Safety

Breaks in cast iron pipes can occur suddenly and are often quite destructive. By contrast, steel and reinforced concrete pipes fail slowly, chiefly by corrosion. However, steel pipelines may collapse under vacuum while they are being drained. With proper operating procedures, this is a rare occurrence. Fiberglass pipe fails suddenly, much like cast iron pipe.

5.8.6 Maintenance

Pipelines of all sizes and kinds must be watched for leakage or loss of pressure—outward signs of failure. There is little choice between materials in this respect. Repairs to precast concrete pipe are perhaps the most difficult, but they are rarely required. Cast iron and small welded-steel pipes can be cleaned by scraping machines and lined in place with cement to restore their capacity. New lines and repaired lines should be disinfected before they are put into service.

5.8.7 Leakage

All pipelines should be tested for tightness as they are constructed. Observed *leakage* is often expressed in gal/day/in. diameter (nominal)/mile of pipe using the US customary units, or in L/day/mm diameter (nominal)/km of pipe using the SI units. The test pressure must naturally be stated. To conduct a leakage test, the line is isolated by closing gates and placing a temporary header or plug at the end of the section to be tested. The pipe is then filled with water and placed under pressure, the water needed to maintain the pressure being measured by an ordinary household meter. Where there is no water, air may be substituted. Losses are assumed to vary with the square root of the pressure, as in orifices.

The allowable leakage of bell-and-spigot cast iron pipe that has been carefully laid and well tested during construction is often set at

$$Q = \frac{nd\sqrt{p}}{1,850} \quad \text{(US customary units)} \tag{5.43a}$$

where Q is the leakage, gal/h; n is the number of joints in the length of line tested; d is the nominal pipe diameter, in.; and p is the average pressure during test, psig. A mile (1.6 km) of 24 in. (600 mm) cast iron pipe laid in 12 ft (3.6 m) lengths and tested under a pressure of 64 psig (444 kPa), for example, can be expected to show a leakage of

$$Q = (5,280/12) \times 24 \times \sqrt{64}/1,850 = 46 \text{ gal/h} \quad (174 \text{ L/h})$$

Considering that the pipe has a carrying capacity of 250,000 gal/h (946,000 L/h) at a velocity of 3 ft/s(0.90 m/s), the expected leakage from joints (46 × 24 gal/day)/24 in. /mi = 46 gal/day/in. /mi, which is relatively small.

The equivalent metric leakage equation using the SI units is as follows:

$$Q = \frac{nd\sqrt{p}}{32,500} \quad \text{(SI units)} \tag{5.43b}$$

where Q is the leakage, L/h; n is the number of joints in the length of line tested; d is the nominal pipe diameter, mm; and p is the average pressure during test, kPa. For the same example introduced above, the expected leakage calculated using SI units is

$$Q = \left[(1.6 \times 1{,}000/3.6) \times 600\sqrt{444}\right]/32{,}500 = 173\ \text{L/h}.$$

The expected leakage from joints is equal to (173 × 24 L/day)/600 mm/1.6 km = 4.32 L/day/mm/km.

5.9 Appurtenances

To isolate and drain pipeline sections for test, inspection, cleaning, and repairs, a number of appurtenances, or auxiliaries, are generally installed in the line (Figs. 5.25 and 5.26).

5.9.1 Gate Valves

Gate valves are usually placed at major summits of pressure conduits because (a) summits identify the sections of line that can be drained by gravity and (b) pressures are least at these points, making for cheaper valves and easier operation. For the sake of economy, valves smaller in diameter than the conduit itself are generally installed together with necessary reducers and increasers. Gates 8 in. (200 mm) in diameter or larger commonly include a 4 or 6 in. (100 or 150 mm) gated bypass. When the larger gate is seated under pressure, water admitted through the bypass can equalize the pressure on both sides and make it easier to lift the main gate.

Gravity conduits are commonly provided with gate chambers (a) at points strategic for the operation of the supply conduit, (b) at the two ends of sag pipes and pressure tunnels, and (c) wherever it is convenient to drain given sections. Sluice gates are normally installed in grade conduits, particularly in large ones. In special situations, needle valves are preferred for fine control of flow, butterfly valves for ease of operation, and cone valves for regulating time of closure and controlling water hammer.

5.9.2 Blowoffs

In pressure conduits, small, gated takeoffs, known as *blowoff* or *scour valves*, are provided at low points in the line. They discharge into natural drainage channels or empty into a sump from which the water can be pumped to waste. There should be no direct connection to sewers or polluted water courses. For safety, two blowoff valves are placed in series. The chance of both failing to close is thus reduced greatly. Their size depends on local circumstances, especially on the time in which a given section of line is to be emptied and on the resulting velocities of flow. Calculations are based on orifice discharge under a falling head, equal to the difference in elevation of the water surface in the conduit and the blowoff, minus the friction head. Frequency of operation depends on the quality of the water carried, especially on silt loads. The drainage gates of gravity conduits are placed in gate chambers.

5.9.3 Air Valves

Rigid pipes and pressure conduits are equipped with *air valves* at all high points. The valves automatically remove (a) air displaced while the line is being filled and (b) air released from the flowing water when the pressure decreases appreciably or summits lie close to the HGL. A manually operated cock or gate can be substituted if the pressure at the summit is high. Little, if any, air will then accumulate, and air needs to escape only while the line is being filled.

Steel and other flexible conduits are equipped with automatic air valves that will also admit air to the line and prevent its collapse under negative pressure (see Fig. 5.26). Pressure differences are generated when a line is being drained on purpose or when water escapes accidentally through a break at a low point. Locations of choice are both sides of gates at summits, the downstream side of other gates, and changes in grade to steeper slopes in sections of line not otherwise protected by air valves.

The required valve size is related to the size of the conduit, and to the velocities at which the line is emptied. The following ratios of air valve size to conduit diameter provide common but rough estimates of needed sizes:

For release of air only: 1:12 or 1 in./ft (83 mm/m).
For admission as well as release of air: 1 : 8 or 1.5 in. /ft (125 mm/m).

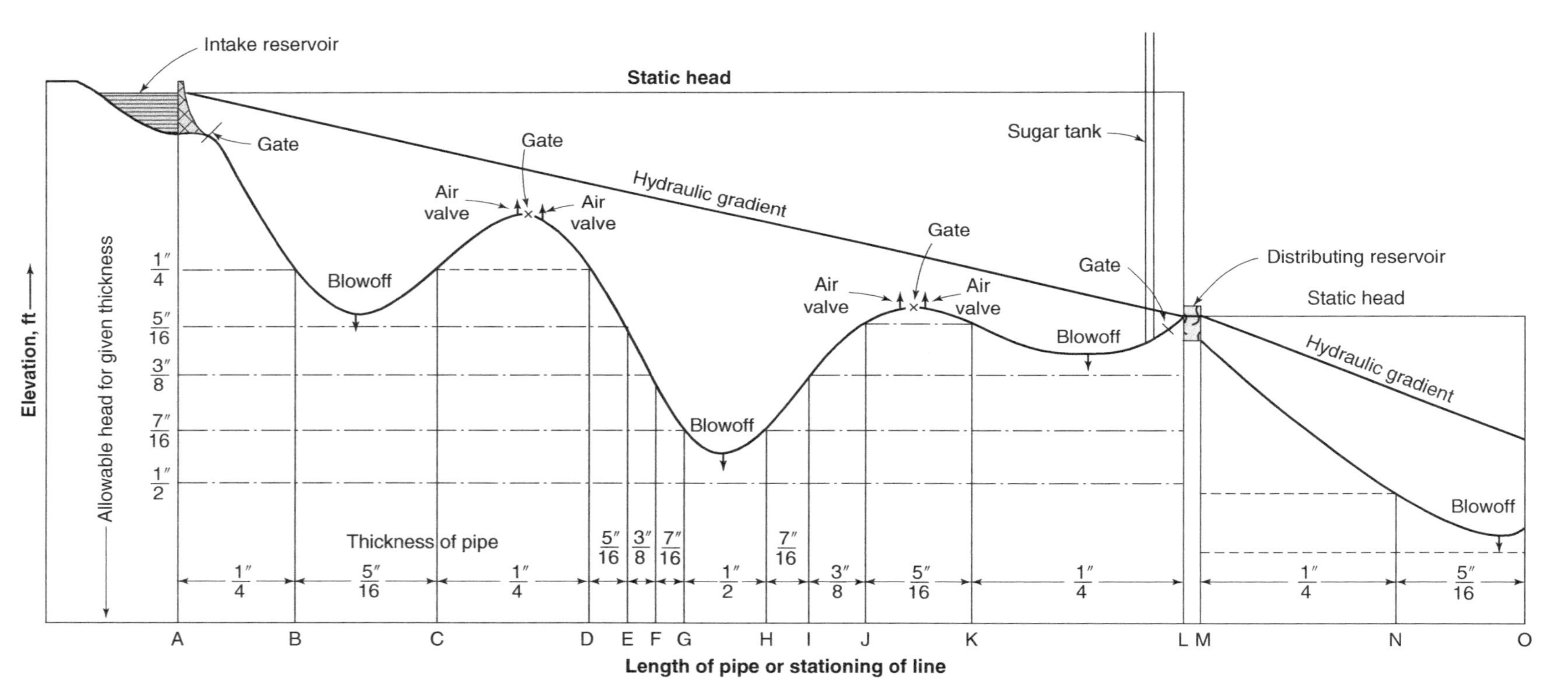

Figure 5.25 Profile of Pipeline Showing Pipe Thickness and Location of Gates, Blowoffs, and Air Valves (Not to Scale, Vertical Scale Magnified). Conversion Factors: 1″ = 1 in. = 25.4 mm; 1′ = 1 ft = 0.3048 m

An approximate calculation will show that under a vacuum of 48 in. (1,220 mm) of water, an automatic air valve, acting as an injection orifice with a coefficient of discharge of 0.5 under a head of $4/(1.3 \times 10^{-3}) = 3{,}080$ ft (939 m) of air of specific gravity 1.3×10, is expected to admit about $0.5\sqrt{2g \times 3{,}080} = 220\ \text{ft}^3/\text{s}$ of air/ft^2 (67.2 m^3/s of air/m^2) of valve. If the diameter ratio is 1:8, the displacement velocity in the conduit can be as high as 220/64 = 3.5 ft/s (1 m/s) without exceeding a vacuum of 48 in. (1,220 mm) of water. A similar calculation will show the rate of release of air. The amounts of air that can be dissolved by water at atmospheric pressure are about 2.9% by volume at 32°F (0°C) and 1.9% at 77°F (25°C), changing in direct proportion to the pressure. Accordingly, they are doubled at 2 atm or 14.7 psig (102 kPa gauge).

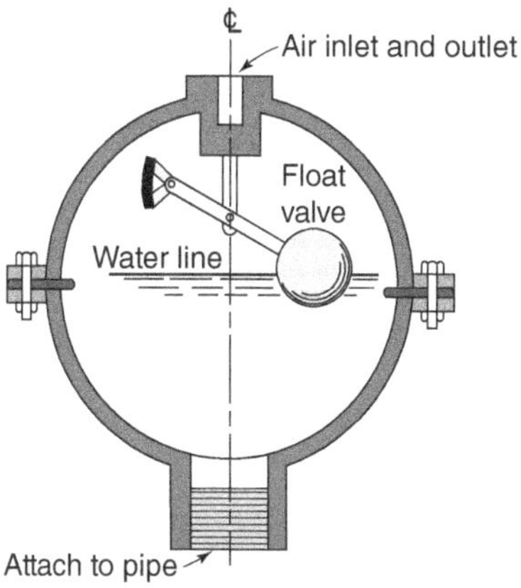

Figure 5.26 Air Inlet and Release Valve

5.9.4 Check Valves

Check valves are used to maintain flow in one direction only by closing when the flow begins to reverse. They are placed on force mains to prevent *backflow* when pumps shut down. When the flow is in the same direction as the specified direction of the check valve, the valve is considered to be fully open.

5.9.5 Pressure-Reducing Valves

Pressure-reducing valves are used to keep pressures at safe levels in low-lying areas. These valves are often used to separate pressure zones in water distribution networks. These valves prevent the pressure downstream from exceeding a specified level, in order to avoid pressures and flows that could otherwise have undesirable effects on the system. A pressure or a hydraulic grade is used to control the operation of the valve.

5.9.6 Pressure-Sustaining Valves

Pressure-sustaining valves maintain a specified pressure upstream of the valve. Similar to the other regulating valves, they are often used to ensure that pressures in the system (upstream, in this case) will not drop to unacceptable levels. A pressure or a hydraulic grade is used to control the operation of a pressure-sustaining valve.

5.9.7 Pressure Breaker Valves

Pressure breaker valves create a specified head loss across the valve and are often used to model components that cannot be easily modeled using standard minor loss elements.

5.9.8 Flow Control Valves

A *flow control valve* limits the flow rate through the valve to a specified value in a specified direction. A flow rate is used to control the operation of a flow control valve. These valves are commonly found in areas where a water district has contracted with another district or a private developer to limit the maximum demand to a value that will not adversely affect the provider's system.

5.9.9 Throttle Control Valves

Throttle control valves simulate minor loss elements whose head loss characteristics change over time. With a throttle control valve, the minor loss K is adjusted based on some other system flow or head.

5.9.10 Manholes

Access manholes are spaced 1,000–2,000 ft (300–600 m) apart on large conduits. They are helpful during construction and serve later for inspection and repairs. They are less common on cast iron lines than on steel and concrete lines.

5.9.11 Insulation Joint

Insulation joints control electrolysis by introducing resistance to the flow of stray electric currents along pipelines. Modern insulation joints make use of rubber gaskets or rings and of rubber-covered sections of pipe sufficiently long to introduce appreciable resistance.

5.9.12 Expansion Joint

The effect of temperature changes is small if pipe joints permit adequate movement. Steel pipe laid with rigid transverse joints must either be allowed to expand at definite points or be rigidly restrained by anchoring the line.

5.9.13 Anchorages

Anchorages are employed (a) to resist the tendency of pipes to pull apart at bends and other points of unbalanced pressure when the resistance of their joints to longitudinal stresses is exceeded, (b) to resist the tendency of pipes laid on steep gradients to pull apart when the resistance of their joints to longitudinal stresses is inadequate, and (c) to restrain or direct the expansion and contraction of rigidly joined pipes under the influence of temperature changes.

Anchorages take many forms as follows:

1) For bends—both horizontal and vertical—concrete buttresses or *kick blocks* resisting the unbalanced pressure by their weight, much as a gravity dam resists the pressure of the water behind it, taking into consideration the resistance offered by the pipe joints themselves, by the friction of the pipe exterior, and by the bearing value of the soil in which the block is buried
2) Steel straps attached to heavy boulders or to bedrock
3) Lugs cast on pipes and fittings to hold tie rods that prevent movement of the pipeline
4) Anchorages of mass concrete on steel pipe to keep it from moving, or to force motion to take place at expansion joints inserted for that purpose—the pipe being well bonded to the anchors, for example, by angle irons welded onto the pipe
5) Gate chambers so designed of steel and concrete that they hold the two ends of steel lines rigidly in place.

In the absence of expansion joints, steel pipe must be anchored at each side of gates and meters in order to prevent their destruction by pipe movement. In the absence of anchors, flanged gates are sometimes bolted on one side to the pipe—usually the upstream side—and on the other side to a cast iron nipple connected to the pipe by means of a sleeve or expansion joint.

5.9.14 Other Appurtenances

Other appurtenances that may be necessary include the following:

1) *Surge tanks* at the end of the line to reduce water hammer created by operation of a valve at the end of the line
2) *Pressure relief valves* or overflow towers on one or more summits to keep the pressure in the line below a given value by letting water discharge to waste when the pressure builds up beyond the design value
3) *Self-acting shutoff* valves triggered to close when the pipe velocity exceeds a predetermined value as a result of an accident
4) *Altitude-control* valves that shut off the inlet to service reservoirs, elevated tanks, and standpipes before overflow levels are reached
5) *Venturi* or other meters and recorders to measure the flows.

5.10 Additional Hydraulics Topics

The following additional hydraulic topics are important to those who plan to take professional engineering (PE) examinations in North America. Several of the included solved examples and homework problems were given previously in PE examinations.

5.10.1 Measurement of Fluid Flow and Hydraulic Coefficient

Various coefficients of hydraulics have been developed and used in civil and environmental engineering practice mainly for measuring fluids flow. The common fluid measurement devices include pitot tubes, nozzles, orifices, venturi meters, and flumes. Application of Bernoulli equation and hydraulic coefficients is important for calibration of fluid measurement devices.

The coefficient of discharge, C_d, is the ratio of actual discharge through a hydraulic device to the ideal discharge. This coefficient is expressed as

$$C_d = \text{actual flow}/\text{ideal flow}$$

$$C_d = Q/A(2gh)^{0.5} \quad (5.44)$$

$$Q = C_d A(2gh)^{0.5} = Av \quad (5.45)$$

$$v = C_d(2gh)^{0.5} \quad (5.46)$$

where

C_d = coefficient of discharge, dimensionless;
Q = actual flow, ft^3/s (m^3/s);
A = cross-sectional area of flow through the hydraulic device, ft^2 (m^2);
g = $32.2\ \text{ft/s}^2$ $(9.81\ \text{m/s}^2)$;
h = total head, ft (m); and
v = ideal velocity, ft/s (m/s).

The coefficient of velocity C_v is the ratio of the actual mean velocity in the cross-section of the flow stream (jet) to the ideal mean velocity, which would occur without friction.

Thus

$$C_v = \text{actual mean velocity}/\text{ideal mean velocity}$$

$$C_v = v/v_i = v/(2gh)^{0.5} \quad (5.47a)$$

$$v = C_v(2gh)^{0.5}, \quad (5.47b)$$

where

C_v = coefficient of velocity, dimensionless;
v = actual mean velocity, ft/s (m/s); and
v_i = ideal or theoretical velocity, ft/s (m/s).

The coefficient of contraction C_c is the ratio of the area of the contracted section of a fluid stream (jet) to the area of the opening through which the fluid flows.

$$C_c = \text{area of stream (jet)}/\text{area of opening}$$

$$C_c = A_{\text{jet}}/A_o = C_d/C_v \quad (5.48a)$$

$$C_d = C_v C_c \quad (5.48b)$$

where

C_c = coefficient of contraction, dimensionless;
A_{jet} = area of jet stream, ft^2 (m^2); and
A_o = area of opening, ft^2 (m^2).

The area of jet, A_{jet}, is measured at the "vena contracta" (v.c.), which is located at ½ orifice diameter downstream from the orifice. Hence, C_c and C_v can be redefined according to v.c.:

$$C_c = \text{area at v.c.}/\text{area of orifice} \quad (5.48c)$$

$$C_v = \text{actual velocity at v.c.}/\text{theoretical velocity at v.c.} \quad (5.47c)$$

$$v_i = (2gh)^{0.5} \quad (5.47d)$$

The head loss, h_f, in hydraulic measuring devices can be expressed as

$$h_f = \left\{\left[(1/(C_v)^2\right] - 1\right\}(v_{\text{jet}})^2/2g \tag{5.49}$$

where

v_{jet} is the jet stream velocity, ft/s or m/s

The coordinates (x, y) of a hydraulic jet can be determined according to the following two kinematic equations:

$$x = vt \tag{5.50}$$

$$y = 1/2gt^2 \tag{5.51}$$

Example 5.54 Determination of Discharge Coefficient

Determine the discharge confident, C_d, for a hydraulic system when water flows through a standard 1-in. (25.4-mm)-diameter orifice under an 18 ft (5.4864 m) head at a rate of 0.111 ft^3/s (0.0031435 m^3/s).

Solution 1 (US Customary System):

$$Q = C_d A(2gh)^{0.5} \tag{5.45}$$

$$0.111 = C_d(\pi/4)(1/12)^2(2 \times 32.2 \times 18)^{0.5}.$$

$$\boldsymbol{C_d = 0.60}.$$

Solution 2 (SI System):

$$Q = C_d A(2gh)^{0.5} \tag{5.45}$$

$$0.0031435 = C_d(\pi/4)(0.0254)^2(2 \times 9.81 \times 5.4864)^{0.5}.$$

$$\boldsymbol{C_d = 0.60}.$$

Example 5.55 Determination of Contraction Coefficient

Water flows through 1-in. (25.4-mm)-diameter orifice under an 18 ft (5.4864 m) head. The jet strikes a wall 5 ft (1.524 m) away and 0.4 ft (0.1219 m) vertically below the center line of the contracted section of the water jet. Determine the coefficient of contraction, C_c, if the coefficient of discharge, C_d, is known to be 0.60 from Example 5.54.

Solution 1 (US Customary System):

$$x = vt \tag{5.50}$$

$$y = 1/2gt^2 \tag{5.51}$$

Eliminating t from the two equations one gets

$$x^2 = (2v^2/g)y$$

$$(5)^2 = (2 \times v^2/32.2)(0.4). \tag{5.47b}$$

$$v = 32.1 \text{ ft/s}.$$

$$v = C_v(2gh)^{0.5}.$$

$$32.1 = C_v(2 \times 32.2 \times 18)^{0.5}.$$

$$C_v = 0.95. \tag{5.48a}$$

$$C_c = C_d/C_v.$$

$$\boldsymbol{C_c} = 0.60/0.95 = \boldsymbol{0.63}.$$

Solution 2 (SI System)

$$x = vt \tag{5.50}$$

$$y = 1/2gt^2 \tag{5.51}$$

Eliminating t from the two equations one gets

$$x^2 = (2v^2/g)y.$$

$$(1.524)^2 = (2v^2/9.81)(0.1219).$$

$$v = 9.67\ \text{m/s}.$$

$$v = C_v(2gh)^{0.5} \tag{5.47b}$$

$$9.67 = C_v(2 \times 9.81 \times 5.4864)^{0.5}.$$

$$C_v = 0.93.$$

$$C_c = C_d/C_v \tag{5.48a}$$

$$\mathbf{C_c} = 0.60/0.93 = \mathbf{0.64}.$$

Example 5.56 Determination of Discharge Coefficient

Determine the coefficient of discharge for an orifice 2 in. (50.8 mm) in diameter, which discharges from a tank with a head of 16 ft (4.876 m). The discharge rate Q is measured at 0.55 ft^3/s (0.0156 m^3/s). The actual velocity at the vena contracta, v.c., is 29.0 ft/s (8.839 m/s).

Solution 1 (US Customary System):

$$v_i = (2gh)^{0.5} \tag{5.47d}$$

$$v_i = \left(2 \times 32.2\ \text{ft/s}^2 \times 16\ \text{ft}\right)^{0.5} = 32.1\ \text{ft/s}.$$

$$C_v = v/v_\text{i} \tag{5.47a}$$

$$C_v = (29\ \text{ft/s})/(32.1\ \text{ft/s}) = 0.90.$$

Area of orifice: $A_\text{o} = (\pi/4)(2/12)^2 = 0.0218\ \text{ft}^2$.

Area at v.c.: $A_\text{jet} = Q/v = \left(0.55\ \text{ft}^3/\text{s}\right)/(29\ \text{ft/s}) = 0.019\ \text{ft}^2$.

$$C_c = A_\text{jet}/A_o = \left(0.019\ \text{ft}^2\right)/\left(0.0218\ \text{ft}^2\right) = 0.87 \tag{5.48a}$$

$$\mathbf{C_d} = C_vC_c = 0.90 \times 0.87 = \mathbf{0.78}. \tag{5.48b}$$

Solution 2 (SI System):

$$v_i = (2gh)^{0.5} \tag{5.47d}$$

$$v_i = \left(2 \times 9.81\ \text{m/s}^2 \times 4.876\ \text{m}\right)^{0.5} = 9.78\ \text{m/s}.$$

$$C_v = v/v_i \tag{5.47a}$$

$$C_v = (8.839\ \text{m/s})/(9.78\ \text{m/s}) = 0.90.$$

Area of orifice: $A_o = (\pi/4)(0.0508)^2 = 0.002\ \text{m}^2$.

Area at v.c.: $A_\text{jet} = Q/v = \left(0.0156\ \text{m}^3/\text{s}\right)/(8.839\ \text{m/s}) = 0.00176\ \text{m}^2$.

$$C_c = A_\text{jet}/A_o = \left(0.00176\ \text{m}^2\right)/\left(0.002\ \text{m}^2\right) = 0.88 \tag{5.48a}$$

$$\mathbf{C_d} = C_vC_\text{c} = 0.90 \times 0.88 = \mathbf{0.79}. \tag{5.48b}$$

5.10.2 Forces Developed by Moving Fluids

Water hammer is caused by the sudden decrease in liquid motion. In a water pipeline, the time of pressure wave to travel upstream and back (round-trip) is given by

$$t = 2L/c \tag{5.52}$$

where

t = round – trip travel time, s;
L = length of pipe, ft (m); and
c = celerity (speed of propagation) of pressure wave, ft/s (m/s).

Increase in water hammer pressure is caused by the sudden closing of a valve or a power outage. The change in pressure is calculated by

$$(\Delta P) = \rho c(\Delta v) \tag{5.53}$$

where

ΔP = change in pressure, lb/ft^2 (N/m^2);
ρ = fluid density, slug/ft^3 (kg/m^3);
c = celerity, ft/s (m/s); and
Δv = change in fluid velocity, ft/s (m/s).

For rigid pipes, such as cast iron pipes, the celerity of the pressure wave is obtained by

$$c = (E_B/\rho)^{0.5} \tag{5.54}$$

where

c = celerity, ft/s (m/s);
E_B = bulk modulus of the fluid, lb/ft^2 (N/m^2); and
ρ = density, slug/ft^3 (kg/m^3).

For non-rigid pipes, such as polyvinyl chloride (PVC) pipes, the expression for celerity is

$$c = \{E_B/\rho[1 + (E_B/E)(D/t_w)]\}^{0.5} \tag{5.55}$$

where

E = modulus of elasticity of pipe wall, lb/ft^2 (N/m^2);
D = inner diameter of pipe, ft (m); and
t_w = pipe wall thickness, ft (m).

The tensile stress, S_t, developed in a moving fluid can be calculated by the following equation:

$$S_t = (\Delta P)r/t_k \tag{5.56}$$

where

S_t = tensile stress of pipe, lb/ft^2 (N/m^2);
ΔP = change in pressure, lb/ft^2 (N/m^2);
r = pipe radius, ft (m); and
t_k = pipe wall thickness, ft (m).

Equation (5.56) can be used for moving fluid as well as non-moving fluid in a pipe or cylindrical liquid storage tank. When the equation is used for a tank, ΔP is the internal–external pressure difference, or the measured gauge pressure (actual pressure = –14.7 psi).

A spherical liquid storage tank is twice as strong as a cylindrical tank with hemispherical ends. In such a case:

$$S_t = 1/2(\Delta P)r/t_k \tag{5.57}$$

where

$$S_t = \text{tensile stress in tank wall, lb/ft}^2\ (\text{N/m}^2);$$
$$\Delta P = \text{internal} - \text{external pressure difference, that is, gauge pressure, lb/ft}^2\ (\text{N/m}^2);$$
$$r = \text{tank radius, in. (mm); and}$$
$$t_k = \text{tank wall thickness, in. (mm).}$$

Example 5.57 Velocity of Pressure Waves
Determine the velocity of pressure waves (celerity) traveling along a rigid pipe containing water at 60°F (15.6°C).

Solution 1 (US Customary System):

From Appendix 3 at 60°F,

$$E_B = 311 \times 10^3\ \text{lb/in}^2 = 4.478 \times 10^7\ \text{lb/ft}^2.$$
$$\rho = 1.938\ \text{slug/ft}^3\ \left(1\ \text{slug} = 1\ \text{lb} - \text{s}^2/\text{ft}\right).$$
$$c = (E_B/\rho)^{0.5} \tag{5.54}$$
$$c = \left(4.478 \times 10^7/1.938\right)^{0.5}.$$
$$c = \mathbf{4{,}807\ ft/s}.$$

Solution 2 (SI System):

From Appendix 3 at 15.6°C,

$$E_B = 214 \times 10^7\ \text{N/m}^2.$$
$$\rho = 999.1\ \text{kg/m}^3.$$
$$c = (E_B/\rho)^{0.5} \tag{5.54}$$
$$c = \left(214 \times 10^7/999.1\right)^{0.5}.$$
$$\mathbf{c = 1{,}463\ m/s}.$$

Example 5.58 Pressure Increase due to Sudden Stop of Flow
Determine the expected increase in pressure if the liquid flow is suddenly stopped assuming the following:

1) 12 in. (0.3048 m) rigid pipe.
2) Liquid in pipe is oil with a specific gravity of 0.8.
3) Flow velocity is 4 ft/s (1.22 m/s).
4) Bulk modulus of oil is 200,000 lb/in.^{2} = 200,000 × 144 lb/ft^2 (138.8 × 10^7 N/m^2).

Solution 1 (US Customary System):

$$c = (E_B/\rho)^{0.5} \tag{5.54}$$

and

$$\rho = \gamma/g.$$
$$c = (200{,}000 \times 144/0.8 \times 62.4/32.2)^{0.5}.$$
$$c = 4{,}310\ \text{ft/s}.$$
$$(\Delta P) = \rho c(\Delta v) \tag{5.53}$$
$$\begin{aligned}(\boldsymbol{\Delta P}) &= (\gamma/g)c(\Delta v)\\ &= (0.8 \times 62.4/32.2)\,(4{,}310)(4-0)\\ &= 26{,}700\ \text{lb/ft}^2\\ &= \mathbf{185\ lb/in.^2}.\end{aligned}$$

Solution 2 (SI System):

$$c = (E_B/\rho)^{0.5}$$
$$c = \left[138.8 \times 10^7/(0.8 \times 1{,}000)\right]^{0.5}. \tag{5.54}$$
$$= 1{,}317\ \text{m/s}.$$

$$\begin{aligned}(\Delta P) &= \rho c(\Delta v)\\ &= (0.8 \times 1{,}000)(1{,}317)(1.22 - 0)\\ &= 1{,}285{,}400\ \text{N/m}^2\\ &= \mathbf{1{,}285.4\ kN/m^2}.\end{aligned} \tag{5.53}$$

Example 5.59 Pipe Tensile Stress due to Sudden Valve Closure

A 48-in. (1,219 mm) steel pipe 3/8 in. (9.525 mm) thick carries water at 60°F (15.6°C) and a velocity of 6.5 ft/s (1.98 m/s). The pipe is 11,000 ft (3,353 m) long. Assume the steel pipe is elastic and non-rigid and its celerity was determined to be 3,155 ft/s (961.6 m/s). Determine

1) The time of travel of the pressure wave upstream and back (round-trip).
2) The increase in water pressure within the pipe.
3) The tensile stress in the pipe, if the valve is closed in 2 s.
4) Discuss how the water hammer can be avoided.

Solution 1 (US Customary System):

1) The time of travel of the pressure wave upstream and back (round-trip):

$$\begin{aligned}t &= 2\,\text{L/c}\\ \boldsymbol{t} &= 2 \times 11{,}000\ \text{ft}/(3{,}155\ \text{ft/s})\\ \boldsymbol{t} &= \mathbf{7.0\ s}.\end{aligned} \tag{5.52}$$

2) The increase in water pressure within the pipe:

$$(\Delta P) = \rho c(\Delta v) \tag{5.53}$$

Since the valve was closed in 2 s, but the time for the wave round-trip is 7 s, this is equivalent to sudden closure.

$$\begin{aligned}(\mathbf{\Delta P}) &= (\gamma/g)c(\Delta v)\\ &= (64.4/32.2)(3{,}155)(6.5 - 0)\\ &= \mathbf{39{,}800\ lb/ft^2}.\end{aligned}$$

3) The tensile stress in the pipe, if the valve is closed in 2 s:

$$\begin{aligned}S_t &= (\Delta P)r/t_k\\ &= \left(39{,}800\ \text{lb/ft}^2\right)(24/12\ \text{ft})/[(3/8)/12\ \text{ft}]\\ &= 25.46 \times 10^5\ \text{lb/ft}^2\\ &= \mathbf{17{,}680\ \ lb/in.^2 (psi\ increase\ in\ pressure)}.\end{aligned} \tag{5.56}$$

4) Discuss how the water hammer can be avoided.

The closure time should be at least 7.0 s, preferably 10 times larger than 2 s, that is, 20 s.

Solution 2 (SI System)

1) The time of travel of the pressure wave upstream and back (round-trip):

$$t = 2\,L/c \tag{5.52}$$

$$\begin{aligned}t &= 2 \times 3{,}353\ \text{m}/(961.6\ \text{m/s}).\\ \boldsymbol{t} &= \mathbf{7.0\ s}.\end{aligned}$$

2) The increase in water pressure within the pipe:

Since the valve was closed in 2 s, but the time for the wave round-trip is 7 s, this is equivalent to sudden closure.

$$\begin{aligned}(\mathbf{\Delta P}) &= \rho c(\Delta v)\\ &= (1{,}000)(961.6)(1.98-0)\\ &= 1{,}905{,}000\ \mathrm{N/m^2}\\ &= \mathbf{1{,}905\ kN/m^2}.\end{aligned} \tag{5.53}$$

3) The tensile stress in the pipe, if the valve is closed in 2 s:

$$\begin{aligned}S_t &= (\Delta P)r/t_k\\ &= (1{,}905{,}000\ \mathrm{N/m^2})(1.2192\ \mathrm{m}/2)/(9.525/1{,}000\ \mathrm{m})\\ &= 1.22\times 10^8\ \mathrm{N/m^2}\\ &= \mathbf{1.22\times 10^5\ kN/m^2\ increase\ in\ pressure}.\end{aligned}$$

4) Discuss how the water hammer can be avoided.

Same as Solution 1.

5.10.3 Impulse–Momentum Principles

Change in direction or magnitude of flow velocity of a liquid leads to change in fluid's momentum. From kinetic mechanics, it is known that linear impulse = change in linear momentum

$$\begin{aligned}\left(\sum F\right)t &= M(\Delta v)\\ &= M(v_2-v_1)\\ \left(\sum F\right) &= \rho Q(v_2-v_1)\\ &= (\gamma/g)Q(v_2-v_1)\end{aligned} \tag{5.58}$$

where

M = mass whose momentum changed in time t, slug (kg);
t = time, s;
$\sum F$ = sum of linear impulse forces, lb (N);
v_2 = exit velocity, ft/s (m/s);
v_1 = entrance velocity, ft/s (m/s);
Q = flow, $\mathrm{ft^3/s}$ $(\mathrm{m^3/s})$;
g = acceleration due to gravity, $32.2\mathrm{ft/s^2}$ $(9.81\ \mathrm{m/s^2})$;
γ = specific weight of fluid, $62.4\ \mathrm{lb/ft^3}$ $(9.8\ \mathrm{kN/m^3})$; and
ρ = density, $\mathrm{slug/ft^3}$ $(\mathrm{kg/m^3})$.

In the horizontal x-direction:

$$(Mv_x)_1 \pm \left(\sum F_x\right)t = (Mv_x)_2 \tag{5.59}$$

In the horizontal x-direction:

$$(Mv_y)_1 \pm \left(\sum F_y\right)t = (Mv_y)_2 \tag{5.60}$$

where

$(Mv_x)_1$ = initial linear momentum in x-direction, slug-ft/s (kg-m/s);
$(Mv_y)_1$ = initial linear momentum in y-direction, slug-ft/s (kg-m/s);
$(Mv_x)_2$ = final linear momentum in x-direction, slug-ft/s (kg-m/s);
$(Mv_y)_2$ = final linear momentum in y-direction, slug-ft/s (kg-m/s);
$\sum F_x$ = sum of linear impulse forces in x-direction, lb (N); and
$\sum F_y$ = sum of linear impulse forces in y-direction, lb (N).

Example 5.60 Impulse–Momentum Analysis
A 24-in. (610 mm) pipe is connected to a 12-in. (305 mm) pipe by a standard pipe reducer fitting. The water flow rate is 31.4 ft^3/s (0.90 m^3/s) and the water pressure in the 24 in. pipe is 40 lb/in.^{2} (278 kN/m^2). Determine the force that is exerted by water on the reducer neglecting the minor head loss.

Solution 1 (US Customary System):

$$Q = A_1v_1 = A_2v_2$$

$$31.4\ \text{ft}^3/\text{s} = \left[(\pi/4)(24/12)^2\text{ft}^2\right]v_1 = \left[(\pi/4)(12/12)^2\text{ft}^2\right]v_2.$$

$$v_1 = 10\ \text{ft/s}.$$

$$v_2 = 40\ \text{ft/s}.$$

$$P_1/\gamma + (v_1)^2/2g + z_1 = P_2/\gamma + (v_2)^2/2g + z_2 + h_f.$$

$$40 \times 144/62.4 + (10)^2/(2 \times 32.2) + 0 = P_2/\gamma + (40)^2/(2 \times 32.2) + 0 + 0.$$

$$P_2/\gamma = 40 \times 144/62.4 + (10)^2/(2 \times 32.2) - (40)^2/(2 \times 32.2) = 69\ \text{ft of water}.$$

$$P_2 = 69\ \text{ft} \times 62.4\ \text{lb/ft}^3 = 4{,}306\ \text{lb/ft}^2 = 30\ \text{psi}.$$

$$F_1 = P_1A_1 = \left(40 \times 144\ \text{lb/ft}^2\right)(\pi/4)(24/12)^2 = 18{,}086\ \text{lb toward the right}.$$

$$F_2 = P_2A_2 = \left(4{,}306\ \text{lb/ft}^2\right)(\pi/4)(12/12)^2 = 3{,}380\ \text{lb toward the left}.$$

$$\left(\sum F\right) = (\gamma/g)Q(v_2 - v_1) \tag{5.58}$$

$$(18{,}086\ \text{lb} - 3{,}380\ \text{lb} - F_x) = (62.4/32.2)(31.4)(40 - 10)$$

$F_x = 14{,}706 - 1{,}825 = 12{,}881$ lb to the left acting on the water.
The force that is exerted by water on the reducer = 12,881 lb to the right.
The vertical component F_y is 0.

Solution 2 (SI System):

$$Q = A_1\,v_1 = A_2v_2$$

$$0.90\ \text{m}^3/\text{s} = \left[(\pi/4)(0.610)^2\ \text{m}^2\right]v_1 = \left[(\pi/4)(0.305)^2\ \text{m}^2\right]v_2.$$

$$v_1 = 3.05\ \text{m/s}.$$

$$v_2 = 12.2\ \text{m/s}.$$

$$P_1/\gamma + (v_1)^2/2g + z_1 = P_2/\gamma + (v_2)^2/2g + z_2 + h_f.$$

$$278/9.8 + (3.05)^2/(2 \times 9.8) + 0 = P_2/\gamma + (12.2)^2/(2 \times 9.8) + 0 + 0.$$

Here $\gamma = 9.8\ \text{kN/m}^3$, and $g = 9.8\ \text{m/s}^2$.

$$P_2/\gamma = 278/9.8 + (3.05)^2/(2 \times 9.8) - (12.2)^2/(2 \times 9.8) = 21\ \text{m of water}.$$

$$P_2 = 21\ \text{m} \times 9.8\ \text{kN/m}^3 = 195\ \text{kN/m}^2.$$

$$F_1 = P_1 A_1 = \left(278\ \text{kN/m}^2\right)(\pi/4)(0.61)^2\ \text{m}^2 = 81\ \text{kN toward the right}$$

$$F_2 = P_2 A_2 = \left(195\ \text{kN/m}^2\right)(\pi/4)(0.305)^2\ \text{m}^2 = 14\ \text{kN toward the left.}$$

$$\left(\sum F\right) = (\gamma/g)Q(v_2 - v_1) \tag{5.58}$$

$$(81 - 14 - F_x) = (9.8/9.8)(0.90)(12.2 - 3.05).$$

$F_x = 67 - 8 = 59$ kN to the left acting on the water.

The force that is exerted by water on the reducer = 59 kN to the right.

The vertical component F_y is 0.

5.10.4 Drag and Lift Forces

Drag is the component of the resultant hydraulic force exerted by a fluid on a subject parallel to the relative motion of the fluid. Lift is the component of the resultant hydraulic force exerted by the fluid on a subject perpendicular to the relative motion of the fluid.

The drag and lift forces are given by the following equations:

$$F_D = C_D \gamma A\left(v^2/2g\right) \tag{5.61}$$

$$F_L = C_L \gamma A\left(v^2/2g\right) \tag{5.62}$$

where

F_D = drag force, lb (N);
F_L = lift force, lb (N);
C_D = drag coefficient, dimensionless;
C_L = lift coefficient, dimensionless;
A = area projected on a plane perpendicular to the relative motion of the fluid, $\text{ft}^2\left(\text{m}^2\right)$;
v = relative velocity of the fluid with respect to the subject, ft/s (m/s);
g = acceleration due to gravity, 32.2 ft/s^2 $\left(9.81\ \text{m/s}^2\right)$; and
γ = specific weight of fluid, 62.4 lb/ft^3 $\left(9.8\ \text{kN/m}^3\right)$.

Example 5.61 Determination of Drag Force

A flat plate 4 ft by 4 ft (1.22 m by 1.22 m) moves at 23 ft/s (7 m/s) normal to its plane. Determine the resistance to the plate moving through water at 60°F (15.6°C). Assume a drag coefficient =1.16 for length/width =1.

Solution 1 (US Customary System):

At 60°F, $\gamma = 62.37\ \text{lb/ft}^3$ (Appendix 3)

$$\begin{aligned} F_D &= C_D \gamma A\left(v^2/2g\right) \\ &= 1.16\left(62.37\ \text{lb/ft}^3\right)\left(4 \times 4\ \text{ft}^2\right)(23\ \text{ft/s})^2/\left(2 \times 32.2\ \text{ft/s}^2\right) \\ &= \mathbf{9{,}513\ lb}. \end{aligned} \tag{5.61}$$

Solution 2 (SI System):

At 15.6°C, $\gamma = 9.798\ \text{kN/m}^3$ (Appendix 3)

$$\begin{aligned} F_D &= C_D \gamma A\left(v^2/2g\right) \\ &= 1.16\left(9.798\ \text{kN/m}^3\right)\left(1.22 \times 1.22\ \text{m}^2\right)(7\ \text{m/s})^2/\left(2 \times 9.8\ \text{m/s}^2\right) \\ &= \mathbf{42.4\ kN}. \end{aligned} \tag{5.61}$$

Problems/Questions

5.1 A water supply system is shown in Fig. 5.27. The values of K (in $h_f = KQ^2$) are indicated in the figure. The total discharge rate from the system is 9 ft^3/s (255 L/s). The residual pressure at point C is to be maintained at 20 psi (139 kPa). Find

a) The flow rate in each pipe
b) The elevation of water surface in the reservoir
c) The pressures at A, B, and E.

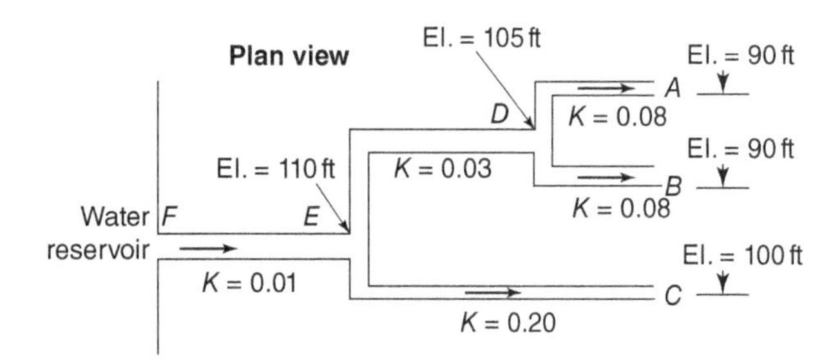

Figure 5.27 Water System for Problem 5.1. Conversion Factor: 1 ft = 0.3048 m

5.2 A village is investigating a new water supply plan. The supply is to consist of a dam impounding water at A, from which it must be pumped over a summit at B. From B it runs by gravity to the village at D (see Fig. 5.28). Since the head available at the village is more than would be used, putting in a hydroelectric plant at C has been suggested. The electric power will be transmitted back to the pumping station and used for pumping water up to B.

Assume the following efficiencies: turbine 80%, generator 90%, transmission line 95%, pump motor 80%, pump 75%. If the rate of water use is to be 6.28 ft^3/s (178 L/s), determine the excess water horsepower still available at C.

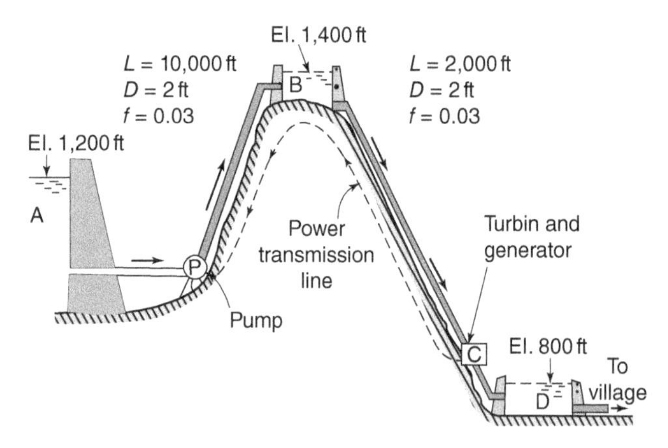

Figure 5.28 Water System for Problem 5.2. Conversion Factor: 1 ft = 0.3048 m

5.3 In the sketch shown in Fig. 5.29, C and D represent points on a university campus that are connected to a common junction B through 100 and 200 m lengths of pipe, respectively. Points B and C are on the lower level of campus at elevation of 12 and 6 m, respectively, above mean sea level (MSL), while point D is on the upper level of the campus at an elevation of 66 m above MSL. The source of supply is at 300 m from the junction at point A. Pipe diameters are 5, 4, and 2.5 cm as indicated in the figure ($g = 9.8$ m/s^2; $\rho = 1$ gm/cm^3; $f = 0.0196$).

a) With C closed off and D freely flowing, determine the total static head required at A for a flow of 2.64 L/s at D.
b) With both points C and D open to the atmosphere, determine the maximum quantity of water that can be provided at point C before any water starts to flow out at D.

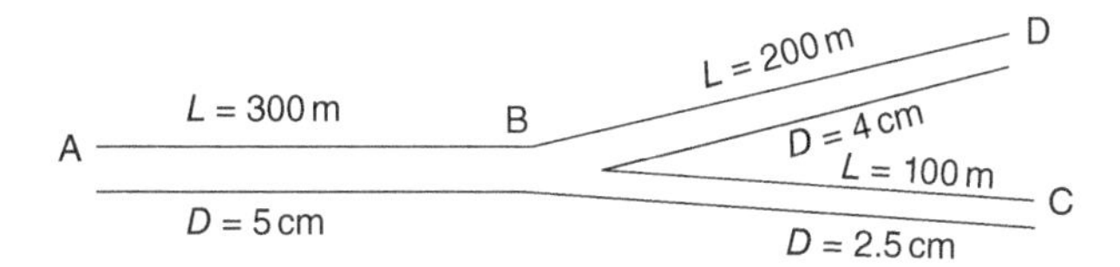

Figure 5.29 Water System for Problem 5.3.

5.4 The elevation of a city distribution reservoir is 400 ft (122 m) and that of the city is 250 ft (76 m). When city water consumption is at the rate of 16 ft^3/s (453 L/s), the pressure in the city is 60 psi (416 kPa).

Assuming that during a fire the minimum required pressure is 20 psi (139 kPa), determine the fire flow available if the coincidental city demand is 14 ft^3/s (396 L/s). (Neglect velocity heads and the variation of f with Reynolds number.)

5.5 Given the system shown in Fig. 5.30 and neglecting minor losses, velocity heads, and the variation of f with Reynolds number:

a) What will be the flow rate from A to D if the pressure at D is to be maintained at 20 psi (139 kPa)?
b) What will the rate of flow be if the water level in reservoir A drops 10 ft (3 m), that is, to El. 170 ft (51.8 m)?
c) What size of ductile iron pipe (f = 0.020) is required to carry an additional 22 ft^3/s (623 L/s) from A to D with the reservoir at 180 ft (55 m) and residual pressure at D to be maintained at 20 psi (139 kPa)?

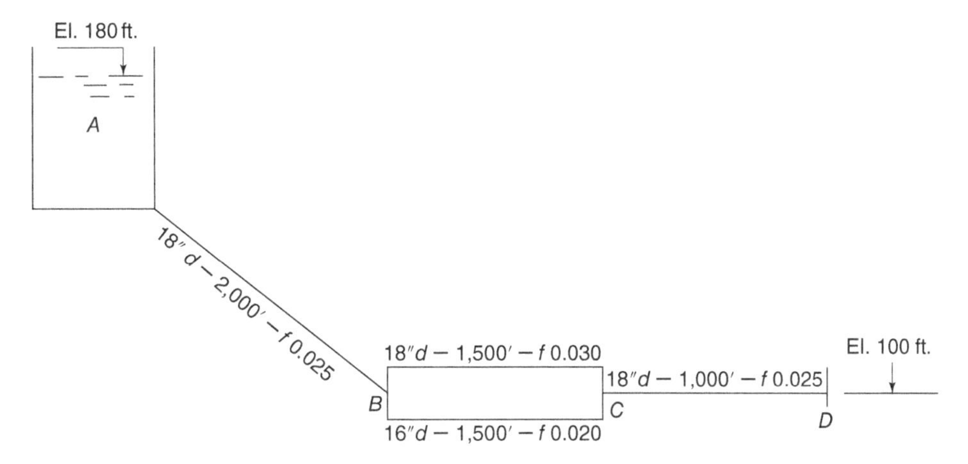

Figure 5.30 Water System for Problem 5.5. Conversion Factors: 1′ = 1 ft = 0.3048 m; 1″ = 1 in. = 25.4 mm

5.6 A pipe 2 ft (600 mm) in diameter and 3 mi (4.83 km) long connects two reservoirs. Water flows through it at a velocity of 5 ft/s (1.5 m/s). To increase the flow, a pipe 1.5 ft (450 mm) in diameter is laid parallel alongside the first pipe for the last mile.

Find the increase in flow between the reservoirs assuming the same value of f for the two pipes. Also assume that the water levels in the reservoirs remain the same. (1 mi = 5,280 ft or 1,610 m.)

5.7 In the water supply system shown in Fig. 5.31, the pumping station is operating at a flow capacity of 125 L/s against a total dynamic head of 50 m. Determine the maximum depth of water that can be stored in the elevated reservoir.

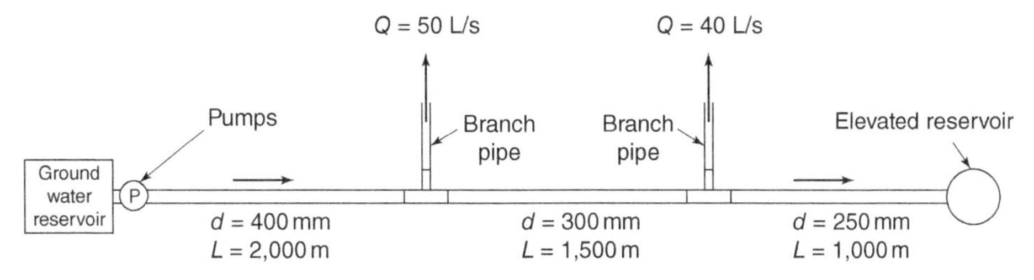

Figure 5.31 Water System for Problem 5.7.

Elevation of water level in underground tank = 500 m.
Elevation of bottom of elevated reservoir = 520 m.
C for all pipes = 100.

5.8 In the water supply system shown in Fig. 5.32, water flows by gravity from reservoir A to points C and D as well as to reservoir E. The elevations of known points are as follows:

Water level A = 650.00 m.
Elevation B = 590.00 m.
Elevation C = 550.00 m.
Elevation D = 570.00 m.

The measured residual pressures are as follows:

Point B = 300 kPa.
Point C = 500 kPa.

The known flows are as follows:

Pipe BD = 30 L/s.
Pipe BE = 75 L/s.

C for all pipes = 100.

a) Calculate the flow in line BC.
b) Find the residual pressure at point D.
c) Compute the maximum water level in reservoir E.
d) Determine the required diameter for pipe AB.

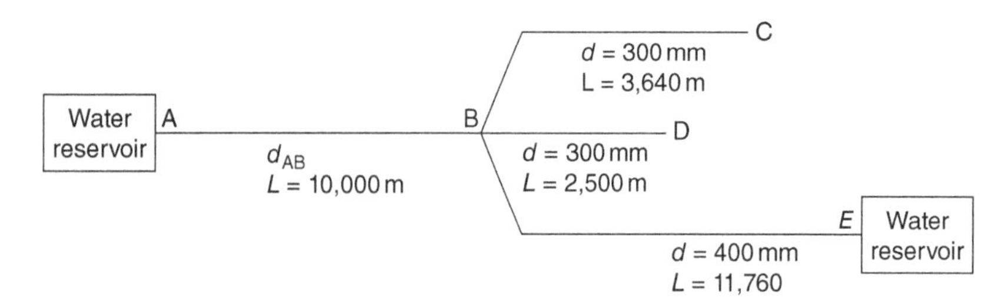

Figure 5.32 Water System for Problem 5.8.

5.9 A water supply system consists of a ground reservoir with lift pumps (A), elevated storage (C), a withdrawal point (B), an equivalent pipelines as given in the plan view in Fig. 5.33. Assume the following elevations:

Water level in ground reservoir (A) = 600.00 m.
Ground level at withdrawal point (B) = 609.10 m.
Ground level at elevated water tank (C) = 612.20 m.
Water level in tank (C) = 30 m above ground.

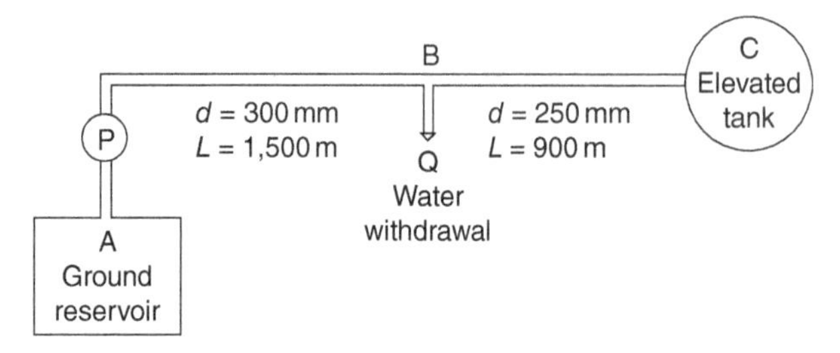

Figure 5.33 Water System for Problem 5.9

The lift pumps provide flow at a discharge pressure of 550 kPa. The value of C for all pipes is 100. Compute the discharge Q in L/s at withdrawal point B for the following conditions:

a) If the residual pressure at point B is 200 kPa
b) If water is neither flowing to, nor from, the elevated storage tank.

5.10 The water supply system shown in Fig. 5.34 is designed to deliver water from the pumping station to city ABC. The water level in the wet well at the pumping station is at 1,000 ft.

The head delivered by the pumps to the water is 100 ft. The value of C for all pipes is 100. Assume the following elevations:

A) 900.00 ft.
B) 915.00 ft.
C) 910.00 ft.

a) Determine the required diameter of main PA, if the pressure at point A is not allowed to drop below 52 psi.
b) Determine the required diameter of pipe BC, if the maximum allowed head loss in the network ABC is 3 ft/1,000 ft.
c) Calculate the actual residual pressures at points A, B, and C.

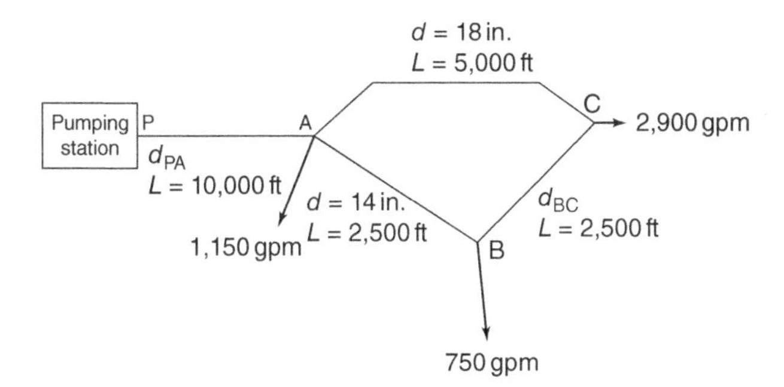

Figure 5.34 Water System for Problem 5.10. Conversion Factors: 1 gpm = 3.785 L/min; 1 ft = 0.3048 m; 1 in. = 25.4 mm.

5.11 Water is pumped from ground reservoir A to the elevated reservoir G through the network of pipes shown in Fig. 5.35.

The following data are given:

All pipes have the same diameter = 300 mm.
C for all pipes = 100.
Water level in ground reservoir = 500 m.
Water flow rate = 80 L/s.
Pump operating head =70 m.

a) Calculate the maximum water level in the elevated reservoir when valves E and F are closed.
b) How many meters will the water level rise in the elevated reservoir if valves E and F were opened?

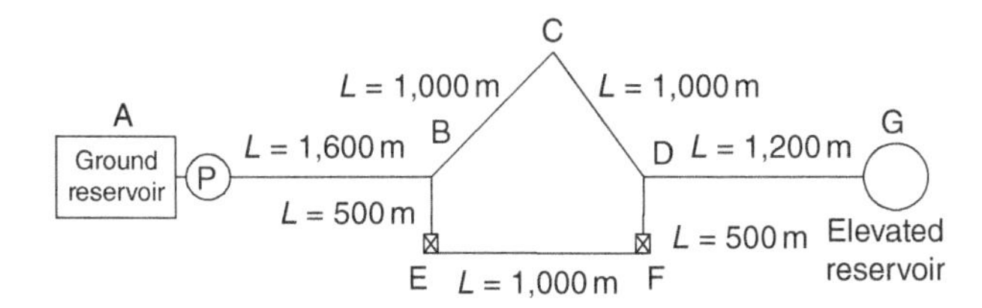

Figure 5.35 Water System for Problem 5.11.

5.12 In the water supply system shown in Fig. 5.36, the pumping station (P) delivers the water from the water treatment plant to city ABCDE. The water head delivered by the pumps is 28 m. The water level in the pumping station wet well is at 600.00 m. The value of C for all pipes is 100.

The following elevations are known:

A) 570.00 m.
B) 575.00 m.
C) 573.00 m.

a) Find the required diameter of the force main PA (d_{PA}), if the pressure at point A is not allowed to drop below 370 kPa.
b) Determine the required diameter of pipe BC (d_{BC}), if the maximum allowed head loss in the network ABC is 3‰.
c) Calculate the actual residual pressures at points A and C.

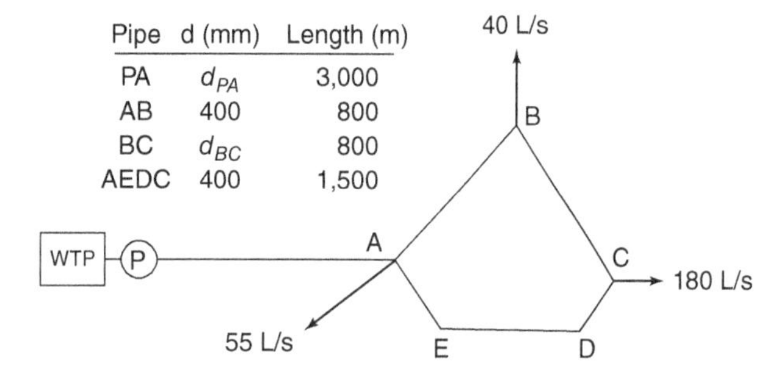

Figure 5.36 Water System for Problem 5.12

5.13 In the water supply system shown in Fig. 5.37, water is pumped from the ground reservoir to the village ABCD (elevation of 120.00 m) through the water main PA. The value of C for all pipes is 100. A water meter located on pipe DC indicates a flow of 20 L/s. The following network data are provided:

Pipe	*d* (mm)	*L* (m)
PA	500	10,000
AB	250	500
AC	200	700
AD	200	500
BC	250	1,000
CD	150	1,000

a) Determine the flow in each pipe as well as the flow delivered to point C (Q_c).
b) Calculate the pumping head, if the residual pressure at point C is not to be less than 150 kPa.
c) Find the residual pressure at point C if a different pump is used, one that can deliver half the flow into the system at the same pumping head above.

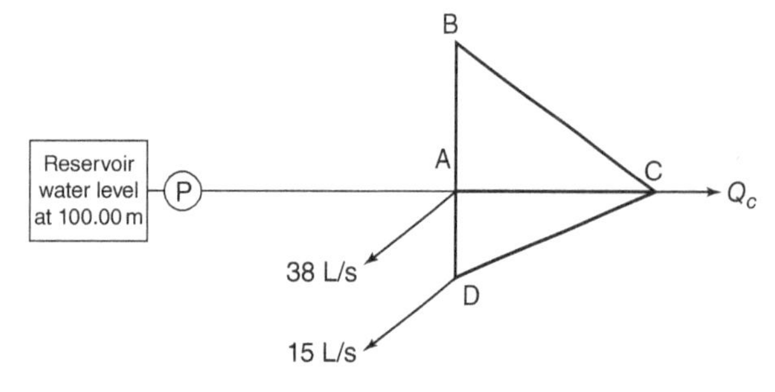

Figure 5.37 Water System for Problem 5.13.

5.14 Consider a vertical rectangular gate ($b \times h$ = 13.12 ft × 6.56 ft = 4 m × 2 m) that is vertically submerged in water so that the top of the gate is 13.12 ft (4 m) below the water surface. Determine the total resultant force on the gate and the location of the center of pressure (see Fig. 5.6).

5.15 In Fig. 5.3 the areas of the plunger A and cylinder B are 6 in.^{2} (38.71 cm^2) and 600 in.^{2} (3,871 cm^2), respectively, and the weight of B is 9,000 lb (4,086 kg). The vessel and the connecting passages are filled with oil of specific gravity 0.75. What force is required for equilibrium, neglecting the weight of A and assuming d = 16 ft (4.88 m).

5.16 Gate EBC shown in Fig. 5.13 is hinged at B and is 4 ft (1.22 m) wide. Neglecting the weight of the gate, determine the unbalanced moment due to water acting on the gate EBC. It is assumed that $\theta = 60°$, d = 8 ft (2.44 m), and a = 3 ft (0.91 m).

5.17 Oil of specific gravity 0.75 flows through the nozzle shown in Fig. 5.1 and deflects the mercury (specific gravity 13.57) in the U-tube gauge. Determine the value of h if the pressure at D is 20 psi (138.8 kN/m^2) and m is 2.75 ft (0.84 m).

5.18 The funnel shown in Fig. 5.38 is full with water. The volume of the upper part is 5.9 ft^3 (0.1671 m^3) and of the lower part is 2.6 ft^3 (0.0736 m^3). The cross-sectional area A_C of the lower part is 0.5 ft^2 (0.0464 m^2). The height of the upper part a = 5 ft (1.524 m), the height of the lower part b = 5 ft (1.524 m), and the water surface area A_W of the upper part is 20 ft^2 (1.8581 m^2). Determine the force tending to push the plug out.

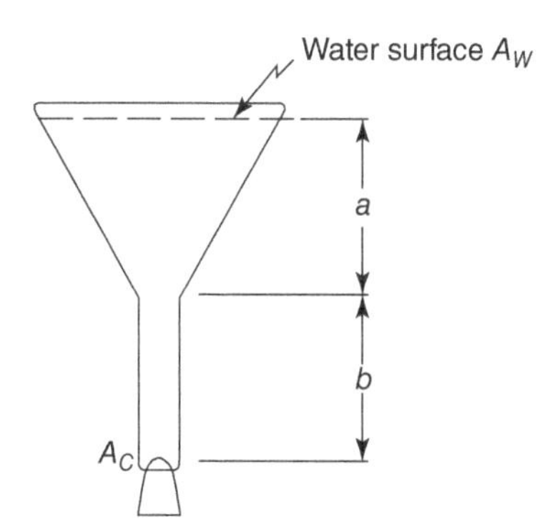

Figure 5.38 Funnel for Problem 5.18

5.19 Determine the components of the forces P_x and P_y due to the water acting on the curved area AB in Fig. 5.39 per unit length ft (m). Also locate the components of these forces P_x and P_y. The radius r is 6.5 ft (1.98 m). The center of gravity of a quadrant of a circle is located at a distance $4r/3\pi$ from either mutually perpendicular radius.

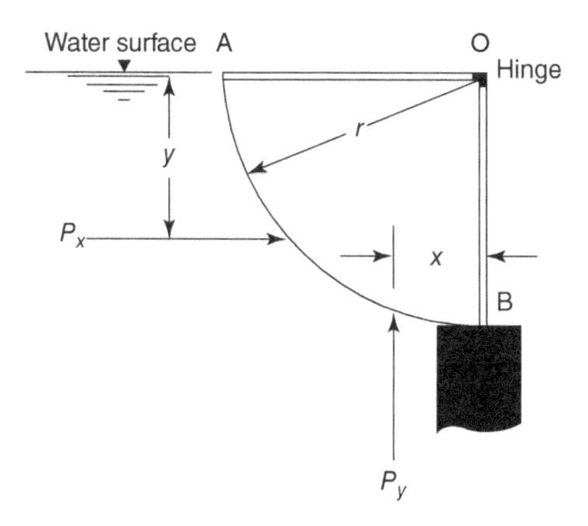

Figure 5.39 Hydrostatic Force Analysis for Problem 5.19

5.20 Determine (1) the total resultant force acting on an inclined rectangular gate shown in Fig. 5.7, and (2) the location of the pressure center. Assume a = 5 ft (1.52 m), b = 6 ft (1.83 m), h = 4 ft (1.22 m), and $\theta = 60°$.

5.21 An object is 1.5 ft (0.4572 m) wide by 2.5 ft (0.762 m) long and 1.5 ft (0.4572 m) high. It weighs 150 lb (68.1 kg) at a water depth of 10 ft (3.048 m).What is the weight of the object in air and what is its specific gravity?

5.22 A prismatic object is 8 in. (20.32 cm) thick, 8 in. (20.32 cm) wide, and 16 in. (40.64 cm) long. It weighs 12 lb (5.45 kg) at a water depth of 20 in. (50.8 cm). What is the weight of the object in air and what is its specific gravity?

5.23 A horizontal pipe carrying oil of specific gravity 0.8 is 6 in. (15.24 cm) in diameter. Near the end E, the pressure is measured to be 13.2 psi (91.6 kPa). After E, the 6-in. (15.24 cm) pipe turns 90° upward for 12 ft (3.66 m) reaching another 90° bend, which connects to an upper horizontal pipe having a diameter of 18 in. (45.72 cm) at point R where the pressure is also measured to be 8.75 psi (60.73 kPa) (Fig. 5.40). Determine

1) The oil velocities in the two horizontal pipes
2) The direction of oil flow
3) The head loss between E and R, if the oil discharge is 5 ft^3/s (0.1415 m^3/s).

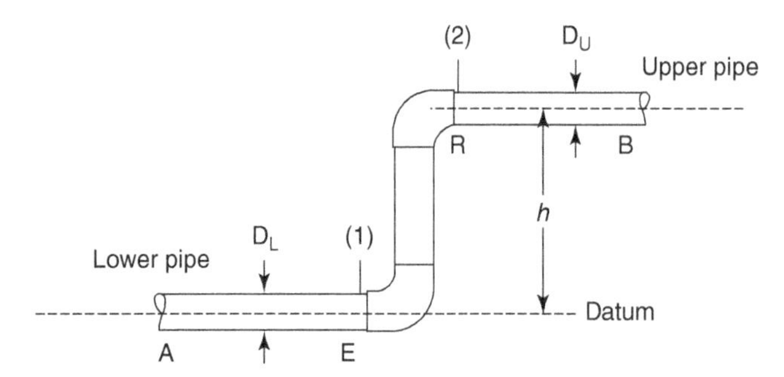

Figure 5.40 Pipeline Analysis for Problem 5.23

5.24 In the sketch shown in Fig. 5.41 a barrel whose empty weight is 60 lb (27.2 kg) contains an unknown quantity of water in which a container holding 200 lb (90.8 kg) of oil (specific gravity = 0.95) floats, and a beaker containing 10 lb (4.54 kg) of mercury (specific gravity = 13.00) floats in the oil. The beaker, the oil container, and the barrel are 6, 24, and 36 in. (0.1524, 0.6096, and 0.9144 m) in diameter, respectively. The total weight indicated by the scale is 860 lb (390.44 kg). Determine the height h of the water in the barrel.

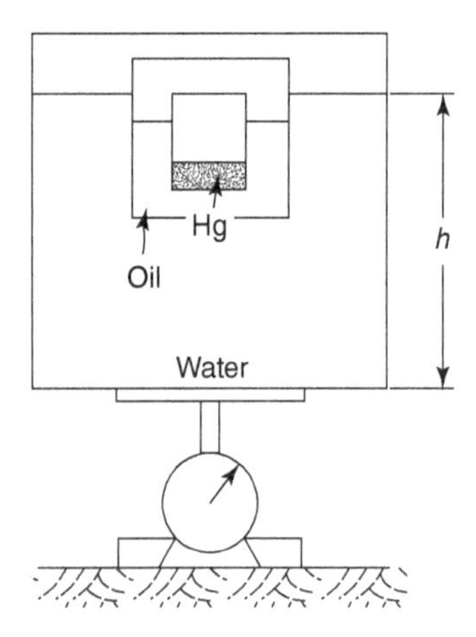

Figure 5.41 Water Barrel, Oil Container, and Mercury Beaker for Problem 5.24

5.25 Determine the force in lb (N) that each connection between the 90° elbow and the 12-in. (0.3048-m)-diameter pipeline in Fig. 5.42 will resist. Assume the water in the system is static (no flow condition) and the pipeline is pressured to 100 psig (694 kPa gauge).

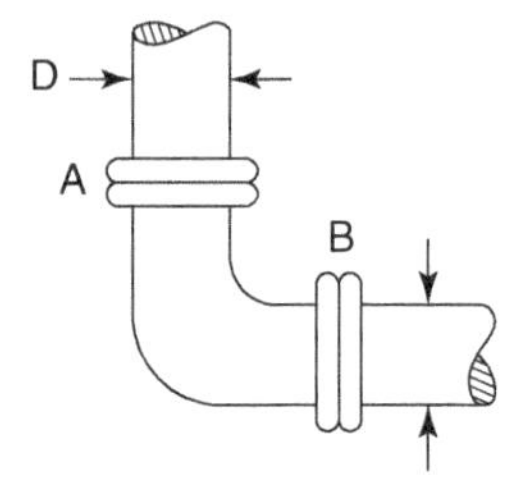

Figure 5.42 Elbow Connecting Two Pipes for Problem 5.25

5.26 A conical reducing section shown in Fig. 5.43 connects an existing 6-in. (152.4 mm) pipeline with a new 4-in. (101.6-mm)-diameter pipeline. Assume the pipeline is at 100 psig (694 kPa = 694 kN/m^2) static pressure under no flow conditions and there is no end restraint from the pipes. Determine the tensile force in lb (kN), which is exerted on the connecting reducer joint.

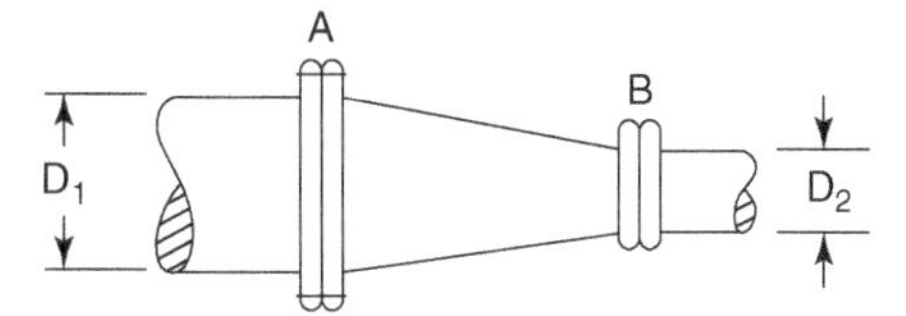

Figure 5.43 Conical Reducer Connecting Two Pipes for Problem 5.26

5.27 A rectangular access port 2 ft by 2 ft (0.6096 m by 0.6096 m) in size, shown in Fig. 5.44, seals an environmental test chamber that is pressurized to 16 psi (111.04 kPa = 111.04 kN/m^2) above external pressure. What force in lb (kN) does the port exert upon its retaining structure?

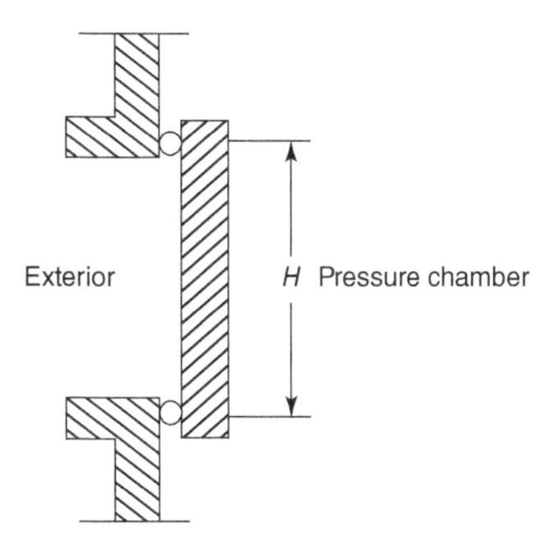

Figure 5.44 Access Port of a Pressure Chamber for Problem 5.27

5.28 Commercial chemical liquid stored in an above-ground spherical steel tank at 102°F (38.89°C) generates a gauge pressure of 80 psig (555.2 kPa = 555.2 kN/m^2). If the chemical storage tank is 6 ft (1.829 m) in diameter and has walls 0.25 in. (6.35 mm) thick, determine the maximum tensile stress in psi (kN/m^2) developed in the steel.

5.29 A steel-reinforced fiber glass cylinder butane storage tank 16 ft (4.88 m) long shown in Fig. 5.45 has hemispherically domed ends 6 ft (1.829 m) in diameter. Butane has a vapor pressure of 46 psig (319.24 kN/m^2 = 319.24 kPa) at 102°F (38.89°C) and the tank walls are 0.25 in. (6.35 mm) thick. Determine the maximum tensile stress in lb/in^2 (kN/m^2) developed in the tank.

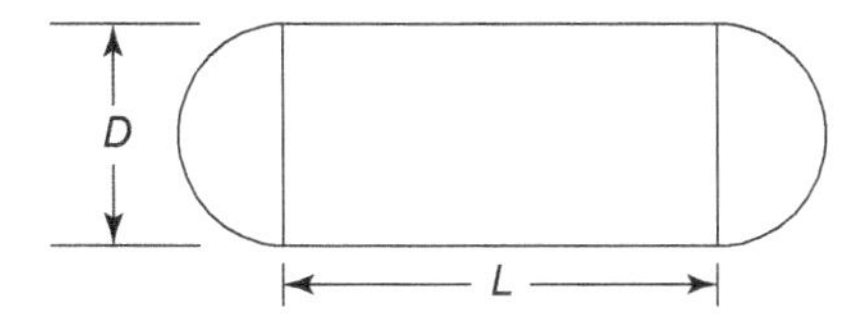

Figure 5.45 Cylindrical Liquid Storage with Hemispherical Domed Ends for Problem 5.29

5.30 Ice in an iceberg has a specific gravity of 0.91. Seawater has a specific gravity of 1.03. Determine the iceberg's percentage volume exposed in air when it is floating in seawater.

5.31 A vertical sliding gate 25 ft (7.62 m) wide and 35 ft (10.668 m) high is submerged in 45 ft (13.76 m) of water. It has a coefficient of friction equals 0.2 between its guides and its edges. The gate weighs 6.5 tons (13,000 lb = 5,902 kg).

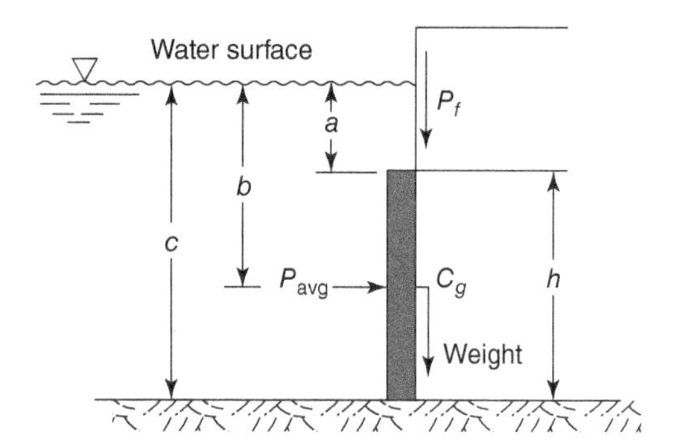

Figure 5.46 Vertical Force to Lift A Sliding Gate for Problem 5.31

Assume friction is due to the normal force of water pressure on the gate and neglect the buoyant force of the water. Determine the vertical force that is required to just lift the gate. The following values for the distances in Fig. 5.46 are $a = 10$ ft (3.048 m), $b = 27.5$ ft (8.382 m), $c = 45$ ft (13.716 m), and $h = 35$ ft (10.668 m).

5.32 A cylinder of cork is floating upright in a container partially filled with water. A vacuum is applied to the container such that the air within the vessel is partially removed. The cork will

1) Rise somewhat in the water
2) Sink somewhat in the water
3) Remain stationary
4) Turn over on its side
5) Sink to the bottom of the container.

Select the correct answer and explain why.

5.33 An incompressible fluid ($\gamma = 52$ lb/ft^3 = 8.175 kN/m^3) enters and leaves a hydraulic system with the following energy in ft-lb/lb (m-kg/kg) of fluid:

1) Potential energy z above datum:

$$\text{Entering} = 6\text{ ft} = 1.8288\text{ m.}$$
$$\text{Leaving} = 16\text{ ft} = 4.8768\text{ m.}$$

2) Kinetic energy, $v^2/2g$:

$$\text{Entering} = 6\text{ ft} = 1.8288\text{ m.}$$
$$\text{Leaving} = 11\text{ ft} = 3.3528\text{ m.}$$

3) Pressure energy, P/γ:

$$\text{Entering} = 32\text{ ft} = 9.7536\text{ m.}$$
$$\text{Leaving} = 152\text{ ft} = 46.3296\text{ m.}$$

4) Total energy:

$$\text{Entering} = 44\text{ ft} = 13.4112\text{ m.}$$
$$\text{Leaving} = 179\text{ ft} = 55.5592\text{ m.}$$

Determine the pressure increase in psi (kN/m^2) between entering and leaving liquid streams.

5.34 Consider the parallel pipe system in Fig. 5.47. The following data are known:

Pipe c is a 10 in. (254 mm) water line.
Pipe d is a 12 in. (304.8 mm) water main.
Pipe a is a 6 in. (152.4 mm) line, 1,000 ft (304.8 m) long.
Pipe b is a 6 in. (152.4 mm) line, 1,440 ft (438.9) long.
Water velocity in pipe b is 10 ft/s (3.048 m/s).

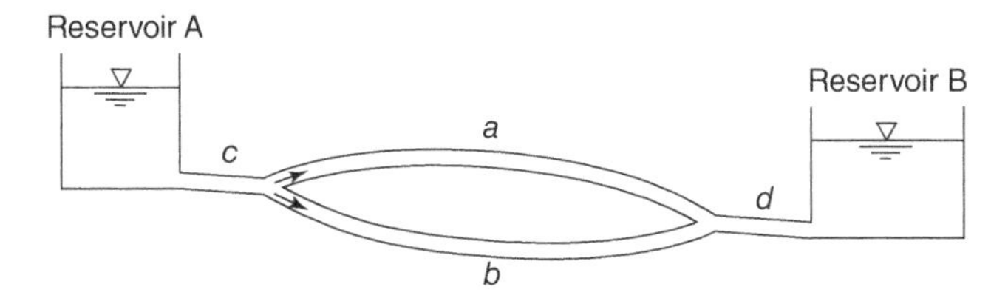

Figure 5.47 Parallel Water Pipes for Problem 5.34

Friction factors in the two pipes a and b are the same and the incidental losses are equal. Determine the water velocity in pipe a.

5.35 The expression for the Reynolds number R for a circular pipe with circular cross-section was given in Eq. (5.11) as follows:

$$\mathrm{R} = v d\rho/\mu = vd/\upsilon \tag{5.11}$$

where d is the pipe diameter, v is the water velocity, μ is the absolute viscosity, $\upsilon = \mu/\rho$ is the kinematic viscosity of the fluid, and ρ is its density. Develop an expression for Reynolds number for an open channel with rectangular cross-section in terms of the hydraulic radius instead of the pipe diameter.

5.36 Summarize the differences between laminar flow and turbulent flow in terms of

1) Motion of fluid particles
2) Energy loss
3) Velocity distribution in pipe
4) Reynolds number

5.37 The vena contracta of a sharp-edged hydraulic orifice usually occurs (select the correct answer)

1) At the geometric center of the orifice
2) At a distance of about 20% of the orifice diameter upstream from the plane of the orifice
3) At a distance equal to about one orifice diameter downstream from the plane of the orifice
4) At a distance equal to about one-half the orifice diameter downstream from the plane of the orifice
5) At a distance within 20% of the orifice diameter upstream from the plane of the orifice

5.38 Determine the coefficients of velocity, discharge, and contraction for a jet of liquid flow through an orifice. Assume the actual velocity in the contracted section of the liquid jet flowing from a 2-in. (50.8-mm)-diameter orifice is 30 ft/s (9.144 m/s), under a head of 16 ft (4.877 m). Actual flow is 0.4 ft^3/s (0.0113 m^3/s).

5.39 A flat plate, 4 ft by 4 ft (1.22 m by1.22 m), moves at 23 ft/s (7.01 m/s) normal to its plane at standard pressure. Determine the resistance of the plate assuming the drag coefficient = 1.16 for length/width ratio equal to 1 and $\gamma_{air} = 0.0752$ lb/ft^3 (0.01181 kN/m^3).

5.40 A standard orifice discharges under a head H as shown in Fig. 5.48. Apply Bernoulli equation from W to J, with datum at J. Assume the head loss of orifice is represented by Eq. (5.49):

$$h_f = \left\{\left[1/(C_v)^2\right] - 1\right\}\left(v_{jet}\right)^2/2g. \tag{5.49}$$

1) Develop the jet velocity (Eq. 5.47b)

$$v_{jet} = v = C_v(2gH)^{0.5} \tag{5.47a}$$

2) Develop the jet flow rate (Eq. 5.45):

$$Q = C_d A(2gH)^{0.5} \tag{5.45}$$

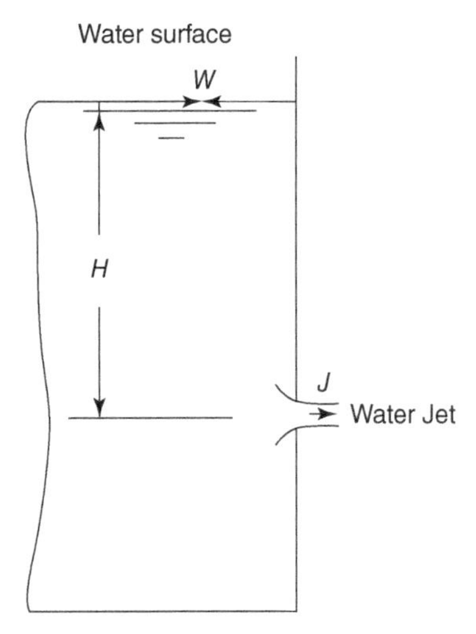

Figure 5.48 Standard Orifice for Problems 5.40 and 5.42

5.41 The pitot tube shown in Fig. 5.49 is used to measure the pressure at a point where the velocity is zero. This point is technically called the stagnation point. The pressure there is called the stagnation (or total) pressure. Assume the tube is shaped and positioned properly; a point of zero velocity is developed at B in front of the open end of the tube. Assume H_A and H_B are known, and there is no head loss. Apply the Bernoulli equation from A to B in Fig. 5.49, datum at B. Develop the equations for the determination of the velocity at A (v_A) and the pressure at B (P_B).

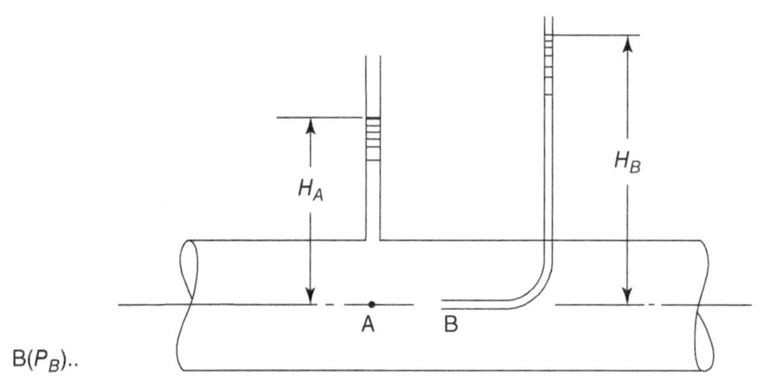

Figure 5.49 Standard Pitot Tube for Problems 5.41 and 5.43

5.42 Determine the flow from a standard orifice J shown in Fig. 5.48. The head above the orifice $H = 20$ ft (6.096 m), the orifice diameter equals 4.5 in. (114.3 mm), and discharge coefficient $C_d = 0.594$.

5.43 The pitot tube shown in Fig 5.49 has a coefficient of 0.9850. It is used to measure the velocity of water at the center of a pipe. Assume a stagnation pressure head H_B of 19 ft (5.7912 m) and the static pressure head H_A in the pipe of 16 ft (4.8768 m). Determine the velocity at A and the pressure at B.

5.44 Given an over-simplified Moody diagram (Fig. 5.50) and the original Moody diagram (Fig. 5.11), explain the applications of curves A, B, C, D, and E.

5.45 Briefly and precisely define

1) The Moody diagram
2) The Bernoulli equation
3) The Darcy–Weisbach formula
4) The Chezy formula
5) Hazen–Williams formula
6) Manning equation

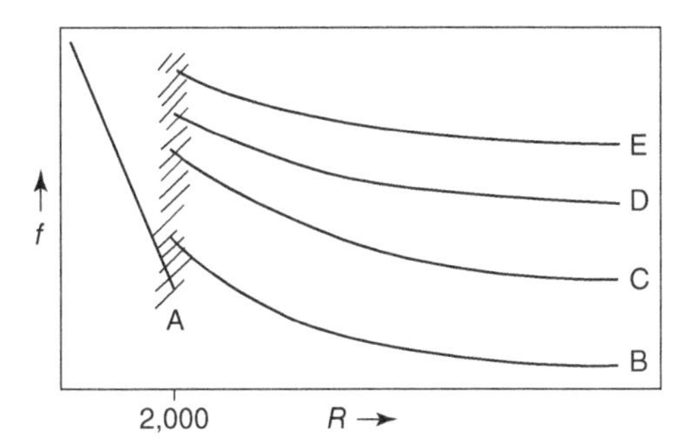

Figure 5.50 Simplified Moody Diagram for Problem 5.44

5.46 Briefly and precisely define

1) Pitot tube
2) Stagnation pressure
3) Hydraulic gradient
4) Energy gradient
5) Friction head loss

5.47 At point A in a pipeline, the elevation is 155 ft (472.44 m); the pressure is 35 psig (242.9 kPa = 242.9 kN/m^2). At point B 5,000 ft (1,524 m) downstream of the pipe, the elevation is 135 ft (41.148 m) and the pressure is 40 psig (277.6 kPa = 277.6 kN/m^2). Determine the head loss between points A and B.

5.48 A 24-in. (609.6 mm), 5,000 ft (1524 m) pipeline carries 1.6 ft^3/s (0.453 m^3/s) of water between points A and B. The head loss between A and B is 8.46 ft (2.56 m). Determine the velocity of flow and the pipe friction factor f.

5.49 A water jet shown in Fig. 5.51 flows vertically upward from a nozzle with a vertical velocity of 16 ft/s (4.8768 m/s) and a flow rate of 0.03 ft^3/s (0.0008496 m^3/s). Above the nozzle at a distance h = 6 in. (152.4 mm), there is a horizontal plate weighing 0.5 lb (0.227 kg). Determine the reaction force F which is required to hold the plate stationary.

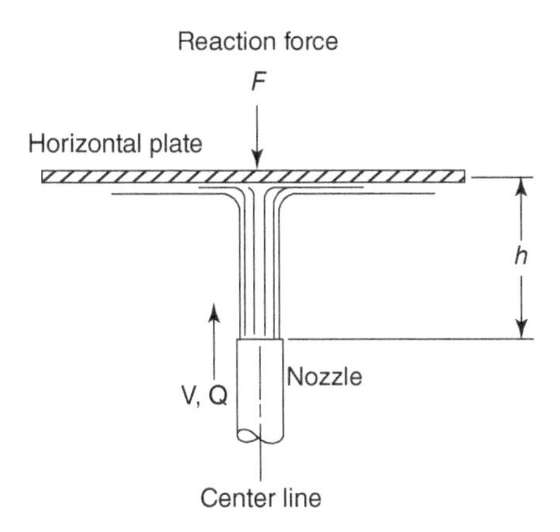

Figure 5.51 Water Jet from A Nozzle for Problem 5.49

5.50 A rectangular wood block floats in water with 6.2 in. (157.48 mm) projecting above the water surface. If the same wood block is placed in an organic solvent of specific gravity 0.81, the wood block projects 4.2 in. (106.68 mm) above the surface of the organic solvent. Determine the specific gravity of the wood block.

Bibliography

Al-Dhowalia, K., and Shammas, N. K., Leak Detection and Quantification of Losses in a Water Network, *Int. J. Water Resour. Develop.*, **7**, 1, 30–38, 1991.

Bradley, J. N., and Thompson, L. R., Friction Factors for Large Conduits Flowing Full, *Bur. Reclamation Eng Monograph 7*, Denver, CO, 1951.

Colebrook, C. F., Turbulent Flow in Pipes, with Particular Reference to the Transition Region Between the Smooth and Rough Pipe Laws, *J. ICE*, **11**, 133–156, 1939.

Colebrook, C. F., and White, C. M., The Reduction of Carrying Capacity of Pipes with Age, *J. ICE*, **7**, 99–118, 1937–1938.

Fair, G. M., Geyer, J. C., and Okun, D. A., *Water and Wastewater Engineering, vol. 1: Water Supply and Wastewater Removal*, John Wiley & Sons, Inc., New York, 1966.

Fair, G. M., Geyer, J. C., and Okun, D. A., *Elements of Water Supply and Wastewater Disposal*, John Wiley & Sons, Inc., New York, 1971.

Flynn, P. J., *Flow of Water in Open Channels, Pipes, Sewers, Conduits, Etc.*, Kessinger Publishing, Whitefish, MT, 2007.

Gile, R., Evett, J., and Liu, C., *Fluid Mechanics and Hydraulics*, 3rd ed., McGraw-Hill Publishing, New York, 1994.

Haestad Methods Water Solutions, *Computer Applications in Hydraulic Engineering*, 7th ed., Bentley Institute Press, Exton, PA, 2007.

Mays, L. W. (Ed.), *Hydraulic Design Handbook*, McGraw-Hill, New York, 1999.

Mays, L. W., *Water Resources Engineering*, 2nd ed., John Wiley & Sons, Inc., New York, 2011.

Munson, B. R., Young, D. F., and Okiishi, T. H., *Fundamentals of Fluid Mechanics*, 3rd ed., John Wiley & Sons, Inc., New York, 1998.

Newnan, D. G., *Engineer-In-Training License Review*, 15th ed., Engineering Press, San Jose, CA, 1998.

Newnan, D. G., *Civil Engineering: License Review*, Engineering Press, San Jose, CA, 2004.

Parmakian, J., Air Inlet Valves for Steel Pipe Lines, *Trans. Am. Soc. Civil Eng.*, **115**, 1, 438–443, 1950.

Prior, J. C., Investigation of Bell-And-Spigot Joints in Cast-Iron Water Pipes, Part I: Pullout Strength; Part II: Bell Strength; Part III: Harness Strength, *Ohio State Uni. Eng. Exp. St. Bull.*, **87**, 1935.

Qasim, S. R., Motley, E. M., and Zhu, G., *Water Works Engineering – Planning, Design and Operation*, Prentice Hall, Upper Saddle River, NJ, 2000.

Rich, G. R., *Hydraulic Transients*, McGraw-Hill, New York, 1951.

Roberson, J. A., Cassidy, J. J., and Chaudhry, M. H., *Hydraulic Engineering*, 2nd ed., John Wiley & Sons, Inc., New York, 1998.

Rouse, H., *Engineering Hydraulics*, John Wiley & Sons, Inc., New York, 1950.

Shammas, N. K., and Al-Dhowalia, K., Effect of Pressure on Leakage Rate in Water Distribution Networks, *J. Eng. Sci.*, **5**, 2, 155–312, 1993.

Shammas, N. K., and Wang, L. K., *Water Supply and Wastewater Removal*, 3rd ed., John Wiley & Sons, Inc., New York, 824 pp, 2011.

Shannon, W. L., Prediction of Frost Penetration, *J. New England Water Works Assoc.*, **59**, 356, 1945.

Sharma, A. K., and Swamee, P. K., *Design of Water Supply Pipe Networks*, John Wiley & Sons, Inc., New York, February 2008.

Swindin, N., *The Flow of Liquids in Pipes*, Kessinger Publishing, Whitefish, MT, 2007.

Tullis, J. P., *Hydraulics of Pipelines: Pumps, Valves, Cavitation, Transients*, John Wiley & Sons, Inc., New York, 1989; online publication December 12, 2007.

Wang, L. K., and Wang, M. H. S., *Standard and Guides of Water Treatment and Water Distribution Systems*, US Dept. of Commerce, National Technical Information Service, Springfield, VA, PB88-177902, 345 pp, 1988.

Williams, S., and Hazen, A., *Hydraulic Tables*, 3rd ed., John Wiley & Sons, Inc., New York, 1933.

6

Water Distribution Systems: Components, Design, and Operation

6.1 Water Supply Systems and Distribution

A complete water supply system consists of (a) raw water supply source facilities from which high-quality raw surface water or groundwater can be obtained; (b) primary water transmission lines or water mains that deliver raw water from its source to a water treatment plant; (c) a water treatment plant or facility that treats the raw water to meet the government required drinking water standards, provides chlorine residue, corrosion control and fire protection pressure, and stores the finished water; (d) secondary water transmission lines that deliver the finished water from water treatment plant to a water distribution grit system; (e) a water distribution grit system (with or without additional water storage for fire protection) that supplies finished water to the homes, businesses, and industries, and provides sufficient water pressure for fire protection. The primary water transmission lines can be either open channels or large diameter pipes (i.e., water mains) for transporting raw water to a facility for treatment. The secondary water transmission lines are normally the water mains with water flowing full under pressure for transmission of the finished water to the customers in the water distribution grit. This chapter introduces the water distribution systems and their components, design, and operation.

Apart from a few scattered taps and takeoffs along their feeder conduits, distribution systems for public water supplies are networks of pipes within networks of streets. Street plan, topography, and location of supply works, together with service storage, determine the type of distribution system and the type of flow through it. Although service reservoirs are often placed along lines of supply, where they may usefully reduce conduit pressures, their principal purpose is to satisfy network requirements. Accordingly, they are, in fact, components of the distribution system, not of the transmission system.

6.1.1 One- and Two-Directional Flow

The type of flow creates four systems, as sketched in Fig. 6.1. Hydraulic grade lines and residual pressures within the areas served, together with the volume of distribution storage, govern the pipe sizes within the network. It is plain that flows from opposite directions increase system capacity. With *two-directional flow* in the main arteries, a *pumped* or *gravity supply,* or a *service reservoir,* feeds into opposite ends of the distribution system or through the system to *elevated storage* in a reservoir, tank, or standpipe situated at the far end of the area of greatest water demand. Volume and location of service storage depend on topography and water needs.

6.1.2 Distribution Patterns

Two distribution patterns emerge from the street plan: (a) a *branching pattern* on the outskirts of the community, in which ribbon development follows the primary arteries of roads (Fig. 6.2a), and (b) a *gridiron pattern* within the built-up portions of the community where streets crisscross and water mains are interconnected (Fig. 6.2b and c). Hydraulically, the gridiron system has the advantage of delivering water to any spot from more than one direction and of avoiding dead-ends. The system is strengthened by substituting for a central feeder a *loop* or belt of feeders that supply water to the *congested,* or *high-value,* district from at least two directions. This more or less doubles the delivery of the grid (Fig. 6.2c). In large systems, feeders are constructed as pressure tunnels, pressure aqueducts, steel pipes, or reinforced-concrete pipes. In smaller communities the entire distribution system may consist of ductile-iron pipes. Ductile iron is, indeed, the most common material for water mains, but plastics, in general, in the case of small supplies, are also important.

Water and Wastewater Engineering: Hydraulics, Hydrology and Management, Volume 1, Fourth Edition.
Lawrence K. Wang, Mu-Hao Sung Wang, and Nazih K. Shammas.

Companion website: www.wiley.com/go/Wang/Waterandwastewater4e

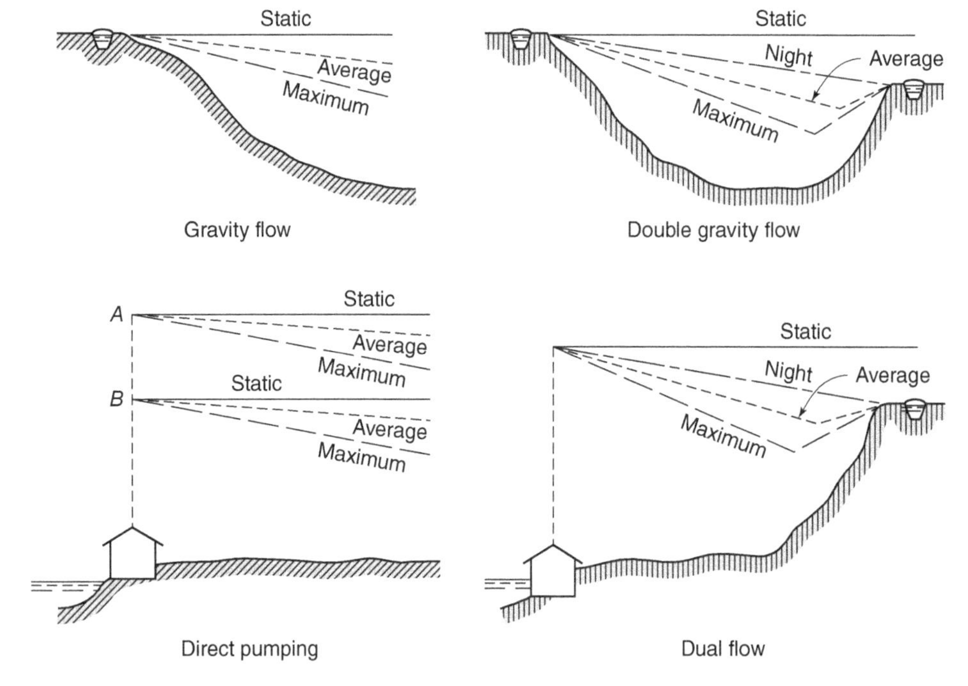

Figure 6.1 One- and Two-Directional Flow in Distribution Systems

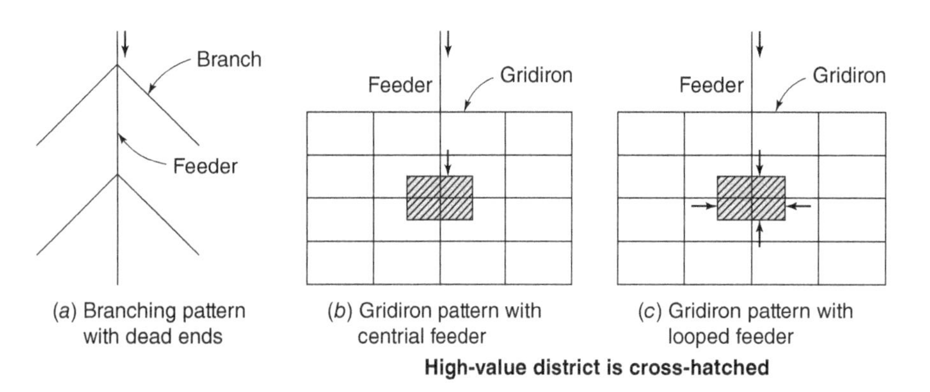

Figure 6.2 Patterns of Water-Distribution Systems. (a) Branching pattern with dead ends, (b) Gridiron pattern with central feeder. (c) Gridiron pattern with looped feeder

6.1.3 Pipe Grids

The gridiron system of pipes stretching over all but the outlying sections of a community (Fig. 6.2) may consist of *single* or *dual mains*. In the Northern Hemisphere, single mains are customarily laid on the north and east sides of streets for protection against freezing. In the Southern Hemisphere, the south and east sides are used. Valves are generally installed as follows: three at crosses, two at tees, and one on single-hydrant branches. In dual-main systems, *service headers* are added on the south (north in Southern Hemisphere) and west sides of streets, and piping is generally placed beneath the sidewalks. Hydraulically, the advantages of dual-main systems over single-main systems are that they permit the arrangement of valves and hydrants in such ways that breaks in mains do not impair the usefulness of hydrants and do not *dead-end* mains.

Dual-main systems must not be confused with dual-water supplies: a high-grade supply for some purposes and a low-grade supply for others.

6.1.4 High and Low Services

Sections of the community too high to be supplied directly from the principal, or *low-service,* works are generally incorporated into separate distribution systems with independent piping and service storage. The resulting *high services* are normally fed by pumps that take water from the main supply and boost its pressure as required. Areas varying widely in elevation may be formed into intermediate districts or zones. Gated connections between the different systems are opened by hand during emergencies or go into operation automatically through pressure-regulating valves. Because high-service areas are commonly small and low-service areas are commonly large, support from high-service storage during breakdowns of the main supply is generally disappointing.

Before the days of high-capacity, high-pressure, motorized fire engines, conflagrations in the congested central, or *high-value,* district of some large cities were fought with water drawn from independent high-pressure systems of pipes and hydrants (Boston, Massachusetts, still maintains a separate fire supply). Large industrial establishments, with heavy investments in plant, equipment, raw materials, and finished products, that are concentrated in a small area are generally equipped with high-pressure fire supplies and distribution networks of their own. When such supplies are drawn from sources of questionable quality, some regulatory agencies enforce rigid separation of private fire supplies and public systems. Others prescribe *protected cross-connections* incorporating *backflow preventers* that are regularly inspected for tightness.

6.1.5 Service to Premises

Water reaches individual premises from the street main through one or more *service pipes* that tap into the distribution system. The building supply between the public main and the takeoffs to the various plumbing fixtures or other points of water use is illustrated in Fig. 6.3. Small services are made of cement-lined iron or steel, brass of varying copper content, copper, and plastics such as polyethylene or polyvinyl chloride (PVC). Because lead and lead-lined pipes may corrode and release lead to the water, they are no longer installed afresh. For large services, coated or lined ductile-iron pipe is often

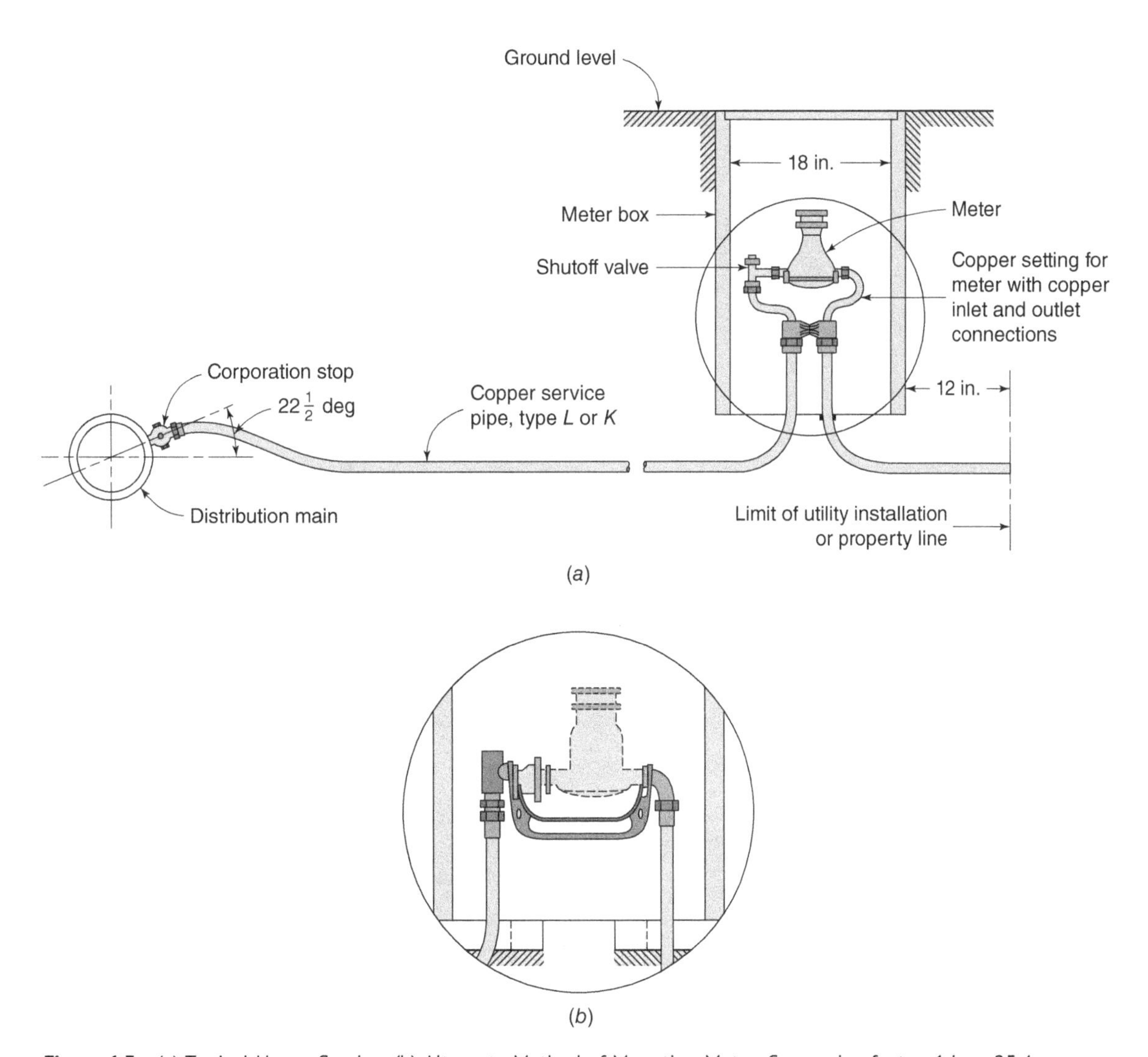

Figure 6.3 (a) Typical House Service; (b) Alternate Method of Mounting Meter. Conversion factor: 1 in. = 25.4 mm

employed. For dwellings and similar buildings, the minimum desirable size of service is 3/4 in. (19 mm). Pipe-tapping machines connect services to the main without shutting off the water. They also make large connections within water distribution systems.

6.2 System Components

Pipes, valves (see Chapter 5), and *hydrants* are the basic elements of reticulation systems (Figs. 6.4–6.7). Their dimensioning and spacing are based on experience that is normally precise enough in its minimum standards to permit roughing in all but the main arteries and feeders. Common standards include the following:

Pipes	
Smallest pipes in gridiron	6 in. (150 mm)
Smallest branching pipes (dead ends)	8 in. (200 mm)
Largest spacing of 6-in. grid (8-in. pipe used beyond this value)	600 ft (180 m)
Smallest pipes in high-value district	8 in. (200 mm)
Smallest pipes on principal streets in central district	12 in. (300 mm)
Largest spacing of supply mains or feeders	2,000 ft (600 m)
Gates	
Largest spacing on long branches	800 ft (240 m)
Largest spacing in high-value district	500 ft (150 m)
Hydrants	
Largest spacing between hydrants, see Table 6.1	
Maximum distance from any point on street/road frontage to a hydrant, see Table 6.1	

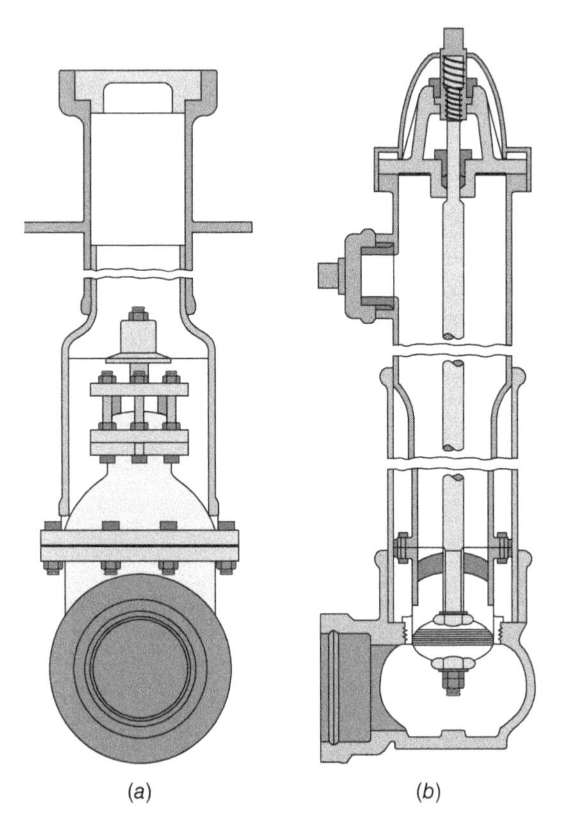

Figure 6.4 (a) Gate Valve and Extendable Valve Box; (b) Post Fire Hydrant with Compression Valve

Figure 6.5 Fire Hydrant (*Source:* http://en.wikipedia.org/wiki/Image:Downtown_Charlottesville_fire_hydrant.jpg)

Figure 6.6 Cast-Iron Gate Valve Being Installed on a New Connection to a Fire Hydrant (*Source:* http://en.wikipedia.org/wiki/Image:Municipal GateValve.JPG)

Figure 6.7 Internals of a Large Butterfly Valve (*Source:* http://upload.wikimedia.org/wikipedia/commons/d/d5/Yagisawa_power_station_inlet_valve.jpg)

Table 6.1 Number and Distribution of Fire Hydrants

Fire-flow Requirement (gpm)	Minimum Number of Hydrants	Average Spacing Between Hydrants[a,b,c] (ft)	Maximum Distance from Any Point on Street or Road Frontage to a Hydrant[d] (ft)
1,750 or less	1	500	250
2,000–2,250	2	450	225
2,500	3	450	225
3,000	3	400	225
3,500–4,000	4	350	210
4,500–5,000	5	300	180
5,500	6	300	180
6,000	6	250	150
6,500–7,000	7	250	150
7,500 or more	8 or more[e]	200	120

Conversion factors: 1 ft = 0.3048 m; 1 gpm = 3.785 L/ min .

[a] Reduce by 100 ft (30 m) for dead-end streets or roads.

[b] Where streets are provided with median dividers, which can be crossed by firefighters pulling hose lines, or where arterial streets are provided with four or more traffic lanes and have a traffic count of more than 30,000 vehicles per day, hydrant spacing shall average 500 ft (150 m) on each side of the street and be arranged on an alternating basis up to a fire-flow requirement of 7,000 gpm (26,500 L/min) and 400 ft (122 m) for higher fire-flow requirements.

[c] Where new water mains are extended along streets where hydrants are not needed for protection of structures or similar fire problems, fire hydrants shall be provided at spacing not to exceed 1,000 ft (3,800 m) to provide for transportation hazards.

[d] Reduce by 50 ft (15 m) for dead-end streets or roads.

[e] One hydrant for each 1,000 gpm (3,800 L/min) or fraction thereof.

The choice of pipe sizes depends on occupancy of the properties along mains (whether residential, commercial, or industrial), their water uses, and the fire risks. Hazards at refineries, chemical plants, and lumber yards require special consideration.

According to the *International Fire Code* (IFC 2006), fire hydrants are required to be provided where a building is more than 400 ft (122 m) from a hydrant, except for buildings equipped with automatic sprinkler systems for which the distance requirement is 600 ft (183 m). A 3-ft (914-mm) clear space is required around the hydrants.

The minimum number of fire hydrants and their average spacing have to be within the requirements listed in Table 6.1. The placement of hydrants should be chosen in such a way that the maximum distance of all points on streets and access roads adjacent to a building is not more than the distance specified by the IFC as shown in Table 6.1.

6.3 System Capacity

The capacity of distribution systems is dictated by domestic, industrial, and other normal water uses and by the *standby* or *ready-to-serve* requirements for firefighting. Pipes should be able to carry the maximum *coincident* draft at velocities that do not produce high pressure drops and surges. Velocities of 2–5 ft/s (0.60–1.50 m/s) and minimum pipe diameters of 6 in. (150 mm) are common in North American municipalities. Capacity to serve is not merely a function of available rate of draft; it is also a function of available pressure. The water must rise to the upper stories of buildings of normal height and must flow from hydrants, directly or through pumpers, to deliver needed fire streams through fire hoses long enough to reach the fire. If there were no fire hazard, the hydraulic capacity of distribution systems would have to equal the maximum demand for domestic, industrial, and other general uses.

The general firefighting requirements according to the IFC are summarized in Chapter 4, Section 4.4.2 and Table 4.13. To these requirements for firefighting must be added a coincident demand of 40–50 gpcd in excess of the average consumption rate for the area under consideration. In small communities or limited parts of large-distribution systems, pipe sizes are controlled by fire demand plus coincident draft. In the case of main feeder lines and other central works in large communities or large sections of metropolitan systems, peak hourly demands may determine the design.

6.4 System Pressure

For normal drafts, water pressure at the street line must be at least 20 psig (140 kPa) to let water rise three stories and overcome the frictional resistance of the house-distribution system, but 40 psig (280 kPa) is more desirable. Business blocks are supplied more satisfactorily at pressures of 60–75 psig (420–520 kPa). To supply their upper stories, tall buildings must boost water to tanks on their roofs or in their towers and, often, also to intermediate floors.

Fire demand is commonly gauged by the *standard fire stream*: 250 gpm (946 L/min) issuing from a $1^1/_8$-in. (28.6 mm) nozzle at a pressure of 45 psig (312.3 kPa) at the base of the tip. When this amount of water flows through $2^1/_2$-in. (63.5 mm) rubber-lined hose, the frictional resistance is about 15 psi per 100 ft of hose (3.42 kPa/m). Adding the hydrant resistance and required nozzle pressure of 45 psig (312.3 kPa) then gives the pressure needs at the hydrant, as shown in Table 6.2. A standard fire stream is effective to a height of 70 ft (21.34 m) and has a horizontal carry of 63 ft (19.20 m).

Because hydrants are normally planned to control areas within a radius of 200 ft (61 m), Table 6.2 shows that direct attachment of fire hose to hydrants (hydrant streams) calls for a residual pressure at the hydrant of about 75 psig (520.5 kPa).

To maintain this pressure during times of fire, system pressures must approach 100 psig (694 kPa). This has its disadvantages, among them danger of breaks and leakage or waste of water approximately in proportion to the square root of the pressure. Minimum hydrant pressures of 50 psig (347 kPa) cannot maintain standard fire streams after passing through as little as 50 ft (15 m) of hose.

Motor pumpers commonly deliver up to 1,500 gpm (5,677 L/min) at adequate pressures. Capacities of 20,000 gpm (75,700 L/min) are in sight, with single streams discharging as much as 1,000 gpm (3,785 L/min) from 2-in.(50-mm) nozzles. To furnish domestic and industrial draft and keep pollution from entering water mains by seepage or failure under a vacuum, fire engines should not lower pressures in the mains to less than 20 psig (140 kPa). For large *hydrant* outlets, the safe limit is

Table 6.2 Hydrant Pressures for Different Lengths of Fire Hose

Length of Hose (ft)	Required Pressure (psig)
100	63
200	77
300	92
400	106
500	121
600	135

Conversion factors: **1 ft = 0.3048 m;1 psig = gage pressure 6.94 kPa.**

sometimes set at 10 psig (70 kPa). In a real way, modern firefighting equipment has eliminated the necessity for pressures much in excess of 60 psig (420 kPa), except in small towns that cannot afford a full-time, well-equipped fire department.

6.5 Field Performance of Existing Systems

The hydraulic performance of existing distribution systems is determined most directly and expeditiously by pressure surveys and hydrant-flow tests. Such tests should cover all typical portions of the community: high-value district, residential neighborhoods and industrial areas of different kinds, outskirts, and high-service zones. If need be, tests can be extended into every block. The results will establish available pressures and flows and existing deficiencies. These can then be made the basis of hydraulic calculations for extensions, reinforcements, and new gridiron layouts. Follow-up tests can show how successful the desired changes have been.

Pressure surveys yield the most rudimentary information about networks; if they are conducted both at night (minimum flow) and during the day (normal demand), they will indicate the hydraulic efficiency of the system in meeting common requirements. However, they will not establish the probable behavior of the system under stress, for example, during a serious conflagration.

Hydrant-flow tests commonly include (a) observation of the pressure at a centrally situated hydrant during the conduct of the test and (b) measurement of the combined flow from a group of neighboring hydrants. Velocity heads in the jets issuing from the hydrants are usually measured by hydrant *pitot tubes*. If the tests are to be significant, (a) the hydrants tested should form a group such as might be called into play in fighting a serious fire in the district under study, (b) water should be drawn at a rate that will drop the pressure enough to keep it from being measurably affected by normal fluctuations in draft within the system, and (c) the time of test should coincide with drafts (domestic, industrial, and the like) in the remainder of the system, reasonably close to *coincident* values.

The requirements of the IFC are valuable aids in planning hydrant-flow tests. A layout of pipes and hydrants in a typical flow test is shown in Fig. 6.8, and observed values are summarized in Table 6.3. This table is more or less self-explanatory. The initial and residual pressures were read from a Bourdon gauge at hydrant 1. Hydrants 2, 3, 4, and 5 were opened in quick succession, and their rates of discharge were measured simultaneously by means of hydrant pitots. A test such as this does not consume more than 5 min, if it is conducted by a well-trained crew.

Necessary hydrant-flow calculations for the flow test may be worked out using Eqs. 6.1 and 6.2 and are recorded as shown in Table 6.3.

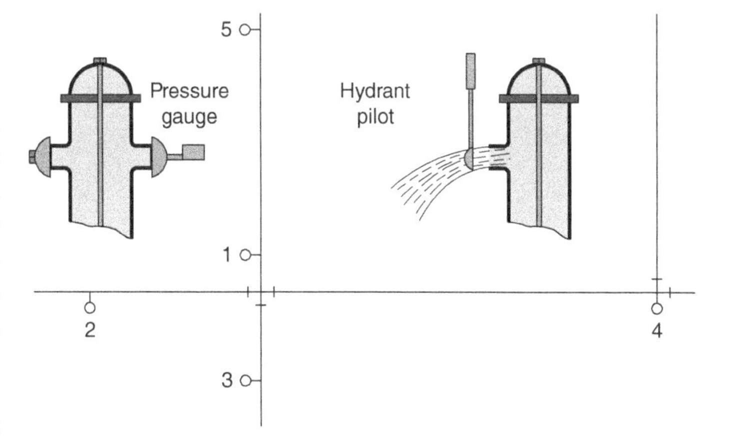

Figure 6.8 Location of Pipes and Hydrants in Flow Test and Use of Hydrant Pitot and Pressure Gauge (see Table 6.3 and Fig. 6.9)

Table 6.3 Record of a Typical Hydrant-Flow Test

Conditions of Test	Observed Pressure at Hydrant 1 psig	Discharge Velocity Head, psig	Calculated Flow (*Q*), gpm	Remarks
All hydrants closed	74	0	0	All hydrant outlets are $2^1/_2$ in. in diameter.
Hydrant 2 opened, 1 outlet	NA	13.2	610	Total Q = 2,980 gpm
Hydrant 3 opened, 2 outlets	NA	9.6	2 × 520	Calculated engine streams = 4,200 gpm
Hydrant 4 opened, 1 outlet	NA	16.8	690	
Hydrant 5 opened, 1 outlet	46	14.5	640	
All hydrants closed	74	0	0	

Conversion factors: **1 psig = gauge pressure 6.94 kPa; 1 gpm = 3.785 L/ min; 1 in. = 25.4 mm**; NA = not available

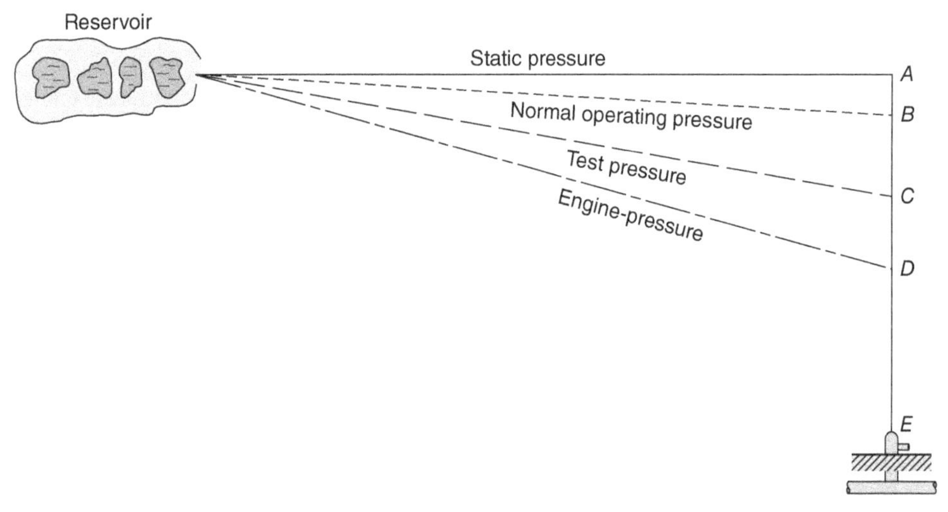

Figure 6.9 Pressure and Discharge Relations Established by Hydrant-Flow Test (see Fig. 6.8 and Table 6.3). A: Static water table. B: No hydrant discharge. Pressure = 74 psig (514 kPa gauge); pressure drop P_0 due to coincident draft Q. C: Hydrant discharge. Pressure = 46 psig (319 kPa gauge); pressure drop P_1 = (74 − 46) = 28 psi (194 kPa) accompanies discharge of Q_1 = 2,980 gpm (908 L/min). D: Engine streams. Pressure 20 psig (140 kPa gauge); pressure drop P_2 = (74 − 20) = 54 psi (375 kPa) accompanies discharge Q_2 = 4,200 gpm (1,280 L/min). E: Hydrant 1, recording residual pressure of hydrant groups shown in Fig. 6.5

For outlets of diameter d (in.), the discharge Q in gpm is

$$Q = 30cd^2\sqrt{p} \quad \text{(US customary units)} \tag{6.1}$$

where p is the pitot reading in psig and c is the coefficient of hydrant discharge.

Equation 6.2 is the hydrant discharge equation using the SI units:

$$Q = 0.0668cd^2\sqrt{p} \quad \text{(SI units)} \tag{6.2}$$

where p is the pitot reading in kPa, c is the hydrant discharge coefficient, d is the outlet diameter in mm, and Q is the hydrant discharge in L/min.

Example 6.1 Hydrant Discharge
Determine the discharge rate of a hydrant using a smooth well rounded 2.5 in. (63.5 mm) outlet at pressures of 13.2, 9.6, 16.8, and 14.5 psig (91.61, 66.62, 116.59, and 100.63 kPa).

Solution 1 (US Customary System):

For smooth well-rounded 2.5 in. outlets, $c = 0.9$
Refer to Table 6.3:

$$Q = 30cd^2\sqrt{p}$$

$$Q = 30 \times 0.9 \times 2.5^2\sqrt{p}$$

$$Q = 169\sqrt{p}$$

For $P = 13.2$ psig, $Q = 169\sqrt{13.2} = 610$ gpm

For $P = 9.6$ psig, $Q = 169\sqrt{9.6} = 520$ gpm

For $P = 16.8$ psig, $Q = 169\sqrt{16.8} = 690$ gpm

For $P = 14.5$ psig, $Q = 169\sqrt{14.5} = 640$ gpm

Solution 2 (SI System):

For smooth well-rounded 63.5 mm outlets, $c = 0.9$
Refer to Table 6.3:

$$Q = 0.0668\ cd^2\sqrt{p}$$

$$Q = 0.0668 \times 0.9 \times 63.5^2\sqrt{p}$$

$$Q = 242.42\ \sqrt{p}$$

For $p = 91.61$ kPa, $Q = 2,320$ L/min

For $p = 66.62$ kPa, $Q = 1,979$ L/min

For $p = 116.59$ kPa, $Q = 2,618$ L/min

For $p = 100.63$ kPa, $Q = 2,432$ L/min

Pressure–discharge relations established in this test are illustrated in Fig. 6.6. If the true static pressure is known, a more exact calculation is possible, although the additional labor involved is seldom justified. In accordance with the common hydraulic analysis of Borda's mouthpiece, a pressure gauge inserted in a hydrant in juxtaposition to the hydrant outlet to be opened will also record the discharge pressure otherwise measured by hydrant pitots.

Hydrant tests are sometimes made to ascertain the capacity of individual hydrants and advertise it to firefighters (particularly to engine companies summoned from neighboring towns) by painting the bonnet a suitable color. The weakness of this practice is its restriction of flow measurements to single hydrants. In firefighting, groups of hydrants are normally brought into action. Tests of individual hydrants may be quite misleading.

6.6 Office Studies of Pipe Networks

No matter how energetically distribution systems are field-tested, needed extensions and reinforcements of old networks and the design of new ones can be adequately identified only by office studies. Necessary analysis presupposes familiarity with processes of hydraulic computation, including high-speed computers. Even without computers, however, the best processes can be so systematized as to make their application a matter of simple arithmetic and pipe-flow tables, diagrams, or slide rules. Useful methods of analysis are

1) Sectioning
2) Relaxation
3) Pipe equivalence
4) Computer programming and electrical analogy.

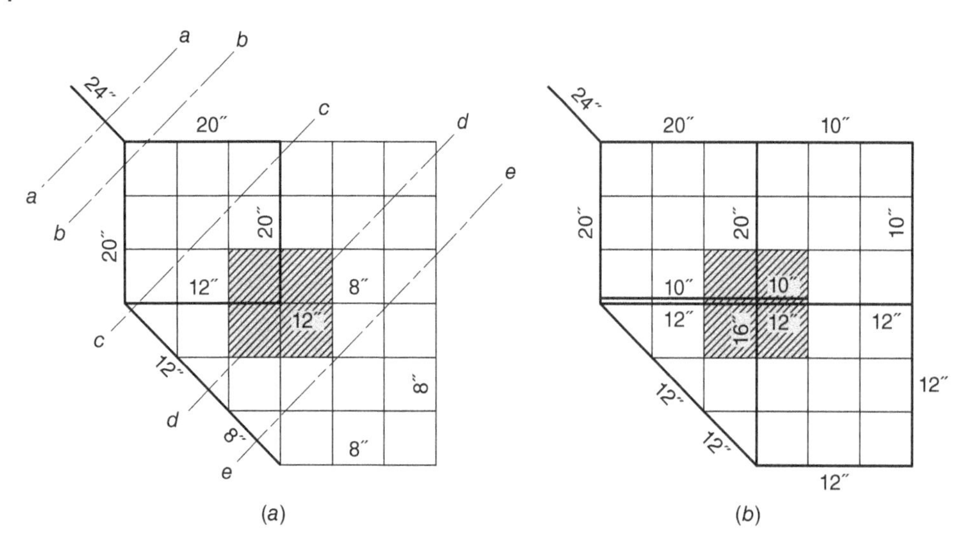

Figure 6.10 Plan of Network Analyzed by Method of Sections (Example 6.2). (a) Existing system; (b) recommended system (unless otherwise indicated, pipe diameters are 6 in. (150 mm). The high-value district is cross-hatched). Conversion factor: 1″ = 1 in. = 25.4 mm

6.6.1 Sectioning

Sectioning is an approximate and, in a sense, *exploratory method,* simple in concept and application and widely useful, provided its limitations are clearly understood. Similar in concept is the circle method, which is usually confined to cutting pipes tributary to a central fire hydrant or group of hydrants at the center of a circle.

Use of the section method is illustrated in Fig. 6.10 and Example 6.2. These are the required steps:

1) Cut the network by a *series of lines*, not necessarily straight or regularly spaced but chosen with due regard to the assumed sources and loads and the estimated location of the piezometric contours. A first series of lines may well cut the distribution piping substantially *at right angles* to the general direction of flow, that is, *perpendicular* to a line drawn from the supply conduit to the high-value district (Fig. 6.7). Further series may be oriented in some other critical direction, for example, horizontally and vertically in Fig. 6.7. For more than one supply conduit, the sections may be curved to intercept the flow from each conduit.
2) Estimate how much water must be supplied to areas *beyond* each section. Base estimates on a knowledge of the population density and the general characteristics of the zone: residential, commercial, and industrial. The water requirements comprise (a) the normal, coincident draft, here called the domestic draft, and (b) the fire demand (Table 4.13). Domestic use decreases progressively from section to section, as population or industry is left behind; fire demand remains the same until the high-value district has been passed, after which it drops to a figure applicable to the type of outskirt area.
3) Estimate the distribution system *capacity* at each section across the piping. To do this, (a) tabulate the number of pipes of each size cut, and count only pipes that deliver water in the general direction of flow; and (b) determine the average available hydraulic gradient or frictional resistance, which depends on the pressure to be maintained in the system and the allowable pipe velocity. Ordinarily, hydraulic gradients lie between 1‰ (per thousand) and 3‰, and velocities range from 2 to 5 ft/s (0.60 to 1.5 m/s). Author note: important correction: 1‰ means one per thousand.
4) For the available, or desirable, *hydraulic gradient,* determine the capacities of existing pipes and sum them for total capacity.
5) Calculate the *deficiency* or difference between required and existing capacity.
6) For the available, or desirable, hydraulic gradient, *select the sizes* and routes of pipes that will offset the deficiency. General familiarity with the community and studies of the network plan will aid judgment. Some existing small pipes may have to be removed to make way for larger mains.
7) Determine the size of the *equivalent pipe* for the reinforced system and calculate the velocity of flow. Excessive velocities may make for dangerous water hammer. They should be avoided, if necessary, by lowering the hydraulic gradients actually called into play.

8) Check important pressure requirements against the plan of the reinforced network.

The method of sections is particularly useful (a) in preliminary studies of large and complicated distribution systems, (b) as a check on other methods of analysis, and (c) as a basis for further investigations and more exact calculations.

Example 6.2 Analysis of a Water Network using the Sections Method
Analyze the network of Fig. 6.10 by sectioning. The hydraulic gradient available within the network proper is estimated to lie close to 2‰. The value of C in the Hazen–Williams formula is assumed to be 100, and the domestic (coincident) draft, in this case, only 150 gpcd (568 Lpcd). The fire demand is taken from Table 4.13. Assume the population for each section as follows: section a–a. 16,000; section b–b. 16,000; section c–c, 14,700; section d–d, 8,000, and section e–e, 3,000. Also assume that the type of building construction in the high-value district is combustible and unprotected and that the maximum total surface area per building is 20,000 ft^2 (1,858 m^2). The area downstream of the high-value district is residential with one- and two-family dwellings having a maximum area of 6,000 ft^2 (557.4 m^2).

Solution:

Calculations are shown only for the first three sections.

1) Section a–a population = 16,000:
   a) Demand: domestic = $16{,}000 \times 150/10^6 = 2.4$ MGD (9.1 MLD); fire (from Table 4.13) = 3,750 gpm = 5.4 MGD (20.4 MLD); total = 7.8 MGD (29.5 MLD).
   b) Existing pipes: one 24 in. (600 mm); capacity = 6.0 MGD (22.7 MLD).
   c) Deficiency: 7.8 − 6.0 = 1.8 MGD or (29.5 − 22.7 = 6.8 MLD).
   d) If no pipes are added, the 24-in. (600-mm) pipe must carry 7.8 MGD (29.5 MLD).
   This it will do with a head loss of 3.2‰ at a velocity of 3.8 ft/s (1.16 m/s).

2) Section b–b population and flow as in section a–a:
   a) Total demand = 7.8 MGD (29.5 MLD).
   b) Existing pipes: two 20-in. (500 mm) at 3.7MGD = 7.4 MGD (28.0 MLD).
   c) Deficiency = 7.8 − 7.4 = 0.4 MGD or (29.5 − 28.0 = 1.5 MLD).
   d) If no pipes are added, existing pipes will carry 7.8 MGD (29.5 MLD) with a loss of head of 2.2‰, at a velocity of 2.8 ft/s (0.85 m/s).

3) Section c–c population = 14,700:
   a) Demands: domestic = $14{,}700 \times 150/10^6 = 2.2$ MGD (8.3 MLD); fire = 5.4 MGD (20.4 MLD); total = 7.6 MGD (28.7 MLD).
   b) Existing pipes: one 20-in. (500 mm) at 3.7 MGD; two 12-in. (300 mm) at 1.0 MGD = 2.0 MGD; five 6-in. (150 mm) at 0.16 MGD = 0.8 MGD; total = 6.5 MGD (24.6 MLD).
   c) Deficiency = 7.6 − 6.5 = 1.1 MGD or (28.7 − 24.6 = 4.1 MLD).
   d) **Pipes added: two 10-in.** (250 mm) at 0.6 MGD = 1.2 MGD (4.5 MLD).
   Pipes removed: one 6-in. at 0.2 MGD (one 150-mm at 0.76 MLD).
   Net added capacity: 1.2 − 0.2 = 1.0 MGD (3.8 MLD).
   Reinforced capacity = 6.5 + 1.0 = 7.5 MGD or (24.6 + 3.8 = 28.4 MLD).
   e) The reinforced system equivalent pipe at 7.5 MGD (28.4 MLD) and a hydraulic gradient of 2‰ is 26.0-in. (650 mm). This will carry 7.6 MGD (28.8 MLD) with a loss of head of 2.1‰.

6.6.2 Relaxation (Hardy Cross)

A method of *relaxation,* or *controlled trial and error,* was introduced by Hardy Cross, whose procedures are followed here with only a few modifications. In applying a method of this kind, calculations become speedier if pipe-flow relationships are expressed by an exponential formula with unvarying capacity coefficient, and notation becomes simpler if the exponential formula is written as follows:

$$H = kQ^n \tag{6.3}$$

where, for a given pipe, k is a numerical constant depending on C, d, and L, and Q is the flow, n being a constant exponent for all pipes. In the Hazen–Williams equation, for example,

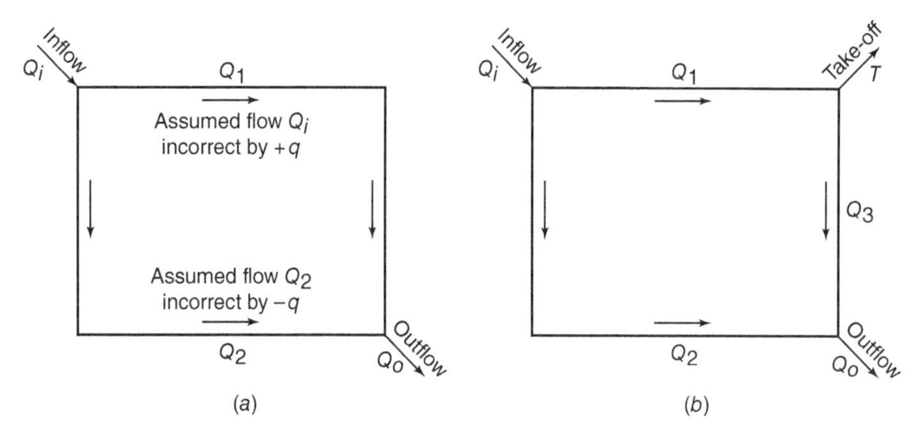

Figure 6.11 Simple Network Illustrating (a) the Derivation of the Hardy Cross Method and (b) the Effect of Changing Flows

$$Q = 405\ Cd^{2.63}s^{0.54} \quad \text{(US customary units)}$$

where Q is the rate of discharge, gpd; d is the diameter of circular conduits, in.; C Hazen–William coefficient, dimensionless; $S = H/L$ = hydraulic gradient, dimensionless; H is the loss of head, ft; L is conduit length, ft.

$$Q = 0.278\ Cd^{2.63}s^{0.54} \quad \text{(SI units)}$$

where Q is the rate of discharge, m^3/s; d is the diameter of circular conduits, m; C is Hazen–Williams coefficient, dimensionless; $s = H/L$ = hydraulic gradient, dimensionless; H is the loss of head, m; L is conduit length, m. For the Hazen–Williams using either the US customary units or the SI units, the following relationship hold true.

$$s = k'Q^{1/0.54} = k'Q^{1.85} \tag{6.4}$$

$$H = sL$$

$$H = kQ^{1.85}$$

Two procedures may be involved, depending on whether (a) the quantities of water entering and leaving the network or (b) the *piezometric levels, pressures,* or *water table* elevations at inlets and outlets are known.

In balancing heads by correcting assumed flows, necessary formulations are made algebraically consistent by arbitrarily assigning *positive signs* to *clockwise flows* and associated head losses, and *negative signs* to *counterclockwise flows* and associated head losses. For the simple network shown in Fig. 6.11a, inflow Q_i and outflow Q_o are equal and known, inflow being split between two branches in such a manner that the sum of the balanced head losses H_1 (clockwise) and $-H_2$ (counterclockwise) or $\Sigma H = H_1 - H_2 = 0$. If the assumed split flows Q_1 and $-Q_2$ are each in error by the same small amount q, then

$$\Sigma H = \Sigma k(Q + q)^n = 0$$

Expanding this binomial and neglecting all but its first two terms, because higher powers of q are presumably very small, we get

$$\Sigma H = \Sigma k(Q + q)^n = \Sigma kQ^n + \Sigma nkqQ^{n-1} = 0,$$

whence

$$q = -\frac{\Sigma kQ^n}{n\Sigma kQ^{n-1}} = \frac{\Sigma H}{n\Sigma H/Q} \tag{6.5}$$

If a takeoff is added to the system as in Fig. 6.11b, both head losses and flows are affected.

In balancing flows by correcting assumed heads, necessary formulations become algebraically consistent when positive signs are arbitrarily assigned to *flows toward junctions* other than inlet and outlet junctions (for which water table elevations are known) and *negative signs to flows away from* these *intermediate junctions,* the sum of the balanced flows at the junctions being zero. If the assumed water table elevation at a junction, such as the takeoff junction in Fig. 6.8b, is in error by a height h, different small errors q are created in the individual flows Q leading to and leaving from the junction. For any one pipe, therefore, $H + h = k(Q + q^n) = kQ^n + h$, where H is the loss of head associated with the flow Q. Moreover, as before,

$$h = nkqQ^{n-1} = nq(H/Q) \text{ and } q = (h/n)(Q/H)$$

Because $\Sigma(Q + q) = 0$ at each junction,

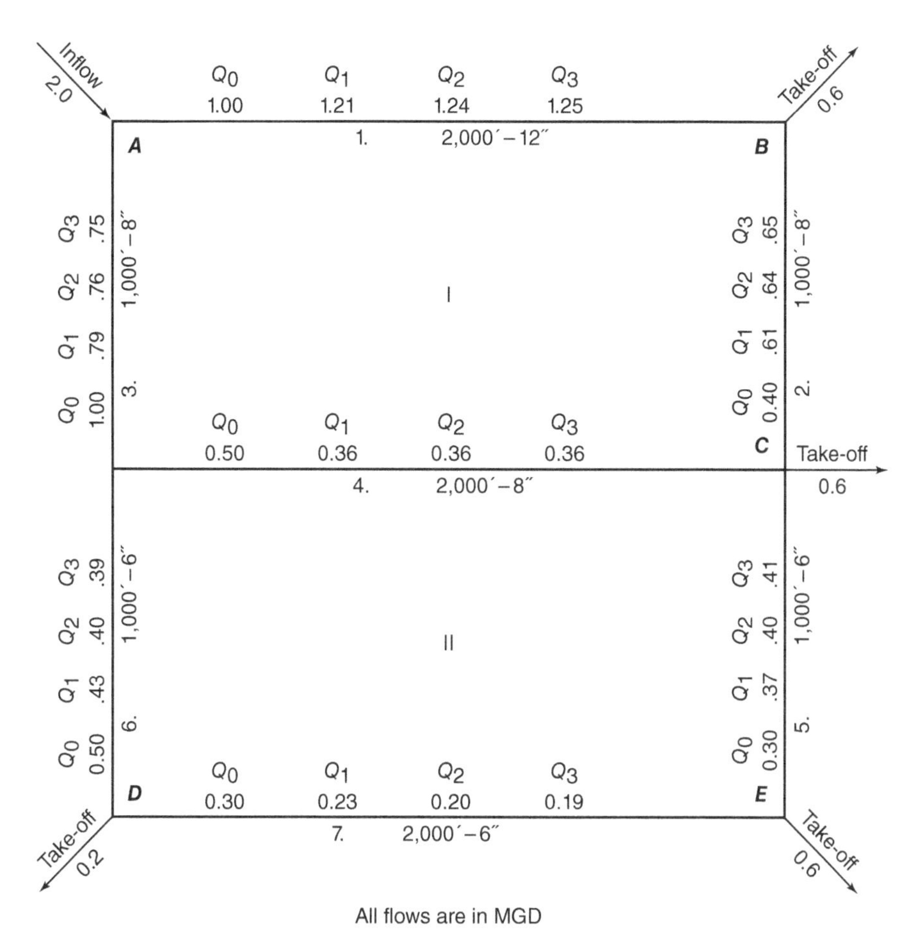

Figure 6.12 Plan of Network Analyzed by the Method of Balancing Heads (Example 6.3). Conversion factors: 1 MGD = 3.785 MLD; 1′ = 1 ft = 0.3048 m; 1″ = 1 in. = 25.4 mm

$$\Sigma Q = -\Sigma q \text{ and}$$
$$\Sigma q = (h/n)\Sigma(Q/H), \text{or}$$
$$\Sigma Q = -(h/n)\Sigma(Q/H); \text{therefore,}$$
$$h = -\frac{n\Sigma Q}{\Sigma(Q/H)} \tag{6.6}$$

The corrections q and h are only approximate. After they have been applied once to the assumed flows, the network is more nearly in balance than it was at the beginning, but the process of correction must be repeated until the balancing operations are perfected. The work involved is straightforward, but it is greatly facilitated by a satisfactory scheme of bookkeeping such as that outlined for the method of balancing heads in Example 6.3 for the network sketched in Fig. 6.12.

Although the network in Example 6.3 is simple, it cannot be solved conveniently by algebraic methods, because it contains two interfering hydraulic constituents: (a) a crossover (pipe 4) involved in more than one circuit and (b) a series of takeoffs representing water used along the pipelines, fire flows through hydrants, or supplies through to neighboring circuits.

Example 6.3 Analysis of a Water Network using the Relaxation Method of Balancing Heads

Solution:

The schedule of calculations (Table 6.4) includes the following:

Columns 1–4 identify the position of the pipes in the network and record their length and diameter. There are two circuits and seven pipes. Pipe 4 is shared by both circuits; “a” indicates this in connection with circuit I; “c” does so with circuit II. This dual pipe function must not be overlooked.

Table 6.4 Analysis of the Network of Fig. 6.12 (Example 6.3) Using the Balancing Heads Method

Network				Assumed Conditions						First Correction					
Circuit No. (1)	Pipe No. (2)	Length, ft (3)	Diameter, in. (4)	Q_0, MGD (5)	s_0, ‰ (6)	H_0, ft (7)	H_0/Q_0 (8)	q_0, MGD (9)		Q_1, MGD (10)	s_1, ‰ (11)	H_1, ft (12)	H_1/Q_1 (13)	q_1, MGD (14)	
I	1	2,000	12	+1.0	2.1	+4.2	4.2	+0.21		+1.21	3.0	+6.0	5.0	+0.03	
	2	1,000	8	+0.4	2.8	+2.8	7.0	+0.21		+0.61	6.1	+6.1	10.0	+0.03	
	3	1,000	8	−1.0	15.1	−15.1	15.1	+0.21		−0.79	9.8	−9.8	12.4	+0.03	
	4[a]	2000	8	−0.5	4.2	−8.4	16.8	+0.21	−0.07[b]	−0.36	2.3	−4.6	12.8	+0.03	−0.03[b]
						$-16.5 \div (43.1 \times 1.85) = -0.21$						$-2.3 \div (40.2 \times 1.85) = -0.03$			
II	4[c]	2,000	8	+0.5	4.2	+8.4	16.8	+0.07	−0.21[b]	+0.36	2.3	+4.6	12.8	+0.03	−0.03[b]
	5	1000	6	+0.3	6.6	+6.6	22.0	+0.07		+0.37	9.8	+9.8	26.5	+0.03	
	6	1,000	6	−0.5	16.9	−16.9	33.8	+0.07		−0.43	12.9	−12.9	30.0	+0.03	
	7	2,000	6	−0.3	6.6	−13.2	44.0	+0.07		−0.23	4.1	−8.2	35.6	+0.03	
						$-15.1 \div (116.6 \times 1.85) = -0.07$						$-6.7 \div (104.9 \times 1.85) = -0.03$			

Network				Second Correction						Result			
Circuit No. (1)	Pipe No. (2)	Length, ft (3)	Diameter, in. (4)	Q_2, MGD (15)	s_2, ‰ (16)	H_2, ft (17)	H_2/Q_2 (18)	q_2, MGD (19)		Q_3, MGD (20)	s_3, ‰ (21)	H_3, ft (22)	Loss of Head *A–E* (23)
I	1	2,000	12	+1.24	3.1	+6.2	5.0	+0.01		+1.25	3.2	+6.4	1. Via pipes 1, 2, 5, 25.0 ft
	2	1,000	8	+0.64	6.6	+6.6	10.3	+0.01		+0.65	6.8	+6.8	2. Via pipes 3, 4, 5, 25.3 ft
	3	1,000	8	−0.76	9.1	−9.1	12.0	−0.01		−0.75	8.9	−8.9	3. Via pipes 3, 6, 7, 25.5 ft
	4[a]	2,000	8	−0.36	2.3	−4.6	12.8	−0.01	−0.01[b]	−0.36	2.3	−4.6	
						$-0.9 \div (40.1 \times 1.85) = -0.01$						−0.3	
II	4[c]	2,000	8	+0.36	2.3	+4.6	12.8	+0.01	−0.01[b]	+0.36	2.6	+4.6	
	5	1,000	6	+0.40	11.3	+11.3	28.2	+0.01		+0.41	11.8	+11.8	
	6	1,000	6	−0.40	11.3	−11.3	28.2	−0.01		−0.39	10.8	−10.8	
	7	2,000	6	−0.20	3.1	−6.2	31.0	−0.01		−0.19	2.9	−5.8	
						$-1.6 \div (100.2 \times 1.85) = -0.01$						−0.2	

[a] Pipe serves more than one circuit; first consideration of this pipe.
[b] Corrections in this column are those calculated for the same pipe in the companion circuit; they are of opposite sign.
[c] Second consideration of this pipe.
Q = flow in MGD; s = slope of hydraulic gradient or friction loss in ft per 1,000 ft (‰) by the Hazen–Williams formula for $C = 100$, H = head lost in pipe (ft).
q = flow correction in MGD; $q = \frac{\sum H}{1.85 \sum (H/Q)}$, $Q_1 = Q_0 + q_0$; $Q_2 = Q_1 + q_1$; $Q_3 = Q_2 + q_2$.
Conversion factors: **1 ft = 0.3048 m; 1 in. = 25.4 mm; 1 MGD = 3.785 MLD.**

Columns 5–9 deal with the assumed flows and the derived flow correction. For purposes of identification the hydraulic elements Q, s, H, and q are given a subscript zero.

Column 5 lists the assumed flows Q_0 in MGD or MLD. They are preceded by positive signs if they are clockwise and by negative signs if they are counterclockwise. The distribution of flows has been purposely misjudged in order to highlight the balancing operation. At each junction the total flow remaining in the system must be accounted for.

Column 6 gives the hydraulic gradients in ft per 1,000 ft (‰) or in m per 1,000 m when the pipe is carrying the quantities Q_0 shown in Column 5. The values of s_0 can be read directly from tables or diagrams of the Hazen–Williams formula.

Column 7 is obtained by multiplying the hydraulic gradients (s_0) by the length of the pipe in 1,000 ft; that is, Column 7 = Column 6 × (Column 3/1,000).The head losses H_0 obtained are preceded by a positive sign if the flow is clockwise and by a negative sign if counterclockwise. The values in Column 7 are totaled for each circuit, with due regard to signs, to obtain ΣH.

Column 8 is found by dividing Column 7 by Column 5. Division makes all signs of H_0/Q_0 positive· This column is totaled for each circuit to obtain $\Sigma(H_0/Q_0)$ in the flow correction formula.

Column 9 contains the calculated flow correction $q_0 = -\Sigma H_0/(1.85 \times \Sigma H_0/Q_0)$. For example, in circuit I, $\Sigma H_0 = -16.5$, $\Sigma(H_0/Q_0) = 43.1$; and $(-16.5)/(1.85 \times 43.1) = -0.21$; or $q_0 = +0.21$. Because pipe 4 operates in both circuits, it draws a correction from each circuit. However, the second correction is of opposite sign. As a part of circuit I, for example, pipe 4 receives a correction of $q = -0.07$ from circuit II in addition to its basic correction of $q = +0.21$ from circuit I.

Columns 10–14 cover the once-corrected flows. Therefore, the hydraulic elements (Q, s, H, and q) are given the subscript 1. Column 10 is obtained by adding, with due regard to sign, Columns 5 and 9; Columns 11, 12, 13, and 14 are then found in the same manner as Columns 6, 7, 8, and 9.

Columns 15–19 record the twice-corrected flows, and the hydraulic elements (Q, s, H, and q) carry the subscript 2. These columns are otherwise like Columns 10–14.

Columns 20–23 present the final result, Columns 20–22 corresponding to Columns 15–18 or 10–12. No further flow corrections are developed because the second flow corrections are of the order of 10,000 gpd (37,850 L/d) for a minimum flow of 200,000 gpd (757,000 L/d), or at most 5%. To test the balance obtained, the losses of head between points A and D in Fig. 6.12 via the three possible routes are given in Column 23. The losses vary from 25.0 to 25.5 ft (7.62 to 7.77 m). The average loss is 25.3 ft (7.71 m) and the variation about 1%.

Example 6.4 Analysis of a Water Network using the Relaxation Method of Balancing Flows

Balance the network of Fig. 6.13 using the balancing flows method. Necessary calculations are given in Table 6.5.

Solution:

The schedule of calculations includes the following:

Columns 1–5 identify the pipes at the three free junctions.

Columns 6 and 7 give the assumed head loss and the derived hydraulic gradient that determines the rate of flow shown in Column 8 and the flow-head ratio recorded in Column 9 = (Column 8 divided by Column 6).

Column 10 contains the head correction h_0 as the negative value of 1.85 times the sum of Column 8 divided by the sum of Column 9, for each junction in accordance with Eq. 6.6:

$$h = \frac{n\Sigma Q}{\Sigma(Q/H)} \tag{6.6}$$

A subsidiary head correction is made for shared pipes as in Example 6.2.

Column 11 gives the corrected head $H_1 = H_0 + h_0$ and provides the basis for the second flow correction by determining s, Q, and Q_1/H_1 in that order.

6.6.3 Pipe Equivalence

In this method, a complex system of pipes is replaced by a single *hydraulically equivalent* line. The method cannot be applied directly to pipe systems containing crossovers or takeoffs. However, it is frequently possible, by judicious skeletonizing of the network, to obtain significant information on the quantity and pressure of water available at important points, or to reduce the number of circuits to be considered. In paring the system down to a workable frame, the analyst can be guided by the fact that pipes contribute little to flow (a) when they are small (6 in. [150 mm] and under in most systems

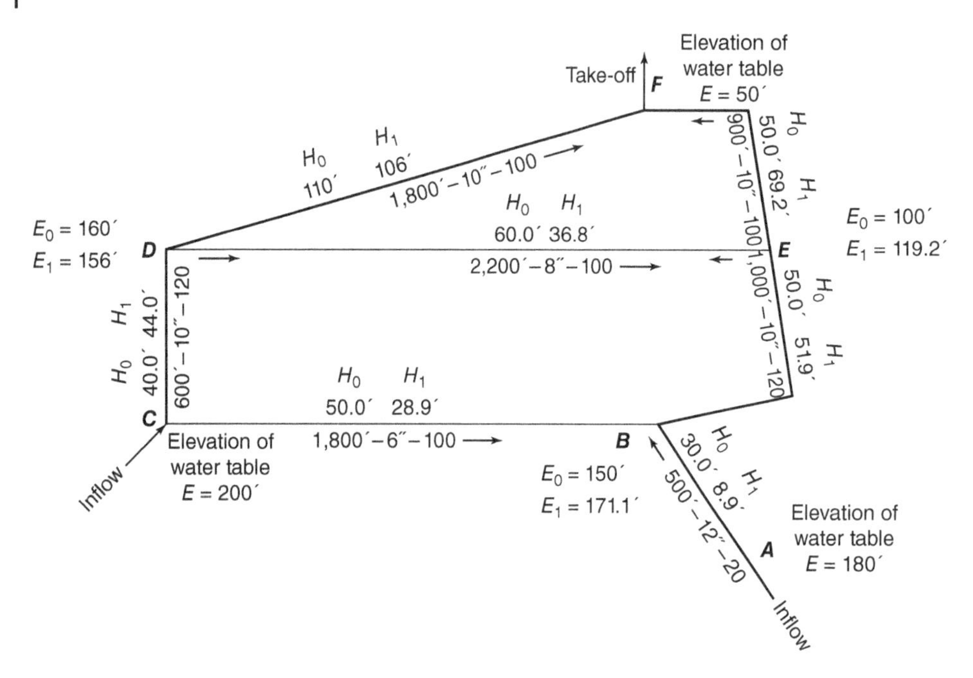

Figure 6.13 Plan of Network Analyzed by the Method of Balancing Flows (Example 6.4). Conversion factors: 1′ = 1 ft = 0.3048 m; 1″ = 1 in. = 25.4 mm

Table 6.5 Analysis of the Network of Fig. 6.13. Using the Balancing Flows Method[a,b] (Example 6.4)

Junction Letter (1)	Pipe (2)	Length ft (3)	Diameter, In. (4)	C (5)	H_0, ft (6)	s_0, % (7)	Q_0, MGD (8)		Q_0/H_0 (9)	h_0, ft (10)		H_1, ft (11)
B	AB	500	12	120	+30	60.0	+7.33		0.244	−21.1		+8.9
	BE	1,000[c]	10	120	−50	50.0	−4.12		0.082	−21.1	+19.2	−51.9
	CB	1,800	6	100	+50	27.8	+0.66		0.013	−21.1		+28.9
						1.85	× (+3.87)	÷	0.339 =	+21.1		
D	CD	600	10	120	+40	66.7	+4.8		0.120	+4.01		+44.0
	DE	2,200[d]	8	100	−60	27.3	−1.37		0.023	+4.01	+19.2	−36.8
	DF	1,800	10	100	−110	61.1	−3.82		0.037	+4.01		−106.0
						1.85	× (−0.39)	÷	0.180 =	−4.01		
E	BE	1,000[e]	10	120	+50	50.0	+4.12		0.082	−19.2	+21.1	+51.9
	DE	2,200[f]	8	100	+60	27.3	+1.37		0.023	−19.2	−4.01	+36.8
	EF	900	10	100	−500	55.6	−3.64		0.073	−19.2		−69.2
						1.85	× (+1.85)	÷	0.178 =	+19.2		

[a] For Example 6.4.
[b] Only the first head correction is calculated for purposes of illustration.
[c] First consideration of pipe BE.
[d] First consideration of pipe DE.
[e] Second consideration of pipe BE.
[f] Second consideration of pipe DE.
Conversion factors: **1 ft = 0.3048 m; 1 in. = 25.4 mm; 1 MGD = 3.785 MLD.**

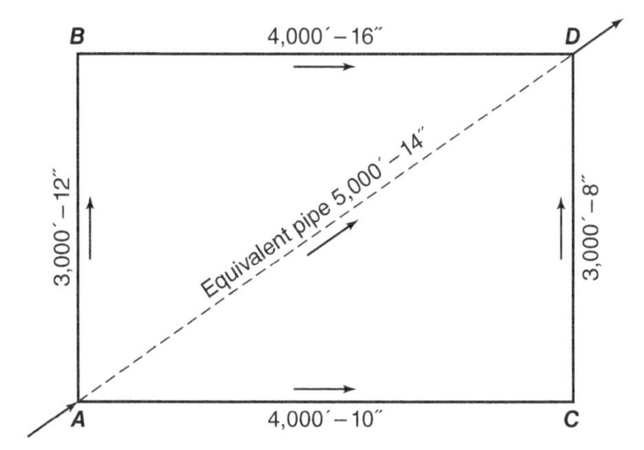

Figure 6.14 Plan of Network Analyzed by the Method of Equivalent Pipes (Example 6.5)

and 8 or 10 in. [200 or 250 mm] in large systems) and (b) when they are at right angles to the general direction of flow and there is no appreciable pressure differential between their junctions in the system.

Pipe equivalence makes use of the two hydraulic axioms:

1) Head losses through *pipes in series,* such as *AB* and *BD* in Fig. 6.14, are additive:

$$H_{AD} = H_{AB} + H_{BD}$$
$$Q_{AB} = Q_{BD}$$
$$H_{AD} = H_{AC} + H_{CD}$$
$$Q_{AC} = Q_{CD}$$

2) Flows through *pipes in parallel,* such as *ABD* and *ACD* in Fig. 6.14, must be so distributed that the head losses are identical:

$$H_{ABD} = H_{ACD}$$

Example 6.5 Analysis of a Water Network Using the Equivalent Pipe Method

Find an equivalent pipe for the network of Fig. 6.14. Express Q in MGD or MLD; s in ‰; H in ft or m; and assume a Hazen–Williams coefficient C of 100.

Solution:

1) Line ABD. Assume $Q = 1$ MGD (3.785 MLD).
   a) Pipe AB, 3,000 ft, 12 in.; $s = 2.1$; $H = 2.1 \times 3 = 6.3$ ft (1.92 m).
   b) Pipe BD, 4,000 ft, 16 in.; $s = 0.52$; $H = 0.52 \times 4 = 2.1$ ft (0.64 m).
   c) Total $H = 6.3 + 2.1 = 8.4$ ft (2.5 6 m).
   d) Equivalent length of 12-in. (300 mm) pipe: $1{,}000 \times 8.4/2.1 = 4{,}000$ ft (1,219 m).

2) Line ACD. Assume $Q = 0.5$ MGD (1.89 MLD).
   a) Pipe AC, 4,000 ft, 10 in.; $s = 1.42$; $H = 1.42 \times 4 = 5.7$ ft (1.74 m).
   b) Pipe CD, 3,000 ft, 8 in.; $s = 4.2$; $H = 4.2 \times 3 = 12.6$ ft (3.84 m).
   c) Total $H = 5.7 + 12.6 = 18.3$ ft (5.58 m).
   d) **Equivalent length of 8-in.** (200 mm) pipe: $1{,}000 \times 18.3/4.2 =$ **4,360 ft(1,329 m**).

3) Equivalent line AD. Assume $H = 8.4$ ft (2.56 m).
   a) Line ABD, 4,000 ft, 12 in.; $s = 8.4/4.00 = 2.1$; $Q = 1.00$ MGD (3.785 MLD).
   b) Line ACD, 4,360 ft, 8 in.; $s = 8.4/4.36 = 1.92$; $Q = 0.33$ MGD (1.25 MLD).
   c) Total $Q = 1 + 0.33 = 1.33$ MGD (5.03 MLD).
   d) Equivalent length of 14-in. (350 mm) pipe: $Q = 1.33$, $s = 1.68$, $1{,}000 \times 8.4/1.68 = 5{,}000$ ft (1,524 m).
   e) Result: **5,000 ft of 14-in. (1,524 m of 350-mm) pipe.**

Necessary calculations are as follows:

1) Because line ABD consists of two pipes in series, the losses of head created by a given flow of water are additive. Find, therefore, from the Hazen–Williams diagram (Chapter 4 and/or Appendix 14), the frictional resistance s for some reasonable flow (1 MGD) or (3.78 MLD) (a) in pipe AB and (b) in pipe BD. Multiply these resistances by the length of pipe to obtain the loss of head H. Add the two losses to find the total loss $H = 8.4$ ft (2.56 m). Line ABD, therefore, must carry 1 MGD (3.78 MLD) with a total loss of head of 8.4 ft (2.56 m). Any pipe that will do this is an equivalent pipe. Because a 12-in. (300-mm) pipe has a resistance s of 2.1% when it carries 1 MGD (3.78 MLD) of water, a 12-in. (300-mm) pipe, to be an equivalent pipe, must be $1{,}000 \times 8.4/2.1 = 400$ ft (122 m) long.
2) Proceed in the same general way with line ACD to find a length of 4,360 ft (1,329 m) for the equivalent 8-in. (200-mm) pipe.
3) Because ABD and ACD together constitute two lines in parallel, the flows through them at a given loss of head are additive. If some convenient loss is assumed, such as the loss already calculated for one of the lines, the missing companion flow can be found from the Hazen–Williams diagram. Assuming a loss of 8.4 ft (2.56 m), which is associated with a flow through ABD of 1 MGD (3.78 MLD), it is only necessary to find from the diagram that the quantity of water that will flow through the equivalent pipe ACD, when the loss of head is 8.4 ft or 2.56 m (or $s = 8.4/4.36 = 1.92$ % o), amounts to 0.33 MGD (1.25 MGD). Add this quantity to the flow through line ABD (1.0 MGD) or 3.78 MLD to obtain 1.33 MGD (5.03 MLD). Line AD, therefore, must carry 1.33 MGD (5.03 MLD) with a loss of head of 8.4 ft (2.56 m). If the equivalent pipe is assumed to be 14 in. (350 mm) in diameter, it will discharge 1.33 MGD (5.03 MLD) with a frictional resistance $s = 1.68$ ‰, and its length must be $1{,}000 \times 8.4/1.68 = 5{,}000$ ft (1, 524 m). Thence, the network can be replaced by a single 14-in. (350-mm) pipe 5,000 ft (1,524 m) long.

No matter what the original assumptions for quantity, diameter, and loss of head, the calculated equivalent pipe will perform hydraulically in the same way as the network it replaces.

Different in principle is the operational replacement of every pipe in a given network by equivalent pipes with identical diameters and capacity coefficients, but variable length. The purpose, in this instance, is to simplify subsequent calculations. For the Hazen–Williams relationship using the US customary units: Q (gpd), d (in.), H (ft), L (ft), C and s (dimensionless), for example,

$$\begin{aligned} Q &= 405\, Cd^{2.63} s^{0.54} \\ s^{0.54} &= Q/405\, Cd^{2.63} s^{0.54} \\ s &= Q^{1.85}/(405)^{1.85}\, C^{1.85} d^{4.87} \\ H/L &= Q^{1.85}/(405)^{1.85}\, C^{1.85} d^{4.87} \\ L &= (405)^{1.85}\, C^{1.85} d^{4.87} H/Q^{1.85} \\ L_e/L &= (100/C)^{1.85} (d_e/d)^{4.87} \end{aligned} \tag{6.7}$$

where Le is the length of a pipe of diameter de; discharge coefficient $C = 100$; and L, d, and C are the corresponding properties of the existing pipe. The desired values for L can be found readily from a logarithmic plot of L_e/L against d_e/d at given values of C, as shown in Fig. 6.15.

The readers are referred to Chapters 5 and 13 for all Hazen–Willliams equations using the SI units.

Example 6.6 Determination of Equivalent Pipe

Find the length of a 24-in. (600-mm) pipe with $C = 100$ that is equivalent to a 12-in. (300-mm) pipe with $C = 130$ and $L = 2{,}000$ ft (610 m).

Solution:

At $C = 130$ read the C-factor 6.2×10^{-1}
At $d_e/d = 2.0$, read the d-factor 2.9×10
Hence $L_e/L = (6.2 \times 10^{-1})\,(2.9 \times 10) = 18$
$L_e = 2{,}000 \times 18 = 36{,}000$ ft (10,973 m)

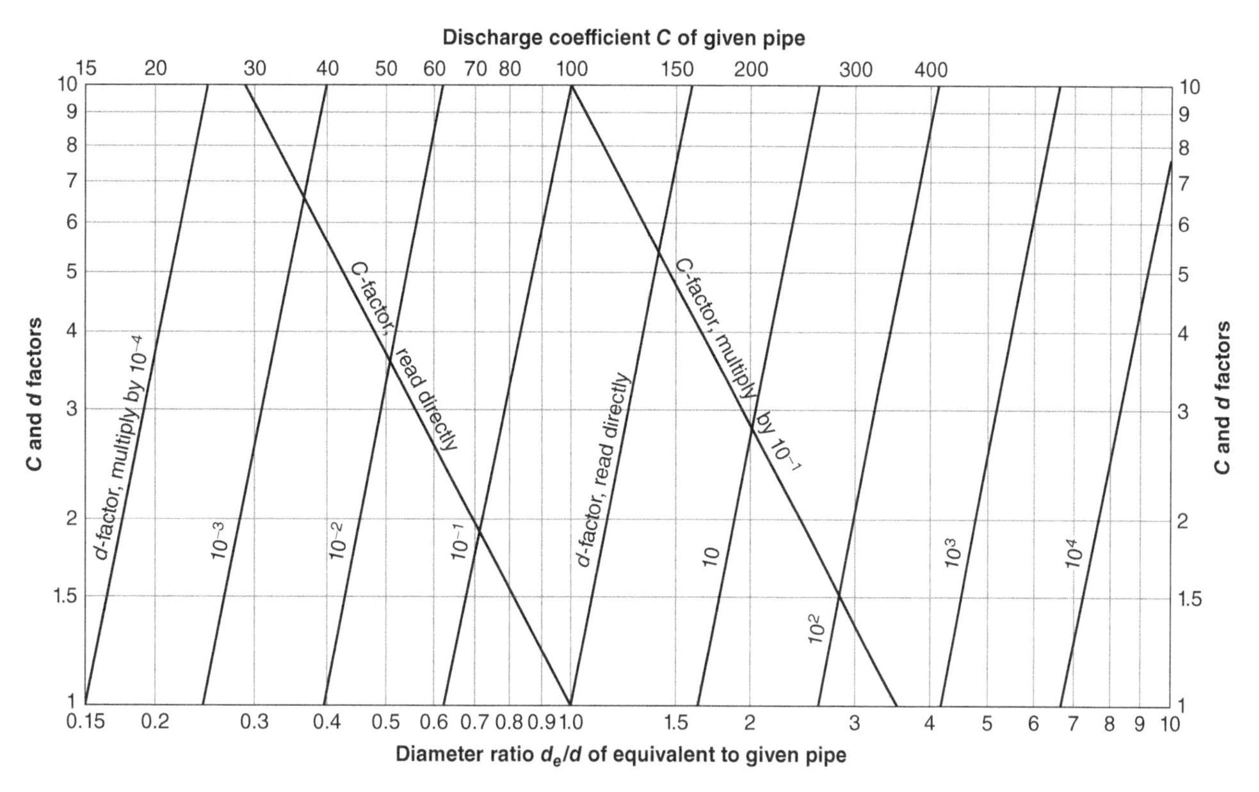

Figure 6.15 Length, Diameter, and Coefficient (L_e, d_e, and C_e = 100) of Pipe Hydraulically Equivalent to an Existing Pipe of Given L, d, and C

6.6.4 Computer Programming

High-speed computers can be programmed to solve network problems in a number of different ways. Convergence formulas need not be introduced as such. Instead, the computer can be assigned the task of adjusting the water table or pressure at each junction not controlled by a service reservoir until the circuit laws discussed in connection with the method of relaxation are satisfied throughout the system. These *laws* can be summarized as follows:

1) At each junction $\Sigma Q_{inflow} = \Sigma Q_{outflow}$
2) In each circuit $\Sigma H = 0$
3) In each pipe $H = kQ^n$ or $Q = (H/k)^{1/n}$.

To program the operation, number each pipe and junction and identify pipe ends by junction numbers. Then tabulate pipe resistances, junction pressures (including assumed values where pressures are unknown), and net inflows at each junction (zero at all but entrance and exit points of the system), and feed the tabulated information into the computer. The computer instructions are then as follows: Calculate by circuit law 3 the total flow into the first junction for which the water table elevation is unknown, adjust the assumed value until the total inflow and outflow are balanced in accordance with *circuit law* 1, proceed in sequence to the remaining junctions, and readjust the first water table elevation. Repeat the cycle of operations until *all circuit laws* are satisfied.

Camp and Hazen (1934) built the first electric analyzer designed specifically for the hydraulic analysis of water distribution systems. Electric analyzers use *nonlinear resistors,* called *fluistors* in the *McIlroy analyzer,* to simulate pipe resistances. For each branch of the system, the pipe equation, $H = kQ^{1.85}$, for example, is replaced by an electrical equation, $V = K_e I^{1.85}$, where V is the voltage drop in the branch, I is the current, and K_e is the nonlinear-resistor coefficient suited to pipe coefficient k for the selected voltage drop (head loss) and amperage (water flow) scale ratios. If the current inputs and takeoffs are made proportional to the water flowing into and out of the system, the head losses will be proportional to the measured voltage drops. Some large, rapidly developing communities have found it economical to acquire electric analyzers suited to their own systems.

Of course, at present, several handy commercial software programs are available for modeling and design of water systems (see Chapter 7). Software is available as a stand-alone interface for Windows or integrated into GIS or CAD systems, for example:

1) Haestad Methods Solutions (Bentley Systems Inc.): Water GEM and WaterCAD
2) MWH Soft: InfoWater
3) Wallingford Software: InfoWorks WS (for water supply).

6.7 Industrial Water Systems

Large industrial establishments, with a heavy investment in plant, equipment, raw materials, and finished products, concentrated in a small area, are generally equipped with high-pressure fire supplies and distribution networks of their own. Because such supplies may be drawn from sources of questionable quality, some regulatory agencies require rigid separation of all private fire supplies and public distribution systems. Others permit the use of *protected cross-connections* and require their regular inspection for tightness. How the two sources of supply can be divorced without denying the protective benefit and general convenience of a dual supply to industry is illustrated in Fig. 6.16. Ground-level storage and pumping are less advantageous.

A widely approved arrangement of double check valves in vaults accessible for inspection and test by the provision of valves, gauges, and bleeders is shown in Fig. 6.17. No outbreak of waterborne disease has been traced to approved and properly supervised cross-connections of this kind. Automatic chlorination of the auxiliary supply can introduce a further safeguard.

6.8 Management, Operation, and Maintenance of Distribution Systems

For intelligent management of distribution storage, reservoir levels must be known at all times of the day and night. Where levels cannot be observed directly by gauges or floats, electrically operated sensors and recorders can transmit the required information to operating headquarters.

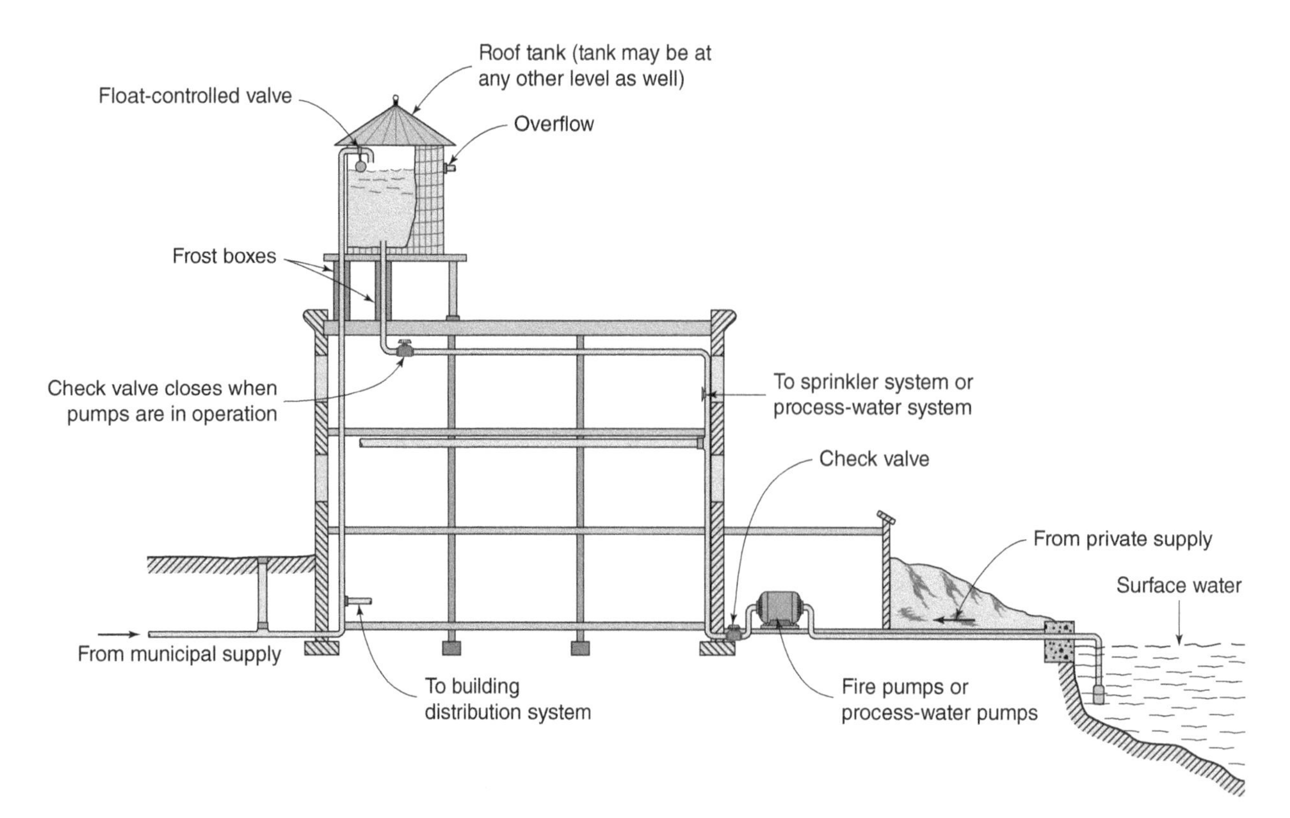

Figure 6.16 Use of Industrial Water Supply without Cross-Connection (After Minnesota State Board of Health)

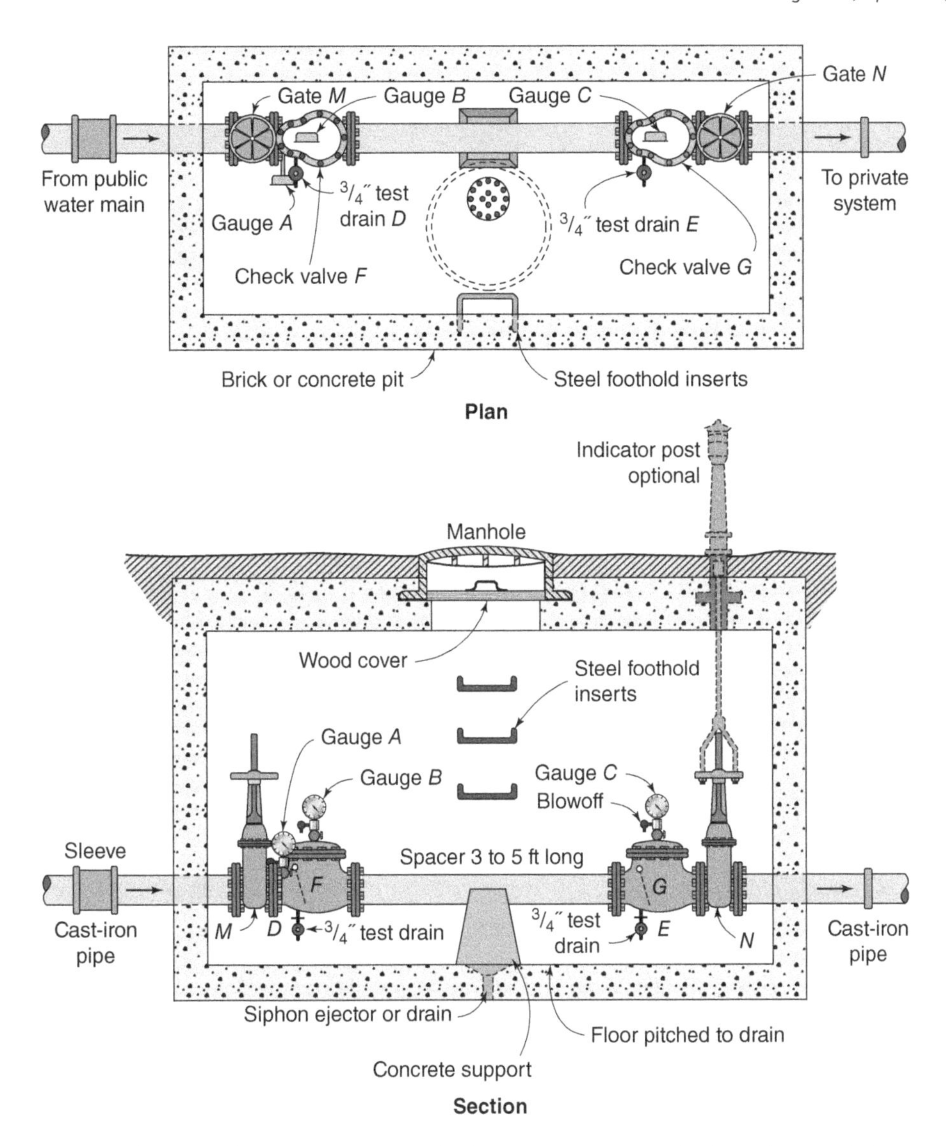

Figure 6.17 Cross-Connection between Municipal Water Supply and Private (Industrial) Water Supply Protected by Double Check-Valve Installation. To test installation: (1) close gates M and N; (2) open test drain B, and observe gauges A and B; (3) open test drain E, and observe gauge C. If check valves F and G are tight, gauge A will drop to zero; gauges B and C will drop slightly owing to compression of rubber gaskets on check valves F and G. Conversion factors: 1″ = 1 in. = 25.4 mm; 1 ft = 0.3048 m.

Well-kept *records* and *maps of pipes* and *appurtenances* are essential to the efficient operation and maintenance of distribution systems. To avoid the occasional discharge of roiled water, piping should be flushed systematically, usually through hydrants. Dead-ends need particular attention; a bleeder on a dead-end will counteract the effects of sluggish water movements. Disinfecting newly laid or newly repaired pipe is important.

There is little flow through service pipes at night, and they may freeze in very cold weather. If water mains themselves are placed at a reasonable depth and enough flow is maintained in the system, they should not freeze. Pipes deprived of adequate cover by the regrading of streets or subjected to protracted and exceptionally cold weather can be protected by drawing water from them through services. Pipes exposed on bridges or similar crossings should be insulated. Large and important lines may be heated where exposure is severe. In very cold climates, water and sewer pipes are often laid in a heated boxlike conduit, known as a *utilidor*.

Frozen pipes are usually thawed by electricity. A transformer connected to an electric power circuit, or a gasoline-driven generator of the electric-welding type, supplies the current: 100–200 A at 3–10 V for small pipes and up to several thousand amperes at 55 or 110 V for large mains. The current applied is varied with the electrical resistance and the melting point of the pipe metals. Nonmetallic jointing and caulking compounds and/or plastic pipes obstruct current flow. Electric grounds on interior water piping, or the piping itself, must be disconnected during thawing operations. Grounds are needed but are an annoyance when they carry high voltages into the pipes and shock workmen. Pipes and hydrants can also be thawed with steam generated in portable boilers and introduced through flexible block-tin tubing.

Loss of water by *leakage* from distribution systems and connected consumer premises should be kept under control by leakage surveys.

Remember that the best way to prevent problems and accidents is to minimize hazards and "design them out" early in the design process (see Chapter 20). Three illustrative examples from the National Institute for Occupational Safety and Health are summarized in the following subsections.

6.8.1 General Maintenance Person Asphyxiated While Attempting to Repair Water Leak

The victim worked as a general maintenance person for a construction company, which employed 13 persons. The construction company provided construction-related maintenance for a local chain of restaurants. A supervisor for the construction company instructed a maintenance person (the victim) to inspect and repair a leaking water valve. The water valve (a screw handle type) controlled the flow of water from the municipal water system to a local restaurant. After the supervisor instructed the victim, he then left the site of the restaurant to check on another job.

Apparently the victim proceeded to the fiberglass water meter pit (14 in. in diameter by 4 ft deep or 350-mm diameter by 1.22 m deep) approximately 25 ft (7.62 m) from the side of the restaurant where the water valve was located. The water meter pit was buried in the ground and the top of the pit was at ground level. A metal cap was attached to the rim of the water meter pit and a water meter with an in-line shutoff valve. A screw handle water valve and the municipal water line were located in the pit. The valves were approximately 36 in. (914 mm) below the top of the pit (or ground level). The victim removed the metal cap covering the pit and placed the cap on the ground next to the pit opening. He then knelt beside the opening on both knees and reached into the pit until his head, both arms, and part of his shoulders were inside the water meter pit. Apparently, the victim became stuck upside down in the opening and could not free himself, causing *asphyxiation* due to *positional deprivation of air.*

Recommendation #1: Supervisory personnel should routinely monitor employee performance to determine if employees have impaired physical and mental capabilities, which may be related to the use of alcohol, illegal or over-the-counter drugs, or prescription medications.

Recommendation #2: Supervisory personnel should identify, evaluate, and address all possible hazards associated with the job site. When employees are expected to work alone at job sites, the area should first be evaluated and all possible hazards identified and addressed by supervisory personnel. The location of the water valve inside the water meter pit required the use of extension tools, thereby eliminating the need to enter the water meter pit (even partially).

6.8.2 Plumber Repairing a Water Line Killed When Struck by a Backhoe Bucket

A 47-year-old male plumber died when he was struck by the bucket of a backhoe while he was repairing a water line in a trench for a shopping center. The victim and an apprentice plumber were at the bottom of the trench removing straight and angled cast-iron pipes and placing them in the backhoe bucket. The operator of the backhoe was an employee of the shopping center. The operator told the police investigator that when he released all controls, the bucket swerved to the left striking the victim on his left side and pinning him against the trench wall.

A trench had to be excavated to reach the pipe. The trench was an oblong shape measuring approximately 31 ft (9.45 m) long. One end was about 5 ft (1.52 m) wide and the other end, where the incident occurred, was approximately 12 ft (3.66 m) wide. The trench depth was 6 ft (1.83 m) at the edges going down to 7 ft (2.13 m) where the pipe was located. About 6–8 in. (150–200 mm) of water was on the bottom of the trench (see Fig. 6.18). There was no shoring, shielding, or sloping of the trench walls.

An employee (hereinafter referred to as "operator") of the shopping center operated the backhoe to dig the trench. He had been with the firm for approximately 13 years and was in charge of the safety of personnel, property, and the public for the mall. He had training on operating backhoes prior to employment with the firm and had operated the same model of backhoe on jobs at the shopping center with the plumbing contractor. However, operating a backhoe was not part of his regular duties. The operator had requested this particular model of backhoe be used, instead of the backhoe the plumbing company

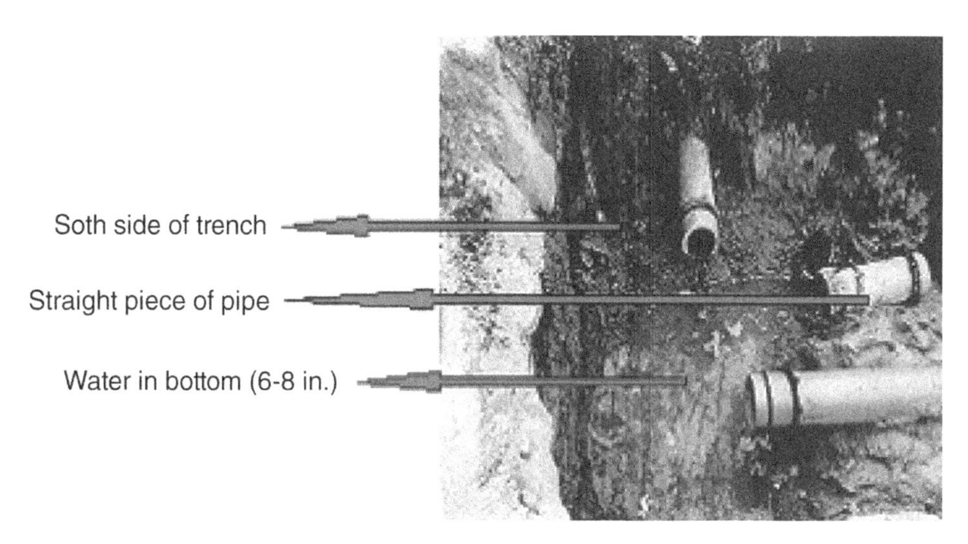

Figure 6.18 Area of Incident

owned. According to the operator, the victim asked him to operate the backhoe because he was more familiar with that particular model.

The backhoe was a farm tractor-type, all-purpose backhoe/loader with rubber tires and a 75-HP (56-kW), diesel, four-cylinder engine. It had a general purpose 18-in. (457 mm) bucket weighing 255 lb. (116 kg) with foot controls to swing it left (left throttle) and right (right throttle). Hand levers raised, lowered, and extended the two-part 430-lb (195-kg) boom with the bucket attached at the end (see Fig. 6.19) to a maximum of 25.5 ft (7.77 m). It had outriggers for stabilization, which were in position on the day of the incident. The backhoe was positioned about 4 ft (1.22 m) from the edge of the trench on the asphalt parking lot behind the curb. The boom with attached bucket was extended and angled over and down into the trench. The operator, if sitting in the seat of the backhoe, could not see activity down in the trench.

To repair the leak, the victim and apprentice plumber had to remove two sections of the old pipe and install new pipe. The pipe sections were cast iron and were 10 in. (250 mm) in diameter with the straight section 3 ft (0.91 m) long and the other section a 90-degree-angle section or "L" joint. The straight piece weighed about 10 lb. (4.54 kg) and the "L" joint weighed between 30 and 40 lb. (14 and 18 kg).

The victim and apprentice plumber were in the trench trying to assemble the 3-ft (0.91-m) section to the 90-degree "L" joint but there was not enough room to work and there was still a large amount (6–8 in. or 150–200 mm) of water in the hole. They decided it would be easier to work on connecting the two pieces outside the trench and then return the assembled unit to the trench to connect to the other pipe lines. To get the pipes out, they decided to place them in the bucket and have them raised out of the trench.

The bucket of the backhoe was positioned about 4 ft (1.4 m) down in the trench close to the pipe line. The victim was facing east about 1 ft directly to the left and in the line of contact with the bucket. The apprentice plumber was also to the left of the bucket but out of its reach and was facing west looking at the victim.

The "L" pipe had just been placed in the bucket by the victim and apprentice plumber when,

Figure 6.19 Bucket and Boom of Backhoe

unexpectedly, the 255-lb (102-kg) bucket with the approximately 40 lb. (18 kg) of pipe swung swiftly to the left and struck the victim in the left arm and upper chest, shoving him about 6 in. and pinning him into the south trench wall. The apprentice plumber yelled at the operator who maneuvered the bucket to the right to release the victim, who then slumped to the ground, groaning.

The operator called emergency personnel, who arrived at the scene in 3 min. The victim had no pulse or respiration. Ambulance personnel removed the victim from the hole and transported him to a hospital where he was pronounced dead of blunt force injuries to the chest. The manner of death was determined to be *accidental*. Hospital personnel told a police officer that death was caused by a *ruptured myocardium* (*heart*).

Recommendation #1: Employers should ensure that backhoe operators shut down backhoes according to the manufacturer's directions and direct workers to remain an adequate distance from operating backhoes. According to the apprentice plumber, the backhoe was running with its bucket positioned about 4 ft (1.22 m) down in the trench directly in line of contact with the victim. If the backhoe had been shut down properly and had workers been directed to remain an adequate distance from the backhoe the fatality may have been averted.

Recommendation #2: Employers should designate a competent person to supervise trench activities. In work involving trenches the Occupational Safety and Health Administration (OSHA) requires a competent person to be responsible for trench activities. A competent trench worker should be able to perform the following duties:

1) View the "big picture" and recognize by proper assessment the present and potential hazards at a site.
2) Classify soil types and changes to the soil composition.
3) Determine the need for sloping, shoring, or shielding.
4) Examine and approve or disapprove material or equipment used for protective systems.
5) Identify utilities located parallel to or crossing the trench.
6) Conduct daily inspections prior to the start of work and as needed. Inspect the adjacent areas and protective systems for indications of failure and hazardous atmospheres and conditions.
7) Inspect the trench after every rainstorm or other hazard-increasing occurrence.
8) Determine suitability of job site for continued work after water accumulation.

Having employees in the trench when the backhoe was running was a hazardous condition for several factors: the vibration of the soil caused by the running machine, the overhead load of the bucket and boom, and the limited visibility by the backhoe operator of activities in the trench. Under any of these conditions, a competent person would remove the exposed employees from the trench until hazards have been controlled.

Recommendation #3: Employer should ensure that workers are protected from cave-ins by an adequate protective system. A protective system designed for the soil conditions found in this excavation could have included a trench shield (also known as a trench box), shoring, or a combination of shoring and shielding. Employers can consult with manufacturers of protective systems to obtain detailed guidance for the appropriate use of these products. In this incident, no protective system had been placed at any point in the 31-ft (9.45-m)-long trench. Although the victim in this incident was not killed because of a trench cave-in, protective systems should always be used in trenches greater than 5 ft (1.52 m) deep.

6.8.3 Welder Killed Following a 100-ft (30-m) Fall from a Water Tower

A welder was killed after falling 100 ft (30 m) from a leg of a municipal water tower. The victim was a 25-year-old male welder who had been working for the company as a tower hand for 4 years. He had previously worked as a welder for a municipality. The incident occurred while the victim and a coworker were welding antenna support brackets onto the leg of the tower. The victim apparently disconnected his fall protection and was climbing the leg of the tower when he fell approximately 100 ft (30 m) to the ground.

The water tower consisted of a 40-ft (12.2-m)-diameter steel water tank mounted on six structural steel legs. Each of the legs was made of welded plate steel with structural steel braces. Although not specifically designed for climbing, these braces were spaced near enough that they could be used as a ladder. The top of the tower was accessed by a steel ladder with a ladder cage that was built on one of the legs. A pump house was located near the bottom of the ladder and was the terminating point for a number of antennas and utility cables leading from the tower.

The day of the incident was clear and sunny. The foreman and three tower hands arrived in the morning and set up the hoist system on the tower leg. The victim was using a lineman's belt that he owned and had modified to make the seat strap more comfortable. Work proceeded uneventfully until the victim and a coworker were finishing the welding of a bracket about 100 ft (30 m) up the tower leg. The coworker stated that they had been up on the leg for a while and that the victim had been tied off to the leg at the same location for about 15 min before the incident. The coworker had completed his weld and handed the welding equipment down to the victim, who was working below him. The victim then finished his weld and began to chip and paint. Noticing that there was a fault in the weld above his head, the victim called down to the foreman to start up the welder so he could redo some spots. He then started to climb up the tower. As the foreman turned to start the machine, he heard the victim yell and saw him fall to the ground. The police and emergency medical services personnel were called and arrived within a few minutes of the fall and began treating the victim, who was still breathing. During this time the shaken coworker was lowered to the ground on the material hoist line. The victim was transported to the local medical center where he was pronounced dead.

Investigators concluded that these *guidelines* should be followed in order to prevent similar incidents in the future:

1) Employers must thoroughly plan all work and perform a job hazard analysis of the site prior to starting work.
2) When practical, employers should provide and require the use of a stable work platform for working at elevated worksites.
3) Employers should provide a system of fall protection that protects employees at all times when working at elevations.
4) Employers should ensure that fall protection equipment is appropriate and maintained in good condition. Employees should inspect fall protection equipment before each use to ensure that all components are in operational order.
5) Employers should ensure that material hoists are not used for raising or lowering employees to or from the worksite.
6) Employers should ensure that electrical safety practices are followed when welding.
7) Owners of water towers and similar structures should design and install a permanent static safety line system on the tower to facilitate the use of fall protection devices.

Problems/Questions

6.1 An elevated water reservoir (water level at 210 m) is to supply a lower groundwater reservoir (water level at 138 m). The reservoirs are connected by 800 m of 400-mm ductile cast pipe ($C = 100$) and 1,200 m of 300-mm ductile cast pipe ($C = 100$) in series.

(a) What will be the discharge delivered from the upper reservoir to the lower one?
(b) If the water demand increases by 50% in the future, determine the required size of pipe that has to be installed in parallel with the existing piping.

6.2 The following data are given for the water supply network shown in Fig. 6.20:

Pipe	*d* (in.)	*L* (ft)
AB	8	2,500
BC	8	500
AD	8	500
DC	12	2,500

The elevation of point A is 100 ft and of point B is 90 ft The C value for all pipes is 100. Find the following:

(a) The length of a single 18-in. pipe that can replace the water network from A to C.
(b) The residual pressure at point C, given that the flow through the network is 4,000 gpm and the pressure at point A is 40 psi.

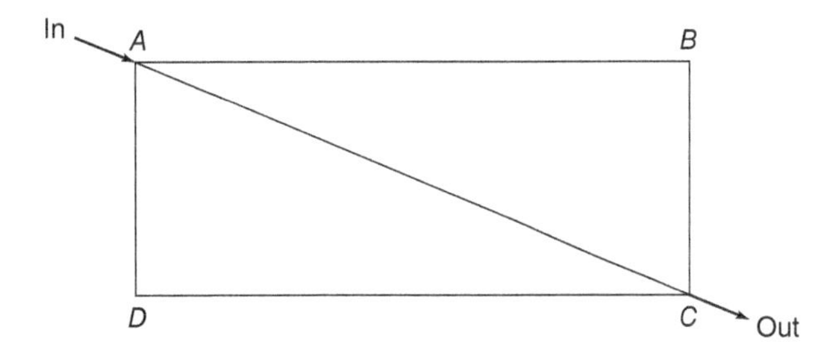

Figure 6.20 Water Network for Problems 6.2 and 6.3

6.3 The following data are given for the water supply network shown in Fig. 6.20:

Pipe	*d* (mm)	*L* (m)
AB	200	800
BC	200	200
AD	200	200
DC	300	800

The elevation of point A is 600 m and of point B is 590 m. The C value for all pipes is 100. Find the following:

(a) The length of a single 400-mm pipe that can replace the water network from A to C.
(b) The residual pressure at point C, given that the flow through the network is 250 L/s and the pressure at A is 300 kPa.

6.4 For the water supply network shown in Fig. 6.21, the following data are given:

Pipe	*d* (mm)	*L* (m)
AB	300	1,200
BD	300	300
AC	200	300
CD	200	1,200
SA	d_{SA}	10,000

Elevation of loop ABCD is 600 m and the value of C for all pipes = 100.

Determine the needed diameter of main SA so that the residual pressure at any point in the network ABCD does not drop below 250 kPa.

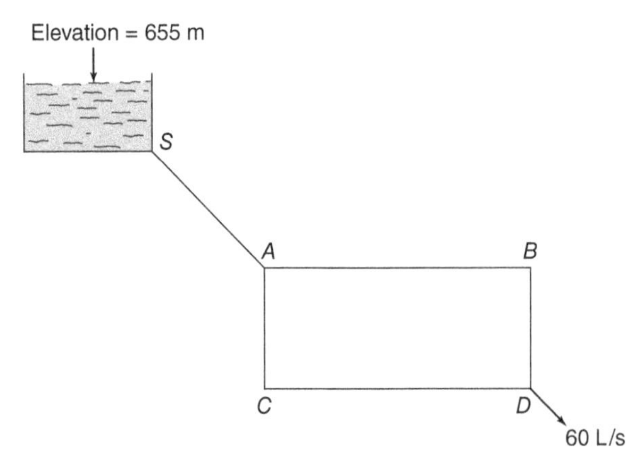

Figure 6.21 Water Network for Problem 6.4

6.5 Potable water is supplied to a city via an 800-mm transmission line at a flow rate of 550 L/s (see Fig. 6.22). A pressure gauge located 500 m upstream from point A registers 6.5 bars at normal operation. The following data are given:

Pipe	*d* (mm)	*L* (m)
PA	800	500
AmnB	500	1,500
AxyB	300	200
BC	800	—

All pipes are made of the same material with a Hazen–Williams coefficient C = 100. The elevations of points P, A, and B are 650, 660, and 635 m, respectively.

(a) Determine the length of an equivalent 500-mm single pipe between points A and B.
(b) Determine the actual flows in the two branches AmnB and AxyB.
(c) Find the residual pressures at both points A and B.

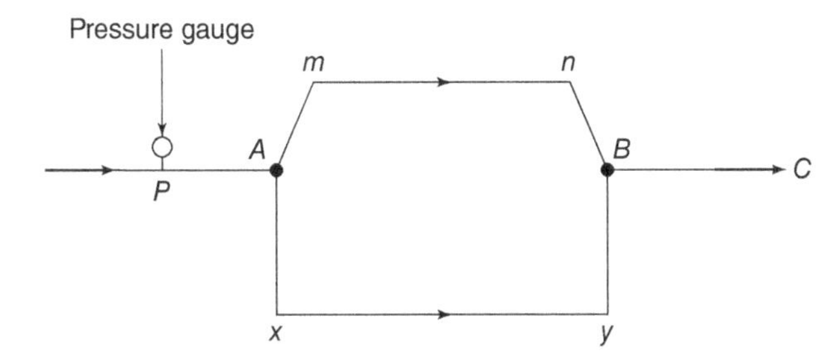

Figure 6.22 Water Network for Problem 6.5

6.6 The layout of a water distribution system is shown in Fig. 6.23. All pipes of the network are made of plastic (C = 150). The maximum flow in the system Q = 5,000 gpm. The following data are given:

Pipe	d (in.)	L (ft)
AB	14	1,000
BC	14	2,500
CD	14	500
DE	14	2,000
AE	10	3,000
AF	12	1,500
FE	10	2,500

(a) Find the length of a single 16-in.-diameter-pipe that can replace the system from A to E.
(b) Find the residual pressure at point E, if the available pressure at point A is 60 psi.

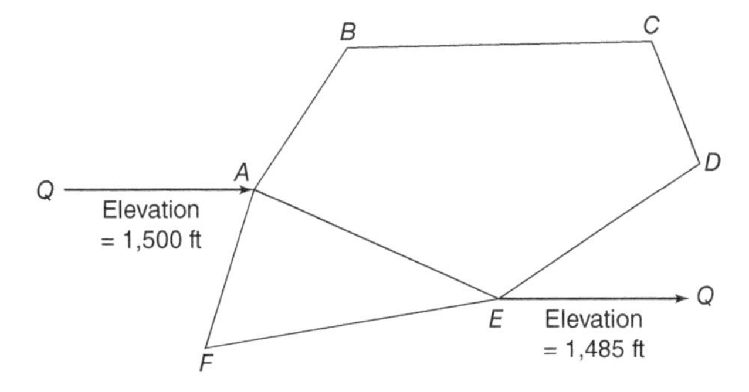

Figure 6.23 Water Network for Problem 6.6

6.7 Check and reinforce (if required) sections d–d and e–e in Example 6.2.

6.8 The layout of a water distribution system is shown in Fig. 6.24. Elevation of water in reservoir = 2,000 ft and elevation of point A = 1, 770 ft. All pipes are made of the same material with a Hazen–Williams coefficient C = 100. The following data are given:

Pipe	*d* (in.)	*L* (ft)
RA	d_{RA}	20,000
AB	18	5,000
BC	8	2,500
BE	6	1,000
CD	8	2,500
DE	6	1,000
AE	10	4,000
AD	d_{AD}	5,000

(a) Determine the required diameter of main from water reservoir to point A, if the minimum required pressure at point A is 60 psi.
(b) Determine the required diameter of pipe AD, assuming that the maximum allowable head loss in the network (ABCDE) is 3 ft/1,000 ft
(c) Determine the required pipe diameter that is to be constructed in parallel with BC assuming that the maximum allowable head loss in the network (ABCDE) is 3 ft/1,000 ft

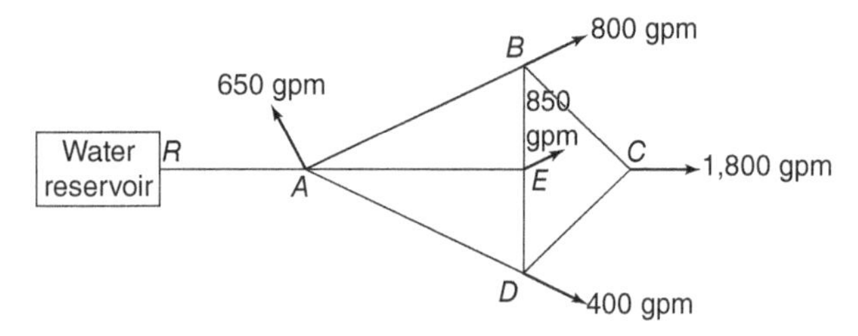

Figure 6.24 Water Network for Problem 6.8

6.9 The layout of a water distribution system is shown in Fig. 6.25. Elevation of water in reservoir = 600 m and elevation of points A and C are 531 and 500 m, respectively. All pipes are made of the same material with a Hazen–Williams coefficient C = 100. The following data are given:

Pipe	*d* (mm)	*L* (m)
RA	d_{RA}	6,000
AB	400	1,500
BC	200	750
BE	150	300
CD	300	750
DE	150	300
AE	250	1,200
AD	d_{AD}	1,500

(a) Determine the required diameter of main from water reservoir to point A, if the minimum required pressure at point A is 450 kPa.
(b) Determine the required diameter of pipe AD; assuming that the maximum allowable head loss in the network (ABCDE) is 3%.
(c) Determine the required pipe diameter that is to be constructed in parallel with BC assuming that the maximum allowable head loss in the network (ABCDE) is 3%.

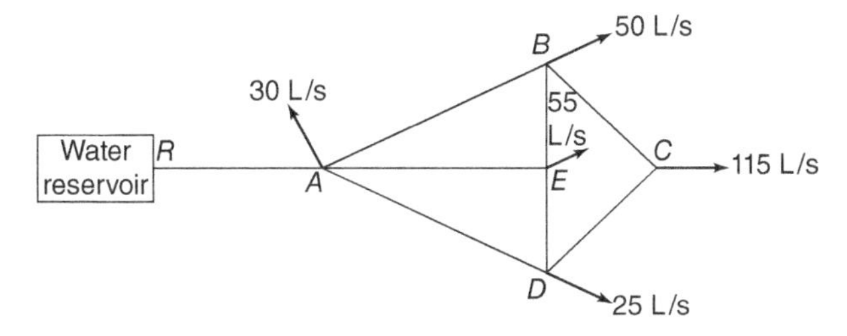

Figure 6.25 Water Network for Problem 6.9

6.10 The water distribution system shown in Fig. 6.26 was designed to serve a small city. Water flows by gravity through a 400-mm water main (ABC) from the water reservoir and feeds into the network at point C. The length of the main, ABC, is 10 km.

At present, the water demand has increased to the values shown in the figure. The water level in the reservoir is at 400 m, and the whole network is located at an elevation of 340 m. Each quadrant in the network is 1,000 m long and 500 m wide. Assume all pipes are ductile iron with C = 100.

(a) If the pressure at point C is not allowed to drop below 300 kPa, determine the required size of an additional ductile pipe to be installed between the reservoir and point B, 7 km away along the line AC.
(b) Determine the required diameters of pipes DE, CK, and JI, assuming that all three pipes will have identical diameters and that the maximum allowable head loss in the network is 3 m/1,000 m (3%).

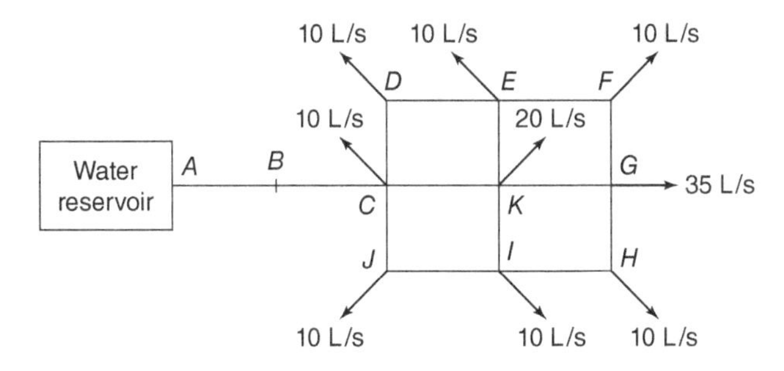

Figure 6.26 Water Network for Problem 6.10

6.11 The main pipe layout of a water distribution system is shown in Fig. 6.27. The water demand for the system (domestic + fire) is 2,300 gpm and is supplied at point A. At point I, the most unfavorable as far as loss of head is concerned, it is assumed that a fire plus the domestic flow requires 900 gpm. At other intersection points it is estimated that the amounts shown will be required to satisfy the normal domestic demand. The elevations at intersections of mains are as follows:

Point	Elevation (ft)	Pipe	*d* (in.)	*L* (ft)
A	95	AB	d_{AB}	500
B	95	BC	d_{BC}	400
C	100	BE	8	1,600
D	95	CF	d_{CF}	1,600
E	100	AD	16	1,600
F	105	DE	6	500
G	100	EF	6	500
H	105	DG	14	2,000
I	109	EH	6	2,000
		FI	d_{FI}	2,000
		GH	12	500
		HI	12	400

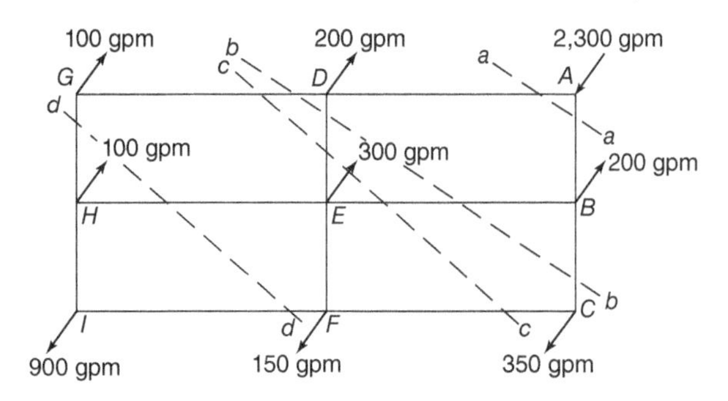

Figure 6.27 Water Network for Problem 6.11

All pipes have a C value of 100.

(a) Determine the required diameters of pipes AB, BC, CF, and FG, assuming that the maximum, allowable head loss in the system is 2%. (*Hint:* Use sections aa, bb, cc, and dd in your analysis.)

(b) What is the minimum possible pressure at point I? Is it acceptable? Assume that a pressure of 30 psi is available at point A.

6.12 The elevations at various points in the water distribution system shown in Fig. 6.28 are as follows:

A, B, and D	528.5 m
C, E, and G	530.0 m
F and H	531.5 m
I	532.5 m

The known lengths and diameters of pipes are as shown below:

Pipe	d (mm)	L (m)
AB	d_{AB}	200
BC	250	200
DE	200	200
EF	150	200
GH	300	200
HI	d_{HI}	200
AD	400	500
BE	200	500
CF	d_{CF}	500
DG	d_{DG}	600
EH	150	600
FI	200	600

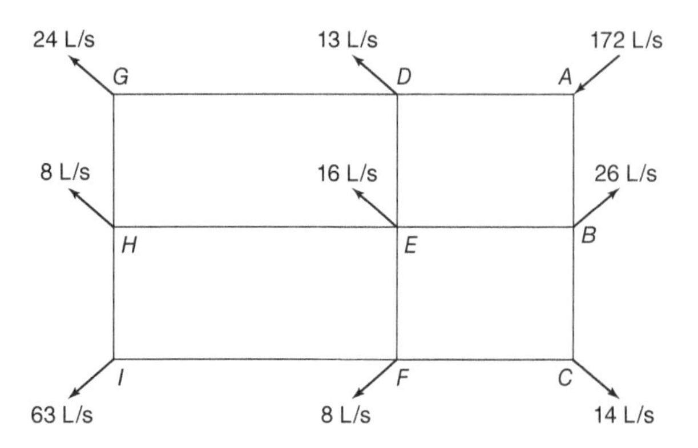

Figure 6.28 Water Network for Problem 6.12

The pressure at point A is 230 kPa and C for all pipes is 100. If the maximum allowable head loss in the system is 2.5‰, find:

(a) Diameter of pipe AB.
(b) Diameter of pipe CF.
(c) Diameter of pipe DG.
(d) Diameter of pipe HI.
(e) The minimum possible pressure at point I.

6.13 Assuming water is to be delivered to a fire through not more than 500 ft of hose, find the water available (in gpm) for a fire in the center of the loop shown in Fig. 6.29. All pipes are located at the same elevation. The pressure in the 12-in. feeder pipes is 40 psi and the residual hydrant pressure is not to be less than 20 psi. C for all pipes is 100.

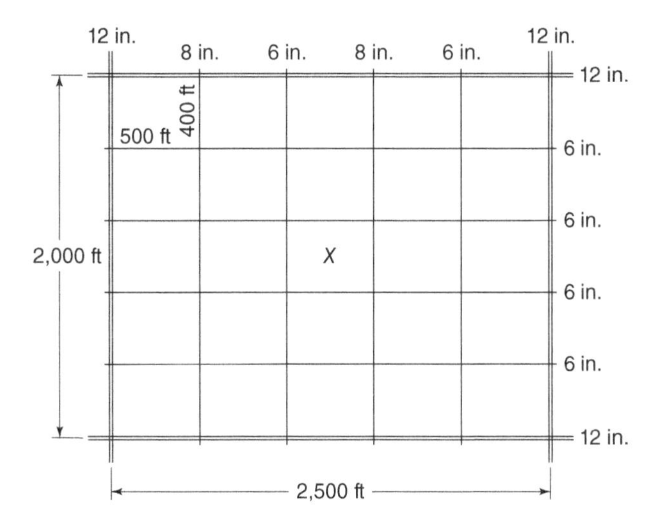

Figure 6.29 Water Network for Problem 6.13

6.14 A gridiron layout of a distribution system located on a flat area is shown in Fig. 6.30. All pipes of the network are made of ductile iron ($C = 100$). The grid serves an area that requires a design fire flow of 10,000 gpm. The pressure in the primary feeder mains ABCDA is 35 psi. All distribution pipes are 8 in. in diameter. The fire hose lines available are 400 ft long.

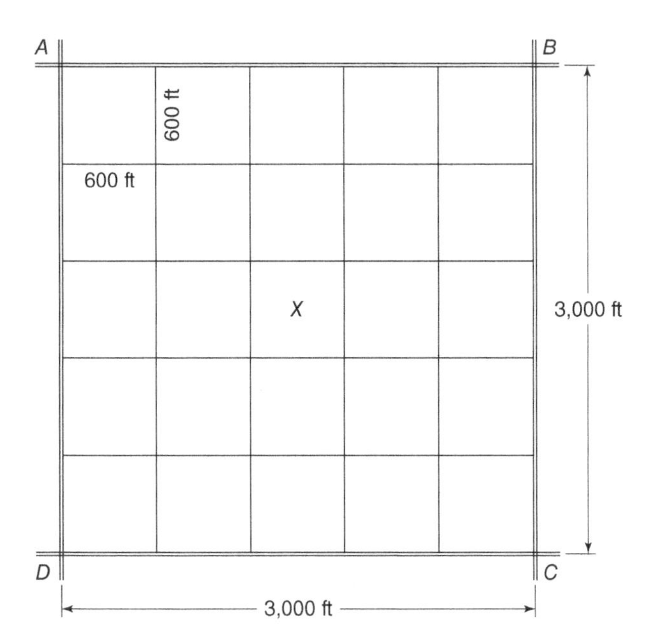

Figure 6.30 Water Network for Problem 6.14

Supposing there is a fire at the middle of the grid (point x):

(a) Find the residual pressure at the hydrants.
(b) How many distribution lines (minimum) should be upgraded from 8 to 10 in. in order to maintain a residual pressure of at least 20 psi at the hydrants?

6.15 The layout of a loop for a city water distribution system located on a flat area is shown in Fig. 6.31. All pipes of the network have a Hazen–Williams C of 100. The fire hose lines available are 150 m long. If the residual pressure in the 300-mm mains is 300 kPa, how much water (Q_F) can be withdrawn for firefighting at the center of loop without lowering the residual pressure at the hydrants below 150 kPa?

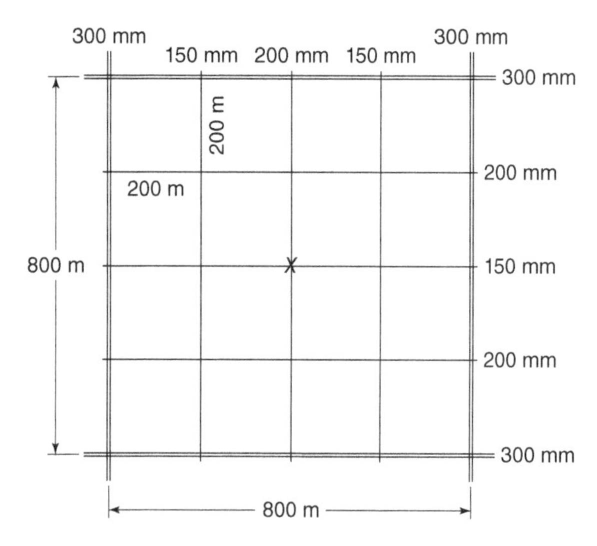

Figure 6.31 Water Network for Problem 6.15

6.16 Mains of a city water distribution loop (ABCDA) are located at an elevation of 635 m (Fig. 6.32). All distribution pipes 1, 2, 3, 4, 5 and 6 have a diameter of 150 mm and a Hazen–Williams C of 100. The fire hose lines available are 150 m long and the required fire flow is 12,200 L/min. The residual pressure in mains ABCDA is 4 bars. Domestic demand can be neglected during firefighting.

(a) Determine the residual pressure at the hydrants (elevation = 645 m) if there is a fire at the center of the loop.
(b) What sizes should lines 2 and 5 be replaced with if the minimum required pressure at the hydrants is 150 kPa?

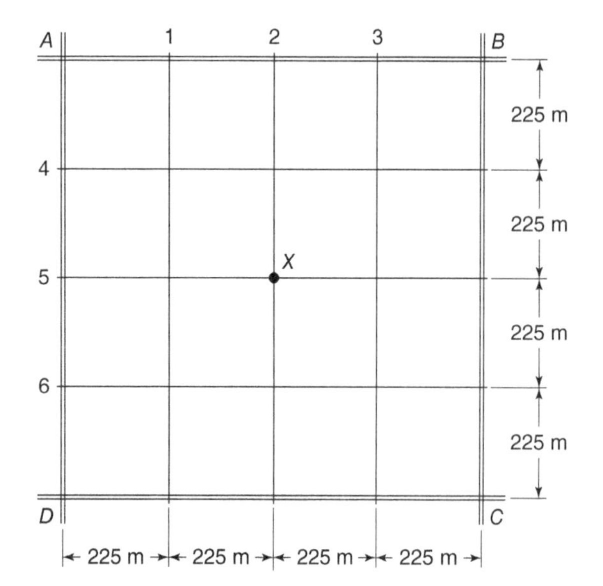

Figure 6.32 Water Network for Problem 6.16

6.17 The layout of a loop in a water distribution system is shown in Fig. 6.33. All pipes of the gridiron network have the same elevation and a Hazen–Williams C of 100. The grid ABCDA serves an area that requires a design fire flow of 9,000 gpm. The pressure in the primary feeders ABCDA is 30 psi, and the fire hose lines available are 410 ft long.
Supposing there is a fire at the middle of the grid (point X),

(a) Find the residual pressure at the hydrants.
(b) How many lines (minimum) should be changed from 8 to 10 in. in order to maintain a residual pressure of at least 20 psi at the hydrants?

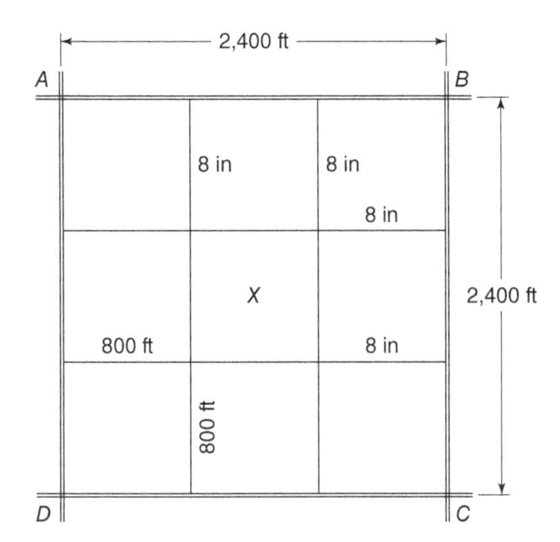

Figure 6.33 Water Network for Problem 6.17

6.18 The main pipes of a city water distribution loop (ABCDA) are located downhill at an elevation of 600 m (Fig. 6.34). All pipes in the network have a Hazen–Williams C of 100. The fire hose lines available are 150 m long. If the residual pressure in the mains is 4 bars, how much water (Q_F in L/s) can be withdrawn for firefighting at point x located at the top of the hill (elevation 619 m) without lowering the residual pressure at the hydrants below 150 kPa?

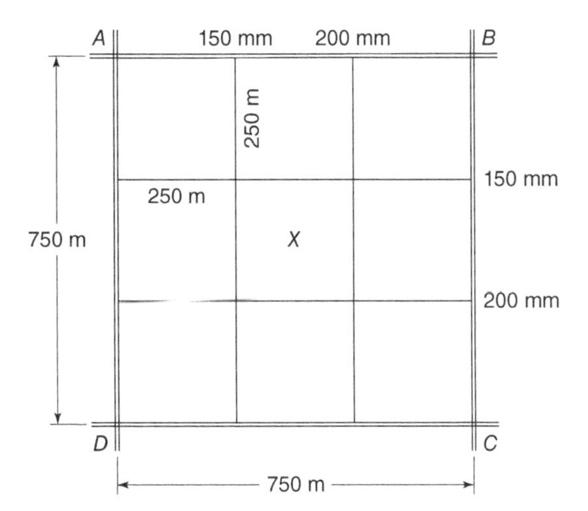

Figure 6.34 Water Network for Problem 6.18

6.19 In the water network shown in Fig. 6.35, the Hazen–Williams coefficient C for all pipes is 100.

(a) Using the Hardy Cross method, determine the flow in each pipe to the nearest 20 gpm
(b) Find the residual pressures at points B and C, given that the pressure at point A is 50 psi.
(c) If the flow input at A is doubled, what would be the residual pressure at C?

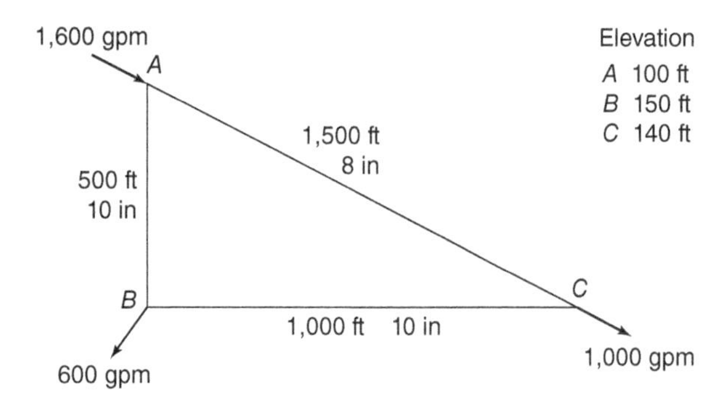

Figure 6.35 Water Network for Problem 6.19

6.20 The water supply system shown in Fig. 6.36 is designed to serve a summer resort. The flow from the reservoir is delivered to network ABC by gravity through a 14-in. main. The elevations of points A, B, and C are as follows:

A 100 ft
B 150 ft
C 140 ft

(a) Find the flow in each pipe.
(b) If the pressure at any point in loop ABC is not allowed to drop below 60 psi, determine the required elevation of the water reservoir.

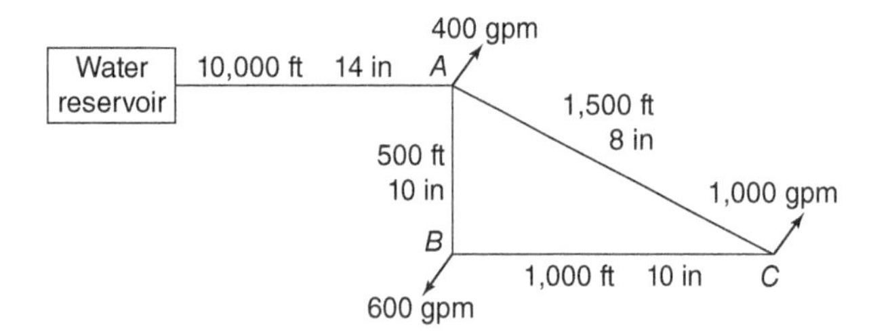

Figure 6.36 Water Network for Problem 6.20

6.21 The water supply system shown in Fig. 6.37 is designed to serve a small town ABDCA. The flow from the reservoir is delivered to the town by gravity through a 400-mm main. All pipes have a Hazen–Williams C of 100. Determine the flow rate in each pipe.

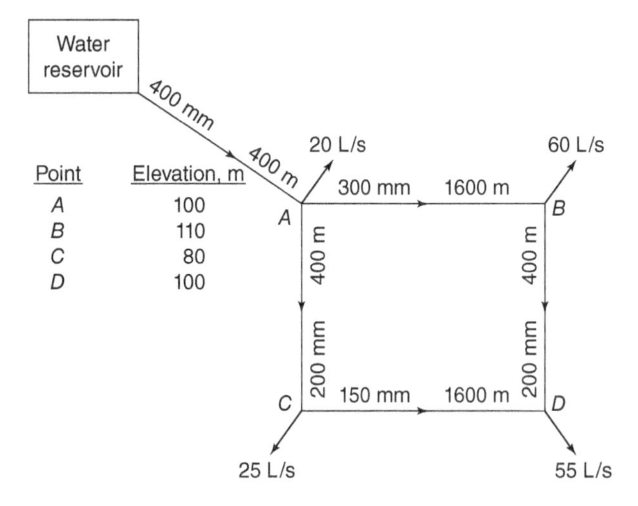

Figure 6.37 Water Network for Problem 6.21

6.22 The water supply system shown in Fig. 6.38 was designed 10 years ago to serve a small town. A water pump P delivers the required flow through a 12-in. water main from the water reservoir to the water distribution loop ABDCA.

At present the water demand has increased to 2,000 gpm, which is assumed to be withdrawn at points A, B, C, and D as shown in the figure. At these drafts the pump can deliver a head of 200 ft. All pipes are made of PVC with a Hazen–Williams coefficient C of 140. Find the flow in each pipe.

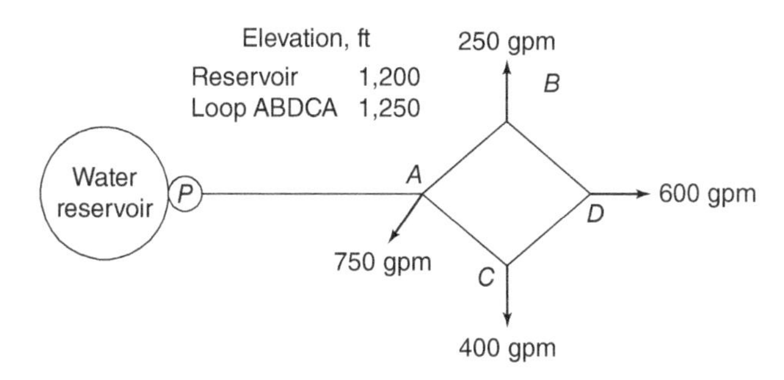

Figure 6.38 Water Network for Problem 6.22

6.23 The regional water supply system shown in Fig. 6.39 is designed to serve four small towns W, X, Y, and Z. Water is delivered from the reservoir to the four towns by pump P. The Hazen–Williams C for all pipes is 130. Pipe diameters and lengths are as follows:

Pipe	*d* (in.)	*L* (ft)
WX	10	10,000
XY	6	12,000
WZ	10	10,000
ZY	6	11,000
PW	16	40,000

Compute the flow in each pipe using an accuracy of 20 gpm.

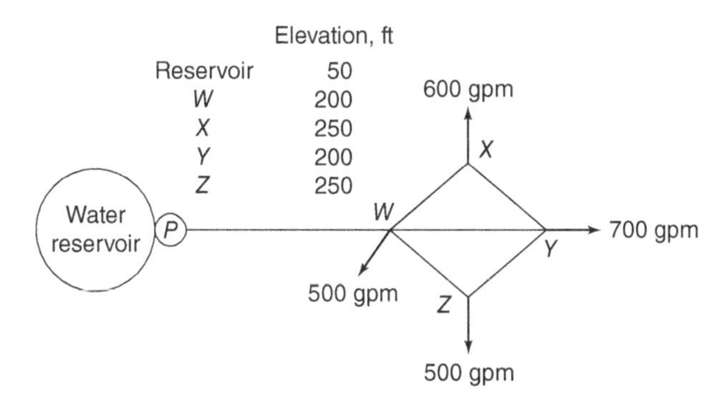

Figure 6.39 Water Network for Problem 6.23

Bibliography

Al-Dhowalia, K. H., and Shammas, N. K., Leak Detection and Quantification of Losses in a Water Network, *Int. J. Water Resour. Develop.*, *7*, 1, 30–38, 1991.

American Water Works Association, *Distribution System Requirements for Fire Protection*, 3rd ed., Manual of Water Supply Practices M31, Denver, CO ed., 1998.

Camp, T. R., and Hazen, H. L., Hydraulic Analysis of Water Distribution Systems by Means of an Electric Network Analyzer, *J. New Eng. Water Works Assoc.*, *48*, 383, 1934.

Cohen, Y. K., *Problems in Water Distribution*, CRC Press, Boca Raton, FL. 240 pp., 2002.

Cross, H., Analysis of Flow in Networks of Conduits or Conductors, *Univ. Illinois Bull.*, *286*, 1936.

Dillingham, J. H., Computer Analysis of Water Distribution Systems, in *Water and Sewage Works series*, January–May, 1967.

Fair, G. M., Geyer, J. C., and Okun, D. A., Water and Wastewater Engineering, in *Water Supply and Wastewater Removal*, vol. *Vol. 1*, John Wiley & Sons, New York, 1966.

Fair, G. M., Geyer, J. C., and Okun, D. A., *Elements of Water Supply and Wastewater Disposal*, John Wiley & Sons, New York, USA, 1971.

Freeman, J. R., Experiments Relating to Hydraulics of Fire Streams, *Trans. Am. Soc. Civil Engrs.*, *21*, 303, 1889.

Haestad Methods Water Solutions, *Computer Applications in Hydraulic Engineering*, 7th ed., Bentley Institute Press, Exton, PA, 2007.

International Code Council, 2006 International Fire Code, Country Club Hills, IL, 2006.

Mays, L. W. (Ed.), *Hydraulic Design Handbook*, McGraw-Hill, New York, 1999.

McIlroy, M. S., Direct-Reading Electric Analyzer for Pipeline Networks, *J. Am. Water Works Ass.*, *42*, 347, 1950.

McIlroy, M. S., Water-Distribution Systems Studied by a Complete Electrical Analogy, *J. New Eng. Water Works Assoc.*, *45*, 299, 1953.

Munson, B. R., Young, D. F., and Okiishi, T. H., *Fundamentals of Fluid Mechanics*, 3rd ed., Wiley, New York, 1998.

National Institute for Occupational Safety and Health, *Fatality Assessment and Control Evaluation (FACE) Program*, Centers for Disease Control and Prevention (CDC), http://www.cdc.gov/niosh/face, 2009.

Roberson, J. A., Cassidy, J. J., and Chaudhry, M. H., *Hydraulic Engineering*, 2nd ed., Wiley, New York, 1998.

Sharma, A. K., and Swamee, P. K., *Design of Water Supply Pipe Networks*, John Wiley & Sons, New York, February 2008.

Swindin, N., *The Flow of Liquids in Pipes*, Kessinger Publishing, Whitefish, MT, 2007.

Tullis, J. P., *Hydraulics of Pipelines: Pumps, Valves, Cavitation, Transients*, John Wiley & Sons, New York, 1989. (Online publication, December 12, 2007.).

7

Water Distribution Systems: Modeling and Computer Applications

7.1 WaterGEMS Software

This chapter deals primarily with the topic of pressure piping as it relates to water distribution systems. If designed correctly, the network of interconnected pipes, storage tanks, pumps, and regulating valves provides adequate pressure, adequate supply, and good water quality throughout the system. If incorrectly designed, some areas may have low pressures, poor fire protection, and even present health risks.

WaterGEMS (Haestad Methods Water Solutions by Bentley) is used in this chapter as a tool to illustrate the application of various available software programs that can help civil and environmental engineers design and analyze water distribution systems. It is also used by water utility managers as a tool to aid in the efficient operation of distribution systems. This software can be used as a stand-alone program, integrated with AutoCAD, or linked to a geographical information system via the GEMS component.

WaterGEMS is used primarily for the modeling and analysis of water distribution systems. Although the emphasis is on water distribution systems, the methodology is applicable to any fluid system with the following characteristics: (a) steady or slowly changing turbulent flow; incompressible, Newtonian, single-phase fluids; and (c) full, closed conduits (pressure system). Examples of systems with these characteristics include potable water systems, sewage force mains, fire protection systems, well pumps, and raw water pumping.

WaterGEMS can analyze complex distribution systems under a variety of conditions. For a typical WaterGEMS project, you may be interested in determining *system pressures* and *flow rates* under average loading, peak *loading*, or fire flow conditions. Extended-period analysis tools also allow you to model the system's response to varying *supply and demand schedules* over a period of time; you can even track *chlorine residuals* or determine the source of the water at any point in the distribution system. In summary, you can use WaterGEMS for

1) Pipe sizing
2) Pump sizing
3) Master planning
4) Construction and operation costs
5) Operational studies
6) Rehabilitation studies
7) Vulnerability studies
8) Water quality studies.

The WaterGEMS program, in addition to other useful software, is available free to users of this book. Educational versions of the software can be accessed online or from the CD that accompanies this textbook.

7.2 Water Demand Patterns

Using a representative diurnal curve for domestic water demand (Fig. 7.1), we see that there is a peak in the *diurnal curve* in the morning as people take showers and prepare breakfast, another slight peak around noon, and a third peak in the evening as people arrive home from work and prepare dinner. Throughout the night, the pattern reflects the relative inactivity of the system, with very low flows compared to the average.

Water and Wastewater Engineering: Hydraulics, Hydrology and Management, Volume 1, Fourth Edition.
Lawrence K. Wang, Mu-Hao Sung Wang, and Nazih K. Shammas.

Companion website: www.wiley.com/go/Wang/Waterandwastewater4e

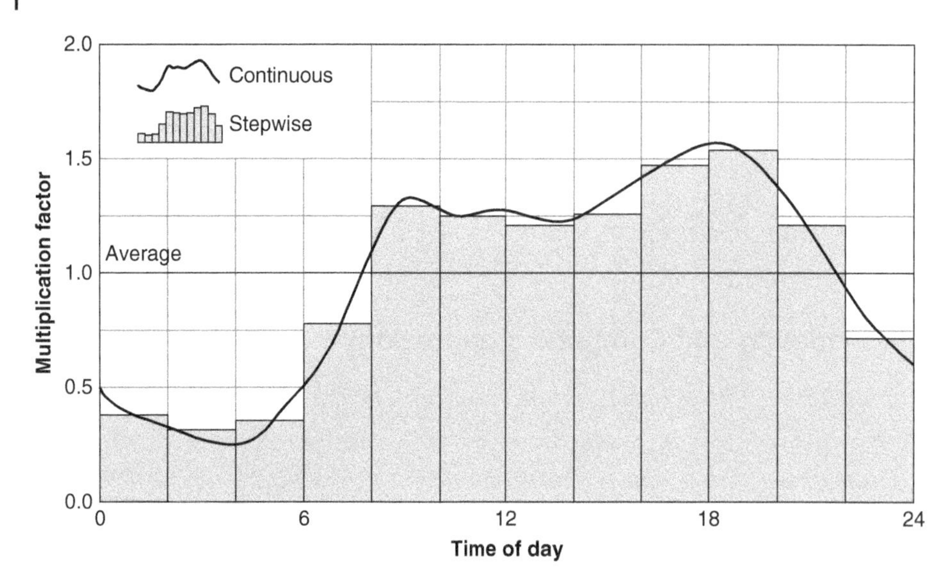

Figure 7.1 Typical Diurnal Curve

Two basic forms are used to represent the patterns of water demand: *stepwise* and *continuous*. A stepwise pattern is one that assumes a constant level of usage over a period of time, and then jumps instantaneously to another level where it again remains steady until the next jump. A continuous pattern is one for which several points in the pattern are known and sections in between are transitional, resulting in a smoother pattern. Notice that, for the continuous pattern in Fig. 7.1, the magnitude and slope of the pattern at the start and end times are the same, a continuity that is recommended for patterns that repeat.

Because of the finite time steps used in the calculations, most computer programs convert continuous patterns into stepwise patterns for use by the algorithms, with the duration of each step equal to the time step of the analysis.

7.3 Energy Losses and Gains

The hydraulic theory behind friction losses is the same for pressure piping as it is for open channel hydraulics. The most commonly used methods for determining head losses in pressure piping systems are the *Hazen–Williams equation* and the *Darcy–Weisbach equation* (see Chapter 5). Many of the general friction loss equations can be simplified and revised because of the following assumptions that can be made for a pressure pipe system:

1) Pressure piping is almost always circular, so the flow area, wetted perimeter, and hydraulic radius can be directly related to diameter.
2) Pressure systems flow full (by definition) throughout the length of a given pipe, so the friction slope is constant for a given flow rate. This means that the energy grade line and the hydraulic grade line (HGL) drop linearly in the direction of flow.
3) Because the flow rate and cross-sectional area are constant, the velocity must also be constant. By definition, then, the energy grade line and HGL are parallel, separated by the constant velocity head ($v^2/2g$).

These simplifications allow for pressure pipe networks to be analyzed much more quickly than systems of open channels or partially full gravity piping. Several hydraulic components that are unique to pressure piping systems, such as regulating valves and pumps, add complexity to the analysis.

Pumps are an integral part of many pressure systems and are an important part of modeling head change in a network. Pumps add energy (head gains) to the flow to counteract head losses and hydraulic grade differentials within the system. Several types of pumps are used for various purposes (see Chapter 8); pressurized water systems typically have centrifugal pumps.

To model the behavior of the pump system, additional information is needed to ascertain the actual point at which the pump will be operating. The *system operating point* is the point at which the pump curve crosses the system curve—the curve representing the *static lift* and *head losses* due to friction and *minor losses*. When these curves are superimposed (as in Fig. 7.2), the operating point is easily located.

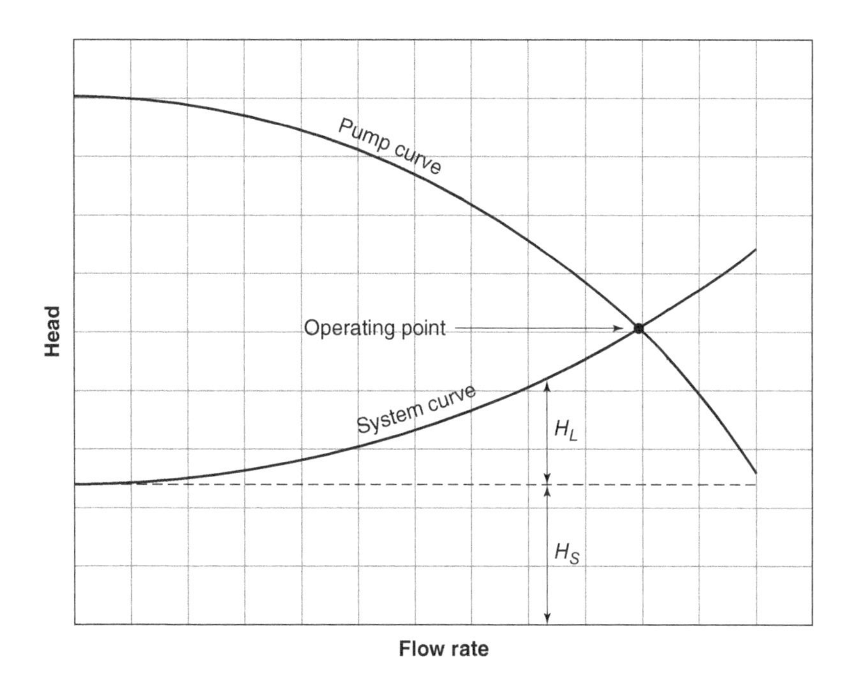

Figure 7.2 System Operating Point

As water surface elevations and demands throughout the system change, the static head and head losses vary. These changes cause the system curve to move around, whereas the pump characteristic curve remains constant. These shifts in the system curve result in a shifting operating point over time (see Chapter 8).

A centrifugal pump's characteristic curve is fixed for a given motor speed and impeller diameter, but can be determined for any speed and any diameter by applying the *affinity laws*. For variable-speed pumps, these affinity laws are presented in Eq. 7.1:

$$\frac{Q_1}{Q_2} = \frac{n_1}{n_2} \text{ and } \frac{H_1}{H_2} = \left(\frac{n_1}{n_2}\right)^2 \tag{7.1}$$

where Q is pump flow rate, m^3/s (ft^3/s); H is pump head, m (ft); and n is pump speed, rpm.

Thus, *pump discharge rate* is proportional to pump speed, and the *pump discharge head* is proportional to the square of the speed. Using this relationship, once the pump curve is known, the curve at another speed can be predicted. Fig. 7.3 illustrates the affinity laws applied to a variable-speed pump. The line labeled "Best Efficiency Point" indicates how the best efficiency point changes at various speeds.

7.4 Pipe Networks

In practice, pipe networks consist not only of pipes, but also of miscellaneous fittings, services, storage tanks, reservoirs, meters, regulating valves, pumps, and electronic and mechanical controls. For modeling purposes, these system elements can be organized into four fundamental categories:

1) *Junction nodes:* Junctions are specific points (nodes) in the system where an event of interest is occurring. Junctions include points where pipes intersect, points where major demands on the system (such as a large industry, a cluster of houses, or a fire hydrant) are located, or critical points in the system where pressures are important for analysis purposes.
2) *Boundary nodes:* Boundaries are nodes in the system where the hydraulic grade is known, and they define the initial hydraulic grades for any computational cycle. They set the HGL used to determine the condition of all other nodes during system operation. Boundary nodes are elements such as tanks, reservoirs, and pressure sources. A model must contain at least one boundary node for the hydraulic grade lines and pressures to be calculated.

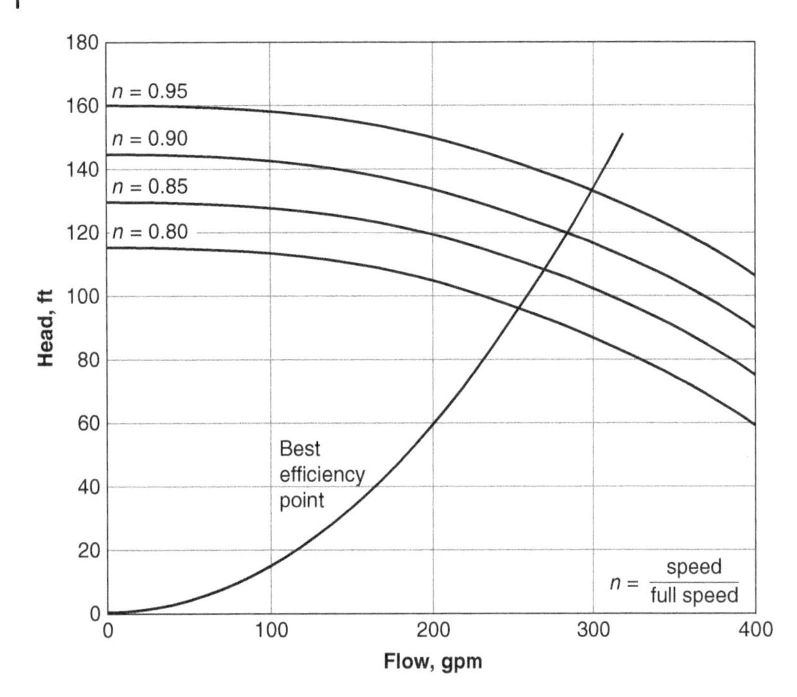

Figure 7.3 Relative Speed Factors for Variable-Speed Pumps. Conversion factors: 1 gpm = 3.785 L/min; 1 ft = 0.3048 m

3) *Links:* Links are system components such as pipes that connect to junctions or boundaries and control the flow rates and energy losses (or gains) between nodes.
4) *Pumps and valves:* Pumps and valves are similar to nodes in that they occupy a single point in space, but they also have link properties because head changes occur across them.

An event or condition at one point in the system can affect all other locations in the system. Although this fact complicates the approach that the engineer must take to find a solution, there are some governing principles that drive the behavior of the network, such as the *conservation of mass* and the *conservation of energy*.

7.4.1 Conservation of Mass

The conservation of mass principle is a simple one. At any node in the system under incompressible flow conditions, the total volumetric or mass flow entering must equal the mass flow leaving (plus the change in storage).

Separating the total volumetric flow into flows from connecting pipes, demands, and storage, we obtain the following equation:

$$\sum Q_{\text{in}} \Delta t = \sum Q_{\text{out}} \Delta t + \Delta s \tag{7.2}$$

where $\sum Q_{\text{in}}$ is total flow into the node, $\sum Q_{\text{out}}$ is total flow out of the node, Δs is change in storage volume, and Δt = change in time.

7.4.2 Conservation of Energy

The principle of conservation of energy dictates that the head losses through the system must balance at each point (Fig. 7.4). For pressure networks, this means that the total head loss between any two nodes in the system must be the same regardless of the path taken between the two points. The head loss must be "sign consistent" with the assumed flow direction (i.e., head loss occurs in the direction of flow, and head gain occurs in the direction opposite that of the flow). For the paths in Fig. 7.4:

Path from A to C: $$H_{L3} = H_{L1} + H_{L2} \tag{7.3}$$

Path from A to B: $$H_{L1} = H_{L3} - H_{L2} \tag{7.4}$$

Loop from A to A: $$0 = H_{L1} + H_{L2} - H_{L3} \tag{7.5}$$

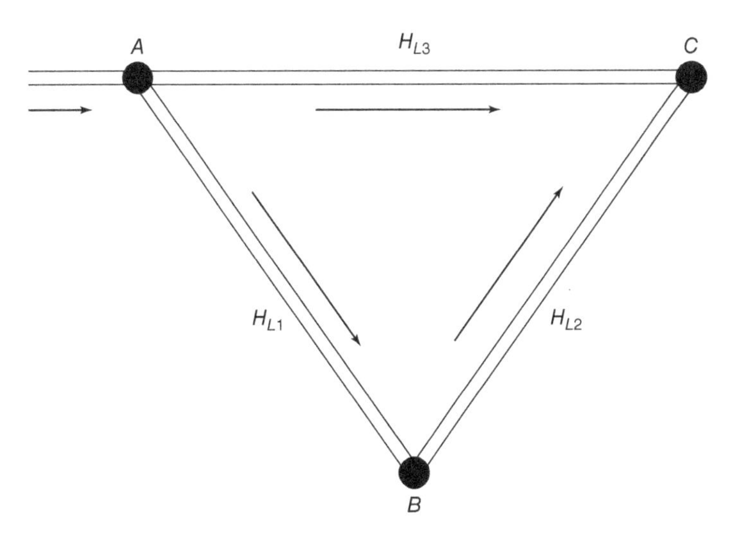

Figure 7.4 Conservation of Energy

Although the equality can become more complicated with minor losses and controlling valves, the same basic principle can be applied to any path between two points. As shown in Fig. 7.4, the combined head loss around a loop must equal zero in order to compute the same hydraulic grade for a given point.

7.5 Network Analysis

7.5.1 Steady-State Network Hydraulics

Steady-state analysis is used to determine the operating behavior of a system at a specific point in time, or under steady-state conditions. This type of analysis can be useful in discovering the short-term effect of fire flows or average demand conditions on the system.

For this type of analysis, the network equations are determined and solved with tanks being treated as fixed-grade boundaries. The results that are obtained from this type of analysis are instantaneous values, and they may not be representative of the values of the system a few hours—or even a few minutes—later in time.

7.5.2 Extended-Period Simulation

An *extended-period simulation* is used to determine the behavior of the system over time. This type of analysis allows the user to model tanks filling and draining, regulating valves opening and closing, and pressures and flow rates changing throughout the system in response to varying demand conditions and automatic control strategies formulated by the modeler.

Whereas a steady-state model may tell the user whether the system has the capability to meet a specific demand, an extended-period simulation indicates whether the system has the ability to provide acceptable levels of service over a period of minutes, hours, or days. Extended-period simulations can also be used for energy consumption and cost studies, as well as for water quality modeling.

Data requirements for an extended-period simulation go beyond what is needed for a steady-state analysis. The user must determine water usage patterns, provide more detailed tank information, and enter operational rules for pumps and valves.

7.6 Water Quality Modeling

In the past, water distribution systems were designed and operated with little consideration of water quality, due in part to the difficulty and expense of analyzing a dynamic system. The cost of extensive sampling and the complex interaction between fluids and constituents makes *numeric modeling* the ideal method for predicting water quality.

To predict water quality parameters, an assumption is made that there is complete mixing across finite distances, such as at a junction node or in a short segment of pipe. Complete mixing is essentially a mass balance given by

$$C_a = \frac{\sum Q_i C_i}{\sum Q_i} \tag{7.6}$$

where C_a is average (mixed) constituent concentration, Q_i represents inflow rates, and C_i represents the constituent concentrations of the inflows.

7.6.1 Age Modeling

Water age provides a general indication of the overall water quality at any given point in the system. Age is typically measured from the time that the water enters the system from a tank or reservoir until it reaches a junction. Along a given link, water age is computed as follows:

$$A_j = A_{j-1} + \frac{x}{v} \tag{7.7}$$

where A_j is the age of water at jth mode, A_{j-1} is the age of water at $j-1$ mode, x is distance from node $j-1$ to node j, and v is velocity from node $j-1$ to node j.

If there are several paths for water to travel to the jth node, the water age is computed as a weighted average using the Eq. 7.8:

$$AA_j = \frac{\sum Q_i \left[AA_i + \left(\frac{x}{v} \right)_i \right]}{\sum Q_i} \tag{7.8}$$

where AA_j is the average age at the node immediately upstream of node j, AA_i is the average age at the node immediately upstream of node i, x_i is the distance from the ith node to the jth node, v_i is the velocity from the ith node to the jth node, and Q_i is the flow rate from the ith node to the jth node.

7.6.2 Trace Modeling

Identifying the origin of flow at a point in the system is referred to as *flow tracking* or *trace modeling*. In systems that receive water from more than one source, trace studies can be used to determine the percentage of flow from each source at each point in the system. These studies can be very useful in delineating the area influenced by an individual source, observing the degree of mixing of water from several sources, and viewing changes in origins over time.

7.6.3 Constituents Modeling

Reactions can occur within pipes that cause the concentration of substances to change as the water travels through the system. Based on *conservation of mass* for a substance within a link (for extended-period simulations only),

$$\frac{\partial c}{\partial t} = v\frac{\partial c}{\partial x} + \theta(c) \tag{7.9}$$

where c is substance concentration as a function of distance and time, t is time increment, v is velocity, x is distance along the link, and $\theta(c)$ is substance rate of reaction within the link.

In some applications, there is an additional term for *dispersion*, but this term is usually negligible (*plug flow* is assumed through the system).

Assuming that complete and instantaneous mixing occurs at all junction nodes, additional equations can be written for each junction node with the following conservation of mass equation:

$$C_k|_{x=0} = \frac{\sum Q_j C_j|_{x=L} + Q_e C_e}{\sum Q_j + Q_e} \tag{7.10}$$

where C_k is concentration at node k, j is pipe flowing into node k, L is the length of pipe j, Q_j is flow in pipe j, C_j is concentration in pipe j, Q_e is external source flow into node k, and C_e is external source concentration into node k.

Once the hydraulic model has solved the network, the velocities and the mixing at the nodes are known. Using this information, the water quality behavior can be derived using a numerical method.

7.6.4 Initial Conditions

Just as a hydraulic simulation starts with some amount of water in each storage tank, initial conditions must be set for a water age, trace, or constituent concentration analysis. These initial water quality conditions are usually unknown, so the modeler must estimate these values from field data, a previous water quality model, or some other source of information.

To overcome the problem of unknown initial conditions at the vast majority of locations within the water distribution model, the duration of the analysis must be long enough for the system to reach *equilibrium conditions*. Note that a constant value does not have to be reached for equilibrium to be achieved; rather, equilibrium conditions are reached when a repeating pattern in age, trace, or constituent concentration is established.

Pipes usually reach equilibrium conditions in a short time, but storage tanks are much slower to show a repeating pattern. For this reason, extra care must be taken when setting a tank's initial conditions, in order to ensure the model's accuracy.

7.6.5 Numerical Methods

Several theoretical approaches are available for solving water quality models. These methods can generally be grouped as either *Eulerian* or *Lagrangian* in nature, depending on the volumetric control approach that is taken. Eulerian models divide the system into fixed pipe segments, and then track the changes that occur as water flows through these segments. Lagrangian models also break the system into control volumes, but then track these water volumes as they travel through the system. This chapter presents two alternative approaches for performing water-quality constituent analyses.

7.6.6 Discrete Volume Method

The *discrete volume method* (DVM) is a Eulerian approach that divides each pipe into equal *segments* with completely mixed volumes (Fig. 7.5). Reactions are calculated within each segment, and the constituents are then transferred to the adjacent downstream segment. At nodes, mass and flow entering from all connecting pipes are combined (assuming total mixing). The resulting concentration is then transported to all adjacent downstream pipe segments. This process is repeated for each water-quality time step until a different hydraulic condition is encountered. When this occurs, the pipes are divided again under the new hydraulic conditions, and the process continues.

7.6.7 Time-driven method

The *time-driven method* (TDM) is an example of a Lagrangian approach (Fig. 7.6). This method also breaks the system into *segments*, but rather than using fixed control volumes as in Eulerian methods, the concentration and size of water *parcels* are tracked as they travel through the pipes. With each time step, the farthest upstream parcel of each pipe elongates as water travels into the pipe, and the farthest downstream parcel shortens as water exits the pipe.

Similar to the DVM, the reactions of a constituent within each parcel are calculated, and the mass and flow entering each node are summed to determine the resulting concentration. If the resulting nodal concentration is significantly different from the concentration of a downstream parcel, a new parcel will be created rather than elongating the existing one. These calculations are repeated for each water-quality time step until the next hydraulic change is encountered and the procedure begins again.

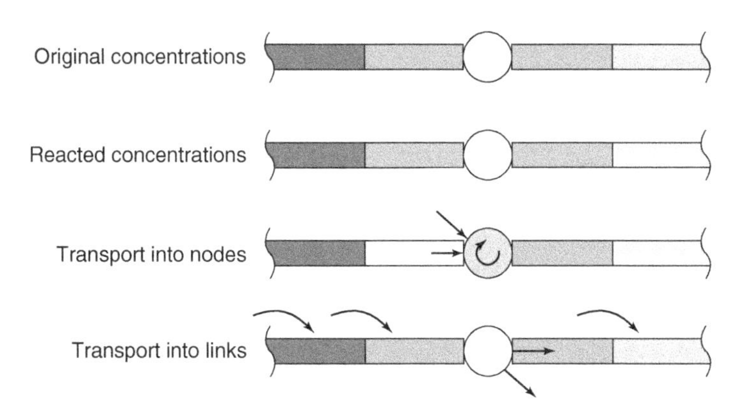

Figure 7.5 Eulerian Discrete Volume Method (DVM)

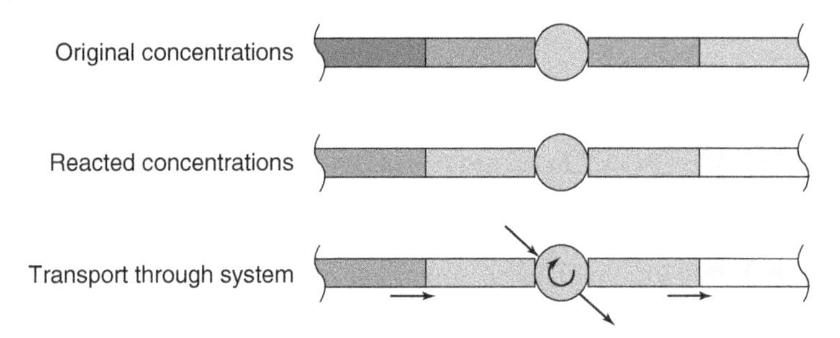

Figure 7.6 Lagrangian Time-Driven Method (TDM)

7.7 Automated Optimization

WaterGEMS has the capability to optimize a model based on *field data* or *design criteria*. Oftentimes, water utility managers will use a model to make design decisions or gather field data to *calibrate a model*. This process is typically a trial-and-error approach in which the modeler will modify a few parameters in a model to either compare design solutions based on cost or benefit, or have the model better predict the real conditions. Because this can be very time-consuming, WaterGEMS has the capability to create many potential solutions and provide a measure of which solution is the "better" solution based on specific *boundary conditions* and *input criteria*.

WaterGEMS employs a *genetic algorithm* search method to find "better" solutions based on the principles of natural selection and biological reproduction. This genetic algorithm program first creates a population of trial solutions based on modeled parameters. The hydraulic solver then simulates each trial solution to predict the HGL and flow rates within the network and compares them to any input criteria. Based on this comparison, a *goodness-to-fit value* is assigned. This information is now used to create a new population of trial solutions. These solutions are then again used to find new solutions. The program compares these solutions to the specific boundary conditions and input criteria until the goodness-to-fit value is optimized. In other words, comparisons are made until no better solution can be generated.

7.7.1 Model Calibration

Model calibration is the process of modifying parameters or values in a model so it better matches what is happening in the real system. The calibration of water distribution models is very complicated. Many values and parameters that are unknown are needed at any one time to reduce the discrepancy between the model and the real system. Oftentimes the pipe roughness value is adjusted to make the model results match the measured or expected values in the real system. However, many other parameters could influence the modeled results. For example, the water demand at junctions and the status of pipes and valves in the system could also be adjusted when calibrating a model.

Calibration of a model relies on accurate field measurement data. Field measurements of pressures in the system, pipe flow rates, water levels in tanks, valve status, and pump operating status and speed are all used to calibrate models. Critical to all of these measurements is the time for which the measurements are made. The times of these measurements must all be synchronized to the time frame of the model. In addition, because the conditions within a real system change throughout the day or year, field data should be collected for many different conditions and times. The calibration process is used to adjust the model to simulate multiple demand loadings and operational boundary conditions. Only then can the modeler be confident that the model is valid for many different conditions.

WaterGEMS has a module called *Darwin Calibrator* that it uses to assist in optimizing the model to match field measurement data. Darwin Calibrator allows the modeler to input field data, then request the software to determine the optimal solutions for pipe roughness values, junction demands or status (on/off). Pipes that have the same hydraulic characteristics where one roughness value is assigned to all pipes can be grouped together. Junctions can also be grouped based on the demand pattern and location. Caution should be used when grouping pipes and junctions because this could greatly affect the model's calibration accuracy.

7.7.2 System Design

The goal of water distribution system design is to maximize the benefits of the system while minimizing the cost. The *optimal solution* is a design that meets all the needs of the system at minimal cost. Some planning is needed to account for additional future needs of the system including potential growth of the system in terms of demand and its location. The modeler must work with the system owner and planning groups to account for both the current and future needs.

Another module in WaterGEMS, called *Darwin Designer*, assists engineers with the planning and design of water distribution networks. Darwin Designer can be used to size new pipe and/or rehabilitate old pipes to minimize cost, maximize benefit, or create a scenario for trading off costs and benefits. The least cost optimization is used to determine the pipe material and size needed to satisfy the design requirements. The maximum benefit optimization is used to determine the most beneficial solution based on a known budget. Darwin Designer will generate a number of solutions that meet the design requirements at minimal cost or maximum benefit. In either case, the best solution for new pipe or rehabilitation of old pipe will be based on the following input hydraulic criteria:

- Minimum and maximum allowable pressures
- Minimum and maximum allowable pipe flow velocity
- Additional demand requirements
- Pipe, pump, tank, valve, etc., status change requirements.

Critical to creating an accurately designed system is time and peak demand requirements. The peak demand and fire flow conditions are used to size pipes since the pipe network must work for all conditions. Using average demand values to size pipe without accurately accounting for peaking factors can create networks that are either undersized and will not deliver the required water needs, or oversized and much more expensive than need be. The daily and seasonal variations can also greatly affect the final design. Demand variations need to be synchronized in the model to accurately reflect what could happen in the real system.

The following examples give step-by-step instructions on how to solve problems and design water systems using WaterGEMS.

Example 7.1 Three Pumps in Parallel

Problem Statement

A pump station is designed to supply water to a small linen factory. The factory, at an elevation of 58.0 m, draws from a circular, constant-area tank (T-1) at a base elevation of 90.0 m with a minimum water elevation of 99.0 m, an initial water elevation of 105.5 m, a maximum water elevation of 106.0 m, and a diameter of 10.0 m.

Three main parallel pumps draw water from a source with a water surface elevation of 58.0 m. Two pumps are set aside for everyday usage, and the third is set aside for emergencies. Each pump has a set of controls that ensure it will run only when the water level in the tank reaches a certain level. Use the Hazen–Williams equation to determine friction losses in the system. The network layout is given in Fig. 7.7; the pipe and pump data are given in Tables 7.1 and 7.2, respectively.

Part 1: Can the pumping station support the factory's 20 L/s demand for a 24-h period?

Part 2: If there were a fire at the linen factory that required an additional 108 L/s of water for hours 0 through 6, would the system with the pump controls given in the problem statement be adequate? Supply the extended-period simulation report describing the system at each time step.

Part 3: How might the system be operated so that the fire flow requirement in part 2 is met?

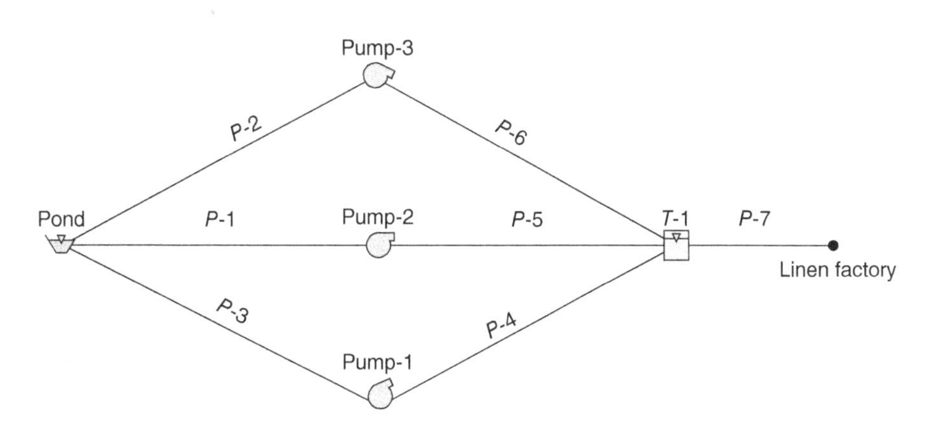

Figure 7.7 Schematic of Example 7.1

Table 7.1 Pipe Information for Example 7.1

Pipe	Length (m)	Diameter (mm)	Material	Roughness
P-1	6	150	Cast iron	90
P-2	6	150	Cast iron	90
P-3	6	150	Cast iron	90
P-4	71	150	Cast iron	90
P-5	72	150	Cast iron	90
P-6	73	150	Cast iron	90
P-7	18	200	Cast iron	90

Table 7.2 Pump Information for Example 7.1

		Pump Curve		
Pump	Elevation (m)	Flow (L/s)	Head (m)	Controls
		0	78.0	
PUMP-1	58.0	32	58.5	On when T-1 is below 105.5 m
		63	0	Off when T-1 is above 106.0 m
		0	78.0	
PUMP-2	58.0	32	58.5	On when T-1 is below 105.2 m
		63	0	Off when T-1 is above 106.0 m
		0	67.0	
PUMP-3	58.0	32	50.5	On when T-1 is below 99.25 m
		63	0	Off when T-1 is above 103.00 m

Solution to Part 1:

- When you start WaterGEMS, you should be prompted with the **Welcome to WaterGEMS** dialog. From this dialog, you can access the tutorials, open existing projects, and create new ones. Select **Create New Project**.
- If the **Welcome to WaterGEMS** dialog does not appear, WaterGEMS is set to **Hide Welcome Page** on startup. To start a new project, select **New** from the **File** menu. You can change from **Hide Welcome Page** mode to **Show Welcome Page** mode in the **Global Options** dialog, which is accessible by selecting **Options** from the **Tools** menu.
- As always when starting a new project, the file should be saved frequently to avoid losing data or simulations. To save a new project, select **Save As** under the **File** menu. Enter the project title **Tutorial 1** and then, at any time, you can save your project by clicking the **Save** button.
- A more descriptive project title and other general information can be entered into the **Project Properties** box found under the **File** menu.
- Before starting, you should set up the general default settings for the project. You can find the default settings in **Options** under the **Tools** menu. In the **Drawing** tab select **Schematic** from the **Drawing Mode** field. This option will allow you to define the pipe lengths and node locations without having to worry about scale and spatial placement on the x–y plane.
- To define the default units, go to the **Units** tab found under **Options** from the **Tools** menu. Select **System International** from the list box in the **Results Default** field if it is not already selected. You can define any of the default label units by clicking the unit field and selecting the desired unit from the list. For example, to change the **Angle** units from radians to degrees, click on **Radians** in the unit field, then select **Degrees** by locating it on the drop-down list of available units.

Laying Out the System

- Begin with the pipeline running horizontally through the center of the system. Because you selected **Schematic** in the **Drawing Mode** field, you do not have to lay out the system exactly as shown in the problem statement. You can roughly sketch the schematic by following the instructions here. You will likely have to rename many of the elements to match the names shown in Fig. 7.7. The steps below will describe how to do this.
- Click the **Pipe Layout** button on the vertical toolbar on the left side of the layout screen.
- Move the cursor to the layout screen and right-click the mouse. Select **Reservoir**. To place the reservoir, simply click the left mouse button.
- Move your mouse horizontally to the right to place a pump. Right-click and select **Pump** from the pop-up menu, then left-click to place the pump.
- Repeat the process for the tank by selecting **Tank** from the pop-up menu.
- Now, place the junction node "Linen Factory." After placing the junction, right-click and select **Done**.
- Continue by entering the remaining two pumps and four pipes in the same way as described previously. To connect a pipe to an object on the layout screen, click the object while in the pipe layout mode. The object should turn red when it is selected.
- Except for the scale, your schematic should look roughly like the one given in the problem statement.
- To exit the pipe layout mode, click the arrow button on the vertical toolbar on the left side of the layout screen.

Entering the Data

- Double-click the reservoir node to open its dialog editor. Change the name to "Pond" in the **Label** field. Enter 58 m in the **Elevation** field. Close the dialog editor.
- Double-click the tank. Enter the given diameter for the circular section and the appropriate elevations from the problem statement. Disregard the inactive volume field. Be sure that **Elevation** is selected in the **Operating Range Type** field. Close the dialog editor.
- Double-click the bottom pump. Change the name to "PUMP-1" in the **Label** field. Enter the appropriate elevation from the pump data table (Table 7.2) into the **Elevation** field. Click the **Pump Definition** field and select **Edit Pump Definitions** to open the Pump Definitions dialog. Add a new pump definition and label it "Pumps 1 and 2." In the **Head** tab select **Standard (3 Point)**. Enter the pump curve data given for PUMP-l. If you need to change the units, right-click on the Flow or Head table headings and open the "Units and Formatting" dialog. Click **Close** to close the Pump Definition dialog. Now select "Pumps 1 and 2" in the **Pump Definitions** field. Close the dialog editor.
- Repeat the above process for the other pumps. When entering the data for PUMP-3, you will have to create a new pump definition titled "Pump 3" for the **Pump Definitions** field.
- Next, enter the pump controls given in the problem statement. Click **Controls** in the **Components** menu.
- Select the **Conditions** tab to enter the five Tank conditions as described from the problem statement information. Enter each condition as **New** and **Simple**. The **Condition Type** is **Element**; select the Tank from the layout screen by clicking the ellipse button ...; select **Hydraulic Grade** as the **Tank Attribute**; the **Operator** and **Hydraulic Grade** are entered based on the problem statement information.
- Select the **Actions** tab to enter whether the pump is on or off. The default setting is generally with the pumps on. Enter the six actions (each pump either on or off) as **New** and **Simple**. For example, to turn off PUMP-1, the **Element** is entered by clicking the ellipse button and selecting PUMP-1 from the layout screen; the **Pump Attribute** would be Pump Status; the **Operator** would be the default "=," then select **Off** for the **Pump Status**.
- Select the **Controls** tab to enter the six controls. The controls are all Simple and entered as If Then statements. For example, click **New** then the **Evaluate as Simple Control** box; in the IF Condition field, select {"Tank" level > 106.00 m}; in the THEN Action field, select {"PUMP-1" pump status = off}. Close the **Controls** dialog.
- Double-click the junction node. Change the name to "Linen Factory." Enter 58 m in the **Elevation** field. Click the **Demand Collection** field to enter a fixed demand of 20 L/s after clicking the ellipse button. Close both dialog editors.
- For the pipes, you can edit the data as you have been by clicking each element individually and then entering the appropriate data. However, this method can be time-consuming, especially as the number of pipe elements increases. It is often easier to edit the data in a tabular format.
- Click the **Flex Tables** button in the toolbar at the top of the screen. Select **Pipe Table** from the available tables.
- The fields highlighted in the **Pipe Table** are output fields. The fields in white are input fields and can be edited as you would edit data in a spreadsheet.

- ***Warning:*** The pipes may not be listed in the table in numerical order. You may want to sort the pipe labels in ascending order. To do this, move the cursor to the top of the table and place it on the **Label** column. Right-click, select **Sort,** and then select **Sort Ascending**. The pipes should then be listed in numerical order.
- Enter the correct pipe lengths into the **Length (User Defined)** column found on the **Pipe Table**. Also enter the pipe diameters and Hazen-Williams *C* value. Close the Pipe Table.
- *Note:* You can customize which columns appear in the **Pipe Table** by clicking the **Edit** button in the toolbar at the top of the table. Table columns can then be added or removed as desired.

Running the Model

- To run the model, first click the **Compute** button on the main toolbar. Arrows should appear on your layout screen indicating the flow direction in each pipe. If you click on any of the objects, you will see the results in the dialog. You could look at the results for all similar objects by opening the **Flex Tables** button. For example, if you want to look at flows in all the pipes, select the **Pipe Table**. To examine the flow through the system over a 24-h period, select the **Calculation Options** under the **Analysis** menu. Double-click the **Base Calculation Options**, then in the **Time Analysis Field** select **EPS**. Set the start time to 12:00:00 a.m. and the duration to 24 h. The **Hydraulic Time Step** of 1 h will provide sufficient output for the purpose of this tutorial. Click the **Compute** button.
- The software provides a couple of different ways to determine whether your model meets the target demand: Scroll through the Calculation Summary and check to see if there are any disconnected node warnings. When the level in tank T-1 drops to the minimum tank elevation of 99 m (tank level of 9 m), the tank closes off, preventing any more water from leaving. This closure will cause the linen factory to be disconnected from the rest of the system (i.e., it will not get the required 20 L/s).

  OR

  Close the Calculation Summary window and select the Linen Factory junction. To create a graph of the pressure at this node, click the **Graphs** button in the main toolbar. Create a **Line-Series Graph** from the **New** button in the Graphs dialog. Select the **Pressure** box in the **Graph Series Options** window, then close the options window. You should see the calculated pressure at the Linen Factory and notice that it never reached zero (no water pressure).

Answer

As you will see for this problem, all the pressures at the linen factory hover around 465 kPa, and no disconnected nodes are detected. Therefore, the pumping station can support the factory's 20 L/s demand for a 24-h period.

Solution to Part 2:

- Add another demand to the Linen Factory node. To do this, double-click the Linen Factory junction and enter into the **Demand Collection** field a second fixed demand of 108 L/s in the row below the 20 L/s demand after clicking the **Ellipse** button. Close the dialog editor.
- Select the **Calculation Options** under the **Analysis** menu, and then double-click the **Base Calculation Options**. You only need to run this model for 6 h, so change 24 to 6 h in the **Duration** field. Click **Compute** to run the model.
- As you scroll through the results, you will see warning messages (yellow or red indicators instead of green) indicating a disconnected node at the linen factory after 3 h. Close the **User Notifications** window. Select the tank and create a Line-Series Graph of the water level in the tank by selecting **Level (Calculated)** in the Graph Series Options window. The graph indicates that the water level in the tank reaches the minimum level of 9 m at 3:37:30 and cannot supply water to the linen factory.

Answer

If there were a fire at the factory, the existing system would **not** be adequate.

Solution to Part 3:

Answer

PUMP-3 could be manually switched on at the beginning of the fire to supply the flow necessary to fight the fire at the linen factory. To do this, delete the pump controls for PUMP-3. Then PUMP-3 will be always on during the model simulation.

Example 7.2 Water Quality

This example demonstrates the use of WaterGEMS to simulate water quality in a water distribution system. The Scenario Manager module is used to facilitate different types of analysis on the same network (within the same project file).

Problem Statement

A local water company is concerned with the water quality in its distribution network. The company wishes to determine the age and chlorine concentration of the water as it exits the system at different junctions. The water surface at the reservoir is 70 m.

Chlorine is injected into the system at the source of flow, R-1 (see Fig. 7.8), at a concentration of 1 mg/L. It has been determined through a series of bottle tests that the average bulk reaction rate of the chlorine in the system (including all pipes and tanks) is approximately 30.5/day.

The network model may be entered in WaterGEMS using the layout in Fig. 7.8 and the data in Tables 7.3 and 7.4.

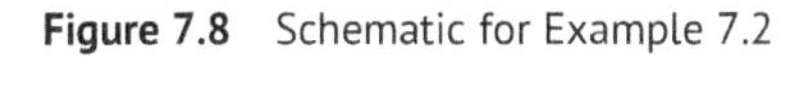

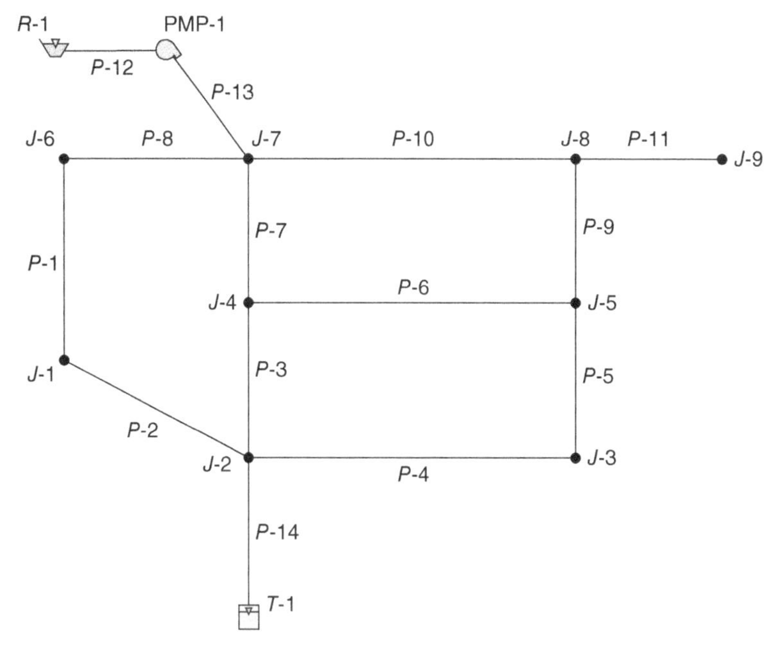

Figure 7.8 Schematic for Example 7.2

Table 7.3 Continuous Demand Pattern Data for Example 7.2

Time from Start (h)	Multiplier	Time from Start (h)	Multiplier
Start	0.80	13	1.30
1	0.60	14	1.40
2	0.50	15	1.50
3	0.50	16	1.60
4	0.55	17	1.80
5	0.60	18	1.80
6	0.80	19	1.40
7	1.10	20	1.20
8	1.50	21	1.00
9	1.40	22	0.90
10	1.30	23	0.80
11	1.40	24	0.80
12	1.40		

Table 7.4 Pump Information for Example 7.2

Flow (L/min)	Head (m)	Controls
0	40	Off of node T-1 above 103.5 m On if node T-1 below 99.5 m
3,000	35	
6,000	24	

The tank is circular with a diameter of 15.0 m. The minimum elevation is 99.0 m. The maximum elevation is 104.0 m, and the initial elevation is 103.4 m. The base elevation is 98.0 m, and the inactive volume is 10.0 m^3. The elevation of the pump is 70.0 m and it is initially on.

Part 1: Perform an age analysis on the system using a duration of 7 days and a time step of 1 h. Determine the youngest and oldest water in the distribution system and the storage tank. Explain why water age varies.

Part 2: Perform a constituent analysis using the same duration and time step as in part 1. Determine the range of concentrations in the system and the storage tank. Explain the behavior of the system with regard to chlorine.

Part 3: Are the simulation results consistent with the known behavior of chlorine?

Part 4: Why is it necessary to run the model for such a long period of time? Do you feel seven days is too long or too short a time period to test the model? Why?

Solution to Parts 1–4:

- Use the same steps as in Example 7.1 to set up the project, lay out the system, and enter the data. Be sure to set units to **System International** and the drawing mode to **Schematic**. Again, you will likely have to rename many of the elements after you draw the general layout to make sure data are correctly entered for each element.
- The demand pattern data can be entered by selecting **Patterns** under the **Components** menu. Right-click **Hydraulic** to select **New**. The Start Time is **12:00:00 AM**, the Starting Multiplier is **0.80**, and the Pattern Format is **Continuous**. Enter the data from the problem statement table under the **Hourly** tab.
- The demand pattern can be assigned to a selected junction by clicking a junction and entering **Hydraulic Pattern—1** into the **Demand Collection** field, or the pattern can be assigned to all junctions as a Global Edit. In this case, assign the demand pattern to all junctions by selecting **Demand Control Center** under the **Tools** menu. Click **Yes** to continue. On the **Junctions** tab, right-click the **Pattern (Demand)** table heading to select **Global Edit**. Select **Hydraulic Pattern—1** in the **Value:** field. Click **OK** and close the Demand Control Center dialog (Tables 7.5 and 7.6).

Table 7.5 Pipe Data for Example 7.2

Pipe	Length (m)	Diameter (mm)	Roughness
P-1	300	200	130
P-2	305	200	130
P-3	225	200	130
P-4	301	200	130
P-5	225	200	130
P-6	301	200	130
P-7	225	200	130
P-8	301	200	130
P-9	200	200	130
P-10	301	200	130
P-11	300	200	130
P-12	1	250	130
P-13	3,000	300	130
P-14	300	300	130

Table 7.6 Junction Data for Example 7.2

Junction	Elevation (m)	Demand (L/min)
J-1	73	151
J-2	67	227
J-3	85	229
J-4	61	212
J-5	82	208
J-6	56	219
J-7	67	215
J-8	73	219
J-9	55	215

Base Scenario

- Run the model for a 24-h period by selecting the **Calculation Options** under the **Analysis** menu. Double-click the **Base Calculation Options,** then in the **Time Analysis Type** select **EPS**. The **Duration (hours)** is 24 h and the **Hydraulic Time Step (hours)** is 1.0 h. Close the Base Calculation Options dialog.
- At the bottom of the Calculation Options window are more tabs. Click on the **Scenarios** tab. Notice that the **Compute** button is in the **Scenario** window toolbar. Click the **Compute** button. WaterGEMS calculates the system parameters for a 24-h simulation period. Details of the calculation can be viewed on the **Calculation Summary** window. Close the **Calculation Summary** and **Scenarios** windows.
- Click on the tank, then the **Graphs** button in the main toolbar. Create a **Line-Series Graph** from the **New** button in the Graphs dialog. Select the **Percent Full** box under "Results" under the **Fields** field in the **Graph Series Options** window. Close the **Graph Series Options** window. Size the Graph window to fit on the layout screen such that you can see most of the layout and the toolbar in the **Graph** window to click the **Play** button . Play the 24-h simulation by clicking the **Play** button in the **Graph** window. The flow in the pipes is indicated by the arrows. Note that that there is no flow in pipes P-12 and P-13 when the pump is not operating. The flow direction reverses in pipes P-1, P-2, P-3, P-4, P-5, P-7, P-8, P-9, and P-14 over the 24-h period. The volume of the water in tank T-1 is indicated by the **Percent Full (%)** on the *y*-axis of the graph.

Age Analysis

- The analysis of the age of water within the network may be performed by defining and running an age analysis scenario. From the **Analysis** menu, select **Scenarios.**
- Create a new **Base Scenario** by clicking the **New** button. Enter "Age Analysis" as the name of the scenario.
- Click on the **Calculation Options** tab at the bottom of the window. Create a new Calculation Option by clicking the **New** button and enter "Age Analysis Calculation Options" as the name. Double-click the calculation options you just created and select **Age** in the **Calculation Type** field. The **Duration** is 168 h (7 days), and the **Hydraulic Time Step** is 1.00 h.
- Go back to the **Scenarios** tab, right-click the Age Analysis scenario, and select **Make Current**. The red check should now be on the Age Analysis scenario. Double-click on the Age Analysis scenario and select the Age Analysis Calculation Options in the **Calculation Options** field.
- Go back to the Scenarios tab and click the **Compute** button.
- Close the **Calculation Summary** window and **Scenarios** dialog to view the layout screen.

Results

The oldest water in the network will be found in tank T-1. Click on the tank then the **Graphs** button in the main toolbar. Create a **Line-Series Graph** from the **New** button in the Graphs dialog. Select the **Age Analysis** box in the **Scenarios** field and **Age**

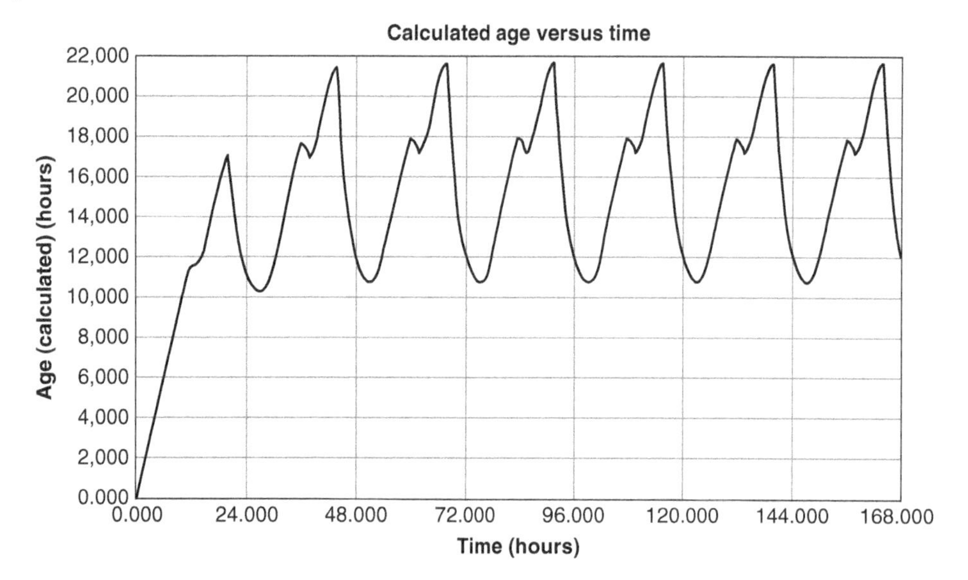

Figure 7.9 Age of Water in Tank T-1 for Example 7.2

(Calculated) found under "Results (Water Quality)" in the **Fields** field in the **Graph Series Options** window. You should also unclick any other selected lines in the **Fields** field. Close the options window. The resulting graph is shown in Fig. 7.9.

Note that the water distribution network reaches dynamic equilibrium after 2 days of the simulation. After 48 h, the maximum age at T-1 is approximately 21.5 h, and the minimum age is approximately 10.5 h.

To view the variation in age in the network, click on the tank, J-2, J-3, J-7, and J-9 while holding the shift on the keyboard to select each object. Then click the **Graphs** button in the main toolbar.

Create a **Line-Series Graph** from the **New** button in the Graphs dialog. Select the **Age Analysis** box in the **Scenarios** field and **Calculated Age**, both under Tank and Junction in the **Fields** field in the **Graph Series Options** window. Close the options window to look at the graph. Notice that the water age in the junctions is much less (2–4 h) than the water in the tank while the pump is on and feeding freshwater into the system. Then the water age in the junctions greatly increases when the system is fed by the tank water after the pump turns off.

Water Quality Analysis

To analyze the behavior of chlorine in the network, the properties of chlorine must be defined in the engineering library.

- From the **Components** menu, select **Engineering Libraries**. Double-click the **Constituent Libraries.** Then right-click on the **ConstituentLibrary.xlm** to select **Add Item**.
- Rename the new constituent by right-clicking it and selecting **Rename**. Change the label to "Chlorine." Click on the **Chlorine** dialog. Enter the **Diffusivity** (1.122e-010 m^2/s). Enter the **Bulk Reaction Order** as 1 and the **Bulk Reaction Rate** as −0.5 $(mg/L)^{1-n}$/day. Because $n = 1$, the units of the rate constant are day^{-1}. Close the **Engineering Libraries.**
- From the **Analysis** menu, select **Scenarios**. Create a new base scenario named "Chlorine Analysis."
- Click on the **Calculation Options** tab at the bottom of the window. Create a new Calculation Option by clicking the **New** button and enter "Chlorine Analysis Calculation Options" as the name. Double-click the calculation options you just created and select **Constituent** in the **Calculation Type field**. The **Duration** is 168 h (7 days), and the **Hydraulic Time Step** is 1.00 h.
- Go back to the **Scenarios** tab and right-click the **Chlorine Analysis** scenario and select **Make Current.** The red check should now be on the Chlorine Analysis scenario. Double-click on the **Chlorine Analysis** scenario and select the **Chlorine Analysis Calculation Options** in the **Calculation Options** field.
- Go to the **Alternatives** tab. Double-click on the **Constituent;** right-click on **Base Constituent Alternative** to select **Open**. Select the **Constituent System Data** tab, then click the ellipse button. Click the **Synchronization Options** button to select **Import from Library**. Select **Chlorine** from the **Constituent Libraries** list. Close the Constituents dialog. Select **Chlorine** in the Constituent field on the **Constituents**: **Base Constituents Alternative** window. Close the **Constituent Alternative** window.

- Double-click the reservoir to define the loading of chlorine. Select **True** in the **Is Constituent Source**? field. The **Constituent Source Type** is **Concentration** and the baseline concentration is 1.0 mg/L. The constituent source pattern is fixed.
- The bulk reaction rate in the pipes can be adjusted using the **Tables** tool. Click the **Flex Table** button, then select the **Pipe Table.** Add the **Bulk Reaction Rate (Local)** and **Specify Local Bulk Reaction Rate**? to the table by clicking the **Edit** button in the toolbar at the top of the table. Scroll to the **Specify Local Bulk Reaction Rate?** column to click the box for the pipe you want to adjust. Now you can enter a reaction value for the pipe in the Bulk Reaction Rate (local) column. In this case we will not change any of the default values so close the Pipe Table.
- Double-click the tank. Set the initial chlorine concentration to 0.000 mg/L, select **True** in the **Specify Local Bulk Rate?** field, then enter the bulk reaction rate of −0.5/day. Close the editor dialog.
- From the **Analysis** menu, select **Scenarios.** Make sure Chlorine Analysis is the current scenario, then run the scenario by clicking the **Compute** button in the top row.
- Open the **EPS Results Browser** window by clicking the icon button. To create a contour map of the chlorine concentration, click the **Contour** button. After clicking the **New** button, contour by **Concentration (Calculated)**; select all elements; set the **Minimum** to 0.0, **Maximum** to 1.0, **Increment** to 0.025, and **Index** to 0.1 mg/L. Click the **Play** button on the **Animation Control** window. The chlorine concentrations for each time step can be viewed through time as shown in Fig. 7.10.
- Save your simulation as "Tutorial 2" since this network will be used in the following tutorials.

Results

To view the variation of the chlorine concentration in the network, click on the tank, J-2, J-3, J-7, and J-9 while holding the shift on the keyboard to select each object. Then click the **Graphs** button in the main toolbar. Create a **Line-Series Graph** from the **New** button in the Graphs dialog. Select the **Chlorine Analysis** box in the **Scenarios** field, and **Concentration (Calculated)** found under "Results (Water Quality)" for both the Tank and Junction in the **Fields** field in the **Graph Series Options** window. Close the options window to look at the graph, which should look like Fig. 7.11. The lowest chlorine concentration is found in tank T-1. Junctions J-2 and J-3 each have similar chlorine concentration values. In addition, the water distribution network reaches dynamic equilibrium during the second day of the simulation. After dynamic equilibrium in achieved, the maximum chlorine concentration at tank T-l is 0.799 mg/L, and the minimum concentration is 0.687 mg/L.

To compare age against chlorine concentration at a selected junction, open the graph that plotted the **Calculated Age** for the tank and junctions. The graphs of age versus time and chlorine concentration should now be open on the desktop.

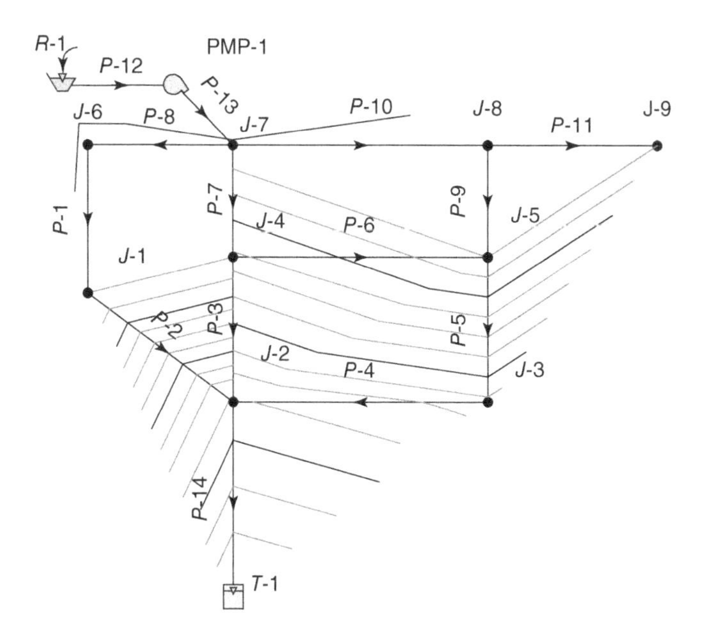

Figure 7.10 Contours of Chlorine Concentration at 42 h for Example 7.2

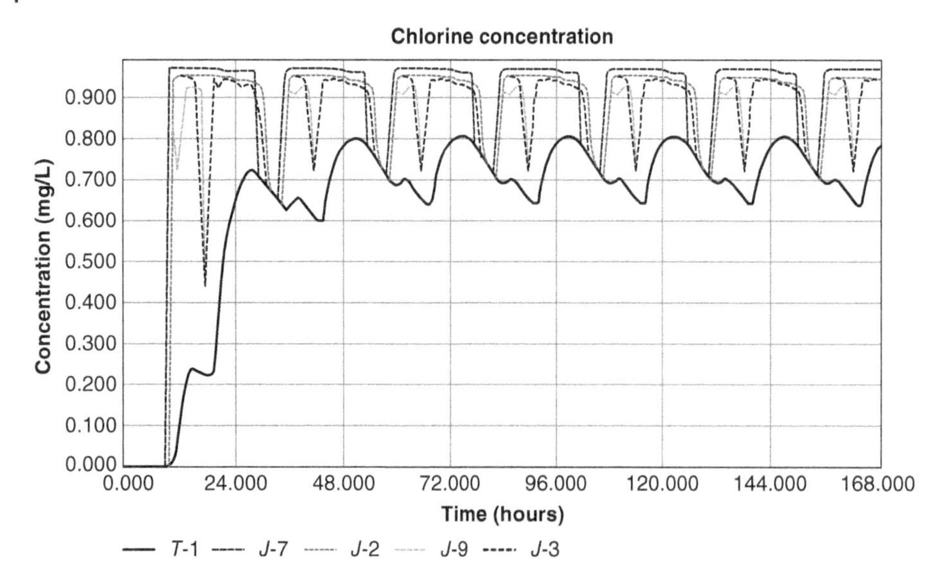

Figure 7.11 Chlorine Concentration in Tank T-1 and Selection Junctions of Example 7.2

Move the graphs so that both are visible and the axes are aligned. Comparison of the two graphs suggests an inverse correlation between age and chlorine concentration.

Answers to Parts 1–4:

Part 1: The oldest water is found in the storage tank. It is far from the source, and incoming water is mixed with the tank's contents. In the distribution system, the oldest water is found at J-9. The newest water is found at J-7 when pump PMP-1 is running.

Part 2: The lowest chlorine concentration is in tank T-1 when it is nearly empty. In the distribution system, the lowest chlorine concentration is found at J-3 when T-1 is emptying. The highest concentration is at J-7 when pump PMP-1 is running.

Part 3: These results are consistent with the fact that chlorine concentration declines over time. For example, during the third day of operation, the minimum chlorine concentration at Junction J-3 is coincident with the maximum age, and the maximum chlorine concentration is coincident with the minimum age.

Part 4: Inspection of the graph of chlorine concentration in tank T-1 suggests that the system stabilizes into a daily pattern on the second day. However, if the initial tank level or the demands are changed, stabilization may take longer. It appears that the 7-day simulation period is adequate for this network.

Example 7.3 Pumping Costs

This example demonstrates the use of WaterGEMS to calculate the energy costs associated with pumping. Calculate the daily electrical costs for the network in Example 7.2 using the following data:

Energy price:	USD 0.10/kWh
Motor efficiency:	90%
Pump efficiency:	50% at 2,000 L/min
	60% at 2,500 L/min
	55% at 3,000 L/min

Solution:

- Open Tutorial 2 (from Example 7.2).
- The first step is to add the pump and motor efficiency data to PMP-1. On the layout view, double-click **PMP-1.** In the **Pump Definition** field, select **Edit Pump Definitions**. The pump definition dialog for PMP-1 appears.

- On the **Efficiency** tab, select the **Multiple Efficiency Points** option for the pump efficiency. In the **Efficiency Points** table, add the efficiency data from the problem statement.
- In the **Motor** section, enter 90% for the motor efficiency.
- Close the Pump Definition tool and the pump dialog.
- The Energy Cost tool is used to calculate energy costs. From the **Analysis** drop-down menu, select **Energy Costs** or click the icon button in the toolbar.
- Click the icon button to open the **Energy Pricing** field dialog.
- Create a **New** label "Energy Pricing-1" to enter the electricity cost. Enter one line in the cost table. The Time from Start is 0.000, and the Energy Price is $0.10/kWh. Close the **Energy Pricing** dialog to return to the **Energy Cost** window.
- Set the scenario to **Chlorine Analysis,** then select "Energy Pricing-1" in the **Energy Pricing** field located on the "Pumps" tab, and then click the **Compute** button.

Answer

On the left panel of the **Energy Cost** window, highlight the **Chlorine Analysis** line. On the right panel, select the **Summary** tab. For the 7-day simulation, the following data were calculated:

Pump energy used	4,030 kWh
Volume pumped	21,220 m^3
Pump cost	USD 403.00
Daily cost	USD 58.60

Example 7.4 Pipe Sizing Using Darwin Designer

WaterGEMS can help size pipes and prepare project cost estimates. In this example, the Darwin Designer with the Minimum Cost function is demonstrated.

Problem Statement

Prepare a minimal cost estimate for the pipe materials and installation portion of the project in Example 7.2. The system pipes should be sized using a demand multiplier of 3.4 (peak flow factor) with a calculated pressure for each junction between 170 and 550 kPa. In addition, the system should supply to an industry located at junction 9 an additional 1,500 L/min with a minimum pressure of 275 kPa. Use the cost data shown in Table 7.7.

Do not consider the cost of the reservoir, tank, pump, or pipes P-12 and P-13.

Solution:

- Open Tutorial 2 (from Example 7.2).
- From the **Analysis** menu, select **Scenarios**. Create a new base scenario named "Designer Analysis."
- Click on the **Calculation Options** tab at the bottom of the window. Create a new Calculation Option by clicking the **New** button and enter "Designer Calculation Options" as the name. Double-click the calculation options you just created. The **Time Analysis Type** is **Steady State**.

Table 7.7 Pipe Material and Cost for Example 7.4

	Pipe Material and Cost	
Material	**Diameter (mm)**	**Cost ($/m)**
Ductile iron	75	40.32
Ductile iron	150	56.64
Ductile iron	200	79.36
Ductile iron	250	114.72
Ductile iron	300	156.16
Ductile iron	350	201.92

- Go back to the **Scenarios** tab, right-click the Designer Analysis scenario, and select **Make Current**. The red check should now be on the Designer Analysis scenario. Double-click on the **Designer Analysis scenario** and select the Designer Calculation Options in the **Calculation Options** field. Click the **Compute** button.
- Click the **Darwin Designer** button or find it under the **Analysis** menu to create a scenario to determine the minimum design cost. Click the **New** button to select **New Designer Study**. Check to make sure the **Scenario** window has **Designer Analysis** selected.
- Enter the design criteria on the **Design Events** tab in the dialog by clicking the **New** button. The top window is used to enter the general design criteria; the bottom window is for entering specific design criteria for any elements (pipe, junction, tank, etc.). Pipe networks are typically designed for high-flow conditions. Scroll across to set the **Demand Multiplier** to 3.4. Criteria are set to maintain a working network to avoid low-flow or low-pressure conditions. The **Minimum Pressure (Global)** to 170 kPa, and **Maximum Pressure (Global)** to 550 kPa. The flow velocity criteria will be the default settings (0–2.44 m/s).
- Enter the design criteria for the industry at junction 9 using the bottom window. In the **Demand Adjustment** tab, use the button to select junction 9 from the drawing by clicking on the J-9 junction, then the green check ✓ in the **Select** dialog. Enter 1,500 L/min into the **Additional Demand** window. In the **Pressure Constraints** tab, again select J-9 from the drawing, then click the box in the **Override Defaults?** window, and then enter 275 kPa and 550 kPa, respectively, for the minimum and maximum pressures.
- Now the software is told which pipes are to be designed. Pipes with similar properties can be grouped together and will be designed the same, or the software can analyze each pipe separately. In this case, the software will analyze each pipe separately. From the problem statement, all pipes will be considered except P-12 and P-13, which are the pipes from the reservoir and pump. On the **Design Groups,** tab click the button to select all the pipes. A table with all the pipes should appear, but if it does not, highlight "<All Available>" in the **Selection Set** window, then click **OK**. In the table, delete pipes P-12 and P-13 to remove them from the analysis.
- The pipe material, properties, and costs to be used in this design scenario are entered in the **Cost/Properties** tab. Open the new pipe table by highlighting **New Pipe** in the window, then select **Design Option Groups** under the New button. Rename "New Pipe-1" to "New Ductile Iron Pipe." Enter the pipe type, diameter, and cost per linear meter from the table in the problem statement. The Hazen–Williams *C* value is 130 for ductile-iron pipe. To change the units for the pipe cost, right-click the column heading **Unit Cost**, then select **Units and Formatting**. In the Unit field, select **$/m**, then click **OK**.
- The objective of this scenario is to size the pipes to deliver the required flow while maintaining reasonable pressures throughout the network at the minimal design cost. To do this, select the **Minimize Cost** criteria in the **Objective Type** window located in the **Design Type** tab.
- After the design criteria have been entered, you can start the simulation by right-clicking on the **New Design Study-1** in the left window to select **New Optimized Design Run**. Now you could perform many different types of design runs by selecting different design events or design groups to be analyzed. Because we have only one design run in this demonstration, you will not have to compare different potential design solutions. In many cases, different solutions will need to be compared to evaluate which would be best for a specific case. The left window helps organize these different solutions.
- On the **Design Events** tab, you can select the design criteria to be evaluated. In this case there is only one choice.
- On the **Design Groups** tab, select the ductile-iron properties and cost that were entered. The data is entered in the **Design Option Group** column. Because all designed pipe will come from the same ductile-iron table, you can enter the data as a global edit. Right-click the **Design Option Group** field and select **Global Edit**. Select the "New Ductile Iron Pipe" in the **Value** window. Click **OK**; all of the Design Option Group fields should automatically fill in.
- The **Options** tab allows the Darwin Designer parameters to be adjusted. In this case the default values will be used.
- To start the run, click the **Compute** button in the Darwin Designer toolbar. When the run is completed, close the Designing ... window. The top three solutions will be listed under the "New Optimized Design Run" in the left window.

Answer

A summary table of the three solutions is shown by clicking the **Solutions** folder. In this example, the minimal pipe cost is USD 217,382. A summary of each solution cost and the design pipe diameters for Solution 1 are shown in Tables 7.8 and 7.9, respectively. The pipe diameters range from 75 to 250 mm and the cost for each pipe is determined based on the expected pipe length. Opening Solution 1 and selecting the **Simulated Results** tab, the calculated pressure at the industry (J-9) is 275.48 kPa, just within the required range.

Note also that this solution uses a number of 75-mm (3-in.) pipes and that the pipe connecting the network to the tank (P-14) is only a 200-mm (8-in.) pipe. If this solution was evaluated for fire flow conditions, it is likely that these pipes would not deliver the required fire flow. Further simulations should be conducted on this solution to ensure that this design can deliver the required flow during a fire event.

Table 7.8 Darwin Designer Solutions for Example 7.4

Darwin Designer Solutions	
Solution	**Total Cost ($)**
Solution 1	217,382
Solution 2	218,687
Solution 3	220,548

Table 7.9 Pipe Diameters and Costs for Solution 1 of Example 7.4

Pipe Diameters and Costs for Solution 1		
Pipe	**Diameter (mm)**	**Cost ($)**
P-3	250	25,812
P-10	200	23,887
P-6	200	23,887
P-14	200	23,808
P-11	200	23,808
P-8	150	17,049
P-4	150	17,049
P-9	200	15,872
P-7	150	12,744
P-2	75	12,298
P-1	75	12,096
P-5	75	9,072

Example 7.5 Model Calibration Using Darwin Calibrator
WaterGEMS has the ability to use measured field data to calibrate a model. In many cases, data that are entered into the model are an approximation or guess. When the model results do not match field data, then parameters in the model are adjusted. Also, using the Darwin Calibrator and field data, we can locate potential differences between the real network and the model that could be caused by problems in the system (blockages, closed valves, etc.).

Adjust the Hazen–Williams *C* factor (roughness factor) for pipes P-2, P-1, and P-8 for the pipe network from Example 7.2. During the field measurements, the tank level was 3.93 m, the pump was off, and a hydrant with a measured flow of 3,400 L/min at junction 7 (J-7) was opened to increase the head loss in the pipe network. The measured field data are shown below:

Junction	Pressure (kPa)
J-7	296.0
J-6	406.5
J-1	263.5
J-2	327.0

Solution:

- Open Tutorial 2 (from Example 7.2).
- From the **Analysis** menu, select **Scenarios**. Create a new base scenario named "Calibrator Analysis."
- Click on the **Calculation Options** tab at the bottom of the window. Create a new Calculation Option by clicking the **New** button and enter "Calibrator Calculation Options" as the name. Double-click the calculation options you just created. The **Time Analysis Type** is **Steady State**.
- Go back to the **Scenarios** tab and right-click the **Calibrator Analysis** scenario and select **Make Current**. The red check should now be on the Calibrator Analysis scenario. Double-click on the **Calibrator Analysis** scenario and select the Calibrator Calculation Options in the **Calculation Options** field. Click the **Compute** button.
- Click the **Darwin Calibrator** button to create a calibration study to determine the pipe roughness factors. Click the **New** button to select **New Calibration Study**. Check to make sure the **Representative Scenario** window has Calibrator Analysis selected.
- Enter the field data in the **Field Data Snapshots** tab. Click the **New** button to enter new data. Enter the measured pressures on the **Observed Target** tab. Select the junctions from the drawing by clicking the **Select** button then clicking junctions J-7 J-6, J-l, and J-2 from the drawing. Click the green check in the **Select** dialog. Select Pressure (kPa) for each junction in the **Attribute** field. Enter the measured pressure for each junction in the **Value** field.
- Enter the tank level in the **Boundary Overrides** tab. Select the tank from the drawing by clicking the **Select** button then clicking the tank. Click the green check in the **Select** dialog. Select **Tank Level (m)** in the **Attribute** field, then enter 3.93 m in the **Value** field.

- To make sure the pump is off, click the **New** button in the **Boundary Overrides** tab. Select the pump by clicking the **Select** button, then clicking the pump. Click the green check in the **Select** dialog. Select **Pump Status** in the **Attribute** field, then select **Off** in the **Value** field.
- Enter the additional demand at junction 7 in the **Demand Adjustments** tab. Select this junction from the drawing by clicking the **Select** button, then clicking **J-7.** Click the green check in the **Select** dialog. Enter 3,400 L/min in the **Value** field.
- Select the pipes where the roughness values are to be determined in the **Roughness Groups** tab. Pipes with similar roughness can be grouped together, or the software can analyze each pipe separately. In this case, the software will determine the pipe roughness values for P-2, P-l, and P-8 separately, and pipes P-3 and P-7 will be grouped since we assume they are both 200-mm (8-in.) ductile-iron pipes, and we do not have any field measurements isolating these pipes. Click the **New** button, then the **Ellipse** button in the **Elements** field. Click the select button, and then pipe P-2 from the drawing. Click the green check in the **Select** dialog. Click **OK** to enter this pipe into the table. Repeat these steps to add pipes P-1 and P-8 to the table. You should have three defined roughness groups in the table on the **Roughness Groups** tab.
- Add the grouped pipes by clicking the **New** button, then the **Ellipse** button in the **Elements** field. Click the select button, then pipes P-3 and P-7 from the drawing. Click the green check in the **Select** dialog. Click **OK** to enter these pipes into the table as a group. There should now be four roughness groups in the table on the **Roughness Groups** tab.
- The calibration method settings are found on the **Calibration Criteria** tab. In this case, these settings will be left as the default settings.
- To use this data to determine the pipe roughness values, create a new run by clicking the **New** button above the left window, then select **New Optimized Run**.
- The pipe roughness values are assumed to have a range between 5 and 140 for each pipe. This is a wide range and any chokes (blockages, partially closed valves, etc.) in a pipe could greatly reduce the pipe roughness value. It is not expected that any roughness value above 140 would not be observed for ductile-iron pipe. On the **Roughness** tab in the **Operation** field, **Set** should be selected; the expected minimum and maximum roughness values are entered in the **Minimum Value** and **Maximum Value** fields. Enter the increment for the software to analyze the roughness values by entering 5 into the **Increment** field. Other increments could be used.
- For this simulation, the top five solutions will be displayed. To do this, click the **Options** tab and enter 5 into the **Solutions to Keep** field.
- To start the run, click the **Compute** button in the Darwin Calibrator toolbar. When the run is completed, close the **Calibration...** window. The top five solutions will be listed under the "New Optimized Run" in the left window.

Answer

The fitness values of the five solutions are shown by clicking the **Solutions** folder. In this example, the fitness values ranged from 0.280 to 0.303 where a lower fitness value indicates a "better" solution. A summary of each solution with its determined roughness values is shown in Table 7.10. To view the determined roughness values for the "best" solution, click the **Solution 1** summary and highlight **Roughness** in the **Adjustments Results** window under the **Solutions** tab. To view the observed and simulated HGL, click the **Simulated Results** tab.

Table 7.10 Darwin Calibrator Solutions for Example 7.5

Darwin Calibration Solutions				
	Roughness Value			
Solution	P-2	P-1	P-8	P-3 and P-7
Solution 1	115	55	140	100
Solution 2	120	55	140	100
Solution 3	100	45	140	100
Solution 4	115	55	125	100
Solution 5	125	55	140	100

These results indicate that pipes P-2 and P-8, and the grouped pipes P-3 and P-7 have a roughness value that is about what would be expected for the installed ductile-iron pipe. However, the results for pipe P-1 show that the roughness value is much lower. This could indicate that a valve is partially closed, that the pipe is blocked, or that the pipe diameter may be smaller than expected. In this case, pipe P-1 should be investigated to determine the cause of this low roughness value. If there is a problem and that problem was fixed, new field measurements should be taken.

These roughness values can be entered into the model and further simulations can be conducted. With enough field data, a model that closely simulates the actual system can be created. Keep in mind that many times the person doing the modeling must decide what values to put into the model. The software can only calculate values based on what is entered. The person doing the modeling must judge how accurate the model is and whether the model can be used to make decisions.

Problems/Questions

7.1 Solve the following problems using the WaterGEMS computer program.

The ductile-iron pipe network shown in Fig. 7.12 carries water at 203°C. Assume that the junctions all have an elevation of 0 m and the reservoir is at 30 m. Use the Hazen–Williams formula ($C = 130$) and the pipe and demand data in Tables 7.11 and 7.12 to perform a steady-state analysis and answer the following questions:

1) Which pipe has the lowest discharge? What is the discharge (in L/min)?
2) Which pipe has the highest velocity? What is the velocity (in m/s)?
3) Calculate the problem using the Darcy–Weisbach equation ($k - 0.26$ mm) and compare the results.
4) What effect would raising the reservoir by 20 m have on the pipe flow rates? What effect would it have on the HGLs at the junctions?

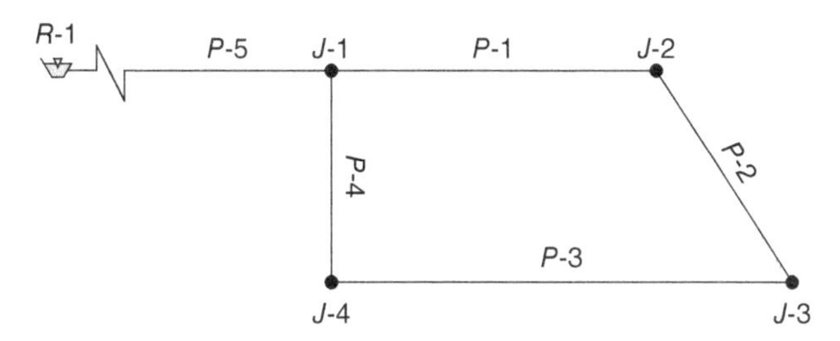

Figure 7.12 Schematic for Problem 7.1

Table 7.11 Pipe Information for Problem 7.1

Pipe	Diameter (mm)	Length (m)
P-1	150	50
P-2	100	25
P-3	100	60
P-4	100	20
P-5	250	760

Table 7.12 Junction Information for Problem 7.1

Junction	Demand (L/min)
J-1	570
J-2	660
J-3	550
J-4	550

7.2 A pressure gauge reading of 288 kPa was taken at J-5 in the pipe network shown in Fig. 7.13. Assuming a reservoir elevation of 100 m, find the appropriate Darcy–Weisbach roughness height (to the hundredths place) to bring the model into agreement with these field records. Use the same roughness value for all pipes. The pipe and junction data are given in Tables 7.13 and 7.14, respectively.

1) What roughness factor yields the best results?
2) What is the calculated pressure at J-5 using this factor?
3) Other than the pipe roughnesses, what other factors could cause the model to disagree with field-recorded values for flow and pressure?

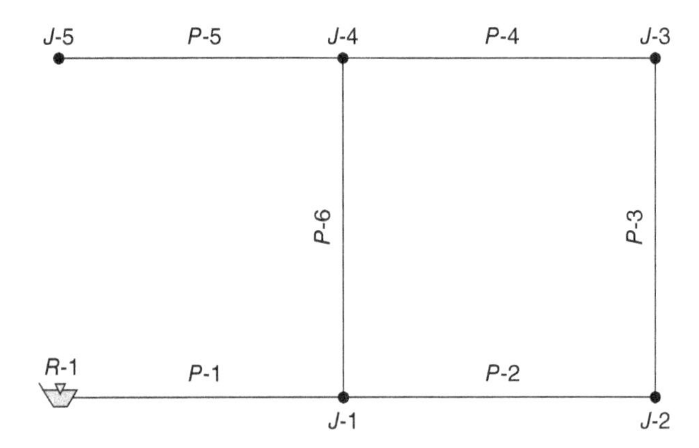

Figure 7.13 Schematic for Problem 7.2

Table 7.13 Pipe Information for Problem 7.2

Pipe	Diameter (mm)	Length (m)
P-1	250	1,525
P-2	150	300
P-3	150	240
P-4	150	275
P-5	150	245
P-6	200	230

Table 7.14 Junction Information for Problem 7.2

Junction	Elevation (m)	Demand (L/min)
J-1	55	950
J-2	49	1,060
J-3	58	1,440
J-4	46	1,175
J-5	44	980

7.3 A distribution system is needed to supply water to a resort development for normal usage and emergency purposes (such as fighting a fire). The proposed system layout is shown in Fig. 7.14. The source of water for the system is a pumped well. The water is treated and placed in a ground-level tank (shown in the figure as a reservoir because of its plentiful supply), which is maintained at a water surface elevation of 210 ft. (64 m). The water is then pumped from this tank into the rest of the system.

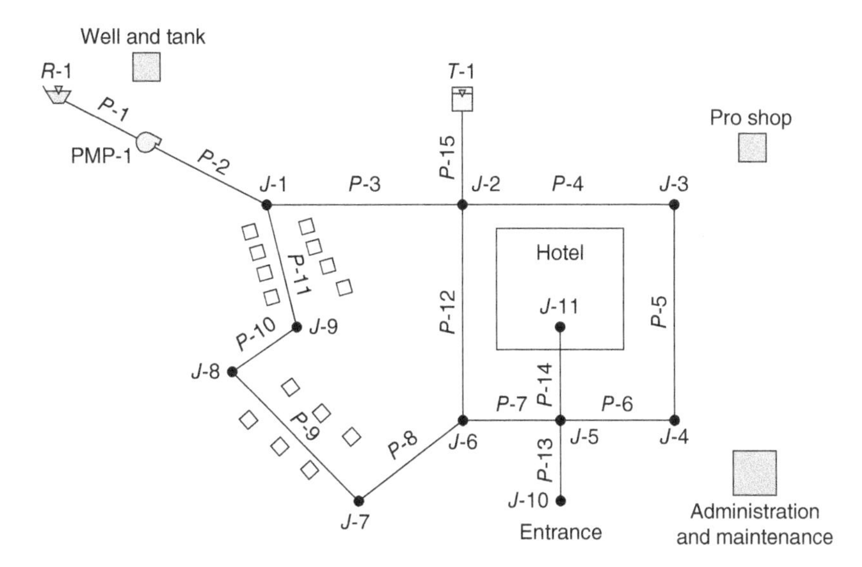

Figure 7.14 Schematic for Problem 7.3

The well system alone cannot efficiently provide the amount of water needed for fire protection, so an elevated storage tank is also needed. The bottom of the tank is at 376 ft. (114.6 m) (high enough to produce 35 psi or 243 kPa at the highest node), and the top is approximately 20 ft. (6.1 m) higher. To avoid the cost of an elevated tank, this 80-ft-diameter (24.4-m-diameter) tank is located on a hillside, 2,000 ft. (610 m) away from the main system. Assume that the tank starts with a water surface elevation of 380 ft. (115.8 m). The pump was originally sized to deliver 300 gpm (1,135 L/min) with enough head to pump against the tank when it is full. Three defining points on the pump curve are as follows: 0 gpm at 200 ft. of head, 300 gpm at 180 ft. of head, and 600 gpm at 150 ft. of head (0 L/min at 610 m of head, 1,135 L/min at 54.9 m of head, and 2,271 L/min at 45.7 m of head). The pump elevation is assumed to be the same as the elevation at J-1, although the precise pump elevation is not crucial to the analysis.

The system is to be analyzed under several demand conditions with minimum and maximum pressure constraints. During normal operations, the junction pressures should be between 35 and 80 psi (243 kPa and 555 kPa). Under fire flow conditions, however, the minimum pressure is allowed to drop to 20 psi (139 kPa). Fire protection is being considered both with and without a sprinkler system.

Demand Alternatives: WaterGEMS enables you to store multiple demand alternatives corresponding to various conditions (such as average day and peak hour). This feature allows you run different scenarios that incorporate various demand conditions within a single project file without losing any input data. For an introduction and more information about scenarios and alternatives, see WaterGEMS's online help system.

Pipe Network: The pipe network consists of the pipes listed in Table 7.15. The diameters shown are based on the preliminary design and may not be adequate for the final design. For all pipes, use ductile iron as the material and a Hazen–Williams C factor of 130. The junction information for this problem is given in Table 7.16.

To help keep track of important system characteristics (like maximum velocity, lowest pressure, etc.), you may find it helpful to keep a table such as Table 7.17.

Another way to quickly determine the performance of the system is to color-code the pipes according to some indicator.

In hydraulic design, a good performance indicator is often the velocity in the pipes. Pipes consistently flowing below 0.5 ft./s (0.15 m/s) may be oversized. Pipes with velocities over 5 ft./s (1.5 m/s) are fairly heavily stressed, and those with velocities above 8 ft./s (2.4 m/s) are usually bottlenecks in the system under that flow pattern. Color-code the system using the ranges given in Table 7.18. After you define the color-coding, place a legend in the drawing (see Table 7.18).

1) Fill in or reproduce the Results Summary table after each run to get a feel for some of the key indicators during various scenarios.
2) For the average day run, what is the pump discharge?
3) If the pump has a best efficiency point at 300 gpm (1,135.5 L/min), what can you say about its performance on an average day?

Table 7.15 Pipe Information for Problem 7.3

Pipe	Diameter		Length	
	(in)	(mm)	(ft)	(m)
P-1	8	200	20	6.1
P-2	8	200	300	91.4
P-3	8	200	600	182.9
P-4	6	150	450	137.2
P-5	6	150	500	152.4
P-6	6	150	300	91.4
P-7	8	200	250	76.2
P-8	6	150	400	121.9
P-9	6	150	400	121.9
P-10	6	150	200	61.0
P-11	6	150	500	152.4
P-12	8	200	500	152.4
P-13	6	150	400	121.9
P-14	6	150	200	61.0
P-15	10	250	2,000	609.6

Table 7.16 Junction Information for Problem 7.3

Junction	Elevation		Average Day		Peak Hour		Minimum Hour		Fire with Sprinkler		Fire Without Sprinkler	
	(ft)	(m)	(gpm)	(L/min)	(gpm)	(L/min)	(gpm)	(L/min)	(gpm)	(L/min)	(gpm)	(L/min)
J-1	250	76.2	0	0	0	0	0	0	0	0	0	0
J-2	260	79.2	0	0	0	0	0	0	0	0	0	0
J-3	262	79.9	20	75.7	50	189.3	2	7.6	520	1,968.2	800	3,028
J-4	262	79.9	20	75.7	50	189.3	2	7.6	520	1,968.2	800	3,028
J-5	270	82.3	0	0	0	0	0	0	0	0	800	3,028
J-6	280	85.3	0	0	0	0	0	0	0	0	800	3,028
J-7	295	89.9	40	151.4	100	378.5	2	7.6	40	151.4	40	151.4
J-8	290	88.4	40	151.4	100	378.5	2	7.6	40	151.4	40	151.4
J-9	285	86.9	0	0	0	0	0	0	0	0	0	0
J-10	280	85.3	0	0	0	0	0	0	360	1,362.6	160	605.6
J-11	270	82.3	160	605.6	400	1,514.0	30	113.6	160	605.6	160	605.6

4) For the peak hour run, the velocities are fairly low. Does this mean you have oversized the pipes? Explain.
5) For the minimum hour run, what was the highest pressure in the system? Why would you expect the highest pressure to occur during the minimum hour demand?
6) Was the system (as currently designed) acceptable for the fire flow case with the sprinkled building? On what did you base this decision?
7) Was the system (as currently designed) acceptable for the fire flow case with all the flow provided by hose streams (no sprinklers)? If not, how would you modify the system so that it will work?

Table 7.17 Results Summary for Problem 7.3

Variable	Average Day	Peak Hour	Minimum Hour	Fire with Sprinkler	Fire Without Sprinkler
Node w/low pressure					
Low pressure (psi)					
Node w/high pressure					
High pressure (psi)					
Pipe w/max. velocity					
Max. velocity (ft/s)					
Tank in/out flow (gpm)					
Pump discharge (gpm)					

Conversion factors: 1 ft/s = 0.3048 m/s; 1 gpm = 3.785 L/ min ; 1 psi = 6.94 kPa.

Table 7.18 Color-Coding Range for Problem 7.3

Max. Velocity		
(ft/s)	(m/s)	Color
0.5	0.15	Magenta
2.5	0.76	Blue
5.0	1.52	Green
8.0	2.44	Yellow
20.0	6.10	Red

7.4 A ductile-iron pipe network ($C = 130$) is shown in Fig. 7.15. Use the Hazen–Williams equation to calculate friction losses in the system. The junctions and pump are at an elevation of 5 ft. (1.52 m) and all pipes are 6 in. (150 mm) in diameter. (*Note:* Use a standard, three-point pump curve. The data for the pump, junctions, and pipes are in Tables 7.19–7.21. The water surface of the reservoir is at an elevation of 30 ft. [9.14 m].)

1) What are the resulting flows and velocities in the pipes?
2) What are the resulting pressures at the junction nodes?
3) Place a check valve on pipe P-3 such that the valve only allows flow from J-3 to J-4. What happens to the flow in pipe P-3? Why does this occur?
4) When the check valve is placed on pipe P-3, what happens to the pressures throughout the system?
5) Remove the check valve on pipe P-3. Place a 6-in. (150-mm) flow control valve (FCV) node at an elevation of 5 ft. (1.52 m) on pipe P-3. The FCV should be set so that it only allows a flow of 100 gpm (378.5 L/min) from J-4 to J-3. (*Hint:* A check valve is a pipe property.) What is the resulting difference in flows in the network? How are the pressures affected?
6) Why does not the pressure at J-1 change when the FCV is added?
7) What happens if you increase the FCV's allowable flow to 2,000 gpm (7,570 L/min). What happens if you reduce the allowable flow to zero?

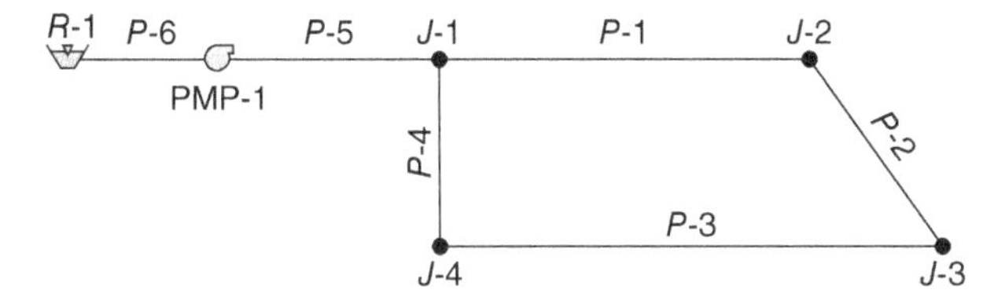

Figure 7.15 Schematic for Problem 7.4

Table 7.19 Pump Information for Problem 7.4

Head (ft)	Head (m)	Flow (gpm)	Flow (L/min)
200	60.96	0	0
175	53.34	1,000	3,785
100	30.48	2,000	7,570

Table 7.20 Junction Information for Problem 7.4

Junction Label	Demand (L/min)	Demand (gpm)
J-1	1,514	400
J-2	2,082	550
J-3	2,082	550
J-4	1,325	350

Table 7.21 Pipe Information for Problem 7.4

Pipe Label	Length (m)	Length (ft)
P-1	23.77	78
P-2	12.19	40
P-3	27.43	90
P-4	11.89	39
P-5	3.05	10
P-6	3.05	10

7.5 A local country club has hired you to design a sprinkler system that will water the greens of their nine-hole golf course. The system must be able to water all nine holes at once. The water supply has a water surface elevation of 10 ft. (3.05 m). All pipes are polyvinyl chloride ($C = 150$; use the Hazen–Williams equation to determine friction losses). Use a standard, three-point pump curve for the pump, which is at an elevation of 5 ft. (1.52 m). The flow at the sprinkler is modeled using an emitter coefficient. The data for the junctions, pipes, and pump curve are given in Tables 7.22–7.24. The initial network layout is shown in Fig. 7.16.

1) Determine the discharge at each hole.
2) What is the operating point of the pump?

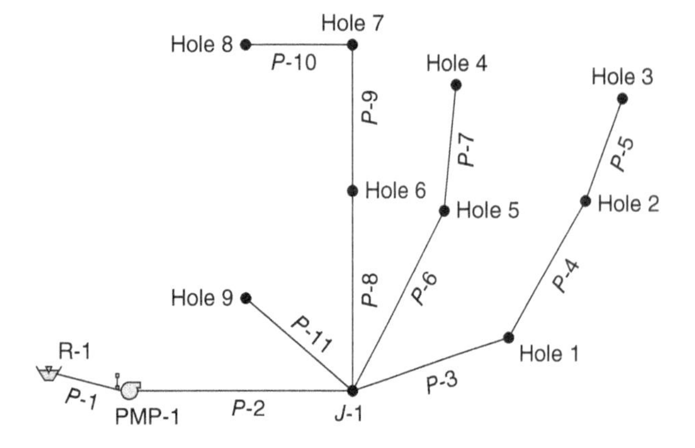

Figure 7.16 Schematic for Problem 7.5

Table 7.22 Junction Information for Problem 7.5

Junction Label	Emitter Coefficient		Elevation	
	(gpm/psi$^{0.5}$)	(L/min/kPa$^{0.5}$)	(ft)	(m)
J-1	–	–	10	3.05
Hole 1	8	11.49	7	2.13
Hole 2	10	14.37	7	2.13
Hole 3	15	21.55	40	12.19
Hole 4	12	17.24	5	1.52
Hole 5	8	11.49	5	1.52
Hole 6	8	11.49	15	4.57
Hole 7	10	14.37	20	6.10
Hole 8	15	21.55	10	3.05
Hole 9	8	11.69	12	3.66

Table 7.23 Pipe Information for Problem 7.5

Pipe Label	Diameter (mm)	Diameter (in)	Length (m)	Length (ft)
P-1	100	4	3.05	10
P-2	100	4	304.8	1,000
P-3	100	4	243.8	800
P-4	76	3	228.6	750
P-5	76	3	152.4	500
P-6	76	3	213.4	700
P-7	50	2	121.9	400
P-8	100	4	243.8	800
P-9	76	3	152.4	500
P-10	50	2	121.9	400
P-11	50	2	152.4	500

Table 7.24 Pump Information for Problem 7.5

Head (ft)	Head (m)	Flow (gpm)	Flow (L/min)
170	51.8	0	0
135	41.1	300	1,135.5
100	30.5	450	1,703.3

7.6 A subdivision of 36 homes is being constructed in a new area of town. Each home will require 1.7 L/s during peak periods. All junction nodes are 192 m in elevation. All pipes are ductile iron ($C = 130$; use the Hazen–Williams Equation to determine the friction losses in the pipe). The current lot and network layout are shown Fig. 7.17. Junction and pipe information are given in Tables 7.25 and 7.26.

Currently, a model of the entire water system does not exist. However, hydrant tests were conducted using hydrants located on two water mains, one in Town Highway #64 and the other in Elm Street. The following data were obtained:

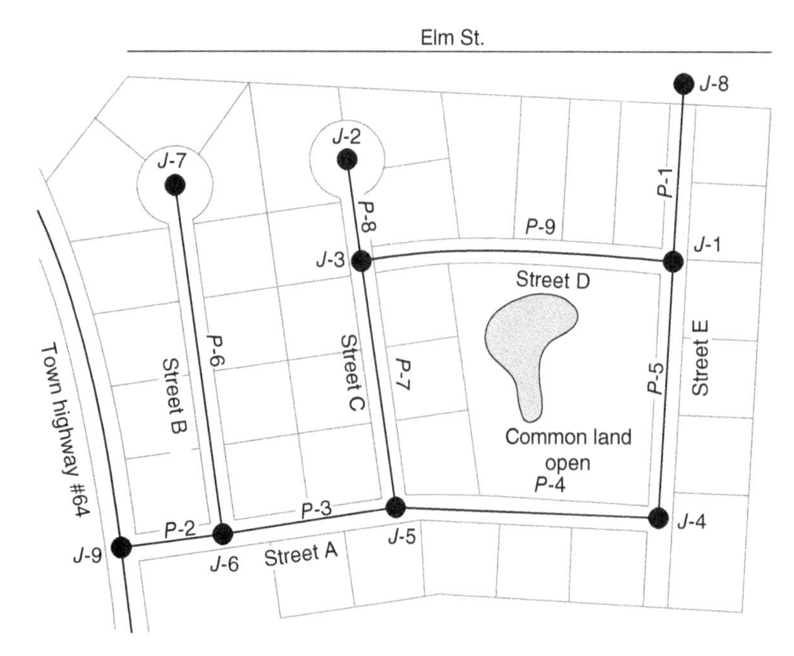

Figure 7.17 Schematic for Problem 7.6

Table 7.25 Junction Information for Problem 7.6

Junction Label	Number of Lots Serviced
J-1	5
J-2	4
J-3	4
J-4	5
J-5	6
J-6	6
J-7	6

Table 7.26 Pipe Information for Problem 7.6

Pipe Label	Length (m)	Diameter (mm)
P-1	60.0	150
P-2	60.0	150
P-3	110.5	150
P-4	164.0	150
P-5	152.5	150
P-6	204.0	100
P-7	148.0	150
P-8	61.0	100
P-9	194.0	150

Town Highway #64 Hydrant Test

Static pressure: 310.3 kPa
Residual pressure: 98.5 kPa at 32 L/s
Elevation of pressure gauge: 190 m

Elm Street Hydrant Test

Static pressure: 413.7 kPa
Residual pressure: 319.3 kPa at 40 L/s
Elevation of pressure gauge: 191.5 m

The subdivision will connect to existing system mains in these streets at nodes J-8 and J-9. (*Hint:* Model the connection to an existing water main with a reservoir and a pump.)

1) What are the demands at each of the junction nodes? What is the total demand?
2) Does the present water distribution system have enough capacity to supply the new subdivision?
3) Which connection to the existing main is supplying more water to the subdivision? Why?
4) Are the proposed pipe sizes adequate to maintain velocities between 0.15 and 2.44 m/s, and pressures of at least 140 kPa?
5) Would the subdivision have enough water if only one connection were used? If so, which one?
6) What do you think are some possible pitfalls of modeling two connections to existing mains within the same system, as opposed to modeling back to the water source?

7.7 Use the pipe sizes given in Table 7.27 for the subdivision in Problem 7.6. City ordinances require that the pressure at the fire flow discharge and at other points in the distribution system cannot fall below 125 kPa during a fire flow of 34 L/s. (*Hint:* The total flow at the fire flow node does not need to include the baseline demand.)

1) If a residential fire occurs at J-7, would the current system be able to meet the fire flow requirements set by the city?
2) If not, what can be done to increase the available flow to provide adequate fire flow to that hydrant?
3) If a fire flow is placed at J-4, does the system meet the requirements with the proposed improvements? Without the proposed improvements?

Table 7.27 Pipe Information for Problem 7.7

Pipe Label	Diameter (mm)
P-1	200
P-2	150
P-3	150
P-4	150
P-5	150
P-6	150
P-7	150
P-8	150
P-9	150

7.8 A local water company is concerned with the water quality within its water distribution network (Fig. 7.18). They want to determine the age and the chlorine concentration of the water as it exits the system at different junctions. The water surface at the reservoir is 70 m.

Chlorine is injected into the system at the source of flow, R-1, at a concentration of 1 mg/L. It has been determined through a series of bottle tests that the average bulk reaction rate of the chlorine in the system (including all pipes and tanks) is approximately −0.5/day. Pump information and stepwise water demand pattern data are as shown in Tables 7.28 and 7.29. Junction and pipe information are given in Tables 7.30 and 7.31.

The cylindrical tank has a diameter of 15 m. The base and minimum elevations are 99 m. The maximum elevation is 104 m, and the initial elevation is 103.4 m.

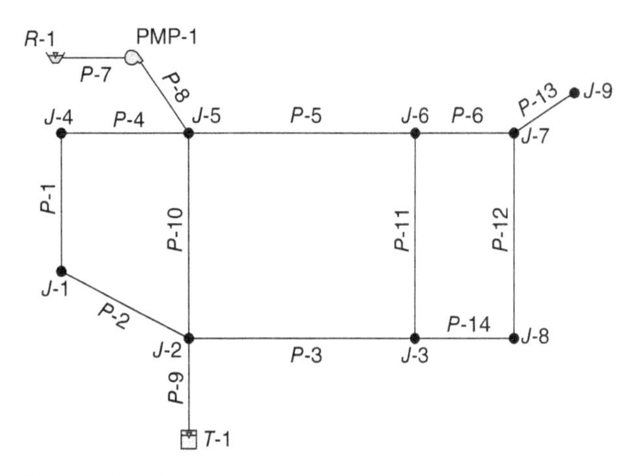

Figure 7.18 Schematic for Problem 7.8

Table 7.28 Pump Information for Problem 7.8

Head (m)	Discharge (L/min)	Controls
40	0	Off if node T-1 above 103.5 m On if node T-1 below 100.5 m
35	3,000	
24	6,000	

Table 7.29 Stepwise Demand Pattern Data for Problem 7.8

Time from Start (h)	Multiplier	Time from Start (h)	Multiplier
0	0.80	13	1.30
1	0.60	14	1.40
2	0.50	15	1.50
3	0.50	16	1.60
4	0.55	17	1.80
5	0.60	18	1.80
6	0.80	19	1.40
7	1.10	20	1.20
8	1.50	21	1.00
9	1.40	22	0.90
10	1.30	23	0.80
11	1.40	24	0.80
12	1.40		

Table 7.30 Junction Data for Problem 7.8

Junction	Elevation (m)	Demand (L/min)
J-1	73	151
J-2	67	227
J-3	81	229
J-4	56	219
J-5	67	215
J-6	73	219
J-7	55	215
J-8	84	180
J-9	88	151

Table 7.31 Pipe Data for Problem 7.8

Pipe	Length (m)	Diameter (mm)	Roughness
P-1	300	200	130
P-2	305	200	130
P-3	300	200	130
P-4	200	200	130
P-5	300	300	130
P-6	200	200	130
P-7	1	300	130
P-8	5,000	300	130
P-9	300	300	130
P-10	500	200	130
P-11	500	200	130
P-12	500	200	130
P-13	150	150	130
P-14	200	200	130

1) Perform an age analysis on the system using a duration of 300 h and a time step of 2 h. Fill in the results in Table 7.32, indicating the maximum water age at each junction and tank after the system reaches equilibrium (a pattern of average water age versus time becomes evident). What point in the system generally has the oldest water? Explain why the water is oldest at this location.
2) Perform a constituent analysis using the same duration and time step as in part 1. Fill in the results in Table 7.32, indicating the minimum chlorine concentration for each junction and tank after the system has reached equilibrium (a pattern of concentration versus time becomes evident). What point in the system has the lowest chlorine concentration? Explain why the chlorine residual is lowest at this location.

Table 7.32 Results Table for Problem 7.8

Junction	J-1	J-2	J-3	J-4	J-5	J-6	J-7	J-8	J-9	T-1
Age (h)										
Chlorine Concentration (mg/L)										

3) From the above table and graphs of demand, age, and concentration versus time generated within WaterGEMS, determine the following correlations:

   (a) Age and chlorine concentration
   (b) Demand and chlorine concentration at a junction
   (c) Demand and water age at a junction.

4) Why is it necessary to run the model for such a long time? Do you feel that 300 h is too long or too short a time period for testing the model? Why?

7.9 A planning commission has indicated a new industry may be connected to the water system described in Problem 7.8. You are to determine the pipe diameters in the network to minimize the installation cost assuming all the pipes are ductile iron. Use the Darwin Designer module to determine the total cost and size of each pipe for each of the following conditions. Use the pipe cost information from Problem 7.4 for the ductile-iron pipe.

I) Size the pipes using a demand multiplier (peaking factor) of 3.2. The pressure must remain between 170 and 550 kPa during peak demand. Exclude pipes P-7 and P-8 in your analysis when determining the pipe sizes. (*Hint:* You will need to specify an additional demand of zero [0 L/min] with the default pressure constraints at the junctions or a fatal error will occur.)

II) It is expected that a new industry with an expected additional demand of 2,000 L/min with a required minimum pressure of 260 kPa will be added to the system. It could be tapped into the network at either junction 6, 7, or 8. Size the pipes for the conditions in part I along with the industry added to all proposed junctions. You will need to analyze the network three times, once for the industry at J-6, again with the industry at J-7, then finally with the industry at J-8.

   1) Indicate which option(s) would work best.
   2) Which junction should the industry be tapped into to be the least costly and what is the expected cost?
   3) What is the size of each pipe for the best solution at the least costly option with the industry added?
   4) What is the calculated minimum pressure at the industry for the best solution?

Bibliography

American Water Works Association, *Introduction to Water Distribution*, Denver, CO, USA, *463* pp. 1986.

American Water Works Association, *Computer Modeling of Water Distribution Systems, Manual of Water Supply Practices M32*, 2nd ed., Denver, CO, USA ed., 2005.

Cesario, A. L., *Modeling Analysis and Design of Water Distribution Systems*, American Water Works Association, Denver, CO, 1995.

Haestad Methods, *WaterGEMS: Geospatial Water Distribution Software*, Haestad Press, Waterbury, CT, 2002.

Haestad Methods, *Advanced Water Distribution Modeling and Management*, Bentley Institute Press, Exton, PA, 2007a.

Haestad Methods, *Computer Applications in Hydraulic Engineering*, 7th ed., Bentley Institute Press, Exton, PA, 2007b.

Males, R. M., Grayman, W. M., and Lark, R. M., Modeling water quality in distribution systems, *J. Water Resour. Plan. Manage., 114*, 197–209, 1988.

Rossman, L. A., and Boulos, P. F., Numerical methods for modeling water quality in distribution systems: a comparison, *J. Water Resour. Plan. Manage., 122*, 2, 137, 1996.

Walski, T. M., *Water System Modeling Using CYBERNET®*, Haestad Press, Waterbury, CT, 1993.

8

Pumping, Storage, Dual Water Systems, and Distribution System Management

8.1 Pumps and Pumping Stations

Pumps and pumping stations (Figs. 8.1 and 8.2) serve the following purposes in water systems:

1) Lifting water from its source (surface or ground), either immediately to the community through *high-lift* installations, or by *low-lift* systems to purification works
2) Boosting water from low-service to high-service areas, to separate fire supplies, and to the upper floors of many-storied buildings
3) Transporting water through treatment works, backwashing filters, draining component settling tanks, and other treatment units; withdrawing deposited solids; and supplying water (especially pressure water) to operating equipment.

Pumps and pumping stations also serve the following purposes in wastewater systems:

1) Lifting wastewater from low-lying basements or low-lying secondary drainage areas into the main drainage system, and from uneconomically deep runs in collecting or intercepting systems into high-lying continuations of the runs or into outfalls
2) Pumping out stormwater detention tanks in combined systems
3) Lifting wastewater to and through treatment works, draining component settling tanks and other treatment units, withdrawing wastewater sludges and transporting them within the works to treatment units, supplying water to treatment units, discharging wastewater and wastewater sludge through outfalls, and pumping chemicals to treatment units.

In addition to *centrifugal* and *propeller pumps*, water and wastewater systems may include (a) *displacement pumps*, ranging in size from hand-operated pitcher pumps to the huge pumping engines of the last century built as steam-driven units; (b) *rotary pumps* equipped with two or more rotors (varying in shape from meshing lobes to gears and often used as small fire pumps); (c) *hydraulic rams* utilizing the impulse of large masses of low-pressure water to drive much smaller masses of water (one-half to one-sixth of the driving water) through the delivery pipe to higher elevations, in synchronization with the pressure waves and sequences induced by water hammer; (d) *jet pumps* or jet ejectors, used in wells and dewatering operations, introducing a high-speed jet of air through a nozzle into a constricted section of pipe; (e) *air lifts* in which air bubbles, released from an upward-directed air pipe, lift water from a well or sump through an *eductor pipe*; and (f) *displacement ejectors* housed in a pressure vessel in which water (especially wastewater) accumulates and from which it is displaced through an eductor pipe when a float-operated valve is tripped by the rising water and admits compressed air to the vessel.

Today most water and wastewater pumping is done by either centrifugal pumps or propeller pumps. These are usually driven by electric motors, less often by steam turbines, internal combustion engines, or hydraulic turbines. How the water is directed through the impeller determines the type of pump. There is (1) *radial flow* in open- or closed-*impeller pumps*, with volute or turbine casings, and single or double suction through the eye of the impeller; (2) *axial flow* in *propeller pumps*; and (3) *diagonal flow* in *mixed-flow, open-impeller pumps*. Propeller pumps are not centrifugal pumps. Both centrifugal pumps and propeller pumps can be referred to as *rotodynamic* pumps.

Open-impeller pumps are less efficient than closed-impeller pumps, but they can pass relatively large debris without being clogged. Accordingly, they are useful in pumping wastewaters and sludges. *Single-stage pumps* have but *one impeller*, and *multistage pumps* have *two or more*, each feeding into the next higher stage. Multistage turbine well pumps may have their motors submerged, or they may be driven by a shaft from the prime mover situated on the floor of the pumping station.

Water and Wastewater Engineering: Hydraulics, Hydrology and Management, Volume 1, Fourth Edition.
Lawrence K. Wang, Mu-Hao Sung Wang, and Nazih K. Shammas.

Companion website: www.wiley.com/go/Wang/Waterandwastewater4e

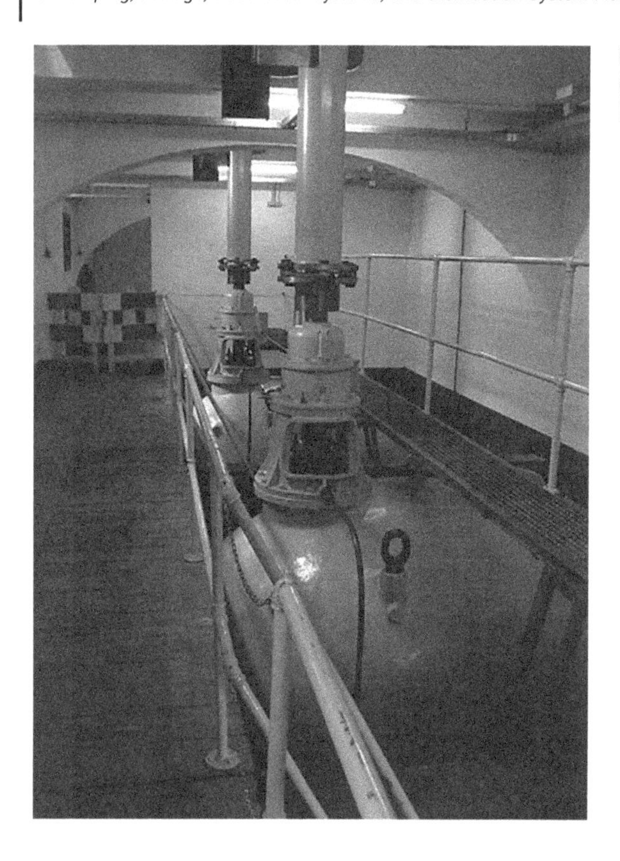

Figure 8.1 Pumping Station Van Sasse in Grave, the Netherlands (*Source:* http://en.wikipedia.org/wiki/Image:Gemaal_van_sasse_interieur.jpg)

Figure 8.2 General View of the Pumping Station Van Sasse in Grave, the Netherlands (*Source:* http://en.wikipedia.org/wiki/Image:Gemaal_van_sasse.jpg)

8.2 Pump Characteristics

A centrifugal pump is defined by its *characteristic curve*, which relates the *pump head* (head added to the system) to the flow rate. Pumping units are chosen in accordance with system heads and pump characteristics. As shown in Fig. 8.3, the system head is the sum of the static and dynamic heads against the pump. As such, it varies with required flows and with changes in storage and *suction levels*. When a distribution system lies between pump and distribution reservoir, the system head also

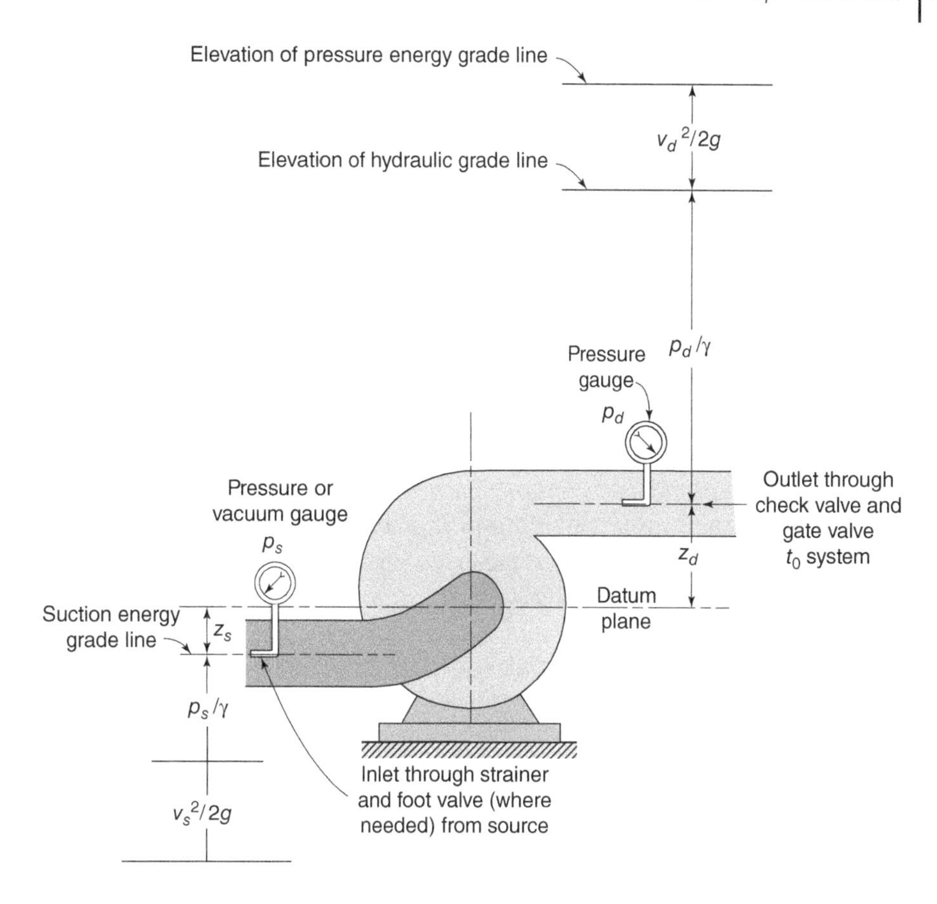

Figure 8.3 Head Relationships in Pumping Systems

responds to fluctuations in demand. Pump characteristics depend on pump size, speed, and design. For a given speed N in revolutions/min, they are determined by the relationships between the rate of discharge Q, usually in gpm (or L/m or m^3/s) and the head H in ft (or m), the efficiency E in %, and the power input P in horsepower (or kilowatt). For purposes of comparison, pumps of given geometrical design are characterized also by their *specific speed* N_s, the hypothetical speed of a homologous (geometrically similar) pump with an impeller diameter D such that it will discharge 1 gpm (3.78 L/m) against a 1-ft (0.30-m) head. Because discharge varies as the product of area and velocity, and velocity varies as $H^{1/2}$, Q varies as $D^2H^{1/2}$. But velocity varies also as $\pi DN/60$. Hence, $H^{1/2}$ varies as DN, or N varies as $H^{3/4}/Q^{1/2}$, and the specific speed becomes

$$N_s = NQ^{1/2}/H^{3/4} \tag{8.1}$$

where N_S is specific speed, rpm; N is speed, rpm; Q is capacity, gpm (m^3/s); and H is head, ft (m).

To obtain the specific speed based on US customary units of head and capacity, multiply the specific speed based on metric units of head and capacity by 52.

Generally speaking, pump efficiencies increase with pump size and capacity. Below specific speeds of 1,000 units, efficiencies drop off rapidly. Radial-flow pumps perform well between specific speeds of 1,000 and 3,500 units; mixed-flow pumps in the range of 3,500–7,500 units; and axial-flow pumps after that up to 12 000 units. As shown in Eq. 8.1, for a given N, high-capacity, low-head pumps have the highest specific speeds. For double-suction pumps, the specific speed is computed for half the capacity. For multistage pumps, the head is distributed between the stages. In accordance with Eq. 8.1, this keeps the specific speed high and with it, the efficiency.

System head II $= (z_d - z_s) + (p_d/\gamma - p_s/\gamma) + (v_d^2/2g - v_s^2/2g)$. For suction lift, p_s/γ and z_s are negative, and $v_s^2/2g$ is positive. For suction pressure, p_s/γ and $v_s^2/2g$ are positive, and z_s is negative.

Changing the speed of a centrifugal pump will change its operating characteristics including the water flow. If the speed of a pump (rpm) is changed, its water flow will change as follows:

$$Q_2 = (Q_1)(N_2/N_1) \tag{8.2}$$

where Q_2 is water flow now, gpm (m^3/s); Q_1 is rated water flow, gpm (m^3/s); N_2 is pump speed now, rpm; and N_1 is rated pump speed, rpm.

Changing the speed of a centrifugal pump will also change its head. If the speed of a pump (rpm) is changed, its head will change as follows:

$$H_2 = (H_1)\,(N_2/N_1)^2 \tag{8.3}$$

where H_2 is head now, ft (m); H_1 is rated head, ft (m); N_2 is pump speed now, rpm; and N_1 is rated pump speed, rpm.

Changing the speed of a centrifugal pump will also change its power requirement. If the speed of a pump (rpm) is changed, its head will change as follows:

$$P_2 = (P_1)\,(N_2/N_1)^3 \tag{8.4}$$

where P_2 is power requirement now, hp. (kW); P_1 is rated power requirement, hp. (kW); N_2 is pump speed now, rpm; and N_1 is rated pump speed, rpm.

Example 8.1 Determination of Pump New Power Requirement

Changing the speed of a centrifugal pump will change its operating characteristics, including the power requirement.

Determine the new power requirement, if

P_1 = rated head = 20 hp. = 14.914 kW
N_2 = pump speed now = 1,200 rpm
N_1 = rated pump speed = 1,425 rpm

Solution 1 (US Customary System):

$$P_2 = (P_1)\,(N_2/N_1)^3$$
$$= (20\text{ hp})\,(1{,}200/1{,}425)^3 = \mathbf{12\ hp}$$

Solution 2 (SI System):

$$P_2 = (P_1)\,(N_2/N_1)^3$$
$$= (14.914\text{ kW})\,(1{,}200/1{,}425)^3 = \mathbf{8.948\ kW}.$$

8.2.1 Power Requirements and Efficiencies of Pumps

Work can be expressed as lifting a weight a certain vertical distance. It is usually defined in terms of ft-lb of work. Power is work per unit of time. These statements are expressed mathematically by the following relationships:

$$\text{Work} = WH \tag{8.5}$$

$$\text{Power} = \text{Work}/t \tag{8.6}$$

where Work is work, ft-lb (m-kg); W is weight, lb. (kg); H is height, ft (m); Power is power, ft-lb/s (m-kg/s); and$t$ is time, s.

If the water flow from a pump is converted to weight of water and multiplied by the vertical distance it is lifted, the amount of work or power done can be represented by the following equation:

$$\text{Power} = \text{Work}/t = \text{WH}/t$$
$$HP = (Q, \text{gpm})\,(8.34\text{ lb/gal})\,(H, \text{ft})\,(\text{min}/60\text{ s})\,(\text{horesepower}/550\text{ ft-lb/s})$$
$$HP = (Q)\,(H)/3{,}957 \tag{8.7}$$

where Q is water flow, gpm; H is head or lift, ft; and HP is power in US customary unit, hp.

In the SI or metric system:

$$MP = \left(Q, \text{m}^3/\text{s}\right)\left(1{,}000\text{ kg/m}^3\right)(H, \text{m})\,(\text{kW}/101.97\text{ m-kg/s})$$
$$MP = 9.8066\,(Q)\,(H) \tag{8.7a}$$

where Q is water flow, m^3/s; H is head or lift, m; and MP is power in metric unit, kW. Here 1 kW = 101.97 m-kg/s = 1,000 watt, 1 watt = 0.102 m-kg/s = 1 J/s, and 1 hp. = 0.746 kW = 550 ft lb./s.

The total power input can be determined using the following equation:

$$INHP = \text{input power in US customary unit} = (V)\,(A)/746 \quad (8.8)$$

$$INMP = \text{input power in metric unit} = 0.001\,(V)\,(A) \quad (8.8a)$$

where V is volt, and A is current in amp.

The wire-to-water efficiency is the efficiency of the total power input to produce water horsepower:

$$E_{ww} = (E_{pump})\,(E_{motor})\,(E_o) \quad (8.9)$$

$$E_{ww} = \text{water horsepower/input horsepower} \quad (8.10)$$

where E_{ww} is the wire-to-water efficiency, E_{pump} is the efficiency of the pump, E_{motor} is the efficiency of the motor driving the pump, and E_o is the efficiency of any other parts in the entire motor-pump-wiring system.

Example 8.2 Determination of Pump Power

Determine the water horsepower, break horsepower, and motor horsepower for a pump operating under the following conditions: Water flow of 620 gpm (0.039 m^3/s = 39 L/s) is to be pumped against a total head of 135 ft (41.15 m); the pump efficiency is 80%; and the motor driving the pump has an efficiency of 90%.

Solution 1 (US Customary System):

$$WHP = \text{water horsepower in US customary unit} = (Q)\,(H)/3{,}957 \quad (8.7)$$
$$= (620)\,(135)/3{,}957 = \mathbf{21.2\ hp}$$

$$BHP = \text{break horsepower in US customary unit} = WHP/E_{pump} \quad (8.11)$$
$$= 21.2/0.80 = \mathbf{26.5\ hp}$$

$$MHP = \text{motor horsepower in US customary unit} = BHP/E_{motor} \quad (8.12)$$
$$= 26.5/0.90 = \mathbf{29.4\ hp}$$

where Q = water flow, gpm; H = head or lift; ft

Solution 2 (SI System):

$$WMP = \text{water power in metric unit} = 9.8066\,(Q)\,(H) \quad (8.7a)$$
$$= 9.8066\,(0.0391)\,(41.15) = 15.7866\ \text{kW}$$

$$BMP = \text{break power in metric unit} = WMP/E_{pump} \quad (8.13)$$
$$= 15.7866/0.80 = 19.733\ \text{kW}$$

$$MMP = \text{motor power in metric unit} = BMP/E_{motor} \quad (8.14)$$
$$= 19.733/0.90 = 21.92\ \text{kW}$$

where Q is water flow, m^3/s; and H is head or lift, m.

Example 8.3 Computation of Total Power Input

Determine the total power input if the electrical input to a motor-pump system is 220 V and 25 amps.

Solution 1 (US Customary System):

$$INHP = \text{input power in US customary unit} = (V)\,(A)/746 \quad (8.8)$$
$$= (220)(25)/746 = \mathbf{7.4\ hp}$$

Solution 2 (SI System):

$$INMP = \text{input power in metric unit} = 0.001\,(V)\,(A) \quad (8.8a)$$
$$= 0.001\,(220)(25)/746 = \mathbf{5.5\ kW}$$

8.2.2 Cavitation

Specific speed is an important criterion, too, of safety against *cavitation*, a phenomenon accompanied by vibration, noise, and rapid destruction of pump impellers. Cavitation occurs when enough potential energy is converted to kinetic energy to reduce the absolute pressure at the impeller surface below the vapor pressure of water at the ambient temperature. Water then vaporizes and forms pockets of vapor that collapse suddenly as they are swept into regions of high pressure. Cavitation occurs when inlet pressures are too low or when pump capacity or speed of rotation is increased without a compensating rise in inlet pressure. Lowering a pump in relation to its water source, therefore, reduces cavitation.

If we replace the head H in Eq. 8.1 by H_{sv}, the *net positive inlet* or *suction head*, namely, the difference between the total inlet head (the absolute head plus the velocity head in the inlet pipe), and the head corresponding to the vapor pressure of the water pumped (table in Appendix 4), we obtain the suction specific speed (S):

$$S = NQ^{0.5}/H_{sv}{}^{0.75} \quad \text{(US customary units)} \tag{8.15}$$

where S is the specific speed, dimensionless; N is the rotative speed, rpm; Q is the capacity, gpm; and H_{sv} is the net positive inlet or suction head, ft

Equation 8.16 is a specific speed equation using the SI or metric units:

$$S = 51.7\,NQ^{0.5}/H_{sv}{}^{0.75} \quad \text{(SI units)} \tag{8.16}$$

where S is the specific speed, dimensionless; N is the rotative speed, rpm; Q is the capacity, m^3/s; and H_{sv} is the net positive inlet or suction head, m.

The specific speed, S, is a number that can be used to compare the performance of specific pump impellers under various conditions of rotative speed, capacity, and head per stage.

The units used to determine specific speed must be consistent within the numerical system for S to become meaningful. Certain general safe limits have been established for S by experiment. The following are examples:

Single-suction pumps with overhung impellers: $S \leq 8{,}000 - 12{,}000$
Single-stage pumps with shaft through eye of impeller: $S \leq 7{,}000 - 11{,}000$
High-pressure, multistage pumps (single suction): $S \leq 5{,}500 - 7{,}500$
High-pressure, multistage pumps with special first-stage impeller (single suction): $S \leq 7{,}500 - 10{,}000$.

$H_{sv} = p_s/\gamma + v_s^2/2g - p_w/\gamma$, where p_s/γ is the absolute pressure, v_s, the velocity of the water in the inlet pipe, and p_w, the vapor pressure of the water pumped, with γ being the specific weight of water and g the gravity constant. The energy grade line is at a distance $h_s = p_a/\gamma - p_s/\gamma + v_s^2/2g$ from the eye of the impeller to the head delivered by the pump, where p_a is the atmospheric pressure. The ratio H_{sv}/h_s, where $h = P_a/\gamma - h_s$, is called the *cavitation parameter.*

8.2.3 Performance Characteristics

Common performance characteristics of a centrifugal pump operating at constant speed are illustrated in Fig. 8.4. Note that the *shut-off head* is a fixed limit and that power consumption is minimum at shut-off. For this reason, centrifugal pumps, after being *primed* or filled with water, are often started with the pump discharge valve closed. As the head falls past the point of maximum efficiency, normal discharge, or rated capacity of the pump (point 1 in Fig. 8.4), the power continues to rise. If a centrifugal pump is operated against too low a head, a motor selected to operate the pump in the head range around maximum efficiency may be overloaded. Pump delivery can be regulated (a) by a valve on the discharge line, (b) by varying the pump speed mechanically or electrically, or (c) by throwing two or more pumps in and out of service to best advantage.

What happens when more than one pumping unit is placed in service is shown in Fig. 8.4 with the help of a curve for the system head. Obviously, pumping units can operate only at the point of intersection of their own head curves with the system head curve. In practice, the system head varies over a considerable range at a given discharge (Fig. 8.5). For example, where a distributing reservoir is part of a system and both the reservoir and the source of water fluctuate in elevation, (a) a *lower* curve identifies head requirements when the reservoir is empty and the water surface of the source is high, and (b) an *upper* curve establishes the system head for a full reservoir and a low water level at the source. The location and magnitude of drafts also influence system heads. Nighttime pressure distributions may be very different from those during the day. How the characteristic curves for twin-unit operation are developed is indicated in Fig. 8.4. Note that

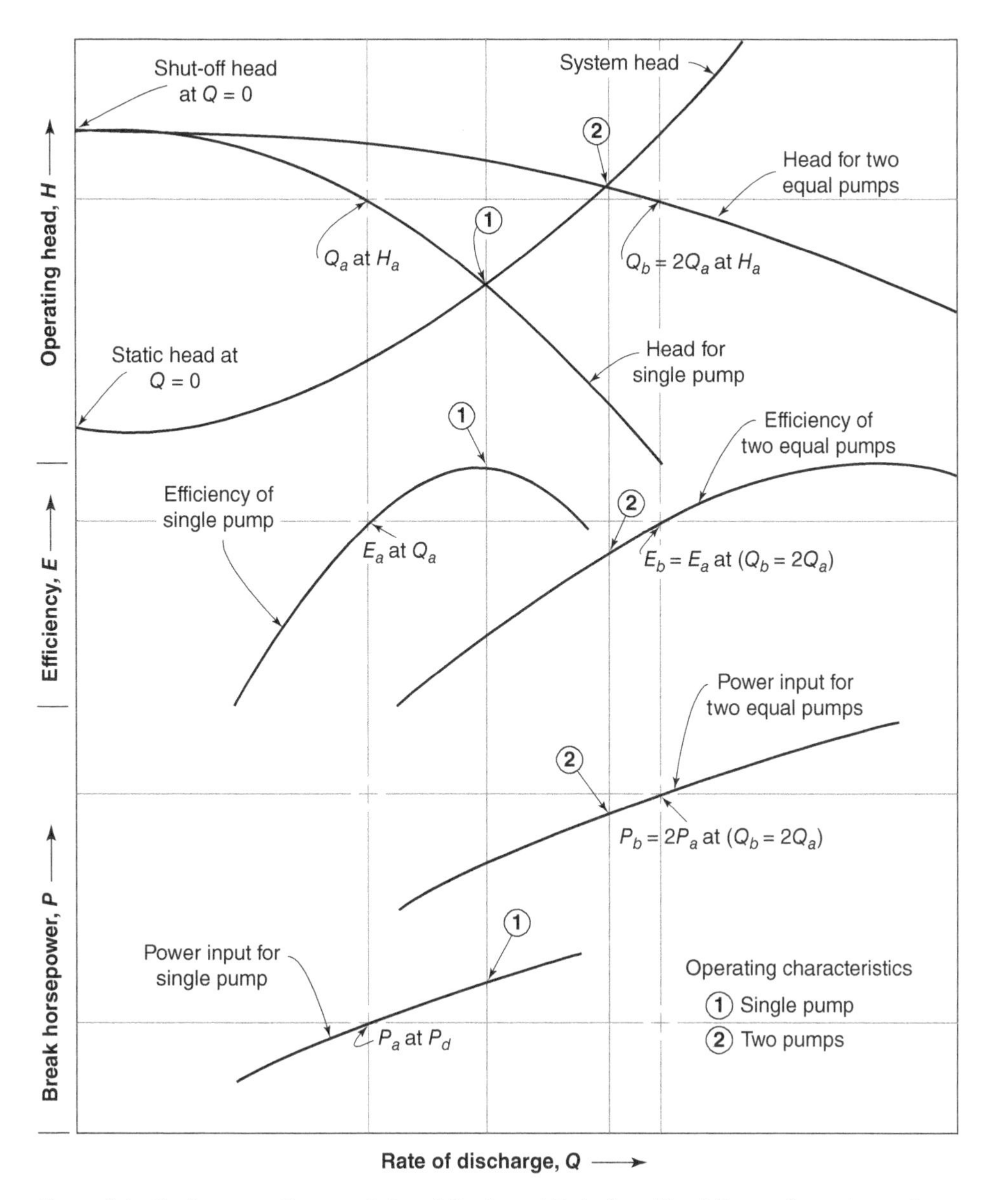

Figure 8.4 Performance Characteristics of Single and Twin Centrifugal Pumps Operating at Constant Speed

the two identical pumping units have not been selected with an eye to highest efficiency of operation in parallel. Characteristic curves for other multiple units are developed in the same way from the known curves of individual units.

Where most of the operating head is static lift—when the water is pumped through relatively short lengths of suction and discharge piping, for example—there is little change in the system head at different rates of flow. In these circumstances, the head curve is nearly horizontal, and the discharge of parallel pumps is substantially additive. This is common in wastewater pumping stations in which the flow is lifted from a lower to an immediately adjacent higher level. Examples are pumping stations along interceptors or at outfalls.

By contrast, friction may control the head on pumps discharging through long force mains, and it may not be feasible to subdivide flows between pumping units with reasonable efficiency. Multispeed motors or different combinations of pumps and motors may then be required.

Because flows from a number of pumps may have to be fed through a different piping system than flows from any single unit, it may be necessary to develop "modified" characteristic curves that account for losses in different combinations of piping.

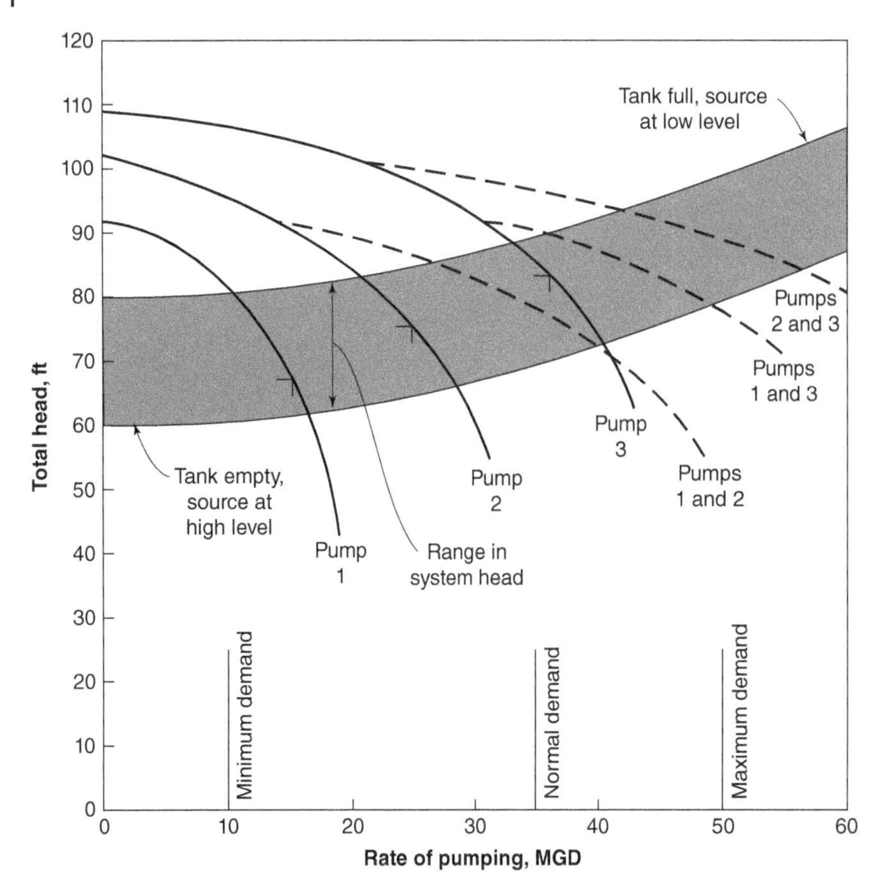

Figure 8.5 Pump Selection in Example 8.4. Conversion factors: 1 ft = 0.3048m; 1 MGD = 3.785 MLD = 0.0438 m^3/s = 43.8 L/s

Centrifugal pumps are normally operated with discharge velocities of 5–15 ft/s (1.5–4.6 m/s). The resulting average outlet diameter of the pump, called the pump size, can be determined by the following two equations:

$$D_{podin} = 0.2\sqrt{Q} \quad \text{(US customary units)} \tag{8.17}$$

where D_{podin} is pump outlet diameter, in.; and Q is the capacity of the pump, gpm; and

$$D_{podcm} = 63.95\sqrt{Q} \quad \text{(SI units)} \tag{8.17a}$$

where Q is the capacity of the pump, m^3/s; and D_{podcm} is the pump outlet diameter, cm. For instance, when the capacity of the pump is 200 gpm (0.01262 m^3/s = 12.62 L/s), the outlet diameter of the pump should be 2.8284 in. (7.1841 cm). An engineer may select 3 in. (75 mm) as the outlet diameter.

Example 8.4 Selection of Pumps Combination to Satisfy Water Demand

A mill supply drawing relatively large quantities of water from a river is to deliver them at a fairly low head. The minimum demand is 10 MGD (0.438 m^3/s), the normal 35 MGD (1.53 m^3/s), and the maximum 50 MGD (2.19 m^3/s). The river fluctuates in level by 5 ft (1.52 m), and the working range of a balancing tank is 15 ft (4.57 m). The vertical distance between the bottom of the tank and the surface of the river at its high stage is 60 ft (18.3 m). The friction head in the pumping station and a 54-in. (76.2-cm) force main rise from a minimum of 1 ft (0.30 m) at the 10-MGD (0.438 m^3/s) rate to a maximum of nearly 20 ft (6.1 m) at the 50-MGD (2.19 m^3/s) rate. Make a study of suitable pumping units, knowing that 1 MGD = 3.785 MLD = 0.0438 m^3/s = 43.8 L/s.

Solution:

The solution to this problem is shown in Fig. 8.5. Three pumps are provided: No. 1 with a capacity of 15 MGD (0.657 m^3/s = 657 L/s) at 66-ft (20.117-m) head, No. 2 with 25 MGD (1.059 m^3/s = 1,059 L/s) at 78-ft (23.774-m) head, and No. 3 with 37 MGD (1.62 m^3/s = 1,620 L/s) at 84-ft (25.6-m) head. Each pump has an efficiency of 89% at the design point. The efficiencies at the top and bottom of the working range are listed in Table 8.1.

Table 8.1 Pumping Characteristics of System in Example 8.4

Pumps in service, No.	1	2	3	1 and 2	1 and 3	2 and 3
Rate of pumping, MGD	10	21	33.5	27	36	42
Head, ft	81	83	88	85	90	93
Efficiency, %	80	88	88	71, 86	35, 87	68, 64
Rate of pumping, MGD	15	25	37	34	43.5	49.5
Heat, ft	66	78	84	80	85	89
Efficiency, %	89	89	89	82, 88	71, 89	79, 87
Rate of pumping, MGD	16.5	28.5	40.5	40	49.5	56.5
Heat, ft	62	66	73	73	79	84
Efficiency, %	88	84	86	88, 88	83, 88	79, 89

Conversion factors: 1 MGD = 0.0438 m^3/s = 43.8 L/s; 1 ft = 0.3048 m.

8.3 Service Storage

The three major components of service storage are as follows:

1) Equalizing, or operating, storage
2) Fire reserve
3) Emergency reserve.

8.3.1 Equalizing, or Operating, Storage

Required *equalizing*, or *operating*, *storage* can be read from a *demand rate curve* or, more satisfactorily, from a *mass diagram*. As shown in Fig. 8.6 for the simple conditions of steady inflow, during 12 and 24 h, respectively, the amount of equalizing, or operating, storage is the sum of the maximum ordinates between the demand and supply lines. To construct such a mass diagram, proceed as follows:

1) From past measurements of flow, determine the *draft* during each hour of the day and night for typical days (maximum, average, and minimum).
2) Calculate the amounts of water drawn up to certain times, that is, the *cumulative draft*
3) Plot the cumulative draft against time.
4) For steady supply during 24 h, draw a straight line diagonally across the diagram, as in Fig. 8.6a. Read the storage required as the sum of the two maximum ordinates between the draft and the supply line.
5) For steady supply during 12 h (by pumping, for example) draw a straight line diagonally from the beginning of the pumping period to its end – for example, from 6 a.m. to 6 p.m., as in Fig. 8.6b. Again read the storage required as the sum of the two maximum ordinates.

Achieving a steady supply at the rate of maximum daily use will ordinarily require an equalizing storage between 15% and 20% of the day's consumption. Limitation of supply to 12 h may raise the operating storage to an amount between 30% and 50% of the day's consumption.

Example 8.5 Determination of Equalizing Storage

Determine the equalizing, or operating, storage for the drafts of water shown in Table 8.2(a) when inflow is uniform during 24 h and (b) when flow is confined to the 12 h from 6 a.m. to 6 p.m.

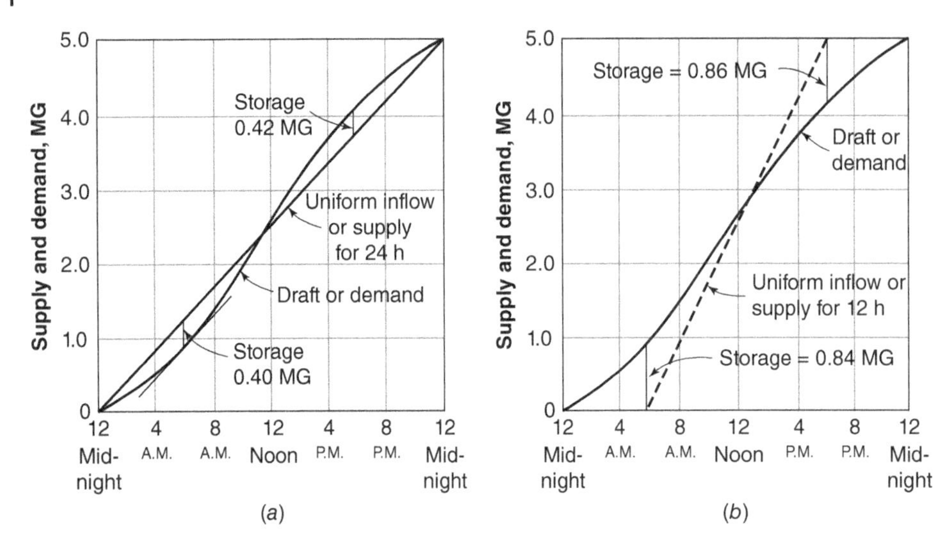

Figure 8.6 Determination of Equalizing, or Operating, Storage by Mass Diagram (See Example 8.5). Uniform Inflow, or Supply, (a) extending over 24 h and (b) Confined to 12 h. (a) Total Storage = 0.40 + 0.42 = 0.82 MG; (b) Total Storage = 0.84 + 0.86 = 1.70 MG. Conversion factor: 1 MG = 1 million gallons = 3.785 ML = 3,785 m^3

Table 8.2 Observed Drafts for Example 8.5

(a) Time	4 a.m.	8 a.m.	Noon	4 p.m.	8 p.m.	Midnight
(b) Draft, MG	0.484	0.874	1.216	1.102	0.818	0.506
(c) Cumulative draft, MG	0.484	1.358	2.574	3.676	4.494	5.000

Conversion factors: 1 MG = 1 million gallons = 3.785 ML = 3785 m^3.

Solution 1 (US Customary System):

a) For steady supply during 24 h, the draft plotted in Fig. 8.6a exceeds the demand by 0.40 MG by 6 a.m. If this excess is stored, it is used up by 11 a.m. In the afternoon, the demand exceeds the supply by 0.42 MG by 6 p.m. and must be drawn from storage that is replenished by midnight. Hence, the required storage is the sum of the morning excess and afternoon deficiency, or **0.82 MG**. This storage volume as a percentage of total draft equals (0.82 MG/5MG) × 100 = 16.4%.

b) For steady supply during the 12-h period from 6 a.m. to 6 p.m., the draft plotted in Fig. 8.6b exceeds the supply by 0.84 MG between midnight and 6 a.m. and must be drawn from storage. In the afternoon, the supply exceeds the demand by 0.86 MG by 6 p.m., but this excess is required to furnish water from storage between 6 p.m. and midnight. Therefore, total storage equals 0.84 + 0.86 = **1.70 MG**, or (1.7 MG/5 MG) × 100 = 34% of the day's consumption.

Solution 2 (SI System):

a) For steady supply during 24 h, the draft plotted in Fig. 8.6a exceeds the demand by 0.40 MG (1,514 m^3) by 6 a.m. If this excess is stored, it is used up by 11 a.m. In the afternoon, the demand exceeds the supply by 0.42 MG (1,590 m^3) by 6 p.m. and must be drawn from storage that is replenished by midnight. Hence, the required storage is the sum of the morning excess and afternoon deficiency, or 0.82 MG **(3,104** m^3). This storage volume as a percentage of total draft equals [(3,104 m^3)/(5 × 3,785 m^3)] × 100 = 16.4%, where 5 × 3,785 m^3 = 5 MG.

b) For steady supply during the 12-h period from 6 a.m. to 6 p.m., the draft plotted in Fig. 8.6b exceeds the supply by 0.84 MG (3,180 m^3) between midnight and 6 a.m. and must be drawn from storage. In the afternoon, the supply exceeds the demand by 0.86 MG (3,255 m^3) by 6 p.m., but this excess is required to furnish water from storage between 6 p.m. and midnight. Therefore, total storage equals 3,180 + 3,255 = **6,435** $\mathbf{m^3}$, or [(6,435)/(5 × 3,785)] × 100 = 34% of the day's consumption.

8.3.2 Fire Reserve

Based on International Fire Code recommendations on observed durations of serious conflagrations, it is recommended that distributing reservoirs be made large enough to supply water for fighting a serious conflagration for (a) 4 h for fire flows of more than 4,000 gpm (252 L/s), (b) 3 h for fire flows of 3,000–3,750 gpm (189–237 L/s), and (c) 2 h for fire flows of 2,750 gpm (174 L/s) and less (see Table 4.13). The resulting fire reserve may not always be economically attainable, and design values may have to be adjusted downward to meet local financial abilities. Changing community patterns, moreover, may make for changing requirements in the future.

8.3.3 Emergency Reserve

The magnitude of this storage component depends on (a) the danger of interruption of reservoir inflow by failure of supply works and (b) the time needed to make repairs. If shutdown of the supply is confined to the time necessary for routine inspections during the hours of minimum draft, the emergency reserve is sometimes made no more than 25% of the total storage capacity, that is, the reservoir is assumed to be drawn down by one-fourth its average depth. If supply lines or equipment are expected to be out of operation for longer times, higher allowances must be made.

8.3.4 Total Storage

The desirable total amount of storage is equal to the sum of the component requirements. In each instance, economic considerations dictate the final choice. In pumped supplies, cost of storage must be balanced against cost of pumping, and attention must be paid to economies affected by operating pumps more uniformly and restricting pumping to a portion of the day only. In all supplies, cost of storage must be balanced against cost of supply lines, increased fire protection, and more uniform pressures in the distribution system.

Storage facilities should have sufficient capacity, as determined from engineering studies, to meet domestic demands and, where fire protection is provided, fire flow demands. The recommended standards for sizing a water storage tank or reservoir as stated in the *Recommended Standards for Water Works, 2007 Edition* (Health Research, 2007; often referred to as the Ten-States Standards) are as follows:

1) Fire flow requirements established by the appropriate state Insurance Services Office should be satisfied where fire protection is provided.
2) The minimum storage capacity (or equivalent capacity) for systems not providing fire protection shall be equal to the average daily consumption. This requirement may be reduced when the source and treatment facilities have sufficient capacity with standby power to supplement peak demands of the system.
3) Excessive storage capacity should be avoided to prevent potential water-quality deterioration problems.

Example 8.6 Total Water Storage Volume

For a steady gravity supply equal to the maximum daily demand, a 4-h fire supply at 8,000 gpm(0.505 m^3/s = 505 L/s), and no particular hazard to the supply works, find the storage to be provided for a city using an average of 7.5 MGD (0.328 m^3/s = 328 L/s) of water.

Solution 1 (US Customary System):

The equalizing storage is 15% of average daily consumption:

$$0.15 \times 7.5\,\text{MG} = 1.13\,\text{MG}$$

The fire reserve is 8,000 gpm for four hours:

$$(8{,}000\,\text{gal/min} \times 4\text{h} \times 60\,\text{min/h})/10^6 = 1.92\,\text{MG}$$

The resulting subtotal is 1.13 + 1.92 = 3.05 MG.

Because the emergency reserve is one-fourth of the total storage, the subtotal is three-fourths (0.75) of the total storage. Therefore,

Total storage = 3.05/0.75 = **4.1 MG**

Solution 2 (SI System):

$$0.328\ \text{m}^3/\text{s} = 0.328 \times 24 \times 60 \times 60\ \text{m}^3/\text{d} = 28{,}387\ \text{m}^3/\text{d}$$

The equalizing storage is 15% of average daily consumption:

$$0.15 \times 28{,}387\ \text{m}^3 = 4{,}258\ \text{m}^3$$

The fire reserve is $0.505\ \text{m}^3/\text{s}$ for four hours:

$$\left(0.505\ \text{m}^3/\text{s}\right)(4 \times 60 \times 60\text{s}) = 7{,}269\ \text{m}^3$$

The resulting subtotal is $4{,}258 + 7{,}269 = 11{,}527\ \text{m}^3$.

Because the emergency reserve is one fourth of the total storage, the subtotal is three fourths (0.75) of the total storage. Therefore,

Total storage = 11,527/0.75 = **15,369 m^3**.

Example 8.7 Total Water Storage Volume Using the Ten-States Standards

For a steady gravity supply equal to the maximum daily demand, a 4-h fire supply at 8,000 gpm ($0.505\ \text{m}^3/\text{s} = 505\ \text{L/s}$), and no particular hazard to the supply works, find the storage to be provided for a city using an average of 7.5 MGD ($0.328\ \text{m}^3/\text{s} = 328\ \text{L/s}$) of water using the Ten-States Standards.

(*Note:* This example uses the same parameters as Example 8.6.)

Solution 1 (US Customary System):

Storage volume for domestic consumption = average daily consumption (Ten-States Standards):

$$= 7.5\ \text{MGD}$$
$$= 7.5\ \text{MG daily}$$

The fire reserve is 8,000 gpm for four hours (similar to Example 8.6):

$$= (8{,}000\ \text{gal/min} \times 4\text{h} \times 60\ \text{min/h})/10^6$$
$$= 1.92\ \text{MG}$$

Total Storage = domestic storage + fire strage = 7.5 + 1.92 = **9.42 MG**

Solution 2 (SI System):

Storage volume for domestic consumption = average daily consumption (Ten-States Standards):

$$= 0.328\ \text{m}^3/\text{s} = 328\ \text{L/s}$$
$$= 28{,}387\ \text{m}^3\ \text{daily}$$

The fire reserve is $0.505\ \text{m}^3/\text{s}$ for four hours (similar to Example 8.6):

$$= \left(0.505\ \text{m}^3/\text{s}\right)(4 \times 60 \times 60\ \text{s})$$
$$= 7{,}269\ \text{m}^3$$

Total Storage = domestic storage + fire strage = 28,387 + 7,269 = **35,656 m^3**.

8.4 Location of Storage

In addition to capacity of service storage, location is an important factor in the control of distribution systems. One MGD (3.785 MLD) of elevated fire reserve, suitably sited in reference to the area to be protected, is equivalent, for example, to the addition of a 12-in. (300-mm) supply main. The underlying reasoning is that, when drawing this volume of water in a 4-h fire, flow is provided at a rate of (24/4) × 1 MGD = 6 MGD or (24/4) (3.785 MLD) = 22.7 MLD. This is the amount of water an 18-in. (450-mm) pipe can carry at a velocity of less than 5 ft/s (1.5 m/s). Why this must be neighborhood storage is explained by the high frictional resistance of more than 8% accompanying such use.

The engineering considerations for deciding the location of water supply storage tanks or reservoirs are as follows:

1) Consideration should be given to maintaining water quality when locating water storage facilities.
2) The bottom of ground-level reservoirs and standpipes should be placed at the normal ground surface and shall be above the 100-year flood level or the highest flood of record.
3) If the bottom elevation of a storage reservoir must be below normal ground surface, it shall be placed above the groundwater table. At least 50% of the water depth should be above grade. Sewers, drains, standing water, and similar sources of possible contamination must be kept at least 50 ft (15 m) from the reservoir. Gravity sewers constructed of water main quality pipe, pressure tested in place without leakage, may be used at distances greater than 20 ft (6 m) but less than 50 ft (15 m).
4) The top of a partially buried storage structure shall not be less than 2 ft (0.6 m) above normal ground surface. Clearwells constructed under filters may be exempted from this requirement when the design provides adequate protection from contamination.

8.5 Elevation of Storage

Storage reservoirs and tanks operate as integral parts of the system of pumps, pipes, and connected loads. In operation all the parts respond to pressure changes as the system follows the diurnal and seasonal demands. Ideally the storage elevation should be such that the reservoir "floats" on the system, neither emptying nor standing continuously full. In systems with inadequate pipes or pumps, or having a storage reservoir that is too high, the hydraulic gradient may at times of peak demand fall below the bottom of the reservoir. When this occurs, the full load falls on the pumps and system pressures deteriorate suddenly.

8.6 Types of Distributing Reservoirs

Where topography and geology permit, service reservoirs are formed by impoundage, balanced excavation and embankment, or masonry construction (Fig. 8.7). To protect the water against chance contamination and against deterioration by algal growths stimulated by sunlight, distributing reservoirs should be covered. Roofs need not be watertight if the reservoir is fenced. Open reservoirs should always be fenced. Where surface runoff might drain into them, they should have a marginal intercepting conduit.

Earthen reservoirs, their bottom sealed by a blanket of clay or rubble masonry and their sides by core walls, were widely employed at one time. Today, lining with concrete slabs is more common. Gunite, a sand-cement-water mixture, discharged from a nozzle or gun through and onto a mat of reinforcing steel, has also been employed to line or reline them. Plastic sheets protected by a layer of earth have also been used to build inexpensive but watertight storage basins. Roofs are made of wood or concrete. Beam and girder, flat-slab, arch, and groined-arch construction have been used. Where concrete roofs can be covered with earth, both roof and water will be protected against extremes of temperature.

Inlets, outlets, and overflows are generally placed in a gate house or two. Circulation to ensure more or less continuous displacement of the water and to provide proper detention of water after chlorination may be controlled by *baffles* or subdivisions between inlet and outlet. Overflow capacity should equal the maximum rate of inflow. *Altitude-control valves* on reservoir inlets (Fig. 8.8) will automatically shut off inflow when the maximum water level is reached. An arrangement that does not interfere with draft from the reservoir includes a bypass with a *swing check valve* seating against the inflow.

Where natural elevation is not high enough, water is stored in concrete or steel *standpipes* and *elevated tanks*. In cold climates, steel is most suitable. Unless the steel in reinforced-concrete tanks is prestressed, vertical cracks, leakage, and freezing will cause rapid deterioration of the structure. Ground-level storage in reinforced-concrete or steel tanks in advance of automatic pumping stations is an alternative.

The useful capacity of standpipes and elevated tanks is confined to the volume of water stored above the level of wanted distribution pressure. In elevated tanks, this level generally coincides with the tank bottom; in standpipes, it may lie much higher. Steel tanks are welded or riveted. Their structural design and erection have become a specialized activity of certain tank manufacturers.

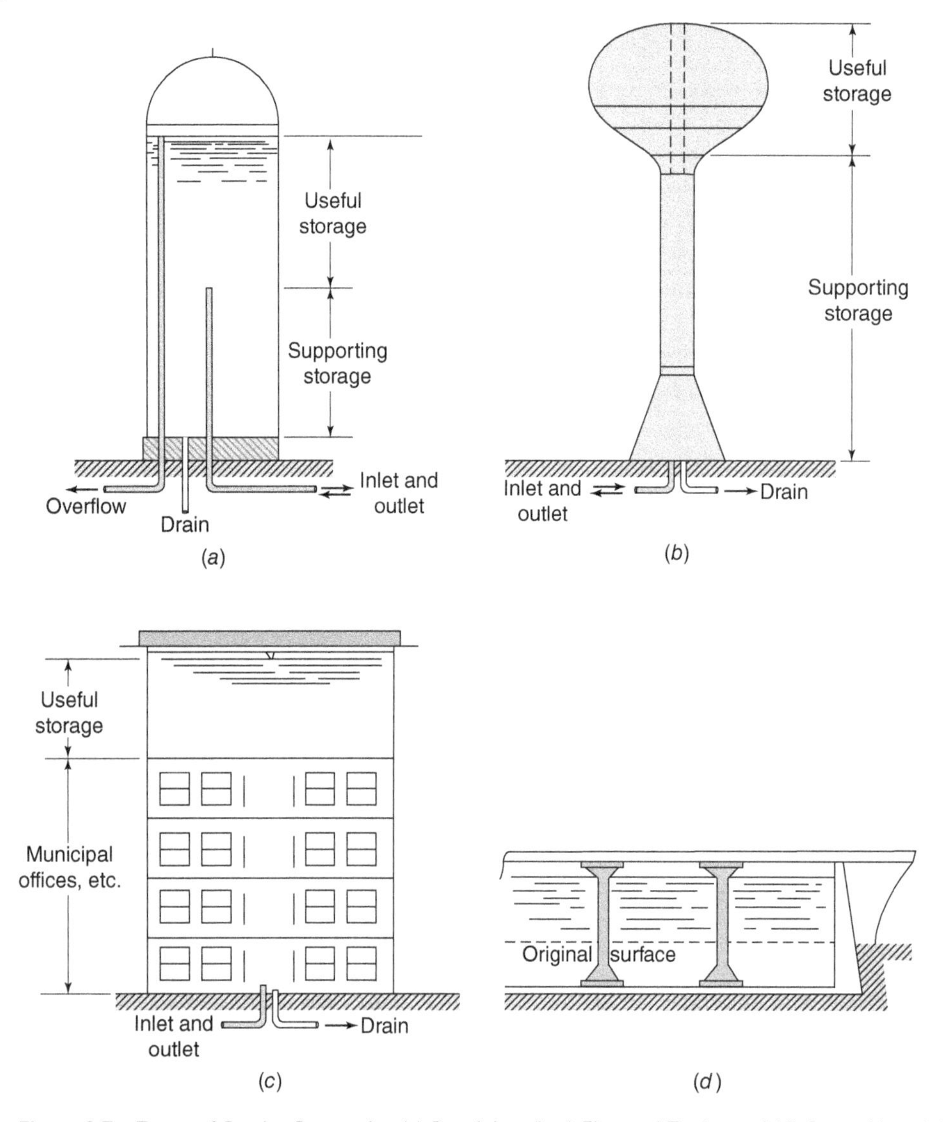

Figure 8.7 Types of Service Reservoirs: (a) Standpipe, (b, c) Elevated Tanks, and (d) Ground-Level Service Reservoir

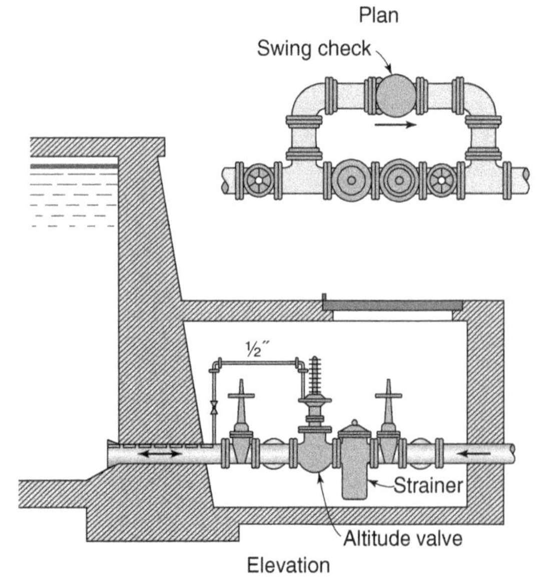

Figure 8.8 Altitude Valve on Supply to Distribution Reservoir

Figure 8.9 Water Storage Tank (*Source:* http://en.wikipedia.org/wiki/Image:Louisville_water_tower.jpg)

Figure 8.10 The Mushroom-Shaped Concrete Water Tower in Helsinki, Finland. It is 52 m high and can hold 12 000 m^3 of water (*Source:* http://en.wikipedia.org/wiki/Water_supply_network)

The function of elevated tanks and *spheroidal tanks* can be expressed to aesthetic advantage in their architecture without resorting to ornamentation (Figs. 8.9 and 8.10). *Standpipes* are simple cylinders. A veneer or outer shell of concrete or masonry may make them attractive. They may be designed as parts of multipurpose structures. The lower level may serve for offices, warehouses, or other functions. At the top, sightseeing or restaurant facilities may convert a potential eyesore into a center of attraction.

Example 8.8 Pumping and Storage System

Draw a sketch, showing a system (including the raw water reservoir, pump station, water transmission lines, elevated water storage tank, hydraulic grade line, elevation in ft or m, etc.) based on the following information: The raw water reservoir of a city is at an elevation of 500 ft (152.4 m). An automatic booster pumping station is proposed, having for its control point a 550 000-gal (2,082 m^3) water storage tank. The design tank water level is at an elevation of 490 ft (149.4 m).

Measurement of the water storage tank level is to be transmitted to the pumping station and is to be the only variable used for the control of the pumps in the pumping station. The head losses of the water transmission line between the raw water reservoir and the elevated water storage tank (including the miscellaneous head losses in the proposed pumping station) computed for the rate of flow of 1,600 gpm (101 L/s), are 40 ft or (12.2 m), which is to be overcome by the pumps. A chlorination system is used for treating the water entering the elevated water storage tank. The chlorinated water from the tank is transmitted to the city's water filtration plant for further treatment prior to domestic consumption (Fig. 8.11).

Example 8.9 Pumps and Water Storage Tanks

The water supply system shown in Example 8.8 provides water to a city at the rates of 2.0 MGD (88 L/s) for average daily water demand, 3.4 MGD (150 L/s) for maximum daily water demand, and 4.8 MGD (210 L/s) for peak hourly water demand. The total head losses are 40 ft (12.2 m) at a flow of 1,600 gpm (101 L/s).

What are the head losses of the transmission line between the raw water reservoir and the elevated water storage tank (including all miscellaneous head losses), which must be overcome by the pumps at 2, 3.4, and 4.8 MGD

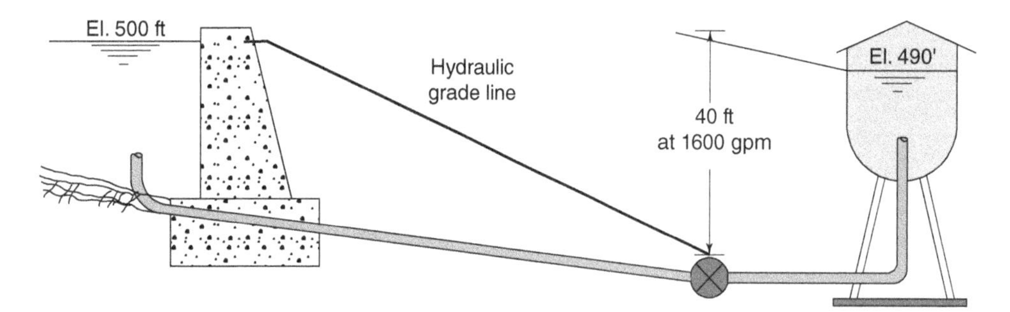

Figure 8.11 Sketch of Water Supply System for Example 8.8. Conversion factors : 1′ = 1 ft = 0.3048 m; 1 gpm = 3.785 L/min = 0.0000631 m^3/s = 0.0631 L/s

(88, 150, and 210 L/s; or 7.6, 12.9, and 18.2 MLD)? Explain which water demand is used for the design of a pumping station between the raw water reservoir and the water storage tank, and why. Is the size of the existing water storage tank sufficient for the city in accordance with the current Ten-States Standards for Water Works?

The following equation shows the relationship between the head loss and the water flow:

$$H_2 = H_1 \left(Q_2/Q_1\right)^2$$

where Q_1 is the first water flow, gpm (m^3/s); Q_2 is the second water flow, gpm (m^3/s); H_1 is head loss at flow of Q_1, ft (m); and H_2 is head loss at flow of Q_2, ft (m).

Solution 1 (US Customary System):

Head loss at average daily water demand of 2 MGD (or 1,388 gpm):

$$H_2 = (40)\ (1,388/1,600)^2 = \mathbf{30\ ft}$$

Head loss at maximum daily water demand of 3.4 MGD (or 2,360 gpm):

$$H_2 = (40)\ (2,360/1,600)^2 = \mathbf{87\ ft}$$

Head loss at maximum hourly water demand of 4.8 MGD (or 3,331 gpm):

$$H_2 = (40)\ (3,331/1,600)^2 = \mathbf{173\ ft}$$

Solution 2 (SI System):

Head loss at average daily water demand of 7.6 MLD (or 0.088 m^3/s, or 88 L/s):

$$H_2 = (12.2)\ (0.088/0.101)^2 = \mathbf{9.36\ m}$$

Head loss at maximum daily water demand of 12.9 MLD (or 0.150 m^3/s, or 150 L/s):

$$H_2 = (12.2)\ (0.150/0.101)^2 = \mathbf{27.00\ m}$$

Head loss at maximum hourly water demand of 18.2 MLD (or 0.210 m^3/s, or 210 L/s):

$$H_2 = (12.2)\ (0.210/0.101)^2 = \mathbf{53.80\ m}.$$

The system will not be economically feasible if the maximum hourly water demand is used for design of a pump station due to its related high head loss (173 ft or 52.73 m). Normally the **maximum daily water demand is used for designing a pump station**. The peak hourly water demand of the city will be provided by the elevated water storage tank.

According to the latest edition of the Ten-States Standards, the minimum water storage capacity for water systems not providing fire protection shall be equal to the average daily consumption, or 2 MG (7,570 m^3). **A new water storage tank (2MG or 7, 570 m^3 + fire demand) is needed** because the existing water storage tank (550 000 gal or 2,082 m^3) is not big enough.

Example 8.10 Design of a Pumping Station

Design a pump station for the water system examined in Examples 8.8 and 8.9. Double-suction, horizontal centrifugal pumps driven by AC electric motors are considered the best application for this project. Select pumps having rated speeds of 1,750, 1,150, or 870 rpm for this pump station using assumed pump characteristics.

Recommend the pump capacity, the number of pumps, and pumping mode (parallel operation or series operation). Calculate the effective head of the selected pump, its water horsepower, brake horsepower (horsepower input to each pump), and motor horsepower assuming the pump efficiency is 80% and the motor efficiency is 90%. Show the assumed capacity-efficiency curve, the head-capacity curve, and the BHP-capacity curve (or the BMP-capacity curve) on a sketch, and then indicate on the sketch the rated points of the pumps. Write brief engineering conclusions stating the following: (a) the number of installed pump units and the type of motor driving each pump unit; (b) the method, if any, of capacity control that an engineer would propose and the probable overall pump station efficiency at the average daily water consumption; and (c) the probable horsepower of the motor driving each unit.

Solution 1 (US Customary System):

Average daily consumption = 2 MGD = 1, 388 gpm
Maximum daily consumption = 3.4 MGD = 2, 360 gpm
Peak hourly demand = 4.8 MGD = 3, 331 gpm

With sufficient water storage capacity available in the future, two pumps of equal capacity should be operating in parallel to supply the maximum daily consumption of 2,360 gpm. Each pump is to supply 1,215 gpm, and both pumps 2,430 gpm.

Head loss at 2, 360 gpm, $H = 40\,(2,360/1,600)^2 = 87$ ft

The effective **head of each pump** should be 87 ft less the difference in water surface elevations.

Difference in water surface elevation = 500 − 490 ft = 10 ft
Effective head of each pump, $H = 87 - 10$ ft = **77 ft**
Water horsepower (WHP) = (QH)/3, 957

$$= (1,215)(77)/3,957 = \mathbf{23.6\,hp}$$

Brake horsepower (BHP) = horsepower input to each pump = WHP/E_{pump}

$$= 23.6/0.8 = \mathbf{29.5\,hp}$$

Motor horsepower (MHP) = horsepower input to motor = $\text{BHP}/E_{\text{motor}}$

$$= 29.5\ \text{hp}/0.9 = \mathbf{32.8\,hp}$$

Solution 2 (SI System):

Average daily consumption = 7.57 MLD = 0.088 m^3/s = 88 L/s
Maximum daily consumption = 12.87 MLD = 0.150 m^3/s = 150 L/s
Peak hourly demand = 18.17 MLD = 0.210 m^3/s.

With sufficient water storage capacity available in the future, two pumps of equal capacity should be operating in parallel to supply the maximum daily consumption of = 12.87 MLD = 0.150 m^3/s = 150 L/s.
Each pump is to supply 0.077 m^3/s, and both pumps 0.153 m^3/s.
Head loss, at 12.87 MLD or 0.150 m^3/s, $H = (12.19\ m)(0.150\ m^3/s/0.101\ m^3/s)^2 = 26.5$ m

The effective **head of each pump** should be 26.5 m (87 ft) less the difference in water surface elevations.

Difference in water surface elevation = 152.4 − 149.3 m = 3.1 m
Effective head of each pump, H = 26.5 − 3.1 m = 23.4 m

Water power (WMP) = 9.8066 (Q H)

$$= 9.8066\,(0.077)\,(23.5) = \mathbf{17.7\,kW}$$

Brake power (BMP) = power input to each pump = WMP/E_{pump}

$$= (17.7)/0.8 = \mathbf{22.2\,kW}$$

Motor power (MMP) = power input to motor = BMP/E_{motor}

$$= 22.2/0.9 = \mathbf{24.6\,kW}$$

The sketch of the selected centrifugal pump will have the pump characteristic curves shown in Fig. 8.12.

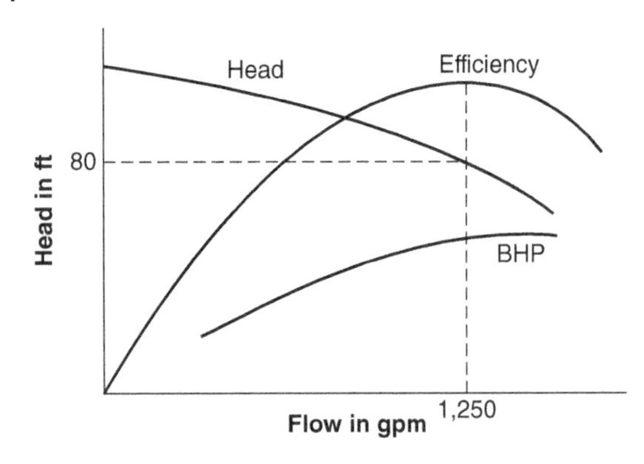

Figure 8.12 Pump Characteristic Curves for Example 8.10. Conversion factors: 1 ft = 0.3048 m; 1 gpm = 3.785 L/min = 0.0000631 m^3/s = 0.0631 L/s

The engineer's recommendations are as follows:

a) Install three pump units, each with a capacity of 1,250 gpm (79 L/s) at a total developed head of 80 ft (24.38 m) at full capacity. Two of the pumps should have standard 60-cycle, AC induction-motor drives, 1,750 rpm. The third pump to be held in reserve in case of electric power failure or shut down of one electric-drive pump for repairs may be driven with a gasoline engine, in case an emergency power generator is not available. If an emergency power generator is available, the third pump can be identical to the selected two pumps.
b) The storage tank capacity (550 000 gal or 2,082 m^3) normally is sufficient to meet the maximum hourly demand and fluctuation during maximum daily demand. However, the existing storage tank capacity is insufficient according to the latest edition (2007) of the Ten-States Standards for water works. Expansion of the water storage tank is recommended if construction funds are available. The quantity of water in the water storage tank should be controlled by a floating control system. If the float drops below a specified level, the control system actuates the starting mechanism of the second pump to put it in parallel with the other pump. The pump station efficiency at average daily consumption will be approximately 50%–60%.
c) Each pump must be driven by a 35-hp (26.1-kW), 1,750-rpm motor.

Example 8.11 Leakage Losses Versus Specific Speed for Double-Suction Pumps

Specific speed S is of significant importance in selecting a pump for a specific application, since there is a relationship and internal leakage within the pump as shown in Table 8.3 for double-suction split-case centrifugal pumps.

Determine and discuss the specific speeds and the expected inherent losses in % of power input for two pumps:

1) Pump A: It has a 356-mm (14-in.) diameter impeller with its optimum efficiency at 0.2145 m^3/s (3,400 gpm) and 54.9 m (180-ft) head when operating at 1,900 rpm.
2) Pump B: It has a 356-mm (14-in.) diameter impeller with its optimum efficiency at 0.0126 m^3/s (200-gpm.) and 79.2 m (260-ft) head when operating at 3,600 rpm

Table 8.3 Specific Speed versus Leakage Losses for Double-Suction Pumps in Example 8.11

Specific Speed S	Leakage Losses in % of Power Input
500	9.5%
1,000	4.3%
1,500	2.6%
2,000	1.8%
2,500	1.4%
3,000	1.2%
3,500	1.0%

Solution 1 (US Customary System):

For Pump A:

$$S = NQ^{0.5}/H_{sv}^{0.75} = 1,900 \times (3,400)^{0.5}/180^{0.75} = \mathbf{2,255}$$

Pump A would have an inherent loss of about **1.6%.**

For Pump B:

$$S = 3,600 \times (200)^{0.5}/260^{0.75} = \mathbf{786}$$

Pump B would have an inherent loss of about **6%**.

Solution 2 (SI System):

For Pump A:

$$S = 51.7\,NQ^{0.5}/H_{sv}^{0.75} = 51.7 \times 1,900 \times (0.2145)^{0.5}/54.9^{0.75} = \mathbf{2,255}$$

Pump A's expected inherent loss would be about **1.6%.**

For Pump B:

$$S = 51.7 \times 3,600 \times (0.0126)^{0.5}/79.2^{0.75} = \mathbf{786}$$

Pump B's expected inherent loss would be about **6%**.

8.7 Dual Water Supply Systems

The term *dual water supply systems* refers to two water supply distribution systems in a city. One is a freshwater system for potable use; the other system is for lower quality water (seawater or untreated raw freshwater, or treated/reclaimed wastewater) for firefighting and toilet flushing purposes. In Hong Kong and Saudi Arabia, dual water supply systems have been in use for more than 40 years.

Water supply networks are sized so that they can provide the large flows needed for firefighting, but this creates potable water quality problems due to the long time water can spend in the network. What is needed, argues Okun (2007), is a switch to dual systems, with potable supplies provided through a smaller bore network using a material such as stainless steel, allowing reclaimed water to be used in existing networks.

Note: This section is based on a timely article written by Dr. Daniel A. Okun a few months before he passed away in 2007. The article appeared in the February 2007 issue of *Water 21*, a magazine of the International Water Association, pages 47–49; reproduced courtesy of IWA Publishing.

8.7.1 Background

In the 18th and early 19th centuries, the demand for protection against fire and the great conflagrations and loss of life that they brought predicated the provision of water distribution systems designed for fire protection. Only later were these distribution systems put into service for commercial and then residential use, which led to the subsequent development of the water closet and sewerage systems.

That our present distribution systems are delivering water of exceedingly poor quality today, almost without regard to the water's source, treatment, or distribution, has been made manifest by the vast literature emerging from every corner of the water supply scene. The American Water Works Association (AWWA) has shown considerable awareness of the problems. Its annual water quality technology conferences have each had more than 100 papers identifying the problems. Some 40 classes of problems are set out by AWWA, with recommendations for individual utilities to assess their own particular problems and find their own answers. But relatively few utilities have the appropriate staff or financial resources to undertake the required studies and address each of the many problems.

The recommended practice of frequent *flushing* of the distribution systems has been widely adopted, but it hardly addresses the problems. Flushing is costly in personnel and extremely wasteful of treated drinking water, which is discharged to stormwater sewerage systems. In addition, frequent flushing is not very effective in keeping the pipes free from *biofilm* growths on pipe walls and maintaining hydraulic capacity

8.7.2 The Nature of the Problems of Current Water Distribution System with Drinking Water Quality

The critical problem is that fire protection requires there to be many hydrants throughout a city, which have to have the capacity to deliver relatively high flow rates at all times and at all locations throughout the area. Pipe sizes were initially a minimum of 6 in. (150 mm), but today this has increased in many communities to a minimum of 8 in. (200 mm).

Because fires are infrequent, the velocity of the water in the network is almost always slow, resulting in residence times of months between when the water is treated and when it arrives at the taps of consumers in the outer regions of the service area. Recent tracer studies by the University of North Carolina in two of the larger cities in the state revealed residence times of more than 10 days. Such times make adequate *chlorine residuals* at the tap unlikely. The inadequacy of *disinfection,* with the resulting risk of microbial exposure at the tap, is not the most troublesome problem arising from ineffective disinfection. In attempting to provide adequate disinfection despite the poor conditions in the pipelines, providers considerably increase the *chlorine dose*, resulting in increased levels of *disinfection by-products (DBPs)* through reactions with both chemical and microbial contaminants in the water. On April 25, 2014, the City of Flint, MI, USA changed their raw water supply source from the Detroit-supplied Lake Huron water to the local Flint River water. The raw water switch caused Flint water distribution pipes to corrode and leach toxic soluble lead and other contaminants into Flint drinking water.

In the United States, the Federal and state regulations require monitoring water within the distribution system under three specific rules to deal with the above problems: (a) Total Coliform Rule, (b) Lead and Copper Rule, and (c) Trihalomethane Rule. Section 8.8 discusses the water quality management of water distribution systems under the US federal and state regulations

8.7.3 The Pipes in the Distribution Systems

Because the pipes in all urban water distribution systems currently need to be a minimum of 6 in. (150 mm) in diameter or larger, they are generally heavy cement-lined ductile-iron pipes, each section 16 ft (5.3 m) in length. These require some 350 joints per mile of pipe, including those needed for fire hydrants. The pipes are laid on soil in trenches, and in time the joints leak and lose water. If the pipes are below the water table, any infiltration of contaminated groundwater would pose a health risk.

Because the pipes are always under pressure, it had been believed that contamination from surrounding groundwater would not be a problem. Recent studies, however, have revealed that sudden changes in the velocity of the water are created by the opening and closing of valves in the lines and the starting and stopping of pumps, events that occur several times a day. This causes negative pressure transients that result in the infiltration of groundwater from the soil in the vicinity of the pipes.

Such transients would, of course, also occur in small pipes, but stainless-steel pipes used for distribution systems that only carry drinking water would not leak because these pipes do not have open joints. In the very small sizes, they can be laid from spools and in the larger sizes the pipe can be welded. Stainless steel is already widely used in Japan for water distribution systems. These pipes have an added advantage over cement-lined pipes because they are not prone to heavy growths of biofilms due to their smooth interior walls.

The fact that leakage from distribution systems increases with time is well recognized, but these losses are generally not considered important. However, recent studies show that leaks have much more serious consequences. The negative pressures that have been found to occur regularly in pipelines encourage infiltration of the water in the soil surrounding the pipes. This is possible because heavy pipes laid on soil tend to subside over time, which opens the joints sufficiently to create two-way leakage.

8.7.4 Biofilms and the Problems They Cause

The poor water quality found in distribution systems today results above all from the considerable growth of biofilms that are attracted to the insides of the pipes because of the pipe materials chosen and the long residence times of the water. The many joints, hydrants, valves, and other appurtenances, along with the cement and other conventional linings, present attractive surfaces for the growth of biota, which deplete the disinfectants. The heavy growths in turn shield the disinfectant from the biota. The biofilms grow thick because of the very slow velocities of the water and the long residence times in the pipes. They build up to the point that, with products of corrosion and tuberculation, they significantly reduce the hydraulic capacity of the pipes. If water consumers were able to see the insides of most distribution systems, they would be encouraged to buy bottled water, now a fast-growing trend. Studies in Britain have compared conventional cement-lined pipe with stainless-steel pipes of various compositions to assess their comparative rates of growth of biofilms. All of the stainless-steel pipes and storage tanks were found to be excellent for preventing corrosion and biofilm growth. The stainless-steel pipes that are available in very small sizes are provided from spools that carry long lengths of product, sufficient to serve the outlying residential areas of larger cities without the problems that are now encountered. Larger stainless-steel pipes and storage tanks are available although the costs are still high. Disinfection is an effective for inactivation or destruction of disease-producing organisms. Section 8.8.5 introduces the proper methods for disinfection of water supply pipes and water storage facilities.

8.7.5 The Proposed System

The professionals engaged in providing drinking water to the public make great efforts to ensure water of high quality, which requires considerable investments in treating the water. However, in the last step, distribution of water to the consumer, the water is permitted to be seriously degraded.

We should be embarrassed by the tremendous efforts and funds we invest in providing high-quality water only to allow it be subjected to the many problems enumerated above, which result in increasing health risks to the public. In our efforts to address each of the problems, we oblige the water utilities to undertake studies and remedies far beyond the capacities of all but the very large utilities.

Just because we have inherited distribution systems created for fighting fires over two centuries, a practice that is responsible for all of the problems, that is no reason for continuing the practice into the future, especially when the only solution offered is frequent flushing of all the pipelines. Flushing is costly in execution, relatively ineffective, and wasteful of the treated drinking water.

A reasonable solution is available: distribution systems designed for drinking water alone. This option has many advantages. The pipe diameters for most of the distribution lengths would be much smaller than the current minimum sizes. Materials such as stainless steel can largely eliminate the bane of our present systems: leaking joints.

One very great advantage is that the size of community water treatment works would be a small fraction of what they now need to be, encouraging the use of membrane treatment, which improves the drinking water quality beyond what is now available in most communities. The result would be purer water at a lower cost. Figure 8.13 illustrates what a new community might do to initiate such an approach to conserving its drinking water, affording high-quality treatment and avoiding degradation in the distribution system.

What makes this approach reasonable today is that dual systems on a large scale began to cover the United States some 40 years ago. Some 2,000 communities in the United States and many abroad, both the largest and the smallest cities, are

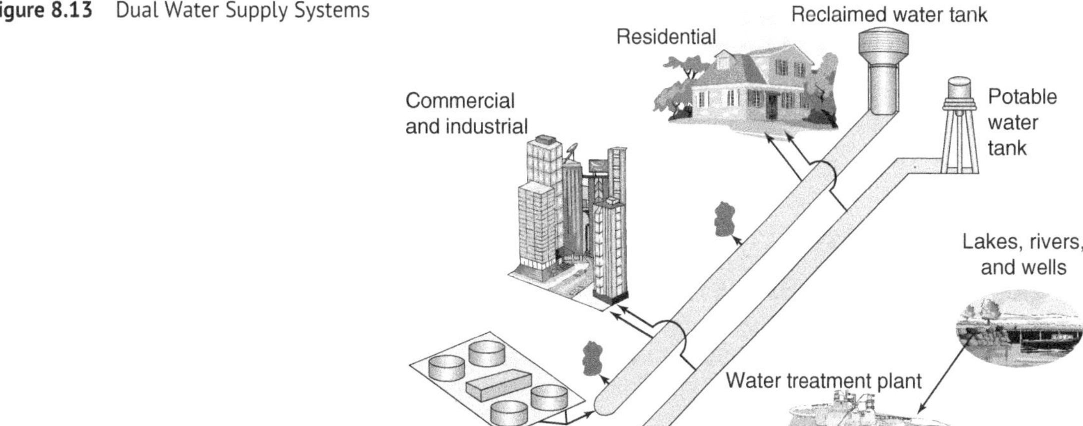

Figure 8.13 Dual Water Supply Systems

adopting dual systems. They began in arid areas but they are present also in Oregon and Washington, the country's wettest states. In Florida, with about 48 in. per year (1,219 mm/yr) of rainfall, some 450 communities have dual systems. Their popularity is based largely on the drinking water supply conservation they provide.

For new communities, a dual system with one system for drinking water only, and the other for all nonpotable purposes, including fire protection, would address not only drinking water supply problems, but water quality problems as well. In addition, such systems would be far less costly than conventional systems. Communities would be able to afford membrane treatment because the amount of drinking water to be treated to high quality would be considerably smaller than overall rates of water to be treated.

One such system was adopted 10 years ago for a new suburb of Sydney, Australia, called Rouse Hill, involving 250 000 people with a first stage for 100 000 residents. It has been operating successfully since. This approach is particularly beneficial for systems that are obliged to take water from sources of poor quality.

The problem is much more difficult to address in existing communities that are growing or retrofitting new systems. All cities are site-specific. Where new distribution systems are being added, installing the stainless-steel pipes gradually would be beneficial but costs would need to be assessed to determine whether they are feasible. The high costs of the conventional pipes and future problems with water quality may justify the higher initial costs.

From a public health perspective we cannot afford to continue our current practices. Dual systems and the water supply conservation they provide, along with distribution systems carrying only well-treated drinking water, dictate the need to study new approaches.

8.8 Monitoring and Management of Water Distribution Systems

8.8.1 Introduction

Water distribution systems with or without problematic lead service line (LSL) will cause water contamination that requires proper systemic monitoring and management. LSL is the drinking water service line that is made of lead, runs underground, and connects a home's plumbing to the water main. Too many LSLs still exist around the world today. LSLs constitute the most significant source of lead contamination in the water distribution systems.

Ordinarily routine samples to analyze for primary and secondary contaminants are drawn from water supply system sources unless it appears there could be some change in constituent levels in the water distribution system. Routine non-source sampling may be performed by utilities and the customers. When water samples are collected from the water distribution system, they normally are taken from representative points throughout the system for water quality control, in turn, for water distribution system management under the US federal and state regulations, namely (a) Total Coliform Rule, (b) Lead and Copper Rule, and (c) Trihalomethane Rule.

Local conditions at a specific sample tap and in the piping connection to the main may make the sample point unrepresentative of the quality of the water being furnished to the consumers. The best and truest evaluation of water quality can be obtained from samples drawn directly from the main at specially designed sampling stations. The water quality management steps under the three rules are discussed in below with emphasis on proper water sampling and analysis for water quality monitoring.

Where tastes, odors, and color may be a problem, samples are often taken at representative or critical points in the distribution system. Such sampling may continue until there is adequate demonstration no problem exists. Critical points would include dead ends and other areas where water movement is slow, and areas where corrosion problems would be expected. Turbidity samples are taken at representative source entry points to the water distribution system.

8.8.2 Distribution System Monitoring for Total Coliform Rule Compliance

This rule controls the microbial water quality aspects by testing for coliform bacteria and chlorine residuals. Bacteriological samples (for coliforms) are taken at points that are representative of the conditions within the water distribution system. Be sure the sample bottles contain sodium thiosulfate to neutralize the chlorine residual. Measure the chlorine residual when the sample is collected.

Chlorine residual samples are frequently collected at the most remote locations of the distribution system to ensure that a chlorine residual exists throughout the entire system. Samples are also collected and tested for chlorine residual throughout the system. Chlorine residual testing is very important. Many utilities attempt to maintain a chlorine residual throughout the distribution system. Chlorine is very effective in biological control and especially in the elimination of coliform bacteria from the finished water. Adequate control of coliform "after growth" is usually obtained only when chlorine residuals are carried to the farthest points of the distribution system. To ensure that this is taking place, make daily chlorine residual tests. A residual of about 0.2 mg/L measured at the extremities of the system is usually a good indication that a free chlorine residual is present in all other parts of the system. This small residual can destroy a small amount of contamination, so a lack of chlorine residual could indicate the potential presence of heavy contamination. If routine checks at a given point show measurable residuals, any sudden absence of a residual at that point should alert the water supplier to the possibility that a potential problem has arisen which needs prompt investigation. Immediate action that can be taken includes retesting for chlorine residual, then checking chlorination equipment, and finally searching for a source of contamination which could cause an increase in the chlorine demand.

8.8.3 Distribution System Monitoring for Lead and Copper Rule Compliance

This rule deals with the corrosivity of water distributed to homes with lead and copper plumbing. Water is tested for lead and copper in the dead ends of water mains and from the drinking water taps of homes when residents first get up in the morning (test stagnated samples at sites with lead and copper plumbing). Other water quality measurements associated with the Lead and Copper Rule include pH, alkalinity, and the residual of any corrosion inhibitor applied to the water.

The Lead and Copper Rule requires that samples be collected at the consumers' taps. The samples must be collected from locations identified as "high-risk," including: (a) homes with lead solder installed after 1982, (b) homes with lead pipes, and (c) homes with LSLs.

Proper lead and copper monitoring and sampling procedures must be followed. The number of samples to be collected for lead and copper analysis is based on the size of the distribution system; the sampling frequency is every six months for the initial monitoring program. There are two monitoring periods each year, January to June, and July to December. If the system is in compliance, either as demonstrated by monitoring or after installation of corrosion control, a reduced monitoring frequency can be initiated. The number of sampling sites required based on system size for initial and reduced monitoring are listed in Table 8.4.

The samples are to be collected as "first draw" samples from the cold water tap in either the kitchen or bathroom, or from a tap routinely used for consumption of water if in a building other than a home. A first draw sample is defined as the first 1 L of water collected from a tap which has not been used for at least 6 h, but preferably unused for no more than 12 h. Faucet aerators should be removed prior to sample collection. The Lead and Copper Rule allows homeowners to collect samples for the utility as long as the proper sample collection instructions have been provided.

To prevent the recurrence of another lead contamination crisis similar to that happened in Flint, MI, USA, the U.S. Environmental Protection Agency (US EPA) issued the Lead and Copper Rule Revisions (LCRR) on December 16, 2021, to strengthen the original 1991 Lead and Copper Rule (LCR) for improving the reliability of lead sampling in the water distribution system and establishing a trigger level to jumpstart mitigation earlier and in more communities. The LCRR retains the lead's "action level" of 15 μg/L while adds a "trigger level" of 10 μg/L, drives more and complete LSL replacements.

Table 8.4 Sampling Sites Required for Lead and Copper Analysis

System Size (Population)	Sampling Sites Required (Base Monitoring)	Sampling Sites Required (Reduced Monitoring)
>100 000	100	50
10 001–100 000	60	30
3,301–10 000	40	20
501–3,300	20	10
101–500	10	5
<100	5	5

Reduced Monitoring: (a) All public water systems that meet the lead and copper action levels or maintain optimal corrosion control treatment for two consecutive six-month monitoring periods may reduce the number of tap water sampling sites as shown in Table 8.3 and their collection frequency to once per year; (b) all public water systems that meet the lead and copper action levels or maintain optimal corrosion control treatment for three consecutive years may reduce the number of tap water sampling sites as shown in Table 8.3 and their collection frequency to once every three years.
Source: US EPA.

8.8.4 Distribution System Monitoring for Trihalomethane Rule Compliance

Trihalomethanes (THMs) and *haloacetic acids* are the only two DBPs that are being regulated by the US EPA, but then with great difficulty. Their *maximum contaminant levels* (MCLs) were epidemiologically uncertain, as indicated by the arbitrary adoption of a THM level of a "round" 0.10 mg/L in 1979. This figure was reduced recently, based in large part on the ability of utilities to reach a lower level.

The DBP problem is much more difficult to manage than is evident from recent research. As shown later, many more contaminants are present in drinking water networks than are recognized today as potential reactants with the chlorine present in the water and therefore there are many more other DBPs that need to be regulated.

Even more concern for the health effects of THMs has arisen because of a study carried out on 50 women in two very different locations (Cobb County, Georgia, and Corpus Christi, Texas in the USA), which have water supplies with very different *THM bromide* concentrations and disinfectant types: *chloroform* in the former and *brominated THMs* in the latter.

Blood samples were taken from women and water samples were taken from their showers in the early morning. It was shown that the THMs in their blood samples rose significantly after showering. The types of THMs in their blood samples matched the type of THM in the water. THM standards are based on lifetime exposures, but recent studies have suggested that THMs pose possible reproductive problems for women that would dictate more rigorous MCLs for DBPs in the future.

This Trihalomethanes Rule is intended to limit chlorinated DBPs. Samples are collected and tested for THMs and chlorine residuals. Sampling within the water distribution system should be done in accordance with accepted collection and transportation procedures. In addition, it will be necessary to test enough samples to provide an optimum amount of information on the quality of the water being supplied. To calculate an average measurement for a time period or a group of measurements or observations, add up all of the measurements and divide by the number of measurements. So, (Average) = (Sum of All Measurements)/(Number of Measurements).

The US Safe Drinking Water Act requires that total THMs be calculated as quarterly averages as shown above. When the running annual average exceeds the MCL of 0.10 mg/L, the value should be reported to the state within 48 hours. The running annual average is calculated by using the quarterly average for the quarter being considered and the three quarters immediately before the one being considered. So, (Annual Running Average) = (Sum of Measurements)/(Number of Measurements).

Running averages are used to "smooth out" data. In the case of THMs, if one quarter was high and exceeded the MCL, the other three quarters could pull the annual running average MCL down and keep the water utility in compliance. For total THM sampling, 25% of the samples are collected at extreme ends of the distribution system and 75% at locations representative of population distribution.

8.8.5 Disinfection of Water Pipes and Storage Facilities

New water supply pipes and water storage facilities and the water facilities that have been repaired, cleaned, or had cathodic protection installed must be disinfected. Always disinfect water storage facilities whenever there has been any opportunity for microbiological contamination. Liquid chlorine (or gas), calcium hypochlorite, and sodium hypochlorite are commonly used as disinfectants. There are two common methods for water storage tank disinfection: (a) high disinfectant concentration method: spraying or brushing water storage tank with 200 mg/L chlorine solution, waiting for 30 minutes, filling the tank with water containing 3 mg/L chlorine residual, and standing for additional 3–6 h; and (b) low disinfectant concentration method: filling the water storage tank with water containing 3 mg/L chlorine residual and providing at least 24 hours of disinfection contact time. The following are the details of the above two disinfection methods.

When the high disinfectant concentration method is used, the interior of the storage facility is sprayed or brushed with 200 mg/L chlorine solution. The chlorine solution can be mixed in a crock or small tank. The application can be accomplished by using garden hoses with nozzles and pressure can be provided by a gasoline-driven pump. Place gasoline engine outside the water storage tank to prevent the buildup of dangerous exhaust vapors. A 200-mg/L chlorine concentration will be uncomfortable to work with inside of a large water storage facility. Provide adequate ventilation whenever a worker is inside a water storage tank. After the tank has been sprayed with the hypochlorite solution, allow it to stand unused for at least 30 minutes before filling. Fill the tank with distribution system water that has been treated with chlorine to provide a chlorine residual of 3 mg/L. Let the water in the tank stand for 3–6 hours. Take a bacterial sample of tank water in a sterile container and test for coliform bacteria. Be sure to add enough sodium thiosulfate to the sampling bottle before it is sterilized to neutralize all chlorine in the water. After a sample is collected, shake the bottle and test to be sure there is no chlorine residual. After bacteriological tests prove negative (no conforms), steps can be taken to put the tank back into service.

When the low disinfectant concentration method is used, the contaminated water storage tank is filled with water having a chlorine residual of 3 mg/L. This method of disinfection is used after the tank has been cleaned and the cathodic protection device has been serviced as necessary. The tank is inspected to be sure that all equipment and tools have been removed. Sufficient chlorine is then applied to the water entering the tank to end up with 3.0 mg/L of chlorine residual. The water in the tank is sampled for bacteriological tests 24 h after the tank has been filled. After bacteriological tests prove negative (no conforms), the tank is put back in service.

Problems/Questions

8.1 Determine the water horsepower, break horsepower, and motor horsepower for a pump operating under the following conditions: Water flow of 490 gpm (31 L/s) is to be pumped against a total head of 110 ft (33.53 m), the pump efficiency is 75%, and the motor driving the pump has an efficiency of 85%.

8.2 A pump is to be located 6 ft (1.83 m) above a wet well and must lift 600 gpm (38 L/s) of water another 52 ft (15.85 m) to a storage reservoir through a piping system consisting of 1,250 ft (381 m) of 6-in. (150-mm) DIP pipe ($C = 110$), two globe valves (open), and two medium sweep elbows. Determine the total dynamic head (TDH) for this water pumping system.

8.3 Using the data from Problem 8.2, assume the following additional data: efficiency of pump = 80%; efficiency of motor = 85%. Determine (a) the motor horsepower in hp. and kWh/day and (b) the daily power cost if the unit power cost is $0.1/kWh.

8.4 Determine the total power input if the electrical input to a motor-pump system is 220 V and 36 amps.

8.5 Determine the wire-to-water efficiency (%) if $Q = 510$ gpm (32 L/s), $TDH = 53.65$ ft (16.35 m), $V = 220$ volts, and $A = 36$ amps.

8.6 Changing the speed of a centrifugal pump will change its operating characteristics, including the water flow. Determine the new flow rate or capacity, Q_2, if

Q_1 = rated water flow = 620 gpm = 39 L/s
N_2 = pump speed now = 1, 320 rpm
N_1 = rated pump speed = 1, 650 rpm.

8.7 Changing the speed of a centrifugal pump will change its operating characteristics, including the head. Determine the new head, if

H_1 = rated head = 120 ft = 36.58 m
N_2 = pump speed now = 1, 320 rpm
N_1 = rated pump speed = 1, 650 rpm.

8.8 Changing the speed of a centrifugal pump will change its operating characteristics, including the power requirement. Determine the new power requirement, if

P_1 = rated head = 16 hp = 12 kW
N_2 = pump speed now = 1, 320 rpm
N_1 = rated pump speed = 1, 650 rpm.

8.9 A water pumping station is designed to raise water from a lake at an elevation of 50 ft (15.24 m) to a reservoir located at an elevation of 140 ft (42.67 m). Water is pumped through a 16-in. (40.64-cm) cast-iron pipe that develops a head loss of

19 ft (5.79 m) of water at a discharge of 1,000 gpm (63 L/s). The pumping station has two pumps, each of which possesses the following characteristics:

Flow (gpm)	Head (ft)
500	195
1,000	180
1,500	150
2,000	110
2,500	50

Flow (L/s)	Head (m)
32	59.44
63	54.86
95	45.72
127	33.53
158	15.24
Shutoff head = 200 ft (60.96 m).	

What will be the discharge of the two pumps when operating in parallel?

8.10 The water supply system shown in Fig. 8.14 is designed to serve city ABCDEFA. Water is treated and collected in a water tank in the water treatment plant (WTP). A pump delivers the water through the main PR1 to an elevated reservoir (water level 2,700 ft, or 822.96 m) at the top of a hill. Pumping is done at a constant rate and only for a period of 16 hours per day from 4 a.m. to 8 p.m. Water flows from the elevated reservoir into the distribution main network ABCDEF through a 24-in. main R_2A.

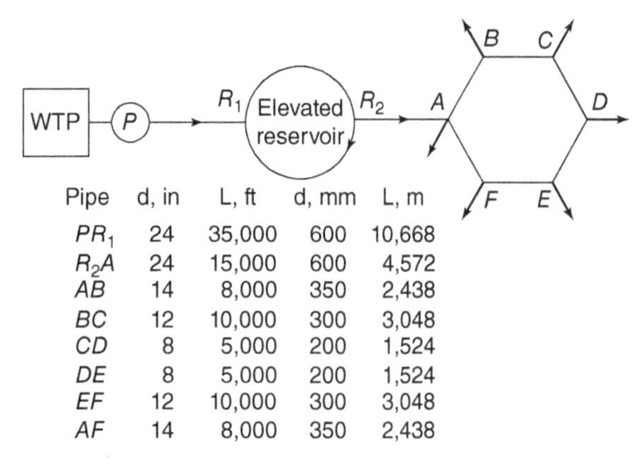

Figure 8.14 Water System for Problem 8.10

The water consumption during the maximum day demand in the city is as follows: Shutoff head = 200 ft (60.96 m).

Period	Flow rate (gpm)
Midnight to 4 a.m.	2,000
4 a.m. to 8 a.m.	4,000
8 a.m. to noon	6,000
Noon to 4 p.m.	5,000
4 p.m. to 8 p.m.	4,000
8 p.m. to midnight	3,000

Period	Flow rate (L/s)
Midnight to 4 a.m.	126
4 a.m. to 8 a.m.	252
8 a.m. to noon	379
Noon to 4 p.m.	315
4 p.m. to 8 p.m.	252
8 p.m. to midnight	189

Assume that the water withdrawal from the network is equally distributed among the points A, B, C, D, E, and F, which are at the same elevation of 2,300 ft (701.04 m) and that all pipes are ductile iron with $C = 100$. Also consider that the pump possesses the following characteristics:

Flow (gpm)	0	2000	4,000	6,000	8,000	12 000
Head (ft)	400	390	370	345	310	225
Flow (L/s)	0	126	252	379	505	757
Head (m)	121.92	118.87	112.78	105.16	94.49	68.58

(a) Calculate the volume of the elevated reservoir needed to balance supply and demand.
(b) Compute the maximum flow in each pipe of the network.
(c) Determine the elevation of the water tank at the treatment plant
(d) At what rate of flow can you pump the water to the elevated reservoir if another identical pump is installed in parallel with the existing pump?

8.11 The water supply system shown in Fig. 8.15 is planned to serve a small village. Treated water from the water treatment plant flowing at a uniform rate throughout the day is collected in a ground tank. Water is then pumped to an elevated reservoir at the top of a hill through the main AB. Water flows from the elevated reservoir into the distribution network through the main CD.

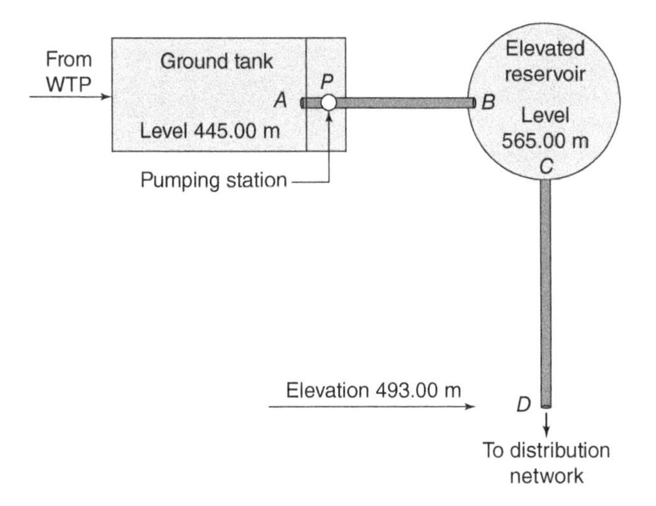

Figure 8.15 Water System for Problem 8.11

Water consumption during the maximum day in the village is as follows:

Period	Flow rate (L/s)
Midnight to 6 a.m.	60
6 a.m. to noon	240
Noon to 6 p.m.	120
6 p.m. to midnight	90

Period	Flow rate (gpm)
Midnight to 6 a.m.	951
6 a.m. to noon	3,804
Noon to 6 p.m.	1,902
6 p.m. to midnight	1,427

The pumping station has three pumps, two of which operate in parallel and the third pump is a standby. Pumping is done at a constant rate and only for a period of 12 hours from 6 a.m. to 6 p.m. Each pump possesses the following characteristics:

Flow (gpm)	0	951	1,427	1,902	2,378	2,853
Head (ft)	54.86	50.29	46.63	41.15	32.00	19.20
Flow (L/s)	0	60	90	120	150	180
Head (m)	180	165	153	135	105	63

The Hazen–Williams coefficient C for all mains is 100. The total equivalent length of main AB is 8,202 ft (2,500 m) and main CD is 29 527 ft (9,000 m).

(a) Determine the volumes needed for the ground tank and elevated reservoir (each separately) to balance supply and demand. (Storage for firefighting and emergencies is not required.)
(b) Determine the minimum standard size needed for the water main AB so that the two operating pumps can deliver the required flow from the ground tank to the elevated reservoir.

8.12 A water pumping station is designed to raise water from a treatment plant ground reservoir at an elevation of 1,968 ft (600 m) to an elevated reservoir located on the opposite side of town at an elevation of 2,165 ft (660 m). The trunk line and the water distribution network connecting the two reservoirs is equivalent to a pipe line ($C = 100$) 16 in. (400 mm) in diameter and 82 020 ft (25 000 m) long.

The pumping station has three pumps, two of which operate in series and the third pump is a standby. Each pump possesses the following characteristics:

Flow (gpm)	0	634	1,268	1,902	2,536	3,170	3,804
Head (ft)	328	321	308	285	259	230	197
Flow (L/s)	0	40	80	120	160	200	240
Head (m)	100	98	94	87	79	70	60

During night hours, when consumption is low, the pumped water is greater than demand and excess water is stored in the elevated reservoir. Considering this mode of operation, determine the flow rate that can be delivered through the system under the following pumping conditions:

(a) Two pumps are operational.
(b) One pump is operational, that is, one pump failed and the standby is under maintenance.

8.13 A water pumping station is designed to raise water from a treatment plant ground reservoir at an elevation of 190 ft (57.91 m) to a downtown elevated reservoir located at an elevation of 415 ft (126.49 m). Water is pumped through a 20-in.(500-mm) pipe ($C = 100$), which is 30,000 ft (9,144 m) long. The pumping station has two pumps, each of which possesses the following characteristics:

Flow (gpm)	15,852	31,704	47,556	63,408	79,260	95,112	110,964
Head (ft)	951	886	804	689	558	410	213
Flow (L/s)	1,000	2,000	3,000	4,000	5,000	6,000	7,000
Head (m)	290	270	245	210	170	125	65
Shutoff head = 300 ft (91.44 m)							

What will be the discharge of the two pumps when operating in series? Also what will be the total head against which the pumps are actually operating?

8.14 An industrial complex utilizes water at a uniform rate of 5,400 gpm (20,439 L/min) during its working hours from 8 a.m. to noon and from 4 p.m. to 8 p.m. This water demand is withdrawn from an elevated water tank located at an elevation of 40 ft (12.19 m) above ground level.

Water is pumped at a uniform rate continuously throughout the day to the tank using two pumps connected in series from a well whose water level is 460 ft (104.21 m) below ground level. The water pipe from the well to the tank is 3,000 ft (914.4 m) long when $C = 100$.

Each pump has the following characteristics:

Flow (gpm)	0	1,000	2,000	3,000	4,000
Head (ft)	350	325	275	175	50
Flow (L/s)	0	63	126	189	252
Head (m)	106.68	99.06	83.82	53.34	15.24

(a) Calculate the water tank volume required to equalize supply and demand.
(b) Determine the size of the pipe delivering water from the well to the water tank.

8.15 A summer resort complex consumes water at the following uniform rates during the indicated periods:

Period	Consumption Rate (gpm)
Midnight to 8 a.m.	5.4
8 a.m. to 4 p.m.	11.4
4 p.m. to midnight	8.2

Period	Consumption Rate (m^3/h)
Midnight to 8 a.m.	20
8 a.m. to 4 p.m.	50
4 p.m. to midnight	30

How large a storage tank would be required to equalize supply and demand for each of the following conditions?

(a) Water is obtained from a nearby city at a uniform rate from 8 a.m. to 4 p.m.
(b) Water is obtained from the same city at a uniform rate over the 24-h period.

8.16 For the water supply of a small town with a daily water requirement of 0.594 MG (2,250 m^3), the construction of a water reservoir has been proposed. The pattern of drawoff is as follows:

Period	% of Daily Demand
7 a.m. to 8 a.m.	30
8 a.m. to 5 p.m.	35
5 p.m. to 6 : 30 p.m.	30
8 : 30 p.m. to 7 a.m.	5

Water is to be supplied to the reservoir at a constant rate for a period of eight hours, 8 a.m. to 4 p.m. Determine the storage capacity of the reservoir needed to balance supply and demand. No storage for fire is required. If pumping is to be done at a constant rate over the 24 h, what will then be the required reservoir size?

8.17 An industrial plant requires 5,000 gpm (316 L/s) of water during its on-shift hours from 6 a.m. to 10 p.m. How large (in gallons or liters) a storage tank would be required to equalize the pumping rate for each of the following conditions:

(a) Water is obtained from a well at a uniform rate over the 24-h period.
(b) Water is obtained from a well during the period from 10 p.m. to 6 a.m., which is the off-peak period of electricity consumption.

8.18 A factory requires 793 gpm (50 L/s) of water during its working hours from 8 a.m. to 4 p.m. How large a storage reservoir would be required to equalize supply and demand for each of the following water supply conditions?

(a) Water is supplied from a well at a uniform rate over the 24-h period.
(b) Water is supplied at a uniform rate from the city network during the off-peak period from 4 p.m. to 8 a.m.

8.19 A residential neighborhood, population 20 000, is supplied with water from an elevated reservoir. The daily water consumption is as follows:

Period	% of Daily Demand
Midnight to 3 a.m.	5
3 a.m. to 6 a.m.	7
6 a.m. to 9 a.m.	13
9 a.m. to noon	20
Noon to 3 p.m.	20
3 p.m. to 6 p.m.	15
6 p.m. to 9 p.m.	10
9 p.m. to midnight	10

The average daily water consumption is 106 gpcd (400 Lpcd) and the maximum daily consumption is 125% of the daily average. Determine the storage volume (in m^3) necessary to balance supply and demand in each of the following cases:

(a) If water is supplied into the reservoir at a constant rate over the 24-h period.
(b) If water is supplied into the reservoir at a constant rate, but only for a period of 12 h from 6 a.m. to 6 p.m.

8.20 A residential complex consists of 60 duplex villas (two housing units in each). Each housing unit can accommodate a maximum of 10 persons. The expected maximum daily water consumption is 106 gal/capita (400 L/capita). Water is supplied from the municipal network at a uniform rate throughout the day and is collected in a ground storage reservoir. Water is then pumped to an elevated reservoir at a constant rate but only for a period of 9 h from 6 a.m. to 3 p.m. Water then flows from the elevated reservoir to the various villas by gravity to satisfy the following daily demands:

Period	% of Daily Demand
Midnight to 3 a.m.	5
3 a.m. to 6 a.m.	5
6 a.m. to 9 a.m.	25
9 a.m. to noon	15
Noon to 3 p.m.	15
3 p.m. to 6 p.m.	10
6 p.m. to 9 p.m.	15
9 p.m. to midnight	10

In order to balance supply and demand, determine:

(a) Required volume of the ground reservoir.
(b) Required volume of the elevated reservoir.

8.21 A small water system in a rural area has a population of 800, and its daily per capita usage is estimated to be 100 gpcd (379 Lpcd). The required fire flow determined by the village engineer is 500 gpm (32 L/s) for a duration of 2 h. There is no particular hazard to the water supply works. Determine the required water storage to be provided for the village using the *Recommended Standards for Water Works, 2007 Edition*. Visit the website or contact the local health department for the latest edition of these Ten-States Standards for water works.

8.22 A city is planning to improve its water supply system (Fig. 8.16). At present the city has a surface water supply reservoir at an elevation of 100 ft (30.48 m) at point A, a pumping station at an elevation of 80 ft (24.38 m) at point B, water storage tank at an elevation of 200 ft (60.96 m) at point C, 500 ft (152.4 m) of suction transmission line between the reservoir and the pumping station, and 10 000 ft (3,048 m) of 10-in. (250-mm) pressure transmission line between the pumping station and the water storage tank. The water at the water storage tank site is treated by UV for disinfection and then discharged to the water distribution system.

A proposal has been made to replace the current pumps with two centrifugal pumps, to construct 10 000 ft (3,048 m) of 16-in. (400-mm) transmission main to parallel the current 10-in. (250-mm) water main, and to provide additional water storage at an elevation of 200 ft (60.96 m) at the current water storage site. The following conditions are assumed: (a) C factor for all pipes = 120, (b) fire flow requirements in the city = 2,000 gpm (126.2 L/s) for 10 h, (c) average daily demand = 3 MGD = 131 L/s, (d) maximum daily demand = 5 MGD = 219 L/s, (e) peak hourly demand = 10 MGD = 438 L/s, (f) pumping station head losses = 500 ft of 16-in. pipe (or 152.4 m of 400 mm pipe) equivalent, and (g) the field-measured total dynamic head delivered by the pump from the raw water supply reservoir to the pump station, then to the water storage tank = 150 ft (45.72 m).

(a) What should the pump rating (gpm or L/min) be for each of the two new pumps?
(b) What should the brake horsepower (BHP or BMP) be for pump selection, and what should the motor horsepower (MHP or MMP) be for motor selection?
(c) Will the two selected pumps be connected in parallel or in series?

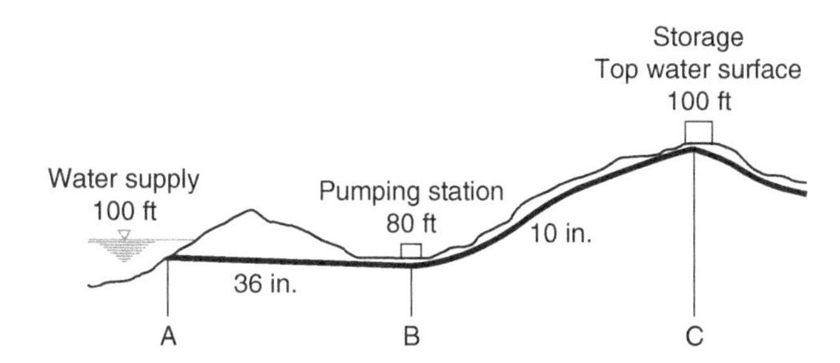

Figure 8.16 Water Supply System for Problem 8.22. Conversion factors: 1 in. = 25.4 mm; 1 ft = 0.3048 m

8.23 Using the same given data in Problem 8.22, explain why the pipe line is designed based on the maximum daily water demand, and determine the total storage capacity of a new water storage tank.

Bibliography

DiGiano, F. A., Zhang, W., and Travaglia, A., Calculation of the Mean Residence Times in Distribution Systems from Tracer Studies and Models, *Aqua, J. Water Supply Res. Technol.*, *54*, 1, 1–14 February 2005.

Fair, G. M., Geyer, J. C., and Okun, D. A., Water and Wastewater Engineering, in *Water Supply and Wastewater Removal*, vol. *vol. 1*, John Wiley & Sons, New York, 1966.

Fair, G. M., Geyer, J. C., and Okun, D. A., *Elements of Water Supply and Wastewater Disposal*, John Wiley & Sons, New York, 1971.

Great Lakes–Upper Mississippi River Board of State and Provincial Public Health and Environmental Managers, *Recommended Standards for Water Works (10 States Standards for Water Works)*, Health Research Inc., Albany, NY, 2022.

Hazen, R., Pumps and Pumping Stations, *J. New England Water Works Assoc.*, *67*, 121, 1953.

Hydraulic Institute, Standards for Rotodynamic (Centrifugal) Pump Applications, ANSI/HI 1.3, 2007.

Institution of Mechanical Engineers, *Centrifugal Pumps: The State of the Art and New Opportunities: IMechE Seminar*, John Wiley & Sons, New York, August 2000.

Miles, A. N., Singei, P. C., Ashley, M. C., Lynberg, M. C., Langlois, R. H., and Nuckols, J. R., Comparisons of Trihalomethanes in Tap Water and Blood, *Environ. Sci. Technol.*, *36*, 8, 1692–1698, April 2002.

Okun, D. A., Distributing Reclaimed Water Through Dual Systems, *J. Am. Water Works Assoc.*, *89*, pp. 62–74 March 1997.

Okun, D. A., Improving Quality and Conserving Resources Using Dual Water Systems, *Water 21*, pp. 47–50 February 2007.

Percival, S. L., Knapp, J. S., Edyvean, R., and Wales, D. S., Biofilm Development on Stainless Steel in Mains Water, *Water Res.*, pp. 243–253, January 1998.

Qasim, S. R., Motley, E. M., and Zhu, G., *Water Works Engineering*, PHI Learning Private Limited, New Delhi, India, 2011. 844 pp.

Ray, M., Flint Water Crisis: Public Health Crisis, Flint, Michigan, United States, in *Britannica*, 2016.

Rich, G. R., *Hydraulic Transients*, McGraw-Hill, New York, 1951.

Rouse, H., *Engineering Hydraulics*, John Wiley & Sons, New York, 1950.

Sharma, A. K., and Swamee, P. K., *Design of Water Supply Pipe Networks*, John Wiley & Sons, New York, February 2008.

Tang, S. L., Yue, D. P. T., and Ku, D. C. C., *Engineering and Costs of Dual Water Supply Systems*, IWA Publishing, London, January 2007.

Tullis, J. P., *Hydraulics of Pipelines: Pumps, Valves, Cavitation, Transients*, John Wiley & Sons, New York, 1989. (Online publication, December 12, 2007.).

Walker, R., *Water Supply, Treatment and Distribution*, Prentice Hall, Inc., Englewood Cliffs, NJ, 1978.

Walski, T. M., Barnard, T. E., Meadows, M. E., and Whitman, B. E., *Computer Applications in Hydraulic Engineering*, 7th ed., Bentley Institute Press–Haestad Methods, 2008.

9

Cross-connection Control

9.1 Introduction

Plumbing *cross-connections*, which are defined as actual or potential connections between a potable and nonpotable water supply, constitute a serious public health hazard. There are numerous, well-documented cases where cross-connections have been responsible for the contamination of drinking water and have resulted in the spread of disease. The problem is a dynamic one because piping systems are continually being installed, altered, or extended.

Control of cross-connections is possible, but only through thorough knowledge and vigilance. Education is essential because even those who are experienced in piping installations fail to recognize cross-connection possibilities and dangers. All municipalities with public water supply systems *should have cross-connection control programs.* Those responsible for institutional or private water supplies should also be familiar with the dangers of cross-connections and should exercise careful surveillance of their systems. The American Water Works Association (AWWA) stated the following in a policy on public water supply matters: *AWWA recognizes water purveyors have the responsibility to supply potable water to their customers. In the exercise of this responsibility, water purveyors or other responsible authorities must implement, administer, and maintain ongoing backflow prevention and cross-connection control programs to protect public water systems from the hazards originating on the premises of their customers and from temporary connections that may impair or alter the water in the public water systems. The return of any water to the public water system after the water has been used for any purposes on the customer's premises or within the customer's piping system is unacceptable and opposed by AWWA. The water purveyor shall assure that effective backflow prevention measures commensurate with the degree of hazard are implemented to ensure continual protection of the water in the public water distribution system. Customers, together with other authorities, are responsible for preventing contamination of the private plumbing system under their control and the associated protection of the public water system. (Reprinted by permission. Copyright © 2009, American Water Works Association.)*

Public health officials have long been concerned about cross-connections and backflow connections in plumbing systems and in public drinking water supply distribution systems. Such cross-connections, which make possible the contamination of potable water, are ever-present dangers. One example of what can happen is an epidemic that occurred in Chicago in 1933. Old, defective, and improperly designed plumbing and fixtures permitted the contamination of drinking water. As a result, 1409 people contracted amebic dysentery, and 98 of them died. This epidemic, and others resulting from contamination introduced into a water supply through improper plumbing, made clear the responsibility of public health officials and water purveyors for exercising control over public water distribution systems and all plumbing systems connected to them. This responsibility includes advising and instructing plumbing installers in the recognition and elimination of cross-connections.

Cross-connections are the links through which it is possible for contaminating materials to enter a potable water supply. The contaminant enters the potable water system when the pressure of the polluted source exceeds the pressure of the potable source. The action may be called *backsiphonage* or *backflow*. Essentially, it is the reversal of a hydraulic gradient, and it can be produced by a variety of circumstances.

One might assume that the steps for detecting and eliminating cross-connections would be elementary and obvious. The reality, however, is that cross-connections can appear in many subtle forms and in unsuspected places. Reversal of pressure in the water may be freakish and unpredictable. The probability of contamination of drinking water through a

Water and Wastewater Engineering: Hydraulics, Hydrology and Management, Volume 1, Fourth Edition.
Lawrence K. Wang, Mu-Hao Sung Wang, and Nazih K. Shammas.

Companion website: www.wiley.com/go/Wang/Waterandwastewater4e

cross-connection occurring within a single plumbing system may seem remote, but considering the multitude of similar systems, the probability is great. Cross-connections exist for these reasons:

1) Plumbing is frequently installed by persons who are unaware of the inherent dangers of cross-connections.
2) Oftentimes, connections are made as a simple matter of convenience without regard to dangerous situations that might be created.
3) Connections are made with reliance on inadequate protection such as a single valve or other mechanical device.

To combat the dangers of cross-connections and backflow connections, education in their recognition and prevention is needed. First, plumbing installers must know that hydraulic and pollution factors may combine to produce a sanitary hazard if a cross-connection is present. Second, they must realize that reliable and simple standard backflow prevention devices and methods are available that can be substituted for convenient but dangerous direct connections. Third, plumbing installers must understand that the hazards resulting from direct connections greatly outweigh the convenience gained from a quick and direct connection.

9.2 Public Health Significance of Cross-connections

Public health officials have long been aware of the threat to public health that cross-connections represent. Because plumbing defects are so frequent and the opportunities for contaminants to invade the public drinking water through cross-connections so general, enteric illnesses caused by drinking water may occur at almost any location and at any time.

The following documented cases of cross-connection problems illustrate and emphasize how cross-connections have compromised water quality and public health.

9.2.1 Human Blood in the Water System

Health department officials cut off the water supply to a funeral home located in a large southern city after it was determined that human blood had contaminated the freshwater supply. The chief plumbing inspector had received a telephone call advising that blood was coming from drinking fountains within the funeral home building. Plumbing and county health department inspectors went to the scene and found evidence that blood had been circulating in the water system within the building. They immediately ordered the building cut off from the water system at the meter. City water and plumbing officials said that they did not think that the blood contamination had spread beyond the building; however, inspectors were sent into the neighborhood to check for possible contamination.

Investigation revealed that the funeral home had been using a hydraulic aspirator to drain fluids from the bodies of human "remains" as part of the embalming process. The aspirator connected directly to the water supply system at a faucet located on a sink in the "preparation" (embalming) room. Water flow through the aspirator created suction that was utilized to draw body fluids through a hose and needle attached to the suction side of the aspirator.

The contamination of the funeral home's potable water supply was caused by a combination of low water pressure in conjunction with the simultaneous use of the aspirator. Instead of the body fluids flowing into the sanitary drain, they were drawn in the opposite direction—into the potable water supply of the funeral home (see Fig. 9.1).

9.2.2 Sodium Hydroxide in the Water Main

A resident of a small town in Alabama jumped in the shower at 5 a.m. one morning, and when he got out his body was covered with tiny blisters. "The more I rubbed it, the worse it got," the 60-year-old resident said. "It looked like someone took a blow torch and singed me."

He and several other residents received medical treatment at the emergency room of the local hospital after the public water system was contaminated with sodium hydroxide, a strong caustic solution.

Other residents claimed that "it [the water] bubbled up and looked like Alka Seltzer. I stuck my hand under the faucet and some blisters came up." One neighbor's head was covered with blisters after she washed her hair and others complained of burned throats or mouths after drinking the water.

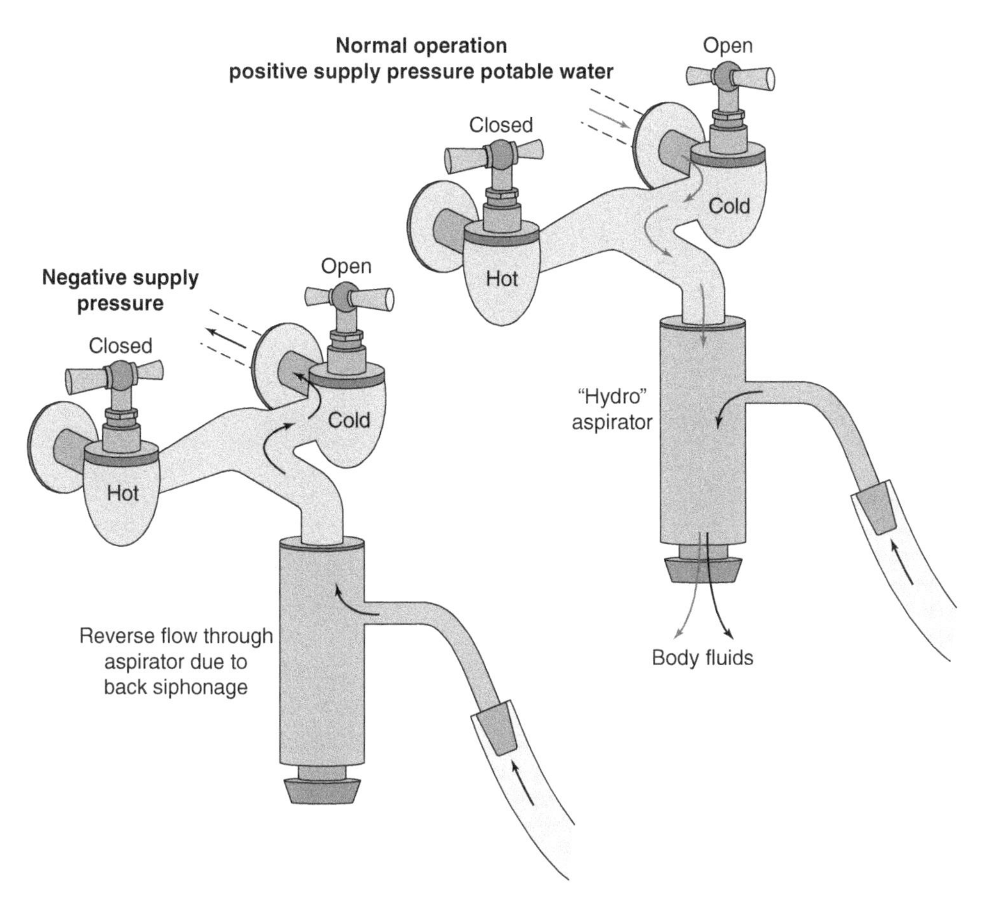

Figure 9.1 Reverse Flow through Aspirator Due to Backsiphonage

The incident began after an 8-in. (200-mm) water main that fed the town broke and was repaired. While repairing the water main, one workman suffered leg burns from a chemical in the water and required medical treatment. Measurements of the pH of the water were as high as 13 in some sections of the pipe.

Investigation into the cause of the problem led to a nearby chemical company that distributes chemicals such as sodium hydroxide as being the possible source of contamination. The sodium hydroxide is brought to the plant in liquid form in bulk tanker trucks, transferred to a holding tank, and then pumped into 55-gal (208-L) drums. When the water main broke, a truck driver was adding the water from the bottom of the tank truck instead of the top, and sodium hydroxide backsiphoned into the water main (see Fig. 9.2).

9.2.3 Heating System Antifreeze in Potable Water

cgBangor, Maine, Water Department employees discovered poisonous antifreeze in a homeowner's heating system and water supply. The incident occurred when they shut off the service line to the home to make repairs. With the flow of water to the house cut off, pressure in the lines in the house dropped and the antifreeze, placed in the heating system to prevent freeze-up of an unused hot water heating system, drained out of the heating system into house water lines and flowed out to the street (see Fig. 9.3). If it had not been noticed, it would have entered the homeowner's drinking water when the water pressure was restored.

9.2.4 Salt Water Pumped into Freshwater Line

A nationally known fast-food restaurant located in the southeastern United States complained to the water department that all of their soft drinks were being rejected by their customers as tasting "salty." This included soda fountain beverages, coffee, orange juice, etc. An investigation revealed that an adjacent water customer complained of salty water occurring simultaneously with the restaurant incident. This second complaint came from a waterfront ship repair facility that was also being

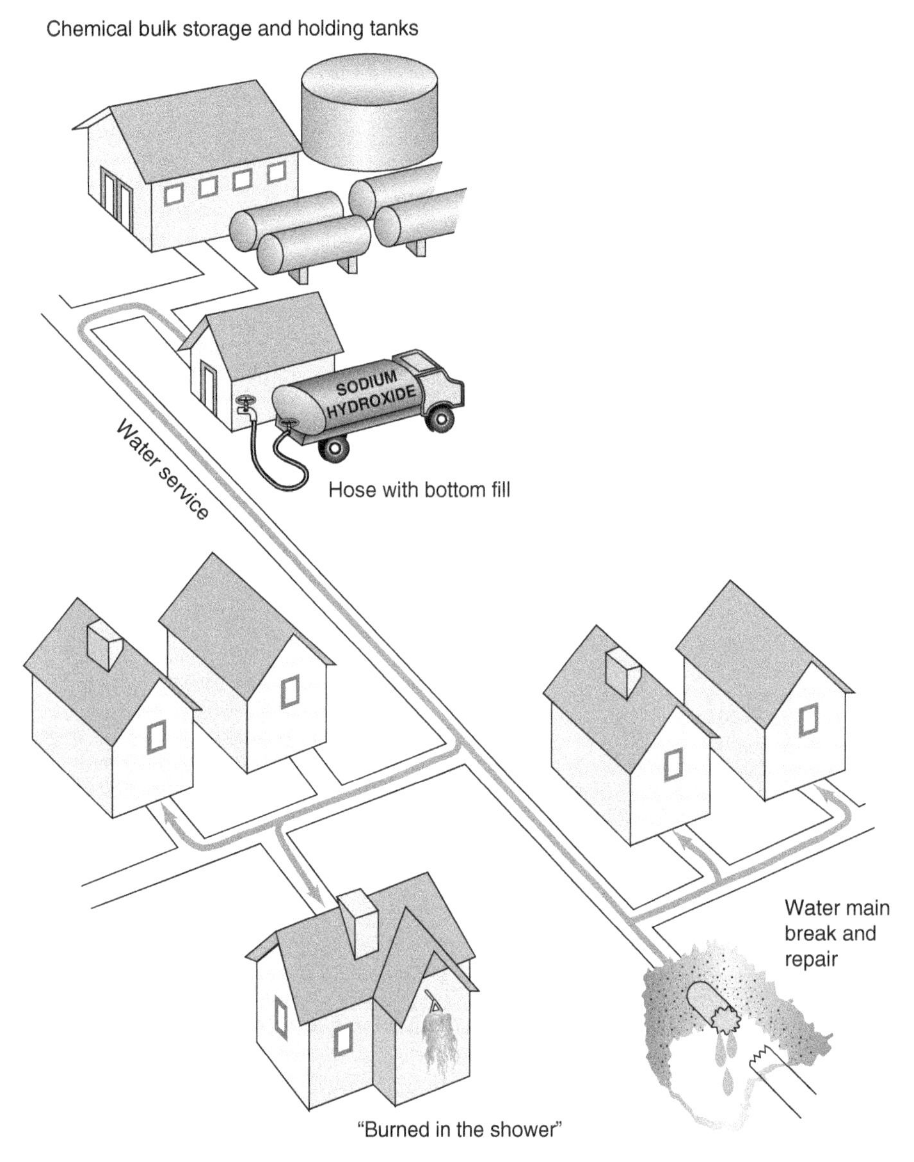

Figure 9.2 Sodium Hydroxide Backsiphoned into the Water Main

served by the same water main lateral. The investigation centered on the ship repair facility and revealed the following (see Fig. 9.4):

- A backflow preventer that had been installed on the service line to the shipyard had frozen and had been replaced with a spool piece sleeve.
- The shipyard fire protection system utilized seawater that was pumped by both electric and diesel-driven pumps.
- The pumps were primed by potable city water.

With the potable priming line left open and the pumps maintaining pressure in the fire lines, raw salt water was pumped through the priming lines, through the spool sleeve piece, to the ship repair facility and the restaurant.

9.2.5 Paraquat in the Water System

"Yellow gushy stuff" poured from some of the faucets in a small town in Maryland, and the state of Maryland placed a ban on drinking the water supply. Residents were warned not to use the water for cooking, bathing, drinking, or any other purpose except for flushing toilets.

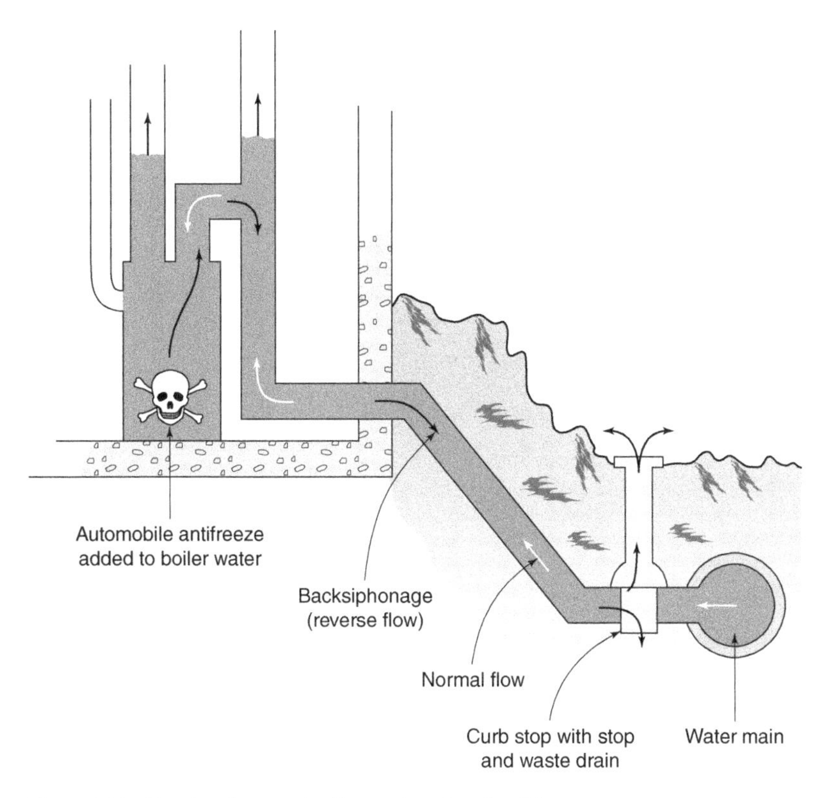

Figure 9.3 Heating System Antifreeze into Potable Water

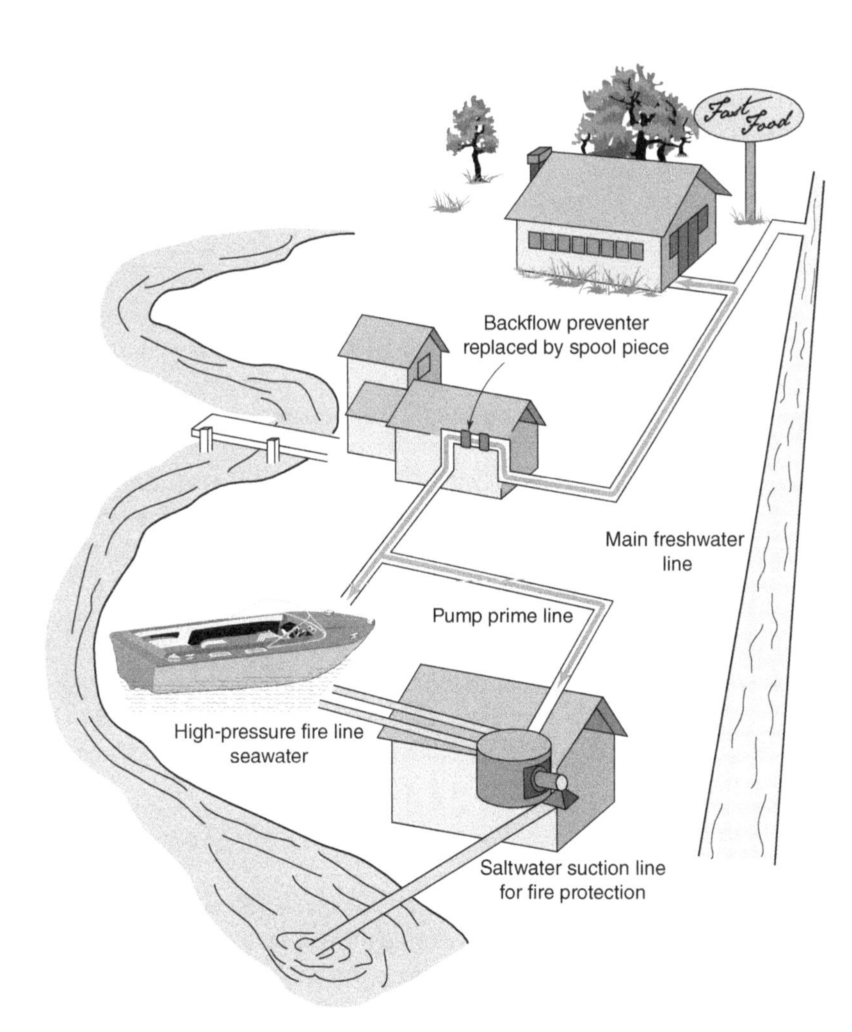

Figure 9.4 Salt Water Pumped into Freshwater Line

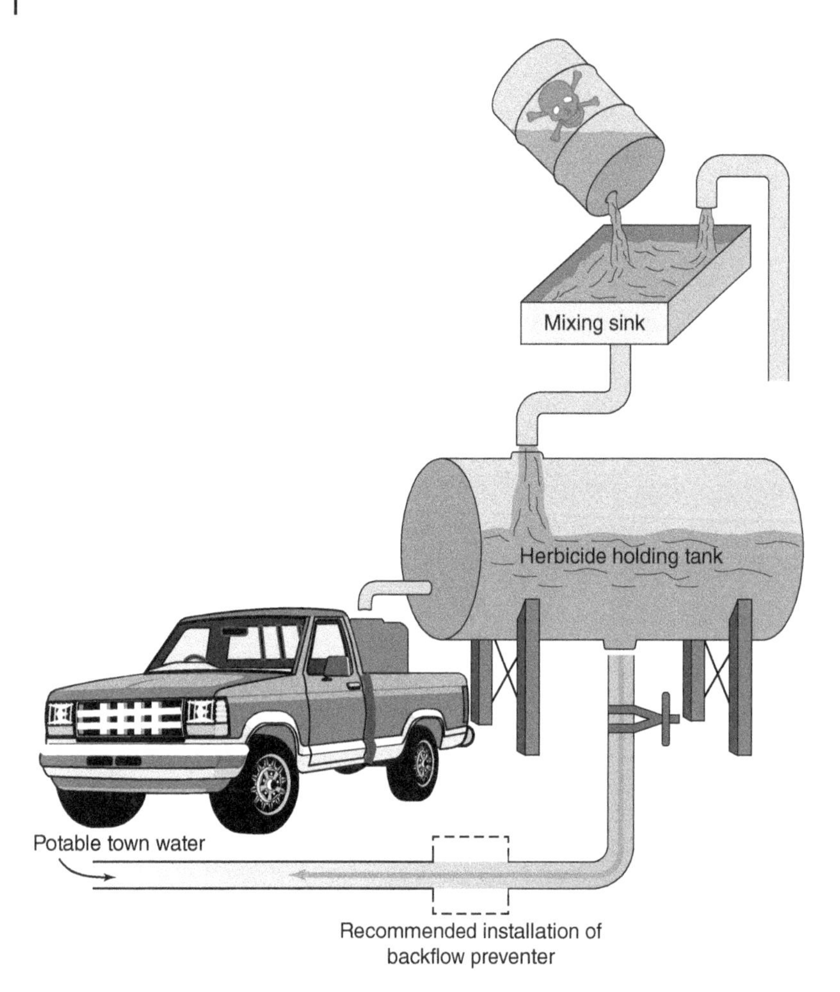

Figure 9.5 Paraquat in the Water System

The incident drew widespread attention and made the local newspapers. In addition, it was the lead story on the ABC news affiliate in Washington, D.C., and virtually all the Washington/Baltimore newspapers that evening. The news media contended that lethal pesticides may have contaminated the water supply and among the contaminants was paraquat, a powerful agricultural herbicide.

The investigation disclosed that the water pressure in the town water mains was temporarily reduced due to a water pump failure in the town water supply pumping system. Coincidentally, a gate valve between an herbicide chemical mixing tank and the town water supply piping had been left open. A lethal cross-connection had been created that permitted the herbicide to flow into the potable water supply system (see Fig. 9.5). On restoration of water pressure, the herbicides flowed into the many faucets and outlets hooked up to the town water distribution system.

This cross-connection, which could have been avoided by the installation of a backflow preventer, created a needless and costly event that fortunately did not result in serious illness or loss of life. Door-to-door public notification, extensive flushing, water sample analysis, and emergency arrangements to provide temporary potable water from tanker trucks all incurred expenses on top of the original problem, all of which would have been unnecessary had the proper precautions been taken.

9.2.6 Propane Gas in the Water Mains

Hundreds of people were evacuated from their homes and businesses in a town in Connecticut as a result of propane entering the city water supply system. Fires were reported in two homes and the town water supply was contaminated. One five-room residence was gutted by a blaze resulting from propane gas "bubbling and hissing" from a bathroom toilet and in another home, a washing machine explosion blew a woman against a wall. Residents throughout the area reported hissing

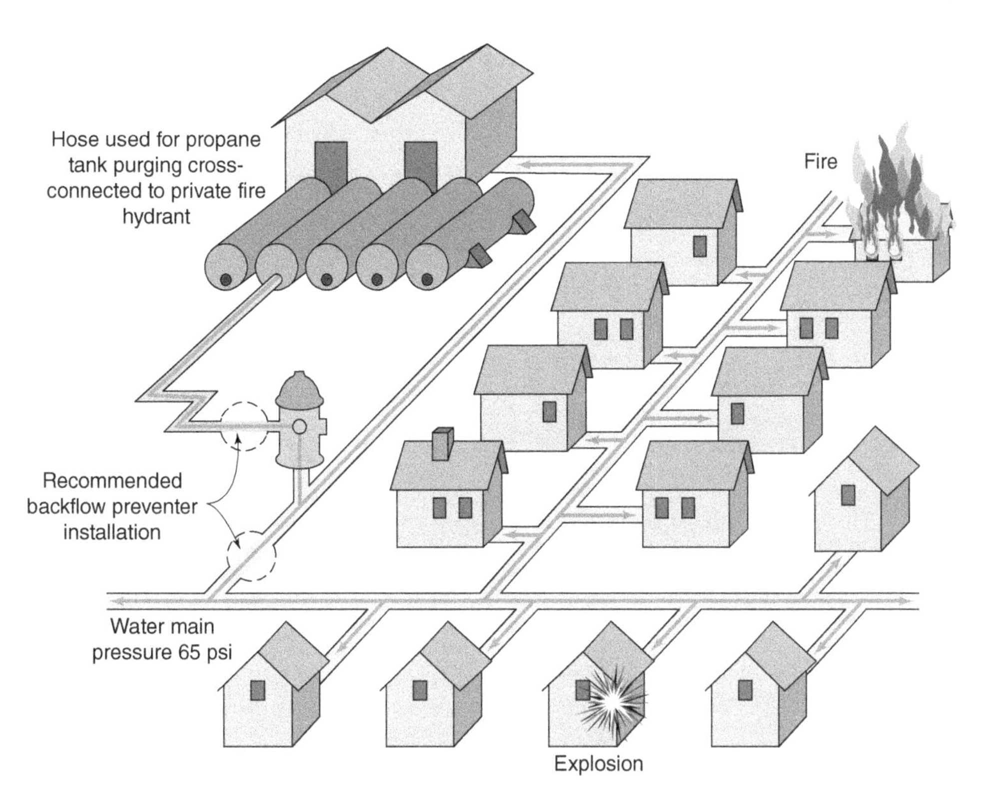

Figure 9.6 Propane Gas in the Water Mains. Conversion factor: 1 psi = 6.94 kPa

and bubbling noises, coming from washing machines, sinks, and toilets. Faucets sputtered out small streams of water mixed with gas, and residents in the area were asked to evacuate their homes.

This near-disaster occurred in one 30 000-gal (113550-L) capacity liquid propane tank when the gas company initiated immediate repair procedures. To start the repair, the tank was "purged" of residual propane by using water from one of two private fire hydrants located on the property. Water purging is the preferred method of purging over the use of carbon dioxide since it is more positive and will float out any sludge as well as any gas vapors. The "purging" consisted of hooking up a hose to one of the private fire hydrants located on the property and initiating flushing procedures (see Fig. 9.6).

Because the vapor pressure of the propane residual in the tank was 85–90 psi (590–625 kPa), and the water pressure was only 65–70 psi (451–486 kPa), propane gas back-pressure backflowed into the water main. It was estimated that the gas flowed into the water mains for about 20 min and that about 2000 ft^3 (57 m^3) of gas was involved. This was approximately enough gas to fill 1 mile of an 8-in. (1.61 km of a 200-mm) water main.

9.2.7 Chlordane and Heptachlor at a Housing Authority

The services to 75 apartments housing approximately 300 people were contaminated with chlordane and heptachlor in a city in Pennsylvania. The insecticides entered the water supply system while an exterminating company was applying them as a preventive measure against termites. While the pesticide contractor was mixing the chemicals in a tank truck with water from a garden hose coming from one of the apartments, a workman was cutting into a 6-in. (150-mm) main line to install a gate valve. The end of the garden hose was submerged in the tank containing the pesticides, and at the same time, the water to the area was shut off and the lines were being drained prior to the installation of the gate valve. When the workman cut the 6-in.(150-mm) line, water started to drain out of the cut, thereby setting up a backsiphonage condition. As a result, the chemicals were siphoned out of the truck, through the garden hose, and into the system, contaminating the 75 apartments (see Fig. 9.7).

Repeated efforts to clean and flush the lines were not satisfactory, and a decision was eventually made to replace the water line and all the plumbing that was affected. There were no reports of illness, but residents of the housing authority were told not to use any tap water for any purpose and they were given water that was trucked into the area by volunteer fire department personnel. They were without their normal water supply for 27 days.

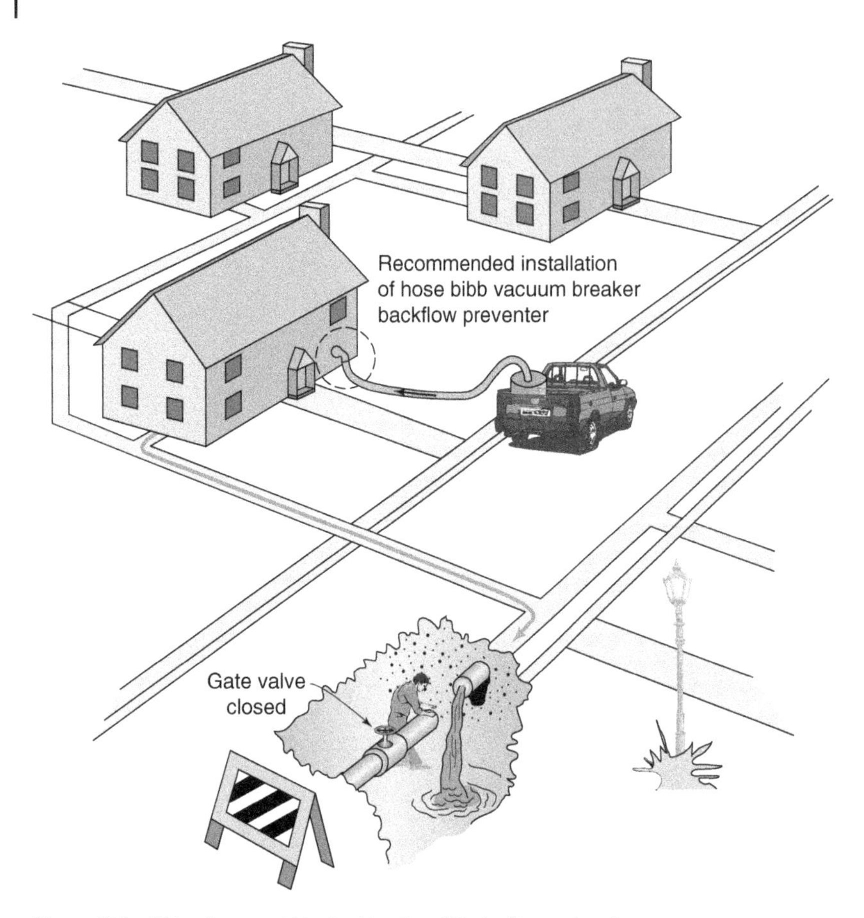

Figure 9.7 Chlordane and Heptachlor in a City in Pennsylvania

9.2.8 Boiler Water Entered High School Drinking Water

A high school in New Mexico was closed for several days when a home economics teacher noticed the water in the potable system was yellow. City chemists determined that samples taken contained levels of chromium as high as 700 mg/L, "astronomically higher than the accepted levels of 0.05 mg/L." The head chemist said that it was miraculous that no one was seriously injured or killed by the high levels of chromium. The chemical was identified as sodium dichromate, a toxic form of chromium used in heating system boilers to inhibit corrosion of the metal parts.

No students or faculty were known to have consumed any of the water; however, area, physicians and hospitals advised that if anyone had consumed those high levels of chromium, the symptoms would be nausea, diarrhea, and burning of the mouth and throat. Fortunately, the home economics teacher, who first saw the discolored water before school started, immediately covered all water fountains with towels so that no one would drink the water.

Investigation disclosed that chromium used in the heating system boilers to inhibit corrosion of metal parts entered the potable water supply system as a result of backflow through leaking check valves on the boiler feed lines (see Fig. 9.8).

9.2.9 Car Wash Water in the Street Water Main

This car wash cross-connection and backpressure incident, which occurred in the state of Washington, resulted in backflow chemical contamination of approximately 100 square blocks of water mains. Prompt response by the water department prevented a potentially hazardous water quality degradation problem without a recorded case of illness.

Numerous complaints of gray-green and "slippery" water were received by the water department coming from the same general area of town. A sample brought to the water department by a customer confirmed the reported problem and preliminary analysis indicated contamination with what appeared to be a detergent solution. While emergency crews initiated flushing operations, further investigation within the contaminated area signaled that the problem was probably caused by a car wash, or laundry, based on the soapy nature of the contaminant. The source was quickly narrowed down to a car wash,

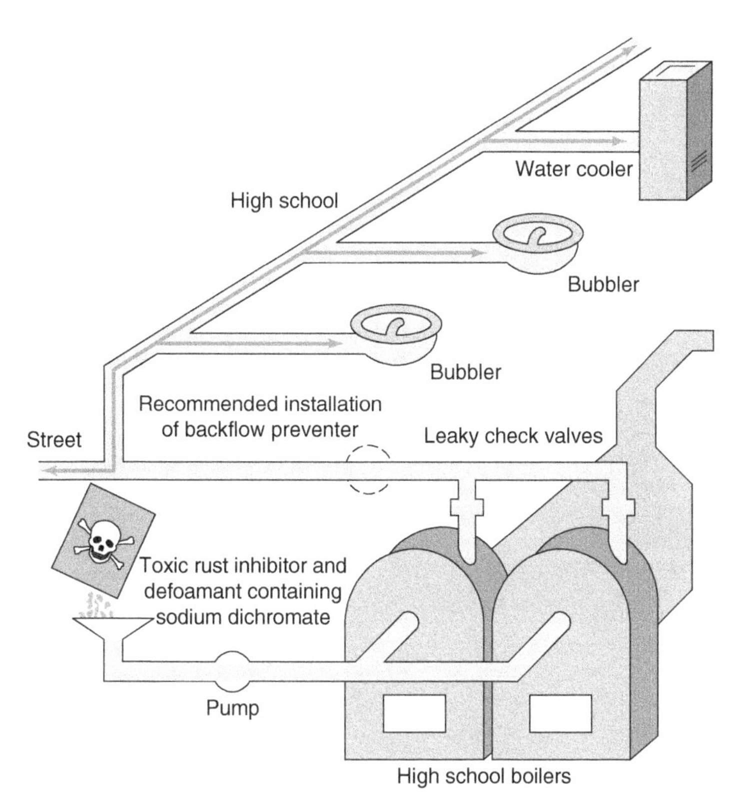

Figure 9.8 Boiler Water Enters High School Drinking Water

and the proprietor was extremely cooperative in admitting to the problem and explaining how it had occurred. The circumstances leading up to the incident were as follows:

- On Saturday, February 10, a high-pressure pump broke down at the car wash. This pump recycled reclaimed wash and rinse water and pumped it to the initial scrubbers of the car wash. No potable plumbing connection is normally made to the car wash's scrubber system.
- After the pump broke down, the car wash owner was able to continue operation by connecting a 2-in. (50-mm) hose section temporarily between the potable supply within the car wash and the scrubber cycle piping.
- On Monday, February 12, the owner repaired the high-pressure pump and resumed normal car wash operations. The 2-in. (50-mm) hose connection (cross-connection) was not removed!
- Because of the cross-connection, the newly repaired high-pressure pump promptly pumped a large quantity of the reclaimed wash/rinse water out of the car wash and into a 12-in. (300-mm) water main in the street. This in turn was delivered to the many residences and commercial establishments connected to the water main (see Fig. 9.9).

Within 24 hours of the incident, the owner of the car wash had installed a 2-in. (50-mm) reduced pressure principle backflow preventer on his water service and all car wash establishments in Seattle that used a wash water reclaim system were notified of the state requirement for backflow prevention.

9.2.10 Health Problems Due to Cross-connection in an Office Building

A cross-connection incident that occurred in a modern seven-story office building located in a large city in New Hampshire resulted in numerous cases of nausea, diarrhea, loss of work time, and employee complaints as to the poor quality of the water.

On a Saturday, a large fire occurred two blocks away from a seven-story office building in a large New Hampshire city. On Sunday, the maintenance crew of the office building arrived to perform the weekly cleaning, and after drinking the water from the drinking fountains, and sampling the coffee from the coffee machines, they noticed that the water smelled rubbery and had a strong bitter taste. On notifying the Manchester Water Company, water samples were taken and preliminary

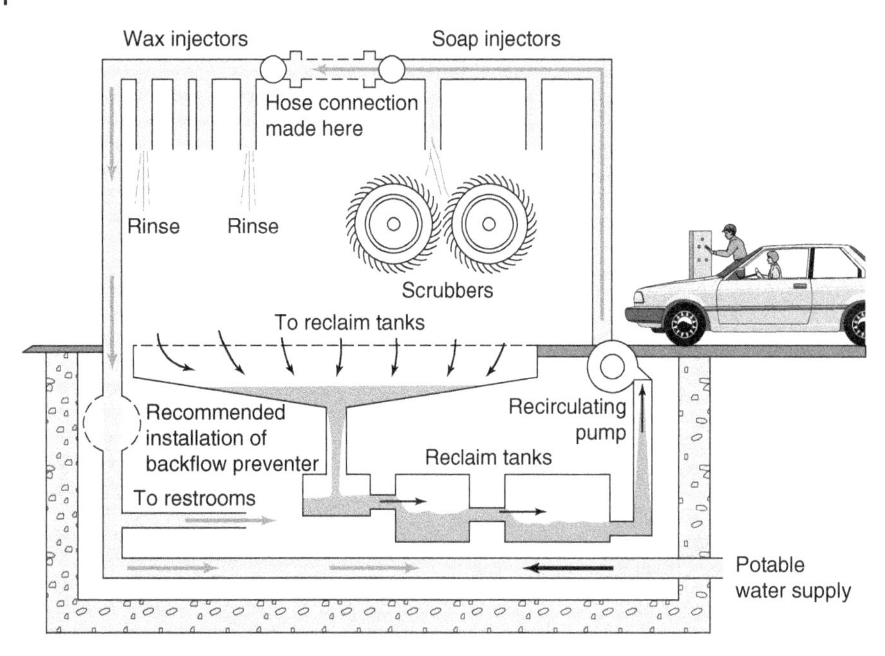

Figure 9.9 Car Wash Water Enters the Street Water Main

analysis disclosed that the contaminants found were not the typical contaminants associated with fire line disturbances. Investigating teams suspected that either the nearby fire could have siphoned contaminants from adjacent buildings into the water mains, or the contamination could have been caused by a plumbing deficiency occurring within the seven-story building itself.

The pH levels of the building water indicated that an injection of chemicals had probably taken place within the seven-story building. Tracing of the water lines within the building pinpointed a 10 000-gal (37850-L) hot-water storage tank that was used for heat storage in the solar heating system. It did not have any backflow protection on the makeup supply line. As the storage tank pressure increased above the supply pressure, as a result of thermal expansion, the potential for backpressure backflow was present. Normally, this would not occur because a boost pump in the supply line would maintain the supply pressure to the storage tank at a greater pressure than the highest tank pressure. The addition of rust-inhibiting chemicals to this tank greatly increased the degree of hazard of the liquid. Unfortunately, at the same time that the fire took place, the pressure in the water mains was reduced to a dangerously low pressure and the low-pressure cutoff switches simultaneously shut off the storage tank booster pumps. This combination allowed the boiler water, together with its chemical contaminants, the opportunity to enter the potable water supply within the building (see Fig. 9.10). When normal pressure was reestablished in the water mains, the booster pumps kicked in, and the contaminated water was delivered throughout the building.

9.3 Theory of Backflow and Backsiphonage

A cross-connection is the link or channel connecting a source of pollution with a potable water supply. The polluting substance, in most cases a liquid, tends to enter the potable supply if the net force acting on the liquid acts in the direction of the potable supply. Two factors are therefore essential for backflow. First, there must be a link between the two systems. Second, the resultant force must be toward the potable supply.

An understanding of the principles of backflow and backsiphonage requires an understanding of the terms frequently used in their discussion. Force, unless completely resisted, will produce motion. Weight is a type of force resulting from gravitational attraction. *Pressure* (P) is a force-per-unit area, such as lb./in.^{2} (psi) or kN/m^2 (kPa). *Atmospheric pressure* is the pressure exerted by the weight of the atmosphere above Earth.

Pressure may be referred to using an absolute scale, lb./in.^{2} absolute (psia), or kN/m^2 absolute (kPa absolute). Pressure may also be referred to using a gauge scale, lb./in.^{2} gauge (psig) or kN/m^2 gauge (kPa gauge). Absolute pressure and gauge

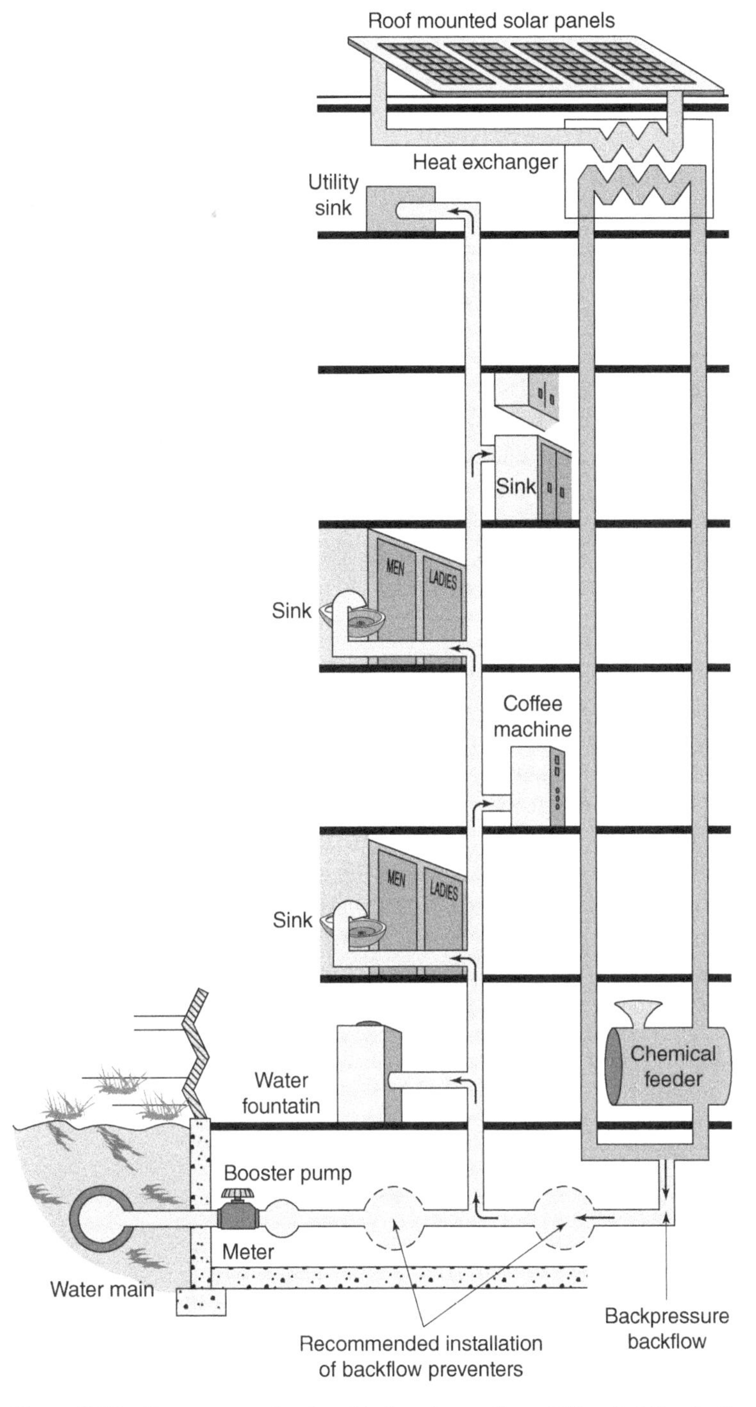

Figure 9.10 Cross-connection in a Modern Seven-Story Office Building in New Hampshire, USA

pressure are related. Absolute pressure is equal to the gauge pressure plus the atmospheric pressure. At sea level, the atmospheric pressure is 14.7 psia using the U.S. customary units. Thus,

$$P_{\text{absolute}} = P_{\text{gauge}} + 14.7\ \text{psi} \quad \text{(U.S. Customary Units)} \tag{9.1}$$

or

$$P_{\text{gauge}} = P_{\text{absolute}} - 14.7\ \text{psi} \quad \text{(U.S. Customary Units)} \tag{9.2}$$

In the SI Units where $P_{absolute}$ and P_{gauge} are in kPa (1 kPa = 1000 Pa = 1000 N/m^2 = 1 kN/m^2) their relationship can be given as follows:

$$P_{absolute} = P_{gauge} + 102\ \text{kPa} \quad \text{(SI Units)} \tag{9.3}$$

or

$$P_{gauge} = P_{absolute} - 102\ \text{kPa} \quad \text{(SI Units)} \tag{9.4}$$

In essence then, absolute pressure is the total pressure. Gauge pressure is simply the pressure read on a gauge. If there is no pressure on the gauge other than atmospheric, the gauge would read zero. Then the absolute pressure would be equal to 14.7 psi (102 kPa), which is the atmospheric pressure.

The term *vacuum* indicates that the absolute pressure is less than the atmospheric pressure and that the gauge pressure is negative. A complete or total vacuum would mean a pressure of 0 psia or −14.7 psig (−102 kPa gauge). Because it is impossible to produce a total vacuum, the term *vacuum*, as used in the text, will mean all degrees of partial vacuum. In a partial vacuum, the pressure would range from slightly less than 14.7 psia (0 psig) to slightly greater than 0 psia (−14.7 psig).

Backsiphonage results in fluid flow in an undesirable or reverse direction. It is caused by atmospheric pressure exerted on a pollutant liquid such that it forces the pollutant toward a potable water supply system that is under a vacuum. Backflow, although literally meaning any type of reversed flow, refers to the flow produced by the differential pressure existing between two systems, both of which are at pressures greater than atmospheric.

9.3.1 Water Pressure

For an understanding of the nature of pressure and its relationship to water depth, consider the pressure exerted on the base of a cubic foot of water at sea level (see Fig. 9.11a). The average weight of a cubic foot of water (62.4 lb./ft^3) will exert a pressure of 62.4 lb./ft^2 (430 kPa) gauge. The base may be subdivided into 144 square inches with each subdivision being subjected to a pressure of 0.433 psig (P_{gauge} = 3 kPa).

Suppose another 1 ft^3 (0.0283 m^3) of water was placed directly on top of the first (see Fig. 9.11b). The pressure on the top surface of the first cube, which was originally atmospheric, or 0 psig, would now be 0.433 psig (P_{gauge} = 3 kPa) as a result of the superimposed cubic foot of water. The pressure of the base of the first cube would also be increased by the same amount of 0.866 psig (P_{gauge} = 6 kPa), or two times the original pressure.

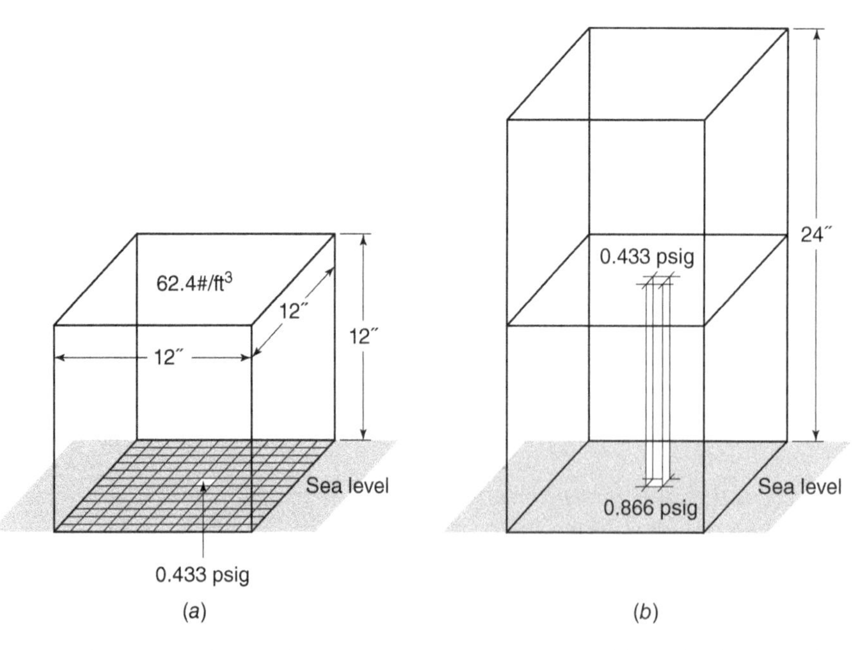

Figure 9.11 Pressures Exerted by (a) 1 ft (0.3048 m) of Water and (b) 2 ft (0.6096 m) of Water at Sea Level. Conversion factors: 1 # = 1b = 0.454 kg; 1″ = 1 in. = 25.4 mm; 1 psig = 6.94 kPa gauge pressure

If this process were repeated with a third cubic foot of water, the pressures at the base of each cube would be 1.299, 0.866, and 0.433 psig (P_{gauge} = 9, 6 and 3 kPa), respectively.

It is evident that pressure varies with depth below a free water surface; in general, each foot (0.3048 m) of elevation change, within a liquid, changes the pressure by an amount equal to the weight-per-unit area of 1 ft (0.3048 m) of the liquid. The rate of increase for water is 0.433 psi/ft (9.84 kPa/m) of depth.

Frequently water pressure is referred to using the terms *pressure head* or just *head*, and is expressed in units of feet of water. One foot (0.3048 m) of head would be equivalent to the pressure produced at the base of a column of water 1 ft (0.3048 m) in depth. One foot (0.3048 m) of head or 1 ft (0.3048 m) of water is equal to 0.433 psig (P_{gauge} = 3 kPa). One hundred feet (30.48 m) of head is equal to 43.3 psig (P_{gauge} = 300 kPa).

9.3.2 Siphon Theory

Figure 9.12a depicts the atmospheric pressure on a water surface at sea level. An open tube is inserted vertically into the water; atmospheric pressure, which is 14.7 psia ($P_{absoute}$ = 102 kPa), acts equally on the surface of the water within the tube and on the outside of the tube.

If, as shown in Fig. 9.12b, the tube is capped and a vacuum pump is used to evacuate all the air from the sealed tube, a vacuum with a pressure of 0 psia ($P_{absoute}$ = 0 kPa) is created within the tube. Because the pressure at any point in a static fluid is dependent on the height of that point above a reference line, such as sea level, it follows that the pressure within the tube at sea level must still be 14.7 psia ($P_{absolute}$ = 102 kPa). This is equivalent to the pressure at the base of a column of water 33.9 ft (10.3 m) high. With the column open at the base, water would rise to fill the column to a depth of 33.9 ft (10.3 m).

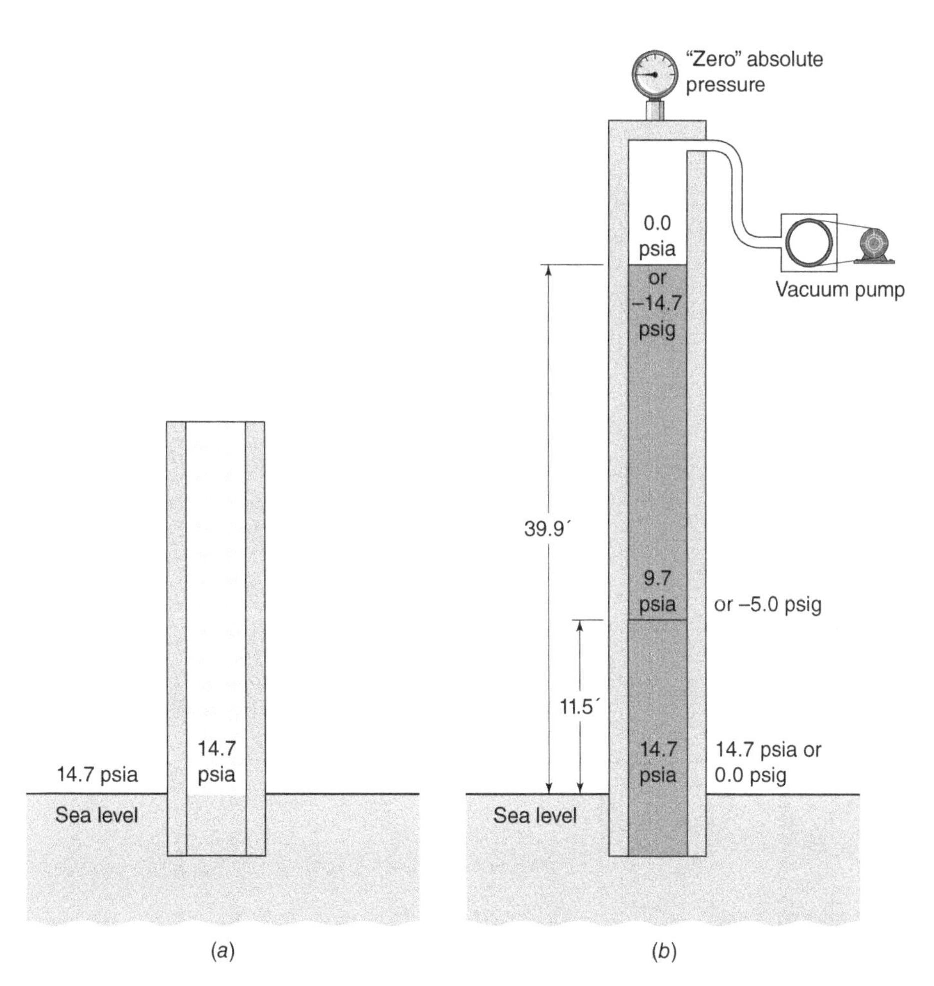

Figure 9.12 (a) Pressure on the Free Surface of a Liquid at Sea Level; (b) Effect of Evacuating Air from a Column. Conversion factors: 1′ = 1 ft = 0.3048 m; 1 psia = 6.94 kPa absolute pressure; 1 psig = 6.94 kPa gauge pressure

In other words, the weight of the atmosphere at sea level exactly balances the weight of a column of water 33.9 ft (10.3 m) in height. The absolute pressure within the column of water in Fig. 9.12b at a height of 11.5 ft (3.5 m) is equal to 9.7 psia ($P_{absolute}$ = 67.3 kPa). This is a partial vacuum with an equivalent gauge pressure of −5.0 psig (P_{gauge} = − 34.7 kPa)

As a practical example, assume the water pressure at a closed faucet on the top of a 100-ft–(30.48-m) high building to be 20 psig (P_{gauge} = 138.8 kpa); the pressure on the ground floor would then be 63.3 psig (P_{gauge} = 439.3 kPa). If the pressure at the ground were to drop suddenly to 33.3 psig (P_{gauge} = 231.1 kPa) due to a heavy fire demand in the area, the pressure at the top would be reduced to −10 psig (P_{gauge} = − 69.4 kPa). If the building water system were airtight, the water would remain at the level of the faucet because of the partial vacuum created by the drop in pressure. If the faucet were opened, however, the vacuum would be broken and the water level would drop to a height of 77 ft (23.47 m) above the ground. Thus, the atmosphere was supporting a column of water 23 ft (7 m) high.

Figure 9.13a is a diagram of an inverted U-tube that has been filled with water and placed in two open containers at sea level. If the open containers are placed so that the liquid levels in each container are at the same height, a static state will exist, and the pressure at any specified level in either leg of the U-tube will be the same. The equilibrium condition is altered by raising one of the containers so that the liquid level in one container is 5 ft (1.52 m) above the level of the other (see Fig. 9.13b). Because both containers are open to the atmosphere, the pressure on the liquid surfaces in each container will remain at 14.7 psia ($P_{absolute}$ = 102 kPa).

If it is assumed that a static state exists, momentarily, within the system shown in Fig. 9.13b, the pressure in the left tube at any height above the free surface in the left container can be calculated. The pressure at the corresponding level in the right tube above the free surface in the right container can also be calculated.

As shown in Fig. 9.13b, the pressure at all levels in the left tube would be less than at corresponding levels in the right tube. In this case, a static condition cannot exist because fluid will flow from the higher pressure to the lower pressure; the flow would be from the right tank to the left tank. This arrangement is referred to as a *siphon*. The crest of a siphon cannot be higher than 33.9 ft (10.3 m) above the upper liquid level because the atmosphere cannot support a column of water greater in height than 33.9 ft (10.3 m).

Figure 9.14 illustrates how this siphon principle can be hazardous in a plumbing system. If the supply valve is closed, the pressure in the line supplying the faucet is less than the pressure in the supply line to the bathtub. Flow will occur, therefore, through siphonage, from the bathtub to the open faucet. This siphon action has been produced by reduced pressures resulting from a difference in the water levels at two separated points within a continuous fluid system.

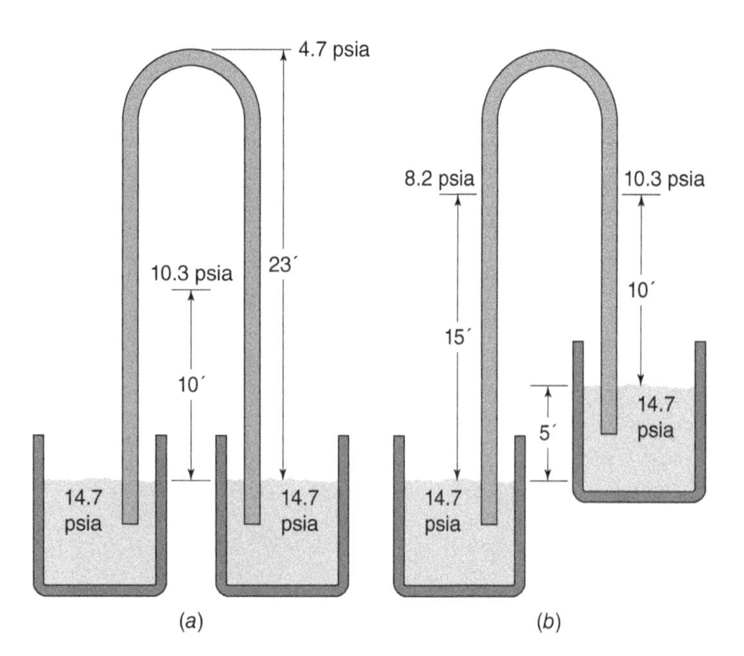

Figure 9.13 Pressure Relationships in (a) Continuous Fluid System at the Same Elevation; (b) at Different Elevations. Conversion factors: 1′ = 1 ft = 0.3048 m; 1 psia = 6.94 kPa absolute pressure

Reduced pressure may also be created within a fluid system as a result of fluid motion. One of the basic principles of fluid mechanics is the *principle of conservation of energy*. Based on this principle, it can be shown that as a fluid accelerates (velocity head increases), as shown in Fig. 9.15 and the following expression, the pressure head (P/γ) is reduced to maintain the same total head:

$$H = z + \frac{P}{\gamma} + \frac{v^2}{2g} \tag{9.5}$$

where

H = total head, ft (m)
Z = elevation, ft (m)
$\frac{P}{\gamma}$ = pressure head, ft (m)
$\frac{v^2}{2g}$ = velocity head, ft (m).

Conversely, it can be shown that as water flows through a constriction ($A_2 < A_1$) such as a converging section of pipe, the velocity of the water increases ($v_2 < v_1$):

$$Q = v_1 A_1 = v_2 A_2 \tag{9.6}$$

where

Q = flow rate, ft^3/s (m^3/s)
v_1 = water velocity at section 1, ft/s (m/s)
v_2 = water velocity at section 2, ft/s (m/s)
A_1 = area of section 1, ft^2 (m^2)
A_2 = area of section 2, ft^2 (m^2).

As a result, the pressure is reduced. Under such conditions, *negative pressures* can develop in a pipe. The simple aspirator is based on this principle. If this point of reduced pressure is linked to a source of pollution, backsiphonage of the pollutant can occur.

One of the common occurrences of dynamically reduced pipe pressures is found on the suction side of a pump. In many cases similar to the one illustrated in Fig. 9.16, the line supplying the booster pump is undersized or does not have sufficient pressure to deliver water at the rate at which the pump normally operates. The rate of flow in the pipe may be increased by a further reduction in pressure at the pump intake. This often results in the creation of negative pressure at the pump intake. This negative pressure may become low enough in some cases to cause vaporization of the water in the line. Actually, in Fig. 9.16 illustration, flow from the source of pollution would occur when pressure on the suction side of the pump is less than the pressure of the pollution source, but this is backflow, which will be discussed below.

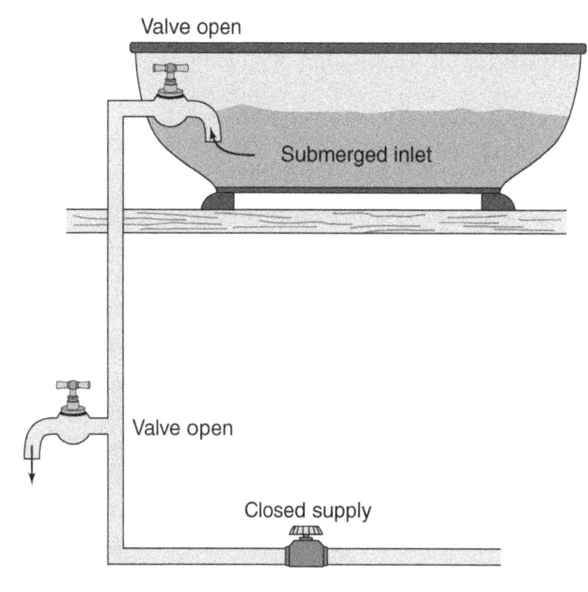

Figure 9.14 Backsiphonage in a Plumbing System

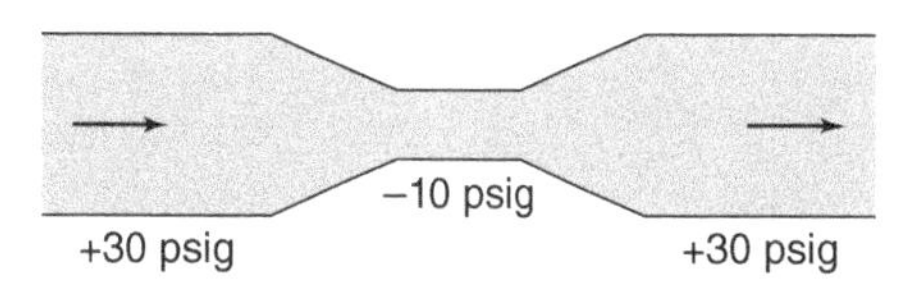

Figure 9.15 Negative Pressure Created by Constricted Flow. Conversion factor: 1 psig = 6.94 kPa gauge pressure

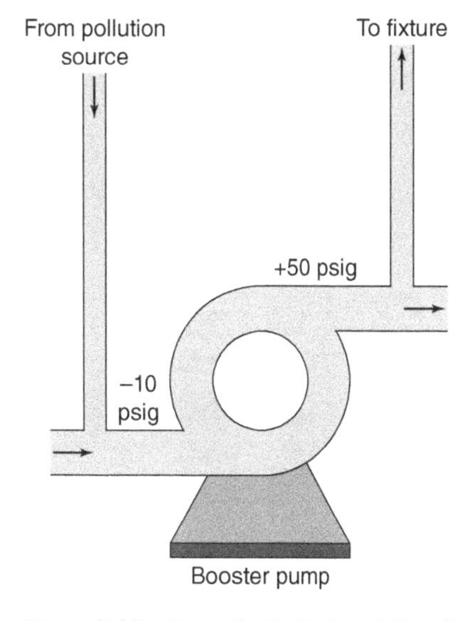

Figure 9.16 Dynamically Reduced Pipe Pressure. Conversion factor: 1 psig = 6.94 kPa gauge pressure

The preceding discussion has described some of the means by which negative pressures may be created and which frequently occur to produce backsiphonage. In addition to the negative pressure or reversed force necessary to cause backsiphonage and backflow, there must also be the cross-connection or connecting link between the potable water supply and the source of pollution. Two basic types of connections are used in piping systems:

1) The solid pipe with a valved connection
2) The submerged inlet

Figure 9.17a and b illustrate solid connections. This type of connection is often installed where it is necessary to lay an auxiliary piping system from the potable source. It is a direct connection of one pipe to another pipe or receptacle. Solid pipe connections are often made to continuous or intermittent waste lines where it is assumed that the flow will be in one direction only. An example of this would be used cooling water from a water jacket or condenser as shown in Fig. 9.17b. This type of connection is usually detectable, but creating concern on the part of the installer about the possibility of reversed flow is often more difficult.

Submerged inlets are found on many common plumbing fixtures and are sometimes necessary features of the fixtures if they are to function properly. Examples of this type of design are siphon-jet urinals or water closets, flushing rim slop sinks, and dental cuspidors. Old-style bathtubs and lavatories had supply inlets below the flood-level rims, but modern sanitary design has minimized or eliminated this hazard in new fixtures. Chemical and industrial process vats sometimes have submerged inlets where the water pressure is used as an aid in diffusion, dispersion, and agitation of the vat contents. Even though the supply pipe may come from the floor above the vat, backsiphonage can occur because the siphon action can raise a liquid such as water almost 34 ft (10.4 m). Some submerged inlets that are difficult to control are those that are not apparent until a significant change in water level occurs or where a supply may be conveniently extended below the liquid surface by means of a hose or auxiliary piping. A submerged inlet may be created in numerous ways, and its detection in some of these subtle forms may be difficult.

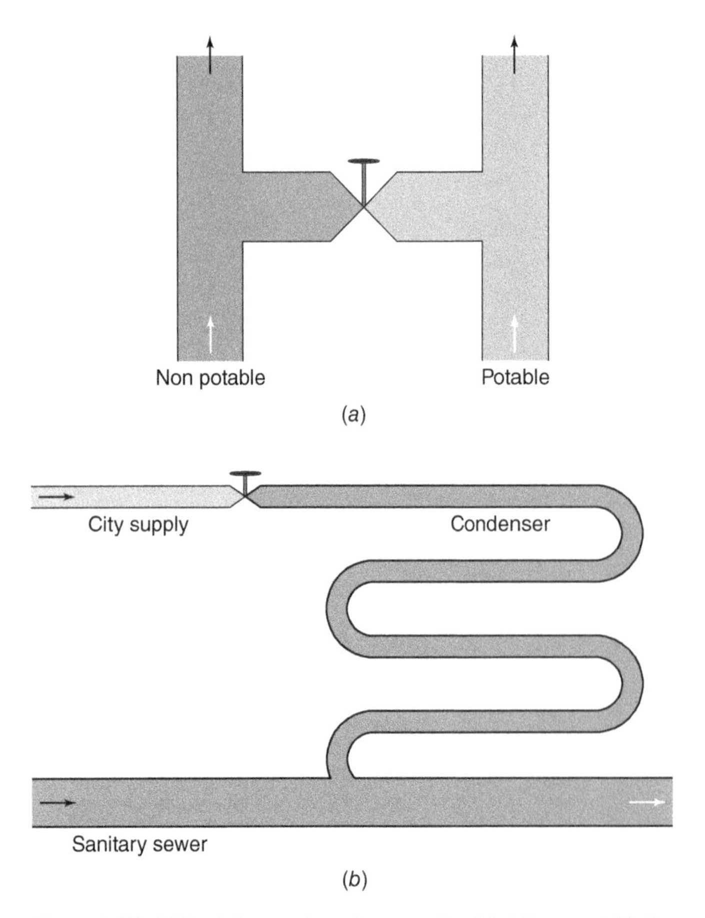

Figure 9.17 Valved Connections Between Potable Water and (a) Nonpotable Fluid and (b) Sanitary Sewer

9.3.3 Backflow

Backflow refers to reversed flow due to backpressure other than siphonic action. Any interconnected fluid systems in which the pressure of one exceeds the pressure of the other may have flow from one to the other as a result of the pressure differential. The flow will occur from the zone of higher pressure to the zone of lower pressure. This type of backflow is of concern in buildings where two or more piping systems are maintained. The potable water supply is usually under pressure directly from the city water main. Occasionally, a booster pump is used. The auxiliary system is often pressurized by a centrifugal pump, although backpressure may be caused by gas or steam pressure from a boiler. A reversal in differential pressure may occur when the pressure in the potable system drops, for some reason, to a pressure lower than that in the system to which the potable water is connected.

The most positive method of avoiding this type of backflow is the total or complete separation of the two systems. Other methods used involve the installation of mechanical devices. All methods require routine inspection and maintenance.

Dual piping systems are often installed for extra protection in the event of an emergency or possible mechanical failure of one of the systems. Fire protection systems are an example. Another example is the use of dual water connections to boilers. These installations are sometimes interconnected, thus creating a health hazard.

9.4 Methods and Devices for the Prevention of Backflow and Backsiphonage

A wide variety of devices are available that can be used to prevent backsiphonage and backpressure from allowing contaminated fluids or gases into a potable water supply system. Generally, the selection of the proper device to use is based on the degree of hazard posed by the cross-connection. Additional considerations are based on piping size, location, and the potential need to periodically test the devices to ensure proper operation.

The six basic types of devices that can be used to correct cross-connections are as follows:

1) Air gaps
2) Barometric loops
3) Vacuum breakers—both atmospheric and pressure type
4) Double-check valves with an intermediate atmospheric vent
5) Double-check valve assemblies
6) Reduced pressure principle devices.

In general, all manufacturers of these devices, with the exception of the barometric loop, produce them to one or more of three basic standards, thus ensuring that dependable devices are being utilized and marketed. The major standards in the industry are devised by the American Society of Sanitary Engineers (ASSE), the American Water Works Association (AWWA), and the University of California Foundation for Cross-Connection Control and Hydraulic Research.

9.4.1 Air Gap

Air gaps are nonmechanical backflow preventers that are very effective devices for use where either backsiphonage or backpressure conditions may exist. Their use is as old as piping and plumbing itself, although their design was standardized relatively recently. In general, the air gap must be twice the supply pipe diameter but never less than 1 in. (see Fig. 9.18a).

An air gap, although an extremely effective backflow preventer when used to prevent backsiphonage and backpressure conditions, does interrupt the piping flow with corresponding loss of pressure for subsequent use. Consequently, air gaps are primarily used at the end of the line service where reservoirs or storage tanks are desired. When contemplating the use of an air gap, here are some other considerations:

1) In a continuous piping system, each air gap requires the added expense of reservoirs and secondary pumping systems.
2) The air gap can be easily defeated in the event that the two-diameter (2D) requirement was purposely or inadvertently compromised. Excessive splash may be encountered in the event that higher than anticipated pressures or flows occur. The splash may be a cosmetic or true potential hazard—the simple solution being to reduce the 2D dimension by thrusting the supply pipe into the receiving funnel. By so doing, the air gap is defeated.

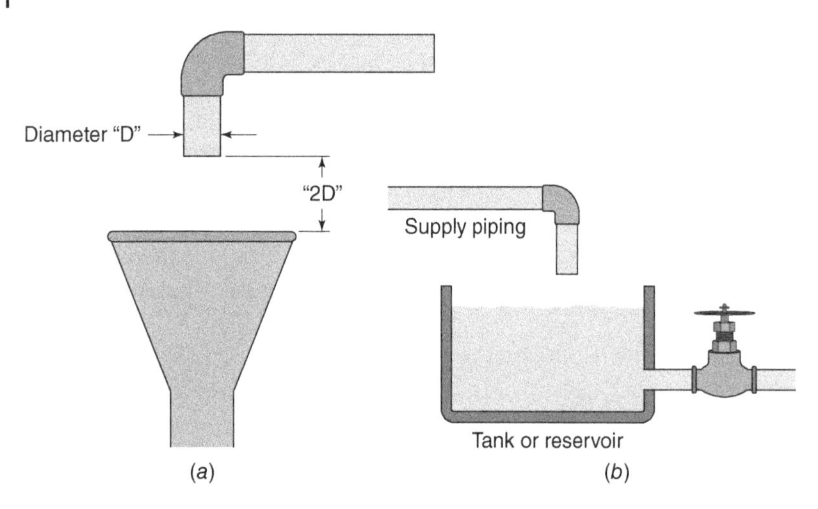

Figure 9.18 Air Gaps

3) At an air gap, we expose the water to the surrounding air with its inherent bacteria, dust particles, and other airborne pollutants or contaminants. In addition, the aspiration effect of the flowing water can drag down surrounding pollutants into the reservoir or holding tank.
4) Free chlorine can come out of treated water as a result of the air gap and the resulting splash and churning effect as the water enters the holding tanks. This reduces the ability of the water to withstand bacteria contamination during long-term storage.
5) For these reasons, air gaps must be inspected as frequently as mechanical backflow preventers. They are not exempt from an in-depth cross-connection control program requiring periodic inspection of all backflow devices.

Air gaps can be fabricated from commercially available plumbing components or purchased as separate units and integrated into plumbing and piping systems. An example of the use of an air gap is shown in Fig. 9.18b.

9.4.2 Barometric Loops

The barometric loop consists of a continuous section of supply piping that abruptly rises to a height of approximately 35 ft (10.6 m) and then returns back down to the originating level. It is a loop in the piping system that effectively protects against backsiphonage. It cannot be used to protect against backpressure.

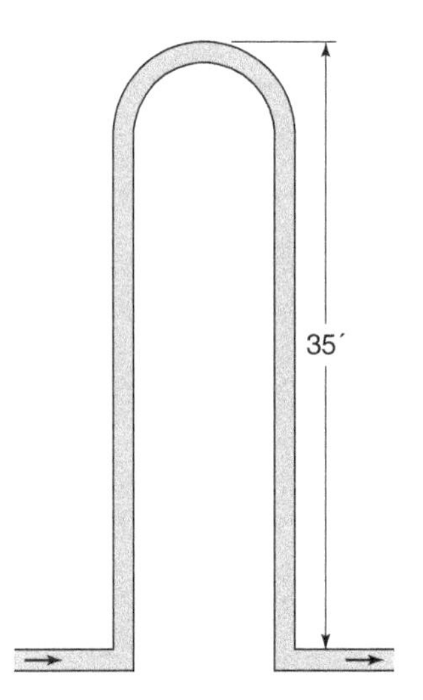

Figure 9.19 Barometric Loop. Conversion factor: 1′ = 1 ft = 0.3048 m

Its operation, in the protection against backsiphonage, is based on the principle that a water column, at sea level pressure, cannot rise above 33.9 ft (10.3 m) (refer to Fig. 9.12b). In general, barometric loops are locally fabricated and are 35 ft (10.6 m) high (see Fig. 9.19).

9.4.3 Atmospheric Vacuum Breakers

These devices are among the simplest and least expensive mechanical types of backflow preventers and, when installed properly, can provide excellent protection against backsiphonage. They must not, however, be utilized to protect against backpressure conditions.

Construction usually consists of a polyethylene float that is free to travel on a shaft and seal in the uppermost position against the atmosphere with an elastomeric disk. Water flow lifts the float, which then causes the disk to seal. Water pressure keeps the float in the upward sealed position. Termination of the water supply will cause the disk to drop down, venting the unit to the atmosphere and thereby opening downstream piping to atmospheric pressure, thus preventing backsiphonage. Figure 9.20a shows a typical atmospheric breaker.

In general, these devices are available in $^1/_2$-in. through 3-in. (12-mm through 75-mm) sizes and must be installed vertically, must not have shutoffs downstream, and must be installed at least 6 in. (152 mm) higher than the final outlet. They cannot be tested once

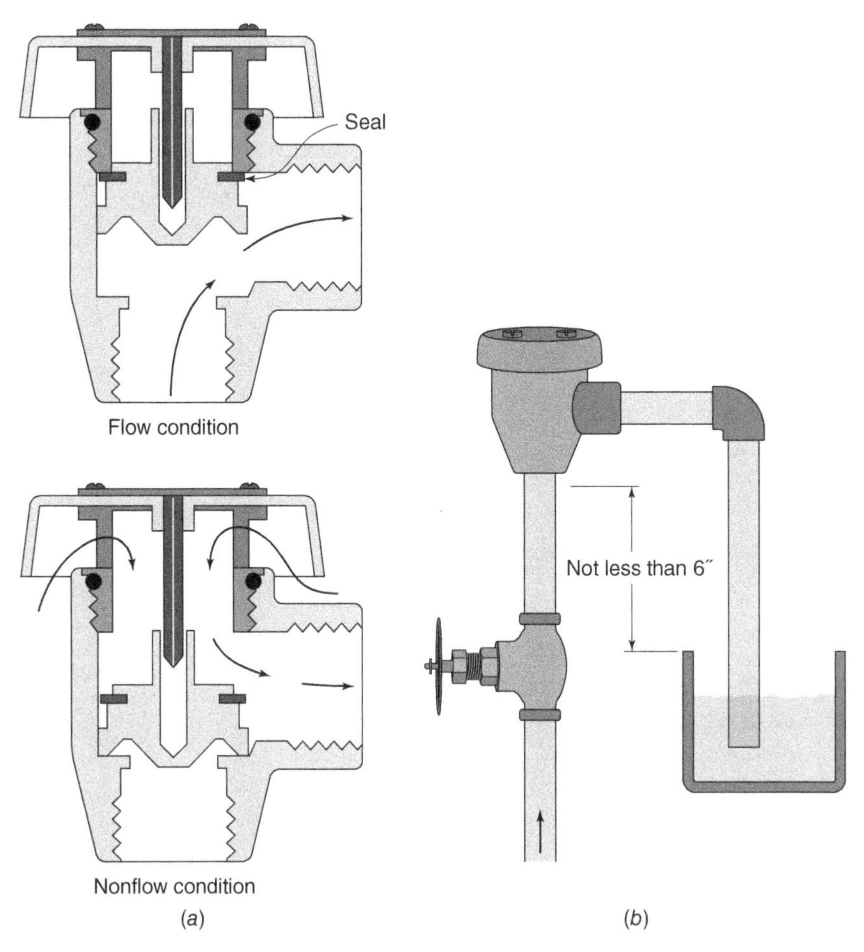

Figure 9.20 Atmospheric Vacuum Breaker (a) Flow and Nonflow Conditions and (b) Typical Installation. Conversion factor: 1″ = 1 in. = 25.4 mm

they are installed in the plumbing system, but are, for the most part, dependable, trouble-free devices for backsiphonage protection.

Figure 9.20b shows the generally accepted installation requirements. Note that no shutoff valve is downstream of the device that would otherwise keep the atmospheric vacuum breaker under constant pressure.

9.4.4 Hose Bib Vacuum Breakers

Hose bib vacuum breakers are small devices that are a specialized application of the atmospheric vacuum breaker. They are generally attached to still cocks and in turn are connected to hose-supplied outlets such as garden hoses, slop sink hoses, and spray outlets. They consist of a spring-loaded check valve that seals against an atmospheric outlet when water supply pressure is turned on. Typical construction is shown in Fig. 9.21.

When the water supply is turned off, the device vents to the atmosphere, thus protecting against backsiphonage conditions. They should not be used as backpressure devices. Manual drain options are available, together with tamper-proof versions.

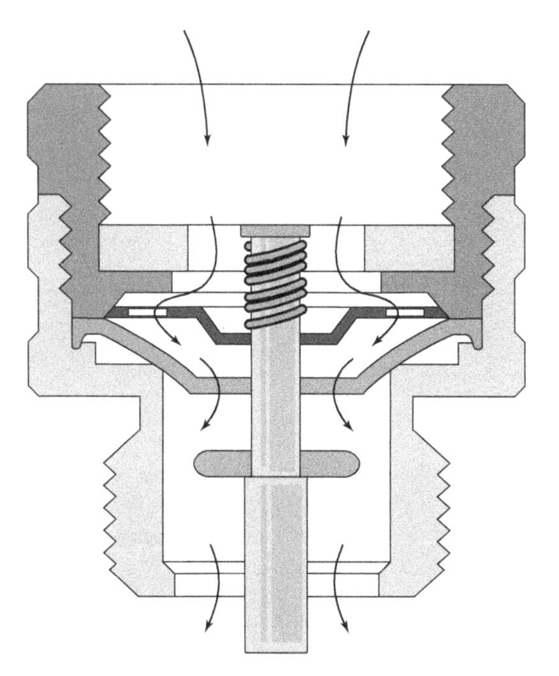

Figure 9.21 Hose Bib Vacuum Breaker

9.4.5 Pressure Vacuum Breakers

This device is an outgrowth of the atmospheric vacuum breaker and evolved in response to a need to have an atmospheric vacuum breaker that could be utilized under constant pressure and that could be tested in line. A spring on top of the disk and float assembly, two added gate valves, test cocks, and an additional first check were required to achieve this device (see Fig. 9.22a).

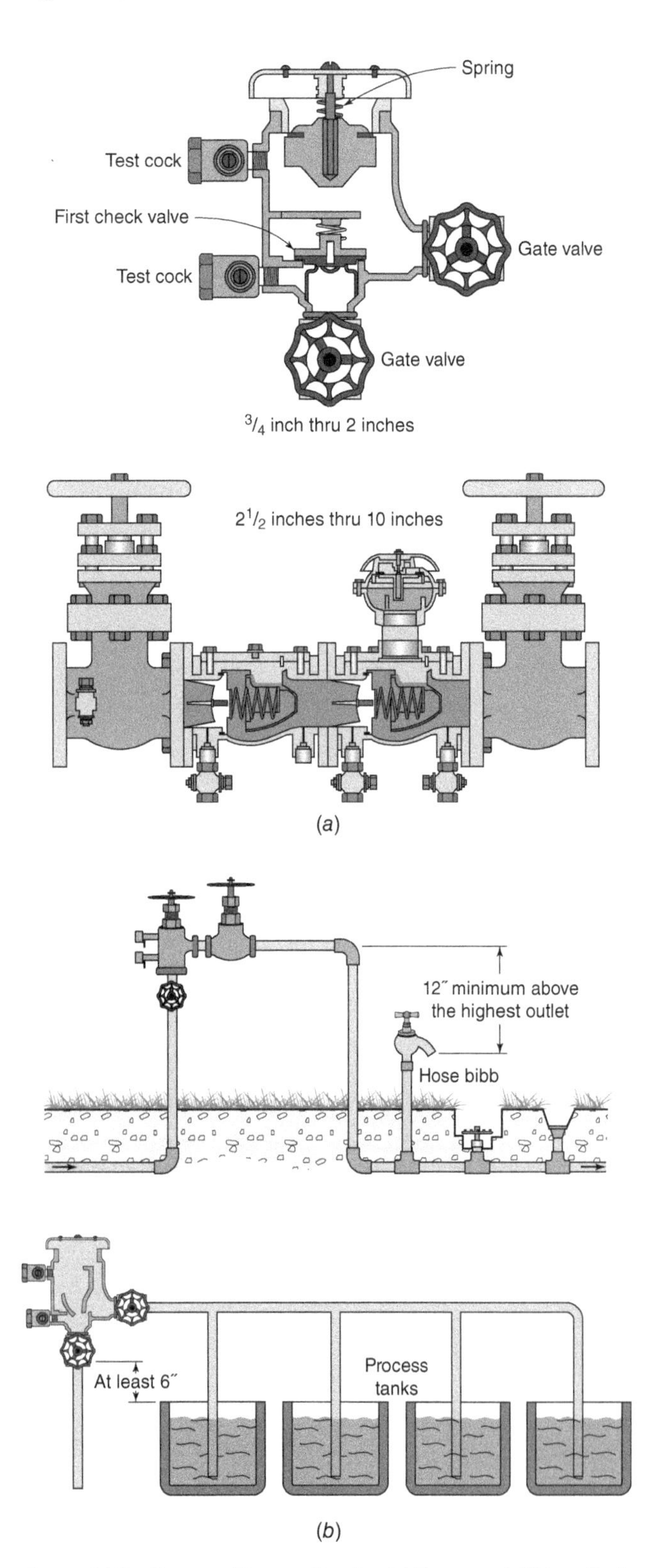

Figure 9.22 Pressure Vacuum Breaker (a) Types and (b) Installations. Conversion factor: 1″ = 1 in. = 25.4 mm

These units are available in $^1/_2$-in. through 10-in. (12-mm through 250-mm) sizes and have broad usage in the agriculture and irrigation market. Typical agricultural and industrial applications are shown in Fig. 9.22b.

Again, these devices may be used under constant pressure, but do not protect against backpressure conditions. As a result, installation must be at least 6–12 in. (150 to 300 mm) higher than the existing outlet.

A spill-resistant pressure vacuum breaker is available that is a modification to the standard pressure vacuum breaker but specifically designed to minimize water spillage. Installation and hydraulic requirements are similar to the standard pressure vacuum breaker and the devices are recommended for internal use.

9.4.6 Double-Check Valves with an Intermediate Atmospheric Vent

There is a need to provide a compact device in $^1/_2$-in. and $^3/_4$-in. (12-mm and 19-mm) pipe sizes that protects against moderate hazards, is capable of being used under constant pressure, and protects against backpressure resulted in this unique backflow preventer. Construction is basically a double-check valve having an atmospheric vent located between the two checks (see Fig. 9.23a).

Line pressure keeps the vent closed, but zero supply pressure or backsiphonage will open the inner chamber to atmosphere. With this device, extra protection is obtained through the atmospheric vent capability. Figure 9.23b shows a typical use of the device on a residential boiler supply line.

9.4.7 Double-Check Valves

A double-check valve is essentially two single-check valves coupled within one body and furnished with test cocks and two tightly closing gaze valves (see Fig. 9.24). The test capability feature gives this device a big advantage over the use of two independent check valves in that it can be readily tested to determine if either or both check valves are inoperative or fouled

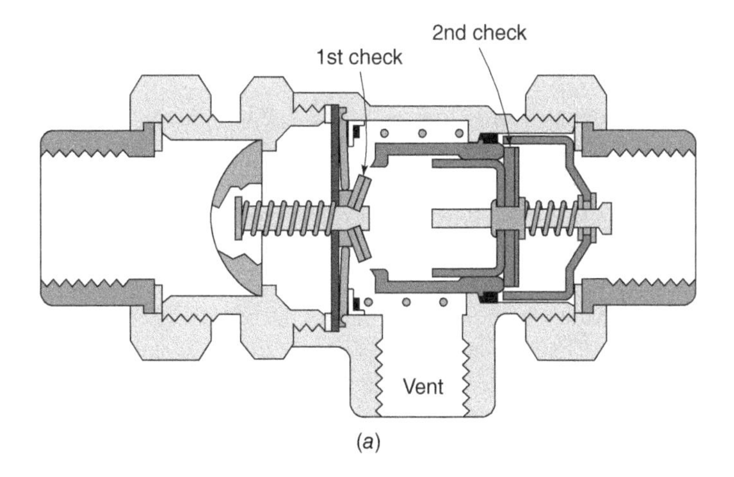

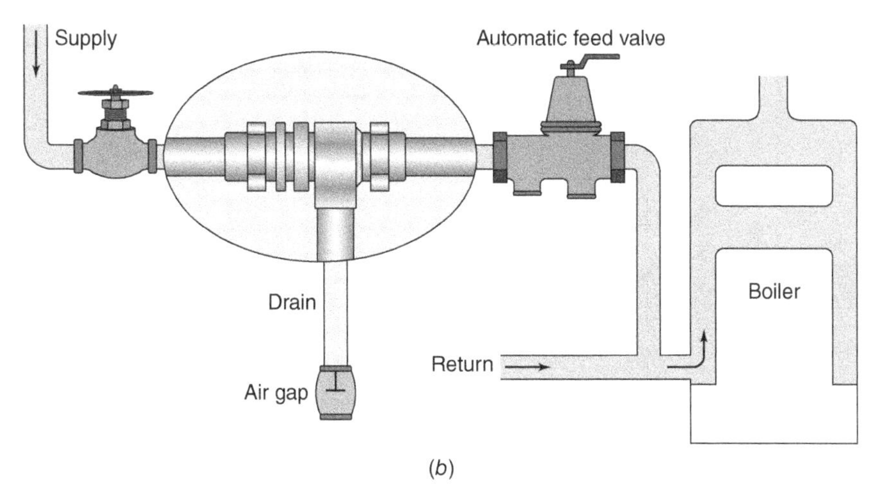

Figure 9.23 (a) Double-Check Valve with Atmospheric Vent and (b) Typical Residential Use of Valve

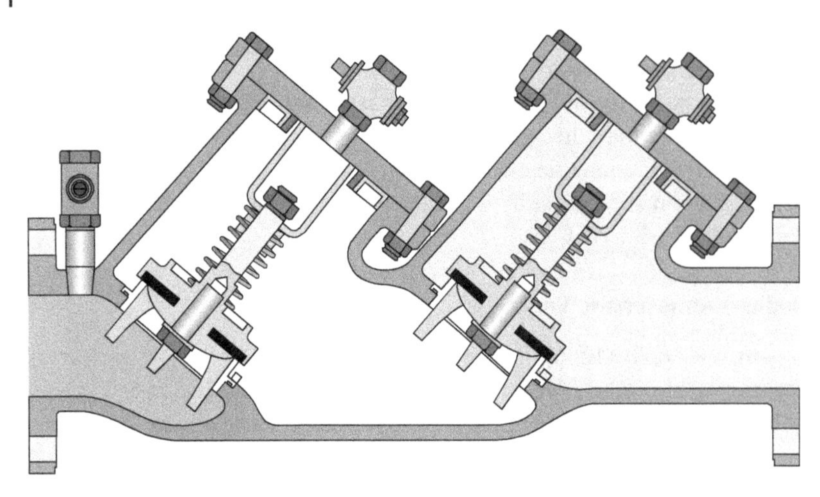

Figure 9.24 Double-Check Valve

by debris. Each check is spring-loaded closed and requires approximately a pound of pressure to open. The spring loading provides the ability to "bite" through small debris and still seal—a protection feature not prevalent in unloaded swing check valves.

Figure 9.24 shows a cross-section of a double-check valve complete with test cocks. Double checks are commonly used to protect against low- to medium-hazard installations such as food processing steam kettles and apartment projects. They may be used under continuous pressure and protect against both backsiphonage and backpressure conditions.

9.4.8 Double-Check Detector Check

This device is an outgrowth of the double-check valve and is primarily utilized in fire line installations. Its purpose is to protect the potable supply line from possible contamination or pollution from fire line chemical additives, booster pump fire line backpressure, stagnant "black water" that sits in fire lines over extended periods of time, the addition of "raw" water through outside fire pumper connections (Siamese outlets), and the detection of any water movement in the fire line water due to fire line leakage or deliberate water theft. It consists of two spring-loaded check valves, a bypass assembly with water meter and double-check valve, and two tightly closing gate valves (see Fig. 9.25).

The addition of test cocks makes the device testable to ensure proper operation of both the primary checks and the bypass check valve. In the event of very low fire line water usage (theft of water), the low pressure drop inherent in the bypass system permits the low flow of water to be metered through the bypass system. In a high flow demand situation, such as that associated with deluge fire capability, the main check valves open, permitting high-volume, low restricted flow through the two large spring-loaded check valves.

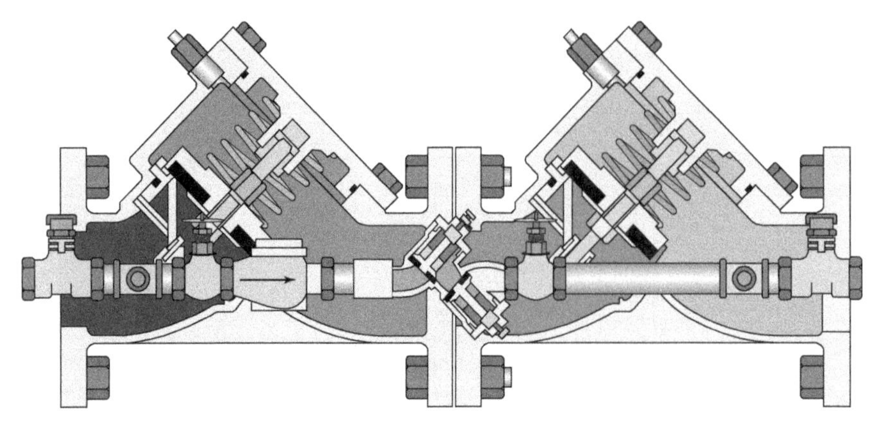

Figure 9.25 Double-Check Detector Check

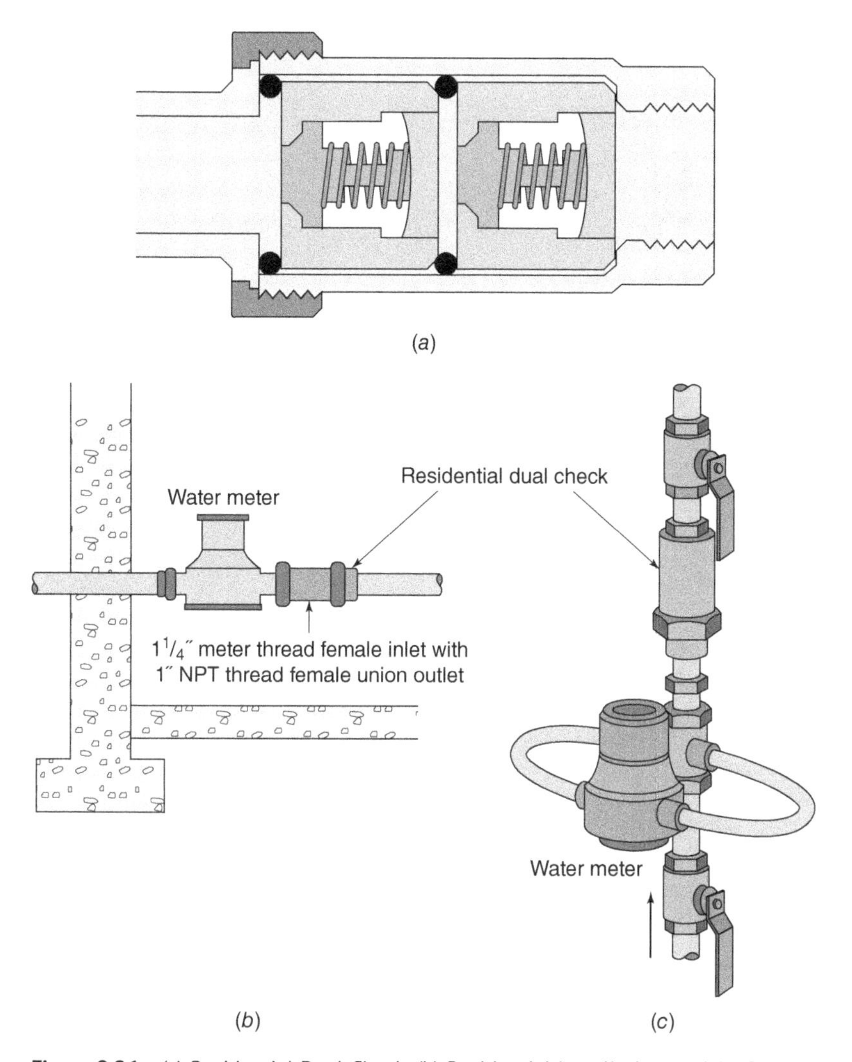

Figure 9.26 (a) Residential Dual Check, (b) Residential Installation, and (c) Copper Horn. Conversion factor: 1″ = 1 in. = 25.4 mm

9.4.9 Residential Dual Check

The need to furnish reliable and inexpensive backsiphonage and backpressure protection for TO individual residences resulted in the debut of the residential dual check. Protection of the main potable supply from household hazards such as home photography chemicals, toxic insect and garden sprays, termite control pesticides used by exterminators, and so forth, reinforced a true need for such a device. Figure 9.26a shows a cutaway of the device.

This device is sized for $^{1}/_{2}$-, $^{3}/_{4}$-, and 1-in (12-, 19-, and 25-mm) service lines and is installed immediately downstream of the water meter. The use of plastic check modules and elimination of test cocks and gate valves keep the cost reasonable while providing good, dependable protection. Typical installations are shown in Fig. 9.26b.

9.5 Reduced Pressure Principle Backflow Preventer

Maximum protection is achieved against backsiphonage and backpressure conditions utilizing reduced pressure principle backflow preventers. These devices are essentially modified double-check valves with an atmospheric vent capability placed between the two checks and designed such that this “zone” between the two checks is always kept at least 2 psi less than the supply pressure (see Fig. 9.27). With this design criterion, the reduced pressure principle backflow preventer can provide protection against backsiphonage and backpressure when both the first and second checks become fouled. They can be used

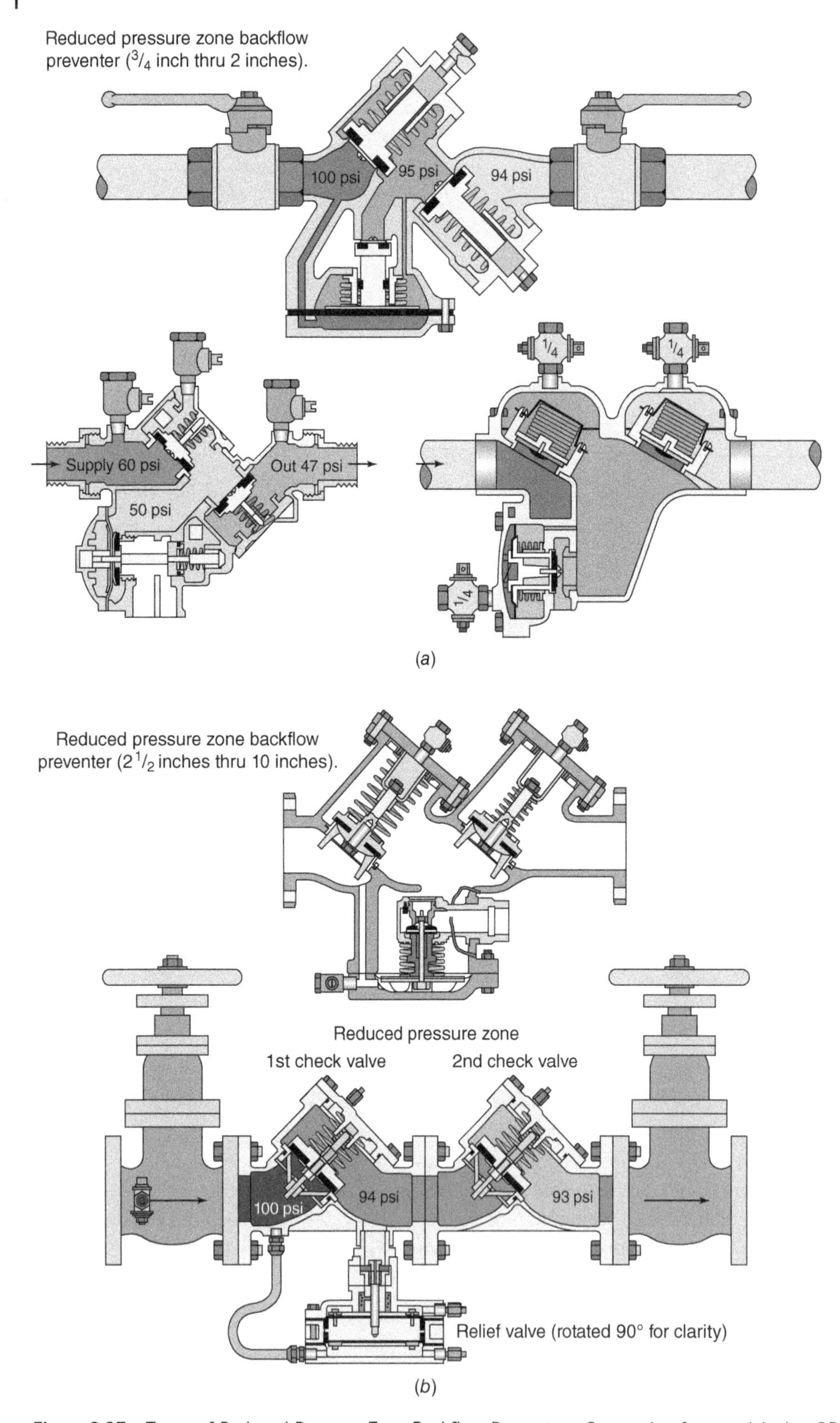

Figure 9.27 Types of Reduced Pressure Zone Backflow Preventers. Conversion factors: 1 inch = 25.4 mm; 1 psi = 6.94 kPa

under constant pressure and at high hazard installations. They are furnished with test cocks and gate valves to enable testing and are available in $^3/_4$-in. through 10-in. (19-mm through 250-mm) sizes.

The principles of operation of a reduced pressure principle backflow preventer are as follows: Flow from the left enters the central chamber against the pressure exerted by the loaded check valve 1. The supply pressure is then reduced by a predetermined amount. The pressure in the central chamber is maintained lower than the incoming supply pressure through the operation of the relief valve, which discharges to the atmosphere whenever the central chamber pressure approaches within a few psi of the inlet pressure. Check valve 2 is lightly loaded to open with a pressure drop of l psi (7 kPa) in the direction of flow and is independent of the pressure required to open the relief valve. In the event that the pressure increases downstream from the device, tending to reverse the direction of flow, check valve 2 closes, preventing backflow. Because all valves may leak as a result of wear or obstruction, the protection provided by the check valves is not considered sufficient. If some obstruction prevents check valve 2 from closing tightly, the leakage back into the central chamber would increase the pressure in this zone, the relief valve would open, and flow would be discharged to the atmosphere.

When the supply pressure drops to the minimum differential required to operate the relief valve, the pressure in the central chamber should be atmospheric; if the inlet pressure should become less than atmospheric pressure, the relief valve should remain fully open to the atmosphere to discharge any backflow water that is the result of backpressure and leakage of check valve 2.

Malfunctioning of one or both of the check valves or relief valve should always be indicated by a discharge of water from the relief port. Under no circumstances should plugging of the relief port be permitted because the device depends on an open port for safe operation. The pressure loss through the device may be expected to average between 10 and 20 psi (69.4 and 138.8 kPa) within the normal range of operation, depending on the size and flow rate of the device.

Reduced pressure principle backflow preventers are commonly installed on high hazard installations such as plating plants, where they protect primarily against backsiphonage potential, carwashes where they protect against backpressure conditions, and funeral parlors, hospital autopsy rooms, etc. The reduced pressure principle backflow preventer forms the backbone of cross-connection control programs. Because it is utilized to protect against backsiphonage from high hazard installations, and because high hazard installations are the first consideration in protecting public health and safety, these devices are installed in large quantities over a broad range of plumbing and water works installations. Figure 9.28 shows typical installations of these devices.

9.6 Administration of a Cross-connection Control Program

9.6.1 Responsibility

Under the provisions of the Safe Drinking Water Act, the federal government has established, through the U.S. Environmental Protection Agency (EPA), national standards for safe drinking water. The states are responsible for the enforcement of these standards as well as the supervision of public water supply systems and the sources of drinking water. The water purveyor (supplier) is held responsible for compliance with the provisions of the Safe Drinking Water Act, including a warranty that water quality provided by its operation is in conformance with the EPA standards at the source, and is delivered to the customer without the quality being compromised as a result of its delivery through the distribution system.

Figure 9.29 depicts several options that are open to a water purveyor when considering cross-connection protection to commercial, industrial, and residential customers. The water supply company may elect to work initially on the "containment" theory. This approach utilizes a minimum of backflow devices and isolates the customer from the water main. It virtually insulates the customer from potentially contaminating or polluting the public water supply system. Although that type of "containment" does not protect the customer within his own building, it does effectively remove him from possible contamination to the public water supply system.

If the water supplier elects to protect its customers on a domestic internal protective basis and/or "fixture outlet protective basis," then cross-connection control protective devices are placed at internal high hazard locations as well as at all locations where cross-connections exist at the "last free-flowing outlet." This approach entails extensive cross-connective survey work on behalf of the water superintendent as well as constant policing of the plumbing within each commercial, industrial, and residential account.

In large water supply systems, fixture outlet protection cross-connection control philosophy, in and of itself, is a virtual impossibility to achieve and police due to the quantity of systems involved, the complexity of the plumbing systems inherent

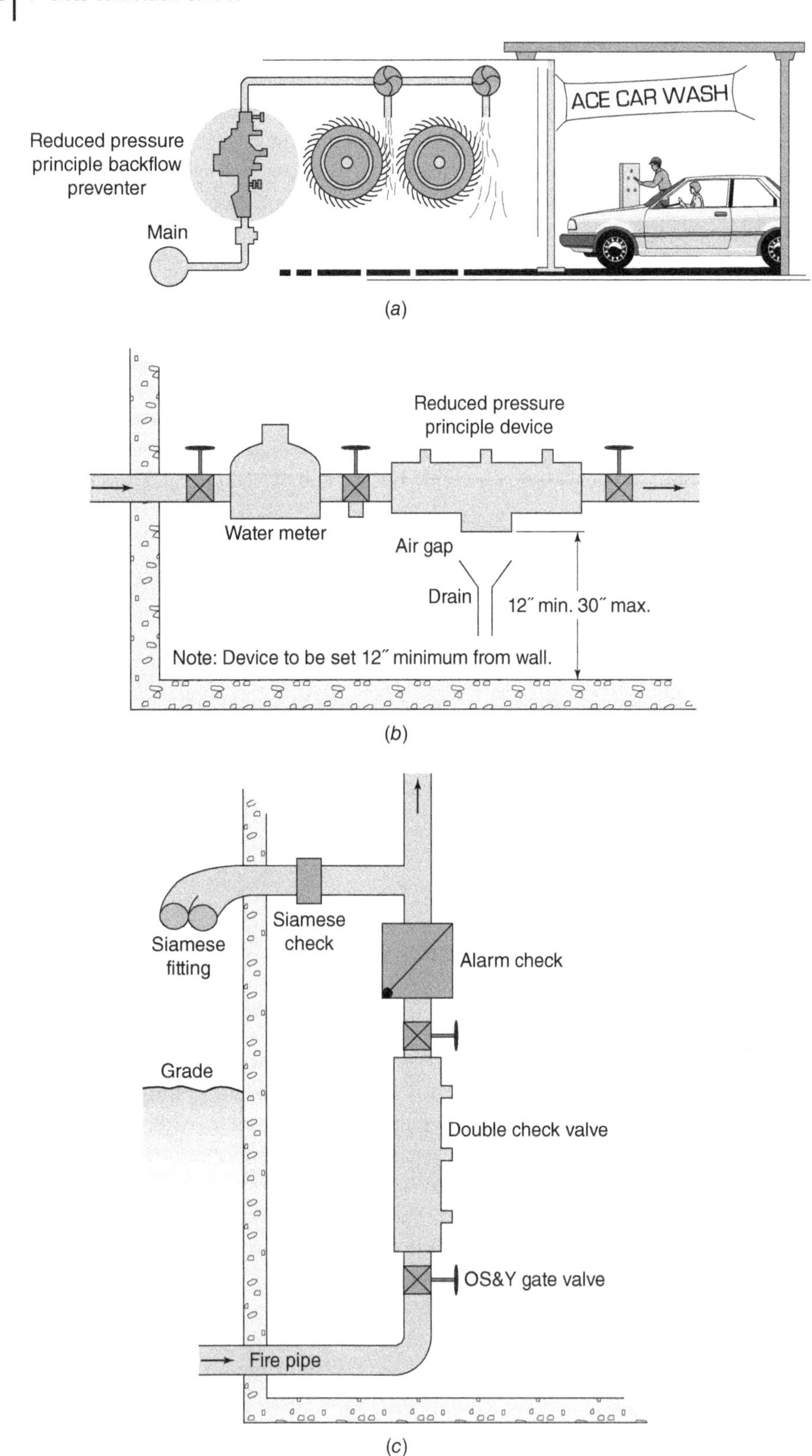

Figure 9.28 Reduced Pressure Device Installations in (a) Car Wash, (b) Water Supply Line, and (c) Fire Line. Conversion factor: 1″ = 1 in. = 25.4 mm

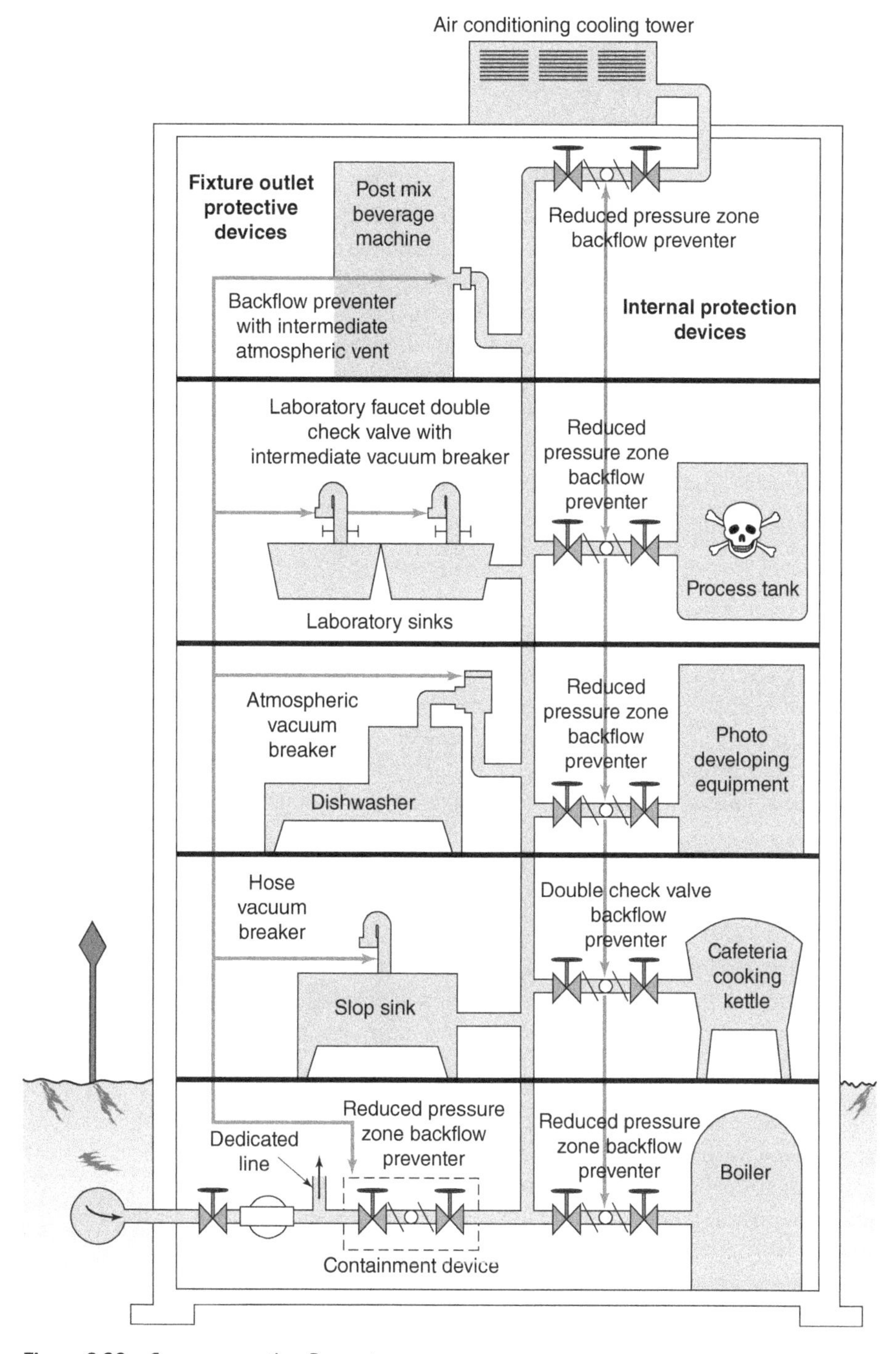

Figure 9.29 Cross-connection Preventers

in many industrial sites, and the fact that many plumbing changes are made within industrial and commercial establishments that do not require the water department to license or otherwise endorse or ratify when contemplated or completed.

In addition, internal plumbing cross-connection control survey work is generally foreign to the average water supplier and is not normally a part of its employees' job descriptions or duties. Although it is admirable for the water supplier to accept and perform survey work, the supplier should be aware that it runs the risk of additional liability in an area that may be in conflict with plumbing inspectors, maintenance personnel, and other public health officials.

Even where extensive "fixture outlet protection," cross-connection control programs are in effect through the efforts of an aggressive and thorough water supply cross-connection control program, the water authorities should also have an active "containment" program in order to address the many plumbing changes that are made and that are inherent within commercial and industrial establishments. In essence, fixture outlet protection becomes an extension of the "containment" program.

Also, in order for the water supplier to provide maximum protection of the water distribution system, consideration should be given to requiring the owner of a premise (commercial, industrial, or residential) to provide, at her own expense, adequate proof that her internal water system complies with the local or state plumbing code(s). In addition, she may be required to install, have tested, and maintain all backflow protection devices that would be required – at her own expense.

The water supplier should have the right of entry to determine degree of hazard and the existence of cross-connections in order to protect the potable water system. By so doing, the water supplier can assess the overall nature of the facility and its potential impact on the water system (determine degree of hazard), personally see actual cross-connections that could contaminate the water system, and take appropriate action to ensure the elimination of the cross-connection or the installation of required backflow devices.

Assisting the water supplier in the total administration of a cross-connection control program requires that all public health officials, plumbing inspectors, building managers, plumbing installers, and maintenance personnel participate and share in the responsibility of protecting the public health and safety of individuals from cross-connections and contamination or pollution of the public water supply system.

9.6.2 Dedicated Line

In addition to the options just discussed, Figure 9.29 also depicts the use of a "dedicated" potable water line. This line initiates immediately downstream of the water meter and is "dedicated" solely for human consumption: drinking fountains, safety showers, eye wash stations, etc. It is very important that this piping be color coded throughout in accordance with local plumbing regulations, flow direction arrows added, and the piping religiously policed to ensure that no cross-connections to other equipment or piping are made that could compromise water quality. In the event that authorities feel the policing of this line cannot be reliably maintained or enforced, the installation of a containment device on this line should be considered.

9.6.3 Method of Action

A complete cross-connection control program requires a carefully planned and executed initial action plan followed by aggressive implementation and constant follow-up. Proper staffing and education of personnel is a requirement to ensure that an effective program is achieved. A recommended plan of action for a cross-connection control program should include the following characteristics:

1) Establish a cross-connection control ordinance at the local level and have it approved by the water commissioners, town manager, etc., and ensure that it is adopted by the town or private water authority as a legally enforceable document.
2) Conduct public information meetings that define the proposed cross-connection control program, review the local cross-connection control ordinance, and answer all questions that may arise concerning the reason for the program, why and how the survey will be conducted, and the potential impact on industrial, commercial, and residential water customers. Have state authorities and members of the local press and radio attend the meeting.
3) Place written notices of the pending cross-connection control program in the local newspaper, and have the local radio station make announcements about the program as a public service notice.
4) Send employees who will administer the program to a course, or courses, on backflow tester certification, backflow survey courses, backflow device repair courses, etc.
5) Equip the water authority with a backflow device test kit.
6) Conduct meeting(s) with the local plumbing inspection people, building inspectors, and licensed plumbers in the area who will be active in the inspection, installation, and repair of backflow devices. Inform them of the intent of the program and the part that they can play in the successful implementation of the program.
7) Prior to initiating a survey of the established commercial and industrial installations, prepare a list of these establishments from existing records, then prioritize the degree of hazard that they present to the water system: plating plants, hospitals, car wash facilities, industrial metal finishing and fabrication, mortuaries, etc. These will be the initial facilities inspected for cross-connections; inspection of less hazardous installations follows.

8) Ensure that any new construction plans are reviewed by the water authority to assess the degree of hazard and ensure that the proper backflow preventer is installed concurrently with the potential degree of hazard that the facility presents.
9) Establish a residential backflow protection program that will automatically ensure that a residential dual check backflow device is installed automatically at every new residence.
10) As water meters are repaired or replaced at residences, ensure that a residential dual check backflow preventer is set with the new or reworked water meter. Be sure to have the owner address thermal expansion provisions.
11) Prepare a listing of all testable backflow devices in the community and ensure that they are tested by certified test personnel at time intervals consistent with the local cross-connection control ordinance.
12) Prepare and submit testing documentation of backflow devices to the state authority responsible for monitoring this data.
13) Survey all commercial and industrial facilities and require appropriate backflow protection based on the containment philosophy and/or internal protection and fixture outlet protection. Follow up to ensure that the recommended devices are installed and tested on both an initial basis and a periodic basis consistent with the cross-connection control ordinance.

The surveys should be conducted by personnel experienced in commercial and industrial processes. The owners or owners' representatives should be questioned as to what the water is being used for in the facility and what hazards the operations may present to the water system (both within the facility and to the water distribution system) in the event that a backsiphonage or backpressure condition were to exist concurrent with a nonprotected cross-connection. In the event that experienced survey personnel are not available within the water authority to conduct the survey, consideration should be given to having a consulting firm perform the survey on behalf of the water department.

Problems/Questions

9.1 What is the definition of a cross-connection?

9.2 Why do we have cross-connections?

9.3 How many types of cross-connections are there?

9.4 What is the difference between backpressure and backsiphonage?

9.5 What types of devices can be used to correct cross-connections?

9.6 Successful control of cross-connection hazards depends not only on inspection for cross-connections by the water system and by water users, but also on an enforceable cross-connection control program. If a community subscribes to a modern plumbing code, such as the National Plumbing Code, its provisions will govern backflow and cross-connections. However, the water system must obtain authority to conduct a community inspection program through an ordinance or other means and carry out a comprehensive program. Please outline the major components of a water system's cross-connection control program.

9.7 Protection against sanitary deficiencies from cross-connections is an important engineering consideration in water and sewer system design. Visit the website www.10statesstandards.com to learn about the latest government regulations regarding the separation distance of drinking water mains crossing sewers.

9.8 How frequently should backflow preventers be tested to ensure their proper functioning? Who is qualified to perform backflow preventer testing? Consult with your local government for the local rules and regulations.

9.9 In addition to the many cross-connections that may exist on the premises of a water system's customers, the water system itself may own or control cross-connections. These potential cross-connections should be subject to the same scrutiny as those that are privately owned. Examples of cross-connections that can pose a risk to water quality and public health can be found in water treatment plants, pumping stations, or in the distribution system. During a sanitary survey for a water treatment system, what are the potential cross-connections that should be identified by an inspector?

9.10 Figure 9.30 shows a chlorine feed system at a water treatment plant. What is required between the chlorine feeding cylinders and the chlorine feed points? The same figure also shows that one chlorine cylinder feeds chlorine to the influent water before chemical flocculation, and another chlorine cylinder feeds chlorine to the effluent from the filter. What are the technical terms of this chlorination system for cross-connection control? Can one chlorine cylinder be used to feed both the influent to the flocculation influent and the filter effluent? Please explain.

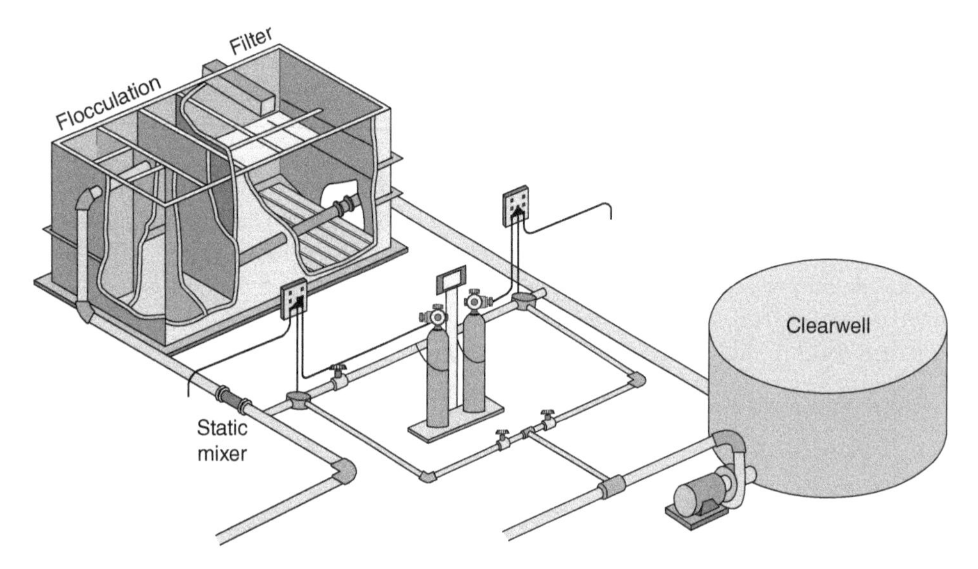

Figure 9.30 For Problem 9.10

9.11 Figure 9.31 shows that an air relief valve is piped directly to a floor drain. Explain the problem and propose a solution.

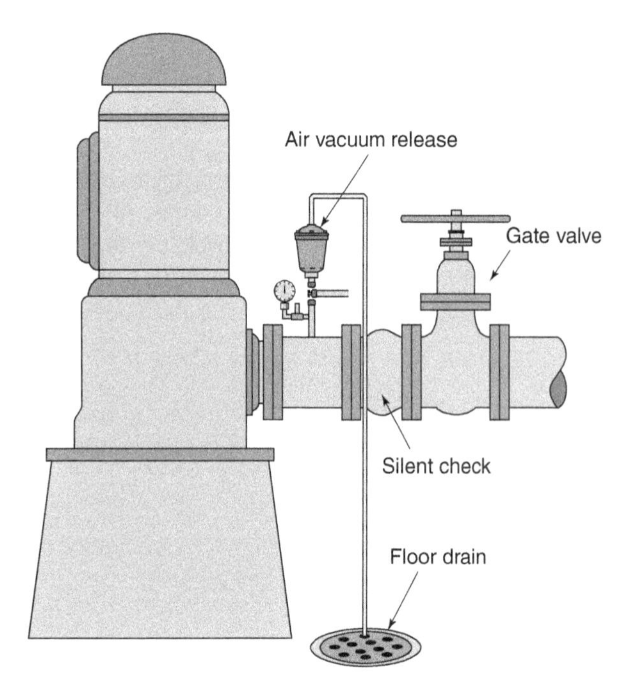

Figure 9.31 For Problem 9.11

9.12 Figure 9.32 shows a direct cross-connection of a potable water supply line and an acid solution feed line. Propose your solution for the cross-connection control.

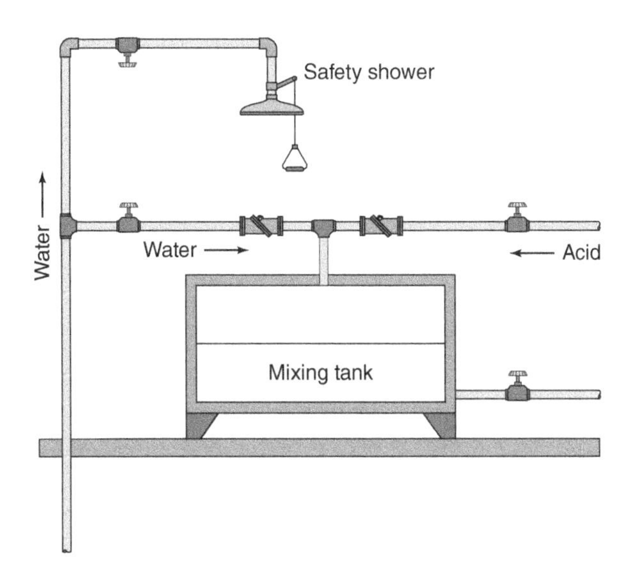

Figure 9.32 For Problem 9.12

9.13 The successful promotion of a cross-connection control and backflow prevention program in a municipality will depend on a legal authority to conduct such a program. Where a community has adopted a modern plumbing code, such as the National Plumbing Code, ASA A40.8–1955, or subsequent revisions thereof, provisions of the code will govern backflow and cross-connections. It then remains to provide an ordinance that will establish a program of inspection for the elimination of cross-connections and backflow connections within the community. A cross-connection control ordinance should have at least three basic parts:

1) Authority for establishment of a program
2) Technical provisions relating to eliminating backflow and cross-connections
3) Penalty provisions for violations.

Review the cross-connection control ordinances of at least two U.S. cities and suggest a model program for adoption by your municipality.

Bibliography

American Water Works Association, Cross-connection control, in *Water Distribution and Transmission*, AWWA, Denver, CO, 1996, pp. 315–355.

American Water Works Association, *Drinking Water Dictionary*, AWWA, Denver, CO, 509 p, 2000.

American Water Works Association, *Recommended Practice for Backflow Prevention and Cross-Connection Control*, Manual of Water Supply Practice Practices M14, Denver, CO, USA, February 2004.

American Water Works Association, *Policy on Public Water Supply*, AWWA Website, 2010.

City of Riverside, Water Rule 13, *Cross-Connections and Pollution of Supplies*, Public Utilities Department, Riverside, CA, 2003. http://www.riversideca.gov/utilities/pdf/water-rules/w_rule13.pdf, retrieved 2010.

Denver Water, *Operating Rules, Chapter 11-Cross-Connections*, Denver, CO, USA, Last Amendments December 1, 2007, http://www.water.denver.co.gov/custserve/commercial/operatingrules/chap11.html, retrieved 2009.

U.S. Environmental Protection Agency, *Cross-Connection Control Manual*, EPA 816-R-03-002, Washington, DC, USA, February 2003.

University of Southern California, *Prevalence of Cross-Connections in Household Plumbing Systems*, February 3, 2004, http://www.usc.edu/dept/fccchr/epa/hhcc.report.pdf, retrieved 2010.

10

Introduction to Wastewater Systems

According to the World Health Organization (WHO), 2.4 billion people lack access to improved sanitation, which represents 42% of the world's population. More than half of those without improved sanitation—nearly 1.5 billion people—live in China and India. In sub-Saharan Africa, sanitation coverage is a mere 36%. Only 31% of the rural inhabitants in developing countries have access to improved sanitation, as opposed 73% of urban dwellers.

About 2 million people die every year due to diarrheal diseases; most of them are children less than five years of age. The most affected are the populations in developing countries who are living in extreme conditions of poverty, normally urban dwellers or rural inhabitants. Among the main problems responsible for this situation are (a) lack of priority given to the sector, (b) lack of financial resources, (c) lack of sustainability of water supply and sanitation services, (d) poor hygiene behaviors, and (e) inadequate sanitation in public places including hospitals, health centers, and schools. Providing access to sufficient quantities of safe water, providing facilities for the sanitary disposal of excreta, and introducing sound hygiene behaviors are of capital importance to reduce the burden of disease caused by these risk factors.

To meet the WHO Millennium Development Goals (MDGs) sanitation target, an additional 370 000 people per day between 2005 and 2015 should gain access to improved sanitation. If this goal is met, the proportion of people without sustainable access to safe drinking water and basic sanitation should be halved by 2015.

This chapter outlines the broad purpose and composition of wastewater systems. With an understanding of the whys and wherefores of needed structures and operations as a whole, we can proceed more fruitfully to a rigorous consideration of details. The practitioner, too, does not move to detailed design until he has settled on a general plan.

Wastewater systems normally comprise (a) collection works, (b) treatment works, and (c) disposal or reuse works. Together, their structures compose a *sewerage* or drainage system. Although individual systems are in a sense unique, they do conform to one of the types outlined in Fig. 10.1. As shown in the figure, wastewaters from households and industries are either collected along with stormwater runoff in the *combined sewers* of a *combined system of sewerage* or are led away by themselves through separate *sanitary sewers*, while stormwaters are emptied by themselves into the separate *storm sewers* of a *separate system* of sewerage. The water-carried wastes from households are called *domestic wastewater*; those from manufacturing establishments are referred to as *industrial* or *trade wastes*; *municipal wastewater* includes both kinds. Combined sewerage systems are common to the older cities of the world, where they generally evolved from existing systems.

The converging conduits of wastewater collection works remove wastewaters from households and industries or stormwater in *free* flow as if it were traveling along branch or tributary streams into the trunk or main channel of an underground river system. Sometimes the main collector of combined systems had, in fact, been a brook at some point that was eventually covered over when pollution made its waters too unsightly, malodorous, and otherwise objectionable. For free or gravity flow, sewers and drains must head continuously downhill, except where pumping stations lift flows through force mains into higher-lying conduits. Pumping is used to either avoid the costly construction of deep conduits in flat country or bad ground or to transfer wastewaters from low-lying subareas to main drainage schemes. Gravity sewers are not intended to flow under pressure. However, some system designs provide comminutors or grinders and pumps for individuals or groups of buildings for the purpose of discharging nonclogging wastewaters through pressurized pipes.

Hydraulically, gravity sewers are designed as *open channels*, flowing partly full or, at most, just filled. Vitrified-clay and plastic pipes are generally the material of choice for small sewers and concrete or plastic reinforced with fiberglass pipes for large ones. Pressurized sewers are smaller in size such that force mains flow full. They are generally laid parallel to the ground surface in a similar fashion to water lines.

Water and Wastewater Engineering: Hydraulics, Hydrology and Management, Volume 1, Fourth Edition.
Lawrence K. Wang, Mu-Hao Sung Wang, and Nazih K. Shammas.

Companion website: www.wiley.com/go/Wang/Waterandwastewater4e

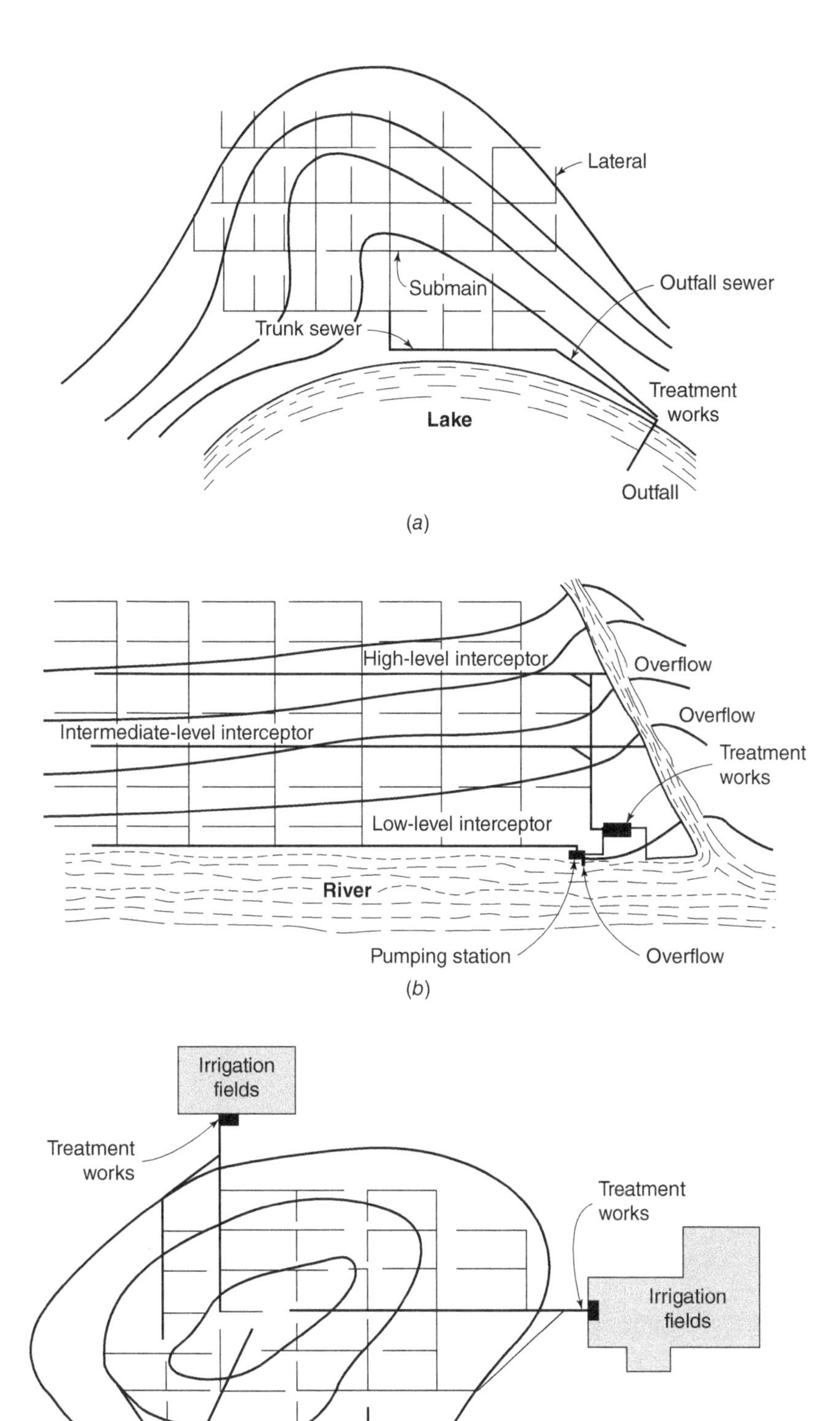

Figure 10.1 Plans of Sewerage Systems, (a) Fan Pattern Sanitary System; (b) Zone Pattern Combined System; and (c) Radial Pattern Sanitary or Combined System of Storm Drains. The sewerage systems of London, Paris (Figs. 10.2 and 10.3), New York, and Boston provide examples of this evolution

Figure 10.2 Sewers in Paris (*Source:* Courtesy Wikipedia. http://en.wikipedia.org/wiki/Image:Paris-Egouts-p1010721.jpg)

Figure 10.3 Wastewater Discharge into a Stream (*Source:* Courtesy Wikipedia. http://upload.wikimedia.org/wikipedia/commons/0/03/Paris-Egouts-p1010760.jpg)

In well-watered regions of the globe, the collected wastewaters are normally discharged into nearby receiving bodies of water after suitable treatment. This is referred to as disposal by *dilution*, but natural purification as well as physical dispersion of pollutants, is involved. In semiarid regions, terminal discharge may be onto land for groundwater recharge or for reuse in *irrigation* and *industry*. Treatment before disposal aims at the removal of unsightly and putrescible matters, stabilization of degradable substances, elimination of organics, removal of nutrients and minerals, and destruction of disease-producing organisms all in a suitable degree. Conservation of water and protection of its quality for other uses are important considerations.

10.1 Sources of Wastewaters

Sanitary wastewater is the spent waters supplied to the community. *Domestic wastewater* is the spent water from the kitchen, bathroom, lavatory, toilet, and laundry. To the mineral and organic matter in the water supplied to the community is added a burden of human excrement, paper, soap, dirt, food wastes, and other substances. Some of the waste matters remain in suspension, while others go into solution or become so finely divided that they acquire the properties of colloidal (dispersed and ultramicroscopic) particles. Many of the waste substances are organic and serve as food for saprophytic microorganisms, that is, organisms living on dead organic matter. Because of this, domestic wastewater is unstable, biodegradable, or putrescible.

Here and there, and from time to time, intestinal pathogens reach domestic and other wastewaters. Accordingly, it is prudent to consider such wastewaters suspect at all times.

The carbon, nitrogen, and phosphorus in most wastewaters are good plant nutrients. They, as well as the nutrients in natural runoff, add to the eutrophication (from Greek *eu*, meaning "well," and *trophein*, "to nourish") of receiving waters and may produce massive algal blooms, especially in lakes. As pollution continues, the depth of heavily eutrophic lakes may be reduced by the benthal buildup of dead cells and other plant debris. In this way, lakes may be turned into bogs in the course of time and eventually completely obliterated.

Industrial wastewaters vary in composition with industrial operations. Some are relatively clean rinse waters; others are heavily laden with organic or mineral matter, or with corrosive, poisonous, flammable, or explosive substances. Some are so objectionable that they should not be admitted to the public sewerage system; others contain so little and such unobjectionable waste matters that it is safe to discharge them into storm drains or directly to natural bodies of water. Fats, lime, hair, and fibers adhere to sewers and clog them; acids and hydrogen sulfide destroy cement, concrete, and metals; hot wastes crack tile and masonry conduits; poisonous chemicals disrupt biological treatment, kill useful aquatic life, and endanger water supplies; fertilizing elements add to the eutrophication of lakes; anthrax and other living organisms are infective to man; flammable or explosive liquors imperil the structures through which they flow; and toxic gases or vapors are hazardous to workmen and operators of wastewater works. Industrial wastewaters are usually pretreated before their discharge into public sewers.

Groundwater may enter gravity sewers through pipe joints. In combined systems and stormwater drains, runoff from rainfall and melting ice and snow adds to the washings from streets, roofs, gardens, parks, and yards. Entering dirt, dust, sand, gravel, and other gritty substances are heavy and inert and form the bed load. Leaves and organic debris are light and degradable and float on or near the water surface. Waters from street flushing, firefighting, and water-main scouring, as well as wastewaters from fountains, wading and swimming pools, and drainage waters from excavations and construction sites swell the tide.

Domestic wastewaters flow through house or building drains directly to the public sewer (Fig. 10.4). Runoff from roofs and paved areas may be directed first to the street gutter or immediately to the storm sewer. In combined systems, water

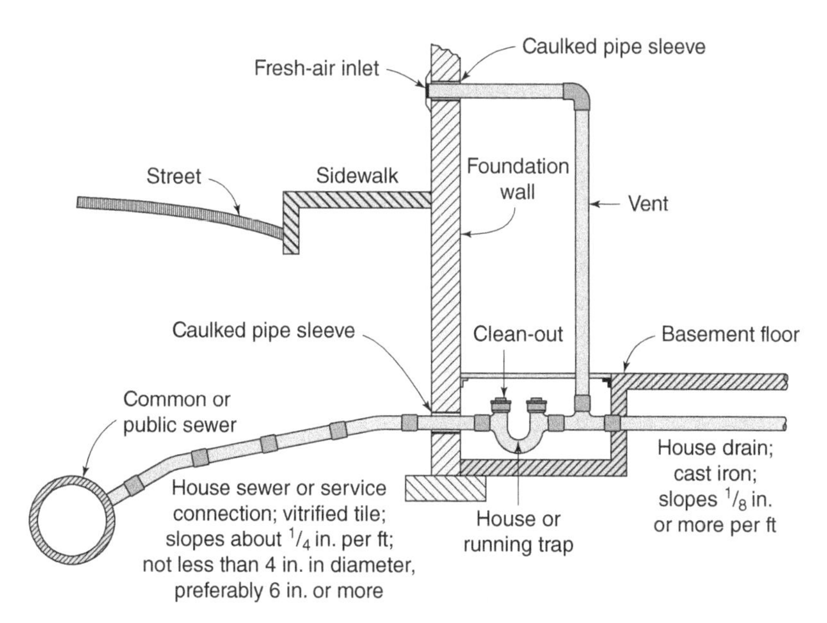

Figure 10.4 Connecting Building Drainage System to Sewer (Trap May Be Installed or Omitted). Conversion factors: 1 in = 25.4 mm; 1 ft = 0.3048 m

from roof and yard areas may be led into the house drain. Other storm runoff travels over the ground until it reaches a street gutter along which it flows until it enters a stormwater inlet or catch basin and is piped to a manhole and to the drainage system. In separate systems, connections to the wrong sewer are commonly in violation of sewer regulations. The dry-weather flows of combined sewers are primarily domestic and industrial wastewaters; the wet-weather flows are predominantly storm runoff. The first flush of stormwater may scour away deposited and stranded solids and increase the discharge of putrescent organic matter through stormwater overflows.

10.2 System Patterns

Among the factors determining the pattern of collecting systems are the following:

1) Type of system (whether separate or combined)
2) Street lines or rights of way
3) Topography, hydrology, and geology of the drainage area
4) Political boundaries
5) Location and nature of treatment and disposal works

Storm drains are naturally made to seek the shortest path to existing surface channels. Combined systems have become rare. They pollute the waters washing the immediate shores of the community and make domestic wastewater treatment difficult. More often, their flows are intercepted before they can spill into waters that need to be protected. If the tributary area is at all large, the capacity of the interceptor should be held to some reasonable multiple of the average dry-weather flow or to the dry-weather flow plus the first flush of stormwater, which, understandably, is most heavily polluted. Rainfall intensities and durations are deciding factors. In North America, rainfalls are so intense that the number of spills cannot be reduced appreciably by raising the capacity of interceptors even to as much as 10 times the dry-weather flow. No more than the maximum dry-weather flow becomes the economically justifiable limit. Domestic wastewater in excess of this amount spills into the receiving water through outlets that antedate interception, or through stormwater overflows constructed for this purpose.

Pumping, so commonly associated with waterfront interceptors, pipe sizes, and difficulties of construction in low-lying and often bad ground, can sometimes be reduced by dividing the drainage area into a series of more or less parallel zones differing in elevation and lending themselves to separate interception. A *zone pattern* is formed as shown in Fig. 10.1b. This pattern is often useful also for sanitary sewers. The *fan pattern* shown in Fig. 10.1a concentrates flows inward from the outskirts of communities and makes for a single outfall. However, its largest sewers quite likely traverse the most congested districts, and it is difficult to increase the capacity of the system, for example, by building *relief sewers* when outlying suburbs grow and add their flows. In the *radial pattern* of Fig. 10.1c, in contrast, wastewater flows outward from the heart of the city, as along the spokes of a wheel. Lines are relatively small and short, but the number of treatment works may be multiplied.

10.3 Collection of Sanitary Wastewater

Because about 70% of the water brought into a community must be removed as *spent water*, the average flow in sanitary sewers is about 100 gpcd (378 Lpcd) in North America. Variations in water use to step up the maximum hourly rate about threefold. Illicit stormwater and groundwater magnify the required capacity still further, and a design value of 400 gpcd (1514 Lpcd) is not uncommon.

Sanitary sewers are fouled by the deposition of waste matters unless they impart self-cleaning velocities of 2–2.5 ft./s (0.60–0.75 m/s). Except in unusually flat country, sewer grades are made steep enough to generate these velocities when the sewers are running reasonably full (half full or more in circular sections because the hydraulic radius of a semicircle equals that of a circle). Nevertheless, there will be deposition of solids. To find and remove them, sewers must be accessible for inspection and cleaning. Except in large sewers, manholes are built at all junctions with other sewers, and at all changes in direction or grade. Straight runs between manholes can then be rodded out effectively if intervening distances are not too great. Maxima of 300 or 400 ft. (90 or 120 m) for pipes less than 24 in. (600 mm) in diameter are generally specified, but effective cleaning is the essential criterion. For larger sewers, distances between manholes may be upped to as much as 600 ft. (180 m). Sewers so large that workmen can enter them for inspection, cleaning,

Figure 10.5 PVC (Polyvinyl Chloride) Plastic Sewer Installation (*Source:* Courtesy Wikipedia http://en.wikipedia.org/wiki/Image:Sansewer.jpg)

and repair are freed from these restrictions, and access manholes are placed quite far apart either above the center line or tangential to one side.

The introduction of flexible cleaning devices has encouraged the construction of curved sewers of all sizes, especially in residential areas. On short runs (<150 ft. or < 45 m) and temporary stubs of sewer lines, terminal cleanouts are sometimes substituted for manholes. They slope to the street surface in a straight run from a Y in the sewer or in a gentle curve that can be rodded out. In very flat country and in other unusual circumstances, sewers are laid on flat grades, in spite of greater operating difficulties, in order to keep depths reasonably small and pumping reasonably infrequent.

The smallest public sewers in North America are normally 8 in. in diameter. Smaller pipes clog quickly and are hard to clean. Vitrified clay or plastic is the material of choice for small sewers (Fig. 10.5); prefabricated concrete or fiberglass for large sewers. Grit or other abrading materials will wear the invert of concrete sewers unless velocities are held below 8–10 ft./s (2.4–3.0 m/s). Very large sewers are built in place, some by tunneling. Hydraulically and structurally, they share the properties of grade aqueducts.

Sewers are laid deep enough (a) to protect them against breakage, by traffic shock, for example; (b) to keep them from freezing; and (c) to permit them to drain the lowest fixture in the premises served. Common depths are 3 ft. (0.90 m) below the basement floor and 11 ft. (3.3 m) below the top of building foundations (12 ft. or 3.7 m, or more for basements in commercial districts), together with an allowance of $^1/_4$ in. per ft. or 2 cm per m (2%) for the slope of the building sewer. In the northern United States, cellar depths range from 6 to 8 ft. (1.8 to 2.4 m) and frost depths from 4 to 6 ft. (1.2 to 1.8 m). A 2-ft (0.60-m) earth cover will cushion most shocks. The deep basements of tall buildings are drained by ejectors or pumps.

Manholes are channeled to improve flow, and the entrance of high-lying laterals is eased by constructing drop manholes rather than lowering the last length of run, a wasteful arrangement. In their upper reaches, most sewers receive so little flow that they are not self-cleaning and must be flushed from time to time. This is done by (a) damming up the flow at a lower manhole and releasing the stored waters after the sewer has almost filled; (b) suddenly pouring a large amount of water into an upstream manhole; (c) providing at the uppermost end of the line a *flushing manhole* that can be filled with water through fire hose attached to a nearby hydrant before a flap valve, shear gate, or similar quick-opening device leading to the sewer is opened; and (d) installing an automatic flush tank that fills slowly and discharges suddenly. Apart from cost and difficulties of maintenance, the danger of backflow from the sewer into the water supply is a bad feature of automatic flush tanks.

Example 10.1 Capacity of Sewer

An 8-in. (200-mm) sewer is to flow full at a velocity of 2.5 ft./s (0.75 m/s).

1) What is its maximum capacity in ft^3/s and MGD using the U.S. Customary Units or m^3/s and MLD using the SI Units?
2) How many **people** can it serve if the maximum per capita flow is 300 gpd (1135 L/day)?
3) How many **acres** (or ha) will it drain if the population density is 50 per acre (124/ha)?

Solution 1 (U.S. Customary System):

1) Capacity = Area × Velocity
$= [(\pi \times 8^2)/(4 \times 144)] \times 2.5\ \text{ft}^3/\text{s}$
$= 0.87\ \text{ft}^3/\text{s}$
$= 0.87 \times 60 \times 60 \times 24 \times 7.48$ gpd
= 562,000 gpd.
= 0.562 MGD
2) People served = 562,000/300 = **1,870**.
3) Area drained = 1,870/50 = **37.4 acres**.

Solution 2 (SI System):

1) Capacity = Area × Velocity
$= [(\pi \times 0.20^2)/4] \times 0.75\ \text{m}^3/\text{s}$
$= 0.0246\ \text{m}^3/\text{s}$
$= 0.0246 \times 1{,}000 \times 1{,}440 \times 60/1000{,}000$ MLD
= 2.13 MLD
2) People served = $2.13 \times 10^6/1{,}135$ = **1,870**.
3) Area drained = 1,870/124 = **15.1 ha**.

10.4 Collection of Stormwater

Much of the suspended load of solids entering storm drains is sand and gravel. Because fine sand is moved along at velocities of 1 ft./s (0.30 m/s) or more and gravel at 2 ft./s (0.60 m/s) or more, recommended minimum velocities are 2.5–3 ft./s (0.75–0.90 m/s), or about 0.5 ft./s (0.15 m/s) more than for sanitary sewers. The following factors determine the capacity of storm drains:

1) Intensity and duration of local rainstorms
2) Size and runoff characteristics of tributary areas
3) Economy of design, determined largely by the opportunity for quick discharge of collected stormwaters into natural water courses.

Rate of storm runoff is ordinarily the governing factor in the hydraulic design of storm drains. To prevent inundation of streets, walks, and yards and flooding of basements and other low-lying structures, together with attendant inconvenience, traffic disruption, and damage to property, storm sewers are made large enough to drain away—rapidly and without becoming surcharged—the runoff from storms shown by experience to be of such intensity and frequency as to be objectionable. The heavier the storm, the greater but less frequent is the potential inconvenience or damage; the higher the property values, the more sizable the possible damage. In a well-balanced system of storm drains, these factors will have received proper recognition for the kind of areas served: residential, mercantile, industrial, and mixed. For example, in high-value mercantile districts with basement stores and stock rooms, storm drains may be made large enough to carry away surface runoff from all but unusual storms, estimated to occur only once in 5, 10, 20, 50, or even 100 years, whereas the drains in suburban residential districts are allowed to be surcharged by all but the 1- or 2-year storm.

Until storm drains have been constructed in a given area and the area itself is developed to its ultimate use, runoff measurements are neither possible nor meaningful. Accordingly, the design of storm sewers is normally based not on analysis of recorded runoff but on (a) analysis of storm rainfalls—their intensity or rate of precipitation, duration, and frequency of occurrence—and (b) estimation of runoff resulting from these rainfalls in the planned development.

Storm sewers are occasionally surcharged and subjected to pressures, but usually no more than their depth below street level. Nevertheless, they are designed for open-channel flow and equipped with manholes in much the same way as sanitary sewers. In North American practice, the minimum size of storm sewers is 12 in. (300 mm) to prevent clogging by trash of one kind or another. Their minimum depth is set by structural requirements rather than basement elevations. Surface runoff enters from street gutters through *street inlets* or *catch basins* (Fig. 10.6) and *property drains*. Size, number, and placement of street inlets govern the degree of freedom from flooding of traffic ways and pedestrian crossings. To permit inspection and cleaning, it is preferable to discharge street inlets directly into manholes. Catch basins are, in a sense, enlarged and trapped street inlets in which debris and heavy solids are held back or settle out. Historically, they antedate street inlets and were devised to protect combined sewerage systems at a time when much sand and gravel were washed from unpaved streets. Historically, too, the air in sewers, called *sewer gas*, was once deemed dangerous to health; this is why catch basins were given water-sealed traps. Catch basins need much maintenance; they should be cleaned after every major storm and may have to be oiled to prevent the production of large crops of mosquitoes. On the whole, there is little reason for continuing their use in modern sewerage systems.

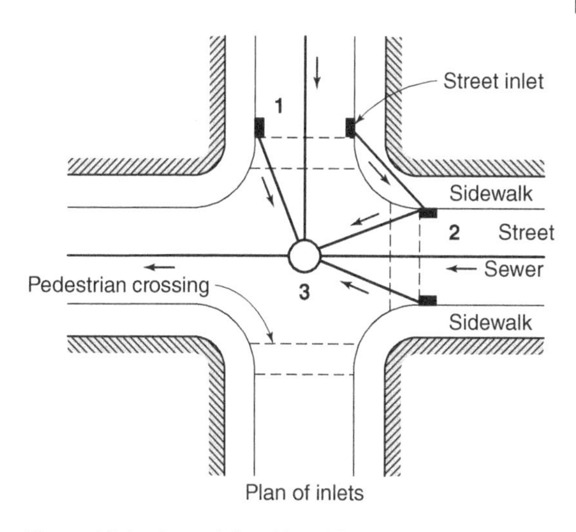

Figure 10.6 Street Inlets (1 and 2) and Their Connection to a Manhole (3)

Example 10.2 Size of storm drain

A storm sewer needs to drain an area of 37.4 acres or 15.1 ha (the area drained by an 8-in. or 200-mm sanitary sewer in Example 10.1). How large must this drain be if it is to carry away rain falling at a rate of 2 in./h (51 mm/h) during 30 min, the time needed for the entire drainage area to become tributary to the sewer? The required velocity of flow is to be 3 ft./s (0.9 m/s), and the ratio of the peak rate of runoff to the rate of rainfall in the area is assumed to be 0.6.

Solution 1 (U.S. Customary System):

1) 1 acre-in./h = 43, 560 $ft^2 \times 1/12$ ft/(60×60) = 1.0083 ft^3/s = 1.0 ft^3/sclosely enough. This is a fact to remember.
2) Rate of runoff = 2 in./h $\times 0.6 \times 37.4$ acre = 45 acre-in./h = 45 ft^3/s.
3) Cross-sectional area of drain = flow/velocity = 45/3.0 = 15 ft^2
4) **Diameter of drain** $= 12\sqrt{4 \times 15/\pi} =$ **53 in.**
5) Ratio of storm runoff to sanitary wastewater (Example 10.1): 45.0 : 0.87 = 52.1; that is, sanitary wastewater, if admitted, would constitute less than 2% of the combined flow.

Solution 2 (SI System):

1) 1 ha-mm/h = [10,000 $m^2 \times 1/1{,}000$ m/(60 min $\times$ 60 s)] $m^3/s \times 1{,}000$ = 2.78 L/s. This is a fact to remember.
2) Rate of runoff = (51 mm/h $\times$ 15.1 ha) mm-ha/h $\times 2.78 \times 0.6/1{,}000$ = 1.25 m^3/s
3) Cross-sectional are of drain = flow/velocity = 1.25/0.90 = 1.39 m^2
4) **Diameter of drain** $= \sqrt{4 \times 1.39/\pi} = 1.33$ m. Use **1400 mm**
5) Ratio of storm runoff to sanitary wastewater (Example 10.1) 1.25 : 0.0246 = 51 : 1; that is, sanitary wastewater, if admitted, would constitute less than 2% of the combined flow.

10.5 Collection of Combined Wastewater

In combined sewerage systems, stormwaters often exceed sanitary wastewater by 50–100 times (Example 10.2), and the accuracy with which rates of surface runoff can be estimated is generally less than the difference between rates of stormwater and combined wastewater flows. Accordingly, most combined sewers are designed to serve principally as storm drains. Understandably, however, they are placed as deep as sanitary sewers. Surcharge and overflow of combined sewers are obviously more objectionable than the backing up of drains that carry nothing but stormwater. Moreover, they are given velocities up to 5.0 ft./s (1.5 m/s) to keep them clean.

The wide range of flows in combined sewers requires the solution of certain special problems, among them the choice of a cross-section that will ensure self-cleaning velocities for both storm and dry-weather flows and provision of stormwater overflows in intercepting systems. Departures from circular cross-sections are prompted by hydraulic as well as structural and economic considerations. Examples are the *egg-shaped* sections and *cunettes* illustrated in Fig. 10.7. Two circular sewers, an underlying sanitary sewer, and an overlying storm drain are fused into a single egg-shaped section. The resulting hydraulic radius is nearly constant at all depths. Cunettes form troughs dimensioned to the dry-weather flow, or sanitary wastewater. *Rectangular sections* are easy to construct and make for economical trenching with low headroom

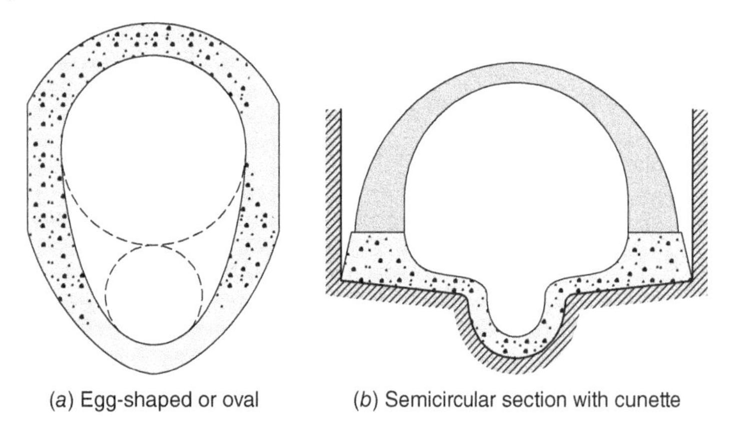

Figure 10.7 Sections of Combined Sewers

requirements. *Horseshoe* sections are structurally very satisfactory; egg-shaped sections are not. Large outfall sewers have been built as pressure tunnels.

The amounts of water entering interceptors at junctions with submains must be controlled. Only as much should be admitted as individual interceptor reaches can carry without being surcharged. Higher flows must be diverted into stormwater overflows. Admission and diversion can be regulated hydraulically or mechanically.

10.6 Choice of Collecting System

Apart from questions of economy, the combined system of sewerage is at best a compromise between two wholly different objectives: water carriage of wastes and removal of flooding runoff. In the life of growing communities, initial economies are offset in the long run (a) by undesirable pollution of natural water courses as a result of stormwater spills and consequent nuisance or, at least, the debased aesthetic and recreational values of the receiving bodies of water; (b) by the increased costs of treating and pumping intercepted wastewater; and (c) by more obnoxious conditions when streets and basements are flooded by combined wastewater instead of stormwater. In the past, small streams, around which parks and other recreational areas could have grown, have been forced into combined wastewater systems because pressing them into service as receiving waters had degraded them into open sewers. By contrast, a separate system of sewerage can exploit natural water courses hydraulically by discharging stormwater into them through short runs of storm drains while preserving their aesthetic and recreational assets. However, they may have to be channelized if they are to perform well.

10.7 Treatment and Disposal or Reuse of Wastewater

The sewerage system is a simple and economical means of removing unsightly, putrescible, and dangerous wastes from households and industries. However, it concentrates potential nuisances and dangers at the terminus of the collecting system. If rivers and canals, ponds and lakes, and tidal estuaries and coastal waters are not to become heavily polluted, the load imposed on the transporting water must be *unloaded* prior to its disposal into the receiving bodies of water. The unloading is assigned to wastewater treatment plants to prevent (a) contamination of water supplies, bathing places, shellfish beds, and ice supplies; (b) pollution of receiving waters that will make them unsightly or malodorous and eutrophication of ponds and lakes; (c) destruction of food fish and other valuable aquatic life; and (d) other impairment of the usefulness of natural waters for recreation, commerce, and industry. The required degree of treatment before disposal depends on the nature and extent of the receiving water and on the regional water economy.

In the treatment of wastewater before reuse in irrigation, full recovery of the *water value* is intended together with as much recovery of *fertilizing value* as is consistent with (a) avoiding the spread of disease by crops grown on wastewater-irrigated lands or animals pastured on them; (b) preventing nuisances such as unsightliness and bad odors around disposal areas; and (c) optimizing, in an economic sense, wastewater disposal costs and agricultural returns. The design of the irrigation areas themselves is based on the nature and size of available lands and the purposes they can serve in the regional agricultural economy.

As countries grow and their waters are used more widely and intensively, the discharge of raw or inadequately treated wastewater into their streams, lakes, and tidal waters becomes intolerable. The daily load of solids imposed on domestic wastewater amounts to about half a pound per person. The resulting mixture of water and waste substances is very dilute—less than 0.1% solid matter by weight when wastewater flows are 100 gpcd (378 Lpcd). Some industrial wastewaters are far more concentrated. Floating and other suspended solids render wastewater and its receiving waters unsightly; settling solids build up sludge banks; organic solids cause wastewater and its receiving waters to putrefy; and pathogenic bacteria and other organisms make them dangerous.

10.7.1 Wastewater Treatment Processes

Waste matter is removed from transporting water in a number of different ways. In municipal wastewater treatment works of fair size, the following processes and devices are common (see Fig. 10.8):

1) Bulky floating and suspended matter is strained out using *racks* and *screens* that produce rakings and screenings. Cutting racks and screens comminute the rakings and screenings in place and return them to the wastewater.
2) Oil and grease are skimmed off after rising during quiescence in *flotation tanks*, producing skimmings.
3) Heavy and coarse suspended matters are allowed to settle to the bottom of stilling chambers, such as *grit chambers*, flotation tanks, *settling tanks*, or *sedimentation basins*, producing grit, detritus, or sludge.
4) Suspended matter and dissolved solids that do not settle are converted into settleable solids and become amenable to sedimentation by flocculation and precipitation with chemicals. *Chemical flocculation* or *precipitation tanks* are used to produce sludge precipitates.
5) Colloidal and dissolved organic matter is metabolized and converted into a settleable cell substance by biological growths or slimes. The hosts of living cells that populate the slimes utilize the waste matter for growth and energy. Their growth

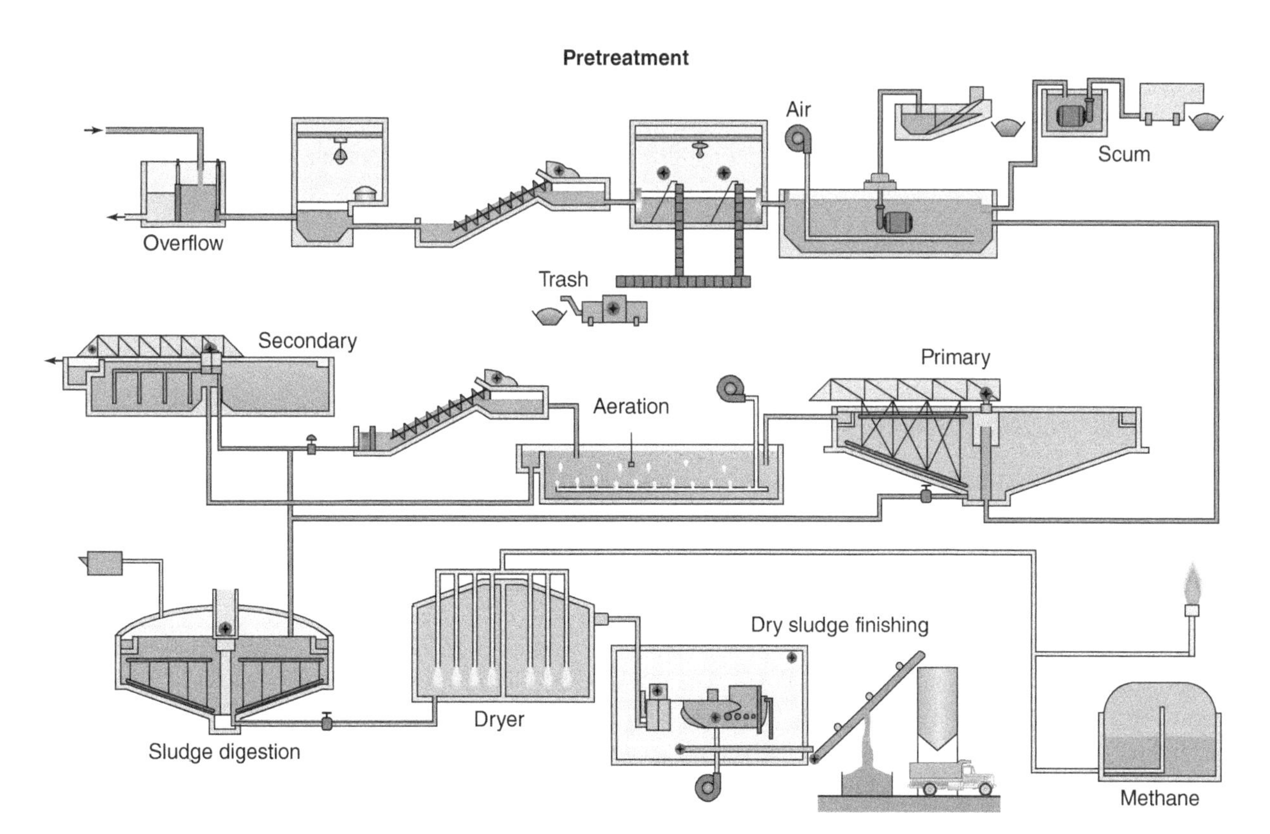

Figure 10.8 Wastewater Treatment Processes Flow Diagram (*Source:* Leonard G, Wikipedia / http://en.wikipedia.org/wiki/Image:ESQUEMPEQUE-EN.jpg, last accessed Sept 16 2023 / CC BY)

unfolds large interfaces at which adsorption, absorption, diffusion, and other interfacial forces or contact phenomena bring about exchanges between wastewater and slimes. To remain active and aerobic, the biomasses are supplied with air. They are either supported on beds of granular material, such as broken stone over which the wastewater trickles more or less continuously, or they are generated in the flowing wastewater, returned to it in wanted amounts, and kept in suspension by agitating the mixed liquor with air. This can also be done mechanically using *trickling filters* and *activated-sludge tanks*, which produce trickling filter humus and excess activated sludge.

6) Excess nitrogen and phosphorous nutrients are usually removed by advanced design and management of the biological treatment processes mentioned in point 5.
7) Some pathogenic bacteria and other organisms are removed from wastewater along with the solids in which they are embedded or to which they cling. Others die because the imposed environment is too unfavorable. Fuller and more direct destruction is accomplished by disinfection with *ozonation* or *chlorination units*.

10.7.2 Sludge Management and Disposal

The solids separated from wastewater in treatment works contain much water and organic matter. Sludge settling from wastewater by plain sedimentation contains about 95% water; activated sludge, 98% or more. If the daily flow of wastewater is millions of gallons, the daily volume of sludge is thousands of gallons. This makes them bulky and putrescible. To simplify handling and reuse/disposal, sludge is dewatered and stabilized in varying degrees. The following ways and devices are common (see Fig. 10.8):

1) Organic matter in sludge stored in tanks is metabolized and converted into relatively stable residues by bacteria and other saprophytic organisms. Continuing hosts of living things use the waste matters in the digesting sludge for growth and energy. Liquefaction and gasification occur. Dissolved oxygen disappears, and the biomass becomes anaerobic. *Sludge-digestion tanks* produce concentrated, stabilized, and rapidly dewatering sludges, sludge liquor, and gases of decomposition, primarily methane and carbon dioxide.
2) Water is removed from sludge (usually digested sludge) and run into beds of sand or other granular materials by evaporation of moisture to the air and percolation of water into and through the beds. This is accomplished with *sludge-drying beds* that produce a spadable sludge cake.
3) Sludge is dewatered by passing it (usually after chemical conditioning) through a centrifuge, filter press, or a filter medium such as cloth or coiled wires wound around a drum to which a vacuum is applied. *Centrifuges*, *presses,* and *vacuum filters* are used for this purpose to produce a sludge cake or paste.
4) Sludge cake or paste is dried by means of heat in *flash driers* that produce commercially dry sludge granules.
5) The biosolids produced by sludge treatment are applied to land as a *soil amendment and fertilizer* for agricultural land.
6) Organic matter in partially dewatered or heat-dried sludge is burned as a fuel in *incinerators,* producing ash.
7) Sludge is thickened in advance of digestion, drying, or dewatering by stirring. *Sludge thickeners* produce a more concentrated sludge.
8) Organic matter in thickened sludge is destroyed by wet combustion in *sludge retorts* operating at high pressures and temperatures and producing readily dewatered, mineralized residues.

The marsh gas or methane released during digestion is a combustible gas of high fuel value and is put to good and varied uses in modern treatment works. After digestion, wastewater solids are no longer recognizable as such. Their colloidal structure has been destroyed and they dry rapidly in the air. Heat-dried sludge is stable and essentially sterile. The final product of incineration is ash. Many of the treatment methods reduce the number of possibly pathogenic organisms concentrated in the sludge to within the standards for safe reuse.

Two flow diagrams of wastewater treatment plants are presented in Fig. 10.9. Numerous other combinations of treatment processes are possible. The plants shown provide so-called primary and secondary treatment of both wastewater and sludge. Partial treatment of either or both is often enough. However, more complete treatment may be needed at critical times of the year; for example, during low summer runoff and high recreational use of receiving waters. Tertiary treatment, also called water renovation, may be provided to remove residual nutrients and toxic, foaming, or otherwise objectionable substances.

The works illustrated in Fig. 10.9 will remove from 80% to 95% or more of the suspended solids, putrescible matter, and bacteria. Effluent chlorination can ensure 99% or higher destruction of bacteria. Partial treatment can achieve removal values between 40% and 70%.

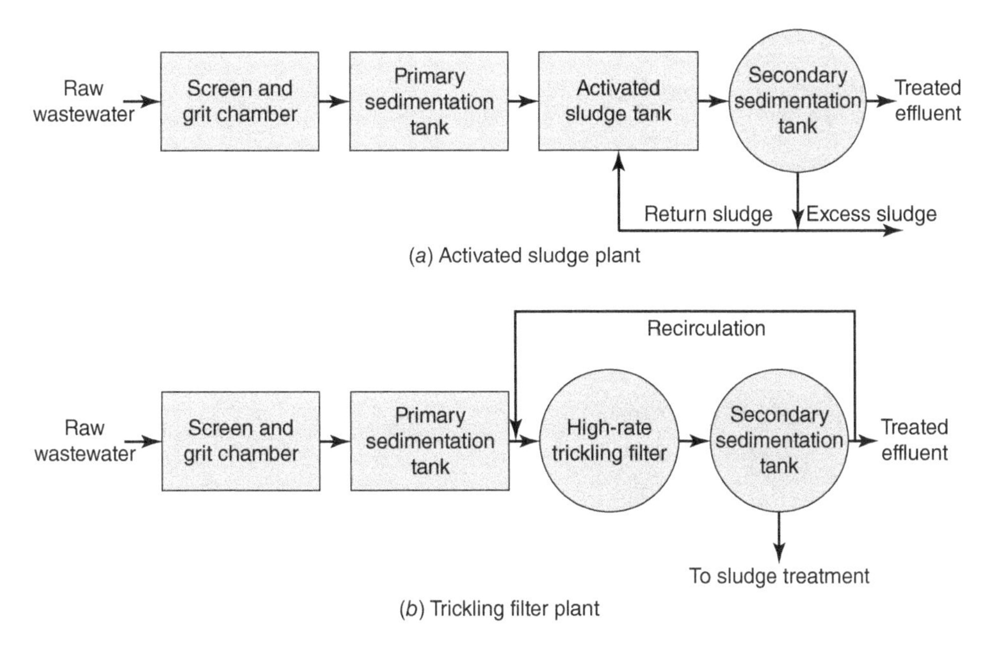

Figure 10.9 Common Types of Wastewater Treatment Plants

10.7.3 Sizing of Wastewater Treatment Works

The design of wastewater treatment works is based on an understanding of (a) treatment processes and devices (process design); (b) factors affecting the flow of wastewater, sludge, and often air, through the structures employed (hydraulic and pneumatic design); (c) behavior of needed structures and mechanisms under load (structural and mechanical design); and (d) treatment costs relative to benefits received (economic design). As a rule of thumb, the following values (normally applicable to domestic wastewater) give some concept of the size of principal structures:

1) Primary settling tanks hold wastewater for about 2 h and are rated at about 900 gpd/ft^2 (36.7 m^3/d/m^2) of water surface. Here primary flotation clarifiers can be used to replace the primary settling tanks when oil and grease (O&G) concentration is high in raw wastewater.
2) Secondary settling tanks, following biological treatment, have detention periods of about 1.5 h and surface loadings up to 1800 gpd/ft^2 (73.4 m^3/d/m^2). Here secondary flotation clarifiers can be used to replace the secondary settling tanks when necessary.
3) Heated separate sludge digestion tanks have a combined capacity of about 2 ft^3/capita (56.6 L/capita) in the northern United States when the digested sludge is to be air-dried. Agitation reduces required detention capacity, and favorable drying conditions as well as mechanical dewatering reduce required storage capacity.
4) Trickling filters (Fig. 10.10) are rated at about 3 MG/acre/d (28 ML/ha/d) in conventional operation and at about 25 MG/acre/d (233 ML/ha/d) in high-rate operation.
5) Activated sludge tanks aerate the wastewater and returned activated sludge, which equals about 25% of the volume of wastewater, for about 6 h in conventional operation; both time of treatment and returned sludge volumes are modified in numerous variations on the conventional process.
6) Open drying beds for digested sludge provide an area of about 1 ft^2/capita (0.093 m^2/capita) in the northern United States. Glass covering and favorable climate lower the required area.

Example 10.3 Capacity of Trickling Filter

Estimate the capacity of the components of a small trickling-filter plant treating 0.8 MGD (3 MLD) of domestic wastewater from 8000 people, assuming the following:

a) Sewage flow production = 100 gpcd (378 Lpcd)
b) Two primary settling tanks with a depth of 10 ft. (3.1 m)

Figure 10.10 Trickling Filter Bed Using Plastic Media (*Source:* Courtesy Wikipedia http://en.wikipedia.org/wiki/Image: Trickling_filter_bed_2_w.JPG)

c) Two sludge digestion tanks with a storage requirement of 2 ft^3 (0.057 m^3) per capita
d) Four sludge drying beds with an area requirement of 1 ft^2 (0.093 m^2) per capita
e) One trickling filter with a loading of 3 MG/acre/d (28 ML/ha/d = 0.0028 ML/m^2/d)

Solution 1 (U.S. Customary System):

1) Primary settling in two tanks, averaging 10 ft. in depth.
   a) Assumed detention period = 2 h.
   b) Effective volume/tank = flow × time = 8,000 × 100 × (2/24)/2 = 33,300 gal = 4,460 ft^3.
   c) Surface area = 4,460/10 = 446 ft^2.
   d) Dimensions such as **10 ft × 45 ft**.
   e) Surface hydraulic loading = (800,000/2)/450 = 890 gpd/ft^2.
2) Sludge digestion in two tanks:
   a) Assumed storage requirement = 2 ft^3/capita.
   b) Effective volume 2 × 8,000/2 = 8,000 ft^3.
   c) Assuming an area equal to settling tanks, depth below settling compartment = 8,000/450 = **18 ft** (plus 2 ft. to keep sludge clear of slots).
3) Sludge-drying beds, four in number:
   a) Assumed area requirement = 1 ft^2/capita.
   b) Effective area 1 × 8,000/4 = 2,000 ft^2.
   c) **Dimensions** such as 20 ft × 100 ft.
4) Trickling filters, one in number:
   a) Assumed loading = 3 MG/acre/d.
   b) Effective area = 0.8/3 = 0.27 acre = 43,560 × 0.27 = 10,500 ft^2.
   c) **Diameter** = $(10{,}500 \times 4/\pi)^{1/2}$ = **116 ft.**

Solution 2 (SI System):

1) Primary settling in two tanks, averaging 3.1 m in depth.
   a) Assumed detention period = 2 h.
   b) Effective volume/tank = flow × time = 8,000 × 378 × (2/24)/2 = 126,000 L = 126 m^3.

c) Surface area = 126 m^3/3.1 m = 40.6 m^2.
d) **Dimensions** such as **3 m × 13.6 m**.
e) Surface hydraulic loading = [(8,000 × 378/2)/1,000 m^3]/(3 m × 13.6 m) = 37.1 m^3/d/m^2.

2) Sludge digestion in two tanks:
a) Assumed storage requirement = 0.057 m^3/capita.
b) Effective volume 0.057 × 8,000/2 = 228 m^3each.
c) Assuming an area equal to settling tanks, **depth below settling compartment** = 228 m^3/3m × 13.6 = **5.6 m(plus 0.6 m** to keep sludge clear of slots).

3) Sludge-drying beds, four in number:
a) Assumed area requirement = 0.093 m^2/capita.
b) Effective area 0.093 × 8,000/4 = 186m^2.
c) **Dimensions** such as **6 m × 31 m**.

4) Trickling filters, one in number:
a) Assumed loading = 0.0028 ML/m^2/d.
b) Effective area = [8,000 × 378/10^6] MLD/[0.0028 MLD/m^2] = 1,080 m^2.
c) **Diameter** = $(1{,}080 \times 4/\pi)^{1/2}$ = **37 m**.

10.7.4 Disposal into Receiving Waters

Outfalls into receiving waters should terminate well below low-water mark. Wastewater or effluent is dispersed effectively when a number of outlets, called *diffusers*, are (a) spaced sufficiently far apart to prevent interference and (b) situated at or near the bottom of the receiving water to keep the generally warmer and lighter wastewater from spreading over the receiving water in a persistent layer (Fig. 10.11). Density differences make themselves felt especially in marine outfalls. Strength, direction, and dimension of prevailing currents and the likelihood of their reaching water works intakes, bathing places, shellfish laying beds, and other important spots are matters for study.

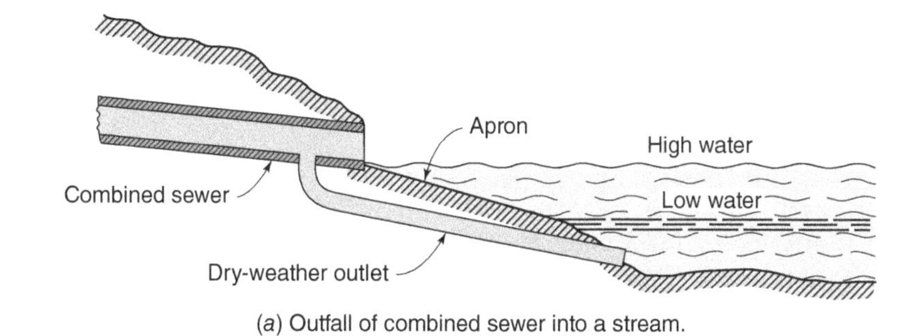

(a) Outfall of combined sewer into a stream.

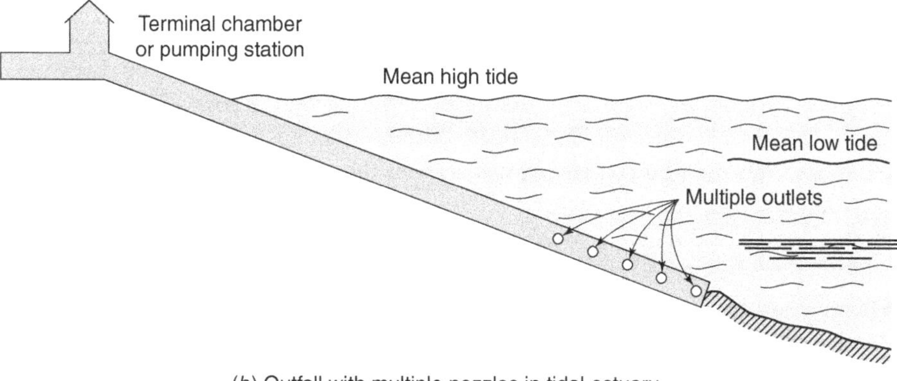

(b) Outfall with multiple nozzles in tidal estuary.

Figure 10.11 Wastewater Outfalls

Whatever the relative dilution, the forces of natural purification, or *self-purification*, inherent in natural bodies of water can, in the course of time and distance, ultimately return the receiving water close to its original state of cleanliness. However, the enrichment of lakes, ponds, and impounding reservoirs with plant nutrients (*eutrophication*) and the resulting ecological changes in receiving waters of this kind are quite another thing. In a sense, natural purification is the prototype of biological treatment. Inherent metabolic activities remain aerobic as long as the rate of oxygen supply is not outbalanced by the rate of oxygen demand. The turbulence of streams usually keeps their running waters aerobic. However, their benthal environment may become anaerobic because oxygen diffuses into bottom deposits only slowly. Anaerobic conditions are most likely to appear in deep-standing waters in which stratification makes for stagnation and consequently poor oxygen transport to the low-water strata. Heavily polluted streams may become, for a time at least, black, unsightly, septic, and malodorous bodies of water in which the normal aerobic, clean-water flora and fauna have given way to a different, generally less acceptable, assemblage of living things.

A rule of thumb formulated by engineers for large American rivers and giving a rough estimate of the amount of dilution that would be required in the absence of treatment if combined wastewater from 1,000 persons was to be discharged without much nuisance is (a) for swift streams, 2.5 ft^3/s (0.071 m^3/s) of diluting water; (b) for normal streams, 6 ft^3/s (0.17 m^3/s); and (c) for sluggish streams, 10 ft^3/s (0.28 m^3/s). For domestic wastewater, treated wastewater, and industrial wastes, *equivalent* populations should be substituted. Combined wastewater averages about 40% stronger than domestic wastewater in an industrial area. Industrial wastewaters may be weaker or stronger. Treated wastewater is weaker in proportion to the amount of putrescible matter removed or destroyed. Where the emphasis is on water supply, recreational enjoyment of water, and conservation of fish and other useful aquatic life, dilution or treatment becomes more urgent.

Example 10.4 Wastewater Treatment and Dilution in Stream Water

The low-water flow of a normally rapid stream draining 2,000 $mile^2$ (5,180 km^2) is 0.1 $ft^3/s/mi^2$ (0.0656 $m^3/min/km^2$). Estimate the extent to which domestic wastewater from a city of 80,000 people must be treated before discharge into the stream if nuisance is to be avoided; also determine the resulting dilution ratio of wastewater to stream water. Assume a per capita flow of wastewater of 100 gpd 378 L/d.

Solution 1 (U.S. Customary System):

1) Low-water flow $= 0.1 \times 2{,}000 = 200\ ft^3/s$.
2) Required flow for disposal of domestic wastewater if it is left untreated $= 6 \times (80{,}000/1{,}000)/1.4 = 340\ ft^3/s$.
3) Percent **removal** of pollution load needed $= 80(340 - 200)/340 = \mathbf{33\,\%}$.
4) **Dilution ratio** $= (80{,}000 \times 100) : (200 \times 7.48 \times 60 \times 60 \times 24) = \mathbf{1:16}$.

Solution 2 (SI System):

1) Low-water flow $= (0.0656\ m^3/min/km^2)(5{,}180\ km^2) = 340\ m^3/min = 5.67\ m^3/s$.
2) Required flow for disposal of domestic wastewater if it is left untreated $= 0.17 \times (80{,}000/1{,}000)/1.4 = 9.71\ m^3/s$.
3) Percent **removal** of pollution load needed $= 80(9.71 - 5.67)/9.71 = \mathbf{33\,\%}$.
4) **Dilution ratio** $= (80{,}000 \times 378/1000) : (5.67 \times 60 \times 60 \times 24) = \mathbf{1:16}$

10.7.5 Disposal onto Land

The objectives of terminal discharge of wastewater onto land or into the soil are safe disposal and, possibly, crop production. In municipal practice, *disposal by irrigation* is viable when the water resources of the region are poor and large tracts of suitable land are available. The extent to which wastewater should be treated before irrigation depends on hygienic considerations and available standards. If wastewater is not sufficiently treated, there is the obvious hazard of contaminating food raised on irrigated soil and infecting animals pastured on irrigated land. By contrast, the discharge of settled wastewater into the ground through agricultural tile pipes in the disposal of wastewater from isolated dwellings, known as subsurface irrigation, can be quite safe.

In one sense, shallow earth basins holding wastewater for a number of days and called *wastewater lagoons* or *stabilization ponds* are purposely inundated or waterlogged irrigation areas producing suspended (algal) rather than rooted crops. There is much evaporation from the ponds and some seepage. In another sense, the displacement of pond waters is often fast enough to approach the natural purification of sluggish receiving streams. Under favorable climatic conditions, pond loadings may be as high as 500 persons per acre (1235 persons per ha) for raw domestic wastewater.

Example 10.5 Wastewater Application on Land for Irrigation
Estimate the daily volume of wastewater that can be applied on an acre of land and the land area required to dispose of the domestic wastewater from a community of 10,000 people by irrigation and through stabilization ponds 4 ft. (1.22 m) deep. Assume that the annual depth of water employed in the irrigation of crops is 10 in. (254 mm) and that the wastewater flow is 80 gpcd (303 Lpcd).

Solution 1 (U.S. Customary System):

1) Rate of irrigation = $[(10/12) \times 7.5 \times 43{,}560]/365 =$ **750 gal/acre/d**.
2) Irrigation area $= 10{,}000 \times 80/750 = 1{,}100$ acres = **1.7 mile^2**.
3) Stabilization-pond area $10{,}000/500 =$ **20 acres**.
4) Detention period in ponds = $20 \times 43{,}560 \times 4 \times 7.48/(10{,}000 \times 80) =$ **33 days**.

Solution 2 (SI System):

1) Rate of irrigation = (254 mm/yr) (1 yr/365 d) (1 m/1,000 mm) (1 m^2/1 m^2) = 0.000696 m^3/m^2/d = 696 m^3/km^2/d = **6.96 m^3/ha/d**.
2) Irrigation area = $(10{,}000 \times 303)$ L/d/696,000 L/km^2/d = **4.35 km^2**.
3) Stabilization-pond area 10,000 persons/(1,235 persons/ha) = **8.1 ha**.
4) Detention period in ponds = $[(8.1 \text{ ha} \times 10{,}000 \text{ m}^2/\text{ha}) \times 1.22 \text{ m} \times 1{,}000 \text{ L/m}^2]/[10{,}000 \times 303]$ L/d = **33 days**.

10.8 Disposal of Industrial Wastewater

Most water-carried industrial wastes can safely be added to municipal wastewater for treatment and disposal. Some wastes, however, are so strong that they damage collection systems and interfere with or overload treatment facilities; pretreatment or separation from the collection system then becomes mandatory. The requisite degree of preparatory treatment depends on the composition, concentration, and condition of the wastes, the nature and capacity of the treatment works, and the nature and capacity of the receiving waters. Shock loads, through sudden release of batches of wastes, are especially objectionable. Holding or storage basins will dampen shocks if they apportion waste discharge to available treatment plant and receiving water capacities.

Wastes rich in carbohydrates, proteins, and fats and not unlike domestic wastewater in their degradability have a valid *population equivalent*. For example, the putrescible matter in the combined wastes from a distillery that processes 1000 bushels of grain a day is equivalent to the wastewater from 3500 people. By contrast, other industrial wastes may persist in water without much change. Some may even interfere with wastewater treatment and natural purification in receiving waters. Copper and other metal wastes are examples. At high concentrations, copper inhibits the anaerobic digestion of settled solids and destroys the biomass in trickling filters and activated sludge units. New synthetic organics may also be quite destructive. Yet it is possible to acclimatize biological slimes to many otherwise toxic organic chemicals, such as phenols and formaldehyde, which are degraded in treatment plants and recciving waters.

Guiding principles in the solution of industrial wastes problems are, in order of preference, (a) recovery of useful materials, (b) waste minimization by improvement of manufacturing processes whereby waste matters and waste waters are reduced in amount, (c) recycling of process waters, and (d) treatment of the waste. Recovery is most successful when the substances recovered are either very valuable or otherwise not so unlike the primary products of manufacture that a separate management organization must be developed. Improvement in manufacturing processes may permit the discharge of remaining waste matter into public sewers.

Satisfactory treatment processes are available for a wide variety of industrial wastes. Most of them are not unlike established wastewater treatment methods, but some chemical wastes require quite a different disposal approach. For example, cyanides from plating industries are most conveniently oxidized to cyanates, chromates from the same source are most conveniently reduced to chromic compounds, and acids and alkalis from many industries are most conveniently neutralized.

10.9 Systems Planning Management and Costs

The planning, design, and construction of water and wastewater systems for metropolitan areas usually bring together sizable groups of engineering practitioners and their consultants, for months and even for years. Under proper leadership, task forces perpetuate themselves in order to attack new problems or deal with old ones in new ways. For publicly owned wastewater systems, studies, plans, specifications, and construction contracts are prepared by engineers normally engaged by the cities and towns or by the wastewater districts to be served. Wastewater engineers may belong to the professional staff of municipal or metropolitan governmental agencies responsible for designing and managing public works, or they may be attached to private works or to firms of consulting engineers. To accomplish large new tasks, permanent staffs may be temporarily expanded. For smaller works, consultant groups may be given most and possibly all of the responsibility. Engineers for manufacturers of wastewater equipment also play a part in systems development. The engineers of construction companies bring the designs into being.

Construction of new wastewater systems, or the upgrading and extension of existing ones, progresses from preliminary investigations or planning through financing, design, and construction to operation, maintenance, and repair. Political and financial procedures are involved, as well as engineering.

The first cost of sanitary sewers lies between USD 223 and USD 744 per capita in the United States. Storm drains and combined sewers, depending on local conditions, cost about three times as much.. The first cost of wastewater treatment works varies with the degree of treatment provided. Depending on plant size, which, for wastewater, is more clearly a function of the population load than the volume of water treated, the per capita cost of conventional wastewater treatment works as of 2022 was as follows:

1) Mechanized settling and heated sludge digestion tanks USD 242 per capita, varying approximately as $1/(P \times 10^{-4})^{1/3}$
2) Activated sludge units as well as primary treatment USD 350 per capita, also varying approximately as $1/(P \times 10^{-4})^{1/4}$
3) Trickling filters as well as primary treatment USD 335 per capita, varying approximately as $1/(P \times 10^{-4})^{2/7}$
4) Stabilization ponds USD 6.70 per capita, varying approximately as $1/(P \times 10^{-4})^{1/4}$.

The approximate annual per capita operating and maintenance cost for the plants themselves, that is, exclusive of central administrative expenses, is as follows:

1) For *primary* plants, USD 20.0, 10.4, 6.8, and 5.0 per capita for communities of 10^3, 10^4, 10^5, and 10^6 people, respectively
2) For activated-sludge plants similarly USD 68.4, 26.0, 14.1, 8.9 and 6.6 per capita for communities of 10^3, 10^4, 10^5, and 10^6 people, respectively
3) For conventional trickling-filter plants USD 26.0, 9.7, and 5.6 per capita for communities of 10^3, 10^4, and 10^5 people, respectively
4) For high-rate trickling-filter plants USD 34.2, 10.4 and 5.5 per capita for communities of 10^3, 10^4, and 10^5 people, respectively.

Including interest and depreciation, as well as charges against operation and management, the removal of domestic wastewater, and its safe disposal costs from USD 372 to USD 744/MG (USD 99.2 to USD 198.4/ML). In comparison with water purification plants, wastewater treatment works are relatively twice as expensive; in comparison with water distribution systems, collection systems for domestic wastewater are about half as expensive. Sewer use charges, like charges for water, can place the cost of sewerage on a value-received basis. Use charges may cover part or all of the cost of the service rendered and are generally related to the water bill as a matter of equity.

Example 10.6 Costs of Sanitary Sewerage and Treatment Plant

Roughly, how much money is invested in the sanitary sewerage system of a city of 100,000 people?

Solution:

1) Assuming the per capita cost of sewers at USD 558, the total first cost is

$P = 100,000 = 10^5$

USD 558 x 100,000 = USD 55,800,000

2) Assuming that the wastewater is treated in an activated sludge plant, the expected cost is

$$= \text{USD } 350 \times P/(P \times 10^{-4})^{1/4}$$
$$= \text{USD } 350 \times 10^5/(10^5 \times 10^{-4})^{1/4}$$
$$= \text{USD } 350 \times 10^5/(10)^{1/4}$$
$$= \text{USD } 19{,}682{,}000$$

No general costs can be assigned to separate treatment of industrial wastewaters. When they are discharged into municipal sewerage systems, treatment costs can be assessed in terms of loads imposed on the municipal works as suspended solids, putrescible matter, or a combination of the two.

10.10 Individual Wastewater Systems

In the absence of public sewerage, wastewaters from individual and rural dwellings and ancillary buildings are normally discharged into the ground. The absorptive capacity of the soil is then of controlling importance. It is greatly increased if settleable waste matter is first removed, for example, by sedimentation combined with digestion and consolidation of the deposited sludge and scum. Sedimentation and digestion are accomplished more often than not either in *leaching cesspools* or in *septic tanks* (or tight cesspools). *Subsurface absorption fields* or *seepage pits* follow. An individual wastewater system as well as a septic tank and absorption field are shown in Fig. 10.12.

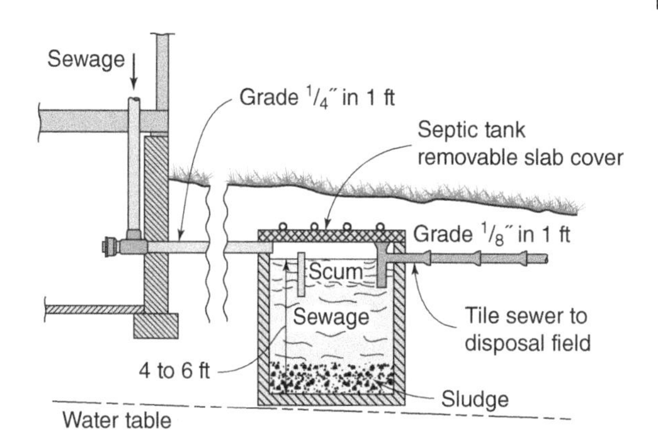

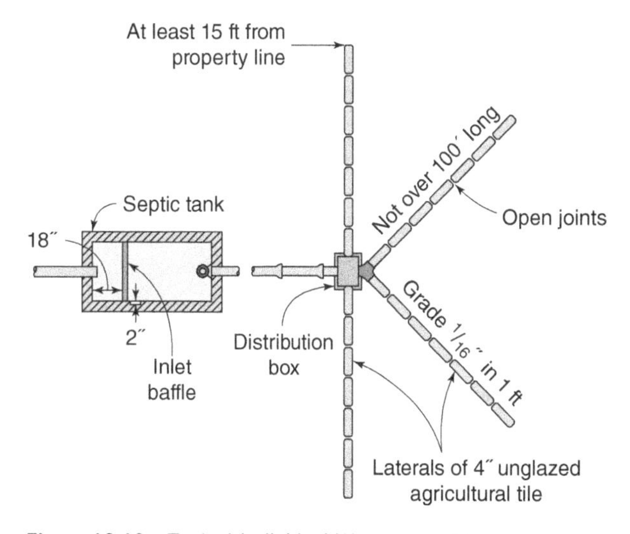

Figure 10.12 Typical Individual Wastewater Disposal System: Septic Tank and Subsurface Absorption Field. Conversion factors: 1″ = 1 in. = 25.4 mm; 1′ = 1 ft = 12 in. = 0.3048 m

Problems/Questions

10.1 A 12-in. sewer is to flow full at a velocity of 3 ft./s. The maximum per capita flow is 350 gpcd. Determine using the US Customary Units (a) the maximum capacity of wastewater flow in ft^3/s and MGD and (b) the number of people the sewer can serve.

10.2 A 300-mm sewer is to flow full at a velocity of 0.90 m/s. The maximum per capita flow is 1300 Lpcd (liter per capita per day). Determine using the SI Units (a) the maximum capacity of wastewater flow in m^3/s and L/day and (b) the number of people the sewer can serve.

10.3 A storm sewer is to drain an area of 40 acres or 16.2 ha (the area drained by a 12-in. or 300-mm sanitary sewer in Examples 10.1 and 10.2). What is the diameter of this storm sewer if it is to carry away rain falling at a rate of 2 in./h (50 mm/h) during 30 min, the time needed for the entire drainage area to become tributary to the sewer? The required velocity of flow is to be 3 ft./s (0.90 m/s), and the ratio of the peak rate of runoff to the rate of rainfall in the area is assumed to be 0.65.

10.4 Estimate the capacity of the components of a small trickling-filter plant treating 1.6 MGD (6 MLD) of domestic wastewater contributed by 16,000 people.

10.5 Estimate the capacity of the components of a small activated sludge plant treating 1.6 MGD (6 MLD) of domestic wastewater contributed by 16,000 people.

10.6 Estimate the following for a community of 12,000 persons. Assume that the annual depth of water employed in the irrigation of crops is 12 in. (300 mm) and that the wastewater flow is 90 gpcd (340 Lpcd).

(a) The daily volume of wastewater that can be disposed of on an acre of land
(b) The land area required to dispose of the domestic wastewater by irrigation
(c) The land area required to dispose of the domestic wastewater through 4-ft (1.22-m) deep stabilization ponds
(d) The detention period in ponds.

10.7 About 2 million people die every year due to diarrheal diseases; most of them are children less than five years of age. The most affected are the populations in developing countries, living in extreme conditions of poverty, normally peri-urban dwellers or rural inhabitants. What are the main problems responsible for this situation?

10.8 What are the three main elements of a wastewater system?

10.9 Differentiate between domestic and municipal wastewaters.

10.10 Explain using sketches of the three patterns that are used in planning a sewerage system.

10.11 Why do we use manholes in a sewerage system? Discuss their purpose and consequently where they have to be located.

10.12 What are the drawbacks of combined sewers?

10.13 What are the principles in the solution of industrial waste problems?

10.14 What is the most common solution to individual and rural wastewater disposal?

Bibliography

Fair, G. M., Geyer, J. C., and Okun, D. A., Water and wastewater engineering, in *Water Supply and Wastewater Removal*, vol. *vol. 1*, John Wiley & Sons, New York, 1966.

Fair, G. M., Geyer, J. C., and Okun, D. A., *Elements of Water Supply and Wastewater Disposal*, John Wiley & Sons, New York, 1971.

Khadam, M., Shammas, N. K., and Al-Feraiheedi, Y., Water losses from municipal utilities and their impacts, *Water Int.*, *13*, 1, 254–261, 1991.

Rowan, P. P., Jenkins, K. L., and Butler, D. W., Sewage treatment construction costs, *J. Water Poll. Control Fed.*, *32*, 594, 1960.

Rowan, P. P., Jenkins, K. L., and Howells, D. H., Estimating sewage treatment operation and maintenance costs, *J. Water Poll. Control Fed.*, *33*, 111, 1961.

Shammas, N. K., The cracking problem in Unplasticized polyvinyl chloride fittings used in Riyadh domestic sewer connections, *J. Pipelines*, *4*, 299–307, 1984.

Shammas, N. K., Wastewater Management and Reuse in Housing Projects, in *Water Reuse Symposium IV, Implementing Water Reuse*, August 2–7, AWWA Research Foundation, Denver, CO, 1987, pp. 1363–1378.

Shammas, N. K., Wastewater Reuse in Riyadh, Saudi Arabia, in *Symposium on Water Reuse, Third North American Chemical Congress, Division of Environmental Chemistry, Toronto, Canada, June 5–11*, 1988.

Shammas, N. K., and El-Rehaili, A., Wastewater engineering, in *General Directorate of Technical Education and Professional Training*, Institute of Technical Superintendents, Riyadh, Kingdom of Saudi Arabia, 1988.

U.S. Army Corps of Engineers, Yearly Average Cost Index for Utilities, in *Civil Works Construction Cost Index System Manual, 110-2-1304*, U.S. Army Corps of Engineers, Washington, DC, 2022.

Wang, L. K., Hung, Y. T., and Shammas, N. K. (Eds.), *Physicochemical Treatment Processes*, Humana Press, Totowa, NJ, 2005.

Wang, L. K., Hung, Y. T., and Shammas, N. K. (Eds.), *Advanced Physicochemical Treatment Processes*, Humana Press, Totowa, NJ, 2006.

Wang, L. K., Hung, Y. T., and Shammas, N. K. (Eds.), *Advanced Physicochemical Treatment Technologies*, Humana Press, Totowa, NJ, 2007a.

Wang, L. K., Shammas, N. K., and Hung, Y. T. (Eds.), *Biosolids Treatment Processes*, Humana Press, Totowa, NJ, 2007b.

Wang, L. K., Shammas, N. K., and Hung, Y. T. (Eds.), *Biosolids Engineering and Management*, Humana Press, Totowa, NJ, 2008.

Wang, L. K., Pereira, N., Hung, Y. T., and Shammas, N. K. (Eds.), *Biological Treatment Processes*, Humana Press, Totowa, NJ, 2009a.

Wang, L. K., Shammas, N. K., and Hung, Y. T. (Eds.), *Advanced Biological Treatment Processes*, Humana Press, Totowa, NJ, 2009b.

Wang, L. K., Ivanov, V., Tay, J. H., and Hung, Y. T. (Eds.), *Environmental Biotechnology*, Humana Press, Totowa, NJ, 2010.

Wang, L. K., Wang, M. H. S., Hung, Y. T., and Shammas, N. K., *Environmental and Natural Resources Engineering*, Springer Nature Switzerland, 512 pp, 2021a.

Wang, L. K., Wang, M. H. S., Shammas, N. K., and Aulenbach, *Environmental Flotation Engineering*, Springer Nature Switzerland, 433 pp., 2021b.

Water Environment Federation, *Water Resource Recovery Facility Design*, WEF, New York, 570 pp., 2015.

World Health Organization, *Water, Sanitation and Health, Facts and Figures*, http://www.who.int/water_sanitation_health/publications/facts2004/en/index.html, November 2004.

World Health Organization, *Water, Sanitation and Health, Water Supply, Sanitation and Hygiene Development*, http://www.who.int/water_sanitation_health/hygiene/en/index.html, 2010.

11

Hydrology: Rainfall and Runoff

11.1 Background

Hydrology is the science of water in nature: its properties, distribution, and behavior. *Statistical hydrology* is the application of statistical methods of analysis to measurable hydrologic events for the purpose of arriving at engineering decisions. By the introduction of suitable statistical techniques, "enormous amounts of quantitative information can often be reduced to a handful of parameters that convey, clearly and incisively, the underlying structure of the original or raw data. These estimates are important for engineers and economists so that proper risk analysis can be performed to influence investment decisions in future infrastructure and to determine the yield reliability characteristics of water supply systems (Fair et al. 1966).

The total water resource of Earth is approximately 330×10^6 mile^3 or 363×10^9 billion gallons (BG) or 1374×10^9 billion liters (BL) about 95% in the oceans and seas and 2% in the polar ice caps. However, the 35 g/L of salt in seawater and the remoteness as well as fundamentally ephemeral nature of the polar ice caps interfere with their use. This leaves as the potential freshwater resource of Earth no more than 10.9×10^9 BG (41.3×10^9 BL) in lakes, streams, permeable soils, and the atmosphere; only about 3% is in the atmosphere, the remainder being split almost equally between surface and ground.

Fortunately, the *hydrosphere* is not static; its waters circulate (Fig. 11.1). Between 110×10^6 and 130×10^6 BG (416×10^6 and 492×10^6 BL) of water fall annually from the skies, about a quarter onto the continents and islands and the remainder onto the seas; 10×10^6 to 11×10^6 BG (37.8×10^6 to 41.6×10^6 BL) return to the oceans as annual runoff. In overall estimates, about a third of the land mass of Earth is classified as well watered, the remainder as semiarid and arid.

Within the contiguous United States, the 100th meridian (it matches the western boundary of Oklahoma and splits North Dakota, South Dakota, Nebraska, Kansas, and Texas) is the general dividing point between annual rainfalls of 20 in. (508 mm) or more to the East and, except for the Pacific slopes, less than 20 in. (508 mm) to the West. The total length of surface streams is about 3×10^6 miles (4.83×10^6 km) and the five Great Lakes, four of them shared with Canada, hold 7.8×10^6 BG (29.5×10^6 BL), and constitute geographically the largest surface storage of freshwater on Earth.

As shown in Fig. 11.1, water is transferred to Earth's atmosphere (a) through the *evaporation* of moisture from land and water surfaces and (b) through the *transpiration* of water from terrestrial and emergent aquatic plants. Solar radiation provides the required energy. Internal or dynamic cooling of rising, moisture-laden air and its exposure to cold at high altitudes eventually lower the temperature of ascending air masses to the *dew point*, condense the moisture, and precipitate it on land and sea. Overland and subsurface flows complete the *hydrologic cycle*.

11.2 Collection of Hydrologic Data

Without adequate quantitative information on Earth's water resource, its use and development become an economic uncertainty and an engineering gamble. The collection of pertinent data is, therefore, an imperative social responsibility that is generally assumed by the government. When and to what extent the government of the United States has accepted this responsibility is summarized in Table 11.1.

Of the different hydrologic parameters, *annual precipitation* is a measure of the maximum annual renewal of the water resource of a given region. About one-fourth to one-third of the water falling on continental areas reaches the oceans as

Water and Wastewater Engineering: Hydraulics, Hydrology and Management, Volume 1, Fourth Edition.
Lawrence K. Wang, Mu-Hao Sung Wang, and Nazih K. Shammas.

Companion website: www.wiley.com/go/Wang/Waterandwastewater4e

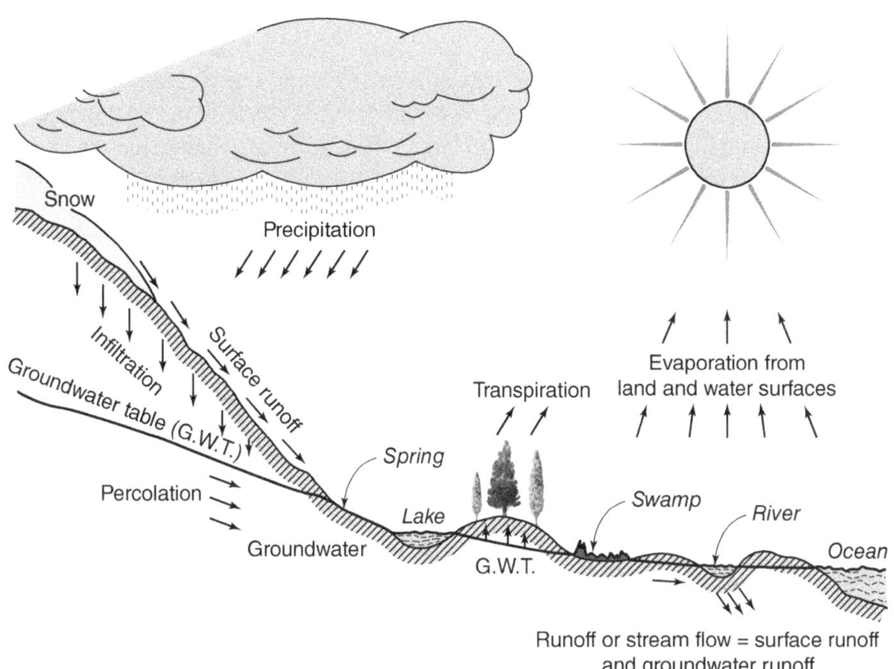

Figure 11.1 The Water Cycle

Table 11.1 Nationwide Collection of Hydrologic Data for the United States

Measurement and Apparatus	Beginning Date	Number of Stations	Areal Density per 100 mile^2
Rain and snow	1870	$>10^5$	4
Nonrecording rain gauges		$^2/_3$ of total	2.7
Recording rain gauges		$^1/_3$ of total	1.3
Storm-tracking radar	1945	—	—
Snow courses	1910	$>10^3$ annually	
Stream flow			
Stream gauges	1890	$>10^4$	0.5–11
Groundwater	1895	—	—
Evaporation		Several hundred	

Conversion factor: 1 mi^2 = 2.59 km^2.

runoff. The balance is returned to the atmosphere by evaporation and transpiration. Where melting of the winter's snows produces a major part of the annual runoff, or where spring thaws cause serious floods, accurate methods for determining the annual snowfall on a watershed are essential requirements. Otherwise, storm rainfalls monitored by recording gauges form the basis for predicting the flood stages of river systems and for estimating the expected rate of runoff from areas yet to be provided with drains or sewers.

Much larger volumes of water are returned to the atmosphere over continents by transpiration than by evaporation. In the well-watered regions of the United States, annual evaporation from free water surfaces more or less equals annual precipitation; in the arid and semiarid portions, it exceeds precipitation manyfold.

About half the annual rainfall normally enters Earth's crust and about half the water stored in its interstices lies within half a mile of the surface and can normally be drawn on for water supply.

11.2.1 Rain Gauging

In the United States, a Weather Service was organized in the Army Signal Corps in 1870, and the Weather Bureau was established in 1891. The bureau maintains more than 10 standard rain gauges, or one for about 250 mi^2 (647.3 km^2). Some two-thirds of the rain gauges are nonrecording, and under cooperative observation. The more informative recording rain gauges have multiplied more rapidly since 1940, with radar storm-tracking stations being added after the Second World War. The average length of record of nonrecording stations is close to 100 years; only a few go back two centuries or more.

11.2.2 Snow Surveys

Snowfall is identified routinely at rain gauging stations. Snow surveys are special undertakings that keep an inventory of the important frozen or winter storage of water in high mountains that is or may be released to lowlands during the summer. Snow surveys began in 1910. They are currently the responsibility of the Soil Conservation Service, which was organized in 1939. More than a thousand snow courses are now traversed annually in cooperation with states and hydraulic or agricultural interests.

11.2.3 Stream Gauging

The yield of upland catchments of moderate size was first measured in the 1860s by cities in need of water. Before that time the Corps of Engineers had gauged only the great rivers of the country. The Geological Survey, created in 1879, established a nationwide stream gauging network in 1894. Its program is supported by state as well as federal funds. Over the years, more than 10,000 stations have been operated. Their density per 1,000 mi^2 (2,590 km^2) varies from 0.5 (Nevada) to 11 (New Jersey) or 0.2 to 4.3 per 1,000 km^2. All too few have been assigned to small, and consequently flashy, streams draining areas of less than 1,000 mi^2 (2,590 km^2). Because the total length of United States surface streams is about 3 million miles (4.8 million km), there is one gauging station in about 400 miles (640 km).

11.2.4 Groundwater Studies

Responsibility for reporting on groundwater was given to the Geological Survey in 1895, but no significant scale of operation was achieved until the 1920s. States share the expense of this program, too. Fluctuations of water levels are measured now in about 10,000 observation wells. It is comforting to know that subsurface water stored at useful depths of less than 2,500 ft (762 m) equals the total recharge of the ground during 16 average years.

11.2.5 Evaporation

The city of Boston made careful measurements of evaporation before 1885, and many thousands of observations have been accumulated since the turn of the last century. Today the Weather Bureau collects information from several hundred land pans, but monthly measurements of Lake Mead (in Nevada and Arizona) evaporation comprise one of the few regular reservoir studies.

11.3 Precipitation

Atmospheric moisture precipitates in large amounts as rain, snow, hail, or sleet; it condenses in small amounts as dew, frost, and rime. The most important causes of precipitation are external and dynamic, or internal, cooling; dynamic cooling implies the reduction in the temperature of the atmosphere accompanying its expansion as air masses rise or are driven to high altitudes. The observed drop in temperature is called the *lapse rate*. Within the troposphere, up to 7 miles (11.26 km) above ground level in the middle latitudes, the lapse rate is about 3°F in 1,000 ft (5.53C in 1,000 m); but it is quite variable within the first 2 or 3 miles (3.22 or 4.83 km) and may, at times, be negative. An increase in temperature with altitude, or negative lapse rate, is called an *inversion*. Adiabatic expansion cools ascending air by about 5.5°F in 1,000 ft (10°C in 1,000 m) if no moisture is precipitated; but if the dew point is reached, the latent heat of evaporation is released, and the rate of the resultant *retarded* or *wet adiabatic* rate of cooling drops to about 3.2°F in 1,000 ft (5.8°C in 1,000 m).

Air is stable and will not ascend by convection when the lapse rate is lower than both the wet and dry adiabatic rates of cooling. Otherwise, its temperature would become lower and its density more than the temperature of its surroundings as it is moved into the higher altitude. If the lapse rate is greater than the dry adiabatic rate, rising air becomes warmer and lighter than the air along its upward path. Hence, it continues to ascend and remain *unstable*. If the lapse rate lies between the wet and dry adiabatic rates, the air remains stable when moisture is not condensing but becomes unstable as soon as precipitation sets in. This *conditional stability* is one of the requirements for successful rainfall stimulation. Dry ice or silver iodide may then provide the nuclei that trigger precipitation and convert stable into unstable air. Nevertheless, rainfall can be heavy only in the presence of a continuing supply of moisture. In other words, cloud seeding becomes favorable only when atmospheric conditions are already conducive to natural precipitation. Accordingly, seeding appears to hold out some, but not much, hope for rainmaking.

Moist air is moved upward principally by (a) convective currents to cause *convective* rainfalls, (b) hills and mountains to produce *orographic* rainfalls, and (c) cyclonic circulation to generate *cyclonic* rainfalls.

11.3.1 Convective Precipitation

Convective precipitation is exemplified by tropical rainstorms. Air masses near Earth's surface absorb heat during the day, expand, and take up increasing amounts of water vapor with a specific gravity near 0.6 relative to dry air. The air mass becomes lighter; almost exclusively vertical currents are induced and they carry the mass to higher altitudes where it is exposed to colder surroundings and expands under reduced pressure. Under both external and dynamic cooling, water vapor is condensed, and precipitation follows.

11.3.2 Orographic Precipitation

Precipitation is orographic when horizontal currents of moist air strike hills or mountain ranges that deflect the currents upward. In North America, the rainfalls of the Pacific Northwest and the Southern Appalachians furnish examples of this type of precipitation.

The condensation as a function of temperature and elevation can be modeled by the following equations:

$$T_c = 60 - 5.5 \times 10^{-3} H_c \quad \text{(U.S. Customary Units)} \tag{11.1}$$

$$D_c = 54 - 1.1 \times 10^{-3} H_c \quad \text{(U.S. Customary Units)} \tag{11.2}$$

where T_c is the air temperature in °F, D_c is the dew-point temperature in °F, and H_c is the elevation in ft.

$$T_c = 15.55 - 10 \times 10^{-3} H_c \quad \text{(SI Units)} \tag{11.3}$$

$$D_c = 12.22 - 2 \times 10^{-3} H_c \quad \text{(SI Units)} \tag{11.4}$$

where T_c is the air temperature in °C, D_c is the dew-point temperature in °C, and H_c is the elevation in m.

The condensation begins when the air temperature, T_c, equals the dew-point temperature, D_c.

Air cools at the dry adiabatic rate below the elevation H_c and at the retarded adiabatic rate above it. The temperature at the top of the mountain above the elevation H_c is

$$T_t = 60 - 5.5 \times 10^{-3} H_c - 3.2 \times 10^{-3}(H_t - H_c) \quad \text{(U.S. Customary Units)} \tag{11.5}$$

Since the descending air warms at the dry adiabatic rate, the temperature on the plain becomes

$$T_p = T_t + 5.5 \times 10^{-3} \Delta H_p \quad \text{(U.S. Customary Units)} \tag{11.6}$$

Where T_t is the air temperature above H_c (such as H_t) in °F, H_t is the elevation above H_c in ft, T_p is the air temperature below H_c (such as H_p) in °F and ΔH_p is the elevation difference below H_t in ft.

The equivalent equations using the SI Units are

$$T_t = 15.55 - 10 \times 10^{-3} H_c - 5.83 \times 10^{-3}(H_t - H_c) \quad \text{(SI Units)} \tag{11.7}$$

$$T_p = T_t + 10 \times 10^{-3} \Delta H_p \quad \text{(SI Units)} \tag{11.8}$$

where T_t is the air temperature above H_c (such as H_t) in °C, H_t is the elevation above H_c in m, T_p is the air temperature below H_c (such as H_p) in °C and ΔH_p is the elevation difference below H_t in m.

The above equations (11.1) through (11.8) can be applied to coastal areas where the dew point is lowered at a rate of about 1.1°F in 1,000 ft (2°C in 1,000 m).

Example 11.1 Condensation as a Function of Temperature and Elevation
Coastal air with a temperature of 60°F (15.6°C) and a dew point of 54°F (12.2°C) is forced over a mountain range that rises 4,000 ft (1,219 m) above sea level. The air then descends to a plain 3,000 ft (914 m) below. If the dew point is lowered at a rate of 1.1°F in 1,000 ft (2°C in 1,000 m), find:

1) The height at which condensation will begin
2) The temperature at the mountain top
3) The temperature on the plain beyond the mountain assuming that condensed moisture precipitates before the air starts downward.

Solution 1 (U.S. Customary System):

1) When condensation starts, $T_c = D_c$, and

$$T_c = 60 - 5.5 \times 10^{-3} H_c$$
$$D_c = 54 - 1.1 \times 10^{-3} H_c$$
$$6 = 4.4 \times 10^{-3} H_c$$

Hence, $H_c = \mathbf{1{,}360\ ft}$

Air cools at the dry adiabatic rate below this elevation and at the retarded adiabatic rate above it. When $H_c =$ 1,360 ft, $T_c = D_c = 52.5°F$

2) The temperature at the top of the mountain is $T_t = \text{Temperature} = 60 - 1{,}360 \times 5.5 \times 10^{-3} - (4{,}000 - 1{,}360) \times 3.2 \times 10^{-3}$
$= \mathbf{44°F}$

3) If the descending air warms at the dry adiabatic rate, the temperature on the plain becomes
$T_p = \text{Temperature} = 44 + 3{,}000 \times 5.5 \times 10^{-3}$
$= \mathbf{60.5°F}$

Solution 2 (SI System):

1) When condensation starts, $T_c = D_c$, and

$$T_c = 15.55 - 10 \times 10^{-3} H_c$$
$$D_c = 12.22 - 2 \times 10^{-3} H_c$$
$$15.55 - 12.22 = (10 - 2) \times 10^{-3} H_c$$

Hence, $H_c = \mathbf{416\ m}$

Air cools at the dry adiabatic rate below this elevation and at the retarded adiabatic rate above it.
When $H_c = 416$ m, $T_c = D_c = 11.4°C$

2) The temperature at the top of the mountain (1219 m) is T_t
$T_t = 15.55 - 10 \times 10^{-3} H_c - 5.83 \times 10^{-3}(H_t - H_c)$
$= 15.55 - (10 \times 10^{-3}) \times 416 - 5.83 \times 10^{-3}(1{,}219 - 416)$
$= \mathbf{6.7°C}$

3) If the descending air warms at the dry adiabatic rate, the temperature on the plain becomes
$T_p = T_t + 10 \times 10^{-3} \Delta H_p$
$= 6.7 + 10 \times 10^{-3} \times (914)$
$= \mathbf{15.8°C}$

11.3.3 Cyclonic Precipitation

Cyclonic precipitation is associated with unequal heating of Earth's surface and buildup of pressure differences that drive air from points of higher to points of lower pressure. Major temperature effects are (a) temperature differences between equator and poles, producing so-called *planetary circulation*, and (b) unequal heating of land and water masses, forming secondary areas of high and low pressure on sea and land and consequent *atmospheric circulation*.

Differences in Earth's relative rotary speed between equator and poles deflect tropical air currents moving toward the poles. This explains the general easterly direction of cyclonic disturbances over the North American continent as well as the rotary or cyclonic motion of horizontal air currents converging at points of low pressure.

In the continuous planetary circulation of the atmosphere between equator and poles, warm, moisture-laden, tropical air masses travel poleward, are cooled, and precipitate their moisture along the way. Ultimately, they are transformed into cold, dry, polar air. A return movement drives polar air masses toward the equator, and heavy precipitation results when tropical and polar air masses collide. The light, warm, tropical air cools and precipitates its moisture as it is forced up and over the heavy polar air. Collisions between tropical and polar air masses normally account for the protracted general rainfalls and accompanying floods of the central and eastern United States. When, for unknown reasons, polar air does not return toward the equator in the usual manner, serious droughts can occur.

The *Bjerkness cyclone model* shown in Fig. 11.2 identifies the movements of warm and cold air masses in the usual type of cyclonic storm. Precipitation is indicated on the plan by shading. The cross-sections suggest the manner in which warm air is forced upward by cold air. At the *cold front,* the colder air wedges itself below the mass of warm air and usually advances southward and eastward in the Northern Hemisphere. At the *warm front,* the warm air is forced to climb over the retreating wedge of cold air and usually advances northward and eastward in the Northern Hemisphere. When there is little or no movement at the boundary of the air masses, the front is called *stationary.* When the cold front overtakes the warm front and lifts all the warmer air above the surface, the front is said to be *occluded.*

Cyclonic storms are eddies in the vast planetary circulation between equator and poles. They are generally several hundred miles in diameter, and their rotational and lateral motions are both relatively slow. Cyclones are distinctly different from the violent whirlwinds of tornadoes or hurricanes. In the central or low-pressure portion of cyclonic disturbances, moisture is precipitated from the rising air, while fair weather usually prevails in surrounding high-pressure areas.

The storms of most importance on the North American continent as a whole originate over the Pacific Ocean, strike the coast of the northern United States or Canada, swing south and east over the central or northern United States, and generally escape through the St. Lawrence Valley to the Atlantic Ocean. Storms born in the Gulf of Mexico drift northward over the continent and out to sea.

Figure 11.2 Bjerkness Cyclone Model

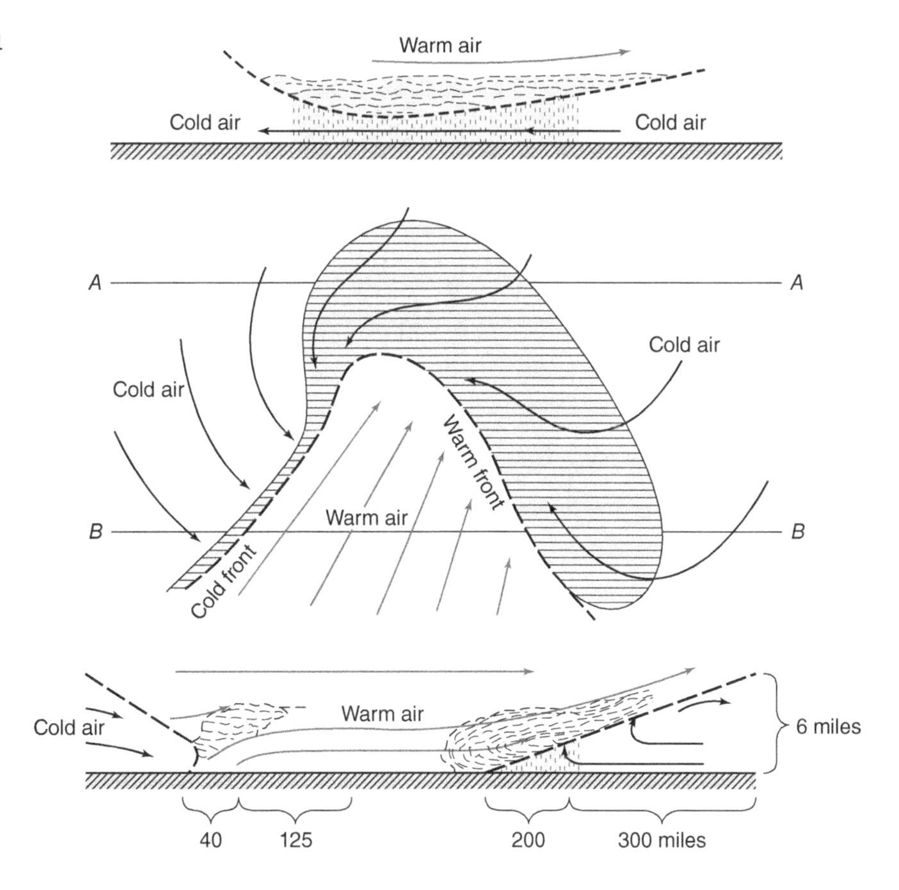

The three types of precipitation—convective, orographic, and cyclonic—seldom occur by themselves. Most precipitation in the temperate regions of Earth results from a combination of two or more causes. Examples are (a) the rainfall of the Pacific slopes, which accompanies cyclonic storms forced upward by the Rocky Mountains, and (b) local thunderstorms, which are both cyclonic and convective in origin.

About a third of the water falling on continental land masses reaches the oceans as *runoff.* The remainder is returned to the atmosphere by evaporation and transpiration. Of the vaporized water, a small part is reprecipitated, but the major part is transported over the oceans. Accordingly, the stream flow from continents represents the net water loss from circulating air masses as they precipitate and reevaporate moisture in their course across the land.

11.3.4 Droughts

Droughts are defined as periods when crops fail to mature for lack of rainfall or when precipitation is insufficient to support established human activities. Drought conditions prevail when annual rainfall is deficient or poorly distributed in time, or when annual precipitation is concentrated in a few heavy rainfalls that drain away rapidly. Droughts impose a critical demand on (a) works designed to furnish a continuous and ample amount of water and (b) streams expected to carry away domestic and industrial treated effluents without nuisance. Statistical studies of past droughts suggest a strong tendency for self-perpetuation. Some dry spells are ultimately broken only by a change in the seasons.

11.3.5 Measuring Precipitation

The *cooperative observer stations* of the United States measure rainfall in standard *cylindrical can gauges,* 2 ft (0.60 m) high and 8 in. (200 mm) in diameter (Fig. 11.3). A funnel-shaped inset receives and discharges the catch to a central measuring tube of such diameter that 1 in. (25.4 mm) on a standard measuring stick inserted in the tube equals 0.1 in. (2.54 mm) of rainfall.

The *official* Weather Bureau stations register rainfall accumulations during short intervals of time. Engineers base their designs of storm drainage systems and forecasts of flood stages for streams on records of this kind. Continuous-recording gauges are in common use. In the *tipping-bucket gauge* (Fig. 11.3b), a twin-compartment bucket is balanced in such a way that the compartments fill, tip, and empty reciprocally. In the *weighing gauge* (Fig. 11.3c), the rain concentrating from a receiving funnel is weighed on recording scales. In both gauges, measurements are traced by movements of a pen on charts operated by clockworks. The screen or shield shown in Fig. 11.3b keeps precipitation from being swept past the gauge by high winds.

Standard gauges will measure snowfalls more satisfactorily if their receiving funnels and collecting tubes are removed and the naked can is inverted and used as a *cookie cutter* to collect samples of snow from undrifted places. The snow is melted

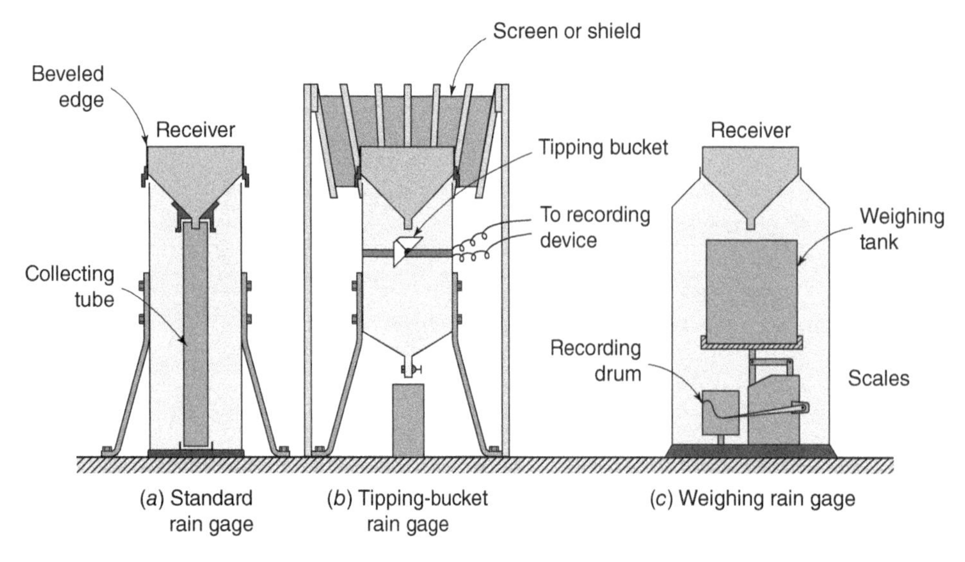

Figure 11.3 Rain Gages: (a) Standard Rain Gage; (b) Tipping-Bucket Gage; (c) Weighing Gage

without loss by adding known amounts of warm water. Volumes in excess of added amounts are recorded as inches of precipitation. Some snows are feathery and dry, others compact and wet. Their water equivalent varies accordingly. An average value to remember is *10 in. of snow to 1 in. of water (10 mm of snow to 1 mm of water).*

The amount of precipitation falling as snow increases rapidly with elevation. Because most rain-gauging stations are situated at relatively low altitudes, more accurate methods are needed for determining snowfall on high-lying watersheds in regions where winter snows provide much of the annual runoff or spring thaws provoke serious floods. In one method, sampling stations are located along a *snow course* traversing the drainage system. An observer walks the course and, at suitable intervals, samples the snow blanket with a hollow-tube collector. The water content of the sample is found by weighing the tube or by melting the collected snow. In another method, a battery of four or five shielded snow gauges is placed 200 or 300 ft (60–90 m) apart in a location typifying average conditions on the watershed. The water equivalent of the snow falling during given periods of time is computed from the weight increment of the containers.

Regional or countrywide rainfall experience is generalized by *isohyetal* (equal rainfall) lines on suitable maps. In the coterminous United States, the 20-in. (500-mm) isohyetal line divides the country into two distinct climatological regions lying roughly to either side of the 100th meridian. In the well-watered east, enough rain falls for normal agriculture and water supply. In the arid west, water development is restricted by the amount that can be collected and stored for use during the dry season.

Errors in precipitation measurements are caused by the following:

1) *Wind eddy currents*—sweeping rain and snow over gauges set on top of the ground. In such situations, less than the true ground-level rainfall is caught. However, it is possible to shield standard rain gauges and to approximate the catch obtained when the rim of the gauge is set flush with the ground.
2) *Obstructions*—trees, bushes, fences, and buildings—cast a rain shadow. To avoid this, gauges are placed in the clear by a distance greater than the height of the tallest obstruction. If gauges must be put on a roof, they should be centered in the largest available flat area.

Engineers should be fully aware of the limitations of precipitation records. Aside from errors of measurement, areal variations in precipitation may be large, even over small stretches of country. In such circumstances, a single gauge cannot yield representative information.

11.4 Evaporation and Transpiration

Evaporation raises the storage requirements of reservoirs and lowers the yield of lakes and ponds. Swamps and other wet surfaces, too, return much water to the atmosphere.

11.4.1 Evaporation from Water Surfaces

Rates of evaporation from open water surfaces vary with the temperature or vapor pressure of the water and the air in contact with it. They also vary with wind speed, barometric pressure, and water quality. Because these factors are by no means independent, individual effects are not clear-cut. In general, evaporation and gas transfer have much in common:

1) For large differences between the maximum vapor pressure at water-surface temperature and the actual pressure of aqueous vapor in the overlying air, evaporation is rapid. For small differences, evaporation is slow; for negative differences, there is condensation.
2) At the temperatures of natural waters, the vapor pressure is almost doubled for every rise of 10°C (18°F). Hence, temperature affects evaporation profoundly, yet the slow warming and cooling of deep bodies of water make for relatively even evaporation.
3) Films of still air above a water surface soon become saturated with moisture, and evaporation practically ceases. Within limits, wind stimulates evaporation by displacing moisture-laden films with relatively dry air.
4) As pressures drop, evaporation rises, but altitude has little effect because of counterbalancing changes in temperature. Fast rates of evaporation at high altitudes are caused, in large measure, by greater wind velocities.
5) Rates of evaporation are decreased slightly when salt concentrations are high.

Rohwer's formula (1931) illustrates the dependence of evaporation on the factors just cited:

$$E = 0.497\left(1 - 1.32 \times 10^{-2} P_a\right)(1 + 0.268w)(P_w - P_d) \quad \text{(U.S. Customary Units)} \tag{11.9}$$

Where E is the evaporation, in in./d; P_a is the barometric pressure, in. Hg; w is the wind velocity, in mi/h; and P_w and P_d are the vapor pressures, in. Hg, at the water temperature and dew-point temperature of the atmosphere, respectively.

The following is an equivalent evaporation equation using the SI Units:

$$E = 0.497\left(1 - 5.2 \times 10^{-4} P_a\right)(1 + 0.167w)(P_w - P_d) \quad \text{(SI Units)} \tag{11.10}$$

Where E is the evaporation, in mm/d; P_a is the barometric pressure, mm Hg; w is the wind velocity, in km/h; and P_w and P_d are the vapor pressures, mm Hg, at the water temperature and dew-point temperature of the atmosphere, respectively. The vapor pressure or vapor tension of water is the maximum gaseous pressure exerted at a given temperature by water vapor in contact with a water surface. The pressure of water vapor in air not saturated with aqueous vapor equals the vapor pressure of water at the dew-point temperature of the air, namely, the temperature at which the air would be saturated by the moisture actually in it. In other words, vapor pressure is the partial pressure exerted by the water vapor in the atmosphere and, in accordance with *Dalton's law*, evaporation is proportional to it.

Thermodynamic formulations relate evaporation to the difference between solar radiation and heat accounted for as (a) back radiation, (b) heat stored in the water, and (c) heat lost in other ways.

Example 11.2 Estimation of Daily Evaporation

Estimate the evaporation for a day during which the following averages are obtained: water temperature = 60°F (15°C); maximum vapor pressure $P_w = 0.52$ in. Hg (13.2 mm Hg); air temperature = 80°F (27°C); relative humidity = 40 %; vapor pressure $P_d = 1.03 \times 0.40 = 0.41$ in. Hg (10.4 mm Hg); wind velocity $w = 8$ mi/h (12.9 km/h); and barometric pressure $p_a = 29.0$ in. Hg (737 mmHg).

Solution 1 (U.S. Customary System):

By Eq. 11.9,

$$\begin{aligned} E &= 0.497(1 - 1.32 \times 10^{-2} P_a)(1 + 0.268w)\,(P_w - P_d) \\ &= 0.497(1 - 1.32 \times 10^{-2} \times 29.0)(1 + 0.268 \times 8)(0.52 - 0.41) \\ &= 0.497(1 - 0.38)(1 + 2.14) \times 0.11 \\ &= 0.497 \times 0.62 \times 3.14 \times 0.11 \\ &= \mathbf{0.10\ in./d} \end{aligned}$$

Solution 2 (SI System):

By Eq. 11.10,

$$\begin{aligned} E &= 0.497(1 - 5.2 \times 10^{-4} P_a)(1 + 0.167w)(P_w - P_d) \\ &= 0.497(1 - 5.2 \times 10^{-4} \times 737)(1 + 0.167 \times 12.9)(13.2 - 10.4) \\ &= 0.497(1 - 0.38)(1 + 2.15) \times 2.8 \\ &= 0.497 \times 0.62 \times 3.15 \times 2.8 \\ &= \mathbf{2.7\ mm/d}. \end{aligned}$$

11.4.2 Evaporation from Land Surfaces

Wet soil loses water rapidly to the atmosphere; moist soil does so more slowly. Immediately after a rainstorm, water, intercepted by vegetation or present in films, pools, or puddles on roofs and pavings, starts to evaporate at about the rate of loss from shallow water. As the free moisture disappears, the rate of evaporation slows. Interstitial moisture is evaporated internally or drawn to the surface by capillarity before it evaporates. Soil characteristics and depth to groundwater are governing factors. Where the water table lies 3 ft (0.90 m) or more down, there is little evaporation after the surface layers have dried out. Because cultivated soils lose less water than uncultivated soils, the practice of loosening soils to conserve moisture prevails.

11.4.3 Transpiration

The amounts of water transpired vary with the kind and maturity of vegetation, conditions of soil moisture, and meteorological factors. On the whole, the continents return more water to the atmosphere by transpiration than by evaporation. The total areal expanse of the leaves in a forest is very great in comparison with land and water exposures. Moreover, some plants can draw water from considerable depths and transpire it. Estimates for the United States place the proportion of precipitation lost to the atmosphere by evaporation and transpiration from forests and land uses and from surface reservoirs and phreatophytes as high as 70%.

11.4.4 Measuring Evaporation and Transpiration

Evaporation from water surfaces is commonly measured by exposing *pans* of water to the atmosphere and recording evaporation losses by systematic measurements or self-registering devices. Both floating and land pans have been used (Fig. 11.4). Neither one is fully satisfactory. The standard (Class A) measuring device of the Weather Bureau is a 4-ft (1.22-m) galvanized iron pan, 10 in. (254 mm) deep, supported on a grid of 2-in. by 4-in. (51 × 103 mm) timbers that raise the pan slightly above the ground to promote air circulation. A hook gauge in a stilling well identifies changes in water level. Temperature, rainfall, and wind speed are also recorded. The anemometer is placed next to and just above the pan. Observed losses are not the same as for floating pans of different materials, color, or depth, nor are they the same as for natural or impounded bodies of water.

Pan evaporation is translated empirically to lake evaporation by the following equation:

$$E = \frac{CE'(P_w - P_d)}{P'_w - P'_d} \quad \text{(U.S. Customary or SI Units)} \tag{11.11}$$

where letters with primes pertain to pan conditions, the others to lake conditions. The coefficient C is said to average 0.7. In most of the United States, mean annual evaporation from water surfaces equals or exceeds mean annual rainfall.

Together, transpiration and evaporation are referred to as *evapotranspiration* or *consumptive use*. They are measured experimentally by agriculturalists in a number of ways. In one of them, evaporation is equated to the difference between rainfall on a plot of ground and water collected by underdrains; in another, soil–water level is held constant in a tank filled with representative materials growing representative plants. If the tank bottom is pervious, the resulting lysimeter measures consumptive use as the difference between (a) water falling on or applied to the surface and (b) water draining from the bottom, corrections being made for changes in soil moisture. Data of this kind are generally of limited use to water supply engineers, but it is good to have some concept of the order of magnitudes involved. For example, the transpiration ratio, or weight of water transpired by a plant per unit weight of dry matter produced exclusive of roots, increases stepwise from about 350 for Indian corn through wheat and rice to almost 800 for flax, and from about 150 for first through oaks to almost 400 for birches.

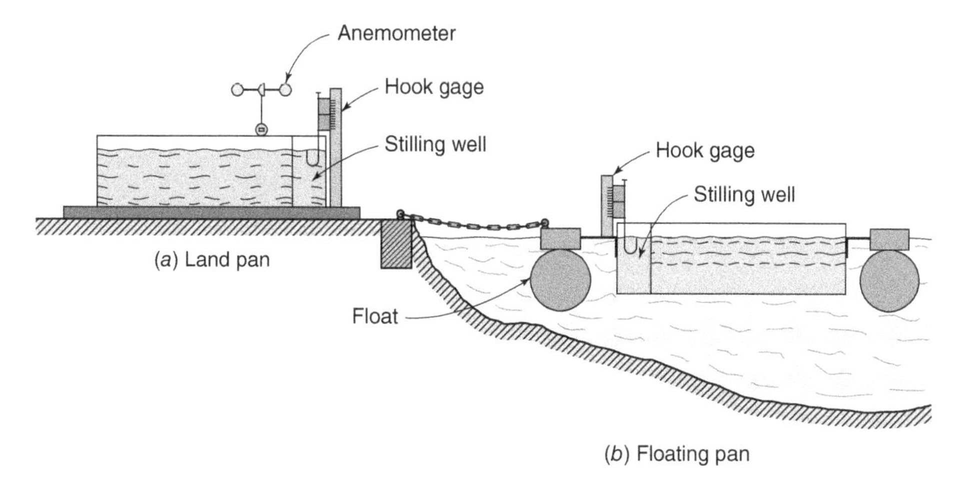

Figure 11.4 Evaporation Pan

Empirical formulations of consumptive use are generally based on temperature records. A relationship proposed by Lowry and Johnson, for example, reads as follows:

$$U = 0.9 + 1.5 \times 10^{-4}\Sigma(T_m - 32) \quad \text{(U.S. Customary Units)} \tag{11.12}$$

where U is the consumptive use, in ft (acre-ft/acre); and T_m is the daily maximum temperature, in °F, for the growing season (last to earliest frost). Hence, $(T_m - 32)$ is the sum of the degree days. When $\Sigma(T_m - 32) = 4 \times 10^4$, for example, U becomes 2.4 ft (0.73 m).

The following is an equivalent consumptive use equation using the SI Units:

$$U = 0.27 + 0.46 \times 10^{-4}\Sigma\, 1.8T_m \quad \text{(SI Units)} \tag{11.13}$$

where U is the consumptive use, in m (ha-m/ha); and T_m is the daily maximum temperature, in °C, for the growing season (last to earliest frost).

Adding the proportion of daylight hours during specific months in the frost-free period, Blaney and Criddle (1957) arrived at the equation:

$$U' = K\,\Sigma\, pT' \quad \text{(U.S. Customary Units)} \tag{11.14a}$$

where the new factors are K, an empirical coefficient for specific crops varying from 0.6 for small vegetables to 1.2 for rice, and p, the monthly percentage of daylight hours in the year. The consumptive use U' is measured in inches for a given period of time, and T' is the mean monthly temperature, °F

An equivalent SI equation using the SI units of U' (mm for a given period time), T' (°C), K (dimensionless), and p (dimensionless) is:

$$U' = 25.4\,K\,\Sigma\, p\,(1.8T + 32) \quad \text{(SI Units)} \tag{11.14b}$$

11.5 Percolation

The term *percolation* is employed to describe the passage of water into, through, and out of the ground. Figure 11.5 shows the conditions in which water occurs below the surface of the ground. Only water in the saturated zone can be withdrawn from subsurface sources, the development of groundwater supplies depending on the yields actually obtainable and their cost. Unwanted entrance of groundwater into manholes and pipes is an important matter in sewerage design.

Groundwater is derived directly or indirectly from precipitation: (a) directly as rainwater and snowmelt that filter into the ground, seep through cracks or solution passages in rock formations, and penetrate deep enough to reach the groundwater table; (b) indirectly as surface water from streams, swamps, ponds, lakes, and reservoirs that filters into the ground through permeable soils when the groundwater table is lower than free water surfaces. Streams that recharge the ground are known as *influent* streams, streams that draw water from the ground as *effluent* streams.

The water table tops out the zone of saturation; the capillary fringe overrides it. The fringe varies in thickness from a foot or so in sand to as much as 10 ft (3 m) in clay. It rises and falls with the water table, lagging behind to become thicker above a falling table and thinner above a rising table. Evaporation is increased when capillarity lifts water to, or close to, the ground surface. Pollution spreads out along the water table and is lifted into the fringe. There it is trapped and destroyed in the course of time. Hydraulically, an aquifer dipping beneath an impervious geologic stratum has a piezometric surface, not a groundwater table.

How much rain filters far enough into the ground to become groundwater is quite uncertain. Governing factors include the following:

1) *Hydraulic permeability*. Permeability, not merely pore space, determines the rate of infiltration of rainfall and its passage to the groundwater table. In winter, the permeability is usually reduced by freezing.
2) *Turbidity*. Suspended matter picked up by erosion of tight soils clogs the pores of open soils.
3) *Rainfall patterns and soil wetness*. Light rainfalls have time to filter into the ground; heavy rainfalls do not. Wet soils are soon saturated; dry soils store water in surface depressions and their own pores. Some stored water may reach the groundwater table eventually. Heavy rains compact soil, and prolonged rains cause it to swell. Both reduce surface openings. Air displacement from soils opposes filtration; sun cracks and biological channels speed it up.

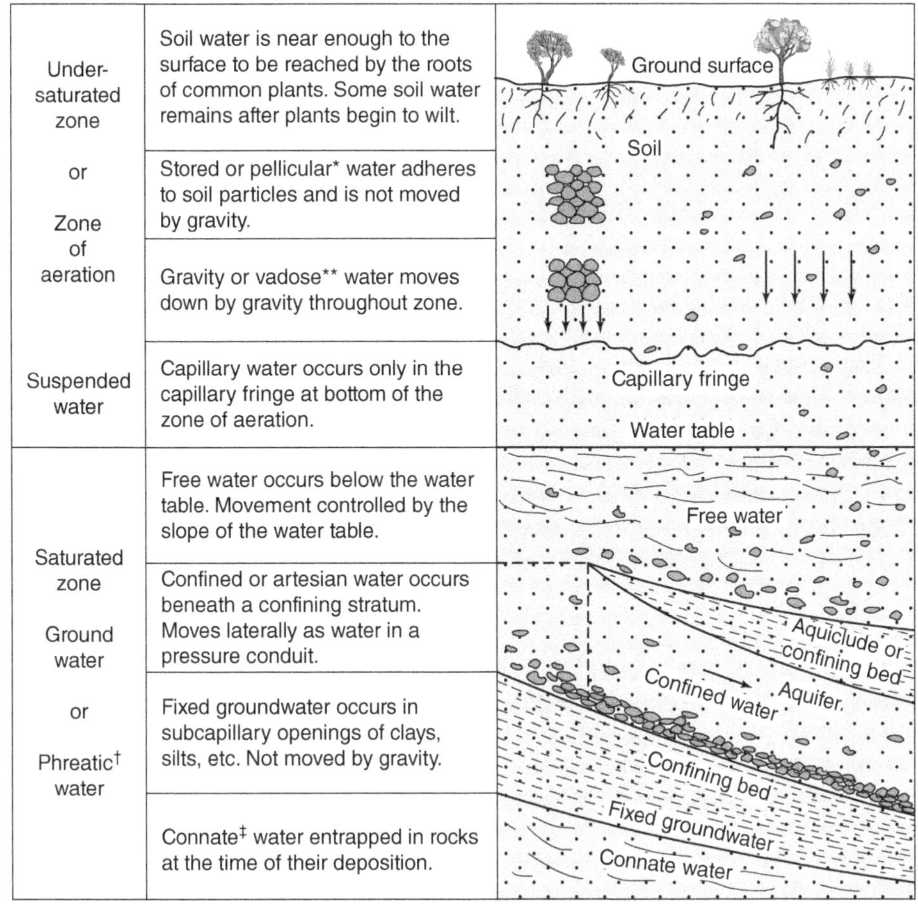

Figure 11.5 Occurrence and Distribution of Subsurface Water

4) *Ground cover.* Vegetation retards runoff and increases surface evaporation as well as retention and transpiration of soil water. Effects such as these are most marked during the growing season.
5) *Geology.* Geologic structure has much to do with infiltration. Examples are (a) lenses of impervious materials, which intercept incoming water and keep it from reaching the groundwater table, and (b) confining layers of tight materials, which direct water into closed-channel flow. Independent zones of saturation above lenses of impervious materials store *perched* water; continuous zones of saturation (aquifers) lying between impervious materials hold *artesian* water.
6) *Surface slope.* Steep slopes hasten surface runoff and reduce infiltration. Earth's crust is porous to depths of 2–8 miles (3.22–12.87 km). Beyond that, pressures are so great that plastic flow closes all interstices.

As explained earlier, infiltration into various soil types supporting different kinds of vegetation is measured in lysimeters. The variables are many, and groundwater flow and yield can be related to rainfall information only in very simple geologic situations such as isolated sand dunes. Nevertheless, it is reasonable, for comparative or bookkeeping purposes, to express annual results in the same units as rainfall. Infiltration of about half the rainfall is not unusual.

11.6 Groundwater Discharge

In nature, subsurface waters are discharged from the ground (a) to the surface through springs and seepage outcrops (hydraulic discharge), and (b) to the atmosphere from the soil or through vegetation (evaporative discharge).

Hydraulic discharge takes place wherever the groundwater table intersects the land surface. Geologic and hydraulic conditions that combine to force the return of groundwater to Earth's surface as springs include the following:

1) Outcroppings of impervious strata covered by pervious soils or other water-bearing formations
2) Overflows of subterranean basins in limestone or lava
3) Leakage from artesian systems through faults that obstruct flow
4) Steep surface slopes that cut into the water table. In humid regions, groundwater may seep into streams throughout their length

Evaporative discharge from soil is commonly confined to the belt of soil water, but it also affects aquifers passing within capillary distance from the land surface. Plants seek moisture at whatever levels their roots can thrive, usually from the belt of soil water or the capillary fringe. Trees and *phreatophytes* may draw water from as far down as the zone of saturation, *xerophytes* only from the zone of aeration. Ways of natural discharge of subsur-face waters are illustrated in Fig. 11.1. Evaporative discharge is normally confined to a few feet in humid climates, and about 20 ft (6.1 m) in dry climates. The roots of phreatophytes may reach downward as far as 50 ft (15.2 m).

Differences between rates of *recharge* and *discharge* are correlated with changes in water stored in the saturation zone. During wet weather, the water table rises; during dry weather, it falls. Because the dry-weather flow of most surface streams is supported by groundwater discharge, correlation between low stream flows and groundwater levels is good, too, and observed coefficients can be used to predict ground storage.

Pumping intrudes into the natural regimen of subsurface waters. Recharge, discharge, and storage are forced to seek new equilibria. Lowering the water table may decrease and even stop natural discharge, but increase natural recharge, especially in soils bordering on surface streams. How much water can be salvaged economically by lowering the water table through pumping depends on the cost of lifting water from increasing depths. If the water table is to remain at a designated level, average rates of withdrawal and recharge must be alike under the conditions generated.

Among other factors affecting groundwater levels are

1) Seasonal variations in evaporation and transpiration
2) Diurnal fluctuations in transpiration
3) Changes in barometric pressure
4) Passage of moving loads, trains for example, over artesian formations
5) Land and ocean tides
6) Earthquakes

Fluctuations in levels registered by a continuous recording gauge on an observation well seldom have one cause only. Records must, indeed, be analyzed with much care if underground hydrologic cycles are to be fully understood.

11.7 Runoff

Water derived directly from precipitation flows over the ground into water courses as surface, storm, or flood runoff. However, the amounts of water actually reaching streams are reduced by infiltration, evaporation, and other losses along the way.

11.7.1 Dry-Weather Flows

Water flowing in streams during dry spells, or when precipitation falls as snow without melting, is known as *dry-weather flow* or *dry-weather runoff*. It is composed of water stored in the ground and impounded in lakes, ponds, swamps, and other backwaters. Accordingly, the dry-weather yield of streams comes both from *natural surface storage* and *natural ground storage*. In some river basins with headwaters at high altitudes, much of the summer runoff is derived from melting snowfields. In the absence of snowmelt or surface storage, streams lying above the groundwater table at all stages of flow are *ephemeral* (short lived); streams lying above the summer groundwater level *intermittent*.

11.7.2 Runoff from Rainfall

Runoff from rainfall is influenced chiefly by (a) the intensity, duration, and distribution of precipitation; (b) the size, shape, cover, and topography of the catchment area; and (c) the nature and condition of the ground. Some of these factors are constant, others vary seasonally. Generally speaking, conditions that tend to promote high surface runoff—high rates of rainfall, steep slopes, frozen or bare and heavy soils, for example—are also conditions that tend to reduce dry-weather flows.

11.7.3 Runoff from Snowmelt

Streams fed by snowfields and glaciers possess certain unique runoff characteristics because of the following subsidiary factors: (a) heat melt, (b) condensation melt, and (c) rainfall melt. Fresh, clean snow reflects about 90% of the incident sunlight. Warm, still air, too, causes little melting, because the thermal conductivity of air is small, but there is condensation of moisture and rapid melt when the vapor pressure of the air is higher than that of the snow. Because the heat of condensation of water vapor is 1,073 Btu/lb at 32°F (596 kg-cal/kg at 0°C) and the heat of fusion of ice is only 144 Btu/lb (80 kg-cal/kg), condensation of 1 in. (25.4 mm) of moisture results in $1,073/144 = 7.5$ in. (190 mm) of melt. Rain falling on snow produces relatively less melting by comparison, namely, $(T - 32)/144$ for each inch of rain (25.4 mm), T being its temperature in °F and equaling the wet-bulb temperature of the air. Even when $T = 60.8$°F (16°C), for example, an inch of rain (25.4 mm) induces only $(60.8 - 32)/144 = 0.2$ in. (5 mm) of melt. In more general terms, each degree day above 32°F (0°C) can result in 0.05–0.15 in. (1.3–3.8 mm) of melt. This is the degree-day factor.

For snow packs on mountainsides, an areally weighted average temperature can be computed for the melting zone. The lower boundary of this zone is drawn by the snow line; the upper boundary by the contour lying above the observation station by about 1000 per three times the difference between the observed temperature and 32°F(0°C). The assumption is made that the temperature of the atmosphere drops by approximately 3°F (1.7°C) for each 1000-ft (305-m) lift in elevation.

11.7.4 Measuring Runoff

No studies of surface-water supplies, storm and combined sewerage systems, and wastewater disposal can advance to the design stage without a thorough evaluation of pertinent runoffs and their magnitude and variability. Stream gauging itself is based on an understanding of open-channel flow. Measuring devices are many: current meters, floats, weirs, and surveying instruments; measuring techniques involve chemicals, radioactive tracers, and persistent dyes. Each of the devices and techniques has its own advantages and disadvantages, and each has its peculiar range of usefulness. Once established at a stable cross-section, and rated by suitable means, gauge height needs to be the sole measure of record, automatic records being traced by vertical movements of a float in a stilling well. Occasional checks of actual runoff will reinforce the validity of the rating curve, which relates measured discharge to measured gauge height.

11.8 Good Records and Their Uses

Collection of representative hydrologic information should be encouraged at every opportunity if the water potential of a river system is to be fully understood and put to use.

11.8.1 Rainfall Records

Although precipitation is the ultimate source of all water supplies, it should be understood at the outset that water supply studies should be based, whenever possible, on direct measurements: runoff records for surface-water supplies and groundwater flows for groundwater supplies. Nonetheless, hydrologists will and should continue to look for useful relations between rainfall and runoff and between rainfall and infiltration, and engineers will and should put these relations to work in the absence of direct measurements. However, the links in the chain of hydrologic sequences still remain too weak to be placed under much stress. If this is kept in mind, rainfall records need not be eschewed by engineers. Instead, they can be put to use to good purpose in the business of water supply and drainage by providing a better idea of the variations to be expected. For example, they illuminate the possibility of the drought that comes but once in a century or the storm that visits a region but once in a score of years; they suggest cycles and trends and identify variations in areal distribution of normal and unusual rainfalls; and they lend form to predictions of runoff in hydraulic systems, such as storm drains and canals still to be built.

11.8.2 Runoff Records

As soon as serious thought is given to the development of a catchment area, actual measurements of runoff should be started. Even the first year's results will tell much because they can be amplified by feedback, for example, by cross-correlation between short-term runoff observations of the river basin under study and long-term records of neighboring or otherwise similar systems. If rainfall records for the watersheds in question are long, correlation of rainfall ratios with runoff ratios for the period of recorded stream flow in each basin provides a reasonable foundation for extending the shorter record. Generally speaking, comparative studies of hydrographs or duration curves are more reliable than correlations of precipitation and other hydrologic information, but there are exceptions.

Records of the intermittent runoff from small areas yet to be drained by storm sewers are neither numerous nor of much use. Rates of flow change markedly when drainage systems are constructed and go into operation, and they are bound to change further as communities grow and age. Accordingly, continuous records supported by imaginative estimates of runoff to be drained away in stormwater systems in a growing urban complex have much to commend them.

Flow measurements in existing sewers are hard to translate into design values for other sewers, unless the new drains are to be situated in similar, normally nearby, catchments. Even then, the uncertainty of future change remains.

Flood-flow estimates must look to unusual as well as normal experience. The range and number of useful observations are best extended by resorting to as much hydrologic information as can be assembled. Fortunately, the labor of correlating pertinent data has been greatly reduced by access to high-speed computers. By combining modern computational methods with modern statistical procedures, it has become possible to harvest information in far greater measure than before and, in a sense, to generate operational sequences of hydrologic information of great significance.

11.9 Hydrologic Frequency Functions

Arrays of many types of hydrologic as well as other measurements trace bell-shaped curves like the *normal frequency distribution* shown in Appendix 5 of this book. Arrays are numerical observations arranged in order of magnitude from the smallest to the largest, or in reverse order. They are transformed into frequency distributions by subdividing an abscissal scale of magnitude into a series of usually equal-size intervals—called class intervals—and counting the number of observations that lie within them. The resulting numbers of observations per class interval, by themselves or in relation to the total number of observations, are called the frequencies of the observations in the individual sequent intervals. To simplify subsequent calculations, frequencies are referenced to the central magnitudes of the sequent class intervals.

11.9.1 Averages

The tendency of a bell-shaped curve to cluster about a central magnitude of an array, that is, its central tendency, is a measure of its average magnitude. The *arithmetic mean* magnitude of an array generally offers the most helpful concept of central tendency. Unlike the *median* magnitude of an array, which is positionally central irrespective of the magnitudes of the individual observations composing the array, and unlike the *mode* of the array, which is the magnitude having the highest frequency of observations composing the array, the arithmetic mean is a collective function of the magnitudes of all component observations.

11.9.2 Variability and Skewness

The tendency of a bell-shaped curve to deviate by larger and larger amounts from the central magnitude of an array with less and less frequency is a measure of its variability, variation, deviation, dispersion, or scatter. The tendency of a bell-shaped curve to be asymmetrical is a measure of its skewness.

Mathematically, the symmetrical bell-shaped *normal* frequency curve of Appendix 5 is described fully by two parameters: (a) the *arithmetic mean, μ*, as a measure of central tendency, and (b) the *standard deviation, σ*, as a measure of dispersion, deviation, or variation. The equation of the curve, which is referred to as the *Gaussian* or *normal probability curve*, is

$$y/n = F(x)/n = f(x) = \left[\frac{1}{\sigma\sqrt{2\pi}}\right] \exp\left\{-\tfrac{1}{2}\left[\frac{x-\sigma}{\sigma}\right]^2\right\} \quad (11.15)$$

where y is the number or frequency of observations of magnitude x deviating from the mean magnitude by$x-\mu$, n is the total number of observations, σ is the standard deviation from the mean, $F(x)$ symbolizes a function of x, and μ is the mathematical constant 3.1416.

As shown in Appendix 5 and calculated from Eq. 11.15, the origin of the coordinate system of the Gaussian curve lies at $x-\mu=0$ where the frequency is $y = n/(\sigma\sqrt{2\pi})$, and the distance from the origin to the points of inflection of the curve is the standard deviation $\sigma = \sqrt{\Sigma(x-\mu)^2/n}$. This presupposes that deviations are referenced to the true mean μ, whereas calculations must be based on the observed mean $\bar{x} = \Sigma x/n$, which is expected to deviate from the true mean by a measurable although small amount. However, it can be shown that when the observed standard deviation is calculated as $s = \sqrt{\Sigma(x-\mu)^2/(n-1)}$, it closely approximates σ. As n becomes large, there is indeed little difference between s and σ as well as between $\bar{x}$ and μ, and the Greek letters are often used even when, strictly speaking, the Latin ones should be. The ratios y/n and $(x-\mu)/\sigma$ generalize the equation of the *normal frequency function*.

Engineers are usually interested in the expected frequencies of observations below or above a given value of x or falling between two given values of x. The magnitudes of areas under the normal curve provide this useful information better than point frequencies. Appendix 5 is a generalized table of these areas. It is referred to as a probability integral table because a given area is a measure of the expected fraction of the total area (as unity) corresponding to different departures from the mean $(x-\mu)$ in terms of the standard deviation σ, that is, $(x-\mu)/\sigma = t$.

11.10 Probability Paper

As suggested by Allen Hazen, the probability integral (Appendix 5) can be used to develop a system of coordinates on which normal frequency distributions plot as straight lines. The companion to the probability scale can be (a) arithmetic for true Gaussian normality, (b) logarithmic for geometric normality, or (c) some other function of the variable for some other functionally Gaussian normality. Observations that have a lower limit at or near zero may be geometrically normal ($\log 0 = -\infty$).

Helpful and useful associations in arithmetic and geometric probability plots can be listed as shown in Table 11.2.

As shown in Fig. 11.6, two or more series of observations can be readily compared by plotting them on the same scales. The ratio of σ to μ is called the coefficient of variation, c_v. It is a useful, dimensionless, analytical measure of the relative variability of different series.

Examination of a series of equally good arrays of information shows that their statistical parameters, their means and standard deviations, for example, themselves form bell-shaped distributions. Their variability, called their *reliability* in such instances, is intuitively a function of the size of the sample. Expressed as a standard deviation, the reliability of the common parameters for normal distributions is shown in Table 11.3.

Table 11.2 Associations in Arithmetic and Geometric Probability Plots

Observed or Derived Frequency (%)	Observed or Derived Magnitude	
	Arithmetic	Geometric
50	μ	μ_g
84.1	$\mu+\sigma$	$\mu_g \times \sigma_g$
15.9	$\mu-\sigma$	μ_g/σ_g

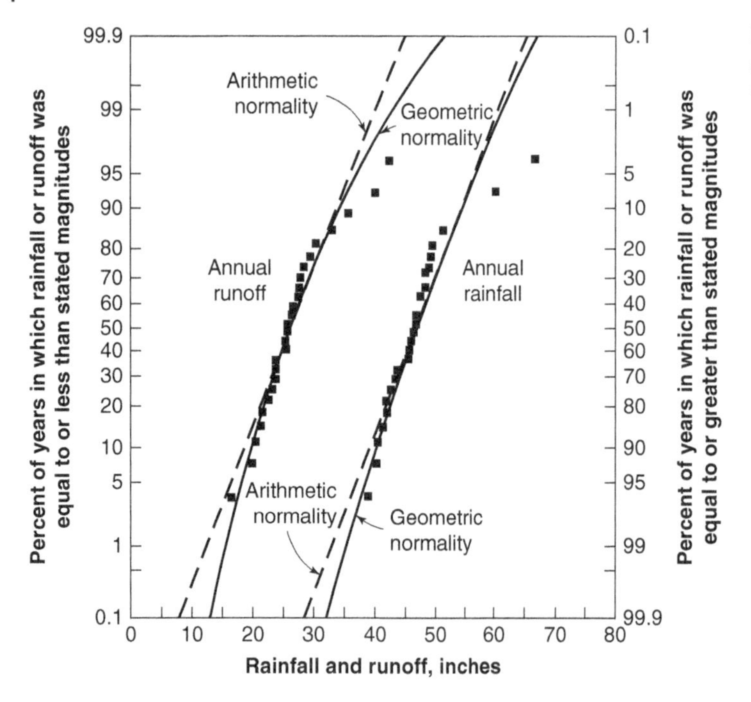

Figure 11.6 Frequency Distribution of Annual Rainfall and Runoff Plotted on Arithmetic-Probability Paper. Conversion factor: 1 inch = 1 in. = 25.4 mm

Table 11.3 The Reliability of Common Parameters for Normal Distributions

Parameter	Computation	Standard Deviation
Arithmetic mean	$\bar{x} = \Sigma x_i/n$	$\sigma/\sqrt{n}$
Median	Midmost observation	$1.25\sigma/\sqrt{n}$
Arithmetic standard deviation	$s = \sqrt{\Sigma(x_i - \bar{x})^2/(n-1)}$	$\sigma/\sqrt{2n} = 0.707\sigma/\sqrt{n}$
Coefficient of variation	$c_v = s/\bar{x}$	$\sqrt{1 + 2c_v^2}/\sqrt{2n}$
Geometric mean	$\log \mu_g = \Sigma \log x_i/n$	$\mathrm{Log}\, \sigma_g/\sqrt{n}$
Geometric standard deviation	$\log \sigma_g = \sqrt{\Sigma \log^2(x_i/\bar{x}_g)/(n-1)}$	$\mathrm{Log}\, \sigma_g/\sqrt{2n}$

11.11 Rainfall and Runoff Analysis

Two types of rainfall and runoff records are analyzed most frequently in the design of water and wastewater works: (a) records of the amounts of water collected by given watersheds in fixed calendar periods such as days, months, and years and (b) records of the intensities and durations of specific rainstorms and flood flows in given drainage areas.

Statistical studies of *annual water yields* provide information on the safe and economic development of surface-water supplies by direct draft and by storage, cast some light on the possible production of groundwater, and are needed in estimates of the pollutional loads that can be tolerated by bodies of water into which wastewaters are discharged. Statistical studies of *rainfall intensities* and *flood runoff* are starting points in the design of stormwater drainage systems, the dimensioning of spillways and diversion conduits for dams and related structures, the location and protection of water and wastewater works that lie in the floodplains of given streams, and the sizing of rainwater collection works.

11.12 Annual Rainfall and Runoff

The presumptive presence of a lower limit of annual rainfall and runoff skews annual rainfall and runoff frequency distributions to the right. The magnitude of this lower limit is generally smaller than the recorded minimum but greater than zero. Although it stands to reason that there must also be an upper limit, its value is less circumscribed and, from the standpoint of water supply, also less crucial. In spite of acknowledging these constraints, most records of annual rainfall and runoff are generalized with fair success as arithmetically normal series and somewhat better as geometrically normal series. Therefore, reasonably accurate comparisons are made in terms of the observed arithmetic or geometric means and the arithmetic or geometric standard deviations. For ordinary purposes, mean annual values and coefficients of variation are employed to indicate the comparative safe yields of water supplies that are developed with and without storage. For comparative purposes, drafts are expressed best as ratios to the mean annual rainfall or runoff, whatever the basis of measurement is.

11.12.1 Rainfall

In those portions of the North American continent in which municipalities have flourished, mean annual rainfalls generally exceed 10 in. (250 mm) and range to almost 80 in. (2,000 mm). In areas with less than 20 in. (500 mm) of annual rainfall and without irrigation, agriculture can be a marginal economic pursuit. For the well-watered regions, the associated coefficients of variation, c_v, are as low as 0.1; for the *arid* regions, they are as high as 0.5, thus implying that a deficiency as great as half the mean annual rainfall is expected to occur in the arid regions as often as a deficiency as great as one-tenth the mean annual rainfall, or less, in the *well-watered regions*. Therefore, high values of c_v are warning signals of low maintainable drafts or high storage requirements.

11.12.2 Runoff

Evaporation and transpiration, together with unrecovered infiltration into the ground, reduce annual runoff below annual rainfall. Seasonally, however, the distribution of rainfall and runoff may vary so widely that it is impossible to establish a direct and meaningful relationship between the two. On the North American continent, the mean annual runoff from catchment areas for water supplies ranges from about 5 to 40 in. (127 to 1,000 mm) and the coefficient of variation of runoff lies between 0.75 and 0.15, respectively. The fact that the mean annual runoff is usually less than half the mean annual rainfall and the variation in stream flow is about half again as great as the variation in precipitation militates against the establishment of direct runoff-rainfall ratios. Storage of winter snows and resulting summer snowmelts offer an important example of conflict between seasonal precipitation and runoff.

Example 11.3 Analysis of Rainfall and Stream Runoff Data

Analyze the 26-year record of a stream (Westfield Little River, which supplies water to Springfield, Massachusetts) in the northeastern United States and of a rain gauge situated in a neighboring valley and covering the identical period of observation (Table 11.4).

Solution:

To plot the data, use the following information: The length of each record is $n = 26$ years; therefore, each year of record spans $100/26 = 3.85\%$ of the experience. However, the arrays are plotted on probability paper in Fig. 11.6 at $100\,k/(n+1) = (100\,k/27)\%$ in order to locate identical points for the left-hand and right-hand probability scales. The resulting plotting observations are shown in Table 11.5.

Necessary calculations are exemplified in Table 11.5 for rainfall and arithmetic normality alone. Calculations for runoff and arithmetic normality would substitute the array of runoff values for that of rainfall values, and assumption of geometric normality would require substitution of the logarithm of the observations for the observations themselves. The calculated statistical parameters can be summarized as shown in Table 11.6.

Examination of the results and the plots shows the following:

1) Both annual rainfall and annual runoff can be fitted approximately by arithmetically normal distributions and somewhat better by geometrically normal distributions.
2) A little more than half the annual rainfall appears as stream flow.
3) Runoff is about 1.7 times as variable as rainfall when measured by c_v.

Table 11.4 Record of Annual Rainfall and Runoff for Example 11.3

Order of Occurrence	Rainfall, in.	Runoff, in.
1	43.6	26.5
2	53.8	35.5
3	40.6	28.3
4	45.3	25.5
5	38.9 (min.)	21.4
6	46.6	25.3
7	46.6	30.1
8	46.1	22.7
9	41.8	20.4
10	51.0	27.6
11	47.1	27.5
12	49.4	21.9
13	40.2	20.1
14	48.9	25.4
15	66.3 (max.)	39.9
16	42.5	23.3
17	47.0	26.4
18	48.0	29.4
19	41.3	25.5
20	48.0	23.7
21	45.5	23.7
22	59.8	41.9 (max.)
23	48.7	32.9
24	43.3	27.7
25	41.8	16.5 (min.)
26	45.7	23.7

Conversion factor: 1 in. = 25.4 mm.

Table 11.5 Calculation of Arithmetic Parameters of Annual Rainfall Frequency for Example 11.3

Magnitude of Observation (1)	Plotting Position % (2)	Deviation from Mean $(x-\mu)$ (3)	$(x-\mu)^2$ (4)
38.9	3.7	−7.9	62.41
40.2	7.4	−6.6	43.56
40.6	11.1	−6.2	38.44
41.3	14.8	−5.5	30.25
41.8	18.5	−5.0	25.00
41.8	22.2	−5.0	25.00
42.5	25.9	−4.3	18.49
43.3	29.6	−3.5	12.25
43.6	33.3	−3.2	10.24
45.3	37.0	−1.5	2.25

Table 11.5 (Continued)

	Magnitude of Observation (1)	Plotting Position % (2)	Deviation from Mean $(x-\mu)$ (3)	$(x-\mu)^2$ (4)
	45.5	40.7	−1.3	1.69
	45.7	44.5	−1.1	1.21
	46.1	48.2	−0.7	0.49
	46.6	51.9	−0.2	0.04
	46.6	55.5	−0.2	0.04
	47.0	59.3	+0.2	0.04
	47.1	63.0	+0.3	0.09
	48.0	66.7	+1.2	1.44
	48.0	70.4	+1.2	1.44
	48.7	74.1	+1.9	3.61
	48.9	77.8	+2.1	4.41
	49.4	81.5	+2.6	6.76
	51.0	85.2	+4.2	17.64
	53.8	88.9	−7.0	49.00
	59.8	92.6	+13.0	169.00
	66.3	96.3	+19.5	380.25
Sum,	1218.8	$n = 26$	0.0	916.04
Mean, $\mu = 46.8$			(Standard deviation)^2, $\sigma^2 = 33.93$	
Median by interpolation or from plot, 46.3				$\sigma = 5.8$
Coefficient of variation,				$c_v = 12.4\%$

Table 11.6 Calculated Statistical Parameters for Example 11.3

	Rainfall	Runoff	Runoff-Rainfall Ratio %
Length of record, n, years	26	26	—
Arithmetic mean, μ, in.	46.8 ± 1.2	26.6 ± 1.1	57
Median, at 50% frequency, in.	46.3 ± 1.5	25.5 ± 1.5	55
Geometric mean, μ_g, in.	$46.5 \overset{\times}{\div} 1.02$	$26.1 \overset{\times}{\div} 1.04$	56
Arithmetic standard deviation, σ, in.	5.8 ± 0.8	5.8 ± 0.8	100
Coefficient of variation, c_v	12.9 ± 1.8	21.8 ± 3.2	169
Geometric standard deviation, σ_g	$1.13 \overset{\times}{\div} 1.02$	$1.24 \overset{\times}{\div} 1.03$	110

4) The probable lower limits of rainfall and runoff are 30 and 10 in., respectively, as judged by a negative deviation from the mean of 36, or a probability of occurrence of $1/(0.5 - 0.4987)$ = once in 700 years.
5) For geometric normality, the magnitudes of the minimum yields expected once in 2, 5, 10, 20, 50, and 100 years, that is, 50, 20, 10, 5, 2, and 1% of the years, are 47, 42, 39, 37, 35, and 33 in. for rainfall and 27, 22, 20, 19, 17, and 15 in. for runoff.

11.13 Storm Rainfall

Storms sweeping over the country precipitate their moisture in fluctuating amounts during given intervals of time and over given areas. For a particular storm, recording rain gauges measure the *point rainfalls* or quantities of precipitation collected during specified intervals of time at the points at which the gauges are situated. Depending on the size of the area of interest, the sweep of the storm, and the number and location of the gauges, the information obtained is generally far from complete. Statistical averaging of experience must then be adduced to counter individual departures from the norm. Given the records of one or more gauges within or reasonably near the area of interest, the rainfall is generally found to vary in intensity (a) during the time of passage or duration of individual storms (time–intensity or intensity–duration), (b) throughout the area covered by individual storms (areal distribution), and (c) from storm to storm (frequency–intensity–duration or distribution in time).

The intensity, or rate, of rainfall is conveniently expressed in in./h (mm/h); and it happens that an inch (25.4 mm) of water falling on an acre in an hour closely equals a ft^3/s (1 in. /h = 1.008 $ft^3/s/acre$). By convention, storm intensities are expressed as the maximum arithmetic mean rates for intervals of specified length, that is, as progressive means for lengthening periods of time. Within each storm, they are highest during short time intervals and decline steadily with the length of interval. Time–intensity calculations are illustrated in Example 11.4.

Example 11.4 Determination of Time–Intensity Relationship from Rain Gauge Records
Given the record of an automatic rain gauge shown in Table 11.7, find the progressive arithmetic mean rates, or intensities, of precipitation for various durations.

Solution:
The records are shown in columns 1 and 2 of Table 11.7, converted into rates in columns 3 and 4, and assembled in order of magnitude for increasing lengths of time in columns 5–7. The maximum rainfall of 5-min duration (0.54 in./5 min = 6.48 in./h) was experienced between the 30th and 35th min.

Table 11.7 Time and Intensity of a Storm Rainfall for Example 11.4[a]

Rain-Gage Record				Time–Intensity Relationship		
Time from Beginning of Storm, min (1)	Cumulative Rainfall, in. (2)	Time Interval, min (3)	Rainfall During Interval, in. (4)	Duration of Rainfall, min (5)	Maximum Total Rainfall, in. (6)[b]	Arithmetic Mean Intensity, in/h (7)[c]
5	0.31	5	0.31	5	0.54	6.48
10	0.62	5	0.31	10	1.07	6.42
15	0.88	5	0.26	15	1.54	6.16
20	1.35	5	0.47	20	1.82	5.46
25	1.63	5	0.28	30	2.55	5.10
30	2.10	5	0.47	45	3.40	4.53
35	2.64	5	0.54	60	3.83	3.83
40	3.17	5	0.53	80	4.15	3.11
45	3.40	5	0.23	100	4.41	2.65
50	3.66	5	0.26	120	4.59	2.30
60	3.83	10	0.17			
80	4.15	20	0.32			
100	4.41	20	0.26			
120	4.59	20	0.18			

[a] Conversion factors: 1 in. = 25.4 mm; 1 in. /h = 25.4 mm/h.
[b] Column 6 records maximum rainfall in consecutive periods. It proceeds out of column 4 by finding the value, or combination of consecutive values, that produces the largest rainfall for the indicated period.
[c] Column 7 = 60 × column 6/column 5.

11.14 Frequency of Intense Storms

The higher the intensity of storms, the rarer their occurrence or the lower their frequency. Roughly the highest intensity of specified duration in a station record of n years has a frequency of once in n years and is called the n-year storm. The next highest intensity of the same duration has a frequency of once in $n/2$ years and is called the $n/2$-year storm.

The recurrence interval I is the calendar period, normally in years, in which the kth highest or lowest values in an array covering n calendar periods are expected to be exceeded statistically. The associated period of time is $100/I$. The fifth value in a 30-year series would, therefore, be calculated to have a recurrence interval $I = n/k = 30/5 = 6$ years, or a chance of $100/6 = 16.7\%$ of being exceeded in any year. If, as in Example 11.1, the frequency of occurrence is calculated as $k/(n+1)$, the recurrence interval is $(n+1)/k = 31/5 = 6.2$ years. Other plotting positions could be chosen instead.

By pooling all observations irrespective of their association with individual storm records, a generalized intensity–duration–frequency relationship is obtained. The following empirical equations are of assistance in weeding out storms of low intensity from the analysis of North American station records:

$$i = 0.6 + 12/t \text{ for the northern U.S} \quad \text{(U.S. Customary Units)} \tag{11.16a}$$

$$i = 1.2 + 18/t \text{ for the southern U.S} \quad \text{(U.S. Customary Units)} \tag{11.17a}$$

Here, i is the rainfall intensity in in./h and t is the duration in min. Two equivalent intensity equations using the SI units of i (mm/h) and t (min) are

$$i = 15.24 + 304.8/t \text{ for the northern U.S.} \quad \text{(SI Units)} \tag{11.16b}$$

$$i = 30.48 + 457.2/t \text{ for the southern U.S.} \quad \text{(SI Units)} \tag{11.17b}$$

For a duration of 10 min, for example, intensities below 3 in./h (76.2 mm/h) in the southern states need not receive attention. The storm recorded in Example 11.4 exhibits double this intensity.

Storm rainfall can be analyzed in many different ways. However, all procedures normally start from a summary of experience such as that shown in Example 11.5. The results may be used directly or after smoothing (generally graphical) operations that generalize the experience. The developed intensity–duration–frequency relationships may be left in tabular form, presented graphically, or fitted by equations.

Example 11.5 Determination of Time–Intensity Relationship for a Given Frequency

The number of storms of varying intensity and duration recorded by a rain gauge in 45 years for New York City is listed in Table 11.8. Determine the time–intensity values for the 5-year storm.

Table 11.8 Record of Intense Rainfalls for Example 11.5

Duration, min	Number of Storms of Stated Intensity (in./h) or More												
	1.0	1.25	1.5	1.75	2.0	2.5	3.0	4.0	5.0	6.0	7.0	8.0	9.0
5							123	47	22	14	4	2	1
10					122	78	48	15	7	4	2	1	
15				100	83	46	21	10	3	2	1	1	
20			98	64	44	18	13	5	2	2			
30	99	72	51	30	21	8	6	3	2				
40	69	50	27	14	11	5	3	1					
50	52	28	17	10	8	4	3						
60	41	19	14	6	4	4	2						
80	18	13	4	2	2	1							
100	13	4	1	1									
120	8	2											

Conversion factors: 1 in. /h = 25.4 mm/h.

Table 11.9 Duration–Intensity Relationship for Example 11.5

Duration, min	5	10	15	20	30	40	50	60	80	100
Intensity, in./h	6.50	4.75	4.14	3.50	2.46	2.17	1.88	1.66	1.36	1.11
Intensity, in./h	1.0	1.25	1.5	1.75	2.0	2.5	3.0	4.0	5.0	6.0
Duration, min	116.0	89.9	70.0	52.5	46.7	29.0	25.7	16.0	9.3	7.5

Conversion factor: 1 in. /h = 25.4 mm/h.

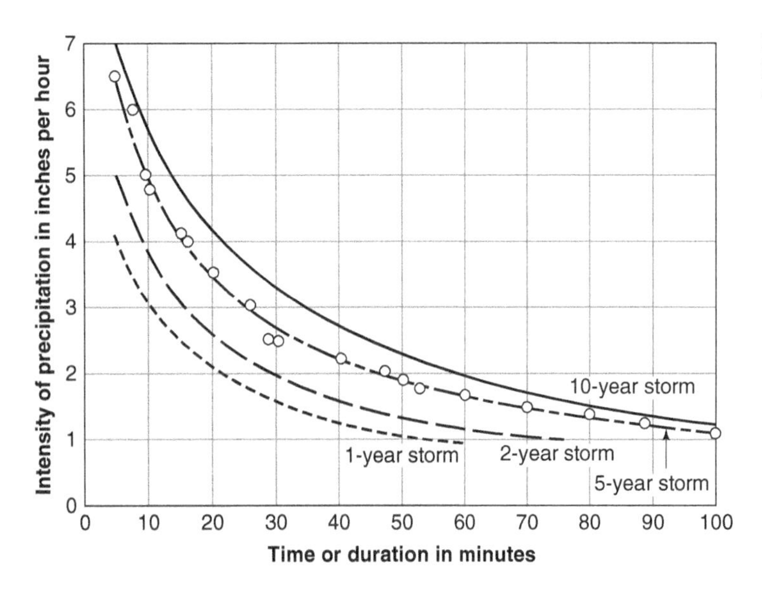

Figure 11.7 Intensity–Duration–Frequency of Intense Rainfalls. Conversion factor: 1 inch/hour = 1 in. /h = 25.4 mm/h

Solution:

If it is assumed that the 5-year storm is equaled or exceeded in intensity 45/5 = 9 times in 45 years, the generalized time–intensity values may be interpolated from Table 11.8 by finding (a) for each specified duration, the intensity equaled or exceeded by nine storms and (b) for each specified intensity the duration equaled or exceeded by nine storms.

The results are as shown in Table 11.9 and are used in constructing Fig. 11.7. Similar calculations for the 1-year, 2-year, and 11-year storms underlie the remaining members of the family of curves in Fig. 11.7.

11.15 Intensity–Duration–Frequency Relationships

Time–intensity curves such as those in Fig. 11.7 are immediately useful in the design of storm drainage systems and in flood-flow analyses. For purposes of comparison as well as further generalization, the curves can be formulated individually for specific frequencies or collectively for the range of frequencies studied.

Good fits are usually obtained by a collective equation of the following form:

$$i = \frac{cT^m}{(t+d)^n} \tag{11.18}$$

where i and t stand for intensity and duration as before; T is the frequency of occurrence, in years; c and d represent a pair of regional coefficients; and m and n represent a pair of regional exponents. Their order of magnitude is about as follows in North American experience: $c = 5$ to 50 and $d = 0$ to 30; $m = 0.1$ to 0.5 and $n = 0.4$ to 1.0.

Equation 11.18 can be fitted to a station record either graphically or by least squares. For storms of specified frequency, the equation reduces to

$$i = \frac{A}{(t + d)^n} \tag{11.19}$$

where

$$A = cT^m \tag{11.20}$$

11.15.1 Graphical Fitting

Equation 11.5 can be transformed to read

$$[\log i] = \log A - n\,[\log\,(t + d)]$$

where the brackets identify the functional scales

$$y = [\log i] \text{ and } x = [\log\,(t + d)]$$

for direct plotting of i against t on double logarithmic paper for individual frequencies. Straight lines are obtained when suitable trial values of d are added to the observed values of t. To meet the requirements of Eq. 11.18 in full, the values of d and of n, the slope of the straight line of best fit, must be the same or averaged to become the same at all frequencies. Values of A can then be read as ordinates at $(t + d) = 1$, if this point lies or can be brought within the plot. To determine c and m, the derived values of A are plotted on double logarithmic paper against T for the frequencies studied. Because $[\log A] = \log c + m\,[\log T]$, the slope of the resulting straight line of best fit equals m, and the value of c is read as the ordinate at $T = 1$.

Example 11.6 Time–Intensity–Frequency Relationship

Fit Eq. 11.18 to the 60-min record of intense rainfalls presented in Example 11.5.

Solution:

Plot the values for the 5-year storm on double logarithmic paper as in Fig. 11.8. Because the high-intensity, short-duration values are seen to bend away from a straight line, bring them into line by adding 2 min to their duration periods, that is,

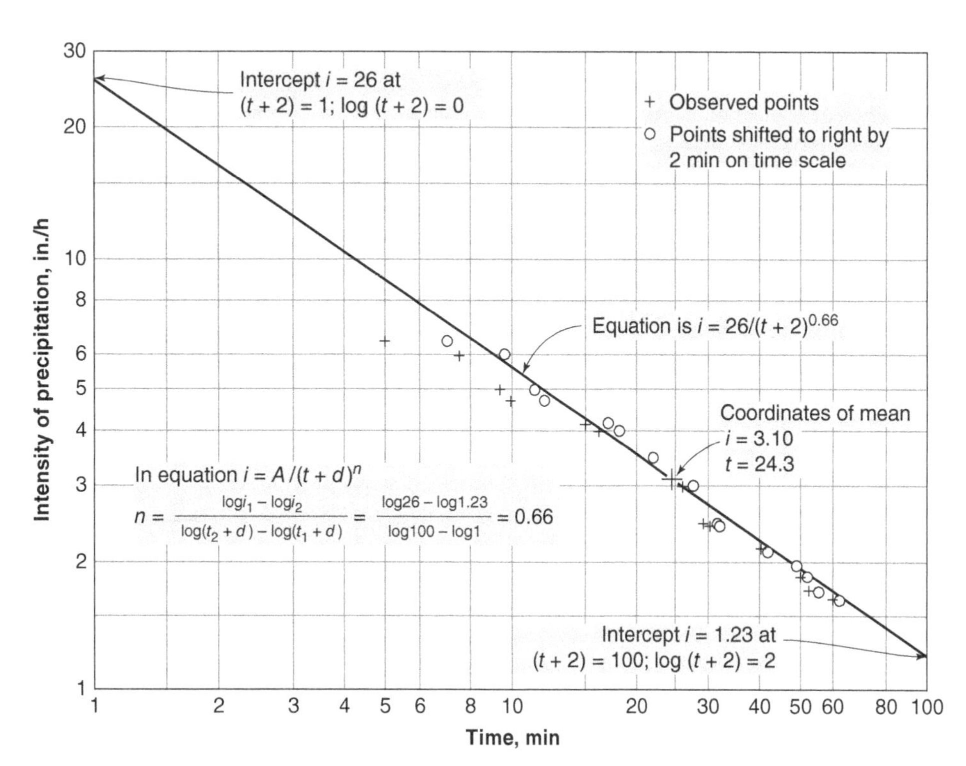

Figure 11.8 Intensity-Duration of 5-Year Rainstorm. Conversion factor: 1 in./h = 25.4 mm/h

$(t+d) = (t+2)$. Derivation of the equation $i = A/(t+d)^n = 26/(t+2)^{0.66}$ is noted on Fig. 11.8. Similar plots for the other storms of Fig. 11.7 would yield parallel lines of good fit. The intercepts A of these lines on the i axis at $(t+d) = 1$ themselves will plot as straight lines on double logarithmic paper against the recurrence interval T. Hence, for $[\log A] = \log c + m\,[\log T]$, find the magnitudes $c = 16$ and $m = 0.31$ to complete the numerical evaluation of the coefficients and with them the equation

$$i = \frac{cT^m}{(t+d)^n} \tag{11.18}$$

$$= \frac{16T^{0.31}}{(t+2)^{0.66}}$$

11.15.2 Intensity–Duration–Frequency Relationships for the United States

Table 11.10 presents metric intensity–duration–frequency relationships, where $n = 1$, for various areas of the United States. These various areas are shown in Fig. 11.9 and represent regions of approximately similar rainfall characteristics. The equations in Table 11.10 for these general areas can be used when no more precise local data are available.

Example 11.7 Intensity–Duration–Frequency Relationships for the United States in English Units
Determine:

1) The intensity–duration–frequency equation for area 1 of the United States using the English system and assuming the frequency of precipitation is 2 years.
2) Calculate the intensity (in./h and mm/h) of a rainstorm for a duration of 20 min.

Table 11.10 Intensity–Duration–Frequency Relationship for the Various Areas Indicated in Fig. 11.9 (Metric system, i = mm/h; t = min)

Frequency, years	Area 1	Area 2	Area 3	Area 4	Area 5	Area 6	Area 7
2	$i=\frac{5{,}230}{t+30}$	$i=\frac{3{,}550}{t+21}$	$i=\frac{2{,}590}{t+17}$	$i=\frac{1{,}780}{t+13}$	$i=\frac{1{,}780}{t+16}$	$i=\frac{1{,}730}{t+14}$	$i=\frac{810}{t+11}$
5	$i=\frac{6{,}270}{t+29}$	$i=\frac{4{,}830}{t+25}$	$i=\frac{3{,}330}{t+19}$	$i=\frac{2{,}460}{t+16}$	$i=\frac{2{,}060}{t+13}$	$i=\frac{1{,}900}{t+12}$	$i=\frac{1{,}220}{t+12}$
10	$i=\frac{7{,}620}{t+36}$	$i=\frac{5{,}840}{t+29}$	$i=\frac{4{,}320}{t+23}$	$i=\frac{2{,}820}{t+16}$	$i=\frac{2{,}820}{t+17}$	$i=\frac{3{,}100}{t+23}$	$i=\frac{1{,}520}{t+13}$
25	$i=\frac{8{,}330}{t+33}$	$i=\frac{6{,}600}{t+32}$	$i=\frac{5{,}840}{t+30}$	$i=\frac{4{,}320}{t+27}$	$i=\frac{3{,}330}{t+17}$	$i=\frac{3{,}940}{t+26}$	$i=\frac{1{,}700}{t+10}$
50	$i=\frac{8{,}000}{t+28}$	$i=\frac{8{,}890}{t+38}$	$i=\frac{6{,}350}{t+27}$	$i=\frac{4{,}750}{t+24}$	$i=\frac{4{,}750}{t+25}$	$i=\frac{4{,}060}{t+21}$	$i=\frac{1{,}650}{t+8}$
100	$i=\frac{9{,}320}{t+33}$	$i=\frac{9{,}520}{t+36}$	$i=\frac{7{,}370}{t+31}$	$i=\frac{5{,}590}{t+28}$	$i=\frac{6{,}100}{t+29}$	$i=\frac{5{,}330}{t+26}$	$i=\frac{1{,}960}{t+10}$

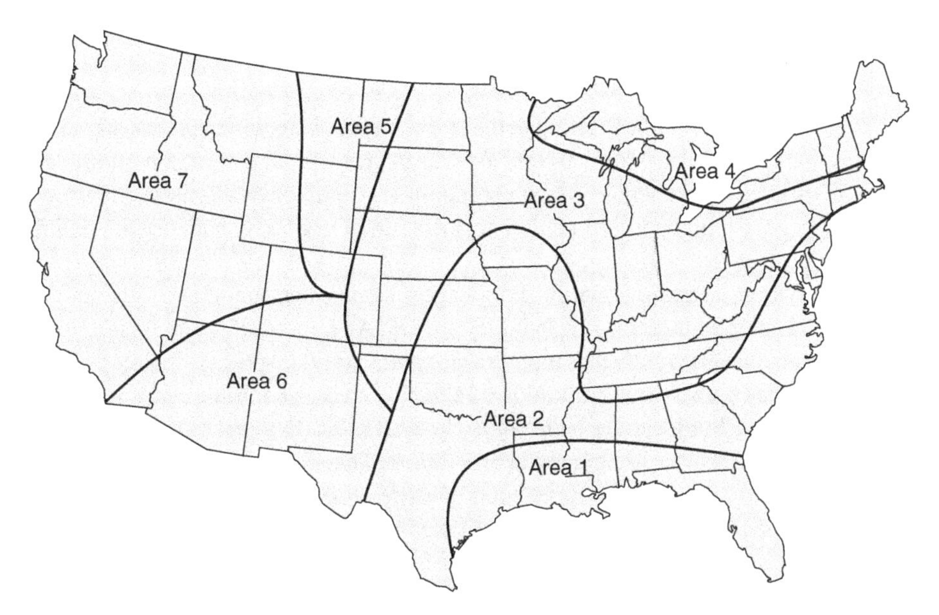

Figure 11.9 Map Showing Areas of Approximately Similar Rainfall Characteristics

Solution 1 (U.S. Customary System):

1) The intensity–duration–frequency from Eq. 11.19 and for $n = 1$ is

$$i = \frac{A}{t + d} \tag{11.21}$$

From Table 11.10 for a frequency of 2 years,

$$i(\text{mm/h}) = \frac{5{,}230}{t + 30} \quad (\text{SI Units}) \tag{11.22}$$

$$i(\text{in./h}) = \frac{\frac{5{,}230}{25.4\ \text{mm/hr}}}{t + 30} \quad (\text{U.S. Customary Units})$$

$$= \frac{206}{t + 30} \quad (\text{U.S. Customary Units}) \tag{11.23}$$

2) The intensity (in./h) of a rainstorm for a duration of 20 min.

$$i(\text{in./h}) = \frac{206}{t + 30}$$

$$= \frac{206}{20 + 30}$$

$$= \mathbf{4.12\ in.\ /h}$$

Solution 2 (SI System):

1) From Table 11.10 for a frequency of 2 years.

$$i(\text{mm/h}) = \frac{5{,}230}{i + 30}$$

2) The intensity is

$$i = \frac{5{,}230}{20 + 30} = 104.6\ \text{mm/h}$$

11.16 Storm Runoff and Flood Flows

The flood flows descending the arterial system of river basins or collecting in the storm drains or combined sewers of municipal drainage districts are derived from rains that fall on the tributary watershed. The degree of their conversion into runoff is affected by many factors, especially in the varied environment of urban communities. Component effects and their relative importance must, therefore, be clearly recognized in the interpretation of storm runoff or flood experience in relation to intense rainfalls.

Flows normally reach flood crest at a given point on a stream or within a drainage scheme when runoff begins to pour in from distant parts of the tributary area. There are exceptions to this rule, but they are few. An important exception is a storm traveling upstream or sweeping across a catchment area so rapidly that runoff from distant points does not reach the point of concentration until long after the central storm has moved on. Diminution of effective area or *retardance* of this kind is rarely taken into consideration in American practice, but it should be in some circumstances. In a given storm, the maximum average rate of rainfall is always highest for the shortest time interval or duration. Therefore, the shorter the elapsed time or *time of concentration* in which distant points are tributary to the *point of concentration*, the larger the flows.

The time of concentration is shortest for small, broad, steep drainage areas with rapidly shedding surfaces. It is lengthened by dry soil, surface inequalities and indentations, and vegetal cover, and by storage in water courses, on floodplains, and in reservoirs. In short intense thunderstorms, peak urban flows often occur when only the impervious or paved areas are shedding water. The volume of runoff from a given storm is reduced by infiltration, freezing, and storage; it is swelled by snow and ice melt, seepage from bank storage, and release of water from impoundages either on purpose or by accident. Maximum rates are obtained when storms move downstream at speeds that bring them to the point of discharge in about the time of concentration, making it possible for the runoff from the most intense rainfall to arrive at the point of discharge at nearly the same instant.

Among the ways devised for estimating storm runoff or flood flows for engineering designs are the following:

1) *Statistical analyses based on observed records of adequate length.* Obviously, these can provide likely answers. Unfortunately, however, recorded information is seldom sufficiently extensive to identify critical magnitudes directly from experience. Information must be generalized to arrive at rational extrapolations for the frequency or recurrence interval of design flows or for the magnitude of flows of design frequency.
2) *Statistical augmentation of available information through cross-correlation* with recorded experience in one or more adjacent and similar basins for which more years of information are available; through correlation between rainfall and runoff when the rainfall record is longer than the runoff record; and through statistical generation of additional values.
3) *Rational estimates of runoff from rainfall.* This is a common procedure in the design of storm and combined sewers that are to drain existing built-up areas and satisfy anticipated change in the course of time, or areas about to be added to existing municipal drainage schemes.
4) *Calculations based on empirical formulations not devised specifically from observations in the design area but reasonably applicable to existing watershed conditions.* Formulations are varied in structure and must be selected with full understanding of the limitations of their derivation. At best, they should be applied only as checks of statistical or rational methods.

   Where failure of important engineering structures is sure to entail loss of life or great damage, every bit of hydrologic information should be adduced to arrive at economical but safe design values. Hydraulic models may also be helpful.

11.17 Estimates of Storm Runoff

Among the various methods used to estimate storm runoff, two are of general interest: the rational method and the unit-hydrograph method.

11.17.1 The Rational Method

This is the method commonly used as a basis for the design of storm drains and combined sewers. Such facilities are expected to carry, without surcharge, the peak runoff expected to be equaled or exceeded on the average once in a period (a recurrence interval) of T years. The interval selected is short, $T = 2$–5 years, when damage due to surcharge and street flooding is small, and is long, $T = 20$–100 years, when the damage is great as it sometimes can be when basements in residential and commercial districts are flooded.

The design peak flow is estimated using the equation

$$Q = \text{cia} \quad \text{(U.S. Customary Units)} \tag{11.24a}$$

where Q is the peak rate of runoff at a specified place, in ft^3/s; a is the tributary area, in acres; and i the rainfall intensity, in in./h (1 in. /h on 1 acre = 1 ft^3/s), for the selected values of T in years and t in min. The rainfall duration, t, is in fact a rainfall-intensity averaging time.

The following are three equivalent Rational Method equations using the SI or metric units:

$$Q = \text{cia} \quad \text{(SI Units)} \tag{11.24b}$$

where the SI units are: Q (m^3/h); i (m/h); a (m^2); c (dimensionless).

$$Q = 238\ \text{cia} \quad \text{(SI Units)} \tag{11.24c}$$

where the SI units are: Q (m^3/d); i (mm/h); a (ha); c (dimensionless).

$$Q = 2.76\ \text{cia} \quad \text{(SI Units)} \tag{11.24d}$$

where the SI units are: Q (L/s); i (mm/h); a (ha); c (dimensionless).

Of the three factors included in Eq. 11.24, the area a is found from a regional map or survey, i is determined for a storm of duration equal to the time of concentration, and c is estimated from the characteristics of the catchment area (see Table 11.11). The time of concentration is found (a) for flood discharge by estimating average velocities of flow in the principal channels of the tributary area and (b) for runoff from sewered areas by estimating the inlet time, or time required for runoff to enter the sewerage system from adjacent surfaces (see Fig. 11.10), and adding to it the time of flow in the sewers or storm drains proper. Because rapid inflow from tributaries generates flood waves in the main stem of a river system, flood velocities are often assumed to be 30–50% higher than normal rates of flow.

When Eq. 11.18 is written in functional terms $Q(T) = ca[i(t, T)]$, it becomes evident that there is implicit in the rational method an assumption that the design peak runoff rate is expected to occur with the same frequency as the rainfall intensity used in the computations. The value computed for $Q(T)$ does not correspond to the peak runoff rate that would be expected from any particular storm. If the peak runoff rates and rainfall intensities for a specified averaging time are analyzed statistically, coefficient c is the ratio of peak unit runoff to average rainfall intensity for any value of T. Examination of the rational method shows that values of c, developed through experience, give reasonably good results even though the assumptions with regard to inlet times and contributing areas are often far from reality.

Peak flows at inlets occur during the very intense part of a storm and are made up of water that comes almost entirely from the paved areas. Lag times between peak rainfall rate and peak runoff rate at inlets are very short, often less than 1 min. For areas of up to 20–50 acres, good values of peak runoff rates can be obtained by estimating the flow to each inlet and

Table 11.11 Runoff Coefficients and Mean Pollutant Concentrations of Runoff Water in the United States

		Pollutant				
Land Use	**C**	**BOD_5**	**SS**	**VS**	**PO_4**	**N**
Residential[a]	0.30	11.8	240	139	0.50	1.9
Commerical[a]	0.70	20.2	140	88	0.48	1.9
Industrial[a]	0.60	8.9	214	105	0.52	2.0
Other Developed Areas[a]	0.10	5.0	119	115	0.44	2.7
Residential[b]	0.30	48.6	989	574	2.04	8.0
Commerical[b]	0.70	83.2	578	364	1.97	7.4
Industrial[b]	0.60	36.7	883	434	2.14	8.4
Other Developed Areas[b]	0.10	20.6	492	473	1.81	11.0

[a] Applies to separate and unsewered drainage areas.
[b] Applies to combined drainage areas.
All units of pollutants are mg/L.
Range of runoff coefficients: 0.3–0.75 for residential areas; 0.5–0.95 for commercial areas; 0.5–1.0 for industrial areas; and 0.05–0.4 for other developed areas. (U.S. EPA, Technical Report EPA-600-/9-76-014, *Areawide Assessment Procedures Manual*).

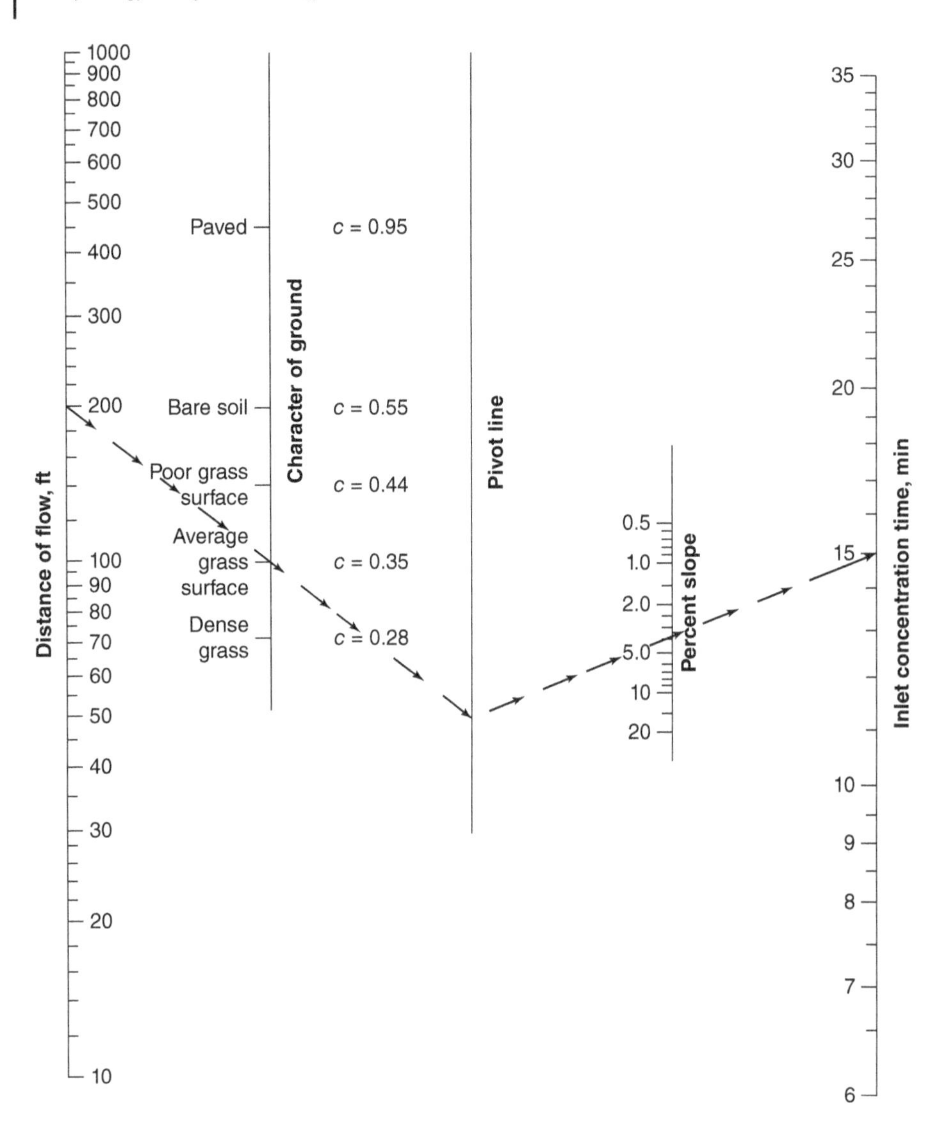

Figure 11.10 Nomogram in U.S. Customary Units for Determining Overland Flow Time (Nomogram in SI System is in Appendix 18). Conversion factor: 1 ft = 0.3048 m

combining attenuated hydrographs to obtain design values. In most intense storms, very little runoff is contributed by the unpaved areas, so good estimates can also be made by considering rainfall on impermeable surfaces only. These and other modifications of the rational method require the use of an appropriate set of coefficients. The selection of suitable values for *c* in estimating runoff from sewered areas is discussed more fully in Chapter 14 in connection with the design of storm drains and combined sewers.

Example 11.8 Determination of Inlet Concentration Time

Determine the inlet concentration time for a land area sloping at 4.5% and covered with average grass with a *c* value of 0.35. The furthest point from the inlet is 60 m (200 ft).

Solution:

Distance of flow = 60 m = 200 ft
Average grass surface, $c = 0.35$
Slope of land = 4.5%
From the nomogram of Fig. 11.10:
$\mathbf{t = 15}$ **min**

Example 11.9 Estimation of Runoff by the Rational Method

Determine the mean runoff flow, in ft^3/s and L/s, for a 5-year recurrence rainfall assuming the runoff coefficient for the semiresidential area in Massachusetts is 0.45, the concentration time is 10 min, and the drainage area a is 10,000 m^2 (2.47 acre; 1 ha).

Solution 1 (U.S. Customary System):

$$Q = \text{cia} \tag{11.24a}$$

$$= (0.45)(i)(2.47 \text{ acre})$$

$$i(in./\text{h}) = \frac{3{,}330/25.4}{t + 19}$$

$$= \frac{3{,}330/25.4}{10 + 19}$$

$$= 4.52 \text{ in. /h}$$

$$Q = (0.45)(4.52)(2.47)$$

$$= \mathbf{5\ ft^3/s}$$

Solution 2 (SI System):

$$Q = \text{cia} \tag{11.24b}$$

$$Q = (0.45)(i)\left(10{,}000 \text{ m}^2\right)$$

From Fig. 11.9, Massachusetts is in Area 3. From Table 11.10, the intensity–duration–frequency for area 3 and a frequency of 5 years is

$$i = \frac{3{,}330}{t + 19}$$

$$= \frac{3{,}330}{10 + 19}$$

$$= 115 \text{ mm/h}$$

$$= 0.115 \text{ m/h}$$

Therefore,

$$Q = (0.45)(0.115 \text{ m/h})(10{,}000 \text{ m}^2) \text{ using Eq. (11.24b)}$$

$$= 518 \text{ m}^3/\text{h}$$

$$= (518 \text{ m}^3/\text{h}) \times (1{,}000 \text{ L/m}^3)/(60 \times 60 \text{ s/h})$$

$$= \mathbf{143\ L/s\ runoff\ rate}$$

$$Q = 2.76\ (0.45)\ (115)\ (1) \text{ using Eq. (11.24d)}$$

$$= \mathbf{143\ L/s}$$

11.17.2 The Unit-Hydrograph Method

In dry weather, or when precipitation is frozen, the residual hydrograph or base flow of a river is determined by water released from storage in the ground or in ponds, lakes, reservoirs, and backwaters of the stream. Immediately after a rainstorm, the rate of discharge rises above base flow by the amount of surface runoff entering the drainage system. That portion of the hydrograph lying above base flow can be isolated from it and is a measure of the true surface runoff (Fig. 11.11). The *unit-hydrograph method* stems from studies of simple geometric properties of the surface-runoff portion of the hydrograph

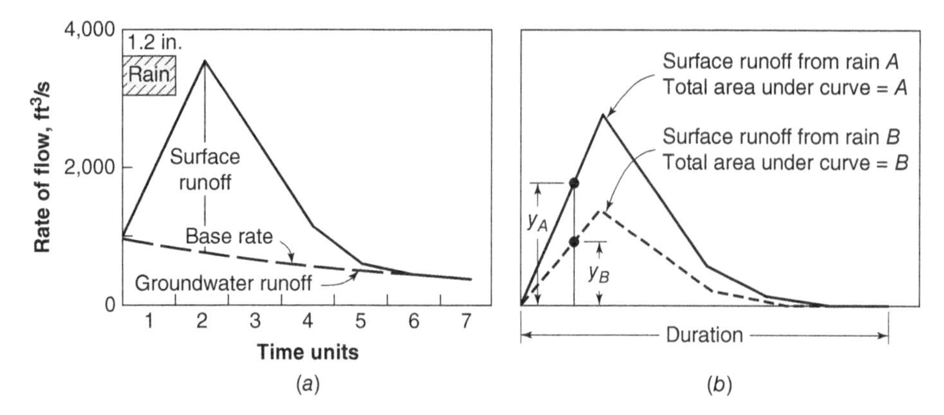

Figure 11.11 Origin and Geometric Properties of the Unit Hydrograph: (a) Hydrograph Resulting from Unit-Time Rain (See Example 11.6). (b) Distribution Graph Showing Geometric Properties of Unit Hydrograph; $y_A : A = y_B : B$; Base Duration Is Constant

in their relation to an *effective rain* that has fallen during a *unit* of time, such as a day or an hour, and that, by definition, has produced surface runoff.

The important geometric properties of the unit hydrograph or surface runoff illustrated in Fig. 11.10 are as follows:

1) The abscissal length measuring time duration above base flow is substantially constant for all unit-time rains.
2) Sequent ordinates, measuring rates of discharge above base flow at the end of each time unit, are proportional to the total runoff from unit-time rains irrespective of their individual magnitudes.
3) Ratios of individual areas to the total area under the hydrograph, measuring the amount of water discharged in a given interval of time, are constant for all unit hydrographs of the same drainage area. These distribution ratios are generally referred to as the *distribution graph*, even when they are not presented in graphical form.
4) Rainstorms extending, with or without interruption, over several time units generate a hydrograph composed of a series of unit hydrographs superimposed in such a manner as to distribute the runoff from each unit-time rain in accordance with the successive distribution ratios derived from unit-time rainfalls. This permits the construction of a hydrograph that might result from not-yet-experienced rainstorms.

These geometric properties do not apply when runoff originates in melting snow or ice, nor when the speed of flood waves in streams is changed appreciably as river stages are varied by fluctuating flows. Time is an important element of this procedure, and rainfall data must be available for unit times shorter than the time of concentration of the drainage area. Unit times as long as a day can be employed successfully only for large watersheds (>1,000 mi^2 or > 2,560 km^2) For sheds of 100–1,000 mi^2 (256–2,560 km^2), it has been suggested that values of 6–12 h be used; for sheds of 20 mi^2 (51 km^2), 2 h; and for very small areas, one-fourth to one-third the time of concentration. The unit-hydrograph method is illustrated in Example 11.10.

The unit-hydrograph method is useful in estimating magnitudes of unusual flood flows, in forecasting flood crests during storms, and in the manipulation of storage on large river systems. It has the important property of (a) tracing the full hydrograph resulting from a storm rather than being confined to a determination of the peak flow alone and (b) producing useful results from short records. For small drainage areas, the method depends on the readings of a recording rain gauge.

Example 11.10 Application of Unit Hydrograph Method

1) Given the rainfall and runoff records of a drainage area of 620 mi^2, determine the generalized distribution of runoff (the distribution graph) from isolated unit-time rainfalls.
2) Apply the average estimate of runoff distribution to the observed rainfall sequence obtained above.

Solution:

This involves first of all a search for records of isolated rainfalls and for records of the resulting surface runoff. The basic data for a typical storm are shown in Table 11.12, together with necessary calculations. Development of this table is straightforward, except for column 4, which records the estimated base flow and can be derived only from a study of the general hydrograph of the stream in combination with all related hydrologic observations of the region.

Table 11.12 Observations and Calculations for Unit Hydrograph for Example 11.10

		Runoff, ft³/s		Estimated Distribution of Surface Runoff		
Sequence of Time, Units	Observed Rainfall, in.	Observed Total	Estimated Base Flow	ft³/s	%	Average Distribution Ratio for 10 Storms, %
(1)	(2)	(3)	(4)	(5) = (3) − (4)	(6) = 100(5)/6, 200	(7)
1	1.20	1,830	870	960	15.5	16
2	0.03	3,590	800	2,790	45.0	46
3	0.00	2,370	690	1,680	27.1	26
4	0.00	1,220	600	620	10.0	10
5	0.00	640	510	130	2.1	1
6	0.00	430	410	20	0.3	1
7	0.00	350	350	0	0.0	0
				—	—	—
Totals	—	. . .	. . .	6,200	100.0	100

Conversion factors: 1 in. = 25.4 mm; 1 ft³/s = 28.3 L/s.

Table 11.13 Application of Unit-Hydrograph Method for Example 11.10

	Rainfall, in				Distribution Runoff for Stated Time Units, in.				Compounded Runoff	
Sequence of Time Units	Observed	Estimated Loss	Net	Average Runoff Distribution Ratio, %	1st	2nd	3rd	5th	in	ft³/s[1]
(1)	(2)	(3)	(4) = (2) − (3)	(5)	(6)	(7)	(8)	(9)	(10)	(11)
1	1.8	1.3	0.5	16	0.08	. . .	. . .	. . .	0.08	1,300
2	2.7	1.6	1.1	46	0.23	0.18	. . .	. . .	0.41	6,900
3	1.6	1.1	0.5	26	0.13	0.50	0.08	. . .	0.71	11,900
4	0.0	0.0	0.0	10	0.05	0.29	0.23	. . .	0.57	9,500
5	1.1	0.2	0.9	1	0.01	0.11	0.13	0.14	0.39	6,500
6	0.0	0.0	0.0	1	0.00	0.01	0.05	0.42	0.48	8,000
7	0.0	0.0	0.0	0	0.00	0.01	0.01	0.23	0.25	4,200
8	0.0	0.0	0.0	0	0.00	0.00	0.00	0.09	0.09	1,500

[1] Rate of runoff in ft³/s = in. × 26.88 × 620 mi² = in. × 16, 700 ft³/s if the time unit is a day. For other time units, multiply by reciprocal ratio of length of time to length of day. Conversion factors: 1 in. = 25.4 mm; 1 ft³/s = 28.3 L/s.

The calculations in Table 11.13 need little explanation except for column 3, the estimated loss of rainfall caused principally by infiltration. This estimate rests on all available information for the region. Column 5 is identical to column 7 of Table 11.10. Column 6 is the net rain of 0.5 in. during the first time unit multiplied by the distribution ratio of column 5. Columns 7–9 are similarly derived for the net rains during the subsequent time units. Column 10 gives the sums of Columns 6–9, and column 11 converts these sums from in. to ft³/s. If the base flow is estimated and added to the surface runoff shown in column 11, the hydrograph becomes complete.

Table 11.14 Examples of Flood-Flow Formulas

Individualized Variable	Region	Formula
None	New England	$Q = Ca^{5/6}$, where $C = 200$ for a in mile^2
Rainfall intensity and slope of watershed	St. Louis, Mo.	$Q = cia^{4/5}s^{1/5}$, where s = slope in % and $c = 0.75$ for a in acres or 480 for a in mile^2
Shape and slope of watershed	Cumberland Plateau	$Q = ca^{7/6}/(l/s^{1/2})$, where l = length of principal waterway in miles, s = slope of waterway in ft/mile, and $c = 1920d$ for a in mile^2, the 10-year peak flood and a factor d relating the basin to the base station at Columbus, Ohio
Shape, slope, and surface storage of watershed	New England	$Q = (0.000036h^{2.4} + 124)a^{0.85}/(rl^{0.7})$, where h = median altitude of drainage basin in ft above the outlet; r = % of lake, pond, and reservoir area; l = average distance in miles to outlet; and a = mile^2
Frequency of flood	Fuller, U.S.A.	$Q = Ca^{0.8}(1 + 0.8 \log T)(1 + 2a^{-0.3})$, where T = number of years in the period considered, and C varies from 25 to 200 for different drainage basins and a in mile^2

Conversion factors: 1 mi^2 = 2.59 km^2; 1 mi = 1.61 km; 1 ft = 0.3048 m; 1 acre = 0.4046 ha; 1 ft/mi = 0.1894 m/km.

11.18 Flood-Flow Formulas

Flood-flow formulas derive from empirical evaluations of drainage-basin characteristics and hydrologic factors falling rationally within the framework of the relation Q = cia. Frequency relations are implied even when they are not expressed in frequency terms. Time–intensity variations, likewise, are included, but (indirectly) as functions of the size of area drained. Equation 11.8 is thereby reduced to the expression $Q = Ca^m$, where m is less than 1. This follows from the relative changes in i and a with t; namely, $i = A/(t + d)^n$ and $a = kt^2$. For d close to zero, therefore, i = constant/$(a^{n/2})$ and, substituting i in Eq. 11.14, $Q = \text{constant}a^{1-n/2} = Ca^m$, where $m = 1 - n/2$. Because n varies from 0.5 to 1.0, m must and does vary in different formulations from 0.8 to 0.5. The value of C embraces the maximum rate of rainfall, the runoff-rainfall ratio of the watershed, and the frequency factor. Table 11.14 shows examples of flood-flow formulas in which certain component variables or their influences on runoff are individualized. The examples shown in Table 11.14 were chosen only as illustrations of forms of flood-flow formulas. They are not necessarily the best forms, nor should they be applied outside the area from which they were derived.

The flood-flow characteristics of U.S. drainage basins have been compared by developing their envelope curve on a Q versus a plot as a function of $\sqrt{a}$ and identifying their Myers rating from the equation

$$Q = 100p\sqrt{a} \quad \text{(U.S. Customary Units)} \tag{11.25a}$$

where Q is the extreme peak flow, in ft^3/s, p is the percentage ratio of Q to a postulated ultimate maximum flood flow of $Q = 10{,}000\sqrt{a}$; and a is the drainage area, which must be 4 mi^2 or more. In the Colorado River Basin, the Myers rating is only 25%; in the northeastern United States, it is seldom more than 50%; in the lower Mississippi basin, it is about 64%.

An equivalent equation using the SI units of Q (L/s), a (km^2) and p (dimensionless) is introduced below:

$$Q = 1{,}760p\sqrt{a} \quad \text{(SI Units)} \tag{11.25b}$$

Problems/Questions

11.1 Coastal air with a temperature of 70°F (21°C) and a dew point of 60°F (16°C) is forced over a mountain range that rises 6,000 ft (1,829 m) above sea level. The air then descends to a plain 4,000 ft (1219 m) below. If the dew point is lowered at a rate of 1.3°F in 1,000 ft (2.47°C in 1, 000 m), find:
(a) The height at which condensation will begin
(b) The temperature at the mountain top
(c) The temperature on the plain beyond the mountain assuming that condensed moisture precipitates before the air starts downward.

11.2 Estimate the daily average evaporation during which the following averages are obtained:

Water temperature = 60°F (16°C)
Maximum vapor pressure, P_w = 0.62 in. Hg (15.8 mm Hg)
Air temperature = 80°F (27°C)
Relative humidity = 50%
Vapor pressure, $P_d = 1.03 \times 0.50 = 0.51$ in. Hg (13 mm Hg)
Wind velocity, w = 15 mi/h (24 km/h)
Barometric pressure, p_a = 28.0 in. Hg (711 mm Hg)

11.3 In the United States, the average annual precipitation is about 30 in. (762 mm), of which about 21 in. (533 mm) is lost to the atmosphere by evaporation and transpiration. The remaining 9 in. (229 mm) becomes runoff into rivers and lakes. Both the precipitation and runoff vary greatly with geography and season. Annual precipitation varies from more than 100 in. (2,540 mm) in parts of the northwest to only 2 or 3 in. (50.8 or 76.2 mm) in parts of the southwest. Determine the annual average runoff from U.S. watersheds in the English units of million gallons per day per square mile (MGD/mi2) and cubic feet per second (ft3/s).

For students who are not living in the United States, please consult with your national government for the information of national average annual precipitation rate and average annual evaporation and transpiration rates.

11.4 In the state of New York, the records of the daily average flows are published annually by the U.S. Geological Survey in its publication *Surface Water Record of New York*. Copies can be obtained by writing to the Water Resources Division, U.S. Geological Survey (USGS), Albany, New York. This year the annual average rainfall is about 45 in. (1,143 mm), of which about 23 in. (584 mm) is lost to the atmosphere by evaporation and transpiration, and only 22 in. (559 mm) becomes surface runoff. Determine the average surface runoff per square kilometer from New York State watersheds in the metric units of cubic meter per day (m^3/d).

For U.S. students who are not living in the state of New York, please consult with your state government or the local USGS office for your state's rainfall and evaporation/transpiration data for an accurate determination for your home state.

11.5 This year the annual average rainfall is about 45 in. (1,143 mm), of which about 23 in. (584 mm) is lost to the atmosphere by evaporation and transpiration, and only 22 in. (559 mm) becomes surface runoff. Determine the average surface runoff per square mile from New York State watersheds in the English units of million gallons per day per square mile (MGD/mi^2) and cubic feet per second (ft^3/s).

11.6 Data from the Hudson River watershed area in the state of New York has been recorded since 1946 at the Green Island Station, near the city of Albany. According to the record, the watershed is 8,090 mi^2 (21,000 km^2), the average annual runoff is 21.9 in. (556 mm), and the average Hudson River flow is 13,060 ft^3/s (370 m^3/s). Obtain the most recent recorded information from the government or the internet and double check the obtained data.

11.7 To be able to withdraw water continuously at or near the average rate of flow from a watershed, water must be stored during periods of high flow. The safe yield of a watershed area is the maximum rate at which the water can be withdrawn continuously over a long period of time. Without storage, the safe yield would be equal to the minimum rate of flow to be expected in the future. It is obvious that over the long period water could not be withdrawn from an area at a rate that exceeded the long-term average flow. To do this, water runoff from the area would have to be stored. As a practical matter, the safe yield of any watershed area will fall somewhere between the minimum expected flow rate and the long-term average. The exact amount depends on the storage capacity of a surface water reservoir. Figure 11.12 shows the relationship of safe yield of surface water supplies versus the reservoir storage capacity for New York State. Determine the safe yield of a reservoir if the following information is known: (a) The reservoir storage capacity is 2 billion gallons or 2,000 MG (7,570 ML), (b) the total contributory watershed area is about 10 mil^2 (26 km^2), and (c) the watershed is in the state of New York.

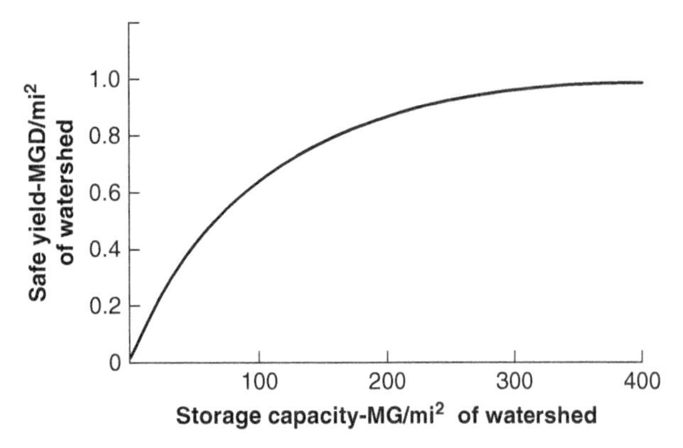

Figure 11.12 Safe Yield versus Reservoir Storage Capacity for Problem 11.5. Conversion factor: 1 MGD/mi^2 = 1.461 MLD/km^2

11.8 Given the record of an automatic rain gauge (Table 11.15), find the progressive arithmetic mean rates, or intensities, of precipitation for various durations.

Table 11.15 Rain Gauge Record for Problem 11.8

Time from Beginning of Storm (min)	Cumulative Rainfall (in.)
5	0.47
10	0.93
15	1.32
20	2.02
25	2.44
30	3.15
35	3.96
40	4.76
45	5.10
50	5.49
60	5.75
80	6.23
100	6.62
120	6.88

Conversion factor: 1 in. = 25.4 mm.

11.9 Determine the mean rainfall intensity (in./h) using the U.S. customary system equation for a 2-year frequency, if the time of concentration, t, is 15 min.

11.10 Determine the mean runoff flow (ft^3/s or m^3/s) assuming the runoff coefficient c for a semiresidential area is 0.35, the mean rainfall intensity i is 4.57 in./h (116 mm/h), and the drainage area a is 24.7 acres (10 ha).

11.11 Determine the mean runoff flow (ft^3/s or m^3/s) assuming the runoff coefficient for a semiresidential area is 0.35, the mean rainfall intensity is 0.55 in./h (14 mm/h), and the drainage area is 131 acres (53 ha).

11.12 Determine the mean runoff flow (ft^3/s) from a drainage area of 120 acres using the English equation for a 2-year frequency i (in. /h) = $206/(t + 30)$. The area is sloping at 5% and is covered with average grass with a c value of 0.35. The furthest point from the inlet is 60 m.

11.13 A residential area with an average estimated runoff coefficient of $c = 0.65$ is shown in Fig. 11.13. The streets have a slope of 5%. The overland and across streets' water flow velocity is 3 m/min. The average flow velocity in the gutters is 2 m/s.

The design rainfall intensity (i in mm/h) as a function of runoff duration (t in min) is given by the following expression:

$$i = \frac{2{,}300}{t + 20}$$

The storm sewer between MH 70 and MH 69 has been previously designed at a flow velocity of 1.85 m/s to serve an upstream drainage area of 72 ha having a concentration time of 70 min.

(a) Determine the total flow reaching each inlet to MH 69.
(b) Determine the design flow for the storm sewer between MH 69 and MH 68

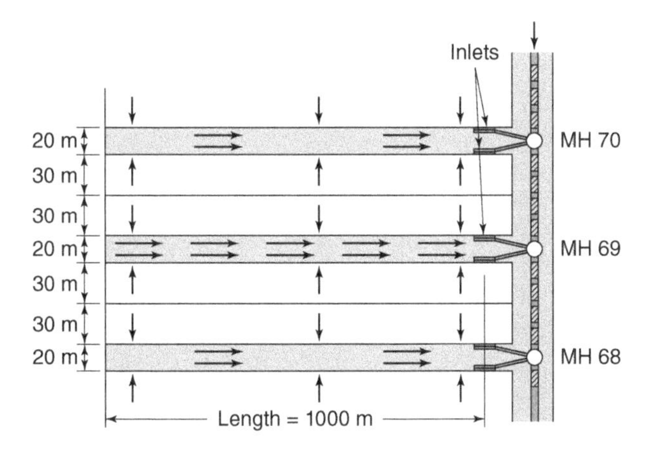

Figure 11.13 Layout of Drainage System for Problem 11.13

11.14 The runoff coefficient indicates the fraction of the storm volume that reaches the conveyance system:

$$V_r = 3{,}630\,(c\,V\,a) \quad \text{(U.S. Customary Units)}$$
$$V_r = 0.00276\,(c\,V\,a) \quad \text{(SI Units)}$$

where
V_r = mean runoff volume (ft^3) (m^3)
V = mean unit rainfall volume (in.) (mm)
c = runoff coefficient
a = drainage area (acre) (ha)
3, 630 = conversion factor to make units consistent (ft^3/acre-in.)
0.00276 = Conversion factor to make units consistent for the SI units.

Determine the mean runoff volume in an unsewered area where the runoff coefficient is 0.35, mean unit rainfall volume is 0.18 in. (4.57 mm), and the drainage area is 131 acres (53 ha).

11.15 Determine the mean runoff volume (V_r) and the mean runoff flow (Q) in a combined sewer area where the runoff coefficient is 0.35, the mean unit rainfall volume is 0.18 in., the drainage area is 565 acres, and the mean rainfall intensity is 0.055 in./h.

11.16 Table 11.11 shows the average pollutant concentration distribution in the United States. Runoff flows may be translated into stormwater loads by multiplying the runoff flows by the appropriate pollutant concentration C_m (mg/L). If storm runoff flows and concentrations are independent, the mean runoff loading rate, W_r, in lb/d, will simply equal the product of the mean concentration, C_m, and the mean runoff flow, Q:

$$W_r = 5.4(C_m)(Q) \quad \text{(U.S. Customary Units)}$$

$$W_r = 0.0864(C_m)(Q) \quad \text{(SI Units)}$$

where
Q = runoff flow (ft^3/s) (L/s)
W_r = mean runoff pollutant loading rate (lb/d) (kg/d)
C_m = mean pollutant concentration (mg/L)
5.4 = a conversion factor for (lb/d)/[(ft^3/s)-(mg/L)].
0.0864 = a conversion factor for SI units.

Determine the mean runoff pollutant loading rate of a separate or unsewered area where the mean biochemical oxygen demand concentration is 20.2 mg/L (from Table 11.11) and the runoff flow is 25 ft^3/s (708 L/s).

11.17 Determine the mean runoff pollutant loading rate of a commercial combined sewer area where the mean nitrogen concentration is 7.7 mg/L (from Table 11.11) and the runoff flow is 11 ft^3/s (311.5 L/s).

Bibliography

Anderson, M. G., and McDonnell, J. J. (Eds.), *Encyclopedia of Hydrological Sciences*, John Wiley & Sons, New York, USA, 2005.

Blaney, H. F., and Criddle, W. D., Quoted by Goodrich, *Trans. Am. Soc. Civil Engrs.*, *122*, 810, 1957.

Chow, V. T., Maidment, D. R., and Mays, L. W., *Applied Hydrology*, McGraw-Hill, New York, 1988.

Davis, S. N., and DeWiest, R.-J. M., *Hydrogeology*, John Wiley and Sons, New York, USA, 1966.

Fair, G. M., Geyer, J. C., and Okun, D. A., Water and wastewater engineering, in *Water Supply and Wastewater Removal*, vol. *vol. 1*, John Wiley & Sons, New York, USA, 1966.

Fair, G. M., Geyer, J. C., and Okun, D. A., *Elements of Water Supply and Wastewater Disposal*, John Wiley & Sons, New York, USA, 1971.

Hazen, A., Storage to be provided in impounding reservoirs, *Trans. Am. Soc. Civil Engrs.*, *77*, 1539, 1914.

Heath, R. C., *Basic Groundwater Hydrology*, USGS Water Supply Paper 2220, U.S. Geological Survey, Denver, CO, 2004.

Kaltenbach, A. B., Storm sewer design by the inlet method, *Public Works*, *94*, 1, 86, 1963.

Kohler, M. A., Lake and pan evaporation, *U.S. Geol. Survey Circ.*, *229*, 127, 1952.

Maidment, D. R., *Handbook of Hydrology*, McGraw-Hill, New York, 1933.

Mays, L. W., *Stormwater Collection Systems Design Handbook*, McGraw-Hill, New York, USA, 2001.

Mays, L. W., *Water Resources Engineering*, John Wiley & Sons, New York, USA, 2005.

McCuen, R. H., *Hydrologic Analysis and Design*, 3rd ed., Prentice Hall, Upper Saddle River, NJ, 2004.

Metcalf & Eddy, Inc., *Wastewater Engineering: Collection and Pumping of Wastewater*, McGraw-Hill, New York, USA, 1981.

Rohwer, C., Evaporation from free water surfaces, *U.S. Dept. Agr. Tech. Bull.*, *271*, 1931.

Schaake, J. C. Jr., Geyer, J. C., and Knapp, J. W., Experimental examination of the rational method, *Proc. Am. Soc. Civil Engrs.*, *93*, HY-6, 353, November, 1967.

Seelye, E. E., Data book for civil engineers, in *Design*, vol. *vol. 1*, John Wiley & Sons, New York, 1960.

U.S. Environmental Protection Agency, *Areawide Assessment Procedures Manual*, Technical Report EPA-600-/9-76-014, Washington, DC, 1976.

Viessman, W., and Lewis, G., *Introduction to Hydrology*, 4th ed., Pearson Higher Education, London, UK, 1996.

Wang, L. K., and Wang, M. H. S., Understanding evaporation, transpiration, evapotranspiration, precipitation and runoff volume for agricultural waste management, *Evolutionary Progress in Science, Technology, Engineering, Arts, and Mathematics*, *4(5)*, Lenox Institute Press, MA, USA, 2022, p. 81. https://doi.org/10.17613/m8tf-zd10.

Wang, L. K., and Yang, C. T., *Modern Water Resources Engineering*, Humana Press, NJ, USA, 866 pp., 2014.

Wang, L. K., Wang, M. H. S., Hung, Y. T., and Shammas, N. K., *Environmental and Natural Resources Engineering*, Springer Nature Switzerland, 512 pp., 2021.

12

Urban Runoff and Combined Sewer Overflow Management

12.1 Hydrologic Impacts of Urbanization

Because stormwater flow quantities are high, reaching many orders of magnitude greater than dry-weather flows, control, whether through flow balancing, multiple uses of facilities, runoff retardation, or combinations thereof, is the focus of cost-effective planning for runoff management.

When precipitation contacts the ground surface, it can take several paths. These include returning to the atmosphere by *evaporation*; *evapotranspiration*, which includes direct evaporation and transpiration from plant surfaces; *infiltration* into the ground surface; *retention* on the ground surface (ponding); and traveling over the ground surface (*runoff*). Altering the surface that precipitation contacts alters the fate and transport of the runoff. Urbanization replaces permeable surfaces with impervious surfaces (e.g., roof tops, roads, sidewalks, and parking lots), which typically are designed to remove rainfall as quickly as possible. As seen in Fig. 12.1, increasing the proportion of paved areas decreases the infiltration and evapotranspiration paths of precipitation, thus increasing the amount of precipitation leaving an area as runoff.

In addition to magnifying the volume of runoff, urban development increases the peak runoff rate and decreases travel time of the runoff. When mechanisms that delay entry of runoff into receiving waters (i.e., vegetation) are replaced with systems designed to remove and convey stormwater from the surface, the stormwater's travel time to the receiving waters is greatly reduced, as is the time required to discharge the stormwater generated by a storm. Figure 12.2 shows an urban area's typical predevelopment and postdevelopment discharge rates over time.

The following changes to hydrology might be expected for a developing watershed:

1) Increased peak discharges (by a factor of 2–5)
2) Increased volume of storm runoff
3) Decreased time for runoff to reach stream
4) Increased frequency and severity of flooding (see Fig. 12.3)
5) Reduced stream flow during periods of prolonged dry weather (loss of base flow)
6) Greater runoff and stream velocity during storm events.

Each of these hydrologic changes can lead to increased pollutant transport and loading to receiving waters. As peak discharge rates increase, erosion and channel scouring become greater problems. Eroded sediments carry nutrients, metals, and other pollutants. In addition, increases in runoff volume result in greater discharges of pollutants. Pollution problems, therefore, multiply with increased urbanization.

Changes in hydrology affect receiving waters through channel widening and subsequent stream bank erosion and deposition, increased stream elevation due to greater discharge rates, and an increased amount of sedimentary material within a stream due to stream bank erosion. The decrease in the ground surface's infiltration capacity and loss of buffering vegetation undermines a significant mechanism for pollutant removal, thereby increasing the load entering the receiving waters. Hydrologic changes can result in more subtle but equally important impacts. Removal or loss of riparian vegetation due to erosion, for example, can increase stream temperature as levels of direct sunlight increase, which can in turn change the biological community structure. With increased sunlight, algae in nutrient-rich receiving waters grow faster and the dominant species changes, which affects the composition of higher organisms. Increased imperviousness and loss of

Water and Wastewater Engineering: Hydraulics, Hydrology and Management, Volume 1, Fourth Edition.
Lawrence K. Wang, Mu-Hao Sung Wang, and Nazih K. Shammas.

Companion website: www.wiley.com/go/Wang/Waterandwastewater4e

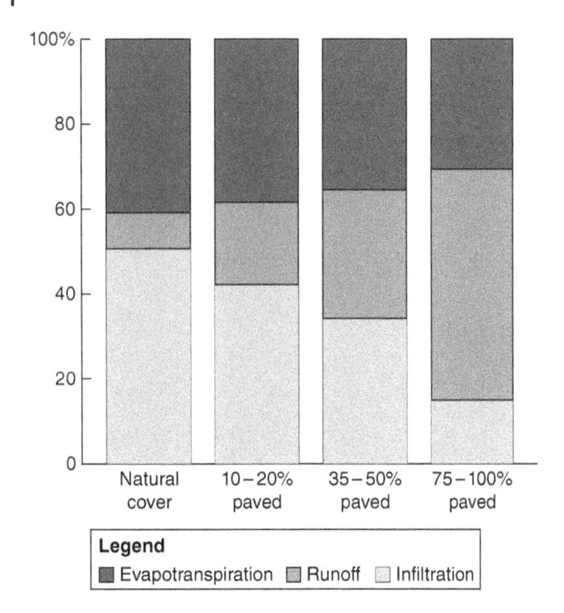

Figure 12.1 Typical Changes in Runoff Flow Resulting from Paved Surfaces (*Source:* U.S. Environmental Protection Agency)

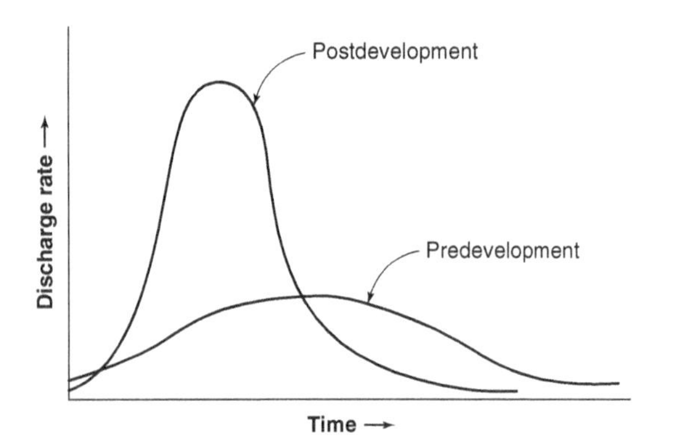

Figure 12.2 Pre- and Postdevelopment Hydraulics (*Source:* U.S. Environmental Protection Agency)

Figure 12.3 2007 Midwest Flooding (*Source:* http://en.wikipedia.org/wiki/Image:Ottawa_OH_2007_Blanchard_River_Flood_Coast_Guard.jpg)

groundwater resupply can lead to more frequent low-flow conditions in perennial streams. The effects of hydrologic changes due to urbanization therefore should be prevented or mitigated to minimize urban runoff pollution.

12.2 Urban Runoff Pollution

Prevention and control of urban runoff pollution requires an understanding of pollutant categories, of the major urban sources of these pollutants, and of the pollutants' effects. When stormwater runoff and municipal wastewaters are carried in the same collector system, the spills of untreated discharges are termed *combined sewer overflow* (CSO). Significantly, 56% of the populations in US cities with 100 000 or more inhabitants are served by such combined or partially combined systems. Also, the separately collected or free-discharging stormwaters alone can produce mass releases of contaminants harmful to receiving waters.

Table 12.1 lists the primary categories of urban runoff pollutants, pollutants associated with each category, typical urban runoff pollutant sources, and potential effects. Table 12.2 summarizes the relative contribution of predominant pollution sources to the degradation of US rivers, lakes, and estuaries. For municipalities, urban storm-generated runoff and construction are the most prevalent sources; outlying agricultural activities also can play a significant role in many urban areas.

The effects of urban runoff pollutants vary for different water resource types. A given municipality's pollutants of concern, therefore, depend on the types of water resources in and downstream of the community and their desired uses.

12.3 The Planning Process

This section outlines the process for developing and initiating urban runoff management plans. It also discusses the establishment and refinement of program goals.

Runoff management and planning elements include the following:

1) Utilization of greenways and detention ponds
2) Utilization of pervious areas for recharge
3) Avoidance of steep slopes for development
4) Maintenance of maximum land area in a natural undisturbed state
5) Prohibitions against developing floodplains
6) Utilization of porous pavements where applicable
7) Utilization of natural drainage features.

Table 12.1 Summary of Urban Runoff Pollutants

Category	Parameters	Possible Sources	Effects
Sediments	Organic and inorganic Total suspended solids (TSS) Turbidity Dissolved solids	Construction sites Urban/agricultural runoff CSOs Landfills, septic fields	Turbidity Habitat alteration Recreational and aesthetic loss Contaminant transport Navigation/hydrology Bank erosion
Nutrients	Nitrate Nitrite Ammonia Organic nitrogen Phosphate Total phosphorus	Urban/agricultural runoff Landfills, septic fields Atmospheric deposition Erosion	Surface waters Algal blooms Ammonia toxicity Groundwater Nitrate toxicity
Pathogens	Total coliforms Fecal coliforms Fecal streptococci Viruses *Escherichia coli* Enterococcus	Urban/agricultural runoff Septic systems Illicit sanitary connections CSOs Boat discharges Domestic/wild animals	Ear/intestinal infections Shellfish bed closure Recreational/aesthetic loss
Organic enrichment	Biochemical oxygen demand (BOD) Chemical oxygen demand (COD) Total organic carbon (TOC) Dissolved oxygen (DO)	Urban/agricultural runoff CSOs Landfills, septic systems	Dissolved oxygen depletion Odors Fish kills
Toxic pollutants	Toxic trace metals Toxic organics	Urban/agricultural runoff Pesticides/herbicides Underground storage tanks Hazardous waste sites Landfills Illegal oil disposal Industrial discharges	Bioaccumulation in food chain organisms and potential toxicity to humans and other organisms
Salts	Sodium chloride	Urban runoff Snowmelt	Vehicular corrosion Contamination of drinking water Harmful to salt-intolerant plants

CSOs = combined sewer overflows.
Source: U.S. Environmental Protection Agency (1977).

Construction controls such as minimizing the area and duration of exposure, protecting the soil with mulch and vegetative cover, increasing infiltration rates, and construction of temporary storage basins or protective dikes to limit storm runoff can significantly reduce receiving water impacts caused by the runoff.

No one single method is a panacea for all combined sewer overflows (CSOs) or storm drain discharge problems. The size and complexity of urban runoff management programs are such that there is a need for an integrated approach to their solution. The type of problems associated with any given community is dependent on a number of variables; as a result, the solution for a community must be developed to fit the needs of that particular urban area. The solution is most often a combination of various *best management practices* (BMPs) and unit process applications.

12.3.1 Description of the Planning Process

When considering stormwater management, the planner is interested in controlling the volume and rate of runoff as well as the pollutional characteristics. The goal is to preserve the initial ecological balance so that expensive downstream treatment facilities can be minimized. Because the size of storm sewer networks and treatment plants is quite sensitive to the flow, quantity, and particularly the peak flow rate, a reduction in total volume or a smoothing out of the peaks will result in lower construction costs.

Table 12.2 Relative Contribution of Nonpoint Source Loading

	Relative Impacts, %		
Source	**Rivers**	**Lakes**	**Estuaries**
Agriculture	55.2	58.2	18.6
Storm sewers/urban runoff[a]	12.5	28.0	38.8
Hydrologic modification	12.9	33.1	4.8
Land disposal	4.4	26.5	27.4
Resource extraction	13.0	4.2	43.2
Construction	6.3	3.3	12.5
Silviculture	8.6	0.9	1.6

Source: U.S. Environmental Protection Agency (1990).
[a] Includes combined sewer overflows.

The planning process for urban runoff management and control programs is based on regulations that require such programs and on technical information about planning approaches. Table 12.3 compares planning approaches required by various regulations. Despite the increasing complexities and uncertainties as one proceeds from left to right in the matrix, the required planning approaches are similar. The process generally consists of the following major components:

1) *Determining existing conditions*: Analyzing existing watershed and water resource data and collecting additional data to fill gaps in existing knowledge.
2) *Quantifying pollution sources and effects*: Utilizing assessment tools and models to determine source flows and contaminant loads, extent of impacts, and level of control needed.
3) *Assessing alternatives*: Determining the optimum mix of prevention and treatment practices to address the problems of concern.
4) *Developing and implementing the recommended plan*: Defining the selected system of prevention and treatment practices for addressing the pollution problems of concern and developing a plan for implementing those practices.

Each regulatory program outlined in Table 12.3 addresses the same components of water quality planning but uses different language to describe the process of each component. For example, as a result of the differing regulatory approaches, municipalities might independently conduct CSO and stormwater planning. Yet since these sources of pollution often exist in the same watersheds and affect the same water resources, this fractured approach is not desirable. To address urban runoff pollution control effectively, communities must consider multiple pollution sources when planning to use a watershed approach.

The planning approach (see Fig. 12.4) is intended to offer municipal officials a systematic approach to developing an urban runoff pollution prevention and control plan. In general, the planning process proceeds as follows:

1) Initiate program.
2) Determine existing conditions.
3) Set site-specific goals.
4) Collect and analyze additional data.
5) Refine site-specific goals.
6) Assess and rank problems.
7) Screen BMPs.
8) Select BMPs.
9) Implement plan.

Although the planning process generally is intended to be followed in sequence, the process can always be altered depending on the specific situation. For example, a municipality might already have begun planning to address certain sources (e.g., stormwater or CSOs). In such cases, starting later in this planning process or integrating other sources into the ongoing planning might be more efficient.

Goal setting and refinement is more appropriately shown as a parallel process rather than a specific step. Only very general goals should be considered at the outset of a program. Existing data should be assessed before setting any site-specific

Table 12.3 Planning Approaches Defined in Regulatory Programs

Project Type	Engineering Facilities	CSO Facilities	Storm Water Management	Nonpoint Source Control	Lake Restoration	Watershed Management
Regulatory basis	National Environmental Policy Act	National CSO Strategy	Storm Water Permit Rule	CWA	CWA	SDWA
Determining existing conditions	Describe existing system	Describe existing conditions	Describe existing conditions	Analyze existing conditions	Describe environmental conditions	Develop watershed description
	Develop planning criteria					
Quantifying pollution sources and water resource impacts	Collect and analyze data	Collect and analyze data	Collect and analyze data	Collect and analyze data	Conduct diagnostic survey	Identify detrimental characteristics
				Identify and rank problems		
Assessing alternatives	Develop alternatives	Develop alternatives	Developing alternatives	Screen BMPs Select BMPs	Conduct feasibility study	Conduct risk assessment
	Assess alternatives	Assess alternatives	Assess alternatives			
Developing and implementing the recommended plan	Develop recommended plan	Develop recommended plan	Develop management plan	Develop recommended plan	Develop recommended plan	Develop detrimental activities control plan
	Develop implementation plan	Develop implementation plan	Develop implementation plan	Develop implementation plan	Develop implementation plan	

CSO = combined sewer overflow; CWA = Clean Water Act; SDWA = Safe Drinking Water Act.
Source: U.S. Environmental Protection Agency (*Urban Runoff Pollution Prevention and Control Planning*, 1993).

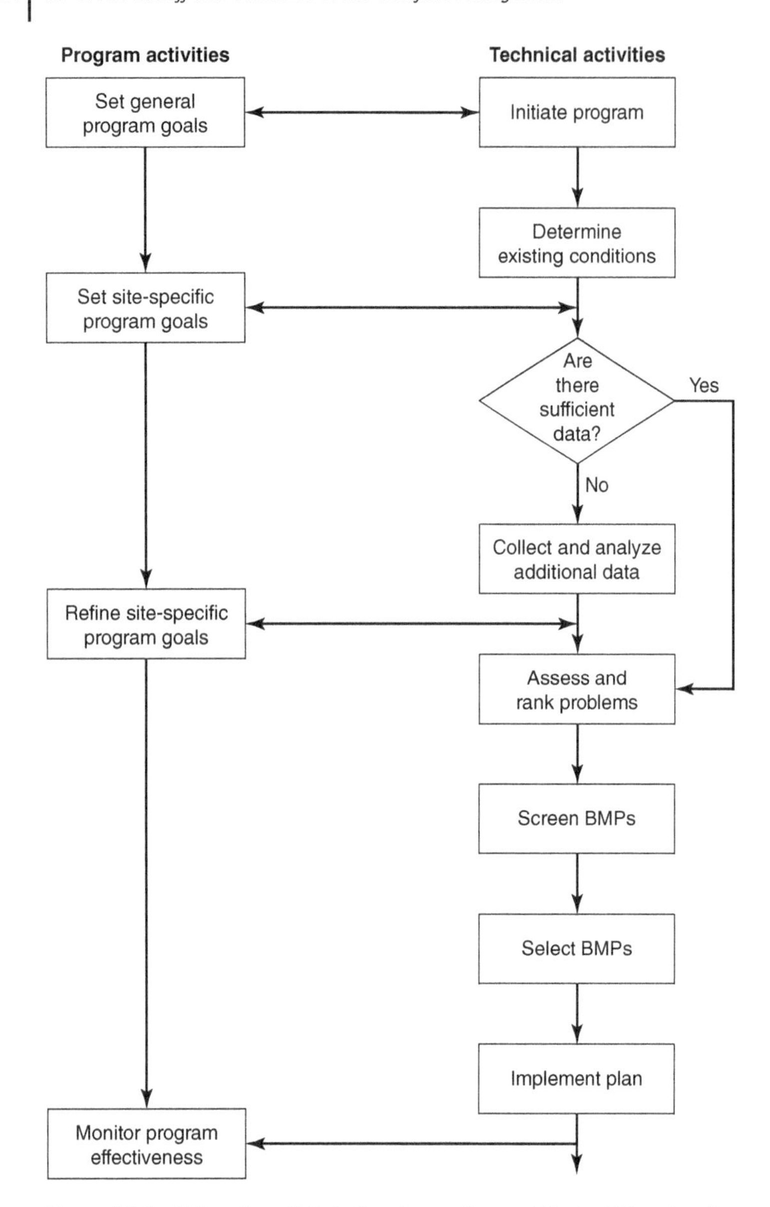

Figure 12.4 Urban Runoff Pollution Prevention and Control Planning Process (*Source:* U.S. Environmental Protection Agency)

goals. As new data are analyzed, new findings and issues are likely to emerge. Program goals therefore must be reevaluated as the planning process progresses. Monitoring the effectiveness of what has been implemented is very important. Because further planning typically will be required, the point of reentry in the planning process needs to be flexible.

12.3.2 Initiate Program

As a first step in the planning process, municipal officials undertaking urban runoff pollution prevention and control planning should develop an overall program structure. Early considerations include organizing a program team; establishing communication, coordination, and control procedures for members of the planning team and other participants; identifying tasks and estimating the number and types of personnel and other resources for each task; and scheduling tasks.

For urban runoff pollution prevention and control programs, the program team should be made up of municipal personnel: public works personnel; conservation officials; engineering personnel; parks personnel; and planning and other officials who regularly deal with or control issues such as utilities, land use and zoning, development review, and environmental issues. The team should be multidisciplinary and able to address the engineering, land use, and environmental issues that will need to be resolved. It is important to involve all entities, including political officials and the public, who have a stake in the program outcome.

12.3.3 Determine Existing Conditions

After initiating the program, the planning team must develop a greater understanding of existing watershed characteristics and water resource conditions in order to

1) Define existing conditions pertinent to the urban runoff management and control program.
2) Identify data gaps.
3) Maximize use of existing available information and data.
4) Organize a diverse set of information in a usable way.

The required research is typically done by gathering existing available watershed information (e.g., environmental, infrastructure, municipal, and pollution source information), as well as receiving-water data (e.g., hydrologic, chemical, and biological data, and water quality standards and criteria). This information can be obtained from various databases; mapping resources; and federal, state, and local agencies. The information can then be used to develop watershed maps; to determine water, sediment, and biological quality; and to establish the current status of streams, rivers, and other natural resources. Once these data are gathered, the program team can organize the information into a coherent description of existing conditions and determine gaps in knowledge. In this way, the existing conditions of the watershed and receiving waters can be defined.

12.3.4 Set Site-Specific Goals

Setting goals is a key aspect of the planning process, and refining goals is an ongoing consideration. Projects that deal with multiple point and nonpoint sources (NPS) require an integrated urban runoff management program, including flood, drainage, and pollution prevention and control. Successful implementation of these programs depends on establishing clear goals and objectives that are quantitative, measurable, and flexible. Setting goals is a process that moves from less to more specificity as additional information on the watershed and water resources is obtained. Fig. 12.4 shows the iterative nature of setting program goals as the planning process proceeds.

The two main types of urban runoff goals are water resource- and technology-based goals. Water resource-based goals are based on receiving-water standards, which consist of designated uses and criteria to protect these uses. For example, water resource-based goals may include reducing the number of oxygen-demanding substances in a lake. In addition, water resource-based goals can place numerical limits on the concentrations of specific pollutants. Further, examples of water resource-based goals include no degradation, no significant degradation, and meeting water quality standards. Applying water resource goals to urban runoff problems, however, might be difficult because water quality standards would need to be assigned to intermittent and variable events.

In contrast, technology-based goals require specific pollution prevention or control measures to address water resource problems. They can be very general, such as "implement the nine minimum technologies for CSO control," or very specific, such as "implementing runoff detention at 50% of the industrial sites in a watershed." A municipality might be able to determine the effectiveness of implementing these goals without conducting future water quality monitoring. With most technology-based goals, implementing the control measures is presumed to be adequate to protect water resources. Monitoring, however, is still essential after implementation to gauge the program's effectiveness and to see if the desired environmental results are being achieved.

The types of goals set by a municipality usually depend on the natural or political forces driving urban runoff control and the public's level of knowledge about the affected water body. Communities might not know or be aware of existing or potential water quality problems. Even under these conditions, however, setting general goals, such as "to meet the requirements of the regulations," is not only possible, but important. Even this general goal directs the program's focus, which then can be made more specific as more information is obtained. In these cases, the municipality typically has to rely on state-mandated goals for the specific water body of concern or general state mandates for the condition of all water bodies. Goal setting will focus the scope of work throughout a program.

12.3.5 Collect and Analyze Additional Data

Even under the best circumstances, municipalities usually will not have all of the required information to describe adequately a program area's existing conditions. The program team, therefore, might have to gather additional information through field investigation and data collection. With this additional information and existing data, the program team can evaluate more fully the existing conditions of the watersheds and water resources of concern. Given the cost and time involved in data gathering, the program team will have to weigh the benefits of additional data collection against using limited funds for plan development and implementation. If the additional data are required, a plan to gather these data must be developed. The plan should include the following:

1) An assessment of available staffing and analytical resources
2) Identification of sampling stations, frequencies, and parameters for sampling and analysis
3) Development of a plan to manage, analyze, and interpret the collected data
4) Analysis of available or needed financial resources.

12.3.6 Refine Site-Specific Goals

Far from static statements, water resource- or technology-based goals should be reassessed as appropriate in the planning process. Once early goals have been stated for a watershed or receiving water, all future actions affecting these resources can be considered against this backdrop and the goals can be reassessed. As more information is gathered, the goals can be maintained, made more specific, or changed completely. By the time the program is defined and ready to be implemented; however, fairly specific goals should exist so that program evaluators can determine whether or not goals have been met.

12.3.7 Assess and Rank Problems

Once sufficient data have been collected and analyzed, the data can then be used to assess and rank the problems. Based on data gathered in earlier steps, the team will need to develop a list of criteria to assess problems. These criteria are used in conjunction with water quality assessment methods and models to determine current impacts and future desired conditions.

Having determined the problems of concern, the project team can rank these problems to set priorities for the selection and implementation of pollution prevention and runoff control measures. The emphasis on ranking of resources and problems is central to the planning strategy. This concept assumes that focusing resources on targeted areas or sources enhances water resource improvement. Further, it assumes that demonstrating water resource benefits increases public support of urban runoff management and control programs as citizens become more closely attuned to overall water quality goals. Also municipalities should investigate water resources within their region to develop priorities so that limited resources can be targeted to areas with the greatest potential for improvement. Various levels of detail can be used in this assessment, ranging from simple unit load methods to complex computer models.

12.3.8 Screen Best Management Practices

Once the water resource problems have been prioritized, specific water resource problems and their sources can be addressed. The program team should compile a list of various runoff management and pollution prevention and treatment practices and review them for their effectiveness in solving the prioritized problems. The next two sections include brief descriptions of various *nonstructural* and *structural* BMPs. Also described is the initial BMP screening step, when potential practices are reviewed for their applicability to the watershed and water resource problems of concern. While the team initially faces a large number of potential practices, obviously inappropriate practices are eliminated in this step based on criteria such as the primary pollutants removed, drainage area served, soil conditions, land requirements, and institutional structure. Following this initial screening, the program team will have a list of potential practices to be evaluated further.

12.3.9 Select Best Management Practices

During this step, the program team investigates the list of potential pollution prevention and treatment practices developed from the previous step to determine which to include in the plan. More specific criteria should be used for analyzing these potential practices than during the initial screening. To make the final selection, the program team must use the analytical tools developed during the ranking and assessment of problems, as well as decision factors such as cost, program goals, environmental effects, and public acceptance. As with the initial screening step, these evaluation criteria depend on established priorities. Generally, the selection process yields a recommended system of various pollution prevention and treatment practices, which together address the pollution sources of concern. Availability of required resources to implement the practices is a major consideration. If needs and resources do not match, the municipality might have to adjust its expectations to what realistically can be accomplished. Both structural and nonstructural practices might be required.

12.3.10 Implement Plan

After choosing the best management and treatment practices, the program team moves from planning to implementation, which often occurs through a phased approach. Inexpensive and well-developed practices can be implemented early in the program as pilot or demonstration studies; and these results might influence further implementation. Given the added requirements of implementation, operation, and maintenance, the original program team might expand to include members with more construction experience. Also, funding sources are needed for initial capital expenses and continuing operation and maintenance costs. Nonstructural practices must be implemented, and the team must arrange for the detailed design and construction of structural practices.

During this step, program responsibilities must be clearly delineated. All involved entities must be familiar with and accept their role in implementing and enforcing the plan. Continuing activities should also be clearly defined and monitoring schedules should be set to determine the program's effectiveness in meeting its goals. Maintenance programs should be developed so that structural practices continue to operate as intended. Finally, the municipality should be aware of available federal and state technical assistance that could help throughout implementation of the plan.

12.4 Best Management Practices

12.4.1 Best Management Practice Overview

Urban runoff problems are more difficult to control than steady-state, dry-weather point source discharges because of the intermittent nature of rainfall and runoff, the number of diffuse discharge points, the large variety of pollutant source types, and the variable nature of the source loadings. Because the expense of constructing facilities to collect and treat urban runoff is often prohibitive, the emphasis of stormwater management should be on developing a least cost approach, which includes nonstructural controls and low-cost structural controls.

Nonstructural controls include regulatory controls that prevent pollution problems by controlling land development and land use. They also include source controls that reduce pollutant buildup or lessen its availability for washoff during rainfall.

Low-cost structural controls include the use of facilities that encourage uptake of pollutants by vegetation, settling, or filtering. Because of the variability of pollutant removal, these controls can be used in series or in parallel combinations. The concept of implementing a "treatment train" might, for example, include initial pretreatment, primary pollutant removal, and final effluent polishing practices to be constructed in series.

Table 12.4 lists several categories of methods that are commonly used as urban runoff and CSO BMPs. The next sections describe methods of BMP screening, then give a brief overview of some of the more important characteristics of these BMPs, including the types of pollutants controlled, the pollution removal mechanisms, limitations on their use, maintenance requirements, and general design considerations.

12.4.2 Best Management Practice Screening

The goal of BMP screening is to reduce the comprehensive list of BMPs to a more manageable list for final selection. Because this step is an initial screening, methods used are generally qualitative and require professional judgment. Although extensive knowledge about specific design criteria is not necessary at this stage in the screening process, understanding each individual BMP's effectiveness and applicability to the program area's problems is crucial.

The BMPs are divided into two general categories: nonstructural and structural. Nonstructural BMPs – which include regulatory practices, such as those that limit impervious area or protect natural resources, and source controls, such as street sweeping or solid waste management – are typically implemented throughout an entire community, watershed, or special area. While structural BMPs, such as detention ponds or infiltration practices, may be designed to address specific pollutants from known sources, they can also be implemented throughout an area. In addition, structural BMPs can be required in new developments or redevelopments.

12.4.2.1 Nonstructural Practices

Since the number of potential nonstructural BMPs to be implemented is very large, initial screening is useful before the final selection process. The regulatory and source control BMP descriptions contained later in this chapter focus on the most commonly implemented practices; other, less commonly used practices, however, could also be considered. In addition, each practice (e.g., solid waste management) can be divided into numerous subpractices (e.g., management of leaf litter,

Table 12.4 Urban Storm Runoff Management BMPs

Urban Runoff Controls	CSO Controls
Regulatory Controls	**Source Controls**
Land use regulations	Water conservation programs
Comprehensive runoff control regulations	Pretreatment programs
Land acquisition	**Collection System Controls**
Source Controls	Sewer separation
Cross-connection identification and removal	Infiltration control
Proper construction activities	Inflow control
Street sweeping	Regulator and system maintenance
Catch basin cleaning	Insystem modifications
Industrial/commercial runoff control	Sewer flushing
Solid waste management	**Storage**
Animal waste removal	Inline storage
Toxic and hazardous pollution prevention	Offline storage
Reduced fertilizer, pesticide, and herbicide use	Flow balance method
Reduced roadway sanding and salting	**Physical Treatment**
Detention Facilities	Bar racks and screens
Extended detention dry ponds	Swirl concentrators/vortex solids separators
Wet ponds	Dissolved air flotation
Constructed wetlands	Fine screens and microstrainers
Infiltration Facilities	Filtration
Infiltration basins	**Chemical Precipitation**
Infiltration trenches/dry wells	**Biological Treatment**
Porous pavement	**Disinfection**
Vegetative Practices	Chlorine treatment
Grassed swales	UV radiation
Filter strips	
Filtration Practices	
Filtration basins	
Sand filters	
Other	
Water quality inlets	

Source: U.S. Environmental Protection Agency, *Stormwater Best Management Practice Design Guide* – Volume 1 – *General Considerations* (2004).

rubbish, garbage, and lawn clippings). In an urban runoff management plan for the Santa Clara Valley, for example, the consultants identified more than 100 separate potential nonstructural BMPs used throughout the country. Municipalities, therefore, have to screen regulatory and source control BMPs based on their particular watershed.

One screening method involves applying screening criteria to each nonstructural practice to determine its applicability to the conditions in the watershed. The screening criteria, which are specific to the watershed and depend on the program goals, include the following:

1) *Pollutant removal*: Because different regulations and source control practices are designed to address different pollutants, the program team should ensure that the screened list of controls includes practices designed to address the pollutants of primary concern. In addition, some practices might not provide sufficient pollutant removal.

2) *Existing government structure*: Some practices implemented throughout the country require a specific government structure. For example, while a strong county government might be important for implementing a specific regulatory control, the role of county governments can vary from one section of the country to another. Practices requiring specific government structures that do not exist in the area of concern therefore could be eliminated from the list.
3) *Legal authority*: For regulatory controls to be effective, the legal authority to implement and enforce the regulations must exist. If municipal boards and officials lack this authority, they could be required to obtain it through local action.
4) *Public or municipal acceptance*: Implementing certain practices could be difficult because of resistance from the public or an involved municipal agency. These practices can be eliminated from the list.
5) *Technical feasibility*: The municipal BMPs that require large expenditures and extensive efforts might not be suitable for small municipalities that lack the required resources.

Another method of screening involves the use of a comparative summary matrix. Such a matrix can be used to screen nonstructural control practices. In the matrix, various regulatory and source control practices are listed and their abilities to meet various criteria are compared. The criteria to be listed include ability to remove specific pollutants, such as nutrients and sediments, maintenance requirements, longevity, community acceptance, secondary environmental impacts, costs, and site requirements. For each practice and criterion, an assessment of effectiveness is indicated; for instance, on a scale of 1 to 5, with 1 indicating the highest effectiveness and 5 the lowest. This type of matrix can provide a basis for an initial assessment of practices and their applicability to the program.

12.4.2.2 Structural Practices

Because structural practices generally are more site-specific and have more restrictions on their use than nonstructural practices, the initial screening step for these practices can be more precise than that for nonstructural practices. Table 12.5 outlines some of the more important criteria for the screening of structural BMPs, including their typical pollutant removal efficiencies, land requirements, the drainage area that each BMP can effectively treat, the desired soil conditions, and the desired groundwater elevation. By using these criteria and the information obtained during data collection and analysis and problem identification and ranking, the program team can narrow the list of BMPs to be further assessed in the BMP selection step.

The initial screening criteria for structural control practices include the following:

1) *Pollutant removal:* The municipality should ensure that the BMPs selected address the primary pollutants of concern to the level of removal desired.
2) *Land requirements:* Large land requirements for some of the aboveground structural BMPs can often restrict their use in highly developed urban areas. Land requirements vary depending on the BMP.
3) *Drainage area:* The structural BMPs listed in Table 12.5 are used primarily to treat runoff from watersheds up to 50 or 60 acres (20 or 25 ha), and the optimum drainage area to be served varies for each practice and according to the land use (connected impervious area, for example). Drainage areas above this size might have to be treated by locating BMPs in subwatersheds.
4) *Soil characteristics:* Structural BMPs have differing requirements for soil conditions. Infiltration facilities generally require permeable soils, while detention BMPs generally require impermeable soils. The municipality must become familiar with soil conditions in the watershed.
5) *Groundwater elevation:* The groundwater elevation in the watershed can be a limiting factor in siting and implementing structural BMPs. Generally, high groundwater elevation can restrict the use of infiltration facilities and filtration practices, but it is necessary for constructed wetlands and may be desirable for detention facilities.
6) *Public acceptance:* Since a municipality could have difficulty implementing a structural BMP without public approval, public acceptance of the BMPs should be considered in the screening step.

Of the screening criteria listed, the pollutant removal, land requirements, and drainage area served are usually absolute restrictions. Soil condition and groundwater elevation, on the other hand, impose restrictions that could be overcome by such means as importing soil or constructing facilities with clay liners to restrict groundwater inflow. Such modifications, however, can add significantly to the BMP costs.

Table 12.5 Structural BMP Initial Screening Criteria

Structural BMPs	Typical Pollutant Removals[a]					Relative Land Requirements	Drainage Area[b]	Desired Soil Conditions	Groundwater Elevation
	Suspended Solids	Nitrogen	Phosphorus	Pathogens	Metals				
Detention Facilities									
Extended detention Dry ponds	Medium	Low-medium	Low-medium	Low	Low-medium	Large	Medium-large	Permeable	Below facility
Wet ponds	Medium-high	Medium	Medium	Low	Medium-high	Large	Medium-large	Impermeable	Near surface
Constructed wetlands	Medium-high	Low	Low-medium	Low	Medium-high	Large	Large	Impermeable	Near surface
Infiltration Facilities									
Infiltration basins	Medium-high	Medium-high	Medium-high	High	Medium-high	Large	Small-medium	Permeable	Below facility
Infiltration trenches/dry wells	Medium-high	Medium-high	Low-medium	High	Medium-high	Small	Small	Permeable	Below facility
Porous pavement	High	High	Medium	High	High	N/A	Small-medium	Permeable	Below facility
Vegetative Practices									
Grassed swales	Medium	Low-medium	Low-medium	Low	Low-medium	Small	Small	Permeable	Below facility
Filter strips	Medium-high	Medium-high	Medium-high	Low	Medium	Varies	Small	Depends on type	Depends on type
Filtration Practices									
Filtration basins	Medium-high	Low	Medium-high	Low	Medium-high	Large	Medium-large	Permeable	Below facility
Sand filters	High	Low-medium	Low	Low	Medium-high	Varies	Low medium	Depends on type	Depends on type
Other									
Water quality inlets	Low-medium	Low	Low	Low	Low	N/A	Small	N/A	N/A

[a] Low = < 30 %; medium = 30 % – 65 %; high = 65 % – 100 %.
[b] Small = < 10 acres; medium = 10 – 40 acres; large = > 40 acres.
Conversion factor: 1 acre = 0.4046 ha; NA: not available.
Source: U.S. Environmental Protection Agency *Stormwater Best Management Practice Design Guide*—Volume 1—*General Considerations* (2004).

12.5 Urban Runoff Control Practices

This section addresses regulatory controls, source controls, and several types of commonly used structural controls.

12.5.1 Regulatory Controls

Urbanization increases the amount of impervious land area, which in turn increases stormwater runoff with its associated pollutants. Municipalities can prevent or reduce many of these pollution problems by implementing regulatory controls to limit the amount of impervious area and to protect valuable resources. These regulatory controls can prevent or limit the quantity of runoff, as well as its pollution load. Regulatory controls typically implemented by municipalities include the following:

1) Land use regulations, such as
   a) zoning ordinances
   b) subdivision regulations
   c) site plan review procedures
   d) natural resource protection
2) Comprehensive runoff control regulations
3) Land acquisition.

Local government regulations can require storm runoff controls, reduce the level of impervious area, require the preservation of natural features, reduce erosion, or require other important practices. The major aspects of stormwater prevention and control – including runoff quantity control, solids control, and other pollution control – are illustrated later on in the chapter in Case Study 12.1 on the regulatory practices implemented by the city of Austin, Texas.

12.5.1.1 Runoff Quantity Control

Regulations addressing runoff quantity control can be used to reduce the effects of land development on watershed hydrology. Hydrologic control in turn results in pollution control, and can be accomplished through requirements such as follows:

- *Open space:* By maintaining specified levels of open space on a development site, the total area of impervious surface is reduced and infiltration of precipitation is increased. This leads to decreases in total pollutant discharge and potential downstream erosion by reducing total and peak runoff flows.
- *Postdevelopment flow control:* Many development regulations require that peak runoff conditions from a site be calculated before and after construction. These requirements specify that conditions after construction must reflect conditions before construction. This control is typically accomplished through the use of detention facilities, which can reduce peak runoff discharge rates, thereby decreasing downstream erosion problems. These regulations specify the desired outcome; the approach for ensuring that outcome, however, is determined by the developer.
- *Runoff recharge:* Regulations may specify that stormwater runoff be recharged on site. Such regulations can reduce the runoff leaving a site, thereby reducing development-induced hydrologic changes and pollutant transport. By directly promoting infiltration, peak and total runoff rates can be decreased and pollutant discharges and downstream erosion can be reduced. Such runoff recharge might also help maintain surficial aquifer levels.

12.5.1.2 Solids Control

Regulations addressing solids control could include requirements for control practices during and after construction, since such activity has been shown to be a major contributor of solids. Construction activities can greatly increase the level of suspended solids in stormwater runoff by removing vegetation and exposing the topsoil to erosion during wet weather. Yet while communities have requirements for implementing erosion control practices on construction sites, fewer communities require erosion control after construction is complete. Because many other land uses can contribute solids loadings, regulatory requirements can cover various types of industrial and commercial activities.

12.5.1.3 Other Pollution Control

Land development increases the concentrations of nutrients, pathogens, oxygen-demanding substances, toxic contaminants, and salt in stormwater runoff. Development regulations, therefore, can be used to address some of these specific pollutants. These regulations can take the form of special requirements for limiting nutrient export in special protection districts or setting performance standards for known problem pollutants.

12.5.1.4 Land Use Regulations

Land use regulations can include zoning ordinances, subdivision and site plan regulations and review requirements, and environmental resource regulations such as wetlands protection. These practices are used as tools to promote development patterns that are compatible with control of urban runoff discharges.

Zoning. Most communities have residential, commercial, industrial, and other zoning districts that specify the types of development allowed and dictate requirements, including the following:

1) Specifying the density and type of development allowed in a given area, thereby maintaining pervious areas
2) Controlling acreage requirements for certain land uses and associated setback, buffer, and lot coverage requirements
3) Directly and indirectly affecting the types of materials that can be stored or used on sites
4) Not allowing potentially damaging uses (e.g., underground chemical storage or pesticide application) in sensitive watersheds.

Subdivision Regulations. Subdivision review deals with land that is divided into separately owned parcels for residential development. Municipalities have the authority to review the plans for such subdivisions and to restrict development options via requirements for drainage, grading, and erosion control, as well as provisions for buffer areas, open spaces, and maintenance. Through this review, municipalities can ensure that proper practices are designed into the development.

Site Plan Review. Site plan review ensures compliance with zoning, environmental, health, and safety requirements. Municipalities can require developers to consider how construction activities will affect drainage on site and to design plans for reducing urban runoff pollution problems. Developers usually are required to submit information to a municipality on the natural drainage characteristics of the site, plans for erosion control, retention and protection of wetlands and water resources, and disposal of construction-related wastes.

Natural Resource Protection. Municipalities can also protect water resources by protecting lands, such as floodplains, wetlands, stream buffers, steep slopes, and wellhead areas. By use of resource overlay zones that restrict high pollution activities in these areas, development can be controlled and the potential for urban runoff pollution can be reduced.

12.5.1.5 Comprehensive Runoff Control Regulations

In addition to strengthening and broadening existing local regulatory control practices, states and municipalities can implement runoff pollution control through comprehensive regulations. Although still relatively rare, comprehensive plans to address urban runoff pollution exist in various states and communities. They are designed to fully address urban runoff pollution problems by identifying specific land use categories and water resources that deserve special attention, and outlining methods for implementing source control and structural BMPs. While the form that these comprehensive regulations take is very specific to the needs of a state or community, reviewing the regulatory approaches that have been tried by others is useful in developing options.

12.5.1.6 Land Acquisition

To protect valuable resources from the effects of development, municipalities can purchase land within the watershed to control land development. Municipalities can acquire land to convert to parks or to maintain as open space; this approach, however, can be very expensive.

12.5.2 Source Control Practices

Source controls include the nonstructural practices designed to reduce the availability of pollutants. Many of these practices tie directly into the U.S. Environmental Protection Agency's (EPA's) pollution prevention strategy, which focuses on preventing pollution sources from entering the system rather than on treatment. Some of the more common practices used by municipalities throughout the country include the following:

1) Cross-connection identification and removal
2) Proper construction activities
3) Street sweeping and catch basin cleaning
4) Industrial/commercial runoff control

5) Solid waste management
6) Animal waste removal
7) Toxic and hazardous waste management
8) Reduced fertilizer, pesticide, and herbicide use
9) Reduced roadway sanding and salting.

12.5.3 Detention Facilities

In its simplest form, detention means capturing stormwater and controlling the release rate to decrease downstream peak flow rates. On-site detention uses simple ponding techniques on open areas where stormwater can be accumulated without damage or interference with essential activities. The design essentials include a contained area that allows the stormwater to pond and a release structure to control the rate at which the runoff is allowed into the drainage system. The release structure is usually a simple construction, such as a small-diameter pipe draining a basin or an orifice plate placed at a sewer inlet. The capacity of the pipe or orifice limits the flow rate to a level acceptable to the downstream system. Where the depth of ponding has to be limited, the release structure will have an automatic overflow to prevent excessive ponding.

One of the most common structural methods for controlling urban runoff and reducing pollution loading is through the construction of ponds (Fig. 12.5) or wetlands to collect runoff, detain it, and release it to receiving waters in a controlled manner. Pollution reduction during the period of temporary runoff storage results primarily from settling of solids. Detention facilities, therefore, are most effective at reducing the concentrations of solids and the pollutants that typically adhere to solids, and less effective at removing dissolved pollutants.

The three types of detention facilities commonly used are extended detention dry ponds, wet ponds, and constructed wetlands; each is discussed next.

12.5.3.1 Extended Detention Dry Ponds

Most municipalities are familiar with the concept of constructing *dry ponds* to control peak runoff. When used as water-quality BMPs, dry ponds are designed with orifices or other structures that restrict the velocity and volume of the discharges (see Fig. 12.6). Dry ponds thereby detain the runoff before discharging it to surface waters.

Pollutant Removal. During the storage period, heavier particles settle out of the runoff, removing suspended solids and pollutants, such as metals, that attach to the particles or precipitate out. Some dry ponds also include vegetated areas that

Figure 12.5 Detention Basin (*Source:* http://en.wikipedia.org/wiki/Image:Trounce_Pond.jpg)

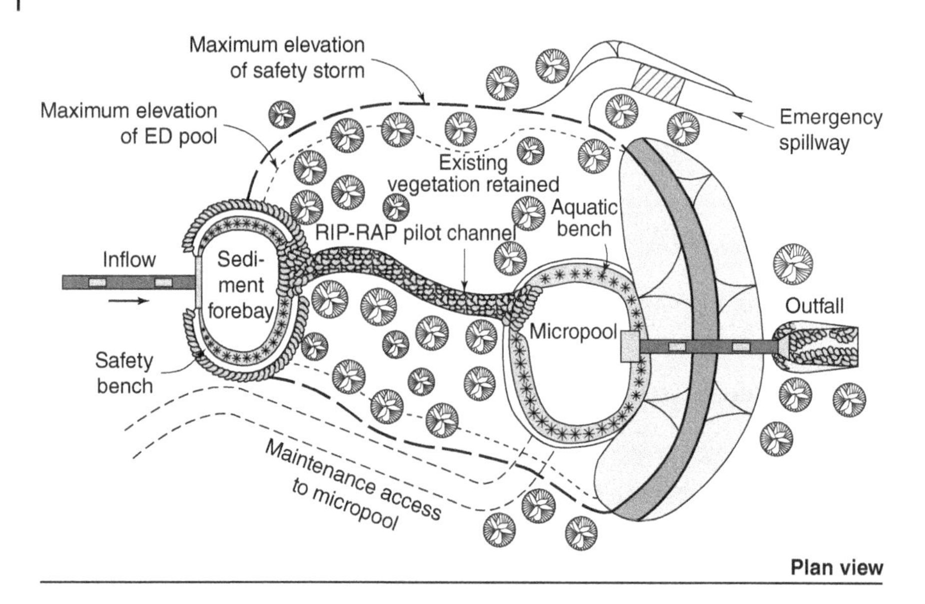

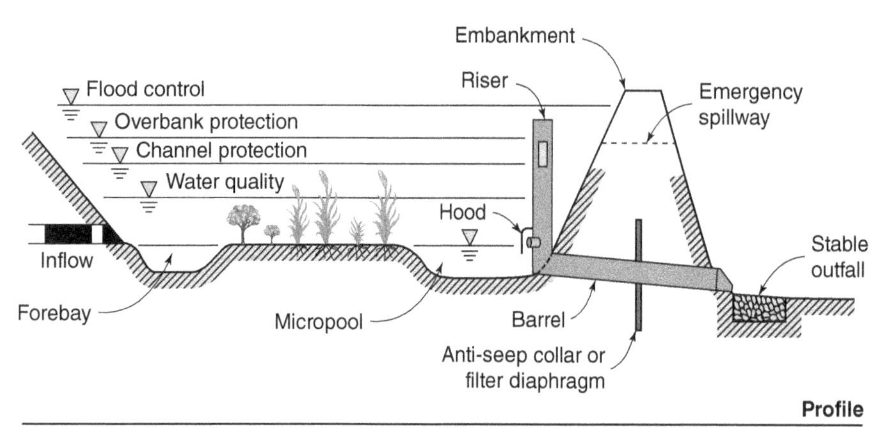

Figure 12.6 Extended Detention Dry Pond (*Source:* New York State Department of Environmental Conservation, USA)

can provide pollutant removal through filtering and vegetative uptake. Overall, the pollutant removal effectiveness of dry ponds has been shown to be less than that for wet ponds and constructed wetlands (see Table 12.5).

Design Considerations. Retrofitting existing dry ponds with new outlet structures can sometimes enhance a municipal flood control structure to increase its pollution control effectiveness. Care must be taken, however, to ensure that the overflow capacity of the pond is maintained, so that it continues to fulfill its original flood control function. Temporary storage can also be provided for runoff from smaller storms by building a small berm around an existing outlet structure. Important design criteria include the desired detention time and the volume of runoff to be detained. These factors dictate the pond's size and affect the pollutant removal efficiency of the structures. Most dry pond sizing criteria specify a certain detention time for a given design storm. For example, the Maryland Water Resources Authority specifies that water quality dry ponds must be large enough to accommodate the runoff volume generated by the 1-year, 24-hour storm to be released over a minimum of 24 hours. In contrast, the Washington State Department of Ecology specifies that dry ponds must be large enough to accommodate the runoff volume generated by the 2-year, 24-hour storm and release it over a period of 40 hours.

Dry ponds should also include some form of low-flow channel designed to reduce erosion; vegetation on the bottom of the pond to promote filtering, sedimentation, and uptake of pollutants; and an outlet structure designed to remove pollutants and withstand clogging. In addition, dry pond designs typically include upstream structures to remove coarse sediments and reduce sedimentation and clogging of the outlet. Also, outlets might be connected to grassed swales (biofilters) to provide additional pollutant removal.

Maintenance Requirements. Maintenance of water quality dry ponds is important. Regular mowing, inspection, erosion control, and debris and litter removal are necessary to prevent significant sediment buildup and vegetative overgrowth. Also, periodic nuisance and pest control could be required. The pond slopes should allow for mowing, and access roads should be provided.

Limitations on Use. Like other stormwater structures used in large watersheds, a physical constraint on the construction of water quality dry ponds is their large land requirements. For this reason, locating dry ponds in new developments is usually more practical than constructing them in already developed areas. Other physical constraints include the topography and the depth to bedrock.

12.5.3.2 Wet Ponds

The design of *wet ponds* is similar to that of dry ponds and constructed wetlands. In wet ponds, stormwater runoff is directed into a constructed pond or enhanced natural pond, in which a permanent pool of water is maintained until being replaced with runoff as shown in Fig. 12.7. Once the capacity of a wet pond is exceeded, collected runoff is discharged through an outlet structure or an emergency spillway.

Pollutant Removal. The primary pollutant removal mechanism in wet ponds is settling. The ponds are designed to collect stormwater runoff during rainfall and to detain it until additional stormwater enters the pond and displaces it. While the

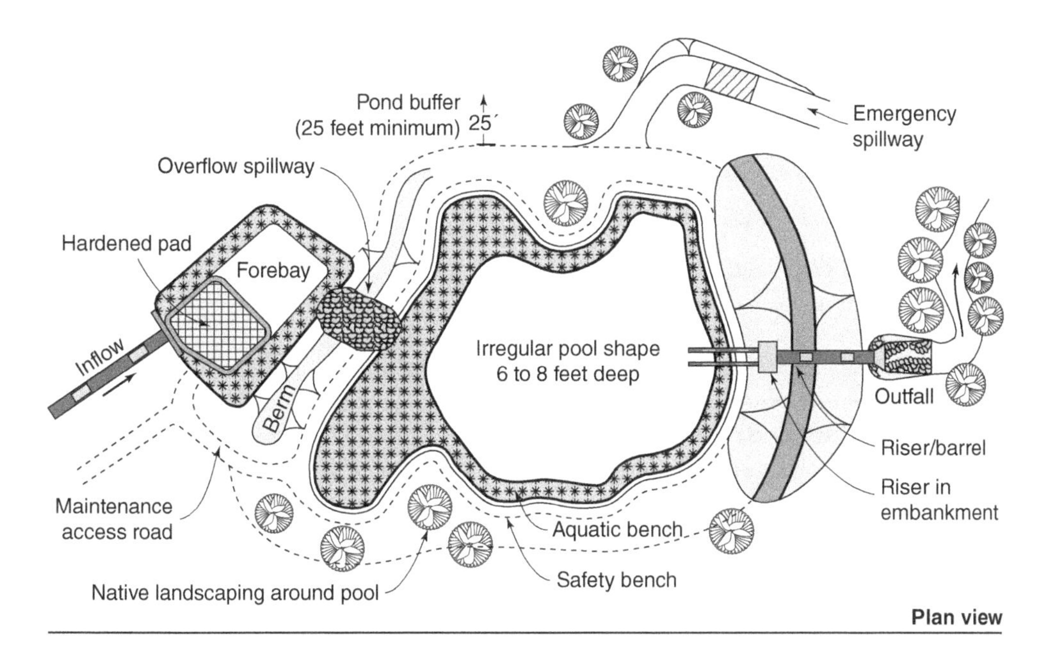

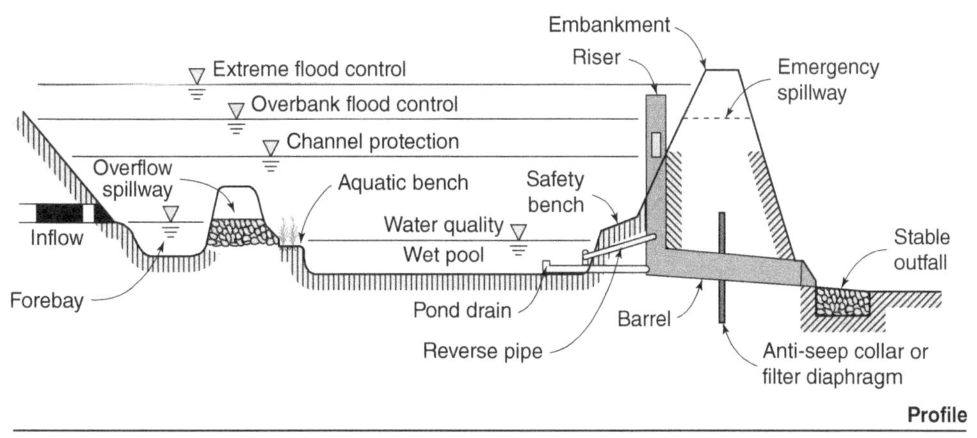

Figure 12.7 Wet Pond. Conversion factor: 1′ = 1 ft = 0.3048 m (*Source:* New York State Department of Environmental Conservation, USA)

runoff is detained, settling of particulates and associated pollutants takes place in the pond. Wet ponds can also remove pollutants from runoff through vegetative uptake. Wet ponds should be vegetated with native emergent aquatic plant species, which can remove dissolved pollutants such as nutrients from the runoff before it is discharged to the receiving water.

Design Considerations. Wet ponds typically are designed with a number of different water levels. One level of the pond has a permanent pool of water. The next level periodically is inundated with water during storms; this area should be vegetated and relatively flat to promote settling and filtering of sediments and vegetative uptake of nutrients. The highest level will be inundated only during extremely heavy rainfall; this area should also be vegetated to prevent soil erosion. At least 30% of the surface area of a wet pond should be a vegetated zone. Typically, this vegetation is concentrated at the outlet as a final "polishing" biofilter. The sizing of wet ponds is similar to that of dry ponds in that a number of different "sizing rules" provide varying levels of pollution control. Generally, these rules specify the volume of runoff to be detained in the wet pond during a storm. For example, the Maryland Water Resources Authority specifies that the permanent pool of a wet pond should be large enough to contain 0.5 in. (13 mm) of runoff distributed over the impervious portion of the contributing watershed. In Florida, storage volume for 1 in. (25.4 mm) of runoff above the normal pool elevation is recommended. This volume must be released at a slow rate; no more than half should be discharged within 60 hours after the event, and all the volume must be released after 120 hours. A hydraulic retention time of 14 days for the permanent pool volume is recommended.

The design of water quality wet ponds must also take into consideration the possibility of large storms. Emergency spillways should be included in the design to prevent flooding difficulties. In addition, the pond's inlet and outlet structures should be separated and constructed at either end of the pond to maximize full mixing when large flows occur and avoid short-circuiting. By separating the inlet from the outlet, the detention time of the pond can also be increased.

Maintenance Requirements. Like many other BMPs, wet ponds require routine maintenance to be effective. Wet ponds are designed to allow for settling of suspended solids; therefore, periodic removal of the accumulated sediment must be performed (perhaps every 10–20 years). In addition, the pond slopes should be regularly mowed to make the sediment removal process easier and to enhance the aesthetic qualities of the area. Inlet and outlet structures should be inspected periodically for damage and accumulated litter, and the pond bottom should be inspected for potential erosion. Erosion of the pond bottom from high-velocity flows can result in increased sediment transport and overall reduction in the pollutant removal capabilities of the pond.

Limitations on Use. Water-quality wet ponds have large land requirements and usually are more suited to new development projects where they can be designed into the site. In addition, wet ponds are not suitable for use in areas with porous soils or low groundwater levels because a pool of water in the bottom is key to their design. Wet ponds should be built into the groundwater with their control elevation set above the level of seasonal high-water tables. Synthetic impermeable materials or clay can be used to prevent seepage. Wet ponds also have physical limitations related to the site topography; since locating wet ponds in areas with extreme slopes is often difficult, relatively flat locations are preferable.

12.5.3.3 Constructed Wetlands

Constructed wetlands are effective in removing many urban stormwater pollutants. Two prevalent types of systems are shallow-constructed wetlands (Fig. 12.8) and wet detention systems (Fig. 12.9). The wet detention system is a wet pond with extensive shoreline shallow wetland areas. Wetland systems combine the pollutant removal capabilities of structural stormwater controls with the flood attenuation provided by natural wetlands. Proper design of constructed wetlands – including their configuration, proper use of pretreatment techniques to remove sediments and petroleum products, and choice of vegetation – is crucial to the functioning of the system.

Pollutant Removal. Constructed wetland systems perform a series of pollutant removal mechanisms including sedimentation, filtration, adsorption, microbial decomposition, and vegetative uptake to remove sediment, nutrients, oil and grease, bacteria, and metals. Wetland systems reduce runoff velocity, thereby promoting settling of suspended solids. Plant uptake accounts for removal of dissolved constituents. In addition, plant material can serve as an effective filter medium, and denitrification in the wetland can remove nitrogen. Specific wetland vegetation species remove specific pollutants from stormwater runoff. Some of the most commonly used wetland vegetation includes cattails, bulrushes, and canary grass.

Design Considerations. General guidelines recognized as important in the design of wetland systems include maximizing the detention time of runoff in the wetland system, maximizing the distance between the inlet and outlet, and providing some form of pretreatment for sediment removal. Travel time can be increased in a wetland by reducing the gradient over which the flow travels or by making the flow travel over a greater distance before being discharged. In either case, some designers recommend a 24-hour detention time during the 1-year, 24-hour storm. Wetland design should also take into

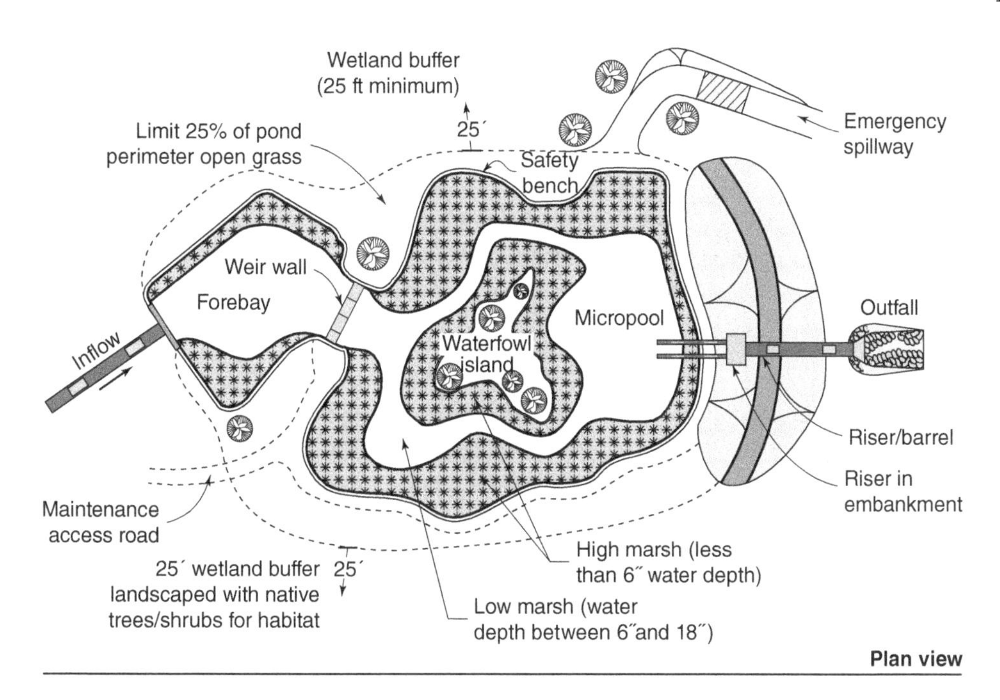

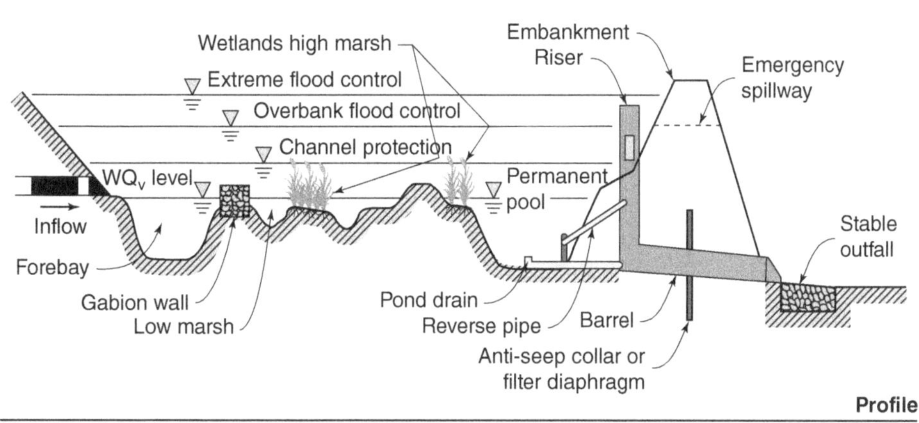

Figure 12.8 Example of a Shallow-Constructed Wetland System Design for Stormwater. Conversion factors: 1″ = 1 in. = 25.4 mm; 1′ = 1 ft = 0.3048 m (*Source:* New York State Department of Environmental Conservation, USA)

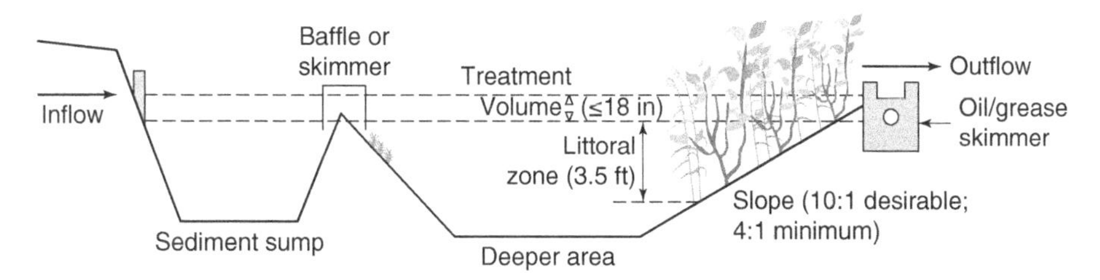

Figure 12.9 Example of a Wet Detention System Design for Stormwater. Conversion factors: 1 in. = 25.4 mm; 1 ft = 0.3048 m (*Source:* Florida State Department of Environmental Regulation, USA)

account that sediment accumulation in wetland systems can greatly shorten their effective life and that some suspended solids should be removed from the runoff before it enters the wetland system. The design should include sloped sides to allow easy removal of accumulated sediments and harvesting of plants.

Maintenance Requirements. Like most stormwater quality controls, constructed wetlands require regular maintenance. In addition to regularly scheduled sediment removal, wetland systems should be periodically cleared of dead vegetation.

Harvesting of plants in the wetland might be appropriate for pollutant removal purposes; if so, disposal of removed material must be planned.

Limitations on Use. While constructed wetland systems can treat stormwater runoff effectively, they do require large areas of undeveloped land, which can make siting of wetland systems difficult especially in urban areas. For this reason, incorporating wetland systems into new development is usually more feasible than retrofitting them into existing developments. Achieving proper soil conditions and groundwater levels can also present difficulties. To maintain a wetland environment, soils must be resistant to infiltration (i.e., have low permeability) and a water supply must be constant. In general, soils in the system must be saturated throughout the growing season so the desired vegetation will survive.

12.5.4 Infiltration Facilities

Unlike detention facilities that capture and eventually release stormwater runoff to a surface water body, infiltration facilities permanently capture runoff so that it soaks into the groundwater (Fig. 12.10). Because they do not release the runoff to surface water, infiltration facilities are sometimes called retention facilities. Pollutant removal in these BMPs occurs primarily through infiltration, which eliminates the runoff volume or lowers it by the capacity of the facility. The three different types of facilities commonly used to promote infiltration and remove pollutants from stormwater runoff are infiltration basins, infiltration trenches/dry wells, and porous pavement (swales, which also promote infiltration, are addressed later under vegetative practices).

12.5.4.1 Infiltration Basins

Infiltration basins are similar to dry ponds, except that infiltration basins have only an emergency spillway and no standard outlet structure. All flow entering an infiltration basin (up to the capacity of the basin) is, therefore, retained and allowed to infiltrate into the soil (see Fig. 12.11).

Pollutant Removal. Infiltration is the major pollutant removal mechanism. Infiltration basins, like dry and wet ponds, receive stormwater runoff from drainage systems and provide storage up to a designed volume. Unlike dry detention ponds, which eventually release stored runoff through a drainage system, or wet ponds, which maintain a permanent pool of water, infiltration basins release stored runoff through the basin's underlying soil. Infiltration basins provide stormwater pollutant removal through volume reduction and filtration and settling. Infiltration basins are particularly effective in removing bacteria, suspended solids, insoluble nutrients, oil and grease, and floating wastes.

Design Considerations. The most important consideration in the design of infiltration basins is calculating the basin's size for the drainage area and the soil type involved. Some designers recommend off-line basins to capture and infiltrate the first 0.5 in. (13 mm) of rainfall from the contributing drainage area. The appropriate amount of flow must be diverted to the system, and soil tests need to be performed to estimate the infiltration rates and appropriately size the basin. Also related

Figure 12.10 Sample Infiltration Basin (*Source:* http://en.wikipedia.org/wiki/Image:Infilt_basin.jpg)

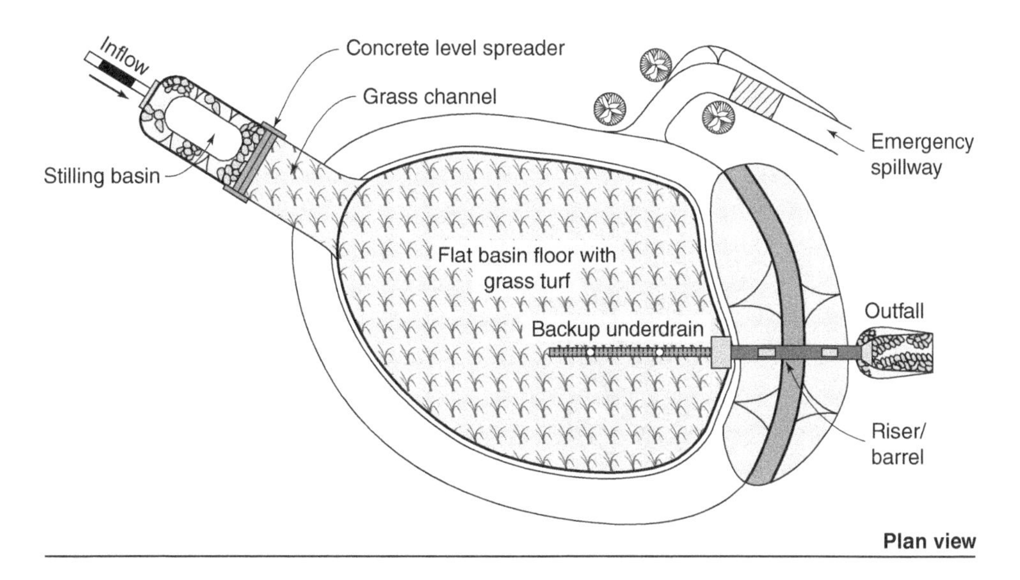

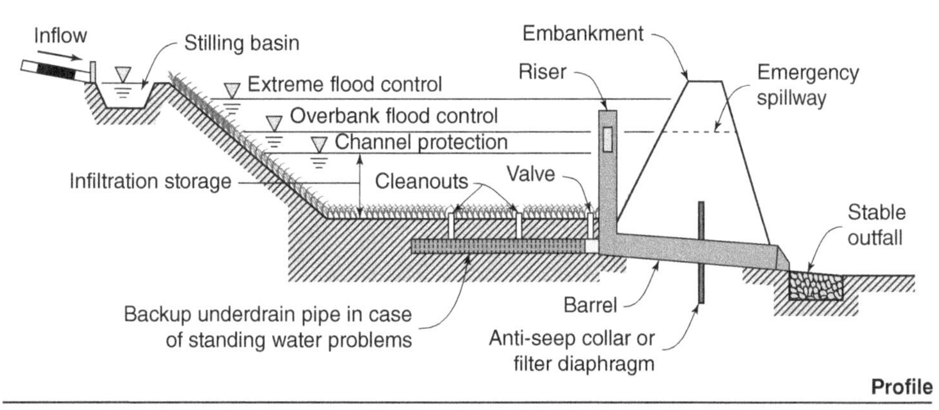

Figure 12.11 Infiltration Basin (*Source:* New York State Department of Environmental Conservation, USA)

to the proper size of infiltration basins is the amount of time necessary for the basin bottom to dry between rainstorms. Designers generally specify that infiltration basins should be designed to be dry for at least three days between storms. This interval allows the soil to dry, thereby increasing its pollutant removal capacity. Basin shape is also important. It should have gently sloping sides to allow for easy access to mow the bottom vegetation. An emergency spillway must also be incorporated into the basin design. Finally, some form of pretreatment is recommended to remove suspended sediments from runoff before it is discharged to the basin. This pretreatment will reduce the need for periodic removal of accumulated sediment, which can clog the soil pores and reduce the level of infiltration.

Maintenance Requirements. Infiltration basins require moderate to high levels of periodic maintenance. Most are designed with vegetated bottoms to provide stabilization and promote some vegetative uptake of nutrients. Periodically, the bottom of the basin must be mowed and accumulated sediments must be removed to maintain desired infiltration rates.

Limitations on Use. Infiltration basins often have relatively large land requirements and are better suited for location in developing areas than in already developed areas. Infiltration basins also require suitable soil to be effective. Accumulating runoff must be able to infiltrate the soil in the bottom of the basin. Typically, sand and loam, with infiltration rates greater than or equal to 0.27 in./h (6.9 mm/h), are the preferred soils for infiltration systems. The use of infiltration basins can be restricted by high groundwater elevations. For infiltration to occur, groundwater levels should be located at least 2–4 ft. (0.60–1.2 m) below the bottom of the basin.

12.5.4.2 Infiltration Trenches/Dry Wells

Subsurface infiltration practices, such as *infiltration trenches* or *dry wells*, force runoff into the soil to recharge groundwater and remove pollutants. These infiltration structures are located below ground and usually must be built off line because of

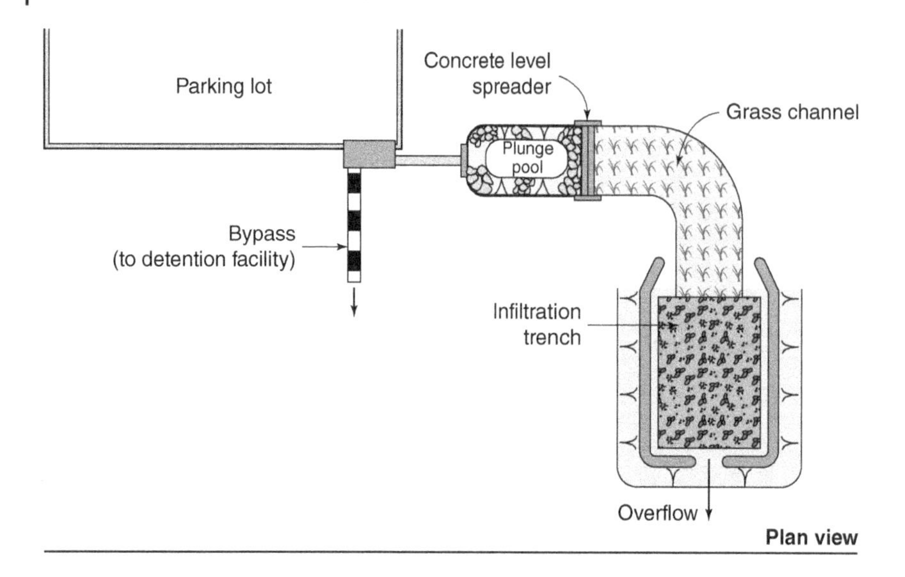

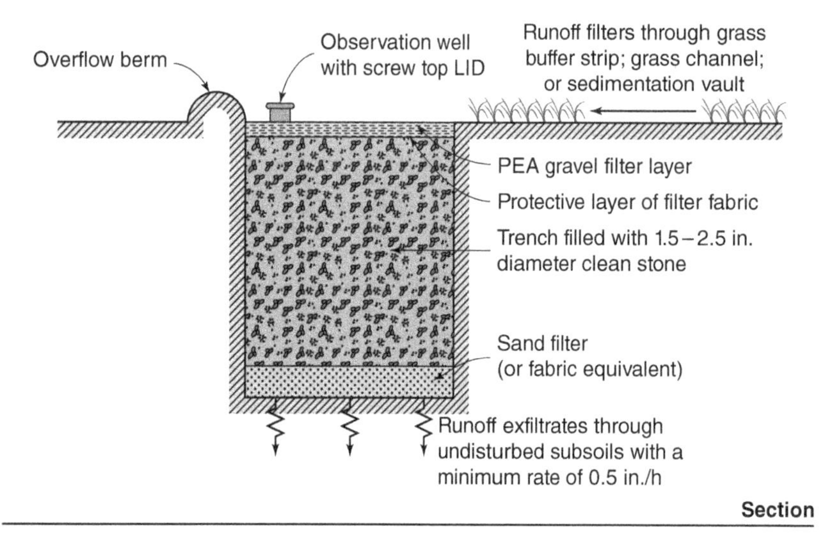

Figure 12.12 Infiltration Trench. Conversion factor: 1 in. = 25.4 mm (*Source:* New York State Department of Environmental Conservation, USA)

their limited storage area (see Fig. 12.12). Subsurface infiltration systems generally consist of precast concrete structures with holes in the sides and bottom surrounded by 2–4 ft. (0.60–1.2 m) of washed stone. Stormwater runoff is directed into these structures and infiltration takes place.

Pollutant Removal. The structural controls described in this section use filtration as the primary pollutant removal mechanism, much like on-site wastewater treatment systems commonly used in many small communities. These controls effectively remove suspended sediments and floating debris, as well as bacteria, which are difficult to remove without disinfection. Infiltration practices are generally less effective at removing dissolved nutrients, such as nitrogen or other soluble contaminants, which can travel through groundwater and be discharged to the receiving water.

Design Considerations. The soil infiltration rate is probably the most important consideration in the design of infiltration structures. The soils underlying the structure must be tested to determine their suitability for infiltration. Some authorities specify the types of soils acceptable for infiltration as noted above for infiltration basins.

Structure size is another primary consideration. The structures must be large enough to handle the desired design storms. Also, the structures must be designed to allow larger storms to bypass them. Because subsurface infiltration structures do not have outlets, they usually have to be designed off line of the regular drainage system. Runoff can then enter the infiltration structure until it is full; additional runoff is directed away from the structure. A diversion structure upstream of the infiltration structure is normally part of the design. The flow entering this structure (which could be a simple manhole) is directed to the subsurface infiltration structure until it is full; then additional flow is directed away from the structure and along the drainage system. A typical sizing rule for subsurface infiltration structures is that they should store the runoff from the first 0.5 in. (13 mm) of rainfall on the site.

Infiltration structures must also be designed to empty in a reasonable length of time. The underlying soils, to remove pollutants from runoff effectively, must be allowed to dry between rainstorms. Most experts specify that infiltration structures should contain a reservoir of runoff for no more than three days after rainfall.

Maintenance Requirements. Infiltration structures require periodic cleaning to remove accumulated sediment and petroleum products. Often the need for this maintenance can be reduced by incorporating into the design a pretreatment structure that removes sediments and petroleum products from the runoff. These pretreatment structures can also minimize the discharge to groundwater of some pollutants, such as solids. Although addressing these issues in the design of infiltration structures can reduce routine maintenance requirements, the design still should include an observation well that allows inspectors to determine sediment deposition.

Limitations on Use. Subsurface infiltration structures can be used for end-of-pipe treatment and can also be located at different points in the drainage system. If located at the downstream end of a drainage system, infiltration structures can have large land requirements. Subsurface infiltration structures, because they are located underground, can be located in areas such as parking lots and access roads.

The primary physical limitation to locating infiltration structures, other than land requirements, is the suitability of soil, which must be neither too impermeable to run off (e.g., clay, silt, or till) nor too rapidly permeated (e.g., sand). Another potential physical limitation is the depth to groundwater. To provide proper treatment and reduce the possibility of groundwater contamination, a distance of at least 2 ft. (0.60 m) should be maintained between the bottom of the infiltration structure and the mean high groundwater elevation.

12.5.4.3 Porous Pavement

Paved roads and parking areas, because they increase watershed imperviousness, are major contributors to stormwater runoff problems in urban areas. Porous pavement, however, allows water to flow through a porous asphalt layer and into an underground gravel bed. Porous concrete pavement or cobblestone paving can also be used (Fig. 12.13). Use of this porous pavement can thereby reduce runoff volume and pollutant discharge. This practice, used in areas with gentle slopes, is generally designed into parking areas that receive light vehicle traffic.

Pollutant Removal. Field studies have shown that porous pavement systems can remove significant levels of both soluble and particulate pollutants. Porous pavement is primarily designed to remove pollutants deposited from the atmosphere, because coarse solids can clog the pavement pores. In these systems, pollutant removal occurs primarily after the runoff has infiltrated the underlying soils. Pollutant removal is accomplished by trapping of sediments, and infiltration through the underlying soils, which can remove pollutants such as bacteria. The removal efficiency depends on the storage volume of the pavement, the basin surface area, and the soil percolation rate.

Figure 12.13 Permeable Cobblestone Paving (*Source:* http://en.wikipedia.org/wiki/Image: Santarem_carfree.JPG)

Design Considerations. Porous asphalt pavement generally is designed with an upper pavement layer 2–4 in. (50–100 mm) thick, a 1–2-in. (25–50-mm) layer of coarse sand, a stone reservoir to provide storage, and a bottom filter fabric as shown in Fig. 12.14. Other types of porous pavement include poured-in-place concrete slabs, precast concrete grids, and modular units of brick or cast concrete. The differences in pavement design result in different ways that the collected runoff is discharged.

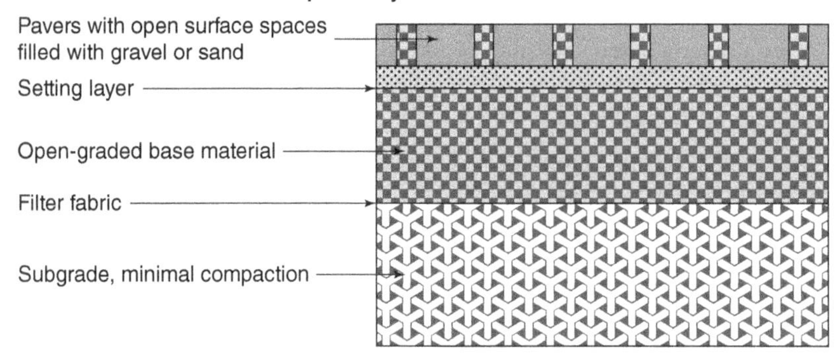

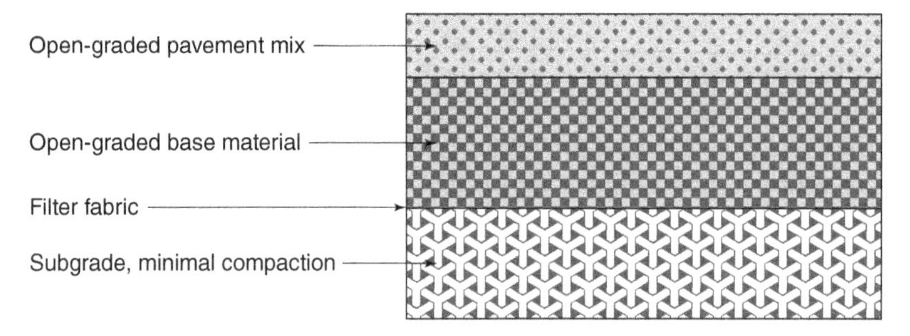

Figure 12.14 Porous Pavement Cross-Section (*Source:* City of Portland Bureau of Environmental Services, USA)

Some systems let all the runoff discharge through the underlying soils and into the groundwater. While these systems provide good pollutant removal, they can result in groundwater contamination. Other systems include perforated pipes to collect the runoff and discharge it directly to surface water; although these systems protect the groundwater below the pavement, they do not provide the same level of pollution removal as the full infiltration systems.

Porous pavement is designed so that a certain amount of runoff is collected and stored in the stone reservoir. The design criteria, therefore, determine the depth of the stone reservoir. The maximum depth of the stone reservoir also is affected by the infiltration rate of the underlying soils. Runoff should be completely drained within a maximum of three days after the maximum design storm event to allow the underlying soils to dry, maintaining aerobic conditions that improve pollutant removal.

Maintenance Requirements. Porous pavement can have extensive maintenance requirements. The pavement must be kept free of coarse particles that can clog the pavement and prevent runoff from collecting. The pavement must, therefore, be regularly inspected and cleaned with a vacuum sweeper and high-pressure jet. The state of Maryland, by reviewing its porous pavement practices, found that after four years of use only 2 of the 13 systems were functioning as designed. The 11 malfunctioning sites were affected primarily by clogging and excessive sediment and debris.

Limitations on Use. Because porous pavement is expensive to replace or repair, it is generally only used on parking areas that receive moderate to low traffic. The area to be paved also should be relatively flat with a depth of 2–4 ft. (0.60–1.2 m) from the bottom of the stone reservoir to the high-water table. In addition, the soils under the pavement must allow for infiltration.

12.5.5 Vegetative Practices

Urbanization results in the elimination of vegetation and increases in impervious area. Vegetative practices (Fig. 12.15) in urban areas decrease the impervious area and promote runoff infiltration and solids capture. These practices generally provide moderate to low pollutant removal and are therefore used as pretreatment for the removal of suspended solids from runoff prior to more intensive treatment by other practices. The two major types of vegetative practices commonly used in urban areas are vegetated swales and filter strips (both sometimes referred to as biofilters). Native vegetation is recommended since it requires less site preparation and maintenance.

12.5.5.1 Vegetated Swales

Vegetated swales are channels covered with vegetation to reduce erosion of soil during storms (Fig. 12.16). They are used to replace conventional catch basin and pipe network systems for transporting runoff to surface waters. Stormwater runoff flows through the swale reducing runoff velocity and promoting the removal of suspended solids.

Pollutant Removal. Infiltration of the runoff and associated pollutants is the most important pollutant removal process accomplished by vegetated swales. Vegetated swales also remove pollutants through filtering by the vegetation and settling of solids in low-flow areas. Because of these pollutant removal mechanisms, swales are most effective at removing suspended solids and associated pollutants, such as metals. The infiltration mechanism also removes bacteria.

Design Considerations. Pollutant removal in vegetated swales can be increased by reducing runoff velocity: reducing the slope, increasing the vegetation density, and installing check dams to promote ponding. Also, the underlying soils should have a high permeability to help promote infiltration.

Figure 12.15 Roof Garden of Rockefeller Center, NY, USA (*Source:* Meathead1962/Wikimedia Commons)

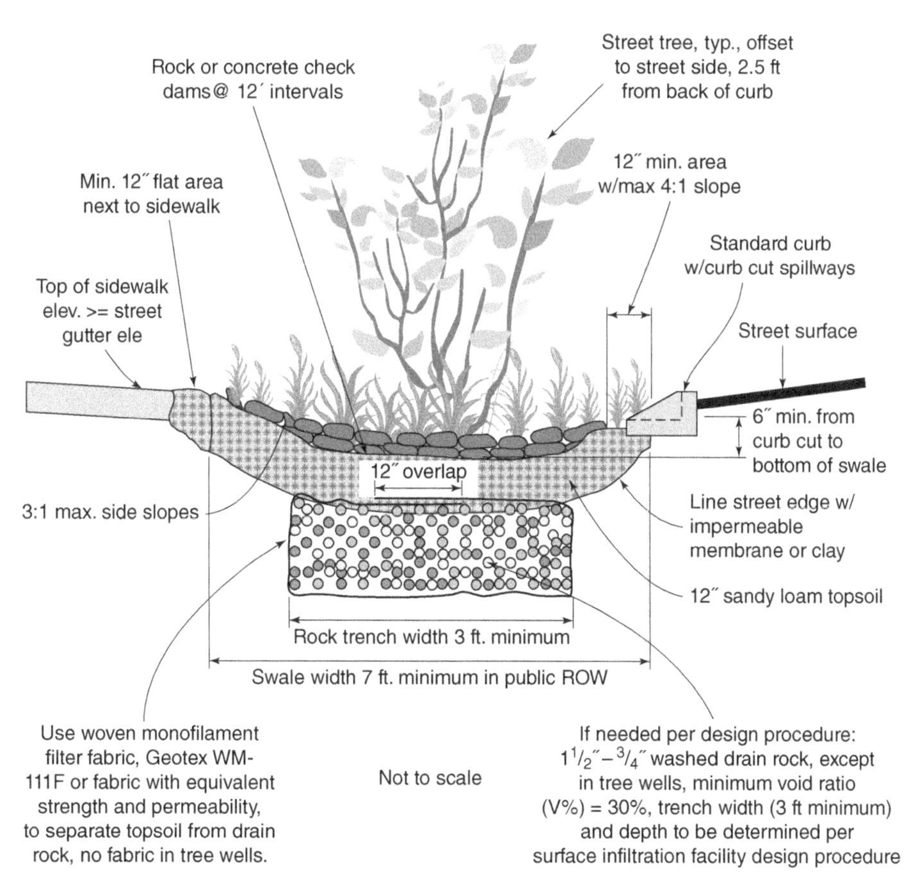

Figure 12.16 Cross-Section of Vegetated Swale. Conversion factors: 1″ = 1 in. = 25.4 mm; 1 ft = 0.3048 m (*Source:* City of Portland Bureau of Environmental Services, USA)

Maintenance Requirements. Maintenance is aimed at preserving dense vegetation and preventing erosion of underlying soils. This maintenance includes regular mowing, weed removal, and watering during drought periods and after initial seeding. In conjunction with mowing, the cut material should be removed.

Limitations on Use. Vegetated swales might be difficult to retrofit in already developed areas. They can replace curb and gutter drainage systems, but work best in low-slope areas with soil that is not susceptible to erosion.

12.5.5.2 Filter Strips

Filter strips, shown in Fig. 12.17, are similar to vegetated swales. Runoff entering these systems, however, generally is sheet flow, is evenly distributed across the filter strip, and flows perpendicular to the filter strip. Because these systems can accept only overland sheet flow, level spreading devices are used so that water is not ponded.

Pollutant Removal. Pollutant removal in filter strips depends on the filter strip's length, size, slope, and soil permeability; the size of the watershed; and the runoff velocity. Filter strips are most effective at removing pollutants such as sediment, organic material, and some trace metals, and less effective at removing dissolved pollutants such as nutrients.

Design Considerations. The major design aspects of filter strips that can be effectively changed are the length, width, slope, and vegetative cover of the strip. Greater pollutant removal is achieved with filter strips that are long and flat. A level spreading device must be incorporated in the design of a filter strip to ensure that concentrated flow does not enter and create a channel. If concentrated flows enter a filter strip, they can cause erosion of the vegetation and soil and reduce the structure's pollutant removal efficiencies. In addition to these considerations, filter strips should be constructed in areas with porous soil to promote infiltration.

Maintenance Requirements. Filter strips must be mowed and weeded regularly – the same maintenance practices as vegetated swales. In addition, the strip must be watered after initial seeding. In some cases, however, large filter strips can be "left on their own" so that large vegetation can grow and create a natural filter strip. This option reduces the level of maintenance required and can enhance the pollution removal of the strip.

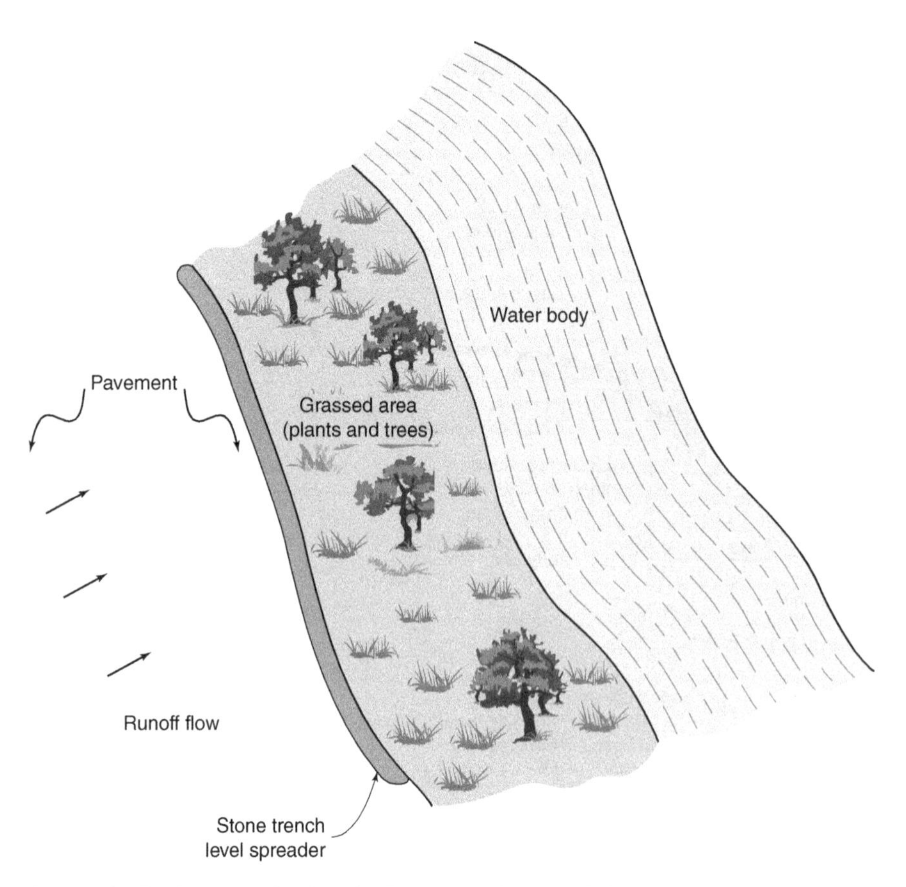

Figure 12.17 Schematic Design of a Filter Strip (*Source:* U.S. Environmental Protection Agency)

Limitations on Use. The major limitation on the use of filter strips is the slope of the land; these strips operate best when placed on flat surfaces that have permeable soils. Also, filter strips treating large watersheds can have large land requirements that preclude their location in urban areas.

12.5.6 Filtration Practices

Filtration practices provide runoff treatment through settling and filtering using a specially placed layer of sand or other filtration medium. Flow enters the structure, ponds for a period of time, and filters through the media to an underdrain that discharges to surface water. These practices attempt to simulate the pollutant removal of infiltration practices using less land area. Two different types of filtration practices in use are filtration basins and sand filters.

12.5.6.1 Filtration Basins

Stormwater runoff diverted to a *filtration basin* can be detained, allowed to percolate through filter media, and collected in perforated pipes as shown in Fig. 12.18. These perforated pipes then transport the filtered runoff to the receiving water. These systems have been used extensively in Austin, Texas, showing good pollutant removal efficiencies and low failure rates. One major question regarding filtration basins is the effect of cold temperature and freezing conditions on the operation of these systems.

Pollutant Removal. Pollutant removal in filtration basins occurs because of settling during the initial ponding time and filtering through the soil media. Removal efficiencies in filtration basins depend on several factors, including the storage volume, detention time, and filter media used. In general, longer detention times increase the system's pollutant removal efficiency. Increasing the detention time usually requires increasing the overall size of the filtration basin. Reducing the size of the perforated pipe, increasing the depth of the filter medium, or decreasing the percolation rate of the filter medium can be used to increase the detention time. Filtration basins primarily use sand as the filtering medium.

Design Considerations. In Austin, Texas, sand filtration basins are typically designed to provide a detention time of 4–6 h and have been used to treat runoff from drainage areas from 3 to 80 acres (1.2 to 32.3 ha). A stormwater sand-peat filtration basin in Montgomery County, Maryland, was designed to store the first 0.5 in. (13 mm) of rainfall from the impervious land in the watershed. In the Maryland area, this sizing criterion results in the treatment of 50%×60% of the annual storm runoff volume. Runoff from larger storms that exceeds the capacity of these filtration systems is diverted away from the filtration basin or discharged through an emergency spillway. To improve the longevity of sand and sand-peat filtration basins, runoff entering the systems is typically pretreated to remove suspended solids. Such pretreatment techniques as the use of a wet pool or water quality inlets can be used in conjunction with filtration basins.

Maintenance Requirements. Stormwater runoff filtration basins require extensive maintenance to remove accumulated sediments and prevent clogging of the filtering medium. Maintenance requirements include inspecting the basin after every major storm event for the first few months after construction and annually thereafter, removing litter and debris, and revegetating eroded areas. In addition, the accumulated sediment should be removed periodically and the filter medium, when clogged with sediment deposits, should be removed and replaced.

Limitations on Use. Filtration basins can often be difficult to locate in highly urbanized areas because of their large land requirements. In addition, high groundwater levels can restrict their use.

12.5.6.2 Sand Fitters

Sand filters are similar to the filtration basins outlined above but can be built underground to reduce the amount of land required. These systems consist of a catch basin for settling of heavy solids and a filtration chamber (see Fig. 12.19). Runoff

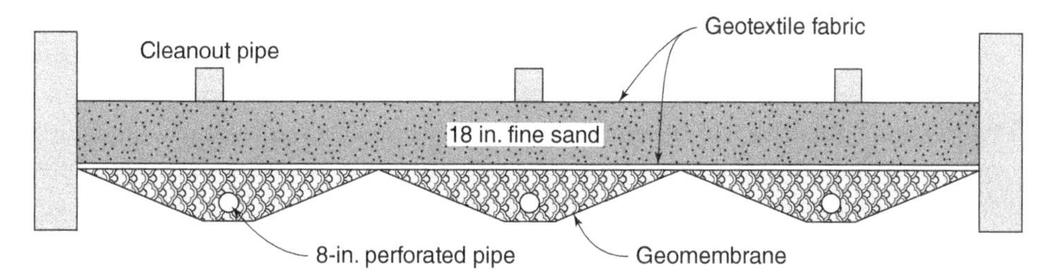

Figure 12.18 Conceptual Design of a Filtration Basin. Conversion factor: 1 in. = 25.4 mm (*Source:* City of Austin, TX, USA)

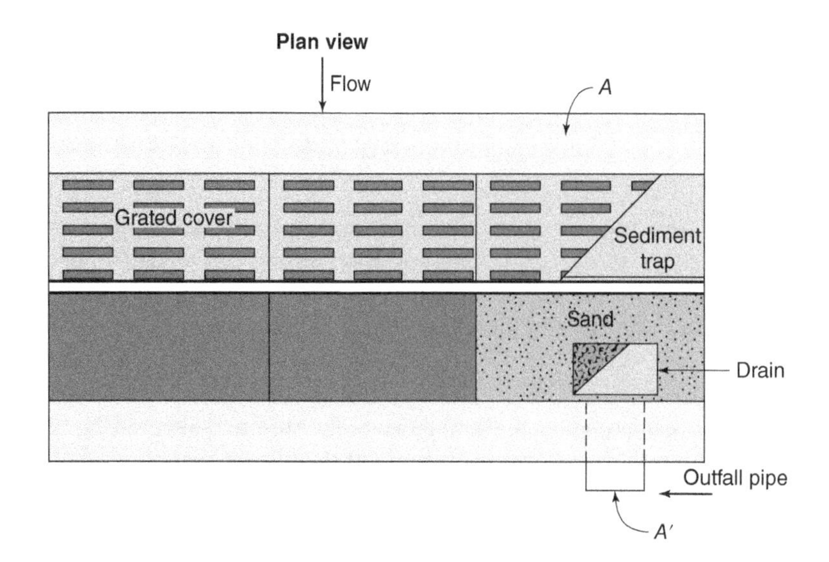

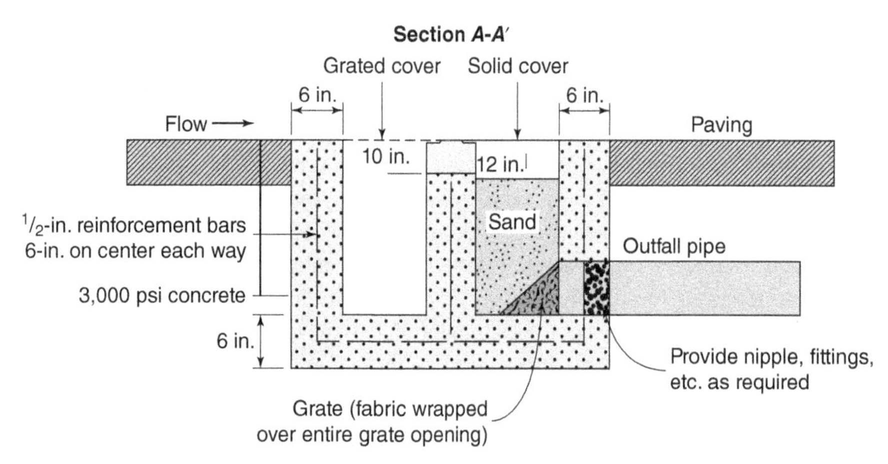

Figure 12.19 Schematic Design of a Sand Filter. Conversion factors: 1 in. = 25.4 mm; 1 psi = 6.94 kPa (*Source:* Delaware State Department of Natural Resources, USA)

enters the catch basin and collects to the basin capacity, overflows into a sand-filled chamber that provides filtration, and is discharged through an outlet pipe in the bottom of the filtration chamber. Other types of systems can be designed in conjunction with wet ponds or other practices, using natural or imported soil banks or bottoms, to increase their pollutant removal capability.

Pollutant Removal. Sand filters use the same pollution removal mechanisms as filtration basins and provide similar pollutant removal. Initial removal of heavy solids occurs through settling in the catch basin and further treatment is provided by filtration through the sand-filled chamber.

Design Considerations. The catch basin section must be designed to provide some sediment removal and to ensure that flow enters the filtration chamber as sheet flow to prevent scouring of the sand. The maximum drainage area that can be treated by a sand filter has been reported to be about 5 acres (2 ha). Sand filters generally are used to treat impervious areas, such as parking lots, so that smaller sediment particles typical of pervious areas will not clog the sand filter.

Maintenance Requirements. Sand fitters require minimal maintenance, consisting of periodically removing accumulated sediment and the top layer of sand from the filtration chamber and removing accumulated sediment and floatables from the catch basin. Regular inspections of the filter system can indicate when this maintenance is required.

Limitations on Use. Because of their small size, sand filters are designed to be used for pretreatment in large watersheds or full treatment in small watersheds. They cannot provide sufficient treatment for large watersheds.

12.5.7 Water Quality Inlets

Water quality inlets, also known as *oil and grit separators*, are similar to septic tanks used for removing floatable wastes in on-site wastewater disposal systems. These inlets provide removal of floatable wastes and suspended solids through the use of a series of settling chambers and separation baffles as shown in Fig. 12.20. Given the limited pollutant removal expected from water quality inlets, they are usually used in conjunction with other BMPs. Fairly effective at removing coarse sediments and floating wastes, water quality inlets can be used to pretreat runoff before it is discharged to infiltration systems or detention facilities. In this way, some of the routine maintenance the other BMPs require (e.g., sediment removal and unclogging of outlet structures) can be reduced. Water quality inlets can also serve to capture petroleum spills that could enter other treatment structures or surface waters.

Pollutant Removal. The primary pollutant removal mechanisms of water quality inlets are separation and settling. The use of three chambers in these inlets serves to increase the detention time of the runoff in the tank, allowing settling to occur. In this way, suspended solids, and the attached pollutants, are removed from the runoff. In addition, the use of baffles and

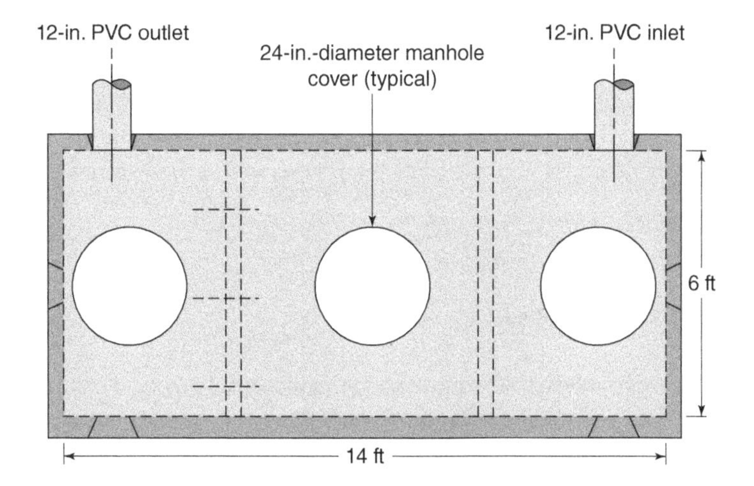

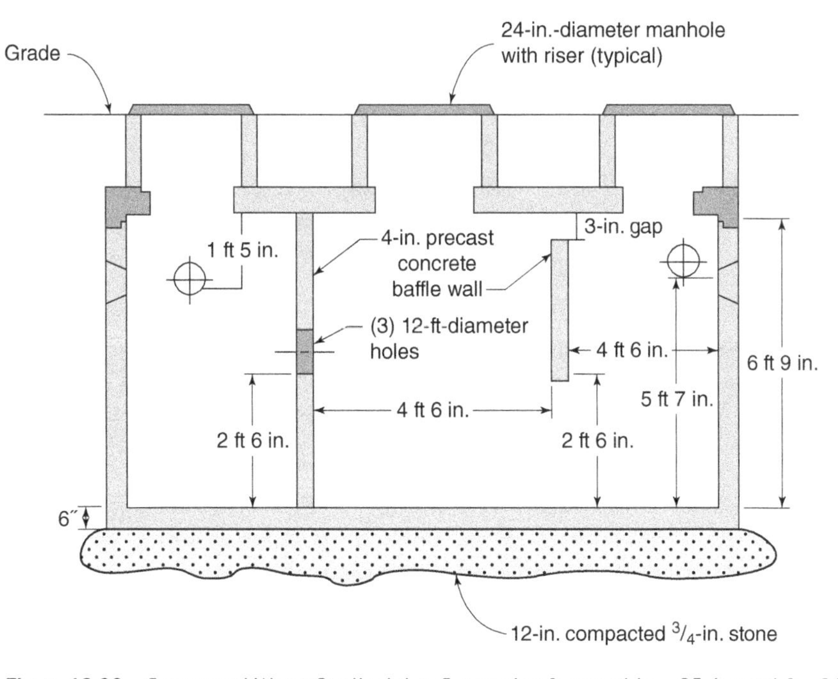

Figure 12.20 Conceptual Water Quality Inlet. Conversion factors: 1 in. = 25.4 mm; 1 ft = 0.3048 m (*Source:* U.S. Environmental Protection Agency)

inverted elbows helps to remove floating litter and petroleum products from the stormwater. The level of removal of these pollutants depends on the volume of water permanently detained in the tank, the velocity of flow through the tank, and the depth of the baffles and inverted elbows in the tank. By increasing detention time and decreasing flow velocity, the level of sediment and floatables expected to be removed from water quality inlets can be improved.

Design Considerations. Water quality inlets design depends on the size of the watershed being treated and the detention time required. Because suggested detention times are usually measured in terms of minutes rather than days, water quality inlets generally do not remove pollutants from stormwater runoff as effectively as some of the more intensive detention facilities. Water quality inlets have the advantage of being relatively small so they can be placed throughout a drainage system rather than just at the downstream end of the system. The flow and velocity of the entering runoff can be hydraulically restricted by limiting the size of the inlet pipe.

Maintenance Requirements. Water quality inlets require periodic maintenance to remove accumulated pollutants; in general, these inlets should be cleaned about twice a year. Cleaning can be performed with a vacuum truck similar to those used to clean catch basins. Periodic inspections between scheduled maintenance are also required to determine the level of accumulated pollutants.

Limitations on Use. There are few physical site limitations on the use of water quality inlets. The inlets are generally designed as belowground structures and do not require large amounts of land. Given their small size, however, large watersheds cannot be drained into a water quality inlet. Removal efficiencies depend on the detention time in the water quality inlets. Their use is usually restricted to small watersheds of less than 2 acres (0.81 ha). Another restriction on the use of water quality inlets is dry-weather base flow. If dry-weather base flow cannot easily be removed from a drainage system, a large water quality inlet and more frequent maintenance are needed to accommodate this flow as well as the flow resulting from a rainfall event.

12.6 Combined Sewer Overflow Control Practices

Some of the urban runoff BMPs discussed above are applicable to CSO control. Additional control practices commonly used for CSO control are described in this section, including a general discussion of each practice's applicability, its pollutant removal effectiveness, and its maintenance requirements. Because CSOs contain sanitary wastewater and other waste streams, the primary pollutants of concern in CSO control are suspended solids, biochemical oxygen demand (BOD), and pathogens. CSOs, however, also contain nutrients, metals, and other toxic substances.

12.6.1 Source Controls

12.6.1.1 Water Conservation Programs

One way to reduce the amount of sanitary wastewater in a combined system is to attempt to control the amount of water used by homes and businesses that is then converted to wastewater. Typical programs and practices for control include the following:

- *Plumbing retrofit:* Using low-flush toilets, flush dams, faucet aerators, and other water-saving devices
- *Plumbing code changes:* Requiring implementation of water-saving devices in new construction or as plumbing is replaced
- *Education programs:* Encouraging water conservation in businesses and homes by providing information on its benefits
- *Technical assistance:* Providing water-use audits or case studies demonstrating potential savings to businesses
- *Rate system modifications:* Adjusting rate systems to promote or reward water savings.

Although these programs might require minor changes in personal habits, they can be cost-effective compared to end-of-pipe treatment. There are limits, however, to the reductions in water use that can be achieved reasonably.

12.6.1.2 Pretreatment Programs

Pretreatment programs are implemented at the local level to control industrial and commercial sources of wastewater discharging to a municipal sewer system. The goals of a local pretreatment program are to stop or prevent industrial and commercial pollutants from passing through a municipal wastewater treatment plant, thereby violating state water quality standards; to stop or prevent disruption of treatment plant operations caused by industrial and commercial pollutants, including the contamination of municipal treatment plant residuals; and to ensure the safety of municipal sewer system and treatment plant workers by minimizing their exposure to potentially dangerous or toxic pollutants.

12.6.2 Collection System Controls

Many collection system controls exist for addressing pollution from CSO discharges. These controls focus on modifying the sewer system to reduce CSO flow, volume, and contaminant load.

12.6.2.1 Sewer Separation

One method for addressing CSO pollution is to convert the combined collection system to separate stormwater and sanitary sewer systems by constructing a new separate sanitary sewer. Sewer laterals from homes and businesses are then connected into the new system. Inappropriate connections to the old system from buildings are plugged. This conversion eliminates the possibility of sanitary wastes entering the drainage system and being discharged to surface water. Sewer separation, however, can be very expensive and disruptive. A municipality implementing this practice likely has to address urban runoff pollution problems. In systems that consist of both combined and separate drainage areas, partial separation (i.e., separation of some combined areas) could be cost-effective.

12.6.2.2 Infiltration Control

Sources of infiltration include groundwater entering the collection system through defective pipe joints, cracked or broken pipes, and manholes, as well as footing drains and springs. Infiltration flow rates tend to be relatively constant and result in lower volumes than inflow contributions. Infiltration problems are usually not isolated and often effect a more general sewer (or drainage) system deterioration. Extensive rehabilitation is typically required to remove infiltration effectively. The rehabilitation effort often must include house laterals, which are normally a significant source.

12.6.2.3 Inflow Control

CSO control can be achieved by diverting some of the surface runoff inflows from the combined sewer system, or by retarding the rate at which these flows are permitted to enter the system. Inflow of surface runoff can be retarded by using special gratings, restricted outlet pipes, or hydrobrakes (or comparable commercial devices) to modify catch basin inlets to restrict the rate at which surface runoff is permitted to enter the conveyance system. Inlet flow restrictions can be designed to produce acceptable levels of temporary ponding on streets or parking lot surfaces, allowing runoff to enter the system eventually at the inflow point, but reducing the peak flow rates that the combined sewer system experiences. Flow detention to delay the entry of runoff into the collection system by storing it temporarily and releasing it at a controlled rate can also be accomplished by rooftop storage under appropriate conditions. Elimination of the direct connection of roof drains to the CSO collection system and causing this runoff to reach the system inlets by overland flow patterns (preferably via unpaved or vegetated areas) is another method of retarding inflows.

12.6.2.4 Regulator and System Maintenance

Malfunctioning regulators are a common problem for combined sewer systems and can result in dry-weather overflows to receiving waters or in system backups and flooding. Static regulators often malfunction because of plugging or interference by debris in the sewer system. Mechanical regulators tend to require frequent maintenance. Municipalities should, therefore, develop an inspection and maintenance program designed to keep these regulators operating as designed.

12.6.2.5 In-System Modifications

These practices are designed to reduce CSO discharges by modifying the system to store more flow and allow it to be carried to the treatment plant. Possible modifications include adjusting regulator control features, such as weir elevation; installing new regulators; or installing new relief conduits.

12.6.2.6 Sewer Flushing

Sewer flushing is an additional practice to address CSO pollution problems. In this practice, water is used to flush deposited solids from the combined system to the treatment plant during dry weather. This practice is typically used in flat areas of the collection system where solids are most likely to settle out. The effectiveness of this practice is site-specific and depends on the flush volume; flush discharge rate; wastewater flow; and sewer length, slope, and diameter.

12.6.3 Storage

CSO discharges occur when the flow in a combined system exceeds the capacity of the sewer system or the treatment plant. Storing all or a portion of the CSO discharges for treatment during dry weather can effectively reduce these overflows. Storage is considered a necessary control alternative because of the high volume and variability associated with storm and CSOs. Storage facilities are frequently used to attenuate peak flows, thereby reducing the size of facilities required for further treatment. Storage techniques include in-line and off-line storage.

In-line storage uses existing capacity in major combined sewer trunk lines or interceptors to store combined flows. During storms, regulators are used to cause flow to back up in the system allowing it to be stored there. Although not all flow can be stored in the sewer system, this practice can reduce overflow volumes during large storms and eliminate overflow volumes during small storms. Care must be taken to ensure that flows do not back up onto streets or into homes.

Off-line storage consists of constructed near-surface or deep-tunnel detention facilities. Near-surface facilities usually consist of concrete tanks or, in some cases, large conduits that convey flow to a treatment facility. Tunnels can provide large storage volumes with relatively minimal disturbance to the ground surface, which can be very beneficial in congested urban areas. Overflows are directed to the storage facility, held during the storm, and pumped to the publicly owned treatment works after the storm, thus reducing the overflow quantity and frequency.

12.6.4 Treatment

Treatment of urban runoff water includes preliminary treatment, flotation, filtration, chemical precipitation, biological treatment and disinfection.

1) *Preliminary treatment*. Most of the urban runoff BMPs previously discussed employ physical processes to reduce pollution. Physical treatment practices can also be used to reduce pollutant discharges from CSOs. These practices use screening technologies to reduce the flow of solids in combined systems. They are typically used as a preliminary treatment step to remove floatables upstream of other processes. Different screens have different size openings to provide various levels of solids removal. Bar racks have the largest openings (typically 1 in. or more) and microstrainers have the smallest openings (typically as small as 15 μm). All of these practices require periodic and regular cleaning to prevent the accumulation of solids.
2) *Dissolved air flotation (DAF)*. DAF removes solids from wastewater by introducing fine air bubbles, which attach to solid particles suspended in the liquid, causing the solids to float to the surface where they can be skimmed off. Because of its relatively high overflow rate and short detention time, DAF does not require as large a facility as conventional sedimentation. Oil and grease are also more readily removed by DAF. The high operating costs for DAF are due to large energy demand; in addition, skilled operators are required for its operation.
3) *Filtration*. Dual-media high-rate filtration has been used for treatment of CSO flows using a two-layer bed, consisting of coarse anthracite particles on top of less coarse sand. After backwash, the less-dense anthracite remains on top of the sand. Filtration rates of 8 gal/ft^2/min (325 L/m^2/min) or more result in substantially smaller area requirements compared with sedimentation.
4) *Chemical precipitation*. Chemical precipitation facilities store and use polymer, alum, or ferric chloride to cause solids to precipitate. Chemical precipitation can increase the pollution removal that generally occurs from other settling practices, thereby allowing for the design of smaller sedimentation tanks. Removal rates for these practices are up to 70% for BOD and 85% for suspended solids. Chemical precipitation generates more sludge buildup and handling can become a major problem.
5) *Biological treatment*. Although biological treatment processes have the potential to provide high-quality effluent, the disadvantages of biological treatment of CSOs include the following:
   - The biomass used to break down the organic material and assimilate nutrients in the combined sewage must be kept alive during dry weather, which can be difficult except at an existing treatment plant; biological processes are subject to upset when exposed to intermittent and highly variable loading conditions.
   - The land requirements for these types of processes can preclude their use in urban areas.
   - Operation and maintenance can be costly and the process requires highly skilled operators.
6) *Disinfection*. Because pathogens are the primary pollutant of concern in CSO control, practices focusing on disinfection are commonly used:
   - Chlorination. Combined flows can be treated with dissolved or gaseous chlorine to reduce the level of pathogens in the flow. Chlorination is typically used in conjunction with upstream solids removal. Dissolved chlorine (hypochlorite) is

currently more commonly used than gaseous chlorine because the equipment is more reliable and storage of the chemicals is safer. Dechlorination might be necessary to minimize the adverse effects of chlorine on aquatic life. With sufficient dosage and mixing, close to 100% destruction of pathogens is possible.

- Ultraviolet (UV) radiation. Introduction of UV radiation to combined wastewater is designed to provide disinfection without the addition of harmful chemicals. This practice uses an UV lamp submerged in a baffled channel located downstream of an effective solids removal process.

Case Study 12.1 City of Austin, Texas, USA, Local Watersheds Ordinances

Austin, a highly urbanized city bisected by the Colorado River, contains a number of high-quality lakes, aquifers, and streams. The major water resources in the area include three lakes – Lake Travis, Lake Austin, and Town Lake – which form a major drinking water reservoir acting as the main water supply for the city; Edwards Aquifer and Barton Springs are the area's other major water resources. These water resources are potentially threatened by urban runoff pollution from urbanized areas; Town Lake has already been affected significantly. To reduce and prevent urban runoff pollution problems in these resources, Austin has developed and passed three major watershed ordinances: the Comprehensive Watersheds Ordinance (CWO) in 1986, the Urban Watersheds Ordinance in 1991, and the Barton Springs Ordinance in 1992.

The primary goal of these ordinances is to protect the water resources of the Austin area from degradation from NPS pollution. Other goals include preventing the loss of recharge to the Edwards Aquifer, preventing adverse impacts from wastewater discharges, and protecting the natural and traditional character of the water resources in the Austin area. In addition, the city has implemented other ordinances that control NPS pollution.

Water pollution problems in the Austin area have been extensively studied since the mid-1970s. In 1981, the city participated in the Nationwide Urban Runoff Program (NURP) and began implementing and monitoring the effectiveness of urban runoff structural controls. The city has been a leader in developing and implementing NPS regulatory controls. The city's first NPS control ordinance, the Lake Austin Watershed Ordinance in 1978, was followed by other watershed ordinances in 1981 and 1984 designed to protect additional sensitive watersheds and upgrade the level of protection. The experience and data gathered as a result of these ordinances led the city to propose and adopt a more complete set of protections for water resources as described in the following summary:

1) *CWO*. The CWO is directed at preventing urban runoff pollution by placing requirements on proposed new developments within a 700-mi^2 (1,800-km^2) area of the city and its extraterritorial jurisdiction. It was developed in 1986 by a task force, appointed by the city council, with representatives from environmental groups, citizens, developers, and a council-appointed environmental board. The ordinance includes requirements for limiting impervious cover, using water quality buffer zones, protecting critical environmental features, limiting the disturbance of natural streams, implementing erosion control practices, constructing sedimentation and filtration basins, and restricting on-site wastewater disposal. The ordinance divides the city into four different watershed categories that each allows for different levels of development intensity: urban, suburban, water supply suburban, and water supply rural. Although urban watersheds were not originally covered by the CWO, they are addressed in the Urban Watersheds Ordinance that is described later. Requirements for all of the applicable watershed categories are shown in Table 12.6.

   The waterways located in each watershed category are classified as minor, intermediate, or major depending on the total drainage area contributory to the waterway (see Table 12.6). Each waterway classification has an associated critical water quality (WQ) zone that encompasses the 100-year floodplain boundary and is located 50–100 ft. (15–30 m) from minor waterways, 100–200 ft. (30–60 m) from intermediate waterways, and 200–400 ft. (60–120 m) from major waterways. No development is allowed in this critical WQ zone.

   Each waterway type also has an associated water quality buffer zone that begins at the end of the critical WQ zone and extends upland for a defined distance as shown in Table 12.6. Development in this zone is restricted by limits on the allowed percent imperviousness of the site. Areas outside the WQ buffer zone are considered upland areas and have less stringent percent imperviousness restrictions. In this zone, the restrictions are tied to the type of development proposed for the site as shown in Table 12.6. Some development restrictions can be reduced if the developer transfers land located in the watershed to the city. In this way, development density can be increased by the developer in exchange for an increase in publicly held lands. For example, a multifamily development in suburban

Table 12.6 Maximum Development Intensity in Austin, Texas

	Waterways			Development Limits				Acceptable Structural Pollution Controls
Watershed Category	Minor	Intermediate	Major	Water Quality Buffer Zone	Area Type	Zone[a] Uplands	Transfer	
Suburban								
						% Impervious Cover		
Drainage area:	320 ac	640 ac	1,280 ac	30% impervious cover	Residential:	50	60	Sedimentation
Critical WQ zone:	100 ft	200 ft	400 ft		Duplex:	55	60	Sedimentation
WQ buffer zone:	None	100 ft	150 ft		Multifamily:	60	70	Filtration
					Commercial:	80	90	Filtration
Water Supply Suburban—Class I								
						% Impervious Cover		
Drainage area:	128 ac	320 ac	640 ac	18% impervious cover; no development over recharge zone	Residential:	30	40	Filtration
Critical WQ zone:	100 ft	200 ft	400 ft		Multifamily:	40	55	Filtration
WQ buffer zone:	100 ft	200 ft	300 ft		Commercial:	40	55	Filtration
Water Supply Suburban—Class II								
						% Impervious Cover		
Drainage area:	128 ac	320 ac	640 ac	30% impervious cover; no development over recharge zone	Residential:	40	55	Filtration
Critical WQ zone:	100 ft	200 ft	400 ft		Multifamily:	60	65	Filtration
WQ buffer zone:	100 ft	200 ft	300 ft		Commercial:	60	70	Filtration
Water Supply Suburban—Class III								
						% Impervious Cover		
Drainage area:	320 ac	640 ac	1,280 ac	30% impervious cover	Residential:	45	50	Filtration
Critical WQ zone:	100 ft	200 ft	400 ft		Duplex:	55	60	Filtration
WQ buffer zone:	100 ft	200 ft	300 ft		Multifamily:	60	65	Filtration
					Commercial:	65	70	Filtration
Water Supply Rural								
						Units/ac		
Drainage area:	64 ac	20 ac	640 ac	One unit per 3 acres; no development over recharge zone	Single-family:	0.5	1.0	–
Critical WQ zone:	100 ft	200 ft	400 ft		Cluster:	1.0	2.0	40% buffer[b]
WQ buffer zone:	100 ft	200 ft	300 ft					
						% Impervious Cover		
					Multifamily:	20	25	40% buffer
					Commercial:	20	25	40% buffer
					Planned:	50	50	40% buffer
					Retail:	50–60[c]	60–70	Filtration

[a] Net site area.
[b] Except in Lake Austin/Lake Travis, where filtration is required.
[c] Only at major intersections.
Conversion factors: ac = acre; 1 acre = 0.4046 ha; 1 ft = 0.3048 m.
Source: U.S. Environmental Protection Agency, *Stormwater Best Management Practice Design Guide*—Volume 1—*General Considerations* (2004).

Class I water supply watershed is restricted to 40% impervious unless the developer is able to use development rights transfers (see Table 12.6). In this case, the development can reach 55% impervious and still meet the requirements of the ordinance.

In addition to the restrictions on site percent imperviousness, developments in these watersheds are required to incorporate structural control practices. The acceptable control practices are sedimentation basins, filtration basins, and vegetative buffers as outlined in Table 12.6. Basins must be designed to capture, isolate, and hold at least the first 0.5 in. (13 mm) of runoff from contributing drainage areas. Also, nonstructural requirements serve to prevent pollution. These include limitations on the depth of cuts and fills, limitations on construction on steep slopes (greater than 15%), and limitations on the disturbance of natural streams including restrictions on the number of stream crossings. Temporary erosion controls, such as silt fences and rock berms, are required during construction.

In Austin, proposed new development plans are reviewed by separate, autonomous environmental review staff from other departments. This allows for a focused review that includes field surveys of projects in sensitive areas. Once the plans are approved, city inspectors monitor construction for compliance with the approved plans. Approximately 50% of the financing for reviews and inspections required by this ordinance comes from development permit fees. The fees vary depending on the development size and are higher in sensitive watersheds because of the increased review requirements. The rest of the expenses are covered through a drainage utility fund that consists of monthly service charges to the residents in the utility service area.

2) *Urban Watersheds Ordinance*. In 1991, the city council approved task force recommendations to include urban area watersheds among those covered by development ordinances. This ordinance, created in response to increased pollution in Town Lake due to urban runoff discharges, focuses on the urban watersheds not previously covered by the CWO. It requires the implementation of structural controls in new developments undergoing site plan review. All new residential, multifamily, commercial, industrial, and civic development in the urban watersheds is required to construct water quality basins (either sedimentation or filtration basins) or provide a cash payment to the city for use in an Urban Watersheds Structural Control Fund. Structural controls must be used to capture the first 0.5 in. (13 mm) of runoff from all contributing areas. The Urban Watersheds Structural Control Fund is used to retrofit and maintain structural controls where required in the urban watersheds. In addition to this requirement, new developments in the urban watersheds are required to provide for removal of floating materials from stormwater runoff through the use of oil/water separators or other practices. Redevelopment projects in the urban watersheds are also included in this ordinance, where structural controls and the removal of floatable materials are required. For redevelopment projects, the city has developed a Cost Recovery Program Fund to provide 75% of the cost of structural controls. These funds are allocated through the drainage utility fund.

   In urban watersheds, the critical WQ zone is the boundary of the 100-year floodplain and is generally located 50–400 ft. (15–120 m) from the waterway. As with the CWO, no development is allowed in the critical WQ zone.

3) *Other NPS control programs*. In addition to the CWO and the Urban Watersheds Ordinance, Austin has developed other ordinances designed to reduce NPS pollution from new developments and redevelopments. One of these, the Barton Springs Zone Ordinance, provides special protection to watersheds contributing to Barton Springs, a widely visited and used natural spring bathing area in Austin. This ordinance, created to be a nondegradation ordinance with specific performance requirements, includes definitions of waterways and development limits similar to those specified in the CWO. Only one- or two-family residential development with a density of 1 unit per 3 acres (1.21 ha) is allowed in the Barton Springs watershed transition zone, which extends up to 300 ft. (90 m) from the water body. In addition, new developments in the Barton Springs watershed must comply with the following requirements: reduced pollutant concentrations compared with the undeveloped conditions and discharge no greater than a specific maximum pollutant concentration after development. The city measures these requirements quarterly on each development through a developer-funded monitoring program. Additional NPS control programs in Austin are as follows:

   - Land Development Code: Enforces landscaping regulations and protects trees and natural areas.
   - Underground Storage Tank Program: Develops guidelines for underground storage of hazardous materials, permitting and inspection of these underground storage tanks, and investigation of problems and response to emergency situations.
   - Water Quality Retrofit Program: Involves engineering and building, with private sector participants, permanent controls for already developed areas and those that are producing stormwater runoff pollution problems for the city's key receiving waters.

- Water Quality Monitoring Program: Monitors and characterizes pollutants from various land uses and structural controls, monitors surface and groundwater quality, and develops water quality models and databases; also conducts specific studies on known NPS problems.
- *Household Chemical Collection Program:* Provides for safe disposal of hazardous materials and other wastes generated from household use. This program is currently located at a permanent site where collection events occur each year.
- Storm Sewer Discharge Permit Program: Involves permitting and regular inspection of industrial and commercial discharges to storm sewers and water courses.
- Emergency and Pollution Incident Response Program: Involves responding to emergency spills, general water pollution incidents, and citizen complaints related to water quality.
- Street Cleaning and Litter Collection Program: Provides regular street cleaning – nightly in the central business district, monthly on other major roads, and bimonthly in residential neighborhoods.
- Integrated Pest Management Program: Encourages application of the most environmentally safe pesticide techniques practicable for pest management in municipal operations.

12.7 Selection of Best Management Practices

Urban runoff problems, because of their diverse nature, need to be addressed through a combination of source control, regulatory, and structural BMPs. The selected combination needs to reflect the program goals and the priorities set during the assessment and ranking of existing problems. This section covers the BMP selection process, which uses the screened list of potentially applicable BMPs to develop and select the BMPs to be implemented.

To select BMPs, the alternatives typically are developed and compared to ensure that all options are considered and that the best possible plan is selected based on a predetermined set of selection criteria. Although a specific problem caused by a specific source might not require development of alternative BMP plans, for most programs that tend to deal with multiple sources and impacts, it is wise to investigate alternatives before selecting a final set of BMPs. This section first addresses development of alternative plans and then the selection of recommended BMPs; at the end of this chapter, Case Study 12.2 is presented on methods of BMP selection.

12.7.1 Alternatives Development

The alternatives can include various combinations of source control, regulatory, and structural BMPs. Source control and regulatory BMPs are often implemented across entire regions or jurisdictions. Structural BMPs can be directed at specific runoff problems/pollutant sources or implemented across geographic areas, including both structural BMPs for new development in currently undeveloped areas or for retrofit in already developed areas. To address fully the urban runoff problems in an area, BMPs from all of these categories are often required.

The method of developing alternatives begins with the screened list of appropriate control measures. Each BMP is then assessed for its ability to address the known and anticipated problems. As an example, preference might generally be given to BMPs that:

1) Address more than one problem or lead to meeting more than one goal.
2) Have lower construction and operating costs.
3) Are most effective in stormwater control.
4) Are more effective in removing pollutants of concern.
5) Emphasize pollution prevention rather than treatment.
6) Are likely to address future problems.
7) Concentrate on addressing the priority problems.

The assessment of individual BMPs results in alternatives based on implementing each BMP throughout the study area. The comparison of alternatives is then in effect a comparison of different BMPs. This approach yields useful data on system-wide implementation of particular BMPs. Although one type of BMP might not address the range of urban runoff problems or goals in a study area, an urban runoff problem might exist that a particular BMP is well suited to control. In this case, implementation of that BMP on a regional basis, with the BMPs strategically located by the municipality, can be more effective and more easily controlled than requiring each developer to implement that BMP for individual developments.

An example of this method of alternative development is the Henrico County, Virginia, regional stormwater detention program. Early in the process of developing a stormwater management plan, it was decided that, given the conditions existing in the watershed, regional detention basins would be used to control runoff pollution. Regional detention basins were chosen because they provide both flood and pollution control, have fewer site restrictions than other pollution control structures, and can be designed to accommodate expected new developments. Therefore, the major remaining decision in the program was the number, location, and size of the detention basins.

The above method leads to the development of alternative plans to address the urban runoff problems of concern. Although the actual contents of each alternative plan are site-specific and depend on the type of alternative evaluation to be conducted, some general guidelines for presenting the alternative plans can help in assessing them. Preliminary sketches, rough cost estimates, expected pollutant removals, and environmental effects can be included for each alternative so comparisons can be made.

12.7.2 BMP Selection Process

After the alternatives have been developed, they are compared using a decision process (Fig. 12.21) that evaluates the relative merits of each plan. Because of the complexity of urban runoff control problems, a number of factors must be considered in assessing alternative plans. These alternatives are represented in Fig. 12.21 as inputs to the decision process, and include analysis tools, design conditions, and decision factors. The analysis tools are those used to assess and rank the existing pollution problems. The design conditions are the set of conditions under which to compare the alternatives. The decision factors are the criteria used to compare the alternatives. All of these inputs are then used to evaluate the alternatives using one or more decision analysis methods. This section first describes each input to the decision analysis and then describes the various decision analysis methodologies that can be used to select BMPs that will comprise the urban runoff pollution prevention and control plan.

12.7.2.1 Analysis Tools

Analysis tools can include watershed models, receiving-water models, and ranking models. The analysis tools are used to project future conditions, given the alternatives being investigated. For example, the total pollutant loads for each alternative can be calculated (whether using a unit load method or complex stormwater management model), yielding one item of input information as the alternatives are compared. Similarly, the impacts to receiving waters can be assessed using these tools, to compare these effects before making a decision.

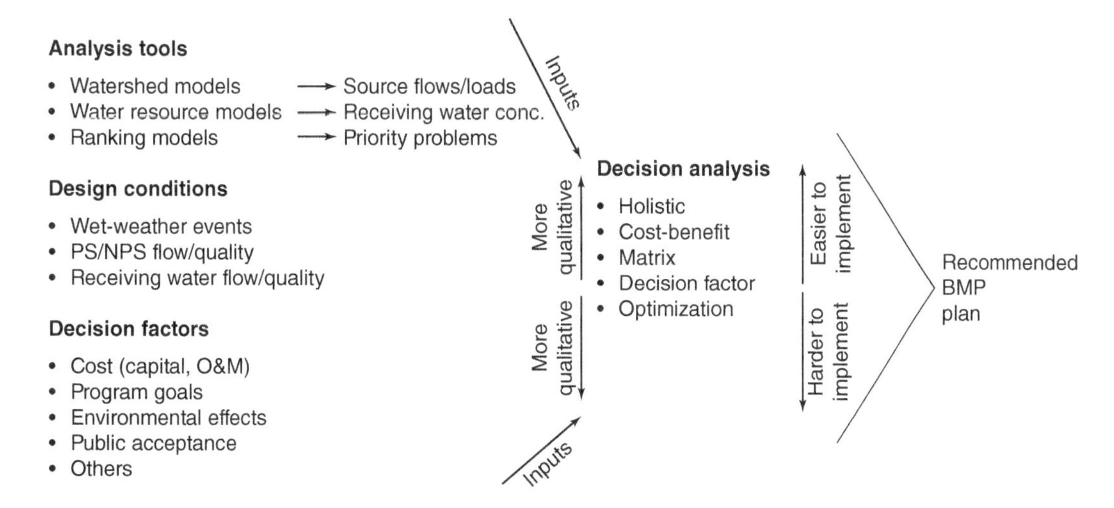

Figure 12.21 Conceptual Diagram of a BMP Selection Method (*Source:* U.S. Environmental Protection Agency)

12.7.2.2 Design Conditions

One major consideration in BMP selection is to determine appropriate conditions under which to compare the alternatives. These so-called design conditions are generally set up to reflect various future conditions, including future no-action conditions, which reflect future expected conditions with no new BMPs. Some important design conditions (each condition defined for specific future planning periods, say, 20 years) to be developed as part of an urban runoff pollution prevention and control plan include the following:

1) Population
2) Land use/expected development
3) Point source/NPS flows/concentrations
4) Background receiving water flows/concentrations.

Part of the comparison involves the selection of worst-case or critical conditions. In the case of a receiving water, this condition could be a summer low-flow period. In the case of urban runoff flow and load estimation, it often involves selection of wet-weather design conditions. These wet-weather conditions are often in the form of design storms. For example, runoff from a new development site might be required to meet preexisting conditions up to a 25-year frequency design storm. A state's CSO policy might require control up to a 1-year, 6-h design storm. Two significant concerns exist when developing wet-weather design conditions. One is distinguishing between wet-weather design criteria used for pollution control and those used for flood control. The second is the use of individual design storms versus multiple storms, continuous simulation, or probabilistic methods.

Historically, design storms have been used to size structures for flood control purposes. These facilities were often sized to control storms of 5-, 10-, or 25-year return periods or even longer. In contrast, BMPs used for wet-weather pollution control can be sized for much smaller storm events (e.g., 1-year storm or less), because most rainfall events (more than 90%) are smaller than a 1-year storm. Thus, a BMP sized for a 1-year storm would control more than 90% of the total runoff volume. Of course, many other BMP design factors are important (e.g., retention time and peak flow capacity), but design criteria appropriate for pollution control should be kept in mind. This also points out the need to consider multiple design conditions for dual-purpose (water quality and flood control) BMPs.

12.7.2.3 Decision Factors

An important step in BMP plan selection is to determine the important decision factors. The selection of these factors is site-specific and needs to be determined by the program team based on the characteristics of the watershed and the financial and personnel resources available. Typical decision factors are listed here:

1) *Cost*. One of the most important decision factors is the relative cost of each alternative (see Table 12.7). In cost assessment, costs of development and implementation of nonstructural BMPs and those for construction and operation of structural BMPs need to be considered. The program benefits such as those associated with restored resources also need to be considered. Costs should generally reflect the life-cycle cost of an alternative over the planning period and are usually easy to derive. The cost benefits associated with the implementation of a control plan, however, are usually more difficult to determine. For example, if an urban runoff control plan is designed to reduce the discharge of fecal coliform to a closed shellfish area, monetary benefits are derived from opening these beds. While analysis of these benefits can be difficult, they should be included in determining total program costs.
2) *Meeting program goals*. Alternatives are also assessed on their ability to meet program goals, including the control of major sources and effects on priority watersheds. Since at this stage in a program, the goals have been reassessed and expanded on a number of times, a large number of specific goals might exist, and each alternative might not meet all of the program goals. Preference generally is given to alternatives that address the most goals or the most important goals. Priority resources and pollution sources should be the focus of the selected alternative.
3) *Operability*. The decision factors included here take into consideration the reliability of structural controls, the reliance of the alternative plan on existing structures, and the number of structures included in the alternative. Operability is a measure of a system's complexity. Complicated systems and plans might be difficult or expensive to implement and operate; these factors are, therefore, taken into consideration in the BMP selection. Typically, this decision factor favors source control and regulatory practices that do not have the level of complexity and possible operational problems of structural controls.

Table 12.7 Capital Cost Functions for Selected Stormwater Runoff Controls in 2022 USD[a]

Management Control Item	Function	Explanatory Variable
Corrugated metal pipe	$C = 0.784\ D^{1.30}L$	D = diameter, cm L = length, m
Reinforced concrete pipe	$C = 0.203\ D^{1.63}L$	D = diameter, cm L = length, m
Manholes	$C = 2170\ H^{0.93}$	H = manhole height, m
Surface storage	$C = 2.26 \times 10^6\ V^{0.83}$	V = volume of storage, ML
Deep tunnels	$C = 3.21 \times 10^6\ V^{0.80}$	V = volume of storage, ML
Detention basins	$C = 3.26 \times 10^4\ V^{0.69}$	V = volume of storage, ML
Retention basins	$C = 3.35 \times 10^4\ V^{0.75}$	V = volume of storage, ML
Infiltration trenches	$C = 2207\ V^{0.63}$	V = volume of voids, m^3
Infiltration basins	$C = 267\ V^{0.69}$	V = volume of basin, m^3
Sand filters	$C = K_1^b A$	A = impervious surface, ha
Grassed swales	$C = K_2^c L$	L = length of swale, m

Source: U.S. Environmental Protection Agency.
[a] None include cost of land acquisition.
[b] K_1 is a constant ranging from 39,680 to 81,840.
[c] K_2 is a constant ranging from 25 to 68.
Conversion factors: 1 acre = 0.4046 ha; 1 ft = 0.3048 m; 1 in. = 2.54 cm = 25.4 mm; 1 MG = 3.785 ML, 1 ft^3 = 0.02832 m^3.

4) *Buildability*. This decision factor is directed primarily at the selection of structural BMPs. Taking into consideration the various aspects of construction, the criteria investigated under this category include the site requirements, extent of disruption, and degree of construction difficulty. When relying on complex structural controls, difficulties inherent in construction and future maintenance might need to be overcome. Although not a consideration in source control and regulatory control practice, this factor can be very important for structural controls.
5) *Environmental effects*. Implementing urban runoff pollution control plans can affect the environment both positively and negatively. The positive effects on resources result from the removal of pollution sources. Resources that can be positively affected include water resources, aquatic animal and plant life, wildlife, and wetlands. The negative environmental effects, which can include aesthetic problems, cross-media contamination, the loss of usable land, wetlands impacts, and many others, must also be considered in the assessment.

   The importance of this decision factor is becoming more widely recognized. There seems to be a shift away from viewing urban runoff control structures only on their pollution control ability. Incorporating structures into new developments or retrofitting them in existing areas can gain wider acceptance if additional aesthetic qualities are considered. For example, unvegetated aboveground infiltration basins or dry ponds are generally not attractive elements of the environment and could serve as insect breeding grounds. Natural-looking wet ponds or vegetated wetlands, however, can be incorporated into the environment and even serve to improve esthetics. These issues can greatly affect public acceptance.
6) *Institutional factors*. This decision factor relates to existing governmental structures, legal authority, and implementation responsibilities. To implement alternatives, the logistical resources must be in place, and the proper authority to pass and enforce regulatory practices must exist. If the proper authority does not exist, an analysis of attaining it must be undertaken. In addition to these considerations, the team should investigate existing urban runoff programs in the community, region, or state. Often, cost savings can be realized and total program efforts can be reduced by taking advantage of material and data compiled during these existing programs.
7) *Public acceptance*. In many instances, the public will be responsible for at least a portion of the funding required to implement the recommended plan. Public reaction to the urban runoff control plan should, therefore, be assessed through the use of public meetings. Measuring public acceptance can be difficult, but can be important to the overall success of a program.

8) *Other decision factors*. Additional decision factors – such as maintainability, level of pollution control, or size requirements – can be included in the assessment of alternative plans if they are more important than those discussed above.

Once the final decision factors have been chosen and applied to the alternative plans, the plans can be assessed by applying a decision analysis tool. Methods for conducting this decision analysis are presented below.

12.7.2.4 Decision Analysis Methods

Assessing alternatives takes into account a variety of factors, both quantitative and qualitative. The type of assessment conducted in these programs, which involves an integration and comparison of these factors, is an example of multiattribute decision making and can be performed with various decision analysis methods.

The following decision analysis methods, which are listed in order from the most qualitative to the most quantitative, can be utilized:

1) Holistic
2) Cost/benefit ratios
3) Matrix comparisons
4) Decision factor analysis
5) Optimization.

Holistic. This approach is qualitative and relies on certain basic facts, intuition, and professional judgment. One key deciding factor (e.g., cost) can guide the process. Given the inherent complexity of assessing alternative urban runoff control plans and the large number of available inputs to the decision, this approach is usually oversimplified. Selecting an appropriate plan from the developed alternatives will generally require an assessment of multiple factors and should be done in as quantitative a manner as is reasonably possible.

Cost/Benefit Ratios. The relative value of different alternatives can be measured using cost/benefit ratios, such as cost per pound of pollutant removed or cost per day of effect on resources. This approach can be used as a tool to determine which BMP should be used first. For example, if it is determined that reducing solids using source control measures costs less per pound than using a structural BMP, then source control measures should be utilized first. Since the unit cost of source control measures increases with the amount of solids eliminated, the cost per pound of solids removed increases with the number of pounds removed. The extent to which source control measures should be used for pollutant removal is then given by the point at which the marginal cost/benefit ratio (i.e., change in cost/change in benefit) becomes larger than that of another alternative.

Another advantage of the cost/benefit ratio approach is that it allows use of the knee-of-the-curve methodology, which seeks to determine the point in the cost/benefit curve where the marginal cost to achieve a marginal benefit becomes significantly higher. This factor is measured by the marginal cost/benefit ratio defined above. Fig. 12.22 shows an example of this methodology where the cost effectiveness drops dramatically as practices are implemented to reduce lake standards exceedances to below 10 days per year.

The cost/benefit ratio approach, however, is limited by the number of cost/benefit ratios that can be conveniently considered simultaneously. To represent the different elements of a complex issue better, where some benefits might be counterbalanced by some detriments, multiple costs and benefits must be considered.

Matrix Comparison. Matrix comparison, a common decision-making method used in facilities planning and siting, is illustrated in Table 12.8. Environmental impacts in Table 12.8 can be divided into short-term construction-related impacts and long-term operational impacts. The matrix comparison approach is also applicable to the assessment of urban runoff control alternatives. This approach involves preparing a matrix that compares alternatives against selected decision factors, both quantitative and qualitative. Where possible, numerical values are given to compare the alternatives and, for qualitative factors, subjective comparisons are used (such as poor, fair, good, and excellent).

Decision Factor Analysis. This is a matrix approach, which further quantifies the decision factors by using weighting methods. In this approach, quantitative factors are used to eliminate the subjective comparisons required in other matrix approaches. These criteria should be

- Nondominant – no criterion should be dominant.
- Complete – no pertinent information should be left out.
- Scorable – criteria cannot be vague, since they must be weighted clearly.
- Independent – criteria should not overlap each other.

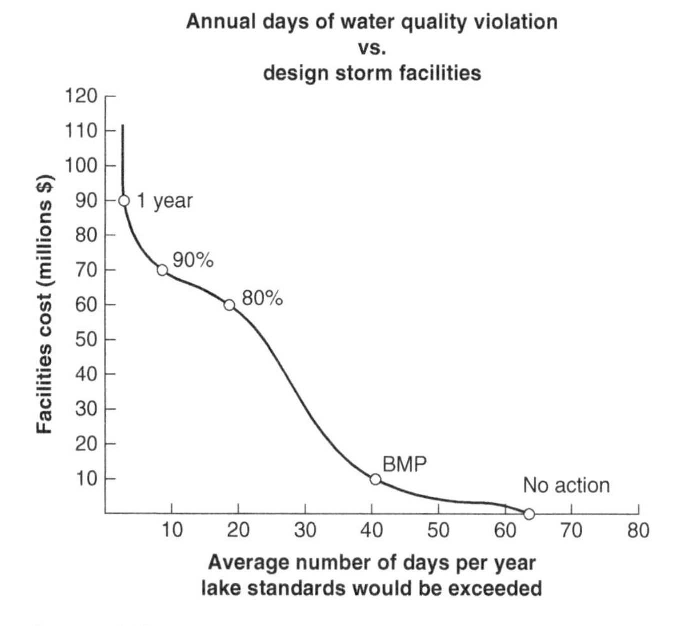

Figure 12.22 Sample Cost/Benefit Ratio Curve (*Source:* U.S. Environmental Protection Agency)

Table 12.8 Example Matrix Comparison

	Alternatives			
Type of Impact	**#1**	**#2**	**#3**	**#4**
Monetary Cost, $				
Capital cost	+ +	–	–	+
Annual O&M cost	–	+	+	+
Cost per household unit	+	+	+	–
Environmental Impact				
Cultural resources	0	+	– –	–
Floodplains and wetlands	+	–	–	–
Agricultural lands	–	+ +	+	–
Coastal zones	+	–	+	+
Wild and scenic rivers	0	+	–	+
Fish and wildlife	+ +	–	+ +	+
Endangered species	+	–	+	+
Air quality	0	0	–	+
Water quality and uses	+	+ +	+	+
Noise, odor, aesthetics	–	+	–	– –
Land use	0	+	–	+
Energy requirements	0	–	0	–
Recreational opportunity	+	+ +	–	–
Reliability	–	–	+	–
Implementability	+	+	– –	+

Legend: + + = significant beneficial impact; + = minimal beneficial impact; 0 = no impact; – = minimal adverse impact; – – = significant adverse impact.
Source: U.S. Environmental Protection Agency. *Stormwater Best Management Practice Design Guide* – Volume 1 – *General Considerations* (2004).

Weights are then generated for each decision factor. These weights must have a common scale, and the relative importance of each factor to the decision should be reflected in the weights. An example is the BMP selection approach in Case Study 12.2 at the end of this chapter. The major difference between this approach and the matrix approach outlined above is that, in this approach, the decision factors must be quantitative. Therefore, subjective comparison terms, such as good or fair, cannot be utilized. The decision factors must be able to be described by values that can be summed. Variations on this type of approach and various decision support software can facilitate the conduct of these analyses.

Optimization. Optimization, a widely used method of quantitative decision making, involves formulating a problem as the maximization (or minimization) of an objective function, subject to a series of constraints. In linear optimization, both the objective function and the constraints must be linear functions of the decision variables. Various methods are available for finding the optimum set of decision variables and several software packages can perform the analyses. These methods are summarized in basic textbooks on optimization.

For plan selection, the objective function can be cost or a more complicated function of cost, benefits, and detriments. Examples of benefits that could be included are gallons of discharge removed, pounds of pollutants removed, and days of beach closure avoided. A multifactor objective function can account for trade-offs among costs, benefits, and detriments by incorporating relative weights for each factor:

$$F = \Sigma\, a_i y_i \tag{12.1}$$

where F is objective function, a_i is weight and conversion factor, and y_i is cost/benefit factor.

All terms in the preceding equation must have the same dimension (e.g., dollars) so that weights also incorporate a conversion factor. The optimization process then consists of maximizing the objective function, by optimally selecting the values of the decision variables on which the different factors depend. Then, each cost/benefit factor, y_i, must be expressed linearly in terms of each of the decision variables x:

$$y_i = \Sigma\, b_i x_i \tag{12.2}$$

where b_i is a different weight or conversion factor.

This relationship is relatively easily established for cost (such as life-cycle cost), but more difficult for other factors, such as pounds of pollutant removed or days of beach closure. For these types of factors, models need to be applied with different values of the decision variable and then straight line fitted to the result. Constraints must also be established as linear functions of the decision variables. Possible constraints are the maximum number of excursions of standards per year or the maximum amount of pollutant reduction achievable given background conditions. Once the objective function and constraints are defined, various algorithms and software packages are available to determine the combination of decision variables maximizing the objective function.

A major problem with this approach is that many relationships pertaining to BMP selections are nonlinear. Qualitative factors are also difficult to incorporate in the process, especially in the form of linear functions of the decision variables. Nonlinear optimization, while accounting for the nonlinear dependence of various factors, is mathematically complex. It also tends to suffer from the same types of drawbacks as linear programming because it is not effective for problems that include qualitative factors.

12.7.2.5 Determination of Appropriate Decision Analysis Approaches

Matrix comparison and decision factor analysis approaches are typically best suited to BMP selection. Such approaches rely on the analytical tools available to analyze the system and on the best professional judgment of those assessing the alternatives. Given specific problems that can be quantified, optimization could be tried. Most BMP selection projects involving urban runoff, however, would be too complex. If the problems being addressed are simple, then the holistic or cost/benefit ratio techniques can be utilized. These simple, qualitative approaches can also be implemented as first approximations for plan assessments whose final results must be made using more complex approaches. In summary, an appropriate decision analysis method or methods must be selected that reflect:

1) The complexity of the problems and the plans to address them
2) The data needs of each method and the ability to obtain the required data
3) The financial and personnel resources available to conduct the assessment.

The selection of BMPs to control urban runoff pollution is difficult and can best be performed by undertaking a systematic assessment process, aided by the use of analytical tools and the selection of appropriate design conditions and decision factors. Because of the qualitative nature of some inputs to the decision, subjective comparisons among the alternative plans typically are necessary. The process outlined in this section is a guide for decision making, but cannot account for all possible circumstances. Professional judgment and care are needed in determining the methods for developing alternatives, the decision factors to be employed, and the decision analysis method to utilize. Once these choices have been made and the BMP plan has been selected, the urban runoff pollution prevention and control plan can be developed in more detail so that it can be implemented.

The following case study provides an example of BMP selection approaches used by the state of Maine for the development and implementation of a major runoff pollution prevention and control plan.

Case Study 12.2 Maine Department of Environmental Protection BMP Selection Matrix

To address stormwater and NPS pollution control in areas of new development, the Maine Department of Environmental Protection (ME DEP), USA, has developed a method to select BMPs. The method, which is presented in a state guidance document, is based on the following information:

- Development land use type and size
- Receiving-water type (e.g., estuary, wetland river, or stream)
- Watershed priority (either priority or nonpriority)
- Erosion and sediment control target or level to achieve
- Stormwater quality control target or level to achieve
- Erosion and sediment control options and treatment level codes
- Stormwater quality control options and treatment level codes.

To implement the BMP selection method, ME DEP developed a series of eight matrices, two matrices for each receiving-water type (i.e., estuary, wetland, river, and stream). One matrix is applied to development in designated priority watersheds and the other is applied to development in nonpriority watersheds. A priority watershed list has been developed by ME DEP based on environmental sensitivity, local support for water quality, and importance of the watershed to the state. Example matrices for priority and nonpriority estuary watersheds are shown in Tables 12.9 and 12.10.

Each matrix has two major components, which are broken down by land use type: an erosion and sediment control level to achieve and a stormwater quality level to achieve. The level to achieve for a given combination of land use and receiving-water category is a relative, qualitative measure of the impact of storm runoff pollution. It ranges from 1 to 5, with 1 being the lowest impact and 5 being the greatest impact. For example, a multihousing development proposed for a priority estuary watershed is given an erosion and sediment level to achieve a level 2 rating and a water quality level to achieve a 3 rating. By comparison, a small residential development in the same priority watershed is given an erosion control level to achieve a level 1 rating and a water quality level to achieve a 1 rating. In all cases, the levels to achieve for priority watersheds are greater than or equal to those for nonpriority watersheds.

Each matrix also addresses the types of BMPs that can be implemented for pollution control. ME DEP selected a number of BMPs and assigned each a treatment level code based on the expected level of pollutant removal. The treatment level code is a relative, qualitative measure designed to indicate the relative pollutant removal expected from various BMPs. Treatment level codes range from 1 to 3, with 1 providing the lowest level of control and 3 providing the greatest level of control. The BMPs and their treatment level codes are shown in Table 12.11. As indicated, various designs for each BMP are given different treatment level codes. For example, a 50-ft (15-m) buffer is given a treatment level code of 1, a 125-ft (38-m) buffer is given a treatment level code of 2, and a 200-ft (61-m) buffer is given a treatment level code of 3.

For a proposed development to be approved, the sum of treatment level codes for the proposed BMPs must be greater than or equal to the level to achieve. For example, if a multihousing unit development is proposed for a priority estuary (erosion level to achieve level 2 and water quality level to achieve level 3), the developer could implement erosion and sediment controls (treatment level 2) and a combination of a swale (treatment level 1) and an infiltration system (treatment level 2). Additional combinations could also be implemented as long as the total treatment level provided is

Table 12.9 Priority Estuary Stormwater Control Matrix

Land Use Category	Erosion and Sediment Level to Achieve	Erosion and Sediment Controls	Water Quality Level to Achieve	Storm Water Controls
Low-density residential, >2 ac/lot	1	Erosion and sediment 1	1	Buffer 1
High-density residential, <2 ac/lot	2	Erosion and sediment 2	3	Buffer 1 or 2 Wet pond 2 Infiltration 1 or 2 Created wetland 2
Commercial, <1 ac disturbed	1	Erosion and sediment 1	1	Buffer 1
Commercial, 1–3 ac disturbed	1	Erosion and sediment 1	2	Buffer 1 or 2 Infiltration 1 Swale 1
Commercial, > 3 ac disturbed	2	Erosion and sediment 2	4	Buffer 1 or 2 Infiltration 1 or 2 Created wetland 2 Wet pond 2 or 3 Fertilizer control 1 Shallow impoundment 1
Intensive-use space (e.g., golf courses, nurseries)	2	Erosion and sediment 2	5	Buffer 1 or 2 Fertilizer control 1 Pesticide control 1 Created wetland 2 or 3 Wet pond 2 or 3
Multihousing users	2	Erosion and sediment 2	3	Buffer 1 or 2 Fertilizer control 1 Pesticide control 1 Created wetland 2 Wet pond 2 Infiltration 1 or 2
Industrial <1 ac disturbed	1	Erosion and sediment 1	1	Buffer 1 Swale 1
Industrial 1–3 ac disturbed	1	Erosion and sediment 1	2	Buffer 1 or 2 Swale 1
Industrial >3 ac disturbed	2	Erosion and sediment 2	5	Buffer 1 or 2 Swale 1 Created wetland 2 or 3 Wet pond 2 or 3

Conversion factor: ac = acre; 1 acre = 0.4046 ha = 0.004046 km^2.
Source: U.S. Environmental Protection Agency (2004).

greater than or equal to the total level to achieve. ME DEP has also recommended that at least one vegetative BMP be implemented unless the site is already 100% impervious. The specified vegetative BMPs are buffers, grassed swales with level spreaders, and swales.

The BMP selection system success depends on the ability to establish levels to achieve that adequately protect water bodies in new developments. It will also depend on the ability of treatment level codes to quantify the effectiveness of the identified control measures. Thus, the system is a technology-based approach for erosion and sediment control, as well as for stormwater pollution control.

This method of BMP selection requires extensive upfront work to develop the matrices and BMP levels of treatment. Once these have been developed, however, this method provides a simple and direct technology-based approach to BMP selection.

Table 12.10 Nonpriority Estuary Stormwater Control Matrix

Land Use Category	Erosion and Sediment Level to Achieve	Erosion and Sediment Controls	Water Quality Level to Achieve	Storm Water Controls
Low-density residential, >2 ac/lot	1	Erosion and sediment 1	1	Buffer 1
High-density residential, <2 ac/lot	2	Erosion and sediment 2	2	Buffer 1 or 2 Infiltration 1
Commercial, <1 ac disturbed	1	Erosion and sediment 1	1	Buffer 1
Commercial, 1–3 ac disturbed	1	Erosion and sediment 1	1	Buffer 1
Commercial, > 3 ac disturbed	2	Erosion and sediment 2	2	Buffer 1 or 2 Infiltration 1 Swale 1 Shallow impoundment 1
Intensive-use space (e.g., golf courses, nurseries)	2	Erosion and sediment 2	3	Buffer 1 or 2 Infiltration 1 or 2 Fertilizer control 1 Created wetland 2 Wet pond 2
Multihousing users	2	Erosion and sediment 2	2	Buffer 1 or 2 Infiltration 1
Industrial <1 ac disturbed	1	Erosion and sediment 1	1	Buffer 1 Swale 1
Industrial 1–3 ac disturbed	1	Erosion and sediment 1	2	Buffer 1 or 2 Swale 1
Industrial >3 ac disturbed	2	Erosion and sediment 2	4	Buffer 1 or 2 Swale 1 or 2 Created wetland 2 or 3 Wet pond 2 or 3

Conversion factor ac = acre; 1 acre = 0.4046 ha = 0.004046 km^2.
Source: U.S. Environmental Protection Agency, (2004).

Table 12.11 Summary of BMP Treatment Level Codes

BMPs	Level of Treatment
Erosion and Sediment Control	
One line of erosion control	1
Two lines of erosion control	2
Nongrassed Buffers	
50 ft	1
125 ft	2
200 ft	3
Infiltration Systems	
Single system	1
Multiple systems	2

(*Continued*)

Table 12.11 (Continued)

BMPs	Level of Treatment
Wet Ponds	
Single-pond system holding 2.5 in. of runoff	2
Double-pond system each pond holding 2.5 in. of runoff	3
Created Wetlands	
Single created wetland	2
Two created wetlands	3
Other BMPs	
Swales	1
Shallow impoundments	1
Street cleaning	1
Fertilizer application control	1
Pesticide use control	1
Grassed swales with level spreaders	1
Reverting land (i.e., allowing currently impervious land to be a vegetative buffer)	1

Conversion factors: 1 in. = 25.4 mm; 1 ft = 0.3048 m.
Source: U.S. Environmental Protection Agency, *Stormwater Best Management Practice Design Guide* – Volume 1 – *General Considerations* (2004).

Problems/Questions

12.1 What is the impact on hydrology expected for developing watershed areas into urbanized areas?

12.2 The development of watershed areas into urbanized areas exerts a large impact on the hydrology. How do the changes in hydrology affect the receiving waters?

12.3 What are the planning elements for runoff management?

12.4 The planning process for runoff management consists of several major components. List and briefly explain these components.

12.5 Briefly describe structural and nonstructural controls for best management of stormwater runoff.

12.6 What are the initial screening criteria for structural control practices?

12.7 What is meant by detention facilities for stormwater runoff? Briefly describe their function.

12.8 Differentiate between detention and retention (infiltration) facilities as stormwater management methods.

12.9 Explain the importance of increasing vegetative cover in urban areas on stormwater runoff.

12.10 What are the aims for the wastewater pretreatment programs?

12.11 An important step in BMP plan selection is to determine the important decision factors. The selection of these factors is site-specific and needs to be determined by the program team based on the characteristics of the watershed and the financial and personnel resources available. Discuss these typical decision factors.

12.12 Assessing alternatives takes into account a variety of factors, both quantitative and qualitative. The type of assessment conducted in these programs, which involves an integration and comparison of these factors, is an example of multiattribute decision making and can be performed with various decision analysis methods. List the decision analysis methods in order from the most qualitative to the most quantitative.

Bibliography

American Society of Civil Engineers, *A Guide for Best Management Practice (BMP) Selection in Urban Developed Areas*, Reston, VA, 2001.

Association of State and Interstate Water Pollution Control Administrators, America's Clean Water: The States' Nonpoint Source Assessment, Washington, DC, 1985.

California Stormwater Quality Association, *California Stormwater Best Management Practice Handbook—New Development and Redevelopment*, 2003

Camp Dresser & McKee, State of California Stormwater Best Management Practice Handbooks, Prepared for California State Water Quality Control Board, 1993.

Center for Watershed Protection, *Impacts of Impervious Cover on Aquatic Systems*, 2003. Retrieved 2010 from http://www.cwp.org/Store/guidance.htm.

City of Austin, Texas, Removal Efficiencies of Stormwater Control Structures, Environmental Resource Management Division, Environmental and Conservation Services Department, Austin, TX, 1990.

City of Portland BES, Stormwater Management Manual, Portland Bureau of Environmental Services, Portland, OR, September 2004.

Delaware Department of Natural Resources, *Sand Filter Design for Water Quality Treatment, Dover, DE*, 1991.

Freedman, P. L., and Marr, J. K., Design Conditions for Wet Weather Controls, in Proceedings Water Environment Federation Specialty Conference, Control of Wet Weather Water Quality Problems, Indianapolis, May 31–June 3, 1992.

George, T. S., and Hartigan, J. P., Regional Detention Planning for Stormwater Management: Model for NPDES management programs, in Proceedings Water Environment Federation Specialty Conference, Control of Wet Weather Water Quality Problems, Indianapolis, May 31–June 3, 1992.

Georgia, Georgia Stormwater Management Manual, Volume 1: Stormwater Policy Handbook, first ed., August, 2001a.

Georgia, *Georgia Stormwater Management Manual, Volume 2: Technical Handbook*, August 2001b

Livingston, E., McCarron, E., Cox, J., and Sanzone, R., The Florida Development Manual: A Guide to Sound Land and Water Management, Florida Department of Environmental Regulation, Stormwater/Nonpoint Source Management Section, 1988.

Maryland Department of Natural Resources, Maryland Water Resources Authority Minimum Water Quality and Planning Guidelines for Infiltration Practices, Massachusetts Department of Environmental Protection, 1986.

Minnesota Pollution Control Agency, Protecting Water Quality in Urban Areas, St. Paul, MN, 1989.

Monks, J. G., Operations Management: Theory and Problems, 3rd ed., McGraw-Hill, New York, 1987.

New York State Department of Environmental Conservation, *New York State Stormwater Management Design Manual, Albany, NY*, August 2003.

Pitt, R., Source Loading and Management Model: An Urban Nonpoint Source Water Quality Model, Volume 1: Model Development and Summary, University of Alabama, Birmingham, 1989.

Schueler, T., Controlling Urban Runoff: A Practical Manual for Planning and Designing Urban BMPs, Metropolitan Washington Council of Governments, 1987.

Schueler, T., 2005, An Integrated Framework to Restore Small Urban Watersheds, Urban Subwatershed Restoration Manual No. 1. Retrieved 2010 from http://www.cwp.org/Store/usrm.htm#3.

Schueler, T., and Holland, H., *The Practice of Watershed Protection: Techniques for Protecting our Nation's Streams, Lakes, Rivers, and Estuaries*, Center for Watershed Protection, 2000. Retrieved 2010 from http://www.cwp.org Store/guidance.htm.

Schueler, T., Hirschman, D., Novotney, M., and Zielinski, J., 2007, Urban Stormwater Retrofit Practices, Urban Subwatershed Restoration Manual No. 3. Retrieved 2010 from http://www.cwp.org/Store/usrm.htm#3.

Stormwater Management, *Volume 2: Stormwater Technical Handbook, Boston, March*, 1997.

Tchobanoglous, G., Burton, F. L., and Stensel, H. D., Wastewater Engineering: Treatment and Reuse, McGraw-Hill, New York, 2002.

U.S. Environmental Protection Agency, *Urban Stormwater Management and Technology: Update and Users' Guide*, EPA-600/8-77-014, Cincinnati, OH, September 1977.

U.S. Environmental Protection Agency, Demonstration of Nonpoint Pollution Abatement through Improved Street Cleaning Practices, EPN600/2-79/161, National Technical Information Service NTIS P880-108988, Washington, DC, 1979.

U.S. Environmental Protection Agency, *Results of the Nationwide Urban Runoff Program (NURP), Vol. 1, Final Report*, Washington, DC; National Technical Information Service, NTIS PB84-185552, Springfield, VA, 1983.

U.S. Environmental Protection Agency, Construction Grants 1985, EPN430109-84/004, Washington, DC, 1985.

U.S. Environmental Protection Agency, *Setting Priorities: The Key to Nonpoint Source Control, EPA Office of Water Regulations and Standards*, Washington, DC, 1987.

U.S. Environmental Protection Agency, *Goals and Objectives for Nonpoint Source Control Projects in an Urban Watershed*, Nonpoint Source Watershed Workshop Seminar, EPA/625/4-1/027, Cincinnati, OH; National Technical Information Service NTIS PB92-137504, Springfield, VA, 1991.

U.S. Environmental Protection Agency, *Stormwater Management for Construction Activities: Developing Pollution Prevention Plans and Best Management Practices*, EPN832/R-92/005, Washington, DC, 1992a.

U.S. Environmental Protection Agency, *Stormwater Management for Industrial Activities: Developing Pollution Prevention Plans and Best Management Practices*, EPN832/R-92/006, Washington, DC, 1992b.

U.S. Environmental Protection Agency, *The Use of Wetlands for Controlling Stormwater Pollution*, Chicago, IL, 1992c.

U.S. Environmental Protection Agency, *Investigation of Inappropriate Pollutant Entries into Storm Drainage Systems. Storm and Combined Sewer Control Program*, 1993a.

U.S. Environmental Protection Agency, *Urban Runoff Pollution Prevention and Control Planning*, EPA, 625/R-93-004, Washington, DC, September 1993b.

U.S. Environmental Protection Agency, *Costs of Urban Stormwater Control*, EPN600/R-02/021, Cincinnati, OH, 2002.

U.S. Environmental Protection Agency, *Stormwater Best Management Practice Design Guide, Volume 1, General Considerations*, EPA/600/R-04/121, Washington, DC, 2004a.

U.S. Environmental Protection Agency, *Stormwater Best Management Practice Design Guide, Volume 2, Vegetative Biofilters*, EPA/600/R-04/121A, Washington, DC, 2004b.

U.S. Environmental Protection Agency, *Stormwater Best Management Practice Design Guide, Volume 3, Basins Best Management Practices*, EPA/600/R-04/121B, Washington, DC, 2004c.

U.S. Environmental Protection Agency, National Water Quality Inventory, Report to Congress, EPN44014-90/003, Washington, DC, 1988, p. 1990.

Walesh, S. G., Urban Surface Water Management, John Wiley & Sons, New York, 1989.

Wang, L. K., Wang, M. H. S., and Shammas, N. K., Treatment of Wastewater, Storm Runoff and Combined Sewer Overflow by Dissolved Air Flotation and Filtration, Handbook of Advanced Industrial and Hazardous Wastes Management, CRC Press, Boca Raton, FL, 2018, pp. 577–610.

Washington State Department of Ecology, *Stormwater Management Manual for the Puget Sound Basin*, Olympia, WA, 1991.

Wisconsin Department of Natural Resources, Stormwater Best Management Practices, Runoff Management, in . Retrieved 2010 from http://dnr.wi.gov/runoff.

Woodward-Clyde Consultants, *Urban Targeting and BMPSelection: An Information and Guidance Manual for State NPS Program Staff Engineers and Managers*, Final Report, Santa Clara Valley, CA, 1990.

13

Hydraulics of Sewer Systems

13.1 Nature of Flow

Hydraulically, wastewater collection differs from water distribution in the following three essentials: (a) sewers, although most of them are circular pipes, normally flow only partially filled and hence as open channels, (b) tributary flows are almost always unsteady and often nonuniform, and (c) sewers are generally required to transport substantial loads of floating, suspended, and soluble substances with little or no deposition, on the one hand, and without erosion of channel surfaces, on the other hand. To meet the third requirement, sewer velocities must be self-cleansing yet nondestructive.

As shown in an earlier chapter, the design period for main collectors, interceptors, and outfalls may have to be as much as 50 years because of the inconvenience and cost of enlarging or replacing hydraulic structures of this nature in busy city streets. The sizing of needed conduits becomes complicated if they are to be self-cleansing at the beginning as well as the end of the design period. Although water distribution systems, too, must meet changing capacity requirements, their hydraulic balance is less delicate; the water must transport only itself, so to speak. It follows that velocities of flow in water distribution systems are important economically rather than functionally and can be allowed to vary over a wide range of magnitudes without markedly affecting system performance. In contrast, performance of wastewater systems is tied, more or less rigidly, to inflexible hydraulic gradients and so becomes functionally and economically important.

13.2 Flow in Filled Sewers

In the absence of precise and conveniently applicable information on how channel roughness can be measured and introduced into theoretical formulations of flow in open channels, engineers continue to base the hydraulic design of sewers, as they do the design of water conduits, on empirical formulations. Equations common in North American practice are the *Kutter–Ganguillett formula* of 1869 and the *Manning formula* of 1890. In principle, these formulations evaluate the velocity or discharge coefficient c in the *Chézy formula* of 1775 in terms of invert slope s (Kutter–Ganguillet only), hydraulic radius r, and a coefficient of roughness n. The resulting expressions for c are as follows:

$$c_{Kutter-Ganguiller} = \frac{(41.65 + 2.81 \times 10^{-3}/s) + 1.811/n}{(41.65 + 2.81 \times 10^{-3}/s)(n/r^{1/2}) + 1} \tag{13.1}$$

$$c_{Meanning} = 1.486\, r^{1/6}/n. \tag{13.2}$$

Of the two, Manning's equation is given preference in these pages, because it satisfies experimental findings fully as well as the mathematically clumsier Kutter–Ganguillet formula. Moreover, it lends itself more satisfactorily to algebraic manipulation and graphical representation.

Introducing Manning's c into the Chézy formula, the complete Manning equation reads:

$$v = (1.49/n)\, r^{2/3} s^{1/2} \quad \text{(US customary units)} \tag{13.3a}$$

Water and Wastewater Engineering: Hydraulics, Hydrology and Management, Volume 1, Fourth Edition.
Lawrence K. Wang, Mu-Hao Sung Wang, and Nazih K. Shammas.

Companion website: www.wiley.com/go/Wang/Waterandwastewater4e

where v is velocity, ft./s; n is the coefficient of roughness, dimensionless; r is hydraulic radius, ft.; and s is the slope of energy gradient, ft./ft. The following is an equivalent Manning formula using the SI units of (m/s), n (dimensionless), r (m), and s (m/m):

$$v = (1/n)\ r^{2/3}s^{1/2} \quad \text{(SI units)} \tag{13.3b}$$

For a pipe flowing full, the hydraulic radius becomes

$$r = A/P_w = \left[(\pi/4)D^2\right]/(\pi D) = D/4 \tag{13.4}$$

where r is hydraulic radius, ft or m; A is cross-section area, ft^2 or m^2; P_w is wetted perimeter, ft. or m; and D is pipe diameter, ft. or m.

Substituting for r into Eqs. 13.3a and 13.3b, the following Manning equations are obtained for practical engineering designs of circular pipes flowing full:

$$v = (0.59/n)(D)^{0.67}(s)^{0.5} \quad \text{(US customary units)} \tag{13.5a}$$

$$Q = (0.46/n)(D)^{2.67}(s)^{0.5} \quad \text{(US customary units)} \tag{13.6a}$$

where Q is flow rate, ft^3/s; v is velocity, ft./s; D is pipe diameter, ft.; s is slope of energy grade line, ft./ft.; and n is roughness coefficient, dimensionless. For SI measurements:

$$v = (0.40/n)(D)^{0.67}(s)^{0.5} \quad \text{(SI units)} \tag{13.5b}$$

$$Q = (0.31/n)(D)^{2.67}(s)^{0.5} \quad \text{(SI units)} \tag{13.6b}$$

where Q is flow rate, m^3/s; v = is velocity, m/s; D is pipe diameter, m; s is slope of energy grade line, m/m; and n is roughness coefficient, dimensionless.

Manning formula is seen to resemble the *Hazen–Williams formula*. Indeed, the Hazen–Williams equation could be used instead. Values of $V_0 = 1.49AR^{2/3}$, $Q_0 = 1.49AR^{2/3}$, and $1/V_0$, the reciprocal of $1.49R^{2/3}$, are listed in Appendix 7 to speed calculations. (Capital letters are chosen here to denote the hydraulic elements of conduits flowing full, and lowercase letters for partially filled sections.) The table is based on a generalization of Manning's formula in terms of the ratio $S^{1/2}/N$, where S/N^2 is, in a sense, the relative slope for varying conduit sizes and roughness coefficients. The ratio $S^{1/2}/N$ appears, too, in formulations for the flow of stormwaters over land and into street inlets.

Choice of a suitable *roughness coefficient* is of utmost importance. The following ranges in values are recommended by the American Society of Civil Engineers and the Water Environment Federation: (a) for vitrified-clay, concrete, asbestos-cement, plastic, and corrugated steel pipe with smooth asphaltic lining, a coefficient ranging from 0.010 to 0.015 for sewage, 0.013 being a common design value for sanitary sewers; (b) for the corrugated steel pipes often used in culverts, a coefficient of 0.018–0.022 when asphalt coatings and paving cover 25% of the invert section, or 0.022–0.026 for uncoated pipe with 1/2-in. corrugations; and (c) for lined open channel, a coefficient of 0.011–0.020. All but corrugated pipes show little difference between values suitable for the Manning and Kutter–Ganguillet equations.

The nomograms included in Fig. 13.1a and b and Appendix 15 show graphical solutions for the Manning equation. Knowing any two parameters for a given value of friction coefficient n, the other parameters can be determined. For example, for a given flow of 20 MGD (31 ft^3/s or 878 L/s) and a pipe to be set at a slope of 1.85%, the required pipe diameter ($n = 0.012$) is 36 in. (900 mm), which when full will have a flow velocity of 4.3 ft./s (1.31 m/s).

Minimum grades S and capacities Q of sewers ($N = 0.013$) up to 24 in. (600 mm) in diameter flowing full at velocities of 2.0, 2.5, 3.0, and 5.0 ft./s (0.60, 0.75, 0.90, and 1.5 m/s) are listed for convenience in Appendix 8.

Example 13.1 Determination of Flow Rate and Velocity in Sewer Using Design Equations

1) Given a 12 in. (304.8 mm) sewer, $N = 0.013$, laid on a grade of 4.05‰ (4.05 ft./1000 ft. or 4.05 m/1000 m), find its velocity of flow and rate of discharge.
2) Given a velocity of 3 ft./s (0.9144 m/s) for this sewer, find its (minimum) gradient for flow at full depth.

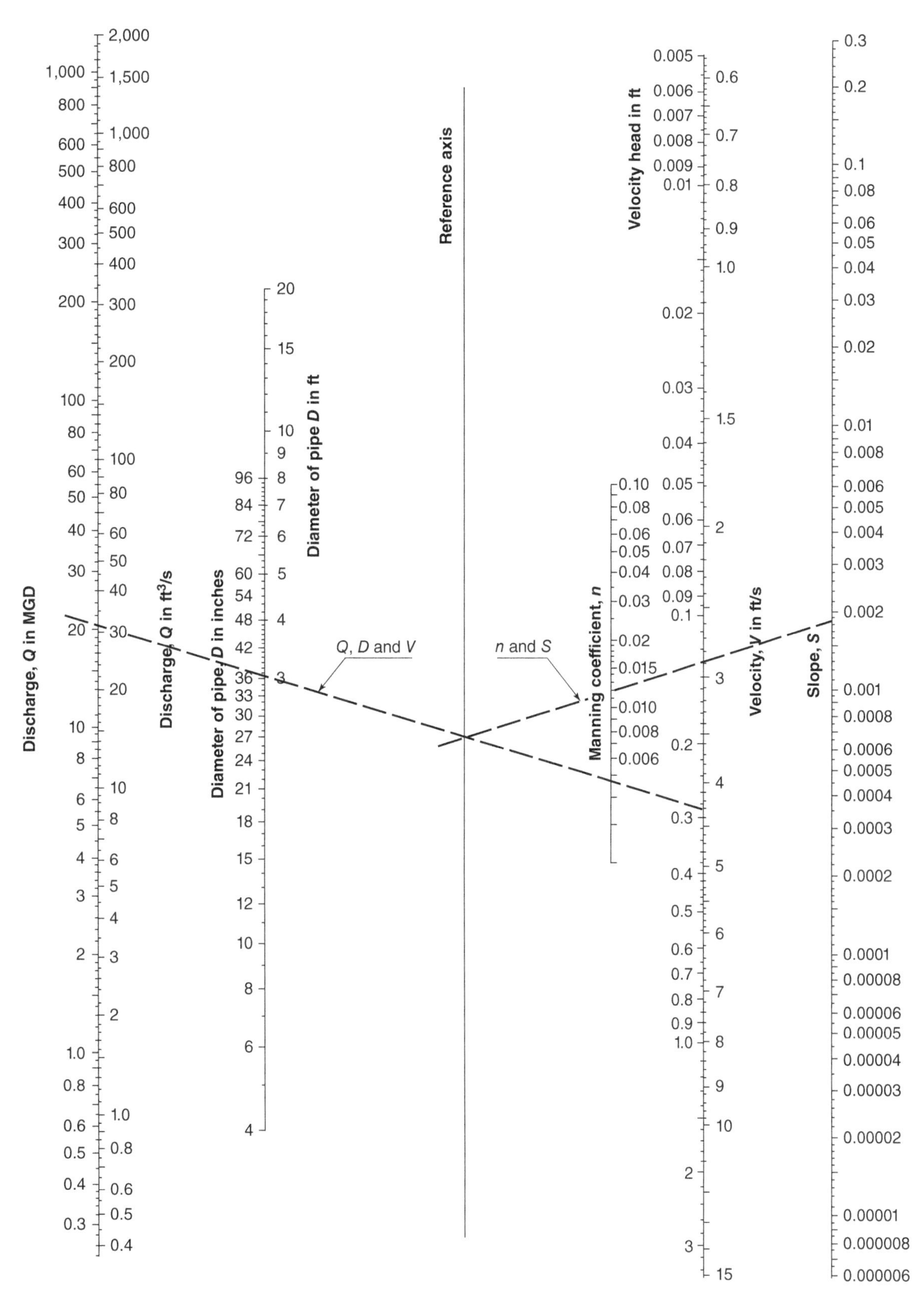

Figure 13.1a Nomogram for Solution of Manning's Equation for Pipes Flowing Full (English Units)

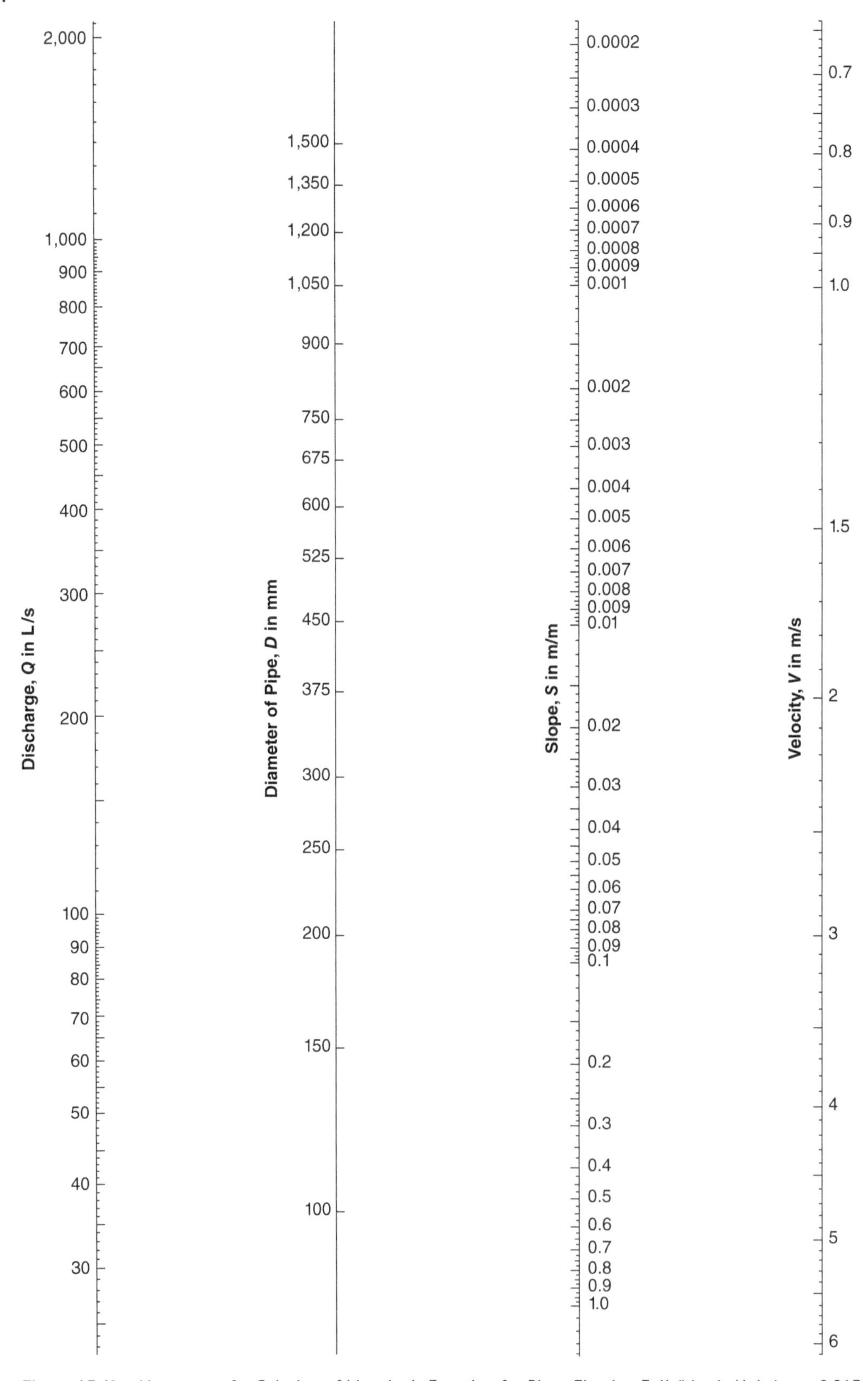

Figure 13.1b Nomogram for Solution of Manning's Equation for Pipes Flowing Full (Metric Units). n = 0.013

Solution 1 (US Customary System):

1) $n = 0.013$
   $D = 12$ in. $= 1$ ft
   $s = 0.00405$
   $v = (0.59/n)(D)^{0.67}(s)^{0.5}$
   $= (0.59/0.013)(1)^{0.67}(0.00405)^{0.5}$
   $=$ **2.89 ft/s**
   $Q = (0.46/n)(D)^{2.67}(s)^{0.5}$
   $= (0.46/0.013)(1)^{2.67}(0.00405)^{0.5}$
   $=$ **2.25 ft³/s**

2) From Eq. 13.5a, the following equation is derived by re-arrangement:
   $s = 2.87\ v^2 \times n^2 \times D^{-1.34}$
   $= 2.87(3)^2(0.013)^2(1)^{-1.34}$
   $= 0.004$ or 4 ‰

Solution 2 (SI System):

1) $n = 0.013$
   $D = 0.3048$ m
   $s = 0.00405$
   $v = (0.40/n)(D)^{0.67}(s)^{0.5}$
   $= (0.40/0.013)(0.3048)^{0.67}(0.00405)^{0.5}$
   $= (0.40/0.013)(0.4511)(0.0636)$
   $=$ **0.88 m/s**
   $Q = (0.31/n)(D)^{2.67}(s)^{0.5}$
   $= (0.31/0.013)(0.3048)^{2.67}(0.00405)^{0.5}$
   $= (0.31/0.013)(0.0419)(0.0636)$
   $= 0.064\ m^3/s$ or **64 L/s**

2) From Eq. 13.5b, the following equation is derived by re-arrangement:
   $s = 6.25\ v^2n^2D^{-1.34}$
   $= 6.25\ (0.9144)^2(0.013)^2(0.3048)^{-1.34}$
   $= 6.25\ (0.8361)(0.000169)(4.9140)$
   $= 0.004$ or 4 ‰

Example 13.2 Determination of Flow Rate, Velocity, or Slope in Sewer Using Nomograms

1) Given a 12-in. (304.8 mm) sewer, $N = 0.013$, laid on a grade of 4.05 ‰ (4.05 ft. per 1000 ft. or 4.05 m/1000 m) and assuming the sewer is flowing fullfind its velocity of flow and rate of discharge from (a) the table in Appendix 7, (b) the nomogram in Fig. 13.1a (US customary system), or Fig. 13.1b (SI system), and (c) the nomogram in Appendix 15
2) Given a velocity of 3 ft./s (0.9 m/s) for this sewer, find its (minimum) gradient for flow at full depth and flow rate, from (a) Appendix 8, (b) the nomogram in Fig. 13.1a (US customary system), or Fig. 13.1b (SI system), and (c) the nomogram in Appendix 15

Solution 1 (US Customary System):

1) First problem : Given sewer diameter $D = 12$ in, roughness coefficient $N = 0.013$, and slope $s = 0.00405$, determine velocity V and flow rate Q:
   a) Using the nomogram in Appendix 7 for solving the first problem: $s^{1/2}/N = (4.05 \times 10^{-3})^{1/2}/1.3 \times 10^{-2} = 4.90$ m the table in Appendix 7, the V and Q values are 0.5897 and 0.4632, respectively for $s^{1/2}/N = 1$; therefore, the real values of V and Q are obtained by multiplication of 4.90:

      $V = 4.90 \times 0.5897 =$ **2.89 ft/s**

      $Q = 4.90 \times 0.4632 =$ **2.27 ft³/s**

b) Using the nomogram of Fig. 13.1a for solving the first problem:
Figure 13.1a can be used for any roughness coefficient N values. Connecting the given values of roughness coefficient N (0.013) and slopes (0.00405) and drawing a straight line will intercept the Reference axis in the middle of Fig. 13.1a, and get an interception point. Connecting the obtained interception point on the reference axis and the given pipe diameter D (12 in.), and drawing another straight line will then obtain the answers of V (2.9 ft/s) and Q (2.3 ft^3/s)

c) Using the nomogram of Appendix 15 ($n = 0.013$) for solving the first problem:
Connecting the given values of slope s (0.00405) and pipe diameter D (12 in.) and drawing a straight line will obtain the answers of V (2.9 ft/s) and Q (2.3 ft^3/s).

2) Second problem : Given sewer diameter $D = 12$ in., roughness coefficient $N = 0.013$, and velocity $V = 3$ ft/s, determine slope s and flow rate Q:

a) Using the nomogram in Appendix 8 for solving the second problem:
Knowing the $N = 0.013$, sewer diameter $D = 12$ in., and velocity $V = 3$ ft/s; the answers from Appendix 8 are: slope $s = 4.37$ ‰, and flow rate $Q = 2.36$ ft^3/s.

b) Using the nomogram of Fig. 13.1a for solving the second problem
Connecting the given values of sewer diameter D (12 in.) and velocity (3 ft/s) , and drawing a straight line will determine the flow rate $Q = 2.36$ ft^3/s , and will intercept the reference axis in the middle of Fig. 13.1a, and get an interception point. Connecting the obtained interception point on the reference axis and the given manning roughness coefficient N (0.013) and drawing another straight line will then obtain the answers of slope s 0.00437 or 4.37 ‰.

c) Using the nomogram of Appendix 15 ($n = 0.013$) for solving the second problem:
Connecting the given values of pipe diameter D (12 in.) and velocity (3 ft/s) and drawing a straight line will obtain the answers of Q (2.36 ft^3/s) and slope s (0.0043 or 4.37 ‰).

$s = 4.4$‰

Solution 2 (SI System):

1) First problem: Given sewer diameter $D = 304.8$ mm, roughness coefficient $N = 0.013$, and slope $s = 0.00405$, determine velocity V and flow rate Q:

a) Using the nomogram in Appendix 7 for solving the first problem:
Appendix 7 is for the US customary system only, therefore, it is not applicable to the SI system determination.

b) Using the nomogram of Fig.13.1b for solving the first problem:
Connecting the given pipe diameter D (304.8 mm) and slope $s = 0.00405$, and drawing a straight line will then obtain the answers of V (0.86 m/s) and Q (60 L/s)

c) Using the nomogram of Appendix 15 ($n = 0.013$) for solving the first problem:
Connecting the given values of slope s (0.00405) and pipe diameter D (304.8 mm) and drawing a straight line will obtain the answers of V (0.86 m/s) and Q (60 L/s)

2) Second problem : Given sewer diameter $D = 304.8$ mm , roughness coefficient $N = 0.013$, and velocity $V = 0.9$ m/s, determine slope s and flow rate Q:

a) Using the nomogram in Appendix 8 for solving the second problem:
Appendix 8 is for the US customary system only, therefore, it is not applicable to the SI system determination.

b) Using the nomogram of Fig. 13.1b for solving the second problem
Connecting the given pipe diameter D (304.8 mm) and and velocity (0.9 m/s), and drawing a straight line will then obtain the answers of slope s (0.0045 or 4.5‰) and Q (62 L/s)

c) Using the nomogram of Appendix 15 ($n = 0.013$) for solving the second problem:Connecting the given values of pipe diameter D (304.8 mm) and velocity (0.9 m/s) and drawing a straight line will obtain the answers of Q (0.063 m^3/s = 63 L/s) and slope s (0.0045 or 4.5‰).

13.3 Limiting Velocities of Flow

Wastes from bathrooms, toilets, laundries, and kitchens are flushed into sanitary sewers through *house* or *building sewers*. Sand, gravel, and debris of many kinds enter storm drains through curb and yard inlets. Combined sewers carry mixtures of the two. Heavy solids are swept down sewer inverts like the *bed load* of streams. Light materials float on the water surface. When velocities fall, heavy solids are left behind as bottom deposits, while light materials strand at the water's edge. When velocities rise again, gritty substances and the flotsam of the sewer are picked up once more and carried along in heavy concentration. There may be erosion. Within reason, all of these happenings should be avoided, insofar as this can be done. Each is a function of the tractive force of the carrying water that should be better known than it commonly is.

Conceptually, the drag exerted by flowing water on a channel is analogous to the friction exerted by a body sliding down an inclined plane. Because the volume of water per unit surface of channel equals the hydraulic radius of the channel,

$$\tau = \gamma rs \tag{13.7a}$$

where τ is the intensity of the tractive force, γ is the specific weight of water at the prevailing temperature, r is the hydraulic radius of the filled section, and s is the slope of the invert or loss of head in a unit length of channel when flow is steady and uniform and the water surface parallels the invert. Substituting $rs = (v/c)^2$ in accordance with the Chézy equation, for example,

$$\tau = \gamma(v/c)^2 \tag{13.7b}$$

and the tractive force intensity is seen to vary as the square of the velocity of flow v and inversely as the square of the Chézy coefficient, c.

13.3.1 Transporting Velocities

The velocity required to transport waterborne solids is derived from Eq. 13.4, with the help of Fig. 13.2. For a layer of sediment of unit width and length, thickness t, and porosity ratio f', the drag force τ exerted by the water at the surface of the sediment and just causing it to slide down the inclined plane equals the frictional resistance $R = W \sin\alpha$, where $W = (\gamma_s - \gamma)t(1 - f')$ is the weight of the sediment in water and α is the friction angle. Accordingly, $\tau = (\gamma_s - \gamma)t(1 - f')\sin\alpha$, and Eq. 13.4 becomes

$$\tau = (\gamma_s - \gamma)t(1 - f')\sin\alpha = k(\gamma_s - \gamma)t. \tag{13.7c}$$

Here, $k = (1 - f')\sin\alpha$ is an important characteristic of the sediment. For single grains, the volume per unit area t becomes a function of the diameter of the grains d as an inverse measure of the surface area of the individual grains exposed to drag or friction. Thus k' replaces k when d replaces t.

It follows from Eqs. 13.7a and 13.7c that the invert slope at which sewers will be self-cleansing is

$$s = (k'/r)[(\gamma_s - \gamma)/\gamma]\,d \tag{13.8a}$$

and that, in accordance with the Chézy equation,

$$v = c[k'd(\gamma_s - \gamma)/\gamma]^{1/2} \tag{13.9a}$$

where the value of c is chosen with full recognition of the presence of deposited or depositing solids and expressed, if so desired, in accordance with any other pertinent capacity or friction factor. Examples include Eq. 13.9b.

$$v = [(8k'/f)gd(\gamma_s - \gamma)/\gamma]^{1/2} \tag{13.9b}$$

which is derived by Camp (1946) from Eq. 13.9a by introducing the *Weisbach–Darcy* friction factor $f = 8\,g/c^2$, and also Eq. 13.10:

$$v = (1.49/n)r^{1/6}[k'd(\gamma_s - \gamma)/\gamma]^{1/2} \tag{13.10a}$$

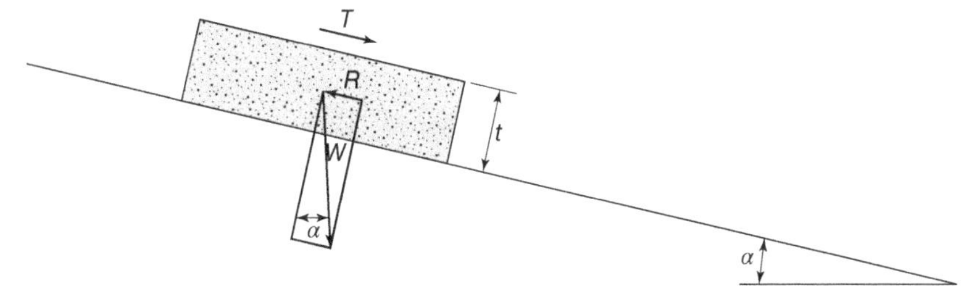

Figure 13.2 Forces Acting on Sediment of Unit Width and Length and Thickness *t*. *T* is the tractive force per unit surface area. *R*, the resisting force, is a function of the weight of the sediment (*W*) and the friction angle (α)

in terms of a Manning evaluation of $c = 1.49(r^{1/6}/n)$, where n is the Manning friction factor. When convenient, the ratio $(\gamma_s - \gamma)/\gamma$ can be replaced by the closely equal term $(s_s - 1)$ where s_s is the specific gravity of the particles (solids) composing the deposit.

Applicable magnitudes of k range from 0.04 for initiating scour of relatively clean grit to 0.8 or more for full removal of sticky grit. Their actual magnitude can be found only by experiment.

Equations 13.8a and 13.10a can be further presented as the following groups of working equations:

$$s = (k'/r)\,(s_s - 1)d \quad \text{(US customary or SI units)} \tag{13.8b}$$

where s is gradient, dimensionless; k' is sediment characteristic factor, dimensionless; r is hydraulic radius, ft. or m; s_s is the specific gravity of particles or solids composing the deposit dimensionless; and d is particle diameter, ft. or m.

$$v = (1.49/n)r^{1/6}[k'd(s_s - 1)]^{1/2} \quad \text{(US customary units)} \tag{13.10b}$$

where v is minimum velocity at which the particle is transported without hindrance, ft./s; n is Manning coefficient, dimensionless; r is hydraulic radius, ft.; k' is sediment characteristic factor, dimensionless; d is particle diameter, ft.; and s_s is specific gravity of the particle, dimensionless.

$$v = (1/n)r^{1/6}[k'd(s_s - 1)]^{1/2} \quad \text{(SI units)} \tag{13.10c}$$

where v is velocity, m/s; n is Manning coefficient, dimensionless; r is hydraulic radius, m; k' is sediment characteristic factor, dimensionless; d is particle diameter, m; and s_s is the specific gravity of the particle, dimensionless.

Example 13.3 Transport of Solid Particles in Sewers

1) Find the minimum velocity and gradient at which coarse quartz sand is transported without hindrance through a 12-in. (304.8-mm)-diameter sewer that is flowing full. Assume the particle diameter is $10^{-1}/30.48$ ft (0.1 cm), the particle specific gravity is 2.65, the sediment characteristic factor is 0.04, and n is 0.013.
2) If flows in storm and combined sewers are given velocities of 3.0 and 5.0 ft./s (0.90 and 1.5 m/s), respectively, find the diameter of sand or gravel moved.

Solution 1 (US Customary System):

1) Introducing a particle diameter d = 0.1 cm ($10^{31}/30.48$ ft), a specific gravity s_s = 2.65, a sediment characteristic k' = 0.04, and a friction factor n = 0.013 into Eqs. 13.10b and 13.8b.

$$v = (1.49/n)r^{1/6}[k'\,d(\gamma_s - \gamma)/\gamma]^{1/2}$$
$$v = (1.49/n)r^{1/6}[k'\,d(s_s - 1)]^{1/2}$$
$$= (1.49/1.3 \times 10^{-2})(1/4)^{1/6}[4 \times 10^{-2}(10^{-1}/30.48)(2.65 - 1.00)]^{1/2}$$
$$= \mathbf{1.34\ ft/s\ (0.41\ m/s)}$$

and

$$s = (k'/r)[(\gamma_s - \gamma)/\gamma]\,d$$
$$s = (k'/r)(d)(s_s - 1)$$
$$= [4 \times 10^{-2}/(1/4)](10^{-1}/30.48)(2.65 - 1.00)$$
$$= \mathbf{0.87 \times 10^{-3}}$$

2) In accordance with Eq. 13.10a,

$$v = (1.49/n)r^{1/6}[k'd(\gamma_s - \gamma)/\gamma]^{1/2}$$

where v varies as $d^{1/2}$. Hence d varies as v^2.

a) For v = 3.0 ft/s, $d = (10^{-1}/30.48\ \text{ft})(3.0/1.34)^2$ = **0.016 ft = 0.19 in.** (small-sized gravel)
b) For v = 5.0 ft/s, $d = (10^{-1}/30.48\ \text{ft})(5.0/1.34)^2$ = **0.046 ft = 0.55 in.** (large gravel).

Solution 2 (SI System):

1) Introducing a particle diameter d = 0.1 cm = 0.001 m, a specific gravity s_s = 2.65, k' = 0.04, and n = 0.013 into Eqs. 13.10c and 13.8b. Here $r = D/4$ = 0.3048 m/4 = 0.0762 m.

$$v = (1/n)r^{1/6}[k'\,d(s_s - 1)]^{1/2}$$

$$= (1/0.013)(0.0762)^{1/6}[0.04 \times 0.001(2.65 - 1)]^{1/2}$$
$$= 0.41 \text{ m/s}$$
$$s = (k'/r)(d)(s_s - 1)$$
$$= (0.04/0.0762)(0.001)(2.65 - 1)$$
$$= 0.87 \times 10^{-3}$$

2) In accordance with Eq. 13.10c,

$$v = (1/n)r^{1/6}[k'd(s_s - 1)]^{1/2}$$

where v varies as $d^{1/2}$. Hence d varies as v^2.

a) For $v = 0.9$ m/s, $d = (0.1 \text{ cm})(0.9 \text{ m/s}/0.41 \text{ m/s})^2 =$ **0.5 cm**, which is small-sized gravel.
b) For $v = 1.5$ m/s, $d = (0.1 \text{ cm})(1.5 \text{ m/s}/0.41 \text{ m/s})^2 =$ **1.4 cm**, which is larger gravel.

13.3.2 Damaging Velocities

Of the materials used in sewers, vitrified tile, glazed brick, and plastic are very resistant to wear; building brick and concrete are less so. Abrasion is greatest at the bottom of conduits, because grit, sand, and gravel are heavy and travel along the invert. The bottom arch or invert of large concrete or brick sewers is often protected by vitrified tile liners, glazed or paving brick, or granite blocks.

Clear water can flow through hard-surfaced channels, such as good concrete conduits, at velocities higher than 40 ft./s (12.2 m/s) without harm. Stormwater runoff, on the other hand, has to be held down to about 10 ft./s (3.1 m/s) in concrete sewers and drains, because it usually contains abrading substances in sufficient quantity to wear away even well-constructed, hard concrete surfaces. The magnitude of the associated tractive force is given by Eq. 13.5.

13.4 Flow in Partially Filled Sewers

In the upper reaches of sanitary sewerage systems, sewers receive relatively little wastewater. Depths of flow are reduced, because minimum pipe sizes (8 in. in North America) are dictated not by flow requirements alone but also by cleaning potentials. The lower portions of the system, too, do not flow full. Even when the end of their design period is reached, they are filled only spasmodically during times of maximum flow. Discharge ratios may vary, indeed, from as little as 4:1 at the end of the design period to as much as 20:1 at the beginning.

Hydraulic performance of the upper reaches of sanitary sewers is improved by steeper-than-normal grades, even though velocities of, for example, 3.0 rather than 2.5 or 2.0 ft./s (0.90 rather than 0.75 or 0.60 m/s) produce still lower depths of flow. Why this is so is shown later. Requisite capacities of storm and combined sewers are even more variable. The hydraulic performance of partially filled and filled sections must, therefore, be well understood, especially in reference to the maintenance of self-cleansing velocities at expected flows.

The variables encompassed by a flow formula such as Manning's, namely, q or v, r or a/p (p = wetted perimeter), s or h/L and n, constitute the *hydraulic elements* of conduits. For a given shape and a fixed coefficient of roughness and invert slope, the elements change in absolute magnitude with the depth d of the filled section. In the case of Manning's formulation, generalization in terms of the ratio of each element of the filled section (indicated here by a lowercase letter) to the corresponding elements of the full section (indicated here by a capital letter) confines all ratios, including velocity and capacity ratios, to ultimate dependency on depth alone. Thus,

$$v/V = (N/n)(r/R)^{2/3} \tag{13.11}$$

and

$$q/Q = (N/n)(a/A)(r/R)^{2/3} \tag{13.12}$$

Of the elements normally included in diagrams or tables, area and hydraulic radius are static, or elements of shape; roughness, velocity, and discharge are dynamic, or elements of flow. Except for roughness, the basis of their computation is explained in Appendix 9. Variation of roughness with depth was observed by Willcox (1924) on 8-in.(200-mm) sewer pipe, by Yarnell and Woodward (1920) on clay and concrete drain tiles 4–12 in. (100–300 mm) in diameter, and by Johnson (1944) in large sewers flowing at low depths. Appendix 10 is a conventional diagram of the basic hydraulic elements of circular sewers. The two sets of curves included for $v > V$ and $q > Q$ mark the influence of a variable ratio of N/n on these dynamic hydraulic elements. It is important to note that velocities in partially filled, circular sections equal or exceed those in full sections whenever sewers flow more than half full and roughness is not considered to vary with depth; moreover, where changes in roughness are taken into account, velocities equal to or greater than those in full sections are confined to the upper 20% of depth only. Nevertheless, sewers flowing between 0.5 and 0.8 full need not be placed on steeper grades to be as self-cleansing as sewers flowing full. The reason is that velocity and discharge are functions of tractive-force intensity, which depends on the friction coefficient and the flow velocity. Needed ratios of v_s/V, q_s/Q, and s_s/S,where the subscript s denotes cleansing equaling that obtained in the full section, can be computed with the help of Eq. 13.4 on the assumption that equality of tractive-force intensity implies equality of cleansing, or $\tau = T = \gamma rs = \gamma RS$; hence,

$$S_s = (R/r)S \tag{13.13}$$

and

$$v_s/V = (N/n)(r/R)^{2/3}(s_s/S)^{1/4} = (N/n)(r/R)^{1/6} \tag{13.14}$$

or

$$q_s/Q = (N/n)(a/A)(r/R)^{1/6} \tag{13.15}$$

What these equations imply is illustrated in Appendix 11 and in Example 13.3.

Example 13.4 Determination of Flow in Partially Filled Sewers Using Design Diagrams

An 8-in. (200-mm) sewer is to flow at 0.3 depth on a grade ensuring a degree of self-cleaning equivalent to that obtained at full depth at a velocity of 2.5 ft./s (0.75 m/s). Find the required grades and associated velocities and rates of discharge at full depth and 0.3 depth. Assume that $N = 0.013$ at full depth.

Solution:

1) From Appendix 8, find for full depth of flow and $V = 2.5$ ft./s (0.61 m/s),

$$Q = 0.873\ \text{ft}^3/\text{s}\ (0.762\ \text{m/s})$$
$$s = 5.20‰$$

2) From Appendix 10 or Appendix 9, find for 0.3 depth,

$$a/A = 0.252$$
$$r/R = 0.684\ (\text{or } R/r = 1.46)$$
$$v/V = 0.776$$
$$q/Q = 0.196$$
$$N/n = 0.78$$

and from Appendix 9, find

$$(r/R)^{1/6} = 0.939$$

Hence, at 0.3 depth and a grade of 5.20‰,

$v = 0.776 \times 2.5 =$ **1.94 ft/s (0.59 m/s) for $n = N$**, or **1.51 ft/s (0.46 m/s) for $N/n = 0.78$** and
$q = 0.196 \times 0.873 =$ **0.171 ft³/s (4.84 L/s) for $n = N$**, or **0.133 ft³/s (3.77 L/s) for $N/n = 0.78$**

For self-cleaning flow, however,

$s = 1.46 \times 5.20‰ =$ **7.6‰** by Eq. 13.13
$v_s = 0.939 \times 2.5 =$ **2.35 ft/s (0.72 m/s) for $n = N$**,
or $0.78 \times 2.35 =$ **1.83 ft/s (0.56 m/s) for $N/n = 0.78$**, by Eq. 13.13

and

$q_s = 0.252 \times 0.939 \times 0.873 =$ **0.207 ft³/s (5.83 L/s) for $n = N$**,
or $0.78 \times 0.207 =$ **0.161 ft³/s (4.56 L/s) for $N/n = 0.78$**, by Eq. 13.15.

Example 13.5 Determination of Diameter of Partially Filled Sewers Using Design Equations
An average sewer pipe with $n = 0.015$ is laid on a slope of 0.0002 and is to carry 83.6 ft^3/s (2.3675 m^3/s) of wastewater when the sewer pipe flows 0.9 full (d/D = 0.9). Determine the diameter of the sewer pipe.

Solution 1 (US Customary System):

$$r = A/P_w = [\text{Circle} - (\text{Sector AOCE} - \text{Traingle AOCD})]/\text{Arc ABC}$$
$$\text{Angle } \theta = \cos^{-1}(0.4D/0.5D) = \cos^{-1} 0.800$$
$$\theta = 36°52'$$
$$\text{Area of Sector AOCE} = [2(36°52')/360°](\pi D^2/4) = 0.1612D^2$$
$$\text{Length of Arc ABC} = \pi D - [2(36°52')/360°](\pi D) = 2.498D$$
$$\text{Area of Triangle AOCD} = 2(0.5 \times 0.4D \times 0.4D \tan 36°52') = 0.12D^2$$
$$r = A/P_w = [0.25\pi D^2 - (0.161D^2 - 0.12D^2)]/(2.498D)$$
$$= (0.7442D^2)/(2.498D)$$
$$= 0.298D$$
$$A = 0.7442D^2 \text{ and } P_w = 2.498D, \text{ then } r = 0.298D$$
$$Q = Av = A[(1.49/n)r^{2/3}s^{1/2}]$$
$$83.6 = (0.7442D^2)[(1.49/0.015)(0.298D)^{2/3}(0.0002)^{1/2}]$$
$$D^{8/3} = 180, \mathbf{D = 7\ ft}$$

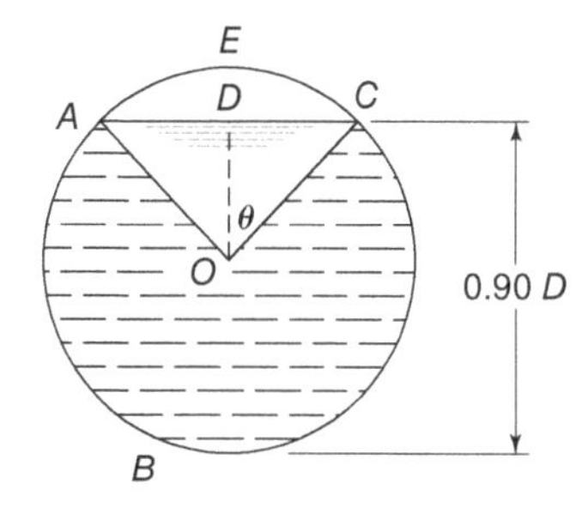

Solution 2 (SI System):

$$r = A/P_w = [\text{Circle} - (\text{Sector AOCE} - \text{Traingle AOCD})]/\text{Arc ABC}$$
$$\text{Angle } \theta = \cos^{-1}(0.4D/0.5D) = \cos^{-1}0.800$$
$$\theta = 36°52'$$
$$\text{Area of Sector AOCE} = [2(36°52')/360°](\pi D^2/4) = 0.1612D^2$$
$$\text{Length of Arc ABC} = \pi D - [2(36°52')/360°](\pi D) = 2.498D$$
$$\text{Area of Triangle AOCD} = 2(0.5 \times 0.4D \times 0.4D \tan 36°52') = 0.12D^2$$
$$r = A/P_w = [0.25\pi D^2 - (0.161D^2 - 0.12D^2)]/(2.498D)$$
$$= (0.7442D^2)/(2.498D)$$
$$= 0.298D$$
$$A = 0.7442D^2 \text{ and } P_w = 2.498D, \text{ then } r = 0.298D$$
$$r = A/P_w = (0.7442\ D^2)/(2.498\ D) = 0.298\ D$$
$$Q = Av = A[(1/n)r^{2/3}s^{1/2}]$$
$$2.3675 = (0.7442\ D^2)[(1/0.015)(0.298\ D)^{2/3}(0.0002)^{1/2}]$$
$$D = 2.13\ \text{m} = \mathbf{2{,}130\ mm}$$

Example 13.6 Determination of Diameter of Partially Filled Sewers Using Design Diagram
Determine the diameter of the sewer pipe described in Example 13.5 assuming $n = 0.015$, $s = 0.002$, $Q = 83.6$ ft^3/s (2.3675 m^3/s), and $d/D = 0.9$.

Solution:

From Appendix 9 "Hydraulic Elements of Circular Conduits," $a/A = 0.949$, $r/R = 1.192$, when $d/D = 0.9$.

Area of the partially filled sewer = $0.949(D^2 \times 0.785) = 0.7442D^2$
Hydraulic radius of the partially filled sewer = $1.192\ (D/4) = 0.298D$

The remaining will be similar to the solution of Example 13.5.

Appendix 11 confirms that minimum grades are enough as long as circular sewers flow more than half full. However, for granular particles, when flows drop to 0.2 depth, grades must be doubled for equal self-cleansing; at 0.1 depth they must be quadrupled. Expressed in terms of the Weisbach–Darcy friction factor f, Eqs. 13.14 and 13.15 become, respectively,

$$v_s/V = (F/f)^{1/2} \tag{13.16}$$

and

$$q_s/Q = (a/A)(F/f)^{1/2} \tag{13.17}$$

Example 13.7 Self-Cleaning Flow in Sewers
An 8-in. (200-mm) sewer is to discharge 0.161 ft^3/s (4.56 L/s) at a velocity as self-cleaning as a sewer flowing full at 2.5 ft./s (0.75 m/s). Find the depth and velocity of flow and the required slope.

Solution:

1) From Example 13.4, $Q = 0.873$ ft^3/s and $S = 5.20$ ‰.
   Hence, $q_s/Q = 0.161/0.873 = 0.185$
2) From Appendix 11, for $N = n$ and $q_s/Q = 0.185$, $d_s/D = 0.25$, $v_s/V = 0.91$, and $s/S = 1.70$.
   Hence, $v_s = 0.91 \times 2.5 =$ **2.28 ft/s (0.69 m/s)**
   $s_s = 1.70 \times 5.20 =$ **8.8‰**
3) From Appendix 11, for N/n and $q_s/Q = 0.185$, $d_s/D = 0.30$, $v_s/V = 0.732$, and $s/S = 1.46$.
   Hence, $v_s = 0.732 \times 2.5 =$ **1.83 ft/s (0.56 m/s)**
   $s_s = 1.46 \times 5.20 =$ **7.6‰**

Egg-shaped sewers and cunettes were introduced, principally in Europe, to provide enough velocity for dry-weather flows in combined sewers. The hydraulic elements of these sewers and of horseshoe-shaped and box sewers that can be charted in the same way as circular sewers are in Appendixes 10 and 11.

13.5 Flow in Sewer Transitions

Although flow in sewers is both *unsteady* (changing in rate of discharge) and *nonuniform* (changing in velocity and depth), these factors are normally taken into account only at sewer transitions. This is so because it is not practicable to identify with needed accuracy the variation in flow with time in all reaches of the sewerage system and because the system is designed for maximum expected flow in any case.

Sewer transitions include (a) changes in size, grade, and volume of flow; (b) free and submerged discharge at the end of sewer lines; (c) passage through measuring and diversion devices; and (d) sewer junctions. Of these, sewer transitions at changes in size or grade are most common. How they affect the profile of the water surface and energy gradient is shown, greatly foreshortened, in Fig. 13.3. Here h_e is the loss in energy or head, h_s the drop in water surface, and h_i the required invert drop. For convenience of formulation, these changes are assumed to be concentrated at the center of the transition.

The energy loss h_e is usually small. In the absence of exact information, it may be considered proportional to the difference, or change, in velocity heads, that is, $he = k\ (h_{v2} - h_{v1}) = k\Delta h_v$. The proportionality factor k may be as low as 0.1 for rising velocities and 0.2 for falling velocities, provided flow is in the *upper alternate stage* (see next section). For the *lower alternate stage* (see next section), k may be expected to increase approximately as the square of the velocity ratios. Camp (1946) has suggested a minimum allowance of 0.02 ft. for the loss of head in a transition of this kind. However, if there is a horizontal curve in the transition, more head will be lost.

The required invert drop h_i follows from the relationships demonstrated in Fig. 13.4. There $h_2 + h_e = h_1 + h_i$, or

$$h_i = (h_2 - h_1) + h_e = \Delta(d + h_v) + kh_v \tag{13.18}$$

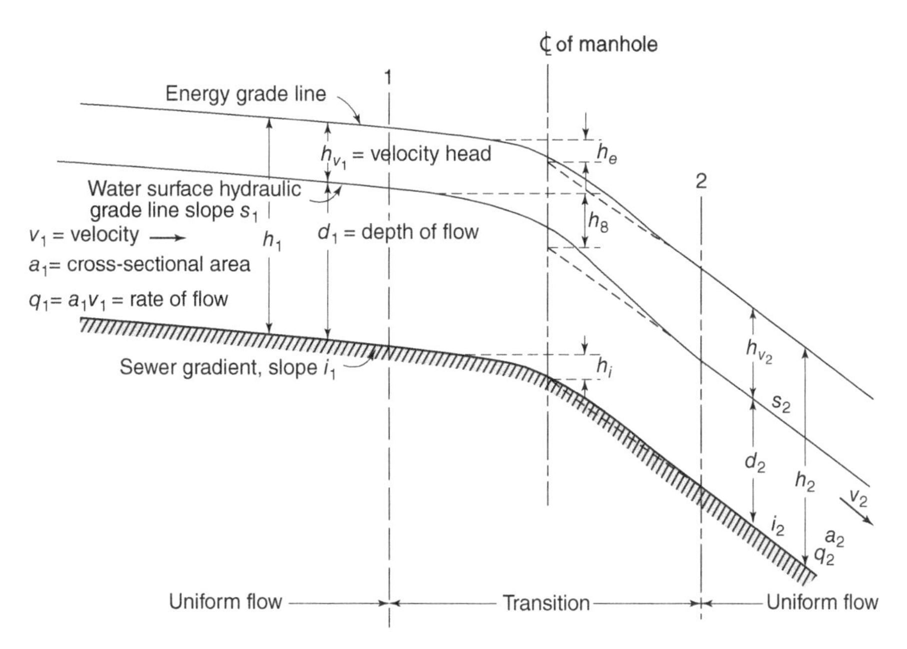

Figure 13.3 Changes in Hydraulic and Energy Grade Lines at a Transition in Size or Grade of Sewer

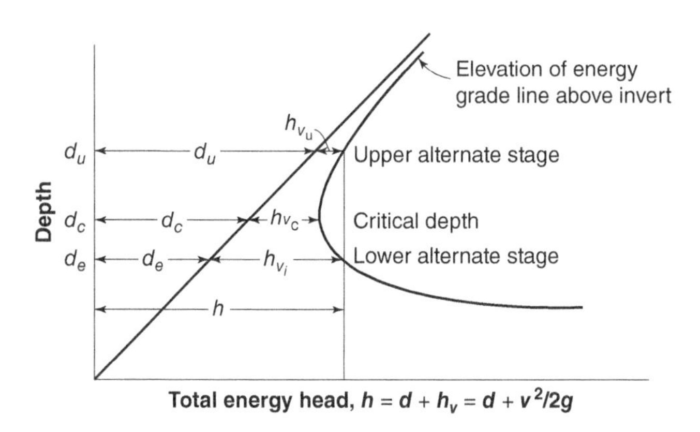

Figure 13.4 Alternate Stages of Flow Total Energy Head at Constant Rate of Discharge in Open-Channel Flow

The calculated change h_i will be positive for increasing gradients and negative for sharply decreasing gradients. A positive value calls for a drop in the invert, a negative value for a rise. However, a rise would obstruct flow, and the invert is actually made continuous. The elevation of the water surface in the downstream sewer is thereby lowered, and the waters in the sewers entering the transition are drawn down toward it.

Rules of thumb sometimes followed by engineers in place of computations reflect average conditions encountered in practice. They may not always be justified by circumstances. Common rules for drops in manholes at changes in size are as follows:

1) Make the invert drop $h_i = \frac{1}{2}(d_2 - d_1)$ for sewers smaller than 24 in. (600 mm), or $h_i = \frac{3}{4}(d_2 - d_1)$ for 24-in. (600-mm) sewers and larger.
2) Allow a drop of 0.1 ft. (0.03 m) in a manhole, 0.2 ft. (0.06 m) in the presence of one lateral or bend, and 0.3 ft. (0.09 m) for two laterals.
3) Keep the 0.8d line continuous on the principle that it is close to the line of maximum velocity.
4) Keep the crowns of sewers continuous.
5) Increase the roughness factor over that in straight runs; $N = 0.015$, for example, instead of $N = 0.013$.

These rules apply only when all sewers entering a manhole will flow full at times of maximum discharge. If entering laterals are partially filled at peak flows, appropriate drops should be provided in water surfaces rather than in sewer barrels. Otherwise solids will deposit from sewage backed up into the laterals.

Example 13.8 Two Sewers Discharging into a Third Sewer
Two 8-in. (200-mm) sanitary sewers, each flowing full and carrying 0.7 ft^3/s (19.8 L/s) at a velocity of 2 ft./s (0.60 m/s) on minimum grade, discharge into a steeper sewer that picks up 0.01 ft^3/s (0.28 L/s) in the course of its next run. The lower sewer can be laid on a grade as low as 10‰, and as high as 14‰. Find the required slope of the lower sewer and the invert drop in the transition.

Solution:

1) From Fig. 13.1 or Appendix 7, an 8-in. (200-mm) sewer flowing full will carry (0.7 + 0.7 + 0.01) = 1.41 ft^3/s (40 L/s) on a slope of 13.6‰ with a velocity of 4.04 ft./s (1.23 m/s) if $N = 0.013$. Pertinent information is, therefore, as follows: $d_1 = 0.67$ ft (0.20 m), $v_1 = 2.00$ ft/s (0.60 m/s), $h_{v1} = 0.062$ ft (0.019 m), $d_1 + h_{v1} = 0.73$ ft (0.22 m); $d_2 = 0.67$ ft (0.20 m), $v_2 = 4.04$ ft/s (1.23 m/s), $h_{v2} = 0.254$ ft (0.077 m), $d_2 + h_{v2} = 0.92$ ft (0.28 m); and $\Delta h_v = 0.19$ ft (0.06 m), $\Delta(d + h_v) = 0.19$ ft (0.06 m). Assuming a loss of head $h_e = 0.2\Delta h_v = 0.038$ ft (0.011 m), Eq. 13.18 gives the required drop in invert, $h_i = 0.19 + 0.04 =$ **0.23 ft (0.07 m)**.
2) A 10-in. (250-mm) sewer laid on a grade of 10‰ has a capacity of 2.19 ft^3/s (62 L/s) and velocity of 4.02 ft/s (1.22 m/s) when flowing full. From Appendix 10, for $N/n \times q/Q = 0.644$, $d/D = 0.65$, and $N/n \times v/V = 0.92$, or $d = 6.5$ in. (165 mm) and $v = 3.69$ ft/s (1.12 m/s). Hence, the upper sewers remaining unchanged, $d_2 = 0.542$ ft (0.165 m), $v_2 = 3.69$ ft/s (1.1 m/s), $h_{v2} = 0.21$ ft (0.06 m), $d_2 + h_{v2} = 0.75$ ft (0.23 m), and $\Delta h_v = 0.15$ ft (0.046 m), $\Delta(d + h_v) = 0.02$ ft (0.006 m). Assuming a loss of head $h_e = 0.2\Delta h_v = 0.036$ ft (0.011 m), Eq. 13.18 states that $h_i = 0.02 + 0.04 =$ **0.06 ft (0.018 m)**.

13.6 Alternate Stages and Critical Depths

In the analysis of transitions, the designer must often find the alternate stage or depths of open-channel flow and their mergence into flow at critical depth. Referred to the sewer invert, the energy grade line shown in Fig. 13.3 is situated at a height

$$h = d + h_v = d + v^2/2g = d + q^2/(2ga^2) \tag{13.19}$$

above this datum, where the cross-sectional area, a, of the conduit is a function of its depth, d. (In terms of the datum, if v is the mean velocity of flow, the kinetic energy head is actually greater than $v^2/2g$ by 10%–20%, depending on the shape and roughness of the channel. But this fact is not ordinarily taken into account in hydraulic computations.) Accordingly, Eq. 13.19 is a cubic equation in terms of d. Two of its roots are positive and, except at critical depth, identify respectively the *upper alternate stage* and the *lower alternate stage* at which a given discharge rate q can be associated with a given energy head h. The two stages fuse into a single critical stage for conditions of maximum discharge at a given total energy head, or minimum total energy head at a given discharge (Fig. 13.4 and Eq. 13.19). For uniform flow, the water surface parallels the invert ($s = i$). Open-channel flow at the near-critical stage is unstable and depth of flow is uncertain and fluctuating.

Equation 13.19 can be generalized by expressing its three components as dimensionless ratios. Bringing q to one side and multiplying both sides by $(l/D)\,(a/A)^2$ give the following straight-line relationship:

$$\left(q/A\sqrt{gD}\right)^2 = 2(a/A)^2(h/D - d/D) \tag{13.20}$$

Again, capital letters denote the hydraulic elements of the full section, and lowercase letters those of the partially filled section. As stated earlier, given the energy head, the maximum rate of discharge is obtained at critical depth d. The associated specific head, h/D, is determined analytically by differentiating Eq. 13.20 with respect to d and equating the result to zero.

For a trapezoidal channel of bottom width b and with side slopes z (horizontal to vertical),

$$h_c/D = d_c/D + \frac{1}{2}(d_c/D)(b + zd_c)/(b + 2zd_c) \tag{13.21}$$

If v is the mean velocity of flow, the kinetic energy head is actually greater than $v^2/2g$ by 10%–20%, depending on the shape and roughness of the channel. But this fact is not ordinarily taken into account in hydraulic computations.

For a rectangular channel ($z = 0$), therefore,

$$h_c/D = \frac{3}{2}d_c/D \text{ or } h_c = \frac{3}{2}d_c \tag{13.22}$$

whence.

$$v_c^2/2g = h = d_c \text{ or } v_c = \sqrt{gd_c} \tag{13.23}$$

For a circular cross-section, finally,

$$\frac{h_c}{D} = \frac{1}{8}\left\{\left[10\left\{\frac{d_c}{D}\right\} - 1\right] + \frac{\frac{1}{4}\pi + \frac{1}{2}\sin^{-1}\left[2\left(\frac{d_c}{d}\right) - 1\right]}{\sqrt{\left(\frac{d_c}{D}\right)\left[1 - \left(\frac{d_c}{D}\right)\right]}}\right\} \tag{13.24}$$

Substituting values of d_c/D varying by tenths from 0.1 to 0.9 yields the numerical results for $h/D, v_c/\sqrt{gD}$, and $[q/(A\sqrt{gD})]^2$shown in Appendix 12. A plot of Eq. 13.20 for circular conduits is shown in Appendix 13.

The critical depth line in a closed conduit is seen to be asymptotic to the line $d/D = 1.0$; that is, there is neither a critical nor an alternate stage for an enclosed conduit flowing full.

Example 13.9 Alternate Stages and Critical Depths

The use of Appendix 13, construction of which is simple and straightforward, can be exemplified as follows. Given a discharge of 60 ft^3/s (1.7 m^3/s) in a 4-ft (1.22-m) circular sewer, find

1) The critical depth
2) The alternate stages for an energy head of 4 ft. (1.22 m)
3) The lower alternate stage associated with an upper alternate stage at 0.8 depth
4) The sewer invert slope that would produce flow at near-critical depth.

Solution:

1) For $q = 60$ ft^3/s and $D = 4$ ft, $[q/(A\sqrt{gD})]^2 = [60 \times 4/(\pi \times 16/\sqrt{4g})]^2 = 0.177$:
   From Appendix 13, read $d_c/D = 0.59$. Hence, $d_c = 0.59 \times 9 =$ **2.36 ft (0.72 m)**.
2) For $h = 4.0$, $h/D = 1.0$, and $[q/(A\sqrt{gD})]^2 = 0.177$ as in part l:
   From Appendix 13, read $d/D = 0.42$ and 0.95, or $d_l = 0.42 \times 4 =$ **1.7 ft (0.52 m)**, and $d_u = 0.95 \times 4 =$ **3.8 ft (1.16 m)**.
3) For $[q/(A\sqrt{gD})]^2 = 0.177$ as in part 1 and $d_u/D = 0.8$, or $d_u = 3.2$ ft (0.98 m):
   From Appendix 13, read $d_l/D = 0.45$, or $d_l =$ **1.8 ft (0.55 m)**.
4) For $N = 0.013$ and $d_c/D = 0.59$:
   From Appendix 10, $q/Q = 0.52$ and $Q = 60/0.52 = 115$ ft^3/s (3.26 m^3/s)
   For $D = 4$ ft, Appendix 7 gives $Q_o = 18.67$ ft^3/s (0.53 m^3/s):
   Since $Q_o = S^{1/2}/N = Q$, $S = (QN/Q_o)^2$. Therefore, $\mathbf{s} = (115 \times 0.013/18.67)^2 = \mathbf{0.64 \times 10^{-2}}$. This is less than a 1% grade. Because many streets have slopes in excess of this, large sewers not infrequently operate at the lower alternate depth, or high-velocity stage.

13.7 Length of Transitions

Transition from one to the other alternate stage carries the flow through the critical depth. Passage from the upper alternate stage (a) to the critical depth or (b) through it, to the lower alternate stage or to free fall, produces nonuniform (accelerating) flow and a *drawdown curve* in the water surface. Passage from the lower to the upper alternate stage creates the hydraulic jump. Reduction in velocity of flow, by discharge into relatively quiet water or by weirs and other flow obstructions, dams up the water and induces nonuniform (decelerating) flow and a *backwater curve* in the water surface. For economy of design, the size of conduit must fit conditions of flow within the range of transient depths and nonuniform flow. If initial and

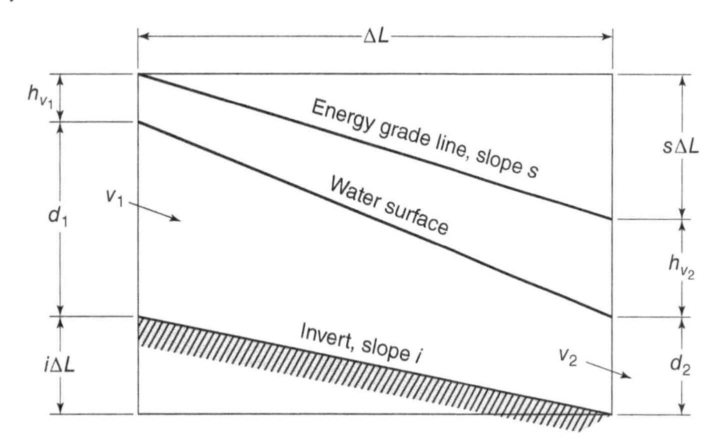

Figure 13.5 Flow Conditions Changing with Increasing Velocity

terminal depths of flow are known, the energy and hydraulic grade lines can be traced either by stepwise calculation or by integration (graphical or analytical). Both stem from the fact that the change in slope of the energy grade line must equal the sum of the changes in the slopes of the invert, the depth of flow, and the velocity head.

For stepwise calculation of the length of conduit between the cross-sections of given depth (Fig. 13.5),

$$s\Delta L = i\Delta L + \Delta(d + h_v)$$

or

$$\Delta L = \Delta(d + h_v)/(s - i) \tag{13.25}$$

Flow being steady, the rate of discharge is constant, and the velocity of flow at given depths is known. For a given invert slope *i*, therefore, only *s* needs to be calculated. This is generally done by introducing the average hydraulic elements of each conduit reach into a convenient flow formula. Averages of choice are ordinarily arithmetic means, but geometric or harmonic means will also give defensible results. Necessary calculations are shown in Example 13.7 for a backwater curve and in Example 13.8 for a drawdown curve.

Beyond the critical depth, the hydraulic drop terminating in free fall is a function of velocity distribution. Flow is supercritical and depth decreases. At the free outfall, pressure on the lower as well as the upper nappe is atmospheric when the nappe is ventilated. Within the conduit, calculated ratios of the terminal depth to the critical depth normally range between $2 > 3$ and $3 > 4$. The critical depth itself lies upstream at a distance of about $4d_c$.

Example 13.10 Tracing the Profile of a Backwater Curve in a Sewer

A 10-ft (3-m) circular sewer laid on a gradient of 0.5‰ discharges 106 ft^3/s (3 m^3/s) into a pump well. The water level in this well rises, at times, 10 ft. (3 m) above the invert elevation of the incoming sewer. Trace the profile of the water surface in the sewer. Assume a coefficient of roughness of 0.012 for the full sewer.

Solution:

A 10-ft sewer ($n = 0.012$) on a grade of 0.0005 has a capacity of 400 ft^3/s (11.3 m^3/s) by Manning's formula (Fig. 13.1).

The value of q/Q, therefore, is $106/400 = 0.265$, and d/D from Appendix 10 is 0.4 for variable N/n.

Hence, the initial depth of flow is $0.40 \times 10 = 4.0$ ft (1.22 m), and the terminal depth 10 ft. (3 m).

The reach in which depths change by chosen amounts is given by Eq. 13.25. Calculations are systematized in Table 13.1. The depth is seen to change from 4.0 to 10.0 ft. (1.22 to 3 m) in 16,580 ft. (5054 m), or slightly over 3 miles (5 km).

Column 1: Depths between initial depth of 4 ft. (1.22 m) and terminal depth of 10 ft. (3 m) are assumed to increase by 2 ft. (0.61 m).

Column 2: Column $1 \div 10$ (the diameter of the sewer).

Columns 3, 4, and 5: a/A, r/R, and N/n, respectively, read from Appendix 10.

Column 6: Column 3×78.5 (the area of the sewer).

Column 7: Column 4×2.50 (the hydraulic radius of the sewer).

Column 8: 106 (the rate of flow)/column 6.

Table 13.1 Calculation of Backwater Curve for Example 13.10

d (1)	*d/D* (2)	*a/A* (3)	*r/R* (4)	*N/n* (5)	*a* (6)	*r* (7)	*v* (8)	$h_v \times 10^2$ (9)	$d + h_v$ (10)
10.0	1.00	1.000	1.000	1.00	78.5	2.50	1.35	2.83	10.028
8.0	0.80	0.858	1.217	0.89	67.5	3.04	1.57	3.83	8.038
6.0	0.60	0.626	1.110	0.82	49.1	2.78	2.16	7.23	6.072
4.0	0.40	0.373	0.857	0.79	29.3	2.14	3.62	20.3	4.203
		Average							
$n \times 10^2$ (11)	$nv \times 10^2$ (12)	*r* (13)	$r^{2/3}$ (14)	$nv \times 10^2$ (15)	$s \times 10^5$ (16)	$(s - i) \times 10^5$ (17)	$\Delta\ (d + h_v)$ (18)	Δl (19)	$\Sigma\Delta l$ (20)
1.20	1.62								0
		2.77	1.97	1.87	4.07	−45.9	−1.990	4330	
1.35	2.12								4330
		2.91	2.04	2.63	7.53	−42.5	−1.966	4620	
1.46	3.15								8950
		2.46	1.82	4.32	25.5	−24.5	−1.869	7630	
1.52	5.50								16,580

Column 9: $v^2/2g$ from column 8.
Column 10: Column 9 + column 1.
Column 11: 0.012 (Manning's *N* for sewer)/column 5.
Column 12: Column 11 × column 8.
Column 13: Arithmetic mean of successive pairs of values in column 7.
Column 14: (Column 13)$^{2/3}$.
Column 15: Arithmetic mean of successive pairs of values in column 12.
Column 16: (Column 15/1.49 × column 14)^2, that is, $s = (n_v/1.49r^{2/3})^2$.
Column 17: Column 16 – i = Col. 16 – 50.
Column 18: Difference between successive pairs of values in column 10.
Column 19: Column 18/column 17 × 10^{-5}, that is, $\Delta l = \Delta(d + hv)/(s - i)$.
Column 20: Cumulative values of column 19.

Example 13.11 Tracing the Profile of a Drawdown Curve in a Sewer
A 10-ft (3-m) circular sewer laid on a gradient of 0.5‰ discharges freely into a water course. Trace the profile of the water surface in the sewer when it is flowing at maximum capacity without surcharge.

Solution:

As shown in Example 13.10, the full capacity of this sewer is 400 ft^3/s (11.3 m^3/s) for $N = 0.012$.

To discharge in free fall, the flow must pass through the critical depth.

Because $[Q/(A\sqrt{gD})]^2 = [400/(78.5\sqrt{10g})]^2 = 8.07 \times 10^{-2}, d_c = 0.47 \times 10 = 4.7$ft (1.43 m), from Appendix 12 or 13.

The reach in which the depth changes from 10 to 4.7 ft. (3 to 1.43 m) is calculated in Table 13.2 in accordance with Eq. 13.25 and as in Example 13.10. The drawdown is seen to extend over a length of 23,600 ft. (7193 m), or over 4 miles (over 7 km), between the full depth of the sewer and a critical depth of 4.7 ft. (1.43 m). There is a further short stretch of flow between the point of critical depth and the end of the sewer. This additional distance is relatively small, namely, $4dc = 4 \times 4.7 = 18.8$ ft (5.73 m).

Table 13.2 Calculation of Drawdown Curve for Example 13.11 (columns as in Table 13.1)

d (1)	*d/D* (2)	*a/A* (3)	*r/R* (4)	*N/n* (5)	*a* (6)	*r* (7)	*v* (8)	h_v (9)	$d + h_v$ (10)
4.7	0.47	0.463	0.960	0.79	36.4	2.40	11.0	1.88	6.58
6.0	0.60	0.626	1.110	0.82	49.1	2.78	8.15	1.02	7.02
8.0	0.80	0.858	1.217	0.89	67.5	3.04	5.93	0.54	8.54
10.0	1.00	1.000	1.000	1.00	78.5	2.50	5.10	0.40	10.40

		Average							
$n \times 10^2$ (11)	$nv \times 10^2$ (12)	*r* (13)	$r^{2/3}$ (14)	$nv \times 10^2$ (15)	$s \times 10^4$ (16)	$(s - i) \times 10^4$ (17)	$\Delta\ (d + h_v)$ (18)	Δl (19)	$\Sigma\Delta l$ (20)
1.52	16.8								0
		2.59	1.88	14.3	26.2	21.2	0.44	210	
1.46	11.9								210
		2.91	2.04	9.9	10.65	5.65	1.52	2690	
1.35	7.99								2900
		2.77	1.97	7.1	5.90	0.90	1.86	20,700	
1.20	6.12								23,600

13.8 Hydraulic Jumps and Discontinuous Surge Fronts

Figure 13.6 Naturally Occurring Hydraulic Jump (*Source:* Wikipedia, http://en.wikipedia.org/wiki/Image:Hydraulic-Jump-on-Upper-Spokane-Falls.jpg)

When a conduit steep enough to discharge at supercritical velocities and depths is followed by a relatively flat channel in which entering velocities and depths cannot be maintained, a more or less abrupt change in velocity and depth takes the form of a *hydraulic jump* (Fig. 13.6). Whereas alternate depths are characterized by equal specific energies ($d + h_v$), sequent depths are characterized by equal pressure plus momentum. In accordance with the *momentum principle*, illustrated in Fig. 13.7, the force producing momentum changes, $1/2(\rho g d_1{}^2 - \rho g d_2{}^2)$, when equated to the momentum change per unit volume, $q\rho(v_2 - v_1)$ and q and v_2 are eliminated by the continuity equation $q = v_1 d_1 = v_2 d_2$, leads directly to the relationship:

$$\left(v_1/\sqrt{gd_1}\right)^2 = \frac{1}{2}(d_2/d_1)[1 + (d_2/d_1)] = \boldsymbol{F}^2 \qquad (13.26)$$

where ρ is the mass density of the water, g is the gravity constant, q is the rate of flow, F is *Froude number*, and $\sqrt{gd_1}$ is the celerity of an elementary gravity wave, or the ratio of the sequent depths d_2 (upper) and d_1 (lower), as determined by

$$d_2/d_1 = \frac{1}{2}\left[\left(1 + 8\boldsymbol{F}^2\right)^{1/2} - 1\right] \qquad (13.27)$$

As shown by Rouse (1950), depths change (a) with substantially no loss of head in a series of undulations when $2 > \boldsymbol{F} > 1$

and (b) with appreciable head loss and a breaking wave when $\boldsymbol{F} > 2$. For cross-sections other than rectangles of unit width, all terms in Eq. 13.27 have numerical coefficients that must be determined experimentally.

As shown in Fig. 13.8, the momentum principle can be adduced to identify also the propagation of discontinuous waves in open-channel flow. Waves of this kind may rush through conduits when a sudden discharge of water from a localized thunderstorm or the quick release of a large volume of industrial wastewater, for example, enters a drainage system. In cases such as these, the volume of water undergoing a change in momentum in unit time and unit channel width is $(v_w - v_1)d_1$. The celerity of propagation, which is the wave velocity or speed or propagation of the surge front, relative to the fluid velocity (equating force to momentum change $1/2gd_2^{\,2} - 1/2gd_1^{\,2} = (v_2 - v_1)(v_w - v_1)d_1$ in a channel of unit width. Continuity of flow, moreover, requires that $v_2d_2 = v_w(d_2 - d_1) + v_1d_1$) being $c = (v_w - v_1)$; it follows that

$$\left(c/\sqrt{gd_1}\right)^2 = 1/2(d_2/d_1)[1 + (d_2/d_1)] \tag{13.28}$$

(US customary or SI units)

where c is the rate of propagation of a discontinuous surge front, ft./s or m/s; g is gravity constant, 32.2 ft./s^2 or 9.806 m/s^2; d_1 is flow depth before surge, ft. or m; and d_2 flow depth raised by the surge, ft. or m.

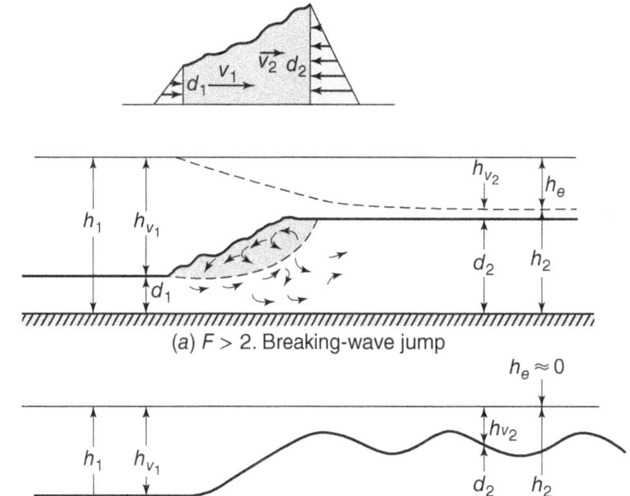

Figure 13.7 Profiles of Hydraulic Jumps: (a) $F > 2$ breaking-wave jump, (b) $2 > F > 1$, undulating jump

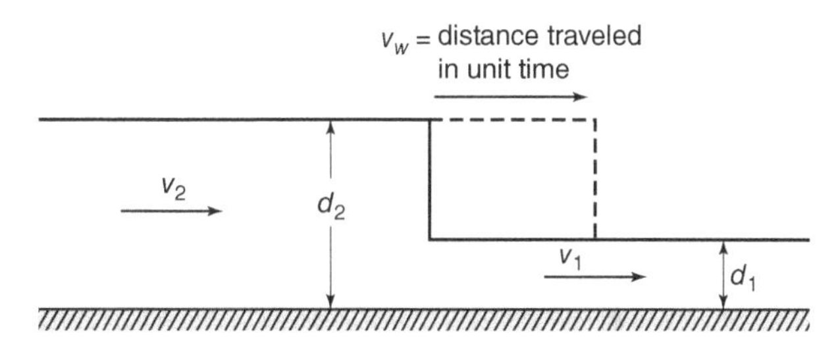

Figure 13.8 Profile of Surge Front

Example 13.12 Hydraulic Jump and Discontinuous Surge Front

Find the rate of propagation of a discontinuous surge front that raises the flow depth from 1 to 2 ft. (0.3048 to 0.6096 m).

Solution 1 (US Customary System):

From Eq. 13.28,

$$\left(c/\sqrt{gd_1}\right)^2 = \frac{1}{2}(d_2/d_1)[1 + (d_2/d_1)]$$

Given $d_1 = 1$ ft (0.3 m) and $d_2 = 2$ ft (0.6 m),

$$c^2 = gd_1[1/2 \times d_2/d_1(1 + d_2/d_1)]$$

$$c = \sqrt{g}[(1/2) \times (2/1)(1 + 2/1)]^{1/2}$$

$$c = \sqrt{3g} = \mathbf{9.8\,ft/s}\ (3\,\mathrm{m/s})$$

Solution 2 (SI System):

From Eq. 13.28,

$$\left(c/\sqrt{gd_1}\right)^2 = \tfrac{1}{2}(d_2/d_1)[1 + (d_2/d_1)]$$

Given $d_1 = 1$ ft (0.3 m) and $d_2 = 2$ ft (0.6 m),

$$c^2 = gd_1\ [0.5 \times d_2/d_1\ (1 + d_2/d_1)] = 0.5gd_2\ (1 + d_2/d_1)$$

$$c = [0.5 \times 9.806 \times 0.6096\ (1 + 0.6096/0.3048)]^{0.5}$$

$$= \mathbf{3\ m/s}$$

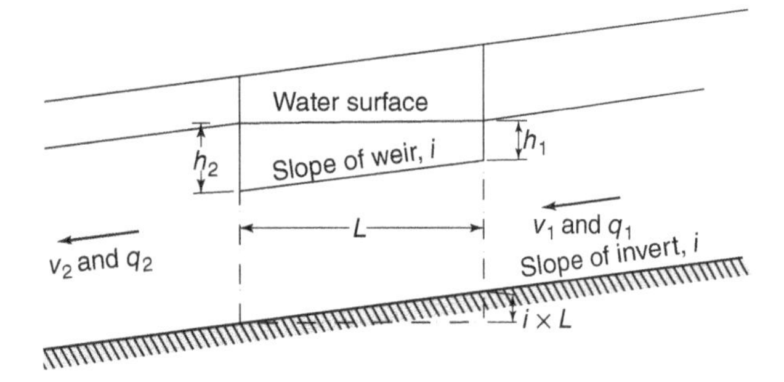

Figure 13.9 Flow Over a Side Weir

13.9 Bends, Junctions, and Overfalls

In common with many other uncertainties of fluid behavior in hydraulic systems, head losses in bends are often approximated as functions of the velocity head $v^2/2g$. For small sewers, the radius of an optimal circular curve of the center line of the sewer is reported to be three to six times the sewer diameter.

The conjoining of the flow patterns of two or more sewers normally involves curvature and impact effects. Generalization of resulting head losses is difficult. Where predictions of surface profiles are important, in the case of large sewers, for example, model studies are advisable. Otherwise, the usual procedure of working upstream from a known point, especially a control point, should identify the water surface profile reasonably well.

Storm flows in excess of interceptor capacity are often diverted into natural drainage channels through overfalls or side weirs. Needed weir lengths depend on the general dimensions and hydraulic characteristics of the sewer and the nature and orientation of the weir itself. Understandably, side weirs paralleling the direction of flow must be longer than weirs at right angles to it.

For the conditions of flow outlined in Fig. 13.9, *Bernoulli's theorem* gives the following relationship when head loss is based on Manning's formula:

$$\left(v_1{}^2/2g\right) + iL + h_1 = \left(v_2{}^2/2g\right) + h_2 + L\left(nv/1.49r^{2/3}\right)^2$$

Hence,

$$h_2 - h_1 = \frac{q_1^2 - q_2^2}{2ga^2} + iL - n^2L\left[\frac{q_1 + q_2}{2 \times 1.49ar^{2/3}}\right]^2 \tag{13.29}$$

The parameters of cross-sectional area a and hydraulic radius r are based on average dimensions of the filled channel, those obtained at the center of the weir, for example. Approximating the flow over the weir, Q, by $cLh^{3/2}$,

$$Q = cL\left[\frac{1}{2}(h_1 + h_2)\right]^{3/2}$$

and

$$h_2 + h_1 = 2[(q_1 - q_2)/(cL)]^{2/3} \tag{13.30}$$

Given q_1, q_2, a, r, i, n, and h_2, values of L and h_1 are then determined by trial, as shown in Example 13.13. A formulation of this kind was first suggested by Forchheimer (1930).

Example 13.13 Storm Flow Diversion Over a Side Weir

Figure 13.9 represents the flow over a side weir. The following data are given: $q_1 = 30$ ft³/s (0.85 m³/s); $q_2 = 16$ ft³/s (0.45 m³/s); $a = 32$ ft² (2.88 m²); $r = 1.6$ ft (0.4877 m); $i = 10^{-4}$; $n = 1.25 \times 10^{-2}$; $h_2 = 0.50$ ft (0.1524 m). Find L and h_1; assume $c = 3.33$.

Solution:

$$h_2 - h_1 = \frac{q_1^2 - q_2^2}{2ga^2} + iL - n^2L\left[\frac{q_1 + q_2}{2 \times 1.49ar^{2/3}}\right]^2 \tag{13.29}$$

$$0.5 - h_1 = \frac{30^2 - 16^2}{2 \times 32.2 \times 32^2} + 10^{-4}L - \left(1.25 \times 10^{-2}\right)^2 L\left[\frac{30 + 16}{2 \times 1.49 \times 32 \times 1.6^{2/3}}\right]^2$$

$$h_1 = 0.490222 - 9.91 \times 10^{-5}L$$

$$h_2 + h_1 = 2[(q_1 - q_2)/(cL)]^{2/3}$$
$$0.5 + h_1 = 2[(30 - 16)/(3.33L)]^{2/3}$$
$$h_1 = 5.21L^{-2/3} - 0.5$$

Hence,

$$0.49022 - 9.91 \times 10^{-5}L = 5.21\ L^{-2/3} - 0.5$$
$$(0.99022 - 9.91 \times 10^{-5}L)L^{2/3} = 5.21$$
$$(12{,}320 - L)L^{2/3} = 64{,}700$$
$$\mathbf{L = 12\ ft\ (3.6575\ m)}$$
$$\mathbf{h_1 = 0.49\ ft\ (0.1494\ m)}$$

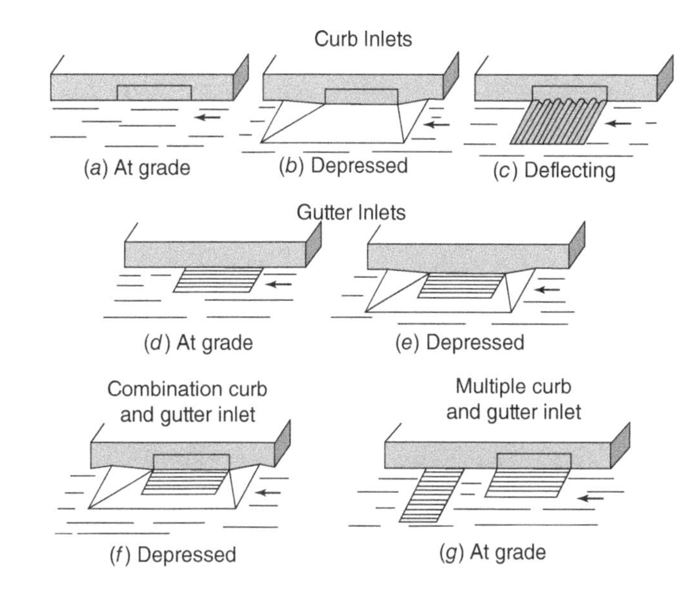

Figure 13.10 Types of Street Inlets. (a) At grade, (b) Depressed, (c) Deflecting, (d) At grade, (e) Depressed, (f) Depressed, (g) At grade

13.10 Street Inlets

Street inlets admitting storm waters to drainage systems are so placed and designed as to concentrate and remove the flow in gutters at minimum cost with minimum interference to both pedestrian and vehicular traffic. The key objective when designing inlets is to minimize the spread of water across a roadway and in the gutter. In stormwater drainage, the gutter is the channel on the side of the road through which stormwater runoff is conveyed to storm sewer inlets. Spread is the top width of the flowing water on the road, measured from the curb. Some features of design improve hydraulic capacity but are costly; other features interfere with traffic. Compromises produce a wide variety of designs.

Inlets are of three general types: *curb inlets, gutter inlets*, and *combination inlets* that combine curb openings with gutter openings (Figs. 13.10 and 13.11). Only where traffic is forced to move relatively slowly may gutter surfaces and gutter inlets be depressed to increase intake capacity.

Figure 13.11 Street Inlet (*Source:* Wikipedia, http://en.wikipedia.org/wiki/Image:Storm_DrainJPG)

The allowable spread length, which is generally determined by local or state regulations, is based on the classification of the road. For example, a road with a higher speed limit should have a smaller allowable spread than a road designed for slower speeds because of the increased risk of hydroplaning. In addition to spread width, roadway classification also dictates the return period of the design storm to use in calculating the spread at a point. Table 13.3, from the Federal Highway Administration (Federal Highway Administration 1996) HEC-22 manual, provides an overview of different road conditions and the design criteria these conditions necessitate.

The incoming surface flow (and spread) observed can be controlled by the efficiency and spacing of the inlets located upstream along the road. One additional factor to consider is whether the inlet is located on grade or in sag, because the design criteria and the equations involved differ. Inlets on grade are located on a slope and intercept a portion of the water as it flows past. Inlets in sag are located at a point where runoff from a given area will ultimately collect, and these inlets are normally designed to capture 100% of the surface flow; otherwise, flooding will occur in the surrounding area.

The intake capacity of inlets, particularly curb inlets, increases with decreasing street grade and increasing crown slope. However, curb inlets with diagonal deflectors in the gutter along the opening become more efficient as grades become steeper. Gutter inlets are more efficient than curb inlets in capturing gutter flow, but clogging by debris is a problem.

Table 13.3 Suggested Minimum Design Frequencies and Spreads for Gutter Sections on Grade

Road Classification		Design Return Period	Design Spreads
High volume or	<70 km/h (45 mph)	10-year	Shoulder + 1 m (3 ft)
Bidirectional	>70 km/h (45 mph)	10-year	Shoulder
	Sag point	50-year	Shoulder + 1 m (3 ft)
	Low volume	10-year	$\frac{1}{2}$ Driving lane
Collector	High volume	10-year	Shoulder
	Sag point	10-year	$\frac{1}{2}$ Driving lane
	Low volume	5-year	$\frac{1}{2}$ Driving lane
Local Streets	High volumes	10-year	$\frac{1}{2}$ Driving lane
	Sag point	10-year	$\frac{1}{2}$ Driving lane

Combination inlets are better still, especially if gratings are placed downstream from curb openings. Debris accumulating on the gratings will then deflect water into the curb inlets. Gratings for gutter inlets are most efficient when their bars parallel the curb. If crossbars are added for structural reasons, they should be kept near the bottom of the longitudinal bars. Depression of inlets, especially curb inlets, enhances their capacity. Long shallow depressions are as effective as short deep ones. If a small flow is allowed to outrun the inlet, the relative intake of water is greatly magnified. Significant economies are effected, therefore, by small carryover flows and their acceptance by downgrade inlets.

13.10.1 Gutter Sections on Grade

The main curb and gutter section types are the uniform section and the composite section, as illustrated in Fig. 13.12 with their defining variables. Uniform gutter sections have a constant slope across the section. Composite gutter sections are defined by a continuous *gutter depression*, a, measured from the bottom of the curb to the projected road cross-slope at the curb. The *frontal flow*, Q_w, is normally defined as the flow in the depressed section of the gutter or the flow over the grate width in the case of a grate inlet.

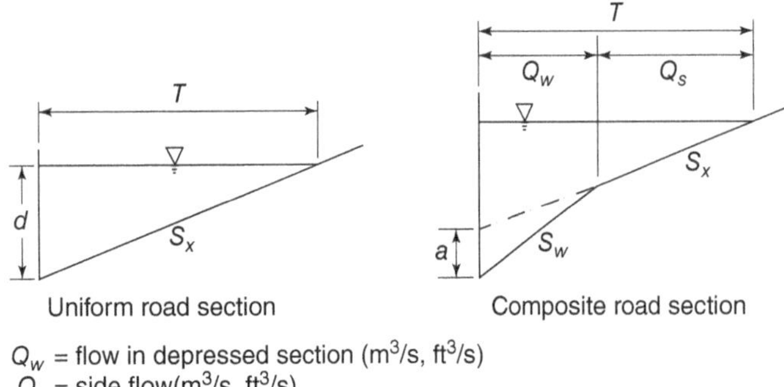

Q_w = flow in depressed section (m^3/s, ft^3/s)
Q_s = side flow(m^3/s, ft^3/s)
S_w = gutter cross-slope (m/m, ft./ft)
S_x = road cross-slope (m/m, ft./ft)
T = total width of flow or spread (m, ft)
W = width of flow in depressed section
d = depth of flow at the curb (mm, in.)
a = continuous gutter depression (mm, in.).

Figure 13.12 Uniform and Composite Gutter Sections

Runoff in gutters on grade is treated as open-channel flow, and Manning's equation is applicable. Because the friction along the curb height is negligible compared to the friction against the pavement width (spread) of flow, Manning's formula was modified, becoming the following equation for calculating flow in uniform gutter sections:

$$Q = \frac{K_e}{n} S_x^{1.67} S_L^{0.5} T^{2.67} \quad \text{(US customary system or SI system)} \tag{13.31}$$

where Q_w is flow in gutter (m^3/s, ft^3/s), S_x is road cross-slope (m/m, ft./ft), T is the total width of flow or spread (m, ft), K_c = 0.376 (0.56 in US customary units), n is Manning's coefficient (usually 0.015), and S_L is road longitudinal slope (m/m, ft./ft).

Using this equation, the *spread*, T, in a uniform gutter section can be explicitly calculated for a given flow rate, Q. However, in the case of composite gutters, T can no longer be expressed as an explicit function of Q. Rather, an iterative process is required to calculate the spread.

On a road with a grade, the spread will be at a maximum just upstream of the inlet. Note that the spread at this location is independent of the inlet's ability to capture flow and its efficiency. The spread for a specific roadway is a function of the discharge in the gutter. To decrease the flow to an inlet, reduce inlet spacings so that they serve as collection points for smaller watersheds.

For *grate inlets* in a uniform section, the variable E_0 is introduced to account for the frontal flow Q_w, which is now defined as the gutter flow contained in the width of the grate. In this case, E_0 becomes

$$E_0 = \left(1 - W_g/T\right)^{2.67} \tag{13.32}$$

where W_g is the width of grate inlet (m, ft).

Drainage inlets typically capture frontal flow more efficiently than side flow. In drainage inlet design, it is good practice to maximize E_0 by increasing the gutter cross-slope or by increasing the width of the gutter depression or the grate extending into the road.

Designs of inlets on grade are based on how much flow will be intercepted for a given total flow (gutter discharge) to the inlet. Inlet design equations solve for this efficiency:

$$E_0 = Q_w/Q \tag{13.33}$$

where E_0 is grate inlet efficiency, Q_w is intercepted frontal flow (m^3/s, ft^3/s), and Q is total gutter flow (m^3/s, ft^3/s).

Flow that is not intercepted by a drainage inlet on grade is bypassed and carried over to another inlet downstream, or is "lost" to a stream or pond, for example.

Example 13.14 Water Spread and Depth of Flow in the Gutter

For a flow of 1.0 ft^3/s (0.0283 m^3/s), a longitudinal street grade of 2.0%, a mean crosswise street grade of 5.6%, and an n coefficient of roughness of 0.015, find

1) the total width of spread, and
2) the maximum depth of flow in the gutter.

Solution 1 (US Customary System):

1) Using Eq. 13.31 and $K_c = 0.56$ in US units.

$$Q = \frac{K_c}{n} S_x^{1.67} S_L^{0.5} T^{2.67}$$
$$1 = (0.56/0.015)(0.056)^{1.67}(0.02)^{0.5}(T^{2.67})$$
$$T^{2.67} = 1/(37.3 \times 0.0081 \times 0.14) = 23.8$$
$$\boldsymbol{T = 23.8^{0.37} = 3.23\ \text{ft}\ (0.98\ \text{m})}\ \text{water spread}$$

2) From Fig. 13.12 and by definition:

$$d/T = S_x$$
$$d = TS_x = 3.23 \times 0.056 = 0.18\ \text{ft}$$
$$= \mathbf{2.2\ in.\ (56\ mm)}\ \text{maximum depth of flow in the gutter}$$

Solution 2 (SI System):

1) Using Eq. 13.31 and $K_c = 0.376$ in SI units.

$$Q = \frac{K_c}{n} S_x^{1.67} S_L^{0.5} T^{2.67}$$

$$0.0283 = \frac{0.376}{0.015}(0.056)^{1.67}(0.02)^{0.5} T^{2.67}$$

$$0.0283 = \frac{0.376 \times 0.0081 \times 0.1417}{0.015} T^{2.67}$$

$$T^{2.67} = 0.98$$

$$T = \mathbf{0.985\ m}$$

2) From Fig. 13.12, and by definition:

$$d/T = S_x$$

$$d = TS_x = (0.985\ \text{m})(0.056) = 0.055\ \text{m}$$

$$= \mathbf{55\ mm}$$

Example 13.15 Grate Inlet

If a 2-ft-wide (0.61-m) grate inlet is introduced in the gutter of Example 13.14, what will be the flow contained in the width of the grate?

Solution 1 (US Customary System):

From Example 13.11, we get $T = 3.23$ ft (0.98 m) and $Q = 1$ ft^3/s (28.3 L/s). From Eq. 13.32:

$$E_0 = 1 - (1 - W_g/T)^{2.67}$$

$$E_0 = 1 - (1 - 2/3.23)^{2.67}$$

$$= 1 - (1 - 0.62)^{2.67}$$

$$= 1 - (0.38)^{2.67}$$

$$= 1 - 0.075$$

$$= 0.925;\ \text{that is, } 92.5\%$$

$$Q_w = QE_0$$

$$Q_w = 1\ \text{ft}^3/\text{s} \times 0.925 = 0.92\ \text{ft}^3/\text{s}\ (26\ \text{L/s})$$

The gutter flow contained in the width of the grate inlet is 0.92 ft^3/s (26 L/s).

Solution 2 (SI System):

From Example 13.11, we get $T = 3.23$ ft (0.98 m) and $Q = 1$ ft^3/s (28.3 L/s). From Eq. 13.32:

$$E_0 = 1 - (1 - W_g/T)^{2.67}$$

$$E_0 = 1 - (1 - 0.61/0.98)^{2.67}$$

$$= 0.925;\ \text{that is, } 92.5\%$$

$$Q_w = QE_0$$

$$Q_w = (0.0283\ \text{m}^3/\text{s}) \times 0.925 = 0.026\ \text{m}^3/\text{s}\ (26\ \text{L/s})$$

The gutter flow contained in the width of the grate inlet is 0.026 m^3/s or 26 L/s.

13.10.2 Grate Inlets on Grade

Grate inlets, as shown in Fig. 13.13, tend to be more efficient than *curb inlets* when on a grade. Two concerns must be addressed in the grate inlet design process. First, the engineer needs to choose a grate type appropriate for the roadway being designed. Bicycle traffic, for example, would limit the engineer to grate types with both longitudinal and transverse

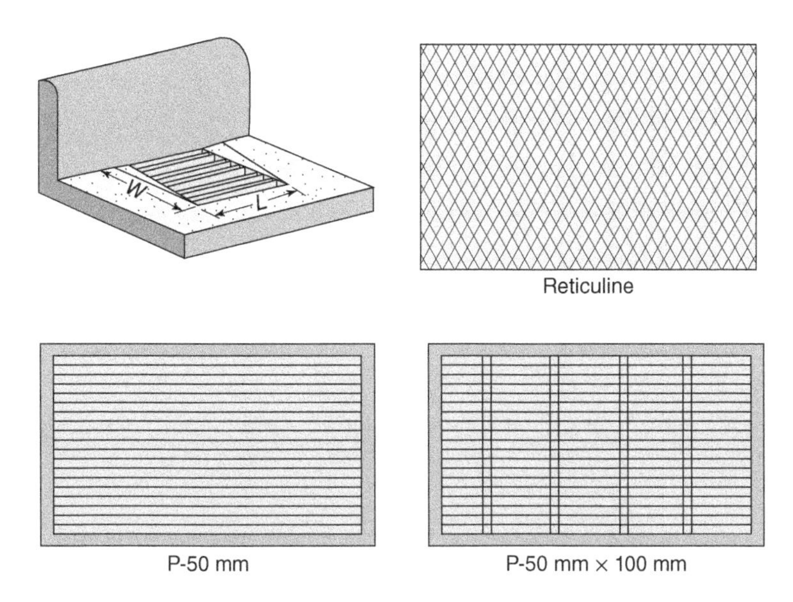

Figure 13.13 Grate Inlet in a Gutter and Some Typical Grate Types

bars in order to prevent bicycle accidents. Second, grate inlets have a higher propensity to clog than other types of inlets. If debris is prevalent in runoff at the point of design, adequate provisions must be made to account for inlet clogging, such as utilizing a combination (grate and curb) inlet at these points.

As shown in Fig. 13.12, the total gutter flow is composed of *frontal flow*, Q_w (the flow in the depressed gutter or over the grate width), and *side flow*, Q_s. The total efficiency of the grate inlet is determined by calculating the grate's ability to capture frontal flow and side flow. The term R_f represents the ratio of intercepted frontal flow to the total frontal flow, and is expressed as follows:

$$R_f = 1 - K_{cf}(V - V_0) \tag{13.34}$$

where $K_{cf} = 0.295$ (0.090 in US customary units), V is average velocity in the gutter at the location of the inlet (m/s, ft./s), and V_0 is splash-over velocity of the inlet (m/s, ft./s).

The *splash-over velocity* is the minimum velocity of the gutter flow capable of inducing enough momentum for some of the flow to skip over the grate opening and be carried over and downstream. The splash-over velocity is a function of the grate type and the grate length and can be found in the HEC-22 manual (Federal Highway Administration 1996) or obtained from the grate manufacturer. If the gutter velocity is less than the splash-over velocity, all frontal flow is intercepted and R_f equals 1.0.

The ratio, R_s, of side flow intercepted to total side flow is expressed as follows:

$$R_s = \left(1 + \frac{K_{cs}V^{1.8}}{S_x L^{2.3}}\right)^{-1} \tag{13.35}$$

where K_{cs} is 0.0828 (0.15 in US customary units) and L is the grate length (m, ft).

The total intercepted flow is expressed as

$$Q_i = Q_w R_f + Q_s R_s \tag{13.36}$$

The total efficiency of the grate inlet on grade is expressed as

$$E = R_f E_0 + R_s(1 - E_0) \tag{13.37}$$

Example 13.16 Intercepted Flow by Grate Inlet

What will be the flow intercepted by the grate inlet in the gutter of Examples 13.14 and 13.15? Assume the grate inlet length is 2 ft. (0.61 m) and its splash-over velocity, V_0, is 3 ft./s (0.91 m/s).

Solution 1 (US Customary System):

From Examples 13.14 and 13.15:

$$Q_g = 1\ \text{ft}^3/\text{s}\ (28.3\ \text{L/s})$$
$$T = 3.23\ \text{ft}\ (0.98\ \text{m})$$
$$W_g = 2\ \text{ft}\ (0.61\ \text{m})$$
$$L = 2\text{ft}\ (0.61\ \text{m})$$
$$d = 2.2\ \text{in} = 0.18\ \text{ft}\ (55\ \text{mm})$$
$$S_L = 2\%$$
$$S_x = 5.6\%$$
$$Q_w = 0.92\ \text{ft}^3/\text{s}\ (26\ \text{L/s})$$
$$V = Q_g/A_g$$
$$A_g = \text{triangular area of gutter flow} = Td/2 = 3.23 \times 0.18/2 = 0.29\ \text{ft}^2\ (0.027\ \text{m}^2)$$
$$V = Q_g/A_g = 1\ \text{ft}^3/\text{s}/0.29\ \text{ft}^2 = 3.45\ \text{ft/s}\ (1.05\ \text{m/s})$$

From Eq. 13.34:

$$R_f = 1 - K_{cf}\,(V - V_0)$$
$$= 1 - 0.090\,(3.45 - 3)$$
$$= 1 - 0.04 = 0.96$$

From Eq. 13.35:

$$R_s = \left(1 + \frac{K_{cs}V^{1.8}}{S_x L^{2.3}}\right)^{-1}$$

$$R_s = 1/\left(1 + \frac{0.15 \times 3.45^{1.8}}{0.056 \times 2^{2.3}}\right)$$

$$= 1/\left(1 + \frac{0.15 \times 9.29}{0.056 \times 2^{2.3}}\right)$$

$$= 0.165$$

From Eq. 13.36:

$$Q_i = Q_w R_f + Q_s R_s$$
$$= 0.92 \times 0.96 + (1 - 0.92) \times 0.165$$
$$= 0.8832 + 0.0132 = 0.8964\ \text{ft}^3/\text{s} = 0.90\ \text{ft}^3/\text{s}\ (25\ \text{L/s})$$

The flow intercepted by the grate inlet = 0.90 ft³/s (25 L/s).

Solution 2 (SI System):

From Examples 13.14 and 13.15:

$$Q_g = 1\ \text{ft}^3/\text{s}\ (28.3\ \text{L/s})$$
$$T = 3.23\ \text{ft}\ (0.98\ \text{m})$$
$$W_g = 2\ \text{ft}\ (0.61\ \text{m})$$
$$L = 2\text{ft}\ (0.61\ \text{m})$$
$$d = 2.2\ \text{in} = 0.18\ \text{ft}\ (55\ \text{mm})$$
$$S_L = 2\%$$

$S_x = 5.6\%$
$Q_w = 0.92\ \text{ft}^3/\text{s}\ (26\ \text{L/s})$
$V = Q_g/A_g$
A_g = triangular area of gutter flow = $Td/2 = 3.23 \times 0.18/2 = 0.29\ \text{ft}^2\ (0.027\ \text{m}^2)$
$V = Q_g/A_g = 1\ \text{ft}^3/\text{s}/0.29\ \text{ft}^2 = 3.45\ \text{ft/s}\ (1.05\ \text{m/s})$

Summary: $Q_g = Q = 28.3\ \text{L/s} = 0.0283\ \text{m}^3/\text{s}$

$T = 0.985\ \text{m}$
$W_g = 0.61\ \text{m}$
$L = 0.61\ \text{m}$ (given)
$d = 55\ \text{mm} = 0.055\ \text{m}$
$S_L = 2\% = 0.02$
$S_x = 5.6\% = 0.056$
$Q_w = 26\ \text{L/s} = 0.026\ \text{m}^3/\text{s}$
A_g = triangular area of gutter flow $= Td/2 = 0.985 \times 0.055/2 = 0.027\ \text{m}^2$
$V = Q_g/A_g = 0.0283/0.027 = 1.05\ \text{m/s}$
Given $V_0 = 0.91\ \text{m/s}$

From Eq. 13.34:

$R_f = 1 - K_{cf}(V - V_0)$
$R_f = 1 - 0.295\ (1.05 - 0.91) = 1 - 0.04 = 0.96$

From Eq. 13.35:

$$R_s = \left(1 + \frac{K_{cs}V^{1.8}}{S_x L^{2.3}}\right)^{-1}$$

$$R_s = \left(1 + \frac{0.0828 \times 1.05^{1.8}}{0.056 \times 0.61^{2.3}}\right)^{-1} = 0.165$$

From Eq. 13.36:

$Q_i = Q_w R_f + Q_s R_s$
$Q_i = (0.026\ \text{m}^3/\text{s}) \times 0.96 + (0.0283 - 0.026\ \text{m}^3/\text{s}) \times 0.165$
$= 0.025\ \text{m}^3/\text{s}$
= 25 L/s

13.10.3 Curb Inlets on Grade

Curb inlets are openings within the curb itself (see Fig. 13.14), and they are used in areas where grate inlets are prone to clogging. The efficiency of a curb inlet is based on the ratio of the actual inlet length to the inlet length necessary to capture 100% of the total runoff. Curb inlets are often inset into the pavement to create a local depression. A local depression, as shown in Fig. 13.15, is a depression of the gutter at the location of the inlet only, as opposed to a gutter depression, which is continuous along the curb.

The curb opening length, L_T, that would be required to intercept 100% of a flow, Q_i, on a roadway section with a uniform cross-slope is defined as

$$L_T = K_c Q^{0.42} S_L^{0.3} \left(\frac{1}{nS_x}\right)^{0.6} \tag{13.38}$$

where K_c is 0.817 (0.60 in US customary units).

To account for local depressions or gutter depressions, an additional composite or equivalent slope, S_e, is necessary:

$$S_e = S_x + S_{w'} E_0 \tag{13.39}$$

where $S_{w'}$ is the gutter cross-slope at the inlet location measured from the pavement cross-slope, S_x (m/m, ft./ft) and E_0 is the ratio of flow in the depressed section to the total gutter flow upstream of the inlet (does not account for local depression).

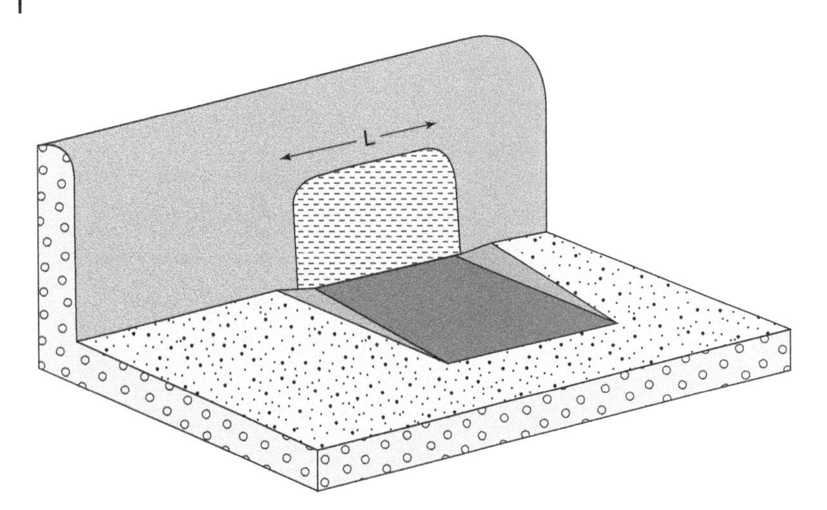

Figure 13.14 Curb Inlet

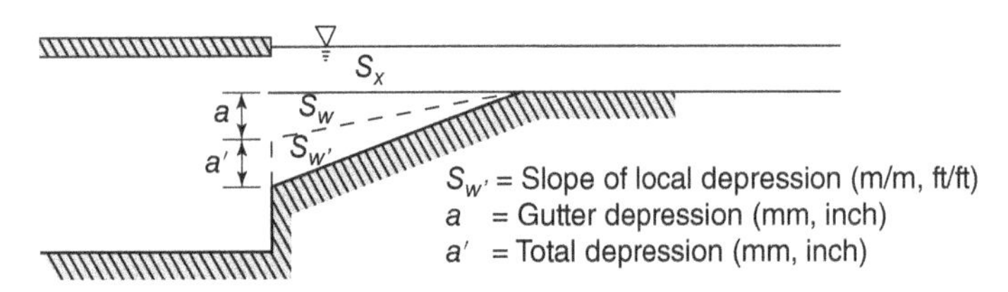

Figure 13.15 Local Depression at a Curb Inlet

To calculate L_T with a composite gutter, replace the road cross-slope S_x with the equivalent slope, S_e, in the equation solving for L_T.

The efficiency of a curb opening on grade shorter than the required length for total interception is

$$E = 1 - \left(1 - \frac{L}{L_T}\right)^{1.8}. \tag{13.40}$$

Example 13.17 Length of an Undepressed Curb Inlet

For a flow of 1.0 ft^3/s (28.3 L/s), a longitudinal street grade of 2.0%, a mean crosswise street grade of 5.6%, and an n coefficient of roughness of 0.015, find

1) the length of an undepressed curb inlet required to capture 100% of the flow, and
2) the length of an undepressed curb inlet required to capture 90% of the flow.

Solution 1 (US Customary System):

1) From Eq. 13.38:

$$L_T = K_c Q^{0.42} S_L^{0.3} \left(\frac{1}{nS_x}\right)^{0.6}$$

$$L_T = (0.60)(1)^{0.42}(0.02)^{0.3}[(1/0.015)(0.056)]^{0.6}$$

$$= (0.60)(0.31)(70) = 13 \text{ ft } (4 \text{ m})$$

The length of curb inlet required to capture 100% of the flow = 13 ft (4 m).

2) From Eq. 13.40:

$$E = 1 - \left(1 - \frac{L}{L_T}\right)^{1.8}$$

$$0.90 = 1 - (1 - L/13)^{1.8}$$

$$(0.10)^{0.55} = 1 - L/13$$

$$L = (1 - 0.28)(13) = 9.4 \text{ ft } (2.87 \text{ m})$$

The length of curb inlet required to capture 90% of the flow = 9.4 ft (2.87 m).

Solution 2 (SI System):

1) From Eq. 13.38:

$$L_T = K_c Q^{0.42} S_L^{0.3} \left(\frac{1}{nS_x}\right)^{0.6}$$

$$L_T = 0.817\ (0.0283)^{0.42}(0.02)^{0.3}(1/0.015 \times 0.056)^{0.6}$$

$$= 0.817 \times 0.2237 \times 0.3092 \times 70.0539$$

$$= 4\ \text{m}$$

The length of curb inlet required to capture 100% of the flow = 4 m.

2) From Eq. 13.40:

$$E = 1 - \left(1 - \frac{L}{L_T}\right)^{1.8}$$

$$0.9 = 1 - (1 - L/4)^{1.8}$$

$$L = 2.87\ \text{m}$$

The length of curb inlet required to capture 90% of the flow = 2.87 m.

13.10.4 Combination Inlets on Grade

Combination inlets, shown in Fig. 13.16, consist of both grate and curb openings. The curb inlet functions as a sweep, removing debris from the runoff before it can clog the grate inlet. If the curb inlet length is equal to the grate inlet length, the flow intercepted by the combination inlet is assumed to be equivalent to that intercepted by the grate inlet alone.

The curb inlet is often extended upstream of the grate for more efficient removal of debris than when the curb opening length equals the grate length. The total flow intercepted by this configuration is calculated as the sum of

1) the flow intercepted by the portion of the curb opening located upstream of the grate, and
2) the flow that bypassed the upstream curb opening and is intercepted by the grate alone (the flow intercepted by the portion of the curb opening adjacent to the grate is neglected).

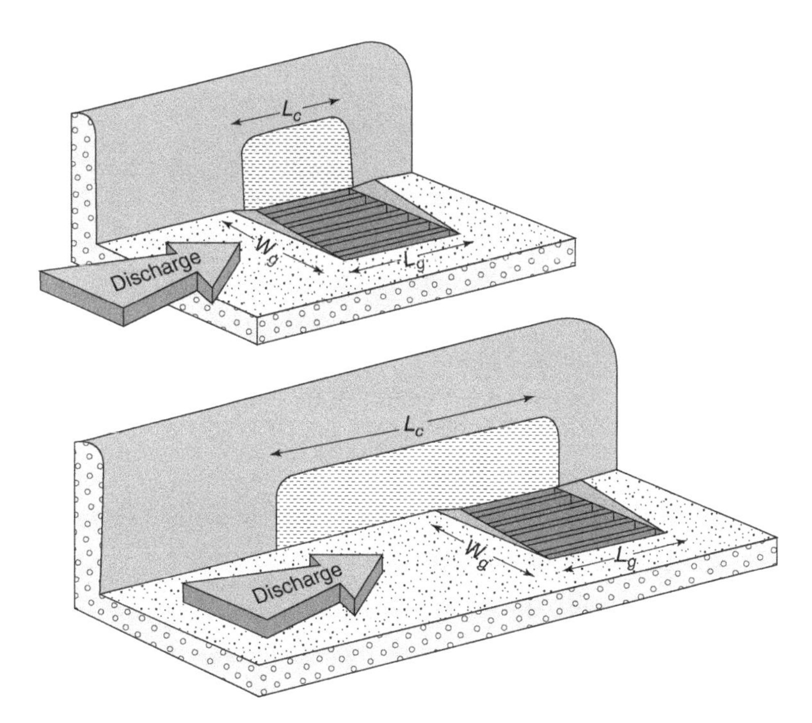

Figure 13.16 Typical Combination Inlets

13.10.5 Inlets in Sag

Inlets located in sag are assumed to capture 100% of flow because, once collected, the runoff in the sag has no other place to go. As opposed to inlets located on a grade, the size and type of inlet directly affect the spread. As shown earlier in Table 13.3, the HEC-22 manual (Federal Highway Administration 1996) typically recommends employing a larger rainfall return period for designing an inlet located in sag than for designing an inlet located on grade.

The computations for calculating the amount of flow intercepted by inlets in sag are based on the principles of weir flow and orifice flow. For an *unsubmerged inlet* operating as a weir, the flow capacity is calculated as

$$Q_{iw} = fC_wPd^{1.5} \tag{13.41}$$

where Q_{iw} is flow intercepted by the inlet operating as a weir (m^3/s, ft^3/s); C_w is weir coefficient, which varies depending on the flow condition and inlet structure; f is a factor; $f = 1$ for the US customary units; $f = 0.55$ for the SI units; P is the perimeter of the inlet (m, ft); and$d$ is flow depth at the curb (m, ft).

Note that for a grate inlet, the perimeter does not include the length along the curb. Also, if the gutter is depressed (locally or continuously), the perimeter, P, of the grate is calculated as

$$P = L + 1.8W \tag{13.42}$$

where L is the grate length (m, ft) and W is the grate width (m, ft).

The depth, d, for both types of inlets is measured from the projected pavement cross-slope. For a curb inlet, the perimeter is equivalent to the length of the inlet.

If the inlet is submerged and is operating as an orifice, its capacity becomes

$$Q_{io} = C_oA(2gd_0)^{0.5} \tag{13.43}$$

where Q_{i0} is flow intercepted by the inlet operating as an orifice (m^3/s, ft^3/s), C_0 is orifice coefficient (varies based on the class of inlet and its configuration), A is area of the opening (m^2, ft^2), g = 9.81 m/s^2 (32.2 ft./s^2 in US customary units), and d_0 is effective head at the orifice (m, ft).

Note that for a grate inlet, the effective head, d_0, is simply the water depth along the curb. For a curb inlet, the effective head is expressed as

$$d_0 = d_i - h/2(\sin\theta) \tag{13.44}$$

where d_i is depth at lip of curb opening (m, ft), h is curb throat opening height (m, ft) (see Fig. 13.17), and θ is the inclination of the curb throat measured from the vertical direction as shown in Fig. 13.17.

Grates alone are not typically recommended for installation in sags because of their propensity to clog and exacerbate ponding during severe weather. A combination inlet may be a better alternative. At low-flow depths, the capacity of a combination inlet for which the grate inlet length equals the curb inlet length is equivalent to the capacity of the grate inlet alone. At higher flow depths for the same type of inlet, both the curb inlet and grate inlet act as orifices. The total intercepted flow is then calculated as the sum of the flows intercepted by the grate and curb openings.

Modeling and computer applications of inlets design will be covered along with sewer networks in Chapter 15.

Example 13.18 Length of an Inlet Located in Sag

For a flow of 1.0 ft^3/s (28.3 L/s), a longitudinal street grade of 2.0%, a mean crosswise street grade of 5.6%, and an n coefficient of roughness of 0.015, find

1) the length of a submerged in sag grate inlet required (assume $C_o = 0.60$) and
2) the required length of unsubmerged in sag grate opening inlet. Assume $C_w = 3$ and $W_g = 1$ ft (0.3 m)

Solution 1 (US Customary System):

1) For 100% interception in sag, $Q_w = 1$ ft^3/s (28.3 L/s) and $C_o = 0.60$. From Example 13.14, $d_o = 2.2$ in = 0.18 ft (55 mm). From Eq. 13.43,

$$Q_{io} = C_oA(2gd_0)^{0.5}$$
$$1 = 0.60 \times A(2 \times 32.2 \times 0.18)^{0.5}$$
$$1 = 0.60 \times A(3.40) = 2.0\ A$$
$$A = 1/2.0 = 0.50\ \text{ft}^2\ (0.046\ \text{m}^2)$$

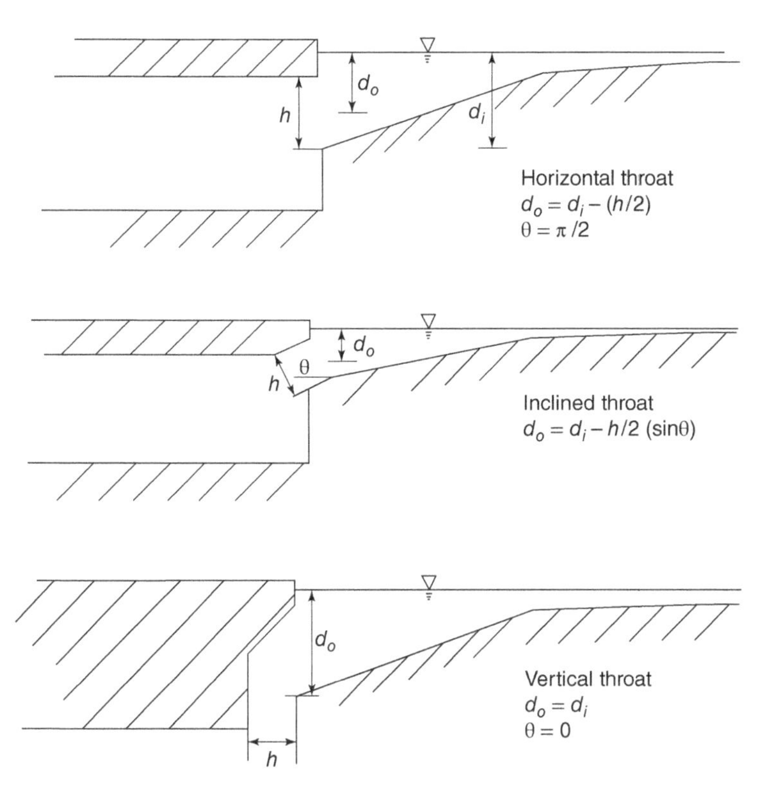

Figure 13.17 Different Curb Inlet Throat Types

Assuming the open area in the grate is 50%,
total grate area $= 2 \times 0.5 = 1$ ft^2 (0.09 m^2)

Using a 1-ft (0.30-m-)-wide grate inlet, the required inlet length will be 1 ft (0.30 m).

Note on the efficiency of submerged grate inlets in sag: Compare the size of this on-sag grate inlet (1 ft^2 in area) intercepting 100% of the gutter flow to the on-grade grate inlet in Example 13.16 (4 ft^2 or 0.37 m^2 in area): four times bigger and intercepts only 92% of the gutter flow.

2) Assume $C_w = 3.0$ and $W_g = 1$ ft. (0.3 m) For 100% interception in sag, $Q_{iw} = 1$ ft^3/s (28.3L/s) and $W_g = 1$ ft (0.3 m). From Example 13.14, $d = 2.2$ in $= 0.18$ ft (55 mm). From Eq. 13.41,

$$Q_{iw} = fC_wPd^{1.5}$$
$$1 = (1)(3.0)\ P\ (0.18)^{1.5}$$
$$1 = 0.23\ P$$
$$P = 1/0.23 = 4.35 \text{ ft } (1.33 \text{ m})$$

Since,

$$P = L + 2W_g \text{ (the perimeter does not include the length along the curb)}$$
$$4.35 = L + 2 \times 1$$
$$L = 2.35 \text{ ft } (0.72 \text{ m})$$

The required inlet length is 2 ft 6 in. (750 mm).

Solution 2 (SI System):

1) For 100% interception in sag, $Q_w = 1$ ft^3/s (28.3 L/s) and $C_o = 0.60$. From Example 13.14, $d_o = 2.2$ in $= 0.18$ ft (55 mm)., From Eq. 13.43:

$$Q_{io} = C_oA(2gd_0)^{0.5}$$
$$0.0283 = 0.60A(2 \times 9.81 \times 0.055)^{0.5}$$
$$A = 0.045 \text{ m}^2$$

Assuming the open area in the grate is 50%, total grate area = $2 \times 0.045 = 0.09\ m^2$.

Using a 0.3-m-wide grate inlet, the required inlet length will be 0.3 m. *Note on the efficiency of submerged grate inlets in sag:* Compare the size of this on-sag grate inlet (0.09 m^2 in area) intercepting 100% of the gutter flow to the on-grade grate inlet in Example 13.16 (0.37 m^2 in area): four times bigger and intercepts only 92% of the gutter flow.

2) Assume $C_w = 3.0$ and $W_g = 0.3$ m. For 100% interception in sag, $Q_{iw} = 0.0283\ m^3/s$, $C_o = 3.0$, and $W_g = 0.3$ m.
   From Example 13.14, $d = 55$ mm = 0.055 m
   From Eq. 13.41:

$$Q_{iw} = fC_w P d^{1.5}$$
$$0.0283 = 0.55\,(3.0)\,P\,(0.055)^{1.5}$$
$$P = 1.33\ \text{m}$$

Since
$$P = L + 2W_g$$
$$1.33 = L + 2 \times 0.3$$
$$L = 0.72\ \text{m}$$

The required inlet length is 750 mm.

Problems/Questions

13.1 Using the US customary system nomogram of Fig. 13.1a:

(a) Given a 10-in. sewer, $N = 0.013$, laid on a grade of 3.0‰ (ft per 1000 ft), find its velocity of flow and rate.
(b) Given a velocity of 3.3 ft./s for this sewer, find its (minimum) gradient for flow at full depth.

13.2 Using the SI System nomogram of Fig. 13.1b:

(a) Given a 250-mm sewer, $N = 0.013$, laid on a grade of 3.0‰ (m per 1000 m), find its velocity of flow and rate of discharge.
(b) Given a velocity of 1 m/s for this sewer, find its (minimum) gradient for flow at full depth.

13.3 The Manning formula can be written as follows:

$$v = (1.49/n)(A/P_w)^{2/3}(s)^{1/2}$$

where v is the velocity of flow (ft/s), n is the coefficient of roughness, A is the cross-sectional area of pipe (ft^2), P_w is wetted perimeter (ft), s is the slope, or rate of grade (dimensionless), and $A/_{wp} = r$ = hydraulic radius (ft).

Derive an alternative Manning equation for a square conduit, with a width of W and its length equals to its width.

13.4 What will be the Manning equation for a rectangular conduit with a dimension of $L \times W$, where L is length in ft. and W is width in ft.?

13.5 What will be the Manning equation for a circular pipe with a diameter of D in ft.?

13.6 Although an environmental engineer may use the table in Appendix 7 and the nomogram in Fig. 13.1, for instance, for engineering design, he or she should also be familiar with the applicable design equations, especially when computer analysis is involved. The Manning equation shown as Eq. 13.3 may be further transformed into the following design equations for computer-aided engineering design assuming circular pipes are used:

$$v = (1.49/n)(0.25D)^{2/3}(s)^{1/2}$$
$$Q = A\,v$$
$$A = 0.785\left(D^2\right)$$

where Q is flow (ft^3/s), A is the cross-sectional area of pipe (ft^2), v is the velocity of flow (ft/s), n is the coefficient of roughness, D is the inside diameter of pipe (ft), s is the slope, or rate of grade (dimensionless)

Given a 12-in. circular sewer, $N = 0.013$, laid on a grade of 4.05‰ (ft per 1000 ft), find its velocity of flow and rate of discharge using the above design equations.

13.7 The Manning equation, Eq. 13.3a, may be further transformed using the known equations in Problem 13.6 and the following equations for calculation of the head loss and hydraulic gradient of a circular sewer when it flows full.

$$
\begin{aligned}
s &= 4.6417\, Q^2\, n^2/\left(D^{5.3334}\right) \\
&= 4.6417\, Q^2\, n^2 D^{-5.333} \\
&= 2.86\, v^2\, n^2 D^{-1.333} \\
H_f &= 4.6417\, Q^2\, n^2 (L)/\left(D^{5.3334}\right) \\
&= 4.6417\, Q^2\, n^2 (L)\left(D^{-5.3334}\right) \\
&= 2.86\, v^2\, n^2 (L)\left(D^{-1.333}\right) \\
s &= H_f/L
\end{aligned}
$$

where $H_f =$ is hydraulic head loss (ft), $L =$ is length of the pipe (ft), Q is flow (ft^3/s), A is the cross-sectional area of pipe (ft^2), v is the velocity of flow (ft/s), n is the coefficient of roughness, D is the inside diameter of pipe (ft), and s is the slope, or rate of grade (dimensionless).

Given a 12-in. sewer, $N = 0.013$, laid on a grade of 4.05‰ (ft per 1000 ft), find (a) its minimum hydraulic gradient for flow at full depth using equations in Problems 13.3 and 13.4, (b) its head loss through 1000 ft. of sewer, and (c) its head loss through 500 ft. of sewer.

13.8 Given a square sewer ($W \times W = 12$ in. $\times$ 12 in.), $n = 0.013$, laid on a grade of 4.05‰ (ft per 1000 ft), find its velocity of flow (v) and rate of discharge (Q) using the design equations from Problem 13.3.

13.9 Given a rectangular sewer ($L \times W = 24$ in. $\times$ 12 in.) $= 4.6417\, L\, (Q\, n)^2\, D^{-5.3334} n = 0.013$, laid on a grade of 4.05‰ (ft per 1000 ft), find the velocity of flow (v) and rate of discharge (Q) using the design equations from Problem 13.4.

13.10 A formula for the required diameter of a circular sewer flowing full, expressed in terms of the required capacity and the planned slope, can be derived by substituting values for $A = 0.785D^2$, $r = 0.25\, D$, and V in the formula $Q = Av$, and solving the resulting equation for the diameter D. The following are the derived equations based on the Manning formula:

$$
\begin{aligned}
D &= \left[2.16\, Q\, n(s)^{-0.5}\right]^{3/8} \\
&= 1.33481 (Q\, n)^{0.375} (s)^{-0.1875} \\
&= 2.2 (v\, n)^{1.5} (s)^{-0.75}
\end{aligned}
$$

where v is the velocity of flow (ft/s), n is the coefficient of roughness, s is the slope, or rate of grade (dimensionless), Q is flow (ft^3/s), and D is the inside diameter of pipe (ft).

The preceding equation is derived assuming the sewer is flowing full. If the sewer is flowing only half full, the Q value must be adjusted to full flow before the above equation can be applied. A sewer pipe is to discharge 4 ft^3/s when laid on a grade of 0.0016 and flowing only half full. Determine the required diameter (D) if the velocity (v) is to be determined by the Manning equation.

13.11 The Manning formula can be rearranged for solving the sewer pipe's slope or rate of grade as follows

$$
\begin{aligned}
s &= \left[v\, n/\left(1.49\, r^{2/3}\right)\right]^2 \\
&= H_f/L \\
Q &= Av
\end{aligned}
$$

When a circular pipe is used, $r = 0.25\, D$, the following working equations can further be derived.

$$s = 2.86(v\,n)^2\,D^{-1.333}$$
$$= 4.6417(Q\,n)^2\,D^{-5.3334}$$
$$H_f = 2.86\,L\,(v\,n)^2\,D^{-1.333}$$
$$= 4.6417\,L\,(Q\,n)^2\,D^{-5.3334}$$

where Q is flow (ft^3/s), A = cross-sectional area of pipe (ft^2), v is the velocity of flow (ft/s), n is the coefficient of roughness, D is the inside diameter of pipe (ft), s is the slope, or rate of grade (dimensionless), H_f is hydraulic head loss (ft), and L is the length of the pipe (ft).

Determine the required velocity (v), the required grade (s), and the total head loss (H_f) of a 48-in. circular pipe when it is flowing full at a discharge rate (Q) of 100 ft^3/s. The coefficient of roughness (n) is 0.015 and the length of the pipe (L) is 1000 ft.

13.12 The following are two general equations for illustration of (a) the relationship between the flow rate in pipe 1 where $n = n_1$ and the flow rate in pipe 2 where $n = n_2$ and (b) the relationship between the velocity in pipe 1 where $n = n_1$ and the velocity in pipe 2 where $n = n_2$.

$$Q_2 = Q_1(n_1/n_2)$$
$$v_2 = v_1(n_1/n_2)$$

where Q_2 is flow rate in pipe 2 (ft^3/s or m^3/s), when the coefficient of roughness = n_2, Q_1 is flow rate in pipe 1 (ft^3/s or m^3/s), when the coefficient of roughness = n_1, v_2 is velocity in pipe 2 (ft/s or m/s), when the coefficient of roughness = n_2, v_1 is velocity in pipe 1 (ft/s or m/s), when the coefficient of roughness = n_1, n_1 is the coefficient of roughness of pipe 1 (dimensionless), and n_2 the coefficient of roughness of pipe 2 (dimensionless).

Determine the flow rate for $n = 0.015$ if the flow rate for $n = 0.013$ is known to be 2.26 ft^3/s. Determine the velocity for $n = 0.015$ if the velocity for $n = 0.013$ is known to be 2.89 ft./s.

13.13 The Manning formula for the SI system can be written as follows:

$$v = (1/n)(A/P_w)^{2/3}(s)^{1/2}$$

where v is the velocity of flow (m/s), n is the coefficient of roughness, A is the cross-sectional area of pipe (m^2), P_w is wetted perimeter (m), s is slope (dimensionless), and $A/P_w = r$ = hydraulic radius (m).

Derive the alternative Manning equations in the SI System for a square conduit, with a width of W and length equals to its width.

13.14 What will be the Manning equations in the SI System for a rectangular conduit with dimensions of $L \times W$, where L is length in m and W is width in m?

13.15 What will be the Manning equation in the SI System for a circular pipe with a diameter of D in m?

13.16 An environmental engineer may use the nomogram in Fig. 13.1, for instance, for engineering design. He or she should, however, also be familiar with the design equations, especially when computer analysis is involved. The Manning equation shown as Eq. 13.3b may be further transformed into the following design equations in the SI System for computer-aided engineering design assuming circular pipes are used:

$$v = (1/n)(0.25D)^{2/3}(s)^{1/2}$$
$$Q = A\,v$$
$$A = 0.785(D^2)$$

where Q is flow (m^3/s), A is the cross-sectional area of pipe (m^2), v is the velocity of flow (m/s), n is the coefficient of roughness, D is the inside diameter of pipe (m), and S is the slope, or rate of grade (dimensionless).

Derive Eq. 13.5b from Eq. 13.3b.

Given a 30.48-cm circular sewer, $n = 0.013$, laid on a grade of 4.05‰ (m per 1000 m), find its velocity of flow and rate of discharge using the preceding design equations.

13.17 The Manning equation, Eq. 13.3b, can be further derived using the known SI System equations in Problem 13.16 and the following equations for calculation of the head loss and hydraulic gradient of a circular sewer when it flows full.

$$s = 10.24557Q^2n^2/\left(D^{5.3334}\right)$$
$$= 10.24557Q^2n^2\, D^{-5.3334}$$
$$= 6.31725v^2n^2\, D^{-1.333}$$
$$H_f = 10.24557\, Q^2n^2(\mathrm{L})/\left(D^{5.3334}\right)$$
$$= 10.24557\, Q^2n^2(\mathrm{L})\left(D^{-5.3334}\right)$$
$$= 6.31725\, v^2n^2(\mathrm{L})\left(D^{-1.333}\right)$$
$$s = H_f/L$$

where H_f is hydraulic head loss (m), L is the length of the pipe (m), Q is flow (m^3/s), A is the cross-sectional area of pipe (m^2), v is the velocity of flow (m/s), n is the coefficient of roughness, D is the inside diameter of pipe (m), and S is the slope, or rate of grade (dimensionless).

Given a 0.3048-m sewer, $n = 0.013$, laid on a grade of 4.05‰ (m per 1000 m), find (a) its minimum hydraulic gradient for flow at full depth using the equations in Problems 13.13 and 13.14, (b) its head loss through 304.8 m of sewer, and (c) its head loss through 152.4 m of sewer.

13.18 Given a square sewer ($W \times W = 0.3048$ m $\times$ 0.3048 m), $n = 0.013$, laid on a grade of 4.05‰ (m per 1000 m), find its velocity of flow (v) and rate of discharge (Q) using the SI System design equations from Problem 13.3.

13.19 Given a rectangular sewer ($L \times W = 0.6096$ m $\times$ 0.3048 m), $n = 0.013$, laid on a grade of 4.05‰ (m per 1000 m), find its velocity of flow (v) and rate of discharge (Q) using the SI System design equations from Problem 13.4.

13.20 A SI System formula for the required diameter of a circular sewer flowing full, expressed in terms of the required capacity and the planned slope, can be derived by substituting values for $A = 0.785\, D^2$, $r = 0.25\, D$, and V in the formula $Q = AV$, and solving the resulting equation for the diameter D. The following are the derived SI System equations based on SI System Manning formula:

$$D = 3.9849\,(v\,n)^{1.5}(s)^{-0.75}$$
$$= 1.5475\,(Q\,n)^{0.375}(s)^{-0.1875}$$

where v is the velocity of flow (m/s), n is the coefficient of roughness, s is the slope, or rate of grade (dimensionless), Q is flow (m^3/s), and D is the inside diameter of pipe (m).

The preceding SI System equation is derived assuming the sewer is flowing full. If the sewer is flowing only half full, the Q value must be adjusted to full flow before the preceding SI System equation can be applied.

A sewer pipe is to discharge 0.11328 m^3/s (or 4 ft^3/s) when laid on a grade of 0.0016 and flowing only half full. Determine the required diameter (D) if the velocity (v) is to be determined by the Manning equation.

13.21 The SI System Manning formula can be rearranged for solving the sewer pipe's slope or rate of grade as follows:

$$s = \left[vn/\left(r^{2/3}\right)\right]^2$$

$$s = H_f/L$$

$$Q = Av$$

When a circular pipe is used, that is, when $r = 0.25\ D$, the following working equations can be derived:

$$s = 6.31725\,(vn)^2 D^{-1.333}$$

$$= 10.24557\,(Qn)^2 D^{-5.3334}$$

$$H_f = 6.31725\,L\,(vn)^2 D^{-1.333}$$

$$= 10.24557\,L\,(Qn)^2 D^{-5.3334}$$

where Q is flow (m^3/s), A is the cross-sectional area of pipe (m^2), v is the velocity of flow (m/s), n is the coefficient of roughness, D is the inside diameter of pipe (m), s is the slope, or rate of grade (dimensionless), H_f is hydraulic head loss (m), and L is the length of the pipe (m).

Determine the required velocity (v), the required grade (s), and the total head loss (H_f) of a 1.2192-m (or 48-in.) circular pipe when it is flowing full at a discharge rate (Q) of 2.832 m^3/s (or 100 ft^3/s). The coefficient of roughness (n) is 0.015 and the length of the pipe (L) is 304.8 m (or 1000 ft).

13.22 For any partially filled circular sewer laid on any grade, the cross-sectional area of the water, the velocity of flow and the water flow rate are laborious, and a chart for partially full circular sewers (Appendix 10) is normally used. For any ratio of depth of flow to diameter of sewer, the curves in Appendix 9 give the ratios of the area, velocity, and flow for that depth to the corresponding values for the circular sewer flowing full. The capacity of a circular sewer flowing full is 7.08 m^3/s (or 250 ft^3/s). Determine

(a) the depth of flow, in terms of the diameter of the sewer, and
(b) the relative velocity when the discharge is 2.832 m^3/s (or 100 ft^3/s).

13.23 A trunk sewer, 1200 mm in diameter ($n = 0.013$), is placed on a slope of 0.003 between two manholes 120 m apart. The average design flow in the line is 667 L/s and the minimum and peak flows are 0.5 and 2.5 of the average flow, respectively.

(a) What would be the depth of flow in the pipe at average flow?
(b) Would the pipe need flushing at regular short intervals? Show your supporting computations.
(c) If the depth of flow is 900 mm, what is the discharge?
(d) What would be the difference in elevation between the inverts at both ends of the pipe?
(e) Compute the maximum population that can be served by this line. Assume that the average per capita contribution is 380 L/day.
(f) Suppose that at the downstream end manhole, this sewer joins with a branch sewer that is 600 mm in diameter, has a slope of 0.005, and a peak flow of 250 L/s. At what height above the invert of the trunk sewer should the invert of the branch sewer be located, so that at design peak flows, there will be no backing up of sewage into the branch sewer?

13.24 The sewage discharges from two neighborhoods A and B have to be disposed of through an existing trunk sewer at MH 100 (see Fig. 13.18). The average sewage flow from the two neighborhoods is 250 Lpcd, and the peaking factor for both areas is 3.0. Neighborhood A has 4300 dwelling units with six persons per unit (n for all sewers is 0.013).

If the sewage level is 626.50 m during peak flow in the existing trunk at MH 100, determine the following:

(a) The maximum number of dwelling units that can be served from neighborhood B.
(b) The size and slope of lines A and B if the velocity at peak flow is not to be less than 1 m/s.
(c) The velocities and depths of flow in lines A and B at average flow condition.

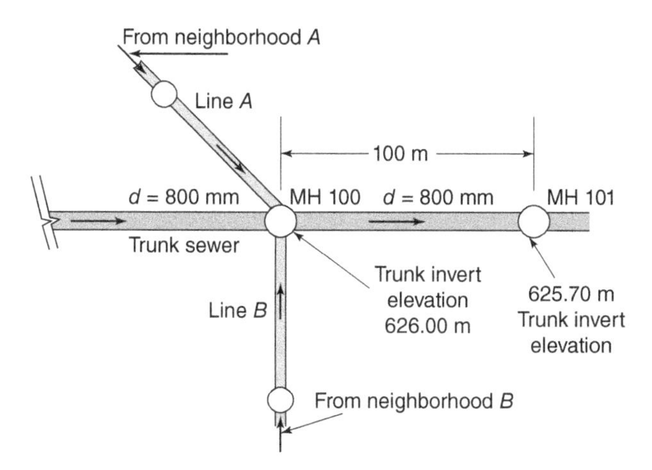

Figure 13.18 Sewer System for Problem 13.24

13.25 Two lateral sewers *zm* and *xm* discharge their sewage flows into the trunk sewer *ymn* at manhole *m* (Fig. 13.19). The average sewage flow in the city is 300 Lpcd and the peaking factor is 2.5. The following information is given:

(a) Trunk sewer *ym* has a diameter of 1000 mm, is constructed at a slope of 0.1%, and has a peak flow of 530 L/s.
(b) Lateral sewer *xm* serves a population of 20,000 persons.
(c) Lateral sewer *zm* serves an area of 150 ha (ha) with a population density of 100 capita/ha.
(d) The minimum and maximum allowable velocities at full flow condition in sewers are 0.75 and 3.0 m/s, respectively.
(e) The Manning *n* for all sewers is 0.013.
(f) The ground surface along the street from MH 3 to *m* and *z* is flat.
(g) The ground slope along the street from *m* to *n* is 0.05%.

Determine the following:

(a) The required size of trunk sewer from MH *m* to MH *n*
(b) The required diameter and slope of laterals *xm* and *zm*
(c) The depth of flow and velocity in sewer *xm* at average flow
(d) The depth of flow and velocity in trunk sewer *ym* at peak flow.

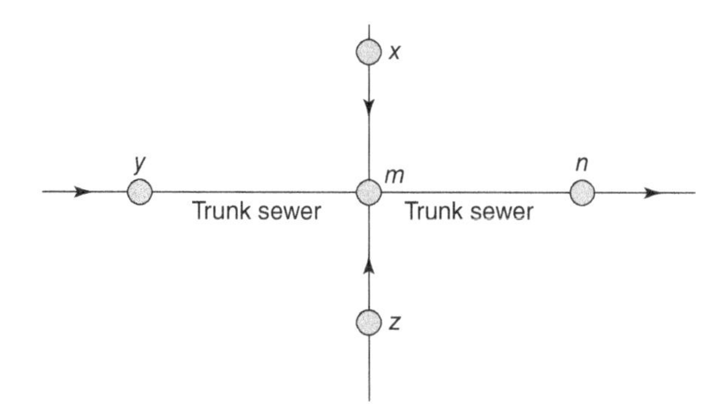

Figure 13.19 Sewer System for Problem 13.25

13.26 Find the minimum velocity and gradient at which particulate material having a diameter of 1.5 mm, a specific gravity *s* of 2.15, and a sediment characteristic *k*′ of 0.04 is transported without hindrance through a sewer 18 in. in diameter that has a Manning *n* of 0.013 flowing full.

13.27 A 12-in. (300-mm) sewer is to discharge 0.82 ft^3/s (0.0232 m^3/s) at a velocity as self-cleaning as a sewer flowing full at 3.0 ft./s (0.91 m/s). Find the depth and velocity of flow and the required slope.

13.28 Find the rate of propagation of a discontinuous surge front that raises the flow depth from 2 to 3 ft. (0.61 to 0.91 m).

13.29 Stormwater is diverted over a side weir as shown earlier in Fig. 13.9. The total flow in the storm sewer, q_1, is 25 ft^3/s (0.71 m^3/s); the diverted stormwater flow is 10 ft^3/s (0.28 m^3/s); the downstream flow in the sewer pipe, q_2, is 15 ft^3/s (0.42 m^3/s); average cross-sectional area, a, is 35 ft^2 (3.25 m^2); average hydraulic radius, r, is 1.7 ft. (0.52 m); slope of weir, i, is 1.5×10^{-4}, Manning's n is 0.013; and water head on the downstream side of the weir, h_2, is 0.60 ft. (0.18 m). Determine the weir length, L, and the head on the upstream side of the weir, h_1; assume $c = 3.33$.

13.30 A stormwater grate inlet is located at the point in the gutter where the water depth in the gutter reaches 6 in. (152 mm). The street has a slope of 0.06 and a transverse slope of 0.04. Manning's n for the gutter is 0.022.

(a) Determine the flow rate in the gutter at the inlet.
(b) If the grate inlet is 1.5 ft. (0.46 m) wide, what will be the flow contained in the width of the grate?

13.31 Grated inlets are located in the gutters of the Main Street where the water depth next to the curb reaches 12.5 cm. All grates are 60 cm wide. The Main Street has a slope of 0.0625 and a transverse slope of 0.050. Manning's n for the gutter is 0.018.

(a) Determine the flow rate in the gutter at the inlet.
(b) Determine the flow contained in the width of the grate.
(c) What will be the flow intercepted by the grate inlet? Assume the grate inlet length is 2 m and its splash-over velocity, V_0, is 1 m/s.

13.32 The layout of a storm water drainage system is shown in Fig. 13.20. The area is residential and it is estimated that the coefficient of runoff, C, will be 0.50 at the time of maximum development.

The street has a slope of 2% and a transverse slope of 4%. The gutters have a Manning coefficient n of 0.016. The overland flow velocity is 6 m/min and the maximum allowable depth of water at the curb is 100 mm. The design rainfall intensity (i in mm/h) as a function of rainfall duration (t in min) for the area is

$$i = \frac{810}{t + 11}$$

(a) Determine the maximum flow, Q_g, that can be handled by the gutter.
(b) Determine the distance, L, at which the first inlets should be installed. (You can assume that the velocity of flow in the gutters is constant.)

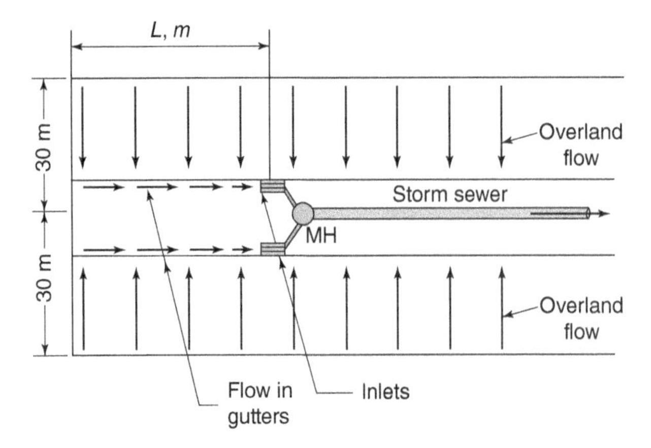

Figure 13.20 Sewer System for Problem 13.32

(c) Find the curb opening length, L_T, that would be required to intercept 100% of the gutter flow.
(d) Determine the length of the undepressed curb inlet if it needs to capture 80% of the flow.

13.33 A residential area with an estimated runoff coefficient $C = 0.65$ is shown in Fig. 13.21. The street has a slope of 6% and a transverse slope of 3%. The gutter has a Manning coefficient of 0.018. The inlet time (overland = gutter) is 16.6 min. The design rainfall intensity (i in mm/h) as a function of rainfall duration (t in min) is given by the following expression:

$$i = \frac{2,000}{t + 25}$$

Determine the following:

(a) The maximum flow in the gutter, if the water depth at the curb is not allowed to exceed 12 cm
(b) The distance × at which the inlet should be installed
(c) The length of submerged in-sag grate inlet required (assume $C_o = 0.60$)
(d) The required length of unsubmerged in-sag grate opening inlet (assume $C_w = 3$ and $W_g = 40$ cm).

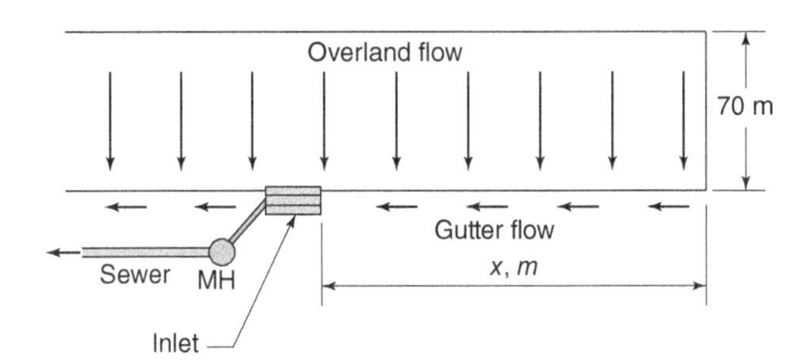

Figure 13.21 Sewer System for Problem 13.33

Bibliography

American Society of Civil Engineers and Water Environment Federation, *Gravity Sanitary Sewer Design and Construction*, 2nd ed., ASCE Manual No. 60 and WEF Manual No. FD-5, Reston, VA ed., 2007.

Camp, T. R., Design of Sewers to Facilitate Flow, *Sewage Work. J., 18*, 3, 1946.

Cassidy, J. J., Generalized Hydraulics of Grade Inlets, *Highway Res. Record, 123*, 36, 1966.

Chanson, H., *The Hydraulics of Open Channel Flow*, 2nd ed., Elsevier Butterworth-Heinemann, Burlington, MA, 2004.

Chaudry, M. H., *Open-Channel Flow*, 2nd ed., Springer-Verlag, New York, 2007.

Chow, V. T., *Open-Channel Hydraulics*, McGraw-Hill, New York, 1959.

Clark, J. W., Viessman, W., and Hammer, M. J., *Water supply and Pollution Control*, Harper & Row Publishers, New York, USA. 857 pp., 1977.

Federal Highway Administration, *Urban Drainage Design Manual*, HEC-22, Washington, DC, 1996.

Flynn, P. J., *Flow of Water in Open Channels, Pipes, Sewers, Conduits, etc.*, Kessinger Publishing, Whitefish, MT, 2007.

Forchheimer, P., *Hydraulik*, B. Teubner, Leipzig, p. 406, 1930.

Haestad Methods, *Computer Applications in Hydraulic Engineering*, 7th ed., Bentley Institute Press, Exton, PA, 2007.

Jain, S. C., *Open-Channel Flow*, Wiley, New York, 2000.

Johnson, C. F., Determination of Kutter's n for Sewers Partly Filled, *Trans. Am. Soc. Civil Engrs., 109*, 240, 1944.

Keifer, G. J., and Chu, H. H., Backwater Functions by Numerical Integration, *Trans. Am. Soc. Civil Engrs., 120*, 429, 1955.

Li, W. H., Geyer, J. C., Benton, G. S., and Sorteberg, K. K., Hydraulic Behavior of Storm-Water Inlets—Parts I and II, *Sewage Ind. Waste., 23*, 34, 722, 1951.

Munson, B. R., Young, D. F., and Okiishi, T. H., *Fundamentals of Fluid Mechanics*, 3rd ed., Wiley, New York, 1998.

Rouse, H. (Ed.), *Engineering Hydraulics*, Wiley, New York, 1950, p. 72.

Von Seggern, M. E., Integrating the Equation of Non-uniform Flow, *Trans. Am. Soc. Civil Engrs., 115*, 71, 1950.

Willcox, E. R., A Comparative Test of the Flow of Water in 8-inch Concrete and Vitrified Clay Sewer Pipe, *Univ. Wash. Eng. Exp. Sta. Ser. Bull, 27*, 1924.

Yarnell, D. L., and Woodward, S. M., The Flow of Water in Drain Tile, *U.S. Dep. Agr. Bull., 854*, 1920.

14

Design of Sewer Systems

14.1 Drainage of Buildings

The water distributed to basins, sinks, tubs, bowls, and other fixtures in dwellings and other buildings, and to tanks and other equipment in industrial establishments, is collected as spent water by the drainage system of the building and run to waste (Fig. 14.1). Plumbing fixtures are arranged singly or in batteries. They empty into substantially *horizontal branches* or *drains* that must not flow full or under pressure if tributary fixtures are to discharge freely and their protecting traps are not to become unsealed. The horizontal drains empty into substantially vertical *stacks*. These, too, must not flow full if wastewaters are not to back up into fixtures on the lower floors. The drainage stacks discharge into the *building drain*, which, 5 ft (1.5 m) outside of the building, becomes the *building sewer* (or *house sewer*) and empties into the *street sewer* (Figs. 14.2–14.4).

Traps are either part of the drainage piping or built into fixtures such as water closets. The traps hold a water seal that obstructs, and essentially prevents, foul odors and noxious gases, as well as insects and other vermin, from passing through the drainage pipes and sewers into the building. Discharging fixtures send water rushing into the drains and tumbling down the stacks; air is dragged along, and air pressures above or below atmospheric within the system might *unseal* the traps were it not for the provision of *vents*. These lead from the traps to the atmosphere and thereby equalize the air pressures in the drainage pipes.

The wastewater from fixtures and floor drains below the level of the public sewer must be lifted by ejectors or pumps (Fig. 14.1). *Sumps* or receiving tanks facilitate automatic operation. Sand and other heavy solids from cellars or yards are kept out of the drainage system by *sand interceptors*, grease by *grease interceptors*, and oil by *oil interceptors*. To act as separators or traps, these generally take the form of small settling, skimming, or holding tanks.

Cast-iron, galvanized-steel or wrought-iron, plastic, brass, and copper piping are employed for drains and vents aboveground; cast iron and plastic are used for drains laid belowground. Building sewers are constructed of vitrified-clay, plastic, or cast-iron pipe. Stormwater from roofs and paved areas taken into a property drain is discharged into the street gutter or directly into the storm sewer. In *combined systems*, roof water may be led into the house drain and water from yard areas into the house sewer. Otherwise, storm runoff travels over the ground, reaches the street gutter, flows along it, enters a stormwater inlet or a catch basin, and is piped to a manhole, from which it then empties into the drainage system.

In *separate systems*, connections to the wrong sewer, in violation of common regulations, carry some stormwater into sanitary sewers and some domestic wastewater into storm drains. The *dry-weather flow* of combined sewers is primarily wastewater and groundwater; the *wet-weather flow* is predominantly storm runoff. The first flush of stormwater will scour away deposited solids, including much putrescent organic matter.

14.2 Gravity Sewers

Sewers are hydraulic conveyance structures that carry wastewater to a treatment plant or other authorized point of discharge. A typical method of conveyance used in sewer systems is to transport wastewater by gravity along a downward-sloping pipe gradient. These sewers, known as *conventional gravity sewers*, are designed so that the slope and size of the pipe are adequate to maintain flow toward the discharge point without surcharging manholes or pressurizing the pipe.

Water and Wastewater Engineering: Hydraulics, Hydrology and Management, Volume 1, Fourth Edition.
Lawrence K. Wang, Mu-Hao Sung Wang, and Nazih K. Shammas.

Companion website: www.wiley.com/go/Wang/Waterandwastewater4e

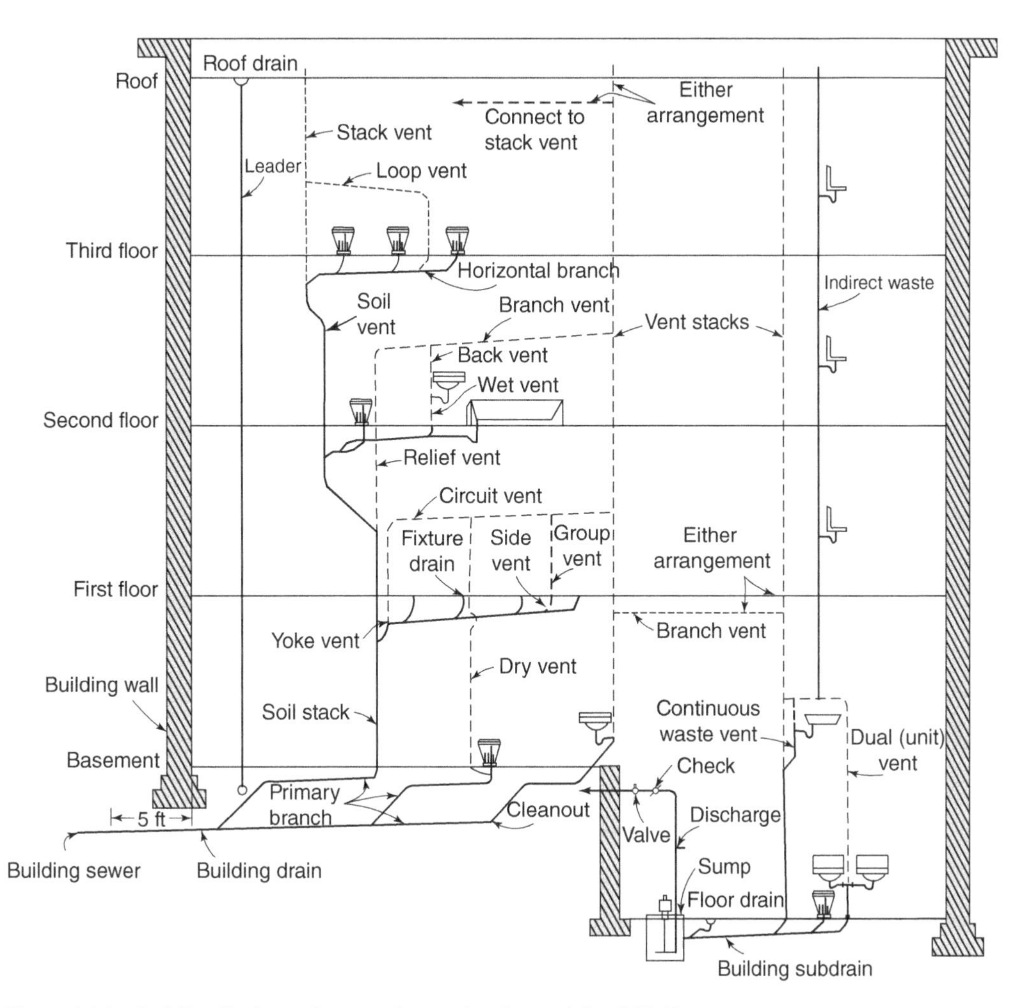

Figure 14.1 Building Drainage System. Conversion factor: 1 ft = 0.3048 m

Figure 14.2 Connecting Building Drainage System to Sewer (House Trap May Be Installed or Omitted). Conversion factor: 1 ft = 0.3048 m; 1 in. = 2.54 cm = 25.4 mm

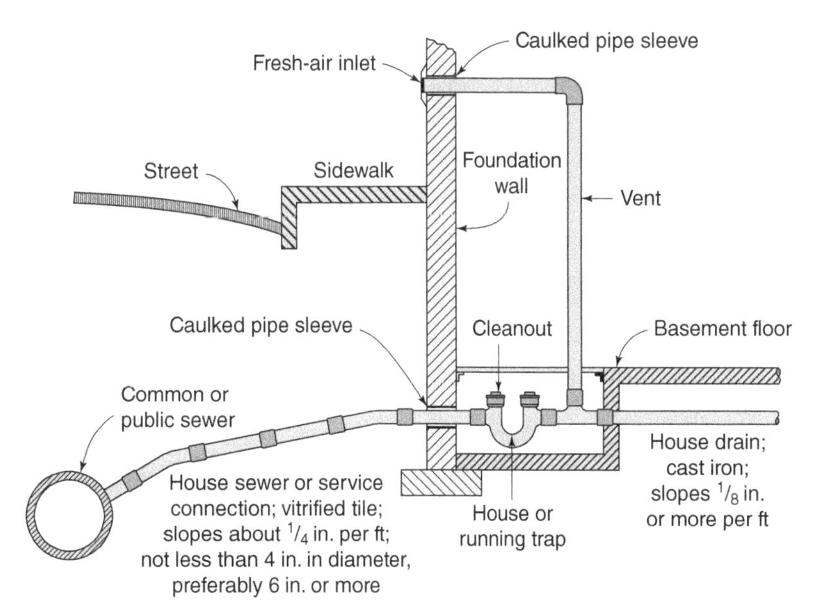

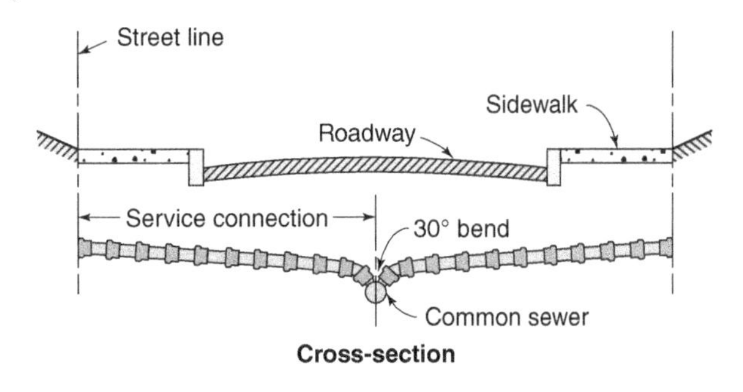

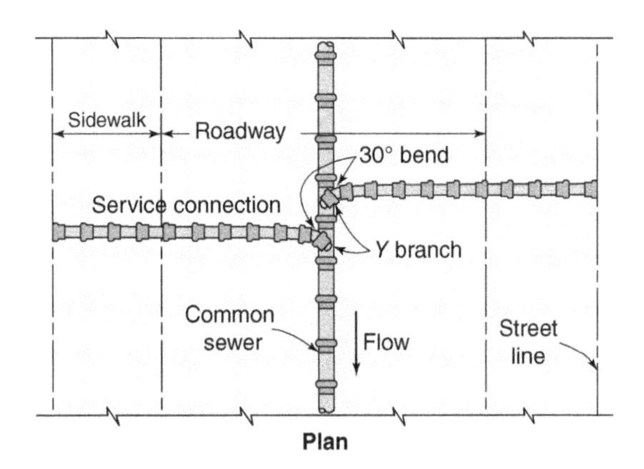

Figure 14.3 Service Connections to Public Sewer at Normal Depth

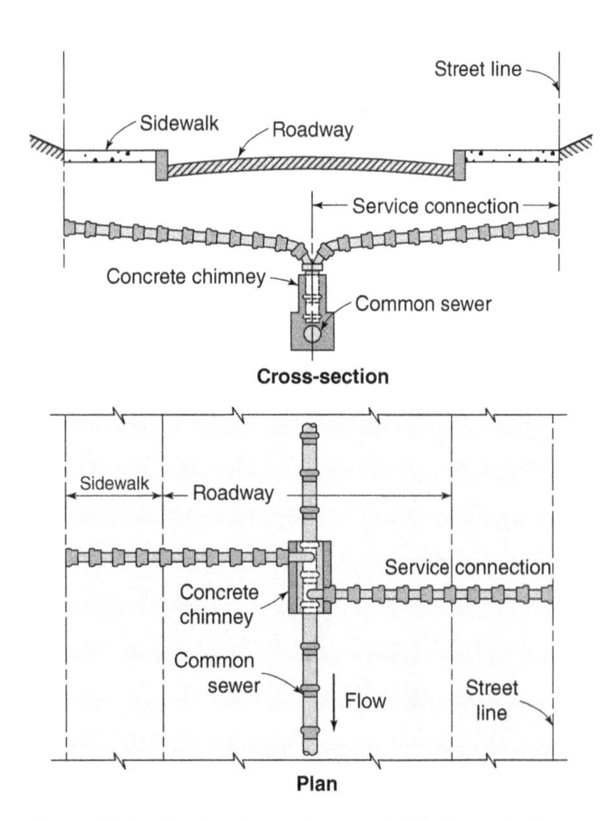

Figure 14.4 Service Connections to Public Sewer in Deep Trench.

Gravity sewers are the oldest and most common wastewater transport system existing. Technology advancement is limited principally to improvement in materials and methods of construction.

Sewers are commonly referred to according to the type of wastewater that each transports. For example, *storm sewers* carry stormwater, *industrial sewers* carry industrial wastes, and *sanitary sewers* carry both domestic sewage and industrial wastes. Another type of sewer, known as a *combined sewer*, is prevalent in older communities, but such systems are no longer constructed. Combined sewers carry domestic sewage, industrial waste, and stormwater.

14.2.1 Applicability

Conventional gravity sewers are typically used in urban areas with consistently sloping ground because excessively hilly or flat areas result in deep excavations and drive up construction costs. Stoppage is frequent due to grease, sedimentation, tree root development, and, in the case of combined sewers, debris. Excessive infiltration and inflow are the most common problems for both old and new systems.

14.2.2 Advantages and Disadvantages

Conventional gravity sewer systems have been used for many years and procedures for their design are well established. When properly designed and constructed, conventional gravity systems perform reliably. Properly designed and constructed conventional gravity sewers provide the following advantages:

1) Can handle grit and solids in sanitary sewage.
2) Can maintain a minimum velocity (at design flow), reducing the production of hydrogen sulfide and methane. This in turn reduces odors, blockages, pipe corrosion, and the potential for explosion.

Disadvantages of conventional gravity sewers include the following:

1) The slope requirements to maintain gravity flow can require deep excavations in hilly or flat terrain, driving up construction costs.
2) Sewage pumping or lift stations may be necessary as a result of the slope requirements for conventional gravity sewers, which result in a system terminus (i.e., low spot) at the tail of the sewer, where sewage collects and must be pumped or lifted to a collection system. Pumping and lift stations substantially increase the cost of the collection system.
3) Manholes associated with conventional gravity sewers are a source of inflow and infiltration, increasing the volume of wastewater to be carried, as well as the size of pipes and lift/pumping stations, and, ultimately, increasing costs.

14.2.3 Collection of Spent Waters

Systems of sanitary sewers are shown in Chapter 10, Fig. 10.1. Because about 70% of the water brought into a community must be removed as spent water, the average flow in sanitary sewers is about 100 gpcd (gallons per capita per day), or 378 Lpcd (liters per capita per day), in North America. Variations in water use step up the maximum hourly rate about threefold. Illicit stormwater and groundwater magnify the required capacity still further, and a design value of 500 gpcd (1,892 Lpcd) is not uncommon.

Sanitary sewers are fouled by the deposition of waste matters unless they impart *self-cleaning velocities* of 2–2.5 ft/s (0.60–0.75 m/s). Except in unusually flat country, sewer grades are made steep enough to generate these velocities when the sewers are running reasonably full (half full or more in circular sections, because the hydraulic radius of a semicircle equals that of a circle). Nevertheless, there will be deposition of solids. To find and remove them, sewers must be accessible for inspection and cleaning. Except in large sewers, *manholes* are built at all junctions with other sewers, and at all changes in direction or grade. Straight runs between manholes can then be rodded out effectively if intervening distances are not too great. Maxima of 300 or 400 ft (91.4 or 121.9 m) for pipes less than 24 in. (600 mm) in diameter are generally specified, but effective cleaning is the essential criterion. For larger sewers, distances between manholes may be upped to as much as 600 ft (183 m). Sewers so large that workmen can enter them for inspection, cleaning, and repair are freed from these restrictions, and access manholes are placed quite far apart either above the center line or tangential to one side. Introduction of flexible cleaning devices has encouraged the construction of curved sewers of all sizes, especially in residential areas.

A plan and profile of a sanitary sewer and its laterals are shown in Figs. 14.5 and 14.6, together with enlarged sections of sewer trenches and manholes. On short runs (<150 ft. or 46 m) and temporary stubs of sewer lines, terminal *cleanouts* are sometimes substituted for manholes. They slope to the street surface in a straight run from a Y in the sewer or in a gentle curve that can be rodded out. To keep depths reasonably small and pumping reasonably infrequent in very flat country and in other unusual circumstances, sewers are laid on flat grades, in spite of greater operating difficulties.

The smallest public sewers in North America are normally 8 in. (200 mm) in diameter. Smaller pipes clog more frequently and are harder to clean. Vitrified clay and plastic are the material of choice for small sewers, and prefabricated lined concrete and fiberglass for large sewers. Vitrified-clay sewer pipe, 4–36 in. (100–900 mm) in diameter, is ordinarily 3–6 ft (0.90–1.83 m) long. Unreinforced-concrete sewer pipe, 4–24 in. (100–600 mm) in diameter, is generally 2–4 ft (0.60–1.22 m) long. Preformed joints made of resilient plastic materials increase the tightness of the system. To reduce the infiltration of groundwater, sewers laid without factory-made joints in wet ground must be undertrained, or made of cast iron, plastic, or other suitable materials. Cast-iron and plastic pipes are long and their joints are tight. Underdrainage is by porous pipes or clay pipes laid with open joints in a bed of gravel or broken stone beneath the sewer. Underdrains may serve during construction only or become permanent adjuncts to the system and discharge freely into natural water courses. Sewage seeping into permanent underdrains may foul receiving waters. As stated before, grit or other abrading materials will wear the *invert* of concrete sewers unless velocities are held below 8–10 ft/s (2.4–3 m/s). Very large sewers are built in place, some by tunneling. Hydraulically and structurally, they share the properties of grade aqueducts.

Sewers are laid deep enough with adequate ground cover to:

1) Protect them against breakage by traffic shock.
2) Keep them from freezing.
3) Permit them to drain the lowest fixture in the premises served.

Common laying depths are 3 ft (0.90 m) below the basement floor and 11 ft (3.35 m) below the top of building foundations (12 ft or more for basements in commercial districts), together with an allowance of 0.3 in. per ft (2.5%) for the slope of the building sewer. At this slope a 6-in. (150-mm) sewer flowing full will discharge about 300 gpm or 40 ft^3/min (1,135 L/min or 1.13 m^3/min) at a velocity of 3.5 ft/s (1.0 m/s). In the northern United States, cellar depths range from 6 to 8 ft and frost depths from 4 to 6 ft (1.22–1.83 m). A 2-ft (0.60-m) earth cover will cushion most shocks. The deep basements of tall buildings are drained by ejectors or pumps.

As shown in Fig. 14.5, manholes are channeled to improve flow, and the entrance of high-lying laterals is eased by constructing drop manholes rather than going to the expense of lowering the last length of run. In their upper reaches, most sewers receive so little flow that they are not self-cleaning and must be flushed from time to time. This is done by

1) Damming up the flow at a lower manhole and releasing the stored waters after the sewer has almost filled
2) Suddenly pouring a large amount of water into an upstream manhole

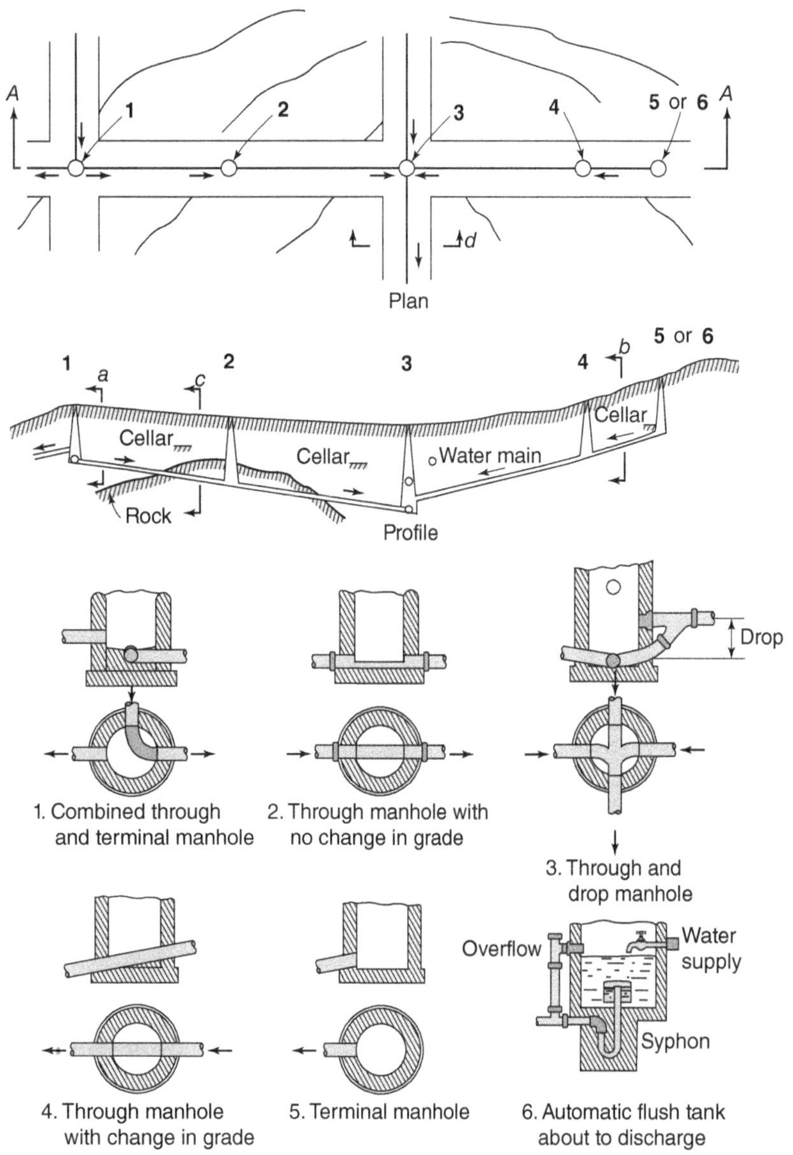

Figure 14.5 Plan, Profile, and Constructional Details of Sanitary Sewers

3) Providing at the uppermost end of the line a *flushing manhole* that can be filled with water through a fire hose attached to a nearby *hydrant* before a flap valve, shear gate, or similar quick-opening device leading to the sewer is opened
4) Installing an automatic *flush tank* that fills slowly and discharges suddenly. Apart from the cost and difficulties of maintenance, the danger of backflow from the sewer into the water supply is a disadvantage of automatic flush tanks.

14.3 Collection of Stormwaters

Much of the suspended load of solids entering storm drains is sand and gravel. Because fine sand is moved along at velocities of 1 ft/s (0.3 m/s) or more and gravel at 2 ft/s (0.60 m/s) or more, recommended minimum velocities are 2.5–3 ft/s (0.75–0.90 m/s), or about 0.5 ft/s (0.15 m/s) more than for sanitary sewers. The following factors determine the capacity of storm drains:

1) *Intensity* and *duration* of local rainstorms
2) Size and runoff characteristics of tributary areas
3) Economy of design, determined largely by the opportunity for quick discharge of collected stormwaters into natural water courses.

Figure 14.6 Sewer Trenches, Access Manholes, and Terminal Cleanouts. Conversion factor: 1 in. = 25.4 mm. Various pipe construction styles in earch trench (a) , in cradle (b), in rock tranch (c), and with underdrain (d) are available

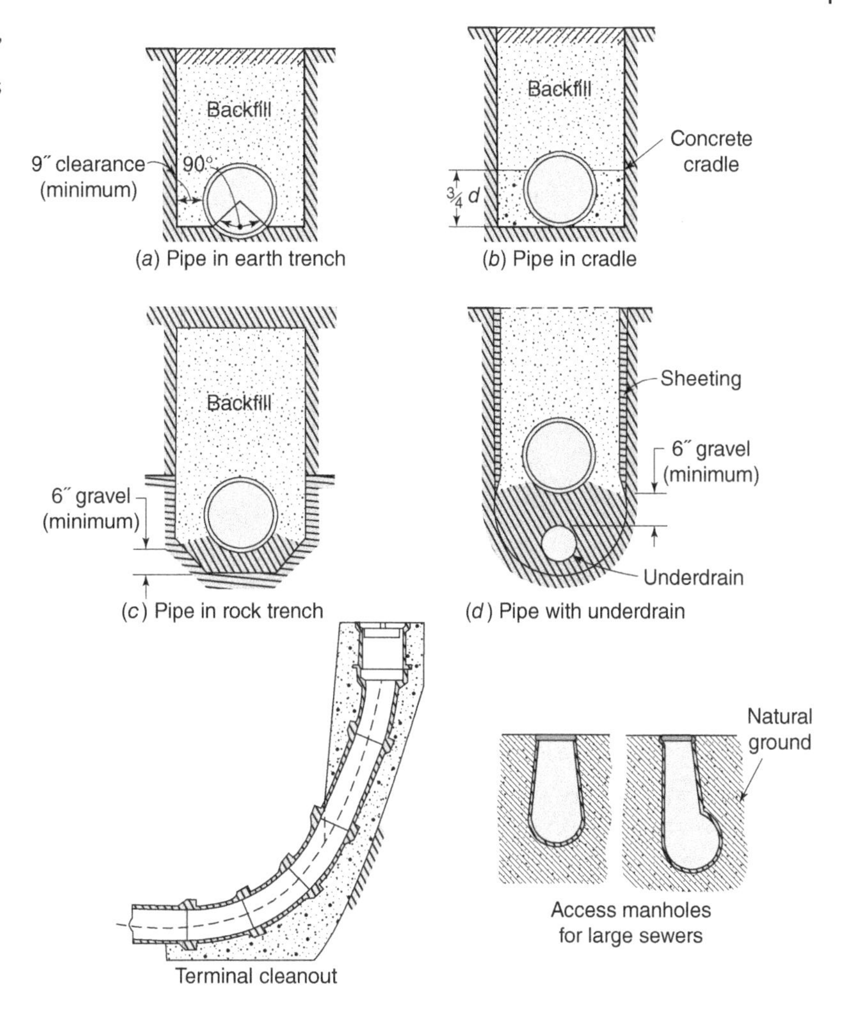

Rate of storm runoff is ordinarily the governing factor in the hydraulic design of storm drains. To prevent inundation of streets, walks, and yards and flooding of basements and other low-lying structures, together with attendant inconvenience, traffic disruption, and damage to property, storm sewers are made large enough to drain away, rapidly and without becoming surcharged, the runoff from storms shown by experience to be of such intensity and frequency as to be objectionable. The heavier the storm, the greater but less frequent the potential inconvenience or damage; the higher the property values, the more sizable is the possible damage. In a well-balanced system of storm drains, these factors will have received proper recognition for the kind of areas served: residential, mercantile, industrial, and mixed. For example, in high-value mercantile districts with basement stores and stockrooms, storm drains may be made large enough to carry away surface runoff from all but unusual storms, estimated to occur only once in 5, 10, 20, 50, or even 100 years, whereas the drains in suburban residential districts are allowed to be surcharged by all storms larger than the one- or two-year storm.

Until there are storm drains in a given area and the area itself is developed to its ultimate use, runoff measurements are neither possible nor meaningful. Accordingly, the design of storm sewers is normally based not on analysis of recorded runoff but on (a) analysis of storm rainfalls—their intensity or rate of precipitation, duration, and frequency of occurrence—and (b) estimation of runoff resulting from these rainfalls in the planned development.

Storm sewers are occasionally surcharged and subjected to pressures, but usually no more than their depth below street level. Nevertheless, they are designed for open-channel flow and equipped with manholes in much the same way as sanitary sewers (Fig. 14.7). In North American practice, the minimum size of storm sewers is 12 in. (300 mm), to prevent clogging by trash of one kind or another. Their minimum depth is set by structural requirements rather than basement elevations.

Figure 14.7 Typical Concrete Storm Sewer Installation (*Source:* Wikipedia, http://en.wikipedia.org/wiki/Image:Stmsewer.jpg.)

Surface runoff enters from *street gutters* through *street inlets* or *catch basins* (Figs. 14.8 and 14.9) and property drains. Size, number, and placement of street inlets govern the degree of freedom from flooding of traffic ways and pedestrian crossings. To permit inspection and cleaning, it is preferable to discharge street inlets directly into manholes. Catch basins are, in a sense, enlarged and trapped street inlets in which debris and heavy solids are held back or settle out. Historically, they antedate street inlets and were devised to protect combined sewerage systems at a time when much sand and gravel were washed from unpaved streets. Historically, too, the air in sewers, called *sewer gas*, was once deemed dangerous to health; this is why catch basins were given water-sealed traps. Catch basins need much maintenance; they should be cleaned after every major storm and may have to be oiled to prevent production of large crops of mosquitoes. On the whole, there is little reason for continuing their use in modern sewerage systems.

14.4 Combined Collection of Wastewaters and Stormwaters

In *combined sewerage systems*, a single set of sewers collects both domestic and industrial wastewater and surface runoff from rainfall (Fig. 10.1). Because stormwaters often exceed sewage flows

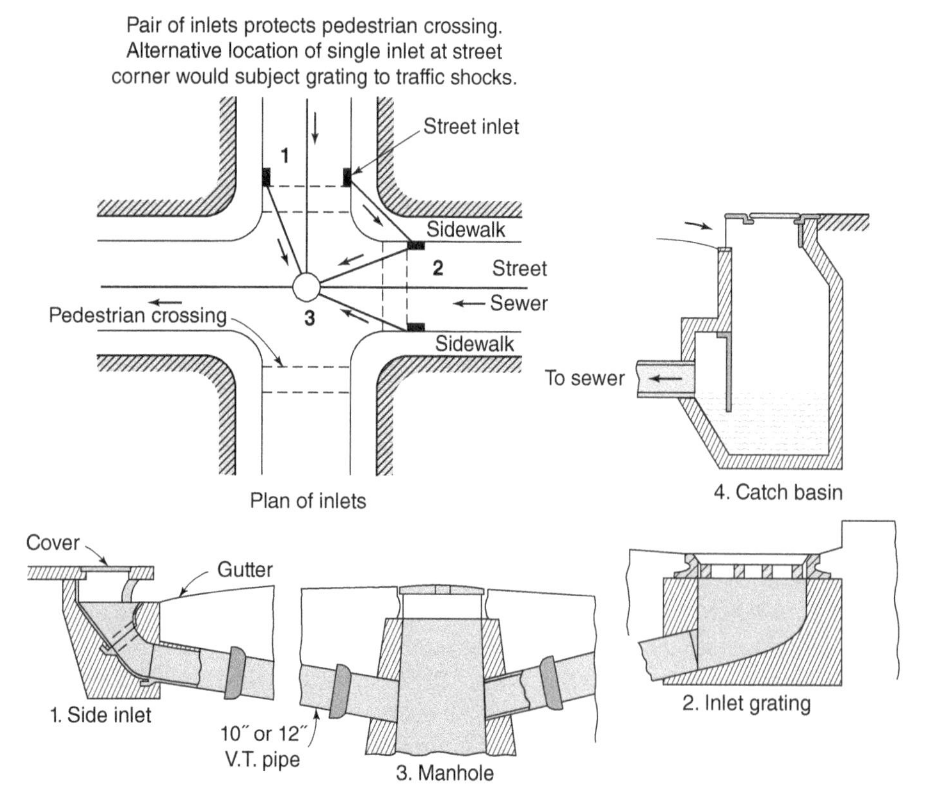

Figure 14.8 Street Inlets and Their Connection to a Manhole. Conversion factor: 1″ = 1 in. = 25.4 mm

Figure 14.9 Typical Catch Basin Connection to Storm Sewer (*Source:* Wikipedia, http://upload.wikimedia.org/wikipedia/en/9/92/Stmsewerandcb.jpg)

by 50–100 times, the accuracy with which rates of surface runoff can be estimated is generally less than the difference between rates of stormwater and combined-sewage flows. Accordingly, most combined sewers are designed to serve principally as storm drains. However, they must be placed as deep as sanitary sewers. The backing up and overflow of combined sewage into basements and streets is obviously more objectionable than the surcharge of drains that carry nothing but stormwater. Combined sewers are given velocities up to 5.0 ft/s (1.5 m/s) to keep them clean.

The wide range of flows in combined sewers requires the solution of certain special problems, among them choice of a cross-section that will ensure self-cleaning velocities for both storm and dry-weather flows; design of self-cleaning inverted siphons—also called sag pipes and depressed sewers—dipping beneath the hydraulic grade line as they carry wastewater across a depression or under an obstruction; and provision of stormwater overflows in intercepting systems.

14.4.1 Cross-Sections

Departures from circular cross-sections are prompted by hydraulic as well as structural and economic considerations. Examples are the egg-shaped sections (Fig. 14.10a), semicircular section with cunettes (Fig. 14.10b), rectangular section (Fig. 14.10c), and horseshoe section (Fig. 14.10d). Two circular sewers, an underlying sanitary sewer and an overlying storm drain are fused into a single *egg-shaped section*. The resulting hydraulic radius is nearly constant at all depths. *Cunettes* form troughs dimensioned to the dry-weather flow. Rectangular sections are easy to construct and make for economical trenching with low head-room requirements. *Horseshoe sections* are structurally very satisfactory; egg-shaped sections are not. Large outfall sewers have been built as pressure tunnels.

14.4.2 Inverted Siphons

Siphons flow full and under pressure, and the velocities in them are relatively much more variable than in open channels, where depth and cross-section change simultaneously with flow. To keep velocities up and clogging by sediments down, two or more parallel pipes are, therefore, thrown in and out of operation as flows rise and fall. The pipes dispatch characteristic flows at self-cleaning velocities of 3 ft/s (0.90 m/s) for pipes carrying sanitary sewage and 5 ft/s (1.5 m/s) for pipes conveying storm or combined sewage. The smallest pipe diameter is 6 in. (150 mm), and the choice of pipe material is adjusted to the hydrostatic head under which it must operate.

Figure 14.11 shows a simple example: Low dry-weather wastewater flows are passed through the central siphon; high dry-weather flows and storm flows spill over weirs into lateral siphons to right and left. The three siphons combine to equal the

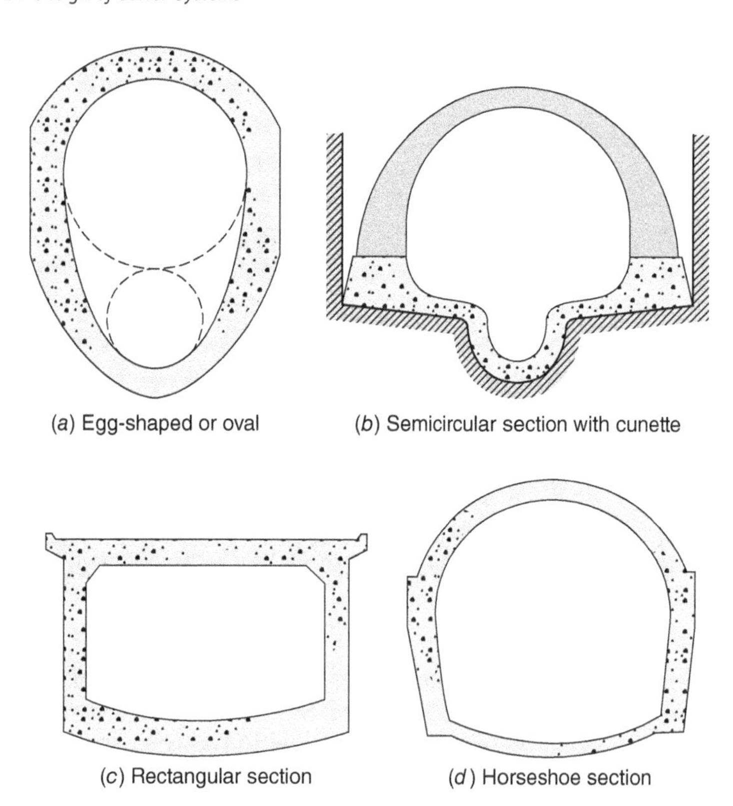

Figure 14.10 Sections of Storm and Combined Sewers

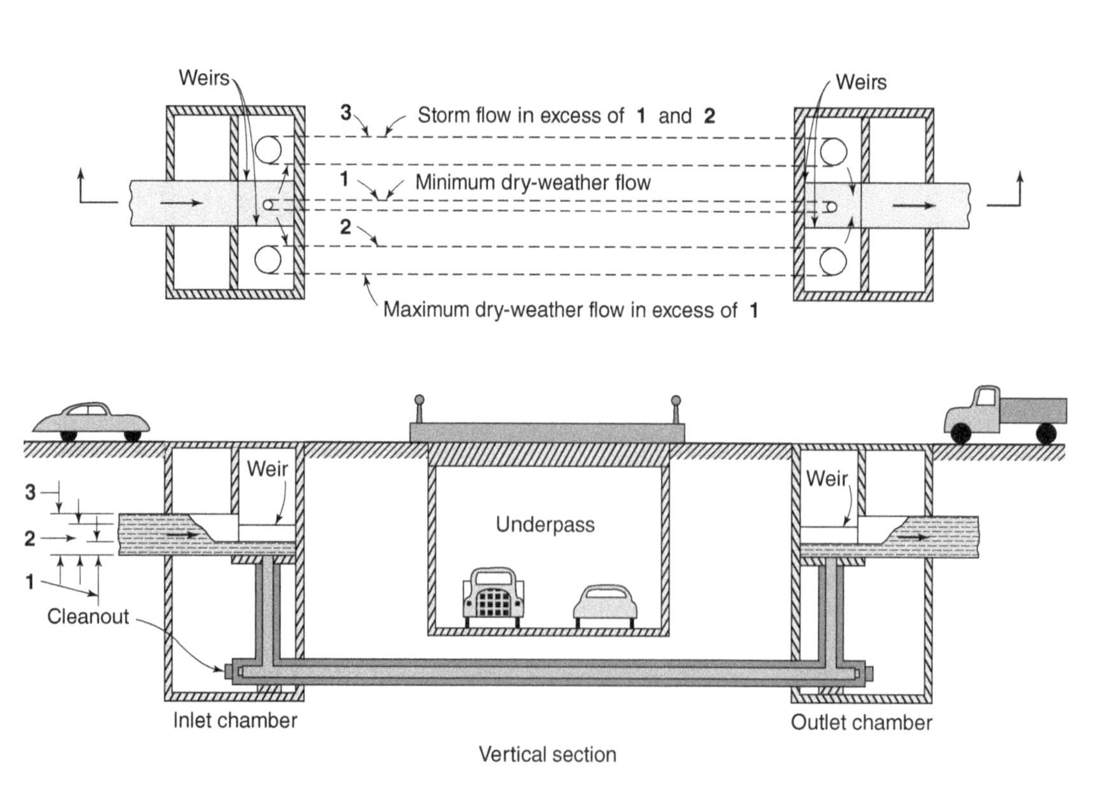

Figure 14.11 Inverted Siphon or Suppressed Sewer for Combined Sewage

capacity of the approach sewer. Weir heights are fixed at depths reached by characteristic flows in the approach sewer and inlet structure. Flows are reunited in a chamber in advance of the outlet sewer.

Example 14.1 Sizing of an Inverted Siphon

A siphon is to carry a minimum dry-weather flow of 1.0 ft^3/s (28.3 L/s or 0.0283 m^3/s), a maximum dry-weather flow of 3.0 ft^3/s (85 L/s or 0.085 m^3/s), and a storm flow of 45.0 ft^3/s (1,274 L/s or 1.274 m^3/s) in three pipes. Select the proper diameters to ensure velocities of 3.0 ft/s (0.90 m/s) in sanitary sewer pipes and 5.0 ft/s (1.5 m/s) in storm sewers.

Solution 1 (U.S. Customary System):

1) For minimum dry-weather flow of 1.0 ft^3/s, the nearest standard diameter of pipe is calculated from:

$$A = Q/V$$
$$\pi D^2/4 = Q/V$$
$$D = \sqrt{4Q/\pi V}$$
$$= 12\sqrt{4 \times 1/\pi \times 3.0} = \mathbf{8\ in.}$$

Actual capacity $= AV = \pi \times (8)^2 \times 3/(4 \times 144) = 1.05\ ft^3/s$

2) For maximum dry-weather flow in excess of the minimum, namely, 3.0 – 1.05 = 2.0 ft^3/s, the nearest standard diameter of pipe is

$$D = 12\sqrt{4 \times 2.0/\pi \times 3.0} = \mathbf{12\ in.}$$

Actual capacity $\pi(1/4) \times 3 = 2.36\ ft^3/s$

3) For storm flows in excess of maximum dry-weather flow, namely, 45 – (2.36 + 1.05) = 41.6 ft^3/s, the next lowest standard diameter of pipe is

$D = 12\sqrt{4 \times 41.6/\pi \times 5.0} = 39$ in.; **use 40 in.,** which will have to flow at a velocity of

$$V = (41.6 \times 4)\ 144/\left(\pi \times 40^2\right) = 4.8\ ft/s$$

Solution 2 (SI System):

1) For minimum dry-weather flow of 0.0283 m^3/s, the nearest standard diameter of pipe is calculated from:

$$A = Q/V$$
$$\pi D^2/4 = Q/V$$
$$D = \sqrt{4Q/\pi V}$$

$D = \sqrt{4 \times 0.0283/\pi \times 0.90} = \mathbf{0.199\ m}$; select 200 mm (or 8 in.)

Actual capacity $Q = AV = (0.785\ D^2)(V) = (0.785 \times 0.200^2)(0.90) = 0.028\ m^3/s = 28\ L/s$

2) For maximum dry-weather flow in excess of the minimum, namely, 0.085 – 0.028 = 0.057 m^3/s, the nearest standard diameter of pipe is

$D = \sqrt{4 \times 0.057/\pi \times 0.90} = \mathbf{0.284\ m}$; **select 300 mm (or 12 in.)**

Actual capacity $Q = AV = (0.785\ D^2)(V) = (0.785 \times 0.300^2)(0.90) = 0.064\ m^3/s = 64\ L/s$

3) For storm flows in excess of maximum dry-weather flow, namely, 1.274 – (0.064 + 0.028) = 1.182 m^3/s, the next lowest standard diameter of pipe is

$D = \sqrt{4 \times 1.182/\pi \times 1.50} = \mathbf{1.002\ m}$; **select 1,000 mm (or 40 in.)**

which will have to flow at a velocity of

$$V = Q/A = \left(1.182\ m^3/s\right)/\left(0.785\ D^2\right) = \left(1.182\ m^3/s\right)/\left(0.785 \times 1.000^2\right)\ m^2 = 1.51\ m/s$$

14.4.3 Interceptors

Intercepting sewers (Fig. 10.1) are generally designed to carry away some multiple of the dry-weather flow in order to bleed off as much stormwater and included wastewater as can be justified by hygienic, aesthetic, and economic considerations. Where rainfalls are intense and sharp, as in most of North America, it is not possible to lead away much stormwater through reasonably proportioned interceptors. Consequently, they are designed to transport not much more than the maximum dry-weather flow, or 250–600 gpcd (946–2,271 Lpcd). A more informative measure of interceptor capacity in excess of average dry-weather flow is the rate of rainfall or runoff they can accept without overflowing. Studies of rainfalls in the hydrologic surroundings of communities in the United States usually lead to the conclusion that most precipitation in excess of 0.1 in. (02.54 mm) is spilled; that spills can occur as frequently as half a dozen times a month; and that interception is not improved greatly by going even to 10 times the dry-weather flow. However, where rains are gentle and long, as in the United Kingdom, six times the dry-weather flow comprises much of the runoff from rainfall and becomes a useful design factor.

The total yearly pollution reaching an interceptor-protected body of water is a significant fraction (3%) of the total annual volume of sanitary sewage. During the periods when spilling occurs, a very high percentage of the sanitary wastewater can be carried through *combined sewer overflows*. If solids have accumulated in the sewer during the interval between rains, these may be washed out also. Thus overflows from combined sewers can be heavily charged with solids. They present a serious pollution problem that will be difficult to correct. Detention, settling, and chlorination are useful, and under some circumstances it may be desirable to work toward full separation of sanitary wastewater from surface runoff.

14.4.4 Retarding Basins

Interception can be improved by introducing into combined systems *retarding devices*—for example, up-system *detention basins* or *equalizing tanks*. Constructed in advance of junctions between submains and interceptors, they store flows in excess of interceptor capacity until they are filled. After that, they continue to retard and equalize flows in a lesser degree, but they do function as settling basins for the removal of gross and unsightly settleable matter. Depending on local conditions, *detention periods* as short as 15 minutes can be quite effective for settling basins and more so for chlorine-contact basins. Operating ranges extend from the dry-weather flowline of the interceptor to the *crown* of the conjoined combined sewer. After storms subside, the tank contents are flushed or lifted into the interceptor, and the accumulated solids eventually reach the treatment works. Where much stormwater is carried as far as the treatment plant—in the British Isles, for example—stormwater standby tanks, which serve as primary sedimentation tanks, become useful adjuncts to the works.

14.4.5 Overflows

The amounts of water entering interceptors at junctions with submains must be controlled. Only as much should be admitted as individual interceptor reaches can carry without being surcharged. Higher flows must be diverted into stormwater overflows. As shown in Fig. 14.12, admission and diversion can be regulated hydraulically or mechanically.

Hydraulic separation of excess flows from dry-weather flows is accomplished by devices such as the following:

1) *Diverting weirs* in the form of side *spillways* leading to overflows, with crest levels and lengths so chosen as to spill excess flows that, figuratively speaking, override the dry-weather flows, which follow their accustomed path of the interceptor (Fig. 14.12a)
2) *Leaping weirs*, essentially gaps in the floor of the channel over which excess flows jump under their own momentum, while dry-weather flows tumble through the gap into the interceptor (Fig. 14.12b)
3) *Siphon spillways* that carry flows in excess of interceptor capacity into the overflow channel (Fig. 14.12c)
4) *Mechanical devices*, in which diversion of stormwater flows is generally regulated by float-operated control valves activated by flow levels in the interceptor (Fig. 14.12d).

14.5 Choice of Collecting System

As explained in a previous chapter, apart from questions of economy, the combined system of sewerage is at best a compromise between two wholly different objectives: water carriage of wastes and removal of flooding runoff. In the life of growing communities, initial economies are offset in the long run (a) by undesirable pollution of natural

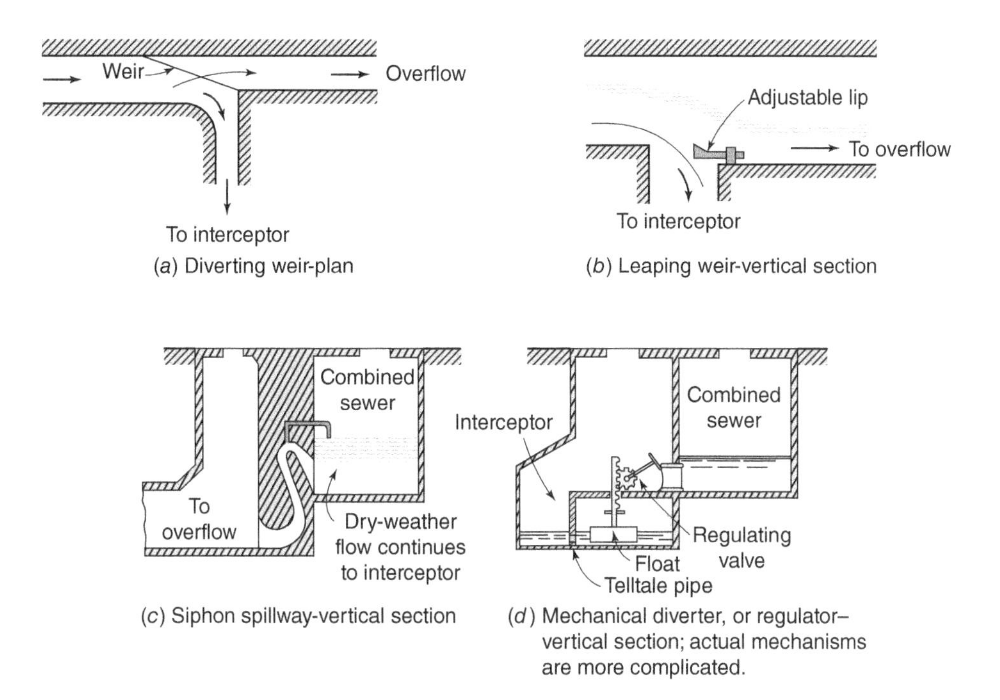

Figure 14.12 Regulation of Stormwater Overflow

water courses through stormwater spills and consequent nuisance or, at least, debased aesthetic and recreational values of receiving bodies of water; (b) by the increased cost of treating and pumping intercepted wastewater; and (c) by more obnoxious conditions when streets and basements are flooded by combined sewage instead of stormwater.

In the past, small streams, around which parks and other recreational areas could have grown, have been forced into combined sewerage systems because pressing them into service as receiving waters had degraded them into open sewers. By contrast, a separate system of sewerage can exploit natural water courses hydraulically by discharging stormwater into them through short runs of storm drains while preserving their aesthetic and recreational assets. However, they may have to be channelized if they are to perform well. In the United States, large sums of money have been and are being expended in sewer separation and related construction in order to protect water courses from combined overflows. The construction of combined sewers was forbidden by most health departments at the turn of the last century.

14.6 Design Information

Much detailed information is required for the design of sewers and drains. Special surveys are generally made to produce needed maps and tables as follows:

1) Detailed plans and profiles of streets to be sewered
2) Plans and contour lines of properties to be drained
3) Sill or cellar elevations of buildings to be connected
4) Location and elevation of existing or projected building drains
5) Location of existing or planned surface and subsurface utilities
6) Kind and location of soils and rock through which sewers and drains must be laid
7) Depth of groundwater table
8) Location of drainage-area divides
9) Nature of street paving
10) Projected changes in street grades
11) Location and availability of sites for pumping stations, treatment works, and outfalls
12) Nature of receiving bodies of water and other disposal facilities.

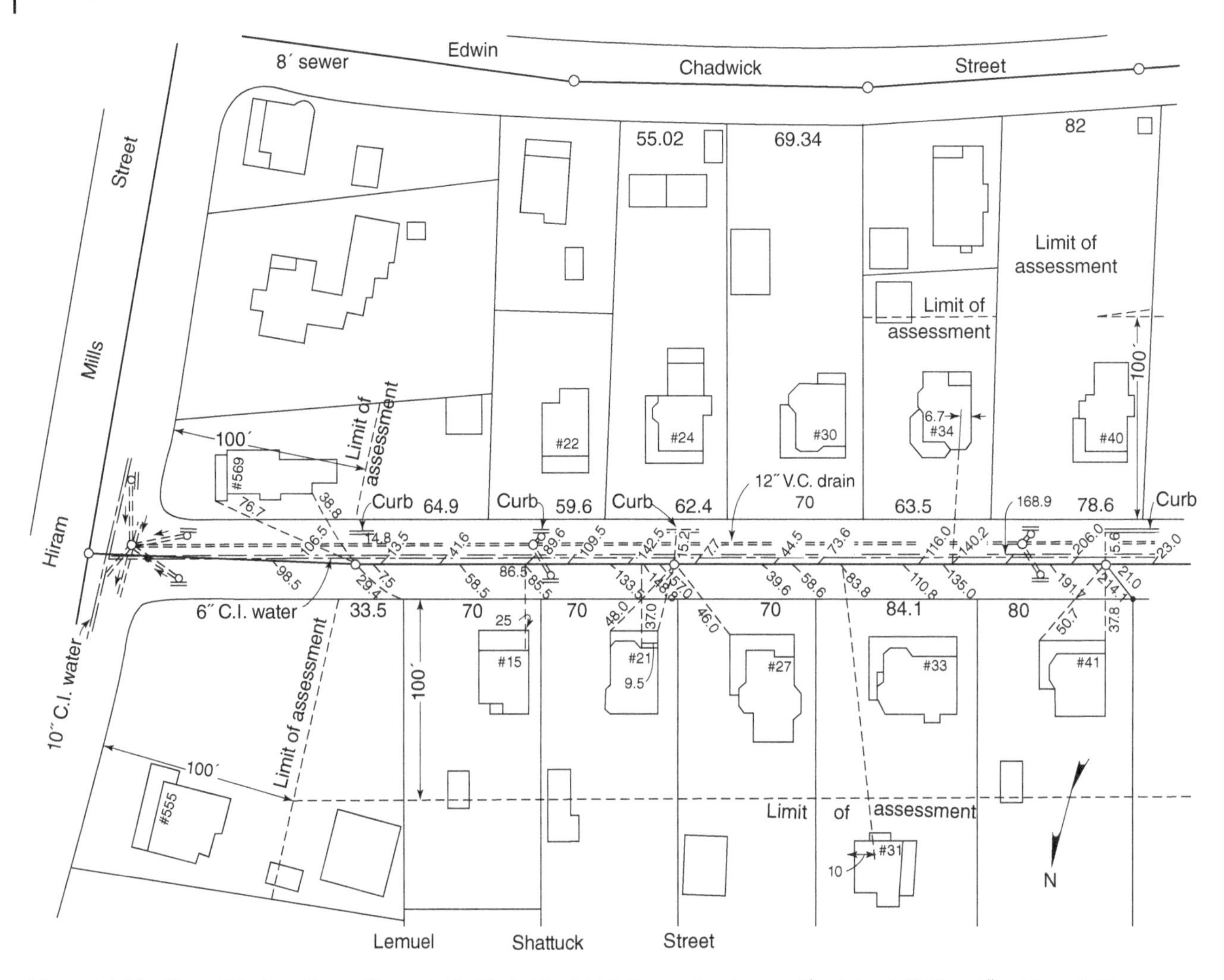

Figure 14.13 Plan of Sanitary Sewer Shown in Profile in Fig. 14.14. Conversion factors: 1′ = 1 ft = 0.3048 m; 1″ = 1 in. = 2.54 cm = 25.4 mm

Much of the topographic information needed is assembled for illustrative purposes in Figs. 14.13 and 14.14 for a single sanitary sewer in a street also containing a storm drain. Aerial maps are useful.

Variations in flow to be handled by sanitary sewers are determined by (a) anticipated population growth and water use during a chosen design period and (b) fluctuations in flow springing from normal water use. Choice of the design period itself will depend on anticipated population increases and interest rates. By contrast, the design period for storm drains and combined sewers is important principally in connection with expected effects of drainage-area development on runoff coefficients and magnitudes of flood damage. Required storm-drain capacity is primarily a matter of probable runoff patterns. Because storms occur at random, adopted values may be reached or exceeded as soon as storm sewers and drains have been laid.

14.6.1 Sanitary Sewers

Although anticipated wastewater volumes and their hourly, daily, and seasonal variations determine design capacity, the system must function properly from the start. Comparative flows are shown in the following schedule:

- *Beginning of design period.* (a) Extreme minimum flow = 1/2 minimum daily flow. Critical for velocities of flow and cleanliness of sewers. (b) Minimum daily flow = 2/3 of average daily flow. Critical for subdivision of units in treatment works.

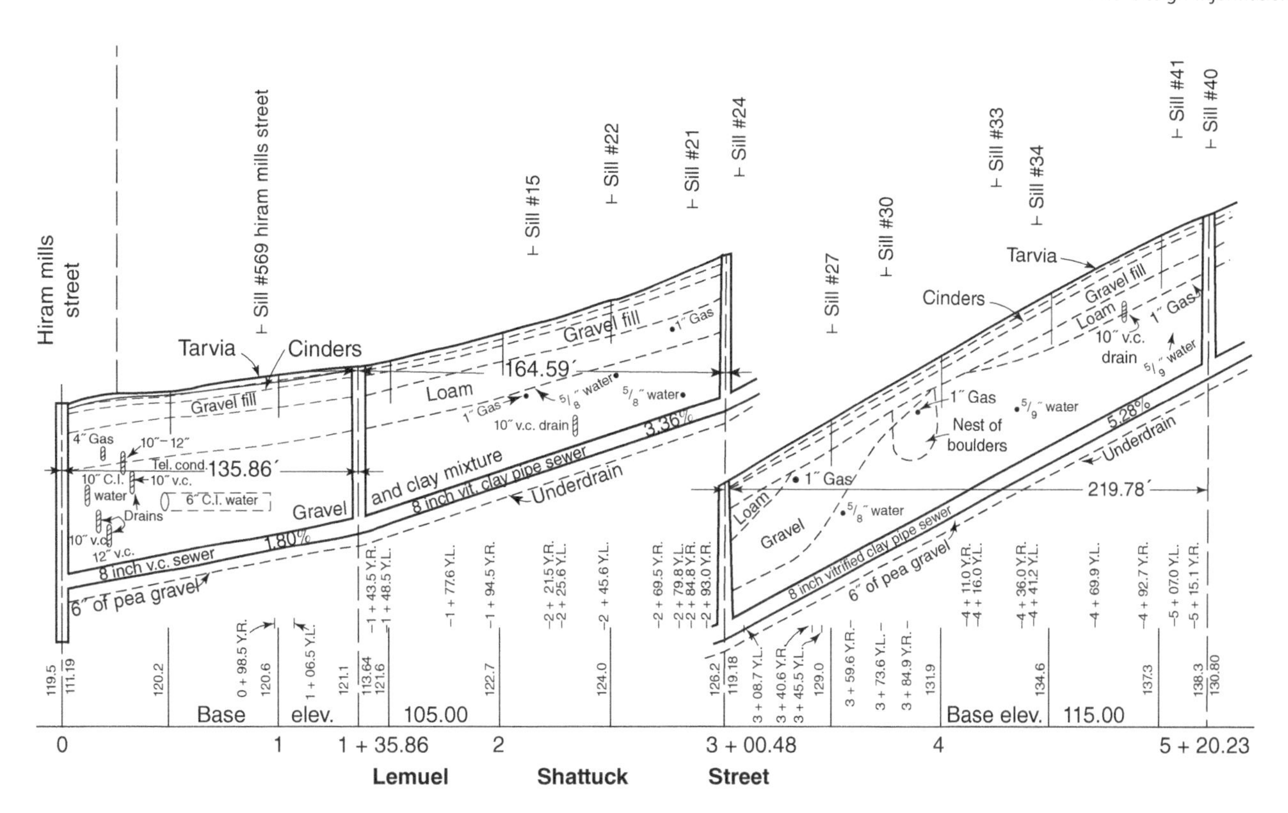

Figure 14.14 Profile of Sanitary Sewer Shown in Plan in Fig. 14.13. Conversion factors: 1′ = 1 ft = 0.3048 m; 1″ = 1 in. = 2.54 cm = 25.4 mm

- *Beginning and end of design period.* (c) Average daily flow at beginning of design period = average daily flow at end of period. Critical for velocities of flow in force mains.
- *End of design period.* (d) Maximum daily flow = 2 × average daily flow. Critical for capacity of treatment works. (e) Extreme maximum flow = 1.5 × maximum daily flow. Critical for capacity of sewers and pumps.

The flow ratios in this outline are suggestive of small sewers and relatively rapidly growing areas, the overall ratio of the extremes being ($2 \times 1.5 \times 2 \times 2 \times 1.5 = 18$) 18 to 1. For large sewers and stationary populations the overall ratio is more, nearly 4 to 1. Important unknowns in necessary calculations are the entering volumes of groundwater and stormwater. Their magnitude depends on construction practices, especially on private property (house or building sewers).

14.6.2 Storm Drains and Combined Sewers

Storm drains are dry much of the time. When rains are gentle, the runoff is relatively clear, and low flows present no serious problem. Flooding runoffs may wash heavy loads of silt and trash into the system. However, most of the drains then flow full or nearly full and so tend to keep themselves clean.

The situation is not as favorable when storm and sanitary flows are combined. A combined sewer designed for a runoff of 1 in./h, for example, will receive a storm flow of 1 ft^3/s or 646,000 gpd (28 L/s) from a single acre (1 acre = 0.4046 ha) of drainage area against an average daily dry-weather contribution of about 10,000 gpd (37,850 Lpd) from a very densely populated acre (0.4046 ha). The resulting ratio, $q/Q = 0.016$, places the depth ratio d/D at only 0.07 and the velocity ratio v/V at only 0.3. This supports the choice of a high design velocity, such as 5.0 ft/s (1.5 m/s) at full depth, for combined sewers. Putrescible solids accumulating in combined systems during dry weather not only create septic conditions and offensive odors, they also increase the escape of sewage solids into receiving waters through storm overflows.

14.7 Common Elements of Sewer Profiles

For specified conditions of minimum velocity, minimum sewer depth, and maximum distance between manholes, a number of situations repeat themselves in general schemes of sewerage wherein street gradient, sewer gradient, size of sewer, and depth of sewer become interrelated elements of design. Some of these recurrent situations are illustrated in Fig. 14.15. Beside a flow formulation such as Manning's, they involve the following simple relationship:

$$h_1 - h_2 = L(g - s) \tag{14.1}$$

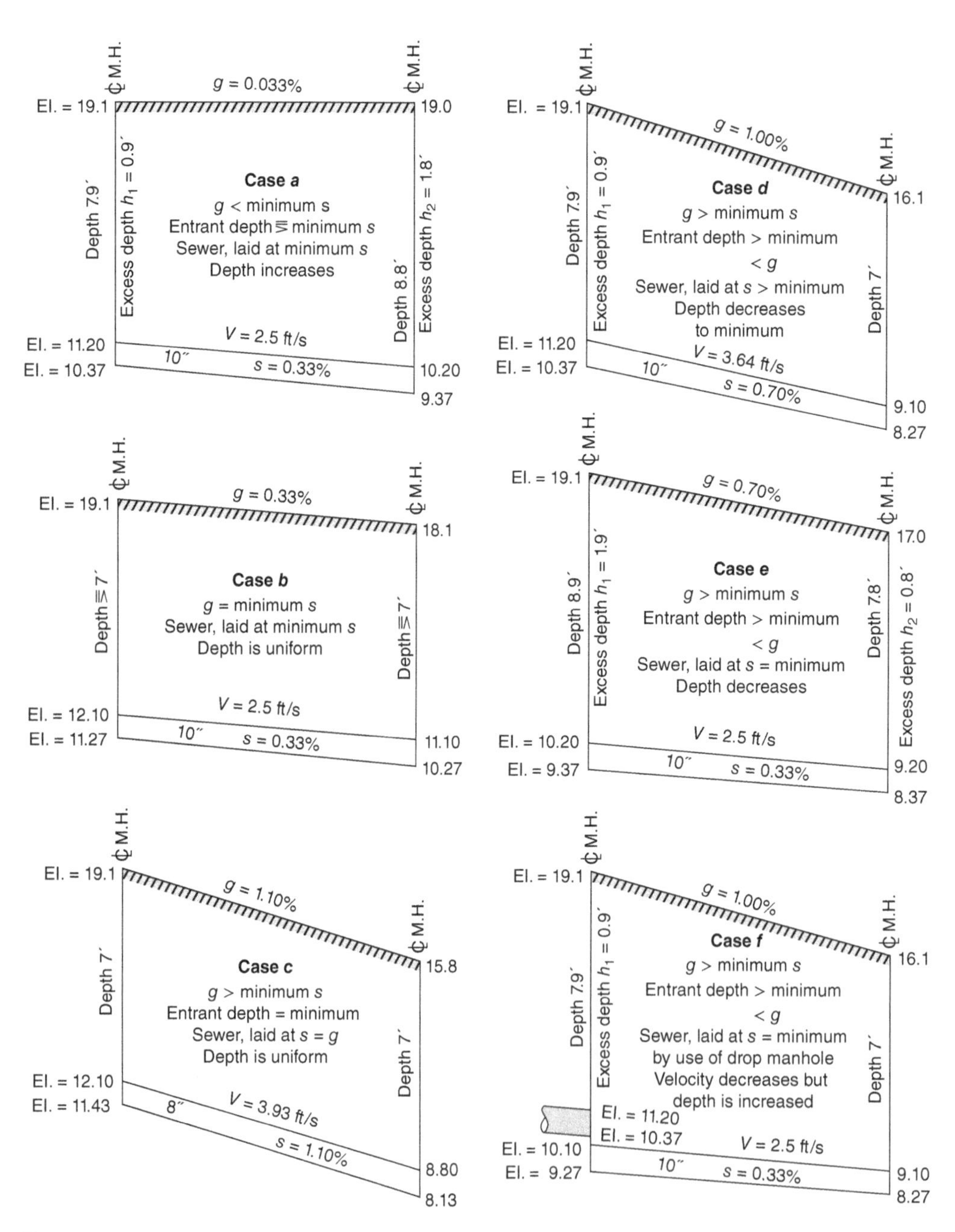

Figure 14.15 Common Elements of Sewer Design
Required: $q = 1.2\ \text{ft}^3/\text{s} = 33.98$ L/s; $Q = 1.2\ \text{ft}^3/\text{s} = 33.98$ L/s; $V = 2.5$ ft/s = 0.76 m/s;
$N = 0.012$; $L = 300$ ft = 91.44 m.
Minimum depth to crown =7.0 ft = 2.13 m.
8-in. (200-mm) sewer: $Q = 1.2\ \text{ft}^3/\text{s} = 33.98$ L/s; $S = 0.84\,\%$; $V = 3.4$ ft/s = 1.04 m/s.
10-in. (250-mm) sewer: $Q = 1.36\ \text{ft}^3/\text{s} = 38.52$ L/s; $S = 0.33\,\%$; $V = 2.5$ ft/s = 0.76 m/s.
Conversion factors: 1″ = 1 in. = 25.4 mm; 1 ft/s = 0.3048 m/s; 1′ = 1 ft = 0.3048 m.

where h_1 and h_2 are sewer depths (ft or m) in excess of minimum requirements, L is the distance (ft or m) between manholes, and g and s are, respectively, the street and sewer grades (dimensionless). Conditions of flow are stated in the legend accompanying Fig. 14.15.

Case *a* from Fig. 14.15 is encountered whenever the required sewer grade is greater than the street grade. Arriving at a depth equal to or greater than the minimum, 7.0 ft (2.13 m) in this instance, the sewer becomes deeper and deeper until it is more economical to lift the sewage by placing a pumping station in the line. Specifically for Case *a* in Fig. 14.15, the sewer grade is held at minimum (0.33%), and $h_2 = h_1 - L(g - s) = 0.9 - 3(0.033 - 0.33) = 1.8$ ft or 0.55 m, the depth increasing by $(1.8 - 0.9) = 0.9$ ft or 0.27 m.

Case *b* is unusual in that the required sewer grade is the same as the street grade. Therefore, the depth of the sewer remains unchanged.

Case *c* introduces a street grade steep enough to provide the required capacity in an 8-in. (200-mm), rather than a 10-in. (250-mm), parallel conduit. Arriving at minimum depth, there is no possibility of utilizing the available fall in part or as a whole to recover minimum depth as in Cases *d* and *e*. The reduced pipe size becomes the sole profit from the steep street grade, provided usually that the upstream sewer is also no greater than 8 in. (200 mm).

Case *d* aims at maximum reduction of excess depth by replacing a 10-in. (250-mm) sewer on minimum grade or, in accordance with Eq. 14.1, $s = g - (h_1 - h_2)/L$. For, $h_2 = 0$, $s = 1.00 - (0.9 - 0.0)/3 = 0.70\,\%$, which is more than the required minimum of 0.33%. Hence the sewer can be brought back to minimum depth, or =7 ft.

Case *e* is like Case *d*, but full reduction to minimum depth is not attainable, because $s = 0.70 - (1.9 - 0.0)/3 = 0.07\%$. This is less than the required minimum of 0.33%. Hence the minimum grade must be used, and $h_2 = 1.9 - 3(0.70 - 0.33) = 0.8$ ft (0.244 m).

Case *f* illustrates how high velocities can be avoided by introducing drop manholes on steep slopes. Case *f* parallels Case *d* but provides a drop of 1.1 ft (0.335 m) to place the sewer on minimum grade and give it minimum velocity. Such action is normal only when grades are extraordinarily steep. Excessive drops and resulting excessive sewer depth can then be avoided by breaking the drops into two or more steps through insertion of intermediate drop manholes.

Because the sewers flow nearly full, no attention needs to be paid to actual velocities and depths of flow in these illustrative cases. Consideration of actual depths and velocities of flow is generally restricted to the upper reaches of sewers flowing less than half full. There, the designer may have to forgo self-cleaning grades for lateral sewers. Otherwise, they might reach the main sewer at an elevation below that of the main itself, and the main would have to be lowered to intercept them. Normally this would be expensive.

Generally speaking, the designer should try for the fullest possible exploitation of the capacity of minimum-sized sewers before joining them to larger sewers. The implications are demonstrated in Fig. 14.16. There Scheme a keeps lateral flows from joining the main conductor until as many as 10 units of flow have accumulated, whereas no lateral carries more than 3 units in Scheme b. Moreover, Scheme b exceeds 10 units in two sections for which the required capacities in Scheme a are still only 6 and 8 units, respectively.

14.8 Design Criteria of Sanitary Sewers

Systems of sanitary sewers receive the waterborne wastes from households, mercantile and industrial establishments, and public buildings and institutions. In addition, groundwater enters by infiltration from the soil and, too often, illicit property drain connections and leaking manhole covers increase the flow. Accordingly, their requisite capacity is determined by the tributary domestic and institutional population, commercial water use, industrial activity, height of groundwater table, tightness of construction, and enforcement of rainwater separation.

It is generally convenient to arrive at unit values of domestic flows on the basis of population density and area served; but it would also be possible to develop figures for the number of people per front foot in districts of varying occupancy and make sewer length rather than area served the criterion of capacity design. Length (sometimes coupled with diameter) of sewer, indeed, offers a perhaps

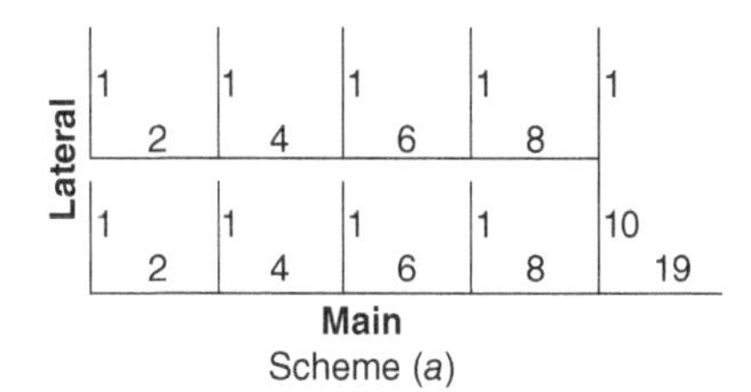

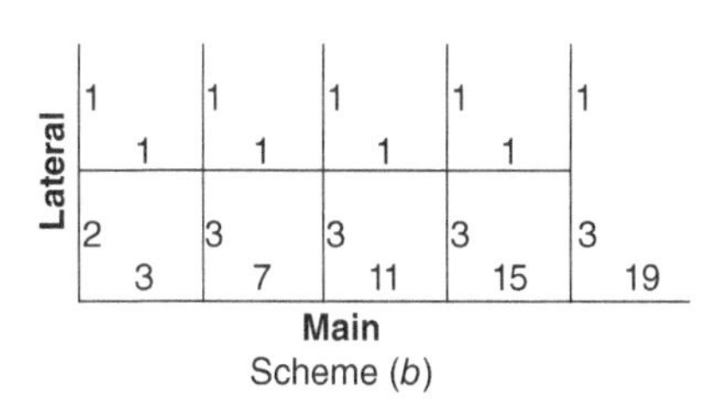

Figure 14.16 Relative Utilization of the Capacity of Lateral Sewers

more rational basis for the estimation of groundwater infiltration. Unit values for flows from commercial districts are generally expressed in terms of the area served. The quantities of wastewaters produced by industrial operations are more logically evaluated in terms of the units of daily production, for example, gallons per barrel of beer, 100 cases of canned goods, 1,000 bushels (35.24 m^3) of grain mashed, 100 lb. (45.36 kg) of live weight of animals slaughtered, or 1,000 lb. (453.6 kg) of raw milk processed.

Peak domestic and commercial flows originate at about the same hour of the day but travel varying distances before they reach a given point in the system. Hence, a reduction in, or damping of, the peak of the cumulative flows must generally be assumed. In a fashion similar to the reduction in flood flows with time of concentration (as represented by the size and shape of drainage area), the lowering of peak flows in sanitary sewers is conveniently related to the volume of flow or to the number of people served, and unit values of design are generally not accumulated in direct proportion to the rate of discharge or to the tributary population.

The design of conventional gravity sewers is based on the following design criteria:

1) *Long-term serviceability*. The design of long-lived sewer infrastructure should consider serviceability factors, such as ease of installation, design period, useful life of the conduit, resistance to infiltration and corrosion, and maintenance requirements. The design period should be based on the ultimate tributary population and usually ranges from 25 to 50 years.
2) *Design flow*. Sanitary sewers are designed to carry peak residential, commercial, institutional, and industrial flows, as well as infiltration and inflow. Gravity sewers are designed to flow full at the design peak flow. Design flows are based on various types of developments. Table 14.1 provides a list of design flows for various development types.
3) *Minimum pipe size*. A minimum pipe size is dictated in gravity sewer design to reduce the possibility of clogging. The minimum pipe diameter recommended by the Ten-States Standards is 8 in. (200 mm). Though the Ten-States Standards have only been adopted by 10 specific states (Illinois, Indiana, Iowa, Michigan, Minnesota, Missouri, New York, Ohio, Pennsylvania, and Wisconsin) and the Province of Ontario, they often provide the basis for other states' standards.
4) *Velocity*. The velocity of wastewater is an important parameter in sewer design. A minimum velocity must be maintained to reduce solids deposition in the sewer, and most states specify a minimum velocity that must be maintained under low-flow conditions. The typical design velocity for low-flow conditions is 1 ft/s (0.30 m/s). During peak dry-weather conditions, the sewer lines must attain a velocity greater than 2 ft/s (0.60 m/s) to ensure that the lines will be self-cleaning

Table 14.1 Average Design Flows for Development Types

Type of Development	Design Flow (gpd)
Residential:	
General	100/person
Single-family	370/residence
Townhouse unit	300/unit
Apartment unit	300/unit
Commercial:	
General	2,000/acre
Motel	130/unit
Office	20/employee
	0.20/net ft^2
Industrial (varies with type of industry):	
General	10,000/acre
Warehouse	600/acre
School site (general)	16/student

Conversion factors: 1 gpd = 3.785 L/d; 1 ft^2 = 0.0929 m^2; 1 acre = 0.4046 ha.

Table 14.2 Minimum Slopes and Capacities for Various Pipe Diameters

Pipe Diameter (Inches)	Minimum Slope for Pipe Velocity of 2 ft/s	8 ft/s	Wastewater Flow, MGD, for Pipe Flowing Full — Velocity, ft/s: 2	3	4	8
6	0.0060	0.075	0.26	0.36	0.47	0.91
8	0.0038	0.045	0.48	0.69	0.91	1.76
10	0.0030	0.035	0.72	1.04	1.37	2.54
12	0.0022	0.026	1.04	1.46	1.94	3.51
15	0.0015	0.020	1.69	2.47	3.25	5.84
18	0.0012	0.016	2.41	3.45	4.42	9.43
21	0.0010	0.013	3.38	4.94	6.37	12.35
24	0.00078	0.011	4.10	6.24	8.13	15.28
27	0.00065	0.0095	5.20	7.80	10.08	18.85
30	0.00058	0.0080	6.50	9.75	13.00	24.05
36	0.00045	0.0060	9.75	14.63	18.20	37.05
42	0.00038	0.0050	13.00	19.50	25.36	48.10
48	0.00032	0.0045	16.25	24.70	31.85	59.80
54	0.00026	0.0039	20.80	31.85	39.65	84.50

Conversion factors: 1 ft/s = 0.3048 m/s; 1 in. = 25.4 mm; 1 MGD = 43.8 L/s.

(i.e., they will be flushed out once or twice a day by a higher velocity). Velocities higher than 10 ft/s (3.0 m/s) should be avoided because they may cause erosion and damage to sewers and manholes.

5) *Slope.* Sewer pipes must be adequately sloped to reduce solids deposition and production of hydrogen sulfide and methane. Table 14.2 presents a list of minimum slopes and capacities for various pipe sizes. If a sewer slope of less than the recommended value must be provided, the responsible review agencies may require depth and velocity computations at minimum, average, and peak flow conditions. The size of the pipe may change if the slope of the pipe is increased or decreased to ensure a proper depth below grade. Velocity and flow depth may also be affected if the slope of the pipe changes. This parameter must receive careful consideration when designing a sewer.
6) *Depth of bury.* Depth of bury affects many aspects of sewer design. Slope requirements may drive the pipe deep into the ground, increasing the amount of excavation required to install the pipe. Sewer depths average 3–6.5 ft (0.90–2.0 m) below ground surface. The proper depth of bury depends on the water table, the lowest point to be served (such as a ground floor or basement), the topography of the ground in the service area, and the depth of the frost line below grade.
7) *Pipe material.* Must meet service application requirements (refer to Chapter 16).
8) *Appurtenances.* Appurtenances include manholes, building connections, junction chambers or boxes, and terminal cleanouts, among others. Regulations for using appurtenances in sewer systems are well documented in municipal design standards and/or public facility manuals. Manholes for small sewers (24 in. or 600 mm in diameter or less) are typically 4 ft (1.22 m) in diameter. Larger sewers require larger manhole bases, but the 4-ft (1.22-m) barrel may still be used. Manhole spacing depends on regulations established by the local municipality. Manholes are typically required when there is a change of sewer direction. However, certain minimum standards are typically required to ensure access to the sewer for maintenance. Typical manhole spacing ranges between 300 and 600 ft (91.44 and 182.88 m) depending on the size of the sewer and available sewer cleaning equipment. For example, one municipality requires that the maximum manhole spacing be at intervals not to exceed 400 ft (120 m) on all sewers 15 in. (400 mm) or less, and not exceeding 500 ft (150 m) on all sewers larger than 15 in. (400 mm) in diameter.

14.9 Layout and Hydraulic Design in Sanitary Sewerage

Before beginning to design individual sewer runs, a preliminary layout is made of the entire system. Sanitary sewers are placed in streets or alleys in proper reference to the buildings served, and terminal manholes for lateral sewers generally lie within the service frontage of the last lot sewered.

Sewers should slope with the ground surface and follow as direct a route to the point of discharge as topography and street layout permit. To this purpose, flow in a well-designed system will normally take the path of surface runoff. Stormwater infiltration through manhole covers is kept down by placing sanitary sewers under the crown of the street.

In communities with alleys, the choice of location will depend on the relative advantages of alleys and streets. Alley location is often preferable in business as well as residential districts.

Instructions for *preliminary layouts* are as follows:

1) Show all sewers as single lines.
2) Insert arrows to indicate flow direction.
3) Show manholes as circles at all changes in directions or grade, at all sewer junctions, and at intermediate points that will keep manhole spacing below the allowable maximum.
4) Number manholes for identification.

Alternate layouts will determine the final design. The avoidance of pumping is not as important as it formerly was, because of the availability of "off-the-shelf" pumping stations equipped with pumps that do not clog readily.

The hydraulic design of a system of sanitary sewers is straightforward and is readily carried to completion in a series of systematic computations, as in Example 14.2.

Example 14.2 Design of Sanitary Sewer System

Determine the required capacity and find the slope, size, and hydraulic characteristics of the system of sanitary sewers shown in Table 14.3, which lists sewer locations, areas and population served, and expected sewage flows.

Solution:

1) Capacity requirements are based on the following assumptions:
   a) Water consumption: average day, 95 gpcd (360 Lpcd); maximum day, 175% of average; maximum hour, 140% of maximum day.
   b) Domestic wastewater: 70% of water consumption; maximum is 285 gpcd (1,080 Lpcd) for 5 acres (2.20 ha), decreasing to 245 gpcd (927 Lpcd) for 100 acres (40.5 ha) or more.
   c) Groundwater: 0.14–0.15 ft^3/s per 100 acres (9.8–10.5 L/s per 100 ha) for 8- to 15-in. (200- to 400-mm) sewers in low land and 0.09–0.11 ft^3/s per 100 acres (6.3–7.7 L/s per 100 ha) in high land. These figures would be lowered by using preformed joints.
   d) Commercial wastewater: 25,000 gal/d/acre = 3.88 ft^3/s per 100 acres = 272 L/s per 100 ha.
   e) Industrial wastewater: Flow in accordance with industry.
2) Hydraulic requirements are as follows:
   a) Minimum velocity in sewers: 2.5 ft/s or 0.75 m/s (actual).
   b) Kutter's coefficient of roughness $N = 0.015$ includes allowances for change in direction and related losses in manholes except for (c) below.
   c) Crowns of sewers are made continuous to prevent surcharge of upstream sewer.
3) Design procedures are as follows (with reference to Table 14.3):

   Columns 1–4 identify the location of the sewer run. The sections are continuous.

   Columns 5–8 list the acreage immediately adjacent to the sewer.

   Column 9 gives the density of the population per domestic acre.

   Column 10 = Column 9 × Column 8.

   Columns 11 to 13 list the accumulated acreage drained by the sewer. For example, in Section b, Column 13 is the sum of column 8 in Sections a and b, or (40 + 27) = 67.

   Column 14 gives the average population density for the total tributary area. For example, in Section b, Column 14 = (40 × 27 + 27 × 19)/(40 + 27) = 23.8.

Table 14.3 Illustrative Computations for the System of Sanitary Sewers for Example 14.2

	Location of Sewer			Adjacent Area						Total Tributary Area				
		Stations or Limits						Population					Population	
Section (1)	Street (2)	From (3)	To (4)	Total Acres (5)	Industrial Acres (6)	Commercial Acres (7)	Domestic Acres (8)	Per Acre (9)	Total (10)	Industrial Acres (11)	Commercial Acres (12)	Domestic Acres (13)	Per Acre (14)	Total (15)
a	A Ave.	B Ave.	C St.	49	5	4	40	27	1,080	5	4	40	27.0	1,080
b	D Ave.	C St.	E St.	37	3	7	27	19	513	8	11	67	23.8	1,593
c	F St.	G St.	H St.	29	8	1	20	25	500	16	12	87	24.1	2,093
d	I St.	J St.	K St.	63	–	10	53	21	1,113	16	22	140	22.9	3,206

	Maximum Volume of Sewage, ft^3/s					Design Profile											
									Velocity, ft/s				Invert Elevation		Cut		
Section (1)	Industrial (16)	Commercial (17)	Domestic (18)	Infiltration (19)	Total (20)	Size, in. (21)	Slopes, % (22)	Capacity, ft^3/s (23)	Full (24)	Actual (25)	Depth of flow, in. (26)	Length, ft (27)	Upper End (28)	Lower End (29)	Upper End (30)	Lower End (31)	Average (32)
a	0.156	0.155	0.440	0.044	0.795	8	0.8	0.82	2.35	2.72	6.37	850	120.00	113.20	7.50	11.50	9.50
b	0.248	0.429	0.650	0.086	1.413	10	0.7	1.42	2.22	3.02	8.1	1,260	113.03	104.21	11.67	8.50	10.08
c	0.496	0.468	0.852	0.115	1.931	12	0.45	2.23	2.45	2.84	9.7	1,880	104.04	95.58	8.67	12.00	10.33
d	0.496	0.858	1.300	0.178	2.832	15	0.3	3.35	2.35	2.72	12.0	1,760	95.33	90.05	12.25	11.00	11.63

Note: Sections of sewers rather than individual runs between manholes are shown in order to include major changes in required capacity and consequent size.
Conversion factors: 1 ft = 0.3048 m; 1 in. = 25.4 mm; 1 acre = 0.4046 ha; 1 ft^3/s = 28.3 L/s.

Column 15 = column 14 × column 13.
Column 16 lists values obtained in a survey of industries in the areas served.
Column 17 = column 12 × 3.88/100.

Column 18 = column 15 × (245 to 285 gpcd) × 1/(7.48 × 24 × 60 × 60) ft^3/s = column 15 × (245 to 285 gpcd) × (1.54×10^{-6}). For example, in Section a, 1080 × 265× (1.54×10^{-6}) = 0.440 ft^3/s or 12.46 L/s
Column 19 = Sum of column 5 times rate of infiltration. For example, in Section a, 49 × 0.09/100 = 0.044 ft^3/s or 1.246 L/s.
Column 20 = Sum of columns 16–19.
Columns 21–29 record the size of sewer for required capacity and available or required grade together with depth and velocity of flow. For example, in Section a, an 8-in. sewer laid on a grade of 6.8/850 = 0.008 or 0.8% will discharge $Q = 0.82$ ft^3/s (23.22 L/s) at a velocity of 2.35 ft/s (0.716 m/s) when it flows full. Hence, for $q/Q = 0.795/0.82 = 0.971$, $d/D = 0.796$, $v/V = 1.16$, or $d = 8 \times 0.796 = 6.37$ in. (160 mm) and $v = 2.35 \times 1.16 = 2.72$ ft/s or 0.83 m/s.
Columns 28 through 31 are taken from profiles of streets and sewers.
Column 28, Section b, shows a drop in the manhole of (113.20 − 113.03) = 0.17 ft (0.052 m) compared with column 28, Section a. This allows for a full drop of 0.17 × 12 = 2 in. (51 mm), thus offsetting the increase in the diameter of the sewer from 8 to 10 in. (203 to 254 mm).
Column 32 = arithmetic mean of columns 30 and 31.

14.10 Capacity Design in Storm Drainage

As shown in a previous chapter, the *rational method* of estimating runoff from rainfall provides a common hydrologic basis for the capacity design of storm-drainage systems. Rainfall rates are normally expressed in terms of in./h or mm/h and are available from the nation's weather bureau. All models of rainfall are empirical approximations based on previous records statistically interrelated, and are represented by the following general hyperbolic equation:

$$i = X/(t + Y) \text{ (US customary units or SI units)} \tag{14.2}$$

where i is the rate of rainfall, in in./h or mm/h; t is the total duration of the rain storm, in minutes; X is a constant determined from previous records; and Y is another constant determined from previous records. Both X and Y vary with location, topography, and relative frequency of storms.

The axiom of surface runoff analysis and design can be represented by the following empirical equation:

$$Q = cia \text{ (US customary units)} \tag{14.3}$$

where Q is runoff, ft^3/s; c is a runoff coefficient; $i = 3$, intensity of rainfall, in./h; and A is drainage area from which runoff takes place, acres. A similar empirical equation using the metric system is shown below:

$$Q = cia \text{ (SI units)} \tag{14.4}$$

where Q is runoff, in L/h; c is a runoff coefficient; i is the intensity of rainfall, in mm/h; and A is drainage area from which runoff takes place, in m^2. Another set of SI units can be: Q, m^3/s; i, m/h; and a, m^2.

The designer must arrive at the best possible estimates of c, the runoff–rainfall ratio, and i, the rainfall intensity, with the area a being determined by measuring tributary surfaces. Because both c and i are variable in time, storm flows reaching a given point in the drainage system are compounded of waters falling within the *time of concentration*.

14.10.1 Time of Concentration

The time of concentration is composed of two parts: (a) the *inlet time*, or time required for runoff to gain entrance to a sewer, and (b) the *time of flow* in the sewerage system.

The *inlet time* is a function of:

1) Surface roughness offering resistance to flow
2) Depression storage delaying runoff and often reducing its total

3) Steepness of areal slope, governing speed of overland flow
4) Size of block or distance from the areal divide to the sewer inlet determining time of travel
5) Degree of direct roof and surface drainage reducing losses and shortening inlet times
6) Spacing of street inlets affecting elapsed times of flow.

In large communities, in which roofs shed water directly to sewers, and runoff from paved yards and streets enters the drainage system through closely spaced street inlets, the time of overland flow is usually less than five minutes. In districts with relatively flat slopes and greater inlet spacing, the time may lengthen to 10–15 minutes.

Except for rains of considerable duration, relatively little of the water entering inlets at the time of peak flow originates from unpaved areas. Therefore, the time between the peak rainfall rate and the peak flow at an inlet is often less than one minute. Applying short inlet times and associated high intensities to paved areas only will sometimes produce the best estimates of inlet flows.

Time of concentration for overland flow has been formulated by Kerby (1959):

$$t = \left[\frac{2}{3}L\left(\frac{n}{\sqrt{s}}\right)\right]^{0.467} \quad \text{(US customary units)} \tag{14.5}$$

where t is the inlet time, in minutes; $L \leq 1{,}200$ ft is the distance to the farthest tributary point, in ft; s is the slope; and n is a retardance coefficient analogous to the coefficient of roughness. Suggested values of n are shown in Table 14.4. For $L = 500$ ft, $s = 1.0\,\%$, and $n = 0.1$, for example,

$$\begin{aligned} t &= \left[\frac{2}{3}500\left(\frac{0.1}{\sqrt{0.001}}\right)\right]^{0.467} \\ &= 1{,}055^{0.467} = 26\ \text{minutes} \end{aligned}$$

The following equation is derived for the SI system:

$$t = \left[(2.187)\,L\left(\frac{n}{\sqrt{s}}\right)\right]^{0.467} \quad \text{(SI units)} \tag{14.6}$$

where t is the inlet time, in minutes; $L \leq 365$ m is the distance to the farthest tributary point, in m; s is the slope; and n is a retardance coefficient analogous to the coefficient of roughness. The same suggested values of n are also shown in Table 14.4. For $L = 500$ ft $= 152.4$ m, $s = 1.0\%$, and $n = 0.1$, for example,

$$\begin{aligned} t &= \left[(2.187)(152.4)\left(\frac{0.1}{\sqrt{0.001}}\right)\right]^{0.467} \\ &= 1{,}055^{0.467} = 26\ \text{minutes} \end{aligned}$$

Table 14.4 Values of Retardance Coefficient, n

Type of Surface	n
Impervious surfaces	0.02
Bare packed soil, smooth	0.10
Bare surfaces, moderately rough	0.20
Poor grass and cultivated row crops	0.20
Pasture or average grass	0.40
Timberland, deciduous trees	0.60
Timberland, deciduous trees, deep litter	0.80
Timberland, conifers	0.80
Dense grass	0.80

As a matter of arithmetic, the time of flow in the system equals the sum of the quotients of the length of constituent sewers and their velocity when flowing full. Ordinarily, neither time increase, as sewers are filled, nor time decrease, as flood waves are generated by rapid discharge of lateral sewers, is taken into account.

14.10.2 Runoff Coefficients

Runoff from storm rainfall is reduced by evaporation, depression storage, surface wetting, and percolation. Losses decrease with rainfall duration. Runoff–rainfall ratios, or shedding characteristics, rise proportionately. The coefficient c may exceed unity because it is the ratio of a peak runoff rate to an average rainfall rate. Ordinarily, however, c is less than 1.0 and approaches unity only when drainage areas are impervious and high-intensity storms last long enough.

The choice of meaningful *runoff coefficients* is difficult. It may be made a complex decision. The runoff coefficient for a particular time of concentration should logically be an average weighted in accordance with the geometric configuration of the area drained, but fundamental evaluations of c and i are generally not sufficiently exact to warrant this refinement.

The choice of a suitable runoff coefficient is complicated not only by existing conditions, but also by the uncertainties of change in evolving urban complexes. Difficult to account for are the variations in runoff–rainfall relations to be expected in given drainage areas along with variations in rainfall intensities in the course of major storms. Fundamental runoff efficiency is least at storm onset and improves as storms progress. A graphic and relationships proposed by different authorities are shown in Fig. 14.17. However, they are not actually very helpful. Weighted average coefficients are calculated for drainage areas composed of districts with different runoff efficiencies.

Least arduous is acceptance of the fact that the degree of imperviousness of a given area is a rough measure of its shedding efficiency. Streets, alleys, side and yard walks, together with house and shed roofs, as the principal impervious components, produce high coefficients; lawns and gardens, as the principal pervious components, produce low coefficients. To arrive at a composite runoff–rainfall ratio, a weighted average is often computed from the information shown in Table 14.5. The resulting overall values for North American communities range between limits not far from those shown in Table 14.6.

14.10.3 Intensity of Rainfall

If the *time–intensity–frequency* analysis of storm rainfalls discussed in a previous chapter is followed, the important engineering decision is not just the selection of a suitable storm but also the pairing of significant values of the runoff coefficient c with the varying rainfall intensities i. Even though c is known to be time- and rainfall-dependent, engineers frequently seek shelter under the umbrella of the mean by selecting an average value of c that will combine reasonably well with varying values of t and i. However, it is possible to avoid poor pairing of c and i values by deriving a runoff hydrograph from the hyetograph of a design storm. How this is done is shown in the following section.

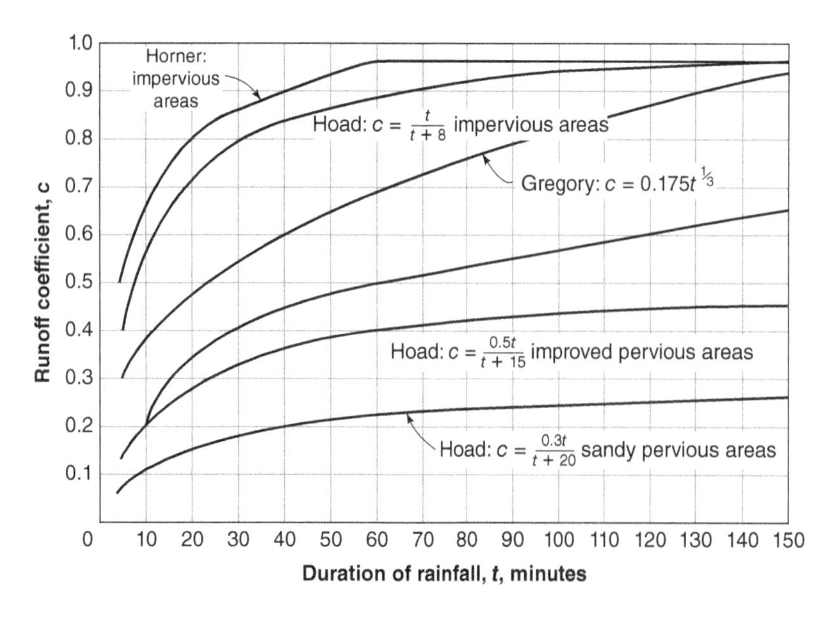

Figure 14.17 Variation in Runoff Coefficients with Duration of Rainfall and Nature of Area Drained

Table 14.5 Values of Runoff Coefficients

Description of Area Component	Runoff Coefficient (%)
Pavement, asphalt and concrete	70–95
Pavement, brick	70–85
Roofs	75–95
Lawns:	
Sandy, flat (2%)	5–10
Sandy, steep (7%)	15–20
Heavy, flat (2%)	13–17
Heavy, steep (7%)	25–35

Table 14.6 Values of Overall Runoff Coefficients for North American Communities

Areas	Coefficient (%)
Business:	
Downtown	70–95
Neighborhood	50–70
Residential:	
Single-family	30–50
Multifamily, detached	40–60
Multifamily, attached	60–75
Suburban	25–40
Apartments	50–70
Industrial:	
Light	50–80
Heavy	60–90
Parks, cemeteries	10–25
Playgrounds	20–35
Unimproved land	10–30

14.11 Storm-Pattern Analysis

Different from the averaging procedures associated in practice with the rational method of runoff analysis is the development of a generalized chronological storm pattern or *hyetograph* and its translation into a design runoff pattern by subtracting rates of (a) *surface infiltration*, (b) *depression storage*, and (c) *surface detention* during overland flow. The runoff hydrograph obtained is routed through overland flow, gutter flow, and flow in building drains, catch basins, and component sewers. Generalization of rainfall information by converting an intensity–duration–frequency curve into a hyetograph is illustrated in Fig. 14.18. An advanced peak in rainfall intensity is assumed in this case at 3/8 the time distance or storm duration from the beginning of appreciable precipitation, selection of a suitable fraction being based on specific rainfall experiences.

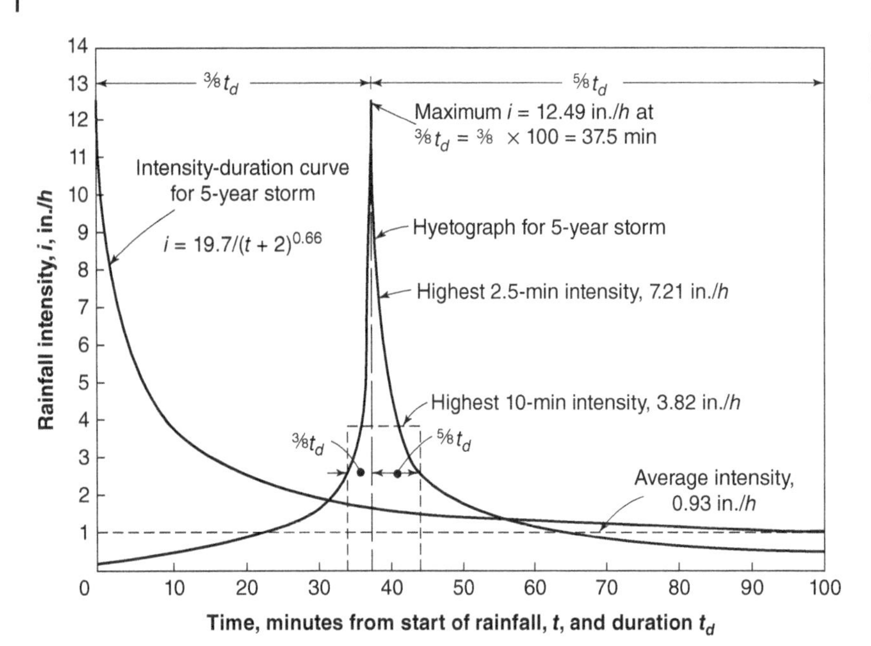

Figure 14.18 Hyetograph Derived from Rainfall Intensity–Duration Frequency Curve. Conversion Factor: 1 in. /h = 25.4 mm/h

Figure 14.18 illustrates results obtained in the application of *conversion* and *routing procedures* developed by Tholin and Keifer (1960) for the city of Chicago. Necessary calculations are based on (a) the infiltration-capacity curves shown in Fig. 14.18; (b) depth of depression storage assumed at 1/4 and 1/2 in. (6.35 and 12.7 mm), for example, and normally distributed about this mean depth, 50% of the area covered by depressions lying within 20% of the mean depth ($\sigma = \pm 14\%$); and (c) surface detention computed by *Izzard's equation*:

$$D = 0.342\left[\left(7 \times 10^{-4} i + c_r\right)/s^{1/3}\right](LQ)^{1/3} \text{ (US customary units)} \tag{14.7}$$

where D is the surface detention, in inches of depth; s is the slope of the ground; L is the distance of overland flow, in ft; Q is the overland supply, in in./h (ft^3/s per acre); i is the intensity of rainfall, in in./h; and c_r is a coefficient of roughness varying downward from 6.0×10^{-2} for pervious areas of turf, through 3.2×10^{-2} for bare, packed pervious areas, and 1.2×10^{-2} for pavements, to 1.7×10^{-2} for flat, gravel roofs. The following equation adopts the SI system:

$$D = 4.39\left[\left(0.276 \times 10^{-4} i + c_r\right)/s^{1/3}\right](LQ)^{1/3} \text{ (SI units)} \tag{14.8}$$

where D is the surface detention, in mm of depth; s is the slope of the ground; L is the distance of overland flow, in m; Q is the overland supply, in mm/h; i is the intensity of rainfall, in mm/h; and c_r is a coefficient of roughness varying downward from 6.0×10^{-2} for pervious areas of turf, through 3.2×10^{-2} for bare, packed pervious areas, and 1.2×10^{-2} for pavements, to 1.7×10^{-2} for flat, gravel roofs.

Variations of c and i with time (time being a function of watershed area a and of other factors such as surface slope s) can be incorporated into overall runoff formulations developed empirically for given localities. An example is *McMath's formula* ($Q = cia^{4/5} s^{1/5}$), for St. Louis, Missouri, which uses the US customary system. If rainfall intensity i and resulting runoff ci are properly determined, overall formulas should yield results comparable with the *rational method*.

As storm drains are actually constructed in a given community, it becomes possible to design stormwater systems for adjacent unsewered areas on the basis of (a) actual runoff measurements conducted in times of heavy rainfall or (b) surcharge experience with recorded storm intensities. In necessary calculations, it is important to identify possible downstream effects on surcharge.

14.12 Layout and Hydraulic Design in Storm Drainage

The layout of storm drains and sanitary sewers follows much the same procedure. Street inlets must be served as well as roof and other property drains connected directly to the storm sewers. How inlets are placed at street intersections to keep pedestrian crossings passable is indicated in Fig. 14.8. To prevent the flooding of gutters or to keep flows within inlet capacities, street inlets may also be constructed between the corners of long blocks. Required inlet capacity is a function of tributary area and its pertinent runoff coefficient and rainfall intensity.

Separate storm drains should proceed by the most direct route to outlets emptying into natural drainage channels. Easements or rights of way across private property may shorten their path. Manholes are included in much the same way and for much the same reasons as for sanitary sewers.

Surface topography determines the area tributary to each inlet. However, it is often assumed that lots drain to adjacent street gutters and thence to the sewers themselves. Direct drainage of roofs and areaways reduces the inlet time and places greater load intensity on the drainage system. Necessary computations are illustrated in Table 14.7, which accompanies Example 14.3.

Example 14.3 Design of a Storm Drains System

Determine the required capacity and find the slope, size, and hydraulic characteristics of the system of storm drains shown in the accompanying tabulation of location, tributary area, and expected storm runoff.

Capacity requirements are based on the rainfall curves included in Fig. 14.18. The area is assumed to be an improved pervious one, and the inlet time is assumed to be 20 minutes. Hydraulic requirements include a value of $N = 0.012$ in Manning's formula and drops in manholes equal to $\Delta(d + h_v) + 0.2\Delta h_v$ for the sewers flowing full.

Solution:

Illustrative computations for the storm drainage system are shown in Table 14.7.

Columns 1 through 4 identify the location of the drains. The runs are continuous.

Column 5 records the area tributary to the street inlets discharging into the manhole at the upper end of the line.

Column 6 indicates the cumulative area tributary to a line. For example, in line 2, column 6 is the sum of column 6, line 1, and column 5, line 2, or $(2.19 + 1.97) = 4.16$.

Columns 7 and 8 record the times of flow to the upper end of the drain and in the drain. For example, the inlet time to manhole 1 is estimated to be 20 minutes, and the time of flow in line 1 is calculated to be $340/(60 \times 3.94) = 1.5$ minutes from (Column 17)/(60 × column 16). Hence, the time to flow to the upper end of line 2 is $(20 + 1.5) = 21.5$ minutes

Column 9 is the mean intensity of rainfall during the inlet time (time of flow to the upper end) read from Fig. 14.18.

Column 10 is the weighted mean runoff coefficient for the tributary area.

Column 11 is (column 9) × (column 10). For example, for line 1, $2.56 \times 0.508 = 1.30$.

Column 12 = (column 11) × (column 6). For example, the runoff entering line 1 is $1.30 \times 2.19 = 2.85$ ft^3/s (81 L/s).

Columns 14, 15, and 16 record the chosen size and resulting capacity and flow velocity of the drains for the tributary runoff and available or required grade. For example, in line 1, a grade of 6.42% and a flow of 2.85 ft^3/s (81 L/s) call for a 12-in. drain. This drain will have a capacity of 3.09 ft^3/s (87.5 L/s) and flow at a velocity of 3.94 ft/s (1.2 m/s).

Columns 17 through 21 identify the profile of the drain. Column 17 is taken from the plan or profile of the street; column 18 = (column 17) × (column 14); column 19 is obtained from Eq. 5.18 [$h_i = \Delta(d + h_v) + kh_v$], the required drop in manhole 2 being $\Delta\ (d + h) + 0.2h = [(1.5 + 0.17) - (1.0 + 0.24)] + 0.2(0.17 - 0.24) = 0.42$ ft; and column 21 = (column 20) − (column 19), column 20 furthermore being (column 21) − (column 19) for the entrant line. For example, for line 2, $(84.28 - 0.42) = 83.86$ and subsequently $(83.86 - 0.92) = 82.94$.

Table 14.7 Illustrative Computations for the Storm Drainage System of Example 14.3

Line Number (1)	Location of Drain			Tributary Area, Acres, *a*		Time of Flow, min		Mean Rainfall Intensity, *i*, in./h (9)	Weighted Mean Runoff Coefficient, *c* (10)
	Street (2)	Manhole Number From (3)	Manhole Number To (4)	Increment (5)	Total (6)	To Upper End (7)	In Drain (8)		
1	A	1	2	2.19	2.19	20.0	1.5	2.56	0.508
2	A	2	3	1.97	4.16	21.5	1.9	2.45	0.518
3	B	3	4	3.05	7.21	23.4	1.3	2.35	0.532

Line Number (1)	Design						Profile				
	Runoff, Q, ft^3/s									Invert Elevation	
	Per acre *ci* (11)	Total (12)	Diameter, in. (13)	Slope, % (14)	Capacity, ft^3/s (15)	Velocity, ft/s (16)	Length, ft (17)	Fall, ft (18)	Drop in MH, ft (19)	Upper End (20)	Lower End (21)
1	1.30	2.85	12	6.42	3.09	3.94	340	2.18	0.00	86.46	84.28
2	1.27	5.28	18	2.71	5.93	3.35	340	0.92	0.42	83.86	82.94
3	1.25	9.02	24	1.50	9.48	3.02	440	0.66	0.46	82.48	81.82

Conversion factors: 1 ft/s = 0.3048 m/s; 1 in. = 25.4 mm; 1 acre = 0.4046 ha; 1 ft^3/s = 28.32 L/s = 0.02832 m^3/s; 1 ft = 0.3048 m; 1 in. /h = 25.4 mm/h.

14.13 Hydraulic Design of Combined Sewers

The capacity design of combined sewers allows for a maximum rate of wastewater flow in addition to stormwater runoff. If entering rainwater is confined to roof water, the wastewater flows are considerable items in required sewer capacities. The resulting system is sometimes called a roof-water system rather than a combined system. If the full runoff from storms of unusual intensity is carried away by the system, wastewater flows become relatively insignificant items in required combined sewer capacity.

14.14 Force Main Sewers

Force mains are pipelines that convey wastewater under pressure from the discharge side of a pump or pneumatic ejector to a discharge point. Pumps or compressors located in a lift station provide the energy for wastewater conveyance in force mains. The key elements of force mains are

1) Pipes
2) Valves
3) Pressure surge control devices
4) Force main cleaning system.

Force mains are constructed from various materials and come in a wide range of diameters. Wastewater quality governs the selection of the most suitable pipe material. Operating pressure and corrosion resistance also impact the choice. Pipeline size and wall thickness are determined by wastewater flow, operating pressure, and trench conditions.

14.14.1 Common Modifications

Force mains may be aerated or the wastewater chlorinated at the pump station to prevent odors and excessive corrosion. Pressure surge control devices are installed to reduce pipeline pressure below a safe operating pressure during lift station start-up and shut-off. Typically, automatically operated valves (cone or ball type) control pressure surges at the pump discharge or pressure surge tanks. Normally, force main cleaning includes running a manufactured "pigging" device through the line and long force mains are typically equipped with "pig" insertion and retrieval stations. In most cases, insertion facilities are located within the lift station and the pig removal station is at the discharge point of the force main. Several launching and retrieval stations are usually provided in long force mains to facilitate cleaning of the pipeline.

14.14.2 Applicability

Force mains are used to convey wastewater from a lower to higher elevation, particularly where the elevation of the source is not sufficient for gravity flow and/or the use of gravity conveyance will result in excessive excavation depths and high sewer pipeline construction costs.

Force mains are very reliable when they are properly designed and maintained. In general, force main reliability and useful life are comparable to that of gravity sewer lines, but pipeline reliability may be compromised by excessive pressure surges, corrosion, or lack of routine maintenance.

14.14.3 Advantages and Disadvantages

Use of force mains can significantly reduce the size and depth of sewer lines and decrease the overall costs of sewer system construction. Typically, when gravity sewers are installed in trenches deeper than 20 ft (6.1 m), the cost of sewer line installation increases significantly because more complex and costly excavation equipment and trench shoring techniques are required. Usually, the diameter of pressurized force mains is one to two sizes smaller than the diameter of gravity sewer lines conveying the same flow, allowing significant pipeline cost reduction. Force main installation is simple because of shallower pipeline trenches and a reduced quantity of earthwork. Installation of force mains is not dependent on site-specific topographic conditions and is not impacted by available terrain slope, which typically limits gravity wastewater conveyance.

While construction of force mains is less expensive than gravity sewer lines for the same flow, force main wastewater conveyance requires the construction and operation of one or more lift stations. Wastewater pumping and use of force mains could be eliminated or reduced by selecting alternative sewer routes, consolidating a proposed lift station with an existing lift station, or extending a gravity sewer using directional drilling or other state-of-the art deep excavation methods.

The dissolved oxygen content of the wastewater is often depleted in the wet well of the lift station, and its subsequent passage through the force main results in the discharge of septic wastewater, which not only lacks oxygen but often contains sulfides. Frequent cleaning and maintenance of force mains is required to remove solids and grease buildup and minimize corrosion due to the high concentration of sulfides.

Pressure surges are abrupt increases in operating pressure in force mains that typically occur during pump start-up and shut-off. Pressure surges may have negative effects on force main integrity but can be reduced by proper pump station and pipeline design.

14.14.4 Design Criteria

Force main design is typically integrated with lift station design. The major factors to consider when analyzing force main materials and hydraulics include the design formula for sizing the pipe, friction losses, pressure surges, and maintenance. The Hazen–Williams formula is recommended for the design of force mains. This formula includes a roughness coefficient C, which accounts for pipeline hydraulic friction characteristics. The roughness coefficient varies with pipe material, size, and age.

Force Main Pipe Materials

Selection criteria for force main pipe materials include

1) Wastewater quantity, quality, and pressure
2) Pipe properties, such as strength, ease of handling, and corrosion resistance
3) Availability of appropriate sizes, wall thickness, and fittings
4) Hydraulic friction characteristics
5) Cost.

Ductile iron pipe offers strength, stiffness, ductility, and a range of sizes and thicknesses and is the typical choice for high-pressure and exposed piping. Plastic pipe is most widely used in short force mains and when smaller diameters are appropriate. Table 14.8 lists the types of pipe recommended for use in a force main system and suggested applications.

Table 14.8 Characteristics of Common Force Main Pipe Materials

Material	Application	Advantages	Disadvantages
Cast or ductile iron: Cement lined	High pressure Available sizes of 4–54 in.	Good resistance to pressure surges	More expensive than concrete and fiberglass
Steel, cement-lined	High pressure All pipe sizes	Excellent resistance to pressure surges	More expensive than concrete and fiberglass
Asbestos cement	Moderate pressure For 36-in.+ pipe sizes	No corrosion Slow grease buildup	Relatively brittle
Fiberglass-reinforced epoxy pipe	Moderate pressure For up to 36-in. pipe sizes	No corrosion Slow grease buildup	350-psi max. pressure
Plastic	Low pressure For up to 36-in. pipe sizes	No corrosion Slow grease buildup	Suitable for small pipe sizes and low pressure only

Conversion factors: 1 in. = 25.4 mm; 1 psi = 6.94 kPa.

Velocity

Force mains from the lift station are typically designed for velocities between 2 and 8 ft/s (0.60 and 2.5 m/s). Such velocities are normally based on the most economical pipe diameters and typical available heads. For shorter force mains (less than 2,000 ft or 610 m) and low lift requirements (less than 30 ft or 9.14 m), the recommended design force main velocity range is 6–9 ft/s (1.83–2.74 m/s). Under certain circumstances, velocities as low as 2 ft/s (0.60 m/s) can be used, provided precautions are used to increase the velocity from time to time. This increased velocity for scouring is commonly obtained by operating the spare and active pump simultaneously on at least a weekly basis. This higher design velocity allows the use of smaller pipe, reducing construction costs. Use of higher velocities increases pipeline friction loss by more than 50%, resulting in increased energy costs. To reduce the velocity, a reducer pipe or a pipe valve can be used. Reducer pipes are often used because of the costly nature of pipe valves. These reducer pipes, which are larger in diameter, help to disperse the flow, therefore reducing the velocity.

The maximum force main velocity at peak conditions is recommended not to exceed 10 ft/s (3 m/s). A peaking factor allowance ranging from 3 for average flows of 1 MGD (44 L/s) and less, to 2 for average flows in excess of 10 MGD (440 L/s), should be used in sizing the force main pipe diameter. If continuous pumping is desired, two or three sizes of pumps may be required, some of which may be constant-speed units and some variable-speed units. Table 14.9 provides examples of force main capacities at various pipeline sizes, materials, and velocities. The flow volumes may vary depending on the pipe material used.

Vertical Alignment

Force mains should be designed so that they are always full and pressure in the pipe is greater than 10 psi (69.5 kPa) to prevent the release of gases. Low and high points in the vertical alignment should be avoided; considerable effort and expense are justified to maintain an uphill slope from the lift station to the discharge point. High points in force mains trap air, which reduces available pipe area, causes nonuniform flow, and creates the potential for sulfide corrosion. Gas relief and vacuum valves are often installed if high points in the alignment of force mains cannot be avoided, and blowoffs are installed at low points.

Pressure Surges

The possibility of sudden changes in pressure (pressure surges) in the force main due to starting and/or stopping pumps (or operation of valves appurtenant to a pump) must be considered during design. The duration of such pressure surges ranges between 2 and 15 seconds. Each surge is site-specific and depends on pipeline profile, flow, change in velocity, inertia of the pumping equipment, valve characteristics, pipeline materials, and pipeline accessories. Critical surges may be caused by power failures. If pressure surge is a concern, the force main should be designed to withstand calculated maximum surge pressures.

Table 14.9 Force Main Capacity

	Velocity = 2 ft/s	Velocity = 4 ft/s	Velocity = 6 ft/s
Diameter (in.)	(gpm)	(gpm)	(gpm)
6	176	362	528
8	313	626	940
10	490	980	1,470
18	1,585	3,170	4,755
24	2,819	5,638	8,457
36	6,342	12,684	19,086

Conversion factors: 1 ft/s = 0.3048 m/s; 1 in. = 25.4 mm; 1 gpm = 3.785 L/ min = 0.0631 L/s.

Valves

Valves are installed to regulate wastewater flow and pressure in the force mains. Valves can be used to stop and start flow, control the flow rate, divert the flow, prevent backflow, and control and relieve the pressure. The number, type, and location of force main valves depend on the operating pressures and potential surge conditions in the pipeline. Although valves have a lot of benefits, the costliness of them prevents them from being used extensively.

Performance

Force main performance is closely tied to the performance of the lift station to which it is connected. Pump-force main performance curves are used to define and compare the operating characteristics of a given pump or set of pumps along with the associated force main. They are also used to identify the best combination of performance characteristics under which the lift station-force main system will operate under typical conditions (flows and pressures). Properly designed pump-force main systems usually allow the lift station pumps to operate at 35–55% efficiency most of the time. Overall pump efficiency depends on the type of pumps, their control system, and the fluctuation of the influent wastewater flow.

14.15 Operation and Maintenance of Drainage Systems

14.15.1 Gravity Sewer System

The principal problem in the operation and maintenance of sewers is the prevention and relief of stoppages. Education and pollution prevention can enhance operation and maintenance programs by informing the public of proper grease disposal methods. Effective operation and hazards control of a conventional gravity sewer begins with *prevention through design* and construction. Tree roots and debris accumulation are the main causes of problems. Important, too, in areas of cohesionless soil is the entrance of sand and gravel through leaky joints and pipe breaks. Cement, mortar, and lime-mortar joints will not keep out roots as effectively as bi-tumastic hot-poured or factory-installed rubber joints. The plastic and other newer jointing materials work well. Understandably, debris is more likely to accumulate in the upper reaches of sewers where flows are low and unsteady. Sharp changes in grade and junctions at grade are danger points. Grease from eating places, oil from service stations, and mud from construction sites, often discharged intermittently and in high concentration, are leading offenders. Well-scheduled sewer flushing is an obvious answer when system design cannot be altered.

Interruptions in sewer service may be avoided by strict enforcement of sewer ordinances and timely maintenance of sewer systems. Regular inspection and maintenance minimize the possibility of damage to private property by sewer stoppages as well as the legal responsibility of the sewer authority for any damages. An operation and maintenance program is necessary and should be developed to ensure the most trouble-free operation of a sanitary sewer system. An effective maintenance program includes enforcement of sewer ordinances, timely sewer cleaning and inspection, and preventive maintenance and repairs. Inspection programs often use closed-circuit television cameras and lamping to assess sewer conditions. Sewer cleaning clears blockages and serves as a preventive maintenance tool. Common sewer cleaning methods include *rodding, flushing, jetting,* and *bailing*. Serious problems may develop without an effective preventive maintenance program. Occasionally, factors beyond the control of the maintenance crew can cause problems. Potential problems include the following:

1) Explosions or severe corrosion due to discharge of industrial wastes without pretreatment.
2) Odors.
3) Corrosion of sewer lines and manholes due to generation of hydrogen sulfide gas.
4) Collapse of the sewer due to overburden or corrosion.
5) Poor construction, workmanship, or earth shifts may cause pipes to break or joints to open up. Excessive infiltration/exfiltration may occur.
6) Protruding taps in the sewers caused by improper workmanship (known as plumber taps or hammer taps). These taps substantially reduce line capacity and contribute to frequent blockages.
7) Excessive settling of solids in the manhole and sewer line may lead to obstruction, blockage, or generation of undesired gases.
8) The diameter of the sewer line may be reduced by accumulation of slime, grease, and viscous materials on the pipe walls, leading to blockage of the line.

9) Faulty, loose, or improperly fitted manhole covers can be a source of noise as well as inflow. Ground shifting may cause cracks in manhole walls or pipe joints at the manhole, which become a source of infiltration or exfiltration. Debris (rags, sand, gravel, sticks, etc.) may collect in the manhole and block the lines. Tree roots may enter manholes through the cracks, joints, or a faulty cover, and cause serious blockages.

In arctic (permafrost) regions, sewers as well as water pipes may have to be placed in utilidors. The transfer of all utilities to such structures is being considered, too, in the rebuilding of old cities and the building of new ones. Effective inclusion of sanitary sewers may call for the introduction of pressurized systems. Storm sewers can probably not be accommodated economically.

Operation and maintenance costs are given in Fig. 14.19. Recent 2022 costs can be obtained by multiplying the costs from the figure by a multiplier of 3.39 (Appendix 16).

14.15.2 Force Main and Lift Station

The operation of force main/lift station systems is usually automated and does not require the presence of a continuous on-site operator. However, annual force-main route inspections are recommended to ensure normal functioning and to identify potential problems.

Special attention is given to the integrity of the force main surface and pipeline connections, unusual noise, vibration, pipe and pipe joint leakage and displacement, valving arrangement and leakage, lift station operation and performance, discharge pump rates and pump speed, and pump suction and discharge pressures. Depending on the overall performance of the force main/lift station system, the extent of grease buildup and the need for pipeline pigging are also assessed.

If there is an excessive increase in pump head and the head loss increase is caused by grease buildup, the pipeline is pigged. Corrosion is rarely a problem since pipes are primarily constructed of ductile iron or plastic, which are highly resistant to corrosion.

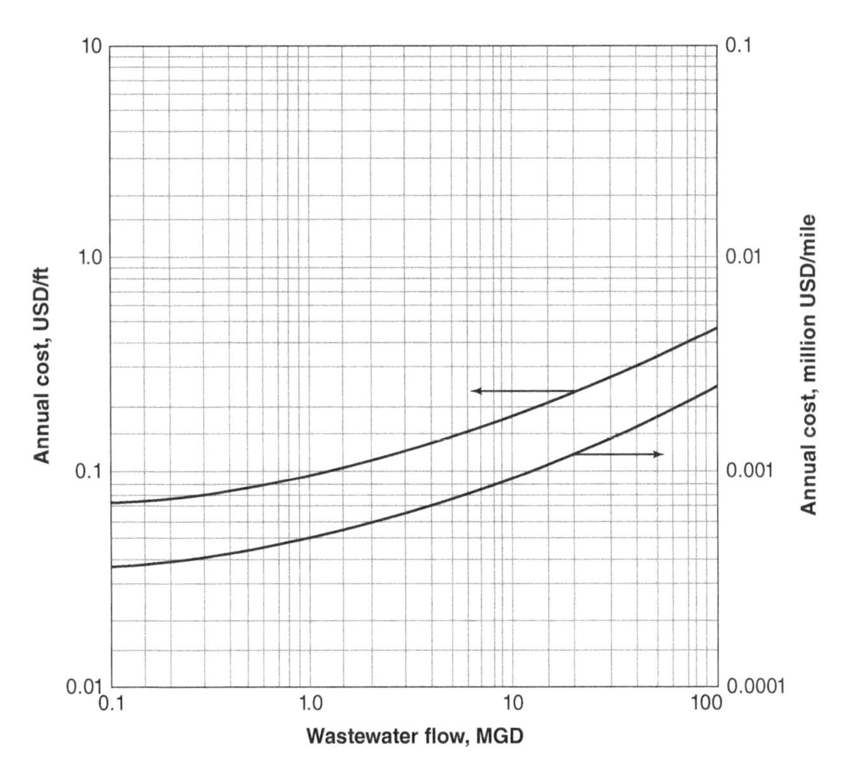

Figure 14.19 Operation and Maintenance Costs for Sewers. Recent 2022 costs can be obtained by multiplying the costs from the figure by a multiplier of 3.39 (Appendix 16). Conversion factors: 1 ft = 0.3048 m; 1 mile = 1.609 km; 1 MGD = 3.785 MLD = 0.0438 m^3/s = 43.8 L/s

14.16 Safety and Hazards Prevention

Safety and hazards prevention during the construction and operation of sewer works can be prevented by taking the appropriate and safety measures starting from the initial design phase. Six case studies are discussed below to illustrate several hazardous conditions and how to avoid them.

14.16.1 Municipal Sewer Maintenance Worker Drowns Inside Sewer Wet Well in Illinois

A 35-year-old male municipal sewer maintenance worker (victim) drowned inside a sewer wet well. The victim was part of a four-man sewer maintenance crew assigned to clean out a sewer wet well, which was 20 ft deep and 6 ft in diameter (6.1 m deep and 1.83 m in diameter). The victim entered through a 24-in.-diameter (600-mm-diameter) manway and climbed down on fixed steel rungs that extended to the bottom. The wet well atmosphere had not been tested nor ventilated before entering. The victim was wearing a full-body harness (secured to a winch cable) and a supplied-air respirator. After descending 8 ft (2.44 m) to a grating platform, the victim installed an inflatable sewer plug into the end of an 18-in.-diameter (450-mm-diameter) inlet sewer pipe 2 ft (0.60 m) below ground level. After inflating the sewer plug, the victim climbed down to the bottom of the wet well and began cleaning out the wet well with an 8-in.-diameter (200-mm-diameter) sewer vacuum hose. Within a few minutes the victim removed the respirator face piece, complaining to a coworker that the respirator was in his way. Approximately 30 minutes later, the sewer plug gave way, causing sewage to flood the wet well.

On hearing the noise, the foreman ran to the manway and yelled for the two workers to get out. Another coworker turned on the winch and began raising the cable. Within 15 seconds the level of sewage inside the wet well was up to the grating. The coworker who was standing on the grating reached down through the opening in the grating and made an unsuccessful attempt to grab the victim who was submerged. During this rescue effort, the winch cable became entangled in the grating support beams. As the sewage level continued to rise, the coworker was forced to climb up further and was ultimately helped out of the wet well by other coworkers. An attempt was also made to start up the pumps inside the lift station. However, the pumps were air-locked and, therefore, would not pump the sewage out.

The fire department rescue squad was notified and arrived within 10 minutes. By this time the sewage was about 2 ft (0.60 m) above the grating. Fire department rescuers entered the wet well, freed the entangled winch cable, removed the victim, and began administering cardiopulmonary resuscitation. The victim revived and was transported to the intensive care unit of a local hospital where he died approximately 11 hours later.

The public works department has a safety policy and confined space entry procedures, but no confined space rescue procedures. The director of the public works department is responsible for the safety program. Public works employees attend monthly safety meetings where job safety issues are discussed and training is occasionally given. The victim had previously attended a one-hour training session on confined space safety, and a one-hour training session on the use of *air respirators* since his employment began with the public works department.

The employer did not ensure that the following sewer plug manufacturer's recommendations were adhered to during the installation of the *plug:* (a) the pipe be cleaned out prior to insertion of the plug, (b) the plug be installed with a backup system (i.e., gate valve), (c) the plug be anchored in place, and (d) the plug be checked to ensure proper inflation to 30 psi (208.2 kPa).

The National Institute for Occupational Safety and Health (NIOSH) investigator concluded that, in order to prevent future similar occurrences, employers and employees must

1) Follow sewer plug manufacturer recommendations on the installation and use of inflatable sewer plugs.
2) Develop, implement, and enforce specific confined space entry procedures.
3) Ensure that appropriate rescue equipment is utilized during confined space entry.
4) Ensure that appropriate personal protective equipment is properly worn.
5) Consider installing self-priming wet-well sewer pumps. Properly installed, a self-priming sewer pump would prevent an air-lock whenever the wet well is manually pumped out below the level of the pump intake. In this incident, if the pumps in the sewage lift station had been self-priming, they could have been immediately activated, possibly preventing the fatality that resulted.

14.16.2 Sewer Worker Dies When Inflatable Sewer Plug Bursts in Washington, D.C.

A sewer maintenance worker died while working inside a sewer gate chamber. An *inflatable sewer plug* downstream from the victim was overinflated and burst, allowing sewage to flood the chamber. The worker was part of a 10-men sewer maintenance crew assigned to divert the flow of sewage in a branched, 6-ft-diameter (1.83-m-diameter) sewer main. The crew lowered an inflatable sewer plug into a diversion gate chamber and anchored it several feet into the right leg of the sewer main. An air line, connected to an air compressor at the surface, was attached to an air valve on the inflatable sewer plug. The victim, who was operating the compressor, left it running unattended and entered the gate chamber to inspect the sewer plug. Within a few minutes the plug burst, forcing water and air into the chamber, fatally injuring the worker.

The rescue squad from the city emergency medical service (EMS) was notified and arrived at the site in five minutes. After a 40-minute search, EMS personnel discovered the body of the victim submerged under the sewage flow, against the bar screen of a sewage pumping station approximately 200 yards (183 m) downstream from the gate chamber. EMS personnel noted that the victim was dead at the scene.

The employer involved is a municipal utility with 1,100 public works employees. Approximately 200 of the employees are sewer maintenance workers and wastewater treatment plant operators. The victim had been employed by the municipality for 23 years as a sewer maintenance worker. The public works department has a full-time safety and health manager and a full-time safety and health specialist. A safety policy exists, but no confined space entry procedures for sewer maintenance workers. However, the victim and other sewer maintenance workers had participated in a 2-hour training session on confined space safety within the past year.

NIOSH investigators concluded that, in order to prevent future similar occurrences, employers should

1) Use *slide gates* instead of, or in conjunction with, inflatable sewer plugs. Slide gates provide a more positive method for diverting/controlling the flow of sewage for maintenance purposes, and should be utilized where possible.
2) Follow sewer plug manufacturers' recommendations and other safety precautions on the installation and use of inflatable sewer plugs. According to the manufacturer, filling the plug with water instead of air when the plug is submerged will greatly reduce the force of a rupture.
3) Develop and implement specific confined space entry and work procedures.

14.16.3 Independent Contractor Dies in Sewer Line Excavation Engulfment, Alaska

A 34-year-old male independent contractor (victim) was killed as a result of traumatic head and neck injuries during a sewer line excavation cave-in.

The sewer pipe leading to a private residence had parted, causing effluent to build up around the point of the break. Sewage was also backing up into the home's garage. The owners of the home contracted with the victim to perform the above procedure. The victim rented a back hoe/front-end loader from a local company to perform the trenching operation. Excavation site and soil conditions on the day of the incident were as follows:

- Approximately 2 in. (50 mm) of blacktop (driveway)
- Three and one-half feet of partially frozen, sandy/silty gravel
- Water-saturated sandy/silty gravel below the semi-frozen gravel.

The victim was repairing the sewer line at the private residence and had excavated a 12-ft-deep by 15-ft-long (3.66-m-deep by 4.57-m-long) trench with 90° sidewalls to access the existing sewer line. The victim then called another contracting service to remove groundwater/sewage that had accumulated in the trench. After this operation was completed, the victim entered the unshored trench via a ladder for another inspection. At this time the walls of the trench collapsed, and concrete slabs (from a sidewalk leading to the private residence) fell into the trench. These slabs struck the victim on the head and neck, and caused him to fall from the ladder. He was subsequently buried by the incoming soil from the trench walls.

Witnesses called 911 and fire rescue personnel responded. They arrived on scene and began to attempt a rescue. The rescuers stood on the "fill" in the trench near the ladder where they began digging with shovels in an attempt to locate the victim. Because of the instability of the trench (rescuers had to abandon the trench on several occasions due to collapsing soil), fire rescue personnel began looking for shoring materials. An attempt was made to shore the walls with 3/8-in. (0.95-cm) plywood, but this proved to be unsuccessful. Because of the elapsed time and the imminent and immediate danger to rescue personnel at the site, Department of Labor officials requested that the rescue attempt be abandoned and the trench be widened and sloped to a safer "angle of repose." The victim's body was located under the concrete slabs approximately three hours after the trench collapse.

The employee was a local independent contractor who specialized in "home improvements." He operated primarily as a "jack of all trades," and had completed jobs in carpentry, electrical installation, and other building trades. He had been in business for five years and worked primarily on rental properties. He was a vocational school graduate and had served in the military. However, little is known about his specific training in trenching and shoring. Most training he received appears to have been "on-the-job" training.

Based on the findings of the epidemiologic investigation, to prevent similar occurrences:

1) Independent contractors should be aware of the potential dangers of trenching or other excavation operations and be knowledgeable about proper techniques of sloping and shoring. The victim failed to use any method of trenching appropriate for the excavation operation. He was apparently either unaware of the dangers associated with collapsing trenches, or failed to correctly evaluate the hazards of his work task.
2) Independent contractors should be aware of the increased potential for excavation collapse due to adverse environmental factors, such as elevated levels of groundwater. The incident described above occurred during the spring "breakup" when large quantities of ice and snow rapidly melt. This phenomenon adds considerable amounts of free water to soil throughout areas of similar geographic and climatic conditions. Soils containing high percentages of silt and gravel can hold significant amounts of groundwater.
3) Independent contractors should be knowledgeable about job safety and always conduct a general hazard assessment prior to beginning any job or work task.
4) EMS and fire rescue personnel should be knowledgeable about proper rescue techniques involving excavation sites and ensure that adequate shoring equipment is on hand at all times. Fire rescue personnel made a number of attempts to rescue the victim, but were impeded by their lack of adequate shoring materials. The 3/8-in. plywood used proved to be insufficient for the task. Fire rescue services should ensure that adequate trenching material is on hand at each station where rescue personnel are housed. Also, emergency shoring devices (e.g., "Quick Shore") are possible solutions for fire rescue stations. These lightweight, narrow devices can be carried by one rescuer, and provide fast, adequate shoring in emergency situations.

14.16.4 Contract Worker Dies While Reinstalling Sewer Line in Wyoming

A 39-year-old male plumber's helper died from injuries incurred when an excavation collapsed while reinstalling a sewer line at a construction site for a new home being built. The victim was in an unshored, vertical-walled excavation replacing a sewer pipe that had been installed nearly a week prior to the incident. While he was shoveling dirt above the pipe to prevent future breakage, a 2- × 3-ft (0.60- × 0.90-m) concrete caisson/footer fell in the excavation along with the surrounding excavation wall, burying the victim to his knees with dirt, striking him in the abdomen, and pinning him to the remaining wall of the trench.

The backhoe operator called to nearby sanitation workers to help, and they found that the victim's legs were pinned by the flat part of the concrete piece. Personnel at the scene had released the victim and pulled him to the surface of the hole by the time emergency services arrived. Fire rescue professionals found the victim balanced on a ground surface area between two excavations, with his feet dangling into one of the holes. The victim was in extreme pain, coherent, breathing well, and answering questions. His lower legs were discolored and he complained of severe pelvic and leg pain. Emergency personnel placed the victim on a backboard for transport 10 ft (3.0 m) to a more stable area, where they applied oxygen and arranged for hospital transfer by ambulance. The victim sustained severe internal injuries to the abdomen and pelvic area and died the following day in the hospital.

Employers may be able to minimize the potential for the occurrence of this type of incident through the following precautions:

1) Shore vertical wall excavations to OSHA standards.
2) Establish quality control procedures to minimize reworks.
3) Improve hiring practices for short-time workers.

14.16.5 Carpenter Dies When 8-ft Trench Wall Collapses During Sewer Pipe Replacement

On October 21, 2004, a 22-year-old male carpenter (victim) died when the walls of an 8-ft (2.44-m) excavation he was working in collapsed and completely covered him (see Fig. 14.20). A homeowner hired the victim's employer to replace a 6-in. (150-mm) clay-tile sewer pipe leading from his home to the alley behind the home and garage. The firm was "threading" a new 4-in. (100-mm) polyvinyl chloride (PVC) pipe through the deteriorating existing clay 6-in. (150-mm) pipe, and leaving the existing 6-in. (150-mm) pipe in place. Prior to the victim's arrival, the employer had excavated an approximately 8-ft-deep (2.4-m-deep) trench from the home's basement to the homeowner's garage. Once beyond the garage the employer dug another 8-ft (2.4-m) excavation from the garage to the alley where the sewer connection was located. The soil conditions in the second excavation were sand/gravel and the angle of repose (maximum permissible slope) for the excavation sides varied from 60° to 80°. To determine how far away the 4-in. (100-mm) PVC pipe was from the sewer line, the victim either kneeled or lay down at the bottom of the excavation. The victim was still kneeling or lying on the ground when the south side of the excavation collapsed, completely burying him and burying his coworker up to his waist. Emergency 911 was called, and at the same time all employees jumped into the excavation to rescue the individuals in the trench. Emergency personnel arrived within minutes, removed the victim, and transported him to a local hospital where he died the next day.

The following recommendations were made:

1) Employers should ensure when employees are working in excavations that require a supporting system that a supporting system is implemented. The excavation was not cut to the proper angle of repose, and no shoring or trench box was used to prevent employee engulfment.
2) Employers should ensure that a qualified person inspects the excavation, adjacent areas, and supporting systems on an ongoing basis and that the qualified person takes the appropriate measures necessary to protect workers. There was only 6 ft (1.83 m) of room between the work area and the adjacent property line, making it difficult to cut the trench to the proper angle of repose for the soil conditions (see Fig. 14.21).
3) Employers should design, develop, and implement a comprehensive safety program.
4) Employers should provide workers with training in the recognition and avoidance of unsafe conditions and the required safe work practices that apply to their work environments. The excavated materials were not stored and retained more than 2 ft (0.60 m) from the excavation edge.
5) Employers should ensure that equipment is moved away from open trenches when not in use.
6) Employers should develop a trench emergency action plan that describes rescue and medical duties and ensure that all employees are knowledgeable of those procedures.

Figure 14.20 Picture of Trench after Cave-in.

Figure 14.21 Overall Scene–Alley View.

14.16.6 Two Confined Space Fatalities Occurred During Construction of a Sewer Line

A city fire department received a report that a man was down at a sewer construction site. When the firefighters arrived on the scene, they learned that two workers were down in the newly constructed sewer. One worker was an employee of the company contracted to construct the sewer. The other worker was a state inspector with the State Department of Transportation. The two workers were removed from the sewer and pronounced dead at the scene. Subsequent autopsy indicated cause of death to be carbon monoxide (CO) poisoning. As a result of the rescue effort, 30 firefighters and 8 construction workers were treated for CO intoxication and/or exhaustion. The synopsis of events is as follows.

In the process of constructing the interstate highway, the contractor had to construct several thousand feet of sanitary sewer line composed of 66-in.-diameter by 16-ft-long (1,550-mm-diameter by 4.88-m-long) sections of concrete pipe. This new line had to tie into an existing line. The upstream portion of the existing line would be abandoned after completion of the new line (see Fig. 14.22).

The existing line had to be kept in service during construction. A bypass line had to be built around the connection point of the new and existing lines. This was done by tapping a 30-in. (76.2-cm) bypass line into the existing line, upstream of the connection point, and tying the bypass line into a newly constructed manhole (No. 1 in Fig. 14.22) at the connection point. To keep sewage from entering the construction area of the connection point, the pipe was diked by sandbags several feet upstream of manhole 1. The dike was left in place for approximately one month while the contractor continued to lay pipe.

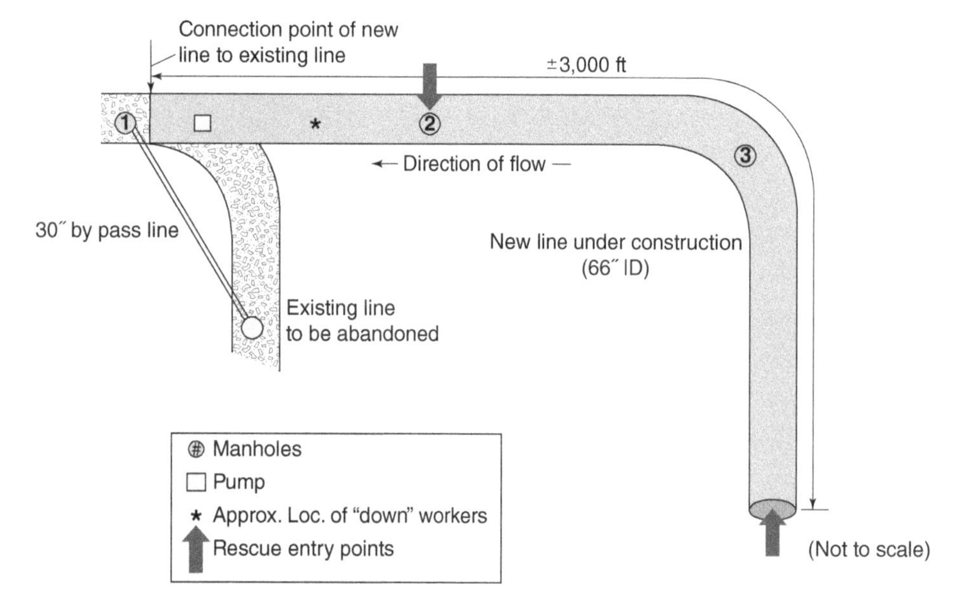

Figure 14.22 Plan for Laying the Sewer. Conversion factors: 1 ft = 0.3048 m; 1″ = 1 in. = 25.4 mm

During this time, sewage seeped/flowed past the dike and extended approximately 480 ft (146.3 m) into the newly constructed line. This sewage had to be removed before the contractor could proceed with grouting the pipe joints.

The contractor replaced the sandbag dike with a steel plug to eliminate further seepage. A gasoline-engine-driven pump was placed upstream of the plug so that the existing sewage could be removed from the pipe. The pumping procedure required a laborer to enter the new line at manhole 2, walk downstream approximately 1,200 ft (365.8 m) to the pump, fuel the gasoline engine, start it, and exit back through manhole 2. This procedure was performed on a three-day cycle. At no time was the atmosphere in the pipe tested prior to entry, nor was there mechanical ventilation to remove air contaminants.

This procedure was not removing the sewage quickly enough and it was decided to increase this cycle to three times per day. On the day of the accident, the labor foreman and one worker (his son) followed the procedure of starting the pump. The two workers returned to manhole 2 to repeat the procedure of refueling the pump. However, manhole 2 had been covered with plywood and framed over in order to have concrete poured the following day. So the two had to enter the pipe from the point of construction. Each carried a flashlight and the worker carried a can of gasoline. They began walking the 3,000-ft (914.4-m) distance to the pump. After passing manhole 3, they took a short break and proceeded past manhole 2 toward the pump. Approximately 750 ft (229 m) past manhole 2, the two came to the board used to mark the water line. While the foreman was moving the board and counting the pipe length to determine how far the water had receded, the worker went on ahead to fuel the pump and start it. After noticing haze in the sewer, the foreman told the worker to keep talking so he could tell if anything was wrong. Shortly the foreman heard the worker attempt to start the pump four times and then say, "I feel dizzy." The foreman ordered the worker out of the pipe. The worker started to leave, dropping his flashlight and stumbling in his unsuccessful attempt. By the time the foreman reached the worker, the worker was down and unresponsive. After failing to carry the worker out, he propped him up out of the water and told him he was going for help. The foreman walked, crawled, and stumbled 3,000 ft (914.4 m) to the outside to report the worker was down near the pump. The only ill effect experienced by the foreman was a severe headache.

Seven workers went into the pipe in an attempt to remove the downed worker. At the same time the state inspector got into his truck and drove to manhole 2, where he removed the plywood cover and entered the sewer. The state inspector proceeded toward the area where the worker had been reported down. The underground superintendent also entered the sewer at manhole 2 but exited after two or three minutes. Six of the seven workers who entered the pipe at the portal exited at manhole 2. The seventh man reached the worker but was unable to remove him. The company safety director entered the sewer at manhole 2 and reported passing the seventh worker and reaching the deceased. Shortly after, the seventh worker and the safety director exited the sewer manhole 2.

At this time three firefighters arrived at the scene and entered manhole 2. The firefighters were equipped with 30-minute *self-contained breathing apparatus (SCBA)*. In addition to the bulkiness of the SCBA, they were hampered by the curved and slick inner surface of the sewer. Initially, the firefighters were told the victims were down approximately 150 ft (45.72 m) into the sewer. However, they had to travel 500–600 ft (152–183 m) to reach the victims. As their air supply decreased, the firefighters placed one SCBA on the victim (the state inspector) who was still breathing, and resorted to buddy breathing to exit. The state inspector was removed through manhole 2. He was pronounced dead at the scene. Subsequent autopsy indicated his carboxyhemoglobin level was 50% and his pO_2 was 0%. The laborer was removed later through manhole 2. He was also pronounced dead at the scene. His carboxyhemoglobin level was 56% and pO_2 level was not available.

Combustible gas measurements and oxygen and carbon monoxide levels were taken 22 hours later at the incident site by an industrial hygienist. Oxygen level was 19% and concentrations of CO were 600 ppmv. The industrial hygienist estimated that concentration of CO next to the pump on the day of the incident was 2,000 ppmv. An air sample taken the following day revealed readings of 19–20% oxygen. Trace amounts of H_2S were also recorded.

Given the industrial hygiene survey results and the toxicologic findings, the cause of death was determined to be exposure to high concentrations of CO, a by-product of the gasoline-powered pump, in an area with no natural ventilation, that is, in a confined space.

While the following list of recommendations is not exhaustive, it does cover some of the salient points which, if implemented, could have prevented this fatal incident:

1) When the existing sewer was activated (passing through manhole 1), no plans were made to prevent the sewage from flowing into the newly constructed sewer. An analysis of the conditions surrounding the connection at manhole 1 should have generated several safe alternatives for an effective temporary barrier in the new sewer, which also considered safe atmospheric conditions.
2) A gasoline-powered pump was installed inside the sewer (a confined space), which was known to have almost no ventilation. Neither workers nor pump could have operated efficiently in the sewer. The rich mixture created by depletion of

O_2 increased the levels of CO. The pump should have been located on the outside of the sewer with a hose running to the sewage via an access hole or an electric-motor-driven pump should have been considered.

3) A static ventilating condition was created when the plug was installed in the new sewer next to manhole 1. Because it was necessary for workmen (either those servicing the pump or those planning to do the grouting) to enter the sewer, adequate ventilation should have been provided. If ventilation could not create a safe atmosphere, the use of SCBA should have been mandatory.
4) Workers were permitted to enter an untested atmosphere of a confined space. The atmosphere should have been tested by a qualified person prior to entry by workers.
5) Both fatal victims lacked experience in working in confined spaces. If workers are expected, as part of their job, to work in confined spaces, they should be given appropriate training.
6) The established corporate safety procedures for work in confined spaces were not implemented. Management, including local supervisors, should comply with approved corporate policy and procedures for confined space entry as well as other rules and regulations approved by the corporate president. The policy and procedure should include entry into confined spaces for rescue efforts.
7) Workers were not able to adequately assess their risk of personal injury of the tasks they were required to perform, much less the additional hazards associated with rescue efforts. Management should develop a safe job procedure for all routine tasks starting with high-risk tasks and specifically establish a policy and procedure regarding rescue efforts.

Problems/Questions

14.1 An inverted siphon is to carry a minimum dry-weather flow of 3.0 ft^3/s, a maximum dry-weather flow of 8.0 ft^3/s, and a wet-weather flow of 130 ft^3/s in three pipes. Select the proper diameters to ensure velocities of 3.0 ft/s in pipes carrying wastewater and 5.0 ft/s in pipes carrying stormwater.

14.2 An inverted siphon is to carry a minimum dry-weather flow of 85 L/s, a maximum dry-weather flow of 226 L/s, and a wet-weather flow of 3,680 L/s in three pipes. Select the proper diameters to ensure velocities of 0.9 m/s in pipes carrying wastewater and 1.5 m/s in pipes carrying stormwater.

14.3 The maximum flow reaching sewer A before the manhole is 850 L/s and lateral B 150 L/s (see Fig. 14.23). If the minimum flow is one-third of the maximum:

(a) Calculate the minimum flow velocities in sewers A and B.
(b) At what minimum height above the invert of sewer A should the invert of lateral B be located so that during maximum flow there will be no backing up of sewage into lateral B?
(c) At a further point downstream, a two-pipe inverted siphon is to be used to cross a valley. One pipe is to handle the minimum flow and the second is to carry the remaining flow at maximum flow condition. Determine the required diameters of the two pipes.

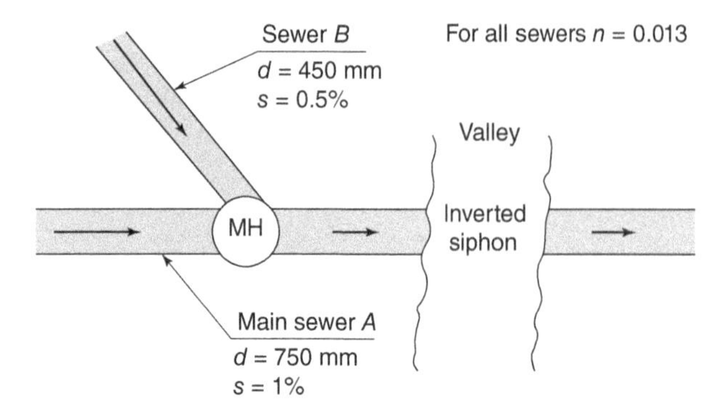

Figure 14.23 Layout of Sanitary Sewers for Problem 14.3

14.4 Part of an existing trunk sewer *ABC* has a diameter of 600 mm (see Fig. 14.24). The length of sewers between successive manholes *A* to *B* and *B* to *C* is 50 m. The following existing conditions are known:

Item	Manhole		
	A	*B*	*C*
Ground elevation (m)	535.50	535.35	535.20
Invert elevation (m)	531.50	531.35	531.20
Peak sewage level (m)	531.80	531.65	531.50

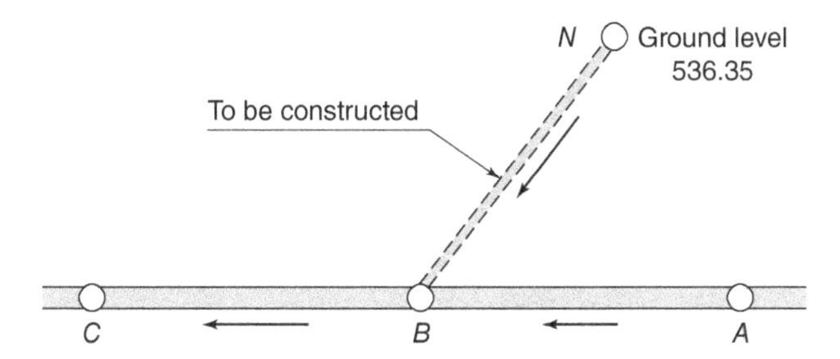

Figure 14.24 Layout of Sanitary Sewers for Problem 14.4

The sewage flow from a new development to the north of the trunk line is planned to be conveyed in a new sewer line *NB* to discharge in manhole *B*. The average flow is estimated to be 300 Lpcd, and the peaking factor is 3. The following limitations are given: $V_{min} = 0.75$ m/s, $V_{max} = 3.0$ m/s, $n = 0.013$, and minimum cover = 1.5 m. Determine the following:

(a) The maximum population that can be served from the new development without exceeding the capacity of the trunk sewer
(b) Diameter, slope, and invert elevations of sewer *NB*.

14.5 The layout of the last part of a sewerage system for a small town is illustrated in Fig. 14.25. Two main sewers *A* and *B* are connected to the trunk sewer at MH 36. After MH 1 the trunk sewer discharges into a wet well from which sewage is pumped into the wastewater treatment plant (WWTP). The following data are known:

- Diameter of sewer A = 800 mm
- Slope of sewer A = 1%
- Diameter of sewer B = 500 mm
- Slope of sewer B = 0.5%
- Trunk sewer between MH 36 and MH 1 has the same diameter and slope of sewer A

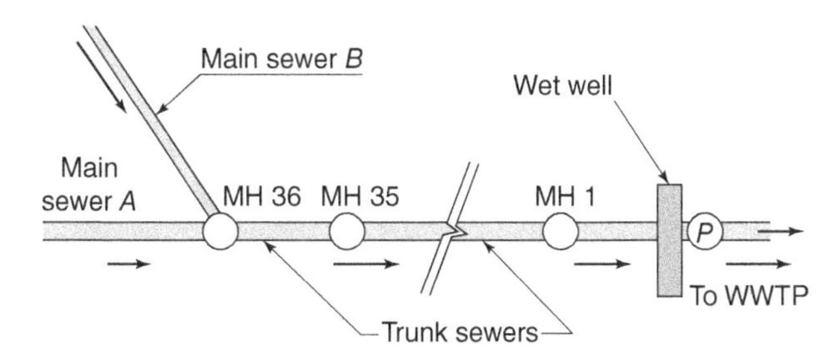

Figure 14.25 Layout of Sanitary Sewers for Problem 14.5

- Peak flow in sewer $A = 850$ L/s
- Peak flow in sewer $B = 150$ L/s
- Invert level of trunk sewer entering MH 1 = 510.40 m
- Elevation of sewage level in wet well = 510.00 m
- Distance between MH 1 and wet well = 100 m.

(a) Calculate the flow velocities in sewers A and B at peak flow conditions.
(b) At what minimum height above the invert of sewer A should the invert of sewer B be located so that during peak flow, there will be no backing up of sewage into sewer B?
(c) Determine the required diameter of trunk sewer between MH 1 and the wet well so that at all times there is free discharge of sewage into the wet well.

14.6 Lateral sewers from MH 5 to MH 1 and from MH 4 to MH 1 intercept the main sewer E (diameter = 400 mm and slope = 0.007) at MH 1. The invert elevation of main sewer E at MH 1 is 609.22 m. All sewers have a Manning coefficient, n, of 0.013 and a minimum cover of 2.00 m. The schematic layout of sewers is shown in Fig. 14.26. Pipe lengths, sewage flows, and ground elevations are provided in Table 14.10.

(a) Determine the total design flow (Q in L/s) in the main sewer F if line E is running full.
(b) Find the required sizes for sewers A, B, C, D, and F.
(c) Determine the invert elevations of all sewers at manholes 1, 2, 3, 4, 5, and 6.
(d) Draw the profile for the sewer line from MH 4 to MH 5 showing all important dimensions and elevations.

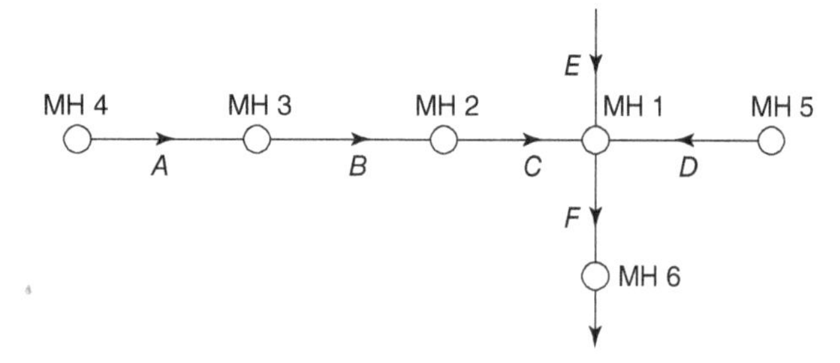

Figure 14.26 Layout of Sanitary Sewers for Problem 14.6

Table 14.10 Data for Problem 14.6

Pipe	Manhole From	Manhole To	Flow (L/s)	Ground Elevation (m) Upper	Ground Elevation (m) Lower	Length (m)
E	–	1	Full	–	612.00	–
A	4	3	20	615.25	613.75	75
B	3	2	40	613.75	611.87	75
C	2	1	60	611.87	612.00	50
D	5	1	17	612.12	612.00	50
F	1	6	Q_F	612.00	610.82	100

Conversion factors: 1 m = 3.2808 ft; 1 L/s = 15.852 gpm.

14.7 Part of the sewerage network of a city is shown in Fig. 14.27. The lateral sewers (lines 1, 2, and 3) discharge their flows into a trunk sewer at manhole T11. The lateral sewers receive sewage at a peak rate of 0.10 L/s per meter length of sewer. The peak flow in the trunk sewer upstream from manhole T11 is 450 L/s. No additional sewage reaches the trunk line between manholes T11 and T10. Manholes are located at 65 m center to center on laterals and at 100-m spacing on trunk lines.

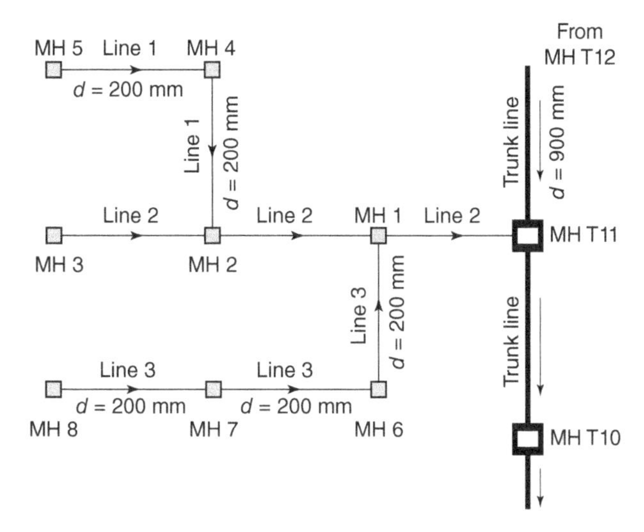

Figure 14.27 Layout of Sanitary Sewers for Problem 14.7

The following elevations are given:
Ground elevations:

MH 3	565.50 m
MH 2	565.00 m
MH 1	564.25 m
MH T11	563.50 m
MH T10	563.00 m

Invert elevations:

Line 1 at MH 2	563.00 m
Line 3 at NH 1	562.55 m
Trunk line T12 to T11 at MH T11	560.60 m

Design line 2 (from MH 3 to MH T11) and the trunk line (between MH T11 and MH T10). Draw a profile starting from manhole 3 and ending at manhole T10. Indicate sewer size and slope as well as ground and invert elevations at each manhole (3, 2, 1, T11, and T10). Consider that the minimum and maximum allowed velocities at peak flows are 0.75 and 3 m/s, respectively, and that the minimum allowed pipe cover is 1.50 m. The peaking factor is assumed to be constant for all sewers.

14.8 Part of a sewerage system and its pertinent data are shown in Fig. 14.28 and Table 14.11. Design sewers MH 2 to MH 1, MH 1 to MH 105 and MH 105 to MH 106. Draw a neat profile for the sewers from MH 2 to MH 106 and indicate all ground and invert levels and pipe diameters.

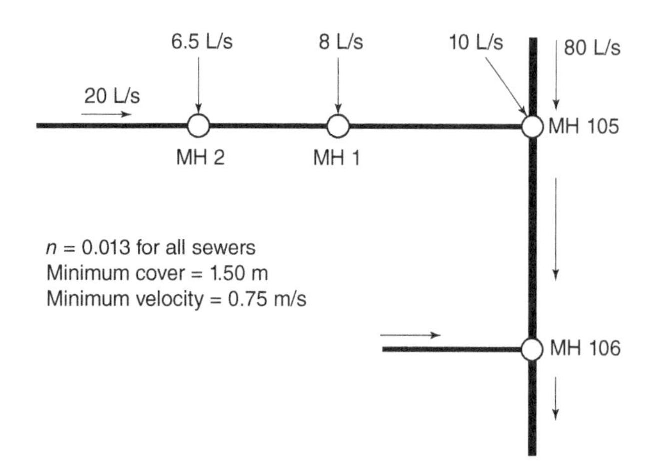

Figure 14.28 Layout of Sanitary Sewers for Problem 14.8. Conversion factors: 1 m = 3.2808 ft; 1 m/s = 3.2808 ft/s; 1 L/s = 15.85 gpm

Table 14.11 Data for Problem 14.8

MH			Ground Level, m		Invert Level, m		
From	**To**	**Length m**	**Upper**	**Lower**	**Upper**	**Lower**	**Pipe Diameter mm**
NA	2	NA	NA	405.50	NA	403.80	200
2	1	65	405.50	404.98	NA	NA	NA
1	105	80	804.98	404.58	NA	NA	NA
NA	105	NA	NA	404.58	NA	402.18	400
105	106	100	404.58	404.43	NA	NA	NA

Conversion factors: 1 m = 3.2808 ft; 1 mm = 0.0394 in.

14.9 The layout of a stretch of a sewer trunk line between MH 102 and MH 104 is shown in Fig. 14.29. Three tributary sewers (400-, 500-, and 300-mm lines) are connected to the trunk line at manholes 102, 103, and 104, respectively. The following design criteria and data are given:

Maximum allowed velocity in sewers	0.80 m/s
Minimum allowed velocity in sewers	3 m/s
Minimum soil cover	1.50 m
Manning coefficient	0.013

Ground elevations

MH 102	649.50 m
MH 103	649.30 m
MH 104	649.00 m

Invert elevations

800 mm trunk line at MH 102	647.20 m
400 mm sewer at MH 102	646.00 m
500 mm sewer at MH 103	647.30 m
300 mm sewer at MH 104	645.83 m

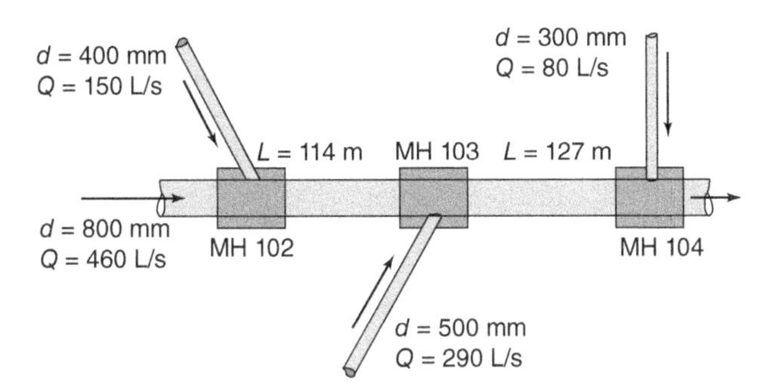

Figure 14.29 Layout of Sanitary Sewers for Problem 14.9. Conversion factors: 1 m = 3.2808 ft; 1 L/s = 15.85 gpm; 1 mm = 0.1 cm = 0.0394 in.

Design the trunk line between MH 102 and MH 104. Draw a neat trunk sewer profile starting at MH 102 and ending at MH 104. Indicate sewer size and slope as well as ground and invert elevations at each manhole.

14.10 Using Kerby's relationship, determine the overland flow time for a bare and moderately rough surface 60 ft long and having a slope of 0.010 that drains toward a storm outlet.

14.11 Calculate the amount of surface detention on a land surface having a ground slope of 0.01, length of overland flow of 100 ft, rainfall intensity of 1 in./h, overland flow of 0.8 in./h, and a coefficient of roughness $=6 \times 10^{-2}$.

14.12 The drainage area shown in Fig. 14.30 contributes its total flow to the inlet and henceforth into the storm sewer. The area is not homogenous and its runoff coefficient c is estimated for the different portions as follows:

% of Area	c
20	0.90
40	0.80
10	0.60
30	0.30

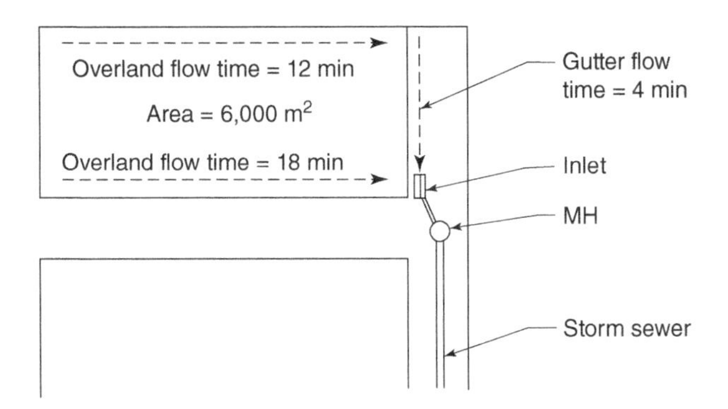

Figure 14.30 Layout of Storm Drainage for Problem 14.12. Conversion factor: 1 m^2 = 10.76 ft^2

The rainfall intensity i (mm/h) as a function of rainfall duration t (minutes) is

$$i = \frac{1,680}{t + 19}$$

(a) Find the overall coefficient c for the drainage area
(b) Find the total flow Q in the storm sewer
(c) Determine the maximum and minimum storm sewer diameter. Assume that the required slope can be economically accommodated in each case.

14.13 Area $ABCD$ contributes all its flow into the gutter BD (see Fig. 14.31). The overland flow time from A to B is 10 minutes, from C to D is 20 minutes, and the flow time in the gutter BD is five minutes.

Inlet D intercepts all the flow from gutter BD and discharges it into the storm sewer. Surface area $ABCD$ is not homogenous, and its runoff coefficient c is estimated for the different portions as follows:

% of Area	c
30	0.50
40	0.80
10	0.60
20	0.40

The rainfall intensity i (mm/h) as a function of rainfall duration t (minutes) is

$$i = \frac{1,220}{t + 12}.$$

Find

(a) The concentration time at inlet D.
(b) The overall runoff coefficient c for the area $ABCD$.
(c) The total flow Q (L/s) entering Inlet D.
(d) What is the maximum and minimum pipe diameter you can use for the storm sewer? Assume that the required slopes can be economically accommodated in each case.

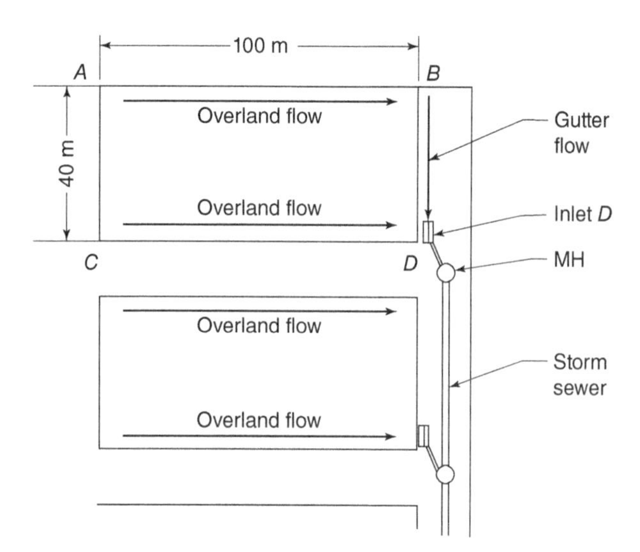

Figure 14.31 Layout of Storm Drainage for Problem 14.13

14.14 Blocks I, II, and III contribute all their runoff into the street *ABC* (see Fig. 14.32). Inlet *C* intercepts all the flow and discharges it into the storm sewer. The drainage areas and their corresponding runoff coefficients are as follows:

Block	Drainage Area (m^2)	Runoff Coefficient
I	10,000	0.80
II	12,000	0.75
III	12,000	0.70

The rainfall intensity i (mm/h) as a function of rainfall duration t (minutes) is

$$i = \frac{1,900}{t + 12}.$$

Find

(a) The total flow at point A (in m^3/min).
(b) The total flow at point B (in m^3/min).
(c) What are the maximum and minimum storm sewer diameters that can be used? Assume that the required slopes can be economically accommodated in all cases.

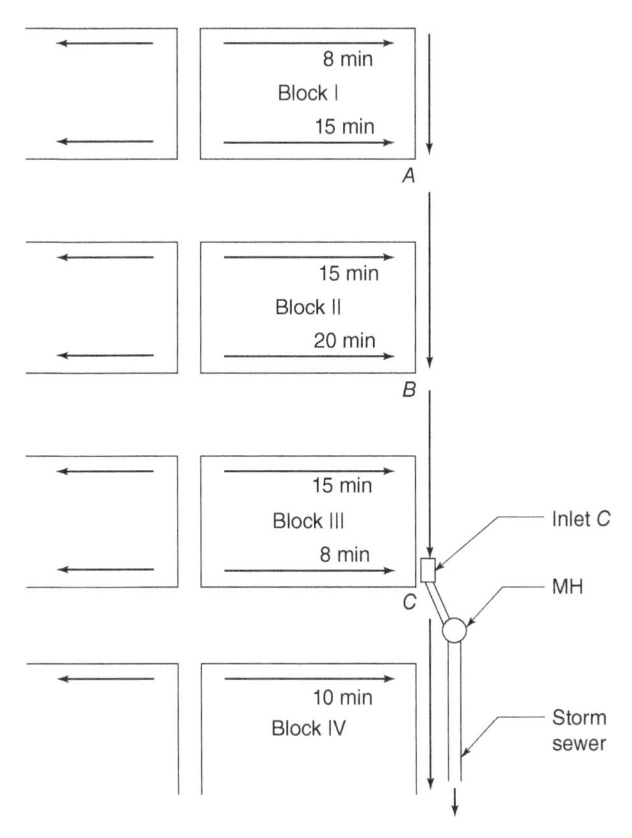

Figure 14.32 Layout of Storm Drainage for Problem 14.14

14.15 The runoff from University Campus is drained through a 2 km long main to a manhole located near University Road at an invert elevation of 610.50 m. From this MH a drainage pipe is to be designed to collect the rainwater runoff and discharge it into the river 1.5 km away. The following data are given:

- Drainage area = 2.5 km^2
- Runoff coefficient = 0.30

- Inlet time to the 2 km-long drain = 30 min .
- Average velocity in the 2 km-long drain = 0.90 m/s
- Manning n for all pipes = 0.013
- Maximum water level in the river = 608.20 m.

Rainfall intensity (mm/h) as a function of rainfall duration (minutes) is

$$i = \frac{1,020}{t + 12}.$$

(a) Determine the required size of the 1.5 km drain from the MH at University Road to the river, so that no back-flow of water from the river into the discharge pipe is possible under any flow condition in the pipe.
(b) What size of pipe will be required, if depth of excavation is to be kept to a minimum? In this particular case assume there is no limitation on depth of cover as long as no pumping shall be used.

14.16 Area *ABC* drains its stormwater into one inlet at point *C* (see Fig. 14.33). The drainage area has a runoff coefficient of 0.60 and a surface overland flow velocity of 1.25 m/s. The rainfall intensity i (mm/h) as a function of rainfall duration t (minutes) is

$$i = \frac{1,120}{t + 12}.$$

(a) Find the total flow Q (m^3/min) entering the inlet at *c*.
(b) What are the maximum and minimum storm sewer pipe diameters (mm) that can be used? Assume that the required grades for the sewer can be economically accommodated in all cases.

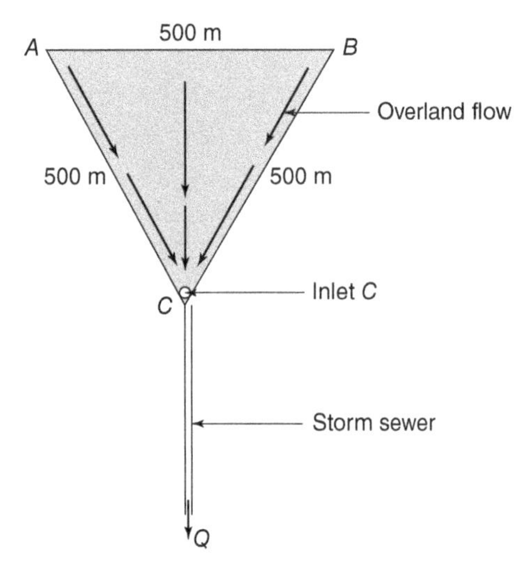

Figure 14.33 Layout of Storm Drainage for Problem 14.16. Conversion factor: 1 m = 3.28 ft

14.17 The drainage system shown in Fig. 14.34 is designed to collect the stormwater from a football field *ABCD* ($c = 0.50$) and a basketball field *EFGH* ($c = 0.90$). The rainfall intensity i (mm/h), as a function of rainfall duration t (minutes) is

$$i = \frac{3,000}{t + 20}.$$

Determine the design flows (m^3/h) for pipes 1, 2, and 3.

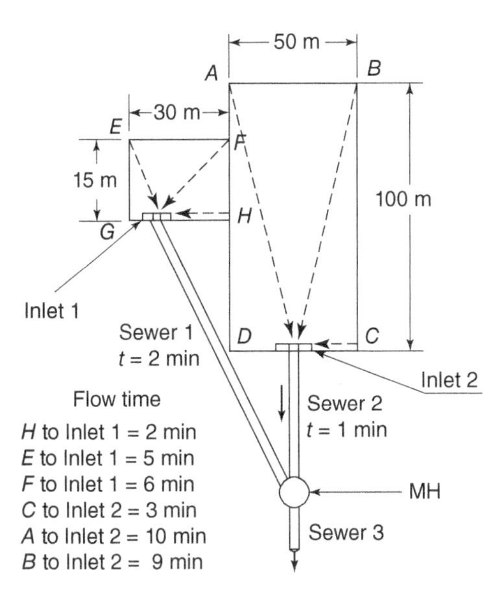

Figure 14.34 Layout of Storm Drainage for Problem 14.17

14.18 The drainage system shown in Fig. 14.35 is designed to collect the stormwater from three areas. The areas' runoff coefficients and overland flow times are as follows:

Area	Area, ha	c	Overland Flow Time, min
I	5	0.6	5
II	15	0.5	10
III	10	0.8	8

The rainfall intensity i (mm/h) as a function of rainfall duration t (minutes) is

$$i = \frac{1,120}{t + 110}.$$

The flow time in sewer 1–2 is two minutes and in sewer 2–3 is 2.5 minutes.

Determine the design flows (m^3/min) for the three sewers: 1–2, 2–3, and 3–4.

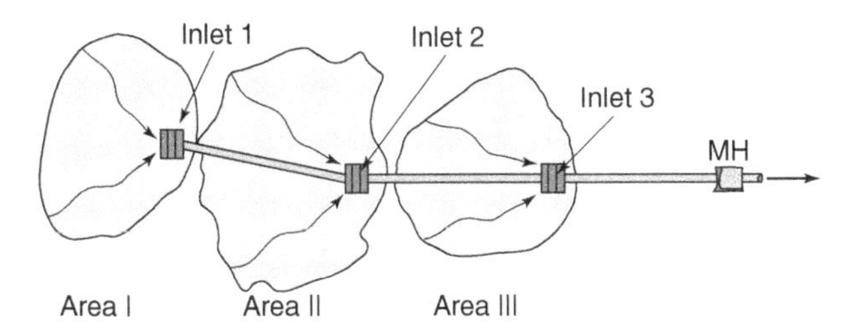

Figure 14.35 Layout of Storm Drainage for Problem 14.18

14.19 The drainage system shown in Fig. 14.36 is designed to collect the stormwater from three areas. The areas, runoff coefficients, and overland flow times are as follows:

Area	Area, m^2	c	Overland Flow Time, min
I	5,000	0.5	7
II	10,000	0.4	5
III	16,000	0.7	5

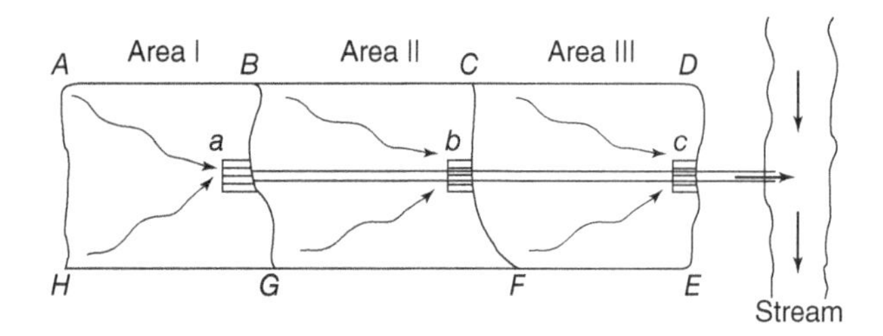

Figure 14.36 Layout of Storm Drainage for Problem 14.19

The rainfall intensity i (mm/h) as a function of rainfall duration t (minutes) is

$$i = \frac{3,360}{t + 17}.$$

The average velocity in sewers = 0.90 m/s. The length of each of the sewers a–b and b–c = 150 m. Calculate the total stormwater discharge (L/s) into the stream from the three areas combined.

14.20 Determine the size of storm sewers 1, 2, 3, and 4 needed to serve the drainage areas shown in Fig. 14.37. The following data are given:

Drainage Area	Area, ha	Overland Flow Time, min	Runoff Coefficient
A1	2.8	20	0.60
A2	2.0	15	0.50
A3	1.5	10	0.40
Park	1.0	15	0.20

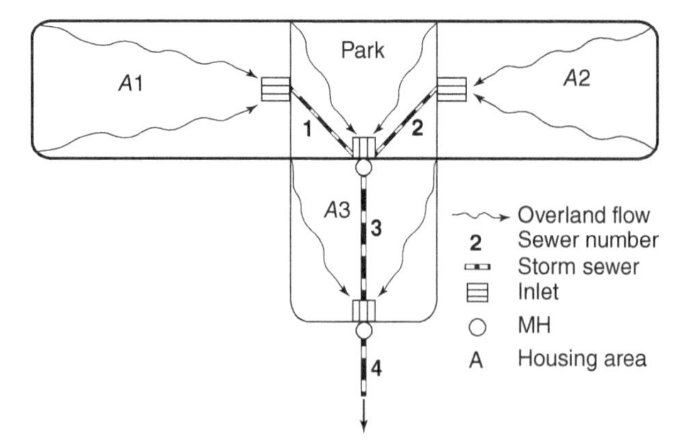

Figure 14.37 Layout of Storm Drainage for Problem 14.20

Storm Sewer	Length, m	Slope	n
1	120	0.0090	0.013
2	120	0.0300	0.013
3	240	0.0040	0.013
4	120	0.0030	0.013

The rainfall intensity (mm/h)-duration (minutes) relationship is

$$i = \frac{3,000}{t + 20}.$$

14.21 Determine the design flows (L/s) for sewers 1, 2, 3, 4, and 5, which serve the drainage blocks A, B, C, D, and E shown in Fig. 14.38. The following data are given:

Block	Area, m^2	Overland Flow Time, min	Runoff Coefficient
A	20,000	11	0.30
B	10,000	8	0.60
C	15,000	12	0.80
D	5,000	5	0.50
E	12,000	18	0.70

Assume that the flow times in storm sewers are as follows:

Storm Sewer	Flow Time, min
1	2
2	1
3	1
4	2
5	4

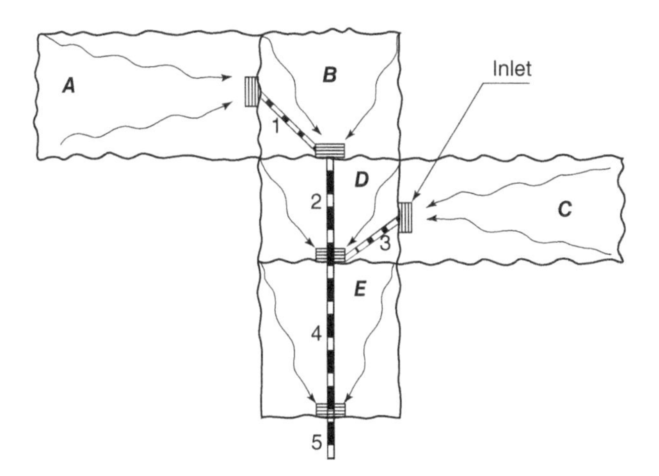

Figure 14.38 Layout of Storm Drainage for Problem 14.21

The rainfall intensity (mm/h)-duration (minutes) relationship is

$$i = \frac{1,640}{t + 17}.$$

14.22 Each of the city blocks contributes its stormwater flow into the street as shown in Fig. 14.39. Each block is 8,000 m^2 in area with a runoff coefficient of 0.65. The rainfall intensity i (mm/h) as a function of rainfall duration t (minutes) is

$$i = \frac{2,510}{t + 14}.$$

Find

(a) The concentration times in minutes at points A, B, C, D, E, and F.
(b) The flows (L/s) at points A, B, C, D, E, and F.
(c) The flows (L/s) in sewers I and II.

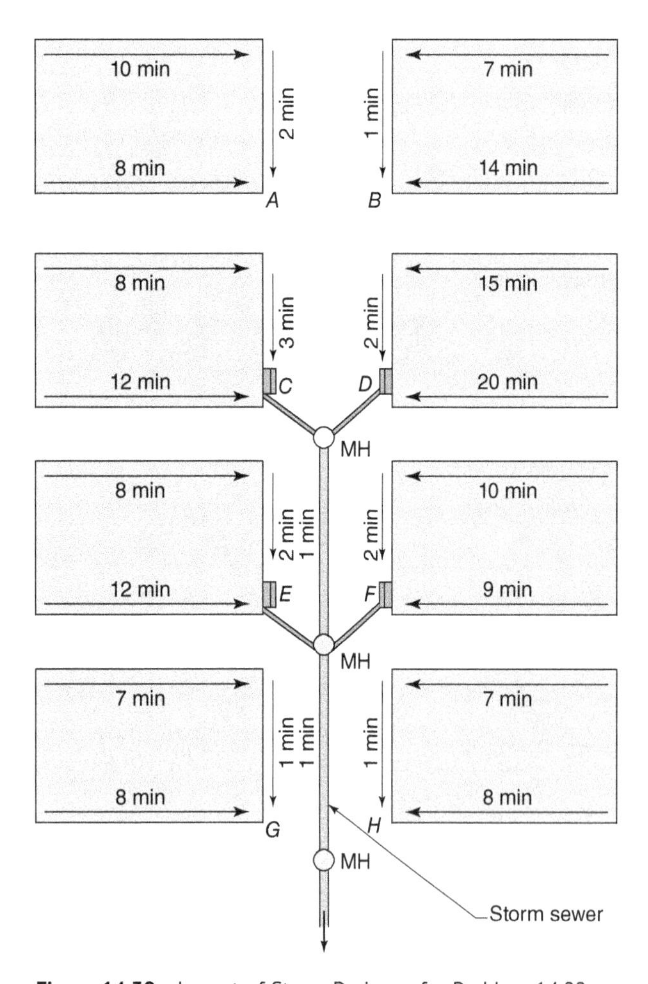

Figure 14.39 Layout of Storm Drainage for Problem 14.22

14.23 Figure 14.40 shows the watershed area that is to be drained through a storm outfall sewer to the stream. Assume the following design criteria:

- The rainfall intensity i (mm/h) as a function of rainfall duration t (minutes) is

$$i = \frac{3,300}{t + 17}.$$

- Runoff coefficient $c = 0.30$
- The velocity in the storm sewers from MH 1 to MH 3 is 0.75 m/s
- Manning coefficient n for sewers = 0.013
- Ground elevation at MH 3 = 625.00 m
- Invert elevation of outfall sewer at MH 3 = 622.00 m
- Ground elevation at edge of stream = 624.00 m
- Bottom elevation of stream = 614.00 m
- Maximum water level in stream = 620.50 m.

(a) Determine the required size of the outfall sewer so that no backflow of water from the stream into the sewer is possible.

(b) What size of outfall sewer will be required if excavation is to be kept to a minimum? In this case assume that there is no minimum limit on depth of cover.

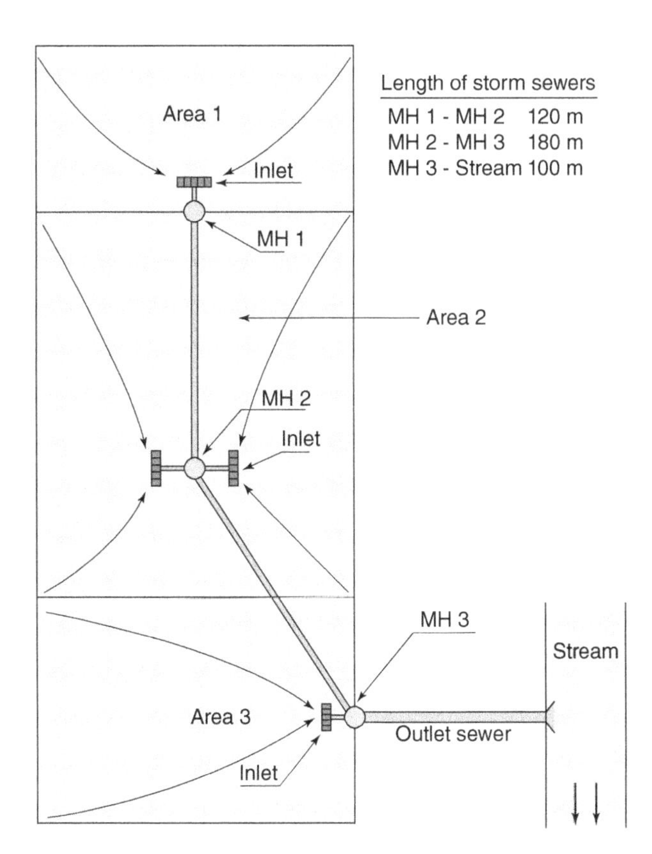

Figure 14.40 Layout of Storm Drainage for Problem 14.23. Conversion factor: 1 m = 3.2808 ft

14.24 Storm sewers 1 and 2 are planned to carry the stormwater runoff contributed by the two areas *A*1 and *A*2 (see Fig. 14.41). Area *A*l is 15,000 m^2 and has a runoff coefficient $c_1 = 0.65$. Area *A*2 is 25,000 m^2 and has a runoff coefficient $c_2 = 0.45$. Flow time in Sewer 1 is two minutes. Inlet times are as follows:

Area	From	To	Inlet Time, min
*A*1	*A*	Inlet 1	5
	C	Inlet 1	4
*A*2	*B*	Inlet 2	10
	C	Inlet 2	12

The rainfall intensity *i* (mm/h) as a function of rainfall duration *t* (minutes) is

$$i = \frac{2,500}{t + 18}.$$

Determine the following:

(a) Overall runoff coefficient c for both areas combined
(b) Concentration times Area 1 at Inlet 1
(c) Concentration time for Area 2 at Inlet 2
(d) Concentration time for Sewer 2
(e) Flow entering Inlet 1
(f) Flow entering Inlet 2
(g) Flow in Sewer 2
(h) Minimum nominal diameter of Sewer 2
(i) Maximum nominal diameter of Sewer 2.

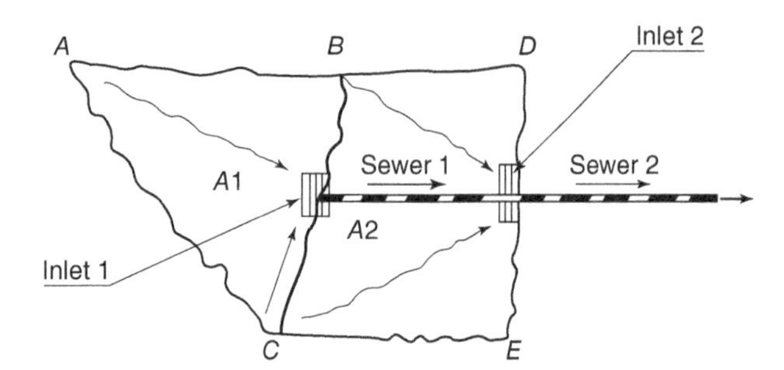

Figure 14.41 Layout of Storm Drainage for Problem 14.24

14.25 Determine the sizes of the three storm sewers needed to serve the 4.8 ha drainage areas shown in Fig. 14.42. Each area has an inlet time of 10 minutes. The distance between manholes is 180 m and all pipes have an $n = 0.013$. The areas and their coefficients of runoff are as follows:

Drainage Area	Area, ha	*c*
1	1.2	0.45
2	2.4	0.45
3	1.2	0.15

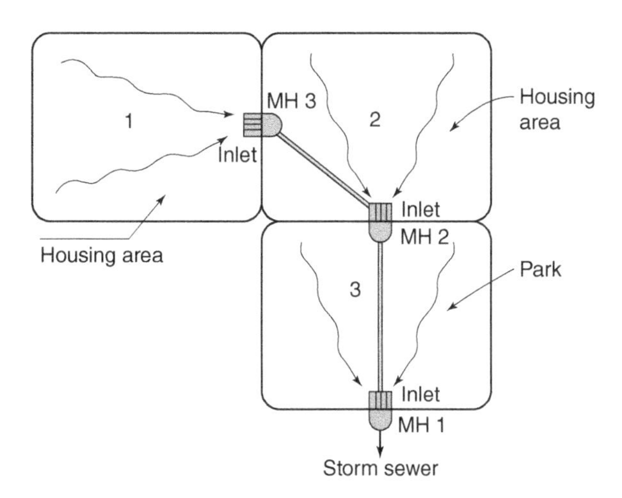

Figure 14.42 Layout of Storm Drainage for Problem 14.25

The storm sewers are set on a slope of 0.0020. The rainfall intensity i (mm/h) as a function of rainfall duration t (minutes) is

$$i = \frac{3,300}{t + 19}.$$

14.26 Design the stormwater drainage system for the area shown in Fig. 14.43. Draw a neat profile along the route of MH 30 – MH 20 – MH 10 – stream edge. On the profile show all pipe sizes, slopes, and ground and invert elevations.

The rainfall intensity (mm/h) as a function of rainfall duration (minutes) is

$$i = \frac{1,000}{t + 50}.$$

The minimum allowed velocity in sewers is 0.80 m/s and the maximum is 3 m/s. The minimum cover over sewers is 1.50 m. The maximum water level in stream is 615.00 m.

The following data concerning the drainage area are known:

Point	Ground Elevation, m
MH 30	625.30
MH 20	624.20
MH 10	622.20
MH 40	622.38
Stream edge	622.50

	Drainage			
Area	Area, ha	c	Draining to MH	Overland Flow Time, min
A	10	0.50	40	10
B	7	0.65	20	12
C	6	0.65	30	5
D	15	0.70	20	15
E	8	0.65	10	8

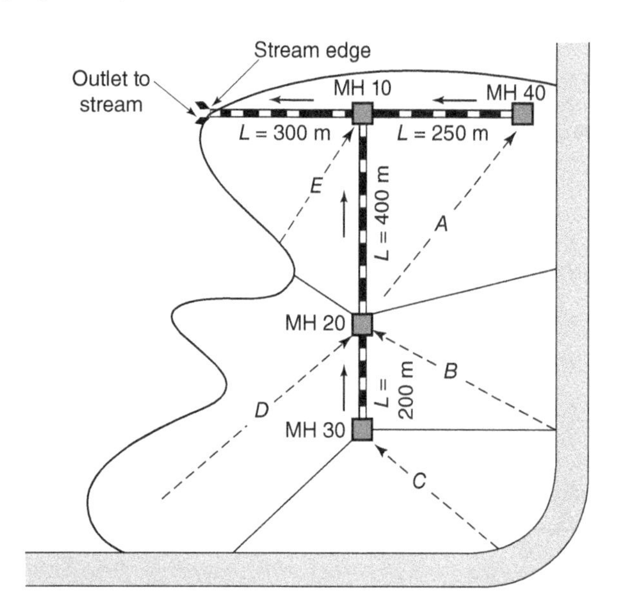

Figure 14.43 Layout of Storm Drainage for Problem 14.26. Conversion factor: 1 m = 3.2808 ft

14.27 Izzard developed the following watershed hydrology equation for calculation of concentration time, t_c:

$$t_c = 41\, L^{1/3} i^{-2/3} \left[(0.0007i + C_r)/s^{1/3} \right]$$

where t_c is the time of concentration, minutes; s is the slope of ground, dimensionless; i is the intensity of rainfall, in./h; C_r is retardance coefficient varying from 6.0×10^{-2} for pervious turf areas, 3.2×10^{-2} for pervious bare, packed areas, 1.2×10^{-2} for pavements, to 1.7×10^{-2} for flat, gravel roofs.

Based on the above equation, develop further relationships for the determination of

(a) overland flow distance, L
(b) retardance coefficient, C_r
(c) slope, s.

14.28 Conduct a university library and an Internet search to find other researchers' mathematical models for determination of concentration time, overland flow distance, retardance coefficient, and slope within a watershed. Write a literature review report.

14.29 A metric Izzard equation using the SI System of units t_c (minutes), L (m), i (mm/h), C_r (dimensionless), and s (dimensionless) has been developed by the authors:

$$t_c = 526\, L^{1/3} i^{-2/3} \left[(0.00002756i + C_r)/s^{1/3} \right]$$

Determine the time of concentration in both US customary and SI units assuming the overland flow distance $L = 1{,}000$ ft (304.8 m), $C_r = 0.06$, slope $s = 0.02$, and the rainfall intensity $i = 1$ in. /h (25.4 mm/h).

14.30 Review Appendix 18, Simplified Overland Flow Time Determination for Urban Drainage. If the distance of water flow is 60 m, what is the time of concentration for a flow passing over an ordinary grass surface with a slope of 4%? Discuss the accuracy of your results.

14.31 The general form of intensity–duration curves is $i = A/(t + B)$. The values of A and B vary from one region to another. The rainfall curves must be derived from historical regional storm records. Assuming you are an environmental engineer in charge of a local drainage project, conduct a search on local and government storm records (such as USGS) and recommend an intensity–duration curve for use in the project. Give a PowerPoint presentation to your class showing your classmates how you reached to your results.

Bibliography

ASCE/WEF, *Gravity Sanitary Sewer, Design and Construction, ASCE Manuals and Reports on Engineering Practice # 60*, WEF Manual of Practice # FD-5, 2nd ed., 2007.

ASCE, Design and Construction of Sanitary and Storm Sewers, *Am. Soc. Civil Eng.*, Manuals of Engineering Practice, *37*, 1969.

Chanson, H., *The Hydraulics of Open Channel Flow*, 2nd ed., Elsevier Butterworth-Heinemann, Burlington, MA, 2004.

Chaudhry, M. H., *Open-Channel Flow*, 2nd ed., Springer-Verlag, New York, NY, December, 2007.

Concrete Pipe and Products Company, Inc, *Technical Manual*, Manassas, VA, 1992.

Crites, R., and Tchobanoglous, G., *Small and Decentralized Wastewater Management Systems*, The McGraw-Hill Companies, New York, NY, 1998.

Fair, G. M., Geyer, J. C., and Okun, D. A., *Elements of Water Supply and Wastewater Disposal*, John Wiley & Sons, New York, NY, 1971.

Fair, G. M., Geyer, J. C., and Okun, D. A., *Water and Wastewater Engineering, Volume 1*, John Wiley & Sons, New York, NY, 1966.

Geyer, J. C., and Lentz, J. J., An Evaluation of the Problems of Sanitary Sewer System Design, in *Final Report of the Residential Sewerage Research Project of the Federal Housing Administration*, Technical Studies Program, 1964.

Horton, A. M., Protective Linings for Ductile Iron Pipe in Wastewater Service, in *Proceedings of the Water Environment Federation. 69th Annual Conference & Exposition, Dallas, TX, Volume 3*, October 1996.

Jain, S. C., *Open-Channel Flow*, Wiley, New York, NY, October, 2000, p. 344.

Kerby, W. S., *Civ. Eng.*, *29*, 174, 1959.

Kerby, W. S., Time of Concentration for Overland Flow, *Civ. Eng.*, *29*, 3, 174, 1959.

Lindeburg, M. R., *Civil Engineering Reference Manual*, Professional Publications, Inc, Belmont, CA, 1986.

Moody, T. C., Optimizing Force Main Odor Control Chemical Dosage: A Tale of Two Systems, in *Proceedings of the Water Environment Federation, 71st Annual Conference, Orlando, FL, Volume 2*, October 1998.

National Institute for Occupational Safety and Health, *Fatality Assessment and Control Evaluation (FACE) Program*, Center for Disease Control and Prevention (CDC) and NIOSH, http://www.cdc.gov/niosh/face, 2009.

Paschke, N. W., Pump Station Basics — Design Considerations for a Reliable Pump Station, *Operations Forum*, *14*, 5, 15–20, May 1997.

Shammas, N. K., Characteristics of the Wastewater of the City of Riyadh, *Int. J. Devel. Technol.*, *2*, 141–150, 1984.

Steel, E. W., and McGhee, T. J., *Water Supply and Sewerage*, McGraw-Hill, New York, 2007, p. 665.

Tholin, A. L., and Keifer, C. J., The Hydrology of Urban Runoff, *Trans. Am. Soc. Civil Eng.*, *125*, 1308, 1960.

Urquhart, L. C., *Civil Engineering Handbook*, McGraw-Hill Book Company, New York, 1962.

U.S. Environmental Protection Agency, *Design Manual: Municipal Wastewater Disinfection*, EPA/625/1-86/021, Cincinnati, OH, 1986.

U.S. Environmental Protection Agency, *Pipe Construction and Materials*, EPA 832-F-00-068, Washington, DC, September 2000.

U.S. Environmental Protection Agency, *Sewers, Conventional Gravity, Collection Systems Technology Fact Sheet*, EPA 832-F-02-007, Washington, DC, September 2002.

U.S. Environmental Protection Agency, *Sewers, Innovative and Alternative Technology Assessment Manual*, EPA 430/9-78-009, Washington, DC, February 1980.

U.S. Environmental Protection Agency, *Sewers, Lift Stations*, EPA 832-F-00-073, Washington, DC, September 2000.

Vu, V. S., and McNown, J. S., Runoff from Impervious Surfaces, *J. Hydraul. Res.*, *2*, 14, 1964.

Workman, G., and Johnson, M. D., Automation Takes Lift Station to New Heights, *Operations Forum*, *11*, 10, 14–16, October 1994.

15

Sewerage Systems: Modeling and Computer Applications

The hydraulics in the gravity portion of sanitary and storm sewers are analyzed with the same techniques used in open-channel flow. When the flow depth is constant, Manning's equation is used. When obstructions or changes in pipe slope exist, the gradually varied flow analysis procedure is applied. The pressure portion of the sanitary sewer is analyzed in the same manner as a water distribution system, as discussed in Chapters 6 and 7. The primary difference is that flow is generally withdrawn from a water distribution network, whereas in a sewer system, flow is injected into force mains.

In this chapter, *SewerCAD* is used as a demonstration for the application of modeling and computer techniques in the sewer design process. More comprehensive reviews of sewer design were introduced in Chapters 13 and 14.

15.1 Extended-Period Simulations

An *extended-period simulation (EPS) model* represents how a sewer network will behave over time. This type of analysis allows the user to model how wet wells fill and drain; pumps toggle on and off; and pressures, hydraulic grades, and flow rates change in response to variable loading conditions and automatic control strategies formulated by the modeler. EPS is a useful tool for assessing the hydraulic performance of alternative pump and wet-well sizes.

The SewerCAD algorithm proceeds in a general downstream direction toward the outfall, following the procedure described here:

1) The analysis begins in the gravity portion of the network. The hydrographs enter the gravity system and are successively routed and summed as the flows approach the downstream wet well or outfall. Ultimately, the total inflow hydrograph to the wet well is determined.
2) Knowing the inflow into the wet well, the pressure calculations for the force main system bounded by the wet well are performed. In addition to flow velocities and pressures, the levels in the wet well over time are determined.
3) The calculation then returns to the gravity portion of the network discussed in step 1. The hydraulics and HGL (hydraulic grade line) profiles are calculated throughout the gravity system for each time step using the known level of the wet well as the boundary condition for the backwater analysis.

The process repeats, continuing through the system downstream of the pressure network until an outlet is reached.

As a hydrograph is routed through a conduit, it undergoes changes in shape and temporal distribution caused by translation and storage effects. SewerCAD uses one of two methods to determine the shape and distribution of a hydrograph routed through a gravity pipe: (a) *convex routing* and (b) *weighted translation routing.*

15.1.1 Convex Routing

The underlying assumptions of the convex routing method are that the routed outflow for a time step is based on the inflow and outflow for the previous time step, and that the flow does not back up in the pipe (i.e., no reverse flow or reduced flow due to tailwater effects exists). Each outflow ordinate is calculated as follows:

$$O_{t+\Delta t} = cI_t + (1-c)O_t \tag{15.1}$$

Water and Wastewater Engineering: Hydraulics, Hydrology and Management, Volume 1, Fourth Edition.
Lawrence K. Wang, Mu-Hao Sung Wang, and Nazih K. Shammas.

Companion website: www.wiley.com/go/Wang/Waterandwastewater4e

where
$O_{t+\Delta t}$ = outflow at time $t + \Delta t$ (L/s, gpm)
t = current time (s)
Δt = hydrologic time step (s)
c = convex routing coefficient
I_t = inflow at time t (L/s, gpm)
O_t = outflow at time t (L/s, gpm).

The convex routing coefficient is essentially a ratio of the hydrologic time step and representative flow travel time through the pipe, and is calculated as follows:

$$c = \Delta t \frac{v}{L} = \frac{\Delta t}{t_t} \tag{15.2}$$

where
Δt hydrologic time step (s)
t_t = travel time (s)
v = velocity established for representative flow (m/s, ft/s)
L = length of pipe (m, ft).

In SewerCAD, the velocity used to calculate the coefficient is either the normal velocity or full velocity generated for a user-specified percentage of the peak of the inflow hydrograph. In other words, if the percentage of the peak flow is greater than the capacity of the pipe, the full-flow velocity is used. If the percentage of the peak flow is less than the pipe capacity, the flow velocity for normal depth is used.

The higher the percentage of flow, the faster the velocity used to calculate the convex routing coefficient, and the closer the routed hydrograph will be to a pure translation of the inflow hydrograph.

The user-specified percentage can be modified in the calculation options. A typical value is around 75%, but this value may be modified for oddly shaped hydrographs with sharp, uncharacteristic peaks or for calibration purposes.

15.1.2 Weighted Translation Routing

The convex routing method is only valid when the convex routing coefficient, c, is less than 1 or when the hydrologic time step is less than the calculated travel time. For certain cases in which the travel time exceeds the hydrologic time step, SewerCAD automatically uses an alternative method of routing called weighted translation routing.

Each ordinate of the outflow hydrograph is derived from a weighted average of the ordinates for the current and previous time steps of the inflow hydrograph. The weights are calculated based on the convex routing coefficient.

Each ordinate of the outflow hydrograph is calculated as follows:

$$O_t = \frac{1}{c} I_{t-\Delta t} + \left(1 - \frac{1}{c}\right) I_t \tag{15.3}$$

where
Q_t = outflow at current time step (L/s, gpm)
c = convex routing coefficient
$I_{t-\Delta t}$ = inflow at previous time step (L/s, gpm)
I_t = inflow at current time step (L/s, gpm).

15.1.3 Hydrologic and Hydraulic Time Steps

SewerCAD uses two distinct time steps when running an extended-period simulation:

1) *Hydrologic time step*: This time step is used to calculate the routed hydrographs and represents the time increment of all hydrographs generated during the analysis. The hydrologic time step is also used as the calculation increment for the pressure calculations.

2) *Hydraulic time step:* This time step represents how often the hydraulic calculations are performed for gravity flow. Flows are interpolated from the previously generated hydrographs using the hydraulic time step and are then used to perform the gradually varied flow analysis for that time step.

The hydrologic time step should be less than or equal to the hydraulic time step, and the hydraulic time step should be a multiple of the hydrologic time step.

15.2 SewerCAD

SewerCAD is an easy-to-use program for the design and analysis of wastewater collection systems. SewerCAD's intuitive graphical interface and data exchange capabilities make it easy to develop and load complex models of combined gravity and pressure networks. Using SewerCAD, you can design and analyze the gravity portion of the system according to either a gradually varied flow calculation or a standard capacity analysis.

The automatic design capabilities of SewerCAD allow you to design the system based on *user-defined constraints* for velocity, cover, and slope. You can also design for partially full pipes, multiple parallel sections, maximum section size, and invert/crown-matching criteria and also allow for drop structures. You can disable the design feature on a pipe-by-pipe basis, to design all or only a portion of your system.

SewerCAD's loading model provides complete support for all types of sanitary and wet-weather loads. Sanitary loads are typically estimated based on a number of contributing units, with a specified average load per unit (such as 260 L/d per apartment resident). The unit is typically a measure of population such as residents or employees; however, loads can also be based on other criteria such as contributing area or user-defined counts of items indicative of loading behavior. See Table 15.1 for a list of standard loading sources and their associated average base loads.

Table 15.1 Typical Unit Sanitary Loads from Different Sources

Unit Sanitary Load	Loading Unit	Base Load (L/d per Loading Unit)
Airport	Passenger	10
Apartment	Resident	260
Automobile Service	Employee	50
Bar	Customer	8
Cabin Resort	Guest	160
Campground	Guest	120
Coffee Shop	Guest	20
Department Store	Toilet Room	2,000
Dormitory	Guest	150
Home (Average)	Resident	280
Home (Better)	Resident	310
Home (Luxury)	Resident	380
Hospital (medical)	Bed	650
Hotel	Employee	40
Prison	Inmate	450
Restaurant	Meal	10
School (Large)	Student	80
School (Medium)	Student	60
School (Small)	Student	40
Swimming Pool	Employee	40
Shopping Center	Parking Space	4

SewerCAD's unit sanitary load library is completely user customizable and supports population, area, discharge, and count-based sanitary loads, as well as hydrographs (flow versus time data) and load patterns. Wet-weather inflow can also be added to the model as instantaneous loads, hydrographs, or base loads with associated patterns. Infiltration can be computed based on pipe length, pipe diameter/length, pipe surface area, or unit count.

Loads are peaked according to a peaking factor or extreme flow factor equations that you select from a user-customizable extreme flow factor library.

You can use SewerCAD to design new systems and analyze the performance of existing systems. SewerCAD's intuitive graphical editor and scenario management capabilities facilitate the process of analyzing a large number of design alternatives and finding potential problems in an existing system. In summary, SewerCAD's modeling techniques enable you to:

1) Design and analyze multiple sanitary sewer networks in a single project.
2) Examine your system using SewerCAD's gradually varied flow algorithms or a standard capacity analysis.
3) Design the system using SewerCAD's automatic, constraint-based design.
4) Load your model based on contributing population, service area, total sanitary flow, or your own loading type.
5) Peak your loads using Babbitt's, Harmon's, or 10-States Standards, or use your own formulas or tables.
6) Calculate infiltration based on pipe length, diameter, surface area, length/diameter, or user-defined data.
7) Generate plan and profile plots of a network.
8) Perform extended-period simulations that include time-variable loads and hydro-logic routing.
9) Animate plans and profiles showing sanitary sewer system performance over time.

SewerCAD can analyze complex sewage collection systems under a variety of loading conditions. SewerCAD's loading capabilities can be used to track system response to an unlimited range of sanitary and wet-weather loading combinations. It can be used for:

1) Pipe sizing
2) Pump sizing
3) Master planning
4) Operational studies
5) Rehabilitation studies.

The theory and background used in SewerCAD are presented in more detail in the *SewerCAD online help system*. The following examples provide step-by-step instructions on how to solve various problems using SewerCAD, which is available free to users of this book. Educational versions of the software can be accessed either online or from the CD that accompanies this textbook.

Example 15.1 Pump Size for Peak Flows

Problem Statement

Figure 15.1 shows a schematic sketch of the sanitary sewer system. A circular, gravity sewer main, P-1, empties into wet-well WW-1. From the wet well, the wastewater is pumped over a hill to a treatment plant at O-1 through force mains FM-2 and FM-3.

Figure 15.1 Schematic of Sewer Used in Example 15.1

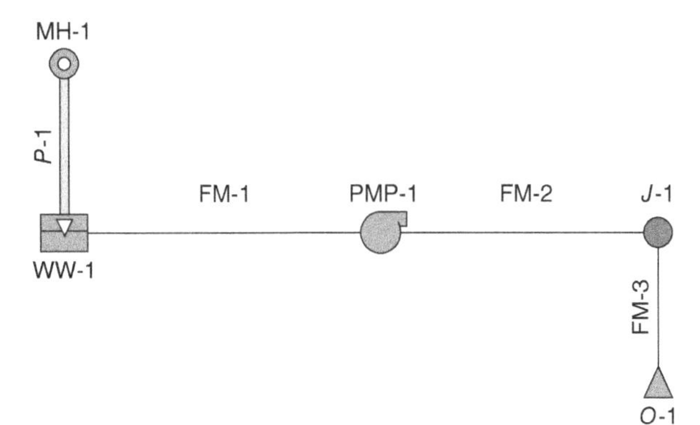

The main line runs through an area with a high water table, and it has been determined that infiltration occurs at a rate of 20 L/d per meter of pipe. It has also been determined that Babbitt's method best correlates the base loads listed in Table 15.2 with the loading peaks in the system. The pipe, pump, and wet-well data are given in Tables 15.3 and 15.4.

Part 1: What is the total sanitary flow (dry-weather flow) exiting the system? What is the total flow?
Part 2: Does the pump have adequate capacity for this system? How can you tell?
Part 3: What is the full-flow capacity of gravity pipe P-1? By how much is this capacity being exceeded?

Table 15.2 Loading Data for Example 15.1

Node	Unit Sanitary Load	Loading Unit Count	Loading Unit
MH-1	Apartment	600	Resident
	Apartment Resort	500	Guest
	Home (Luxury)	200	Resident
WW-1	Home (Average)	2,000	Resident

Table 15.3 Node Data for Example 15.1

Element	Type	Element Characteristics		
MH-1	Manhole	Ground Elevation	101.0 m	
		Rim Elevation	101.0 m	
		Sump Elevation	99.3 m	
		Structure Diameter	1.0 m	
WW-1	Wet Well	Ground Elevation	100.0 m	
		Section Type	Constant Area–Circular	
		Maximum Elevation	100.0 m	
		Alarm Elevation	99.0 m	
		Initial Elevation	96.5 m	
		Minimum Elevation	96.0 m	
		Base Elevation	96.0 m	
		Diameter	4.0 m	
PMP-1	Pump	Elevation	96.25 m	
		Pump Type	Standard (3-Point)	
		Shutoff Head/Discharge	22.0 m	0 m^3/min
		Design Head/Discharge	15.0 m	2.5 m^3/min
		Max Operating Head/Discharge	9.0 m	4.0 m^3/min
J-1	Pressure Junction	Elevation	105.0 m	
O-1	Outlet	Ground Elevation	107.5 m	
		Rim Elevation	107.5 m	
		Sump Elevation	106.0 m	
		Tailwater Condition	Free Outfall	

Table 15.4 Pipe Data for Example 15.1

Pipe	Pipe Type	Length (m)	Diameter (mm)	Material	Upstream Invert (m)	Downstream Invert (m)
P-1	Gravity	150	200	Concrete $n = 0.012$	99.35	97.50
FM-1	Force Main	1	200	PVC $C = 150$	96.00	96.25
FM-2	Force Main	215	200	PVC $C = 150$	96.25	105.00
FM-3	Force Main	200	200	PVC $C = 150$	105.00	107.00

Solution:

Project Setup Wizard

- When you start SewerCAD, you should be prompted with the Welcome to SewerCAD dialog. From this dialog, you can access the tutorials, open existing projects, and create new ones. Select **Create New Project**, provide a filename, and click **Save**.

If the Welcome to SewerCAD dialog does not appear, SewerCAD is set to **Hide Welcome Dialog on startup**. To start a new project, select **New** from the **File** menu, enter a filename, and click **Save**. (You can change from Hide Welcome Dialog mode to Show Welcome Dialog mode in the Global Options dialog, which is accessible by selecting **Options** from the **Tools** menu.)

- When you provide a filename, the **Project Setup Wizard** appears. Add a project title and any comments, and then click **Next**.
- In the second screen of the Project Setup Wizard, select **Manning's Formula** as the Gravity Friction Method and select **Hazen-Williams Formula** as the Pressure Friction Method. Click **Next**.
- In the third screen of the Project Setup Wizard, select **Schematic** as the Drawing Scale for the project. This option allows you to define the lengths for the pipes without having to worry about scale and spatial placement on the *x-y* plane. Click **Next**.

  Note: If this screen does not show SI units, click **Cancel** to exit the Project Setup wizard and reset the units to **SI**. Select **Options** from the **Tools** menu on the main drawing screen and, if necessary, change the **Unit System** field (under the **Global** tab) to **System International**. Click **OK** to exit the **Global Options** dialog. On changing the unit, click **Yes** when prompted to confirm the change. Begin the Project Setup wizard again using the instructions above for starting a new project (see first entry in this list).
- Click **Finished**.

Laying Out the System

- You do not have to lay out the sanitary sewer network exactly as shown in the problem statement. For now, just roughly sketch the schematic by following the instructions below.
- Click the **Pipe Layout** button in the vertical toolbar on the left side of the screen.
- To place manhole MH-1, click the left mouse button in the workspace.
- To place wet-well WW-1, move the cursor below the manhole and click the right mouse button. Select **Wet Well** from the pop-up menu, and then click the left mouse button. Notice that the cursor has changed to the wet-well icon. Position the cursor where you want to place the wet well and click the left mouse button to insert it. Gravity pipe P-1 will automatically be placed between MH-1 and WW-1.

- Place pump PMP-1 by right-clicking, selecting **Pump** from the pop-up menu, and left-clicking to place it in the model. Note that force main FM-1 was created automatically between WW-1 and PMP-1. SewerCAD will automatically create either a gravity pipe or a force main (pressure pipe), depending on the type of end nodes that are present.
- Lay out pressure junction J-1 and outlet O-1 in exactly the same way.
- Except for the scale, your schematic should look roughly like that of Fig. 15.1.

Data Entry

- Double-click manhole MH-1. Under the **General** tab, enter the ground elevation, sump elevation, and the structure diameter as given in the problem statement. Note that you do not have to enter the rim elevation if the box entitled **Set Rim to Ground Elevation** has been checked.
- Click the **Loading** tab at the top of the Manhole dialog. Select the **Add** button beside the Sanitary (Dry-Weather) Flow section. Select **Unit Load – Unit Type & Count** for the Load Definition and click **OK.** Select **Apartment** from the **Unit Sanitary (Dry Weather) Load** pull-down menu and enter **600** as the number of apartment residents to the Loading Unit Count. Click **OK**. Insert data for the remaining two unit loads for manhole MH-1 in the same way.
- Double-click wet-well **WW-1**. Under the General tab, enter the ground elevation. Under the Section tab, select **Constant Area** from the pull-down menu in the Section field. Select **Circular** from the pull-down menu in the Cross Section field. Enter a wet-well diameter of **4 m**. Finally, enter the maximum, alarm, initial, minimum, and base elevations as given in the problem statement. Enter these as elevations, not levels.
- Select the **Loading** tab at the top of the Wet Well dialog. Enter in the loading data as you did for manhole MH-1. Click **OK** to exit the dialog.
- Double-click pump PMP-1. Under the General tab, enter the pump's elevation of **96.25 m**. Next, select **Standard (3 Point)** from the pull-down menu in the Pump Type field. Then, enter the pump curve data as given in the problem statement. Click **OK** to exit the dialog.
- Double-click pressure junction **J-1** to edit its properties. Enter the junction's elevation and click **OK** to exit the dialog.
- Double-click outlet **O-1.** Enter the ground elevation and sump elevation. Also, select **Free Outfall** from the list of available tailwater conditions. As with the manhole, if the box entitled **Set Rim to Ground Elevation** has been checked; you do not have to enter the rim elevation because it is automatically set to the ground elevation. Click **OK** to exit the dialog.
- Double-click gravity pipe P-1. Enter the upstream and downstream inverts in the appropriate fields. Enter the length in the **Length** field. Select **Circular** from the pull-down menu in the **Section Shape** field. Select **Concrete** from the pull-down menu in the **Material** field. Change the Manning's n value from 0.013 to 0.012. Finally, select **200 mm** from the **Section Size** field.
- Select the **Infiltration** tab at the top of the Gravity Pipe dialog. Set the **Infiltration Type** to **Pipe Length.** Select an **Infiltration Loading Unit** of **m**. Enter the appropriate infiltration rate per unit length per day in the **Infiltration Rate per Loading Unit** field. Click **OK** to exit the dialog.
- Double-click force main **FM-1** to edit its properties. Enter the diameter and select the material type as given in the problem statement. Enter the upstream invert and the length. If the **Set Invert to Upstream Structure** box is checked, you will not have to enter the upstream invert for this problem. Also note that invert elevations are not editable if the upstream or downstream element is a pump or a pressure junction. In such cases, the invert elevation is set equal to the pump or pressure junction elevation. Click **OK.**
- Enter the data for FM-2 and FM-3 in the same way.
- To apply a peaking factor method for each of the different unit sanitary loads used in the project, select **Extreme Flow** from the **Analysis** menu at the top of the screen. Highlight the extreme flow setup labeled Base Extreme Flow Setup and click **Edit**. Apply the appropriate extreme flow method as dictated by the problem by selecting the method from the cell under the Extreme Flow Method column for each of the unit sanitary loads in the list. (*Hint:* Right-click the column heading and choose **Global Edit** to change all of the rows simultaneously.) Click **OK** to exit the dialog. Click **OK** to exit the Extreme Flow Setup Manager.
- Now that all input data has been entered, save the project file to disk by choosing **Save** from the **File** menu. It is a good idea to periodically save your work in this way.

Running the Model

- Click the **GO** button in the toolbar at the top of the SewerCAD window.

- Under the Calculation tab, make sure that the Calculation Type is set to **Steady State**. Also, make sure that the **Design** box is unchecked, so that SewerCAD does not automatically design the system. Select **Base Extreme Flow Setup** from the list box in the **Extreme Flow Setup** field.
- Before running the model, click the **Check Data** button. The input data are checked for any problems that will prevent the model from running. There should not be any problems, so click **OK** to return to the dialog. If there are problems, you can view them by selecting the **Results** tab.
- Click **GO** to run the model. A dialog box displaying the results should appear with a green light displayed on the **Results** tab. The green light indicates a successful run. Yellow and red lights indicate warnings and problems, respectively. Close the **Scenario: Base** dialogue window.

Answer to Part 1

To check the total sanitary flow exiting the system, double-click the outlet O-1. In the **Flow Summary** section of the dialog, the Total Sanitary Flow (dry weather only) is 3 552 007 L/d. This is the peak sanitary flow. The Total Flow, which includes wet-weather flow, is 3 555 007 L/d. The difference between the two values is caused by the infiltration along pipe P-1.

Answer to Part 2

Double-click **PMP-1.** At 2.75 m^3/min, the pump is operating near its design point of 2.50 m^3/min, which is greater than the total flow to the wet well of 2.47 m^3/min. The pump capacity is adequate. Note that a pump station would typically consist of redundant pumps to allow for servicing.

Answer to Part 3

Double-click gravity pipe **P-1**. In the Hydraulic Summary section, find that the pipe has a full capacity of 3 556 744 L/d. The excess full capacity is 1 931 159 L/d.

Example 15.2 24-Hour Simulation of Dry-Weather Flow

In this example, the EPS feature of SewerCAD will be used to simulate flows through the sewer over a 24-h period. The Scenario Manager will be used to specify different loadings to the network from Example 15.1.

Problem Statement

The sewer system designed in Example 15.1 is used for this problem. The 24-h diurnal pattern shown in Table 15.5 is applied to the sanitary flows. The multiplier is used to convert average daily flows to hourly flows.
Part 1: During an average day, how many times does the pump turn on and off?
Part 2: What is the maximum rate of pumping?

Solution:

Set Pump Controls

- The user must specify how the pump operates. This is done by specifying the levels in WW-1 that trigger the pump to turn on and off. Select the pump by double-clicking **PMP-1.** Select the **Controls** tab.

Table 15.5 Data for the Stepwise Diurnal Sanitary Loads of Example 15.2

Hour	Multiplier	Hour	Multiplier	Hour	Multiplier	Hour	Multiplier
1	0.85	7	0.87	13	1.21	19	1.05
2	0.72	8	1.08	14	1.13	20	1.11
3	0.56	9	1.32	15	1.06	21	1.19
4	0.52	10	1.42	16	0.95	22	1.05
5	0.43	11	1.61	17	0.84	23	1.06
6	0.52	12	1.49	18	0.88	24	1.00

- To add a pump "on" setting, click the **Add** button, set the Control Type to **Status** and the Pump Status to **On**. The Condition should be set to **Node**, the Node should be set to **WW-1**, the Comparison to **Above**, and the Elevation to **98.0 m**. Click **OK**.
- To add a pump "off" setting, click the **Add** button, set the Control Type to **Status**, and set the Pump Status to **Off**. The Condition should be set to **Node**, the Node should be set to **WW-1**, the Comparison to **Below**, and the Elevation to 96.5. Click **OK** twice to close the pump dialog.
- To apply a peaking factor method for each of the different unit sanitary loads used in the project, select **Extreme Flow** from the **Analysis** menu at the top of the screen. Highlight the extreme flow setup labeled Base Extreme Flow Setup and click **Edit**. Apply the appropriate extreme flow method as dictated by the problem by selecting the method from the cell under the Extreme Flow Method column for each of the unit sanitary loads in the list. (*Hint:* Right-click the column heading and choose **Global Edit** to change all the rows simultaneously.) Click **OK** to exit the dialog. Click **OK** to exit the Extreme Flow Setup Manager.
- Now that all input data have been entered, save the project file to disk by choosing **Save** from the **File** menu. It is a good idea to periodically save your work in this way.

Running the Model

- Click the **GO** button in the toolbar at the top of the SewerCAD window.
- Under the Calculation tab, make sure that the Calculation Type is set to **Steady State**. Also, make sure that the **Design** box is unchecked, so that SewerCAD does not automatically design the system. Select **Base Extreme Flow Setup** from the list box in the **Extreme Flow Setup** field.
- Before running the model, click the **Check Data** button. The input data are checked for any problems that will prevent the model from running. There should not be any problems, so click **OK** to return to the dialog. If there are problems, you can view them by selecting the **Results** tab.
- Click **GO** to run the model. A dialog box displaying the results should appear with a green light displayed on the **Results** tab. The green light indicates a successful run. Yellow and red lights indicate warnings and problems, respectively. Close the **Scenario: Base** dialogue window.

Set Up the EPS Scenario

- Open Scenario Manager by selecting **Scenarios** from the **Analysis** drop-down menu.
- Click the **Scenario Wizard** button and enter **24-h Cycle** as the name of the scenario. Then click **Next**.
- Select **Base** as the scenario to serve as the basis for the 24-h Cycle scenario. Then click **Next**.
- The Calculation Type is **Extended Period**. Use the default settings for Duration (**24 hours**), **Hydrologic Time Step** (0.10 hour), and Hydraulic Time Step (1.0 hour). For the Pattern Setup option, click the ellipsis button (. . .). The **Pattern Setup Manager** will appear. Click **Add** and enter **24-hour pattern** for the name of the pattern.
- In the Pattern Setup table, add the new diurnal pattern to each of the unit loads. On the Apartment row in the Diurnal Pattern column, click once and you will see an ellipsis button. Click this box to enter the Pattern Manager. Select **Add** to enter the pattern. The Label is **24-hour diurnal**, the **Start Time** is 12:00, and the Starting Multiplier is 1.00. On the right side of the box, enter the 24-hour pattern given in the problem statement beginning with hour 1 and continuing to hour 24. Click the **Stepwise** radio button. Click **OK** twice to return to the **Pattern Setup** window. For each **Unit Load**, set the diurnal pattern to **24-hour diurnal.** Click **OK** to return to the Scenario Wizard, set the Pattern Setup field to "24-hour diurnal," and then click **Next**.
- Select the **Alternative** that is to be changed for this scenario. Click on the check box next to **Sanitary (Dry Weather) Loading**. Click Next.
- Select **Create New Sanitary (Dry Weather) Loading Alternative**.
- Click the **Next** button again to review the scenario. Note that all of the alternatives are inherited with the exception of the sanitary loads. Click **Finished**.

Run the Scenario

- In the Scenario Manager window, select **Go Batch Run** from the left column.

- Check the box next to **24 hr Cycle** and click **Batch**. SewerCAD will ask "Run 1 scenarios as a batch?" Select **Yes** to run the scenario. When the run is completed, click **OK.**
- Close the Scenario Manager window.

View EPS Results

- Notice that above the network view, there is a toolbar as shown here:

- This area contains the controls for displaying results of the EPS. Clicking the arrow on the left begins the display of the results at each time step. This button changes to a red square when the scenario is playing. The down arrow is used to specify the delay at each time step. The numbers in the left window indicate the time step. The left and right double arrows may be used to manually step through the simulation. The other windows indicate the time step increment and the name of the scenario.
- Double-click **O-1** to view information on the sewer outlet. Note that the EPS controls are located in the lower right corner of the outlet window. As you step through time, the total flow (in the flow summary section) changes.
- Select the **Report** button at the bottom of the outlet window, and then select the **Hydrograph**. In the Graph Setup box, select the **MH-1**, **O-1**, and **WW-1** elements. Then click **OK.** Figure 15.2 shows the 24-hour flows at the three locations.

Answer to Part 1

The graph in Fig. 15.2 shows the flow at MH-1, WW-l, and at the outlet. The outlet peaks correspond to pumping cycles. Note that the pump was on for 36 intervals during the 24-hour simulation period.

Answer to Part 2

The maximum discharge through the pump is 2.97 m^3/min.

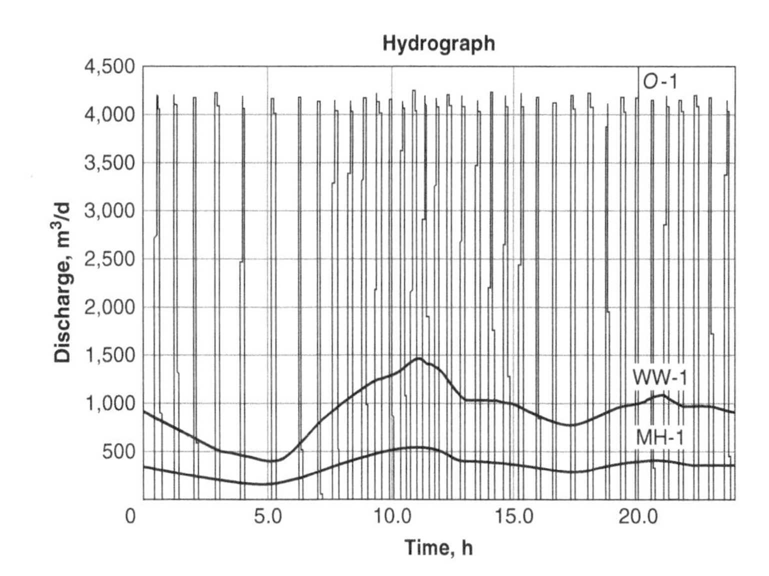

Figure 15.2 Example 15.2 Discharge Rates at MH-1, WW-l, and O-1 for 24-hour Cycle

Example 15.3 Constructing a Profile

Problem Statement

In the third example, a profile of a sewer is to be constructed. The system used in the first two examples will be used again. The profile is to show all structures, ground elevations, and the hydraulic profile.

Part 1: What is the maximum hydraulic head produced by the pump?
Part 2: Does the water in the wet well back up into pipe P-1?

Solution:

Constructing the Profile

- The Profile Manager is located in the first row of tools. Open it by clicking its icon.
- Select **Profile Management** and then **Add**. Enter **Example Profile** as the profile name and click **OK**.
- The **Profile Wizard** walks the user through the profile construction process. Click the **Select From Drawing** button. Select the pipe elements you wish to profile by clicking once on each pipe (the solid lines will turn into dashed lines). Then, right-click and select **Done** to indicate that the selection is complete. Click the **Next** button.
- Select the **Standard** template and click the **Next** button.
- Set the **Horizontal Increment** to 0350 and the **Vertical Increment** to 2 m. Exit the Profile Wizard by clicking **Finished**. The profile will appear.

Customizing the Profile

- Annotations can be moved to the desired location by dragging and dropping them.
- Annotation text or properties can be changed right-clicking it. Annotations can also be rotated.
- Use the **Options** button in the profile window to change the range and the increments of the horizontal and vertical axes and the text height, as well as to define the annotation for each element type.
- When the annotation is complete, select **Print Preview** to preview a hardcopy of the profile. Select **Print** to send the file to your printer.

Using the Hydraulic Grade Line

- For this run, changes in the hydraulic grade line are best viewed with a small time increment. While in the profile view, click the **Scenario Manager** button. Highlight the 24-hour EPS scenario, right-click, and select **Edit**. Click the **Calculation** tab and set both the Hydraulic and Hydrologic time steps to **0.05 h**. Close the Scenario edit dialog. Click **GO Batch Run**, recalculate the scenario, and then close out of Scenario Manager.
- In the profile view, set the Increment to **0.05 h** and click the **Play** button to animate.

Answer to Part 1

Note that when the pump is not running, the hydraulic grade line in the force main is level with the crown of the outlet (107.2 m). When the pump is running, the hydraulic grade line rises at the pump, and then declines as flow approaches the outlet.

Answer to Part 2

Note that the level in the wet well varies over the course of the simulation. The profile in Fig. 15.3 shows the simulation at hour 12. The level in the wet well is at 98 m, which causes pipe P-1 to be full at the downstream end. The zone of full flow in P-1 extends upstream approximately 25 m.

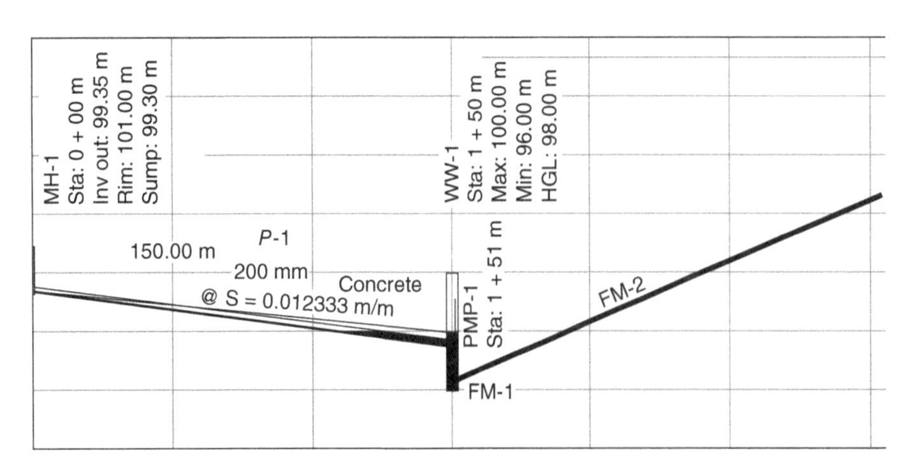

Figure 15.3 Example 15.3 Profile of Sewer Network at Hour 12

Example 15.4 Capital Cost Estimating

SewerCAD has many tools for preparing project cost estimates. In this exercise, the Unit Cost function and the Cost Table features will be demonstrated.

Problem Statement

Prepare a cost estimate for the materials and installation portion of the project from Example 15.2. Use the cost data in Tables 15.6 and 15.7. The coefficients a, b, c, and d will be applied to the cost function formula form (unit cost) $= d + a(x - c)^b$.

Table 15.6 Cost Functions for the Pipes and Gravity Sewer of Example 15.4

Structure	Unit Cost Unit	Unit Cost Function Attribute	a	b	c	d
Force Main	\$/m	Diameter (mm)	0.15	1.3	150	60
Gravity Sewer	\$/m	Rise (mm)	0.15	1.1	175	55
Sewer Manhole	\$	Depth (m)	200	1.8	0	1,200

Table 15.7 Fixed Costs for the Other System Components of Example 15.4

Pipe Junction	\$1,000
Pump Station	
Pump	\$15,000
Installation	\$4,000
Valves and Piping	\$750
Wet Well	\$50,000
Outlet	\$2,500

Solution:

- From the Analysis pull-down menu, select **Capital Costs**, and then select **Unit Cost Functions**.
- Select the **Pressure Pipe** tab and click the **Add** button. Select **Formula Unit Cost Function** as the Unit Cost Function. Type and **Diameter** in the Unit Cost Attribute. Click **OK**. Enter the data for the force main from the preceding tables. To view a plot of the function, set the **Valid Cost Data Range to** 150 to 300 and choose **Plot**. Close the plot viewer. Click **OK** to close the Cost Function dialog box and return to the Unit Cost Functions dialog.
- Select the **Gravity Pipe** tab and click the **Add** button. Select **Formula Unit Cost Function** as the Unit Cost Function Type and **Rise** in the Unit Cost Attribute field. Click **OK**. Enter the data for gravity sewer pipe from Tables 15.6 and 15.7. Set the **Valid Cost Data Range** to 175–300. Click **OK** to close the Cost Function dialog and return to the unit cost functions dialog.
- Select the **Manhole** tab and click the **Add** button. Select **Formula Unit Cost Function** as the Unit Cost Function Type and **Structure Depth** as the Unit Cost Attribute. Click **OK**. Enter the data for manholes from the table above. Set the **Valid Cost Data Range** to 1–5. Click **OK** to close the Cost Function dialog and return to the Unit Cost Functions dialog. Click **Close** twice to close the dialog and exit the Cost Manager.
- Click the **Tabular Reports** button in the toolbar at the top of the screen. Select **Pipe Cost Report** from the Available Tables list and click **OK**.
- Check the box next to **Include in Capital Cost Calculation?** for FM-1, FM-2, and FM-3.

- Right-click the header of the **Total Cost** ($) column. Select the **Global** Edit option. Enter **Materials and Installation** in the first line of the label column. Click the **Advanced** button. Check the box next to **Set Quantity Equal to Pipe Length**, and then select **Diameter Cost Function** for the Unit Cost Function. Use the same procedure to apply the **Rise Cost Function** to the gravity pipe P-1. Click **OK** twice and close the pipe cost report.
- Double left-click **MH-1** and select the **Capital Cost** tab. Check the box next to **Include in Capital Cost Calculation?** Enter **Materials and Installation** in the first line of the label column. Click the **Advanced** button. Select **Structure Depth Cost Function** for the Unit Cost Function. Click **OK** to close the MH-1 window.
- Double left-click **WW-1** and select the **Capital Cost tab**. Check the box next to **Include in Capital Cost Calculation?** Enter **Materials and Installation** in the first line of the label column. Enter **50,000** in the **Unit Cost ($)** column. Click on **OK** twice to close the window. Define the costs for O-1 and J-1 using the same procedure.
- Double left-click **PMP-1** and select the **Capital Cost** tab. Check the box next to **Include in Capital Cost Calculation?** Enter **Pump** in the first line of the label column. Enter **15,000** in the **Unit Cost ($)** column. Add the remaining cost data for installation and valves and piping on the next two lines. Click **OK** to close the window.
- You are now ready to calculate the total project costs. On the Analysis drop-down menu, select **Capital Costs** to open Cost Manager. The cost will be calculated for the base scenario. Click the **Cost Reports** button and select **Detailed Report** from the drop-down menu. The Detailed Cost Report table will be activated. To obtain a hardcopy, select **Print Review** and then **Print**.

It should always be noted that the costs is a function of time due to inflation. By the time the construction costs of an engineering project are estimated even in a timely manner, the costs have changed. Therefore, the estimated costs must be dated with a specific Month-Year. Appendix 16 shows the U.S. Army Corps of Engineers (USACE) civil works construction yearly average cost index for utilities from 1967 to 2022. USACE publishes the cost index for utilities every month; therefore, the calculated costs can be updated to the future months and years anytime when necessary. USACE cost index is only an example. Other cost index, such as Engineering News Record (ENR) cost index, etc. can also be used. ENR publishes three pages of construction economics data, in PDF format each week. Building material prices in 20 major U.S. cities, labor cost data, etc. are reported. For engineering project negotiation, the cost index (USACE, ENR, etc.) for updating the future costs must be legally specified.

Answer

- From the **Analysis** drop-down menu, select **Capital Costs**. Note that costs may be calculated for any scenario. In this example, the physical layout is identical for both scenarios. Click the **Report** button next to "Scenario Costs." A report like that shown in Table 15.8 is produced.

Table 15.8 Reported Results for Example 15.4

Label	Construction Cost ($)	Nonconstruction Costs ($)	Total Cost ($)
Wet Well	50,000	0	50,000
Manhole	1,720	0	1,720
Pressure Pipe	35,049	0	35,049
Outlet	2,500	0	2,500
Gravity Pipe	9,136	0	9,136
Pump	19,750	0	19,750
Pressure Junction	1,000	0	1,000
Total			119,155

15.3 Storm Sewer Applications

Storm sewer analysis occurs in two basic calculation sequences:

1) *Hydrology*: Watersheds are analyzed and flows are accumulated from upstream inlets toward the system outlet.
2) *Hydraulics:* A tailwater condition is assumed at the outlet, and the flow values (from the hydrology calculations) are used to compute hydraulic grades from the outlet toward the upstream inlets.

15.3.1 Hydrology Model

As the runoff from a storm event travels through a storm sewer, it combines with other flows, and the resulting flows are based on the overall watershed characteristics. As with a single watershed, the peak flow at any location within the storm sewer is assumed to occur when all parts of the watershed are contributing to the flow. Therefore, the rainfall intensity that produces the largest peak flow at a given location is based on the controlling system time at that same location.

The controlling system time is the larger of the local time of concentration (to a single inlet) and the total upstream system time (including pipe travel times). The controlling time is used for computing the intensity (and therefore the flow) in the combined system.

Example 15.5 Flow Accumulation
A storm sewer inlet has a local time of concentration of 8 min for a watershed with a weighted CA (that is, the weighted runoff coefficient times the total drainage area) of 1.23 acres (0.50 ha). This inlet discharges through a pipe to another storm sewer inlet with a weighted CA of 0.84 acres (0.34 ha) and a local time of concentration of 9 min. If the travel time in the pipe is 2 min, what is the overall system CA and corresponding storm duration?

Solution:

The total CA can be found by simply summing the CA values from the two inlets. The storm duration, however, must be found by comparing the local time of concentration at the second inlet to the total time for flow from the upstream inlet to reach the downstream inlet:

$$\text{Total CA} = 1.23\ \text{acres} + 0.84\ \text{acre} = 2.07\ \text{acres}\ (0.84\ \text{ha})$$

$$\text{Upstream time} = 8\ \text{min} + 2\ \text{min} = 10\ \text{min}$$

The total upstream flow time of 10 minutes is greater than the local time of concentration at the downstream inlet (9 minutes). The 10-minute value is therefore the controlling time, and should be used as the duration of the storm event.

A storm sewer system may have a number of flow sources other than direct watershed inflow. Flow may be piped into an inlet from an external connection, or flow could be entering an inlet that is the carryover (bypass) flow from another storm sewer inlet.

Design practices for handling carryover flows vary from jurisdiction to jurisdiction. Some local regulations may require that pipes be sized to include all flow that arrives at an inlet, whereas other locales may specify that pipes be sized to accommodate only the flow that is actually intercepted by the upstream inlets. It is the responsibility of the design engineer to ensure that the design is in agreement with the local criteria and policies.

15.3.2 Hydraulics Model

The hydraulics for a storm sewer is computed as described previously for hydraulics of sanitary sewers. Normal depth may occur in portions of the system, whereas other areas may experience pressure (submerged) conditions. Gradually varied flow and rapidly varied flow may also occur.

Computations start at the system outlet, where a tailwater condition is assumed. Four basic assumptions about tailwater conditions are made:

1) *Normal depth:* The depth at the outfall of the farthest downstream pipe is assumed to be equal to normal depth.
2) *Critical depth:* The depth at the pipe outfall is assumed to be critical depth, as in subcritical flow to a free discharge.
3) *Crown elevation:* The depth is set to the crown (top) of the pipe for free outfall.

4) *User-specified tailwater:* A fixed tailwater depth can also be used, for instance, when there is a known pond or river water surface elevation at the outfall of the storm sewer.

Care should be taken to choose an accurate tailwater condition because this value can affect the hydraulics of much of the system. The designer of a storm sewer system should consider the tailwater depth during storm conditions. An outlet may be above the receiving stream during dry weather, but may be submerged during the design storm event.

15.4 StormCAD

StormCAD from Bentley is a user-friendly program that helps civil engineers design and analyze storm sewer systems. Just draw your network on the screen by using the tool palette, double-click any element to enter data, and click the **GO** button to calculate the network.

Rainfall information is calculated using *rainfall tables, equations,* or the *National Weather Service's HYDRO-35 data*. StormCAD also plots intensity-duration-frequency (IDF) curves. You have a choice of conveyance elements that include circular pipes, pipe arches, boxes, and more. Flow calculations handle pressure and varied flow situations including hydraulic jumps, backwater, and drawdown curves. StormCAD's flexible reporting features allow you to customize and print the design and analysis results in report format or as a graphical plot.

StormCAD is so flexible you can use it for all phases of your project, from the feasibility report to the final design drawings and analysis of existing networks. During the feasibility phase, you can use StormCAD to create several different system layouts with an AutoCAD® or MicroStation drawing as the background. For the final design, StormCAD lets you complete detailed drawings with notes that can be used to develop construction plans. In summary, you can use StormCAD for the following tasks:

1) Design multiple storm sewer systems with constraint-based design.
2) Design/analyze inlets based on HEC-22 methodology.
3) Use AASHTO, HEC-22 energy, standard, absolute, or user-specified ("generic") methods to compute structure losses.
4) Analyze various design scenarios for storm sewer systems.
5) Import and export AutoCAD and MicroStation's DXF files.
6) Predict rainfall runoff rates.
7) Generate professional-looking reports for clients and review agencies.
8) Generate plan and profile plots of the network.

StormCAD's automatic design feature allows you to design a whole or part of a storm sewer system based on a set of user-defined design constraints. These constraints include minimum/maximum velocity, slope, and cover; choice of pipe invert or crown matching at structures; inlet efficiency; and gutter spread and depth. StormCAD will automatically design the invert elevations and diameters of pipes, as well as the size of a drainage inlet necessary to maintain a given spread (for inlets in sag) or capture efficiency (for inlets on grade).

StormCAD also includes an option to automatically generate storm sewer profiles as longitudinal plots of the storm sewer. Profiles allow the design engineer, the reviewing agency, the contractor, and others to visualize the storm system. They are useful for viewing the HGL and determining if the proposed storm sewer is in conflict with other existing or proposed underground utilities.

The theory and background used in StormCAD are presented in more detail in the StormCAD Bentley online help system. The following examples provide step-by-step instructions on how to solve various problems using StormCAD (included with the CD that accompanies this textbook and also available online).

15.5 Inlet Design and Analysis Using Computer Modeling

This section provides a quick overview of the features necessary to solve inlet problems using StormCAD software, which is used here to illustrate how to model inlets and gutter in storm sewer networks.

StormCAD calculates and designs drainage inlets by applying the HEC-22 methodology. StormCAD will design an entire gutter network. The gutter network dictates what happens to flow on the surface when it bypasses an inlet and is routed downstream to another inlet.

To edit an inlet's properties, first double-click on that inlet in the drawing pane and select the **Inlet** tab at the top of the dialog. In the **Inlet** field, select an inlet from the pull-down menu, which lists all of the inlets currently defined in the **Inlet Library**. To examine the characteristics of the inlet selected, click the ellipsis button (. . .) next to this field. The **Inlet Library** dialog will appear. The Inlet Library is a separately saved file and editable library that allows the user to customize for local regulations. Select the inlet of interest and click **Edit** . . . to examine or change its properties, or click the **Insert** button to create a new inlet. To exit an inlet's properties dialog in the library, click **OK** or **Cancel**. To exit the **Inlet Library**, click **Close**.

In the **Inlet** tab of an **Inlet Editor** dialog, the **Inlet Opening** section displays variables for the type of inlet chosen. Gutter characteristics can be entered under **Inlet Section.**

Finally, you have to specify an inlet location by clicking either the **On Grade** or **In Sag** radio button. If the inlet is on grade, fill in the **Longitudinal Slope** of the road and the **Manning's n** value for the pavement. You must also select a **Bypass Target** to establish the connectivity of the gutter network. The bypass target is another inlet or outlet in the same project (the target element does not have to be an element in the same pipe network) where the flows bypassing the current inlet will be carried. If no bypass target is chosen, all bypass flow will automatically carry over to the outlet of the sewer network that the inlet is part of (in this case, the bypassed flow is not accounted for in the sewer network). Click **OK** to exit the dialog. To view a graphical representation of the gutter network in plan view, select **Gutter Network** from the **Tools** pull-down menu.

Example 15.6 Design of a Network with StormCAD's Auto Design Feature

Problem Statement

A small residential road has three houses. Each house has one storm drain to collect its runoff. The outfall for the system is across the street and drains into a pond. Pipe P-3 leading to the outfall is a 10-m-long, concrete box culvert. Pipes P-1 and P-2 are 20-m-long, circular concrete pipes. Each structure has a head loss coefficient of 0.5 (using the standard head loss method).

Use Manning's equation to calculate the friction losses through the pipe. The rainfall data and hydrologic data are given in Tables 15.9 and 15.10. All inlets are Generic Default 100%, which are assumed to capture 100% of the surface flow.

1) Use the "Auto Design" feature to find pipe sizes and invert elevations for a 25-year storm if the elevation of the drainage pond's water surface is 262.8 m. The inlet and rainfall information, along with design constraints, are given in Fig. 15.4. Additional design constraints are given in Table 15.11.
2) What would the effect on the system be if the pond's water surface elevation were 0.1 m below the ground elevation at the outfall?
3) What will happen to the system if the pond's water surface elevation is 262.8 m and an additional flow of 0.5 m^3/s is added to I-2?

Table 15.9 Inlet Information for Example 15.6

Inlet	Ground Elevation (m)	Impervious Area $C = 0.9$ (ha)	Pervious Area $C = 0.3$ (ha)	Time of Concentration (min)
I-1	264.8	0.16	0.18	5
I-2	264.5	0.13	0.15	5
I-3	265.0	0.17	0.13	6
Outlet	264.6	–	–	–

Table 15.10 Rainfall Data for Example 15.6

	Rainfall Intensity (mm/h)		
Duration (min)	**5 Year**	**10 Year**	**25 Year**
5	165	181	205
10	142	156	178
15	123	135	154
30	91	103	120
60	61	70	80

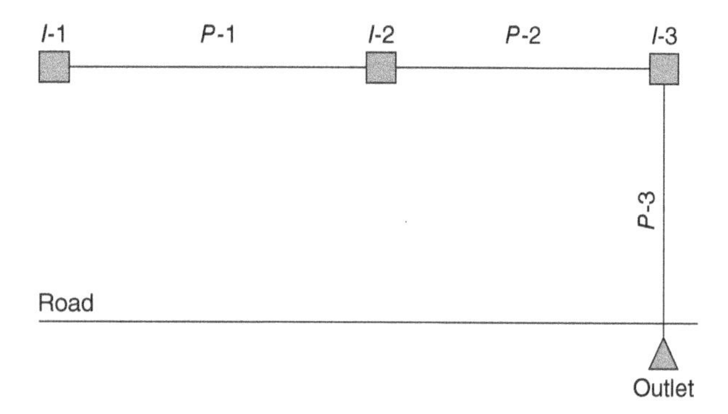

Figure 15.4 Schematic for Example 15.6

Table 15.11 Design Constraints for Example 15.6

Constraint	Minimum	Maximum
Velocity	1 m/s	3 m/s
Cover	1 m[a]	–
Pipe Slope	0.5%	10%
	Match Inverts	

[a] Cover constraints may be relaxed at the outfall, where invert location requirements will typically govern.

Solution to Part 1:

Project Setup Wizard

- When you start StormCAD, you should be prompted with the **Welcome to StormCAD** dialog. From this dialog you can access the tutorials, open existing projects, and create new ones. Select **Create New Project**, provide a filename, and click **Save**.
- If the **Welcome to StormCAD** dialog does not appear, StormCAD is set to **Hide Welcome Dialog on startup**. To start a new project, select **New** from the **File** menu, enter a filename, and click **Save**. (You can change from **Hide Welcome Dialog** mode to **Show Welcome Dialog** mode in the **Global Options** dialog, which is accessible by selecting **Options** from the **Tools** menu.)
- The **Project Setup Wizard** will appear. Add a project title and any appropriate comments, and click **Next**.
- In the second screen of the **Project Setup Wizard**, select **Manning's Formula** from the pull-down menu in the **Friction Method** field and click **Next**.

- In the third screen of the **Project Setup Wizard**, click the **Schematic** radio button in the **Drawing Scale** section of the dialog. This action will allow you to define the lengths of the pipes without having to worry about scale and spatial placement on the *x-y* plane. Click **Next**, and then click **Finished**.

Laying Out the System

- Because the units are given mostly in SI, you can simplify data entry by specifying a global unit system for the model. From the **Tools** menu, select **Options**. Then under the **Global** tab, select **System International** from the pull-down list in the **Unit System** field if it has not already been selected. Click **OK**. When changing the unit system, you may be prompted to confirm the switch. If you are prompted, click **Yes** to confirm the change.
- You do not have to lay out the system exactly as shown in the problem statement. For now, just roughly sketch the schematic following the instructions below.
- Click the **Pipe Layout** button on the vertical toolbar on the left side of the screen.
- To place inlet I-1, click the left mouse button in the workspace.
- To place inlet I-2, move the cursor to the right and click the left mouse button in the location you would like to drop the inlet. Pipe P-1 will be placed automatically.
- Place inlet I-3 in the same manner.
- Click the right mouse button and select the **Outlet** by highlighting it and left-clicking. Click the left mouse button to place the outlet.
- Except for the scale, your schematic should look roughly like the one given in Fig. 15.4.

Data Entry

- From the **Analysis** pull-down menu, select **Rainfall Data** and then select **Table**.
- You need to edit the duration rows and the return period columns in the table to match those given in the problem statement. To do this, click the **Edit Return Periods** button and select either **Delete** or **Add Return Period** to delete the 2-year return period and add the 25-year return period. Edit the duration rows in the table the same way using the **Edit Durations** button. Enter the data for rainfall intensity as you would enter data into a spreadsheet and click **OK**.
- Double-click inlet I-1 to edit the dialog.
- Enter **264.8 m** in the **Ground Elevation** field on the **General** tab.
- Click the **Catchment** tab to enter the rational method data for the watershed draining to inlet I-1.
- Enter the impervious area, **0.16 ha**, and its Rational coefficient C, **0.9**, into the table in the **Subwatershed Information** section. Enter the pervious area and its rational coefficient C in the second row. The model will compute the composite C coefficient.
- Enter the 5-minute T_c in the **Time of Concentration** field.
- Click the **Headlosses** tab. Select **Standard** from the list of **Headloss Methods**. Enter **0.5** in the **Headloss Coefficient** field.
- Edit the data for inlets I-2 and I-3 in the same way as for inlet I-1 using the data found in the tables, making sure to enter the standard head loss coefficients under the **Headlosses** tab.
- Double-click the outlet to edit its dialog. Enter the ground elevation in the appropriate field.
- A tailwater depth of 262.8 m is given in the problem statement. Therefore, select **User Specified** from the **Tailwater Conditions** field of the dialog. Enter **262.8 m** (the elevation of the drainage pond surface) for the tailwater elevation and click **OK**.
- Double-click pipe P-3 to edit its dialog.
- Input 10 m in the **Length** field.
- Select **Box** from the **Section Shape** pull-down menu.
- Select the appropriate material from the **Material** pull-down menu.
- You do not need to enter the upstream and downstream invert elevations under the **General** tab because StormCAD will design these in this example.
- Edit pipes P-1 and P-2 in the same manner. Make sure that both are 20-m-long, circular, concrete pipes. You do not need to enter any new diameters or invert elevations because StormCAD will design these, ignoring any current values.

Running the Model

- Select **Default Design Constraints** from the **Analysis** pull-down menu.
- In the **Default Design Constraints** dialog, select the **Gravity Pipe** tab and enter the minimum and maximum velocities, ground covers, and pipe slopes as shown in Fig. 15.5.
- Select the **Node** tab and select **Inverts** from the **Pipe Matching** field. Click **Close** to exit the dialog.
- Click the **GO** button on the toolbar at the top of the screen. Select **25 year** from the **Return Period** pull-down menu.
- Click the **Design** radio button to design pipe sizes and inverts.
- Click **GO** to design the network. The program will ask you if you want to save the designed data as a new physical alternative, which would give you access to the initial data and the final design within the same model. In this case, click **No**. Click **Close** to exit the dialog.

Answer to Part 1

- Click the **Tabular Reports** button to examine the newly designed pipe characteristics. Select **Pipe Report** by highlighting it in the **Available Tables** list and click **OK**.
- The length, size, slope, and invert elevations for each pipe can be easily compared using this table. You can also examine the data for any individual pipe under the different tabs of that pipe's dialog.

Solution to Part 2:

- Double-click on the outlet and change the tailwater elevation as dictated by the problem. Close out of this dialog.
- To calculate the model, click the **GO** button in the top toolbar. Switch the calculation type from **Design** to **Analysis**, and click the **GO** button.
- Under the **Results** tab, notice that a warning is displayed, as indicated by a yellow light (if there are no warnings, a green light appears). The warning indicates that flooding occurs at inlet I-2 and that P-3 fails the minimum velocity constraint.
- You can also examine conditions by double-clicking the inlet and clicking the **Message** tab in the **Inlet** dialog box. Another way to look at data is to select **Profiling** from the **Tools** menu. Follow the steps in the **Profile Wizard** to create a profile of the hydraulic grade line within the storm sewer system.

Answer to Part 2

Flooding will occur at inlet I-2.

Solution to Part 3:

- Double-click the outlet and change the tailwater elevation back to its original level. Click **OK** to accept the change and exit the dialog.
- Double-click inlet I-2. Click the **Flows** tab. Enter **0.5 m^3/s** in the **Additional Flow** field. Click **OK** to accept the change and exit the dialog.
- Calculate the network in analysis mode as you did in part 2.

Answer to Part 3

Flooding will occur at inlet I-2.

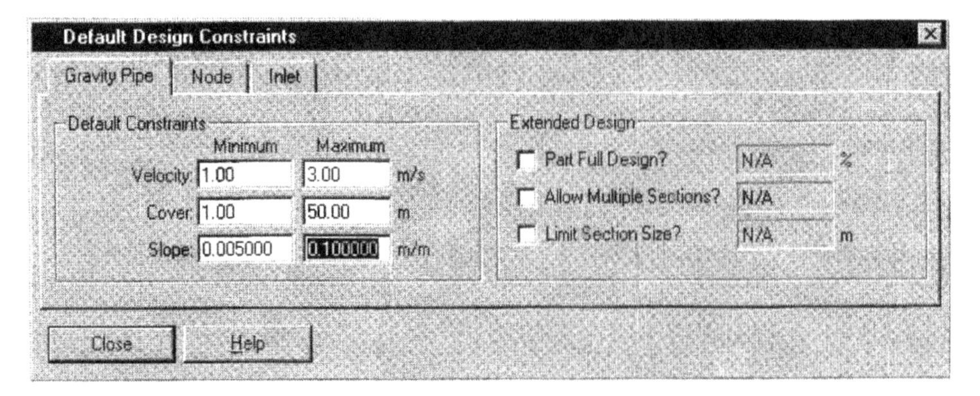

Figure 15.5 Default Pipe Design Constraints in StormCAD

Example 15.7 Alternatives Analysis with StormCAD's Scenario Manager
The Scenario Manager tool of StormCAD is a versatile utility for analyzing alternative designs and loadings without having to reenter data. In this example, Scenario Manager is used to construct an alternative design to convey additional flows.

Problem Statement

Under an alternative scheme to Example 15.6, additional flow will enter the network at I-3 and be conveyed under the road via pipe P-3. This scheme adds 1.0 m^3/s of flow at inlet I-3. Determine the required size of the box culvert under this alternative.

Solution:

Reset the conditions to the original design by resetting the tailwater at O-1 to 262.8 and removing the 0.5 m^3/s additional flow at I-2.

- On the **Analysis** drop-down menu, select **Scenarios** and then click **Scenario Wizard**.
- Use the default name of Scenario 2 and click **Next**.
- Scenario 2 will be based on the Base scenario. Click **Next**.
- Set the calculation type to **Design**. Make sure that the Return Period is set to 25 years. Click **Next**.
- The Alternatives that will be changed are the **System Flows**. Check this box and then click **Next**.
- Select the "Create New System Flows Alternative" and then click **Next**. Click **Finish** to return to Scenario Manger.
- Note that with the Scenario section set to Scenario 2, the System Flows alternative is local. All other alternatives are inherited from the Base scenario. Double-click the **System Flow Alternative** to open the dialog window. On the I-3 line, enter **1.0** in the additional flow column, and then click **Close**.
- With the Scenarios section set to Scenario 2, click on the **Go** – **Batch Run** button. In the Batch Run window, select **Scenario 2** and then click on the **Batch** button. Confirm that one batch run will be performed and click on **Yes** to allow StormCAD to create a new Physical Properties alternative.

Answer to Part 2

Return to the network layout view. Note that the Scenario setting can be used to toggle between Base and Scenario 2. With the Scenario set to Scenario 2, double-click P-3 and examine its properties. Note that that box culvert has changed from a 610–3,610-mm size in the Base scenario to a 1,220–3,610-mm box in Scenario 2. All other physical properties are unchanged.

Example 15.8 Cost Estimating
StormCAD provides many tools for preparing project cost estimates. This example demonstrates the Unit Cost function and the Cost Table features.

Problem Statement

Prepare a cost estimate for materials and installation of the network designed in Example 15.7. The outlet structure has a fixed cost of \$1,500, and the box culvert has a unit cost of \$125/m. Table 15.12 outlines the cost functions for pipes and inlets.

Table 15.12 Data for the Cost Functions of Example 15.8

			Cost Function Coefficients			
Label	Unit Cost Unit	Unit Cost Function Attribute	*a*	*b*	*c*	*d*
Circular Pipe	\$/m	Rise (mm)	0.15	1.1	250	55
Inlet	\$ea.	Depth (m)	200	1.8	0.0	1,000

Solution

- From the **Analysis** pull-down menu, select **Capital Costs** and then select **Unit Cost Functions**.
- Select the **Pipe** tab and then click the **Add** button. Select **Formula Unit Cost Function** as the Unit Cost Function Type, select **Rise** for Unit Cost Function Attribute, and click **OK**. Specify the unit cost, unit, local unit, and enter the data for circular pipe. To view a plot of the function, click on the **Initialize Range** button and then on **Plot**. Close the Plot viewer. Click on **OK** to close the Cost Function dialog.
- To define the cost for the box culvert, click **Add** to add a second cost function. Select **Tabular Unit Cost Function** as the Function Type and click **OK**. Enter **Box Culvert** in the label. In the Unit Cost Data section, enter **610** as the rise (length) and $125 in the unit cost. Then click **OK**.
- To add a cost function for the inlets, select the **Inlet** tab. Click **Add** and select **Formula Unit Cost Function** as the Unit Cost Function Type, select **Depth** for Unit Cost Function Attribute, and click **OK**. Click the **Initialize Range** button. Enter **Inlets** as the label. Enter the data for inlets from the cost table. Click **OK** and then close the Cost Function dialog box.
- The next step is to assign the cost data to the structures. Close the Cost Manager and return to the network layout view.
- Double-click **I-1** to open the inlet dialog window. Select the **Capital Cost** tab. Check the **Include in Construction Cost** box and then click the **Insert** button to add the first line of the construction cost table. In the Label column, enter **Material and Installation** and then click the **Advanced** button. In the Cost Item pop-up, select **Inlets**. Click **OK** twice to exit the I-1 dialog. Repeat the same procedure for I-2 and I-3.
- Double-click on **P-1** to open the pipe dialog window. Select the **Capital Cost** tab. Check the **Include in Construction Cost** box and then the insert button to add the first line of the construction cost table. In the Label column, enter *Material and Installation*, and then click the **Advanced** button. In the Cost Item pop-up, select **Circular pipe**. Click **OK** twice to exit the I-1 dialog. Repeat the same procedure for P-2. For P-3 repeat these steps except that unit cost function is **Box Culvert**.
- Double-click on **O-1** to open the Outlet dialog. Select the **Capital Cost** tab. Check the **Include in Construction Cost** box and click the **Insert** button to add the first line of the construction cost table. In the label column, enter **Material and Installation** and then enter **$1,500** in the Unit Cost column. Click on **OK**.
- You are now ready to calculate the total project costs. On the **Analysis** drop-down menu, select **Capital Costs** to open the Cost Manager. The cost will be calculated for the Base Scenario. Click the **Cost Reports** button and select **Detailed Report** from the drop-down menu. The Detailed Cost Report table will appear. To obtain a hardcopy, select **Print Review** and then **Print**.

Answer

The estimated costs for materials and installation are:

Inlets	4,449
Outlet	1,500
Pipe	6,446
TOTAL	$12,395

Problems/Questions

15.1 In the sanitary sewer represented by the schematic in Fig. 15.6, a load from development of high-rise apartments (at wet well WW-1) with 10,000 residents is pumped to the top of a hill, where it is then transported via a circular gravity line to a treatment plant (represented by O-1). At manhole MH-1 is a resort apartment with 300 guests. The load generated here flows down a circular gravity pipe to the same gravity line mentioned above. A bar that serves, on average, 50 people per day and two large cafeterias with 20 employees each are located near manhole MH-2. The pipe and node data for this problem are shown in Tables 15.13 and 15.14, respectively.

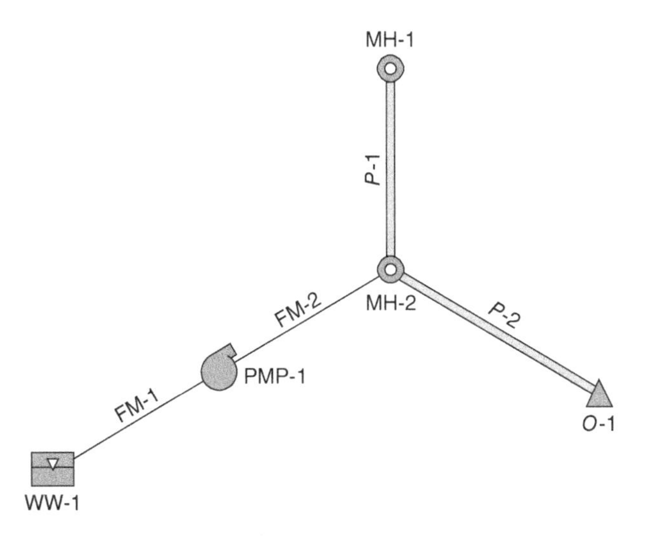

Figure 15.6 Schematic for Problem 15.1

Table 15.13 Pipe Data for Problem 15.1

Pipe	Pipe Type	Material	Diameter (mm)	Length (m)	Upstream Invert (m)	Downstream Invert (m)
P-1	Gravity	PVC $n = 0.010$	200	100	98.90	97.62
P-2	Gravity	PVC $n = 0.010$	375	200	97.62	96.62
FM-1	Pressure	Ductile Iron $C = 130$	300	1	70.00	70.00
FM-2	Pressure	Ductile Iron $C = 130$	350	1,000	70.00	97.62

Table 15.14 Node Data for Problem 15.1

Node	Ground Elevation (m)	Rim Elevation (m)	Sump Elevation (m)	Structure Diameter (m)
MH-1	100.30	100.30	98.90	1.00
MH-2	99.10	99.10	97.62	1.00
O-1	100.00	100.00	96.62	N/A
WW-1	74.00	–	–	–

The base and minimum elevations for the wet well are both 70 m. The initial elevation is 73 m, and the maximum elevation is 73.5 m. The diameter of the circular wet well is 3 m.

The tailwater elevation at outlet O-1 is 98 m.

The three defining points of the pump curve are 0 m^3/min at 33.33 m, 5.00 m^3/min at 25.00 m, and 10 m^3/min at 0 m. The pump's elevation is 70 m.

1) What is the total peak sanitary outflow if no peaking factor method is applied to the four-unit sanitary loads mentioned above? If Babbitt's peaking factor method is applied? If Harmon is applied? Which peaking factor method is the most conservative?
2) With the Harmon peaking factor applied to each of the four unit dry loads, what is the hydraulic grade at MH-2? How does this peaking factor change the hydraulic load and the flow velocity of pipe P-1 from when no peaking factor was applied?
3) Identify and describe any problems for each of the three scenarios from part 1

15.2 The estimated infiltration rate for each of the concrete gravity pipes in the proposed sanitary sewer system represented in Fig. 15.7 is 20 Lpd per mm-km. The estimated inflow into each of the manholes during a 5-year storm event is approximately 75 Lpd. The ground elevations for MH-1, MH-2, MH-3, JC-l, and O-1 are 12.10, 12.10, 11.20, 11.80, and 10.25 m, respectively. The top of junction chamber JC-1 is 11.8 m. The tailwater condition is a free outfall.

Use SewerCAD's Automatic Design feature to design the inverts and sizes of concrete pipes in the proposed sanitary sewer. The pipe lengths and design constraints are given in Tables 15.15 and 15.16. Apply both the wet-weather (given above) and dry-weather sanitary loads (given in Table 15.17). Use Federov's equation to calculate the peaking factor for each of the unit sanitary loads.

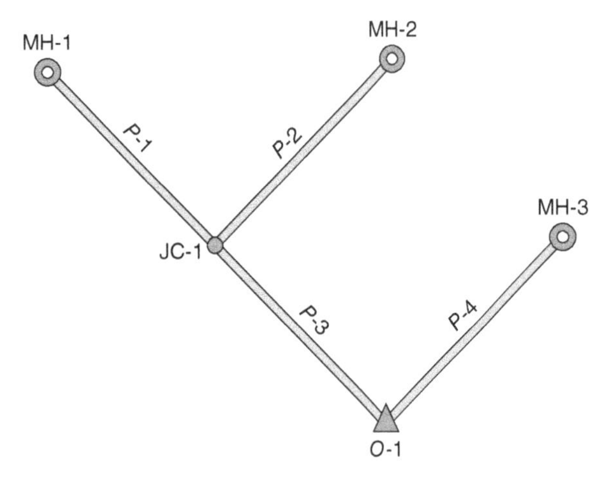

Figure 15.7 Schematic for Problem 15.2

Table 15.15 Pipe Data for Problem 15.2

Pipe	Length (m)
P-1	200
P-2	150
P-3	200
P-4	300

Table 15.16 Constraint Data for Problem 15.2

Constraint	Minimum	Maximum
Velocity (m/s)	0.5	3.0
Cover (m)	1.0	3.0
Slope (m/m)	0.005	0.1

Table 15.17 Node Data for Problem 15.2

Node	Unit Sanitary Load	Loading Unit	Unit Count
MH-1	Hospital (Medical) per Bed	Bed	400
	Cafeteria per Employee	Employee	20
	Apartment Resort	Guest	100
MH-2	Apartment	Resident	400
	Shopping Center per Employee	Employee	100
	Laundromat per Wash	Wash	200
MH-3	School (Boarding)	Student	500

List the diameter and slope for each newly designed pipe. Are there any problems with the designed system?

1) What percentage of the total flow is wet-weather flow? Which pipe has the most infiltration?
2) For a larger magnitude storm, the inflow rate into each manhole is estimated at 100 Lpd. Analyze the model using the previously designed system and apply the larger wet-weather loading. What percentage of the total flow is wet-weather loading?
3) Do you consider the amount of wet-weather flow into the system significant? What are some methods for alleviating infiltration?

15.3 The network in Fig. 15.8 is an initial design for a system of force mains in a sanitary sewer. All pipes are ductile iron (Hazen–Williams coefficient $C = 130$). The pump, PMP-l, is at an elevation of 691 m. Enter the system data given in Tables 15.18–15.21 and answer the following questions. The ground elevation at the outlet is 715 m and the sump elevation is 712 m.

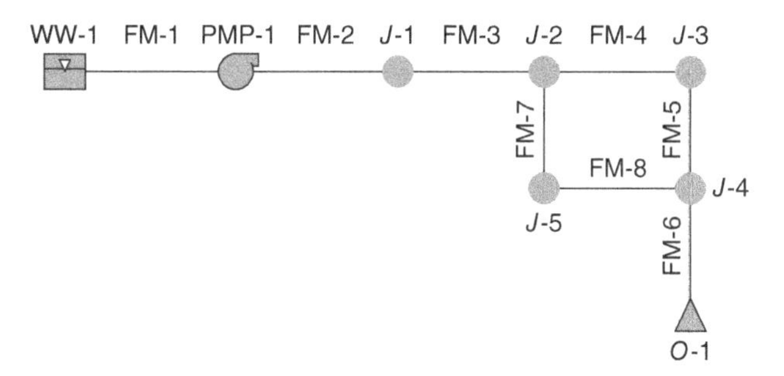

Figure 15.8 Schematic for Problem 15.3

Table 15.18 Pump Data for Problem 15.3

Pump Curve Point	Head (m)	Flow (m^3/min)
Shutoff	30	0
Design	20	4
Max Operating	5	6

Table 15.19 Wet Well Data for Problem 15.3

Ground Elevation (m)	694.7
Maximum Elevation (m)	694.0
Initial Elevation (m)	691.5
Minimum Elevation (m)	688.5
Base Elevation (m)	687.0
Diameter (m)	3.0

Table 15.20 Junction Data for Problem 15.3

Pressure Junction	Elevation (m)
J-1	698
J-2	701
J-3	703
J-4	705
J-5	703

Table 15.21 Pipe Data for Problem 15.3

Pipe	Diameter (mm)	Length (m)	Upstream Invert (m)	Downstream Invert (m)
FM-1	400	2	687.00	691.00
FM-2	250	58	691.00	698.00
FM-3	350	45	698.00	701.00
FM-4	350	88	701.00	703.00
FM-5	350	71	703.00	705.00
FM-6	450	67	705.00	712.00
FM-7	450	39	701.00	703.00
FM-8	450	68	703.00	705.00

Hints: Make sure that the **Fixed Level in Steady State** box is checked under the Section tab of the Wet Well dialog. In addition, before running the model, make the following modification to the calculation options: Within the **GO** dialog, click the **Options** button, scroll to the right, and in the **Pressure Hydraulics** tab check the **Use Pump Loads** box.

1) What is the head loss across the entire system?
2) Why is there more flow in FM-7 than FM-4?
3) Fill in Table 15.22.
4) If the minimum velocity required in the force main to keep particles from settling is 0.6 m/s, which areas are going to have problems?
5) What are some possible changes to the design to fix the problem portions of the system?
6) Implement one of the solutions you suggested in part 5. Describe the fix(es) and fill in Table 15.23.

Table 15.22 Answer Table for Part 3 of Problem 15.3

Pipe	Flow (m^3/s)	Velocity (m/s)	Head Loss (m)
FM-2			
FM-3			
FM-4			
FM-5			
FM-6			
FM-7			
FM-8			

Table 15.23 Answer Table for Part 6 of Problem 15.3

Pipe	Diameter (mm)	Flow (m^3/s)	Velocity (m/s)	Head Loss (m)
FM-2				
FM-3				
FM-4				
FM-5				
FM-6				
FM-7				
FM-8				

15.4 A major interceptor along a river collects laterals from subdivisions. The lower residential area loads are collected in a wet well and pumped to the major interceptor on the other side of a hill. The layout of the system is shown in Fig. 15.9. All pipes shown as double lines are circular, concrete, gravity-flow pipes ($n = 0.013$). The two pressure pipes (FM-1 and FM-2) are ductile iron (Hazen–Williams coefficient $C = 130$). There is an overflow diversion at MH-5. All input data is given in Tables 15.24–15.26.

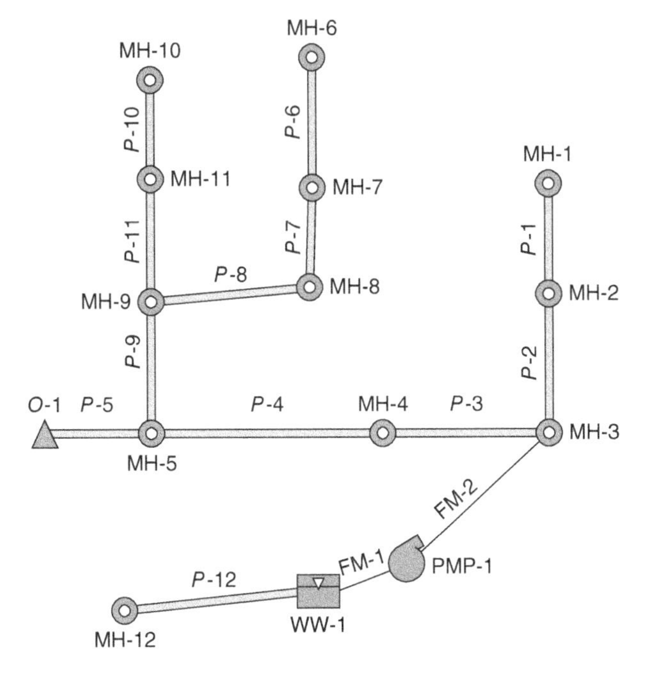

Figure 15.9 Schematic for Problem 15.4

Table 15.24 Pipe Data for Problem 15.4

Pipe	Section Size (in)	Upstream Invert Elevation (ft)	Downstream Invert Elevation (ft)	Length (ft)
P-1	18	146.50	141.50	400
P-2	18	141.50	136.00	400
P-3	24	136.00	119.00	400
P-4	36	119.00	114.00	400
P-5	36	114.00	94.00	400
P-6	12	157.00	146.50	400
P-7	18	146.50	136.50	400
P-8	18	136.50	126.00	400
P-9	24	126.00	114.00	400
P-10	12	147.00	136.50	400
P-11	18	136.50	126.00	400
P-12	18	74.50	45.00	400
FM-1	24	45.00	45.00	2
FM-2	24	45.00	136.00	60

Conversion factors: 1 in. = 25.4 mm; 1 ft = 0.3048 m.

Table 15.25 Node Data for Problem 15.4

Node	Ground Elevation (ft)	Sump Elevation (ft)
MH-1	150.00	146.50
MH-2	145.00	141.50
MH-3	140.00	136.00
MH-4	129.00	119.00
MH-5	124.00	114.00
MH-6	160.00	157.00
MH-7	150.00	146.50
MH-8	140.00	136.50
MH-9	130.00	126.00
MH-10	150.00	147.00
MH-11	140.00	136.50
MH-12	80.00	74.50
Outlet	100.00	94.00
PMP-1	45.00	N/A
WW-1	60.00	N/A

Conversion factor: 1 ft = 0.3048 m.

Table 15.26 Overflow Data for MH-5 in Problem 15.4

System Flow (gpm)	Diverted Flow (gpm)
0	0
20,000	0
30,000	10,000

Conversion factors: 1 gpm = 0.0631 L/s; 1 in. = 25.4 mm; 1 ft = 0.3048 m.

To determine the performance of the system, set up and run three scenarios:

1) A steady-state analysis of the average (base) sanitary loading only
2) An extended-period analysis of the sanitary loading only
3) An extended-period analysis of both the sanitary and wet-weather loading.

For the extended-period analyses, use a 24-hour duration with a 1-hour hydraulic time step and a 0.1-hour hydrologic time step.

Hint: For this problem, it is only necessary to create one Sanitary Loading Alternative that will contain the base and pattern loads. During a steady-state analysis, SewerCAD will ignore the time-based pattern. However, it will be necessary to create two infiltration and inflow loading alternatives (one without the wet-weather loads and the other with the wet-weather loads) because Scenario 2 should not consider the wet-weather loads. Define the wet-weather loading pattern in the manhole prototype before laying out the network. All manholes have a sanitary base load of 100 gpm and a continuous diurnal pattern applied to them as defined in Table 15.27. Each manhole also has wet-weather loading given by the hydrograph in Table 15.28.

The pump turns on when the elevation in the wet well rises to 57.0 ft, and shuts off when the elevation drops to 45.0 ft. The data for the wet well and pump are given in Tables 15.29 and 15.30

To verify important information such as minimum and maximum velocities, it will be helpful to keep a table similar to Table 15.31.

1) Is there surcharging or flooding in the system? Explain the difference between surcharging and flooding. Fill in Table 15.31 for each scenario.
2) Does the diversion divert any flow? If so, for which scenario?
3) If the pump has a best efficiency point at 4,000 gpm, what can you say about its performance for the dry-and wet-weather scenarios?
4) Plot profiles from MH-1 to the outlet for all three scenarios and compare the HGLs.

Table 15.27 Domestic Pattern Data for Problem 15.4

Time (h)	Multiplier
Starting	0.4
3	1.0
6	1.4
9	1.2
12	1.4
15	0.9
21	0.6
24	0.4

Table 15.28 Wet-Weather Loading Data for Problem 15.4

Time (h)	Flow (gpm)
0.00	0
3.00	0
4.00	100
5.00	300
6.00	900
7.00	1,500
8.00	1,800
9.00	1,600
10.00	1,000
11.00	600
12.00	300
13.00	100
14.00	0

Conversion factor: 1 gpm = 0.0631 L/s.

Table 15.29 Wet-Well Data for Problem 15.4

Wet Well	Ground Elevation (ft)	Base Elevation (ft)	Minimum Elevation (ft)	Initial Elevation (ft)	Alarm Elevation (ft)	Maximum Elevation (ft)	Wet-Well Diameter (ft)
WW-1	60.00	45.00	45.00	55.00	59.00	60.00	20.00

Conversion factor: 1 ft = 0.3048 m.

Table 15.30 Pump Data for Problem 15.4

Pump Curve Point	Head (ft)	Flow (gpm)
Shutoff	100	0
Design	80	4,000
Max Operating	40	8,000

Table 15.31 Answer Table for Part 1 of Problem 15.4

Variable	Steady State	EPS Sanitary Only	EPS Dry and Wet
Maximum Flow at Outlet (gpm)			
Time of Max. Flow at Outlet (h)	N/A		
Maximum Velocity in System at This Time (ft/s)			
Pipe with Max. Velocity at This Time			
Minimum Flow at Outlet (gpm)	N/A		
Time of Min. Flow at Outlet (h)	N/A		
Min. Velocity in System at This Time (ft/s)			
Pipe with Min. Velocity at This Time			
Maximum Diverted Flow (gpm)			
At What Hour?			

Conversion factors: 1 gpm = 0.0631 L/s; 1 ft = 0.3048 m; 1 ft/s = 0.3048 m/s.

15.5 Lay out the storm sewer system shown in Fig. 15.10 in StormCAD and enter the data for the network from Tables 15.32 to Table 15.36. Calculate the results using Manning's equation and the 10-year storm event. Inlet I-1 is on grade with a longitudinal slope of 3%, whereas inlet I-2 is in sag.

The tailwater condition at the outlet is free outfall. Assume that there is no clogging of the inlets.

Fill out an answer table similar to Table 15.37 for each of the following situations:

1) Assume uniform gutters with a slope of 0.02 m/m.
2) Assume continuously depressed gutters with a road cross-slope of 0.02 m/m, a gutter cross-slope of 0.04 m/m, and a gutter width of 0.8 m.
3) Assume continuously depressed gutters as described in part 2 along with inlet lengths that are increased from 1.1 to 1.6 m.
4) Explain the reasons for the differences between the three resulting tables.

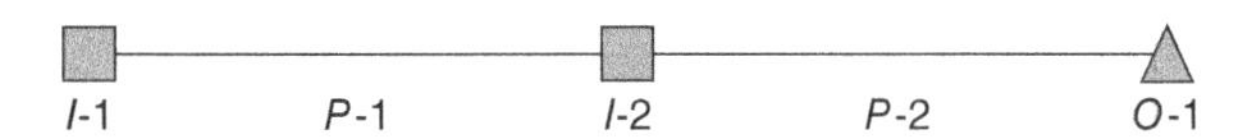

Figure 15.10 Schematic for Problem 15.5

Table 15.32 Pipe Data for Problem 15.5

Pipe	Circular Section Size (mm)	Upstream Invert Elevation (m)	Downstream Invert Elevation (m)	Length (m)	Pipe Material (*n*-value)
P-1	300	115.40	115.10	46	Concrete (0.013)
P-2	300	115.05	114.60	61	Concrete (0.013)

Table 15.33 Node Data for Problem 15.5

Nodes	Ground Elevation (m)	Sump Elevation (m)	Head Loss Method	Head Loss Coefficient
I-1	117.40	115.40	Standard	0.5
I-2	117.15	115.05	Standard	0.5
O-1	118.00	114.60	–	–

Table 15.34 Inlet Catchment Data for Problem 15.5

Inlet	Time of Concentration (min)	Additional Carryover (m^3/s)	Area (ha)	Inlet *C*
I-1	6.3	0	0.45	0.75
I-2	5.2	0	0.22	0.80

Table 15.35 Inlet Data for Problem 15.5

Inlet	Inlet Type	Grate Length (m)	Road Cross-Slope (m/m)	Bypass Target	Manning's *n*
I-1	Grate DI-1	1.1	0.02	I-2	0.012
I-2	Grate DI-1	1.1	0.02	–	–

Note: Info for inlet type grate DI-1 is provided in case it is missing from the inlet library. Structure width = 0.67 m; structure length = 0.67 m; grate type P - 50 mm × 100 mm; width = 0.76 m standard, length = 0.76 m.

Table 15.36 Rainfall Data for Problem 15.5

	Rainfall Intensities (mm/h)		
Duration (min)	5 Year	10 Year	50 Year
5	69.8	78.7	99.6
10	54.6	61.0	77.5
20	40.6	45.2	56.6
30	31.7	36.2	45.7

Table 15.37 Answer Table for Problem 15.5

Inlet	Gutter Spread (m)	Total Flow to Inlet (m^3/s)	Intercepted Inlet Flow (m^3/s)	Bypassed Inlet Flow (m^3/s)	Efficiency (%)
I-1					
I-2					

15.6 Enter and calculate the storm sewer network shown in Fig. 15.11 using Manning's equation and the rainfall data used in Problem 15.5. The tailwater condition is free outfall. Assume no clogging of the inlets. The data for the network are given in Tables 15.38–15.41. Answer the questions that follow.

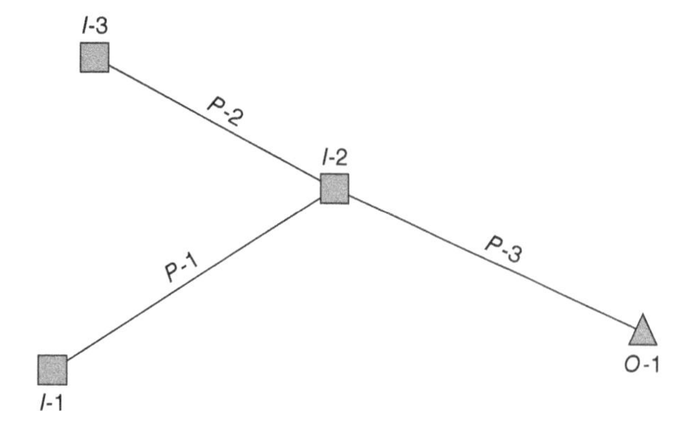

Figure 15.11 Schematic for Problem 15.6

Table 15.38 Pipe Data for Problem 15.6

Pipe	Circular Section Size (mm)	Upstream Invert Elevation (m)	Downstream Invert Elevation (m)	Length (m)	Pipe Material (*n*-value)
P-1	300	425.40	424.70	45	Concrete (0.013)
P-2	450	424.70	424.00	40	Concrete (0.013)
P-3	300	425.00	424.70	54	Concrete (0.013)

Table 15.39 Node Data for Problem 15.6

Node	Ground Elevation (m)	Sump Elevation (m)
I-1	428.00	425.40
I-2	427.50	424.70
I-3	428.60	425.00
O-1	427.10	424.00

Table 15.40 Inlet Data for Problem 15.6

Inlet	Inlet Type	Inlet Length (m)	Road Cross-Slope (m/m)	Inlet Location	Bypass Target	Long. Slope (m/m)	Manning's *n*
I-1	Curb DI-3A	1	0.02	On Grade	I-2	0.01	0.012
I-2	Curb DI-3A	1	0.02	In Sag	–	–	–
I-3	Curb DI-3A	1	0.02	In Sag	–	–	–

Table 15.41 Inlet Catchment Data for Problem 15.6

Inlet	Time of Concentration (min)	Additional Carryover (m^3/s)	Area (ha)	Inlet *C*
I-1	5.5	0	0.67	0.90
I-2	5.0	0	0.70	0.80
I-3	6.0	0	0.80	0.90

1) Given the design criteria shown in Table 13.3 (Chapter 13), what minimum gutter cross-slope should the 3-m-wide shoulders have (within 1%) on this high-volume, bidirectional road if the speed limit is 90 km/h?
2) As an alternative, set all gutter cross-slopes to 4% and replace the curb inlet at I-2 with a grate inlet called **Test**. You will need to add this new inlet to the program's inlet library. The inlet has the following characteristics: structure width and length 1.2 m, grate type P-50 mm, and grate width 1 m. What is the minimum grate length for I-2 that would satisfy the 50-year design criteria shown in Table 13.3, assuming no clogging?
3) Using the same data and inlet designed in part 2, calculate the spread at I-2 assuming 50% clogging.

15.7 The data shown in Tables 15.42–15.44 describe the existing storm sewer system shown in Fig. 15.12. For runoff calculations, assume $C = 0.3$ for pervious land cover and $C = 0.9$ for impervious cover. The ground elevation at the system discharge point is 17.0 m. All pipes are concrete ($n = 0.0\ 13$) and circular. Apply a standard head loss coefficient of 0.5 to inlet I-3. Assume all inlets are **Generic Default 100%**, which means that they are assumed to capture 100% of the surface flow.

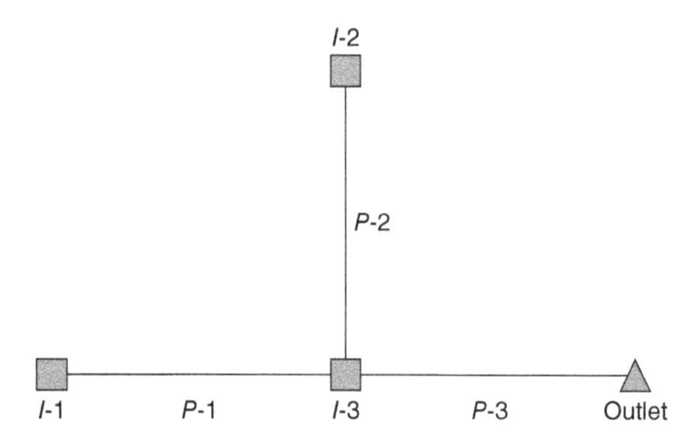

Figure 15.12 Schematic for Problem 15.7

Table 15.42 Rainfall Data for Problem 15.7

	Rainfall Intensity (mm/h)		
Duration (min)	5 Year	10 Year	25 Year
5	165	181	205
10	142	156	178
15	123	135	154
30	91	103	120
60	61	70	80

Table 15.43 Inlet Data for Problem 15.7

Inlet	Ground Elevation (m)	Impervious Area (ha)	Pervious Area (ha)	Time of Concentration (min)
I-1	17.9	0.13	0.32	6.0
I-2	18.0	0.15	0.58	5.0
I-3	17.6	0.08	0.36	5.0

Table 15.44 Pipe Data for Problem 15.7

Pipe	Upstream Invert (m)	Downstream Invert (m)	Diameter (mm)	Length (m)
P-1	16.7	16.15	300	56
P-2	16.8	16.1	375	46
P-3	16.1	15.3	375	54

1) Analyze the system for a design return period of 10 years. Assume a free outfall condition. Provide output tables summarizing pipe flow conditions and hydraulic grade lines at the inlets. How is this system performing?
2) Increase the size of pipe P-3 to 450-mm. Rerun the analysis and present the results. How does the system perform with this improvement?
3) Local design regulations require that storm sewer systems handle 25-year return periods without flooding. Rerun the analysis for the improved system in (2). Does the system meet this performance requirement?
4) The above analyses are run using a default Manning's n of 0.013. Many drainage design manuals propose a less conservative design roughness of 0.012. Reanalyze the improved system under 25-year flows using $n = 0.012$. How does this change influence the predicted performance of the system?

15.8 You have been asked by the lead project engineer for a water supply utility to design the stormwater collection system for the proposed ground storage tank and pump station facility shown in Fig. 15.13. Pipe lengths for P-1, P-2, P-3, and P-4 are 88, 92, 185, and 46 ft (26.82, 28.04, 56.39, and 14.02 m), respectively. The data for the system layout are given in Tables 15.45 and 15.46. Assume $C = 0.3$ for pervious areas and $C = 0.9$ for impervious areas. Assume all inlets are **Generic Default 100%**, which means that they are assumed to capture 100% of the surface flow.

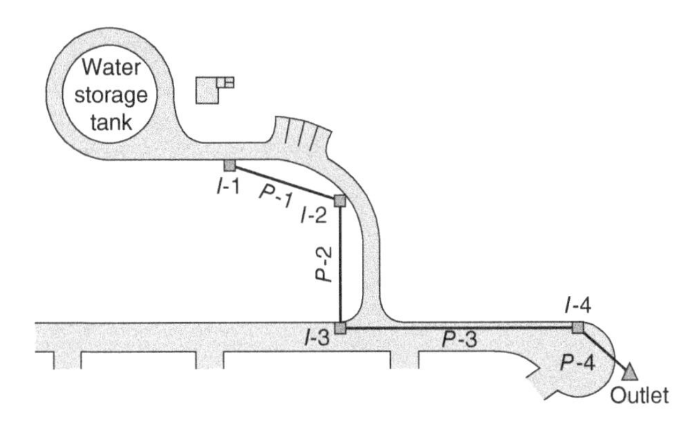

Figure 15.13 CAD System Layout for Problem 15.8

Table 15.45 Inlet Data for Problem 15.8

Inlet	Ground Elevation (ft)	Impervious Area (acre)	Pervious Area (acre)	Time of Concentration (min)
I-1	848.9	0.25	0.25	6.0
I-2	848.3	0.09	–	5.0
I-3	847.0	0.20	0.49	5.0
I-4	846.5	0.18	0.5	7.5

Conversion factors: 1 ft = 0.3048 m; 1 acre = 0.4046 ha.

Table 15.46 Design Constraints for Problem 15.8

Constraint	Minimum	Maximum
Velocity	2 ft/s	15 ft/s
Cover	4 ft[a]	–
Pipe Slope	0.005 ft/ft	0.100 ft/ft

[a] Cover constraints may be relaxed at the outfall, where invert elevation requirements will typically govern.

Conversion factors: 1 ft/ft = 1 m/m; 1 ft = 0.3048 m; 1 ft/s = 0.3048 m/s.

1) Using the StormCAD program's Automatic Design feature, size the system using the following design data. Use concrete pipe ($n = 0.013$) and the 25-year intensity-duration-frequency data provided in Problem 15.7 (*Hint:* StormCAD can mix SI and U.S. customary units.) The top of bank elevation at the outfall ditch is 846.1 ft (257.89 m). The outfall pipe invert must be located at or above elevation 838.0 ft (255.42 m). Assume that the water surface elevation at the outfall is 842.0 ft (256.64 m) and that the pipes should have matching soffit (crown) elevations at every structure. Present your design in tabular form and provide a profile plot of your design.
2) During agency review, the county engineer requests that the water utility and the county work cooperatively to accommodate the planned construction of an elementary school nearby by increasing the size of the proposed storm system so that it can handle the design runoff from the school. The county engineer performs his own calculations and asks that you increase the size of pipes P-3 and P-4 to handle an external CA of 9.5 acres (3.84 ha) with a time of concentration of 12 minutes. Using StormCAD, introduce the additional flow at inlet I-3 and revise the facility design using the Automatic Design functionality of the program. Are all the design constraints met? What can you say about the flow conditions in pipe P-3 and pipe P-4? In the rest of the pipes?
3) If necessary, manually fine-tune and revise the design to meet all design criteria. Document your design as in part 1.

15.9 An inlet with a ground elevation of 260 m is connected to an outlet with a ground elevation of 259 m and a tailwater elevation of 256 m. The pipe connecting the inlet and outlet is a 20-m-long, concrete, circular pipe with a diameter of 525 mm and an upstream invert/sump elevation of 257 m. Assume all inlets are **Generic Default 100%**, which means that they are assumed to capture 100% of the surface flow. Print and compare pipe slope, average velocity, pipe flow time, capacity, energy slope, hydraulic grade line upstream, and flow profiles for each of the following conditions:

Scenario 1:
- Inlet inflow = 0.3 m^3/s
- Downstream invert/sump elevation = 256.9 m

Scenario 2:
- Inlet inflow = 0.5 m^3/s
- Downstream invert/sump elevation = 256.9 m

Scenario 3:
- Inlet inflow = 0.5 m^3/s
- Downstream invert/sump elevation = 256.7 m

Scenario 4:
- Inlet inflow = 0.5 m^3/s
- Downstream invert/sump elevation = 256.2 m

What are the differences between the four scenarios, and why do they occur?

15.10 The system described in Fig. 15.14 drains into a river. Because the water surface elevation of the river varies, analyze the following system with different tailwater elevations. Use the Darcy-Weisbach equation because it is more appropriate for pressure flow than Manning's equation. All pipes are concrete (roughness 3 0.122 mm), circular pipes. The rainfall, pipe, and inlet data are shown in Tables 15.47–15.49. The ground elevation at the outlet is 312.0 m. Assume all inlets are **Generic Default 100%**, which means that they are assumed to capture 100% of the surface flow.

What is the hydraulic grade at the entrance of inlet I-2 under the following tailwater conditions for a 10-year storm event?

1) Tailwater elevation = 310.0 m
2) Tailwater elevation = 311.0 m
3) Tailwater elevation = 313.0 m.

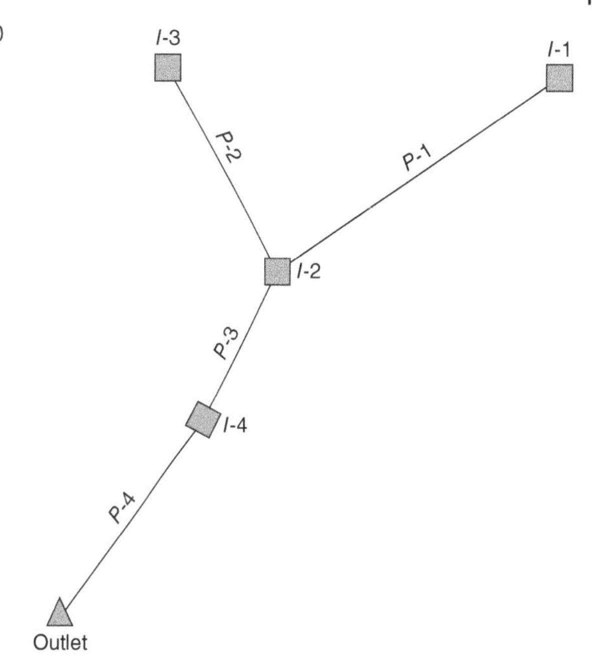

Figure 15.14 Schematic for Problem 15.10

Table 15.47 Rainfall Data for Problem 15.10

	Rainfall Intensity (mm/h)		
Duration (min)	**2 Year**	**10 Year**	**100 Year**
5	80	130	210
15	65	97	145
30	45	68	100
60	25	37.5	55

Table 15.48 Pipe Data for Problem 15.10

Pipe	Length (m)	Diameter (mm)	Upstream Invert Elevation (m)	Downstream Invert Elevation (m)
P-1	23	300	312.4	311.5
P-2	25	300	311.5	311.0
P-3	32	525	311.0	310.5
P-4	17	525	310.5	310.4

Table 15.49 Inlet Data for Problem 15.10

Inlet	$Area_1$ (ha)	C_1	$Area_2$ (ha)	C_2	t_c (min)	Ground Elev. (m)	Head Loss Coefficient (Standard Method)	Sump Elev. (m)
I-1	0.2	0.7	0.35	0.6	5.0	315.0	0.5	312.4
I-2	0.3	0.8	0.18	0.6	5.0	314.5	0.7	311.0
I-3	0.8	0.4	0.3	0.7	6.0	314.3	0.5	311.5
I-4	0.9	0.4	0.3	0.5	8.0	313.0	0.5	310.5

15.11 A schematic of an existing storm system for a residential subdivision is shown in Fig. 15.15. The rainfall, pipe, inlet, and hydrologic information data are provided in Tables 15.50–15.53. All pipes are circular, PVC pipes with Manning's $n = 0.010$. Assume all inlets are **Generic Default 100%**, which means that they are assumed to capture 100% of the surface flow.

1) Analyze the system for a 25-year storm event. Assume a free outfall (critical depth) condition. Provide output tables summarizing pipe flow, velocity, and hydraulic grade at the upstream end of each pipe. How is this system performing?
2) If the existing system were constructed using *corrugated metal pipe* (CMP; $n = 0.024$) instead of PVC, how would the system perform in a 25-year storm event?
3) the outfall were discharging to a pond that had a water surface elevation equal to that of the ground (using the original PVC pipes); how would the system perform in a 25-year storm? A 50-year storm?
4) Using the tailwater conditions specified in part 1, if the park were paved (with a $C = 0.9$ and time of concentration = 8 min), how would the system perform in a 25-year storm? In a 50-year storm?

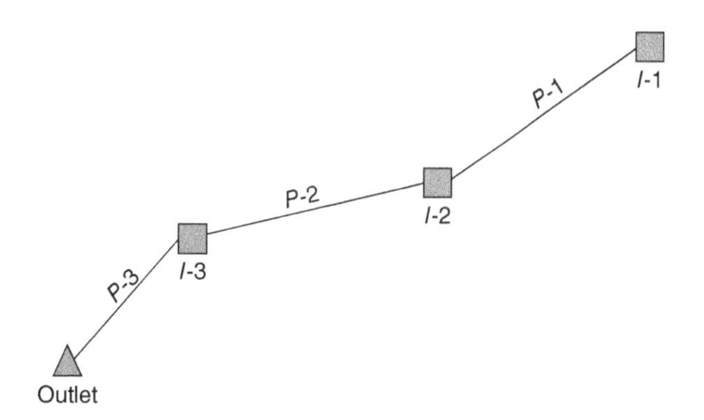

Figure 15.15 Schematic for Problem 15.11

Table 15.50 Rainfall Data for Problem 15.11

	Rainfall Intensity (in/h)	
Duration (min)	25 Year	50 Year
5	4.00	4.60
15	2.80	3.20
30	1.80	2.00
60	0.75	0.90
100	0.55	0.65

Conversion factor: 1 in. /h = 25.4 mm/h.

Table 15.51 Pipe Data for Problem 15.11

Pipe	Length (ft)	Diameter (in.)	Upstream Invert Elevation (ft)	Downstream Invert Elevation (ft)
Pipe-1	70	18	166.70	165.30
Pipe-2	40	18	165.20	164.80
Pipe-3	53	24	164.80	164.40

Conversion factors: 1 in. = 25.4 mm; 1 ft = 0.3048 m; 1 ft/s = 0.3048 m/s.

Table 15.52 Inlet Data for Problem 15.11

Inlet	Ground Elevation (ft)	Rim Elevation (ft)	Sum Elevation (ft)	Head Loss Coefficient (Standard Method)
1	172.00	172.00	166.50	0.5
2	172.00	172.00	165.00	0.5
3	168.00	168.00	164.50	0.5
Outfall	166.50	166.50	164.40	–

Conversion factor: 1 ft = 0.3048 m.

Table 15.53 Inlet Hydrologic Data for Problem 15.11

Inlet	Area (acre)	Area Type	Description	C	(min)
	8	Residential	Single Family	0.5	
1	1	Pavement	Asphalt/Concrete	0.7	12
	0.04	Misc.	Playground	0.2	
2	1.2	Pavement	Asphalt/Concrete	0.7	5
3	10	Misc.	Park	0.2	15

Conversion factor: 1 acre = 0.4046 ha.

15.12 Use the same network and data provided in part 1 of Problem 15.11. During a 25-year storm, it was determined that Inlet 1 is not capable of capturing all of the flow produced by the runoff for its catch basin. A flow of 5 ft^3/s (141.6 L/s) is bypassed to inlet 2, which is in sag (all gutter flow is captured). Assume all inlets are **Generic Default 100%**, which means that they are assumed to capture 100% of the surface flow. (*Hint:* You can apply negative and positive additional carryovers to Inlets 1 and 2 under the **Catchment** tab.)

1) What is the gutter flow captured by Inlet 1?
2) What is the gutter flow captured by Inlet 2?
3) How is the system performing?
4) A flow monitoring study was performed and data were collected (see Table 15.54):

Table 15.54 Collected Data from the Flow Monitoring Study of Problem 15.12

Pipe	25-Year Storm (ft^3/s)	50-Year Storm (ft^3/s)
Pipe-1	10	11.5
Pipe-2	17.3	19.9
Pipe-3	21.0	25.0

Conversion factor: 1 ft^3/s = 28.32 L/s.

Analyze the system using these known flow values for both the 25- and 50-year storms. How is the system performing? Report the flows, upstream velocities, and hydraulic grade lines for each pipe.

Bibliography

American Society of Civil Engineers and Water Environment Federation, Gravity Sanitary Sewer Design and Construction, 2nd ed., ASCE Manual No. 60 and WEF Manual No ed., FD-5, Reston, VA, 2007.

Brown, S. A., Stein, S. M., and Warner, J. C., *Urban Drainage Design Manual*, 2nd ed., Hydraulic Engineering Circular No. 22 ed., U.S. Department of Transportation, Federal Highway Administration, Washington, DC, 2001.

Chanson, H., *The Hydraulics of Open Channel Flow*, 2nd ed., Elsevier Butterworth-Heinemann, Burlington, MA, 2004.

Chaudhry, M. H., *Open-Channel Flow*, 2nd ed., Springer-Verlag, New York, 2007.

Engineering News-Record, *Construction Economics*, ENR, New York. https://www.enr.com/ext/resources/static_pages/Economics/2022/0110_CE_WK1_combined.pdf, 2022.

Flynn, P. J., *Flow of Water in Open Channels, Pipes, Sewers, Conduits, etc.*, Kessinger Publishing, Whitefish, MT, 2007.

Haestad Methods, *Computer Applications in Hydraulic Engineering*, 7th ed., Bentley Institute Press, Exton, PA, 2007.

Haestad Methods, *SewerCAD: Sanitary Sewer Modeling Software*, Waterbury, CT, 2001.

Haestad Methods, *StormCAD: Storm Sewer Design and Analysis Software*, Waterbury, CT, 2000.

Haestad Methods, *Stormwater Conveyance Modeling and Design*, Bentley Institute Press, Exton, PA, 2007.

Haestad Methods, *Wastewater Collection System Modeling and Design*, Bentley Institute Press, Exton, PA, 2007.

Mays, L. W., *Stormwater Collection Systems Design Handbook*, McGraw-Hill, New York, 2007.

U.S. Army Corps of Engineers, *Yearly Average Cost Index for Utilities, Civil Works Construction Cost Index System Manual, 110-2-1304*, U.S. Army Corps of Engineers, Washington, DC, 2022.

16

Sewer Material, Appurtenances, and Maintenance

16.1 Gravity Sewer Pipe Material

Several different pipe materials are available for wastewater collection systems, each with a unique characteristic that is applicable to different conditions. Pipes have life cycles ranging from 15 to more than 100 years depending on the type of material and the environment. Looking at pipe, the material used can be a greater indicator of failure than age. The four commonly used pipe materials are ductile-iron, concrete, plastic, and vitrified clay.

Pipe material selection considerations include trench conditions (geologic conditions), corrosion, temperature, safety requirements, and cost. Key pipe characteristics are corrosion resistance (interior and exterior), the scouring factor, leak tightness, and hydraulic characteristics.

Pipe manufacturers follow requirements set by the American Society of Testing Materials (ASTM) or American Water Works Association (AWWA) for specific pipe materials. Specification standards cover the manufacture of pipes and specify parameters such as internal diameter, loadings (classes), and wall thickness (schedule). The methods of pipe construction vary greatly with the pipe materials.

Some new pipe materials and construction methods use the basic materials of concrete pipes with modifications (i.e., coatings). Other pipe manufacturing methods use newly developed resins that offer improvements in strength, flexibility, and resistance to certain chemicals. Construction methods may also allow for field modifications to adapt to unique conditions (river crossings, rocky trenches, etc.) or may allow for special, custom-ordered diameters and lengths.

16.1.1 Ductile-Iron Pipe

Ductile-iron pipe (DIP) is an outgrowth of the cast-iron pipe industry. Improvements in the metallurgy of cast iron in the 1940s increased the strength of cast-iron pipe and added ductility, an ability to slightly deform without cracking. This was a major advantage and ductile-iron pipe quickly became the standard pipe material for high-pressure service for various uses (water, gas, etc.).

16.1.2 Concrete Pipe

Concrete represents an important sewer pipe material, particularly for large trunk sewers. Concrete can be fortified against acid attack by using calcareous aggregate, increasing the cement content, or both, which provides additional alkalinity and acid-neutralizing capacity. PVC liners can be employed where severely corrosive conditions are anticipated.

Two types of concrete pipe commonly used today are *prestressed concrete cylinder pipe* (PCCP) and *reinforced concrete pipe* (RCP). PCCP is used for force mains, while RCP (Fig. 16.1) is used primarily for gravity lines. PCCP may be of either *embedded-cylinder* (EC) or *lined-cylinder* (LC) construction. The process for both LC and EC construction begins with the casting of a concrete core in a steel cylinder. This single process produces the LC pipe. Once the cylinder cures, it is wrapped with a prestressed steel wire and coated with cement slurry and a dense mortar or concrete coating to produce the EC pipe. The manufacturing process for reinforced concrete cylinder pipe (RCCP) is similar to that for EC pipe; however, a reinforcing cage and the steel cylinder are positioned within a reusable vertical form and the concrete is cast instead of using the prestressed wire. RCCP can be cured by using either water or steam.

Water and Wastewater Engineering: Hydraulics, Hydrology and Management, Volume 1, Fourth Edition.
Lawrence K. Wang, Mu-Hao Sung Wang, and Nazih K. Shammas.

Companion website: www.wiley.com/go/Wang/Waterandwastewater4e

Figure 16.1 Concrete Sewer Pipe. (*Source:* Wikipedia, http://upload.wikimedia.org/wikipedia/en/9/98/WaterPipeLarge.JPG.)

16.1.3 Plastic Pipe

Plastic pipe is made from either *thermoplastic* or *thermoset plastics*. Characteristics and construction vary, but new materials offer high strength and good rigidity. *Fluorocarbon plastics* are the most resistant to attack from acids, alkalies, and organic compounds, but other plastics also have high chemical resistance. Plastic pipe design must include stiffness, loading, and hydrostatic design stress requirements for pressure piping.

Thermoplastics are plastic materials that change shape when they are heated. Common plastics used in pipe manufacturing include polyvinyl chloride (PVC), polyethylene (PE or HDPE for high-density PE), acrylonitrile-butadiene-styrene (ABS), and polybutylene (PB). HDPE is commonly used with pipe bursting. PVC is strong, lightweight, and somewhat flexible. PVC pipe is the most widely used plastic pipe material. Other plastic pipes or composites with plastics and other materials may be more rigid.

Thermoset plastics are rigid after they have been manufactured and are not able to be re-formed. Thermoset plastic pipes are composed of epoxy, polyester, and phenolic resins, and are usually reinforced with fiberglass. Resins may contain fillers to extend the resin and to provide specific characteristics to the final material. The glass fibers may be wound around the pipes spirally, in woven configurations, or they may be incorporated into the resin material as short strands. The pipes may be centrifugally cast. Stiffness may also be added in construction as external ribs or windings. *Reinforced plastic mortar* (RPM) and *reinforced thermosetting resin* (RTR) are the two basic classes of these pipes. Another name is *glass-reinforced plastic* (GRP), a composite material made of plastic reinforced by fine fibers made of glass. Thermoset pipes are often manufactured according to the specific buyer requirements and may include liners of different composition for specific chemical uses.

For plastic pipes, resins composed of polymerized molecules are mixed with lubricants, stabilizers, fillers, and pigments, to produce mixtures with different characteristics. Plastic pipes are generally produced by extrusion. Plastic pipe may be used for sliplining or for rehabilitating existing pipes by inserting or pulling them through a smaller diameter pipe. HDPE pipes may also be used for bursting and upgrading. The smaller diameter pipe may be anchored into place with mortar or grout.

16.1.4 Vitrified Clay Pipe

Vitrified clay pipes (VCPs) are composed of crushed and blended clay that are formed into pipes, then dried and fired in a succession of temperatures. The final firing gives the pipes a glassy finish. VCPs have been used for hundreds of years and are strong and resistant to chemical corrosion, internal abrasion, and external chemical attack. They are also heat resistant. It has been shown that the thermal expansion of VCPs is less than that of many other types (such as DIP and PVC). In older VCP lines, cement mortar joints expanded and sometimes broke the bells. Other types of gasketed joints now in general use avoid this problem. However, VCP is brittle and, hence, requires special installation practice and care in handling and transport.

16.2 Trenchless Rehabilitation Pipe Material

Three types of material stand out in terms of their excellent all-around characteristics for use in trenchless rehabilitation:

1) *Glass-reinforced plastic pipes* have a high strength-to-weight ratio and a long design life (more than 100 years). However, they can be used in only slightly curving pipes. Commercially available diameters are 12–96 in. (300–2,500 mm).
2) *Polyolefin pipes* range in diameter from 6 to 24 in. (150–600 mm), have excellent corrosion resistance, and have a design life of 50 years. These pipes can be welded together at the surface in a continuous length before insertion, or they can be threaded together in small sections to allow insertion from smaller access pits. Polyolefin pipes include polyethylene, polypropylene, and polybutylene.
3) *Ductile-iron pipes* are very strong and reliable. They can overcome friction and obstructions more easily than plastic pipes because they can withstand the high stresses applied by rams and winches. However, ductile-iron is more susceptible to corrosion than plastic.

16.3 Force Main Pipe Material

Ductile-iron and PVC are the most frequently used materials for wastewater force mains. Ductile-iron pipe has particular advantages in wastewater collection systems due to its high strength and high flow capacity with greater than nominal inside diameters and tight joints. For special corrosive conditions and extremely high flow characteristics, polyethylene-lined ductile-iron pipe and fittings are widely used.

Cast-iron pipe with a glass lining is available in standard pipe sizes, with most joints in lengths up to 20 ft (6 m). Corrosion-resistant, plastic-lined piping systems are used for certain waste-carrying applications. Polyethylene-lined ductile-iron pipe and fittings known as "poly-bond-lined" pipe are widely used for force mains conveying highly corrosive industrial or municipal wastewater.

The types of thermoplastic pipe materials used for force main service are PVC, ABS, and PE. The corrosion resistance, light weight, and low hydraulic friction characteristics of these materials offer certain advantages for different force main applications, including resistance to microbial attack. Typically, PVC pipes are available in standard diameters of 4–36 in. (100–900 mm), and their laying lengths normally range from 10 to 20 ft (3–6 m). The use of composite material pipes, such as fiberglass-reinforced mortar pipe ("truss pipe"), is increasing in the construction of force mains. A truss pipe is constructed on concentric ABS cylinders with annular space filled with cement. Pipe fabricated of fiberglass-reinforced epoxy resin is almost as strong as steel, as well as corrosion and abrasion resistant.

Certain types of asbestos-cement pipe are applicable in construction of wastewater force mains. The advantage of asbestos-cement pipes in sewer applications is their low hydraulic friction. These pipes are relatively lightweight, allowing long laying lengths in long lines. Asbestos-cement pipes are also highly corrosion resistant. At one time, it was thought that many asbestos-containing products (including asbestos-cement pipe) would be banned by the U.S. Environmental Protection Agency. However, a court ruling overturned this ban and this pipe is available and still used for wastewater force main applications.

16.4 Application of Pipe Material

The applicability of different pipe materials varies with each site and the system requirements. The pipe material must be compatible with the soil and groundwater chemistry. The pipe material also must be compatible with the soil structure and topography of the site, which affects the pipe location and depth, the supports necessary for the pipe fill material, and the required strength of the pipe material. The following list shows background information to be used in determining what type of pipe best fits a particular situation:

1) Maximum pressure conditions (force mains)
2) Overburden, dynamic, and static loading
3) Lengths of pipe available
4) Soil conditions, soil chemistry, water table, and stability

5) Joining materials required
6) Installation equipment required
7) Chemical and physical properties of the wastewater
8) Joint tightness/thrust control
9) Size range requirements
10) Field and shop fabrication considerations
11) Compatibility with existing systems
12) Manholes, pits, sumps, and other required structures to be included
13) Valves (number, size, and cost)
14) Corrosion/cathodic protection requirements
15) Maintenance requirements.

16.4.1 Advantages and Disadvantages

The advantages and disadvantages of specific pipe materials are listed in Table 16.1. The primary advantages and disadvantages to consider for pipes used in sewer applications include those that are related to construction requirements, pressure requirements (force mains), depth of cover, and cost.

Table 16.1 Advantages and Disadvantages of Pipe Materials

Advantages	Disadvantages
Ductile-Iron	
Good corrosion resistance when coated	Heavy
High strength	
Concrete	
Good corrosion resistance	Requires careful installation to avoid cracking
Widespread availability	Heavy
High strength	Susceptible to attack by H_2S and acids when pipes are not coated
Good load-supporting capacity	
Vitrified Clay	
Very resistant to acids and most chemicals	Joints are susceptible to chemical attack
Strong	Brittle (may crack); requires careful installation
	Short length and numerous joints make it prone to infiltration and more costly to install
Thermoplastics (PVC, PE, HDPE, ABS)	
Very lightweight	Susceptible to chemical attack, particularly by solvents
Easy to install	Strength affected by sunlight unless UV protected
Economical	Requires special bedding
Good corrosion resistance	
Smooth surface reduces friction losses	
Long pipe sections reduce infiltration potential	
Flexible	
Thermosets (FRP)	
High strength	High material cost
Lightweight	Brittle (may crack); requires careful installation
Corrosion resistant	High installation cost

16.4.2 Design Criteria

Design requirements may vary greatly. Pipe design is approached differently for both materials and construction methods. The mechanics of the soil that will surround the pipes is a fundamental design aspect for the support characteristics, especially for flexible pipes. The soil type, density, and the moisture content are important characteristics.

16.4.3 Costs of Gravity Sewer

Costs for piping comparisons should include both the costs of the materials as well as the construction costs. The pipe cost is usually given in dollars per unit length, traditionally in USD/linear ft (or USD/linear m), plus the costs of the fittings, connections, and joints.

Construction costs will depend on the type of digging necessary, special field equipment requirements, and an allowance for in-field adjustments to the system. Access to pipe systems will also be a relevant cost factor, because manhole spacing is dependent on pipe size.

Sanitary sewer construction costs depend on several variables, including depth, type of soil, presence of rock, type of bedding material, location (rural versus urban areas) clearing costs, and other factors.

Table 16.2 summarizes unit costs for PVC pipe, trenching, bedding, and backfill. Typical current costs for construction of sanitary sewers (excluding service connections and manholes) are provided in Table 16.3. All costs are given in 2022 USD using U.S. Army Corps of Engineers' *Cost Index for Utilities*.

Table 16.2 Unit Costs for Sanitary Sewers in 2022 USD

Item	Description	Cost (USD/unit)
PVC pipe (not including excavation and backfill)	8-in. diameter	6.58/linear ft
	10-in. diameter	10.20/linear ft
	15-in. diameter	20.72/linear ft
	Brick, 4 ft in diameter, 4 ft deep	1,240 each
Catch basins or manholes (including footing and excavation; not including frame or cover)	Concrete, 4 ft × 4 ft, 8 in. thick, 4 ft deep	1,116 each
Trenching	4 ft wide, 6 ft deep, 1/2 yd^3 bucket	31.56/linear ft
Pipe bedding	Side slope 0–1, 4 ft wide	5.93/linear ft
Backfill	Spread dump material by dozer, no compaction	2.15/yd^3

Conversion factors: 1 in. = 25.4 mm; 1 ft = 0.3048 m; 1 USD/linear ft = 3.28 USD/linear m; 1 yd^3 = 0.76456 m^3.

Table 16.3 Construction Cost of Pipe in 2022 USD/Linear ft

	Pipe Diameter (in.)							
Pipe Material	**2**	**4**	**6**	**8**	**12**	**15**	**18**	**24**
VCP	—	—	37	43	56	74	96	161
DIP	—	—	—	56	73	—	110	161
RCP	—	—	—	—	16	25	34	46
PVC	22	27	34	37	49	56	74	109
PE	—	10	17	21	14[a]	—	24[a]	—
FRP	31	43	62	87	—	—	—	—
ABS	16	—	—	—	—	—	—	—

Conversion factors: 1 in. = 25.4 mm; 1 USD/linear ft = 3.28 USD/linear m.
[a] Corrugated.

The cost per linear foot in Table 16.3 is based on an average trench depth of 8 ft (2.5 m) and excludes service connections and manholes. The following is not included in the cost per linear foot (or cost per linear meter):

1) Asphalt and gravel driveway repair
2) Boring and jacking
3) Concrete encasement of pipe at stream crossings or other locations
4) Erosion control
5) Relocation of other utilities

Soil material is assumed to be silt, clay, or other soil mixtures with no requirement for shoring, rock removal, or dewatering.

The capital costs for gravity sewers are given in Fig. 16.2. Capital costs include the following:

1) Construction costs for in-place sewer pipe
2) Appurtenances and nonpipe costs
3) Nonconstruction costs.

Construction costs for in-place pipe include material and labor. Appurtenances and nonpipe costs include manholes, thoroughfare crossings, pavement removal and replacement, rock excavation (minimal), special pipe bedding, and miscellaneous appurtenances. Nonconstruction costs include administrative/legal costs; land, structures, and right-of-way costs; architect/engineer fees; bond interest; and contingency, indirect, and miscellaneous costs.

Typical capital costs in Fig. 16.2 do not include costs for special or site-specific requirements such as extensive rock excavation, dewatering, or shoring.

Recent 2022 costs are obtained by multiplying the costs from Fig. 16.2 by a multiplier of 3.39. Future costs can be updated using the future USACE Construction Cost Index (Appendix 16 is an example), or the future Engineering News-Record (ENR) Cost Index.

16.4.4 Costs of Force Main

Force main costs depend on many factors:

1) Conveyed wastewater quantity and quality
2) Force main length
3) Operating pressure
4) Soil properties and underground conditions

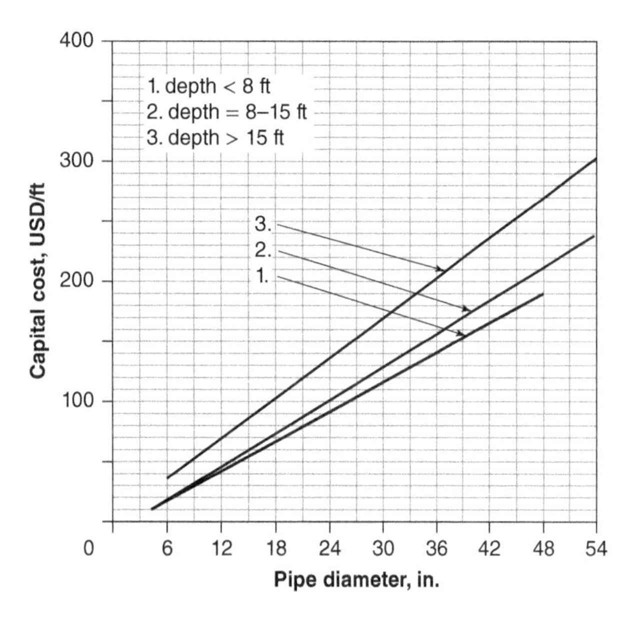

Figure 16.2 Capital Costs for Sewers. Recent March 2022 costs can be obtained by multiplying the costs from Fig. 16.2 by a multiplier of 3.39. Conversion factors: 1 in. = 25.4 mm; 1 ft = 0.3048 m; 1 USD/ft = 3.28 USD/m

Table 16.4 Construction Costs for Ductile-Iron and Plastic Force Mains in 2022 USD

Pipe Diameter (in.)	Ductile-Iron Pipe (USD/Linear ft)	PVC Pressure Pipe (USD/Linear ft)
8	35	22
10	43	30
12	53	38
14	68	50
16	79	61
18	98	72
20	107	83
24	124	97
30	211	134
36	283	201

Conversion factors: 1 in. = 25.4 mm; 1 USD/linear ft = 3.28 USD/linear m.

5) Pipeline trench depth
6) Appurtenances such as valves and blowoffs
7) Community impacts.

These site- and system-specific factors must be examined and incorporated in the preparation of force main cost estimates.

Unit force main construction costs are usually expressed in USD/linear ft (or cost/linear m) of installed pipeline, and costs typically include labor and the equipment and materials required for pipeline installation. Table 16.4 lists unit pipeline construction costs for ductile-iron and plastic (PVC) pipes used for force main construction. All unit pipeline costs are adjusted to 2022 USD. These costs are base installation costs and do not include the following:

1) General contractor overhead and profit
2) Engineering and construction management
3) Land or right-of-way acquisition
4) Legal, fiscal, and administrative costs
5) Interest during construction
6) Community impacts.

Force main operation and maintenance costs include labor and maintenance requirements. Typically, labor costs account for 85–95% of total operation and maintenance costs and are dependent on the force main length. The maintenance costs usually vary from 3.1 to 8.7 USD/linear ft (10.2–28.5 USD/linear m), depending on the size and number of appurtenances installed on the force main. An internal inspection using TV equipment can be completed, if visual inspection is not sufficient. TV inspection can be costly, ranging from 1,490 to 17,000 USD/mile (925–10,570 USD/km) with an average cost of 6,820 USD/mile (4,241 USD/km).

16.5 Manholes

Manholes (or maintenance holes) are underground vaults that provide an access point for making connections or performing maintenance on underground and buried sewers and storm drains. Manholes (MHs) are usually outfitted with metal or polypropylene steps installed in the inner side of the wall to allow easy descent into the manhole.

Manholes are required at junctions of gravity sewers and at each change in pipe direction, size, or slope. The distance between manholes must not exceed 400 ft (123 m) in sewers of less than 18 in. (450 mm) in diameter. For sewers 18 in.

(450 mm) in diameter and larger, and for outfalls from wastewater treatment facilities, a spacing of up to 600 ft (183 m) is allowed provided the velocity is sufficient to prevent sedimentation of solids.

The invert of the outlet pipe from a manhole will be on line with or below the invert of the inlet pipe. When the outlet pipe from a manhole is larger than the largest inlet pipe, the crown of the outlet pipe is to be no higher than the lowest inlet pipe crown. Where the invert of the inlet pipe would be more than 24 in. (600 mm) above the manhole floor, a drop connection will be provided. Typical manholes are shown in Figs. 16.3 and 16.4.

The primary construction materials to be used for manhole structures are precast concrete sections; prefabricated GRP units; cast-in-place, reinforced, or unreinforced concrete; and, if necessary, brick masonry. In the past, most manholes were built of brick masonry, which are now frequently the source of significant volumes of groundwater infiltration. More recently, in attempts to alleviate this problem, precast concrete and fiberglass manholes have been utilized. In certain

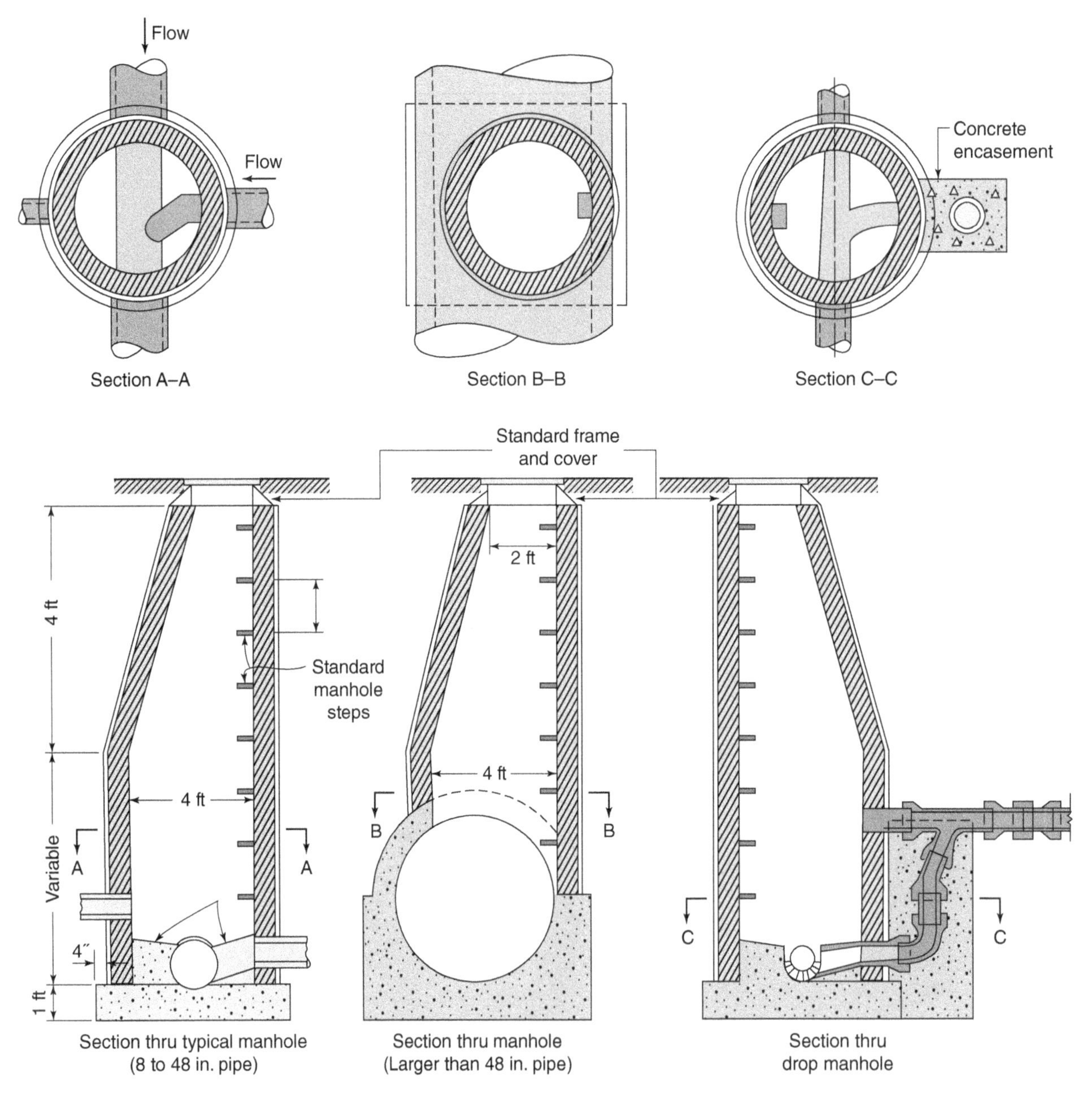

Figure 16.3 Sections through Manholes. Conversion factors: 1 in. = 25.4 mm; 1 ft = 0.3048 m

Figure 16.4 Concrete Sewer Manhole. (*Source:* Wikipedia, http://en.wikipedia.org/wiki/Image:Manhole.jpg)

Figure 16.5 Manhole Cover. (*Source:* Wikipedia, http://en.wikipedia.org/wiki/Manhole_cover) and ponding. Frames and covers must be sufficient to withstand impact from wheel loads where subject to vehicular traffic

situations, precast units will not be suitable, and cast-in-place reinforced concrete will be required. Cast-in-place construction permits greater flexibility in the configuration of elements and, by varying the reinforcing, the strength of similar-sized structures can be adjusted to meet requirements.

Smooth flow channels are formed in the manhole bottom. Laying half tile through the manhole or laying full pipe with the top of the pipe being broken out later are acceptable alternatives.

Manholes are protected by manhole covers designed to prevent accidental or unauthorized access to manholes. A manhole cover is a removable plate forming the lid over the opening of a manhole (Fig. 16.5). Manhole covers usually weigh more than 100 lb. (roughly 45 kg), partly because the weight keeps them in place when traffic passes over them, and partly because they are often made out of cast iron, sometimes with infills of concrete. This makes them inexpensive and strong but heavy. A manhole cover sits on a metal base, with a smaller inset rim that fits the cover.

The covers usually feature "pick holes," in which a hook handle is inserted to lift them. Pick holes can be concealed for a more watertight lid or can allow light to shine through. A manhole pick or hook is typically used to lift them, though other tools can be used as well. Manhole top elevations are set to avoid submergence of the cover by surface-water runoff.

16.6 Building Connections

Building connections are designed to eliminate as many bends as practical and provide convenience in rodding. Generally, connections to other sewers will be made directly to the pipe with standard fittings rather than through manholes. However, a manhole must be used if the connection is more than 100 ft (30 m) from the building cleanout.

16.7 Cleanouts

Cleanouts must be installed on all sewer building connections to provide a means for inserting cleaning rods into the underground pipe. An acceptable cleanout consists of an upturned pipe terminating at, or slightly above, final grade with a plug or cap. Preferably, the cleanout pipe will be of the same diameter as the building sewer, and never smaller than 6 in. (150 mm).

16.8 Lift Stations

Wastewater lift stations are facilities designed to move wastewater from lower to higher elevation through pipes. Key elements of lift stations include a wastewater receiving well (wet well), often equipped with a screen or grinder to remove coarse materials, pumps and piping with associated valves, motors, a power supply system, an equipment control and alarm system, and an odor control system and ventilation system.

Pumps and pumping stations serve the following purposes in wastewater systems:

1) Lifting wastewater from low-lying basements or low-lying secondary drainage areas into the main drainage system, and from uneconomically deep runs in collecting or intercepting systems into high-lying continuations of the runs or into outfalls
2) Pumping out stormwater detention tanks in combined systems
3) Lifting wastewater to and through treatment works, draining component settling tanks and other treatment units, withdrawing wastewater sludges and transporting them within the works to treatment units, supplying water to treatment units, discharging wastewater and wastewater sludge through outfalls, and pumping chemicals to treatment units.

Lift station equipment and systems are often installed in an enclosed structure. They can be constructed on-site (custom designed) or prefabricated. Lift station capacities range from 20 gpm (1.26 L/s) to more than 100,000 gpm (6,310 L/s). Prefabricated lift stations generally have capacities of up to 10,000 gpm (631 L/s). Centrifugal pumps are commonly used in lift stations. A trapped air column, or bubbler system, that senses pressure and level is commonly used for pump station control. Other control alternatives include electrodes placed at cutoff levels, floats, mechanical clutches, and floating mercury switches. A more sophisticated control operation involves the use of variable-speed drives.

Lift stations are typically provided with equipment for easy pump removal. Floor access hatches or openings above the pump room and an overhead monorail beam, bridge crane, or portable hoist are commonly used.

The two most common types of lift stations are the *dry-pit* or *dry-well* and *submersible lift stations*. In dry-well lift stations, pumps and valves are housed in a pump room (dry pit or dry well) that is easily accessible. The wet well is a separate chamber attached or located adjacent to the dry-well (pump room) structure. Figures 16.6 and 16.7 illustrate the two types of pumps.

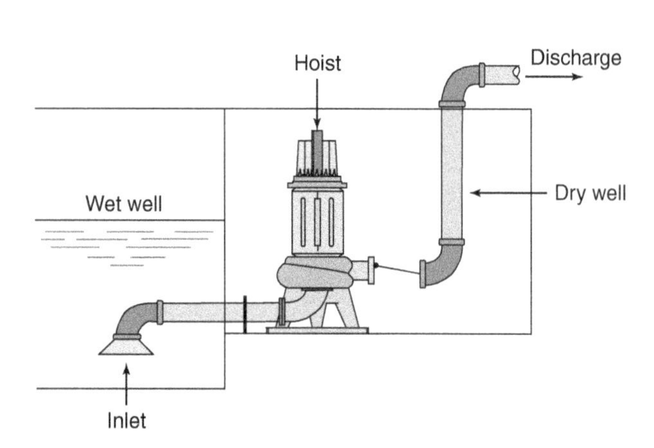

Figure 16.6 Dry-Well Pump

Submersible lift stations do not have a separate pump room; the lift station header piping, associated valves, and flow meters are located in a separate dry vault at grade for easy access. Submersible lift stations include sealed pumps that operate submerged in the wet well. These are removed to the surface periodically and reinstalled using guide rails and a hoist. A key advantage of dry-well lift stations is that they allow easy access for routine visual inspection and maintenance. In general, they are easier to repair than submersible pumps. An advantage of submersible lift stations is that they typically cost less than dry-well stations and operate without frequent pump maintenance. Submersible lift stations do not usually include large aboveground structures and tend to blend in with their surrounding environment in residential areas. They require less space and are easier and less expensive to construct for wastewater flow capacities of 10,000 gpm (630 L/s) or less.

16.8.1 Applicability

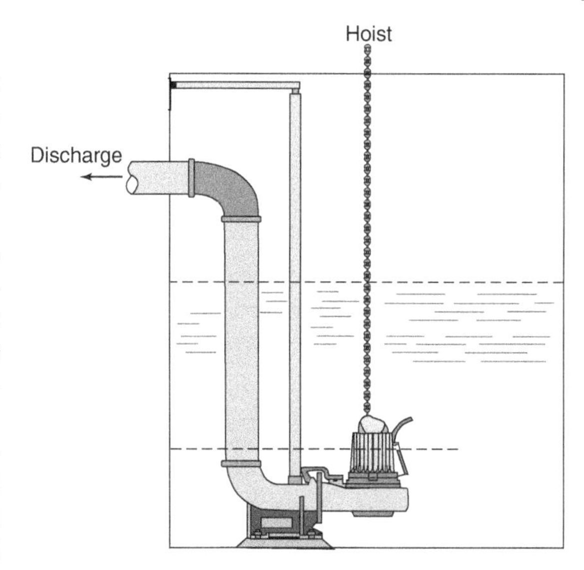

Figure 16.7 Wet-Well Submersible Pump

Lift stations are used to move wastewater from a lower to a higher elevation, particularly where the elevation of the source is not sufficient for gravity flow and/or when the use of gravity conveyance will result in excessive excavation depths and high sewer construction costs.

Lift stations are widely used in wastewater conveyance systems. Dry-well lift stations have been used in the industry for many years. However, the current industry-wide trend is to replace dry-well lift stations of small and medium size (typically <6,500 gpm or 410 L/s) with submersible lift stations mainly because of lower costs, a smaller footprint, and simplified operation and maintenance.

Variable-speed pumping is often used to optimize pump performance and minimize power use. Several types of variable-speed pumping equipment are available, including variable-voltage and variable-frequency drives, eddy current couplings, and mechanical variable-speed drives. Variable-speed pumping can reduce the size and cost of the wet well and allows the pumps to operate at maximum efficiency under a variety of flow conditions. Because variable-speed pumping allows lift station discharge to match inflow, only nominal wet-well storage volume is required and the well water level is maintained at a near-constant elevation. Variable-speed pumping may allow a given flow range to be achieved with fewer pumps than a constant-speed alternative. Variable-speed stations also minimize the number of pump starts and stops, reducing mechanical wear. Although the potential energy savings are significant for stations with large friction losses, they may not justify the additional capital costs unless the cost of power is relatively high. Variable-speed equipment also requires more room within the lift station and may produce more noise and heat than constant-speed pumps.

Lift stations are complex facilities with many auxiliary systems. Therefore, they are less reliable than gravity wastewater conveyance. However, lift station reliability can be significantly improved by providing standby equipment (pumps and controls) and emergency power supply systems. In addition, lift station reliability is improved by using non-clogging pumps suitable for the particular wastewater quality and by applying emergency alarm and automatic control systems.

16.8.2 Advantages and Disadvantages

Lift stations are used to reduce the capital cost of sewer system construction. When gravity sewers are installed in trenches deeper than 10 ft (3 m), the cost of sewer line installation increases significantly because of the more complex and costly excavation equipment and trench shoring techniques required. The size of the gravity sewer lines is dependent on the minimum pipe slope and flow. Pumping wastewater can convey the same flow using smaller pipeline size at shallower depth, thereby reducing pipeline costs.

Compared to sewer lines where gravity drives wastewater flow, lift stations require a source of electric power. If the power supply is interrupted, flow conveyance is discontinued and can result in flooding upstream of the lift station. It can also interrupt the normal operation of the downstream wastewater conveyance and treatment facilities. This limitation is typically addressed by providing an emergency power supply.

Key disadvantages of lift stations include the high cost to construct and maintain and the potential for odors and noise. Lift stations also require a significant amount of power, are sometimes expensive to upgrade, and may create public concerns and negative public reaction.

The low cost of gravity wastewater conveyance and the higher costs of building, operating, and maintaining lift stations means that wastewater pumping should be avoided, if possible and technically feasible. Wastewater pumping can be eliminated or reduced by selecting alternative sewer routes or extending a gravity sewer using direction drilling or other state-of-the-art deep excavation methods. If such alternatives are viable, a cost/benefit analysis can determine if a lift station is the most viable choice.

16.8.3 Design Criteria

Cost-effective lift stations are designed to accomplish the following:

1) Match pump capacity, type, and configuration with wastewater quantity and quality.
2) Provide reliable and uninterruptible operation.
3) Allow for easy operation and maintenance of the installed equipment.
4) Accommodate future capacity expansion.
5) Avoid septic conditions and excessive release of odors in the collection system and at the lift station.
6) Minimize environmental and landscape impacts on the surrounding residential and commercial developments.
7) Avoid flooding of the lift station and the surrounding areas.

16.8.3.1 Wet Well

Wet-well design depends on the type of lift station configuration (submersible or dry well) and the type of pump controls (constant or variable speed). Wet wells are typically designed large enough to prevent rapid pump cycling but small enough to prevent a long detention time and associated odor release.

Wet-well maximum detention time with constant-speed pumps is typically 20–30 minutes. Use of variable-frequency drives for pump speed control allows wet-well detention time reduction to 5–15 minutes. The minimum recommended wet-well bottom slope is to 2:1 to allow self-cleaning and minimum deposit of debris. Effective volume of the wet well may include sewer pipelines, especially when variable-speed drives are used. Wet wells should always hold some level of sewage to minimize odor release. Bar screens or grinders are often installed in or upstream of the wet well to minimize pump clogging problems.

16.8.3.2 Wastewater Pumps

The number of wastewater pumps and associated capacity should be selected to provide head-capacity characteristics that correspond as nearly as possible to wastewater quantity fluctuations. This can be accomplished by preparing pump/pipeline system head-capacity curves showing all conditions of head (elevation of a free surface of water) and capacity under which the pumps will be required to operate.

The number of pumps to be installed in a lift station depends on the station capacity, the range of flow, and the regulations. In small stations, with maximum inflows of <700 gpm (44 L/s), two pumps are customarily installed, with each unit able to meet the maximum influent rate. For larger lift stations, the size and number of pumps should be selected so that the range of influent flow rates can be met without starting and stopping pumps too frequently and without excessive wet-well storage.

Depending on the system, the pumps are designed to run at a reduced rate. The pumps may also alternate to equalize wear and tear. Additional pumps may provide intermediate capacities better matched to typical daily flows. An alternative option is to provide flow flexibility with variable-speed pumps.

For pump stations with high head losses, the single-pump flow approach is usually the most suitable. Parallel pumping is not as effective for such stations because two pumps operating together yield only slightly higher flows than one pump. If the peak flow is to be achieved with multiple pumps in parallel, the lift station must be equipped with at least three pumps: two duty pumps that together provide peak flow and one standby pump for emergency backup. Parallel peak pumping is typically used in large lift stations with relatively flat system head curves. Such curves allow multiple pumps to deliver substantially more flow than a single pump. The use of multiple pumps in parallel provides more flexibility.

Several types of centrifugal pumps are used in wastewater lift stations. In the straight-flow centrifugal pumps, wastewater does not change direction as it passes through the pumps and into the discharge pipe. These pumps are well suited for low-flow/high-head conditions. In angle-flow pumps, wastewater enters the impeller axially and passes through the volute casing at 90° to its original direction (Figs. 16.8 and 16.9). This type of pump is appropriate for pumping against low or moderate heads. Mixed-flow pumps are most viable for pumping large quantities of wastewater at low head. In these pumps, the outside diameter of the impeller is less than an ordinary centrifugal pump, increasing flow volume.

Example 16.1 Head Designation For Pumps

Head is commonly used to represent the vertical height of a static column of liquid corresponding to the pressure of a fluid at the point in question. Head can also be considered as the amount of work necessary to move a liquid from its

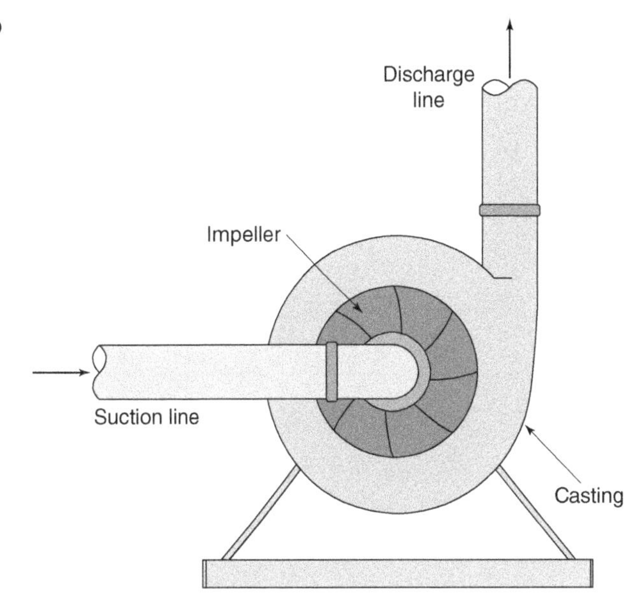

Figure 16.8 Centrifugal Angle-Flow Pump

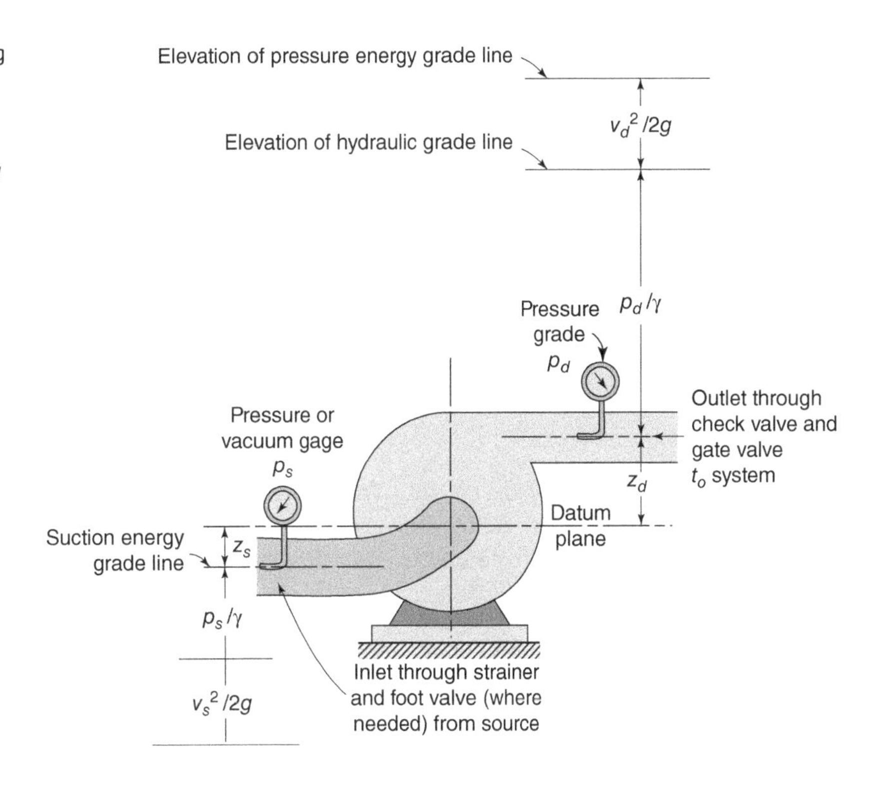

Figure 16.9 Head Relationships in Pumping Systems. System head H = $(z_d - z_s)$ + $(p_d/\gamma - p_s/\gamma)$ + $(v_d^2/2g - v_s^2/2g)$; for suction lift, p_s/γ and z_s are negative, and $v_s^2/2g$ is positive; for suction pressure, p_s/γ and $v_s^2/2g$ are positive, and z_s is negative

origin to the required delivery position. This includes the extra work necessary for a pump to overcome the resistance to flow in the pipe line. Figure 16.10 shows two common pumping situations: (a) a pump with suction head and (b) a pump with suction lift.

Define the two pumping situations and the terms for the static suction lift, the static suction head, total head, friction head loss in the discharge piping, friction head loss in the suction piping, and velocity head shown in Fig. 16.10.

Solution:

The pump with suction lift is a pump operational condition when the fluid source is below the pump. The vertical distance from the free surface of liquid to pump datum is called static suction lift. The net suction head, in this case, is the sum of static suction lift (h_s) plus friction losses (f_s) (negative net suction lift).

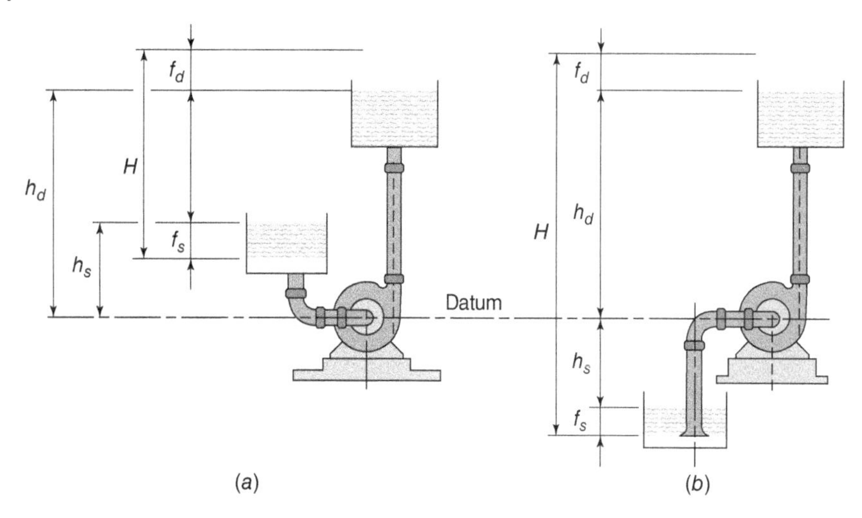

Figure 16.10 Pump with (a) Suction Head and (b) Pump with Suction Lift for Example 16.1. (*Source:* American Society of Plumbing Engineers)

The pump with suction head is a pump operational condition when the fluid is above the pump. Vertical distance from the free surface of liquid to pump datum is called static suction head (h_s). The net suction head is the sum of the static suction head (h_s) minus friction losses (f_s) (either positive or negative).

The static discharge head (h_d) equals the vertical distance between the pump datum and the surface of liquid in the discharge reservoir.

The static suction head or lift (h_s) equals the vertical distance from the liquid surface to the pump datum. This value is positive when operating with a suction lift and negative when operating with a suction head.

Friction head loss in the discharge piping (f_d) is caused by the pipe and fittings on the pump's discharge side.

Friction head loss in the suction piping (f_d) is caused by the pipe and fittings on the pump's suction side.

Total head (H) is developed by the pump, which can be expressed by one of the following equations:

Pump with suction lift: $$H = h_d + h_s + f_d + f_s + (v^2/2g) \tag{16.1}$$

Pump with suction head: $$H = h_d - h_s + f_d + f_s + (v^2/2g) \tag{16.2}$$

The velocity head is represented by $v^2/2g$. For a vertical turbine and submersible pumps, the velocity head is measured at the discharge flange. However, for booster pumps and centrifugal pumps, the velocity head developed by the pump is the difference between the velocity head at the discharge flange and the velocity head at the suction flange.

16.8.3.3 Ventilation

Ventilation and heating are required if the lift station includes an area routinely entered by personnel. Ventilation is particularly important to prevent the collection of toxic and/or explosive gases. According to the Nation Fire Protection Association (NFPA), all continuous ventilation systems should be fitted with flow detection devices connected to alarm systems to indicate ventilation system failure. Dry-well ventilation codes typically require six continuous air changes per hour or 30 intermittent air changes per hour. Wet wells typically require 12 continuous air changes per hour or 60 intermittent air changes per hour. Motor control center (MCC) rooms should have a ventilation system adequate to provide six air changes per hour and should be air-conditioned to between 55° and 90 °F (12° and 32 °C). If the control room is combined with an MCC room, the temperature should not exceed 85 °F (30 °C). All other spaces should be designed for 12 air changes per hour. The minimum temperature should be 55 °F (12 °C) whenever chemicals are stored or used.

16.8.3.4 Odor Control

Odor control is frequently required for lift stations. A relatively simple and widely used odor control alternative is minimizing wet-well turbulence. More effective options include collection of odors generated at the lift station and treating them in scrubbers or biofilters or the addition of odor control chemicals to the sewer upstream of the lift station. Chemicals typically used for odor control include chlorine, hydrogen peroxide, metal salts (ferric chloride and ferrous sulfate) oxygen, air, and potassium permanganate. Chemicals should be closely monitored to avoid affecting downstream treatment processes, such as extended aeration.

16.8.3.5 Power Supply

The reliability of power for the pump motor drives is a basic design consideration. Commonly used methods of emergency power supply include electric power feed from two independent power distribution lines, an on-site standby generator, an adequate portable generator with quick connection, a standby engine-driven pump, ready access to a suitable portable pumping unit and appropriate connections, and availability of an adequate holding facility for wastewater storage upstream of the lift station.

16.8.4 Performance

The overall performance of a lift station depends on the performance of the pumps. All pumps have four common performance characteristics: capacity, head, power, and overall efficiency. Capacity (flow rate) is the quantity of liquid pumped per unit of time, typically measured as gpm or MGD (L/min or MLD). Head is the energy supplied to the wastewater per unit weight, typically expressed as feet of water. Power is the energy consumed by a pump per unit time, typically measured as kilowatt-hours. Overall efficiency is the ratio of useful hydraulic work performed to actual work input. Efficiency reflects the pump's relative power losses and is usually measured as a percentage of applied power.

Common performance characteristics of a centrifugal pump operating at constant speed are illustrated in Figs. 16.11 and 16.12. Note that the *shutoff head* is a fixed limit and that power consumption is minimum at shutoff. For this reason, centrifugal pumps, after being *primed* or filled with water, are often started with the pump discharge valve closed. As the head falls past the point of maximum efficiency, normal discharge, or rated capacity of the pump (point 1 in Fig. 16.11), the power continues to rise. If a centrifugal pump is operated against a head that is too low, a motor selected to operate the pump in the

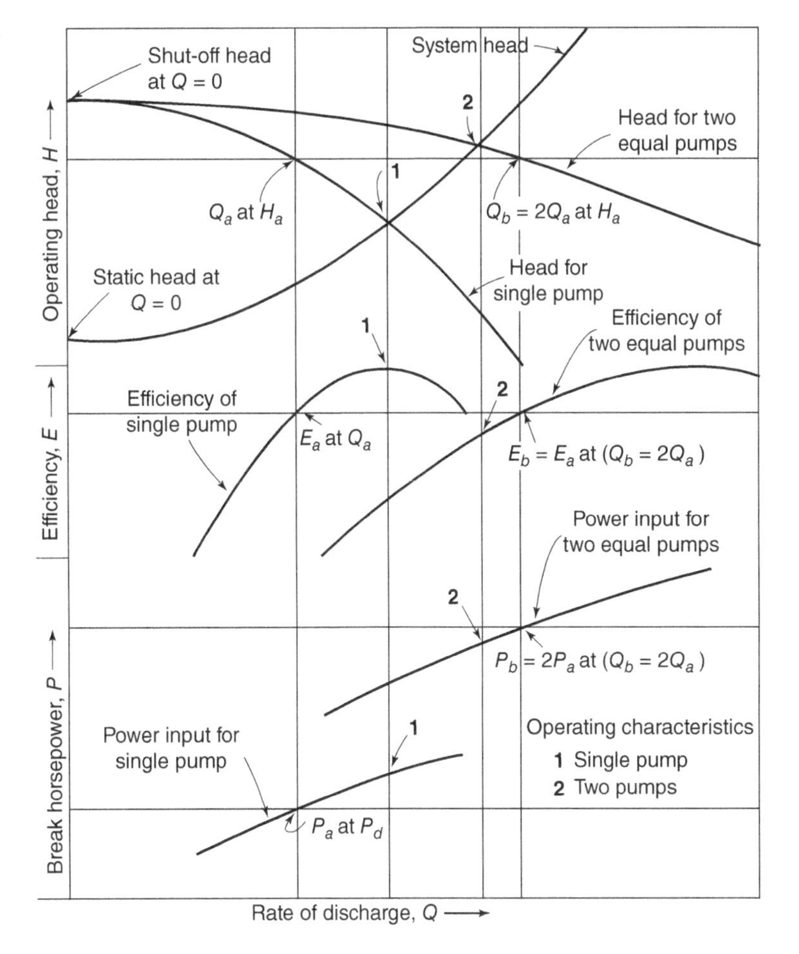

Figure 16.11 Performance Characteristics of Single and Twin Centrifugal Pumps Operating at Constant Speed

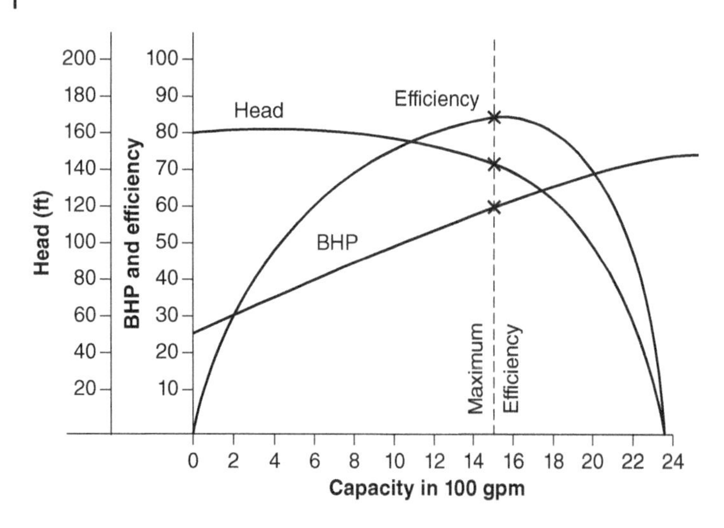

Figure 16.12 Pump Characteristic Curves. Conversion factors: 1 gpm = 0.0631 L/s; 1 ft = 0.3048 m. (*Source:* California Department of Health Services and U.S. Environmental Protection Agency, *Small Water System Operation and Maintenance*, 2002)

head range around maximum efficiency might become overloaded. Pump delivery can be regulated (a) by a valve on the discharge line, (b) by varying the pump speed mechanically or electrically, or (c) by throwing two or more pumps in and out of service to the best advantage.

What happens when more than one pumping unit is placed in service is shown in Fig. 16.11 with the help of a curve for the system head. Obviously, pumping units can operate only at the point of intersection of their own head curves with the system head curve. In practice, the system head varies over a considerable range at a given discharge (see Fig. 16.13 in Example 16.2). For example, where a detention tank is part of a system and both the tank and the source of wastewater fluctuate in elevation, (a) a lower curve identifies head requirements when the tank is empty and the wastewater surface of the source (the wet well in this instance) is high, and (b) an upper curve establishes the system head for a full tank and a low wastewater level in the wet well. How the characteristic curves for twin-unit operation are developed is indicated in Fig. 16.11. Note that the two identical pumping units have not been selected with an eye to the highest efficiency of operation in parallel. Characteristic curves for other multiple units are developed in the same way from the known curves of individual units.

Where most of the operating head is a static lift—when the wastewater is pumped through relatively short lengths of suction and discharge piping, for example—there is little change in the system head at different rates of flow. In these

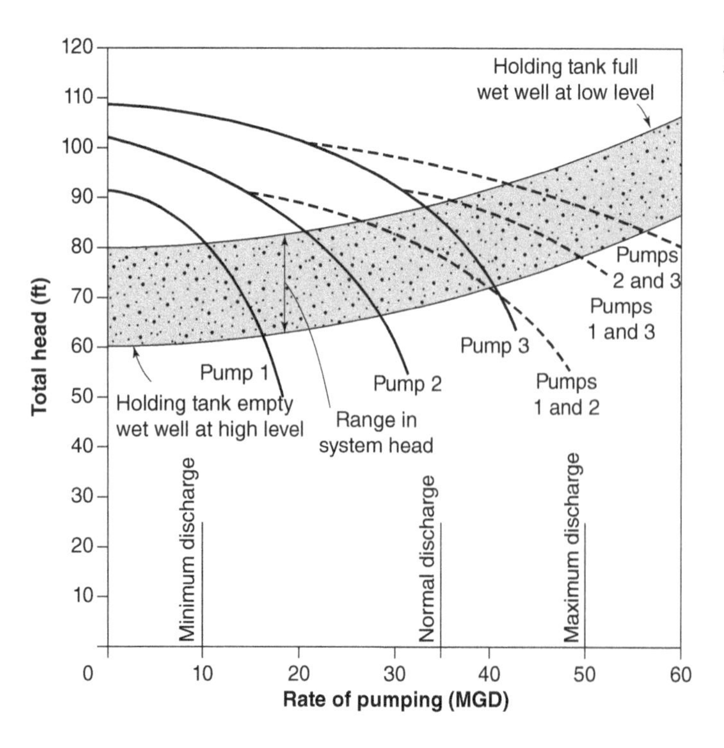

Figure 16.13 Pump Selection in Example 16.2. Conversion factors: 1 MGD = 3.785 MLD = 43.8 L/s; 1 ft = 0.3048 m

circumstances, the head curve is nearly horizontal, and the discharge of parallel pumps is substantially additive. This is common in wastewater pumping stations in which the flow is lifted from a lower to an immediately adjacent higher level. Examples are pumping stations along interceptors or at outfalls.

By contrast, friction may control the head on pumps discharging through long force mains, and it may not be feasible to subdivide flows between pumping units with reasonable efficiency. Multispeed motors or different combinations of pumps and motors may then be required.

Because flows from a number of pumps may have to be fed through a different piping system than flows from any single unit, it may be necessary to develop "modified" characteristic curves that account for losses in different combinations of piping.

Performance optimization strategies focus on different ways to match pump operational characteristics with system flow and head requirements. They may include the following options: adjusting system flow paths, installing variable-speed drives, using parallel pumps, installing pumps of different sizes, trimming a pump impeller, or putting a two-speed motor on one or more pumps in a lift station. Optimizing system performance may yield significant electrical energy savings.

Example 16.2 Choice of Pumps in a Stormwater Lift Station

A city is pumping a relatively large quantity of stormwater from a lift station wet well to deliver it at a fairly low head to a holding tank at the wastewater treatment plant located on a hill overlooking the city. The minimum flow is 10 MGD (37.85 MLD), the normal 35 MGD (132.48 MLD), and the maximum 50 MGD (189.25 MLD). The stormwater level in the wet well fluctuates in level by 5 ft (1.52 m), and the working range of the stormwater holding tank is 15 ft (4.57 m). The vertical distance between the bottom of the holding tank and the water surface of the wet well at the high stage is 60 ft (18.29 m). The friction head in the pumping station and a 54-in. (1,400 mm), force main rises from a minimum of 1 ft (0.30 m) at the 10-MGD (37.85 MLD) rate to a maximum of nearly 20 ft (6 m) at the 50-MGD (189.25 MLD) rate. Make a study of suitable pumping units.

Solution:

The solution for this example is shown in Fig. 16.13. Three pumps are provided: No. 1 with a capacity of 15 MGD (56.78 MLD) at a 66-ft (20.12 m) head, No. 2 with 25 MGD (94.63 MLD) at a 78-ft (23.77 m) head, and No. 3 with 37 MGD (140 MLD) at an 84-ft (25.60 m) head. Each pump has an efficiency of 89% at the design point. The efficiencies at the top and bottom of the working range are listed in Table 16.5.

Centrifugal pumps are normally operated with discharge velocities of 5–15 ft/s (1.50–4.60 m/s). The resulting average outlet diameter (in inches) of the pump, called the pump size, can be represented by the following equation:

$$d = 0.2\sqrt{Q} \quad \text{(U.S. Customary Units)} \tag{16.3}$$

where d is the average outlet diameter, in inches, and Q is the capacity of the pump, in gpm in the U.S. customary system. The following equation is for the metric system:

$$d = 20.22\sqrt{Q} \quad \text{(SI Units)} \tag{16.4}$$

where d is the average outlet diameter, in mm, and Q is the capacity of the pump, in L/s.

Table 16.5 Pumping Characteristics of System in Example 16.2

Pumps in Service, No.	1	2	3	1 & 2	1 & 3	2 & 3
Rate of Pumping, MGD	10	21	33.5	27	36	42
Head, ft	81	83	88	85	90	93
Efficiency (%)	80	88	88	71, 86	35, 87	68, 84
Rate of Pumping, MGD	15	25	37	34	43.5	49.5
Head, ft	66	78	84	80	85	89
Efficiency (%)	89	89	89	82, 88	71, 89	79, 87
Rate of Pumping, MGD	16.5	28.5	40.5	40	49.5	56.5
Head, ft	62	66	73	73	79	84
Efficiency (%)	88	84	86	88, 88	83, 88	79, 89

Conversion factors: 1 MGD = 3.785 MLD = 43.8 L/s; 1 ft = 0.3048 m.

Example 16.3 Determination of the Average Outlet Diameter of a Pump
Determine the average outlet diameter of a pump if the pump capacity is 600 gpm (37.86 L/s).

Solution 1 (U.S. Customary System):

$Q =$ capacity of the pump $= 600$ gpm

$d =$ average outlet diameter $= 0.2\sqrt{Q} = 0.2(600)^{0.5} = 4.90$ in.

Select $d = 6$ in. (or 150 mm).

Solution 2 (SI System):

$Q =$ capacity of the pump $= 37.86$ L/s
$d =$ average outlet diameter $= 20.22\sqrt{Q} = 20.22\sqrt{37.86} = 124$ mm

Select $d = 150$ mm (or 6 in.).

16.8.5 Operation and Maintenance

Lift station operation is usually automated and does not require continuous on-site operator presence. However, frequent inspections are recommended to ensure normal functioning and to identify potential problems. Lift station inspection typically includes observation of pumps, motors, and drives for unusual noise, vibration, heating, and leakage; checking of pump suction and discharge lines for valve arrangement and leakage; checking of control panel switches for proper position; monitoring of discharge pump rates and pump speeds; and monitoring of the pump suction and discharge pressure. Weekly inspections are typically conducted, although the frequency really depends on the size of the lift station.

If a lift station is equipped with grinder bar screens to remove coarse materials from the wastewater, these materials are collected in containers and disposed of in a sanitary landfill site as needed. If the lift station has a scrubber system for odor control, chemicals are supplied and replenished typically every three months. If chemicals are added for odor control ahead of the lift station, the chemical feed stations should be inspected weekly and chemicals replenished as needed.

The most labor-intensive task for lift stations is routine preventive maintenance. A well-planned maintenance program for lift station pumps prevents unnecessary equipment wear and downtime. Lift station operators must maintain an inventory of critical spare parts. The number of spare parts in the inventory depends on the critical needs of the unit, the rate at which the part normally fails, and the availability of the part. The operator should tabulate each pumping element in the system and its recommended spare parts. This information is typically available from the operation and maintenance manuals provided with the lift station.

16.8.6 Costs

Lift station costs depend on many factors, including these:

1) Wastewater quality, quantity, and projections
2) Zoning and land use planning of the area where the lift station will be located
3) Alternatives for standby power sources
4) Operation and maintenance needs and support
5) Soil properties and underground conditions
6) Required lift to the receiving (discharge) sewer line
7) The severity of the impact of an accidental sewage spill on the local area
8) The need for an odor control system.

These site- and system-specific factors must be examined and incorporated in preparing a lift station cost estimate.

16.8.6.1 Construction Costs

The most important factors influencing cost are the design lift station capacity and the installed pump power. Another cost factor is the lift station complexity. Factors that classify a lift station as complex include two or more of the following: (a) extent of excavation; (b) congested site and/or restricted access; (c) rock excavation; (d) extensive dewatering requirements, such as cofferdams; (e) site conflicts, including modification or removal of existing facilities; (f) special foundations,

Table 16.6 Lift Station Construction Costs

Lift Station	Design Flow Rate (MGD)	Construction Costs (2022 USD)
Cost curve data	0.5	200,136
Cost curve data	1	367,040
Cost curve data	3	584,040
Valencia, CA	6	2,070,800
Sunnymead, CA	12	4,935,200
Sunset/Heathfield, CA	14	3,868,800
Terry Street Pumping Station, Springfield, OR	20	8,134,400
Detroit, MI	750	191,704,000

Conversion factors: 1 MGD = 3.785 MLD = 43.8 L/s.

including piling; (g) dual power supply and on-site switch stations and emergency power generator; and (h) high pumping heads (design heads in excess of 200 ft or 61 m).

Mechanical, electrical, and control equipment delivered to a pumping station construction site typically accounts for 15–30% of total construction costs. Lift station construction has a significant economy of scale. Typically, if the capacity of a lift station is increased 100%, the construction cost would increase only 50–55%. An important consideration is that two identical lift stations will cost 25–30% more than a single station of the same combined capacity. Usually, complex lift stations cost two to three times more than more simple lift stations with no construction complications. Table 16.6 provides examples of complex lift stations and associated construction costs. Costs are given in 2022 USD using U.S. Army Corps of Engineers' *Cost Index for Utilities*.

16.8.6.2 Operation and Maintenance Costs

Lift station operation and maintenance costs include power, labor, maintenance, and chemicals (if used for odor control). Usually, the costs for solids disposal are minimal, but are included if the lift station is equipped with bar screens to remove coarse materials from the wastewater. Typically, power costs account for 85–95% of the total operation and maintenance costs and are directly proportional to the unit cost of power and the actual power used by the lift station pumps. Labor costs average 1–2% of total costs. Annual maintenance costs vary, depending on the complexity of the equipment and instrumentation. Figure 16.14 shows the operation and maintenance costs for lift stations as a function of wastewater flow. Recent 2022 costs are obtained by multiplying the costs from Fig. 16.14 by a multiplier of 3.39.

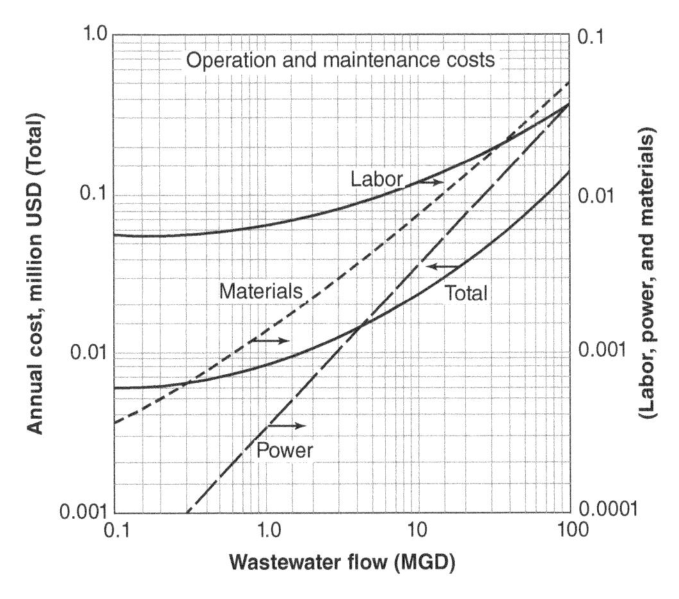

Figure 16.14 Operation and Maintenance Costs for Lift Stations. Recent 2022 costs are obtained by multiplying the costs from Fig. 16.14 by a multiplier of 3.39. Conversion factor: 1 MGD = 3.785 MLD = 43.8 L/s

16.9 Evaluation and Maintenance of Sewers

The mission and responsibility of sewerage authorities is to provide environmentally sensitive sewerage systems and to share in the responsibility of protecting water resources now and for future generations. In conjunction with this mission, the sewerage utility operation and maintenance unit needs to perform continual evaluation and maintenance of the public wastewater system. A comprehensive program of ongoing investigation and preventive maintenance maximizes system performance and is more economical than expensive and inconvenient emergency repairs.

Wastewater discharged from a home or business enters the system through *service lines* of size 12 in. (300 mm) or smaller. These smaller pipes are known as *collectors*. The waste-water then travels to larger trunk lines consisting of pipes 12 in. (300 mm) or larger. Finally, the wastewater reaches the largest lines, the *interceptors*, which vary in size from 18 to 60 in. (450–1,500 mm). These interceptors carry the wastewater directly to the treatment plants.

Pumping stations are an integral part of the collection and conveyance system. Collection systems are built to utilize the natural flow of gravity when possible. When wastewater cannot travel through the lines by gravity, pumps are used to force the waste-water through the force mains until it can again be transported by gravity flow.

16.10 Preliminary Analysis of Sewer Systems

It is essential to conduct a sewer system analysis to determine quickly and easily if there are serious *infiltration/inflow (I/I)* problems, evaluate the extent of these problems, and select the approach for further analysis and investigation. Occurrences that indicate the need for a sewer system analysis include the following:

1) Greater than anticipated flows measured at the wastewater treatment plant
2) Flooded basements during periods of intensive rainfall
3) Lift station overflows
4) Sewer system overflows or bypasses
5) Excessive power costs for pumping stations
6) Overtaxing of lift station facilities, often resulting in frequent electric motor replacements
7) Hydraulic overloading of treatment plant facilities
8) Excessive costs of wastewater treatment including meter charges levied by sanitary districts or other jurisdictional authorities
9) Aesthetic and water quality problems associated with bypassing of raw waste-water
10) Surcharging of manholes, resulting in a loss of pipe overburden through defective pipe joints and eventual settlement or collapse
11) Odor complaints
12) Structural failure
13) Corrosion.

In particular, evaluation of sewer systems occurs when required by regulatory requirements in order to receive federal funding or when dealing with capacity limitations, structural failure, or indirect evidence of excessive I/I in the overall system. I/I problems are often abated by the construction of relief sewers, larger lift stations, and treatment plants and also by the use of wastewater bypasses throughout the system. This last approach, however, often results in untreated wastewater flows being discharged into rivers, streams, lakes, and open ditches, which is no longer an acceptable solution. An effective sewer system evaluation and rehabilitation plan is required for effective protection of the infrastructure in nearly all cases regardless of the initial reasons for the evaluation.

16.10.1 Regulatory Requirements

Regulations promulgated as a result of Public Law 92-500 require that any engineer or public official concerned with the design of improvements to existing sewer system infrastructure components or wastewater treatment plants become familiar with and follow certain procedures to ensure that excessive I/I is not present in order to become eligible for U.S. Environmental Protection Agency (U.S. EPA) grant funding.

16.10.2 Structural Failure

Wastewater collection system structural failures often occur due to H_2S crown corrosion, natural aging, and factors such as defective design, excessive overburden, soil settlement, and earthquakes. The historical method for repairing structural problems in sewer systems was to excavate and replace the pipe. With the advent of new technologies, such as *trench-less sewer construction*, rehabilitation of wastewater collection lines has become more cost effective and can often be accomplished without extensive excavation and replacement.

16.10.3 Capacity Limitations

Given the natural increase in population and industrial growth in cities, the capacity of the wastewater pipes often becomes insufficient. Sewer collection lines and treatment plants become inadequate to handle the increase in sanitary flows. Without the correction of excessive I/I, existing sewer lines are unable to carry the increased flows, thus prohibiting expansion and growth within the existing tributary area.

16.10.4 Citizens' Complaints

Citizens' complaints are often reported during periods of extensive rainfall because sewers surcharge and cause local, area, and residential flooding. When such phenomena occur on a regular basis, an analysis of the sewer system is necessary because these complaints indicate that the sewer lines exhibit excessive amounts of I/I during periods of rainfall.

16.10.5 Need to Enlarge Service Area

Traditional planning of sewer systems has included allowances for growth and expansion within specific *drainage basins* or within specific geographical or political subdivisions of communities. As existing systems continue to expand, however, the demands on the existing sewer infrastructure continue to grow, and the capacity and condition of existing interceptor sewers, lift stations, and appurtenant structures must be continually evaluated. During these planning activities, it often becomes apparent that existing facilities have experienced deterioration and require rehabilitation or replacement to remain serviceable and to accommodate the flow of expanding service areas.

Evaluation of many existing systems as a part of federally funded I/I and *sewer system evaluation survey* (SSES) investigations has often shown that severe deterioration has occurred, thus creating additional financial pressures for future sewer system planning and expansion. Because sewer systems are designed for service lifetimes of 30–50 years or more and because the planning of these systems does not normally include replacement financing, future expansion, and development planning must take into account the cost of replacement. The continued expansion of existing collection systems normally continues until the capacity of the critical components of existing collection and treatment systems is reached. Because of the high cost of increasing interceptor and collection system capacity, especially in fully developed areas, it is important for I/I to be minimized and for the necessary investment to be made over the lifetime of existing facilities to preserve their condition and capacity. It is for this reason that the major federal funding sources for sewer construction have emphasized the importance of I/I control and protection of systems from major deterioration due to corrosion.

At any given point in time within a sewered community, there is a continuing need to recognize the following:

1) Value of the existing sewer infrastructure
2) Condition of the system
3) Rate of deterioration
4) Cost of mitigation of deterioration
5) Estimated remaining service lifetime
6) Ultimate system capacity.

A realistic evaluation of these factors is a crucial element of sound public works management and a fundamental requirement for effective financial planning of sewer system infrastructure improvements.

16.10.6 Budgetary Planning Needs

Sewer system budgetary planning normally includes the following major cost categories:

1) Legal and administrative
2) Long-term and short-term debt
3) Short-term capital financing
4) Operations and maintenance labor
5) Operations materials and utilities
6) Contingency or reserve funds.

Evaluation of the age and condition of existing sewer systems allows inclusion of the total system needs into the sewer system operations budget. A well-planned sewer system survey will provide information such as this:

1) Sewer line manhole (other structure) replacement needs and costs
2) Lift station equipment needs
3) Extent of corrosion of lift station equipment and structures, force mains, and downstream receiving sewers
4) Immediate and longer-term rehabilitation needs
5) Long-term and short-term maintenance needs.

Although all needs cannot be met by annual operating budgets, the budgeting and expenditure of funds annually for repair, maintenance rehabilitation, and replacement of critical sewer system components in many cases can eliminate or reduce the need for major capital expenses at a later date. For example, early identification of deterioration due to corrosion may save more than 60% of the cost of eventual repair or replacement.

16.11 Preventive Maintenance Methods and Procedures

Despite proactive efforts to maintain the system, operational complications may sometimes occur. Some of the common causes of system problems include using the drain to dispose of items such as grease, paper, garbage, or household hazardous waste; vandalizing the system by putting foreign or incompatible materials in the system or down manholes; and blocking of the system by tree or shrub roots seeking water and entering the system.

To maintain the wastewater collection system and to keep it functioning properly, the following preventive maintenance procedures are utilized:

1) *Visual inspection*. Maintenance crews periodically check manholes, frames, and covers to look out for cracks, breaks, or missing parts that may prevent them from maintaining airtight integrity. Replacement and maintenance are scheduled as necessary.
2) *TV inspection of lines*. Sewer lines are inspected internally with a special closed-circuit TV camera that is lowered into a manhole and pulled through the line. Testing and repair equipment used in conjunction with the camera will determine if there are areas of weakness in the joints and pipes and look for leakage. If the line is in poor condition and cannot be repaired, it will be scheduled for replacement.
3) *Smoke testing*. By blowing smoke into a sewer line, crews can determine areas of breaks, improper connections, and other system problems, which can then be scheduled for repair or replacement. This procedure sometimes identifies problems on the property owner's side of the system. In these cases, the property owners are notified and advised to make the appropriate repairs.
4) *Chemical root treatment*. In some areas, workers may find it helpful to use a foaming chemical root treatment. This foam is pumped into selected sewer mains to kill existing roots and to inhibit their regrowth.
5) *Jet washing and root cutting*. Sewer lines are often rodded to remove roots or other material and then cleaned with high-pressure water by using a combination jet vacuum system.
6) *Sewer main relining program*. Some sewer lines can be rehabilitated by installing PVC plastic liners. The lining extends the structural life of the pipes, inhibits root growth, and reduces groundwater leakage into the sewer pipes.
7) *Sewer service replacement*. Sewer "services," which are the lines that run between the property line and the main line, are periodically checked for structural and operational soundness. If they are found to be in poor condition, the pipes are replaced (Figs. 16.15 and 16.16).

16.12 Sanitary Sewer Overflows

Sanitary sewer overflows (SSOs) are releases of untreated sewage into the environment. They have always been illegal under the Clean Water Act, but the U.S. EPA's SSO Control Rule further clarifies the prohibition and provides a program for helping municipalities track and report activities undertaken to control SSOs.

16.12.1 Need to Control Sanitary Sewer Overflows

Sewer infrastructure represents an enormous public asset that accounts for trillions of dollars worth of local, state, and federal investment during the past century. Most collection system projects were spurred by strong public demand for relief from unsanitary, unsightly, and smelly sewage problems that plagued many areas of the country, contaminating water and causing deadly disease outbreaks.

Much of the nation's sewage collection infrastructure is between 30 and 100 years old, placing the infrastructure's various elements at increased risk for leaks, blockages, and malfunctions due to deterioration. The longer sewer collection system problems go unresolved, the more serious they become, placing vital public assets at risk of further degradation, posing an unacceptable risk to human health and the environment, damaging public and private property, and impacting state and local economies.

For some communities, implementation of the SSO rule requires significant additional investment to replace, repair, or expand parts of the system. For others, better operation and maintenance practices will resolve many of the problems that lead to SSOs.

Figure 16.15 Failure in PVC House Connection Pipe

16.12.2 Causes of Sanitary Sewer Overflows

When an SSO occurs, the cause is usually listed as a recent, immediately traceable condition, such as a pipe break or pump failure. However, merely repairing the ruptured pipe without understanding the underlying cause of its failure may not protect against future SSOs. The major causes leading to SSOs include age, lack of maintenance, poor operational procedures, and inadequate flow capacity.

Many sewer system failures are attributable to natural aging processes, such as the following:

1) Years of wear and tear on system equipment such as pumps, lift stations, check valves, and other equipment with movable parts that can lead to mechanical or electrical failure
2) Freeze/thaw cycles, groundwater flow, and subsurface seismic activity that can result in pipe movement, warping, brittleness, misalignment, and breakage
3) Deterioration of pipes and joints due to exposure to saltwater or other corrosive substances.

Figure 16.16 Crack in PVC Elbow

Lack of maintenance exacerbates age-related deterioration. Systems that are not routinely cleaned and repaired experience more frequent clogging and collapsing of lines due to root growth and accumulation of debris, sediment, oil, and grease. Similarly, the condition and function of mechanical equipment depreciate much faster without regular maintenance. Regular inspection and cleaning can eliminate many of these problems and keep the system functioning smoothly.

Not only do collection system bottlenecks and pipe breaks lead directly to SSOs, but they also exert hydraulic stress on other parts of the system, resulting in an expanding web of failures.

Operational procedures that lead to SSOs include mistakes, such as accidentally activating a pump without ensuring that all necessary check valves are in position, and disregarding or disconnecting available warning mechanisms, such as warning bells and lights.

Rapid development has also caused sewage flows to exceed system capacity in a number of communities. Some have tried to be proactive, imposing growth restrictions or building moratoriums until sewer capacity catches up. Many more are uncertain of the actual design capacity of their sewer systems, or do not adequately consider it in their planning process. Capacity-constrained areas may need additional miles of sewer pipe, bigger interceptors, more underground storage, or additional treatment capacity to control their overflows.

Any of the above causes, by itself or in combination, can set the stage for an SSO.

16.12.3 Size of the Problem

There are about 19,500 sewer systems nationwide designed to handle an average daily flow of roughly 50 billion gallons (BG) or 189 billion liters (BL) of raw sewage.

SSO reporting requirements vary from state to state, and many go unreported. However, based on a sampling of news reports during 2000, 59 SSOs in 18 states resulted in the release of an estimated 1.2 BG (4.54 BL) of sewage. Of these reported SSOs, one of the most serious was an estimated 72 million gallons (MG) or 272.5 million liters (ML) release into Florida's Indian River that resulted in drinking water advisories and beach closures throughout most of the state.

A number of states did participate in voluntary monitoring and reporting of SSOs. In 1999, 122 separate SSO events were reported in 13 states and the U.S. Virgin Islands. In another 36 events, SSO was listed as one of the elements leading to beach contamination and closure. Several states, including California and Texas, have passed laws mandating reporting of SSOs.

16.12.4 Impacts of Sanitary Sewer Overflows

16.12.4.1 Human Health

Raw sewage contains disease-causing pathogens, including viruses, bacteria, worms, and protozoa. Diseases resulting from enteric pathogens range from stomach flu and upper respiratory infections to potentially life-threatening illnesses such as cholera, dysentery, hepatitis B, and cryptosporidiosis. Children, the elderly, and people with suppressed immune systems face added risk of contracting serious illnesses. When SSOs contaminate public places and waters of the United States, people can be at risk of exposure to the untreated sewage in these circumstances:

1) *Swimming in open water.* Between 1997 and 1998, the Centers for Disease Control and Prevention (CDC) recorded 1,387 cases of enteric illness contracted during nine outbreaks among swimmers in lakes, ponds, rivers, and canals. Although the source of the pathogens wasn't listed in the CDC survey, the disease-causing organisms were consistent with those found in human sewage, including *E. coli*, *Cryptosporidium*, and a Norwalk-like virus. Health professionals suspect that the actual number of outbreaks resulting from open-water swimming is many times this number, but most cases go unreported.
2) *Drinking from a contaminated community water supply.* In June 1998, in Austin, Texas, 1,300 people fell ill with cryptosporidiosis after an SSO in Brushy Creek flowed through underground fissures into an aquifer supplying five municipal wells. In September 2000, drinking water alerts were issued to residents of Springfield, Missouri, and several neighboring communities when a million-gallon SSO entered Goodwin Hollow Creek, an underground stream that feeds several springs and private water wells.
3) *Eating contaminated fish or shellfish.* Shellfish are bottom-dwelling filter feeders that pass large quantities of water through their systems. They accumulate diseases, bacteria, and biotoxins and pass them on to humans who eat them. Fish that prey on contaminated shellfish or contract diseases themselves can also make people ill.

16.12.4.2 Recreation

Every year, vacationers take 1.8 billion trips to a public waterfront. About a fourth of them, or 45 million, come to boat, swim, or fish—activities that can include primary and secondary contact with the water. Each year, tourism dollars are lost because hundreds of coastal beaches are closed due to SSO contamination, often repeatedly or for extended periods. In 1999, 7,214 beach closings and advisories were issued, 20% due to SSOs. Beach closings took place in 13 states, affecting the Pacific, Atlantic, and Gulf Coasts; the Great Lakes; and many smaller inland streams, lakes, and reservoirs.

Sport fishing, which also benefits local economies, is impacted by SSOs when fisheries become less productive due to poor water quality. Nearly 17 million marine sport anglers took 68 million fishing trips in 1997 and caught 366 million fish, 50% of which were released. In some fisheries, recreational anglers harvest as much, if not more, fish than commercial fisherman.

16.12.4.3 Natural Resource Impacts

According to U.S. EPA's National Water Quality Inventory, 40% of U.S. waterways monitored by states during 1998 were found to be impaired. In rivers, streams, and estuaries, the major contaminants contributing to the impairment were pathogens, nutrients, and metals—all contaminants typically found in sewage. Although it is hard to gauge the importance of SSOs in the overall problem, they are suspected of being a contributing factor.

The environmental impacts of sewage include hypoxia, harmful algal blooms, habitat degradation, floating debris, and impacts to threatened or endangered species. According to the U.S. Fish and Wildlife Service, more than 50% of threatened and endangered species are water dependent.

16.12.4.4 Public and Private Property Damage

An untold number of private basement backups occur each year. In addition to the problem of human exposure, these spills can cause structural damage to building frames and foundations as well as water damage to electrical and gas appliances that are typically located in the basement. They can also damage or destroy floor and wall coverings and personal property. The 2022 cost of cleaning up a sewage spill has been estimated at between USD 875 and 5,000. SSOs frequently spill into homeowner yards, damaging landscaping, driveways, and outside possessions.

Municipal property damage from a major SSO can be severe. Communities pay billions per year to clean up and repair overflow damage to sewer infrastructure, roads and other transportation assets, parks and recreation areas, and municipal water supplies and treatment facilities.

16.12.4.5 Economic Impact on Shellfish Beds and Fisheries

In 1995, 6.7 million acres (2.71 million ha) of shellfish beds were restricted, 72% of them due to water pollution. The primary basis for harvest restriction is the concentration of bacteria typically found in sewage. Bed closures can have a devastating impact on local economies that rely on commercial shellfishing. Demand for shellfish has roughly doubled since 1966 and continues to grow, placing pressure on coastal states to improve water quality to open up more shellfish nurseries to harvest.

Commercial and recreational fishing suffers when SSOs impact fishing waters. Polluted water creates lowered fishery productivity, reduced and more costly harvests, and weakened consumer confidence. These impacts on the fishing industry also impact local economies in coastal regions. Each year, commercial fishing enterprises spend millions on boats, motors, docking fees, fuel, etc. Industry cutbacks mean loss of income to local service providers in small fishing communities that have grown up around the fishing industry and have few replacement options. They also lead to a reduction in the food supply, which leads to higher prices for consumers and more imports.

16.12.4.6 Economic Impact on Manufacturing

Manufacturers need access to adequate wastewater collection and treatment facilities to sustain or increase production. In communities that are experiencing capacity problems and/or SSOs, manufacturers may not be able to obtain the increased POTW (Publicly Owned Treatment Works) discharge limits needed to expand, forcing them to relocate. Loss of major manufacturers can cripple a local economy.

16.12.4.7 Economic Impact on Property Values

Property with access to surface water is worth more to homeowners and businesses if the water is perceived to be of high quality. Conversely, neighborhoods that experience chronic SSOs or perceived impairments to water quality drop in value.

16.13 Exfiltration

Many municipalities throughout the United States have sewerage systems (separate and combined) that may experience exfiltration of untreated wastewater from both sanitary and combined sewers. Sanitary sewer systems are designed to collect and transport to wastewater treatment facilities the municipal and industrial wastewaters from residences, commercial buildings, industrial plants, and institutions, together with minor or insignificant quantities of groundwater, stormwater, and surface waters that inadvertently enter the system. Over the years, many of these systems have experienced major infrastructure deterioration due to inadequate preventive maintenance programs and insufficient planned system rehabilitation and replacement programs. These conditions have resulted in deteriorated pipes, manholes, and pump stations that allow sewage to exit the systems (exfiltration) and contaminate adjacent ground and surface waters and enter storm sewers. Exfiltration is different from sanitary sewer overflows. SSOs are overflows from sanitary sewer systems usually caused by infiltration and inflow (I/I) leading to surcharged pipe conditions. SSOs can be in the form of direct overflows to receiving water, street flooding, and basement flooding, whereas exfiltration is not necessarily caused by excess I/I and is merely caused by a sewer leaking its contents to the surroundings outside.

Untreated sewage from exfiltration often contains high levels of suspended solids, pathogenic microorganisms, toxic pollutants, floatables, nutrients, oxygen-demanding organic compounds, oil and grease, and other pollutants. Exfiltration can result in discharges of pathogens into residential areas, cause exceedances of water quality standards (WQS), and pose risks to the health of the people living adjacent to the impacted streams, lakes, groundwater, sanitary sewers, and storm sewers; threaten aquatic life and its habitat; and impair the use and enjoyment of the nation's waterways.

Although it is suspected that significant exfiltration of sewage from wastewater collection systems occurs nationally, there is little published evidence of the problem and no known attempts to quantify or evaluate it on a national basis. Accordingly, the objectives of this chapter are to identify the factors that cause and contribute to the problem and to document the current approaches for correcting the problem, including costs.

16.13.1 Causative Factors

There is no information on exfiltration-specific and unique/causative factors because most factors that cause inflow/infiltration are identical to those associated with exfiltration (i.e., they both occur through leaks in pipes, depending on the relative depth of the groundwater).

The following factors contribute to exfiltration:

1) Size of sewer lines
2) Age of sewer lines
3) Materials of construction (sewer pipe, point/fitting material, etc.)
4) Type and quality of construction (joints, fittings, bedding, and backfill)
5) Depth of flow in the sewer.

Geologic conditions that contribute to exfiltration include these:

1) Groundwater depth (in relation to sewer line/depth of flow of sewage)
2) Type of soil
3) Faults.

Climate conditions that influence exfiltration include the following:

1) Average frost line in relation to sewer depth
2) Average rainfall, which helps determine groundwater depth.

In a typical exfiltrating sanitary sewer system, with the groundwater level below the sewage flow surface, exfiltration can occur in several areas. Figure 16.17 schematically represents these exfiltration sources, including defective joints and cracks in the service laterals, local mains, and trunk/interceptor sewers. The level of groundwater and the depth of flow in the sewer will influence the extent of exfiltration rates, since the pressure differential between the hydraulic head in the sewer and the groundwater hydraulic head will force water out of the sewer apertures into the surrounding soil material.

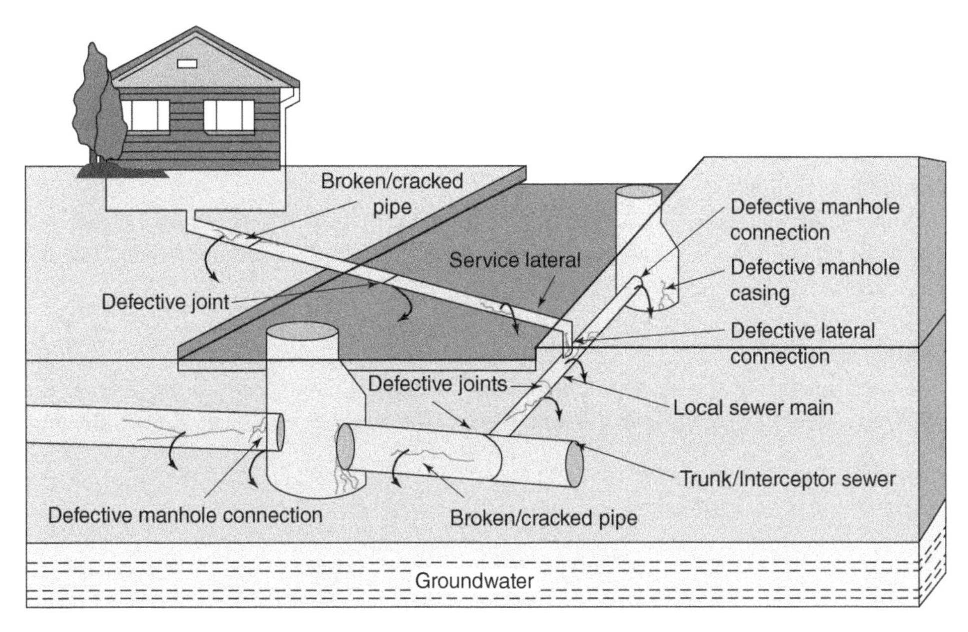

Figure 16.17 Sanitary Sewer System Components and Exfiltration Sources

16.13.2 Health and Environmental Impacts

This section addresses the potential health impacts of exfiltration on groundwater, drinking water distribution systems, and surface water.

16.13.2.1 Groundwater

Several studies have indicated widespread pollution of groundwater in urban areas arising from the general leakiness of sewers, including bacteria and ammonium.

Transport of the sewage and pollutants leaking into the subsurface/groundwater depends on a variety of factors, including but not limited to the difference in hydraulic head between the sewage surface and the groundwater table level, the substrate physical/chemical/biological characteristics (which determines attenuation potential), and the sewage pollutants and their concentrations. Fecal bacteria contamination is the most serious health risk associated with domestic sewage exfiltration. Contamination by viruses, protozoa, and other microorganisms is also a concern. Increased concentrations of total organic carbon, nitrate, chloride, and sulfate, however, can also make the water unfit for consumption. Phosphate and boron are good indicators of sewage pollution since they are not naturally occurring in groundwater.

16.13.2.2 Water Supply Distribution Systems

Because of minimum separation requirements for potable water supply distribution systems and sanitary sewers and vigilant application of cross-connection control programs, the opportunity for sewer exfiltration to contaminate drinking water supplies is theoretically rather limited. Sewage from exfiltration can enter a distribution system through a broken water main or, under reduced pressure conditions, through a hole that leaks drinking water out under normal positive-pressure conditions. Situations that could allow infiltration of the sewage through a lowering of water main pressure primarily involve backflow and surges.

16.13.2.3 Surface-Water Pollution

The occurrence of exfiltration is limited to those areas where sewer elevations lie above the groundwater table. Because groundwater elevations near surface-water bodies are typically near the ground surface, sewers near surface-water bodies are generally below the groundwater table, and infiltration (rather than exfiltration) will dominate the mode of sewer leakage in these areas. In areas of steep topographic conditions, where sewers are located near surface waters and at elevations that lie above the surface water, exfiltration impacts may be possible.

16.13.3 Corrective Measures

The proper selection of corrective or rehabilitation methods and materials depends on a complete understanding of the problems to be corrected, as well as the potential impacts associated with the selection of each rehabilitation method. Pipe rehabilitation methods to reduce exfiltration (and simultaneously infiltration) fall into one of two categories: external rehabilitation methods or internal rehabilitation methods.

Certain conditions of the host pipeline influence the selection of the rehabilitation method. It is therefore necessary to assess these factors to prepare the pipe for rehabilitation. Rehabilitation is preceded by surface preparation by cleaning the pipe to remove scale, tuberculation, corrosion, and other foreign matter.

16.13.3.1 External Sewer Rehabilitation Methods

External rehabilitation methods are performed from the aboveground surface by excavating adjacent to the pipe, or the external region of the pipe is treated from inside the pipe through the wall. Some of the methods used include the following:

1) External point repairs
2) Chemical grouting:
   - Acrylamide base gel
   - Acrylic base gel
3) Cement grouting
   - Cement
   - Microfine cement
   - Compaction.

16.13.3.2 Internal Sewer Rehabilitation Methods

Internal sewer rehabilitation methods include the following:

1) Internal *grouting* is the most commonly used method for sealing leaking joints in structurally sound sewer pipes. Chemical grouts do not stop leaks by filling cracks; they are forced through cracks and joints, and gel with surrounding soil, forming a waterproof collar around leaking pipes. This method is accomplished by sealing off an area with a "packer," air testing the segment, and pressure injecting a chemical grout for all segments that fail the air test. The three major types of chemical grout are:

   - Acrylic
   - Acrylate
   - Urethane.

2) Replacement or additional parallel sewer.
3) Trenchless rehabilitation (see next chapter).

16.13.4 Rates of Exfiltration from Sewers and Cost of Correction

The exfiltration susceptibility of sewer systems is defined by the relative depths of the sewers and groundwater table. In cases where sewer depths are generally shallower than the surrounding groundwater, the potential to exfiltrate exists (because the direction of the hydraulic gradient is toward the exterior of the sewer); these sewers can therefore be considered exfiltration susceptible. The most significant types of sewer damage that permit exfiltration are leaking service junctions, leaking sewer joints, pipe cracks, and pipe fractures.

Three basic approaches have been used to quantify sewer exfiltration rates:

1) Direct measurement of flow in isolated sewer segments
2) Theoretical estimates using Darcy's law and related hydraulic theory
3) Water balance between drinking water produced/delivered and wastewater collected/treated.

Sewer exfiltration rates are reported in units of (a) gallons per inch diameter per mile length per day (gal/in./mile/d) or (b) liters per millimeter diameter per kilometer length per day (L/mm/km/d), as shown in Table 16.7.

Most of the urban areas in the northeastern, southeastern, and coastal areas of the United States have relatively shallow groundwater tables (<15 ft or 4.57 m). In these areas, where a significant portion of the population (and therefore sewer systems) exists, relatively few exfiltration-susceptible sewer systems are expected. One caveat is exfiltration from service laterals. Even in the areas mentioned, many shallow service laterals may exist above groundwater tables. However, the hydraulic head available to drive exfiltration in these service lines is generally very low (typically only 1–2 in. or 25–50 mm) and intermittent.

Table 16.7 Reported Average Exfiltration Rates

Location	Exfiltration	
	(gal/in./mi/d)	(L/mm/km/d)
Berkeley, CA	5,730	531
Santa Cruz, CA	23,760	2,200
Lexington, KY	15,690	1,453
Riyadh, Saudi Arabia	29,280	2,711

Conversion factor: 1 gal/in. /mile/d = 0.0926 L/mm/km/d.

Based on a review of the depth-to-groundwater map, it is expected that widespread exfiltration is probably limited to a relatively small portion of the total U.S. population because relatively few large urban areas in the United States are located in these deeper groundwater areas. Cities such as Albuquerque, Phoenix, and Tucson are among the larger urban areas where significant exfiltration potential exists.

Areas with extremely deep groundwater tables probably experience relatively less risk associated with exfiltration due to the long subsurface travel times and distances of the exfiltrated sewage from the sewer to the groundwater table. Areas with significant portions of the system above, but in proximity to, the groundwater table are probably at greatest risk. There is an increased risk in the relatively few areas with significant exfiltration potential when there is, for example, a thin soil and fractured rock hydrogeologic setting that allows pathogens and other contaminants from the sewage to reach the groundwater quickly and with minimal attenuation. However, because public water supplies are treated with chlorination, ozonation, or other systems to kill fecal bacterial contamination, an added measure of protection is provided.

A greater potential problem, albeit isolated, may be exfiltration from sewers carrying industrial wastewater. Organic and inorganic constituents of industrial wastewater can be much more persistent than those of domestic sewage, and therefore much more likely to reach the groundwater in areas of significant exfiltration potential. The disposition of industrial wastewater contaminants that reach groundwater used for drinking water supplies may not be the same as that from domestic sewage. Untreated well water in some rural, small community, commercial, and private-owner drinking water systems does not enjoy the added protection offered by treatment. However, these systems are not typically in proximity to large municipalities and their associated sewer systems/exfiltration potential.

In summary, exfiltration appears to be a problem in certain cities in the United States—mainly those located west of the Mississippi River and those along the Appalachian Mountains. Exfiltration may be a regional or, more likely, local problem where the groundwater table lies closely under the sewage flow surface. Situations where the exfiltrate can reach even deep groundwater through a thin soil/fractured rock hydrogeologic setting, especially where persistent, potentially toxic contaminants are present (such as those often associated with industrial wastewater), also pose a problem.

Corrective actions to address exfiltration in those situations where local-level evaluation calls for such action will generally be accomplished with technologies similar to those used to address infiltration. Table 16.8 provides an example of those costs assuming the use of cured-in-place lining as the method of sewer rehabilitation.

Table 16.8 Costs for Exfiltration Corrective Action

Sewer Diameter (in.)	Cost (2022 USD/Linear ft)
8	87
10	105
12	112
15	186
18	236
21	329
24	434
27	459
30	781
36	868

Conversion factors: 1 in. = 25.4 mm; 1 USD/linear ft = 3.28 USD/linear m.

16.14 Corrosion Analysis and Control

Structural problems in wastewater collection systems can sometimes occur as a result of corrosion; thus, it is important to consider corrosion when designing, rehabilitating, or analyzing sewer systems. This section discusses the types of sewer corrosion, explains how a corrosion survey is conducted, and describes methods for controlling corrosion. Major emphasis is placed on hydrogen sulfide (H_2S) corrosion because it is the most prevalent form of corrosion in sewer systems.

Internal corrosion in sewer systems is normally related to the characteristics of the wastes being transported and is caused by chemical, electrochemical, and biochemical reactions. External corrosion is primarily caused by thermal, physical, structural, or electrochemical stresses.

16.14.1 Types and Mechanisms of Corrosion

Internal pipe corrosion in sewer systems is primarily caused by two mechanisms:

1) Direct attack by corrosive gases released from the wastewater, such as H_2S and SO_2
2) Bacterial oxidation of H_2S to sulfuric acid in the unsubmerged portions of pipes (crown corrosion).

In their gaseous forms, H_2S and SO_2 are directly corrosive to metals. Note that the presence of H_2S raises concerns for safety because H_2S gas is toxic to humans. H_2S represents an imminent life threat at a concentration of 300 ppmv (part per million by volume) in air. OSHA recommends a time-weighted average exposure during an 8-hour period of <10 ppmv.

Corrosive wastes (e.g., industrial acidic wastes) discharged to the sewer can cause direct corrosion in submerged portions of the sewer. Furthermore, certain chemicals that are used in wastewater treatment and collection systems can be corrosive.

Electrochemical corrosion may when electrical currents are created between dissimilar metals or when an electrolytic waste removes one or more metals from an alloy.

Hydrogenation occurs when hydrogen ions react with metal pipes. This occurs, however, under high-temperature, high-pressure, stress, and anaerobic conditions and is not commonly found in sewer systems.

Fatigue corrosion and stress corrosion are similar in that both are caused when external stresses are applied to pipes and both occur inside the pipes. Fatigue corrosion occurs when pipes are exposed to repeated stresses.

Filiform corrosion occurs on metal piping with organic coatings. It is characterized by filament-like corrosion in the metal surface originating at pinpoint penetrations of the surface. Note that corrosion is often the result of more than one mechanism. For example, if iron pipe is already experiencing hydrogen sulfide corrosion, it becomes more brittle and is more prone to cracking when stress is applied.

16.14.2 Hydrogen Sulfide in Sewers

Sulfide generation is a bacterially mediated process occurring in the submerged portion of sewers. Fresh domestic wastewater entering a wastewater collection system is usually free of sulfide. However, a dissolved form of sulfide soon appears as a result of low dissolved oxygen (DO) content, high-strength wastewater, low-flow velocity and long detention time in the collection system, elevated wastewater temperature, and extensive pumping.

The root cause of odor and corrosion in collection systems is sulfide, which is produced from sulfate by bacteria residing in a slime layer on the submerged portion of sewer pipes and structures. Once released from the wastewater as H_2S gas, odor and corrosion problems begin. Another type of bacteria utilizes H_2S gas to produce sulfuric acid (H_2SO_4), which causes the destruction of wastewater piping and facilities. Operation and maintenance (O&M) expenditures are required to correct the resulting damage caused by this H_2SO_4. In severe instances, pipe failure, disruption of service, and uncontrolled releases of wastewater can occur.

The first step in this bacterially mediated process is the establishment of a slime layer below the water level in a sewer. This slime layer is composed of bacteria and inert solids held together by a biologically secreted protein "glue" or film called *zooglea*. When this biofilm becomes thick enough to prevent DO from penetrating it, an anoxic zone develops within it. Approximately two weeks is required to establish a fully productive slime layer or zoogleal film in pipes. Within this slime layer, sulfate-reducing bacteria use the sulfate ion (SO_4^{2-}), a common component of wastewater, as an oxygen source for the assimilation of organic matter in the same way DO is used by aerobic bacteria. The SO_4^{2-} concentrations are almost never limiting in normal domestic wastewaters. When SO_4^{2-} is utilized by these bacteria, sulfide (S^{2-}) is the by-product. The rate at which S^{2-} is produced by the slime layer depends on a variety of environmental conditions including the concentration of organic food source or biochemical oxygen demand (BOD), DO concentration, temperature, wastewater velocity, and the area of the normally wetted surface of the pipe.

As SO_4^{2-} is consumed, the S^{2-} by-product is released back into the wastewater stream, where it immediately establishes a *dynamic chemical equilibrium* between four forms of sulfide, the sulfide ion (S^{2-}), the bisulfide or hydrosulfide ion (HS^-), dissolved or aqueous H_2S [$H_2S(aq)$], and H_2S gas [$H_2S(g)$]:

1) *Sulfide ion (S^{2-}).* The sulfide ion is a colorless ion in solution and cannot leave wastewater in this form. It does not contribute to odors in the ionic form.
2) *Bisulfide ion (HS^-).* The HS^- (or hydrosulfide) ion is a colorless, odorless ion that can only exist in solution. It also does not contribute to odors.
3) *H_2S(aq).* H_2S can exist as a gas dissolved in water. The polar nature of the H_2S molecule makes it soluble in water. In the aqueous form, H_2S does not cause odors; however, this is the only sulfide species that can leave the aqueous phase to exist as a free gas. The rate at which H_2S leaves the aqueous phase is governed by Henry's law, the amount of turbulence of the wastewater, and the pH of the solution.
4) *H_2S(g).* Once H_2S leaves the dissolved phase and enters the gas phase, it can cause odor and corrosion. H_2S(g) is a colorless but extremely odorous gas that can be detected by the human sense of smell in very low concentrations. In high concentrations, it is also very hazardous to humans. In concentrations as low as 10 ppmv, it can cause nausea, headache, and conjunctivitis of the eyes. Above 100 ppmv, it can cause serious breathing problems and loss of the sense of smell along with burning of the eyes and respiratory tract. Above 300 ppmv, death can occur within a few minutes.

Due to the continuous production of S^{2-} in wastewater, H_2S(g) rarely, if ever, reenters the liquid phase. S^{2-} continuously produced by the slime layer replaces that which is lost to the atmosphere as H_2S(g) in the collection system. In addition, once the H_2S(g) is released, it usually disperses throughout the sewer environment and never reaches a high enough concentration to be forced back into solution. The four sulfide chemical species are related according to the following equilibrium:

$$\begin{array}{l} H_2S(g) \leftrightarrow H_2S(aq) \leftrightarrow HS^- \leftrightarrow S^{2-} \\ \qquad \text{pKa} = 6.9 \qquad\qquad \text{pKa} = 14 \end{array}$$

where pK_a is defined as the *ionization constant*; the higher the pK_a value, the weaker the acid.

As indicated by the equilibrium equations, once H_2S is released into the gas phase, HS^- is immediately transformed into more $H_2S(aq)$ to replace that which is lost. Concurrently, S^{2-} is transformed into HS^- to replace that lost to $H_2S(aq)$. Through this type of continuously shifting equilibrium, it would be possible to completely remove all S^{2-} from wastewater as H_2S(g) through stripping. This is generally not recommended or advantageous due to odor releases and the accelerated corrosion that can take place.

The quantitative relationship between the four sulfide species is controlled by the pH of the wastewater. The S^{2-} does not exist below a pH of approximately 12 and, as indicated by the pK_a, is in a 50/50 proportional relationship with the HS^- at a pH of 14. Because the normal pH of wastewater is far lower, the S^{2-} is rarely experienced. The pK_a of much greater importance is the one controlling the proportional relationship between the HS^- and $H_2S(aq)$. Most domestic wastewater has a pH near 6.9. This means that at the pH of normal wastewater, half of all S^{2-} present exists as the HS^- and the other half exists as $H_2S(aq)$ (a dissolved gas). Since the concentration of dissolved gases in solution is primarily controlled by the specific Henry's law coefficient for that gas, they can be released from solution to exist as the free gas form. Once subjected to turbulence or aeration, wastewater can release the dissolved gas as free H_2S(g), and more HS^- is transformed into the dissolved gas form to replace that lost to the atmosphere.

16.14.3 Factors Affecting Sulfide Concentration

16.14.3.1 Settleable Solid

Periods of low flow in the collection system correlate to lower average wastewater velocities. Low-flow velocities allow material, usually grit and large organics, to settle in the collection system piping. This increases the mass and surface area of material in the collection system on which SO_4^{2-}-reducing bacteria (slime layer) can grow and can lead to an increased conversion of SO_4^{2-} to S^{2-}.

Collection systems with sedimentation problems can experience S^{2-} concentration spikes during the historically high-flow, cool temperature months. This phenomenon occurs when significant sand or grit accumulations exist and the particles are covered by an anaerobic slime layer that contains SO_4^{2-}-reducing bacteria. Only the bacteria on the surface of the grit pile receive a continuous supply of SO_4^{2-} because they are exposed to the wastewater. The buried SO_4^{2-}-reducing bacteria are not exposed to a continuous supply of SO_4^{2-}. This forces them to exist in a semidormant, anaerobic state with very low cell activity (but they are not dead). When a high-flow event occurs, with sufficient velocity and shear force to resuspend the sediment, this enormous surface area of SO_4^{2-}-reducing bacteria is suddenly exposed to ample SO_4^{2-} and the bacteria rapidly convert it to dissolved sulfide. This causes a relatively short-duration, high-S^{2-} event with resulting H_2S(g) release, odor, and corrosion.

The grit particles and their attached SO_4^{2-}-reducing bacteria that were semidormant are suspended and exposed to a tremendous quantity of SO_4^{2-} and quickly begin producing S^{2-}. The interaction between a large quantity of bacteria and an almost unlimited food source will create dissolved S^{2-} spikes that are subsequently released in areas of high turbulence. This trend is common and well documented in many cities with similar grit deposition problems such as Boston, Los Angeles, St. Louis, and Houston.

16.14.3.2 Temperature

In addition to the factors described earlier, summer conditions result in an increase in wastewater temperatures. Greater wastewater temperatures increase the metabolic activity of the SO_4^{2-}-reducing organisms, causing faster conversion of SO_4^{2-} to S^{2-} and increased dissolved S^{2-} concentrations. It has been estimated that each incremental 7 °C (12.5 °F) increase in wastewater temperature doubles the production of S^{2-}.

16.14.3.3 Flow Turbulence

Turbulence is a critical parameter to consider in preventing H_2S(g) release from wastewater. The effects of H_2S(g) odor and corrosion are increased by orders of magnitude at points of turbulence. Henry's law governs the concentration of gas over a liquid containing the dissolved form of the gas. *Henry's law* states, in effect, that the concentration of a gas over a liquid containing the dissolved form of the gas is controlled by the partial pressure of that gas and the mole fraction of the dissolved gas in solution. Because this law governs the relationship between the dissolved form and gaseous form of sulfide over a given surface area, any action that serves to increase the surface area of the liquid also increases the driving force from the liquid to the gas phase.

The most common form of increased surface area is turbulence. In turbulent areas, small droplets are temporarily formed. When this happens, the forces governing Henry's law (partial pressure) quickly try to reach equilibrium between the liquid and atmospheric phases of the gas. The result is often a dramatic release of sulfide from the dissolved to the gaseous form. Structures causing turbulence should be identified and measures should be taken to protect and/or control the subsequent H_2S(g) releases. This same release mechanism is exhibited whenever wastewater containing dissolved sulfide is aerated.

16.14.4 Structural Corrosion

Thiobacillus aerobic bacteria, which commonly colonize pipe crowns, walls, and other surfaces above the waterline in wastewater pipes and structures, have the ability to consume H_2S(g) and oxidize it to H_2SO_4. This process can only take place where there is an adequate supply of H_2S(g) (32.0 ppmv), high relative humidity, and atmospheric oxygen. These conditions exist in the majority of wastewater collection systems for some portion of the year. A pH of 0.5 (which is approximately equivalent to a 7% H_2SO_4 concentration) has been measured on surfaces (crown of pipes) exposed to severe H_2S(g) environments (>50 ppmv in air).

The effect of H_2SO_4 on concrete surfaces exposed to the sewer environment can be devastating. Sections of collection interceptors and entire pump stations have been known to collapse due to loss of structural stability from corrosion. The process of concrete corrosion, however, is a stepwise process that can sometimes give misleading impressions. The following briefly describes the general process of concrete corrosion in the presence of a sewer atmosphere.

Freshly placed concrete has a pH of approximately 11 or 12, depending on the composition of mixed aggregates. This high pH is the result of the formation of calcium hydroxide [$Ca(OH)_2$] as a by-product of the hydration of cement. $Ca(OH)_2$ is a very caustic crystalline compound that can occupy as much as 25% of the volume of concrete. A surface pH of 11 or 12 will not allow the growth of any bacteria; however, the pH of the concrete is slowly lowered over time by the effect of carbon dioxide (CO_2) and H_2S(g). These gases are both known as "acid" gases because they form relatively weak acid solutions when dissolved in water. CO_2 produces carbonic acid and H_2S produces thiosulfuric and polythionic acid. These gases dissolve into the water on the moist surfaces above the wastewater flow and react with the $Ca(OH)_2$ to reduce the pH of the surface. Eventually, the surface pH is reduced to a level that can support the growth of bacteria (pH 9–9.5).

The time it takes to reduce the pH is a function of the concentration of CO_2 and H_2S(g) in the sewer atmosphere. It can sometimes take years to lower the pH of concrete from 12 to 9; however, in some severe situations, it can be accomplished in a few months. Once the pH of the concrete is reduced to around pH 9, biological colonization can occur. More than 60 different species of bacteria are known to regularly colonize wastewater pipelines and structures above the water line. Most species of bacteria in the genus *Thiobacillus* have the unique ability to convert H_2S(g) to H_2SO_4 in the presence of oxygen.

Because each species of bacteria can only survive under a specific set of environmental conditions, the particular species inhabiting the colonies changes with time. Because the production of H_2SO_4 from H_2S is an aerobic biological process, it can only occur on surfaces exposed to atmospheric oxygen.

16.14.5 Sulfuric Acid Production

As a simplified example, one species of *Thiobacillus* only grows well on surfaces with a pH between 9 and 6.5. However, when the H_2SO_4 waste product they excrete decreases the pH of the surface below 6.5, they die off and another species that can withstand lower pH ranges takes up residence. The succeeding species grows well on surfaces with a pH between 6.5 and 4. When the acid produced by these species drops the pH below 4, a new species takes over. The process of successive colonization continues until species that can survive in extremely low pH conditions take over. One such species is *Thiobacillus thiooxidans*, which is sometimes known by its common name, *Thiobacillus concretivorous*, which is Latin for "eats concrete." This organism has been known to grow well in the laboratory while exposed to a 7% solution of H_2SO_4. This is equivalent to a pH of approximately 0.5.

Sulfuric acid attacks the matrix of the concrete, which is commonly composed of calcium silicate (CaSi) hydrate gel, calcium carbonate ($CaCO_3$) from aggregates (when present), and unreacted $Ca(OH)_2$. Although the reaction products are complex and result in the formation of many different compounds, the process can be generally illustrated by the following reactions:

$$H_2SO_4 + CaSi \leftrightarrow CaSO_4 + Si + 2H^+$$

$$H_2SO_4 + CaCO_3 \leftrightarrow CaSO_4 + H_2CO_3$$

$$H_2SO_4 + Ca(OH)_2 \leftrightarrow CaSO_4 + 2H_2O$$

The primary product of concrete decomposition by SO_4^{2-} is calcium sulfate ($CaSO_4$), more commonly known by its mineral name, gypsum. From experience with this material in its more common form of drywall board, we know that it does not provide much structural support, especially when wet. It is usually present in sewers and structures as a pasty white mass on concrete surfaces above the water line. In areas where diurnal or other high flows intermittently scour the walls above the water line, concrete loss can occur rapidly. The surface coating of gypsum paste can protect underlying sound concrete by providing a buffer zone through which freshly produced H_2SO_4 must penetrate. Because *Thiobacillus* bacteria are aerobic, they require free atmospheric oxygen to survive. Therefore, they can only live on the thin outer covering of any surface. This means that acid produced on the surface must migrate through any existing gypsum paste to reach sound concrete. When the gypsum is washed off, fresh surfaces are exposed to acid attack and this accelerates the corrosion.

The color of corroded concrete surfaces can be various shades of yellow caused by the direct oxidation of H_2S to elemental sulfur. This only occurs where a continuous high-concentration supply of atmospheric oxygen or other oxidants is available. The upper portions of manholes and junction boxes exposed to high H_2S concentrations are often yellow because of the higher oxygen content there. This same phenomenon can be observed around the outlets of odor scrubbers using hypochlorite solutions to treat high concentrations of $H_2S(g)$.

Another damaging effect of H_2SO_4 corrosion of concrete is the formation of a mineral called *ettringite*. The chemical name for ettringite is calcium sulfoaluminate hydrate. It is produced by a reaction between $CaSO_4$ and alumina, which is found in virtually all cements. It forms at the boundary line between the soft $CaSO_4$ layer and the sound, uncorroded concrete surface. Ettringite is damaging because it is an expansive compound that occupies more space than its constituents. When ettringite forms, it lifts the corroded concrete away from the sound concrete and causes a faster corrosion by continually exposing new surfaces to acid attack. Although the rate of concrete loss is dependent on a number of factors, including ettringite formation, it is not uncommon to see concrete loss of 1 in./year in heavy sulfide environments.

16.14.6 Conducting a Corrosion Survey

As a part of a sewer system evaluation survey (SSES), sewer systems should be examined for certain characteristics that encourage corrosion. Corrosion is more likely to occur when the DO is below 0.5 mg/L because this condition favors the anaerobic bacteria that convert sulfate to sulfide. In gravity sewers, DO levels are likely to decrease when the wastewater velocity decreases because the lower velocity (a) decreases the scouring of the microbial slime growing on the submerged pipe walls and invert, (b) promotes solids deposition, and (c) increases the residence time. In force mains, inverted siphons, and surcharged sewers, anaerobic conditions often exist since the pipe is full, thereby precluding surface aeration by oxygen addition from the sewer atmosphere.

The depth of flow in sewers, the amount of exposed surface area, BOD of the waste, and pipe slope also affect the DO. Turbulence (which is found at pipe junctions and places where the pipe changes direction or slope) can add oxygen to the water, preventing sulfide generation. However, if sulfide is present, turbulence has an overall negative effect because it allows H_2S gas to be released from the wastewater, making it available for corrosion above the water level.

16.14.6.1 Identifying Likely Locations for Corrosion

The first step in a corrosion survey is to interview people involved in design, cleaning, inspection, and repair of the sewer system to identify locations where corrosion may have been observed. Collection system maps should be available that include sizes and types of pipes, slopes of lines, flows, manhole locations, frequency of pumping operations, and locations of force main discharges and surcharged sewers. Maintenance records, odor complaint files, and TV inspection logs can also be informative. Likely field locations to check include these:

1) Locations of low velocities or solids deposits
2) Force main discharge points
3) Transition manholes
4) Sewage lift stations
5) Areas of high turbulence
6) Sewers with flat slopes and long detention times
7) Inverted siphon discharges
8) Headworks of wastewater treatment plants
9) Junction chambers and metering stations.

16.14.6.2 Performing Visual Inspections

A visual inspection of the condition of manholes, metering stations, wet wells, headworks, and other structures as a part of the SSES physical survey is essential to identify corrosion problems. Areas that are accessible can be entered and inspected. However, hazardous atmospheres can exist in such confined spaces, and proper safety procedures for confined space entry must be strictly followed. These items should be noted in a visual inspection:

1) Condition of ladder rungs, bolts, conduit, and other metal components
2) Presence of protruding concrete aggregate
3) Presence of exposed reinforcing steel
4) Development of black coating (copper sulfate) on copper pipes and electrical contacts
5) Loss of concrete from pipe crown or walls
6) Soundness of concrete
7) Depth of penetration, using screwdriver or a sharp tool to expose uncorroded material.

A quick method of inspecting the general condition of sewers can be performed with a telescoping rod onto which are attached a halogen light and adjustable mirror at one end, and a low-magnification (e.g.,4×) sight scope at the other end. The rod is inserted into a manhole, and by slightly tilting the rod and flashing the light beam down the sewer, its condition can be observed. This procedure is useful when small-diameter sewers are involved. Also, because entry into a confined space is not required, there is little risk of being overcome by potentially harmful sewer gas.

16.14.6.3 Collecting Data

Useful data that can be collected to assess the presence of or the potential for corrosion include the following:

1) Concentration of gaseous H_2S in manholes and sewer atmospheres
2) Wastewater pH
3) Total and dissolved sulfide in wastewater
4) DO and oxidation-reduction potential (ORP)
5) Surface pH on manhole and sewer walls
6) Total and soluble BOD
7) Temperature
8) Depth of corrosion penetration.

One of the most useful early warning indicators of potential H_2S corrosion problems is the pH of the pipe crown or structure wall. This is a simple test using color-sensitive pH paper, which is applied to the moist crown of the pipe. New concrete pipe has a pH of 10–11. After aging, the pH of the crown under noncorrosive conditions may drop to near neutral. Pipe experiencing severe H_2S corrosion may have a pH of 2 or lower. pH levels below 4 are generally indicative of corrosion problems.

If it is possible to estimate the amount of concrete lost from sewers or manholes and their age is known, the rate of corrosion can be approximated. Estimates of the remaining useful life of a structure (e.g., to exposure of reinforcing steel) can then be used to prioritize sewer segments or structures for further action.

Predicting Sulfide Corrosion

Models have been developed that allow for the prediction of the rate of sulfide accumulation in sewers as well as the rate of hydrogen sulfide corrosion of concrete pipe. The predictive equations can be found in several of the entries in the bibliography at the end of this chapter.

16.14.7 Rehabilitating Corroded Sewers

If it is determined that a sewer is severely corroded and will require rehabilitation, an appropriate rehabilitation method must be chosen. Table 16.9 lists common sewer problems involving corrosion and provides applicable rehabilitation methods for each. Rehabilitation techniques are discussed in detail in the following chapter.

16.14.8 Sulfide Corrosion Control

H_2S corrosion can be controlled by reducing the levels of dissolved sulfide in the wastewater. Common control techniques include oxygenation, oxidation, precipitation, and pH elevation.

If the concentration of DO in the wastewater exceeds 1.0 mg/L, sulfides will not be generated. Therefore, maintaining a high DO concentration is an effective method of sulfide corrosion control. Common methods of H_2S control in sewer systems are summarized next.

16.14.8.1 Aeration

Aeration can be a cost-effective method for controlling sulfide generation, but unless air is introduced by passive means, such as the presence of turbulent conditions in the system, equipment must be provided to compress the air and to introduce it into the wastewater. An advantage of using air injection can be the simplicity of equipment when compared to other

Table 16.9 Common Sewer Corrosion Problems and Applicable Rehabilitation Methods

Problem	Rehabilitation Method
1. Severe corrosion and poor structural integrity	a. Excavation and replacement b. Sliplining c. Some specialty concretes
2. Severe corrosion; minor structural reinforcement needed	a. Cured-in-place inversion lining b. Sliplining c. Some specialty concretes d. Fold and formed pipe
3. Corrosion in structurally sound pipes with diameters 76 cm (2.5 ft) or greater	a. PVC or other corrosion-resistant liners b. Sliplining c. Cured-in-place inversion lining d. Some specialty concrete e. Fold and formed pipe
4. Corrosion in noncircular pipes	a. Cured-in-place inversion lining b. Some specialty concretes
5. Corroded pipes under busy streets	a. Cured-in-place inversion lining b. Fold and formed c. Sliplining (may be applicable)

sulfide control methods. Often, compressed air is added to the discharge side of the wastewater pumps at the upstream end of a force main. The major disadvantages of this approach are (a) the potential for gas pocket formation and air binding and (b) the relatively short duration for which aerobic conditions can be maintained due to DO uptake by bacteria.

16.14.8.2 Pure Oxygen

Because pure oxygen is five times more soluble in water than in air, it is possible to achieve higher DO levels in wastewater by injecting pure oxygen into wastewater instead of air. As with air injection, use of pure oxygen as a sulfide control measure is particularly advantageous in pressurized systems because dissolution of oxygen is greater at higher pressures. However, because less oxygen gas is required than air to achieve the desired DO levels, the potential for gas pocket generation in force mains is substantially reduced. Maintaining the DO above 1 mg/L is usually sufficient to prevent sulfate reduction.

16.14.8.3 Hydrogen Peroxide

When hydrogen peroxide (H_2O_2) is added to wastewater, it oxidizes dissolved sulfide. Excess H_2O_2 decomposes to water and oxygen. Common dosage rates are 1–5 lb. H_2O_2/lb. H_2S (1–5 kg H_2O_2/kg H_2S), depending on the degree of control desired, wastewater characteristics, sulfide levels, and length of time involved between the injection and sulfide control point. Equipment used for H_2O_2 addition is relatively simple, consisting mainly of a storage vessel and metering pumps. Materials for storage and feed equipment must be compatible with H_2O_2.

16.14.8.4 Potassium Permanganate

Potassium permanganate ($KMnO_4$) is a strong oxidizing agent that has a reaction similar to that of H_2O_2. It is normally supplied in a dry state and is fed as a 6% solution in water. Therefore, equipment for dissolving and feeding it must be supplied. Because of its high costs, it is not commonly used for sulfide control in wastewater collection systems.

16.14.8.5 Chlorine

Chlorine will oxidize sulfide to sulfate or to elemental sulfur, depending on pH. It is commonly added at a dosage rate of 10–15 lb. Cl_2/lb. H_2S removed (10–15 kg Cl_2/kg H_2S removed). It may be added as sodium hypochlorite or a chlorine solution using equipment similar to that installed in wastewater treatment plants for effluent disinfection. A disadvantage of using chlorine is the formation of dangerous byproducts: chlorinated organic compounds.

16.14.8.6 Iron Salts

Iron salts react with sulfide to produce insoluble precipitates and are added after the generation of sulfides has occurred to tie up the dissolved sulfide and prevent the release of H_2S into the sewer atmosphere. Dosages are usually dependent on initial sulfide levels and the targeted level of control, but will generally be 4–15 lb. Fe/lb. H_2S (4–15 kg Fe/kg H_2S). Iron salts may be purchased as dry chemicals and dissolved in water for ease of injection, but are commonly purchased as a solution. Ferrous chloride and ferrous sulfate are often purchased in bulk, usually as a 40% solution, and being acidic in nature, they must be handled in corrosion-resistant materials. As with other sulfide control chemicals (such as hydrogen peroxide), a typical feed system involves feeding the iron solution at multiple rates in relation to diurnal fluctuations in dissolved sulfide and flow rate.

16.14.8.7 Sodium Hydroxide

H_2S corrosion may also be controlled by inactivating the sulfate-reducing bacteria. Sewer systems in Los Angeles, as well as other cities, have used pH elevation to control sulfide corrosion. A caustic solution is added as a shock dosage for 20–30 minutes, raising the pH in the sewer to 12–13. Soon after the high-pH slug has passed, the sulfate-reducing bacteria will become reestablished so that dosing with caustic must be repeated when sulfide levels begin to increase. Typically, the caustic will be added at intervals that vary from several days to two weeks.

With this approach, caution must be taken to avoid upsetting the biological treatment system with the slug of high-pH wastewater. The slug may be diluted as it passes through the sewer system, in which case there is no problem. If this is not the case, the slug can be diluted by directing it to spare tankage and slowly adding it to the treatment plant influent.

Although a pH above 8.0 will result in lower levels of dissolved H_2S gas, continuously adding caustic to maintain a high pH is not generally practical.

16.14.8.8 Sodium Nitrate

The addition of sodium nitrate to sulfate-containing wastewaters will suppress the generation of H_2S. This occurs because bacteria will preferentially reduce nitrates before sulfates. Thus, no sulfides will be produced until the sodium nitrate has all been reduced.

16.14.8.9 Designing to Avoid Corrosion

Sulfide corrosion can be minimized through careful design of the sewer system. The corrosion resistance of the materials used is an important factor. The most important considerations when designing for sulfide control are:

1) Minimizing the occurrence of force mains, siphons, and surcharged sewers
2) Designing for velocities that are sufficient to prevent the accumulation of solids and that provide surface aeration of the wastewater
3) Prohibiting the direct addition of sulfides from any source.

Design practices can affect the degree of corrosion found in a sewer system. If pipes are oversized, wastewater flow rates will be reduced and organic materials may accumulate. This condition is favorable for the production of sulfide and can lead to corrosion problems. O&M practices also affect corrosion. Because accumulation of organic solids is favorable to initiation of sulfide generation, regular cleaning of oversized segments will help prevent these problems. Prolonged surcharging, or other conditions that result in full pipes, should be avoided because lack of air in the pipes can increase sulfide production. Short periods of surcharging, however, can be beneficial because the increased flow rates may wash out accumulated solids. Lift station and force main designs that allow long pipeline detention times increase the pressure of H_2S.

16.14.9 Control of Other Forms of Corrosion

Where industrial wastes contribute to H_2S generation, industries that discharge to the sewer should be required to make that discharge meet some type of pretreatment standards. If the wastewater in the sewers is found to be significantly acidic or basic, the pH should be adjusted.

Moisture must be present for most types of corrosion to occur. For example, a moist atmosphere allows corrosive vapors to condense on pipe walls and on other fixtures in the sewer system. Ventilation systems have been used to reduce moisture and corrosion in sewers.

When designing a new system or making replacements in an old system, materials should be chosen that are resistant to the types of waste present. Guidelines for selecting resistant materials were provided in Chapter 15.

If electrochemical corrosion of iron or steel is the problem, cathodic protection may be appropriate. One type of cathodic protection involves impressing an electrical current on the corroding surface. The metal surface that is to be protected is electrically connected to the negative terminal of a current source and a sacrificial anode is connected to the positive terminal. The sacrificial anode must be in the electrolytic wastewater and must be constructed of a material that has a higher electrical potential than the protected material. Cathodic protection does not always work well, so an experienced corrosion control engineer should be consulted before such a system is attempted.

16.15 Sewer Sediment Control

Generally, if sediments are left to accumulate in pipes, hydraulic restrictions can result in blockages at flow line discontinuities. Otherwise, the bed level reaches an equilibrium level. Conventional sewer cleaning techniques include rodding, balling, flushing, and the use of poly pigs or bucket machines. These methods are used to clear blockages once they have formed but also serve as preventive maintenance tools to minimize future problems. With the exception of flushing, these methods are generally used in a "reactive" mode to prevent or clear up hydraulic restrictions.

Flushing of sewers has been a concern dating back to the Romans. The concept of sewer flushing involves inducing an unsteady wave by either rapidly adding external water or creating a "dam break" effect by quickly opening a restraining gate. This aim is to resuspend, scour, and transport deposited pollutants to the wastewater treatment plant or to displace solids deposited in the upper reaches of large collection systems closer to the system outlet. The control idea is either to reduce depositing pollutants that may be resuspended and subsequently overflow during high flows or to decrease the time of concentration of solids transport within the collection system. During wet-weather events in combined sewers, these accumulated loads may then be more quickly displaced to the treatment headworks before overflows occur or be more efficiently captured by wet-weather "first-flush" capture storage facilities.

Manual flushing methods usually involve discharge from a fire hydrant or quick opening valve from a tank truck to introduce a heavy flow of water into the line at a manhole. Equipment for automated cleansing of sediments in both sewer pipes and CSO tanks is available. These include flushing gates, which are considered the most efficient method for flushing large-diameter, low-slope sewers. The tipping flusher and flushing gate technology are the most cost-effective means for flushing solids and debris from CSO storage tanks.

Problems/Questions

16.1 What are the four most commonly used pipe materials in sewers?

16.2 What considerations should you take into account in the choice of sewer pipe material? What pipe characteristics should be taken into consideration?

16.3 List and describe the types of concrete pipe used in sewers.

16.4 Describe the types of plastic pipe used in sewers.

16.5 What are the three types of pipe material used in trenchless rehabilitation? Briefly describe their most important properties.

16.6 Discuss the applicability of different pipe materials and list the background information needed in determining what type of pipe best fits a particular situation.

16.7 List the factors that affect sewer construction costs.

16.8 Where are manholes required?

16.9 What are the purposes of pumps and pumping stations in wastewater systems?

16.10 What are the two types of wastewater pumping stations? Discuss their advantages and disadvantages.

16.11 Discuss the four common performance characteristics of lift station pumps.

16.12 List the factors that affect lift station costs.

16.13 A pumping station consists of a wet well and a dry well. The wet well receives the influent wastewater, while the dry well holds the pumps and other equipment and fittings. The pump's flow rate and efficiency are evaluated, in part, by measuring the time for the pump to fill or empty a portion of a wet well or diversion box when all inflow is blocked off. Determine the pumping rate, in gpm or L/s, if (a) the size of the wet well is $L \times W \times D = 12$ ft $\times$ 12 ft $\times$ 6 ft = 3.66 m $\times$ 3.66 m $\times$ 1.83 m, (b) the time for wastewater to drop 4.5 ft (1.37 m) in the wet well is 9 minutes and 30 seconds, and (c) the engineer measures the time it takes to lower the wet-well water level a distance of 4.5 ft (1.37 m) in depth.

16.14 The suction lift head, discharge head, and total head of a pump can be determined by the installation of a vacuum/pressure gauge immediately on the suction side and a pressure/vacuum gauge immediately on the discharge side. Assume (a) the vacuum/pressure gauge on the suction side shows 2.5 in Hg vacuum, (b) the pressure gauge on the discharge side shows a positive 22-psi (152.7-kPa) pressure, (c) the two gauges are at the same height, and (d) the pipe diameters of both the suction side and the discharge side are the same. Determine the suction lift head, discharge head, and total head.

16.15 The efficiency of a pump can be determined in the field by determining the (a) actual pump capacity when a pump is withdrawing wastewater from a wet well, (b) actual suction lift head on the pump's suction side and actual discharge head on the pump's discharge side during pumping operation, (c) actual kilowatts drawn by the pump motor using a kilowatt meter, and (d) the electric motor's efficiency from the manufacturer. Determine the pump efficiency and draw a horsepower balance diagram for the motor-pump system assuming the following are known:

1) Actual pump capacity = 510 gpm (32.18 L/s).
2) Actual suction lift head on the pump's suction side during pumping operation = 2.83 ft (0.86 m).
3) Actual discharge head on the pump's discharge side during pumping operation = 50.82 ft. (15.49 m).
4) Actual kilowatts drawn by the pump motor using a kilowatts meter = 8 kW.
5) Electric motor's efficiency per manufacturer = 80 % .

16.16 A pump's characteristic curves, such as those shown in Fig. 16.12, are provided by the manufacturer. The head-capacity curve shows that the head of a centrifugal pump normally reduces as the capacity is increased. The pump's efficiency-capacity curve shows that the pump efficiency begins from zero at no pump discharge, increases to a maximum value, and then decreases as the capacity is increased. The brake horsepower-capacity curve shows that the brake horsepower (or the power input to the pump) increases with increasing capacity until it reaches a maximum, and then it normally reduces slightly. Analyze the pump characteristic curves at (a) the maximum efficiency point and (b) at the intersection point of the head-capacity curve and the efficiency-capacity curve.

16.17 A pumping station lifts wastewater from a collection tank A (wastewater surface elevation = 600 ft = 182.88 m above the datum of the pump) to treatment plant tank B (wastewater surface elevation = 700 ft = 213.36 m above the datum of the pump) through a force main (1,000 ft long and 1 ft in diameter, or 304.8 m long and 304.8 mm in diameter). The pump has the head-capacity characteristic curve shown in Fig. 16.18. What will be the discharge rate (gpm or L/s) of this pump and its operating head (ft or m)?

Assume the following conditions:

1) Reservoir A's contraction head loss = $Kv^2/2g$, where $K = 0.5$ and g = 32.2.
2) Reservoir B's expansion head loss = $Kv^2/2g$, where $K = 1.0$.
3) 90° smooth bend's head loss = $Kv^2/2g$, where $K = 0.35$.
4) Pipe friction loss = $f(L/D)v^2/2g$ where fraction factor $f = 0.015$ (*Note:* Hazen-Williams formula can also be used.)

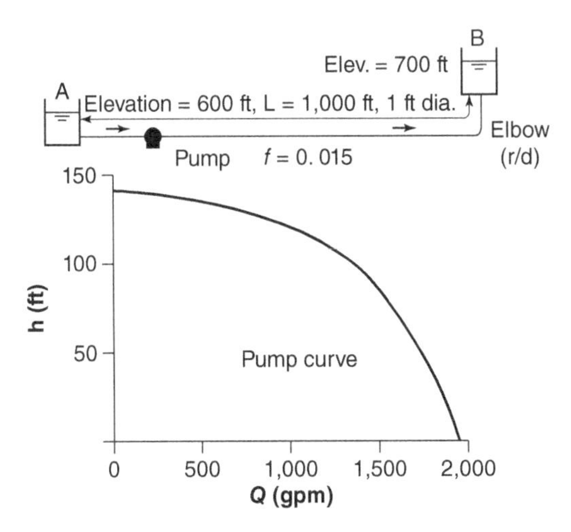

Figure 16.18 Pumping System and Pump Head-Capacity Characteristic Curve for Problem 16.17. Conversion factors: 1 gpm = 0.0631 L/s; 1 ft = 0.3048 m. (*Source:* Adapted from American Society of Plumbing Engineers)

16.18 Two centrifugal pumps have head-capacity characteristics and efficiencies as shown in Table 16.10. Pump 1 has a speed of 960 rpm and pump 2 has a speed of 1,150 rpm. What would be the combined discharge when both pumps work in parallel against a total dynamic head of 40 ft (12.19 m)?

Table 16.10 Head-Capacity Characteristics for Pumps of Problem 16.18

Flow (gpm)	TDH-1 (ft)	Eff-1 (%)	TDH-2 (ft)	Eff-2 (%)
0	50	60	—	—
25	46	20	57	20
50	43	50	54	40
75	39	70	50	60
100	35	75	45	75
125	29	70	40	80
150	20	50	34	78
175	0	25	27	71
200	0	—	20	60
225	0	—	11	40
250	0	—	0	—

Conversion factors: 1 gpm = 0.0631 L/s; 1 ft = 0.3048 m.

16.19 Considering the given data and the pump characteristic curves in Problem 16.18, what would be the combined discharge when both pumps work in parallel against a total dynamic head of 20 ft (6.10 m)?

16.20 Considering the given data and the pump characteristic curves in Problem 16.18, against what total dynamic head could the pumps deliver 75 gpm (4.73 L/s) when working in series?

16.21 The sewerage authority needs to conduct a comprehensive program of ongoing investigation and preventive maintenance. Why?

16.22 Sewer system analyses must be conducted to determine quickly and easily if there are serious infiltration/inflow (I/I) problems, evaluate the extent of these problems, and select the approach for further analysis and investigation. Name at least five occurrences that indicate the need for a sewer system analysis.

16.23 What are the possible causes of wastewater collection system structural failures?

16.24 To maintain a wastewater collection system and keep it functioning properly, certain preventive maintenance procedures are necessary. List and briefly explain such preventive procedures.

16.25 What are the major causes of sanitary sewer overflows? Discuss each cause and give examples.

16.26 What are the impacts of sanitary sewer overflows?

16.27 What factors contribute to exfiltration from sewers?

16.28 Discuss the problem of exfiltration from sanitary sewers and the approaches used to quantify their rates.

16.29 Discuss the types of sewer corrosion.

16.30 Identify the most likely locations for corrosion in sewers.

16.31 Explain the method of controlling sulfide corrosion in sewers by the addition of iron salts.

16.32 Discuss sewer sediment control.

Bibliography

American Concrete Pipe Association, Irving, TX, USA, http://www.concrete-pipe.org, accessed 2010, 2010.

American Concrete Pipe Association, *Sulfide and Corrosion Prediction and Control*, American Concrete Pipe Association, Vienna, VA, USA, 1984.

American Society of Chemical Engineers, *Sulfide in Wastewater Collection and Treatment Systems,* ASCE Manual and Reports on Engineering Practice No. 69, New York, USA, 1989.

Camp, Dresser & McKee Inc., *City of Albuquerque—Water Construction Program Evaluation—Sanitary Sewer Exfiltration Analysis*, September 1998.

Casada, D., Pump Optimization for Changing Needs, *Oper. Forum*, *9*, 5, 14–18, 1998.

Centers for Disease Control and Prevention, *Surveillance Summaries: Surveillance for Waterborne Disease Outbreaks—United States, 1997–1998*, Centers for Disease Control and Prevention, Atlanta, GA, 2000.

Decker, J., Environmental Hazard by Leaking Sewers, in *Proceedings of Water Quality International, IAWQ 17th Biennial International Conference, Budapest, Hungary, July 24–29*, 1994.

Department of Public Works, *Sewer Line Maintenance*, http://www, Department of Public Works, Anne Arundel County, Annapolis, MD, 2008. http://aacounty.org/DPW/Utilities/sewerLine.cfm.

Ductile-Iron Pipe Research Association, Birmingham, AL, http://www.dipra.org/ductile.html, accessed 2009, 2009.

Gravette, B. R., Benefits of Dry-Pit Submersible Pump Stations, in *Proceedings of the Water Environment Federation, 68th Annual Conference*, Miami Beach, FL, USA, Vol. *3*, pp. 187–196, October 1995.

Hoffman, M., and Lerner, D., Leak Free Sewers, *Water and Waste Treat.*, *35*, 8, 1992.

Hydraulic Institute, *Standards for Rotodynamic (Centrifugal) Pump Applications*, ANSI/HI 1.3, Parsippany, NJ, 2007.

Institution of Mechanical Engineers, *Centrifugal Pumps: The State of the Art and New Opportunities, IMechE Seminar*, John Wiley & Sons, New York, USA, August 2000.

Khadam, M., Shammas, N., and Al-Feraiheedi, Y., Water Losses from Municipal Utilities and Their Impacts, *Water Int.*, *13*, 1, 254–261, 1991.

Lamit, L. G., *Pipe Fitting and Piping Handbook*, Prentice-Hall, Upper Saddle River, NJ, USA, 1984.

Lerner, D. N., and Halliday, D., The Impact of Sewers on Groundwater Quality, in *Groundwater Problems in Urban Areas*, Thomas Telford, London, 1994.

Moser, A. P., *Buried Pipe Design*, McGraw-Hill, New York, USA, 1990.

National Clay Pipe Institute, Lake Geneva, WI, http://ncpi.org, accessed October 2008, 2008.

Paschke N. W., Pump Station Basics—Design Considerations for a Reliable Pump Station, *Oper. Forum, 14,* 5, 15–20 May 1997.

Rauch, W., and Stegner, T., The Culmination of Leaks in Sewer Systems during Dry Weather Flow, *Water Sci. Technol.*, *1*, 30, 205–210, 1994.

Sanks, R. L., Tchobanoglous, G., Newton, D., Bosserman, B. E., and Jones, G. M., *Pump Station Design*, Butterworth, Boston, MA, USA, 1998.

Seelye, E. E., *Data Book for Civil Engineers, Vol. 1, Design*, John Wiley & Sons, New York, USA, 1960.

Shammas, N. K., The Cracking Problem in Unplasticized Polyvinyl Chloride Fittings Used in Riyadh Domestic Sewer Connections, *J. Pipelines*, *4*, 299–307, 1984.

Tullis, J. P., *Hydraulics of Pipelines: Pumps, Valves, Cavitation, Transients*, John Wiley & Sons, New York, USA, 1989, online publication, December 12, 2007.

U.S. Army Corps of Engineers, Yearly Average Cost Index for Utilities, in *Civil Works Construction Cost Index System Manual, 110-2-1304*, U.S. Army Corps of Engineers, Washington, DC, USA, March, 2022.

U.S. Environmental Protection Agency, *Benefits of Measures to Abate Sanitary Sewer Overflows (SSOs)*, Contract No. 68-C6-0001, U.S. Environmental Protection Agency, Washington, DC, August 1999.

U.S. Environmental Protection Agency, *Design Manual: Odor and Corrosion Control in Sanitary Sewerage Systems and Treatment Plants, EPA/625/1-85/018*, U.S. Environmental Protection Agency, Washington, DC, 1985.

U.S. Environmental Protection Agency, *Detection, Control, and Correction of Hydrogen Sulfide Corrosion in Existing Wastewater Systems, EPA/430/9–91 /019*, U.S. Environmental Protection Agency, Washington, DC, 1991.

U.S. Environmental Protection Agency, *Exfiltration in Sewer Systems, EPA/600/R-01/034*, U.S. Environmental Protection Agency, Cincinnati, OH, December 2000.

U.S. Environmental Protection Agency, *Pipe Construction and Materials, Wastewater Technology Fact Sheet*, EPA 832-F-00-068, U.S. Environmental Protection Agency, Washington, DC, September 2000.

U.S. Environmental Protection Agency, *Rehabilitation of Wastewater Collection Systems, Science Brief, EPA/600/F-07/012*, U.S. Environmental Protection Agency, Washington, DC, September 2007.

U.S. Environmental Protection Agency, *Report to Congress: Hydrogen Sulfide Corrosion in Wastewater Collection and Treatment Systems*, EPN 430/9-91/009, U.S. Environmental Protection Agency, Washington, DC, 1991.

U.S. Environmental Protection Agency, *Sewer and Tank Flushing for Corrosion and Pollution Control, EPA/600/ J-01/120*, U.S. Environmental Protection Agency, Washington, DC, 2001.

U.S. Environmental Protection Agency, *Sewer System Infrastructure Analysis and Rehabilitation, EPA/625/6-91/030*, U.S. Environmental Protection Agency, Cincinnati, OH, October 1991.

U.S. Environmental Protection Agency, *Sewers Lift Station, Collection Systems Technology Fact Sheet*, EPA 832-F-00-073, U.S. Environmental Protection Agency, Washington, DC, September 2000.

U.S. Environmental Protection Agency, *Sewers, Innovative and Alternative Technology Assessment Manual*, EPA 430/9-78-009, U.S. Environmental Protection Agency, Washington, DC, February 1980.

U.S. Environmental Protection Agency, *SSO Fact Sheet: Why Control Sanitary Sewer Overflows*, U.S. Environmental Protection Agency, Washington, DC, http://www.epa.gov/npdes/sso/control, 2008.

U.S. Environmental Protection Agency, *Water Quality Conditions in the United States: A Profile from the 1998 National Water Quality Inventory Report to Congress*, EPA 841-F-00-006, U.S. Environmental Protection Agency, Washington, DC, June 2000.

Walski, T. M., Barnard, T. E., Meadows, M. E., and Whitman, B. E., *Computer Applications in Hydraulic Engineering*, 7th ed., Bentley Institute Press, Exton, PA, USA, 2008.

Water Infrastructure Network, *Clean Safe Water for the 21st Century: A Renewed National Commitment to Water and Wastewater Infrastructure*, Water Infrastructure Network, Washington, DC, USA, August 2000.

17

Trenchless Technology and Sewer System Rehabilitation

System *rehabilitation* is the application of infrastructure repair, renewal, and replacement technologies in order to reinstate the functionality of a wastewater system or subsystem. As the infrastructure in the United States ages, increasing importance is being placed on rehabilitating the nation's wastewater treatment collection systems. Cracks, settling, tree root intrusion, and other disturbances that develop over time deteriorate pipelines and other conveyance structures that comprise wastewater collection systems. These deteriorating conditions can increase the amount of *inflow* and *infiltration* (I/I) entering the system, especially during periods of wet weather. Increased I/I levels create an additional hydraulic load on the system and thereby decrease its overall capacity. In addition to I/I flow, stormwater may enter the wastewater collection system through illegal connections such as downspouts and sump pumps. If the combination of wastewater, infiltration, and illegal stormwater connections entering the wastewater treatment plant (WWTP) exceeds the capacity of the system at any point, untreated wastewater may be released into the receiving water. This bypass of untreated wastewater, known as a *sanitary sewer overflow* (SSO), may adversely affect human health as well as impair the usage and degrade the water quality of the receiving water.

Under the traditional method of sewer relief, a replacement or additional parallel sewer line is constructed by digging along the entire length of the existing pipeline. These traditional methods of sewer rehabilitation require unearthing and replacement of the deficient pipe (the *dig-and-replace method*), *trenchless methods of rehabilitation* use the existing pipe as a host for a new pipe or liner. Trenchless sewer rehabilitation techniques offer a method of correcting pipe deficiencies that requires less restoration and causes less disturbance and environmental degradation than the traditional dig-and-replace method.

Trenchless sewer rehabilitation methods include the following alternative techniques:

1) Pipe bursting, or in-line expansion
2) Sliplining
3) Cured-in-place pipe (CIPP)
4) Modified cross-section liner
5) Spot (point) repair.

17.1 Excavation and Replacement

17.1.1 Description

Replacement of deteriorated pipelines was once the most common rehabilitation practice, but is becoming more limited due to the availability of trenchless technologies. Excavation and replacement of defective pipe segments are normally undertaken under the following conditions:

1) When the structural integrity of the pipe has deteriorated severely; for example, when pieces of pipe are missing, pipe is crushed or collapsed, or the pipe has large cracks, especially longitudinal cracks
2) When the pipe is significantly misaligned
3) When additional pipeline capacity is also needed
4) When trenchless rehabilitation methods that would be adequate to restore pipeline structural integrity would produce an unacceptable reduction in service capacity

Water and Wastewater Engineering: Hydraulics, Hydrology and Management, Volume 1, Fourth Edition.
Lawrence K. Wang, Mu-Hao Sung Wang, and Nazih K. Shammas.

Companion website: www.wiley.com/go/Wang/Waterandwastewater4e

5) For point repair where short lengths of pipeline are too seriously damaged to be effectively rehabilitated by any other means
6) Where entire reaches of pipeline are too seriously damaged to be rehabilitated
7) Where removal and replacement are less costly than other rehabilitation methods.

The following are the disadvantages of pipeline removal and replacement as a method of sewer line rehabilitation:

1) Removal and replacement is usually more expensive than other rehabilitation methods.
2) Removal and replacement construction causes considerably greater and longer lasting traffic and urban disruption than does rehabilitation.
3) Removal and replacement construction involves a greater threat of damage to, or interruption of, other utilities than does pipeline rehabilitation.

17.1.2 Procedures and Equipment

Sewer pipe rehabilitation through pipeline replacement can be carried out in the following two general forms: (a) excavation and replacement, in which the existing pipeline is removed and a new pipeline is placed in the same alignment, and (b) abandonment and parallel replacement, in which the existing pipeline is abandoned in place and replaced by a new pipeline in either physically parallel alignment adjacent to the existing line or in a functionally parallel alignment along a different route

Pipeline replacement materials include traditional materials such as reinforced concrete, clay, ductile iron, and a variety of plastics.

Removal of an existing pipeline and replacement with new pipe involves all of the problems that occur in construction of new pipelines in new alignments, plus special problems that are unique to removal and replacement. The problems unique to removal and replacement are as follows:

1) Maintaining tributary system and/or service flows during construction
2) Removing and disposing of old pipes
3) Filling up old abandoned pipes with structurally sound material to prevent potential collapse
4) Working through utilities overlying or closely parallel to the pipe. These special problems usually result in added construction costs, which often do not occur in new construction along a new pipe alignment.

Problems that occur in both removal and replacement and in new construction on a new alignment are as follows:

1) Disruption of street traffic
2) Disruption of access to residential, commercial, and industrial properties
3) Temporary loss of street parking
4) Trench shoring in deep construction involving unstable soil
5) Trench dewatering in areas with high groundwater levels.

17.1.3 Costs

The cost of pipeline excavation and replacement is specific to individual job conditions. Cost factors include the following:

1) Old pipe removal and disposal
2) Manhole removal and replacement
3) Trench shoring
4) New pipe materials installation
5) Service reconnections to the sanitary sewer
6) Street inlet reconnections to storm drains
7) Upstream flow diversion during construction
8) Maintenance of local sanitary service
9) Traffic control systems and scheduling
10) Pavement restoration
11) Interference with other utilities.

Table 17.1 presents the costs for the excavation and replacement method of sewer rehabilitation. The costs in this table were adjusted to the recent 2022 prices through the use of the U.S. Army Corps of Engineers' *Cost Index for Utilities*.

Table 17.1 Rehabilitation Costs for Excavation and Replacement

Pipe Diameter (in.)	Cost (2022 USD/linear ft)[a]
6	81–124
8	87–130
10	93–149
12	112–167
14	124–186
16	130–192
18	146–211
20	167–254
30	236–360
40	341–422
48	409–496
54	422–515
60	484–592
72	639–787
90	812–967
102	930–1,128

Conversion factors : 1 in. = 2.54 cm = 25.4 mm; 1 ft = 0.3048 m; 1 USD/ft = 3.28 USD/m.
[a] USD = US dollars (March 2022).

17.2 Chemical Grouting

Chemical grouting of sewer lines is mainly used to seal leaking joints and circumferential cracks. Small holes and radial cracks may also be sealed by chemical grouting. Chemical grouts can be applied to pipeline joints, manhole walls, wet wells in pump stations, and other leaking structures using special tools and techniques. One grouting procedure is illustrated in Fig. 17.1. All chemical grouts are applied under pressure after appropriate cleaning and testing of the joint.

Chemical grouting is used in precast concrete brick, vitrified-clay sewers, and other pipe materials to fill voids in backfill outside the sewer wall. Such backfill voids can reduce lateral support of the wall and allow outward movement, resulting in the rapid deterioration of the structural integrity as the arch of the top of the pipe loses its support. Chemical grouting adds no external structural properties to the pipe where joints or circumferential cracking problems are due to ongoing settlement or shifting of the pipelines. It is not effective to use chemical grouting to seal longitudinal cracks or to seal joints where the pipe near the joints is longitudinally cracked. Grouting is a joint sealing technique to be used for each joint in a pipeline segment that fails the initial leakage test. Chemical grouting is normally undertaken to control groundwater infiltration in nonpressure pipelines when these are caused by leaking pipe joints or circumferential cracking of pipe walls.

Chemical grouting is applied internally within a pipe, and thus does not damage or interfere with other underground utilities. It does not require excavation or surface restoration, such as pavement or sidewalk replacement and ground cover reseeding.

Chemical grouting does not improve the structural strength of the pipeline and thus should not be considered when the pipe is severely cracked, crushed, or badly broken. Chemical grouts may also dehydrate and shrink if the groundwater drops below the pipeline and the moisture content of the surrounding soil is reduced significantly. Large joints and cracks may be difficult to seal because large quantities of grout may be required. Large cracks, badly offset joints, and misaligned pipes may not be sealable. Offset joints may prevent the inflatable rubber sleeves of the sealing unit from seating properly against the walls of the pipe, making it impossible to isolate and seal the joint.

The most common chemical grouts are *acrylamide gel*, *acrylic gel*, *acrylate gel*, *urethane gel*, and *polyurethane foam*. The use of acrylamide gel as a grout has been banned by the U.S. Environmental Protection Agency (U.S. EPA) because it is suspected that this grout causes health problems for the application workers. The basic characteristics of foam and gel grouts are described next.

Gel grouts are resistant to most chemicals found in sewer lines but they may produce a gel-soil mixture that is susceptible to dehydration and shrinkage cracking. When using gel grouts, the grouting contractor and/or the grout supplier should be required to submit data supporting the nonshrink characteristics of the grout. Urethane gel uses water as the catalyst. No significant water contamination of the urethane grout should be permitted prior to its injection. Gel grouts are not recommended where there are large voids outside the pipeline joints.

Foam grouts consist of liquid urethane prepolymers, which are catalyzed by water during injection. The foaming reaction of the grout and water expands the materials into the joint cavity, thereby sealing the crack. The foam grouts are capable of expanding 8–12 times their initial volumes. Foam grouts are usually difficult to apply and are more expensive than gel grouts.

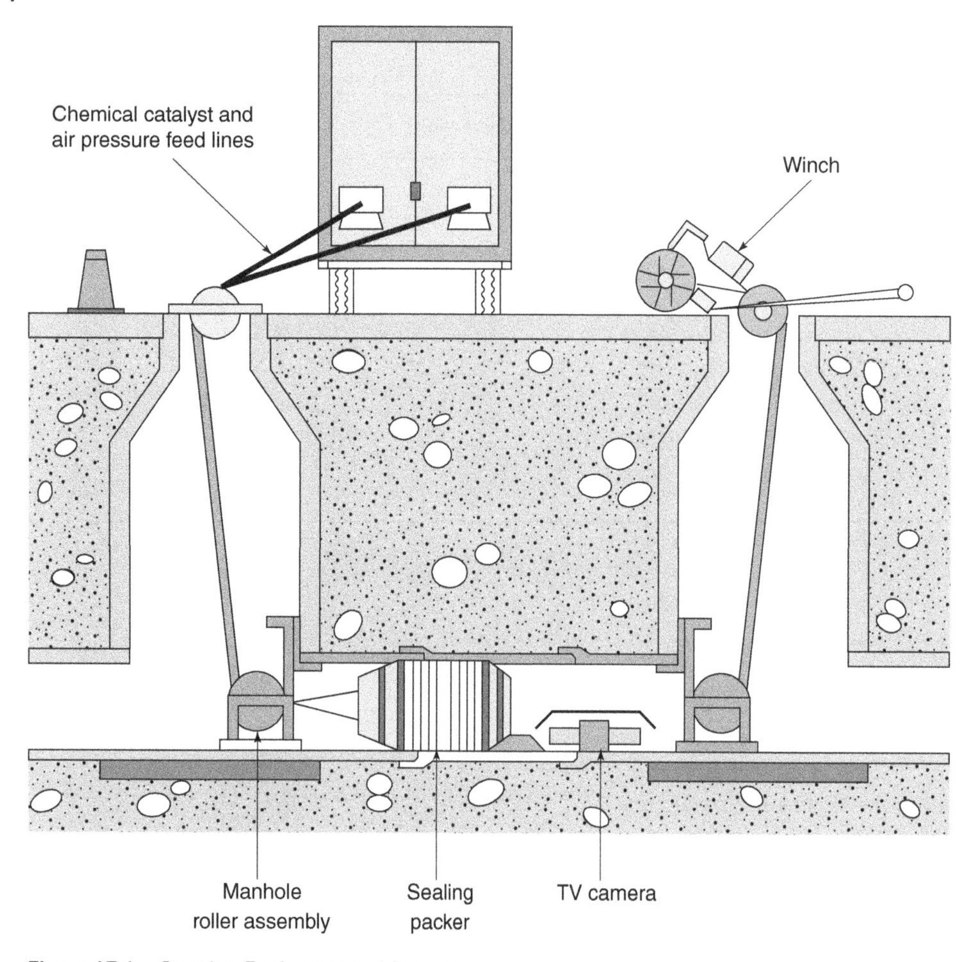

Figure 17.1 Grouting Equipment and Procedures

Before choosing to use grouting for joint rehabilitation, the pipeline should be inspected for the following:

1) Determine pregrouting cleaning needs and extent of root intrusions. For effective grouting, the pipeline must be relatively free of sand, sediment, and other deposits. Cleaning should occur just prior to grouting.
2) Identify crushed or broken sections that must be replaced. Deformed and longitudinally cracked pipe sections should not be grouted.

Joints and circumferential cracks in small- and medium-sized pipes of about 6–42 in. (150–1,100 mm) in diameter can be remotely tested and grouted using a *packer system* monitored by *closed-circuit TV*.

Pipe size, joint spacing, and the percentage of joints requiring sealing are factors to consider in determining the cost of chemically sealing a line. The larger the pipe, the higher the cost because of increased manpower, equipment, and materials. Chemical grouting requires rerouting of wastewater flow around the section being grouted until the grout is cured.

17.3 Pipe Bursting or In-line Expansion

Pipe bursting is a method of replacing existing sewers by fragmenting the existing pipe and replacing the pipe in the void. The *Pipebursting™ method*, patented by the British Gas Company in 1980, was successfully applied by the gas pipeline industry before its applicability was identified by other underground utility agencies. During the past two decades, other methods of in-line expansion have been patented as well. During in-line expansion, the existing pipe is used as a guide for inserting the *expansion head* (part of the bursting tool). The expansion head, typically pulled by a cable rod and winch, increases the area available for the new pipe by pushing the existing pipe radially outward until it cracks. The bursting device pulls the new pipeline behind itself.

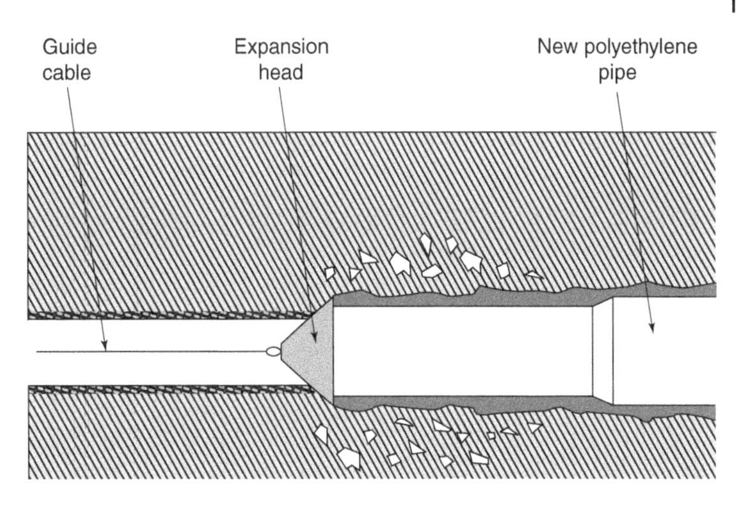

Figure 17.2 Pipe Bursting Process

The pipe bursting process is illustrated in Fig. 17.2. Various types of expansion heads, categorized as static or dynamic, can be used on the bursting tool to expand the existing pipeline. *Static heads*, which have no moving internal parts, expand the existing pipe only through the pulling action of the bursting tool. Unlike static heads, *dynamic heads* provide additional pneumatic or hydraulic forces at the point of impact. Pneumatic heads pulsate internal air pressure within the bursting tool, whereas hydraulic heads expand and collapse the head. While the dynamic head pulsates or expands and contracts, the bursting device is pulled through the existing pipeline and breaks up the existing pipe, replacing it with the new pipe directly behind it. Dynamic heads are often required to penetrate difficult pipe materials and soils. However, because dynamic heads can cause movement of the surrounding soils—resulting in additional pressure and ground settlement—static heads are preferred where pipe and soil conditions permit.

During the pipe bursting process, the rehabilitated pipe segment must be taken out of service by rerouting flows around it. After the pipe bursting is completed, laterals are reconnected, typically with robotic cutting devices.

17.4 Sliplining or Insertion

17.4.1 Description

Sliplining is a well-established method of trenchless rehabilitation. During the sliplining process, a new liner of smaller diameter is inserted inside the existing pipe. This is done by pulling or pushing new pipe into a deteriorated pipeline. The liner forms a continuous, watertight length within the existing pipe after installation. This is followed by reconnecting the service connection to the new liner. The annular space, or area between the existing pipe and the new pipe, is typically grouted to prevent leaks and to provide structural integrity.

Pipe insertion techniques can be used to rehabilitate sewer, water, and other pipelines that may have severe structural problems such as extensive cracks, lines in unstable soil conditions, deteriorated pipes in corrosive environments, pipes with massive and destructive root intrusion problems, and pipes with relatively flat grades.

In most sliplining applications, manholes cannot function as proper access points to perform the rehabilitation. In these situations, an insertion pit must be dug for each pipeline segment. Because of this requirement, in most applications, sliplining is not a completely trenchless technique. However, the excavation required is considerably less than that for the traditional dig-and-replace method. System and site conditions will dictate the amount of excavation spared.

The advantages and disadvantages of sliplining as a method of rehabilitation are listed in Table 17.2.

The most popular materials used to slipline sewer lines are *polyolefins*, *fiberglass-reinforced polyesters* (FRPs), *reinforced thermosetting resins* (RTR), *polyvinyl chloride* (PVC), and *ductile iron* (cement-lined and polyvinyl-lined).

Polyethylene (PE) is the most common polyolefin material used and is available in low density, medium density, and high density. *High-density PE* (HDPE) compounds are best suited for rehabilitation applications because they have good stiffness and are hard, strong, tough, and corrosion resistant. PE pipe is manufactured as either *extruded* or *corewall*. Extruded pipe has smooth inner and outer surfaces and has structural characteristics that are determined by wall thickness. Corewall pipe

Table 17.2 Advantages and Disadvantages of Sliplining

Advantages	Disadvantages
• Minimal disruption to traffic and urban activities (as compared to replacement).	• Possible reduction in pipe capacity.
• Minimal disturbances to other underground utilities; affects only those in the vicinity of access pits.	• Requires excavation of an access pit.
• Significantly less costly than replacement	• Less applicable to sewers with numerous curves or bends, since multiple pits would be required.
• Quick installation time.	• Requires obstruction removal of internal obstructions prior to sliplining.
• Good protection against acid corrosion	• Installation difficulties may be encountered during grouting of annular space.
• Does not require bypassing.	
• Wide range of pipe sizes (i.e., 3–144 in.)	
• Can be used to rehabilitate pipelines with severe corrosion.	

Conversion factors: 1 in. = 2.54 cm = 25.4 mm.

has an exterior hollow rib that gives structural integrity to the pipe and minimizes pipe weight. Extruded PE pipe is manufactured in diameters of 2–48 in. (50–1,200 mm) and corewall PE pipe 12–144 in. (300–3,600 mm).

Polybutylene (PB), another polyolefin, is similar to medium-density PE pipe in stiffness and chemical resistance but has better continued stress loading characteristics. It also has good temperature resistance. PB pipe is manufactured as extruded pipe in diameters of 3–42 in. (80–1,100 mm).

Centrifugally cast FRP pipe was originally manufactured in Switzerland in the 1960s and is now being widely used in the United States. It is a composite of resin made from fiberglass and sand that is formed within a revolving mold. It is frequently specified as an acceptable alternative to PE. FRP pipe has very good chemical and corrosion resistance and is suitable for use over a pH of 1–10. FRP pipe is manufactured in standard 20-ft (6-m) lengths and is available in diameters of 17–96 in. (450–2,400 mm).

RTR pipe is a composite of fibers and resins that are either spiral wound on a rotating mandrel or composited on the inside of a rotating drum, similar to FRP pipe. RTR pipe has good axial and longitudinal strength, enabling it to be pushed into an existing pipe without buckling. Friction losses are low due to smooth interior surfaces. This pipe has high strength and high modules of elasticity. It also has good corrosion, erosion, and abrasion resistance. Manufactured sizes include lengths of 20–80 ft. (6–24 m) and diameters of 4–142 in. (100–3,600 mm).

Flexible PVC has been used successfully as a sliplining material. PVC is highly resistant to acid attack and is very smooth, exhibiting good hydraulics. Grouting of the annular space is required to give strength to the PVC.

17.4.2 Procedures and Equipment

Prior to sliplining, the sewer should first be thoroughly cleaned and inspected by closed-circuit TV to identify all obstructions such as displaced joints, crushed pipes, and protruding service laterals. The inspection also should locate all service connections that will need to be connected to the new liner pipe. The pipe must be thoroughly cleaned. It may at times be necessary to proof-test the existing pipe by pulling a short piece of liner through the sewer section.

Sliplining is performed by either a *push* or a *pull* technique; both methods are illustrated in Fig. 17.3. In the pull method a pulling head is attached to the end of the pipe. A cable is run from the termination point to the access point and connected to the pulling head. The sliplining pipe is then pulled through the existing pipe with the cable by a track-mounted winch assembly. Pulling is used for extruded PE and PB pipe that is heat fused into a continuous length. The push method can be performed by either a backhoe or a jacking machine. A backhoe can be used for extruded heat-fused PE or PB pipe, corewall PE pipe, FRP pipe, and RTR pipe. However, the backhoe is limited to small-diameter and/or short-length pipe. The jacking machine is used for larger diameter, sleeve-coupled or bell spigot jointed pipes (e.g., corewall PE, FRP, and RTR) but it is not used for heat-fused polyolefin pipe.

For most insertion projects it is not necessary to eliminate the entire flow stream within the existing pipe structure. Actually, some amount of flow can assist positioning of the liner by providing a lubricant along the liner length as it moves

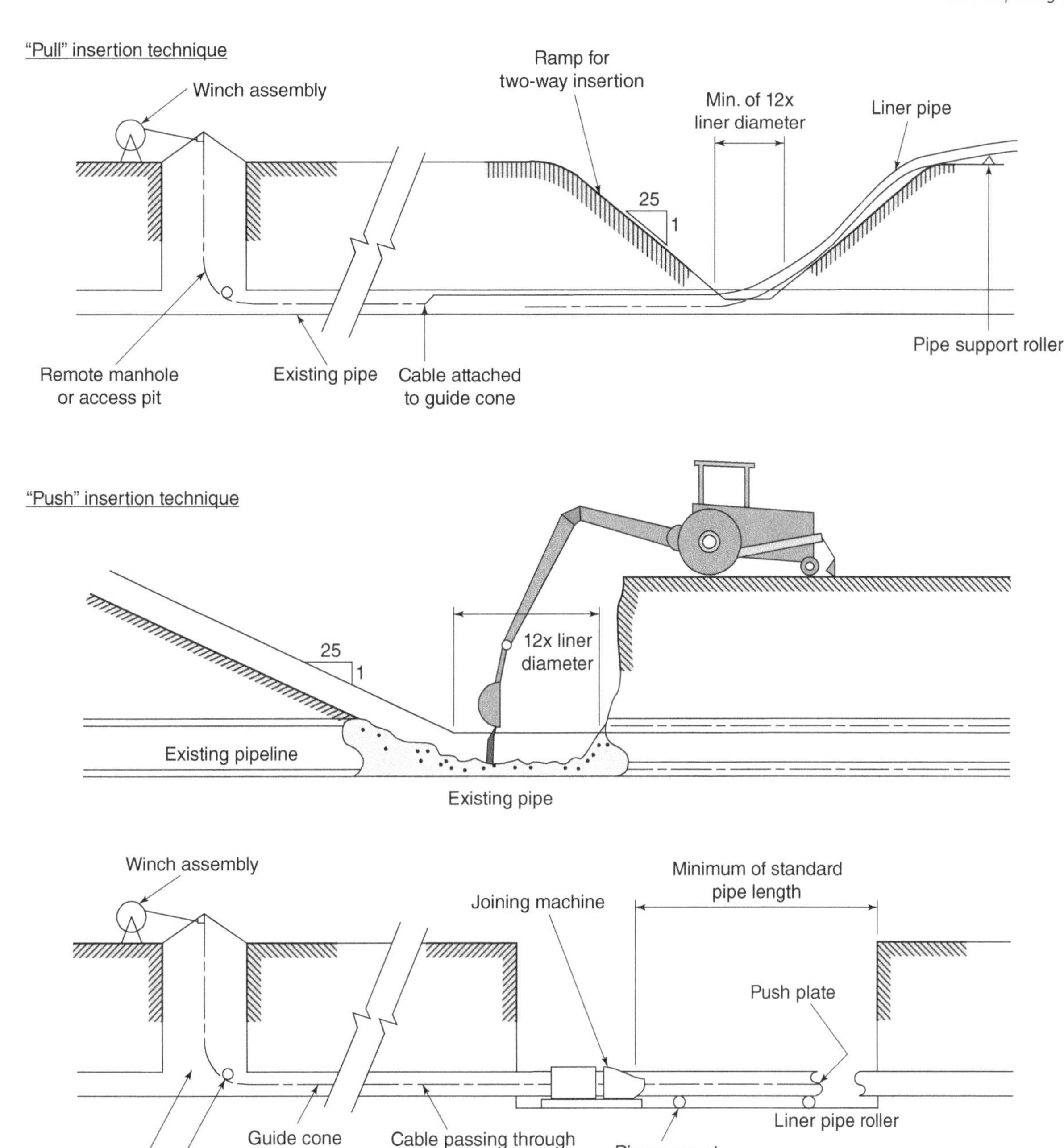

Figure 17.3 Insertion Methods

through the deteriorated pipe structure. Excessive flows can inhibit the insertion process; however, the insertion procedure should be timed to take advantage of cyclic periods of low flows that occur during the operation of most gravity piping systems. During the insertion process, often a period of 30 minutes or less, the annular space between the sliplining pipe and the existing pipe will probably carry sufficient flow to maintain a safe level in the operating section of the system being rehabilitated. Flow can then be diverted into the liner upon final positioning of the liner. During periods of extensive flow blockage, the upstream piping system should be monitored, and provisions for bypassing provided in order to avoid unexpected flooding and drainage areas.

Once the sliplining pipe has been pulled through the existing pipe, it is grouted in place. Grouting at manhole connection is required, but grouting of the entire length of the pipe is not required if the liner is strong enough to support loads in the event of collapse of the original pipe. It must be determined on a site-specific basis whether or not to grout after evaluating the severity of structural deterioration and anticipated hydrostatic and structural loadings. Grouting provides the following advantages:

1) Provides structural integrity.
2) Increases hydrostatic and structural loading capabilities.
3) Prevents liner from moving.
4) Locks in service connections.
5) Extends the service life of the pipe.
6) Provides increased temperature resistance.
7) Provides support to liner when cleaning.

Methods of sliplining include *continuous*, *segmental*, and *spiral* wound. All three methods require laterals to be reconnected by excavation or by a remote cutter. In continuous sliplining, the new pipe, joined to form a continuous segment, is inserted into the host pipe at strategic locations. The installation access point, such as a manhole or insertion pit, must be able to handle the bending of the continuous pipe section. Continuous pipe material can be either PE or polypropylene.

Installation by the segmental method involves assembling pipe segments at the access point. Sliplining by the segment method can be accomplished without rerouting the existing flow. In many applications, the existing flow reduces frictional resistance and thereby aids in the installation process. Generally, this method is used on larger sized pipe and forced into the host pipe. Pipe material can be any of the following:

1) PE
2) PVC
3) Reinforced plastic mortar
4) Fiberglass-reinforced plastic
5) Ductile iron
6) Steel.

Spiral-wound sliplining is performed within a manhole or access point by using interlocking edges on the ends of the pipe segments to connect the segments. This involves winding strips of PVC in a helical pattern to form a continuous liner on the inside of the existing pipe. The liner is then strengthened and supported with grout that is injected into the annular void between the existing pipe and the liner. A modified spiral method is also available that winds the liner pipe into a smaller diameter than the existing pipe. The spiral wound pipe is then inserted into the existing pipe as illustrated in Fig. 17.4.

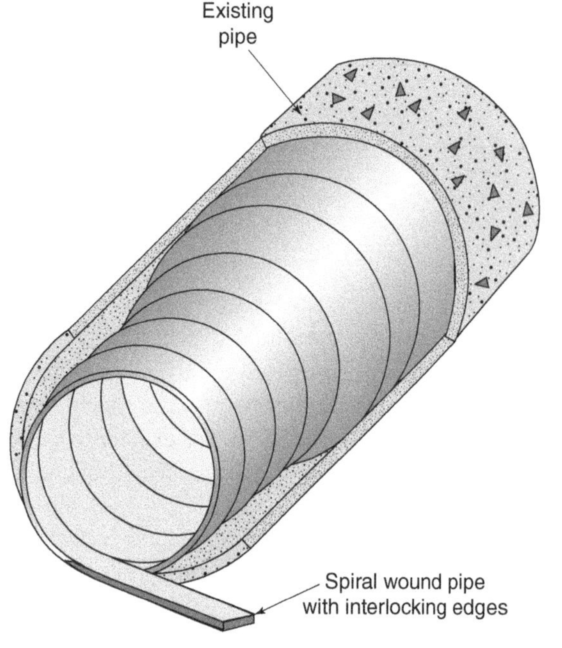

Figure 17.4 Spiral-Wound Sliplining Process

17.5 Cured-In-Place Pipe Lining

17.5.1 Description

Inversion lining is formed by inserting a resin-impregnated felt tube into a pipe, which is inverted against the inner wall of the pipe, and allowing it to cure. After the lining system has been installed and cured, a special cutting device is used with a closed-circuit TV camera to reopen service connections, which are located with the camera before the liner is installed. The pliable nature of the resin-saturated felt prior to curing allows installation around curves, filling of cracks, bridging of gaps, and maneuvering through pipe defects. After installation, the fabric cures to form a new pipe of slightly smaller diameter, but of the same shape as the original pipe. The new pipe has no joints or seams and has a very smooth interior surface, which may actually improve flow capacity despite the slight decrease in diameter.

Two resin types (*polyester* and *epoxy*) are widely used in this method of pipe rehabilitation. Both of these resins are liquid thermosetting resins and have excellent resistance to domestic wastewater. Chemical resistance tests should be specified for CIPP for other than domestic wastewater. *Vinylester* resins may be used where superior corrosion resistance is required at high temperatures. Epoxy resins are used where adhesion to the existing pipeline is desired.

Inversion lining is successful in dealing with a number of structural problems, particularly in sewers needing minor structural reinforcement. Caution must be used, however, in the application of this method to any structural problems involving major loss of pipe wall. Inversion lining can be accomplished relatively quickly and without excavation, and thus this method of pipeline rehabilitation is particularly well suited for repairing pipelines located under existing structures, large trees, or busy streets or highways where traffic disruption must be minimized. Inversion lining produces minor reductions in pipe cross-sections. It is applicable to noncircular pipes and pipes with irregular cross-sections. This method is also effective in correcting corrosion problems and can be used for misaligned pipelines or in pipelines with bends where realignment or additional access is not required. See Table 17.3 for a summary of the advantages and disadvantages of sewer rehabilitation by *cured-in-place lining.*

17.5.2 Procedures and Equipment

Installation of cured-in-place inversion lining is carried out by inserting the resin-impregnated fabric tube (turned inside out) into the existing pipeline and inverting it as it progresses inside the pipe. It is then cured in place through the use of heated water or air steam. Prior to the installation of the liner, the pipeline section must be cleaned to remove loose debris, roots, protruding service connections, and excessive solids. A typical CIPP process by the *water inversion* method is illustrated is Fig. 17.5. The pipeline segment must be isolated from the system by bypassing flows during the installation of the

Table 17.3 Advantages and Disadvantages of Cured-in-Place Lining

Advantages	Disadvantages
• Applicable to all shapes • Rapid installation • Minimum traffic disruption	• Bypass required during installation • Post-installation remote camera inspection required • Maximum effluent temperature 82 °C (180 °F) using specially formulated resins
• Excavation normally not required • In-line lateral reconnections • Improved hydraulics • Bridges gaps and misaligned joints • Special resins are available to provide acid resistance • Custom-designed wall thickness to aid in structural strength • Only 50%–70% of replacement costs • Adds some structural integrity • Does not interfere with or damage other utilities • No pavement repairs • Safer than some other rehabilitation methods	

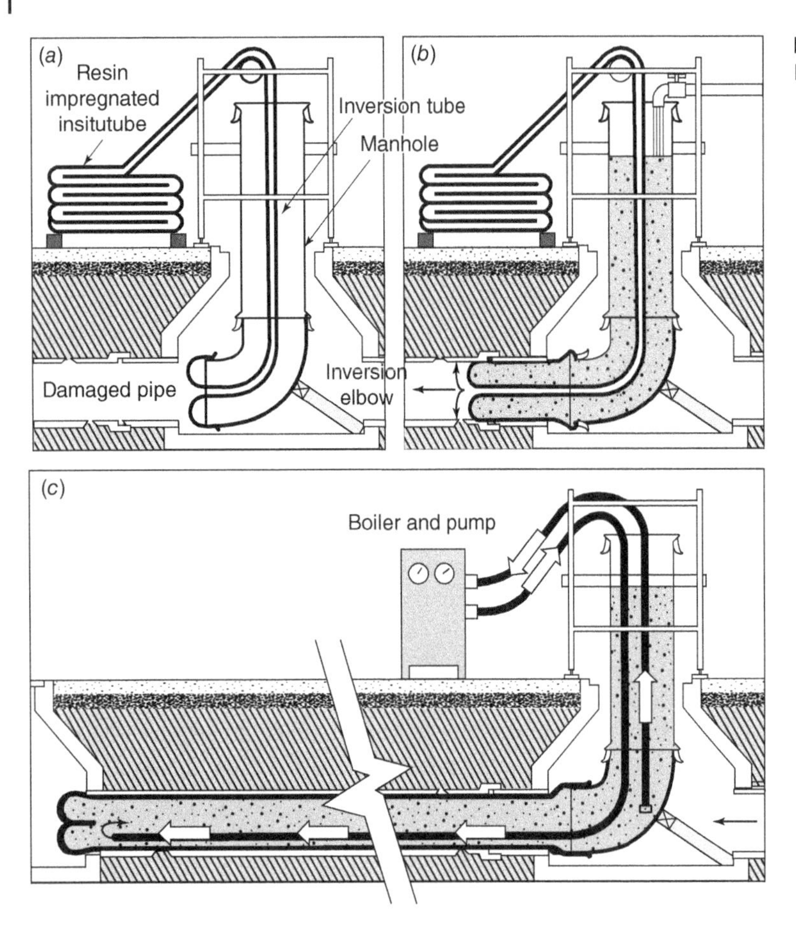

Figure 17.5 A Typical Cured-in-Place Pipe Installation Procedure

inversion lining. The inversion felt tube liner is usually inserted from existing manholes. Following curing of the liner, the ends are cut and sealed and service connections are restored.

17.6 Fold and Formed (Modified Cross-Section Lining)

The fold-and-form process uses a folded thermoplastic (PE or PVC) pipe that is pulled into place and then rounded to conform to the internal diameter of the existing pipe. This method of pipe rehabilitation can be considered as an improved version of sliplining. Excavation is not required for installation when there are existing manhole access points, and lateral reinstatement is accomplished internally. The finished pipe has no joints and produces a moderately tight fit to the existing pipe wall. This method of pipe rehabilitation is less versatile than cured-in-place methods in terms of diameter range and installation length. Only slight offsets and bends can be negotiated.

The fold-and-form method of rehabilitation does not require a long curing process in terms of speed of installation. Twofold-and-form processes are commercially available in the United States: *U-Liner* and *NuPipe*. The fold-and-form method of pipeline rehabilitation is suitable for pipe diameters of 100–400 mm (4–15 in.) with typical lengths of installation of 210 linear m (700 ft). Butt-fused U-Liner pipe can be used for lengths up to the stress-resistant pull force of the material.

The selection of the appropriate wall thickness for the U-liner will depend on the particular loading conditions from project to project. U-Liner is extruded as round pipe and then, through a combination of heat and pressure, is deformed into the "U" shape. It is then wound onto spools ready for installation. This technology is currently applicable for pipe sizes of 4–16 in. (100–400 mm).

The installation of U-Liner is basically a three-step process. After cleaning and TV inspection and analysis to identify defects and to determine the applicability of U-Liner, the first step includes winching of a preengineered seamless coil of pipe of a precut length into place. The pipe is pulled off the spool at ambient temperature, fed through an existing

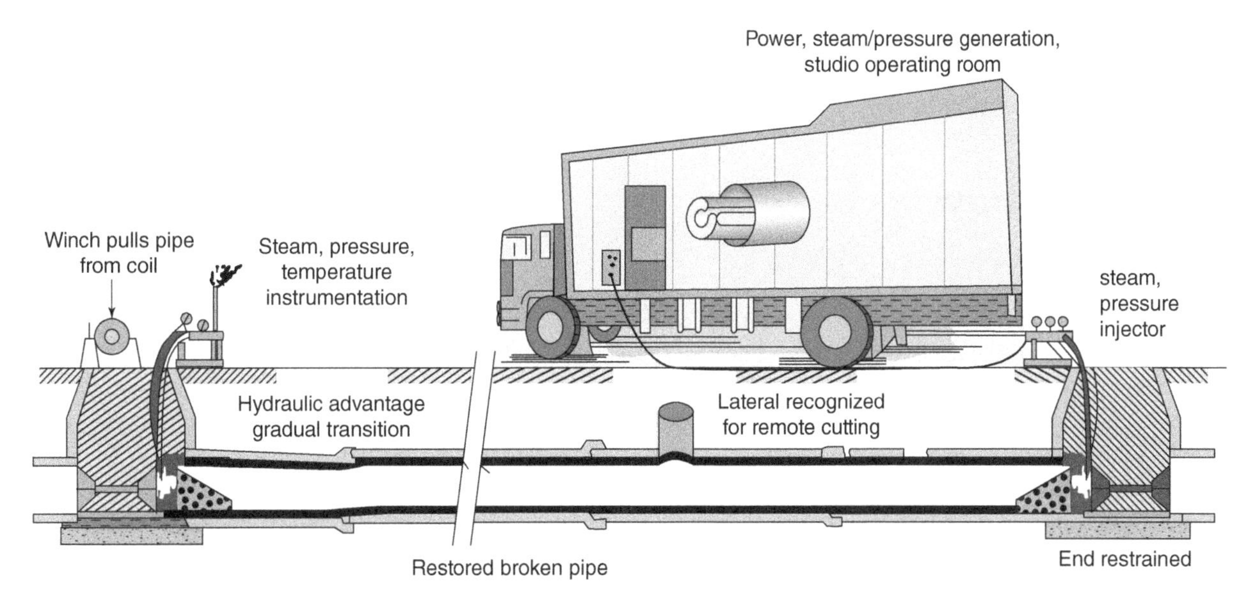

Figure 17.6 U-Liner Installation Method

manhole, and winched through the existing pipe to the terminal point. Once it is in place, steam is fed through the inside of the folded pipe, softening the plastic to allow for the reforming process. Temperatures of 235°–270 °F (110°–130 °C) are used to soften the plastic. Diagrams depicting the U-Liner installation procedure are presented in Figs. 17.6 and 17.7. After the plastic has been heated, pressure is used to *reround* the pipe. The pressure required to reround the pipe will vary, depending on the wall thickness, between 25 and 35 psi (170 and 240 kPa).

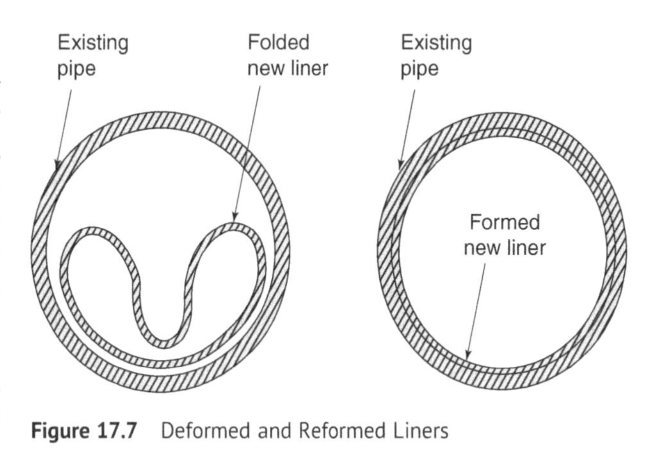

Figure 17.7 Deformed and Reformed Liners

Due to the relatively high coefficient of thermal expansion of PE, combined with the extreme temperature changes associated with the process, sufficient time should be allowed for the system to stabilize before laterals are reinstated and end treatment is finished. Laterals are recut utilizing a remote-controlled cutter head, in conjunction with a TV camera.

The NuPipe process was developed in the United States and has been commercially available since 1990. NuPipe is manufactured from PVC and is extruded in a folded shape. While it is still pliable, the folded PVC is wound onto spools. It is available in 6–12 in. (150–300 mm). The resin composition conforms to the standard specification for direct-bury PVC pipe. As with PE, the corrosion resistance of PVC is excellent.

The installation of NuPipe includes cleaning of existing host pipe along with a TV inspection to determine the extent of deterioration and to verify the applicability of NuPipe. A flexible reinforced liner called the *heat containment tube* (HCT) is inserted into the host pipe. The HCT provides a closed environment in which the NuPipe is installed and processed. The folded PVC is heated while on the spool and is pulled through the host pipe. Heating the plastic reduces the forces required to pull the pipe in place. Once the folded NuPipe reaches the termination point, steam is introduced into the system both through the interior and around the exterior of the folded pipe. The use of the HCT also allows for heating both sides of the plastic to provide complete heat transfer through the pipe wall, which minimizes the effect of infiltrating water. After the PVC becomes pliable, a rounding device is introduced into one end of the pipe, which is then propelled through the folded pipe. The rounding device progressively rounds the NuPipe, moving standing water out of the way while also expanding the plastic tightly against the host pipe, creating a mechanical lock at joints and laterals. Approximately 5–10 psi (35–70 kPa) is needed to propel the rounding device. Cold water is then injected into the NuPipe, effectively quenching the plastic.

17.7 Specialty Concrete

17.7.1 Description

Specialty concretes containing sulfate-resistant additives such as potassium silicate and calcium aluminate have shown greater resistance than typical concrete to acidic attack on sewer pipes and manhole structures.

Specialty concrete is used to reinforce weakened concrete pipes and structures by applying an acid-resistant coating over the original surface. Specialty concretes are unique in that their matrix is not formed by a hydration reaction. Rather, they are the result of the reaction of an acid reagent with an alkaline solution of a ceramic polymer of potassium silicate. Portland cement releases calcium hydroxide during hardening, whereas the specialty cements do not release calcium hydroxide. Specialty cements can resist attack by many substances including mineral salts, mild solutions of organic and mineral acids, sugar solutions, fats, and oils. In some cases, the acid reagents also act as effective bactericides.

Applicability of specialty concrete depends on the degree of corrosion-related deterioration and the structural integrity of the sewer. Thin-film specialty concrete is applicable to mildly deteriorated pipes or structures, whereas an elastic membrane concrete system is applicable to all cases. After curing, the specialty concrete bonds firmly to the original surface. The new acid-resistant layer, if applied and cured properly, extends the useful life of the structure. Advantages and disadvantages of specialty concretes are listed in Table 17.4.

Table 17.4 Advantages and Disadvantages of Specialty Concretes

Advantages	Disadvantages
Cement mortar	
• Minimal service interruption	• Excavation required for sharp ends or curves
• Improved structural integrity	• Cannot be done in winter if freezing potential exists
• Applicable for wide range of pipe sizes	• Bypass required
	• Extensive surface preparation required (e.g., chipping, sandblasting)
	• Access holes required every 150–210 m (500–700 ft)
	• Concrete shelf life must be tracked for the project duration
	• Transportation cost higher for concrete
	• May not provide adequate corrosion resistance
Shotcrete	
• Minimal excavation required	• Extensive surface preparation required
• Access in through manholes	• Extended downtime period of three to seven days or longer required for cleaning
• Can restore structural integrity to a pipe that would otherwise require replacement	• Some reduction in hydraulic capacity
• Minimum traffic interruptions	• Limited to man-entry size structures
• Applicable to all shapes of man-entry size pipes	• Concrete shelf life to be tracked for project duration
	• Transportation cost high for concrete
	• May not provide adequate corrosion resistance.
Cast concrete	
• Established procedure	• Cleaning required
• Simple to design	• Bypass required
• Applicable to all shapes of pipes	• Seldom applicable to pipes less than 1.2 m (4 ft) in diameter
	• Concrete shelf life to be tracked for project duration
	• Transportation cost high for concrete
	• May not provide adequate corrosion resistance

17.7.2 Procedures and Equipment

Specialty concretes are available in three types: *cement mortar*, *shotcrete*, and *cast concrete*. Acid-resistant mortars have been used in industry as linings in tanks or as mortar bricks. Development of mechanical in-line application methods (centrifugal and mandrel) has established mortar lining as a successful and viable rehabilitation technique for sewer lines, manholes, and other structures.

Mortar lining is applied using a centrifugal lining machine. The machine has a revolving, mortar-dispensing head with trowels on the back to smooth the mortar immediately after application. In smaller diameter pipes a variable-speed winch pulls the lining machine through a supply hose. Reinforcement can also be added to the mortar with a reinforcing spiral-wound rod. The reinforcing rod is inserted into the fresh mortar and a second coat is applied over it. For man-entry structures the mortar can be applied manually with a trowel.

Shotcrete, sometimes referred to as *gunite*, is a low-moisture, high-density mixture of fine aggregate (particle size of 0.75 in. [19 mm] and smaller), cement, and water; solids-to-liquid mix ratios are typically 5 : 1. Well-placed shotcrete has high modules of elasticity (greater than 4 million psi) and a coefficient of thermal expansion similar to that of low-carbon steel. Bonding with the original surface is usually stronger than the base material itself, with better adhesion occurring with the more deteriorated and irregular existing pipe. Shotcrete is applied to a minimum thickness of 2 in. (50 mm). Shotcrete is used in man-entry size sewers [32 in. (800 mm) or greater] and manholes. Prior to shotcreting, reinforcing steel is set into place. The shotcrete lining machine is self-propelled and controlled by a person riding it. Mortar is supplied to an electrically driven supply cart that conveys mortar from the access hole to the feeder, which is attached to the lining machine. The dry specialty cement and aggregate are mixed with water in a specially designed spray nozzle. Hydration occurs, and the resulting mixture is shot into place under pressure. Curing occurs under moist conditions for the first 24 hours and an additional six days at a temperature above 40 °F (4 °C).

Cast concretes are potassium-silicate-bonded, poured or cast-in-place structural concretes. They typically have half of the in-place density or strength value of shotcrete. The solids-to-liquid mix ratios are generally 2 : 1, similar to that of cement mortar.

Cast concrete is pored over prefabricated or hand-built interior pipe forms that can be removed and reused section by section. Reinforcing steel is added between the original surface and the form, setting within the cured thickness.

Each of the three application techniques requires prior cleaning to remove oils, greases, foreign objects, and loose materials, as well as wastewater bypass during application and initial curing.

17.8 Liners

17.8.1 Description

Rehabilitation techniques using liners include the installation of prefabricated panels or flexible sheets on the existing structure usually with anchor bolts or concrete penetrating nails shot into place. The following liner types are available for rehabilitation purposes:

1) PVC liners
2) PE liners
3) Segmented, fiberglass-reinforced plastic liners
4) Segmented, fiberglass-reinforced cement liners.

PVC liners are manufactured from acid-resistant, rigid unplasticized PVC, which has excellent resistance to acids and also is initially conductive to better hydraulics than concrete. The liner is composed of high-molecular-weight vinyl chloride resins combined with chemical-resistant plasticizers. The completely inert mixture is extruded under pressure and temperature into a liner plate with a minimum thickness of 1.6 mm (0.065 in.). The liners are pinhole-free, forming an effective barrier to gaseous penetration.

PE liners are similar to PVC liners but are made of PE resins. These liners are tough, rigid, acid-resistant, smooth, and inexpensive.

Fiberglass-reinforced plastic liners are manufactured in a range of wall thicknesses and consist of a composite of fiberglass and acid-resistant resin. The resins are specified according to the degree of acid resistance required. The composite has high

Table 17.5 Advantages and Disadvantages of Liners

Advantages	Disadvantages
• Material cost inexpensive	• Applicable only to man-entry size sewers (i.e., 76 cm [2.5 ft] or greater)
• Liner materials have very good acid resistance	• Susceptible to leakage due to number of joints
• No disruptions to traffic as installation is performed entirely in-line	• Timely to install. Thus total project cost may be uneconomical
• Smooth surfaces provide good hydraulics	• Prolonged bypass required
	• Surface preparation required
	• PE can crack in areas of turbulent flow

mechanical and impact strength and good abrasion resistance. These liners can be manufactured to a wide variety of shapes and are applicable to sewers more than 42 in. (1,100 mm) in diameter.

These liners do not provide any structural support but they do provide an adequate corrosive barrier and smooth lining for structurally sound sewers. These liners have little absorption capability and no apparent permeability.

Fiberglass-reinforced cement liners consist of cement and glass fibers. They usually are 9.8 mm (0.385 in.) thick and are in thin panel form. They have high mechanical and impact strength and good acid and alkaline resistance. They are also highly resistant to abrasion, with negligible absorption and permeability. These liners are not designed to support earth loads and should be used only in structurally sound sewers. The liners can be easily assembled to fit variations in grades, slopes, and cross-section. The smooth interior surface improves hydraulic capabilities. These liners can be used in circular, oval, rectangular, and other sewer shapes above 42 in. (1,100 mm) in normal size and can be segmented to fit the diameter required.

The advantages and disadvantages of liners as a sewer rehabilitation technique are listed in Table 17.5.

17.8.2 Procedures and Equipment

Segmented plastic and fiberglass-reinforced cement liners are installed so they overlap at the joints and are then attached to the concrete surface by anchor bolts or nails. Space is left between the existing surface and the liners for grouting purposes. After a thorough line cleaning and dewatering, the segments are installed in 4-ft (2.4-m) lengths, which overlap at the joints, and the flanges on the segments may be predrilled for filling with screws or by means of an impact nail gun. Joints are coated with an adhesive to better connect panels and are sealed with an acid-resistant resin. After all the panels have been set in place, the entire section is cement-pressure-grouted in place to prevent sagging and deformation. Some liners are installed in conjunction with the casting of concrete by placing the liners against the inner surface of the form prior to pouring. Anchors become embedded in the concrete during curing, thereby securing the liner in place (see Fig. 17.8). Alternate installation procedures for flexible PVC liners or ribbed sheets involve only grouting, without anchor bolts or impact nails for attachment. Installation of another type of liner requires an access pit to fit the special winding machine that joins a male and female PVC strip. This process is applicable to pipes up to 2.5 ft. (750 mm) in diameter. The latest liner installation method uses acid-resistant mastic to fasten the sheets directly to the sewer concrete surface. This technique does not require grouting but requires thorough cleaning prior to installation.

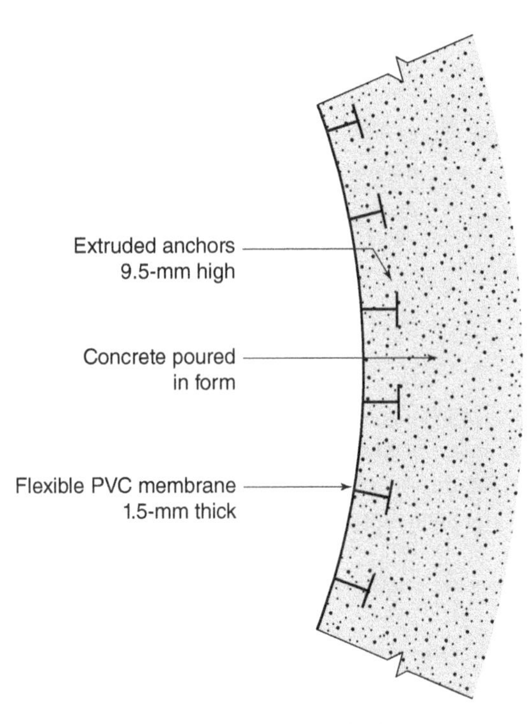

Figure 17.8 Detail of Liners with Anchors

Liner failures can occur due to leaking joints that allow H_2S gas or sulfuric acid to penetrate the liner materials and attack the concrete substrate beneath. Liner failures have also been reported in areas of high turbulence where cracks developed. Cracking has been identified in PE pipes and thus PE is not recommended in areas of turbulent flows.

17.9 Coatings

17.9.1 Description

Coatings include a myriad of proprietary materials including *coal tar epoxy, concrete sealers, epoxy, polyester, silicone, urethane*, and *vinylester*, which can be applied by spray machines or brushed onto a concrete surface. They are intended to form an acid-resistant layer that protects the substrate concrete from corrosion. Some of the advantages and disadvantages involved with the use of coatings are listed in Table 17.6.

17.9.2 Procedures and Equipment

Application of coatings usually includes the following procedures:

1) Bypass wastewater.
2) Prepare/clean concrete surface.
3) Allow concrete surface to dry.
4) Apply coating by brush or spray (more than one coat is usually necessary).
5) Allow coating to cure.
6) Remove bypass.

Most coatings can be brush- or spray-applied. Spray application requires 3,000 psi (20 820 kPa), which is double the pressure used for conventional airless spraying. Spraying is excellent for coating uneven surfaces and is much faster than brush application methods for some products.

17.10 Spot (Point) Repair

Point repairs are used to correct isolated problems in a pipe. Sometimes they are used as the initial step in the use of other rehabilitation methods. Point repairs include

1) Robotic repair
2) Grouting/sealing
3) Special sleeves
4) Point CIPP.

Trenchless sewer rehabilitation methods are now routinely applied to wastewater collection system improvement projects in the United States and many other countries. Trenchless sewer rehabilitation has been successfully applied by large municipalities such as New York City, Los Angeles, Boston, Miami, and Houston, and by smaller municipalities such as Baton Rouge, Louisiana; Madison, Wisconsin; and Amarillo, Texas. The market value for sewer and pressure pipe rehabilitation projects exceeded USD 5 billion worldwide at the start of the twenty-first century.

In many municipalities, sewer rehabilitation projects are an essential part of operation and maintenance (O&M) programs for the collection system. For example, as part of an O&M program focused on proactive maintenance, Fairfax

Table 17.6 Advantages and Disadvantages of Coatings

Advantages	Disadvantages
• Economical • No disruption to traffic or other utilities • Most are fast curing, some cure in less than one hour • Quick to apply • Can be applied to uneven surfaces	• Applicable only to man-entry sewers and manholes • Surface imperfections—pinholes, blowholes • Poor bonding to vertical or overhead surfaces • Bypass required • Surface preparation required • Few contractors have experience with products • Surface repairs often required prior to application • Still a developing technology

County, Virginia, has identified two older sewer sheds for rehabilitation. All trunk and main lines within each sewer shed are television-inspected. Results of the TV inspection are used to prioritize cleaning needs and to help determine appropriate rehabilitation measures. Projects within the targeted sewer sheds have utilized the CIPP and fold-and-form rehabilitation methods. In an effort to monitor the effectiveness of the rehabilitation efforts, the department installed permanent and temporary meters in these two sewer sheds. Fairfax County's focused approach to maintenance has reduced average flows to the WWTP despite several years of above-normal rainfall.

17.11 Advantages and Disadvantages of Trenchless Technology

By reducing *I/I* levels in the collection system, trenchless rehabilitation projects can assist communities in complying with the EPA's Clean Water Act and thereby protect the aquatic integrity of receiving water bodies from potentially high pollutant concentrations by reducing SSOs. In addition to potential improvements in receiving water bodies, trenchless sewer rehabilitation requires substantially less construction work than traditional dig-and-replace methods. In wetland areas and areas with established vegetation, construction influences can be especially harmful to the plant and aquatic habitat. Underground utility construction can disrupt citizens living and working in areas near the construction zone. Trenchless sewer rehabilitation, with the potential to reduce surface disturbance over traditional dig-and-replace methods, can reduce the number of traffic and pedestrian detours, spare tree removal, decrease construction noise, and reduce air pollution from construction equipment. In addition to these benefits, reducing the amount of underground construction labor and surface construction zone area confines the work zones to a limited number of access points, reducing the area where safety concerns must be identified and secured.

Rehabilitation techniques should be selected based on site constraints, system characteristics, and project objectives. A comparison of economic, cultural, and social costs of sewer rehabilitation with those of traditional dig-and-replace methods can help determine whether or not trenchless sewer rehabilitation is suitable and economically feasible for a particular site. Because some digging may be required for point repairs, construction limitations should be evaluated when deciding whether trenchless sewer rehabilitation techniques can be applied. If there are major changes in cross-section between manholes or if the existing alignment, slope, or pipe bedding material must be changed, each line must be rehabilitated as an independent segment, necessitating even more digging.

Specific limitations of each trenchless rehabilitation method are listed in Table 17.7. As can be seen, the sliplining, modified cross-section and CIPP methods will reduce the pipe diameter, tending to decrease the hydraulic capacity of the sewer.

Table 17.7 Limitations of Trenchless Sewer Rehabilitation

Method	Limitations
Pipe bursting	Bypass or diversion of flow required Insertion pit required Percussive action can cause significant ground movement May not be suitable for all materials
Sliplining	Insertion pit required Reduces pipe diameter Not well suited for small-diameter pipes
CIPP	Bypass or diversion of flow required Curing can be difficult for long pipe segments Must allow adequate curing time Defective installation may be difficult to rectify Resin may clump together on bottom of pipe Reduces pipe diameter
Modified cross-section	Bypass or diversion of flow required Cross-section may shrink or unfold after expansion Reduces pipe diameter Infiltration may occur between liner and host pipe unless sealed Liner may not provide adequate structural support

The rehabilitated pipeline, however, may be less rough than the original; the roughness coefficient depends on the liner material. New high-performance plastic materials tend to reduce pipe roughness against aged concrete materials. Additionally, the hydraulic capacity may be modified during rehabilitation as groundwater intrusion is inadvertently redirected to unlined side sewers. An evaluation may be performed to determine whether the change in pipe friction and groundwater redirection will offset the decrease in pipe diameter and meet project objectives for an increase in peak flow and/or reduction in SSOs.

Most trenchless rehabilitation applications require laterals to be shut down for a 24-h period. Coordinating shutdowns with property owners can be a difficult and unpopular task. Unforeseen conditions can increase construction time and increase the risk and responsibility to the client and contractor. For example, during a rehabilitation project in Norfolk, Virginia, pipe bursting had to be coordinated with the relocation of a nearby electrical substation and the rerouting of flow from a sanitary force main found in a manhole where an insertion pit was to be located. In addition to these issues, numerous abandoned underground utilities that were not indicated on city or private utility records were encountered during the project. Such underground conditions are found in many other urban environments around the United States. When trenchless rehabilitation is planned, public works projects and utility work by other agencies should be coordinated with sewer rehabilitation projects.

17.12 Applicability of Trenchless Technology

Trenchless techniques can be applied to rehabilitate existing pipelines in a variety of conditions. They are particularly valuable, however, in urban environments where construction impacts are particularly disruptive to businesses, homeowners, and automotive and pedestrian traffic. Other underground utilities and existing infrastructure are obstacle in the traditional dig-and-replace method, and trenchless techniques are widely applied where these are present. Most trenchless techniques are applicable to both gravity and pressure pipelines. Many trenchless methods are capable of performing spot repairs as well as manhole-to-manhole lining.

For most applications, trenchless sewer rehabilitation techniques require less installation time—and therefore less pump-around time—than traditional dig-and-replace methods. Installation time can be critical in deciding between trenchless sewer rehabilitation methods and dig-and-replace methods. For example, when considering sewer repair or replacement options for a critical force main crossing the Elbe River in Heidenau, Germany, city officials determined that the line could not be out of service for more than 12 days. As a result of this time constraint, as well as reduced disruption to riverboat traffic, city officials chose to rehabilitate the sewer using the *Swagelining™ process*. The successfully rehabilitated sewer was out of service only eight days.

Trenchless sewer rehabilitation can be performed to increase the hydraulic capacity of the collection system. While pipe bursting typically yields the largest increase in hydraulic capacity, rehabilitation by other trenchless methods may also increase hydraulic capacity, by reducing friction. A hydraulic analysis of the pre- and post-rehabilitation conditions can be performed to evaluate the impact on collection system capacity. In general, the hydraulic analysis is performed by municipal engineers and/or consultants who prepare specifications for contractors.

Each of the trenchless rehabilitation methods described has been used for various applications over a range of pipe sizes and lengths. A comparison of trenchless techniques is shown in Table 17.8.

17.13 Performance of Trenchless Technology

The performance of trenchless techniques in reducing the I/I can be determined through flow measurements taken before and after the rehabilitation. Effectiveness is typically calculated by correlating flow measurements with precipitation data to determine the peak rate and volume of I/I entering the collection system. Another method of calculating I/I is to isolate the rehabilitated line and measure flows both before and after the rehabilitation.

The performance of sewer rehabilitation projects in various communities has been documented. Results of pre- and post-monitoring within these communities indicate that I/I reductions of 49%, 65%, and 82% were achieved. The Washington Suburban Sanitary Commission uses the isolation and measurement method to assess the performance of rehabilitation projects. An analysis of 98 rehabilitated sewer mains indicates that I/I flow was reduced by 70% in the rehabilitated sewers.

Table 17.8 Comparison of Various Sewer Rehabilitation Techniques

Method		Diameter Range [mm (in.)]	Maximum Installation [m (ft)]	Liner Material
In-line expansion	Pipe bursting	100–600 (4–24)	230 (750)	PE, PP, PVC, GRP
Sliplining	Segmental	100–4,000 (4–158)	300 (1,000)	PE, PP, PVC, GRP
	Continuous	100–1,600 (4–63)	300 (1,000)	PE, PP, PE/EPDM, PVC
	Spiral wound	150–2,500 (6–100)	300 (1,000)	PE, PVC, PP, PVDF
Cured-in-place product linings	Inverted-in-place	100–2,700 (4–108)	900 (3,000)	Thermoset resin/fabric composite
	Winched-in-place	100–1,400 (4–54)	150 (500)	Thermoset resin/fabric composite
	Spray-on-linings	76–4,500 (3–180)	150 (500)	Epoxy resins/cement mortar
Modified cross-section methods	Fold-and-form	100–400 (4–15)	210 (700)	PVC (thermoplastics)
	Deformed/re-formed	100–400 (4–15)	800 (2,500)	HDPE (thermoplastics)
	Drawdown	62–600 (3–24)	300 (1,000)	HDPE, MDPE
	Rolldown	62–600 (3–24)	300 (1,000)	HDPE, MDPE
	Thin-walled lining	500–1,100 (20–46)	960 (3,000)	HDPE
Internal point repair	Robotic repair	200–760 (8–30)	N/A	Epoxy resins cement mortar
	Grouting/sealing and spray-on	N/A	N/A	Chemical grouting
	Link seal	100–600 (4–24)	N/A	Special sleeves
	Point CIPP	100–600 (4–24)	15 (50)	Fiberglass/polyester, etc.

Note: Spiral-wound sliplining, robotic repair, and point CIPP can be used only with gravity pipeline. All other methods can be used with both gravity and pressure pipelines.
EPDM = ethylene propylene diene monomer; GRP = glassfiber-reinforced polyester; HDPE = high-density polyethylene; MDPE = medium-density polyethylene; PE = polyethylene; PP = polypropylene; PVC = polyvinyl chloride; PVDF = polyvinylidene chloride; CIPP = cured-in-place pipe; N/A = not available.

The Miami-Dade Water and Sewer Department (MDWASD) has completed one of the country's largest I/I reduction programs. The program, aimed at reducing I/I throughout the system, utilizes the fold-and-form, CIPP, pipe bursting, and sliplining rehabilitation techniques in conjunction with point and robotic repairs. MDWASD has experienced success with this program; an average I/I reduction of 19% (20 MGD or 876 L/s) was achieved after the first three years of rehabilitation based on comparing plant flow and billed flow.

These studies should only be used as an indicator of potential I/I removal. Removal rates will vary depending on the material and condition of the pipe, local soil type, groundwater flow, and other site-specific conditions.

17.14 Costs of Trenchless Technology

A cost comparison of trenchless and traditional sewer rehabilitation methods must consider the condition and site characteristics of the existing pipeline. Factors influencing the cost of a trenchless sewer rehabilitation project include the following:

1) The diameter of the pipe
2) The amount of pipe to be rehabilitated
3) Specific defects in the pipe (such as joint offsets, root intrusions, severe cracking, or other defects)
4) The depth of the pipe to be replaced and changes in grade over the pipe length
5) The locations of access manholes

6) The number of additional access points that need to be excavated
7) The location of other utilities that have to be avoided during construction
8) Provisions for flow bypass
9) The number of service connections that need to be reinstated
10) The number of directional changes at access manholes.

Cost comparisons for trenchless rehabilitation of typically sized sewer mains are provided in Tables 17.9 and 17.10. The costs were adjusted to recent prices (2022) through the use of the U.S. Army Corps of Engineers (2022)' *Cost Index for Utilities* (see Appendix

Table 17.9 Cost Comparisons of Trenchless Rehabilitation Technologies in 2022 USD/ft^2

Method	Cost (2022 USD/ft^2)
Coatings	19–37
Sliplinings	50–87
Cured-in-place pipe	50–68
Fold-and-form pipe	43–62
In situ liners	31–50

Conversion factors: 1 ft = 0.3048 m; 1 USD/ft^2 = 10.76 USD/m^2.

Table 17.10 Cost Comparisons of Trenchless Rehabilitation Technologies in 2022 USD/Linear ft

		Sliplining					
Pipe Diameter (in.)	Grouting	High-Density Polyethylene (HDPE)	Polyethylene (PE)	Reinforced Thermosetting Resin (RTR)	Cured-in-Place Inversion Lining	Cement Mortar Lining	Reinforced Shotcrete Lining
6	37–50	–	62–105	–	56–81	–	–
8	43–62	25–62	68–112	43–50	–	–	–
10	50–74	–	81–124	–	105–130	–	–
12	56–87	43–81	87–130	68–99	–	25–43	–
14	–	–	99–143	–	130–217	–	–
16	–	50-99	–	93–112	–	–	–
18	68–105	–	112–143	–	155–254	–	–
20	–	62–124	–	124–149	–	–	–
22	–	–	130–198	–	–	–	–
24	130–198	62–155	–	149–155	192–335	25–56	–
26	–	–	155–229	–	–	–	–
28	–	93–192	186–273	–	–	–	–
30	167–248	–	–	174–229	236–397	–	–
32	–	124–229	211–316	–	–	–	–
33	186–285	136–272	273–335	–	248–434	–	–
39	229–347	–	–	–	–	–	–
40	–	155–316	–	–	–	–	–
42	–	174–322	335–403	254–304	285–533	–	–
48	–	186–403	384–484	298–347	322–614	37–87	198–254
52	–	–	434–508	–	–	–	–
54	–	–	–	329–372	360–682	–	236–310
58	–	–	508–620	–	–	–	–
60	–	–	–	372–446	372–756	43–105	248–360
63	–	254–515	–	–	–	–	–
66	–	–	–	372–477	–	–	279–372
68	–	–	577–701	–	–	–	–
72	–	–	657–806	–	–	–	304–446
80	–	–	750–930	–	–	–	–
84	–	–	–	–	–	–	360–508
88	–	–	849–1,166	–	–	–	–
92	–	–	942–1,153	–	–	–	–
96	–	–	–	–	–	–	403–577
100	–	–	1,042–1,265	–	–	–	–

Conversion factor: 1 USD/ft = 3.28 USD/m.

16). These costs include a standard cleaning of the sewer line (major blockages and point repairs increase the cost) and inspection of the sewer line before and after the sewer is rehabilitated. Sewer rehabilitation by both trenchless and traditional dig-and-replace methods can reduce treatment and O&M costs at the receiving treatment plant by potentially eliminating I/I flows to the plant. In addition to treatment cost savings, energy costs for transporting flows to the treatment plant could also be reduced due to the reduced flow volume.

In general, the less the amount of excavation required for a rehabilitation operation, the more cost-effective trenchless sewer rehabilitation becomes as compared with the traditional dig-and-replace method (compared to Table 17.1). In addition to excavation and installation costs, sewer cleaning and inspection are typically required before sewer rehabilitation.

17.15 All Techniques for Manholes

17.15.1 Description of Materials, Equipment, and Products

Sewer manholes require rehabilitation to prevent surface water inflow and groundwater infiltration, to repair structural damage, and to protect surfaces from damage by corrosive substances. When rehabilitation methods will not solve the problems cost-effectively, manhole replacement should be considered. Selection of a particular rehabilitation method should consider the type of problems, physical characteristics of the structure, location, condition, age, and type of original construction. The extent of successful manhole rehabilitation experiences and cost should also be considered.

Manhole rehabilitation methods are directed at either (a) the frame and cover or (b) the sidewall and base. The following sections provide a summary of manhole and base rehabilitation methods. The advantages and disadvantages of these rehabilitation methods are presented in Table 17.11.

Table 17.11 Advantages and Disadvantages of Manhole and Sump Rehabilitation Methods

Method	Advantages	Disadvantages
Frame and cover		
Stainless steel, and neoprene washersor corks in holes in covers	Simple to install	Restricts natural venting
Prefabricated lid insert	When installed properly it prevents surface water, sand, and grit from entering manhole through or around cover	Requires perfect fit for success
Joint sealing tape	Simple to install	Short service life
Hydraulic cement	Provides strong waterproof seal to stop infiltration	Labor-intensive; freeze-thaw cycle may reduce patch life
Raise frame above grade	Minimizes inflow through cover and frame	Limited to areas outside of street right-of-way
Sidewall and base		
Epoxy or polyurethane coatings oninterior infiltration	Protects interior walls against corrosion	Requires structurally sound and dry manhole surface walls must be very clean prior to application; short service life
Chemical grout	Can be very inexpensive method for stopping infiltration	Short service life; cannot predict amount of grout required to eliminate infiltration
Structural liner	Provides structural restoration; manholes require less disruption of traffic and utilities than replacement; longer service life than coatings	Complex and costly installation
Hydraulic cement	Seals manhole frame in place. Prevents infiltration between frame and cone section	
Raise frame	Prevents surface-water inflow through manhole cover	
Sidewall and base		
Epoxy or polyurethane coatings on interior walls	Protects wall from corrosion and infiltration on structurally sound manholes	

17.15.2 Description of Procedures

Manhole frame and cover rehabilitation prevents surface water (stormwater runoff) from flowing into the manholes. Surface water from storm runoff and other sources can often flow into the manholes through the holes in the cover lid, through the annular space around the lid and the framed cover and under the frame if it is improperly sealed. Manhole frames and covers can be rehabilitated by the following techniques:

1) By installing stainless steel bolts with caulking compound and neoprene washers or corks to plug holes in the cover.
2) By installing a prefabricated lid insert between the frame and the cover. These plastic lids are resistant to corrosion and damage by sulfuric acid or road oils. The lids come with gas relief and vacuum relief valves to allow gas escape. They prevent water, sand, and grit from entering the manhole. The lids are easy to install, can fit any manhole, and require periodic maintenance to function properly.
3) By installing a resin-based joint sealing tape between metal frame and cover. The sealing tape provides flexibility to seal imperfectly fitting surfaces and to move with ground shifting. These sealing tapes can be used for all types of manholes.
4) For cracks and openings on the existing manhole/frame, seals are applied with hydraulic cement and waterproofing epoxy
5) By raising the manhole frames to minimize flows through the frame covers.

Manhole sidewall and base rehabilitation is primarily done to prevent infiltration of groundwater. Casting or patching can be used to rehabilitate structurally sound sidewalls. Complete replacement should be carried out for severely deteriorated manholes and bases.

Manhole steps also deteriorate frequently and should be replaced. Manhole sidewall and base rehabilitation can be carried out using these procedures:

1) By applying *epoxy, acrylic*, or *polyurethane-based coatings* to the interior wall of the manhole. These waterproof and corrosion-resistant coatings can be applied to brick, block, and precast concrete manholes and bases. The coatings are applied by towel brush or sprayer. Prior to coating application, the surfaces of the manhole walls should be cleaned and all leaks plugged using patching or grouting materials.
2) By applying *chemical grout* from interior walls to exterior walls to stop infiltration through cracks and holes.
3) By inserting *structural liners* inside existing manholes. These liners are typically fiberglass of the reinforced polyester mortar type.

17.15.3 Costs

Manhole rehabilitation costs are provided in Table 17.12. The costs were adjusted to recent (2022) prices through the use of U.S. Army Corps of Engineers (2022)' *Cost Index for Utilities* (see Appendix 16).

Table 17.12 Rehabilitation Cost of Manholes

Item	Cost (2022 USD/manhole)
Chemical grouting	942–1,457
Seal frames to corbels	688–725
Chemically seal and plaster walls	688–750
Raise manhole to grade	1,128–1,916
Replace frame	725–1,128
Insert structural liners	8,060–24,174
Manhole replacement	2,096–4,185
Manhole repair	211–2,108
Raise manhole frame and cover	422–632
Manhole cover replacement	211–422

17.16 Service Lateral Techniques

17.16.1 Description

Service laterals are the pipes that connect building sewers to the public sewer main. The service laterals usually range in size from 3 to 6 in. (75 to 150 mm) and are often laid at a uniform slope from the building to the immediate vicinity of the main sewer. They can enter the sewer at angles of 0–90° from horizontal. For many years the effects of leaking service connections were considered insignificant because it was assumed that most service connections were above the water table and therefore subject to leakage only during periods of excessive rainfall or high groundwater levels. Recent studies indicate that a significant percent of infiltration in any collection system is the result of service connection defects such as cracked, broken, or open-jointed pipes. Service connections may also transport water from inflow sources such as roof drains, cellar and foundation drains, basement or subcellar sump pumps, and stormwater flows from commercial and industrial properties. In a national survey carried out by state and local agencies, the estimated percentage of total system infiltration from service laterals was found to range from 30% to as high as 95% in some cases.

17.16.2 Procedures and Equipment

Following are the procedures and equipment used for rehabilitating service laterals:

17.16.2.1 Chemical Grouting

The following chemical grouting methods are utilized:

1) *Pump-full method:* Chemical grout is injected through a conventional sealing packer from a sewer main into the service connection to be grouted. The forced grout surrounds the pipe and a seal is formed after the gel has set. Excessive grout is augured from the building sewer and the sewer is returned to service after the sealing has been accomplished.
2) *Sewer sausage method:* This method is similar to the pump-full method except that a tube is inverted into the service connection before sealing to reduce the quantity of grout to be used and to minimize the amount of cleaning required after the sealing has been completed.
3) *Camera-packer method:* This method utilizes a miniature TV camera and a specialized sealing packer that is pulled out while it is simultaneously repairing faults that are seen through the TV camera. The equipment is removed and the service connection returned to service after the repairs are completed.

17.16.2.2 Inversion Lining

This technique is similar to the sewer main installation in that it involves the insertion of a resin-impregnated flexible polyester felt liner into the service line. No annular space is created between the liner and the pipe that might result in infiltration migration. No prior excavations are required to correct slight offsets. An access point is always needed on the upstream side of the service connection line. A variation from the sewer main installation is the use of a special pressure chamber to provide the needed pressure to invert the fabric materials through the service pipeline. After the completion of the curing process, the downstream end of the liner is cut manually or via a remotely controlled cutting device placed in the sewer main. The upstream end is trimmed at the access point, restoring the sewer service.

Problems/Questions

17.1 As the infrastructure in the United States ages, increasing importance is being placed on rehabilitating the nation's wastewater treatment collection systems. Cracks, settling, tree root intrusion, and other disturbances that develop over time deteriorate pipelines and other conveyance structures that comprise wastewater collection systems. These deteriorating conditions can increase the amount of inflow and infiltration (I/I) entering the system, especially during periods of wet weather. Discuss the effects of these deteriorating conditions.

17.2 With the growing concern over groundwater contamination and our polluted waterways, shorelines, and rivers, governments around the world are spending billions of dollars to repair a crumbling water and wastewater infrastructure. Poor materials in the ground, root intrusion, corrosive chemicals, and ground shifting are damaging our drain systems. The U.S. Environmental Protection Agency estimates 70%–80% of the increased pollution comes from the I/I of residential lateral pipes. Trenchless technology is the "new" solution for the municipalities and business property owners. Give an overview of (a) the trenchless technology known as cured-in-place (CIP) technology and (b) applications of the CIP technology.

17.3 Name and briefly describe the various trenchless sewer rehabilitation methods.

17.4 What are the disadvantages of pipeline removal and replacement as a method of sewer line rehabilitation?

17.5 Explain the pipe bursting or in-line expansion technique for sewer rehabilitation and compare the use of static and dynamic heads.

17.6 Sliplining is performed by either a push or a pull technique. Explain each method.

17.7 Discuss the advantages and disadvantages of the cured-in-place pipe (CIPP) trenchless technology in comparison with conventional sewer repair technology.

17.8 Search the Internet for "trenchless technology" and visit some of the websites you find. Comment on the design considerations and specifications of the technology

17.9 Conduct an internet and/or university library search and report on the thermal cure method for installation of trenchless liners.

17.10 Conduct an internet and/or university library search and report on the ambient cure method for installation of trenchless liners.

17.11 List and discuss the sewer rehabilitation techniques that use liners.

17.12 Discuss the pros and cons of trenchless technology.

17.13. What factors influence the cost of a trenchless sewer rehabilitation project?

17.14 Give the reasons why manholes need rehabilitation and list the factors to be considered in the selection of a particular rehabilitation method.

17.15 Explain the three procedures for manhole sidewall and base rehabilitation.

17.16 List and explain the three chemical grouting procedures used in the rehabilitation of service laterals.

Bibliography

American Public Works Association, *Sewer System Evaluation, Rehabilitation and New Construction: A Manual of Practice*, EPA/600/2-77/017d, United States Environmental Protection Agency, Cincinnati, OH, December 1977.

American Society of Civil Engineers and Water Pollution Control Federation, 1983. *Existing Sewer System Evaluation and Rehabilitation*. ASCE Manuals of Reports on Engineering Practice No. 62, WPCF Manual of Practice FD-6.

Brown and Caldwell, *Utility Infrastructure Rehabilitation*, NTIS No. PB86-114642,, National Technical Information Service, Washington, DC, 1984.

Fehnel, S. K., Dorward, M., and Mansour, S., New Orleans Sewer System Evaluation and Rehabilitation Program Success Through Vision and Innovation, in *Proceedings of the Water Environment Federation, Collection Systems 2005*, 2005, pp. 333–350.

Fernandez, R. B., Sewer Rehab Using a New Subarea Method, *Water/Eng. Manage.*, *133*, 28–30, 1986.

Goumas, J., Tri-Villages of Greater Chicago Reduce I/I with CIPP, *Trenchless Technol.*, *4*, 3, 70–71, 1995.

Kung'u, F., Excavation and Elimination: No-Dig Solutions to Sewer Problems, *Civil Eng. News*, *10*, 7, 45–49, 1998.

Kutz, G. E., Predicting I/I Reduction for Planning Sewer Rehabilitation, in Osborn, L. E. (Ed.), *Trenchless Pipeline Projects: Practical Applications*, American Society of Chemical Engineers, New York, 1997, pp. 103–110.

Najafi, M., and Gokhale, S. B., *Trenchless Technology: Pipeline and Utility Design, Construction, and Renewal*, McGraw-Hill, New York, 2004.

Sabnis, G. M., and Shook, W. E., Rehabilitation of Concrete Sewer System Using Anti-Bacterial Admixture, in *Proceedings of the 27th Conference on our World in Concrete and Structure, Keynote Address, Singapore, August 29–30*, vol. *21*, 2002.

Saccogna, L. L., Swagelining Renews Force Main under the Elbe River, *Trenchless Technol. Int.*, *2*, 1, 28–29, 1998.

Stepkes, H., Zimmermann, J., Müller, K., Siekmann, M., and Pinnekamp, J., Economical Rehabilitation of Sewer Systems by Ground Penetration Radar Investigations, in *Leading-Edge Asset Management Conference, October 17–19*, International Water Association, Lisbon, 2007.

U.S Army Corps of Engineers, Application of Trenchless Technology at Army Installations, in *Public Works Technical Bulletin 420-49-10*, U.S. Army Corps of Engineers, Washington, DC, February 1999.

U.S. Army Corps of Engineers, 2022. Yearly Average Cost Index for Utilities, in *Civil Works Construction Cost Index System Manual, 110-2-1304*, Washington, DC., USA.

U.S. Environmental Protection Agency, *Hydrogen Sulfide Corrosion in Wastewater Collection and Treatment Systems*, EPA/ 430/9-91/009, Washington, DC, 1991.

U.S. Environmental Protection Agency, *Sewer System Infrastructure Analysis and Rehabilitation*, EPA/625/6-91/030, Cincinnati, OH, October 1991

U.S. Environmental Protection Agency, *Exfiltration in Sewer Systems*, EPA/600/R-01/034, Cincinnati, OH, December 2000.

U.S. Environmental Protection Agency, *Pipe Bursting, Water Technology Fact Sheet*, EPA 832-F-06-030, Washington, DC, September 2006

U.S. Environmental Protection Agency, *Rehabilitation of Wastewater Collection Systems*, Science Brief, EPA/600/F-07/012, Washington, DC, September 2007.

U.S. Environmental Protection Agency, *Sustainable Infrastructure for Water and Wastewater*, http://www.epa.gov/waterinfrastructure/index.html, 2008.

18

Alternative Wastewater Collection Systems

Conventional collection systems transport wastewater from homes or other sources by gravity flow through buried piping systems to a central treatment facility. These systems are usually reliable and consume no power. However, the slope requirements to maintain adequate flow by gravity may require deep excavations in hilly or flat terrain, as well as the addition of wastewater pump stations, which can significantly increase the cost of conventional collection systems. Manholes and other sewer appurtenances also add substantial costs to conventional collection systems.

In the late 1960s, the cost of conventional gravity collection systems in rural communities was found to dwarf the cost of treatment and disposal. In response to this condition, efforts were initiated throughout the United States to develop low-cost sewer systems to serve the needs of rural communities, which constituted more than 80% of demand for centralized collection and treatment. In developing alternative collection systems (ACS) for these small communities, engineers turned to concepts that had theretofore been either forgotten or ignored by the profession.

Pressure sewers had been conceived toward the middle of the 20th century as a means of separating combined sewers in large cities by the late Professor Gordon M. Fair of Harvard University. Vacuum sewers had been around since the 19th century, but had not been seriously considered for widespread use until then. Small-diameter gravity sewers (SDGSs) also found 19th-century roots in the United States, but the principles had been all but forgotten in the rush to codify urban civil engineering technology. These systems returned to the United States from Australia where they had been employed successfully for several years.

18.1 Alternative Sewer Systems

After initial demonstration projects had been underwritten by the U.S. Environmental Protection Agency (EPA) and the Farmers Home Administration, these technologies were given special status under the innovative and alternative (I&A) technology provisions of the Clean Water Act of 1977. Thus stimulated, these technologies flourished in small communities, which were able to secure grants under this program. More than 500 alternative sewer systems were installed under the I&A provisions, and a significant number were also constructed with state, local, and private funding during the 1970s and 1980s.

Alternative wastewater collection systems can be cost-effective for homes in areas where traditional collection systems are too expensive to install and operate. Alternative sewers are used in sparsely populated or suburban areas in which conventional collection systems would be expensive. These systems generally use smaller diameter pipes with a slight slope or follow the surface contour of the land, reducing excavation and construction costs.

Alternative sewers differ from conventional gravity collection systems because they break down large solids in the pumping station before they are transported through the collection system. Their watertight design and the absence of manholes eliminate extraneous flows into the system. Thus, alternative sewer systems may be preferred in areas that have high groundwater that could seep into the sewer, increasing the amount of wastewater to be treated. They also protect groundwater sources by keeping wastewater in the sewer. In general, the disadvantages of alternative wastewater systems include increased energy demands, higher maintenance requirements, and greater on-lot costs. However, the grinder-pump-driven low-pressure sewer (LPS) system can, under certain circumstances, require less energy demands and lower operation and maintenance (O&M) requirements than conventional gravity sewers. In areas with varying terrain and population density, it may prove beneficial to install a combination of sewer types.

Water and Wastewater Engineering: Hydraulics, Hydrology and Management, Volume 1, Fourth Edition.
Lawrence K. Wang, Mu-Hao Sung Wang, and Nazih K. Shammas.

Companion website: www.wiley.com/go/Wang/Waterandwastewater4e

This chapter discusses the three types of alternative sewer systems:

1) LPS system
2) SDGS system
3) Vacuum sewer system.

18.1.1 Common Features of Alternative Collection Systems

Although each of the alternative sewer technologies uses very different motive forces, they share many similarities. All use lightweight plastic pipe buried at shallow depths, with fewer joints compared to conventional gravity sewers due to their increased individual pipe lengths. Each has the ability to save significantly on capital investment if properly designed and installed in rural areas where their inherent advantages can be exploited. All have suffered from some misuse and misapplication in early installations, as have all new technologies.

A common need of all ACS is proper administration and management. Because the needs of these technologies are different from conventional sewers and treatment facilities, O&M staff members must be properly trained in the particular needs of the type of system employed.

A common concern with all ACS types is the shallow burial depth, which increases the potential for damage from the ground surface, for example, from excavation projects. Good management and design can minimize this problem by inclusion of marking tape and toning wires in the trench and surface markers, which point excavators in the direction of the O&M staff for assistance in locating facilities. Quality as-built drawings and, possibly, the use of geographic information systems software will prove invaluable for all of these systems.

The other concern is the need for more on-lot activity than is normally experienced with conventional sewers. Homeowner involvement in the planning process is a requirement for success with any ACS project to minimize the potential for subsequent damage to public relations and to maximize the potential support of the homeowners for the project. Similarly, the system staff representatives will be considered the embodiment of management and must be able to relate positively to the public.

18.1.2 Evaluation Issues

All of the alternative sewer systems can be considered for municipalities of 10,000 people or less. Those communities of 3,500–10,000 population can likely handle all ACS technologies with proper training. Small communities under 1,000 population are probably the most restrictive in terms of available O&M capability. Anything more mechanical than a SDGS with no lift station should be given another level of scrutiny for most of these locations. Arrangements with county government, private management entities, or other larger utilities may eliminate this O&M barrier for even the smallest communities, permitting an unconstrained choice of the optimum technology for each community.

The most common combination is that of SDGS with *septic tank effluent pumping* (STEP) sewers. This combination system is sometimes referred to as an effluent sewer system, because both employ septic tank pretreatment. *Grinder-pump* (GP) pressure systems are commonly used with conventional gravity sewers to reduce total costs. Combinations of ACS, other than "effluent sewers," however, are rare. Theoretically, both SDGS and STEP could feed into a vacuum or a GP sewer, but the reverse could have repercussions since the former two are designed to carry wastewater that does not contain heavy solids and grease.

Conservative engineers have designed ACS treatment systems that are identical to those for conventional wastewater. Given the rural nature of these systems, most treatment facilities have been stabilization ponds that are somewhat insensitive to the wastewater characteristic variance between ACS types. Some mechanical treatment systems (extended aeration and oxidation ditches) have been used without significant difficulty for larger ACS installations. Also, some subsurface soil absorption systems have been successfully employed for some smaller ACS sites.

18.2 Low-Pressure Sewer System

The LPS system is emerging as one of the most popular and successful of the collection system alternatives. A pressure sewer is a small-diameter pipeline, shallowly buried, that follows the profile of the ground. Typical main diameters are 2–6 in. (50–150 mm). Polyvinyl chloride (PVC) is the usual piping material. Burial depths usually are below the frost line, or a 30-in. (750-mm) minimum, whichever is greater. In northern areas insulated and heat-traced piping offer relief from these criteria.

Pressure sewers are particularly adaptable for rural or semirural communities where public contact with effluent from failing drain fields presents a substantial health concern. Because the mains for pressure sewers are, by design, watertight, the pipe connections ensure minimal leakage of sewage. This can be an important consideration in areas subject to groundwater contamination. Two major types of pressure sewer systems are the STEP system and the GP system mentioned earlier. Neither requires any modification to plumbing inside the house.

In STEP systems, wastewater flows into a conventional septic tank to capture solids. The liquid effluent flows to a holding tank containing a pump and control devices. The effluent is then pumped and transferred for treatment. Retrofitting existing septic tanks in areas served by septic tank/drain field systems would seem to present an opportunity for cost savings, but a large number (often a majority) must be replaced or expanded over the life of the system because of insufficient capacity, deterioration of concrete tanks, or leaks. In a GP system, sewage flows to a vault where a GP grinds the solids and discharges the sewage into a pressurized pipe system. GP systems do not require a septic tank but may require more horsepower than STEP systems because of the grinding action. A GP system can result in significant capital cost savings for new areas that have no septic tanks or in older areas where many tanks must be replaced or repaired. Figure 18.1 shows a typical septic tank effluent pump system, while Fig. 18.2 shows a typical GP used in residential wastewater treatment.

The choice between GP and STEP systems depends on three main factors:

1) *Cost*: On-lot facilities, including pumps and tanks, will account for more than 75% of total costs and may even run as high as 90%. Thus, there is a strong motivation to use a system with the least expensive on-lot facilities. STEP systems may lower on-lot costs because they allow some gravity service connections due to the continued use of a septic tank. In addition, a GP must be more rugged than a STEP pump to handle the added task of grinding and, consequently, the GP system is more expensive. If many septic tanks must be replaced, costs will be significantly higher for a STEP system than a GP system, and overall life-cycle cost of the GP system should be lower as pump life is longer than the effluent pumps traditionally used in STEP systems.
2) *Downstream treatment*: GP systems produce a larger amount of *total suspended solids* (TSS) that may or may not be acceptable at a downstream treatment facility.

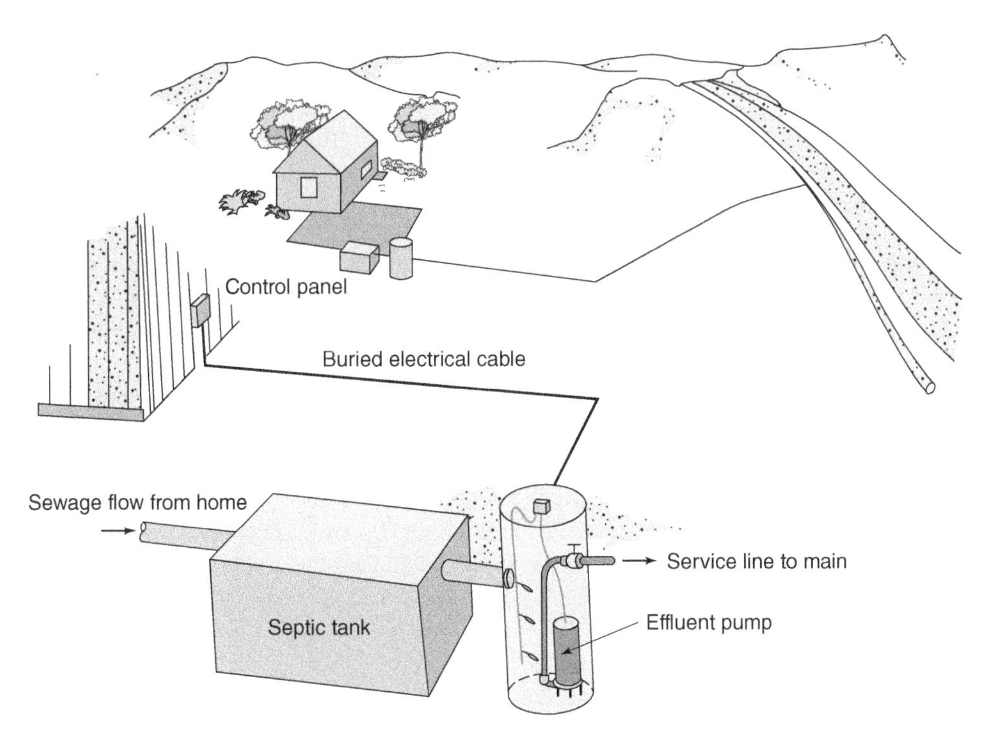

Figure 18.1 Septic Tank Effluent Pump (STEP) System

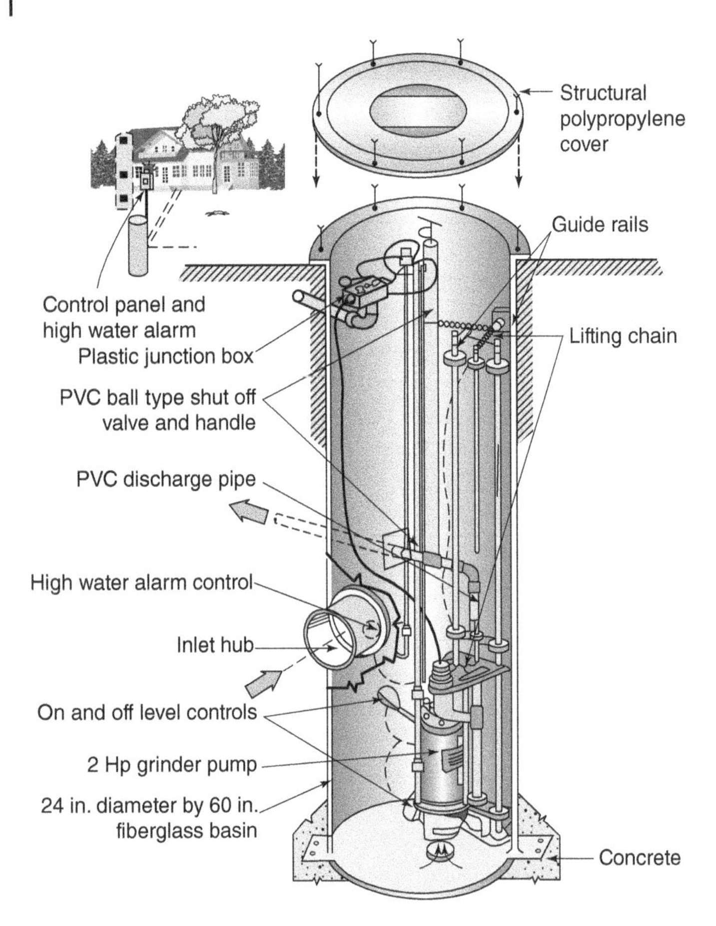

Figure 18.2 Grinder Pump (GP). Conversion factors: 1 in. = 25.4 mm; 1 Hp = 0.7457 kW

3) *Low-flow conditions*: STEP systems will better tolerate low-flow conditions that occur in areas with highly fluctuating seasonal occupancy and those with slow build-out from a small initial population to the ultimate design population. Thus, STEP systems may be better choices in these areas than GP systems.

18.2.1 Applicability

The primary reason for the use of pressure sewers is economic, but in some cases the decisions are environmentally motivated. In areas where rock is encountered when excavating to install mainline sewers, pressure sewers can be cost-effective. The deep, wide trenches required to install conventional sewers are expensive to construct. Pressure sewers require only shallow, narrow trenches.

Where groundwater is high, the deep excavations for conventional sewers may enter that groundwater. In some cases dewatering is not achievable. When conventional sewers are installed under these conditions, the cost is high and the quality of installation is questionable. Shoring can also add considerably to the cost of conventional sewers.

Some topography does not favor gravity collection. One typical example is around lakes where the homes are built fronting the lake. The road serving these homes, and often the only practical location for the sewer, may be upslope from the homes. The profile of that route may go up and down as it circles the lake; numerous and costly pumping stations would be necessary if conventional sewers were used.

Conventional sewers have a high cost per foot of sewer installed. Where homes are sparse, the resulting cost can be exorbitant. Pressure sewers can be installed less expensively on a per-foot basis.

Extremely flat terrain poses a problem for gravity sewer installations since the gravity sewer must continually slope downward. This causes the sewer to become increasingly deep until a lift station is necessary. Both the deep excavations and the lift stations are expensive, and the latter represents a considerable O&M expense.

Damage consequential to the installation of deep sewers is a factor. In some cases blasting is required to install sewers. This may cause upheaval of the road, damage to nearby buried utilities and homes, and disruption to the community. Deeply buried conventional sewers may intercept and drain groundwater. In many cases the groundwater will enter the gravity sewer as unwanted infiltration.

Developments experiencing slow growth find pressure sewers economically attractive. The front-end infrastructure (mainline) is inexpensively provided. Directional boring for GP sewer lines results in essentially zero environmental disruption. The cost of the pumping units is deferred until the homes are built and occupied. The cost for the pumping units may also be financed with the home. Time value of money considerations makes this feature particularly attractive.

Pressure sewer equipment is also used in conjunction with conventional systems. Where a low-lying home or basement is too low to allow gravity flow into a fronting conventional sewer, a GP or pressure-sewer-type solids-handling pump may be used at that home to discharge to the sewer. Similarly, STEP units are used to discharge to high-lying drain fields, sand filters, mounds, and other forms of on-site wastewater disposal.

18.2.2 Advantages

Pressure sewer systems that connect several residences to a "cluster" pump station can be less expensive than conventional gravity systems. On-property facilities represent a major portion of the capital cost of the entire system and are shared in a cluster arrangement. This can be an economic advantage because on-property components are not required until a house is constructed and are borne by the homeowner. Low front-end investment makes the present-value cost of the entire system lower than that of conventional gravity sewerage, especially in new development areas where homes are built over many years.

Because wastewater is pumped under pressure, gravity flow is not necessary and the strict alignment and slope restrictions for conventional gravity sewers can be relaxed. Network layout does not depend on ground contours; pipes can be laid in any location and extensions can be made in the street right-of-way at a relatively small cost without damage to existing structures. Other advantages of pressure sewers include the following:

1) Material and trenching costs are significantly lower because pipe size and depth requirements are reduced.
2) Low-cost cleanouts and valve assemblies are used rather than manholes and may be spaced further apart than manholes in a conventional system.
3) Infiltration is eliminated, resulting in reductions in pipe size, as well as the plant size for wastewater treatment.
4) The user pays for the electricity to operate the pump unit. The resulting increase in electric bills is small and may replace municipality or community bills for central pumping eliminated by the pressure system.
5) Final treatment may be substantially reduced in hydraulic and organic loading in STEP systems. Hydraulic loadings are also reduced for GP systems.
6) Because sewage is transported under pressure, more flexibility is allowed in siting final treatment facilities and may help reduce the length of outfall lines or treatment plant construction costs.

18.2.3 Disadvantages

The system has the following disadvantages:

1) It requires a lot of institutional involvement because the pressure system has many mechanical components throughout the service area.
2) The O&M costs for a pressure system are often higher than for a conventional gravity system due to the high number of pumps in use. However, lift stations in a conventional gravity sewer can reverse this situation.
3) Annual preventive maintenance calls are usually scheduled for the GP components of pressure sewers that employ float-level sensors. However, some GPs manufactured by Environment One Corporation do not employ floats and are subsequently preventative—maintenance-free. STEP systems also require pump out of septic tanks at 2- to 3-year intervals.
4) Public education is necessary so the user knows how to deal with emergencies and how to avoid blockages or other maintenance problems.
5) The number of pumps that can share the same downstream force main is limited.
6) Extended Power outages can result in overflows if standby generators are not available.

7) Life cycle replacement costs are expected to be higher because pressure sewers have a lower life expectancy than do conventional systems.

Odors and corrosion are potential problems for both alternative and conventional sewers because the wastewater in the collection sewers is usually septic. Proper ventilation and odor control must be provided in the design and noncorrosive components should be used. Air release valves are often vented to soil beds to minimize odor problems, and special discharge and treatment designs are required to avoid terminal discharge problems.

18.2.4 Design Criteria

Many different design flows can be used in pressure systems. When positive displacement GP units are used, the design flow is obtained by multiplying the pump discharge by the maximum number of pumps expected to be operating simultaneously. When centrifugal pumps are used, the equation, using US customary units, is

$$Q = 20 + 0.5N \quad \text{(US customary units)} \tag{18.1}$$

where Q is the flow in gpm and N is the number of homes served. The operation of the system under various assumed conditions should be simulated by computer to check design adequacy. No allowances for infiltration and inflow (I/I) are required. No minimum velocity is generally used in design, but GP systems must attain 3–5 ft/s (0.90–1.50 m/s) at least once per day. A Hazen–Williams coefficient C of 130–140 is suggested for hydraulic analysis.

An equivalent equation using metric units is as follows:

$$Q = 1.262 + 0.032\,N \quad \text{(SI units)} \tag{18.2}$$

where Q is the flow in L/s and N is the number of homes served.

The service line leading from the pumping unit to the main is usually 1–1.5 in. (25–38 mm) in diameter and made of PVC. A check valve on the service line prevents backflow. Prevention is ensured with a redundant check valve at the pumping unit. If a malfunction occurs, a high-liquid-level alarm is activated. This may be a light mounted on the outside wall of the home, or it may be an audible alarm that can be silenced by the resident. The resident then notifies the sewer service district, which responds to make the necessary repair.

Pressure mains generally use 2-in. (50-mm) or larger PVC pipe and rubber-ring joints or solvent welding to assemble the pipe joints. *High-density polyethylene* (HDPE) pipe with fused joints is widely used in the United States, Canada, and European countries. The required pipe size is a function of the number of homes as follows:

Pipe Size (in.)	Pipe Size (mm)	Number of Homes
2	50	6
3	75	60
4	100	120
6	150	240
8	200	560

Electrical requirements, especially for GP systems, may necessitate rewiring and electrical service upgrading in the service area. Pipes are generally buried to at least the winter frost penetration depth; in far northern sites, insulated and heat-traced pipes are generally buried at a minimal depth. GP and STEP pumps are sized to accommodate the hydraulic grade requirements of the system.

GPs to serve individual homes are usually 1 hp (0.75 kW) in size, but 2 hp (1.49 kW) units are also used. Some installations use 3- to 5-hp (2.24- to 3.73-kW) motors, but these are usually used when serving several homes with one pumping unit. STEP pumps are usually a fractional horsepower. Discharge points must use drop inlets to minimize odors and corrosion. Air release valves are placed at high points in the sewer and often are vented to soil beds. Both STEP and GP systems can be assumed to be anaerobic and potentially odorous if subjected to turbulence (stripping of gases such as H_2S).

Example 18.1 Determination of Head Loss in Low-Pressure Sewer System

Determine the friction head loss (ft/100 ft or m/100 m) of a housing development zone where seven GPs are operating simultaneously at the rate of 11 gpm/pump (0.69 L/s/pump). Assume that 2,200 ft of 3-in. (670 m of 75-mm)-diameter PVC Class SDR 21 pipe (nominal size; $C = 150$) is to be used for transporting the sewage in the LPS system.

Solution 1 (US Customary System):

Two modified Hazen–Williams formula are presented here for the convenience of designing an LPS system:

$$H_f = 0.2083\left[(100/C)^{1.85}\left(Q_{gpm}\right)^{1.85} D_i^{-4.86}\right] \quad (18.3a)$$

$$v = 0.3208\, Q_{gpm}/A \quad (18.3b)$$

where H_f is friction head loss (ft/100 ft of pipe); C is coefficient, which depends on the type and condition of the pipe; Q_{gpm} is water flow (gpm); D_i is inside diameter of pipe (in.) = average outside diameter – (2 × minimum wall thickness); v is velocity (ft/s); and A is area (in.^{2}).

Conversion factors: 1 ft/100 ft = 1 m/100 m; 1 gpm = 0.0631 L/s; 1 in. = 25.4 mm.

Using the above modified Hazen–Williams formula and the design criteria (11 gpm/pump or 0.69 L/s/pump), and the PVC pipe manufacturer's data (regarding the actual inside diameter of the 3-in. [75-mm] diameter [nominal size] PVC Class SDR 21 pipe), an engineering design table is developed as shown in Table 18.1. In this case $C = 150$.

From Table 18.1 (or from the equations in this problem), the maximum water velocity (v) and the friction loss (H_f) are determined to be **3.14 ft/s and 1.12 ft/100 ft**, respectively (0.96 m/s and 1.12 m/100 m, respectively).

The same results can be obtained by engineering calculation:

$$c = 150 \text{ and outside pipe diameter} = 3.5 \text{ in.}$$

$$D_i = 3.5\text{ in.} - 2 \times \text{thickness} = 3.146 \text{ in inside diameter (given)}$$

$$Q_{gpm} = (11\text{ gpm/pump}) \times (7\text{ pump}) = 77\text{ gpm (Note: } N = 7 \text{ when using Table 18.1)}$$

$$H_f = 0.2083\left[(100/C)^{1.85}\left(Q_{gpm}\right)^{1.85} Di^{-4.86}\right]$$

$$= 0.2083\left[(100/150)^{1.85}(77)^{1.85}(3.146)^{-4.86}\right]$$

$$= 0.2083[0.4723 \times 3090.35 \times 0.0038]$$

$$= 1.16 \text{ ft/100 ft (Note: from Table 18.1, } Hf = 1.12 \text{ ft/100 ft)}$$

$$A = D_i^2\pi/4 = (3.146)^2 \times 3.14/4 = 7.77 \text{ in.}^2$$

$$v = 0.3208\, Q_{gpm}/A = 0.3208 \times 77/7.77 = 3.17 \text{ ft/s (Note: from Table 18.1, } v = 3.14 \text{ ft/s)}$$

Alternatively the following two modified Hazen–Williams equations can also be used to obtain same results:

$$h_f = 4.72\left(Q_{\text{ft}^3/\text{s}}/C\right)^{1.85} L/D^{4.87} \quad (5.37)$$

$$v = 0.55\, CD^{0.63}s^{0.54} \quad (5.34)$$

where the common US customary units are used: h_f(ft) $Q_{\text{ft}^3/\text{s}}(\text{ft}^3/\text{s})$, C (dimensionless), L (ft), D (ft), v (ft/s), and s (dimensionless).

Because the length (L) of the pipe is 2,200 ft (670 m), the net friction loss in this zone is calculated to be 2,200 ft × (1.16 ft/100 ft) = **25.52 ft of net head loss**, or (670 m) × (1.16 m/100 m) = 7.77 m of net head loss.

Solution 2 (SI System):

Given data are as follows:

1) Circular PVC pipe SDR 21 inside diameter = 3.146 in. = 79.9 mm = 0.0799 m
2) C = 150
3) $Q = 7 \times 11$ gpm = 77 gpm = 77×0.0000631 m^3/s = 0.00486 m^3/s
4) L = 670 m

Table 18.1 Flow Velocity and Friction Head Loss Versus Pumps in Simultaneous Operation for PVC Pipe Class SDR 21 (Example 18.1)

	1 1/4 in.		1 1/2 in.		2 in.		2 1/2 in.		3 in.		4 in.		5 in.		6 in.	
N	v	H_f	v	H_f	v	H_f	v	H_f	v	H_f	v	H_f	v	H_f	v	H_f
1	1.99	**1.15**	1.52	**0.60**												
2	3.99	**4.16**	3.04	**2.15**	1.95	**0.73**										
3	5.98	**8.82**	4.56	**4.56**	2.92	**1.54**	1.99	**0.61**								
4	7.97	**15.02**	6.08	**7.77**	3.89	**2.63**	2.66	**1.04**	1.79	**0.40**						
5					4.87	**3.97**	3.32	**1.57**	2.24	**0.60**						
6					5.84	**5.57**	3.99	**2.20**	2.69	**0.85**						
7					6.81	**7.41**	4.65	**2.93**	3.14	**1.12**	1.90	**0.33**				
8							5.32	**3.75**	3.59	**1.44**	2.17	**0.42**				
9							5.98	**4.66**	4.04	**1.79**	2.44	**0.53**				
10							6.64	**5.67**	4.49	**2.18**	2.71	**0.64**				
11									4.93	**2.60**	2.98	**0.76**	1.95	**0.27**		
12									5.38	**3.05**	3.25	**0.90**	2.13	**0.32**		
13									5.83	**3.54**	3.52	**1.04**	2.31	**0.37**		
14									6.28	**4.06**	3.80	**1.19**	2.48	**0.43**		
15											4.07	**1.36**	2.66	**0.48**	1.88	**0.21**
16											4.34	**1.53**	2.84	**0.55**	2.00	**0.23**
17											4.61	**1.71**	3.02	**0.61**	2.13	**0.26**
18											4.88	**1.90**	3.19	**0.68**	2.25	**0.29**
19											5.15	**2.10**	3.37	**0.75**	2.38	**0.32**
20											5.42	**2.31**	3.55	**0.82**	2.50	**0.35**
21											5.69	**2.53**	3.73	**0.90**	2.63	**0.39**
22											5.96	**2.76**	3.90	**0.98**	2.75	**0.42**
23											6.24	**2.99**	4.08	**1.07**	2.88	**0.46**

Note: N = number of GPs; v = velocity, ft/s; H_f = friction head loss, ft/100 ft of pipe.
Conversion factors: 1 in. = 25.4 mm; 1 ft/s = 0.3048 m/s.
Source: Environment One Corp, Niskayuna, New York, USA.
Example: 7 grinder pumps = $N = 7$; Di = 3 in; $v = 3.14$ ft/s; $H_f = 1.12$ ft/100 ft.

SI system Hazen–Williams equations are as follows:

$$h_f = 10.67\left(Q_{m^3/s}/C\right)^{1.85} L/D^{4.87} \tag{5.37}$$

$$s = h_f/L$$

$$v = 0.3545\, CD^{0.63} s^{0.54} \tag{5.34}$$

Therefore,

$$h_f = 10.67\,(0.00486/150)^{1.85}\,(670)/(0.0799)^{4.87}$$
$$= 10.67(5 \times 10^{-9})(670)/0.000004523$$
$$= 7.9 \text{ m } (25.9 \text{ ft}) \textbf{ net head loss}$$

$$s = h_f/L = 7.9\, m/670\, m = 0.01179$$
$$= 1.179 \text{ m}/100 \text{ m } (1.179 \text{ ft}/100 \text{ ft})$$

$$v = 0.3545 \times 150 \times (0.0799)^{0.63}\,(0.01179)^{0.54}$$
$$= 0.3545 \times 150 \times 0.2035 \times 0.0909$$
$$= 0.9836 \text{ m/s } (3.2 \text{ ft/s})$$

A design table similar to Table 18.1 can also be established using the SI units for rapid LPS and GP selections.

18.2.5 Wastewater Force Main Design

The procedures for an LPS system are discussed here. The preliminary LPS system design is straightforward because it only involves (a) design, selection, and installation of GPs; and (b) design, selection, and installation of force main pipes. This is primarily a result of two characteristics of new semipositive displacement pumps: near-constant flow over the entire range of operating pressures and the ability of the pump to handle transient overpressures.

This section outlines a systematic approach to LPS system design, going from pump model and pipe selection to a detailed zone and system analysis.

18.2.5.1 Information Required for LPS System Design

The following information should be assembled prior to initiation of the LPS system design:

1) Topography map
2) Soil conditions
3) Climatic conditions (frost depth, low temperature, and duration)
4) Water table
5) Applicable codes
6) Discharge location
7) Lot layout (with structures shown, if available)
8) Total number of lots
9) Dwelling type(s)
10) Use and flow factors (seasonal occupancy or year-round, appliances, water supply sources)
11) Area development sequence and timetable
12) LPS system design manual and design software (*Note*: available from Environment One Corporation, or from the instructor who adopts this book as the textbook).

18.2.5.2 Grinder Pump Station Size Selection

Use the manufacturer's recommendations to select GP models for the types of occupancy to be served. For instance, the smallest GP may have a flow capacity of up to 700 gpd (2,649.5 L/d), which is adequate for managing the flow from one average single-family home, and up to two average, single-family homes where codes allow and with consent of the factory, while the largest GP may have a capacity of up to 6,000 gpd (22,710 L/d), which is adequate for managing the flow from up to nine average single-family homes.

Design considerations for the pump sizing, selection, and installation include the following:

1) Wet-well and discharge piping must be protected from freezing.
2) Model and basin size must be appropriate for incoming flows, including peak flows.
3) An appropriate alarm device must be used.
4) The location must be suitable.

Daily flows above those recommended may exceed the tank's peak flow holding capacity and/or shorten the interval between pump overhauls. The company should be consulted if higher inflows are expected. The final selection will have to be determined by the engineer on the basis of actual measurements or best estimates of the expected sewage flow.

18.2.5.3 Grinder Pump Placement

The most economical location for installation of the GP station is in the basement of the building it will serve. However, due consideration must be given when choosing an indoor location. If there is a risk of damage to items located in the basement level, other provisions should be made during basement installation or an outdoor unit should be considered.

Considerations such as ownership of the pumps by a municipality or private organization or the need for outdoor accessibility frequently dictate outdoor, in-ground installations. For outdoor installations, some GP models are available with HDPE integral access ways ranging in height up to 10 ft (3.048 m). By keeping the unit as close as possible to the building, the lengths of gravity sewer and wiring will be minimized, keeping installation costs lower while reducing the chances of infiltration in the gravity flow section.

AC power from the building being served should be used for the GP. Separate power sources add to installation and O&M costs, decrease overall reliability, and frequently present an aesthetic issue. When two dwellings are to be served by a single unit, the station is usually placed in a position requiring the shortest gravity drains from each home. With multifamily buildings, more than one GP may be required.

18.2.5.4 LPS System Pipe Selection

The final determination of the type of pipe to be used is the responsibility of the consulting engineer. In addition, the requirements of local codes, soil, terrain, water, and weather conditions that prevail will guide this decision.

Although pipe fabricated from any approved material may be used, most LPS systems have been built with PVC and HDPE pipes. Continuous coils of small-diameter HDPE pipe can be installed with automatic trenching machines and horizontal drilling machines to sewer areas at lower cost.

Table 18.2 compares the water capacity of two types of commonly used PVC pipe, SDR 21 and Sch 40, and one type of HDPE, SDR 11. All three have adequate pressure ratings for LPS service. Although both types of PVC pipes are suitable, the three parameters compared in Table 18.3 illustrate why SDR 21 is suggested as a good compromise between capacity, strength, friction loss characteristics, and cost.

18.2.5.5 LPS System Layout

A preliminary sketch of the entire pressure sewer system should be prepared (Fig. 18.3). Pump models should be selected and their location (elevation) should be noted. The location and direction of flow of each lateral, zone, and main and the point of discharge should be shown. The system should be designed to give the shortest runs and the fewest abrupt changes

Table 18.2 Recommended Pipe Water Capacities (gal/100 ft of pipe length)

Nominal Pipe Size (in.)	Sch 40 PVC	SDR 21 PVC	SDR 11 HDPE
1¼	7.8	9.2	7.4
1½	10.6	12.1	9.9
2	17.4	18.8	15.4
2½	23.9	27.6	—
3	38.4	40.9	33.5
4	66.1	67.5	55.3
5	103.7	103.1	84.5
6	150.0	146.0	119.9
8	260.0	249.0	203.2

Conversion factors: 1 gal = 3.785 L; 100 ft = 30.48 m; 1 in. = 25.4 mm; 1 gal/100 ft = 12.418 L/100 m.
Source: Environment One Corp, Niskayuna, New York, USA.

Table 18.3 PVC Pipe Comparisons (Nominal Pipe Size =2 in.)

Parameter	Sch 40	SDR 21
Wall thickness, in.	0.154	0.113
Inside diameter, in.	2.067	2.149
50 gpm friction loss, ft/100 ft	4.16	3.44

Conversion factors: 1 in. = 25.4 mm; 1 ft/100 ft = 1 m/100 m; 1 gpm = 3.785 L/min.
Source: Environment One Corp, Niskayuna, New York, USA.

in direction. "Loops" in the system must be avoided because they lead to unpredictable and uneven distribution of flow. Although not shown in Fig. 18.3, the elevation of the shutoff valve of the lowest lying pump in each zone should be recorded and used in the final determination of static head loss. Because Environment One GPs are semipositive displacement and relatively insensitive to changes in head, precisely surveyed profiles are unnecessary.

Air/vacuum valves, air release valves, and combination air valves serve to prevent the concentration of air at high points within a system. This is accomplished by exhausting large quantities of air as the system is filled and also by releasing pockets of air as they accumulate while the system is in operation and under pressure. Air/vacuum valves and combination air valves also serve to prevent a potentially destructive vacuum from forming.

Air/vacuum valves should be installed at all system high points and significant changes in grade. Combination air valves should be installed at those high points where air pockets can form. Air release valves should be installed at intervals of 2,000–2,500 ft (610–762 m) on all long horizontal runs that lack a clearly defined high point.

Air relief valves should be installed at the beginning of each downward leg in the system that exhibits a 30 ft (9.10 m) or more drop. Trapped pockets of air in the system not only add static head, but also increase friction losses by reducing the cross-sectional area available for flow. Air will accumulate in downhill runs preceded by an uphill run.

Long ascending or descending lines require air and vacuum or dual-function valves placed at approximately 2,000-ft (610-m) intervals. Long horizontal runs require dual-function valves also placed at approximately 2000-ft (610-m) intervals.

Pressure air release valves allow air and gas to continuously and automatically be released from a pressurized liquid system. If air or gas pockets collect at high points in a pumped system, those pressurized air pockets can begin to displace usable pipe cross-section. As the cross-section of the pipe artificially decreases, the pump sees this situation as increased resistance to its ability to force the liquid through the pipe.

Air relief valves at high points may be necessary, depending on total system head, flow velocity, and the particular profile. The engineer should consult Environment One in cases where trapped air is considered a potential problem.

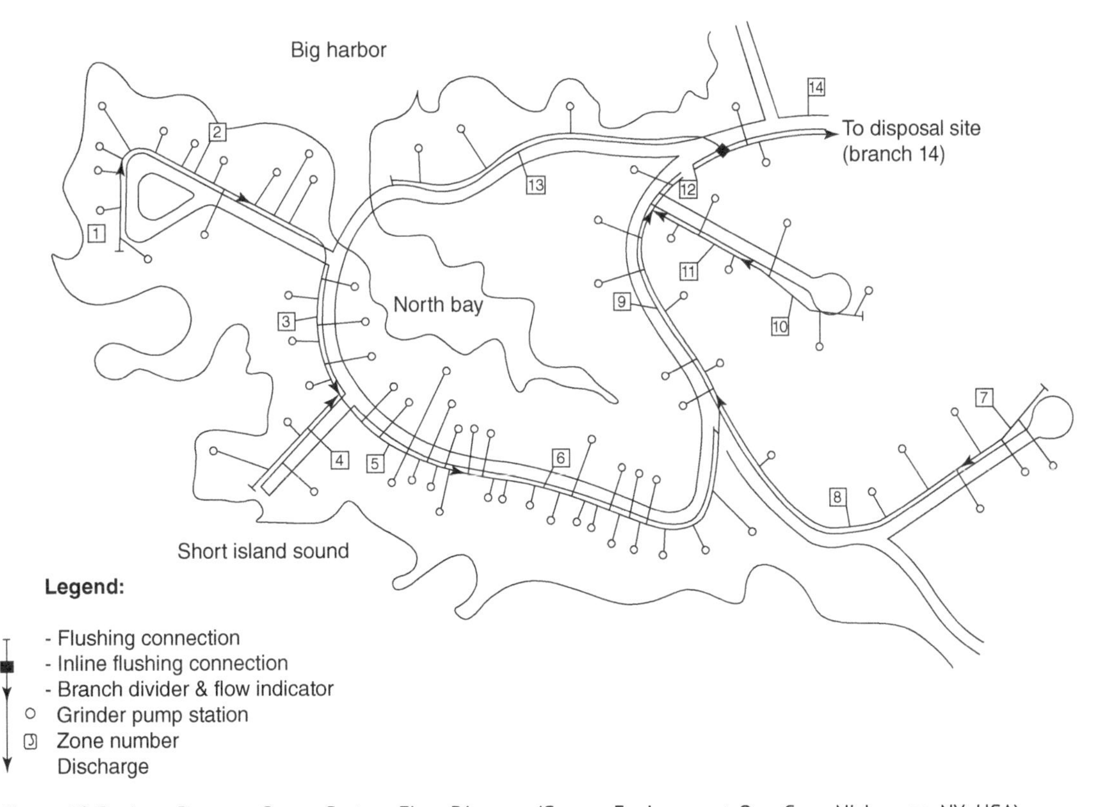

Figure 18.3 Low-Pressure Sewer System Flow Diagram (*Source:* Environment One Corp, Niskayuna, NY, USA)

Cleanout and flushing stations should be incorporated into the pipe layout. In general, cleanouts should be installed at the terminal end of each main, every 1,000–1,500 ft (305–457 m) on straight runs of pipe, and whenever two or more mains come together and feed into another main.

18.2.5.6 LPS System Zone Designations

The LPS system illustrated in Fig. 18.3, for instance, contains 72 pumps and is divided into 14 individually numbered zones. Division into zones facilitates final selection of pipe sizes, which are appropriate in relation to the requirements that flow velocity in the system be adequate and that both static and dynamic head losses be within design criteria. Assignment of individual zones follows from the relationship between the accumulating total numbers of pumps in a system and the predicted number that will periodically operate simultaneously (Table 18.4). This table was initially developed after careful analysis of more than 58,000 pump events in a 307-day period during the demonstration project in Albany, New York. It was extended for larger systems by application of probability theory. The validity of this table has since been confirmed by actual operating experience with thousands of large and small LPS systems during a 34-year period.

Table 18.4 Maximum Number of Grinder Pump Cores Operating Daily

Number of Grinder Pump Cores Connected	Maximum Daily Number of Grinder Pump Cores Operating Simultaneously
1	1
2–3	2
4–9	3
10–18	4
19–30	5
31–50	6
51–80	7
81–113	8
114–146	9
147–179	10
180–212	11
213–245	12
246–278	13
279–311	14
312–344	15
345–377	16
378–410	17
411–443	18
444–476	19
477–509	20
510–542	21
543–575	22
576–608	23
609–641	24
642–674	25
675–707	26

Table 18.4 (Continued)

Number of Grinder Pump Cores Connected	Maximum Daily Number of Grinder Pump Cores Operating Simultaneously
708–740	27
741–773	28
774–806	29
807–839	30
840–872	31
873–905	32
906–938	33
939–971	34
972–1,004	35

Source: Environment One Corp, Niskayuna, New York, USA.

Using Fig. 18.3, the actual exercise of assigning zones is largely mechanical. The single pump farthest from the discharge point in any main or lateral constitutes a zone. This and downstream pumps along the main are accumulated until their aggregate number is sufficient to increase the number of pumps in simultaneous operations by one, that is, until the predicted maximum flow increases by 11 gpm (41.64 L/min or 0.69 L/s).

Fig. 18.3 shows that zones 1, 2, and 3 end when the number of pumps connected total three, six, and nine, and the number of pumps in daily simultaneous operation are two, three, and four, respectively.

Any place where two or more sections of main join, or where the outfall is reached, also determines the end of a zone. This design rule takes precedence over the procedure stated above, as seen in zones 3, 4, 6, 8, 9, 11, 12, 13, and 14.

Example 18.2 Design of a Low-Pressure Sewer System

The new residential realty subdivision plan shown in Fig. 18.3 has divided the subdivision area into 14 housing zones. The chief engineer chose PVC pipe, Class SDR 21, and an LPS system. The manufacturer's design criteria of (a) 11 gpm/pump or 41.64 L/min/pump, and (b) the chart of maximum number of GP cores operating daily (Table 18.4) will be followed. Table 18.5 indicates the given data for all 14 zones. Fill in the data in columns 7–14 and columns 17–18 of Table 18.6 in turn, to complete the preliminary engineering design.

Solution

The first step is to complete the pipe schedule and zone analysis (Table 18.6). The data recorded on the system flow diagram (Fig. 18.3) is then transferred to Table 18.6. For clarification, Table 18.6's column 2 is "Connects to Zone," column 4 is "Accumulated Pumps in Zone," and the unit for column 11 is feet.

Column 4 is completed by referring to Table 18.4, where the maximum number of pumps in simultaneous operation is given as a function of the number of pumps upstream from the end of the particular zone.

The output of each zone will vary slightly with head requirements, but under typical conditions, the flow is approximately 11 gpm (41.64 L/min). Calculate the maximum anticipated flow for each zone by multiplying the number of simultaneous operations in column 7 by 11 gpm and record the results in column 8.

To complete columns 9, 10, 12, and 13, refer to the flow velocity and friction head loss table, Table 18.1, for the type of pipe selected in this case, commercial PVC pipe, Class SDR 21. The Hazen–Williams equation can also be used for calculations. Note that the engineer will frequently be presented with more than one option when selecting pipe size. Sometimes a compromise in pipe size will be required to meet present needs and planned future development. As a general rule, pipe sizes should be selected to minimize friction losses while keeping velocity near or above 2 ft/s (0.60 m/s).

For example, zone 1 has a maximum of two pumps running (column 7). Table 18.1 offers a choice of 1.25-, 1.5-, or 2-in. (32-38- or 50-mm) pipe. The 1.5-in. (38-mm) pipe is selected because flow velocity equals 3.04 ft/s (0.93 m/s) and friction loss

Table 18.5 Preliminary LPS System Pipe Schedule and Zone Analysis for Example 18.2

1	2	3	4	5	6
Zone No.	**Conn. to Zone**	**No. of Pumps in Zone**	**Accum. Pumps in Zone**	**Flow Per Core, gpd**	**Max. Flow Per Core, gpm**
1	2	3	3	200	11
2	3	6	9	200	11
3	5	9	18	200	11
4	5	3	3	200	11
5	6	9	30	200	11
6	9	17	47	200	11
7	8	3	3	200	11
8	9	4	7	200	11
9	12	6	60	200	11
10	11	3	3	200	11
11	12	3	6	200	11
12	14	1	67	200	11
13	14	3	3	200	11
14	14	2	72	200	11

Conversion factors: 11 gpm = 41.635 L/ min; 1 gpd = 3.785 L/d.
Source: Environment One Corp, Niskayuna, New York, USA.

equals 2.15 ft/100 ft (2.15 m/100 m). Since the zone is 205 ft (62.48 m) in length (column 11), the total friction loss (column 13) is (2.15 ft/100 ft) (205 ft) = 4.41 ft or 1.34 m.

For zone 14, with 72 upstream pumps, it is seen that a maximum of 7 pumps can be running simultaneously. Table 18.6 provides options of:

$$\text{3-in.(75-mm) pipe: } v = 3.14\ \text{ft/s} = 0.96\ \text{m/s}; H_f = 1.12\ \text{ft}/100\ \text{ft} = 1.12\ \text{m}/100\ \text{m}$$

or

$$\text{4-in.(100-mm) pipe: } v = 1.90\ \text{ft/s} = 0.589\ \text{m/s}; H_f = 0.33\ \text{ft}/100\ \text{ft} = 0.33\ \text{m}/100\ \text{m}$$

The smaller diameter 3-in. (75-mm) pipe is selected because of the increased velocities, especially with the TDH (total dynamic head) below 138 ft (42.06 m). A choice of 3-in. (75-mm) pipe would lead to a friction loss in this zone of:

US customary units: H_f = (1.12 ft/100 ft) (2,200 ft) = 24.75 ft

Metric system: H_f = (1.12 m/100 m) (670.56 m) = 7.51 m

Accumulated friction loss (column 14) for each zone is next determined by adding the friction loss for each zone from the system outfall (zone 14) to the zone in question. Thus, from Fig. 18.3 we can see that the accumulated friction loss for zone 1 is

Zone 14: Friction loss =24.75 ft = 7.54 m
Zone 12: Friction loss =2.7 ft = 0.82 m
Zone 9: Friction loss =5.85 ft = 1.78 m
Zone 6: Friction loss =8.46 ft = 2.58 m
Zone 5: Friction loss =4.83 ft = 1.47 m
Zone 3: Friction loss =16.56 ft = 5.05 m
Zone 2: Friction loss =5.86 ft = 1.79 m
Zone 1: Friction loss =4.41 ft = 1.34 m.

Accumulated friction head loss for zone 1 = 73.41 ft (22.38 m). The same summation is completed for each zone.

Table 18.6 Filled Preliminary Design of an LPS System: Pipe Schedule and Zone Analysis for Example 18.2

1	2	3	4	5	6	7	8	9	10	11	12	13	14	15	16	17	18
Zone No.	Conn. to Zone	No. of Pumps in Zone	Accum. Pumps in Zone	Flow Per Core (gpd)	Max. Flow Per Core (gpm)	Max. Sim Ops	Max. Flow (gpm)	Pipe Size (in.)	Max. Velocity (ft/s)	Length of Main This Zone	Friction Loss Factor (ft/ 100 ft)	Friction Loss This Zone (ft)	Accum. Friction Loss (ft)	Max. Main Elev. (ft)	Min. Pump Elev. (ft)	Static Head (ft)	Total Dynamic Head (ft)
1	2	3	3	200	11	2	22	1.5	3.04	205	2.15	4.41	73.41	40	10	30	103.41
2	3	6	9	200	11	3	33	2.0	2.92	380	1.54	5.86	69.00	40	10	30	99.00
3	5	9	18	200	11	4	44	2.0	3.89	630	2.63	16.56	63.14	40	5	35	98.14
4	5	3	3	200	11	2	22	1.5	3.04	310	2.15	8.46	53.25	40	5	35	88.25
5	6	9	30	200	11	5	55	3.0	2.24	800	0.60	4.83	46.58	40	5	35	81.58
6	9	17	47	200	11	6	66	3.0	2.69	1,000	0.85	8.46	41.75	40	5	35	76.75
7	8	3	3	200	11	2	22	1.5	3.04	175	2.15	3.77	49.56	40	5	35	84.56
8	9	4	7	200	11	3	33	2.0	2.92	810	1.54	12.50	45.80	40	30	10	55.80
9	12	6	60	200	11	7	77	3.0	3.14	520	1.12	8.85	33.30	40	10	30	63.30
10	11	3	3	200	11	2	22	1.5	3.04	230	2.15	4.95	37.03	40	10	30	67.03
11	12	3	6	200	11	3	33	2.0	2.92	300	1.54	4.63	32.08	40	10	30	62.08
12	14	1	67	200	11	7	77	3.0	3.14	240	1.12	2.70	27.45	40	10	30	57.45
13	14	3	3	200	11	2	22	1.5	3.04	985	2.15	21.19	45.94	40	5	35	80.94
14	14	2	72	200	11	7	77	3.0	3.14	2,200	1.12	24.75	24.75	40	30	10	34.75

Conversion factors: 1 gpd = 1 gal/day = 3.785 L/d; 1 gpm = 3.785 Lpm; 1 ft/s = 0.3048 m/s; 1 ft/100 ft = 1 m/100 m; 1 ft = 0.3048 m; 1 in. = 25.4 mm.
Source: Environment One Corp, Niskayuna, New York, USA.

To complete the hydraulic analysis of Table 18.6, refer to the drawing contours and record in column 15 the maximum line elevation between the point of discharge and the zone under consideration. In column 16 record the elevation of the lowest pump in the zone. Subtract the values in column 16 from those in column 15 and record only positive elevation differentials in column 17. Add the values in column 14 to those in column 17 and record the total in column 18 to show the maximum combination of friction and static head a pump will experience at any given point in the system.

The accumulated data in Table 18.6 represents the preliminary LPS system design, which should then be reviewed for conformity with the criteria of flow velocity greater than or equal to 2.0 ft/s (0.60 m/s) and total design head less than or equal to 138 ft (42 m). If the system pressure exceeds 92 ft (28 m), the number of cores operating will remain the same and the flow from each pump will be reduced from 11 to 9 gpm (42 to 34 L/min). Data should be reviewed to determine whether system improvements could result from construction modifications. As an example, deeper burial of pipe in one or two critical high-elevation zones might bring the entire system into compliance with design criteria. The manufacturer (in this case Environment One) should be consulted in marginal cases or when odor control issues, frost protection issues, excessive static head conditions, excessive total dynamic head conditions, and unusual applications are of concern.

18.2.6 STEP Performance

When properly installed, septic tanks typically remove about 50% of *biochemical oxygen demand* (BOD), 75% of TSS, virtually all grit, and about 90% of grease, reducing the likelihood of clogging. Also, wastewater reaching the treatment plant will be weaker than raw sewage. Typical average values of BOD and TSS are 110 and 50 mg/L, respectively. On the other hand, septic tank effluent has virtually zero dissolved oxygen.

Primary sedimentation is not required to treat septic tank effluent. The effluent responds well to aerobic treatment, but odor control at the headworks of the treatment plant should receive extra attention.

The small community of High Island, Texas, was concerned that septic tank failures were damaging a local area frequented by migratory birds. Funds and materials were secured from the U.S. EPA, several state agencies, and the Audubon Society to replace the undersized septic tanks with larger ones equipped with STEP units and an LPS system that ultimately discharged to a constructed wetland. This system is designed to achieve an effluent quality of less than 20 mg/L each of BOD and TSS, less than 8 mg/L of ammonia, and greater than 4 mg/L of dissolved oxygen.

In 1996, the village of Browns, Illinois, replaced a failing septic tank system with a STEP system discharging to LPSs and ultimately to a recirculating gravel filter. Cost was a major concern to the residents of the village, who were used to average monthly sewer bills of 2008 USD 23. Conditions in the village were poor for conventional sewer systems, making them prohibitively expensive. An alternative low-pressure/STEP system averaged only USD 22 per month per resident and eliminated the public health hazard caused by the failed septic tanks.

18.2.7 GP Performance

The wastewater reaching the treatment plant will typically be stronger than that from conventional systems because infiltration is not possible. Typical design average concentrations of both BOD and TSS are 350 mg/L.

GP-LPS systems have replaced failing septic tanks in Lake Worth, Texas; Beach Drive in Kitsap County, Washington; and Cuyler, New York and hundreds of communities in the United States. Each of these communities chose alternative GP-LPS or SDGS systems over conventional gravity sewer systems based on lower costs and better suitability to local soil conditions.

18.2.8 Operation and Maintenance

Routine O&M requirements for both STEP and GP systems are minimal. Small systems that serve 300 or fewer homes do not usually require a full-time staff. Service can be performed by personnel from the municipal public works or highway department. Most system maintenance activities involve responding to homeowner service calls usually for electrical control problems or pump blockages. STEP systems also require pumping every 2–3 years.

The inherent septic nature of wastewater in pressure sewers requires that system personnel take appropriate safety precautions when performing maintenance to minimize exposure to toxic gases, such as hydrogen sulfide, which may be present in the sewer lines, pump vaults, or septic tanks. Odor problems may develop in pressure sewer systems because of improper house venting. The addition of strong oxidizing agents, such as chlorine or hydrogen peroxide, may be necessary to control odor where venting is not the cause of the problem.

Generally, it is in the best interest of the municipality and the homeowners to have the municipality or sewer utility be responsible for maintaining all system components. General easement agreements are needed to permit access to on-site components, such as septic tanks, STEP units, or GP units on private property.

18.2.9 Extent of Use in the United States

The pressure sewer market size is large and growing. No comprehensive lists have been kept to document pressure sewer projects, but hundreds of systems are known to exist throughout the United States. Locations of major projects range from Florida to Alaska and from Texas to New York. Many of these systems serve 50–200 homes. A few systems serve more than 1,000 homes each, and some systems have been designed and built that will serve more than 10,000 homes.

The use of pressure sewer components to serve low-lying homes fronting gravity sewers is substantial. Pressure sewer components used with on-site disposal practices has become commonplace. Canada also uses pressure sewers, as do several European and Asian countries.

Pressure sewers should not be glamorized as a panacea. The endeavor should be to use the technology appropriate for the setting. If the appropriate technology is the use of conventional sewers, or the use of septic tank/drain field systems, that should be used. Conventional practices are mature, well understood, and well accepted. Because of particular grant funding conditions, in some cases alternative technologies have been used inappropriately.

It has been a common error for people to learn of excellent operating performance from especially well-designed and well-built pressure sewer systems and, oddly, to expect the same performance from a shoddy installation. Too often engineers inexperienced with the technology have been employed. Components have often been chosen without their having demonstrated competence. Inspection has frequently been inadequate. The results are a poor system that is likely to be replaced in the near future and a poor reputation for that particular pressure sewer concept that is not deserved.

Experience specific to pressure sewers is vital to provide a good installation. A small system should be built first, preferably with guidance from experienced people. Then, performance of the operating system should be closely observed to close the loop between planning, design, construction, and long-term O&M.

The attitude and talent of the district owning and operating the system are major factors. If the maintenance forces or the management reluctantly accept pressure sewers, or do not have the ability to work with new concepts, the project will probably be a failure.

A frequently held misunderstanding is that pressure sewers are inherently maintenance-intensive. Experience has not supported that opinion. Well-designed pressure sewers, made easy to maintain by design and attended by qualified personnel, have been relatively easy to maintain. However, they are not tolerant of withheld maintenance, and O&M can be worse.

The engineer and the district must be willing to interface closely with the homeowners, and personnel assigned to the task must be knowledgeable and skillfully diplomatic. Each installation causes disruption to the homeowners' yard and inconvenience to them personally. The time required for public relations is usually poorly conceived and underestimated.

18.2.10 Costs

Pressure sewers are generally more cost-effective than conventional gravity sewers in rural areas because capital costs for pressure sewers are generally lower than for gravity sewers. While capital cost savings of 90% have been achieved, no universal statement of savings is possible because each site and system is unique. Table 18.7 presents a generic comparison of

Table 18.7 Relative Characteristics of Alternative Sewers

Sewer Type	Slope Requirement	Construction Cost in Rocky, High Groundwater Sites	Operation and Maintenance Requirements	Ideal Power Requirements
Conventional	Downhill	High	Moderate	None[a]
GP-LPS	None	Low	Moderate–High	Low
STEP	None	Low	Moderate–High	Moderate

[a] Power may be required for lift stations.

Table 18.8 Installed Unit Costs for Pressure Sewer Mains and Appurtenances.

Item	Cost (2022 USD/Linear ft)
Mains (in.):	
2	13.27
3	14.14
4	16.00
6	22.32
8	24.80
Extra for mains in asphalt concrete pavement	8.93
Item	**Cost (2022 USD/Item)**
Isolation valves (in.):	
2	446
3	484
4	620
6	707
8	1,017
Individual grinder pump	2,133
Single (simplex) package pump system	7,266
Package installation	880-2,654
Automatic air release stations	1,773

Conversion factor: 1 USD/ft = 3.28 USD/m.; 1 in. = 2.54 cm.

common characteristics of sanitary sewer systems that should be considered in the initial decision-making process on whether to use pressure sewer systems or conventional gravity sewer systems.

Table 18.8 presents data from evaluations of the costs of pressure sewer mains and appurtenances (essentially the same for GP and STEP), including items specific to each type of pressure sewer. All costs were updated in terms of 2022 USD employing the U.S. Army Corps of Engineers' *Cost Index for Utilities*. Purchasing pumping stations in volume may reduce costs by up to 50%. The linear cost of mains can vary by a factor of 2 to 3, depending on the type of trenching equipment and local costs of high-quality backfill and pipe. The local geology and utility systems will impact the installation cost of either system.

The homeowner is responsible for energy costs, which will vary from 2022 USD 1.43 to 3.53/month for GP systems, depending on the horsepower of the unit. STEP units generally cost less than USD 1.43/month.

No preventative maintenance needs to be performed on GPs without float-level sensors. However, preventive maintenance should be performed annually for GP with float-level sensors, with monthly maintenance of other mechanical components. STEP systems require periodic pumping of septic tanks. Total O&M costs average USD 124–248/yr/unit (2022 costs) and include costs for troubleshooting, inspection of new installations, and responding to problems.

Mean time between service calls data vary greatly, but values of 4–10 years for both GP and STEP units are reasonable estimates for quality installations.

18.3 Small-Diameter Gravity Sewer System

SDGS systems convey effluent by gravity from an interceptor tank (or septic tank) to a centralized treatment location or pump station for transfer to another collection system or treatment facility. A typical SDGS system and interceptor tank are depicted in Figs. 18.4 and 18.5, respectively.

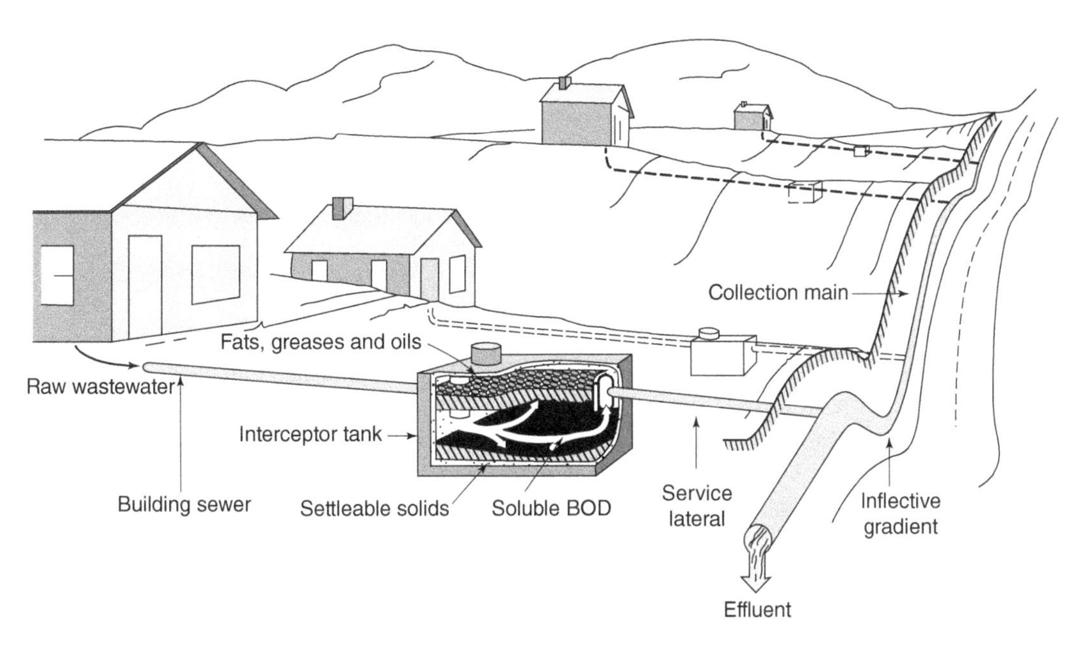

Figure 18.4 SDGS System

Most suspended solids are removed from the wastewater by septic (interceptor) tanks, reducing the potential for clogging to occur and allowing for smaller diameter piping both downstream of the septic tank in the lateral and in the sewer main. The interceptor tanks are an integral part of the system. They are typically located on private property, but usually owned or maintained by the utility districts so that regular pumping to remove the accumulated solids for safe disposal is ensured. Cleanouts are used to provide access for flushing; manholes are rarely used. Air release risers are required at or slightly downstream of summits in the sewer profile. Odor control is important at all access points because the SDGS carries odorous septic tank effluent. Because of the small diameters and flexible slope and alignment of the SDGS, excavation depths and volumes are typically much smaller than with conventional sewers. Minimum pipe diameters can be 3 in. (76 mm).

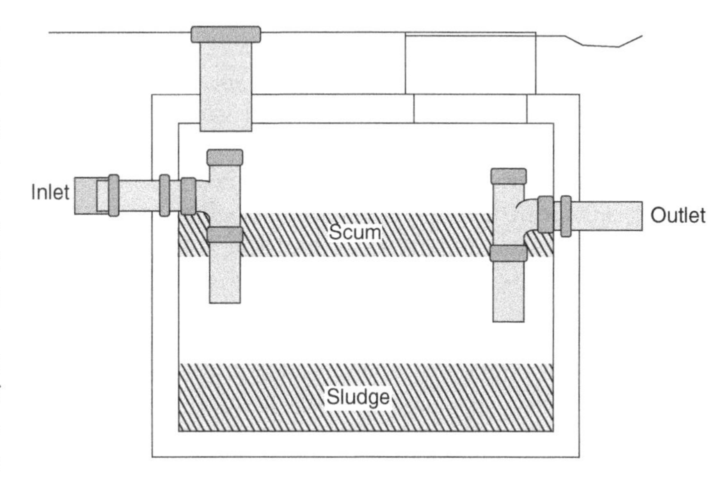

Figure 18.5 Precast Concrete Interceptor Tank

The required size and shape of the mains is dictated primarily by hydraulics rather than solids-carrying capabilities as with conventional gravity sewers. Plastic pipe is typically used because it is economical in small sizes and resists corrosion.

SDGS were first constructed in Australia in the 1960s. They were used to provide a more cost-effective solution than conventional gravity sewers to correct problems with failing septic tank systems in densely developed urban fringe areas. The SDGS were designed to collect the effluent from existing septic tanks. Because the tanks would remove the suspended solids that might settle or otherwise cause obstructions in the mains, smaller collector mains 4 in. (100 mm) in diameter, laid on a uniform gradient sufficient to maintain only a 1.5 ft/s (0.46 m/s) flow velocity, were permitted. This alternative has been estimated to reduce construction costs by 30%–65%. Routine maintenance also proved to be low in cost. As a result, by 1986 more than 80 systems had been constructed with up to 4000 connections per system.

In the United States, SDGSs were not introduced until the mid-1970s. The first systems were small demonstration systems with 13–90 connections. As knowledge of the success of these systems spread, SDGS began to gain acceptance and by the mid-1980s, more than 100 systems had been constructed. The designs of most of the systems constructed prior to 1990 followed the Australian guidelines, but as experience has been gained, engineers are finding that the guidelines can be relaxed without sacrificing performance or increasing maintenance costs. Variable-grade systems in which the sewers are allowed to operate in a surcharged condition are becoming more common. Minimum flow velocities are no longer considered as a design criterion. Instead, the design is based on the system's capacity to carry the expected peak flows without raising the hydraulic grade line above the interceptor tank outlet inverts for extended periods of time. Inflective gradients are allowed such that sections of the mains are depressed below the static hydraulic grade line. Despite these significant changes from the Australian guidelines, O&M costs have not increased.

SDGS systems consist of these components:

1) *House connections* are made at the inlet to the interceptor tank. All household wastewaters enter the system at this point.
2) *Interceptor tanks* are buried, watertight tanks with baffled inlets and outlets. They are designed to remove both floating and settleable solids from the waste stream through quiescent settling over a period of 12–24 h. Ample volume is also provided for storage of the solids, which must be periodically removed through an access port. Typically, a single-chamber septic tank, vented through the house plumbing stack vent, is used as an interceptor tank.
3) *Service laterals* connect the interceptor tank with the collector main. Typically, they are 3–4 in. (75–100 mm) in diameter, but should be no larger than the collector main to which they are connected. They may include a check valve or other backflow prevention device near the connection to the main.
4) *Collector mains* are small-diameter plastic pipes with typical minimum diameters of 3–4 in. (75–100 mm). The mains are trenched into the ground at a depth sufficient to collect the settled wastewater from most connections by gravity. Unlike conventional gravity sewers, SDGSs are not necessarily laid on a uniform gradient with straight alignment between cleanouts or manholes. In places, the mains may be depressed below the hydraulic grade line. Also, the alignment may be curvilinear between manholes and cleanouts to avoid obstacles in the path of the sewers.
5) *Cleanouts,* manholes, and vents provide access to the collector mains for inspection and maintenance. In most circumstances, cleanouts are preferable to manholes because they are less costly and can be more tightly sealed to eliminate most infiltration and grit that commonly enter through manholes. Vents are necessary to maintain free-flowing conditions in the mains. Vents in the household plumbing are sufficient except where depressed sewer sections exist. In such cases, air release valves or ventilated cleanouts may be necessary at the high points of the main.
6) *Lift stations* are necessary where elevation differences do not permit gravity flow. Either STEP units (see pressure sewer systems) or mainline lift stations may be used. STEP units are small lift stations installed to pump wastewater from one or a small cluster of connections to the collector main, while a mainline lift station is used to service all connections in a larger drainage basin.

Although the term *small-diameter gravity sewers* has become commonly accepted, it is not an accurate description of the system, since the mains need not be small in diameter (the size is determined by hydraulic considerations) nor are they "sewers" in the sense that they carry wastewater solids. The most significant feature of small-diameter sewers is that primary pretreatment is provided in interceptor tanks upstream of each connection. With the settleable solids removed, it is not necessary to design the collector mains to maintain minimum self-cleansing velocities. Without the requirement for minimum velocities, the pipe gradients may be reduced and, as a result, the depths of excavation. The need for manholes at all junctions, changes in grade and alignment, and at regular intervals is eliminated. The interceptor tank also attenuates the wastewater flow rate from each connection, which reduces the peak-to-average flow ratio below what is typically used for establishing design flows for conventional gravity sewers. Yet, except for the need to evacuate the accumulated solids in the interceptor tanks periodically, SDGS systems operate similarly to conventional sewers.

18.3.1 Applicability

Approximately 250 SDGS systems have been financed in the United States by the U.S. EPA's Construction Grants Program. Many more have been financed with private or local funding. These systems were introduced in the United States in the mid-1970s, but have been used in Australia since the 1860s.

SDGS systems can be most cost-effective where housing density is low, the terrain has undulations of low relief, and the elevation of the system terminus is lower than all or nearly all of the service area. They can also be effective where the terrain is too flat for conventional gravity sewers without deep excavation, where the soil is rocky or unstable, or where the groundwater level is high.

SDGS systems do not have the large excess capacity typical of conventional gravity sewers and should be designed with an adequate allowance for future growth.

18.3.2 Advantages

SDGS systems have these advantages:

1) Construction is fast, requiring less time to provide service.
2) Unskilled personnel can operate and maintain the system.
3) Elimination of manholes reduces a source of inflow, further reducing the size of pipes, lift/pumping stations, and final treatment, ultimately reducing cost.
4) Excavation costs are reduced because trenches for SDGS pipelines are typically narrower and shallower than those used for conventional sewers.
5) Material costs are reduced because SDGS pipelines are smaller than conventional sewers, reducing pipe and trenching costs.
6) Final treatment requirements are scaled down in terms of organic loading since partial removal is performed in the septic tank.
7) The shallower depth required for the mains reduces construction costs due to high groundwater or rocky conditions.

18.3.3 Disadvantages

Though not necessarily a disadvantage, limited experience with SDGS technology has yielded some situations where systems have performed inadequately. This is usually more a function of poor design and construction than the ability of a properly designed and constructed SDGS system to perform adequately. Although SDGS systems have no major disadvantages specific to temperate climates, some restrictions may limit their application:

1) SDGS systems cannot handle commercial wastewater with high grit or settleable solids levels. Restaurants may be hooked up if they are equipped with effective grease traps. Laundromats may be a constraining factor for SDGS systems in small communities.
2) In addition to corrosion within the pipe from the wastewater, corrosion outside the pipe has been a problem in some SDGS systems in the United States where piping is installed in highly corrosive soil. If the piping will be exposed to a corrosive environment, noncorrosive materials must be incorporated in the design.
3) Disposing of collected septage from septic tanks is probably the most complex aspect of the SDGS system and should be carried out by local authorities. However, many tanks are installed on private property requiring easement agreements for local authorities to gain access. Contracting to carry out these functions is an option, as long as the local authorities retain enforceable power for hygiene control.
4) Odors are the most common problem. Many early systems used an on-lot balancing tank that promoted stripping of hydrogen sulfide from the interceptor (septic) tank effluent. Other odor problems are caused by inadequate house ventilation systems and mainline manholes or venting structures. Appropriate engineering can control odor problems.
5) SDGS systems must be buried deep enough so that they will not freeze. Excavation may be substantial in areas where there is a deep frostline.

18.3.4 Design Criteria

Peak flows are based on Eq. 18.1 when using US customary units or Eq. 18.2 when using metric units. Whenever possible, it is desirable to use actual flow data for design purposes. However, if this is not available, peak flows are calculated. Each segment of the sewer is analyzed by the Hazen–Williams or Manning equations to determine if the pipe is of adequate size and slope to handle the peak design flow. No minimum velocity is required and PVC pipe is commonly used for gravity segments. Stronger pipe may be dictated where STEP units feed the system. Check valves may also be used in flooded sections or where backup (surcharging) from the main may occur. These valves are installed downstream of mainline cleanouts.

Typical pipe diameters for SDGS are 3 in. (75 mm) or more, but the minimum recommended pipe size is 4 in. (100 mm) because 3-in. (75-mm) pipes are not readily available and need to be special ordered. The slope of the pipe should be adequate to carry peak hourly flows. SDGS systems do not need to meet a minimum velocity because solids settling is not a design parameter in them. The depth of the piping should be the minimum necessary to prevent damage from anticipated earth and truck loadings and freezing. If no heavy earth or truck loadings are anticipated, a depth of 24–30 in. (600–750 mm) is typical.

All components must be corrosion-resistant and all discharges (e.g., to a conventional gravity interception or treatment facility) should be made through drop inlets below the liquid level to minimize odors. The system is ventilated through service-connection house vent stacks. Other atmospheric openings should be directed to soil beds for odor control, unless they are located away from the populace.

Septic tanks are generally sized based on local plumbing codes. STEP units used for below-grade services are covered in the previous section on pressure sewers. It is essential to ensure that on-lot I/I are eliminated through proper testing and repair, if required, of building sewers, as well as preinstallation testing of septic tanks.

Mainline cleanouts are generally spaced 400–1,000 ft (120–300 m) apart. Treatment is normally by stabilization pond or subsurface infiltration. Effluent may also be directed to a pump station or treatment facility.

A well-operated and maintained septic tank will typically remove up to 50% of BOD, 75% of TSS, virtually all grit, and about 90% of grease. Clogging is not normally a problem. Also, wastewater reaching the treatment plant will typically be more dilute than raw sewage. Typical average values of BOD and TSS are 110 and 50 mg/L, respectively.

Primary sedimentation is not required to treat septic tank effluent. Sand filters are effective in treatment. Effluent responds well to aerobic treatment, but odor control at the headworks of the treatment plant should receive extra attention.

18.3.5 Operation and Maintenance

O&M requirements for SDGS systems are usually low, especially if there are no STEP units or lift stations. Periodic flushing of low-velocity segments of the collector mains may be required. The septic tanks must be pumped periodically to prevent solids from entering the collector mains. It is generally recommended that pumping be performed every 3–5 years. However, the actual operating experience of SDGS systems indicates that once every 7–10 years is adequate. Where lift stations are used, such as in low-lying areas where waste is collected from multiple sources, they should be checked on a daily or weekly basis. A daily log should be kept on all operating checks, maintenance performed, and service calls. Regular flow monitoring is useful to evaluate whether inflow and infiltration problems are developing.

The municipality or sewer utility should be responsible for O&M of all of the SDGS system components to ensure a high degree of system reliability. General easement agreements are needed to permit access to components such as septic tanks or STEP units on private property.

18.3.6 Costs

The installed costs of the collector mains and laterals and the interceptor tanks constitute more than 50% of total construction costs. Average unit costs for 12 projects (adjusted to 2022 USD) were 4-in. (100-mm) mainline, USD 21.32/ft (USD 69.91/m); cleanouts, USD 508 each; and service connections, USD 15.89/ft (USD 50.06/m). Average unit cost for a 1,000-gal (3,785-L) septic tank was USD 2,300. The average cost per connection was USD 9,362 and the major O&M requirement for SDGS systems is the pumping of the tanks. Other O&M activities include gravity line repairs from excavation damage, supervision of new connections, and inspection and repair of mechanical components and lift stations. Most SDGS system users pay USD 17–35 per month for management, including O&M and administrative costs.

18.4 Vacuum Sewer System

18.4.1 System Types

Vacuum sewer collection systems were patented in the United States in 1888, when Adrian LeMarquand invented a system of wastewater collection by barometric depression. The first commercial applications of such systems were by the Liljendahl Corporation (now known as Electrolux) of Sweden in 1959. Since that time, three other companies have been active in this market: Colt-Envirovac, Vac-Q-Tec, and AIRVAC. Significant differences exist among these systems in terms of design concepts. The major differences lie in the extent to which the systems use separate black (toilet) and gray (the balance) water collection mains. Electrolux uses a separate system for these sources; Colt-Envirovac uses vacuum toilets and one main; and

Vac-Q-Tec and AIRVAC take the normal household combined wastes. Other differences relate to the location of the gravity/vacuum interface and to the design of pumps, valves, lines, etc.

The Liljendahl-Electrolux system (Fig. 18.6) was first used in the Bahamas in the 1960s. In this concept, separate black and gray water collection mains are used. The black water is discharged to one of the vacuum mains through a vacuum toilet (Fig. 18.7), while the gray water enters the other through the use of a specially designed vacuum valve. The separate vacuum mains are connected to the vacuum station. For critically water-short areas, such as the Bahamas, the reduction in toilet wastewater volume was a definite factor in the selection of vacuum transport.

The Colt-Envirovac system is the direct descendent of the Liljendahl-Electrolux system (Fig. 18.8). The houses have separate black and gray water plumbing. The black water piping from the vacuum toilet joins the gray water piping immediately downstream of the gray water valve. A single pipe with the combined contents transports the wastewater to the vacuum station.

The Vac-Q-Tec system was the first residential vacuum collection system used in the United States. This system uses concepts of the Liljendahl system but has many important differences. The Vac-Q-Tec system requires no inside vacuum toilets or vacuum plumbing. This system employs a single combined black and gray water collection main. Large (750-gal or 2839-L) storage tanks are required at each residence. Finally, an external power source is required for each valve because they are electrically operated. Several Vac-Q-Tec residential systems have been used by private developers.

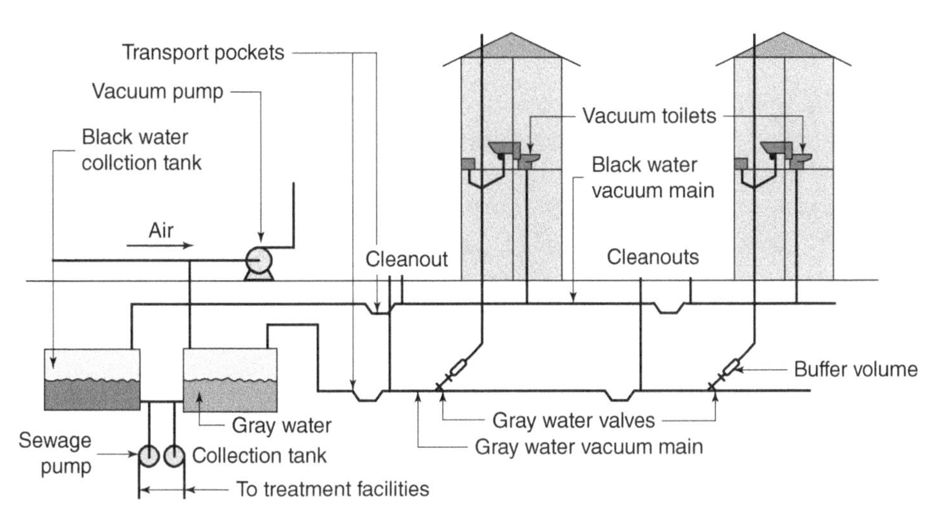

Figure 18.6 Liljendhahi-Electrolux Vacuum Sewer System

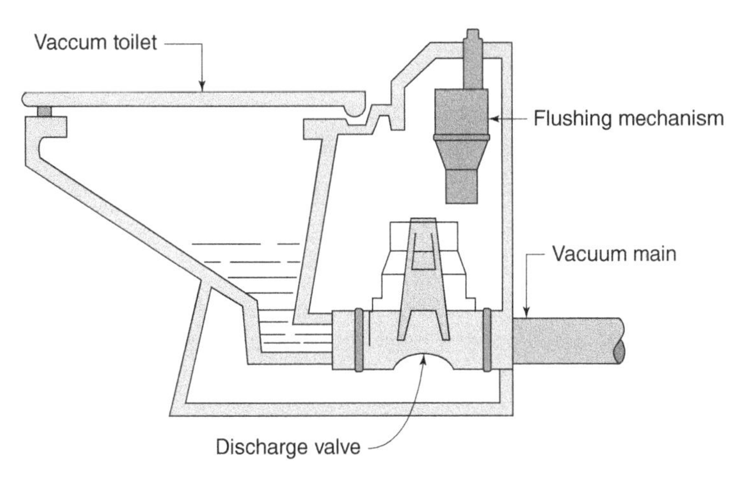

Figure 18.7 Vacuum Toilet

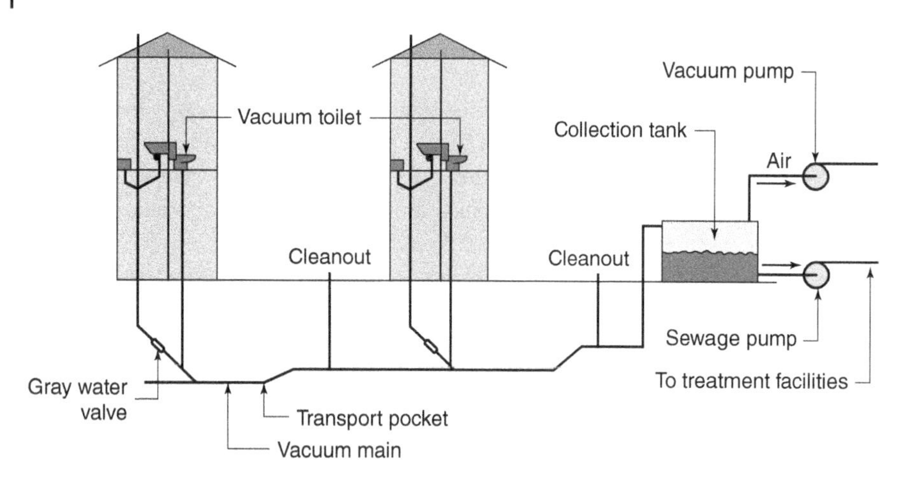

Figure 18.8 Colt-Envirovac Vacuum Sewer System

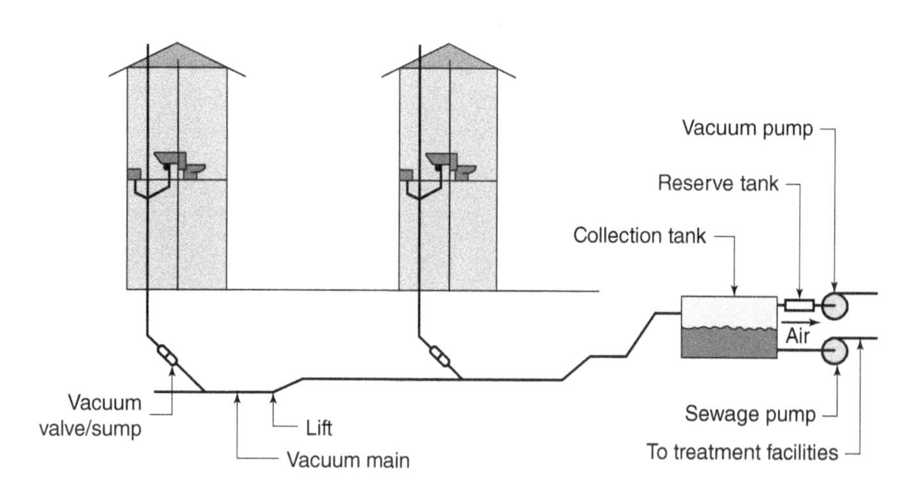

Figure 18.9 AIRVAC Vacuum Sewer System

AIRVAC markets a pneumatically controlled and operated vacuum valve, which is used for combined gray and black water systems (Fig. 18.9). The AIRVAC system allows for use of conventional plumbing in the house, with the wastewater flowing by gravity to a combined sump/valve pit. The valve starts its cycle when it senses that approximately 10 gal (38 L) has accumulated in the sump. It opens for a few seconds, which is enough to evacuate the contents of the sump as well as to allow atmospheric air to enter the system. The wastewater/air mixture then travels to the vacuum station.

AIRVAC's first system was installed in Virginia, in 1970. Currently AIRVAC has more than tens of additional systems operating in the United States with many more being planned, designed, or in construction. AIRVAC has also been very active in the foreign market with operating systems in Australia, Canada, Japan, Holland, and some other European countries.

18.4.2 System Comparison

Each of the four systems has unique design features. The major differences between these systems are shown in Table 18.9. The water-saving feature of the Electrolux and Colt systems is reported to be as much as 27% of the total in a domestic application with the use of vacuum toilets. The AIRVAC and Vac-Q-Tec systems can be altered to accommodate these and other water-saving devices.

Vacuum valves operate automatically, based on the volume of wastewater behind the valve. Provided that sufficient vacuum is available in the main, the valves will open after a predetermined volume of wastewater has accumulated. Wastewater enters the mains through these valves, followed by a volume of atmospheric air. The valve is actuated by a pneumatic controller in all systems except the Vac-Q-Tec system.

Table 18.9 Vacuum Collection System Parameters

System Type	House Piping	Valve Type	Piping Profile	Collection Line
Electrolux	Black and gray separate	Black: vacuum toilets; gray: pneumatic valves	Set configuration with traps	Black: 1-1/2″ and 2″; gray: 2″ and 3″; PVC solvent weld
Colt-Envirovac	Black and gray separate	Black: vacuum toilets; gray: pneumatic valves	Set configuration with traps	Single main, 3″, 4″, and 6″; PVC, special "O" ring
Vac-Q-Tec	Conventional plumbing	Electrically actuated, pneumatic valve	Parallels terrain with traps	Single main, 4″; PVC solvent weld
AIRVAC	Conventional plumbing	Pneumatic valve	Set configuration with profile changes	Single main, 4″; 6″, and 8″; PVC, solvent weld, or "O" ring

Conversion: 1″ = 1 in. = 25.4 mm.

The Vac-Q-Tec's gravity/vacuum interface valve assembly is unique in that it requires an external power source. The valve can be monitored and operated from the vacuum station through an extra set of contacts in the controller. A separate cycling mode, called AutoScan, can be added, which offers flexibility to the Vac-Q-Tec system. This mode locks out the accumulated volume-cycle command from each valve and subsequently operates each valve during low-flow periods. This flexibility allows the system to store flows during peak periods and release them later during low-flow periods. All of the other systems must be designed to handle peak flows. This feature does, however, add costs to the base system. Also, additional operating and skilled electronics technicians are required to maintain these complex systems.

Depending on the manufacturer, the amount of water entering the system with each valve operation varies. The vacuum toilet admits approximately 0.3–0.4 gal/flush (1.13–1.57 L/flush), whereas the pneumatically controlled vacuum valves admit 12–15 gal/cycle (45.4–56.8 L/cycle).

Good transport characteristics are found with sufficient inlet air and small enough slug loadings for the available pressure differential to overcome the liquid's inertia. This results in rapid slug breakdown, reestablishing vacuum quickly at upstream valves.

Piping profiles differ, depending on uphill, downhill, or level terrain. The pipe profiles recommended by each manufacturer also differ. Only AIRVAC offers a complete piping design program. All systems use PVC pipe. Both solvent-weld and gasketed O-ring pipe have successfully been used.

Vacuum stations, sometimes referred to as collection stations, vary from manufacturer to manufacturer. Table 18.10 shows the varying design parameters of each type. Electrolux and Colt vary their use of vacuum reserve tanks with each installation, whereas Vac-Q-Tec and AIRVAC always use reservoir tanks between the collection tank and the vacuum pumps.

In summary, four manufacturers have played a major role in the development of vacuum sewer systems. There are significant differences in overall system philosophy, design concepts, system components, and marketing approaches (Tables 18.9–18.11). While all four were active 35 years ago in the United States, only AIRVAC has continued to place residential systems into operation on a regular basis. Some of the early systems of Colt-Envirovac and Vac-Q-Tec have been retrofitted with AIRVAC valves.

Table 18.10 Vacuum Station Parameters

System Type	Receiving Tank	Receiving Tank Evacuation Device	Valve Monitoring and Control Capability
Electrolux	Separate black and gray water vessels. Reserve tank use varies by installation	Sewage pumps	No
Colt-Envirovac	Common receiving vessel Reserve tank use varies	Sewage pumps	No
Vac-Q-Tec	One receiving vessel plus reserve tank	Pneumatic ejectors	Yes
AIRVAC	One receiving vessel plus reserve tank	Sewage pumps	No

Table 18.11 Summary of Vacuum System Types

System Type	Target Market	Current Status	US Systems	Design Approach
Electrolux	Residential	Sold license to Colt in 1970s	0	In-house No design manual
Colt-Envirovac	Shipbuilder Industrial	Now a subsidiary of Evak	Few	In-house No design manual
Vac-Q-Tec	Residential	Ceased operation	Few	In-house No design manual
AIRVAC	Residential	Active	Many	Published design manual

18.4.3 System Description

Figure 18.10 shows the basic vacuum sewer system layout, including the major components. This layout is based on an AIRVAC type of system since it is the most common. A vacuum sewer system consists of three major components: the vacuum station, the collection piping, and the services. Each is described in the sections that follow.

Wastewater flows by gravity from one or more homes into a 30-gal (113 L) holding tank. As the wastewater level rises in the sump, air is compressed in a sensor tube that is connected to the valve controller. At a preset point, the sensor signals for the vacuum valve to open. The valve stays open for an adjustable period of time and then closes. During the open cycle, the holding tank contents are evacuated. The timing cycle is field adjusted between 3 and 30 s. This time is usually set to hold the valve open for a total time equal to twice the time required to admit the wastewater. In this manner, air at atmospheric pressure is allowed to enter the system behind the wastewater. The time setting is dependent on the valve location since the vacuum available will vary throughout the system, thereby governing the rate of wastewater flow.

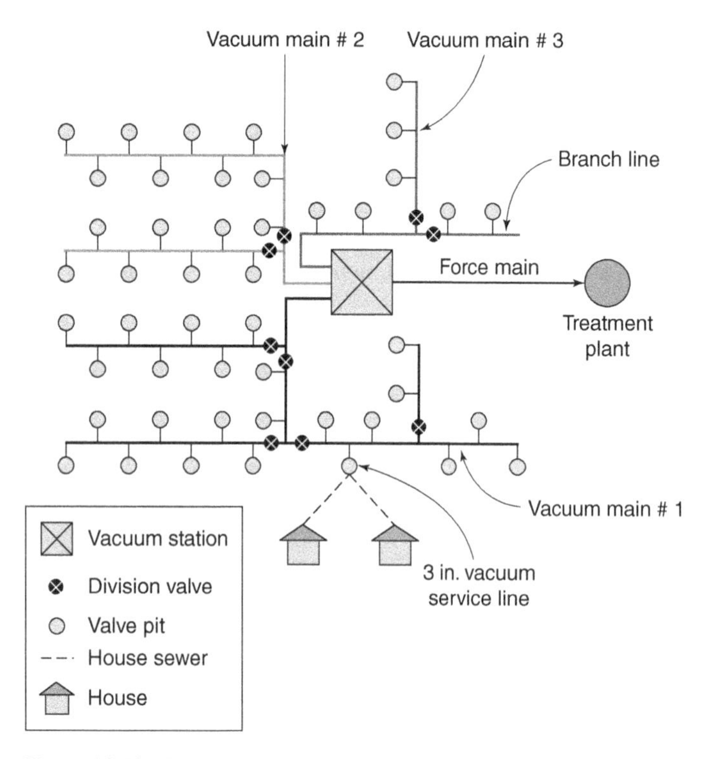

Figure 18.10 Major Components of a Vacuum Sewer System. Conversion factor: 1 in. = 25.4 mm

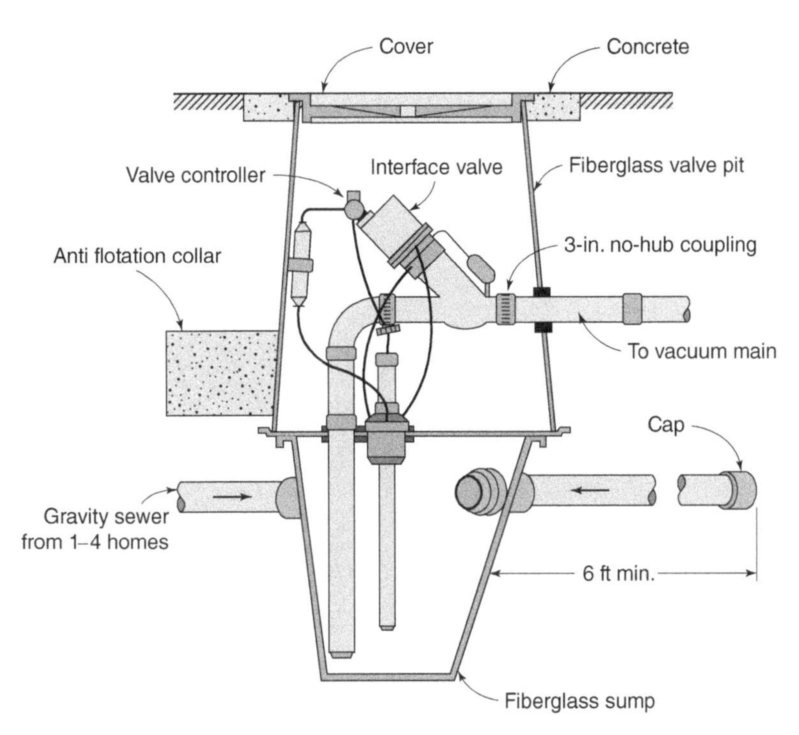

Figure 18.11 AIRVAC Valve/Pit/Sump Arrangement. Conversion factors: 1 in. = 25.4 mm; 1 ft = 0.3048 m

The valve pit typically is located along a property line. AIRVAC's valve pit/holding tank arrangement (Fig. 18.11) is usually made of fiberglass, although modified concrete manhole sections have been used for special situations (deep basements, large user, pressure/ vacuum interface, etc.). Where traffic is not an issue, for instance, in yard installations, a lightweight aluminum or cast iron lid is available. Where the installation will be subjected to vehicular loading, a flush-mounted cast iron lid is used. An anti-flotation collar may be required in some cases.

18.4.3.1 Collection Piping

The vacuum collection piping usually consists of 4- and 6-in. (100- and 150-mm) mains, although more recent installations also include 10-in.(250-mm) mains in some cases. Smaller 3-in. (75-mm) mains used in early vacuum systems are no longer recommended, because the cost savings of using 3- to 4-in. (75- to 100-mm) mains are considered to be insignificant.

Both solvent-welded PVC pipe and rubber-gasketed pipe have been used, although past experience indicates that solvent welding should be avoided when possible. Where rubber gaskets are used, they must be certified by the manufacturer as being suitable for vacuum service. The mains are generally laid to the same slope as the ground with a minimum slope of 0.2%. For uphill transport, lifts are placed to minimize excavation depth (Fig. 18.12). There are no manholes in the system; however, access can be gained at each valve pit or at the end of a line where an access pit may be installed. Installation of the pipe and fittings follows water distribution system practices. Division valves are installed on branches and periodically on the mains to allow for isolation when troubleshooting or when making repairs. Plug valve and resilient wedge gate valves have been used.

18.4.3.2 Vacuum Station

The vacuum station is the heart of the vacuum sewer system. It is similar to a conventional wastewater pumping station. These stations are typically two-story concrete and block buildings approximately 25 × 30 ft (7.62–9.14 m) in floor plan. Equipment in the station includes a collection tank, a vacuum reservoir tank, vacuum pumps, wastewater pumps, and pump controls (Fig. 18.13). In addition, an emergency generator is standard equipment, whether it is located within the station or outside the station in an enclosure or is of the portable, truck-mounted variety.

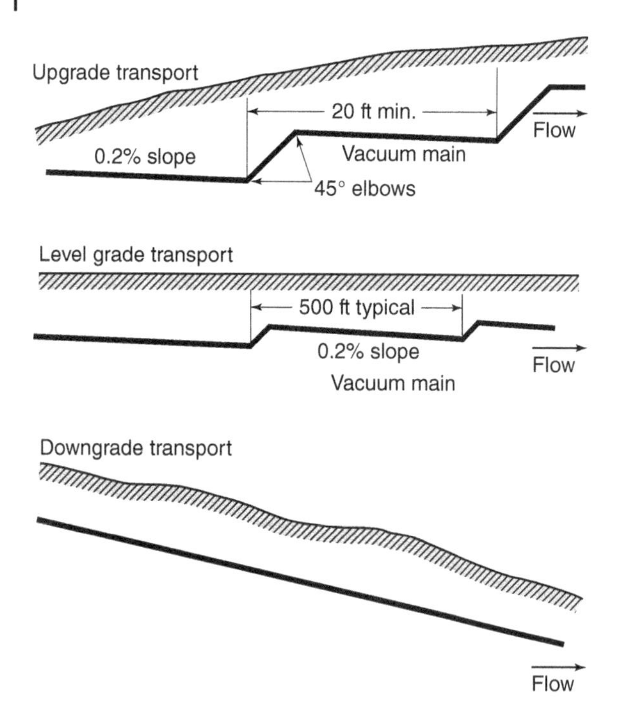

Figure 18.12 Typical Vertical Profile of Vacuum Sewers. Conversion factor: 1 ft = 0.3048 m

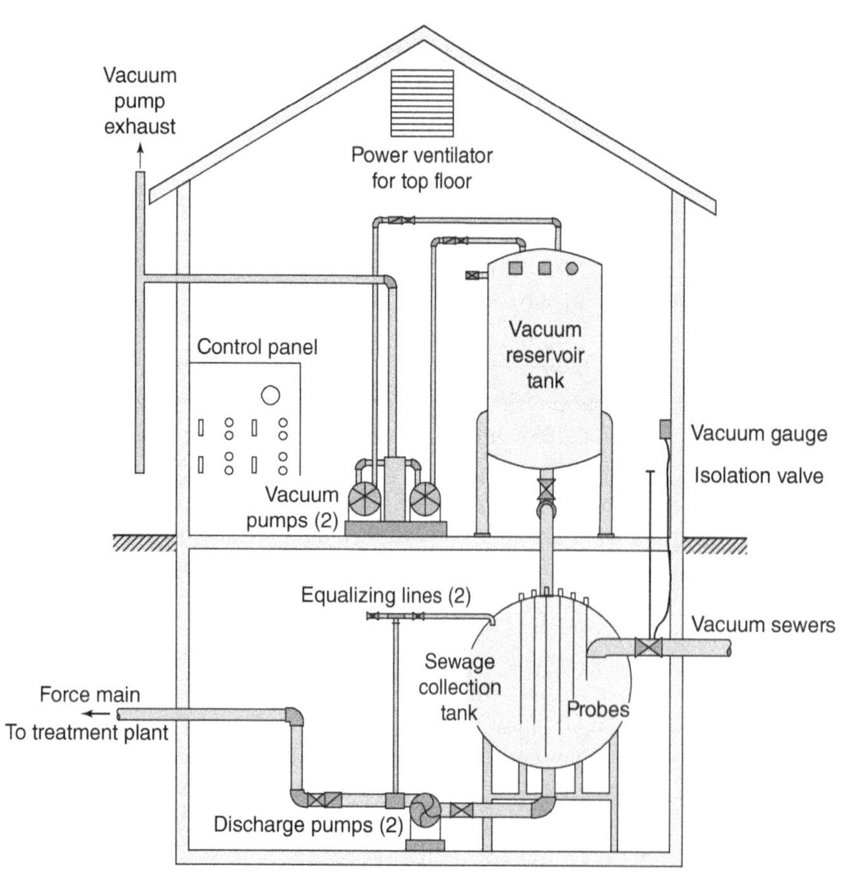

Figure 18.13 Diagram of a Typical Vacuum Station

The collection tank, made of either steel or fiberglass, is the equivalent of a wet well in a conventional pumping station. The vacuum reservoir tank is connected directly to the collection tank to prevent droplet carryover and to reduce the frequency of vacuum pump starts and thereby extend their life. The vacuum pumps can be either the liquid ring or sliding vane type. These pumps are usually sized for 3–5 h/d run times. The wastewater discharge pumps are nonclogging pumps with sufficient net-positive suction head to overcome tank vacuum. Level control probes are installed in the collection tank to regulate the wastewater pumps. Vacuum switches on the reservoir tank regulate the vacuum pumps. A fault monitoring system alerts the system operator should a low-vacuum or high-wastewater level condition occur. Average sized vacuum stations contain 20-hp (14.91-kW) vacuum pumps.

18.4.4 Operation

Vacuum or negative-pressure sewer systems use vacuum pumps at central collection stations to evacuate air from the lines, thus creating a pressure differential. In negative-pressure systems, a pneumatically operated valve serves as the interface between the gravity system from the individual user and the vacuum pipelines. Pressure sensors in a wastewater holding tank open and close the interface valve to control the flow of wastewater and air into the vacuum system.

The normal sequence of operation is as follows:

1) Wastewater from the individual service flows by gravity to a holding tank.
2) As the level in the holding tank continues to rise, air is compressed in a small-diameter sensor tube. This air pressure is transmitted through a tube to the controller/sensor unit mounted on top of the valve. The air pressure actuates the unit and its integral three-way valve, which allows vacuum from the sewer main to be applied to the valve operator. This opens the interface valve and activates a field adjustable timer in the controller/sensor. After a set time period has expired, the interface valve closes. This happens as a result of the vacuum being shut off, allowing the piston to close by spring pressure.
3) The wastewater within the vacuum sewer approximates the form of a spiral rotating hollow cylinder traveling at 15–18 ft/s (4.57–5.49 m/s). Eventually, the cylinder disintegrates from pipe friction, and the liquid flows to low points (bottom of lifts) in the pipeline.
4) The next liquid cylinder and the air behind it will carry the liquid from the previously disintegrated cylinders up over the sawtooth lifts designed into the system. In this manner, the wastewater is transported over a series of lifts to the vacuum station.

The principle concept of operation was that of liquid plug flow. In this concept, it was assumed that a wastewater plug completely sealed the pipe bore during static conditions. The movement of the plug through the pipe bore was attributed to the pressure differential behind and in front of the plug. Pipe friction would cause the plug to disintegrate, thus breaking the vacuum. With this being the situation, reformer pockets were located in the vacuum sewer to allow the plug to reform and thus restore the pressure differential (Fig. 18.14a). In this concept, the reestablishment of the pressure differential for each disintegrated plug was a major design consideration.

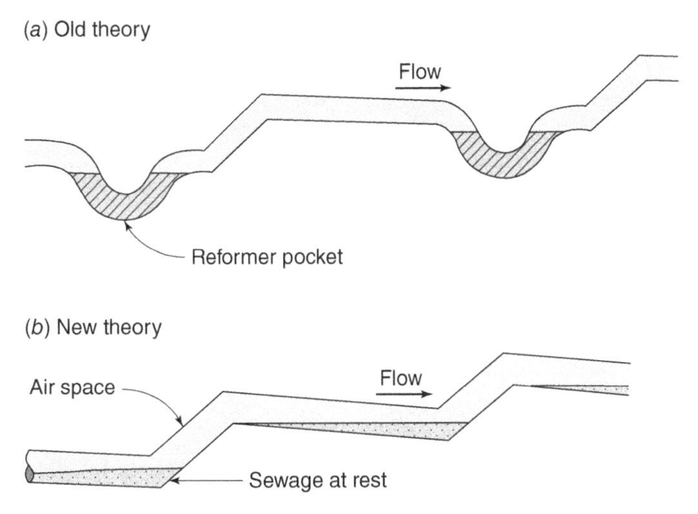

Figure 18.14 Design Concepts of Vacuum Sewers Operation

In the current design concept, the reformer pockets are eliminated so that the wastewater does not completely fill or "seal" the pipe bore. Air flows above the liquid, thus maintaining a high vacuum condition throughout the length of the pipeline (Fig. 18.14b). In this concept, the liquid is assumed to take the form of a spiral, rotating, hollow cylinder. The momentum of the wastewater and the air carries the previously disintegrated cylinders over the downstream sawtooth lifts. The momentum of each subsequent air/liquid slug and its contribution to the progressive movement of the liquid component of the previous slugs are the major design considerations.

Both of the above design concepts are approximations and oversimplifications of a complex, two-phase flow system. The character of the flow within the vacuum sewer varies considerably. The plug flow concept is probably a reasonable approximation of the flow as it enters the system, whereas the progressive movement concept is probably a better approximation of the flow throughout the vacuum main.

The significance of the air as a driving force cannot be overemphasized. The atmospheric air expands within the vacuum sewer, thus driving the liquid forward. The air affects not only the liquid in the associated air/liquid slug, but also the liquid downstream.

18.4.5 System Design

States have their own standards for the design of vacuum sewer systems. In Massachusetts and other states in the New England region of the United States, the design is based on the *Guides for the Design of Wastewater Treatment Works*, published by the New England Interstate Water Pollution Control Commission (1988).

18.4.5.1 Piping Design

Vacuum sewers should be designed to handle the peak sewage flow from dwellings. Various formulas exist for determining peak flow rates, but in general, a peak flow factor of 3.5 times the daily flow rate should be used:

$$Q_{peak} = 3.5\, Q_{avg} \tag{18.4}$$

where Q_{peak} is peak sewage flow from dwellings, in gpm (L/s); and Q_{avg} is average sewage flow from dwellings, in gpm (L/s).

The pipe is sized using the Hazen–Williams formula for full-bore flow and a C factor of 150 for PVC pipes. Within the system, the flow consists of 2 parts of air to 1 part of liquid. It is recommended to use the equivalent of Class 200, SDR 21 PVC piping or better to provide the necessary working pressure rating for the system and to provide durability during installation. Piping should be deep enough to prevent freezing.

18.4.5.2 Collection Station Design

The following minimum *vacuum pump* sizes (Q_{vp}) are recommended:

For sewers length up to 3,000 ft (914 m) long:

$$Q_{vp} = (5\, Q_{max}/K_2) + (K_1)\,(N_v) \tag{18.5}$$

For sewers more than 3,000 ft (914 m) long:

$$Q_{vp} = (6\, Q_{max}/K_2) + (K_1)\,(N_v) \tag{18.6}$$

where Q_{vp} is vacuum pump capacity (ft^3/min or L/s); Q_{max} is collection station peak flow (gpm or L/s); $K_1 = 0.15$ ft^3/min using US customary units = 0.07 L/s using metric units; K_2 = conversion factor = 7.5 gal/ft^3 using US customary units = 1 using metric units; and N_v is the number of valves connected to the sewage collection station.

These formulas are provided for guidance. Other factors to consider include type of vacuum pump, vacuum pump efficiency, temperature of sewage, type and temperature of service liquid (if applicable), and altitude at which vacuum pumps are to operate.

Each sewage *discharge pump* capacity is 20% greater than the design peak flow:

$$Q_{dp} = 1.2\,(Q_{max}) \tag{18.7}$$

where Q_{dp} is the discharge pump capacity, in gpm (L/s).

The motors are sized using the procedure for force mains. However, 25 ft (7.6 m) of additional head is required to pump against the vacuum in the collection tank.

The operating volume of the *collection tank* is equal to the sewage accumulation required to restart the discharge pump. The operating volume should be sized so that at minimum design flow, the pump will operate once every 15 min as represented by the following equation:

$$V_o = T\left(Q_{dp} - Q_{min}\right)\left(Q_{min}\right)/\left(Q_{dp}\right) \tag{18.8}$$

$$Q_{\text{min}} = Q_{\text{avg}}/Z \tag{18.9}$$

where V_o is the operating volume of the collection tank (gal or L); Q_{min} is station minimum flow (gpm or L/s); Q_{avg} is station average flow (gpm or L/s); Z is the ratio of average station flow to minimum flow; and $T = 15$ min using US customary units =900 s using metric units.

The total volume of the collection tank (V_t) is 3.0 times the operating volume with a minimum size of 400 gal (1500 L). After fixing the operating volume, the designer should check to ensure that an excessive number of pump-starts per hour will not occur:

$$V_t = 3\,V_o \tag{18.10}$$

where V_o is the operating volume of the collection tank, in gal (L); and V_t is the actual volume of the collection tank, in gal (L).

Standby power should be provided for use during emergency conditions. One hundred percent standby power is required. The lengths of the collection lines are governed by two main factors: static lift and losses. Total available head loss is 13 ft (4.0 m). Therefore, the sum total of static lifts and friction lost must not exceed this figure.

Example 18.3 Design of a Vacuum Sewer System

Design a vacuum sewer system assuming the following design conditions:

N_v = number of valves connected to the sewage collection station = 60
$K_1 = 0.15$ ft^3/min using US customary units = 0.07 L/s using metric units
K_2 = a conversion factor = 7.5 gal/ft^3 using US customary units = 1 using metric units
Q_{avg} = station average flow = 2,880 gpm = 181.73 L/s
Vacuum length = 1,500 feet = 450 m
Q_{max} = collection station peak flow = Q_{peak} = 10,080 gpm = 636.05 L/s
Z = ratio of average station flow to minimum flow = 3.

Solution 1 (US Customary System):

Q_{min} = station minimum flow = Q_{avg}/Z = 2,880/3 = 960 gpm

For sewer length up to 3,000 ft long:

Q_{vp} = vacuum pump capacity, ft^3/min
= $(5\,Q_{max}/K_2) + (K_1)\,(N_v)$
= [(5 × 10,080 gpm)/(7.5 gal/ft^3)] + (0.15 ft^3/min) (60)
= 6,729 ft^3/min
= 50,468 gpm vacuum pump capacity

Q_{dp} = discharge pump capacity, gpm
= $1.2(Q_{max})$
= 1.2 (10,080 gpm)
= 12,096 gpm discharge pump capacity

V_o = operating volume of the collection tank, gal
= $T(Q_{dp} - Q_{min})\,(Q_{min})/(Q_{dp})$
= 15(12,096 − 960) (960)/(12,096)
= 13,257 gal

V_t = total volume of the collection tank =3 V_o
= 3 × 13,257 = 39,771 gal

Select a total volume for the collection tank = 40,000 gal.

Solution 2 (SI System):

Q_{min} = station minimum flow = Q_{avg}/Z = 181.73/3 = 61 L/s

For sewer length up to 914 m long:

Q_{vp} = vacuum pump capacity, L/s
$= (5\ Q_{max}/K_2) + (K_1)\ (N_v)$
$= [(5 \times 636.05\ \text{L/s})/(1)] + (0.07\ \text{L/s})\ (60)$
= 3,184 L/s vacuum pump capacity
Q_{dp} = discharge pump capacity, L/s
$= 1.2\ (Q_{max})$
= 1.2(636.05 L/s)
= 763 L/s discharge pump capacity
V_o = operating volume of the collection tank, L
$= T(Q_{dp} - Q_{min})\ (Q_{min})/(Q_{dp})$
= 900 (763.26 − 60.58)(60.58)/(763.26)
= 50,195 L
V_t = total volume of the collection tank $= 3\ V_o$
$= 3 \times 50{,}195 = 150{,}585$ L

Select a total volume for the collection tank = 160,000 L.

18.4.6 Potential Applications

The following general conditions are conducive to the selection of vacuum sewers:

1) Unstable soils
2) Flat terrain
3) Rolling terrain with small elevation changes
4) High water table
5) Restricted construction conditions
6) Rock
7) Urban development in rural areas.

Experience has shown that for vacuum systems to be cost-effective, a minimum of 75–100 customers is needed per custom vacuum station. Package vacuum stations have proven to be cost-effective for service areas of 25–150 customers. The average number of customers per station in systems currently in operation is about 200–300. A few systems have fewer than 50 and some have as many as 2000 per station. Some communities have multiple vacuum stations, each serving hundreds of customers.

Hydraulically speaking, vacuum systems are limited somewhat by topography. The vacuum produced by a vacuum station is capable of lifting wastewater 15–20 ft (5–6 m), depending on the operating level of the system. This amount of lift many times is sufficient to allow the designer to avoid the lift station(s) that would be required in a conventional gravity system.

Many myths exist concerning vacuum sewer systems. In reality, a vacuum system is not unlike a conventional gravity system. Wastewater flows from the individual homes and utilizes gravity to reach the point of connection to the sewer main. The equipment used in the vacuum station is similar in mechanical complexity to that used in a conventional lift station.

Vacuum valves can fail in either the open or closed position. One failing in the closed position will result in backups. This would be analogous to a blockage or surcharging of a gravity sewer. Fortunately, failure in this mode is rare. Almost all valve failures happen in the open position. This means that the vacuum continues to try to evacuate the contents of the pit. The vacuum pumps usually run continuously to keep up, because this failure simulates a line break. In these cases, a telephone dialer feature available in vacuum stations notifies the operator of this condition. Correction of the problem can generally be made in less than an hour after the operators arrive at the station.

In short, many of the major objections to the use of vacuum systems are not well founded. These systems have been acceptable in a variety of applications and locations. Any hypothetical or abstract difficulty that can be applied to the vacuum system can also be applied to the more conventional systems. In any event, the vacuum system offers the same convenience as any other type of public sewer system with reference to the actual discharge from the home and meeting the needs of the particular locality.

18.4.7 System Costs

Certain site conditions contribute to the high costs of a conventional sewer installation. These include unstable soils, rock, and a high water table. In addition, restricted construction zones or areas that are flat may also result in high construction costs. By using vacuum sewers, the construction costs for these difficult conditions can be reduced. Smaller pipe sizes installed at shallower depths are the prime reasons for this. The uphill transport capability, even when used only to a very small degree, may save many dollars in installation costs. One other major advantage is the extent to which unforeseen subsurface obstacles can be avoided. Each of these considered individually results in lower costs. Considered collectively, they may result in substantial cost savings.

Many factors affect construction cost bids. Material surpluses or shortages, prevailing wage rates, the local bidding climate, geographic area, time of year, soundness of the design documents, and the design engineer's reputation are examples of these factors. Funding and regulatory requirements also play a part in project cost estimating to the extent that the regulations may be a help or hindrance to the contractor, the client, and the engineer. Because of the many variables, accurate cost-estimating guidelines are beyond the scope of this book. The following general guidelines, however, will be of help to the estimator.

Pipe installation prices are a function of pipe diameter, pipe material, trench depth, and soil conditions. Other factors, such as shoring, dewatering, and restricted construction area conditions will have a large effect on the price.

Piping systems are best estimated using guidance from water system projects built in the same area, if similar materials and specifications are used. Table 18.12 provides estimating data for planning purposes. Pipe prices include furnishing and installing the pipe, excavation, bedding, backfilling, compaction, vacuum testing, cleanup, and similar requirements. Not included are allowances for such items as rock excavation, engineering, and administration.

Valve pit prices will vary depending on the type of valve, type of pit, and depth of pit. Table 18.13 gives average installed unit prices for various valve pit settings and appurtenances. The prices include furnishing and installing the valve pit, excavation, bedding, backfill, compaction, vacuum testing, and surface restoration.

The price of a custom-designed vacuum station depends very much on the equipment selected, the type of structure it is housed in, and the amount of excavation required. Equipment cost varies with the capacity required for each component. Table 18.14 gives average installed prices for both custom-designed stations and package stations. The prices include the equipment (including the generator for all stations), station piping, electrical, excavation, site restoration, and labor.

Table 18.12 Installed Unit Costs for Vacuum Sewer Mains and Appurtenances

Item	Unit Cost (2022 USD)
Mains (in.):	(per linear ft)
4	20
6	25
8	30
Division valve (in.):	(Each)
4	620
6	707
8	992
Gauge tap	87
Lifts	87

Conversion factors: 1 USD/ft = 3.28 USD/m; 1 in. = 25.4 mm; 1 ft = 0.3048 m.

Table 18.13 Installed Unit Costs for Valve Pits and Appurtenances

Item	Unit Cost (2022 USD)
Standard setting (30 in. deep)	4,092
Deep setting (54 in. deep)	4,464
Single buffer tank	5,332
Dual buffer tank	7,068
Extra for anti-flotation collar	174
4-in. auxiliary vent	87
External breather	87
Optional cycle counter	223

Conversion factor: 1 in. = 25.4 mm.

Table 18.14 Installed Cost for Vacuum Station

Station	Number of Customers	Total Cost (2022 USD)
Package	10–25	168,700
Package	25–50	230,600
Package	50–150	285,200
Custom	100–300	372,000
Custom	300–500	443,900
Custom	>500	567,900

18.5 Comparison with Conventional Collection

18.5.1 Population Density

Conventional sewers are typically costly on a lineal foot basis. Where housing is sparse, resulting in long reaches between services, the cost of providing conventional sewers is often prohibitive.

Pressure sewers, SDGSs, and vacuum sewers are typically less costly on a lineal foot basis, so they often prove to be more cost-effective when serving sparse populations.

Conversely, where the required length of sewer between service connections is comparatively short, the cost of providing conventional sewers is usually affordable unless some other obstacle is present, such as adverse slopes, high water table, or rock excavation.

18.5.2 Growth Inducement

The minimum allowed size of pipe in conventional sewers is generally 8 in. (200 mm) in diameter, to accommodate sewer cleaning equipment. Being comparatively large in diameter (and capacity), conventional sewers are often seen as being growth inducing. This is especially true if assessment costs to fronting properties are high, which prompts property owners to develop the property for housing. This is often the most profitable alternative for the property owner and as such provides the greatest financial relief from the assessment.

Pressure sewer, SDGS, and vacuum sewer system mains may be intentionally downsized to limit growth. Innovation in assessment rates can reduce the need for property development as a means of escaping imposed charges. However, designer/owners may wish to allow for some level of growth. This can be incorporated with alternative sewers, as well as conventional ones.

18.5.3 Ground Slopes

Where the ground profile over the main slopes continuously downward in the direction of flow, conventional or small-diameter gravity sewers are normally preferred. If intermittent rises in the profile occur, the cost of conventional sewers may become prohibitively deep. The variable-grade gravity sewer variation of SDGSs, by use of inflective gradients and in conjunction with STEP pressure sewer connections, can be economically applied. Vacuum sewers may be particularly adaptable to this topographic condition, as long as head requirements are within the limits of available vacuum.

In flat terrain conventional sewers become deep due to the continuous downward slope of the main, requiring frequent use of lift stations. Both the deep excavation and the lift stations are expensive. SDGS are buried less deep, owing to the flatter gradients permitted. Pressure sewers and vacuum sewers are often found to be practical in flat areas, as ground slope is of little concern. In areas higher than the service population, pressure sewers and vacuum sewers are generally preferred, but should be evaluated against SDGS systems with lift stations.

18.5.4 Subsurface Obstacles

Where rock excavation is encountered, the shallow burial depth of alternative sewer mains reduces the amount of rock to be excavated. Deep excavations required of conventional sewers sometimes encounter groundwater. Depending on severity, dewatering can be expensive and difficult to accomplish.

18.5.5 Discharge to Gravity Sewers

Where homes are in proximity to a conventional gravity sewer, but where conventional service is impractical, alternatives may often be used. GPs or solids-handling pumps are used at individual homes to discharge from low-lying homes to the conventional sewer. Vacuum sewers are used to serve large enough groups of homes to justify the cost of the vacuum station.

STEP pressure sewers are commonly used in conjunction with SDGS. Such hybrid installations are more common than strictly STEP or strictly SDGS, even though these sewers are usually classified as one of these types. Their discharge into some conventional sewers may be feasible, but the discharge of sulfides to the sewer must be evaluated when such discharges are large enough to constitute a significant portion of the total flow.

18.5.6 Discharge to Subsurface Disposal Fields

STEP pressure sewer equipment is commonly used to discharge septic tank effluent to subsurface disposal fields that are distant or located at higher elevations than the homes served. SDGS may also be used for conveyance of effluent to subsurface disposal facilities as long as ground slopes are favorable for gravity flow.

Problems/Questions

18.1 Discuss the difference between alternative and conventional gravity sewers. List the types of alternative sewers.

18.2 List and describe the factors on which the choice between GP and STEP systems depends.

18.3 List the advantages of small-diameter gravity systems.

18.4 List the disadvantages of small-diameter gravity systems.

18.5 Briefly compare vacuum sewer systems with pressure sewer systems, and discuss their cost factors and applications.

18.6 The low-pressure sewer system, which is one type of wastewater force main system, is commonly used for new, small residential and commercial housing developments where the use of gravity sewer systems is unfeasible. An LPS system is not subject to infiltration from groundwater or from surface stormwater entering through leaking pipe joints and manholes. Grinder pumps of approved design accomplish all pumping and sewage-grinding processes for small-diameter LPS systems. Normally the Hazen–Williams equation is used for LPS system design:

$$v = 1.318\, C\, (r)^{0.63} (s)^{0.54}$$

where v is the velocity of flow (ft/s); C is Hazen–Williams coefficient, which depends on the type and condition of the conduit; r is hydraulic radius (ft); and s is the slope of hydraulic gradient, or loss of head in feet per feet of length.

Conversion factors: 1 ft/s = 0.3048 m/s; 1 ft = 0.3048 m; 1 ft/ft = 1 m/m; 1 gpm = 3.785 L/min.

Determine the velocity of flow (v) and the friction head loss (s), if new 2-in. (50-mm) PVC pipe with a C value of 150 is to be used for transporting 33 gpm (124.9 L/min) in a new LPS system.

18.7 A low-pressure sewer system is to be designed for a new, small community. If the 630-ft (192-m), 2-in. (50-mm) PVC sewer pipe (Class SDR 21) is used for transporting 44 gpm (167 L/min) of sewage in one housing development zone under low pressure and the head loss was calculated to be 0.0263 ft/ft (0.0263 m/m), determine the friction head loss in this zone.

18.8 In a low-pressure sewer system, the total accumulated friction loss that must be overcome by a grinder pump is 63.14 ft (19.25 m). The maximum force main elevation and the minimum grinder pump elevation are 40 and 5 ft (12 and 1.5 m), respectively. Determine the total dynamic head that must be overcome by the selected pump.

18.9 The subdivision development shown in Fig. 18.3, which uses a low-pressure sewer system, is to be analyzed separately for zone 13 and zone 14, which are connected in the numerical order from 13 to 14. Zone 13 is the first or beginning zone, whereas zone 14 is the last zone to discharge to the wastewater treatment plant. Each zone has its own pipe friction head loss during sewage collection in its zone. What will be the accumulated friction loss of each zone that must be overcome by their grinder pumps? The individual friction head loss (ft or m) values of the two zones in the order of zone 13 and zone 14 are 21.19 and 24.75 ft (6.46 and 7.54 m), respectively.

18.10 Design a vacuum sewer system assuming the following design conditions:

N_v = number of valves connected to the sewage collection station = 60
$K_1 = 0.15\ \text{ft}^3/\text{min} = 0.07\ \text{L/s}$
K_2 = a conversion factor = 7.5 gal/ft^3 using US customary units and = 1 using metric units
Q_a = station average flow = 2,880 gpm = 10,901 L/ min = 181.68 L/s
Vacuum sewer length = 3,005 ft = 916 m
Q_{max} = collection station peak flow = 10,080 gpm = 38,153 L/ min = 635.88 L/s
Z = ratio of average station flow to minimum flow = 3.

18.11 What is the definition of SDR for a PVC pipe? How can you find out the outside diameter, inside diameter, minimum wall thickness, and nominal weight of the SDR PVC pipes? Discuss the meaning of SDR in piping design.

18.12 Conduct an Internet search to find the manufacturers and designers of low-pressure sewer (LPS) systems and grinder pumps (GPs). Contact the manufacturers and/or the consulting engineers involved in the LPS and GP projects, asking them for their technical bulletins, reports, design manuals, papers, downloadable design software, and case histories for LPS and GP. Contact the manufacturers or the installing engineering firm for arrangement of a field trip to witness the installations and the products. Finally write a report entitled "Recent Advances in Low-Pressure Sewer Systems and Grinder Pumps."

18.13 Conduct the same investigations outlined in Problem 18.12 for the small-diameter gravity sewer (SDGS) systems instead of LPS and GP, and write a report entitled "Recent Advances in Small-Diameter Gravity Sewer Technology."

18.14 Conduct the same investigations outlined in Problem 18.12 for the Vacuum Collection Systems instead of LPS and GP, and write a report entitled "Recent Advances in Vacuum Collection Systems."

Bibliography

AIRVAC, *Vacuum Sewerage Systems, AIRVAC Design Manual*, Rochester, IN, USA, 1989.

Averil, D. W., and Heinke, G. W., *Vacuum Sewer Systems*, Report prepared for the Northern Science Group of the Canadian Department of Indian Affairs and Northern Development, 1973.

Barrett, M. E., and Malina, J. F. Jr., *Wastewater Treatment Systems for Small Communities: A Guide for Local Government Officials*, University of Texas at Austin, September 1991.

Cooper, I. A., and Rezek, J. W., Vacuum Sewer System Overview, in *Presented at the 49th Annual Water Pollution Control Federation Conference, Minneapolis, MN, USA, October 3–8*, 1976.

Crites, R., and Tchobanoglous, G., *Small and Decentralized Wastewater Management Systems*, WCB McGraw-Hill, Boston, USA, 1998.

Environment One Corporation, *Low Pressure Sewer Design Manual*, Environment One Corporation, Niskayuna, NY, www.eone.com/wastewater, 2010.

Environment One Corporation, *E/One Pressure Sewer Systems: Environmentally Sensitive, Economically Sensible*, www.eone.com/downloads/tech_reprints/Grinder-Pump-Driven-Pressure-Sewer.html, 2011.

Head, L. A., Mayhall, M. R., Tucker, A. R., and Caffey, J. E., *Low Pressure Sewer System Replaces Septic System in Lake Community*, http://www.eone.com/sewer/resources/resource01/content.html, 2010.

Illinois Community Action Association, *Alternative Wastewater Systems in Illinois*, http://www.icaanet.com/rcap/aw_pamphlet.htm, 2010.

New England Interstate Water Pollution Control Commission, *Guides for the Design of Wastewater Treatment Works, TR-16 Manual*, New England Interstate Environmental Training Center, South Portland, ME, 1988.

Parker, M. A., *Step Pressure Sewer Technology Package*, National Small Flows Clearinghouse, Morgantown, WV, 1997.

Saunders, M., O&M considerations for STEP systems, *Water Environ. Technol.*, *21*, 3, 22–27, March 2009.47

U.S. Army Corps of Engineers, Yearly Average Cost Index for Utilities, in *Civil Works Construction Cost Index System Manual, 110-2-1304*, Washington, DC, USA, 2022.

U.S. Environmental Protection Agency, *Alternatives for Small Wastewater Treatment Systems, EPA/625/4-77/011*, Cincinnati, OH, 1977.

U.S. Environmental Protection Agency, *Innovative and Alternative Technology Assessment Manual*, EPA 430/9-78-009, Washington, DC, USA, February 1980.

U.S. Environmental Protection Agency, *Design Manual: Alternative Wastewater Collection Systems*, EPA 625/1-91/024, Cincinnati, OH, USA, October 1991.

U.S. Environmental Protection Agency, *Design Manual: Wastewater Treatment and Disposal for Small Communities*, EPA 625/R-92/005, Cincinnati, OH, USA, September 1992.

U.S. Environmental Protection Agency, *Small Diameter Gravity Sewers, Decentralized Systems Technology Fact Sheet*, EPA 832-F-00-038, Washington, DC, USA, September 2000.

U.S. Environmental Protection Agency, *Pressure Sewers, Wastewater Technology Fact Sheet*, EPA 832-F-02-006, Washington, DC, USA, September 2002.

19

Engineering Projects Management

19.1 Role of Engineers

The planning, design, and construction of water and wastewater systems for metropolitan areas usually bring together a sizable and varied group of engineering practitioners and their consultants, not for months but for years, in a bold and busy venture. As their work is completed, elements of their most important and powerful membership move on to new enterprises; older engineers drop out; younger engineers move in; and offshoots accept parallel, usually smaller, but nevertheless important, assignments. Under proper leadership, task forces perpetuate themselves to attack new problems or deal with old ones in new ways. The science and practice of water supply and wastewater collection and treatment are preserved and promoted in this way.

Because the systems are generally in public ownership in the United States, studies, plans, specifications, and contracts for the construction of water and wastewater works are prepared by engineers normally engaged by the cities and towns or the water or wastewater districts to be served. Private water companies are increasing in number, while private sewerage corporations are still rare institutions. The engineers may belong to the professional staff of the municipal or metropolitan governmental agencies responsible for designing and managing public works, or they may be members of a firm of consulting engineers. For very large undertakings, governmental and consulting staff may be expanded for the duration of the enterprise, as suggested in the first paragraph of this chapter. For smaller undertakings, this is seldom true. Consultant groups are given most and possibly all of the responsibility. Engineers for manufacturers of water and wastewater equipment also have a part. The engineers of contractors or construction companies bring the design into being.

19.2 Steps in Project Development

Community action leading to the study, design, construction, and operation of new or enlarged water or wastewater systems and the engineering response elicited by community action are conveniently listed in sequence in the next section, with special reference to projects for which engineering consultants are engaged.

19.2.1 Community Action

Community actions expected for water and wastewater works project development are summarized in Fig. 19.1 and include the following sequential steps:

1) Invitation to consulting engineers to submit proposals for preparing an engineering report or appointment of a consultant to the regular or expanded engineering staff of the community.
2) Engagement of a consultant on evidence of his or her qualifications and not by competitive bidding.
3) Examination of the consulting engineering report and its acceptance or rejection. If the report is not accepted, step 1 may have to be repeated.
4) Authorization of the preparation of plans and specifications. The consultant responsible for the report, and other engineers, too, may be asked to submit proposals for doing the work under the stipulations of step 2.

Water and Wastewater Engineering: Hydraulics, Hydrology and Management, Volume 1, Fourth Edition.
Lawrence K. Wang, Mu-Hao Sung Wang, and Nazih K. Shammas.

Companion website: www.wiley.com/go/Wang/Waterandwastewater4e

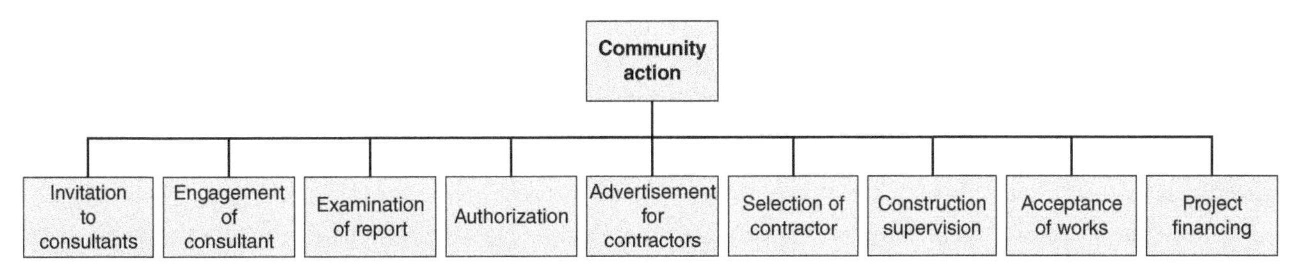

Figure 19.1 Steps for Community Action

5) Advertisement for construction contracts after the plans and specifications have been accepted. Local newspapers and national engineering journals are used for this purpose.
6) Selection of the contractor, generally on the basis of the lowest bid received from a technically qualified and financially responsible construction company.
7) Construction under supervision of a resident engineer, employed by the owner (the municipality or district) or the consulting engineer.
8) Acceptance of the completed works on the recommendation of the engineer. The consultant may be retained to advise and assist in putting the system into operation.
9) Examination and adoption of project financing. This is done at a suitable stage of project development, often with the assistance of the engineering consultant or a financial consultant.

19.2.2 Engineering Response

The engineering response for water and wastewater works project development is summarized in Fig. 19.2 and in the following list:

1) Collection and evaluation of available and required basic information: demographic, hydrologic, geologic, topographic, and industrial. The advice of local engineering practitioners who know the community and region may be useful.
2) Preparation of a preliminary or feasibility report, or of a final engineering report.
3) Preparation of plans and specifications, if the report is accepted and the detailed design authorized. Basing her or his decision on needed and available materials, the engineer will normally discuss the ideas with qualified manufacturers and suppliers.
4) Preparation of contract documents, including an estimate of construction costs after the plans and specifications have been accepted.
5) Assistance in advertising for bids and selecting the successful bidder.
6) Supervision of construction to make sure that the contractor performs the work in accordance with the plans and specifications. Approval of necessary shop drawings supplied by the contractor.
7) Authorization of payment when the work has been completed and preparation of as-built plans to record the construction work.
8) Preparation of an operating manual for the system and supervision of its operation during its early years, most often when the community does not have experienced operating personnel of its own.
9) Assistance in setting up accounting procedures and establishing appropriate rates for service.

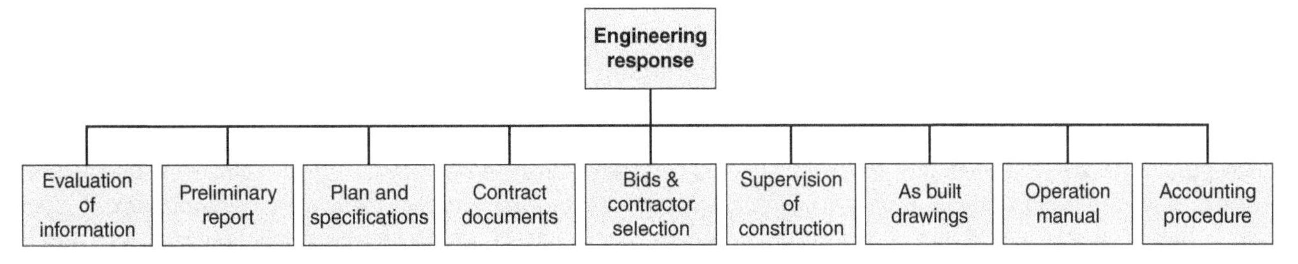

Figure 19.2 Steps for Engineering Response

19.3 The Engineering Report

Engineers are known by the quality of their reports. They are judged by the performance of completed designs. Good reports have opened up new channels of learning. Good designs and constructions have created new technologies. Of the many outstanding examples, some are allied to consultants, others to special commissions, and others to governmental agencies at all levels.

Regardless of whether the community does its own engineering work or engages a consultant to do it, its water and wastewater schemes should receive the benefit of well-documented studies of feasible projects before detailed designs are authorized. This maxim of sound engineering practice and good governmental procedure should be honored.

Engineering reports commissioned for the purpose of identifying the need for new or expanded water and wastewater constructions and offering acceptable proposals for their development are expected to state their mission clearly; analyze and summarize available and needed data; assess the technical, economic, legal, and political feasibility of projected works; offer alternative answers to the questions asked and point out the one or two most suitable replies; estimate costs; investigate methods of financing; and, as a result of the preceding studies, lay a firm base for the recommendations made and the execution of a feasible scheme.

If the report is well written, its purpose will be understood by lay boards and by fellow engineers. If its findings, conclusions, and recommendations are carefully worded, they can be quoted verbatim in news releases and give the public the information to which it is entitled. Bonds funding the proposed works can then receive justifiable support. If the report is imaginative and exhaustive, it should become the document of reference for future studies and further planning and development of the water resource. If the report is accepted, it should allow for the orderly and economic acquisition of needed rights and properties, the preparation and enactment of required legislation, and the exploration of obtainable financial support.

Most engineering reports are scheduled to contain the information itemized in the form of a table of contents as shown in Fig. 19.3.

19.4 Feasibility Studies

Engineering feasibility implies both technical and economic practicability. Technical practicability is readily demonstrated in engineering studies such as those described in earlier chapters of this book. Although economic practicability can find expression in optimization procedures of benefits in relation to costs, either as ratios or as differences, it may not be possible to do so in full measure and in all circumstances. Resource developments competing for funds in developing countries and the national, regional, or basin-wide comprehensive planning of industrially advanced societies, for example, may create so many constraints or introduce so many decision variables that true optimization is left in suspense. Moreover, it is difficult to attach monetary values to intangible benefits. Recreation is a notable example.

Engineering feasibility may also be encumbered by legal and political restrictions. The fact that surface waters are common bounds between states or provinces and that large rivers may cross-national and international boundaries in their course toward the sea may add political and legal constraints of much significance. The Congress of the United States may be asked to lift some of them through interstate and international compacts. However, legal contests between states are not unknown. The federal courts, in which interstate cases are tried, have handed down far-reaching decisions. In the United States, the *common law* or *riparian doctrine* governing water use in the eastern portion of the country and the *doctrine of prior use* or *prior appropriation* prevailing in the western portion may add legal constraints. They, too, may require interpretation by the courts.

19.5 Alternatives

There is no single solution to a given water resource problem, and development of the best solution is not a matter of objective optimization alone. Community and regional needs and wishes must be taken into account. To be meaningful, decisions must identify comparative advantages and disadvantages of promising alternative schemes. The community is then free to make its own choice. The most immediately economical system may not be the most acceptable system. A recurrent

Table of Contents (TOC)

- Letter of transmittal, addressed to the responsible agency of government.
- Letter of authorization, from the agency to the engineer.
- Summary of findings, conclusions, and recommendations.
- Detailed report:
  1. Purpose and scope of the report.
  2. The community, its geography and people, its history and expectations.
  3. Existing water or wastewater works and their historical development.
  4. Population: past, present, and probable future population and population density.
  5. Report on water supply system:
     a. Water use—domestic, fire, mercantile, and industrial, by area served and expected total.
     b. Available water sources, source development, pumping, transmission, treatment, distribution, and service storage.
     c. Project comparisons, including construction and operating costs.
     d. Recommended project.
  6. Report on wastewater system:
     a. For sanitary sewerage: wastewater production—domestic, infiltration, mercantile, and industrial by kind of area served and expected total.
     b. For stormwater drainage: rainfall and runoff—design storm and its recurrence interval by area served.
     c. Proposals for reuse/disposal, including interception of sewage from combined systems; stormwater management plans, wastewater treatment works, and outfalls.
     d. Effect of treated effluents including stormwater on receiving bodies of water.
     e. Project comparisons, including capital and operating costs.
     f. Recommended project.
  7. Financing.
  8. Rates.
- Appendix:
  Charts and tables of basic data.

Figure 19.3 Information Itemized in Engineering Reports

example is the common preference for naturally clean water rather than water made clean by treatment. In many instances, upland supplies reaching the city by gravity meet the specification of naturally clean waters; supplies pumped from polluted rivers coursing past the city and purified in treatment works before delivery to its distribution system are descriptive of the second. Of the two, upland water supplies are usually more costly to develop, but their cost of operation, maintenance, and repair may be smaller because of differences in power and treatment requirements. Aesthetic imponderables become decision variables in cases such as this. Quite different is the example of large industrial users who must make their choice between tying into a public supply and developing one of their own. Entering into their decisions are the economy of lower interest rates available to public bodies, the economy of scale, and possible advantages of more useful water temperatures where surface and groundwater temperatures are in competition.

19.6 Plans and Specifications

More expensive and time-consuming than an engineering report, but proportionately more remunerative to the engineer, are the detailed planning and specification of works to be built. To ensure a meeting of the minds of the owner, the designer, and the contractor, plans and specifications must be comprehensive as well as precise. Vague and conflicting documents create confusion and increase bids as well as actual costs. Indeed, they may be responsible for unsatisfactory constructions.

In the course of their careers, most engineers develop specific interests and capacities. If these are given recognition in design offices—and it is generally advantageous to do so—design assignments can be based on competencies in basic engineering, structural, mechanical, and electrical elements, for instance, or on competencies in specific system components such as pipelines, pumps, filters, and other treatment units. For major projects, a team leader carries the responsibility and should be given the discretion of work assignments and the proper timing and coordination of effort by members of the team. Specifications may be compiled from individual statements, or they may be written by a separate group.

Preliminary reports, by contrast, are usually tasks performed by one or two senior engineers who generally tackle the essential subjects in sequence.

19.7 Sources of Information

A busy office must be supported by a good library. A sizable engineering organization may employ a professional librarian. Professional organizations develop standards.

The shelf list of a working library usually includes the references shown in Fig. 19.4.

19.8 Standards

Undoubtedly, standards have done much to improve the performance of water and wastewater works, but the underlying philosophy continues to be questioned. Standardization of pipes and other equipment for water and wastewater systems normally reduces the cost of the standard items. Together with standardization of materials of construction, it simplifies not only design but also construction and the procurement of materials that meet minimum criteria and ensure compatibility. Standards of water quality set a goal to be reached in the protection of waters for the many purposes they serve. Design standards of regulatory agencies are generally written for the protection of the communities. Their aim is to promote

Contents of Shelf Library

- Engineering manuals, texts, and serial publications.
- Standards and specifications of professional organizations. Examples are standards published by:
  - The American Water Works Association
  - The American Society for Testing and Materials
- Manuals of engineering practice and design, among them those of:
  - The American Society of Civil Engineers
  - The American Water Works Association
  - The Water Environment Federation
  - North American Society for Trenchless Technology
- United Nations, national, state, and local building and electrical codes.
- Handbooks of associations, manufacturers, and book companies:
  - The Ductile Iron Pipe Research Association
  - The American Concrete Pipe Association
  - The Clay Sewer Pipe Association
  - The Hydraulic Institute
  - The Portland Cement Association
  - John Wiley & Sons, Inc. (Water and Wastewater Engineering Series)
  - Humana Press/Springer (Handbook of Environmental Engineering Series)
- Catalogs of equipment manufacturers
- Reference annuals published by trade magazines

Figure 19.4 Shelf List of References in a Working Library

the successful construction of water and sewerage works. To this purpose, they may specify minimum sizes and strengths of water and wastewater pipes and minimum velocities of flow in sewers, for example.

The standards of small water and sewerage works should not be applied to works of all sizes. Otherwise they may impede the development and adoption of new processes and fruitful introduction of new ideas. The engineering profession should see to it that new enterprise is not obstructed because it happens to be in conflict with existing standards. Opportunities for large-scale experimentation should be kept open. Examples of obsolescent requirements are (a) rules and regulations against curved sewers of small diameter in residential developments and (b) rules and regulations governing the maximum allowable spacing of manholes in sanitary sewers.

There can be no quarrel with the importance of health and safety standards. Standards of this kind protect water from gathering ground to points of use, maintain adequate capacity and pressures, and introduce standby power and water reserves for use in emergencies. In similar fashion, standards of this kind introduce safeguards into wastewater systems from points of collection to their disposal grounds, prevent surcharge of sewers and resulting flooding of basements and low-lying areas, and protect receiving waters.

19.9 Design Specifications

Two extremes for specifying the work to be performed by a contractor are exemplified by (a) the *turnkey project* or *performance specification* and (b) the *descriptive specification*. Where performance is specified, the contractor becomes responsible for both design and construction. This is common practice in industry and in many developing countries. Where descriptive specifications underlie design, the owner's engineer specifies what the work is to be, and constructions are expected to perform properly if the specifications are met. As equipment becomes more complex, there is a tendency to shift to performance specifications. The engineer then becomes a coordinator of devices and relinquishes some of his responsibility in design.

Performance specification has a place when mass production is of benefit to the purchaser of the product. It is justified, too, when performance can be pretested, or when it is possible to test equipment after installation and to replace it if it does not meet specified performance. Pumps are examples of equipment that is generally selected on the basis of performance specifications. However, pump performance specifications are usually supplemented by descriptive specifications that provide protection against overload and ensure compatibility with other elements of the system. Performance specifications for a pumping station rather than for pumps only, and for an entire treatment plant or even for individual units of a treatment plant, are rarely appropriate. The claim that performance specifications save money because the engineer need not prepare detailed designs is valid only when standardized shelf items can be incorporated in the projected system. When this is not so, the owner, in fact, bears not only the direct cost of design but also the hidden cost of turnkey or other projects prepared for bids that were not successful.

When a descriptive specification interferes with competitive bidding by suppliers of material or equipment that will serve the design purpose equally well, the engineers normally prepare two or more alternative designs and invite bids on them.

19.10 Project Construction

Public construction projects are generally required by law to go to the low bidder. Awards can be withheld only on evidence that a contractor does not possess the qualifications and financial backing to undertake and complete the project successfully.

Documents formulated by the engineer and used by the contractor in preparing bids are given below.

19.10.1 Notice to Bidders

The notice to bidders advertises the project and tells where copies of plans and specifications can be inspected and how they can be obtained.

19.10.2 General Conditions

The general conditions document is a statement of the conditions under which the contract is to be performed. Normally, it includes information on the following matters:

1) *Proposal requirements and conditions.* These specify the conditions under which proposals will be received, requirements for bond, necessary qualifications of bidders, the basis for disqualifying bidders, and conditions for the employment of subcontractors.
2) *Award of contract.* In the case of public agencies, a statement says that if the contract is awarded, it will go to the lowest responsible bidder.
3) *Contract terms.* To ensure a meeting of the minds of the contracting parties, the terms used in the contract document are defined, and the authority of the resident engineer, management of extra work orders, and general rules for interpreting plans and specifications are stated.
4) *Bonds and insurance.* Bonding requirements and guarantees against defective workmanship are specified together with types of insurance demanded of the contractor.
5) *Responsibilities and rights of the contractor.* The specific responsibilities of the contractor on the project are stipulated. As a rule the contractor must give assurance to the owner that he does not become liable for patent infringement when patented equipment is installed.
6) *Responsibilities and rights of the owner.* The owner furnishes property surveys and, through his engineers, necessary baselines and benchmarks for the work to be done. He specifies the conditions under which she or he or the engineers have the right to inspect the operation and the mechanics of issuing changes in work. Although the engineer is employed by the owner, she or he is made the final judge in disputes between owner and contractor. However, the contractor is given a right of appeal for arbitration.
7) *Workmanship and materials.* The general basis for controlling the quality of materials and equipment is stated. The contractor must submit shop drawings, lists of materials, and other required information in ample time for review by the engineer prior to their incorporation into the construction. The contractor is generally required to field-test equipment after its installation to ensure its proper operation. In turn, the contractor may require the equipment supplier to direct the installation of specialized equipment and to supervise its initial operation.
8) *Prosecution of work.* Conditions governing the time for completion of the work are covered. Except in unusual cases there is no penalty for late compliance with the contract. However, associated costs of engineering and inspection may have to be reimbursed by the contractor. Specification of damages for delays generally raises the bid prices for contracts.
9) *Payments.* The contractor is paid periodically for work actually performed and materials brought to the construction site. A small share of the total, normally 10%, is held against project completion and final acceptance. Methods of payment for extra work and work omitted are stipulated.

19.10.3 Special Provisions

To apply to a particular project, the special provisions may include requirements for the continuation of existing services while construction is under way, for particular methods of construction the engineers believe should be followed in executing the work, and for the specific scope of work.

19.10.4 Detailed Specifications

The detailed specifications comprise the bulk and most-used section of the document. Although some specifications are not changed significantly from project to project (e.g., specifications for concrete, steel, and certain kinds of pipe), most of the detailed specifications have reference only to the project for which they have been written. They are essential companions of the plans. Neither plans nor detailed specifications are self-sufficient.

The detailed specifications include such items as site preparation, demolition of existing structures, earth excavation, fill and backfill, rock excavation, preparation of foundations, embankments, paving, concrete, reinforcing steel, piping, drain piping, gates, valves, meters, specific items of equipment, metal work, painting, plumbing, heating and ventilating, electrical work, fencing, final grading and surfacing or seeding, and planting.

19.10.5 The Proposal

The contractor's proposal, accompanied by a certain percentage of the total bid price, constitutes his bid. Ordinarily he quotes a price for each category of materials or equipment detailed in the specifications. Some of them are unit prices, for example, dollars per acre for clearing and grubbing; per cubic yard for concrete, excavation, and fill; per linear foot for piping; per square foot for paving; and *lump sum* for items of equipment. Where unit prices are requested, the engineer lists the estimated quantities. In a conflict between actual quantities and quantities listed in the proposal, the actual quantity is paid for. The more precise the engineer can make the estimates and the more extensively he or she uses unit prices, the lower the total bid is likely to be.

In arriving at cost estimates, both engineer and contractor spend much time in measuring and calculating quantities—the engineer to make sure that the bid is compatible with the budget of the client, and the contractor to decide on the bids. Itemized estimates cover bulk and finished concrete, brick, painting, steel, pipe, paving, trim, planting, and many others in seemingly endless flow. Bid prices normally reflect the local situation of construction and employment.

Written too tightly, specifications may lose the advantages of competitive bidding among equipment suppliers. Written too loosely, they may allow unsatisfactory equipment to be installed. To protect the owner, the engineer may either require the bidder to identify the equipment on which the bid is based, or exclude equipment from price competition. This lets the successful bidder and the engineer select the equipment after the contract has been awarded.

With the proposal, the bidder is asked to supply information on his or her experience, equipment available for constructing the project, names and qualifications of persons who will have responsible charge of the work, and personal financial resources.

19.10.6 The Contract

The contract is the agreement signed by representatives of the owner and contractor. A contract bond serves as a guarantee of performance, quality of materials, and workmanship. Normally it remains in force for 12 months beyond final acceptance of the project.

19.11 Project Financing

Funds for the construction of major water or wastewater systems are usually borrowed. Loans normally stipulate how funds will be obtained for their repayment and for meeting other continuing obligations of the enterprise. Interest payments and operation, maintenance, and replacement (OMR) costs are examples (see Fig. 19.5).

1) *Capital costs* are the costs of the project from its beginning to the time the works are placed in operation. Included are (a) the purchase of property and rights-of-way, (b) payments for equipment and construction and for engineering and legal services, and (c) interest charges during construction. For this phase of the undertaking, money must be borrowed on short-term bond anticipation notes.
2) *Fixed charges* are the annual charges made to repay capital costs, both interest and principal, together with applicable taxes.
3) *Amortization* is the serial repayment of principal.
4) *Principal* (P) is the amount borrowed. *Repayment* (R) of principal is a part of the fixed charges.
5) *Interest* (i) is the cost of borrowing money. It is a function of the unrepaid principal and is expressed as a percentage per year. Like repayment, it is part of the fixed charges.
6) *OMR costs* include the expenditures for operation of the works, their maintenance and repair, the replacement of equipment in the normal course of operation, and minor normal extensions.

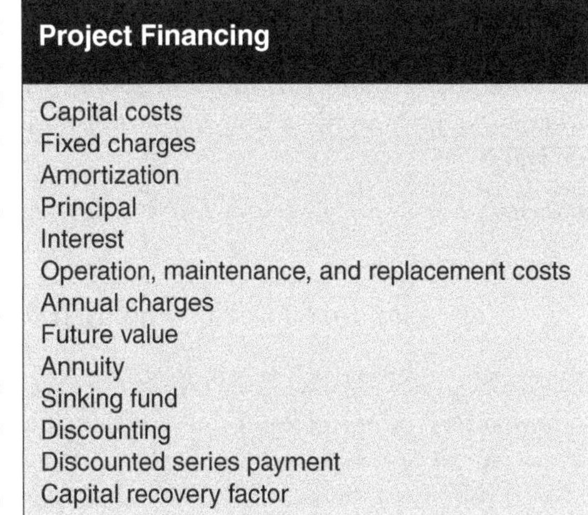

Figure 19.5 Obtaining Funds and Their Repayment

7) *Annual charges* comprehend the sum of fixed charges and OMR costs. Two or more dissimilar alternative projects are often compared on the basis of annual charges because each one must be paid for from taxes, special assessments, service charges, or commodity rates.
8) *Future value* is a function of P, i, and n (number of years for the loan). Calling FV the value of a single payment after n years, $FV = P(1 + i)^n$. A payment or loan of USD 1,000 accumulating interest at 4% annually, for instance, has a future value or repayment requirement of USD 1,480 after 10 years.
9) *Annuity* or uniform series future worth is a function of R, i, and n. If A is the value of the annuity or series of n annual end-of-the-year investments, $A = (R/i)[(1 + i)^n - 1]$. For example, USD 1,000 invested each year for 10 years at an interest rate of 4% compounded annually has a future value of USD 12,000.
10) *Sinking fund* is a fund established to retire a debt in a series of equal payments R to provide an amount A in n years. It is a function of A, i, and n, namely, $R = Ai/[(1 + i)^n - 1]$. For example, a sinking fund of USD 12,000 after a 10-year period at 4% interest compounded annually is built up by an annual investment of USD 1,000.
11) *Discounting* describes the practice of reducing future costs of benefits to an equivalent present value PV. As a function of A, i, and n, it provides common ground for estimating alternative projects by bringing them to a common present date, that is, $PV = A/(1 + i)^n$. The present worth of USD 1,480 paid at the end of 10 years is USD 1,000, for instance, if the interest rate is 4% compounded annually.
12) *Discounted series payment* determines the present value, or discounts the value of a series of equal annual future payments in terms of R, i, and n, or $PV = (R/i)[(1 + i)^n - 1]/(1 + i)^n$. Thus the present worth of USD 1,000 to be repaid each year for 10 years with interest at 4% is USD 8,111; that is, this amount invested at 4% interest compounded annually over a 10-year period would produce an income or payments of USD 1,000 a year for each of the 10 years.
13) *Capital recovery factor* is the annual payment, including both principal and interest, necessary to amortize debt A in n years at an interest rate of i, that is, $R = Ai(1 + i)^n/[(1 + i)^n - 1]$. To repay a loan of USD 8,111 by a series of payments each year over a period of 10 years, for example, requires equal annual payments of USD 1,000, if the annual interest rate is 4%.

Tables of these factors for various periods of time and different rates of interest are found in most engineering handbooks.

Example 19.1 Future Value Calculation

What is the future value or a single repayment requirement after five years for a USD 20,000 loan (such as a municipal bond) with a 3% annual interest rate?

Solution:

$$FV = P(1 + i)^n$$

$$FV = 20{,}000(1 + 0.03)^5 = \text{USD } 23{,}180.55$$

Example 19.2 Annuity of Uniform Series

What is the future worth of USD 20,000 invested by a city each year for five years at an interest rate of 3% compounded annually?

Solution:

$$A = (R/i)[(1 + i)^n - 1]$$

$$A = (20{,}000/0.03)[(1 + 0.03)^5 - 1] = \text{USD } 106{,}180.27$$

Example 19.3 Sinking Fund to Pay a Consultant

With a mutual agreement, a city is to retire a debt to a consulting engineer by establishing a sinking funds of USD 24,000, for a 3-year period at 4% interest rate compounded annually. What is the annual amount the city must put aside to retire the debt after three years?

Solution:

$R = Ai/[(1+i)^n - 1]$
$R = 24{,}000 \times 0.04/[(0.04)^3 - 1] = \text{USD } 7{,}688.36$

The annual amount that the city must put aside is USD 7,688.36.

Example 19.4 Present Value for Cost Comparison
Two equally qualified consulting engineering firms, A and B, are bidding for a city's water works project, and both agree to be paid in full after the design project is completed. The engineering firm A can complete the project in two years and will charge USD 29,000. The engineering B firm can complete the project in one year and six months, and will charge USD 31,000. Which engineering firm's service charge is cheaper in terms of present values, assuming the city has plenty of time to wait, so the completion time is not a factor, and the steady interest rate is 7%?

Solution:

The present value of firm A's service charge:

$$PV = A/(1+i)^n = 29{,}000/(1+0.07)^2 = \text{USD } 25{,}329$$

The present value of firm B's service charge:

$$PV = 31{,}000/(1+0.07)^{1.5} = \text{USD } 28{,}008$$

Assuming the completion time is not a factor, firm A's service is lower than that of firm B.

Example 19.5 Discounted Series Payment
Determine the present worth amount, which is invested at 4% interest compounded annually over a 10-year period and will produce an income or payments of USD 50,000 a year for each of the 10 years.

Solution:

$PV = (R/i)[(1+i)^n - 1]/(1+i)^n$
$PV = (50{,}000/0.04)[(1+0.04)^{10} - 1]/(1+0.04)^{10}$
$= \text{USD } 405{,}400$

Example 19.6 Capital Recovery Factor
The city has obtained a federal loan of USD 500,000 for a water works renovation project. Determine the equal annual payments that will be necessary to amortize and repay the debt in 10 years at an interest rate of 4%.

Solution:

$R = Ai(1+i)^{10}/[(1+i)^n - 1]$
$= 500{,}000\ (1+0.04)^{10}/[(1+0.04)^{10} - 1]$
$= \text{USD } 61{,}666$

The city must repay the federal government **USD 61,666 annually.**

19.12 Methods of Borrowing

For relatively simple and straightforward projects, the engineer advises on the most suitable methods of borrowing needed funds. Where funds are to be derived from several sources or for large projects, special financial advice may be sought. The methods of borrowing depend on the resources of the borrower, the sources from which funds can be borrowed, regulations of appropriate government agencies, and the nature of repayment arrangements. Borrowing is arranged through the sale of bonds. In the United States, the income from municipal bonds is not taxable. Bonds are of three general types, as discussed in the following sections.

19.12.1 General Obligation Bonds

General obligation bonds generally carry the lowest interest rates because they are backed by the full faith and credit of the community, with income generally derived from ad valorem taxes on property. When the bonds are to be repaid over a period of years, they are called serial bonds.

19.12.2 Revenue Bonds

Revenue bonds are based on repayments earned from the sale of water or from sewer service charges (sometimes called sewer rentals). The revenue bonds of an enterprise with a history of good management may carry as low an interest rate as general obligation bonds. New projects or projects for which the quality of management is uncertain may have to pay high interest rates.

19.12.3 Special Assessment Bonds

Special assessment bonds, like general obligation bonds, are backed by the value of the property they serve. They are generally short-term bonds that are normally designed to permit borrowing for a specific project serving only part of the community. Examples are lateral sewers or booster pumping stations. Only rarely are two or three types of bonds issued in combination. Reasons for combining bonds may be the allocation of charges for service in some relation to benefits received or, more pragmatically, the avoidance of legal limits on bonded indebtedness.

19.13 Rate Making

The principle adopted as a guide in financing and rate-making schemes has been stated by American Society of Civil Engineers (1951) usefully as follows: "The needed total annual revenue of a water or sewage works shall be contributed by users and nonusers (or by users and properties) for whose use, need, and benefit the facilities of the work are provided, approximately in proportion to the cost of providing the use and the benefits of the works."

On the one hand, payments for water supply and wastewater removal obtained solely through general taxation would be inequitable, because property owners would be assessed for these goods and services in full and irrespective of use. On the other hand, payments derived solely from the sale of water and through it from sewer-use charges would also be inequitable because properties benefiting from the availability of water or drainage facilities would not contribute to any part of system costs. Even if the cost of delivering water or collecting wastewater could be ascertained exactly for each water consumer or owner of property, it would be impractical to base a rate structure on such changeable information. Instead, the rate maker generally resorts to the mean; that is, he arrives at an approximation by averaging costs within categories of users of the same general kind. Should inequity be proved for one or more of the categories, imbalance can be righted when rates are next adjusted.

Understandably, rates must bring in sufficient income to cover fixed charges, normal OMR costs, and the cost of reasonable improvements. In some instances they may be designed to also provide a modest reserve for normal expansion of the system. Too large a reserve would place an unfair burden on current users; too small a reserve would entail frequent and expensive bond issues.

19.13.1 Water Rates

Water rates are normally structured according to the classes of consumers served and their water uses. Common classifications include manufacturing or wholesale, commercial, or intermediate, and residential or domestic users, with each category being subdivided according to rate of draft. A first or *minimum block* of charges covers the cost of metering and meter reading and of billing and collecting; it is independent of the quantity of water drawn. A second or *wholesale block* of charges are assessed in direct proportion to the cost of supplying water. Rates are obtained by dividing the system costs by the volume of water delivered. Water furnished to a neighboring community that resells it to its residents is usually charged for in this way. A third block of charges allows for the addition of incremental costs to the wholesale cost; the block is divided into subclasses with individual rates.

In terms of cost alone, unit prices generally decrease for large users. However, where water is in short supply and each incremental use adds higher costs, it is not unreasonable to increase unit prices for larger users.

19.13.2 Fire Protection

Water service for fire protection bears no relationship to the amount of water used. For this reason, charges for fire service should be subtracted from the total cost of water. Fire service is charged for in various ways. If revenue is to be proportioned according to use and benefits, for example, the cost of each element of the system must be isolated and apportioned in relation to its contribution to fire protection and to general water service. Costs of a transmission main sized to carry water for fire protection plus the coincident draft, for instance, would then be divided in proportion to these flows. By contrast, no charges for water purification would be allocated to fire protection, because a negligible quantity of water is actually consumed in firefighting.

If the primary purpose of a water system is to be a water supply service, only the incremental costs necessary to equip the system for fire protection would be so assessed. To find the cost, a hypothetical system not providing fire protection would be laid out and its cost estimated. Fire protection charges would then be based on the difference in cost between the hypothetical system (without fire service) and the actual system (with fire service).

When the cost of fire service is met from general tax funds, as it usually is, the benefits of fire protection are assumed to be proportional to property values. Actually, costs of fire service range from 10% of the total cost of water for large communities to as much as 30% for small communities.

19.13.3 Peak-Flow Demands

Peak-flow costs are exemplified by larger capacities of pipelines furnishing seasonal peak-flow demands for lawn sprinkling. There is no easy way of adjusting income received to these flows. Charges are normally the same as for other water uses. In the electrical industry, by contrast, peak costs are based on demand-meter readings, and charges per unit of electricity are related to the consumer's peak-demand rate or to demand during peak periods. Although it would probably not pay to install demand meters on residential water services, their use may be justified for wholesale customers—neighboring communities, for instance. An incidental advantage would be the resulting encouragement of wholesale customers to put in service storage and equalize the system demand.

19.13.4 Sewer Service Charges

For historical reasons, sewerage systems were paid for from general taxation. However, as communities have turned to easier, although sometimes more costly, revenue financing, and as wastewater treatment has become more common, sewer service charges have been introduced with the example of water rates in mind. Inasmuch as the wastewater released to the system is a more or less uniform fraction of the water used, the service charge is often made a fixed percentage of the water bill.

However, it can be argued that well-balanced financing of a sewerage system should actually be composed of special assessments for lateral sewers, general taxation for storm sewers, or the stormwater portion of combined sewers, as well as service charges based on water-meter readings for domestic wastewater collection and disposal.

Although metering the wastewater from households may be neither practical nor necessary where the water is metered, it may be both practical and equitable to meter industrial wastewaters. Basic charges may be determined in much the same way as for water. Surcharges may be imposed when admission of the wastewater increases the cost of treatment out of proportion to their quantity. An incidental advantage to be gained from surcharges is the possible inducement of industries to reduce and alter their waste discharges by recirculation of process water and modification of manufacturing processes. Pretreatment of wastewaters before their discharge to the sewer could be either an option or a requirement depending on the industrial wastewater quality.

19.14 Systems Planning and Management

Each municipal water or wastewater engineering program begins with an analysis of a county or multi-county area to define efficient water districts or sewer districts and regional piping systems. This section chooses formation of a sewer district based on natural drainage areas as a typical example. By law, only treated effluents can be discharged into a surface receiving water.

At initial gross regional level, the actual siting and design of treatment plants do not receive detailed study beyond a feasibility analysis. Major land forms, regional land uses, future growth of the area served by the plant, and engineering criteria will be the significant parameters studied. These parameters, all related to a planned receiving water body, will define the individual districts and allow more detailed design studies to progress in later stages. It is the requirement of the US Environmental Protection Agency (USEPA) that all publicly owned treatment works needed to maintain acceptable water quality over a 20-year period (Source: Areawide Waste Treatment Management Planning, 1974).

The planned sewer district is based on the delineation of the trunk sewer lines and the analysis of alternate interceptor systems. To arrive at the best location for these elements, a licensed professional engineer makes detailed studies of natural factors and current urban development, and then recommends (a) the type of wastewater treatment process system required, (b) a number of potential plant sites, and (c) one of these plant sites for the conceptual construction of the wastewater treatment plant. This initial conceptual design is used as a check on engineering feasibility and as a basis for the economic evaluation of alternates.

The components of the professional engineer's study carried to a point just prior to the recommendation of a site will include the receiving water, gravity sewer trunks, interceptor sewer, corridor of potential plant sites, proposed alternate plant sites, and proposed sewer district boundary. The optimum lines for the gravity sewer trunks and the interceptors are defined in relation to a receiving water body, prior to identification of a corridor of potential plant sites. The recommendation of a particular site within this corridor will depend greatly on the analysis of the engineer, but it was the conclusion of the conference that the expertise of other professions is essential for the study of non-engineering factors. (Source: Siting & Design of Municipal Treatment Plants, Hudson River Valley Commission 1970)

Making the final choice of the wastewater treatment plant location within the "Corridor of Potential Plant Sites" depends on a detailed study of natural factors, a thorough understanding of ecological conditions, a clear picture of existing and projected land-use, and research into the physical needs of the Community. Most engineering reports do not provide this type of detail. Such studies should involve the considerations of an ecologist, a landscape architect, an urban planner, and an architect. As spokesmen for their communities, elected officials should be included at all stages in the process of site selection. Combining the talents of a number of professions in a decision making process is called a team approach. By using this approach, the local communities will get plans and structures reflecting a desire for pure waters and an appreciation for the natural and man-made environment.

At the site selection phase an interdisciplinary team would explore environmental factors within the potential site corridor. For example, in a natural corridor the following would be considered: (a) The natural features of the site can best accommodate a water resources recovery facility (i.e. wastewater treatment plant) with little or no disturbance; (b) the site will not present excessive problems of drainage, run-off, slope, soil conditions, or geologic sub-surface disturbances; (c) the site will have a nearby water body of sufficient size and quality to accommodate treated effluent while still maintaining high water quality for other uses; (d) the site offers multiple use opportunities; and (e) the interceptor right-of-way near the site may become part of regional open space network.

If a potential site corridor is urbanized, the following factors would be studied by the professional planning team. A site should be selected considering the following situations: (a) the site will least impair environmental and land use values when it is developed; (b) the site can be expanded to meet future sewage treatment demand, without causing excessive problems; (c) the site is compatible with existing land uses and the comprehensive development plan; (d) there is possibility of locating certain existing facilities in the urban area near the proposed site; (e) the site may offer multiple use opportunities with other facilities; and (f) the site's interceptor right-of-way will enhance pedestrian access to riverfront.

Selecting a final water resource recovery facility (WRRF) site within the "corridor" entails comparisons among several alternatives. The conference participants indicated that a variety of factors need to be considered as general design guidelines for locating and orienting the plant on the site selected.

The following engineering situations should be considered: (a) WRRF site development should express the design relationship between the WRRF function and existing site character; (b) WRRF layout should respect the existing site character, topography, site features and shoreline; (c) WRRF site development should take advantage of the existing site topography to either emphasize or diminish the visual impact of the facility within the landscape, depending on design goals; and (d) the long axis of large or tall WRRF structures should be perpendicular to natural water bodies in order to avoid blocking views and access.

WRRF site planning considers the following additional planning criteria: (a) The long axis of WRRF structures should be parallel to significant lines of sight; (b) WRRF structures should be located as far back from the water edge as possible to offer opportunities for locating other types of compatible land uses, particularly recreation and open space. It is understood that this principle may not apply if the plant structure is to be physically depressed to permit recreational development above;

(c) site development should alter existing naturally stabilized site contours and drainage patterns as little as possible; (d) plant layout should be flexible and adjust to existing site conditions. Rigid layouts can require expensive and damaging earth moving; (e) the area needed for future plant expansion should also be planned at this time to avoid evolution of the sprawled unordered treatment facility in undeveloped areas or conflicts with the surrounding community in developed urban areas; and (f) when landscaping is utilized it should reflect the character of the surrounding area, and consider multiple use opportunities with any of the following compatible facilities such as water purification plants, WRRF, water storage facilities, power generating plants, municipal incineration facilities, composting facilities, public works garages, equipment storage yards, salt storage yards, power transfer stations, transportation centers, recreational open space, and water-based recreation uses.

A good system is a flexible system. However, if ways of taking advantage of built-in flexibility are not clearly understood and put to use by operating personnel, the advantages of flexibility are lost. Only if it is operated effectively and efficiently does an otherwise well-conceived, well-designed, and well-constructed system become a credit to the community and to the participants in the project. To meet their responsibility to society in full measure, engineers should see to it that the systems they have designed accomplish their mission. Accordingly, they must be prepared to assist the community in the operation of projects as effectively as they did in their design and construction. To this purpose, consultants may be engaged by communities for introductory or continuing surveillance of systems operations and management. A manual describing the purpose and operations of each unit and the required sequence of operations for the works as a whole may be useful. So may schematic diagrams that outline available methods of control, as well as record forms, data collection sheets, and equipment and maintenance cards.

State regulatory agencies, professional organizations, and educational institutions often assist in training plant personnel and other officials in the management of water and wastewater facilities. Both technical and fiscal operations may be covered to advantage.

Problems/Questions

19.1 Outline the role of the community in the development of water and wastewater projects.

19.2 Outline the steps of the engineering response in the development of water and wastewater projects.

19.3 Discuss the basic content of an engineering report for identifying the need for the development of new water and wastewater projects.

19.4 List the items in a table of contents for an engineering report.

19.5 Engineering feasibility implies both technical and economic practicability. Discuss the problems that could be encountered when creating engineering economic feasibility studies.

19.6 An engineering firm's office must be supported by a good library. What type of references does a typical engineering firm's library include?

19.7 The design standards of regulatory agencies are generally written for the protection of small communities. Their aim is to promote the successful construction of small works. To this purpose, they may specify minimum sizes and strengths of water and wastewater pipes and minimum velocities of flow in sewers, for example. Standards of this kind should not be applied to works of all sizes. Why?

19.8 Two extremes for specifying the work to be performed by a contractor are exemplified by (a) the turnkey project or performance specification and (b) the descriptive specification. Differentiate between the two types and discuss their pros and cons.

19.9 The document on general conditions is a statement under which the contract is to be performed. What information does this statement include?

19.10 Detailed specifications are essential companions of the plans. Neither plans nor detailed specifications are self-sufficient. What items do the detailed specifications include?

19.11 Funds for the construction of major water or wastewater systems are usually borrowed. Loans normally stipulate how funds will be obtained for their repayment and for meeting other continuing obligations of the enterprise. Give 10 examples of such costs.

19.12 The methods of borrowing depend on the resources of the borrower, the sources from which funds can be borrowed, regulations of appropriate government agencies, and the nature of repayment arrangements. Borrowing is arranged through the sale of bonds. Name and explain the general types of bonds.

19.13 Water rates are normally structured according to the classes of consumers served and their water uses. What are the three common categories of consumers?

19.14 Discuss the steps and criteria for selecting a right and qualified engineering firm to plan, design, and build future municipal water and wastewater works.

Bibliography

American Public Works Association, *Innovative Funding—Getting to the End of the Rainbow*, Washington, DC, 2008.

American Society of Civil Engineers, *Fundamental Consideration in Rates and Rate Structure for Water and Sewage Works*, ASCE Bulletin No. 2, reprinted from Ohio Law J., Spring 1951.

American Society of Civil Engineers, *How to Work Effectively with Consulting Engineers*, rev. ed., ASME Manual on Engineering Practice No. 45, Reston, VA, 2003.

American Water Works Association, *Developing Rates for Small Systems (M54)*, Denver, CO, 2004.

American Water Works Association, *Water Utility Management (M5)*, 2nd ed., Denver, CO ed., 2005.

American Water Works Association, *Fundamentals of Water Utility Capital Financing (M29)*, 3rd ed., Denver, CO ed., 2008.

American Water Works Association (Ed.), *Principles of Water Rates, Fees, and Charges (M1)*, 5th ed., American Water Works Association, Denver, CO, p. 2000.

Beard, J. L., Wundram, E. C., and Loulakis, M. C., *Design-Build: Planning through Development*, McGraw-Hill, New York, 2001.

Grant, E. L., and Ireson, W. G., *Principles of Engineering Economy*, Ronald Press, New York, 1960.

Great Lakes–Upper Mississippi River Board of State and Provincial Public Health and Environmental Managers, *Recommended Standards for Individual Sewage Systems (10 States Standards for Individual Sewage Systems)*, Health Research Inc., Albany, NY, 1980.

Great Lakes–Upper Mississippi River Board of State and Provincial Public Health and Environmental Managers, *Recommended Standards for Wastewater Facilities (10 States Standards for Wastewater Facilities)*, Health Research Inc., Albany NY, 2014.

Great Lakes–Upper Mississippi River Board of State and Provincial Public Health and Environmental Managers, *Recommended Standards for Water Works (10 States Standards for Water Works)*, Health Research Inc., Albany, NY, 2022.

Hajek, V. J., *Management of Engineering Projects*, 3rd ed., McGraw-Hill, New York, 1984.

Hudson River Valley Commission, *Siting & Design of Municipal Treatment Works*, 1970.

Insurance Services Office, Inc., *Guide for Determination of Needed Fire Flow*, ISO, Jersey City, NJ, 2010.

Kneese, A., *The Economics of Regional Water Quality Management*, Johns Hopkins Press, Baltimore, MD, 1964.

New York State Department of Environmental Conservation, *Design Standards for Wastewater Treatment Works*, NYSDEC, Albany, NY, 1988.

Smith, N. J., *Engineering Project Management*, 3rd ed., Wiley-Blackwell, New York, 2008.

Stanley, C. M., *The Consulting Engineer*, John Wiley & Sons, New York, 1961.

U.S. Environmental Protection Agency, *Areawide Waste Treatment Management Planning*, US EPA Office of Water, Washington DC, 1974.

U.S. Environmental Protection Agency, *New or Repaired Water Mains*, US EPA Office of Water, Washington DC, Aug 2002.

U.S. Environmental Protection Agency, *Effluent Guidelines*, US EPA Office of Water, Washington DC, 2023. https://www.epa.gov/eg.

Wang, L. K., *The State-of-the-Art Technologies for Drinking Water Treatment and Management*, United Nations Industrial Development Organization, Vienna, Austria. UNIDO Manual 8-8-95,, Aug 1995.

Water Environment Federation, *Financing & Charges for Wastewater Systems (MOP 27)*, WEF Press, Alexandria, VA, 2004.

20

Prevention Through Design and System Safety

Prevention through Design (PtD) is a collaborative initiative in the United States based on the belief that the best way to prevent work-related injuries and illnesses is by anticipating and "designing out" potential hazards and risks at the "drawing board" or as early as possible in the design phase of new processes, structures, facilities, equipment, or tools, and organizing work to take into consideration construction, maintenance, production, and decommissioning operations. The PtD initiative is a positive catalyst to the creation and dissemination of business tools, case studies, demonstration projects, and good engineering practices centered on design solutions that reduce worker injuries, illnesses, and costs.

Many practicing engineers view the ideas behind PtD as a necessity in today's highly competitive business environment, and PtD concepts are maturing in a number of businesses worldwide. These leaders understand that it is less costly to anticipate and minimize workplace hazards and risks early in the design process than to retrofit changes after workers get hurt. Besides the cost of retrofitting, not designing for prevention can lead to increased chemical exposures, ergonomic hazards, explosions, fires, falls, amputations, etc. PtD helps plan for success by integrating occupational safety and health solutions with key business processes. It also increases efficiency and cost-effectiveness.

The PtD concept is not new. The engineering profession has long recognized the importance of preventing safety and health problems with the designs its members create. Although the PtD initiative today is broad in scope, one of the building blocks that led to its recognized value for businesses is the development of a process called *system safety*. The remainder of this chapter focuses on system safety and how it applies to water and wastewater engineering. Key examples are introduced here and in other chapters of the book where applicable.

20.1 Introduction to System Safety

Systems are analyzed to identify their hazards and those hazards are assessed as to their risks for a single reason: to support management decision making. Management must decide whether system risk is acceptable. If that risk is not acceptable, then management must decide what is to be done, by whom, by when, and at what cost.

Management decision making must balance the interests of all *stakeholders*: employees at all levels of the company, customers, suppliers, the public, and the stockholders. Management decision making must also support the multiple goals of the enterprise and protect all of its resources: human, equipment, facility, product quality, inventory, production capability, financial, market position, and reputation.

System safety originated in the aircraft and aerospace industries. *Systems engineering* was developed shortly after World War II. It found application in US nuclear weapons programs because of the complexity of these programs and the perceived costs (risks) of nonattainment of nuclear superiority. Systems engineering seeks to understand the integrated whole rather than merely the component parts of a system, with an aim toward optimizing the system to meet multiple objectives. During the early 1950s, the RAND Corporation developed *systems analysis methodology* as an aid to economic and strategic decision making. These two disciplines were used in the aerospace and nuclear weapons programs for several reasons: (1) Schedule delays for these programs were costly (and perceived as a matter of national security); (2) the systems were complex and involved many contractors and subcontractors; (3) they enabled the selection of a final design from among various competing designs; and (4) there was intense scrutiny on the part of the public and the funding agencies. Over the years, the distinction between systems engineering and systems analysis has blurred. Together, they form the philosophical foundation for system safety. That is, safety can – and should – be managed in the same manner as any other design or operational parameter.

Water and Wastewater Engineering: Hydraulics, Hydrology and Management, Volume 1, Fourth Edition.
Lawrence K. Wang, Mu-Hao Sung Wang, and Nazih K. Shammas.

Companion website: www.wiley.com/go/Wang/Waterandwastewater4e

System safety was first practiced by the U.S. Air Force (USAF). Historically, most aircraft crashes were blamed on pilot error. Similarly, in industry, accidents were most commonly blamed on an unsafe act. To attribute an aircraft crash to pilot error or an industrial accident to an unsafe act places very little intellectual burden on the investigator to delve into the design of the system with which the operator (pilot or worker) was forced to coexist. When the USAF began developing intercontinental ballistic missiles (ICBMs) in the 1950s, there were no pilots to blame when the missiles blew up during testing.

Because of the pressure to field these weapon systems as quickly as possible, the USAF adopted a concurrent engineering approach. This meant that the training of operations and maintenance personnel occurred simultaneously with the development of the missiles and their launch facilities. Remember that these weapon systems were far more complex than had ever been attempted and that many newly developed technologies were incorporated into these designs. Safety was not handled in a systematic mariner. Instead, during these early days, safety responsibility was assigned to each subsystem designer, engineer, and manager. Thus safety was compartmentalized, and when these subsystems were finally integrated, interface problems were detected too late.

The USAF describes one incident in a design manual: An ICBM silo was destroyed because the counterweights used to balance the silo elevator on the way up and down in the silo were designed with consideration only to raising a fueled missile to the surface for firing. There was no consideration that, when you were just testing and not firing in anger, you had to bring a fueled missile back down to defuel. The first operation with a fueled missile was nearly successful. The drive mechanism held it for all but the last 5 ft (1.5 m), at which point gravity took over and the missile dropped down. Very suddenly, the 40-ft (12.2-m)-diameter silo was altered to about a 100-ft (30.5-m) diameter.

The investigations of these losses uncovered deficiencies in management, design, and operations. The USAF realized that the traditional (reiterative) "fly–crash–fix–fly" approach could not produce acceptable results (because of cost and geopolitical ramifications). This realization led the USAF to adopt a system safety approach, which had the goal of preventing accidents before their first occurrence.

Beyond mere regulatory compliance, companies and utilities are realizing that waiting for accidents to occur and then identifying and eliminating their causes is simply too expensive, whether measured in terms of the costs of modification, retrofit, liability, lost market share, or tarnished reputation.

The principal advantage of a system safety program – compared with a conventional or traditional industrial safety program – is that early in the design stage, the forward-looking system safety program considers the hazards that will be encountered during the entire life cycle. The industrial safety program usually considers only the hazards that arise during the operational phases of the product or manufacturing system.

The system safety techniques allow the analysis of hazards at any time during the life cycle of a system, but the real advantage is that the techniques can be used to detect hazards in the early part of the life cycle, when problems are relatively inexpensive to correct. System safety stresses the importance of designing safety into the system, rather than adding it on to a completed design. Most of the design decisions that have an impact on the hazards posed by a system must be made relatively early in the life cycle. System safety's early-on approach leads to more effective, less costly control or elimination of hazards. System safety looks at a broader range of losses than is typically considered by the traditional industrial safety practitioner. It allows the analyst (and management) to gauge the impact of various hazards on potential "targets" or "resources," including workers, the public, product quality, productivity, environment, facilities, and equipment.

System safety relies on analysis, and not solely on past experience and standards. When designing a new product, no information may be available concerning previous mishaps; a review of history will have little value to the designer. Because standards writing is a slow process relative to the development of new technology, a search for – and review of – relevant standards may not uncover all of the potential hazards posed by the new technology.

System safety is broader than *reliability*. Reliability asks the question, "Does the component or system continue to meet its specification, and for how long?" System safety asks the broader question, "Was the specification correct, and what happens if the component meets (or doesn't meet) the specification?" Reliability focuses on the failure of a component; system safety recognizes that not all hazards are attributable to failures and that all failures do not necessarily cause hazards. System safety also analyzes the interactions among the components in a system and between the system and its environment, including human operators.

When new processes are developed, the designer seldom begins with a blank canvas. Rather, there is a mixture of retained knowledge, combined with new technology that is fashioned into the new design. The retained knowledge (lessons learned) and new technology drive the safety program planning; hazard identification and analyses; and the safety criteria, requirements, and constraints. The designer's "upstream" knowledge of the safety issues allows for the cost-effective integration of

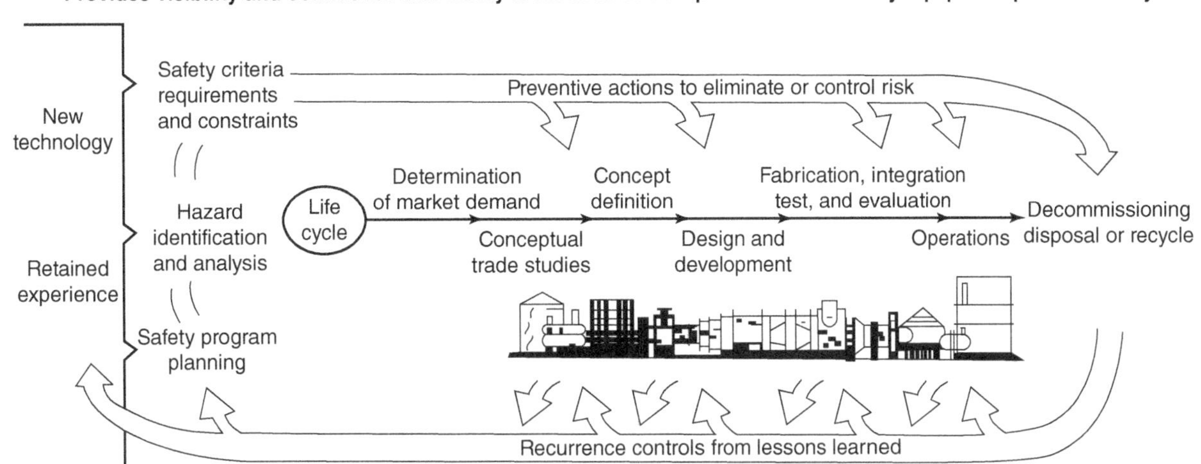

Figure 20.1 Schematic of System Safety Approach

safety, health, and environmental considerations at all points of the product life cycle. Knowledge will be gained as the product/process life cycle moves forward. This knowledge or "lessons learned" can be applied at earlier stages of the product life cycle, leading to changes in design, materials, manufacturing methods, inspection, etc. This approach to continuous process improvement is shown graphically in Fig. 20.1.

Figure 20.1 provides a schematic description of the system safety approach as it is successfully used in various settings, including the design of utilities, semiconductor manufacturing facilities, chemical and food processing plants, air and ground transportation systems, and consumer products. Many modern systems are software-controlled. This has resulted in increasing recognition of the importance of integrating software safety efforts within the system safety program.

20.2 Nature and Magnitude of Safety and Health Problems

Many graduates of civil engineering programs frequently find themselves in responsible positions as employees of or consultants to construction project participants, having little or no background or training in safety and health. Their knowledge on this important topic may be less than adequate for providing the optimal professional service in design, management, or supervision. Traditionally, the emphasis in civil engineering curricula has been on the safety of a facility as designed or constructed and little attention has been given to public safety and health, and occupational safety and health during the construction, operation, and maintenance processes. Most of the knowledge on work site safety and health has been gained, and to a large extent continues to be acquired, through experience on the job. This chapter is designed to fill an important gap in the civil engineering curriculum, affording opportunities for teaching and learning the significance of and the principles and practice of construction, operation, and maintenance safety and health.

The construction industry in the United States is a $780 billion (in 2008 USD) per year industry. About 90% of the total is new construction work, while 10% is maintenance and repairs. Over the years, construction has been a major contributor to the national economy. Its share of the gross national product is about 5%. The construction industry employs more than 5 million people annually (3 million full-time jobs). This number represents more than 5% of the national labor force. There are also more than a million contractors and subcontractors engaged in construction work.

Construction projects are conceived, planned, designed, and built by a team usually consisting of the owner, designer-architect/engineer, and constructor (general contractor and subcontractors). A fourth entity, the construction manager, may also be involved. In addition, support roles are played by material and equipment suppliers, financiers, insurers, regulators, and consultants. Workers carry out the construction tasks.

The construction industry is large but diffuse and fragmented, generally consisting of small units. A number of very large construction firms also exist, each doing more than USD 2 billion business annually, often for very large corporate clients whose construction bills run into the billions of dollars. The industry undertakes diverse projects ranging in size from single-family residential units to multibillion-dollar power and treatment plants. The work may encompass general building construction (residential, nonresidential, institutional, industrial), heavy construction (highways, public works infrastructure, utilities), and industrial construction (major industrial complexes such as refineries and power plants).

The construction industry is uniquely different from manufacturing and other industries, in that

1) Its products are site-produced and one-of-a-kind. With the exception of perhaps the housing sector, each project can be considered a "first, full-scale prototype."
2) The construction project is highly dependent on local conditions: site, climate, local codes and regulations, and so forth.
3) The work is very labor-intensive, involving large forces and energies. The tasks are very variable; mass production and standardization are very difficult to achieve.
4) The industry brings together diverse entities such as designers, engineers, planners, developers, skilled craftspeople, laborers, suppliers, distributors, managers, lawyers, and regulators. These entities are joined together for a single project usually under complex arrangements and are then dispersed on completion of the work. The workforce is transient in most cases.

Because of the inherent diversity and complexity of construction projects, the industry has traditionally suffered low productivity, a high degree of disputes, litigation and business failures, and a poor record of site safety. The national and state statistics clearly point out that construction work is dangerous, and significant problems need to be addressed concerning safety and health in the construction industry. Although some improvement in certain indicators is apparent, the situation is not where it should be. The National Forum on Construction Safety and Health Priorities identified seven critical issues that need attention from the industry:

1) Accident surveillance
2) Construction team interface
3) Engineering needs
4) Legal/legislative aspects
5) Occupational health
6) Primary safety exposures
7) Education and training.

Concerns about *accident surveillance* include proper and meaningful data collection, accident reporting requirements and procedures, data analysis, and contractual aspects, including liability issues. A vastly improved system of accident surveillance is needed to help contractors and regulatory agencies acquire the data significant to the understanding and solution of the problems. Training of the pertinent personnel on the effective implementation of all surveillance-related tasks is essential.

With respect to the *construction team interface*, the overriding problem revolves around the lack of a real understanding of the roles and responsibilities for safety and health by the project participants, that is, the owner, designer, consultant, contractor, and workers. Safety and health aspects in the construction industry are not managed in the same manner as cost, quality, schedule, and productivity, mainly due to the lack of appropriate education and training in this area. The industry needs to fully understand the economic benefits of effective safety and health programs, and owners should screen contractors based on their safety and health records.

Engineering needs deal with the application of engineering principles and skills to accident prevention, particularly with respect to construction erection systems, the construction process, equipment design, and accident data. Many construction accidents and failures can be prevented by proper engineering. A sound structural design rationale is needed for temporary erection systems and criteria for acceptable risk in various situations and environments. Layout controls and operating procedures for construction equipment should be standardized based on ergonomics principles and on an effective accident data collection and analysis system.

The role and effectiveness of *the Occupational Safety and Health Administration* (OSHA) are recognized as the single most important area of concern in regard to legal and legislative aspects. It is suggested that OSHA is not particularly effective in accident prevention due to its focus on regulatory compliance and penalties, and not on providing the appropriate education, training, and incentives for the contractor to work safely. This issue is being rectified through OSHA's PtD initiative.

In terms of *occupational health*, the primary concern is over the applicability of the OSHA hazard communication standard to the construction industry, although its benefits are readily acknowledged. There is a need for a medical screening and monitoring system tailored to the requirements and potential exposures of the job. The need for effective safety and health programs with clearly assigned responsibilities emphasizing the health aspects and hazard control aspects of a particular project requires that a model safety and health program be developed at the project level.

In terms of *primary safety exposures,* the root causes of accidents include (a) lack of company programs and written rules; (b) lack of basic, specific, and continuing training; (c) lack of planning, organization, motivation, communication, and control; (d) lack of a sound value system, for example, attitude, loyalty, and sensitivity; (e) poor work practices and improper procedures; and (f) lack of accountability and responsibility. The key factors for solutions to these problems can be identified as effective education and training, and the implementation and enforcement of well-designed safety and health programs.

Education and training are, by far, the most important issue and need in the construction industry at all levels, from owners, architects, design engineers, and construction management personnel, all the way through to the site supervisors and foremen, and to the workers. Lack of adequate education and training is a part of many problems related to construction safety and health, and the need for it is strongly emphasized for devising effective solutions to the problems. An awareness of safety and health needs has to be fostered by training and must be phased into the educational system very early, such as in grade schools and vocational training centers. The earlier mentioned national forum particularly recommended that safety and health education becomes a required part of the engineering curriculum in the United States.

20.3 Risk Assessment Matrix

20.3.1 Description

A *risk assessment matrix* is a tool that is used to conduct subjective risk assessments for use in hazard analysis. The definition of risk and the principle of the iso-risk contour are the basis for this technique.

The risk posed by a given hazard to an exposed resource can be expressed in terms of an expectation of loss, the combined severity and probability of loss, or the long-term rate of loss. *Risk* (R) is the product of *severity* (S) and *probability* (P) (loss events per unit time or activity). Note that the probability component of risk must be attached to an exposure time interval.

The severity and probability dimensions of risk define a risk plane. As shown in Fig. 20.2, *iso-risk contours* depict constant risk within the plane. The concept of the iso-risk contour is useful to provide guides, conventions, and acceptance limits for risk assessments (see Fig. 20.3).

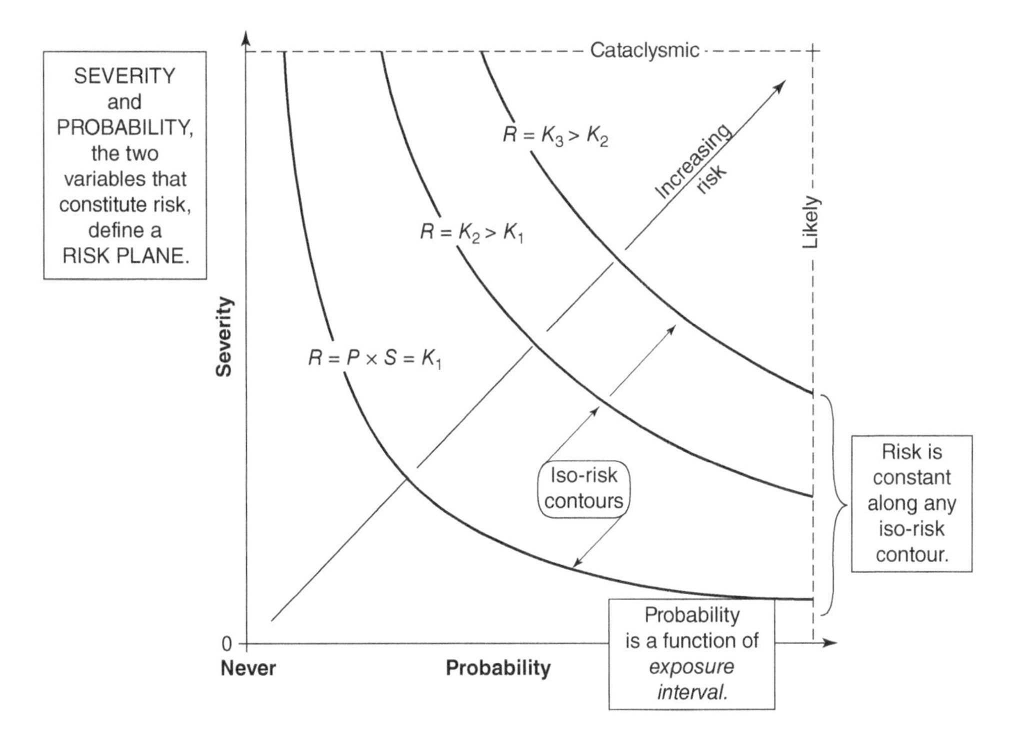

Figure 20.2 Iso-Risk Contours

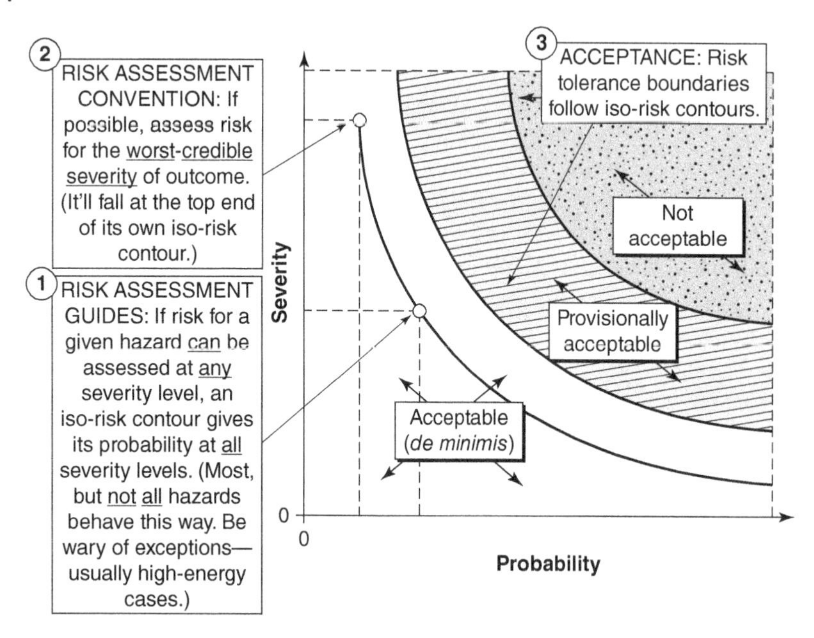

Figure 20.3 Iso-Risk Contours Usage

Risk should be evaluated for the worst credible case, not worst conceivable case, or conditions. Failure to assume credible (even if conceivable is substituted) cases may result in an optimistic analysis, which is a nonviable analysis.

The risk assessment matrix is typically used during the design and development phase, but may also be used in the conceptual trade studies phase. This technique is used as a predetermined guide or criterion to evaluate identified hazards. These risks are expressed in terms of severity and probability. Use of this tool allows an organization or firm to institute and standardize its approach for performing hazard analyses.

20.3.2 Procedures

The procedures for developing a risk assessment matrix are as follows:

1) Categorize and scale the subjective probability levels for all targets or resources, such as frequent, probable, occasional, remote, improbable, and impossible. Note that a *target* or *resource* is defined as the "what" that is at risk. One typical breakout of targets or resources is personnel, equipment, downtime, product loss, and environmental effects.
2) Categorize and scale the subjective severity levels for each target or resource, such as catastrophic, critical, marginal, and negligible.
3) Create a matrix of consequence severity versus the probability of the mishap (the event capable of producing loss). Approximate the continuous, iso-risk contour functions in the risk plane with matrix cells (see Fig. 20.4). These matrix cells fix the limits of risk tolerance zones. Note that management – not the analyst – establishes and approves the risk tolerance boundaries. Management will consider social, legal, and financial impacts when setting risk tolerance
4) The following hints are helpful for creating the matrix:
   a) Increase adjacent probability steps by orders of magnitude. The lowest step, "impossible," is an exception (see Fig. 20.5).
   b) Avoid creating too many matrix cells. Because the assessment is subjective, too many steps add confusion with no additional resolution (see Fig. 20.6).

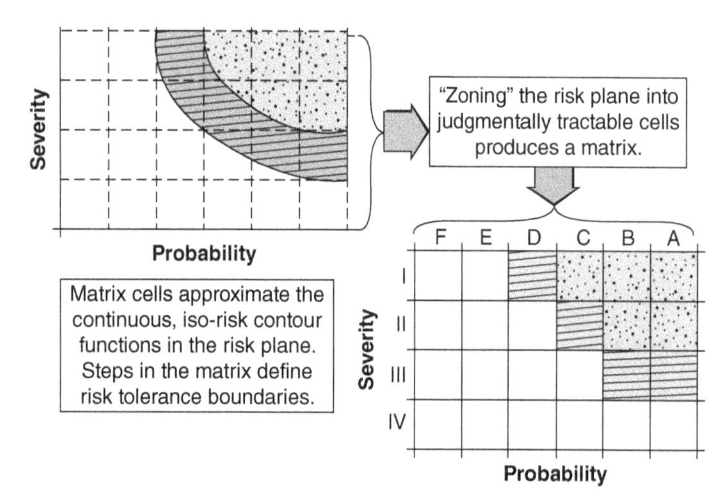

Figure 20.4 Risk Contours to Risk Matrix Transformation

c) Avoid discontinuities in establishing the risk zones, that is, make sure every one-step path does not pass through more than one zone (see Fig. 20.7).
d) Establish only as many risk zones as there are desired categories of resolution to risk issues, that is, (1) unacceptable, (2) accepted by waiver, and (3) routinely accepted (see Fig. 20.8).
e) Link the risk matrix to a stated *exposure period.* When evaluating exposures, a consistent exposure interval must be selected, otherwise risk acceptance will be variable. An event for which the probability of occurrence is judged as remote during an exposure period of three months may be judged as frequent if the exposure period is extended to 30 years. For occupational applications, the exposure period is typically 25 years. All stakeholders (management or the client) who participate in establishing the risk acceptance matrix must be informed of any changes to the exposure interval for which the matrix was calibrated.

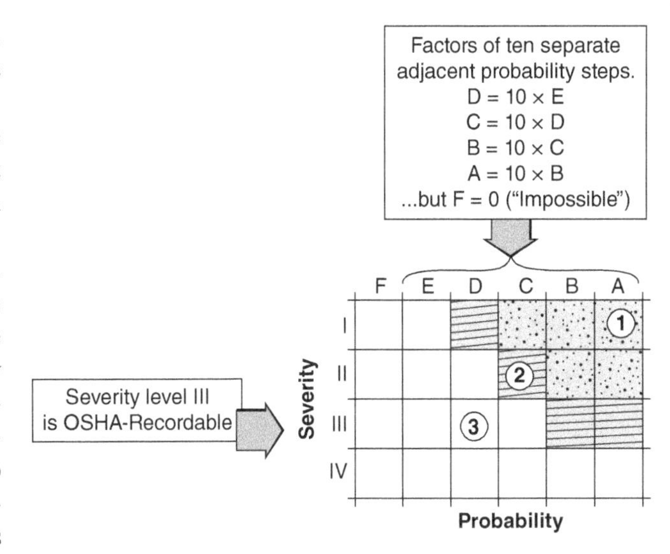

Figure 20.5 Useful Conventions

Figure 20.6 Creation of Matrix Cells

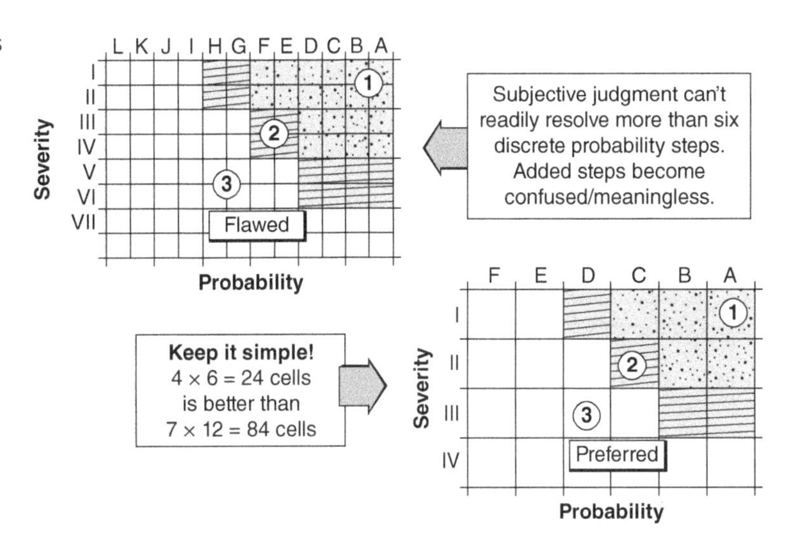

Figure 20.7 Avoidance of Discontinuities

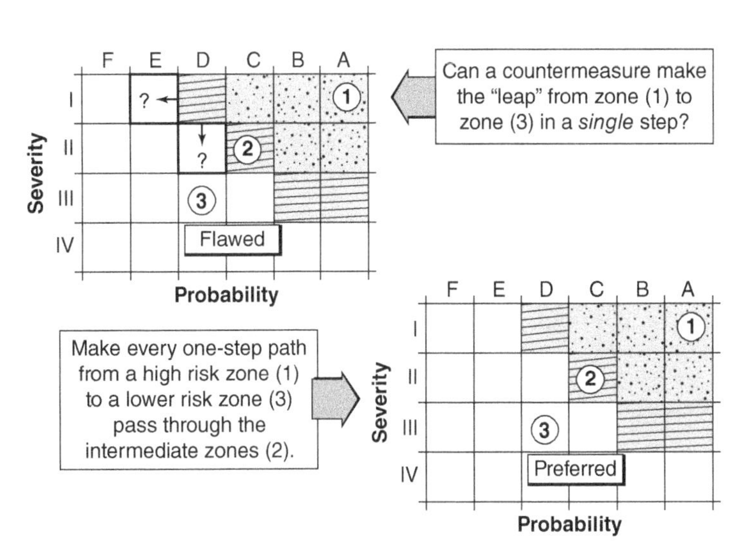

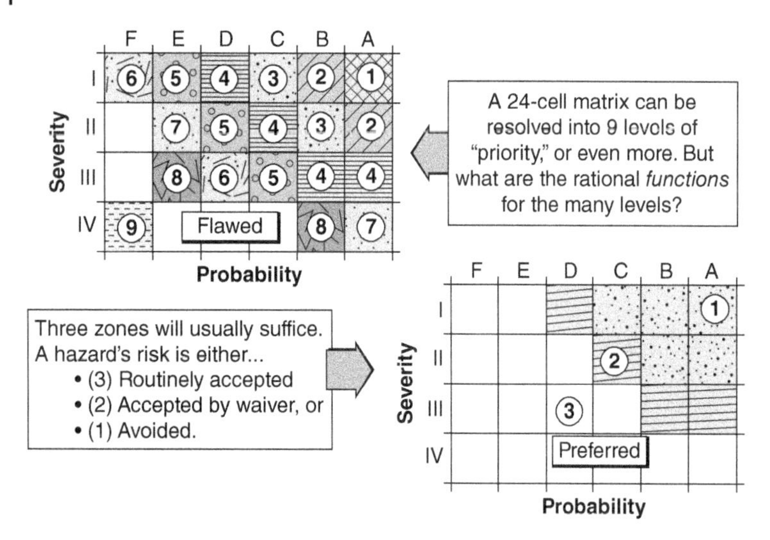

Figure 20.8 Number of Risk Zones

5) Calibrate the risk matrix by selecting a cell and attaching a practical hazard scenario to it. The scenario should be familiar to potential analysts or characterize a tolerable perceivable threat. Assign its risk to the highest level severity cell just inside the acceptable risk zone. This calibration point should be used as a *benchmark* to aid in evaluating other, less familiar risk.
6) An example for developing a risk assessment matrix is given in Example 20.1. Examples of strategies to manage harmful energy flow are listed in Table 20.1.

Table 20.1 Examples of Strategies to Manage Harmful Energy Flow

Strategy	Examples
Eliminate energy concentrations	• Control/limit floor loading • Disconnect/remove energy source from system • Remove combustibles from welding site • Change to nonflammable solvent
Limit quantity and/or level of energy	• Store heavy loads on ground floor • Lower dam height • Reduce system design voltage/operating pressure • Use small(er) electrical capacitors/pressure accumulators • Reduce/control vehicle speed • Monitor/limit radiation exposure • Substitute less energetic chemicals
Prevent energy release	• Heavy-wall pipe or vessels • Interlocks • Tagout-lockouts • Double-walled tankers • Wheel chocks
Modify rate of energy release	• Flow restrictors in discharge lines • Resistors in discharge circuits • Fuses/circuit interrupters
Separate energy from target in time and/or space	• Evacuate explosive test areas • Impose explosives quantity-distance rules • Install traffic signals • Use yellow no-passing lines on highways • Control hazardous operations remotely

Table 20.1 (Continued)

Strategy	Examples
Isolate by imposing a barrier	• Guard rails • Toe boards • Hard hats • Face shields • Machine tool guards • Dikes • Grounded appliance frames/housing • Safety goggles
Modify target contact surface or basic structure	• Cushioned dashboard • Fluted stacks • Padded rocket motor test cell interior • Whipple plate meteorite shielding • Breakaway highway sign supports • Foamed runways
Strengthen potential target	• Select superior material • Substitute forged part for cast part • "Harden" control room bunker • Cross-brace transmission line tower
Control improper energy input	• Use coded keyed electrical connectors • Use match-threaded piping connectors • Use backflow preventors

Example 20.1 Development of a Risk Assessment Matrix

Figure 20.9 shows a typical risk assessment matrix. Interpret the severity and probability steps for this matrix.

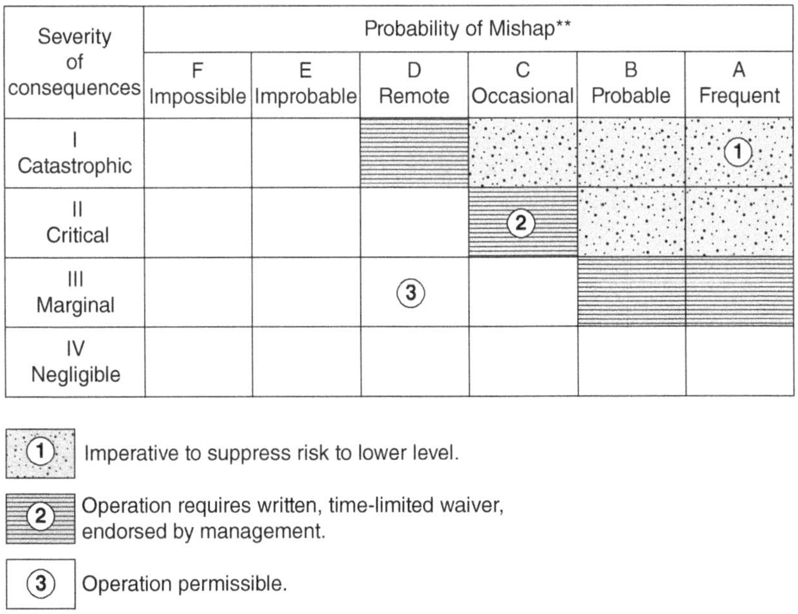

Severity of consequences	Probability of Mishap**					
	F Impossible	E Improbable	D Remote	C Occasional	B Probable	A Frequent
I Catastrophic						①
II Critical				②		
III Marginal			③			
IV Negligible						

① Imperative to suppress risk to lower level.

② Operation requires written, time-limited waiver, endorsed by management.

③ Operation permissible.

Note: Personnel must not be exposed to hazards in risk zones 1 and 2.

Figure 20.9 Typical Risk Assessment Matrix (Life Cycle = 25 yr)

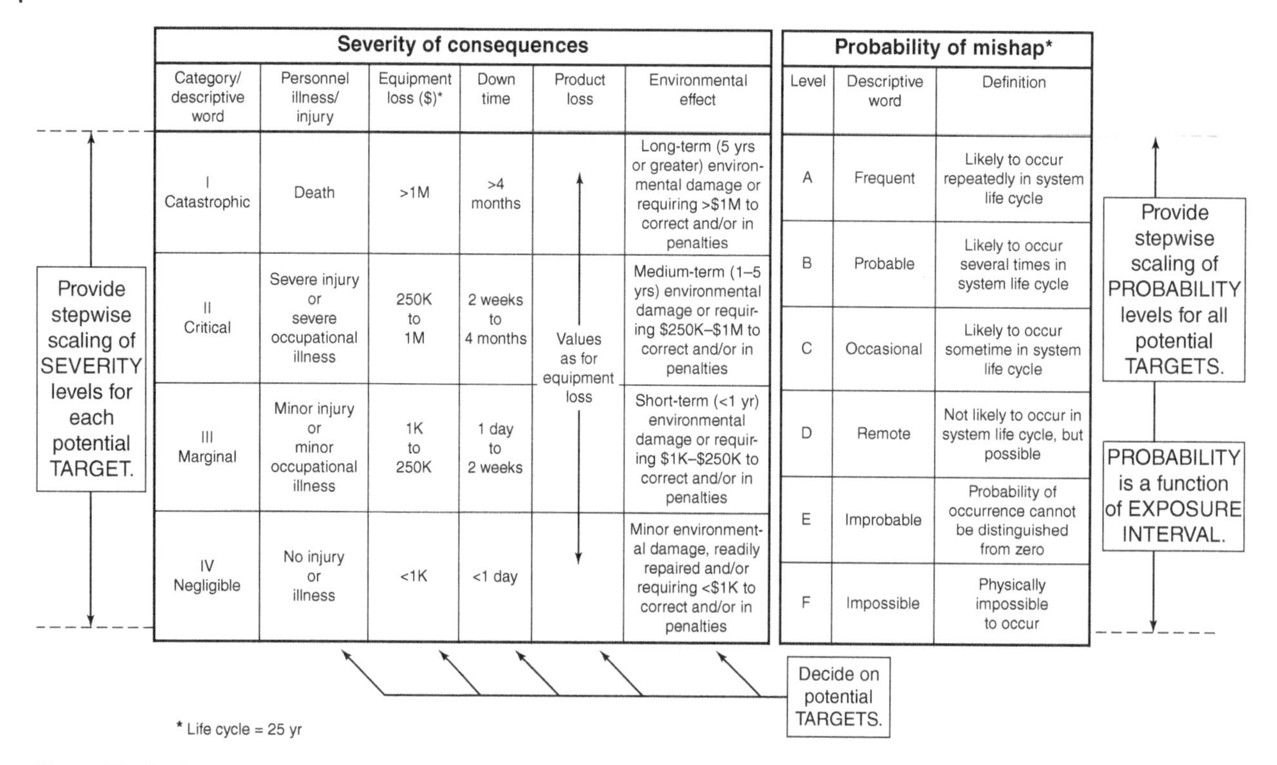

Figure 20.10 Severity and Probability Interpretations

Solution:

Figure 20.10 shows sample interpretations of the severity and probability steps for this matrix.

20.3.3 Advantages and Limitations of the Risk Matrix

The risk assessment matrix has certain advantages:

1) The risk matrix provides a useful guide for prudent engineering.
2) The risk matrix provides a standard tool for treating the relationship between severity and probability in assessing risk for a hazard.
3) Subjective risk assessment avoids unknowingly accepting intolerable and senseless risk, allows operating decisions to be made, and improves resource distribution for mitigation of loss resources.

The risk assessment matrix has the following limitations:

1) The risk assessment matrix can only be used if hazards are already identified; this tool does not assist the analyst in identifying hazards.
2) Without data, this method is subjective and is a comparative analysis only.

20.4 Failure Modes, Effects, and Criticality Analysis

20.4.1 Description

A *failure modes and effects analysis* (FMEA) is a forward logic (bottom-up), tabular technique that explores the ways or modes in which each system element can fail. It also assesses the consequences of each of these failures. In its practical application, its use is often guided by top-down screening to establish the limit of analytical resolution. A *failure modes, effects, and criticality analysis* (FMECA) also addresses the criticality or risk of individual failures. Countermeasures can be defined for each failure mode, and consequent reductions in risk can be evaluated. FMEA and FMECA are useful tools for cost/benefit studies to implement effective risk mitigation and countermeasures.

It is important to remember that the analytical techniques discussed in this chapter complement (rather than supplant) each other. It has long been sought, but there is no "Swiss army knife" technique that answers all questions and is suitable for all situations.

20.4.2 Application

A FMEA can call attention to system vulnerability to failures of individual components. Single-point failures can be identified. This tool can be used to provide reassurance that the cause, effect, and associated risk (FMECA) of component failures have been appropriately addressed. These tools are applicable within systems or at the system/subsystem interfaces and can be applied at the system, subsystem, component, or part levels.

These failure mode analyses are typically performed during the design and development phase. The vulnerable points identified in the analyses can aid management in making decisions to allocate resources in order to reduce vulnerability.

20.4.3 Procedures

Procedures for preparing and performing FMECAs are presented next. Procedures for preparing a FMEA are the same with the exception of steps 8 through 12 being omitted.

1) Define the scope and boundaries of the system to be assessed. Gather pertinent information relating to the system, such as requirement specifications, descriptions, drawings, components and parts lists. Establish the mission phases to be considered in the analysis.
2) Partition and categorize the system into convenient and logical elements to be analyzed. These system elements include subsystems, assemblies, subassemblies, components, and piece parts.
3) Develop a numerical coding system that corresponds to the system breakdown (see Fig. 20.11).
4) Identify resources of value to be protected, such as personnel, facilities, equipment, productivity, mission or test objectives, and environment. These resources are potential targets.
5) Identify and observe the levels of acceptable risk that have been predetermined and approved by management or the client. These limits may be the risk matrix boundaries defined in a risk assessment matrix.
6) By answering the following questions, the scope and resources required to perform a classic FMEA can be reduced, without loss of benefit:
   - Will failure of the *system* render an unacceptable or unwanted loss?

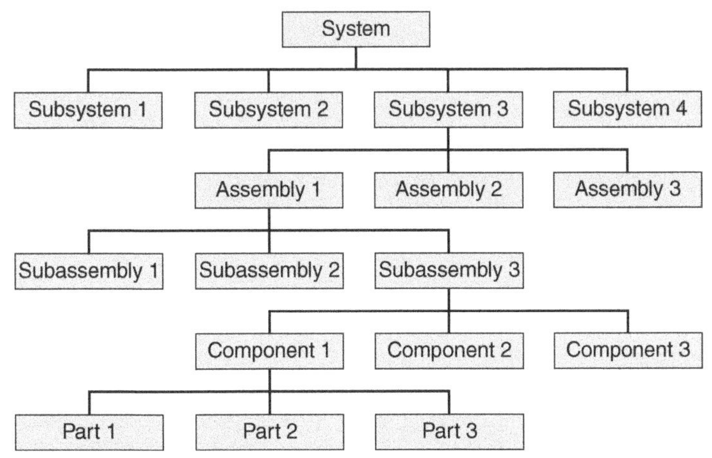

Figure 20.11 Example of System Breakdown and Numerical Coding

If the answer is no, the analysis is complete. Document the results. (This has the additional benefit of providing visibility of non-value-added systems, or it may correct incomplete criteria used for the FMEA.) If the answer is yes, ask the following question for each subsystem identified in step 2:

- Will failure of this *subsystem* render an unacceptable or unwanted loss?

  If the answer for each subsystem is no, the analysis is complete. Document the results. If the answer is yes for any subsystem, ask the following question for each assembly of those subsystems identified in step 2:
- Will failure of this *assembly* render an unacceptable or unwanted loss?

  If the answer for each assembly is no, the analysis is complete. Document the results. If the answer is yes for any assembly, ask the following question for each component of those assemblies identified in step 2:
- Will failure of this *subassembly* render an unacceptable or unwanted loss?

  If the answer for each subassembly is no, the analysis is complete. Document the results. If the answer is yes for any subassembly, ask the following question for each component of those subassemblies identified in step 2:
- Will failure of this *component* render an unacceptable or unwanted loss?

  If the answer for each component is no, the analysis is complete. Document the results. If the answer is yes for any component, ask the following question for each part of those components identified in step 2:
- Will failure of this part render an unacceptable or unwanted loss?

7) For each element (system, subsystem, assembly, subassembly, component, or part) for which failure would render an unacceptable or unwanted loss, ask and answer the following questions:
   - What are the failure modes for this element?
   - What are the effects (or consequence) of each failure mode on each target?
8) Assess worst credible case (not the worst conceivable case) severity and probability for each failure mode, effect, and target combination.
9) Assess the risk of each failure mode using a risk assessment matrix. The matrix should be consistent with the established probability interval and force or fleet size for this assessment.
10) Categorize each identified risk as acceptable or unacceptable.
11) If the risk is unacceptable, then develop countermeasures to mitigate it.
12) Then reevaluate the risk with the new countermeasure installed.
13) If countermeasures are developed, determine if they introduce new hazards or intolerable or diminished system performance. If added hazards or degraded performance are unacceptable, develop new countermeasures and reevaluate the risk.
14) Document your completed analysis on a FMEA or FMECA worksheet. The contents and formats of these worksheets vary among organizations. Countermeasures may or may not be listed.

Figure 20.12 represents a flowchart for FMEA or FMECA. Table 20.2 presents a sample FMEA worksheet.

Example 20.2 Assessment of Risk
Assess the risk for an automated mountain climbing rig.

Solution:

A sample FMECA is illustrated in Fig. 20.13. The system being assessed is an automated mountain climbing rig. Table 20.3 illustrates the breakdown and coding of the system into subsystem, assembly, and subassembly elements. A FMECA worksheet for the control subsystem is presented in Table 20.4.

20.4.4 Advantages and Limitations

Performing FMEA and FMCEA has the following advantages:

1) Provides an exhaustive, thorough mechanism to identify potential single-point failures and their consequences. A FMECA provides risk assessments of these failures.
2) Results can be used to optimize reliability, optimize designs, incorporate "fail-safe" features into the system design, obtain satisfactory operation using equipment of "low reliability," and guide in component and manufacturer selection.
3) A FMECA can be a very thorough analysis suitable for prioritizing resources to higher risk areas if it can be performed early enough in the design phase.

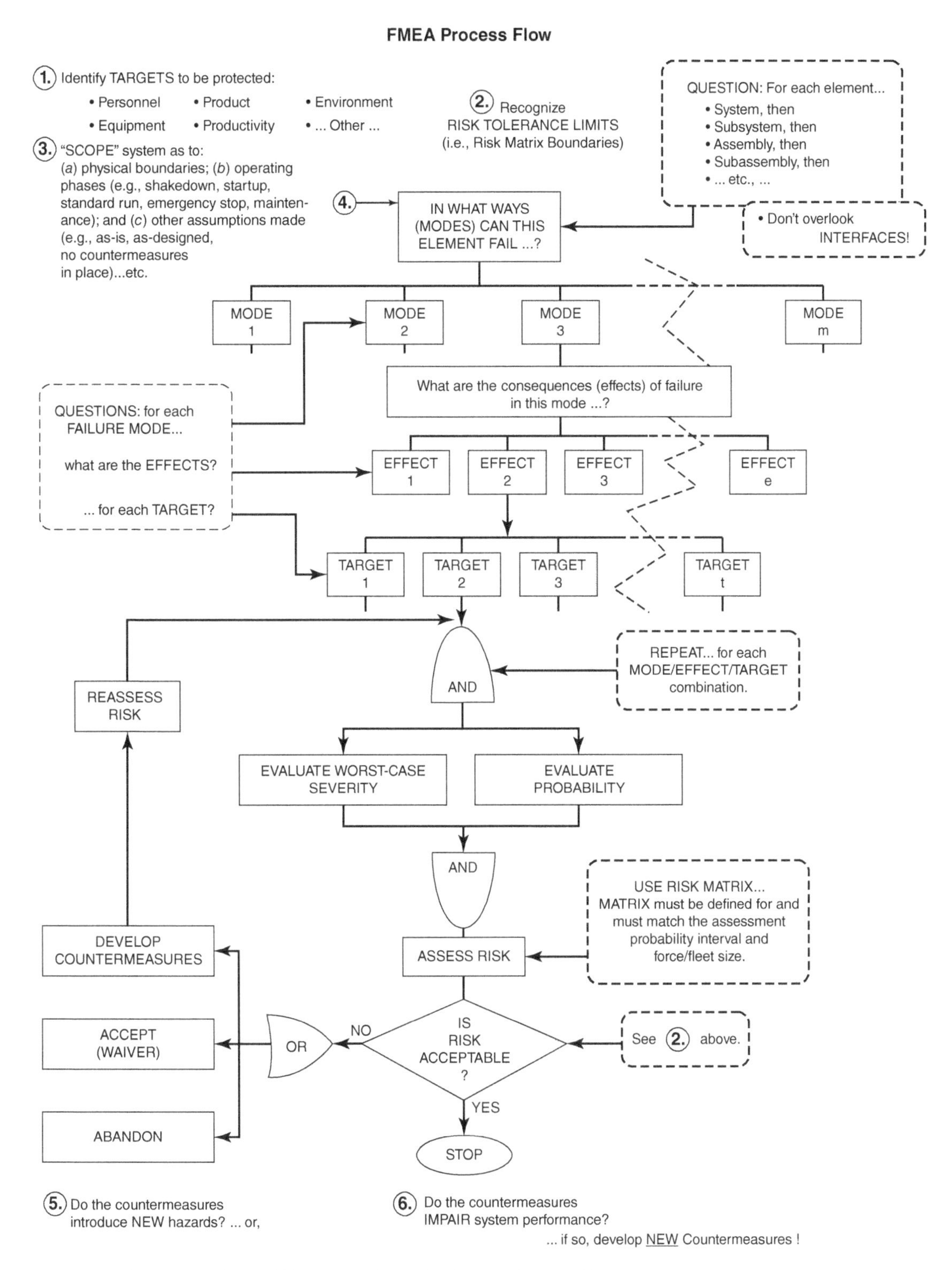

Figure 20.12 Failure Modes, Effects, and Criticality Analysis Process Flowchart

Table 20.2 Typical Failure Modes, Effects, and Criticality Analysis Worksheet

FMEA NO.: ______ Project No.: ______ Subsystem No.: ______ System No.: ______ PROB. Interval: ______	Failure Modes, Effects, And Criticality Analysis Worksheet	Sheet ______ of ______ Date ______ Prepared By: ______ Reviewed By: ______ Approved By: ______
Target/Resource Code: P-Personnel/E-Equipment/ T-Downtime/R-Product/D-Data/V-Environment		

Id. No.	Item/ Functional Identification	Failure Mode	Failure Cause	Failure Event	Target	Risk Assessment			Action Required/ Comments
						Sev	Prob	Risk Code	

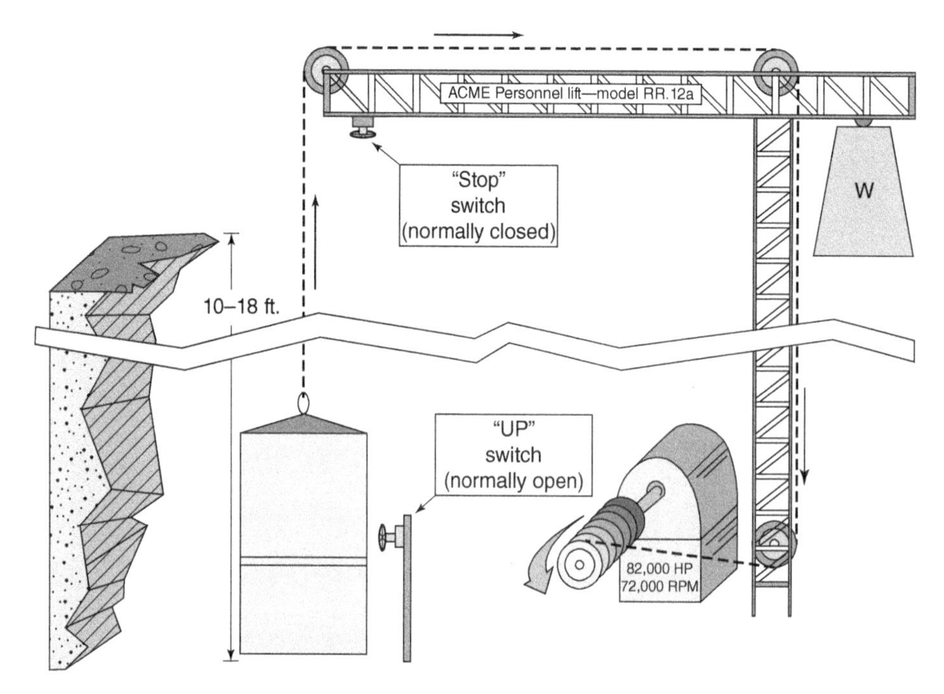

Figure 20.13 System for Example 20.2

Table 20.3 System Breakdown and Coding for Example 20.2

Subsystem	Assembly	Subassembly
Hoist (A)	Motor (A-01)	Windings (A-01-a) Inboard bearing (A-01-b) Outboard bearing (A-01-c) Rotor (A-01-d) Stator (A-01-e) Frame (A-01-f) Mounting plate (A-01-g) Wiring terminals (A-01-h)
	Drum (A-02)	
External power source (B)		
Cage (C)	Frame (C-01) Lifting lug (C-02)	
Cabling (D)	Cable (D-01) Hook (D-02) Pulleys (D-03)	
Controls (E)	Electrical (E-01)	START Switch (E-01-a) FULL UP LIMIT Switch (E-01-b) Wiring (E-01-c)
	Operator (E-02)	

Table 20.4 Failure Modes, Effects, and Criticality Analysis for Example 20.2

FMEA No.:________ Project No.:________ Subsystem No.: Controls System No.: Mountain Climbing Rig Prob. Interval: 30 years	Failure Modes, Effects, and Criticality Analysis Worksheet	Sheet ____ of ____ Date________ Prepared By:______ Reviewed By:______ Approved By:______

Target/Resource Code: P-Personnel/E-Equipment/T-Downtime/R-Product/D-Data/V-Environment

Id. No.	Item/ Functional Identification	Failure Mode	Failure Cause	Failure Event	Target	Risk Assessment			Action Required/ Comments
						Sev	Prob	Risk Code	
E-01-a	Start Switch	Switch fails closed.	Mechanical failure or corrosion.	Cage will not move.	PET	IV IV IV	C C C	3 3 3	
E-01-b	Full Up Switch	Switch fails open.	Mechanical failure or corrosion.	Cage does not stop.	P	II	A	1	
E-02	Wiring	Cut, disconnected.	Varmint invasion, faulty assembly	No response in a switch. Start switch fails open. Stop switch fails closed. Cage staysin safeposition.	PET	IV IV IV	D D D	3 3 3	

The following limitations are imposed when performing FMEAs and FMECAs:

1) These analyses are costly in terms of man-hour resources, especially when performed at the parts-count level within large, complex systems.
2) Probabilities or the consequences of system failures induced by coexisting, multiple-element faults or failures within the system are not addressed or evaluated.
3) Although systematic, and guidelines/checksheets are available for assistance, no check methodology exists to evaluate the degree of completeness of the analyses.
4) These analyses depend heavily on the ability and expertise of the analyst for finding all necessary modes.
5) Human error and hostile environments frequently are overlooked.
6) Failure probability data are often difficult to obtain for a FMECA.
7) If too much emphasis is placed on identifying and eliminating single-point failures, then focus on more severe system threats (posed by coexisting failures/faults) may be overlooked.

20.5 Engineering and Design for Safe Construction

20.5.1 Construction Failures

Tragic construction failures, especially those that involve fatalities and substantial economic losses, often receive wide media coverage and bring public outcry for preventive actions. Such incidences have frequently led to federal investigations, and controversy and debate within the civil engineering and construction community, and have prompted legislative initiatives. The following are a few of the most infamous examples of construction failures, which have caused great losses and generated significant public and professional interest.

20.5.1.1 L'Ambiance Plaza Condominium Collapse

On April 23, 1987, L'Ambiance Plaza, a 16-story apartment building in Bridgeport, Connecticut, collapsed during construction. The lift slab method was being used in this project to erect the structure. The collapse resulted in 28 fatalities and 18 injuries, which made L'Ambiance the most catastrophic construction accident of the 1980s. An investigation was performed by the National Bureau of Standards (NBS; now called NIST, the National Institute for Standards and Technology) to determine the most probable cause of this collapse. This investigation concluded that the most probable technical cause of the collapse was the failure of some lifting assembly components, designed and erected by the lift slab contractor and its subcontractor.

Several subsequent forensic investigations were also conducted on this accident, identifying design deficiencies and suggesting various alternative failure mechanisms. Many experts involved in this work believe that the fragmentation of responsibility, the delegation of structural design to the contractors, and poor communication between the parties were the underlying reasons that led to the collapse of L'Ambiance Plaza.

20.5.1.2 Willow Island Cooling Tower Collapse

In April 1978, the hyperbolic cooling tower of the Pleasants Power Station near Willow Island, West Virginia, partially collapsed during construction, resulting in the death of 51 construction workers. The patented jump form scaffolding, which was supported by the partially completed concrete shell, collapsed as the connecting anchors pulled out of the partially matured concrete, 1 day after the pouring. This accident brought to the nation's attention the potential hazards of temporary structures and the potential limitations of construction materials.

20.5.1.3 Kansas City Hyatt Regency Walkway Collapse

Although this failure did not occur during construction, an important lesson may be learned from the errors that led to it. On July 17, 1981, the second- and fourth-floor hanging walkways spanning across the atrium of the Hyatt Regency Hotel in Kansas City, Missouri, collapsed and fell to the ground floor during a social event. Described as "the most devastating structural collapse ever to take place in the United States," this failure took the lives of 114 people and injured 200 more. According to NBS, which performed an investigation on the failure, the cause of the collapse was inadequate connection details between the steel box beams and the hanging rods.

It was found that without the consent of the engineer of record, these critical details were altered by the steel fabricators in such a way that the main source of support for the second-floor walkway was transferred from the roof to the fourth-floor walkway, creating an overload situation. This flawed revision was not noticed by the engineer of record, who checked the shop drawings only for conformance to the original design. This failure raised the question of responsibility for the shop drawings and the design of structural details, causing the engineering profession to start reconsidering the traditional practices with regard to structural details, particularly those that do not conform to the American Institute for Steel Construction standards.

20.5.2 Causes of Construction Failures

Prevention of construction failures, first and foremost, requires an understanding of the causes. Many efforts have been directed toward this end by engineers, forensic scientists, the legal community, and legislators. In the early 1980s, this problem was addressed by the Investigations Subcommittee of the House Committee on Science and Technology through a study aimed at determining the underlying causes of the construction failures, and the possible related shortcomings in technology and education. This investigation was based on extensive interviews with construction engineers focusing on the understanding of the actions that take place during construction. The findings of this study were summarized as follows:

1) The organization of a construction project is unique since it brings together many parties for a limited period of time.
2) There is little evidence of the presence of consistent methods of maintaining overall project quality.
3) The structural engineer in charge of design does not perform all of the design work; important details are implemented by steel fabricators.
4) The structural engineer, who is the logical person to inspect a structure, is discouraged from doing so.
5) Recent engineering graduates are well trained in mathematics and computers, but are relatively unsophisticated about the behavior of structures.
6) Long-span structures require careful quality control, and there is a need for new approaches emphasizing dynamic behavior analysis.

Eldukair and Ayyub (1991) conducted a comprehensive study of the causes of construction/structural failures. They reviewed 604 cases that occurred in the United States between 1975 and 1986. They discovered the following:

1) Nearly 57% of failures resulted in collapse and 4% resulted in other unsafe conditions.
2) A critical percentage (44%) of the structural failures occurred during the construction phase.
3) Errors in design and construction phases of the projects were the most predominant sources of failures. Design errors occurred in 51% of the total failure cases; 57% of the total failure cases recorded deficiencies in construction procedures.
4) Structural designers were responsible for the errors in 40% of the cases. Contractors' site staff were accountable for errors in 60% of the total failure cases.
5) Deficiencies in the construction procedures primarily involved inadequate construction methods, ineffective evaluation of laws and safety regulations, and inadequate planning and supervision.
6) Errors in management practices had a tremendous effect on the performance of project activities, the schedule, and safety. These included deficiencies in work responsibilities, deficiencies in the communication process, and insufficient work cooperation. Errors in defining work responsibilities were dominant (30%) as part of the deficiencies in management practices.
7) A great majority of the cases (86%) indicated that reinforced concrete elements were involved in the failures, and the highest number of failed members (34%) was slabs and plates.
8) The primary causes of most failures were poor erection procedures and inadequate connection elements leading to inadequate load behavior, followed by unclear contract information, and contravention of information. A small percentage of failures were attributed to unforeseen events.
9) Secondary causes of failures were led by environmental effects (mostly poor weather conditions), followed by poor supervision and control, poor communications, poor material and equipment usage, poor workmanship, and other minor technical and nontechnical factors.

The study found that the cases examined caused 416 total deaths and 2,515 total injuries. The total direct costs of damage associated with the studied structural and construction failures in 1986 were estimated at USD 3.5 billion (5.6 billion in 2008 USD). The hidden indirect costs were excluded from the estimate because of the difficulty of determining them.

20.5.3 Classification of Causal Factors

The above-mentioned studies and others have shed considerable light on some of the common causes of construction failures. Based on these considerations, it is possible to summarize the causal factors governing construction failures into five major classifications:

1) Natural disasters
2) Engineering and design considerations
3) Material properties and performance
4) Construction methods and procedures
5) Contract administration and project management.

20.5.3.1 Natural Disasters

Natural phenomena such as earthquakes, hurricanes, violent storms, and tornadoes often cause enormous destruction to buildings and other structures during construction or service, leading to loss of life and property damage of significant magnitude. Although most structures can be designed and constructed to withstand such overwhelming forces of nature, some damage is often unavoidable (hence the term "act of God"). However, a natural disaster may also reveal a latent deficiency in design, construction techniques, or code requirements.

An example from 1992 is Hurricane Andrew, which resulted in the costliest disaster in US history. This violent storm with gusty winds exceeding 150 mph (240 km/h) devastated many buildings, while revealing critical deficiencies in construction practices in the South Florida area. This fact became particularly evident in residential construction, where roof failures caused by inadequate connections resulted in the total destruction of many wooden frame houses. Experts pointed to cost-cutting measures by contractors as a contributing factor for the great damage caused by Hurricane Andrew.

Such disasters may expose critical deficiencies in structures, raising concerns for public safety. The engineering community should recognize the threat of natural disasters in geographic areas susceptible to them, and propose corrective measures through code revisions to eliminate or mitigate the potential failures.

20.5.3.2 Engineering and Design Considerations

Engineering and design for construction involve complex processes, which blend technical expertise with skillful project organization and management. In addition, with respect to design and construction knowledge, an understanding of government regulations and building codes is necessary, along with effective communications with various groups involved in the construction process.

An important prerequisite to safe design and engineering is the consideration of *constructability*. The constructability task force of the Construction Industry Institute (CII) has defined constructability as the optimum integration of construction knowledge and experience in planning, design, procurement, and field operations to achieve overall project objectives. This approach will lead to greater productivity, lower costs, and improved quality and safety.

According to CII, the design process must integrate construction needs from the conceptual stage to final delivery of contract documents. Constructability analysis of the design can identify conflicts and shortcomings that may complicate construction. It is therefore beneficial to integrate such an analysis into the project execution plans starting with the preliminary design stage. Meaningful input from the constructors, construction managers, or constructability consultants can be very beneficial in the early stages of a project. Several critical decisions are made during the schematic design stage that will determine many subsequent construction activities. Factors such as availability of raw materials and skilled labor, costs, and the overall project schedule can be incorporated into these decisions. It might be assumed that the project schedule is construction-sensitive, and the general construction methods might consider the impact of seasonal weather conditions, site characteristics, drainage, and soil conditions.

Constructability reviews can be applied in more depth during the design development and construction document preparation phases by incorporating the following objectives and principles:

- Design configurations should enable efficient construction and eliminate spatial constraints, which complicate proper assembly, especially for large structural elements.
- Sequence of design should follow construction logic and take into account feasible construction time. For example, builders need to realize that certain slabs that are designed for service loads may not support the construction loads.

- Complicated construction details should be minimized, especially if they require extensive on-site assembly. Standardization and preassembled design should be maximized.
- Construction task interdependencies that would bring many trades together should be minimized.
- Flexibility for field adjustments during implementation should be allowed, and tight tolerances should be eliminated. An example of this would be certain structural members that have adequate depth to support the service loads not allowing for the necessary connections.
- The site layout should consider accessibility, and the contract documents should contain provisions for safe access for workers and storage materials and equipment.
- The construction team should provide input and advice on access needs, major equipment placement, and storage requirements.
- Potential conflicts between underground work and heavy equipment movement should be eliminated.
- The design and erection sequence should consider the stability of incomplete structural frames during construction.
- If the project needs to be completed during adverse weather conditions, design procurement should facilitate an early enclosure of the structure to minimize the exposure to harsh weather. Specifications should promote the use of weather-compatible materials and limit field work by using preassembled modules.

These principles imply that the consideration of constructability issues during design can eliminate difficulties in execution of construction tasks and avoid unsafe conditions on the job site. Also, all document conflicts and discrepancies of information caused by the possible lack of coordination between different sections of the construction documents should be eliminated. Such conflicts and gaps may go unnoticed and create constructability problems (such as incompatibilities between structural members and mechanical elements) when the project has reached an advanced stage.

20.5.3.3 Material Properties and Performance

Many structural/construction failures are caused by material deficiencies. Materials, in and of themselves, may not be defective; however, their inappropriate selection or utilization in the construction process may lead to failures. Understanding the nature and properties of the materials, their structural behavior, and their strength and durability performance in the service environment is vital to the successful performance of the structure and the avoidance of failures. For instance, freeze/thaw effects and shrinkage problems in concrete are well known. However, strength development (maturity) is somewhat less well understood. The collapse of the Willow Island cooling tower mentioned earlier was primarily attributable to incomplete curing of the concrete, resulting in insufficient material strength at the time of formwork removal.

Building products have been continually evolving and their changing attributes may present a challenge to the designers. New materials, as well as technological advancements in traditional materials, create the possibility of not fully understanding their potential limitations. For instance, the performance of plastics, which are gaining acceptance in the construction industry, has not been fully tested for all service conditions. Improved properties of steel, concrete, and wood can enhance structural performance, but may also present some problems. For instance, higher strength steel facilitates greater space availability, but it may offer reduced ductility.

The effects of environmental conditions on materials are another important concern. For example, metals under certain conditions are subject to corrosion, despite conventionally adequate safeguards and quality control. Galvanic action between certain metals is a common cause of corrosion. Metal fatigue may create cracks that weaken the material and facilitate chemical reactions promoting corrosion. Contrary to common belief, concrete does not always provide adequate protection to reinforcing steel. Moisture can reach the steel through the cracks (and sometimes pores) in concrete causing corrosion, which can lead to total deterioration. Chlorides also promote corrosion in steel. They can come from deicing salts, or from chloride additives used in high-bond mortar.

Incompatibility of materials that come in contact in the constructed form may also cause deterioration. The different materials that form the composite members of a structural member may have incompatible tolerances, strengths, and ductilities. These differences can produce unanticipated stress redistributions. Incompatibility may also lead to differential movements resulting from temperature and moisture gradients. Different materials have varying coefficients of expansion. For instance, brick, limestone, and concrete have uniquely different coefficients, and under the same temperature gradient, they may expand and contract at different rates, leading to cracks.

Some of the most important decisions regarding design and constructability reviews involve material selection. Engineers and architects must specify the materials, approve substitutions (as may be required by value engineering analysis), and test the quality of the material during construction. These tasks must take into account all exposures, compatibility, and the limits of performance. In the selection of materials, manufacturer's design and performance data must be scrutinized, and qualitative information should be clearly documented.

20.5.3.4 Construction Methods and Procedures

In present practice, the construction methods and procedures by which a facility or project is erected are considered to be the responsibility of the contractor. In a great majority of construction projects, constructors are proficient and possess sufficient knowledge and experience in their work. However, some construction situations may require sophisticated design decisions at the time of erection, and some constructors may attempt to make such decisions on their own despite a possible lack of capacity to do so. This may adversely affect the project's safety.

Temporary structures cover those structures temporarily erected to assist in the construction operations and are central to shoring, formwork and slip forms, scaffolding, ramps and platforms, bracing and guying, cofferdams, etc. Contract documents prepared by the engineers may specify the design of the temporary structures, or may require that the contractor develop the necessary temporary structures. In the latter case, the contractor generally assumes responsibility for their integrity.

Scaffold failure is a common type of accident associated with temporary structures as demonstrated by many case histories. A scaffold resting on concrete blocks collapsed in Chicago killing 12 workers. Insufficient strength of the concrete blocks was determined as the cause of collapse. Fourteen injuries were reported on another accident in New Jersey when a scaffold collapsed pulling over a 12-ft (3.7-m) high masonry wall to which it was tied. Overloading with construction materials, coupled with poor base support of the planks, was the apparent cause of failure. In a gymnasium remodeling project in Illinois, the roof collapsed when the scaffold failed to support the roof after the supporting beams were removed. The court decided that the designers were at fault in that they did not determine the scaffold's ability to support the roof since the scaffold was intended to be used as a shoring system. These cases underscore the importance of sound design and construction of scaffolds.

In a legal dispute, opposing parties may present expert witnesses (often engineers) who may testify about the competence of services as compared to that of another reasonable/prudent engineer in the same community. It is the expert's duty to perform the research necessary to establish the facts about the case and be totally impartial, although he or she is paid by a plaintiff or defendant. Thus, when experts disagree about given issues, as is commonly the case, the disagreement should stem from differing judgments and not from their acting as advocates for their clients.

Privity indicates direct contractual relationships. In the past, lack of privity protected design engineers from liability for third-party claims (mainly for injuries from construction accidents). It was understood that engineers were not obligated to protect the workers from safety hazards since they had no contract with them. However, increasingly, the engineer's liability is determined by the tort concept of foreseeability. In this context, if a party is negligent in the performance of its duties, it may be held liable to any party who suffered harm that was foreseeable at the time duties were undertaken.

The doctrine of *strict liability* generally applies to manufacturers of a product. It is far more stringent than the doctrine of professional liability, since negligence does not have to be proved. However, the following must be shown:

- There was a defect in the product.
- The defect existed at the time the product was transferred to the purchaser/user.
- The defect caused or contributed to an injury.
- The product failed in normal use.

Claimants frequently assert that the engineer should also be held strictly liable in tort for injuries caused by allegedly defective designs of inherently dangerous structures. However, several courts have rejected the strict liability doctrine in application to engineers, viewing the science of engineering and construction as inexact, in that engineers are not expected to guarantee the satisfactory results of their services.

20.5.3.5 Contract Administration and Project Management

Contract administration encompasses all activities from the beginning to the end of a construction project relating to the execution of the project tasks as provided for by the contract documents. Shop drawings and change orders are the two most important aspects that may impact construction safety. They are discussed in the following paragraphs.

Shop drawings are documents prepared by various subcontractors who will perform specific construction tasks. They usually show the specific components of the constructed system and the details of the erection procedure. Shop drawings have two critical purposes. First, they constitute the contractor's interpretation of the design documents. Secondly, they provide the freedom to contractors (especially in complex projects) to propose construction details. This is particularly necessary where it is not economically feasible to include all of the necessary details on the contract documents. Note that, in many cases, suppliers and subcontractors may have greater expertise in materials, construction methods, and familiarity with some new technologies than the designer. This freedom generally results in a competitive bid since the contractor is not forced to comply with specific methods in which he may not be totally proficient. Shop drawings are reviewed by both the contractors and the designers. They are reviewed more critically by the responsible designers if the case concerns structural details. Usually shop drawings receive precedence over contract documents and, thus, become the sole source of information and guidance for the construction team. Any errors or omissions on shop drawings can lead to inadequate construction, which may result in failures.

Deficiencies in shop drawing submittal and processing practices can jeopardize safe and efficient construction. Three crucial factors must be observed to minimize such problems, namely

1) Efficient planning and time management
2) Adequate shop drawing reviews
3) Good communication between the designer and the contractor.

Effective communication is perhaps the most critical factor in preventing failures. In practice, there is often some overlap of design responsibility between the engineer/design professional and the contractor/subcontractor. This is especially common in certain areas like steel connections or precast concrete construction. It is important that the designers develop excellent communications with the contractors, and clearly express their intentions. Also, the contract documents must clearly define the extent of the contractors' responsibilities to eliminate any misunderstandings.

The Hyatt Regency walkway collapse generated considerable debate with regard to the inherent dangers of the traditional practices concerning shop drawing reviews. In approving the shop drawings, design engineers generally do not assume responsibility for the accuracy of the work represented in the drawings. They simply verify that the information submitted is in conformance with the design. In fact, the reviewed shop drawings usually contain a statement that defines the limits of the engineer's professional liability in terms of the scope of evaluation and the resulting action expected from the review. On the other hand, some contractors view the engineer's approval as assumption of responsibility for the outcome of the shop drawing. The fact that structural engineers for the Hyatt Regency walkways were found liable indicates that the engineer's role and responsibility go beyond just reviewing the shop drawings to check for conformance to the design concept.

A change order is essentially a communication tool that is issued by designers to the contractors/subcontractors during the construction phase to alter the course of a project with owner's approval. The necessity for a change order may arise from unexpected site conditions, unavailability of materials or equipment, potential design flaws, changes in owner's requirements and other relevant factors.

Change orders may have profound impacts on the project budget and schedule, often imposing additional demands and liability for cost of the related design and construction. This fact, coupled with time constraints and pressures during construction, can cause the designers to pay insufficient attention to the preparation of change orders. Yet, change orders become a part of the contract documents, and any errors or omissions on them actually can become design flaws, which might have an impact on the safety of the constructed facility and the site personnel. Change orders should contain clearly written instructions supported by complete and accurate drawings to avoid confusion and poor guidance on the project changes.

20.6 Construction Safety and Health Management

Managers responsible for construction projects must recognize that safety and health must be managed in the same way as cost, schedule, quality, and productivity for effective loss control. Management of safety and health deals with managerial decisions and actions taken at all levels to produce an organizational setting in which employees are trained, motivated, and supervised to perform work free from injury and illness. Development and implementation of a company safety and health program are key to good management. Safety and health programs are most effective when they are designed to meet the specific and individual needs of each company. It may even be useful to develop one for each project.

20.6.1 Safety and Health Program Elements

Depending on the sophistication of the contracting firm, the nature of the project, and owner requirements, a safety and health program can be relatively plain and simple, or highly detailed. In this regard, there is not a unique set of elements for a company safety and health program. The key elements believed to be common to successful programs are listed and summarized here.

Research has shown that safety starts at the top. Companies in which the chief executive has made a strong commitment to safety and health and communicates this concern to employees by word and deed have better safety records than companies for which this is not true. Although chief executives do not normally supervise construction workers on the jobsite, the image they project and the behavior they display on and off the jobsite can have a great impact on the company's safety performance. This can be achieved by

- Creating an organizational culture in which safety and health are top priority
- Holding line managers accountable for the safety and health of their subordinates
- Focusing on and providing staff support and other necessary resources to help line managers meet their safety and health goals.

A written safety and health policy statement issued by top management is the best way to show the commitment to safety and health, to start the communication of goals and expectations, and to set the ground rules for the delivery of a successful program.

Clear goals for the safety and health program need to be set at the corporate and the project level. These goals need to be communicated to all employees. Goals may include zero fatalities and permanent disability causing injuries, prevention of major fires, vehicle accidents, etc., and full compliance with OSHA standards. The goals set for the program should be measurable to assess whether they are being successfully met.

An organizational and administrative structure should be established to achieve the best results from a safety and health program through line management and support staff. This is possible through setting an effective chain of communication between all levels. There is no set rule as to the proper location of safety and health responsibility in an organization; it is company-specific, depending on the need. It is common, however, to have a safety and health department at both the home office of a company, as well as at the individual project level. Usually, the home office serves a staff function, while the field office provides a line function under the project superintendent.

It is important to recognize that safety on the project is everybody's responsibility. Following the commitment of top management comes the assignment of safety duties. Employees at all levels must share in these duties because each employee has a safety duty to every other employee. Safety assignments must be understood by all employees. The right of each employee to a safe workplace must be explained and delineated, along with the duty to help keep it that way. It is highly desirable to put the assignments in writing. These might include individual responsibilities for program accomplishment and/or compliance for the company's president, management, job site superintendents, safety staff, foremen, craft personnel, subcontractors, suppliers, architect/engineers, owner personnel, and visitors. Accountability should also be addressed stating how incentives are distributed or disciplinary actions will be taken.

Maintaining an effective communication network at the job site is key to successful safety and health management. Managers can accomplish this by the chain-of-command system, direct contacts, and group meetings. The chain-of-command system alone has drawbacks since it can lead to distorted or "filtered" information through the hierarchy, which may provide misleading data on which to act. Managers who walk through the job frequently and talk directly with the site personnel can create a medium of effective two-way communication and convey their safety and health priorities. It is better when they share information with workers and foremen and listen to their feedback, rather than just issuing orders. This way the integrity of the chain-of-command system is not undermined. It is also important for follow-up action to be taken to address the issues discussed with the site personnel. Otherwise, the interest in participation may dwindle.

The safety supervisor/director plays an important role in the implementation of the safety program. These professionals are usually in staff positions rather than line positions, because safety is a staff responsibility. They provide support to the line managers who actually control safety and health performance. Among their duties are

- Serving as consultants to management on technical and organizational aspects of safety and health
- Developing and participating in the training and worker orientation programs
- Assisting in safety and health planning, monitoring (job site supervision), and record keeping
- Keeping the organization up to date on safety and health regulations and related matters.

20.6.2 Project Safety Rules

All company safety rules should be published in a written form. Written rules are more easily enforced than unwritten ones, and provide readily available guidance for operational safety. These rules should include the penalties for noncompliance. Each current employee, as well as any new employee, should be furnished a copy of the safety rules.

Each contractor should have rules for the basic types of operations it performs. In some cases there should be rules for the different company divisions or crews. Also, these rules should be modified as site conditions or owner requirements dictate. These work practices may be simple, one-page handouts for specific operations such as erecting scaffolds, excavating trenches, or operating specific pieces of equipment. Depending on the situation, the safety rules may also be very involved based on a hazard analysis for critical tasks. In any event, there should be a list of general rules for everyone, such as those shown in Table 20.5. Rules must be concise and easy to understand. They are instructions to field personnel for safe working procedures; therefore, they must be in a format that can be easily implemented in the field.

20.6.3 Training and Worker Orientation

Training and worker orientation are necessary elements of an effective safety and health program. There are two types of training:

1) Training for specific tasks
2) General training in accident avoidance and prevention.

It is desirable to integrate these efforts and train employees in safety and health as part of craft training. There are several ways to provide employee training. Discussions at weekly toolbox meetings are an effective training method. Many insurance companies and trade associations, such as the Associated General Contractors, provide "toolbox talks" to assist employers in the selection and delivery of safety topics. Private subscription services are also available to provide new topics on a regular basis.

Many insurance companies conduct safety seminars on the jobsite as part of their services associated with the workers' compensation coverage. This type of training can be utilized effectively before employees start work or on specific hazardous operations. Local safety councils and trade associations may provide free or low-cost training programs. These programs are tailor-made to meet specific training requirements. Many academic institutions also offer extensive safety courses. Finally, safety consultants provide both training and training program evaluation services. All training efforts must be carried out by qualified persons who are capable of stimulating and motivating the trainees.

20.6.4 New Worker Orientation

A significant component of the company safety and health training program is new worker orientation. New construction workers, and workers starting on an entirely new activity without previous experience, are particularly susceptible to being injured. The objectives of the orientation program are to relieve beginner anxieties, indoctrinate the workers to the company's safety program, and teach safe work practices, which include the use of personal protective equipment. Each employee should be handed a set of company safety rules during orientation and should be encouraged to read and understand them thoroughly.

20.6.5 Accident Investigation and Record Keeping

Accident investigations can highlight problem areas, and help detect patterns of unsafe acts or conditions, which should be addressed in preventive efforts. In this regard, accident reports make excellent training tools. Every accident, including those without injury should be investigated as part of the company safety and health program. The record-keeping requirements are also an important element of the safety and health program. A thorough and continuous record-keeping system for recordable cases, accident investigations, safety training programs, test results, etc., provides managers with significant information and management capability for good decision making.

20.6.6 Safety Budget and Audits

A company truly interested in safety will include a budget covering safety program expenses such as costs of safety personnel, training, protective equipment, and first-aid station maintenance. An advanced program will include these safety costs showing them as part of direct labor. An enhanced program will be able to demonstrate these costs as part of operational or

Table 20.5 General Safety Rules

All of our safety rules must be obeyed. **Failure to do so will result in strict disciplinary action being taken.**
1. Keep your mind on your work at all times. No horseplay on the job. Injury or termination, or both, can be the result.
2. Personal safety equipment must be worn as prescribed for each job, such as safety glasses for eye protection, hard hats at all times within the confines of the construction area, and gloves when handling materials. Safety shoes are highly recommended for protection against foot injuries.
3. Shirts and long-legged pants must be worn to prevent sunburn and to protect against acid burns, steam burns, weld splatter, and cuts. Minimum clothing for the upper body is a T-shirt.
4. If any part of your body should come in contact with an acid or caustic substance, rush to the nearest water available and flush over the affected part. Secure medical aid immediately.
5. Watch where you are walking. Do not run.
6. The use of illegal drugs or alcohol or being under the influence of same on the project shall be cause for termination. If you take or are given strong prescription drugs that warn against driving or using machinery, let your supervisor know about them.
7. Do not distract the attention of fellow workers. To do so may cause injury.
8. Sanitation facilities have been or will be provided for your use. Defacing or damaging these facilities is forbidden.
9. A good job is a clean job and a clean job is a safe one. So keep your working area free from rubbish and debris.
10. Do not use a compressor to blow dust or dirt from your clothes, hair, face, or hands.
11. Never work aloft if you are afraid to do so, are subject to dizzy spells, or if you are apt to be nervous or sick.
12. Never move an injured person unless it is absolutely necessary. Further injury may result. Keep the injured person as comfortable as possible and utilize job site first-aid facilities until a doctor arrives.
13. Know where firefighting equipment is located and learn how to use it.
14. Learn to lift correctly—with the legs not the back. If the load is too heavy, GET HELP. Twenty percent of all construction-related injuries result from lifting materials.
15. Riding on loads, fenders, running boards, sideboards, and gates or with your legs hanging over the ends or sides of trucks will not be tolerated.
16. Do not use power tools and equipment until you have been properly instructed in safe work methods and become authorized to use them.
17. Be sure that all guards are in place. Do not remove, displace, damage, or destroy any safety device or safeguard furnished or provided for use on the job, nor interfere with the use thereof.
18. Do not enter an area that has been roped off or barricaded.
19. If you must work around power shovels, cranes, trucks, and dozers, make sure operators can always see you.
20. Never oil, lubricate, or fuel equipment while it is running or in motion.
21. Rope off barricade danger areas.
22. Keep away from the edge of cuts, embankments, trenches, holes, and pits.
23. Trenches must be shored or sloped to comply with the most stringent requirements. Keep out of trenches or cuts that have not been properly sloped or shored. Excavated or other material shall not be stored nearer than 3 ft (1 m) from the edge of any excavation.
24. Use the "four and one" rule when using a ladder. One foot of base for every four feet of height (25 cm of base for every meter of height).
25. Always secure the bottom of the ladder with cleats and/or safety feet. Lash off the top of ladder to avoid shifting.
26. Ladders must extend three feet (one meter) above a landing for proper use.
27. Defective ladders must be properly tagged and removed from service.
28. Keep ladder base free of debris, hoses, wires, material, etc.
29. Build scaffolds according to manufacturers' recommendations.
30. Scaffold planks must be cleated or secured to prevent them from sliding.
31. Use only extension cords of the three-prong type. Check the electrical grounding system daily.
32. The use of safety belts with safety lines when working from unprotected high places is mandatory. Always keep your line as tight as possible.
33. Tar kettles must be kept at least 25 ft (7.6 m) from buildings or structures and never on roofs.
34. Open fires are prohibited.
35. Know what emergency procedures have been established for your job site (location of emergency phone, first-aid kit, stretcher location, fire extinguisher locations, evacuation plan, etc.).
36. Notify your supervisor of unlabeled or suspect toxic substances immediately and avoid contact.

task performance, such as a percentage of each cubic yard of concrete poured. Contractors should be required to show that they are prepared to allocate resources for safety supervision, joint consultation, training, safety equipment, and other necessary safety program assets in relation to the size and nature of the project.

Safety and health audits measure the program's effectiveness. Audits include both field inspections and overall program evaluation. One of the better tools for enhancing safety programs is a system of frequent on-site inspections. These inspections, or field audits, provide immediate and tangible evidence of the performance level of a safety program. They also provide for direct observation of any present or developing site hazards. On-site inspections keep safety at the forefront and ensure a high degree of compliance with the rules. Safety audits should be an ongoing activity throughout the project.

Inspections may be performed by owner representatives, site superintendents, foremen, or the company safety staff. The advantages of management personnel performing the audits are as follows: (a) Management is primarily responsible for safety, (b) managers control work performance, and (c) supervisors should know the hazards on the job site. The main disadvantage is the possibility of a conflict of interest. There may be a tendency to overlook or minimize some situations that might imply negligence on management's part. This problem can be alleviated when safety professionals perform the audits.

Numerous items can be targeted for inspection and jobsite observation. From a hazard standpoint, these may include housekeeping and sanitation, fire prevention, electrical installations, hand and power tools, ladders, scaffolding, cranes, hoists and derricks, heavy equipment, motor vehicles, barricades, handling and storage of materials, excavation and shoring, demolition, hazardous materials, welding and cutting, personal protective equipment, concrete and masonry construction, and steel erection.

A comprehensive safety and health program audit may include the following key elements:

- Degree of management commitment
- Presence and effectiveness of the policy statement
- Program goal setting
- Definition of safety responsibilities
- Experience modification rating (over two to three year)
- Management supervisory meetings
- Preplanning for job site safety
- Effectiveness of training and orientation programs
- Accident investigation activities
- Record keeping
- Substance abuse policy
- Safety budget
- Field performance audits.

Incorporation of safety and health in construction contracts can make a significant and positive impact on the project's safety and health performance. One effective way of doing this is to establish target safety and health standards and criteria for the project.

Inclusion of safety and health in construction contracts places the required emphasis on their significance and encourages the contractor to pay systematic attention to safety and health. However, safety is usually not a pay item in the contract, and its costs are lumped into the "cost of doing business," or the project overhead. At the National Forum on Construction Safety and Health Priorities, it was strongly suggested that safety and health be included in the contracts as a bid item just like excavation, borrow material, concrete, etc. This will make owners pay for this item separately and it will provide the resources to contractors to invest in safety and health appropriately, delivering improved performance.

20.7 Requirements for Safety in Construction Projects

20.7.1 Falls

Falls may occur near excavations or manholes when workers are entering or leaving the area, or when they simply may have not noticed the opening. In sewers, fall hazards may lead to drowning if the person experiences unconsciousness. The most effective means of fall protection on such locations are guard rails with toe boards, covers for openings, and hazard communication signs. Toe boards and screens are to be used to safeguard lower elevations from falling objects. Materials should not be placed near edges, where such protection is not provided. Prohibiting signs can warn workers against entering a fall hazard zone.

20.7.2 Excavation and Trenching

Excavation and trenching operations are essential to many types of construction projects. Foundations, drains, sewers, and underground utilities are part of a great majority of construction projects, requiring excavation or trenching. Excavation is the removal of soil and rock from the original location. The behavior of soil during excavation is dependent on its composition and the environmental conditions. Soil composition varies, from sand, which flows easily, through silt to clay, which is cohesive. Water is often present in soil to some degree and affects soil behavior. Many types of soils cannot support their own weight during excavation. Therefore, some form of support is often required. The exact method of support will depend on the soil type, as well as site and groundwater conditions. In shallow trenches, or excavations with depths less than 5 ft (1.5 m), sloping or shoring may not be required if the soil is cohesive. For deeper trenches, support will be required as decided by a qualified person.

20.7.2.1 Excavation and Trenching Hazards

Detailed information on excavation and trenching hazards has been presented by the Center for Excellence in Construction Safety (CECS). The primary hazard in excavation operations is the possibility of *cave-ins*, or earth slides, burying workers in the trenches. Many of these accidents result in death. Persons buried beneath the soil become unable to breath and ultimately suffocate. Complete burial is, in fact, not necessary to cause death. The pressure of surrounding soil will force air from lungs and prevent further breathing. The likelihood of rescuing buried workers in such cases is usually not very high because heavy equipment cannot be used to uncover victims for fear of causing further injuries. Furthermore, sending rescuers inside a caved-in trench exposes more persons to the same danger.

The four important factors that contribute to cave-ins are as follows:

1) *Weight*: Weight is generated by the soil itself, construction equipment and vehicles, or other objects.
2) *Amount of water in soil*: Excessive moisture or lack of moisture can weaken the soil leading to collapse.
3) *Vibrations*: Vibrations are caused by activities such as vehicular traffic, blasting, and pile driving.
4) *Soil composition*: The composition and structure of the soil affects its stability during the excavation process. Mechanical failure due to removal of lateral support is also a critical factor in cave-ins.

Water, either by rain or percolation, can affect the lateral resistance of soil. Water can enter the dry soil through cracks and voids and cause softening and sudden loss of strength. In cold climates freezing and thawing can affect soil strength. Water can also increase the weight of soil, and act as a lubricant contributing to slippage. Vehicle traffic, especially large trucks and heavy construction equipment, can cause soil displacement and instability. Other hazardous conditions are created by placing excavated earth, pipes, and shoring equipment too close to the trench to form additional (surcharge) loads. Impacts during unloading of these materials can lead to cave-ins as well.

Certain types of excavation hazards may be particular to a specific location, such as a trenching operation at or near previously excavated ground; excavation reduces the inherent strength of the soil. Therefore, starting a new excavation on previously excavated ground increases the possibility of a cave-in. Where new excavation is planned close or parallel to a previously excavated trench, it must be recognized that the ground between them may be particularly susceptible to collapse. Hazards are also caused by differing strata within a soil cut, or by pockets of weak soil at the site. Existing vegetation such as large trees extracts the soil moisture through the roots and may contribute to loss of soil strength.

One of the greatest hazards encountered during excavation is the collapse of adjacent structures, which can result in catastrophes and fatal injuries. In construction projects involving proposed additions and renovations, a new trench is commonly excavated close to an existing foundation wall. If the soil is removed to a depth that is equal to or greater than the depth of the existing footing, and proper shoring is not provided to support the existing structure, the loss of bearing support of the excavated soil may cause a partial or full collapse of the existing structure.

Although cave-in failures are generally the major concern in excavation operations, other hazards may also be encountered. Excavations are performed for a variety of purposes and they involve a variety of activities. Physical injury from falls or falling objects, exposure to buried electrical cables, and hazards associated with confined spaces are also of major concern and should not be overlooked.

Excavation cave-ins are preventable if certain precautionary measures are adopted. It is important to recognize that visual inspection of soil stability is not sufficient to ensure the long-term safety of an excavation because one cannot anticipate the changes in the weather, rainfall, or future unsafe practices. Therefore, soil analyses must be conducted by a competent person, for example, a geotechnical engineer, who will determine the adequate support system depending on the nature of the excavation, the soil, and groundwater conditions.

20.7.2.2 Shoring

Shoring is the support system designed to prevent the lateral movement of soil, which can lead to cave-ins. It is commonly used for deep excavations, those that are 5 ft (1.5 m) or deeper. Various shoring systems are available to prevent excavation cave-ins. Aluminum hydraulic shoring systems have been developed as an economic method to ensure the safety of workers. Their lightweight and adaptability to varying conditions make them very attractive. Other shoring systems include vertical shores for compact soils ("skip shoring"), stringer systems providing horizontal support for intermittent or solid sheathing for relatively unstable soils, and manhole braces.

Shoring systems must incorporate certain precautionary features to ensure the safety of workers:

- Providing effective means of access and egress, such as securely fixed ladders in trenches that are 4 ft (1.2 m) or deeper
- Erection of temporary barriers to guard against falls
- Construction of diversion dikes and ditches to provide drainage of the adjacent area, which will prevent water from entering a trench.

20.7.2.3 Sloping

Sloping involves cutting the sidewalls of an excavation to form a safe angle of repose, which represents the angle at which the soil settles to a natural state with no tendency to further settle to a shallower form. The angle of repose varies with the specific soil conditions; however, it cannot exceed 45%. Sloping is not a practical approach for deep excavations because it requires excessive removal and replacement of soil. However, sloping can be effectively used in conjunction with shoring. Typically, the top of the trench is sloped in such cases to allow for easy installation of the shoring system at lower depths.

20.7.2.4 Trench Shields

Trench shields (boxes) are essentially used as "personnel protectors." Such devices are commonly placed in an already excavated, but unshored, trench or pit. Normally, the trench walls remain intact long enough to complete the construction. If they collapse, however, a properly designed shield must be capable of withstanding the maximum anticipated lateral soil stress at a given depth. This minimizes the potential for worker injury. Trench shields are most often used in open areas away from existing utilities, streets, and buildings that may require a support system. If ground conditions are favorable and the trench zone requires no direct support, the use of a trench shield may be an excellent choice. However, when the job calls for strict compaction and replacement requirements, one must be cautious. If unstable soils slough off as the shield is pulled up the trench line, it could make proper construction a difficult and costly task.

20.7.2.5 Inspections

Inspection of excavations by a competent person can assist in assessing the soil stability and the quality of the shoring system. Daily inspections should be performed. Attention should be directed to the following during the inspections:

- Bulges on trench walls, cracks near the edges and walls
- Accumulation of loose rocks fallen from the trench wall
- Water in the excavation area
- Soundness and quality of the shoring materials (e.g., timber that has decayed)
- Improper connections in the shoring system and other signs of distress.

Other safety measures adopted to protect workers against excavation hazards are as follows:

- Keep the excavated soil pile at least 2 ft (0.6 m) from any opening to avoid excessive surcharge pressure.
- Keep all vehicles and equipment away from the trench.
- Be aware of vibrations, overhead utilities, and underground utilities.
- Be aware of toxic fumes. Toxic fumes can seep through the soil and accumulate at the bottom of an excavation. Clean air should be provided in sufficient volume to dissipate the toxic gases.
- While excavating the soil, clear all unnecessary workers from the area.
- Protect the public by installing guard rails, barricades, and warning signs around the excavation area.

20.7.3 Confined Space Entry

Confined spaces are enclosed areas having limited access or egress. They can be storage tanks, boilers, bins, silos, etc., which are accessible through a manhole. Confined spaces commonly encountered on construction projects include basements, trenches, shafts, bore holes, ducts, pipelines, drains, and sewers. Entry into confined spaces can be hazardous and requires special care. Fatal accidents in confined spaces are common.

The atmosphere in confined spaces may be hazardous due to a lack of oxygen, the presence of toxic agents, or the presence of flammable and explosive gases. Other hazards in confined spaces include moving parts of machines, extreme temperatures, collapses and cave-ins, and sudden flooding. The difficulty of access and egress and the difficulty of moving in confined spaces amplify the impact of these hazards. Some specific hazards are discussed next.

20.7.3.1 Oxygen Deficiency

Small drops in oxygen level can cause loss of balance and concentration, breathing difficulties, and fatigue; a continued drop in the oxygen level may result in unconsciousness and death by suffocation. According to the National Institute for Occupational Safety and Health (NIOSH), an oxygen-deficient atmosphere has less than 19.5% available oxygen (O_2), requiring an approved self-contained breathing apparatus. Atmospheres with less than 16% O_2 cause faulty judgment and rapid fatigue. Atmospheres below a 6% O_2 level create breathing difficulties and cause death in minutes. Oxygen deficiency is most commonly caused by gas leakage into the confined space, oxidation due to corrosion or bacterial growth, and exhaustion of the oxygen supply due to inhalation and combustion.

20.7.3.2 Other Atmospheric Hazards

Presence of toxic gases in sufficient quantities is another source of hazard. Hydrogen sulfide is a toxic gas commonly produced by the decomposition of organic matter and is often present in the sewer systems. Hydrogen sulfide is both toxic and explosive. Another highly explosive gas, methane, is also often present in sewer lines. Other lethal gases are carbon dioxide, which is naturally present in soil, and carbon monoxide, which is usually produced by internal combustion engines. Leakage can cause chemical vapors to poison a confined space atmosphere; ammonia and chlorine are typical toxic fumes that may be encountered. Nitrous fumes may be left in a confined space immediately after explosive operations. Very small quantities of certain vapors and gases can cause fires and explosions. Whereas methane and hydrogen sulfide are explosive gases, petroleum and vapors of solvents such as acetone and toluene can cause fire.

20.7.3.3 Hazard Control in Confined Spaces

Good advance planning and adequate training in safe methods of work in confined spaces will enable workers to enter these areas safely. Confined spaces training will vary according to the nature of the activity and the confined space. The training program should consider the role of supervisors, workers, persons outside the confined spaces, and rescue personnel. A permit-to-work system, in which each step is planned and authorized, is required for confined space entry. Other alternatives to entry should be considered for confined space operations wherever possible. If entry to the space is unavoidable, then appropriate breathing apparatus should be considered. The use of mechanical or forced ventilation is the first choice. Before entry into a confined space, the space should be withdrawn from service and be isolated from electrical and mechanical sources.

20.7.3.4 Monitoring, Protective Equipment, and Communications

Personnel planning to work in confined spaces should check for leaks and test the atmosphere for oxygen and flammable, explosive, and toxic gases. Once begun, the operation should be constantly monitored. Personal protective equipment such as breathing apparatus, harnesses, lifelines, and rescue equipment must be provided and used as necessary. Personnel must be informed about the details of activities and the required communications. A trained person should be in attendance and in continual verbal communication with the workers in confined spaces throughout the operation.

20.7.3.5 Rescue Operations

Rescue operations may be needed when a person is injured, or collapses in a confined space, becoming unconscious. Rescuers must wear breathing apparatus and safety harnesses that are attached to a lifeline before entering the space. The rescue equipment must include lifting tools such as tripods and winches to carry the unconscious worker.

Example 20.3 Confined Spaces

Discuss different types of confined spaces that the water and wastewater engineers and operators must all understand.

Solution:

Any *confined space* means: (a) a space that is large enough and so configured that an employee can bodily enter and perform assigned work, (b) a space that has limited or restricted means for entry or exit (for example, tanks, vessels, silos, storage bins, hoppers, vaults, and pits are spaces that may have limited means of entry), and (c) a space that is not designed for continuous employee occupancy. (Definition from the Code of Federal Regulations [CFR] Title 29 Part 1910.146.)

A *non-permit confined space* is a confined space that does not contain or, with respect to atmospheric hazards, have the potential to contain any hazard capable of causing death or serious physical harm.

A *permit-required confined space (permit space)* has one or more of the following characteristics: (a) contains or has a potential to contain a hazardous atmosphere, (b) contains a material that has the potential for engulfing an entrant, (c) has an internal configuration such that an entrant could be trapped or asphyxiated by inwardly converging walls or by a floor which slopes downward and tapers to a smaller cross section, or (d) contains any other recognized serious safety or health hazard. (Definition from the Code of Federal Regulations [CFR] Title 29 Part 1910.146.)

Example 20.4 Competent Person, Danger, Dangerous Air Contamination, Oxygen Deficiency Oxygen Deficiency, and Oxygen Enrichment

Legally define the terms of competent person, danger, dangerous air contamination, oxygen deficiency oxygen deficiency, and oxygen enrichment.

Solution:

A competent person is defined by OSHA as a person capable of identifying existing and predictable hazards in the surroundings, or working conditions which are unsanitary, hazardous or dangerous to employees, and who has authorization to take prompt corrective measures to eliminate the hazards.

The word *danger* is used where an immediate hazard presents a threat of death or serious injury to employees.

Dangerous air contamination occurs when an atmosphere presents a threat of causing death, injury, acute illness, or disablement due to the presence of flammable and/or explosive, toxic, or otherwise injurious or incapacitating substances. There are three kinds of dangerous air contaminations: (a) Dangerous air contamination due to the flammability of a gas or vapor is defined as an atmosphere containing the gas or vapor at a concentration greater than 10% of its lower explosive (lower flammable) limit, (b) dangerous air contamination due to a combustible particulate is defined as a concentration greater than 10% of the minimum explosive concentration of the particulate, and (c) dangerous air contamination due to the toxicity of a substance is defined as the atmospheric concentration immediately hazardous to life or health.

Oxygen deficiency occurs when an atmosphere containing oxygen at a concentration of less than 19.5% by volume. Oxygen enrichment means that an atmosphere containing oxygen at a concentration of more than 23.5% by volume. Either one may be a cause of *danger* to a person's life.

Example 20.5 Permit-Required Confine Space Monitoring and Operation

There are many locations in a water or wastewater utility where it is necessary to monitor the atmosphere before entering. Some of these specific locations may be underground regulator vaults, solution vaults, manholes, tanks, trenches, and other *confined spaces*. There are a variety of devices that are used to monitor (check) the available oxygen and also combustible and toxic gases. Regardless of what type of monitoring device or devices are used, discuss the monitoring and operation of a permit-required confine space monitoring and operation.

Solution:

A competent person should always test for dangerous air contamination and/or oxygen deficiency/oxygen enrichment with an approved device immediately prior to an operator entering a permit-required confined space and at intervals frequent enough to ensure a safe atmosphere during the time an operator is in the structure. Dangerous air contamination includes explosive conditions and toxic gases. Toxic gases include hydrogen sulfide and carbon monoxide. A record of the test must be kept at the job site for the duration of the work. The operator's life will be in jeopardy if the air in a confined space is not tested before entering. The following five steps should be performed before entering a confined space: (a) Calibrate the gas detection device; (b) barely open the confined space; (c) test for oxygen deficiency/enrichment, combustible gases, and toxic gases using a probe or tube to collect the sample; (d) record the results of these tests in your gas log; and (e) ventilate the confined space.

If a hazardous atmospheric condition is discovered, the competent person must repeat the tests for oxygen deficiency/enrichment and combustible and toxic gases while ventilating and again record the results.

On rare occasions, unusual conditions may exist or be created even though the required testing procedures have been followed. Air currents through duct lines can easily change if other vaults in the same system are opened. Toxic or explosive gases from broken gas lines or decayed vegetation may then flow through the ducts into a previously gas-free vault. Therefore, it is important that adequate ventilation be maintained and that the space be rechecked for hazardous atmospheres periodically while operators are working in such locations. Use portable fans or ventilators to provide fresh air.

At least one competent person must stay outside the confined work area with another person standing by. This backup person should check continuously on the status of personnel working in the confined space. Should a person working in the confined space be rendered unconscious due to asphyxiation or lack of oxygen, the backup person at the surface could descend into the space by: (a) securing the help of an additional backup rescue person and (b) putting on the correct type of respiratory protective equipment. Self-contained breathing equipment should be nearby and used as necessary.

20.7.4 Heavy Construction Equipment

Heavy construction equipment is an integral part of construction operations. The machinery is used for many demanding tasks that are beyond human capability. Earth-moving equipment such as bulldozers, scrapers, motor graders, and front-end loaders are typical of the heavy construction equipment that performs cutting, transporting, and grading tasks.

Bulldozers are used to strip soil in layers. They move and grade earth/rock material for distances less than 300 ft (91.4 m). The various types include bulldozers and angle dozers. Scrapers also strip in layers. However, they also load the earth into bowls, and haul, spread, and partially compact it. Types are crawler-tractor-pulled scrapers and wheel-tractor-pulled scrapers. They represent a compromise between the best backing and best hauling machines. Motor graders cut, shape, and grade. They have a blade that can be set at different angles. They can be quite versatile when fitted with rippers, backslopers, snow plows, etc. Front-end loaders are used to transport bulk materials, to load trucks, to excavate earth, etc. Types are crawler and wheel loaders. Other types of heavy construction equipment such as hydraulic excavators, trenching machines, tractors, and rollers are also available.

20.7.4.1 Heavy Equipment Safety Hazards

Many of the hazards associated with heavy construction equipment are common to all:

- Poor repairs and service
- Obstructed view during backing
- Striking people and collision with other equipment
- Traveling empty at excessive speeds
- Pinch points between equipment and objects
- Riders falling from equipment or bucket
- Overturning of the equipment
- Unexpected electrical shock
- Failure of lifting mechanisms/operational failures
- Injuries to operators due to ingress/egress difficulties
- Runaway machines (e.g., if wheels were not blocked upon parking or if operator was unable to control the machinery)
- Being struck by limbs of trees or other overhead obstructions, and moving equipment.

20.7.4.2 General Safety Precautions

Recommendations for general safety precautions associated with heavy construction equipment are as follows:

- Management should carefully select competent operators.
- Rules for operation should be clearly stated.
- All equipment should be kept in good working condition.
- All new machines should be equipped with rollover protection.
- Existing machines with rollover protection should be maintained in good working condition.
- Adequate illumination should be provided for night operations.
- Dust must be kept down on all roads.
- Operators must be given clear and specific instructions.

- All personnel should be clear of the work area.
- Unauthorized riding on the equipment should be prohibited.
- Speeds should be consistent with job conditions and OSHA requirements.
- Use a signal person in busy areas.
- Equipment should have an audible reverse signal alarm system that operates automatically with backward movement.
- Repairs should be done when the machine is not running.
- During refueling operations, all personnel in the vicinity should stop all motors and refrain from smoking.
- Personal protective equipment (hard hat, steel-toed shoes, gloves, safety glasses, respirators, etc.) should be used at all times.

20.7.4.3 Maintenance and Training

Maintenance is extremely important in securing the safety of machinery and equipment. OSHA standards require inspection of all machinery before each use. Results of tests and inspections should be recorded. Frequent inspections should be made on wire ropes and guys, hoists and trolley cables, jib and counterweight jib guy lines, hoist rope anchorage on winding drum, foundations, and structural connections.

Training of workers on all cases is essential, and must include a review of potential hazards and familiarization with the equipment.

20.8 Occupational Diseases

In the past decade, a greater public awareness of occupational diseases and their consequences has created a broad-based involvement with issues related to worker health and safety. Studies related to occupational diseases are no longer restricted to health care professionals, and the field of industrial hygiene contributed significantly to this change.

As scientists have increased their ability to analyze minute concentrations, those who study occupational disease have established that materials, biological agents, or energies existing in very low concentrations may be harmful to humans. Exposure to materials that exist as minute contaminants of other industrial raw materials may also represent a risk to worker health.

Engineers who, as a profession, are often responsible for the design and construction of municipal and industrial facilities and the protocols relating to their operation must now become more involved with worker health-related issues. New materials or new or improved processes, or both, that are constantly being introduced require greater awareness of their health effects at the design stage. Engineers should not relegate health considerations to retrofitting practices. Increased public awareness has also had positive results, for example, greater worker awareness of health-related issues, setting new standards of exposure and adjustment of old standards to reflect a more up-to-date understanding of occupational diseases, and recognition of the need to instruct professionals such as engineers in health-related issues. By greater understanding of these issues, engineers will be able to contribute to improving the conditions that influence the well-being of workers.

Occupational diseases are preventable. An understanding of the nature of occupational diseases (identifying the causal agents, quantitatively assessing the human exposure, and learning about the interactions between hazardous agents and the human body) should result in a safer work environment. In this enlightened situation, arrived at partly through the actions of engineers, health effects associated with a specific operation will be considered in the overall engineering design. Being aware of the interaction of a worker with his or her work environment will affect process design, including engineering controls, decisions about safe operating practices, and the use of protective equipment.

20.8.1 System Approach

When investigating the development of an occupational disease, one is concerned with the interface and the interaction between the human body and the work environment. An additional and important concern is the nature of the disease process, including the structural and functional changes occurring in the human body that result in or are a result of a disease process (the study of human pathophysiology). Together, all of these disciplines contribute to an understanding of occupational diseases. In this type of work, a systems approach is used to investigate the behavior of the human body in an occupational setting. This requires a description of the interaction of the human body with potential disease-causing chemical, physical, or biological agents in the environment. Matters are complicated by the fact that only a fraction of a worker's time may be spent in an occupational environment and that exposure to potential disease-causing agents (including chemical agents, physical energies, or biological entities) may take place in the home environment or as a result of nonoccupational activities.

A thorough study of occupational disease requires a multidisciplinary approach. Physical sciences and engineering analyses are useful to describe the interface of the human with the work environment. Life sciences information, including biochemical, biomechanical, toxicological, and pathological descriptions, is needed to assess the physiological effects of occupational exposure. In addition, the social patterns of the workers (i.e., lifestyle considerations) may influence the outcome of exposure.

20.8.2 Complexity of the Issues

The central focus in discussing occupational diseases is the human worker and his or her body – a complex system of physiological and psychological (behavioral) components. Just as no two people will behave exactly alike in a work environment, they will also not have the same responses to stimuli. Not only can the range of human responses to physical, chemical, or biological agents be dramatic, so can the range of idiosyncratic behavior and personal hygiene. Cigarette smoking and willingness of a worker to wear protective equipment are two examples of personal habits that may influence the risk of occupational diseases.

A person's work history may also influence his or her potential for developing an occupational disease. Past exposure to disease-causing agents may have resulted in the bodily accumulation of agents that are slow to be removed. Body burdens (the amounts of these agents in various body compartments) may stress normal physiological systems. Alternatively, previous work exposure may have affected homeostasis (the stabilizing tendency in the human body) and, thus, rendered the individual more susceptible to certain occupational diseases. Assessing the effect of work histories on the potential for developing an occupational disease is difficult and is a subject of ongoing research.

Generally, the differences in individual susceptibility to occupational disease can be related to general health, age, sex, race, diet, and heredity. These differences, as well as the limited information relating exposure and human response, significantly complicate the task of setting "safe" levels for potentially harmful agents. For these reasons, limits for occupational exposure cannot be considered as absolute levels of protection. Although limits are established, this is not assurance that individual workers may not show deleterious effects if they have unusual susceptibility.

20.8.3 Scientific Factors

Many dilemmas exist concerning the role of scientific "information" and occupational disease. Most scientific information that relates exposure to diseases comes from three qualitatively different types of investigation:

1) Case studies of accidents or disasters
2) Epidemiological studies
3) Basic and applied scientific research.

Accidents and disasters such as occurred in Chernobyl and Hiroshima serve as unparalleled sources of information concerning human exposure to radiation. The human and environmental exposure as a result of the industrial accident in Bhopal, India, is and will remain under investigation for many years. While the moral, ethical, and legal implications of these accidents are being explored, scientists are gathering data from medical records and environmental measurements to assess human and environmental responses to ionizing radiation and widespread isocyanate exposure. Sometimes information from accidents and disasters is claimed to be "gained from experience." Nonetheless, perhaps the most useful outcome of these accidents/disasters is the corrective changes that diminish the probability of a recurrence.

Epidemiological research attempts to relate exposure to harmful agents and the occurrence and distribution of disease or injury in segments of society. Unlike clinical medicine, where the focus is on the diagnosis and treatment of an individual, epidemiology emphasizes the patterns of disease or injury in groups of individuals in order to identify causality. Interpreting causal relationships is made difficult because of systematic errors (called biases) introduced into a study or because individuals are exposed to factors other than a particular environmental agent – factors that may influence the outcome of a disease (a *confounding factor*). The epidemiologist's raw data may consist of monitored or estimated patterns of exposure, job titles and death records, or medical history records. The results of a study may be a description of an incidence rate for a particular disease, and it may be standardized to the general population or expressed in a proportional manner. The statistically determined increased risk of developing a disease based on exposure to a specific agent is often sought in these studies. Results of epidemiological investigations have been extremely beneficial in assessing causal relationships. They are, however, often clouded by imprecision because of faults with the data sets and the presence of confounding agents and biases in the studies.

Scientific laboratory research aimed at better understanding the interaction between exposure to an agent and a disease is limited primarily to animal experimentation. For example, laboratory tests seek to identify harmful agents that interfere with reproduction in particular, agents that interfere with transferring genetic information from parent to offspring or that may produce physical defects in the offspring. The great complexity of and physiological differences between human and other animal reproductive systems is such that extrapolating laboratory information from one species to the other should be approached cautiously.

Because conducting animal experiments is expensive in terms of time and money, the usual approach is to subject the animals (usually rats and mice or other small mammals) to very high levels of the substance being investigated. This way a sufficient number of positive (disease-producing) results can be obtained. The extrapolation procedure is subject to criticism, and some of the issues relate to

1) Differences between laboratory test animals and humans
2) Well-controlled administration of test agents in the laboratory versus uncontrolled occupational exposure of humans
3) The often several orders of magnitude difference per mass of body weight of laboratory versus occupational exposure
4) The short-term laboratory exposure versus the chronic (long-term) occupational exposure.

Much of what we know about exposure to potentially harmful agents and the development of occupational diseases is obtained from laboratory experiments on animals. Regulatory policy relies heavily on such scientific endeavors. The drawbacks and weaknesses of this approach, however, continue to create a dilemma for medical professionals, other scientists, and policy makers.

20.8.4 Occupational Disease as a Process

Occupational diseases develop out of the complex circumstances involving the pathophysiological response of a human exposed to harmful agents. Despite recent, rapid advances in science and medicine, much of the information describing the etiology (the study of factors that cause diseases) of occupational disease either does not exist or is poorly understood. The large variability among humans in their responses also contributes to this unclear situation. Still other factors have hindered the understanding of occupational diseases: the long latency period of some occupational diseases, the difficulty in establishing causality between exposure and development of a disease, the lack of emphasis on occupational diseases during medical training, and the multifactorial nature of real-life exposure.

From an engineering point of view, the development of an occupational disease may be considered a complex, incompletely understood process involving a contaminated environment (forcing function) acting on a physiological entity (the worker or physiological transfer function) with the potential of producing a pathological condition (a perturbed state). Occupational diseases are usually studied according to functional physiological units (e.g., organs) or according to hazardous agents (or materials). The approach here differs from the standard health-care-oriented approach; the emphasis is on system behavior including interaction with the local environment.

Exposure to some potentially harmful agent (a chemical substance, a biological material, or energy) becomes an input to the human body (dose) when it crosses the hypothetical surface separating the occupational environment from the internal milieu of the body. This dose now elicits the possibility of two initial responses: normal or altered body processes, each with their inherent defense mechanisms and elimination. The altered body process route may be in response either to previous exposure or to genetic characteristics that predisposed the body to be hyper- or hyposusceptible to a stimulus. The final state of the body that results from the input and physiological and biochemical processing steps is (a) homeostasis, in which normal conditions are attained; (b) acclimation, in which changes occur that do not impede normal function; (c) disease that differs qualitatively from acclimation only in that the changes that occur are greater and in that normal function is disrupted; or (d) death.

20.8.5 Potential Hazards

Chemical agents (including particulates, gases, vapors, liquids, and combined forms of these) probably represent the largest category of industrial hazards to health. In addition to these chemical substances, energies (including various forms of radiant energies, mechanical energies associated with physical labor, and biomechanical operation of the human body) may be hazardous. Climatic conditions such as temperature or humidity represent potential hazards. Biological agents (ranging from microorganisms to insects) may also be deleterious. In a negative sense, the deprivation of any normal input required to maintain normal human function (e.g., oxygen deprivation) is itself a hazard. Finally, physiological and sociological stimuli are potential hazards.

20.8.6 Modes of Entry

Any of the potential hazards or circumstances above can affect the internal human body via inhalation, ingestion, irradiation, or information exchange (information that may affect mental health). Where an energy transfer takes place between the worker and the work environment, the form of exchange may result not only from environmental stresses such as climate, light, sound, and heat but from the physical aspects of labor. The physical aspects of labor are part of the worker–job interaction, that is, the study of ergonomics.

Hazardous agents usually enter the body via *inhalation*. Our respiratory system is a highly organized mass transfer apparatus; its primary function is oxygen exchange into the blood and carbon dioxide exchange from the blood to the outside environment. This physiological system is effective in removing particulates, gases, and vapors from inspired air. The specific effects and the site of action of these airborne materials are determined by the organization of the respiratory system, including the dynamics of breathing, as well as by the physical, chemical, and physiological properties of the agents being captured.

Ingestion is not considered a major pathway by which foreign agents enter the body; nevertheless, it is worth noting. Either accidently (e.g., poor hygiene) or as a consequence of a contaminated environment (exacerbated by mouth breathing), foreign agents may be ingested. The ingested agents usually wind up in the stomach where they are exposed to an active physiological system characterized by low pH and high digestive enzyme activity.

Transport of agents across the skin to the internal body environment can occur via several distinct pathways. Transcutaneous absorption will depend on the barrier function of the skin and the physicochemical properties of the absorbing agent. Adsorption onto the skin may be a first step that results in the material eventually being transferred to the internal milieu (milieu interne). Abraded or punctured skin opens a direct entry route to the internal environment for hazardous agents. The direct route may provide an entry several orders of magnitude greater than that of adsorption. Biological agents such as ticks or microorganisms can also penetrate the skin and gain entry to the internal environment.

Irradiation (the exposure of the body to either ionizing or nonionizing radiation) may affect the skin surface or may penetrate to deeper layers. This form of energy exchange could be from sunshine or from exposure to radioactive agents.

Information is a psychosocial form of exchange with an environment that is very difficult to quantify. A conscious or subconscious input is received by the worker, and his or her mental health depends in part on information.

The amount of a hazardous agent that reaches the body's internal environment is referred to as the *dose*. For most agents, the notion of a dose may be easy to quantify (e.g., chemical agents); for others, the concept of a dose may be almost meaningless (e.g., as in the case of information). A dose may be referenced to a specific target organ or to a site of exposure, or it may be generalized to represent systemic contamination.

20.8.7 Body Processes and Defenses

Once a dose of some agent has entered, a normal biological response indicates that no altered sensitivities or hyper- or hyporesponse is initiated immediately following the dose. As part of the immediate and subsequent response to foreign agents or energies, defense mechanisms and eliminations are involved. These are related intimately to normal body processes.

Normal body behavior or normal physiology is beyond the scope of this book, but many texts are available dealing with this subject. The normal biological processes following exposure to some potentially hazardous agent may depend on the quantity of the dose, the particular agent or energies involved, and the temporal pattern of the dose. As a result of either acute (short-term) or chronic (long-term) exposure to harmful agents, the state of health of the worker may change. The normal body process should be considered as a starting point or as an initial condition in this dynamic system. That is, the "normal" processes can be changed as a result of inputs.

The human body is well equipped for defense against most foreign agents. For each mode of entry of agents or energies, the body has specific defenses or protective mechanisms. Examples of these are nasal filtration of particulates and the barrier function of the skin. Once an exposure has resulted in a dose of an agent, the defense mechanism associated with the internal biological environment may be summoned. For a thorough description, refer to appropriate biological or medical sources.

The *immune system*, for example, is a complex defensive network that relies on specific cells to recognize foreign agents and to mobilize other cells (e.g., lymphocytes) to "attack" the foreign matter (biological or chemical). Concurrently, circulating antibodies (antagonists to specific materials) are synthesized, after recognition of the foreign agent, to enter the fray to destroy the foreign agents. As part of this system, a highly specific memory is developed from previous exposures to some agents. With the aid of this recall, the response time can be shortened and the degree of response increased for subsequent exposure.

Cellular defenders, other than lymphocytes as discussed above, can be found systemically in body fluids (e.g., blood or tissue fluids) or they can be associated with a particular tissue. The manner by which cells can defend against the presence of foreign agents, includes engulfing or ingesting foreign agents (phagocytosis) and thereafter exposing these agents to internal cellular materials such as enzymes, which may chemically digest them; and producing and secreting substances, which aggressively coat or attack the foreign agent.

Inflammation, broadly defined, is a defensive reaction of tissue to injury characterized by redness, heat, swelling, and tenderness in the affected area. This reaction usually occurs along with other defensive mechanisms. Inflammation involves microvascular, cellular, body-fluid-associated, and systemic components. The term *inflammation* is somewhat arbitrary because, in reality, it is part of a process with ill-defined stages. Inflammation usually results from biological agents (e.g., bacteria), but energies such as those associated with sunburn, frostbite, or physical labor may produce inflammations. Inflammation is usually classified as acute or chronic. With an acute inflammation, the return to a normal condition can be expected in a matter of days as a specific population of defensive cells migrates to the inflamed area. Chronic inflammation is characterized by persistence over a long period of time, a cellular population in the injured area that is very different from that of the normal and the acute condition, and the excessive production of tissue structural materials (e.g., collagen). Inflammation, even though it is involved in tissue response to injury and the subsequent repair process, may itself have harmful effects. Inflammation with excessive pulmonary edema (accumulation of fluid in the lungs) may be life-threatening.

The body can also transform agents foreign to its internal environment. This biotransformation results from the action of enzymes associated primarily with liver function and is classified into breakdown reactions (oxidation, reduction, or hydrolysis) and synthesis-reactions (conjugation). Generally, biotransformations convert materials to other forms (metabolites) that are more easily eliminated from the body, or that are less toxic, or both. These effects are achieved generally by alterations in solubility and/or in chemical activity of the xenobiotics.

In addition to the primary defense mechanisms discussed earlier, protective characteristics can be attributed to the normal homeostatic (preservation of the stability of the system) mechanisms and to alterations in growth patterns. Altogether, the body can muster impressive defensive possibilities.

20.8.8 Elimination

Elimination is part of the dynamic and interacting system that determines the intake, distribution, and fate of agents that are potentially hazardous. As such, it can be considered an aspect of normal processing as well as a specific defense mechanism. The major excretion pathways are via urine and feces. Materials can also be eliminated in sweat or during normal surface cell turnover (exfoliation), including loss of hair and nails, and in exhaled gases.

The pathway of altered body processes acknowledges the dramatically different response of an individual with altered susceptibility to a dose of some agent. Hypersensitivity can be found in a subpopulation of individuals whose immediate response to certain agents involves immune-mediated events not characteristic of the immediate response by the normal population. The relative response by the subpopulation of hypersensitive individuals can be extreme when compared with the response of normal persons. Also, the susceptibility of the otherwise normal population can vary significantly. Some individuals can exhibit extreme responsiveness to certain agents in the absence of action by the immune system. Both the hypersensitive and normal but hypersusceptible groups display altered body processes.

Altered body processes can involve a target organ, such as the skin or the lungs, or they can be systemic. The onset of the physiological response to a foreign agent may be rapid and exaggerated or, under specific conditions, it may be delayed. In any event, the eventual long-term outcome may be very similar to that result obtained following the normal body process route.

20.9 Ergonomics

20.9.1 The Worker and Work

Occupational disease may result from biomechanical, psychological, or environmental stresses associated with work. *Ergonomics* is the interdisciplinary study of the interaction of the worker with the work environment. The objective of ergonomic studies is to make the human work experience more efficient while promoting the well-being of the worker. This involves establishing and maintaining compatibility between equipment, tools, tasks, and environmental factors on one hand, and human anatomical and biomechanic considerations and perceptual and behavioral characteristics on the other. Its interdisciplinary nature is a critical feature of ergonomics, which combines physiology, psychology, engineering, and anthropometry (measurements with respect to the human body). The nonpsychological aspects of ergonomics are sometimes referred to as human factors engineering. The range of subjects that can be studied from an ergonomic point of view is quite diverse and includes analyses of static and dynamic human body biomechanics, metabolic and physical work requirements, use of tools, repetitive motion tasks, climatic and other environmental effects, lighting, equipment and process design, job demands, and mental and cognitive demands, to name a few.

Ergonomic considerations include aspects of safety program and managerial components as well as engineering, equipment, equipment maintenance, medicine, and training. To investigate the possibility of potential problems in a consistent, rational manner, ergonomic checklists have been designed. These devices are used to systematically examine the wide range of factors that may be involved in a system composed of a human working in an occupational environment. Common to many of the checklists are the following:

1) Analysis of human capabilities with respect to work station design and layout
2) Investigation of equipment design regarding the reliability and ease of equipment use
3) Analysis of the risk associated with physical workloads
4) Information handling and decision making
5) A survey of environmental factors including illumination, noise, vibration, and climatic conditions
6) Consideration for work schedules.

NIOSH has proposed prevention strategies for the 10 leading work-related diseases and injuries. Included in this list are musculoskeletal injuries (including back injuries), traumatic injuries including death, noise-induced hearing loss, and psychological disorders. The incidence of these injuries or diseases can be reduced by effectively applying ergonomic considerations.

20.9.2 Adverse Effects Caused by Workplace Conditions

Examples of adverse effects on a worker serve to illustrate the scope of ergonomic problems. Cumulative traumas can result in disorders to

1) The nervous system, such as damage to the peripheral nervous system
2) The tendons and tendon sheaths, such as in carpal tunnel syndrome or in epicondylitis (tennis elbow), which may result from repeated hand/wrist movements
3) The lower back, the site of some of the most costly occupational injuries based on number of injuries and associated medical expenses
4) Joints, such as in bursitis and degenerative joint diseases.

The specific region associated with the symptoms of cumulative trauma is usually in the upper part of the body in either soft tissue or joints.

Workers who use vibratory equipment, such as jackhammers or chain saws, sometimes register such complaints as numbness and blanching of fingers, pain, loss of muscular control, or reduced sensitivity to heat and cold. These are the symptoms associated with Raynaud's or vibrational white finger disease. This condition can arise after prolonged and repeated minor insults (cumulative trauma) to the body such as from vibrations or from being struck by objects.

Noise-induced hearing loss can be either temporary or permanent. It can result from physical interference with the transmission of sound or it can result from neuropathologies. The causative agent is the sound power level or the level of impact noise. Industry abounds with equipment and operations that produce excessive sound pressure and noise impact levels.

Table 20.6 Conditions Caused by Biomechanical and Environmental Stresses

Injury/Symptoms	Commonly Affected Workers
Back problems	Material handlers
Carpal tunnel syndrome or tendonitis	Clerical workers, assembly line workers, check-out workers, stamping job workers
Raynaud's syndrome	Forestry workers, construction workers
Degenerative joint diseases	Material handlers, forestry workers
Eye strain resulting in fatigue	Clerical workers, foundry workers, high precision assembly and inspection workers
Hearing impairment	Furnace operators, truck drivers, machinery operators, construction workers
Segmented vibratory diseases	Chainsaw chipper and jackhammer operators
Loss of strength, problems with hand–eye coordination, decreased mental capacity, fatigue	Most workers

Another example of an occupational disorder preventable by ergonomic considerations is that of fatigue. Though fatigue is not technically an occupational disease, it is so prevalent that it warrants mention. *Fatigue* is defined operationally as impairment in the ability of a person to perform efficiently because of prolonged or excessive physical or mental exertion. Causes of fatigue include monotony, work intensity, and psychological and environmental factors. Fortunately, fatigue can be cured by rest. If insufficient rest follows bouts of fatigue, then complete recovery cannot be ensured and further relapse into unproductive work may result.

Table 20.6 is a selected list of occupational injuries or symptoms and the workers that fall within the field of study of ergonomics. As the average age of the workforce increases, so too will the potential for problems requiring ergonomic solutions.

Problems/Questions

20.1 What is the principal advantage of a system safety program compared with a conventional or traditional industrial safety program?

20.2 How is the construction industry different from manufacturing and other industries?

20.3 List and discuss the seven critical issues that were established as needing attention at the National Forum on Construction Safety and Health Priorities. Which one of these issues do you think is the most important? Why?

20.4 What are the root causes of accidents in the construction industry?

20.5 Contrast the perspective of the reliability engineer with that of the systems safety engineer.

20.6 Who in an enterprise establishes risk tolerance levels?

20.7 Why is it important to establish an exposure interval when evaluating risk?

20.8 What are the advantages and limitations of a risk assessment matrix?

20.9 What are the differences between a FMEA and a FMECA?

20.10 What are the five major classes of causal factors of construction failures?

20.11 Describe the role of natural disasters in causing or contributing to failures.

20.12 Define constructability. What should designers know about constructability?

20.13 What causes material failures? How do material failures lead to construction failures?

20.14 How do shop drawings and change orders impact construction site safety?

20.15 What is meant by the term *temporary structures*? What is their significance with regard to construction safety?

20.16 What should go into a policy statement prepared by top management?

20.17 Why should all projects have a set of project safety rules?

20.18 What are the most effective means of fall protection?

20.19 What is meant by the term *cave-ins* and what are the factors contributing to cave-ins during excavation and trenching operations?

20.20 Name some excavation and trenching hazards other than cave-ins.

20.21 Discuss hazard control in excavation and trenching by addressing the shoring systems, sloping, and trench shields.

20.22 Define the term *confined space*. What types of hazards are present in confined spaces? Describe the hazard control principles associated with confined space entry.

20.23 What are the principal safety hazards of operating heavy construction equipment such as bulldozers, scrapers, graders, and loaders? How can these hazards be prevented or mitigated?

20.24 What are the main modes of hazardous agents' entry into the human body?

20.25 Give four examples of adverse effects on a worker that illustrate the scope of ergonomic problems.

Bibliography

Air Force Space Division, *System Safety Handbook for the Acquisition Manager*, SDP 127–1, 1987.

Bert, J. L., *Occupational Diseases*, U.S. Department of Health and Human Services, National Institute for Occupational Safety and Health, Cincinnati, OH, 1991.

Center for Excellence in Construction Safety, *Newsletter, 2, No. 3*, West Virginia University, Morgantown, WV, 1989.

Clemens, P. L., *Working with the Risk Assessment Matrix: Lecture Notes*, 2nd ed., Sverdrup Technology, Tullahoma, 1993.

Clemens, P. L., and Simmons, R. J., *System Safety and Risk Management: A Guide for Engineering Educators*, U.S. Department of Health and Human Services, National Institute for Occupational Safety and Health, Cincinnati, OH, 1998.

Construction Industry Institute, *Constructability: A Primer*, The University of Texas at Austin, Austin, TX, 1986.

Eldukair, Z.A., and Ayyub, B. M., Analysis of Recent U.S. Structural and Construction Failures, *J. Perform. Constr. Facil.*, *5*, 1, 57–73 February 1991.

Mohr, R. R., *Failure Modes and Effects Analysis: Lecture Presentation*, 6th ed., Sverdrup Technology, Tullahoma, TN, 1992.

National Bureau of Standards, *Investigation of L'Ambiance Plaza Building Collapse*, NBS Technical Report PB88-112438, 1987.

National Institute for Occupational Safety and Health, *Engineering Education in Occupational Safety and Health*, http://www.cdc.gov/niosh/topics/SHAPE, 2008.

National Institute for Occupational Safety and Health, *Prevention through Design*, http://www.cdc.gov/niosh/topics/ptd, 2009.

National Institute for Occupational Safety and Health, *Program Portfolio, Prevention through Design*, http://www.cdc.gov/niosh/programs/PtDesign, 2009.

Pettit, T., and Linn, H., *A Guide to Safety in Confined Spaces*, Department of Health and Human Services, National Institute for Occupational Safety and Health, Cincinnati, OH, 1987.

Pfrang, E. O., and Marshall, R. M., Collapse of the Hyatt Regency Walkways, *Civ. Eng.*, *52*, 65–68, 1982.

Occupational Safety and Health Administration Alliance Program, *PtD Concept: Prevention through Design—Design for Construction Safety*, http://www.designfor-constructionsafety.org/concept.shtml, 2009.

Roland, H. E., and Moriarty, B., *System Safety Engineering and Management*, 2nd ed., Wiley Interscience, New York, 1990.

Schulte, P. A., Rinehart, R., Okun, A., Geraci, C. L., and Heidel, D. S., National Prevention through Design PtD Initiative, *J. Safety Res.*, *39*, 115–121, 2008.

Stephans, R. A., and Talso, W. W. (Eds.), *System Safety Analysis Handbook*, 2nd ed., New Mexico Chapter, System Safety Society, Albuquerque, NM, 1997.

Usmen, M. A., *Construction Safety and Health for Civil Engineers, Instructional Module*, Department of Health and Human Services, National Institute for Occupational Safety and Health, Cincinnati, OH, 1994.

U.S. Department of Defense, *Procedures for Performing a Failure Modes, Effects, and Criticality Analysis*, MIL-STD-1629A, Washington, DC, 1980.

U.S. Department of Defense, *System Safety Program Requirements*, MIL-STD-882C, Washington, DC, 1993.

U.S. House of Representatives, *Structural Failures in Public Facilities*, House Committee on Science and Technology, Report 98-621, 1984.

Wang, L. K., In-plant management and disposal of industrial hazardous substances, Wang, L. K., Hung, Y. T., Lo, H. H., and Yapijakis, C. (Eds.), *Handbook of Industrial and Hazardous Wastes Treatment*, Marcel Dekker, CRC Press, NY, 2004, pp. 515–584.

Wang, L. K., Hung, Y. T., Lo, H. H., and Yapijakis, C., *Hazardous Industrial Waste Treatment*, CRC Press, NY, 516, p. 2006.

Appendixes 1 to 19

Appendix 1

Nomenclature and Abbreviations

A	angstrom
A	Ampere
$A_{leakage}$	actual leakage rate of asbestos-cement or ductile iron pipes, gpd/mi-in
$A_{leakage}$	actual leakage rate of asbestos-cement or ductile iron pipes, Lpd/km-cm
ABS	acrylonitrile butadiene styrene
ac	acre
a.m.	before noon
ASCE	American Society of Civil Engineers
atm	atmosphere
bbl	barrel
BMP	best management practice
BOD	biochemical oxygen demand
Btu	British thermal unit
c	curie
°C	degree centigrade
cal	calorie
CECS	Center for Excellence in Construction Safety
Chap.	chapter
Chaps.	chapters
CII	Construction Industry Institute
cm	centimeter
CMP	corrugated metal pipe
COD	chemical oxygen demand
d	day, diameter
D	darcy
DO	dissolved oxygen
DWF	dry-weather flow
emf	electromotive force
EPS	extended-period simulation
Eq.	equation

Water and Wastewater Engineering: Hydraulics, Hydrology and Management, Volume 1, Fourth Edition.
Lawrence K. Wang, Mu-Hao Sung Wang, and Nazih K. Shammas.

Companion website: www.wiley.com/go/Wang/Waterandwastewater4e

Eqs.	equations
°F	degree Fahrenheit
Fig.	figure
Figs.	figures
ft	foot
ft^2	square foot
ft^3	cubic foot
g	gram
gal	gallon
gpcd	gallon per capita per day
gpd	gallons per day
gph	gallons per hour
gpud	gallons per unit per day
gpm	gallons per minute
h	hour
ha	hectare
HGL	hydraulic grade line
hp	horsepower
in.	inch
I_t	inflow at time *t*
J	Joule
J_{100}	each 100 plastic pipe joints
K	hydraulic conductivity
kg	kilogram
kWh	kilowatt-hour
L	liter
Lpcd	liter per capita per day
Lpd	liter per day
Lph	liter per hour
Lpm	liter per minute
lb	pound
m	meter
m^2	square meter
m^3	cubic meter
MAF	mean annual flow
mD	millidarcy
me	milliequivalent
mg	milligram
MG	million gallons
MGD	million gallons per day
min	minute

mL	milliliter
ML	million liters
MLD	million liters per day
MLSS	mixed liquor suspended solids
MLVSS	mixed liquor volatile suspended solids
mm	millimeter
mol	mole (gram-molecular weight)
mol wt	molecular weight
mph	miles per hour
MPN	most probable number
N	Newton
NBS	National Bureau of Standards
NIOSH	National Institute for Occupational Safety and Health
O	outflow or discharge
O_t	outflow at time *t*
O&M	operation and maintenance
OSHA	Occupational Safety and Health Administration
p.	Page
$P_{leakage}$	actual leakage rate of plastic pipes, gph/100 joints
$P_{leakage}$	actual leakage rate of plastic pipes, Lph/100 joints
PF	peaking factor
p.m.	afternoon
ppmv	parts per million, volume
psi	pound per square inch
psia	pound per square inch, absolute
psig	pound per square inch, gauge
PtD	Prevention through Design
PVC	polyvinyl chloride
$Q_{leakage}$	measured leak rate, gph
$Q_{leakage}$	measured leak rate, Lph
R, R_g	gas constant
rpm	revolutions per minute
s	second
SCBA	self-contained breathing apparatus
SDGS	Small-diameter gravity sewer
SDI	sludge density index
SG	specific gravity
SS	suspended solids
SSO	sanitary sewer overflow
STEP	septic tank effluent pump

SVI	sludge volume index
t	ton (English)
T	tonne (metric)
U.S.	United States
USD	United States Dollar
U.S. DOD	U.S. Department of Defense
U.S. EPA	United States Environmental Protection Agency
UV	ultraviolet
W	watt
WQS	water quality standards
$W(u)$	well function of u
yd	yard
yd^2	square yard
yd^3	cubic yard
yr	year
μ	micron
μc	micro curie
μg	microgram
μm	micrometer
Δt	hydrologic time step

Appendix 2

Units

a	area
a	continuous gutter depression
A	area
A	absorbance at about 254 nm
A	debt
A_j	age of water at jth node
AA_j	average age at the node immediately upstream of node j
avg	average
b	breadth, width, thickness
B	resistance characteristics of a formation
BG	billion gallons
BHP	break horsepower
BL	billion liters
c	coefficient, Chézy coefficient
c	runoff coefficient
c	rate of surge propagation
c	convex routing coefficient
c	substance concentration as a function of distance and time
c	the killing concentration of HOCl, mg/L
C	concentration of the disinfectant, mg/L
C	Hazen–Williams coefficient
C	runoff coefficient
C_a	average (mixed) constituent concentration
C_e	external source concentration into node k
C_i	constituent concentrations of the inflows
C_k	concentration at node k
C_0	orifice coefficient
C_w	weir coefficient
$C1$	first concentration of liquid disinfectant, %
$C2$	second concentration of liquid disinfectant, %

Water and Wastewater Engineering: Hydraulics, Hydrology and Management, Volume 1, Fourth Edition.
Lawrence K. Wang, Mu-Hao Sung Wang, and Nazih K. Shammas.

Companion website: www.wiley.com/go/Wang/Waterandwastewater4e

CD	chemical disinfectant dose, mg/L (lb/MG)
Ct_{calc}	calculated Ct value, mg/L-min
d	distance or depth
d_c	critical depth
d_i	depth at lip of curb opening
d_0	effective head at the orifice
D	diameter
D	draft
D	hydraulic diffusivity
D	housing density
D	surface detention, in. of depth
D	number of homes served
D	UV dosage, mJ/cm^2 or $mW\text{-}s/cm^2$
D_c	dew-point temperature
D_i	inside diameter of pipe
D_w	well drawdown
deg	degree
DD	disinfectant demand, mg/L
DR	disinfectant residual (chlorine residual), mg/L
E	rate of evaporation, efficiency
E	activation energy, cal
E_0	efficiency
E_{pump}	efficiency of the pump
E_{motor}	efficiency of the motor driving the pump
E_{ww}	wire-to-water efficiency
f	dimensionless friction factor
f'	porosity ratio
f_d	friction head loss in the discharge piping
f_s	friction head loss in the suction piping
$\mathbf{F}$	Froude number
F	freezing index
F	frequency of occurrence
g	acceleration of gravity
h	head, height
h	curb throat opening height
h_c	critical head
h_{cg}	height of center of gravity
h_d	static discharge head
h_f	head loss
hs	static suction head or lift

H	head, height
HP	power in US customary unit
H_A	energy at Section A
H_B	energy at Section B
H_a	energy added
H_f	friction head loss or energy loss
H_{sv}	suction head
i	intensity
i	slope
i	interest rate
I	hydraulic gradient
I	inflow or recharge
I	recurrence interval
I	UV light intensity in the bulk solution, mW/cm^2
I_{cg}	Moment of inertia of an area about its center of gravity axis
$INHP$	input power in US customary unit
$INMP$	input power in metric unit
I_p	intensity of pressure at the center of gravity of an area
I_t	inflow at time t
$It\text{-}\Delta t$	inflow at previous time step
j	pipe flowing into node k
k	constant, flow multiplier for time of day
k	coefficient of proportionality or the rate constant, 1/min
k′	$-k \log e$, 1/min
k′	constant
K	permeability
K	hydraulic conductivity
K	constant
K_c	0.376 (0.56 in US units) in gutter equation
K_c	0.817 (0.60 in US units) in curb inlet equation
K_{cs}	0.0828 (0.15 in US units)
K_{cf}	0.295 (0.090 in US units)
K_h	hydrolysis constant, mg/L
K_i	ionization constant, mg/L
K_w	ratio of length of screen to saturated thickness of an aquifer
K_1	$0.15\ ft^3/min = 0.07\ L/s$
K_2	conversion factor, $7.48\ gal/ft^3$
l, ℓ	length
L	length, saturation or maximum population
L_T	length of curb inlet required to capture 100% of the flow

MHP	motor horsepower
MP	power in metric unit
n	number
n	number of years considered
n	number of cycles
n	coefficient of dilution or a measure of the order of the reaction
n	coefficient of retardance
n	constant
n	coefficient of roughness, Manning's coefficient
n	pump speed, rpm
N	number of organisms remaining, #
N	pump speed, rpm
N	Kutter's coefficient of roughness
N_O	initial number of organisms, #
N_s	specific speed
N_v	number of valves connected to the sewage collection station
O	outflow or discharge
p	monthly percentage of daylight hours in the year
P	pressure
P	population in thousands
P	total force
P	perimeter of the inlet
p	percent concentration of disinfectant, %
p	fractional part of a cycle in well pumping
P_a	atmospheric pressure
P_a	absolute pressure
P_{atm}	atmospheric pressure
P_d	vapor pressure at dew-point temperature
P_{gauge}	gauge pressure
P_k	design year population in thousands
P_n	future population
P_o	present population
P_p	past population
P_s	absolute pressure
p_t	percentage of residual ozone at time t, %
P_w	vapor pressure at a given water temperature
q, qs	flow rate
Q	flow rate
$Q1$	first flow of liquid disinfectant, L/min or ft^3/min
$Q2$	second flow of liquid disinfectant, L/min or ft^3/min

Q_{19}	the useful ratio t_1/t_2 for $T_2 - T_1 = 19$
Q_a	pumping station average flow
Q_{avg}	average daily water demand
Q_d	disinfectant feed rate, mg/d or lb/d
Q_{dp}	discharge pump capacity
Q_e	external source flow into node k
Q_i	flow rate to the jth node from the ith node
$Q_{ins\ peak}$	instantaneous peak water demand, gpm
Q_i	intercepted flow by a street stormwater inlet
Q_{i0}	flow intercepted by the inlet operating as an orifice
Q_{iw}	flow intercepted by the inlet operating as a weir
Q_j	flow in pipe j
Q_{max}	collection station peak flow
Q_{min}	pumping station minimum flow
Q_s	side flow at street inlet
Q_t	flow at time t
Q_{vp}	vacuum pump capacity
Q_w	flow in depressed section of gutter
Q_w	water flow rate, L/d or MG/d
r	radius
r	hydraulic radius
r	radial distance
r	the proportionate efficiency of OCl^- ions relative to HOCl
r_c	permissible distance between production and disposal wells
r_w	effective well radius
R	rate of rainfall
$\mathbf{R}$	Reynolds number
R	residuals
R	hydraulic radius
R	frictional resistance
R	probable rate of population increase per year
R	the gas constant, 1.99 cal/(mol-K)
R	capital recovery factor
R_f	the ratio of intercepted frontal flow to the total frontal flow in the gutter
R_g	gas constant
R_s	the ratio of side flow intercepted to total side flow in the gutter
R_T	total concentration of chlorine required to produce a given% of kill, mg/L
s	drawdown, hydraulic gradient
s_s	specific gravity of particles
S	storage coefficient
S'	unit storage

S_e	composite or equivalent slope in the gutter
S_L	road longitudinal slope
S_n	drawdown in the pumped well
S_w	gutter cross-slope
Sw'	gutter cross-slope at the inlet location measured from the pavement cross-slope, S_x
S_x	road cross-slope
t	time, thickness
t_c	time of concentration
t_p	time required to effect a constant percentage kill of the organisms, min
t_t	travel time
T	temperature
T	absolute temperature, K
T	transmissivity
$\% T$	percent UV transmittance, %
T	total width of flow or spread in gutter
T'	mean monthly temperature
T_a	absolute temperature
T_c	time of concentration
T_m	daily maximum temperature
TL_m	median tolerance limit
U	water consumptive use
v, v_s	velocity in a partially full section
v_c	critical velocity
V	velocity in full section
V	volume
V	volt
V_0	splash-over velocity of the inlet
V_o	operating volume of the collection tank
V_s	specific volume
V_s	volume of silt
V_t	total volume of the collection tank
w	wind velocity
W	weight
W	width
W	width of flow in gutter depressed section
WHP	water horsepower
W_g	width of grate inlet
W_p	wetted perimeter
x	distance
y	number of organisms destroyed in unit time, #/min

y	population
y_{cg}	moment of inertia of the area about its center of gravity axis
y_{cp}	distance of the center of pressure measured along the plane from an axis located at the center of gravity
y_m	midyear population
y_{cg}	distance of center of gravity
Z	elevation
Z	ratio of average pumping station flow to minimum flow
α	coefficient
α	friction angle
α	vertical compressibility of aquifer material
β	coefficient, compressibility
γ	specific weight of water
γ_s	specific weight of sediment or solids
ε	a measure of absolute roughness
Δs	change in drawdown
Δs	change in storage volume
ΔS	change in storage
Δt	change in time
Δt	hydrologic time step
υ	kinematic viscosity
P	specific density
θ	porosity, angle
θ	inclination of the curb throat measured from the vertical direction as shown in Fig. 13.17
θ (c)	substance rate of reaction within the link
μ	viscosity, measure of central tendency
σ	standard deviation
$\sum$	sum
$\sum Q_{in}$	total flow into the node
$\sum Q_{out}$	total flow out of the node
τ	drag or tractive force
Ω	specific weight of water

Appendix 3

Physical Properties of Water

Viscosity and Density of Water

Temperature (°C)	Density p, γ (grams/cm^3), also s[c]	Absolute Viscosity μ, centipoises[a]	Kinematic Viscosity υ, centistokes[b]	Temperature (°F)
0	0.99987	1.7921	1.7923	32.0
2	0.99997	1.6740	1.6741	35.6
4	1.00000	1.5676	1.5675	39.2
6	0.99997	1.4726	1.4726	42.8
8	0.99988	1.3872	1.3874	46.4
10	0.99973	1.3097	1.3101	50.0
12	0.99952	1.2390	1.2396	53.6
14	0.99927	1.1748	1.1756	57.2
16	0.99897	1.1156	1.1168	60.8
18	0.99862	1.0603	1.0618	64.4
20	0.99823	1.0087	1.0105	68.0
22	0.99780	0.9608	0.9629	71.6
24	0.99733	0.9161	0.9186	75.2
26	0.99681	0.8746	0.8774	78.8
28	0.99626	0.8386	0.8394	82.4
30	0.99568	0.8004	0.8039	86.0

[a] 1 centipoise = 10^{-2} g mass/cm.s
[b] 1 centistoke = 10^{-2} cm^2/s
[c] 1 lb/ft^3 = 0.016 g/cm^3
Conversion factors:

1 stoke = 1 cm^2/s = 100 centistokes = 1.0763×10^{-3} ft^3/s
1 poise = 1 gram/cm.s = 100 centipoise = 0.1 N^{-s}/m^2 2.088×10^{-3} lb-s/ft^3

Water and Wastewater Engineering: Hydraulics, Hydrology and Management, Volume 1, Fourth Edition.
Lawrence K. Wang, Mu-Hao Sung Wang, and Nazih K. Shammas.

Companion website: www.wiley.com/go/Wang/Waterandwastewater4e

Physical Properties of Water in US Customary Units

Temp. (°F)	Specific weight, γ (lb/ft^3)	Density, ρ (slugs/ft^3)	Absolute Viscosity, $10^{-5}\,\mu$ (lb · s/ft^2)	Kinematic viscosity, $10^{-5}\,\upsilon$ (ft^2/s)	Surface tension, 100 σ (lb/ft)	Vapor-pressure head, P_v/γ (ft)	Bulk modulus of elasticity, $10^7\,\beta$ (lb/in^2)
32	62.42	1.940	3.746	1.931	0.518	0.20	293
40	62.43	1.940	3.229	1.664	0.514	0.28	294
50	62.41	1.940	2.735	1.410	0.509	0.41	305
60	62.37	1.938	2.359	1.217	0.504	0.59	311
70	62.30	1.936	2.050	1.059	0.500	0.84	320
80	62.22	1.934	1.799	0.930	0.492	1.17	322
90	62.11	1.931	1.595	0.826	0.486	1.61	323
100	62.00	1.927	1.424	0.739	0.480	2.19	327
110	61.86	1.923	1.284	0.667	0.473	2.95	331
120	61.71	1.918	1.168	0.609	0.465	3.91	333
130	61.55	1.913	1.069	0.558	0.460	5.13	334
140	61.38	1.908	0.981	0.514	0.454	6.67	330
150	61.20	1.902	0.905	0.476	0.447	8.58	328
160	61.00	1.896	0.838	0.442	0.441	10.95	326
170	60.80	1.890	0.780	0.413	0.433	13.83	322
180	60.58	1.883	0.726	0.385	0.426	17.33	313
190	60.36	1.876	0.678	0.362	0.419	21.55	313
200	60.12	1.868	0.637	0.341	0.412	26.59	308
212	59.83	1.860	0.593	0.319	0.404	33.90	300

Source: Water Engineering: Hydraulics, Distribution and Treatment, First Edition. Nazih K. Shammas and Lawrence K. Wang. © 2016 John Wiley & Sons, Inc. Published 2016 by John Wiley & Sons, Inc.

Conversion factors: 1 slug = 32.174 lb = 14.59 kg; 1 lb/ft^3 = 0.016 g/cm^3; 1 slug/ft^3 = 32.174 lb/ft^3; 1 g/cm^3 = 62.4176 lb/ft^3; 1 N = 1 Newton = 0.10197 kg = 0.22481 lb; 1 lb/in^2 = 6940 N/m^2 = 6.94 kPa

Physical Properties of Water in SI Units

Temp. (°C)	Specific weight, γ (N/m^3)	Density, ρ (kg/m^3)	Absolute Viscosity, $10^{-3}\ \mu$ ($N \cdot s/m^2$)	Kinematic viscosity, $10^{-6}\ \upsilon$ (m^2/s)	Surface tension, 100 σ (N/m)	Vapor-pressure head, P_v/γ (m)	Bulk modulus of elasticity, $10^7\ \beta$ (N/m^2)
0	9,805	999.9	1.792	1.792	7.62	0.06	204
5	9,806	1,000.0	1.519	1.519	7.54	0.09	206
10	9,803	999.7	1.308	1.308	7.48	0.12	211
15	9,798	999.1	1.140	1.141	7.41	0.17	214
20	9,789	998.2	1.005	1.007	7.36	0.25	220
25	9,779	997.1	0.894	0.897	7.26	0.33	222
30	9,767	995.7	0.801	0.804	7.18	0.44	223
35	9,752	994.1	0.723	0.727	7.10	0.58	224
40	9,737	992.2	0.656	0.661	7.01	0.76	227
45	9,720	990.2	0.599	0.605	6.92	0.98	229
50	9,697	988.1	0.549	0.556	6.82	1.26	230
55	9,679	985.7	0.506	0.513	6.74	1.61	231
60	9,658	983.2	0.469	0.477	6.68	2.03	228
65	9,635	980.6	0.436	0.444	6.58	2.56	226
70	9,600	977.8	0.406	0.415	6.50	3.20	225
75	9,589	974.9	0.380	0.390	6.40	3.96	223
80	9,557	971.8	0.357	0.367	6.30	4.86	221
85	9,529	968.6	0.336	0.347	6.20	5.93	217
90	9,499	965.3	0.317	0.328	6.12	7.18	216
95	9,469	961.9	0.299	0.311	6.02	8.62	211
100	9,438	958.4	0.284	0.296	5.94	10.33	207

Note: Mathematical models of all above water properties can be found in the following U.S. government report: Wang, L. K., Wang, M. H. S., and Terranova, D., Development of Water Property Models for Water Quality Control, U.S. Dept, of Commerce, National Technical Information Services, Springfield, VA, PB80-153224, 1980, 53p.

Conversion factors:

1 psi = 1 lb/in^2 = 6940 N/m^2 = kPa;
1 N/m = 0.0678 lb/ft;
1 $lb\text{-}s/ft^2$ = 47.8377 $N\text{-}s/m^2$;
1 lb/ft^3 = 157.0811 N/m^3

Appendix 4

Vapor Pressure and Surface Tension of Water in Contact with Air

Temperature (T), °C	0	5	10	15	20	25	30
Vapor pressure (p_w), mm Hg[a]	4.58	6.54	9.21	12.8	17.5	23.8	31.8
Surface tension (σ), dyne/cm[b]	75.6	74.9	74.2	73.5	72.8	72.0	71.2

[a] To convert to in. Hg divide by 25.4.

[b] To convert to (lb force)/ft divide by 14.9.

$p_w = 4.571512 + 0.352142\,T + 0.007386\,T^2 + 0.000371\,T^3$

$p_w = 4.571512$ mm Hg when $T = 0°C$

$p_w = 6.563247$ mm Hg when $T = 5°C$

$p_w = 12.767617$ mm Hg when $T = 15°C$

$p_w = 23.78818$ mm Hg when $T = 25°C$

Source: Wang, L. K., Shammas, N. K., Selke, W. A., and Aulenbach, D. B. (eds.). Flotation Technology, Humana Press-Springer Science, NYC, NY. 49–84, 2010.

Water and Wastewater Engineering: Hydraulics, Hydrology and Management, Volume 1, Fourth Edition.
Lawrence K. Wang, Mu-Hao Sung Wang, and Nazih K. Shammas.

Companion website: www.wiley.com/go/Wang/Waterandwastewater4e

Appendix 5

Area Under the Normal Probability Curve

Fractional parts of the total area (1.0000) corresponding to distances between the mean (μ) and given value (x) in terms of the standard deviation (σ), i. e., $(x - \mu)/\sigma = t$.

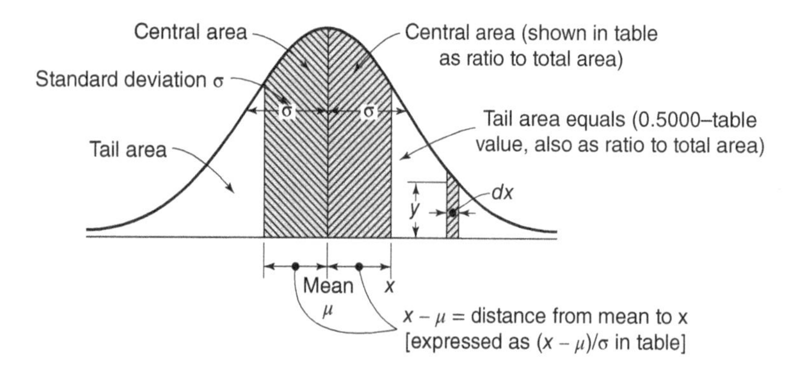

Water and Wastewater Engineering: Hydraulics, Hydrology and Management, Volume 1, Fourth Edition.
Lawrence K. Wang, Mu-Hao Sung Wang, and Nazih K. Shammas.

Companion website: www.wiley.com/go/Wang/Waterandwastewater4e

t	0.00	0.01	0.02	0.03	0.04	0.05	0.06	0.07	0.08	0.09
0.0	0.0000	0.0040	0.0080	0.0120	0.0160	0.0199	0.0239	0.0279	0.0319	0.0359
0.1	0.0398	0.0438	0.0478	0.0517	0.0557	0.0596	0.0636	0.0675	0.0714	0.0753
0.2	0.0793	0.0832	0.0871	0.0910	0.0948	0.0987	0.1026	0.1064	0.1103	0.1141
0.3	0.1179	0.1217	0.1255	0.1293	0.1331	0.1368	0.1406	0.1443	0.1480	0.1517
0.4	0.1554	0.1591	0.1628	0.1664	0.1700	0.1736	0.1772	0.1808	0.1844	0.1879
0.5	0.1915	0.1950	0.1985	0.2019	0.2054	0.2088	0.2123	0.2157	0.2190	0.2224
0.6	0.2257	0.2291	0.2324	0.2357	0.2389	0.2422	0.2454	0.2486	0.2517	0.2549
0.7	0.2580	0.2611	0.2642	0.2673	0.2704	0.2734	0.2764	0.2794	0.2823	0.2852
0.8	0.2881	0.2910	0.2939	0.2967	0.2995	0.3023	0.3051	0.3078	0.3106	0.3133
0.9	0.3159	0.3186	0.3212	0.3238	0.3264	0.3289	0.3315	0.3340	0.3365	0.3389
1.0	0.3413	0.3438	0.3461	0.3485	0.3508	0.3531	0.3554	0.3577	0.3599	0.3621
1.1	0.3643	0.3665	0.3686	0.3708	0.3729	0.3749	0.3770	0.3790	0.3810	0.3830
1.2	0.3849	0.3869	0.3888	0.3907	0.3925	0.3944	0.3962	0.3980	0.3997	0.4015
1.3	0.4032	0.4049	0.4066	0.4083	0.4099	0.4115	0.4131	0.4147	0.4162	0.4177
1.4	0.4192	0.4207	0.4222	0.4236	0.4251	0.4265	0.4279	0.4292	0.4306	0.4319
1.5	0.4332	0.4345	0.4357	0.4370	0.4382	0.4394	0.4406	0.4418	0.4429	0.4441
1.6	0.4452	0.4463	0.4474	0.4484	0.4495	0.4505	0.4515	0.4525	0.4535	0.4545
1.7	0.4554	0.4564	0.4573	0.4582	0.4591	0.4599	0.4608	0.4616	0.4625	0.4633
1.8	0.4641	0.4649	0.4656	0.4664	0.4671	0.4678	0.4686	0.4693	0.4699	0.4706
1.9	0.4713	0.4719	0.4726	0.4732	0.4738	0.4744	0.4750	0.4758	0.4761	0.4767
2.0	0.4772	0.4778	0.4782	0.4788	0.4793	0.4798	0.4803	0.4808	0.4812	0.4817
2.1	0.4821	0.4826	0.4830	0.4834	0.4838	0.4842	0.4846	0.4850	0.4854	0.4857
2.2	0.4861	0.4864	0.4868	0.4871	0.4875	0.4878	0.4881	0.4884	0.4887	0.4890
2.3	0.4893	0.4896	0.4898	0.4901	0.4904	0.4906	0.4909	0.4911	0.4913	0.4916
2.4	0.4918	0.4920	0.4922	0.4925	0.4927	0.4929	0.4931	0.4932	0.4934	0.4936
2.5	0.4938	0.4940	0.4941	0.4943	0.4945	0.4946	0.4948	0.4949	0.4951	0.4952
2.6	0.4953	0.4955	0.4956	0.4957	0.4959	0.4960	0.4961	0.4962	0.4963	0.4964
2.7	0.4965	0.4966	0.4967	0.4968	0.4969	0.4970	0.4971	0.4972	0.4973	0.4974
2.8	0.4974	0.4975	0.4976	0.4977	0.4977	0.4978	0.4979	0.4979	0.4980	0.4981
2.9	0.4981	0.4982	0.4982	0.4983	0.4984	0.4984	0.4985	0.4985	0.4986	0.4986
3.0	0.4987	0.4987	0.4987	0.4988	0.4988	0.4989	0.4989	0.4989	0.4990	0.4990
3.5	0.499367		4.0	0.499968		4.5	0.499997		5.0	0.4999997

Example: For $x = 10.8$, $\mu = 9.0$, $\sigma = 2.0$, $(x{-}\mu)/\sigma = 0.90$, and $0.3159 = 31.59\%$ of the area is included between $x = 10.8$ and $\mu = 9.0$.

Appendix 6

Flow Velocity and Discharge Rate for Pipes Flowing Full

Velocity of Flow and Rate of Discharge for Pipes Flowing Full When Frictional Resistance is 2 ft per 1,000 ft or 2 m per 1,000 m (2‰) and C is 100 in Hazen–Williams Formula

Diameter d (in.)	Area A (ft^2)	Velocity v (ft/s)	Discharge Q (1000 gpd)
(1)	(2)	(3)	(4)
4	0.0873	0.96	54.3
5	0.137	1.11	97.5
6	0.196	1.24	157
8	0.349	1.49	336
10	0.546	1.71	602
12	0.785	1.92	971
14	1.07	2.12	1,380
16	1.40	2.29	2,080
18	1.77	2.48	2,830
20	2.18	2.64	3,760
24	3.14	2.97	6,060
30	4.91	3.42	10,800
36	7.07	3.83	17,500
42	9.62	4.32	26,200
48	12.57	4.60	37,300
54	15.90	4.93	50,900
60	19.64	5.29	67,200

Conversion factors: 1 in. = 25.4 mm; 1 ft^2 = 0.0929 m^2; 1 ft/s = 0.3048 m/s; 1 gpd = 3.785 L/d = $3.785 \times 10^{-3} m^3/d$.

Water and Wastewater Engineering: Hydraulics, Hydrology and Management, Volume 1, Fourth Edition.
Lawrence K. Wang, Mu-Hao Sung Wang, and Nazih K. Shammas.

Companion website: www.wiley.com/go/Wang/Waterandwastewater4e

Appendix 7

Flow Velocity and Discharge Rate for Pipes Flowing Full

Velocity of Flow and Rate of Discharge for Pipes Flowing Full for $\sqrt{S}/N = 1$ in Manning's Formula[a]

Diameter D (in.)	Area A (ft^2)	Velocity V (ft/s)	Discharge Q (ft^3/s)	Reciprocal of Velocity (1/V)
(1)	(2)	(3)	(4)	(5)
6	0.1963	0.3715	0.07293	2.6921
8	0.3491	0.4500	0.1571	2.2222
10	0.5455	0.5222	0.2848	1.9158
12	0.7852	0.5897	0.4632	1.6958
15	1.2272	0.6843	0.8398	1.4613
18	1.7671	0.7728	1.366	1.2940
21	2.4053	0.8564	2.060	1.1677
24	3.1416	0.9361	2.941	1.0683
27	3.9761	1.0116	4.026	0.9885
30	4.9087	1.0863	5.332	0.9206
36	7.0686	1.2267	8.671	0.8152
42	9.6211	1.3594	13.08	0.7356
48	12.5664	1.4860	18.67	0.6729
54	15.9043	1.6074	25.56	0.6221
60	19.6350	1.7244	33.86	0.5799

[a] To find V or Q for given values of S and N, multiply column (3) or (4) by $\sqrt{S}/N$. To find S for given values of V or Q and N, multiply NV or NQ/A, respectively, by column (5) and square the product.
Conversion factors: 1 in. = 25.4 mm; 1 ft^2 = 0.0929 m^2; 1 ft/s = 0.3048 m/s; 1 ft^3/s = 28.32 L/s = 0.02832 m^3/s.

Water and Wastewater Engineering: Hydraulics, Hydrology and Management, Volume 1, Fourth Edition.
Lawrence K. Wang, Mu-Hao Sung Wang, and Nazih K. Shammas.

Companion website: www.wiley.com/go/Wang/Waterandwastewater4e

Appendix 8

Minimum Grades and Capacities of Circular Conduits Flowing Full

Minimum Grades and Capacities of Circular Conduits Flowing Full when N is 0.013 in the Manning Formula

		Diameter (in.)							
Velocity (ft/s)		6	8	10	12	15	18	21	24
2.0	S (‰)	4.89	3.33	2.48	1.94	1.44	1.13	0.923	0.775
	Q (ft^3/s)	0.393	0.698	1.09	1.57	2.45	3.53	4.81	6.28
2.5	S (‰)	7.64	5.20	3.88	3.06	2.25	1.74	1.44	1.21
	Q (ft^3/s)	0.491	0.873	1.36	1.96	3.07	4.42	6.01	7.85
3.0	S (‰)	11.0	7.50	5.58	4.37	3.24	2.54	2.08	1.74
	Q (ft^3/s)	0.589	1.05	1.64	2.36	3.68	5.30	7.22	9.42
5.0	S (‰)	30.5	20.8	16.1	12.2	9.00	6.96	5.76	4.84
	Q (ft^3/s)	0.982	1.75	2.73	3.93	6.14	8.84	12.0	15.7

Conversion factors: 1 in. = 25.4 mm; 1 ft/s = 0.3048 m/s; 1 ft^3/s = 28.32 L/s = 0.02832 m^3/s.

Water and Wastewater Engineering: Hydraulics, Hydrology and Management, Volume 1, Fourth Edition.
Lawrence K. Wang, Mu-Hao Sung Wang, and Nazih K. Shammas.

Companion website: www.wiley.com/go/Wang/Waterandwastewater4e

Appendix 9

Hydraulic Elements of Circular Conduits

Central angle: $\cos \frac{1}{2}\theta = 1 - 2d/D$

Area: $\frac{D^2}{4}\left(\frac{\pi\theta}{360} - \frac{\sin\theta}{2}\right)$

Wetted perimeter: $\pi D\theta/360$

Hydraulic radius:

$\frac{D}{4}\left(1 - \frac{360\sin\theta}{2\pi\theta}\right)$

Velocity: Manning's formula

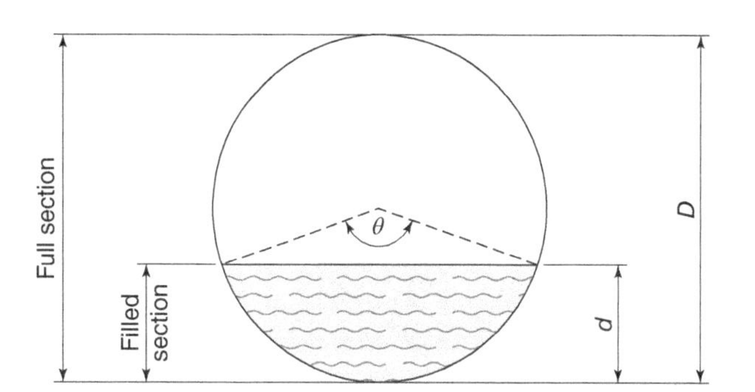

Depth d/D (1)	Area a/A (2)	Hydraulic Radius r/R (3)	R/r (4)	$(r/R)^{1/6}$ (5)	Velocity v/V for $N/n = 1.0$ (6)[a]	Discharge q/Q (7)[a]	Roughness N/n (8)
1.000	1.000	1.000	1.000	1.000	1.000	1.000	1.00
0.900	0.949	1.192	0.839	1.030	1.124	1.066	0.94
0.800	0.858	1.217	0.822	1.033	1.140	0.988	0.88
0.700	0.748	1.185	0.843	1.029	1.120	0.838	0.85
0.600	0.626	1.110	0.900	1.018	1.072	0.671	0.83
0.500	0.500	1.000	1.00	1.000	1.000	0.500	0.81
0.400	0.373	0.857	1.17	0.975	0.902	0.337	0.79
0.300	0.252	0.684	1.46	0.939	0.776	0.196	0.78
0.200	0.143	0.482	2.07	0.886	0.615	0.088	0.79
0.100	0.052	0.254	3.94	0.796	0.401	0.021	0.82
0.000	0.000	...	...	...	...	0.000	...

[a] For values corrected for variations in roughness with depth multiply by roughness ratio N/n in column 8.

Water and Wastewater Engineering: Hydraulics, Hydrology and Management, Volume 1, Fourth Edition.
Lawrence K. Wang, Mu-Hao Sung Wang, and Nazih K. Shammas.

Companion website: www.wiley.com/go/Wang/Waterandwastewater4e

Appendix 10

Basic Hydraulic Elements of Circular Conduits for All Values of Roughness and Slope

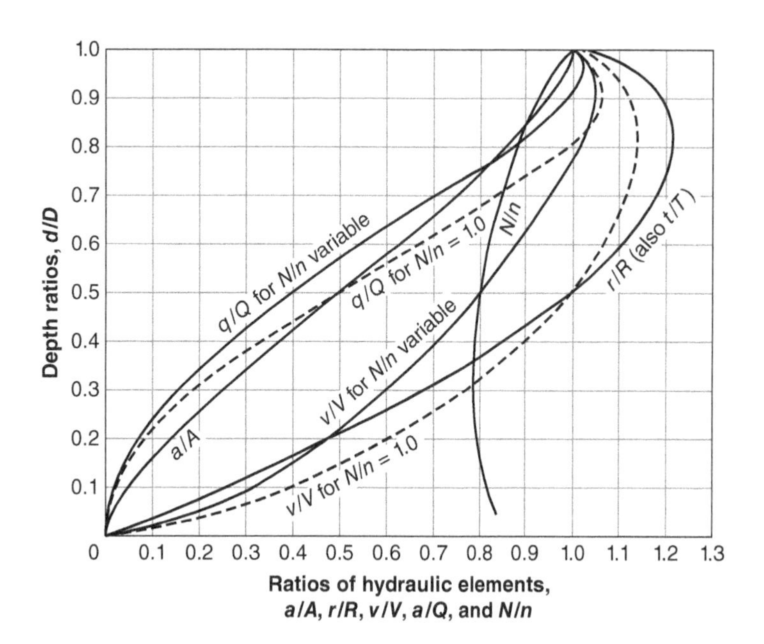

Water and Wastewater Engineering: Hydraulics, Hydrology and Management, Volume 1, Fourth Edition.
Lawrence K. Wang, Mu-Hao Sung Wang, and Nazih K. Shammas.

Companion website: www.wiley.com/go/Wang/Waterandwastewater4e

Appendix 11

Hydraulic Elements of Circular Conduits with Equal Self-Cleansing Properties at All Depths

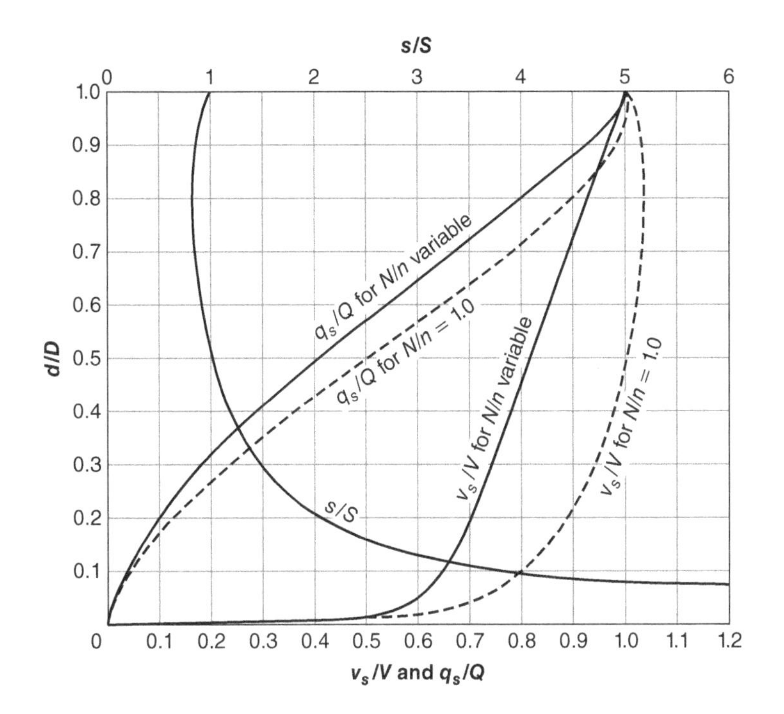

Water and Wastewater Engineering: Hydraulics, Hydrology and Management, Volume 1, Fourth Edition.
Lawrence K. Wang, Mu-Hao Sung Wang, and Nazih K. Shammas.

Companion website: www.wiley.com/go/Wang/Waterandwastewater4e

Appendix 12

Values of Hydraulic Elements in Circular Conduits

Values of $h/D, V_c/\sqrt{gD}$, and $\left[q/\left(A\sqrt{gD}\right)\right]^2$ for Varying Values of d_c/D in Circular Conduits

d_c/D	h/D	$V_c/\sqrt{gD}$	$[q/(A\sqrt{gD})]^2$
0.1	0.134	0.261	1.184×10^{-4}
0.2	0.270	0.378	2.86×10^{-3}
0.3	0.408	0.465	1.37×10^{-2}
0.4	0.550	0.553	4.18×10^{-2}
0.5	0.696	0.626	9.80×10^{-2}
0.6	0.851	0.709	1.97×10^{-1}
0.7	1.020	0.800	3.58×10^{-1}
0.8	1.222	0.919	6.23×10^{-1}
0.9	1.521	1.11	1.12

Water and Wastewater Engineering: Hydraulics, Hydrology and Management, Volume 1, Fourth Edition.
Lawrence K. Wang, Mu-Hao Sung Wang, and Nazih K. Shammas.

Companion website: www.wiley.com/go/Wang/Waterandwastewater4e

Appendix 13

Alternate Stages and Critical Depths of Flow in Circular Conduits

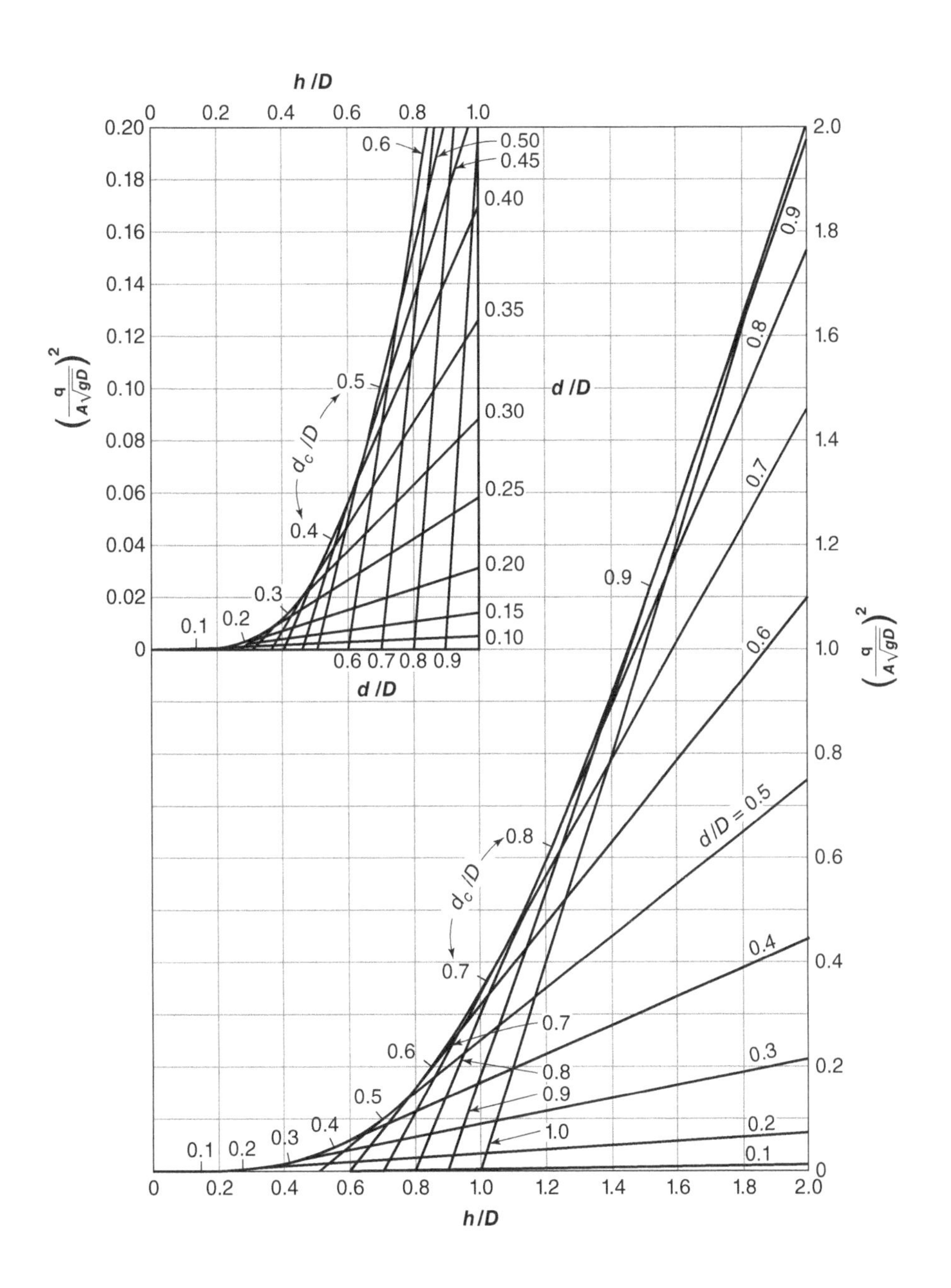

Water and Wastewater Engineering: Hydraulics, Hydrology and Management, Volume 1, Fourth Edition.
Lawrence K. Wang, Mu-Hao Sung Wang, and Nazih K. Shammas.

Companion website: www.wiley.com/go/Wang/Waterandwastewater4e

Appendix 14

Nomograms for Solution of Hazen–Williams Pipe Flow Equation

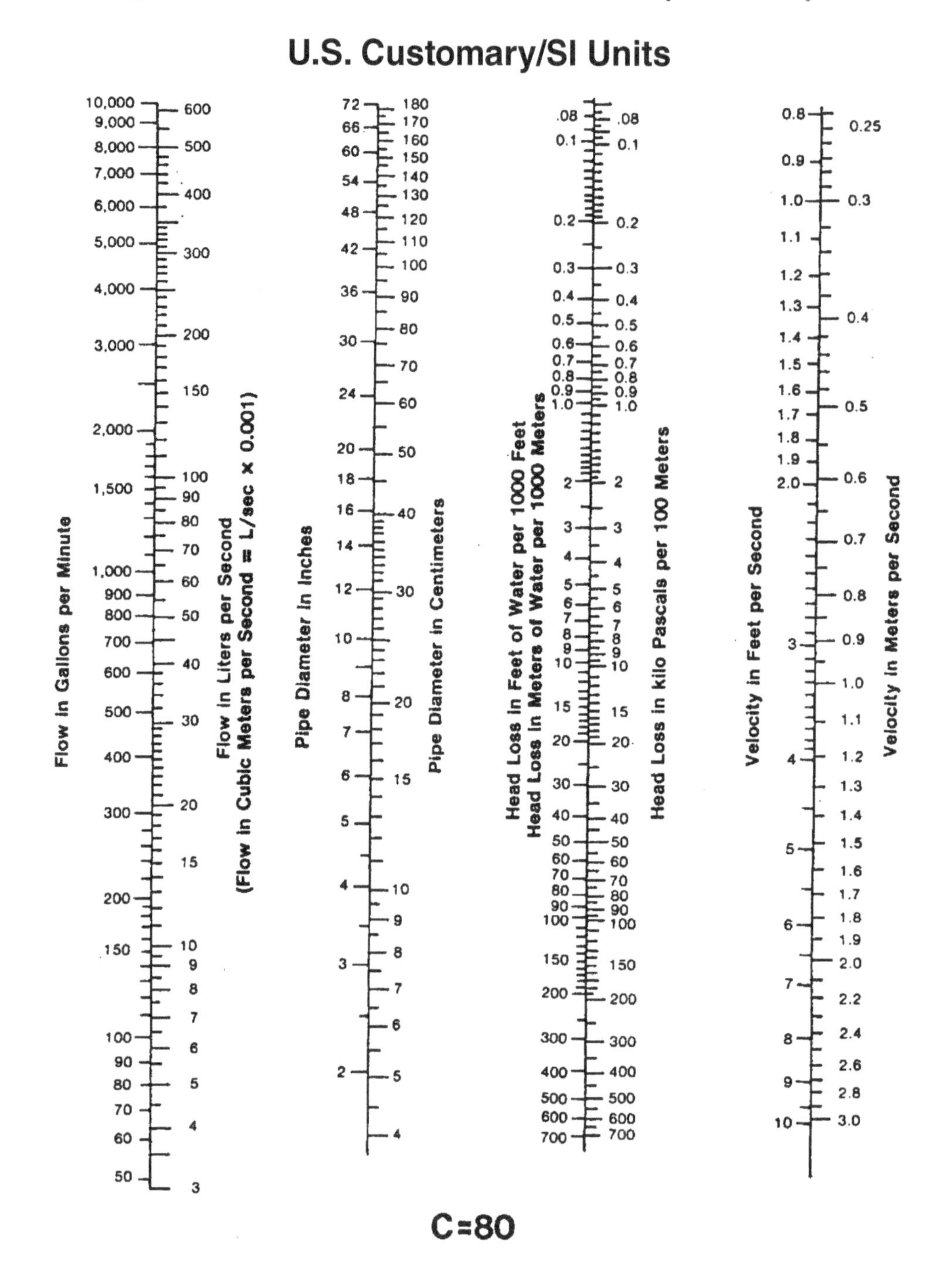

Water and Wastewater Engineering: Hydraulics, Hydrology and Management, Volume 1, Fourth Edition.
Lawrence K. Wang, Mu-Hao Sung Wang, and Nazih K. Shammas.

Companion website: www.wiley.com/go/Wang/Waterandwastewater4e

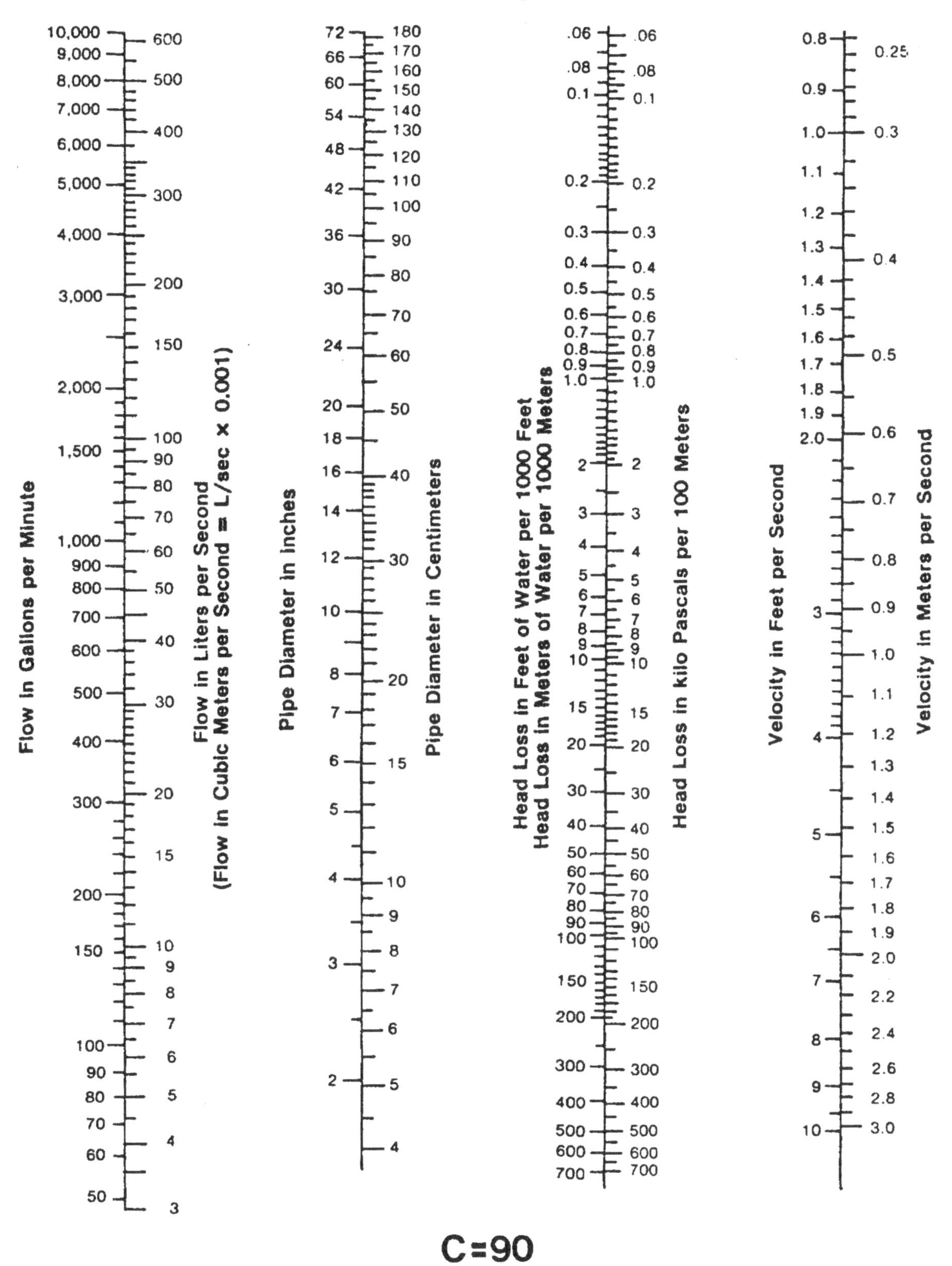
U.S. Customary/SI Units
Flow in Gallons per Minute
Flow in Liters per Second
(Flow in Cubic Meters per Second = L/sec × 0.001)
Pipe Diameter in Inches
Pipe Diameter in Centimeters
Head Loss in Feet of Water per 1000 Feet
Head Loss in Meters of Water per 1000 Meters
Head Loss in kilo Pascals per 100 Meters
Velocity in Feet per Second
Velocity in Meters per Second
C=90

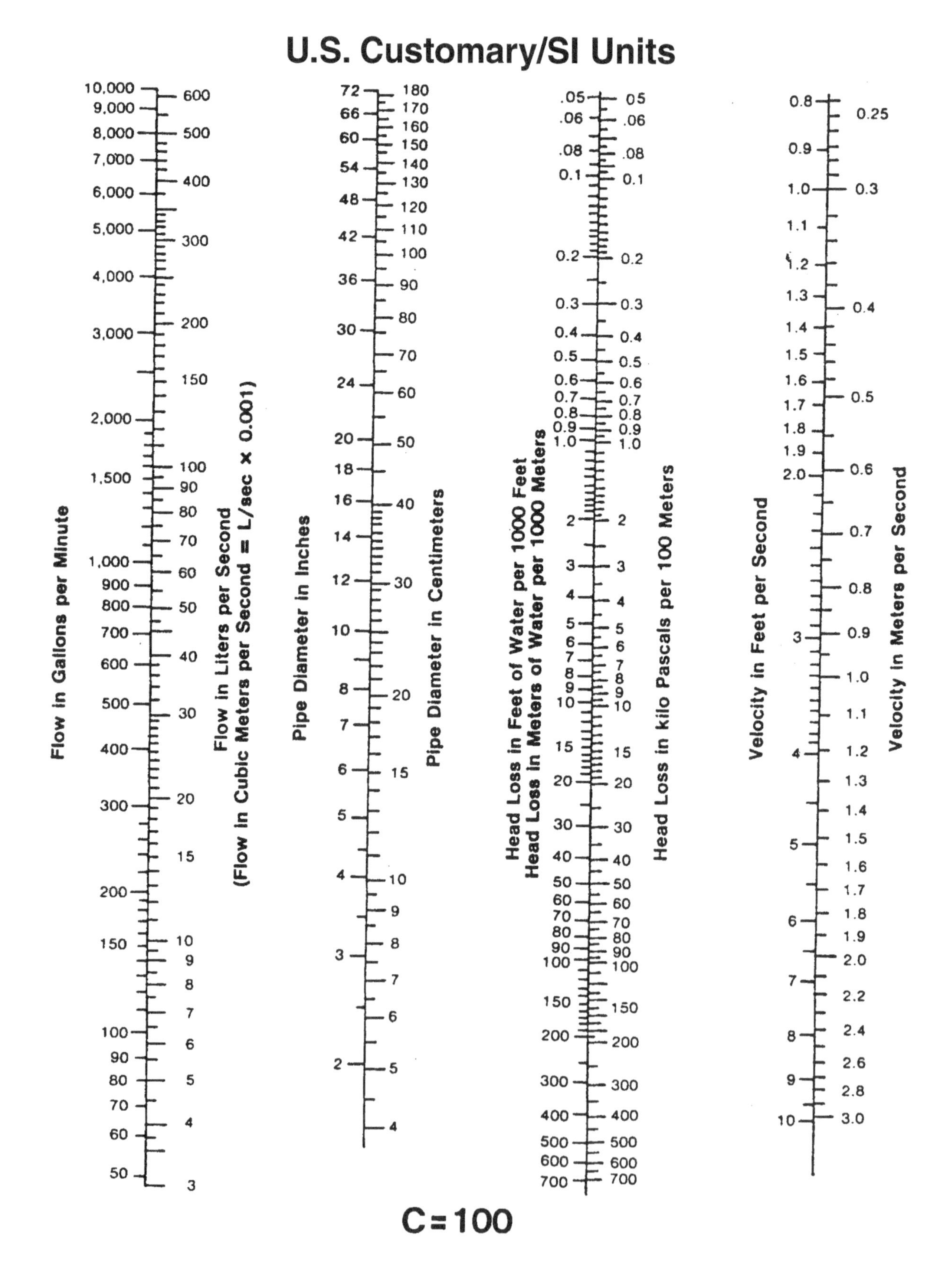
U.S. Customary/SI Units
Flow in Gallons per Minute
Flow in Liters per Second
(Flow in Cubic Meters per Second = L/sec × 0.001)
Pipe Diameter in Inches
Pipe Diameter in Centimeters
Head Loss in Feet of Water per 1000 Feet
Head Loss in Meters of Water per 1000 Meters
Head Loss in kilo Pascals per 100 Meters
Velocity in Feet per Second
Velocity in Meters per Second
C=100

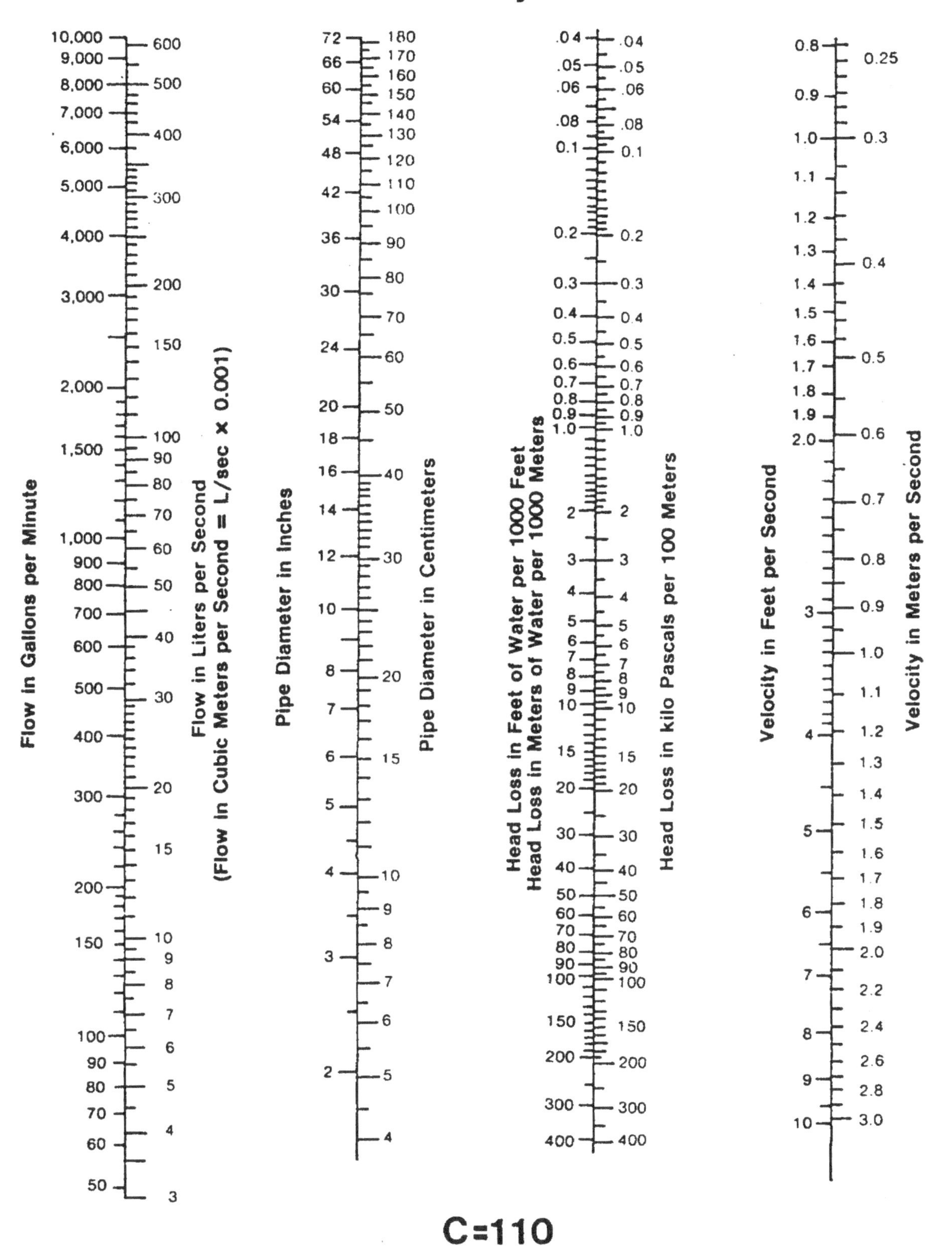
U.S. Customary/SI Units
Flow in Gallons per Minute
Flow in Liters per Second
(Flow in Cubic Meters per Second = L/sec × 0.001)
Pipe Diameter in Inches
Pipe Diameter in Centimeters
Head Loss in Feet of Water per 1000 Feet
Head Loss in Meters of Water per 1000 Meters
Head Loss in kilo Pascals per 100 Meters
Velocity in Feet per Second
Velocity in Meters per Second
C=110

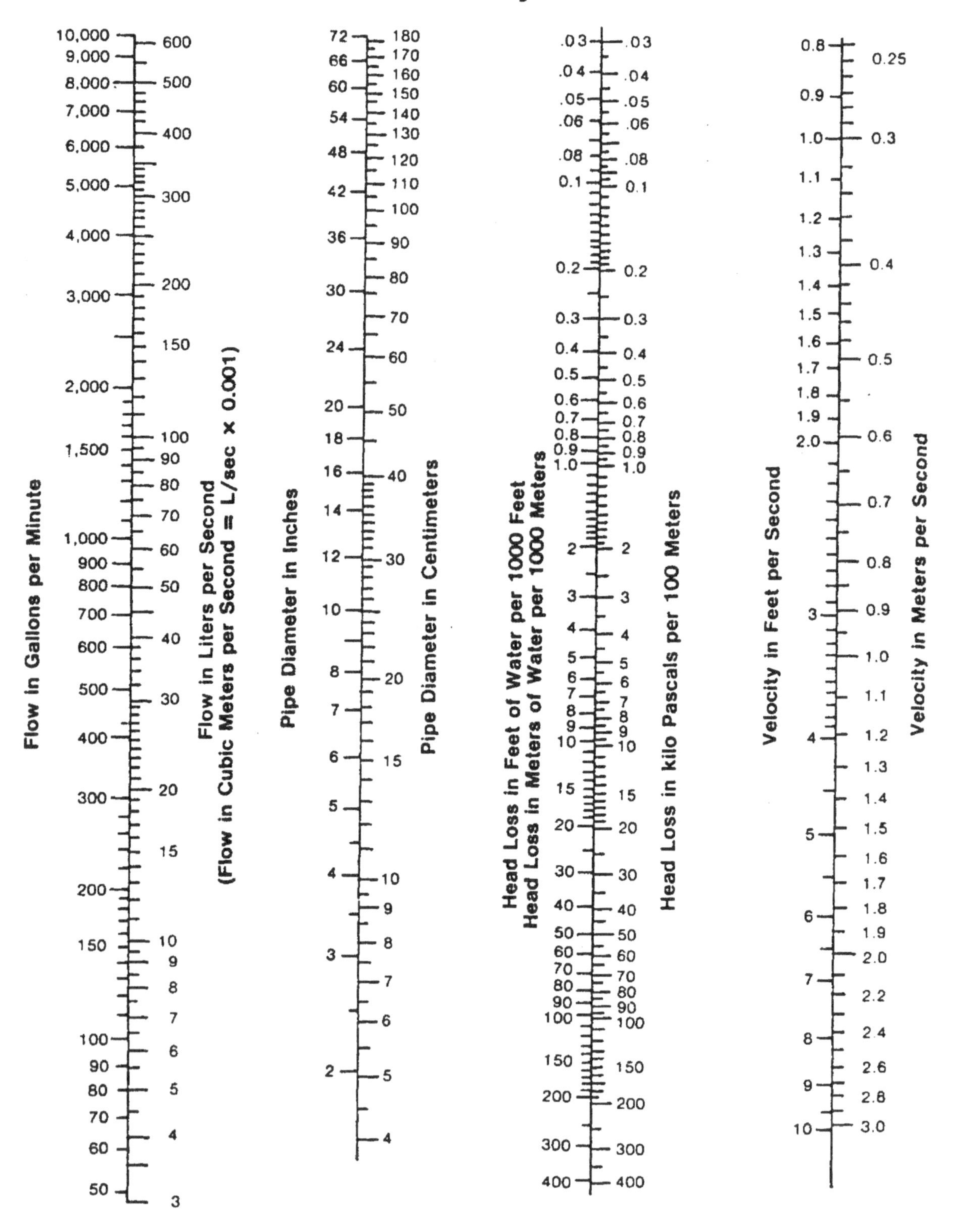
U.S. Customary/SI Units
Flow in Gallons per Minute
Flow in Liters per Second
(Flow in Cubic Meters per Second = L/sec × 0.001)
Pipe Diameter in Inches
Pipe Diameter in Centimeters
Head Loss in Feet of Water per 1000 Feet
Head Loss in Meters of Water per 1000 Meters
Head Loss in kilo Pascals per 100 Meters
Velocity in Feet per Second
Velocity in Meters per Second
C=130

U.S. Customary/SI Units

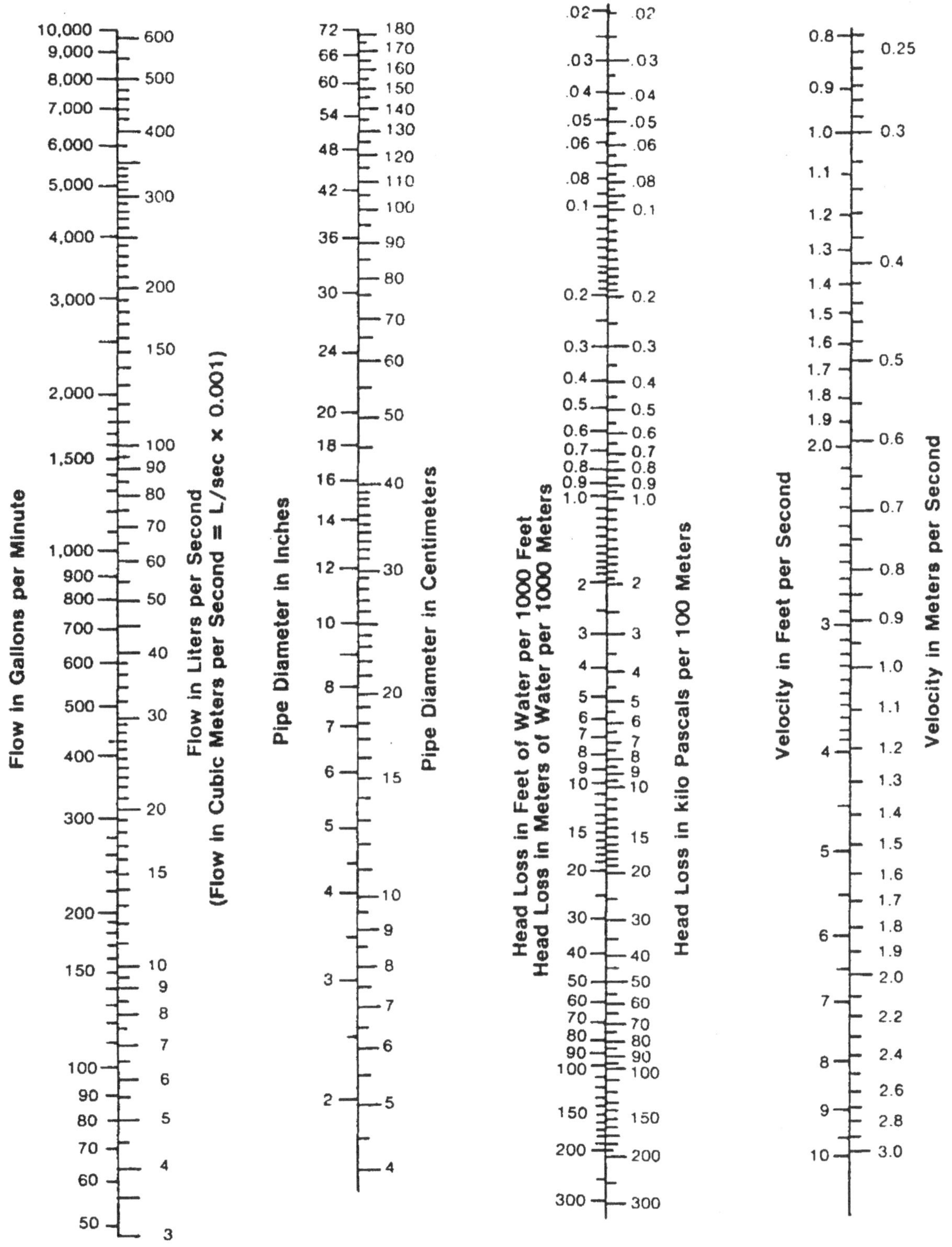

Flow in Gallons per Minute

Flow in Liters per Second
(Flow in Cubic Meters per Second = L/sec × 0.001)

Pipe Diameter in Inches

Pipe Diameter in Centimeters

Head Loss in Feet of Water per 1000 Feet
Head Loss in Meters of Water per 1000 Meters

Head Loss in kilo Pascals per 100 Meters

Velocity in Feet per Second

Velocity in Meters per Second

C=150

Appendix 15

Nomograms for Solution of Manning's Equation for Pipes Flowing Full

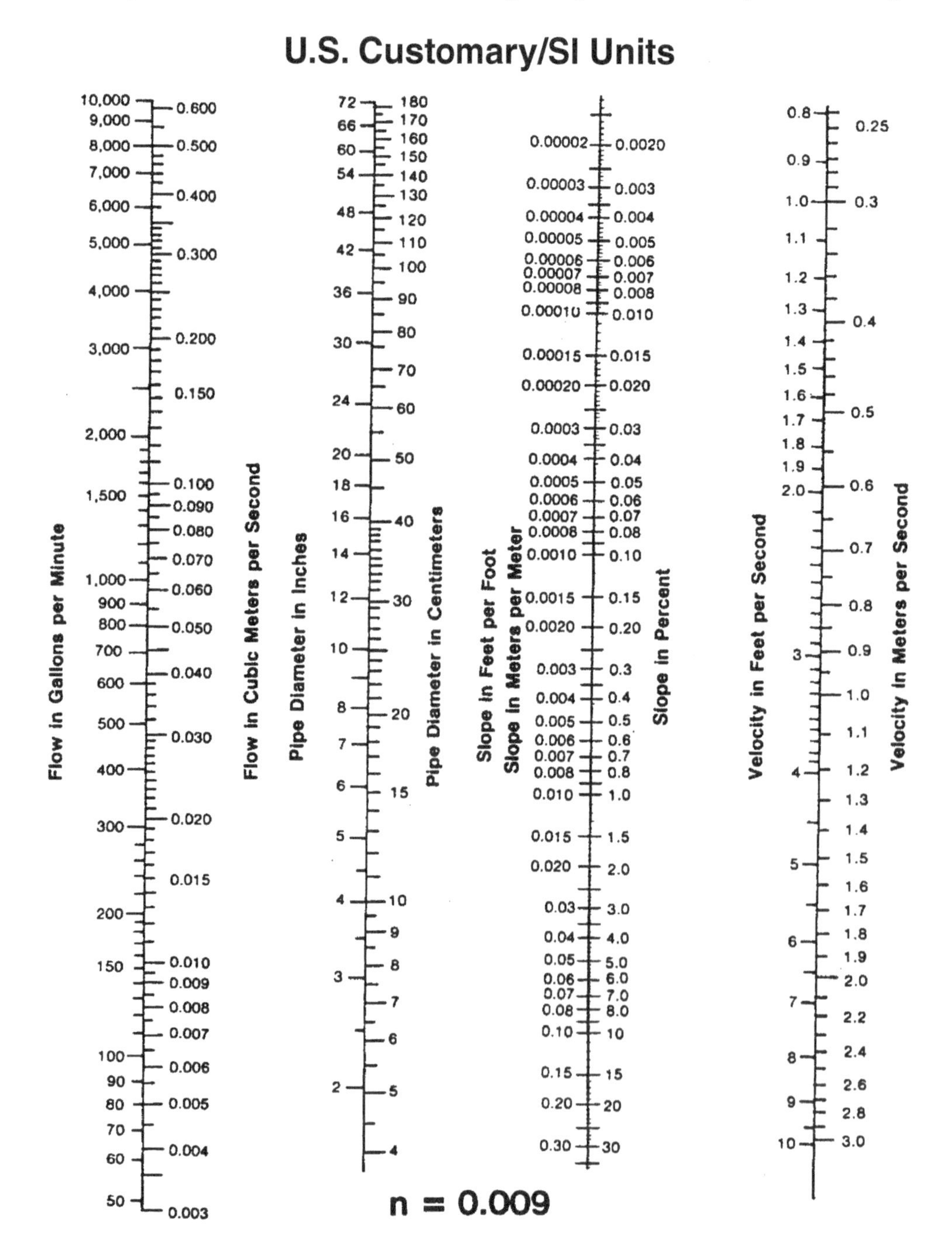

Water and Wastewater Engineering: Hydraulics, Hydrology and Management, Volume 1, Fourth Edition.
Lawrence K. Wang, Mu-Hao Sung Wang, and Nazih K. Shammas.

Companion website: www.wiley.com/go/Wang/Waterandwastewater4e

U.S. Customary/SI Units

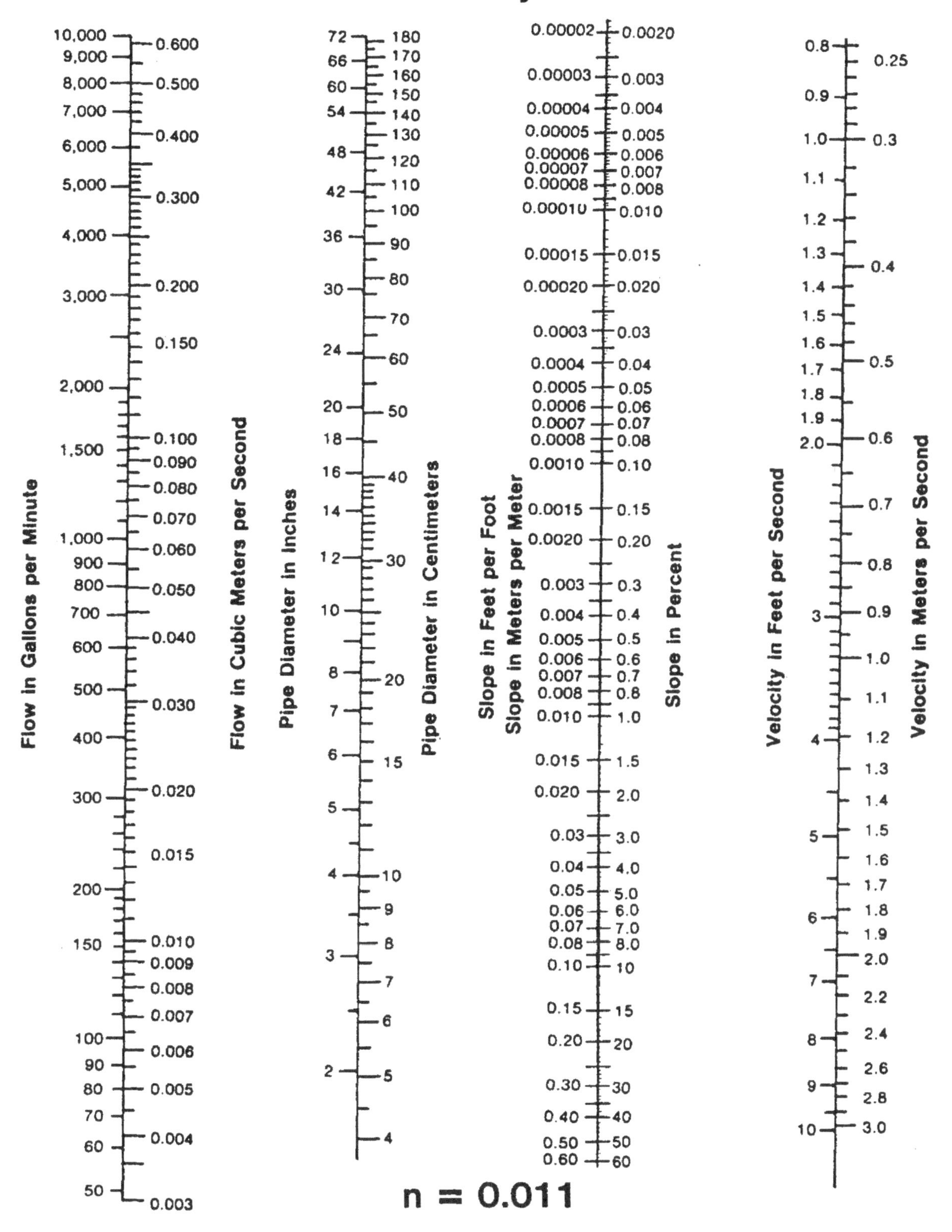

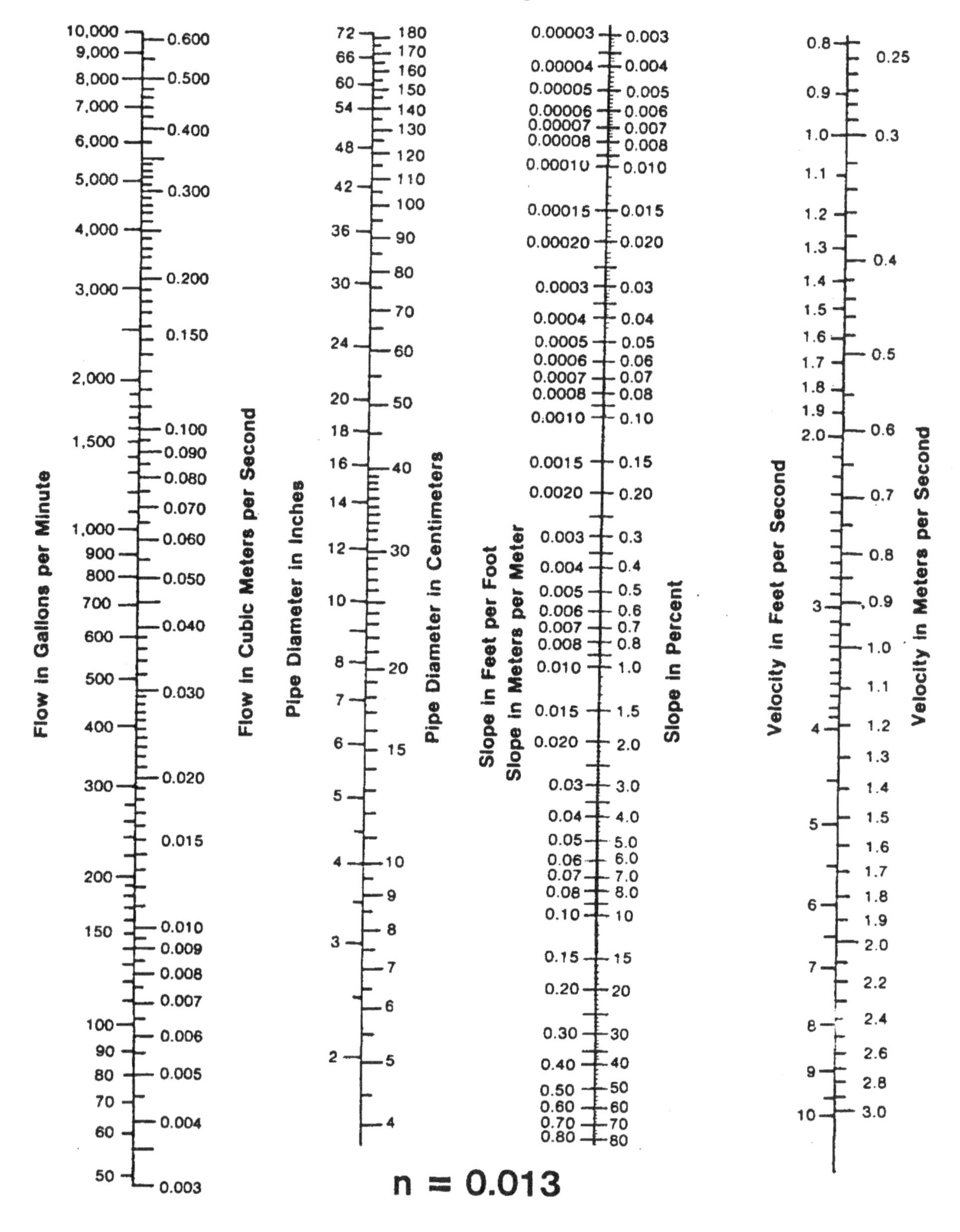
U.S. Customary/SI Units
Flow in Gallons per Minute
Flow in Cubic Meters per Second
Pipe Diameter in Inches
Pipe Diameter in Centimeters
Slope in Feet per Foot
Slope in Meters per Meter
Slope in Percent
Velocity in Feet per Second
Velocity in Meters per Second
n = 0.013

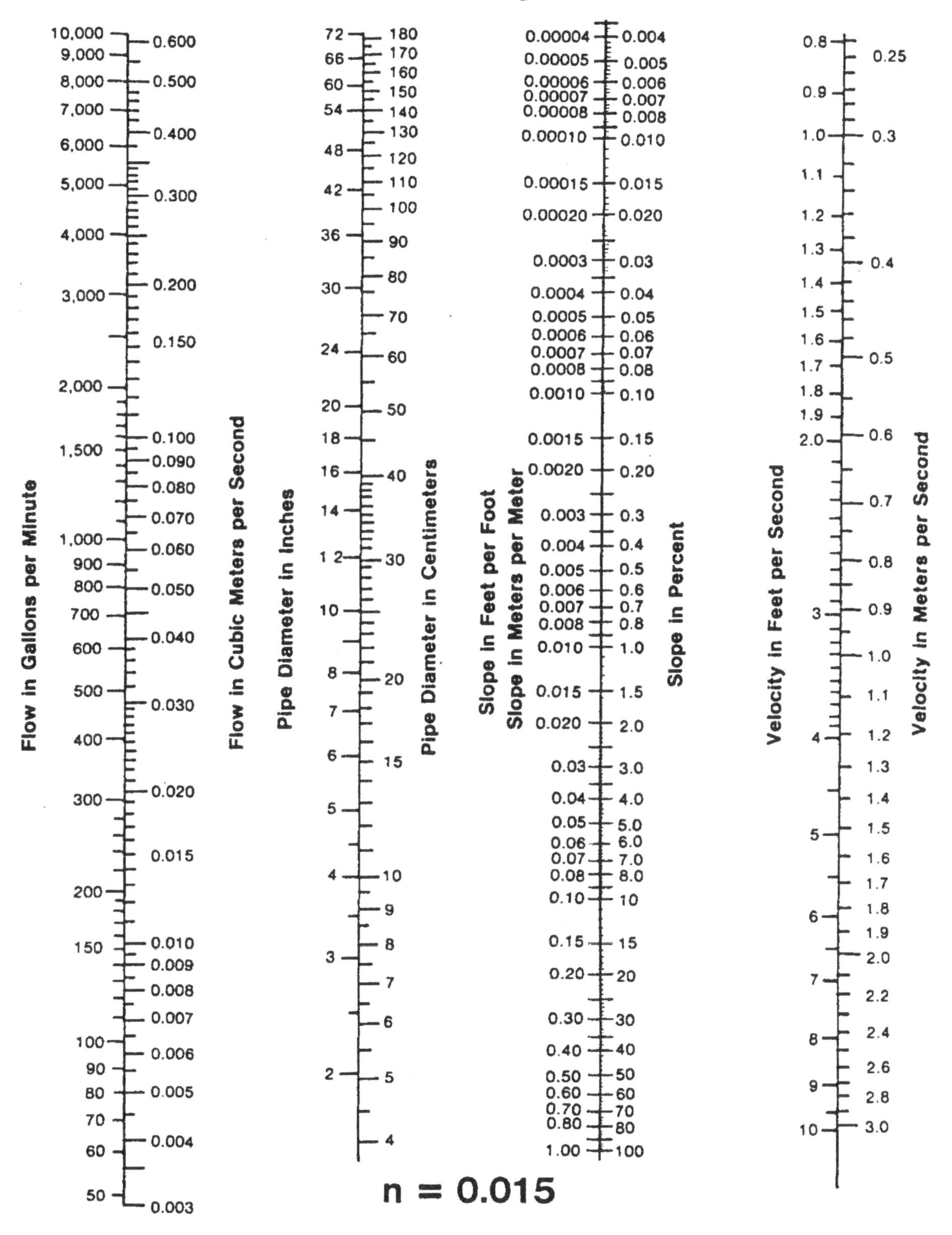
U.S. Customary/SI Units
Flow in Gallons per Minute
Flow in Cubic Meters per Second
Pipe Diameter in Inches
Pipe Diameter in Centimeters
Slope in Feet per Foot
Slope in Meters per Meter
Slope in Percent
Velocity in Feet per Second
Velocity in Meters per Second
n = 0.015

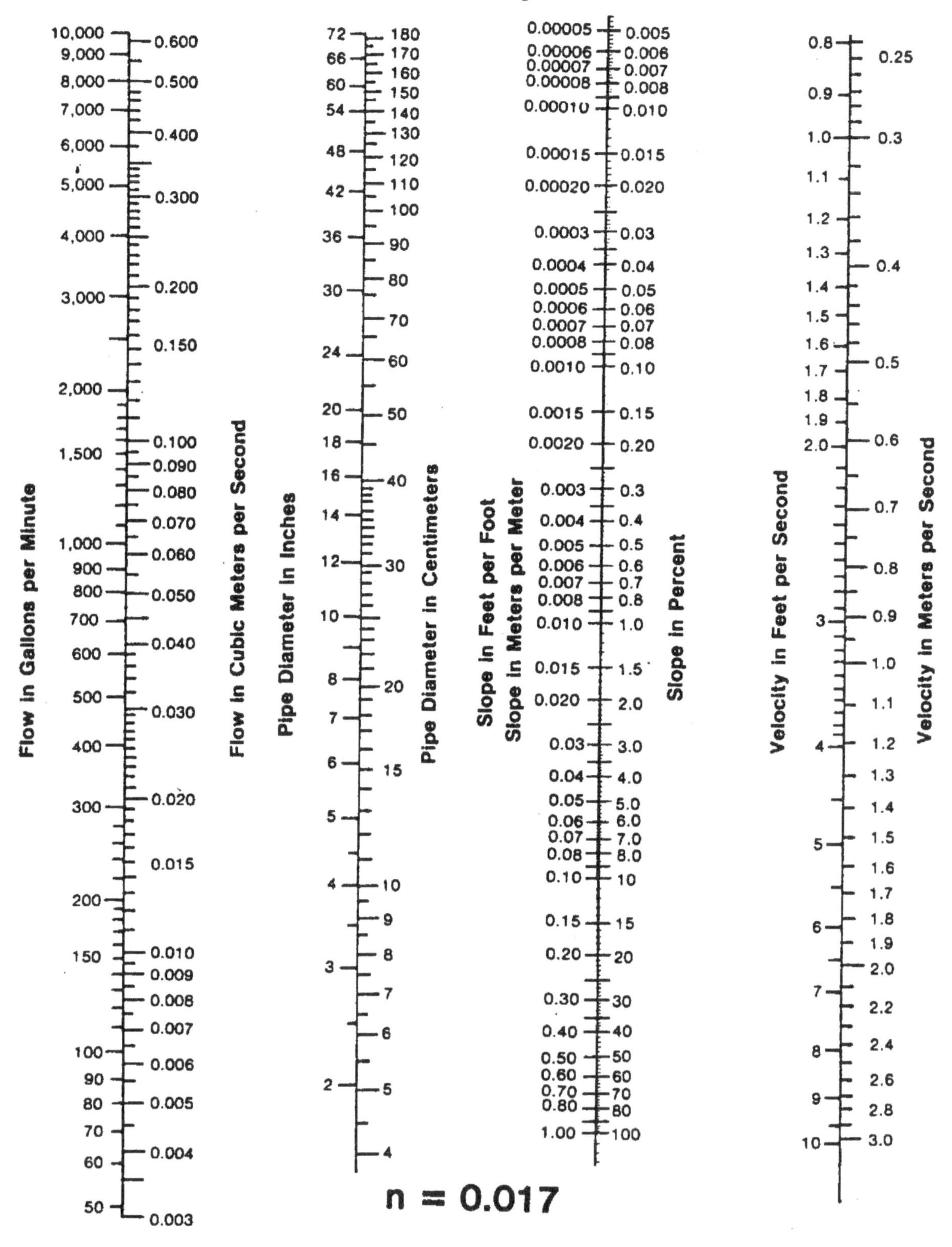
U.S. Customary/SI Units
Flow in Gallons per Minute
Flow in Cubic Meters per Second
Pipe Diameter in Inches
Pipe Diameter in Centimeters
Slope in Feet per Foot
Slope in Meters per Meter
Slope in Percent
Velocity in Feet per Second
Velocity in Meters per Second
n = 0.017

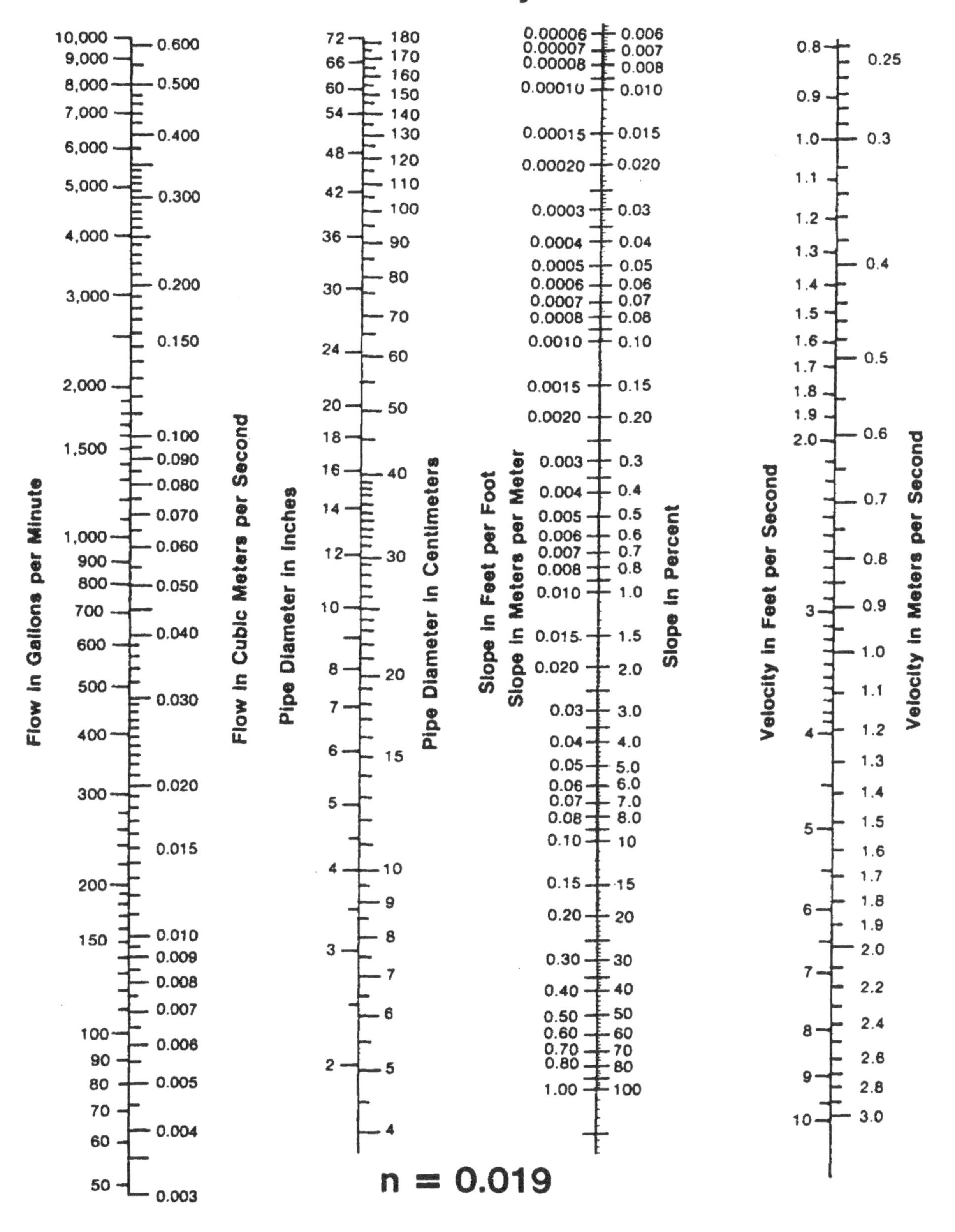
U.S. Customary/SI Units
Flow in Gallons per Minute
Flow in Cubic Meters per Second
Pipe Diameter in Inches
Pipe Diameter in Centimeters
Slope in Feet per Foot
Slope in Meters per Meter
Slope in Percent
Velocity in Feet per Second
Velocity in Meters per Second
n = 0.019

Appendix 16

U.S. Army Corps of Engineers Civil Works Construction Yearly Average Cost Index for Utilities

Year	Index	Year	Index
1967	100	1995	439.72
1968	104.83	1996	445.58
1969	112.17	1997	454.99
1970	119.75	1998	459.40
1971	131.73	1999	460.16
1972	141.94	2000	468.05
1973	149.36	2001	472.18
1974	170.45	2002	486.16
1975	190.49	2003	497.40
1976	202.61	2004	563.78
1977	215.84	2005	605.47
1978	235.78	2006	645.52
1979	257.20	2007	681.88
1980	277.60	2008	741.36
1981	302.25	2009	699.70
1982	320.13	2010	720.80
1983	330.82	2011	758.79
1984	341.06	2012	769.30
1985	346.12	2013	776.44
1986	347.33	2014	791.59
1987	353.35	2015	786.32
1988	369.45	2016	782.46
1989	383.14	2017	803.93
1990	386.75	2018	841.84
1991	392.35	2019	866.18
1992	399.07	2020	867.71
1993	410.63	2021	893.02[a]
1994	424.91	2022	918.91[a]

[a]US ACE. Yearly Average Cost Index for Utilities. In: *Civil Works Construction Cost Index System Manual*, 110-2-1304, U.S. Army Corps of Engineers, Washington, DC, pp. 44. (2020—Tables Revised 31 March).

Water and Wastewater Engineering: Hydraulics, Hydrology and Management, Volume 1, Fourth Edition.
Lawrence K. Wang, Mu-Hao Sung Wang, and Nazih K. Shammas.

Companion website: www.wiley.com/go/Wang/Waterandwastewater4e

Appendix 17

Equivalent Pipe Length to Head Loss in Fittings

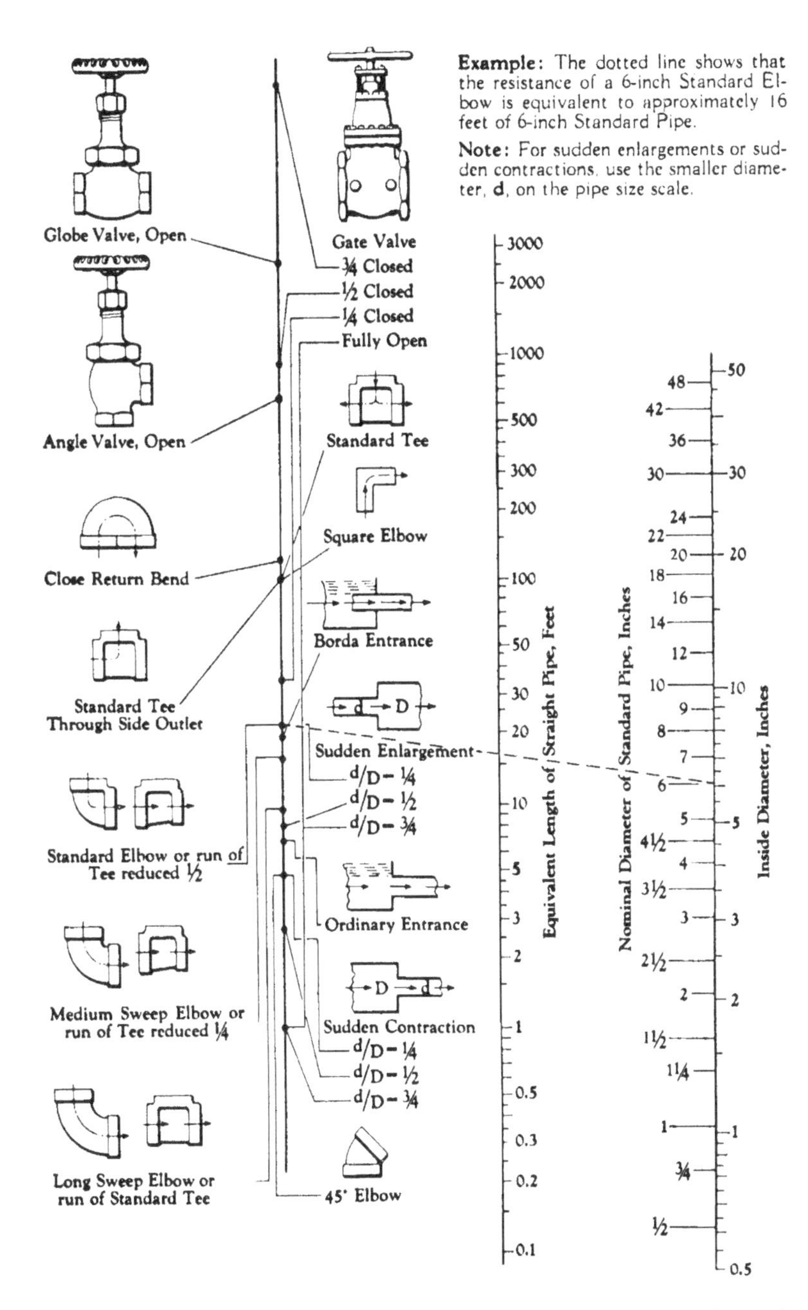

After Seelye, E. E. *Data Book for Civil Engineers*, Volume 1, Design, John Wiley & Sons, New York, NY, 1960.

Water and Wastewater Engineering: Hydraulics, Hydrology and Management, Volume 1, Fourth Edition.
Lawrence K. Wang, Mu-Hao Sung Wang, and Nazih K. Shammas.

Companion website: www.wiley.com/go/Wang/Waterandwastewater4e

Appendix 18

Simplified Overland Flow Time Determination for Urban Drainage

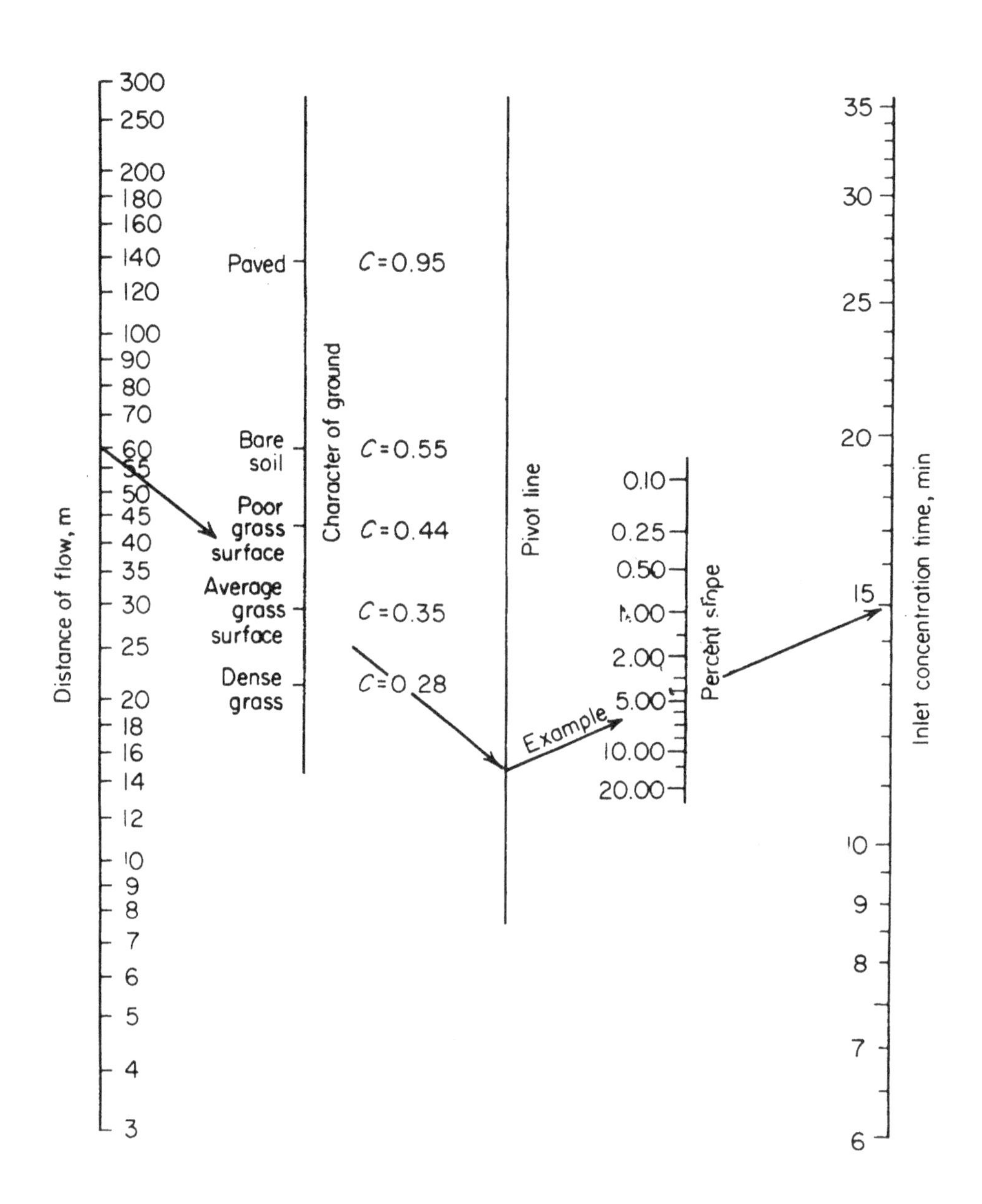

Water and Wastewater Engineering: Hydraulics, Hydrology and Management, Volume 1, Fourth Edition.
Lawrence K. Wang, Mu-Hao Sung Wang, and Nazih K. Shammas.

Companion website: www.wiley.com/go/Wang/Waterandwastewater4e

Appendix 19

Loss Coefficients for Transitions and Fittings

Description	Sketch	Additional data	K	
Pipe entrance	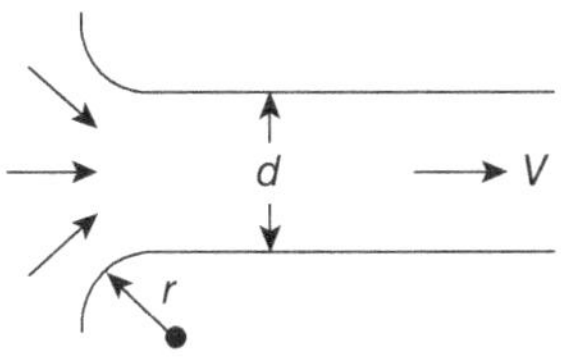	r/d	K_e	
		0.0	0.50	
		0.1	0.12	
		>0.2	0.03	
$h_f = K_e V^2/2g$				
Contraction	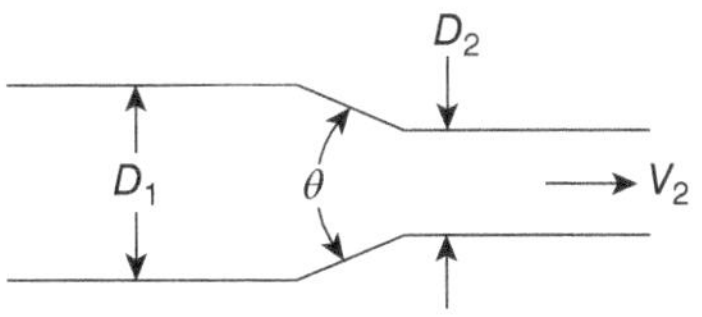		K_c	K_c
		D_2/D_1	$\theta = 60°$	$\theta = 180°$
		0.0	0.08	0.50
		0.20	0.08	0.49
		0.40	0.07	0.42
		0.60	0.06	0.32
$h_f = K_c V_2^2/2g$		0.80	0.05	0.18
		0.90	0.04	0.10
Expansion	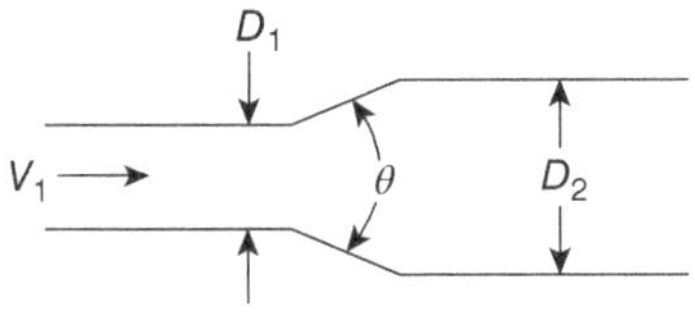		K_E	K_E
		D_1/D_2	$\theta = 10°$	$\theta = 180°$
		0.0		1.00
		0.20	0.13	0.92
		0.40	0.11	0.72
$h_f = K_E V_1^2/2g$		0.60	0.06	0.42
		0.80	0.03	0.16
90° miter bend	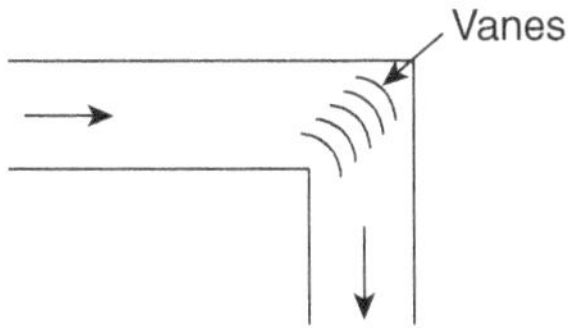	Without vanes		$K_b = 1.1$
		With vanes		$K_b = 0.2$
$h_f = K_b V^2/2g$				

(*Continued*)

Water and Wastewater Engineering: Hydraulics, Hydrology and Management, Volume 1, Fourth Edition.
Lawrence K. Wang, Mu-Hao Sung Wang, and Nazih K. Shammas.

Companion website: www.wiley.com/go/Wang/Waterandwastewater4e

Description	Sketch	Additional data	K	
Smooth bend	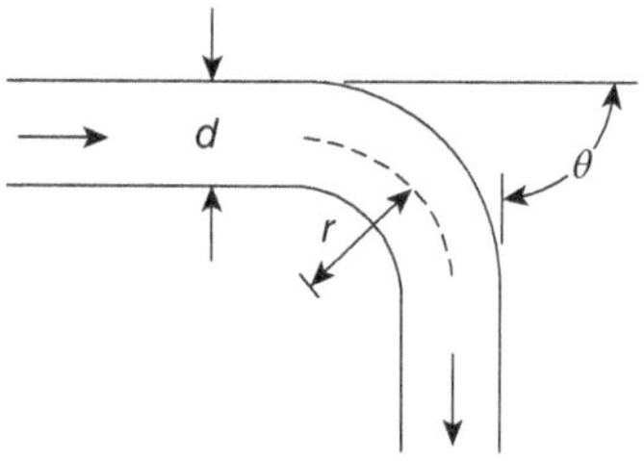	r/d	K_b $\theta = 45°$	K_b $\theta = 90°$
		1	0.10	0.35
		2	0.09	0.19
		4	0.10	0.16
		6	0.12	0.21
$h_f = K_b V^2/2g$				
Threaded pipe fittings	Globe valve-wide open			$K_v = 10.0$
	Angle valve-wide open			$K_v = 5.0$
	Gate valve-wide open			$K_v = 0.2$
	Gate valve-half open			$K_v = 5.6$
	Return bend			$K_b = 2.2$
	Tee			$K_t = 1.8$
	90° elbow			$K_e = 0.9$
	45° elbow			$K_e = 0.4$

Index

Water and Wastewater Engineering: Hydraulics, Hydrology and Management, Volume 1, Fourth Edition.
Lawrence K. Wang, Mu-Hao Sung Wang, and Nazih K. Shammas.

Companion website: www.wiley.com/go/Wang/Waterandwastewater4e

d

e

t

u

v